A TEXTBOOK OF THERMAL ENGINEERING

[A Textbook for the Students of B.Sc. Engg., UPSC (Engg. Services), Section 'B' of AMIE (I) and Diploma Courses]

(S.I. UNITS)

R.S. KHURMI
J.K. GUPTA

S. CHAND
PUBLISHING

S Chand And Company Limited
(ISO 9001 Certified Company)

S Chand And Company Limited

(ISO 9001 Certified Company)

Head Office: D-92, Sector–2, Noida – 201301, U.P. (India), Ph. 91-120-4682700

Registered Office: A-27, 2nd Floor, Mohan Co-operative Industrial Estate, New Delhi – 110 044, Phone: 011-49731800

www.**schandpublishing.com**; e-mail: **info@schandpublishing.com**

Marketing Offices:

Chennai : Ph: 23632120; chennai@schandpublishing.com
Guwahati : Ph: 2738811, 2735640; guwahati@schandpublishing.com
Hyderabad : Ph: 40186018; hyderabad@schandpublishing.com
Jalandhar : Ph: 4645630; jalandhar@schandpublishing.com
Kolkata : Ph: 23357458, 23353914; kolkata@schandpublishing.com
Lucknow : Ph: 4003633; lucknow@schandpublishing.com
Mumbai : Ph: 25000297; mumbai@schandpublishing.com
Patna : Ph: 2260011; patna@schandpublishing.com

First Edition 1978
Subsequent Editions and Reprints 1978, 79, 80, 81, 83, 84, 85, 86, 87, 88, 89, 91, 93, 94, 95, 96, 97, 98, 99, 2000, 2001, 2002, 2003, 2004, 2005, 2006, 2007, 2008, 2009 (Twice), 2010 (Twice), 2011, 2012, 2013, 2014, 2015, 2016, 2018 (Twice), 2019, 2020

Low Priced Students' Paperback Edition (LPSPE) 2022; Reprint 2022

Reprint 2023

ISBN: 978-93-550-1054-4 **Product Code:** H3THE62MECH10ENAA21L

PRINTED IN INDIA

By Vikas Publishing House Private Limited, Plot 20/4, Site-IV, Industrial Area Sahibabad, Ghaziabad – 201 010 and Published by S Chand And Company Limited, A-27, 2nd Floor, Mohan Co-operative Industrial Estate, New Delhi – 110 044.

PREFACE TO THE FIFTEENTH EDITION

We feel satisfied in presenting the new Edition of this standard treatise named as *A Textbook of Thermal Engineering*. The warm reception which the previous editions and reprints of this book have enjoyed all over India and abroad, is a matter of great satisfaction for us.

The present Edition of this treatise has been thoroughly revised and brought up-to-date. Two new chapters on *General Thermodynamic Relations* and *Variable Specific Heat* have been added. The mistakes which had crept in have been eliminated. We wish to express our sincere thanks to numerous professors and students, both at home and abroad, for sending their valuable suggestions and also for recommending the book to their students and friends. We hope that they will continue to patronise this book in the future also.

Any errors, omissions and suggestions, for the improvement of this volume brought to our notice, will be acknowledged and incorporated in the Next Edition.

R.S. KHURMI
J.K. GUPTA

CONTENTS

1. Introduction ... 1
2. Properties of Perfect Gases ... 30
3. Thermodynamic Processes of Perfect Gases ... 50
4. Entropy of Perfect Gases ... 103
5. Kinetic Theory of Gases ... 140
6. Thermodynamic Air Cycles ... 153
7. Formation and Properties of Steam ... 199
8. Entropy of Steam ... 219
9. Thermodynamic Processes of Vapour ... 230
10. Thermodynamic Vapour Cycles ... 264
11. Fuels ... 289
12. Combustion of Fuels ... 301
13. Steam Boilers ... 323
14. Boiler Mountings and Accessories ... 334
15. Performance of Steam Boilers ... 345
16. Boiler Draught ... 360
17. Simple Steam Engines ... 374
18. Compound Steam Engines ... 394
19. Performance of Steam Engines ... 422
20. Steam Condensers ... 446
21. Steam Nozzles ... 469
22. Impulse Turbines ... 501
23. Reaction Turbines ... 521
24. Performance of Steam Turbines ... 535
25. Modern Steam Turbines ... 560
26. Internal Combustion Engines ... 582
27. Testing of Internal Combustion Engines ... 611
28. Reciprocating Air Compressors ... 637
29. Rotary Air Compressors .. 665
30. Performance of Air Compressors ... 682
31. Air Motors ... 700
32. Gas Turbines ... 707
33. Performance of Gas Turbines ... 721
34. Introduction to Heat Transfer ... 736
35. Air Refrigeration Cycles ... 751
36. Vapour Compression Refrigeration Systems ... 771
37. Psychrometry ... 798
38. Air Conditioning Systems ... 824
39. General Thermodynamic Relations ... 835
40. Variable Specific Heat ... 859

Index ... 887

1

Introduction

1. Definition. 2. Fundamental Units. 3. Derived Units. 4. Systems of Units. 5. C.G.S. Units. 6. F.P.S. Units. 7. M.K.S. Units. 8. S.I. Units (International System of Units). 9. Metre. 10. Kilogram. 11. Second. 12. Kelvin. 13. Presentation of Units and their Values. 14.Rules for S.I. Units. 15. Newton's Laws of Motion. 16. Mass and Weight. 17 Force. 18. Absolute and Gravitational Units of Force. 19. Thermodynamic Systems. 20. Classification of Thermodynamic Systems. 21. Properties of a System. 22. Classification of Properties of a System. 23. State of a System. 24. Path of Change of State. 25. Thermodynamic Process. 26. Thermodynamic Cycle or Cyclic Process. 27. Quasi-static or Quasi-equilibrium Process. 28. Temperature. 29. Absolute Temperature. 30. Thermodynamic Equilibrium. 31. Equality of Temperature. 32. Pressure. 33. Absolute Pressure and Gauge Pressure. 34. Normal Temperature and Pressure (N.T.P.). 35. Standard Temperature and Pressure (S.T.P.). 36. Energy. 37. Types of Stored Energy. 38. Law of Conservation of Energy. 39. Heat. 40. Specific Heat. 41. Thermal Capacity or Heat Capacity. 42. Water Equivalent. 43. Mechanical Equivalent of Heat. 44. Work. 45. Heat and Work-A Path Function. 46. Comparison of Heat and Work. 47. Power. 48. Laws of Thermodynamics. 49. Zeroth law of Thermodynamics. 50. First law of Thermodynamics. 51. Limitations of First Law of Thermodynamics. 52. Second Law of Thermodynamics. 53. Equivalence of Kelvin-Planck and Clausius Statements.

1.1. Definition

The field of science, which deals with the energies possessed by gases and vapours, is known as *Thermodynamics*. It also includes the conversion of these energies in terms of heat and mechanical work and their relationship with properties of the system. A machine, which converts heat into mechanical work or vice versa, is known as *Heat Engine*. The field of engineering science, which deals with the applications of thermodynamics and its laws to work producing and work absorbing devices, in order to understand their functions and improve their performance, is known as *Thermal Engineering*.

The heat is, usually, generated by the combustion of fuel which may be solid, liquid or gas. It is supplied to the working substance (a source of conveying heat to the heat engine for doing work in the engine cylinder) at a higher temperature. A part of the heat energy is converted into mechanical work by expanding the working substance, within the engine cylinder. The remaining heat energy is rejected at a lower temperature.

The working substances, widely used in the heat engines, are fluids in the gaseous or liquid state. A mixture of air and fuel is used as a working substance in internal combustion engines, and water vapour (steam) in the steam engines or steam turbines.

1.2. Fundamental Units

The measurement of physical quantities is one of the most important operations in engineering. Every quantity is measured in terms of some arbitrary, but internationally accepted units, called *fundamental units*.

1.3. Derived Units

Some units are expressed in terms of other units, which are derived from fundamental units are known as *derived units e.g.* the unit of area, velocity, acceleration, pressure etc.

1.4. Systems of Units

There are only four systems of units, which are commonly used and universally recognised. These are known as :

1. C.G.S. units, 2. F.P.S. units, 3. M.K.S. units, and 4. S.I. units.

1.5. C.G.S. Units

In this system, the fundamental units of length, mass and time are *centimetre*, *gram* and *second* respectively. The C.G.S. units are known as absolute units or physicist's units.

1.6. F.P.S. Units

In this system, the fundamental units of length, mass and time are *foot*, *pound* and *second* respectively.

1.7. M.K.S. Units

In this system, the fundamental units of length, mass and time are *metre*, *kilogram* and *second* respectively. The M.K.S. units are known as gravitational units or engineer's units.

1.8. S.I. Units (International System of Units)

The 11th General Conference* of Weights and Measures have recommended a unified and systematically constituted system of fundamental and derived units for international use. This system is now being used in many countries. In India, the standards of Weights and Measures Act, 1956 (vide which we switched over to M.K.S. units) has been revised to recognise all the S.I. units in industry and commerce.

In this system of units, there are seven fundamental units and two supplementary units, which cover the entire field of science and engineering. These units are shown in the following table.

Table 1.1. Fundamental and supplementary units.

S.No.	*Physical quantity*	*Unit*
	Fundamental units	
1.	Length (l)	Metre (m)
2.	Mass (m)	Kilogram (kg)
3.	Time (t)	Second (s)
4.	Temperature (T)	Kelvin (K)
5.	Electric current (I)	Ampere (A)
6.	Luminous intensity (I_v)	Candela (cd)
7.	Amount of substance (n)	Mole (mol)
	Supplementary units	
1.	Plane angle ($\alpha, \beta, \theta, \phi$)	Radian (rad)
2.	Solid angle (Ω)	Steradian (sr)

* It is known as General Conference of Weights and Measures (C.G.P.M.). It is an international organisation, of which most of the advanced and developing countries (including India) are members. The conference has been entrusted with the task of prescribing definitions for various units of weights and measures, which are the very basic of science and technology today.

The derived units, which will be commonly used in this book, are given in the following table :

Table 1.2. Derived units.

S.No.	*Quantity*	*Symbol*	*Unit*
1.	Linear velocity	V	m/s
2.	Linear acceleration	a	m/s^2
3.	Angular velocity	ω	rad/s
4.	Angular acceleration	α	rad/s^2
5.	Mass density	ρ	kg/m^3
6.	Force, Weight	F, W	N ; 1N = 1kg-m/s^2
7.	Pressure	p	N/m^2
8.	Work, Energy, Enthalpy	W, E, H	J ; 1J = 1N-m
9.	Power	P	W ; 1W = 1J/s
10.	Absolute or dynamic viscosity	μ	N-s/m^2
11.	Kinematic viscosity	ν	m^2/s
12.	Frequency	f	Hz ; 1Hz = 1 cycle/s
13.	Gas constant	R	J/kg K
14.	Thermal conductance	h	W/m^2 K
15.	Thermal conductivity	k	W/m K
16.	Specific heat	c	J/kg K
17.	Molar mass or Molecular mass	M	kg/mol

1.9. Metre

The metre is defined as the length equal to 1 650 763.73 wavelengths in vacuum of the radiation corresponding to the transition between the levels 2 p_{10} and 5 d_5 of the Krypton - 86 atom.

1.10. Kilogram

The kilogram is defined as the mass of the international prototype (standard block of platinum - iridium alloy) of the kilogram, kept at the international Bureau of Weights and Measures at Sevres, near Paris.

1.11. Second

The second is defined as the duration of 9 192 631 770 periods of the radiation corresponding to the transition between the two hyperfine levels of the ground state of the caesium - 133 atom.

1.12. Kelvin

The kelvin is defined as the fraction 1/273.16 of the thermodynamic temperature of the triple point of water.

Note. The triple point of water is taken as a fundamental fixed point having a temperature 273.16 K.

1.13. Presentation of Units and their Values

The frequent changes in the present day life are facilitated by an international body known as International Standard Organisation (ISO) which makes recommendations regarding international standard procedures. The implementation of ISO recommendation, in a country, is assisted by its organisation appointed for the purpose. In India, Bureau of Indian Standards (BIS) previously known as Indian Standards Institution (ISI) has been created for this purpose. We have already discussed that the fundamental units in M.K.S. and S.I. units for length, mass and time is metre, kilogram and second respectively. But in actual practice, it is not necessary to express all lengths in metres, all masses in kilograms and all times in seconds. We shall, sometimes, use the convenient units, which are multiples or divisions of our basic units in tens. As a typical example, although the metre is the unit of length, yet a small length of one-thousandth of a metre proves to be more convenient unit, especially in the dimensioning of drawings. Such convenient units are formed by using a prefix in front of the basic units to indicate the multiplier. The full list of these prefixes is given in the following table.

Table 1.3. Prefixes used in basic units.

Factor by which the units is multiplied	*Standard form*	*Prefix*	*Abbreviation*
1 000 000 000 000	10^{12}	tera	T
1 000 000 000	10^{9}	giga	G
1 000 000	10^{6}	mega	M
1 000	10^{3}	kilo	k
100	10^{2}	hecto*	h
10	10^{1}	deca*	da
0.1	10^{-1}	deci*	d
0.01	10^{-2}	centi*	c
0.001	10^{-3}	milli	m
0.000 001	10^{-6}	micro	μ
0.000 000 001	10^{-9}	nano	n
0.000 000 000 001	10^{-12}	pico	p

1.14. Rules for S.I. Units

The eleventh General Conference of Weights and Measures recommended only the fundamental and derived units for S.I. system. But it did not elaborate the rules for the usage of the units. Later on many scientists and engineers held a number of meetings for the style and usage of S.I. units. Some of the decisions of the meetings are as follows :

1. For numbers having five or more digits, the digits should be placed in groups of three separated by spaces** (instead of commas) counting both to the left and right to the decimal point.

2. In a four digit number,*** the space is not required unless the four digit number is used in a column of numbers with five or more digits.

3. A dash is to be used to separate units that are multiplied together. For example, newton × metre is written as N-m. It should not be confused with mN, which stands for millinewton.

4. Plurals are never used with symbols. For example, metre or metres are written as m.

5. All symbols are written in small letters except the symbols derived from the proper names. For example, N for newton and W for watt.

---

* These prefixes are generally becoming obsolete probably due to possible confusion. Moreover it is becoming a conventional practice to use only those powers of ten which conform to 10^{3x}, where x is a positive or negative whole number.

** In certain countries, comma is still used as the decimal mark.

*** In certain countries, a space is used even in a four digit number.

6. The units with names of scientists should not start with capital letter when written in full. For example, 90 newton and not 90 Newton.

At the time of writing this book, the authors sought the advice of various international authorities, regarding the use of units and their values. Keeping in view the international reputation of the authors, as well as international popularity of their books, it was decided to present units* and their values as per recommendations of ISO and BIS. It was decided to use :

4500	not	4 500	or	4,500
7 589 000	not	7589000	or	7,58,90,00
0.012 55	not	0.01255	or	.01255
30×10^6	not	3,00,00,000	or	3×10^7

The above mentioned figures are meant for numerical values only. Now let us discuss about the units. We know that the fundamental units in S.I. system of units for length, mass and time are metre, kilogram and second respectively. While expressing these quantities, we find it time consuming to write the units such as metres, kilograms and seconds, in full, every time we use them. As a result of this, we find it quite convenient to use some standard abbreviations.

We shall use :

m	for metre or metres
km	for kilometre or kilometres
kg	for kilogram or kilograms
t	for tonne or tonnes
s	for second or seconds
min	for minute or minutes
N-m	for newton × metres (*e.g.* work done)
kN-m	for kilonewton × metres
rev	for revolution or revolutions
rad	for radian or radians

1.15. Newton's Laws of Motion

Newton has formulated three laws of motion, which are the basic postulates or assumptions on which the whole system of dynamics is based. Like other scientific laws, these are also justified as the results, so obtained, agree with the actual observations. These three laws of motion are as follows :

1. *Newton's First Law of Motion.* It states, "*Every body continues in its state of rest or of uniform motion in a straight line, unless it is acted upon by some external force.*" *This is also known as Law of inertia.*

The inertia is that property of a matter, by virtue of which a body cannot move of itself, nor change the motion imparted to it.

2. *Newton's Second Law of Motion.* It states, "*The rate of change of momentum is directly proportional to the impressed force and takes place in the same direction in which the force acts.*"

3. *Newton's Third Law of Motion.* It states "*To every action, there is always an equal and opposite reaction.*"

---

* In some of the question papers of the universities and other examining bodies, standard values are not used. The authors have tried to avoid such questions in the text of the book. However, at certain places the questions with sub-standard values have to be included, keeping in view the merits of the question from the reader's angle.

1.16. Mass and Weight

Sometimes much confusion and misunderstanding is created, while using the various systems of units in the measurement of force and mass. This happens, because of the lack of clear under-standing of the difference between mass and weight. The following definitions of mass and weight should be clearly understood.

1. *Mass.* It is the amount of matter contained in a given body, and does not vary with the change in its position on the earth's surface. The mass of a body is measured by direct comparison with a standard mass by using a lever balance.

2. *Weight.* It is the amount of pull, which the earth exerts upon a given body. Since the pull varies with the distance of the body from the centre of the earth, therefore weight of the body will also vary with its position on the earth's surface (say latitude and elevation). It is thus obvious, that the weight is a *force.*

The earth's pull in metric units, at sea level and 45° latitude, has been adopted as one force unit and named one kilogram of force. Thus it is a definite amount of force. But, unfortunately, it has the same name as the unit of mass. The weight of a body is measured by the use of a spring balance, which indicates the varying tension in the spring as the body is moved from place to place.

Note. The confusion in the units of mass and weight is eliminated, to a great extent, in S.I. units. In this system, mass is taken in kg and weight in newtons. The relation between the mass (m) and the weight (W) of a body is

$$W = m\,g \quad \text{or} \quad m = W/g$$

where W is in newtons, m is in kg and g is the acceleration due to gravity in m/s^2.

1.17. Force

It is an important factor in the field of Engineering science, which may be defined as an agent which produces or tends to produce, destroy or tends to destroy the motion. According to Newton's Second Law of Motion, the applied force or impressed force is directly proportional to the rate of change of momentum. We know that

$$\text{Momentum} = \text{Mass} \times \text{Velocity}$$

Let

m = Mass of the body,

u = Initial velocity of the body,

v = Final velocity of the body,

a = Constant acceleration, and

t = Time required to change the velocity from u to v.

$$\therefore \text{ Change of momentum} = m\,v - m\,u$$

and rate of change of momentum $= \dfrac{m\,v - m\,u}{t} = \dfrac{m\,(v-u)}{t} = m\,a \qquad \ldots \left(\because \dfrac{v-u}{t} = a\right)$

or $\qquad$ Force, $F \propto m\,a$ or $F = k\,m\,a$

where k is a constant of proportionality.

For the sake of convenience, the unit of force adopted is such that it produces a unit acceleration to a body of unit mass.

$$\therefore \quad F = m\,a = \text{Mass} \times \text{Acceleration}$$

In S.I. system of units, the unit of force is called newton (briefly written as N). A newton may be defined as the force while acting upon a mass of one kg produces an acceleration of 1 m/s^2 in the direction of which it acts. Thus

$$1\text{ N} = 1\text{ kg} \times 1\text{ m/s}^2 = 1\text{ kg-m/s}^2$$

1.18. Absolute and Gravitational Units of Force

We have already discussed that when a body of mass 1 kg is moving with an acceleration of 1 m/s^2, the force acting on the body is 1 newton (briefly written as 1 N). Therefore when the same body is moving with an acceleration of 9.81 m/s^2, the force acting on the body is 9.81 N. But we denote 1 kg mass attracted towards the earth with an acceleration of 9.81 m/s^2 as 1 kilogram-force (briefly written as kgf) or 1 kilogram-weight (briefly written as kg-wt). It is thus obvious, that

$$1 \text{ kgf} = 1 \text{ kg} \times 9.81 \text{ m/s}^2 = 9.81 \text{ kg-m/s}^2 \doteq 9.81 \text{ N} \quad \ldots (\because 1\text{N} = 1 \text{ kg-m/s}^2)$$

The above unit of force *i.e.* kilogram force (kgf) is called *gravitational or engineer's units of force*, whereas newton is the *absolute* or *scientific* or *S.I. units of force*. It is thus obvious, that the gravitational or engineer's units of force are '*g*' times greater than the unit of force in the absolute or S.I. units.

It will be interesting to know that the *mass of the body in absolute units is numerically equal to the weight of the same body in gravitational units*. For example, consider a body whose mass,

$$m = 100 \text{ kg}$$

Therefore the force, with which the body will be attracted towards the centre of the earth,

$$F = m\,a = m\,g = 100 \times 9.81 = 981 \text{ N}$$

Now, as per definition, we know that the weight of a body is the force, by which it is attracted towards the centre of the earth. Therefore weight of the body,

$$W = 981 \text{ N} = 981/9.81 = 100 \text{ kgf} \quad \ldots (\because 1 \text{ kgf} = 9.81 \text{ N})$$

In brief, the weight of a body of mass *m* kg at a place where gravitational acceleration is '*g*' m/s^2 is *m.g* newtons.

1.19. Thermodynamic Systems

The *thermodynamic system* (or simply known as *system*) may be broadly defined as a *definite area* or *a space* where some thermodynamic process* is taking place. It is a region where our attention is focussed for studying a thermodynamic process. A little observation will show that a thermodynamic system has its boundaries and anything outside the boundaries is called its *surroundings* as shown in Fig. 1.1. These boundaries may be *fixed* like that of a tank enclosing a certain mass of compressed gas, or *movable* like boundary of a certain volume of liquid in a pipe line.

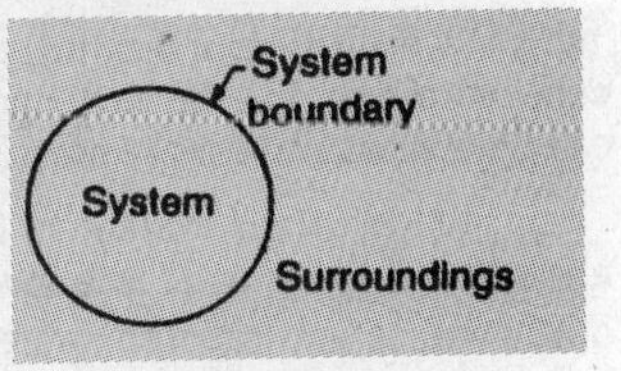

Fig. 1.1. Thermodynamic system.

1.20. Classification of Thermodynamic Systems

The thermodynamic systems may be classified into the following three groups :

1. Closed system ; 2. Open system ; and 3. Isolated system.

These systems are discussed, in detail, as follows :

1. *Closed system.* This is a system of fixed mass and identity whose boundaries are determined by the space of the matter (working substance) occupied in it.

A closed system is shown in Fig. 1.2. The gas in the cylinder is considered as a system. If heat is supplied to the cylinder from some external source, the temperature of the gas will increase and the piston will rise.

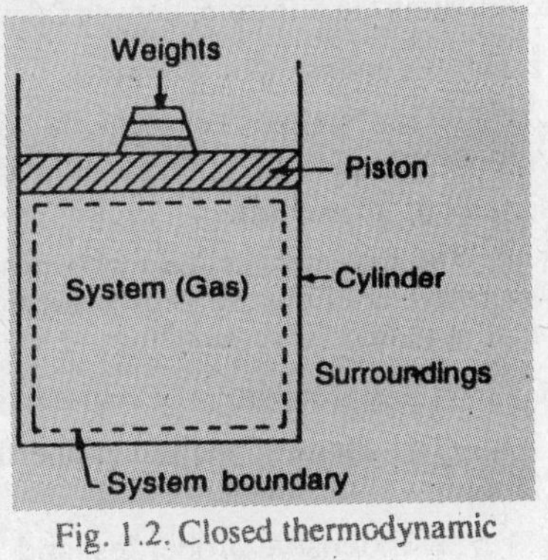

Fig. 1.2. Closed thermodynamic system.

---

* Refer Art. 1.25

As the piston rises, the boundary of the system moves. In other words, the heat and work energy crosses the boundary of the system during this process, but there is no addition or loss of the original mass of the working substance. It is thus obvious, that the mass of the working substance, which comprises the system, is fixed.

Thus, a closed system does not permit any mass transfer across its boundary, but it permits transfer of energy (heat and work).

2. *Open system.* In this system, the mass of the working substance crosses the boundary of the system. Heat and work may also cross the boundary. Fig. 1.3 shows the diagram of an air compressor which illustrates an open system.

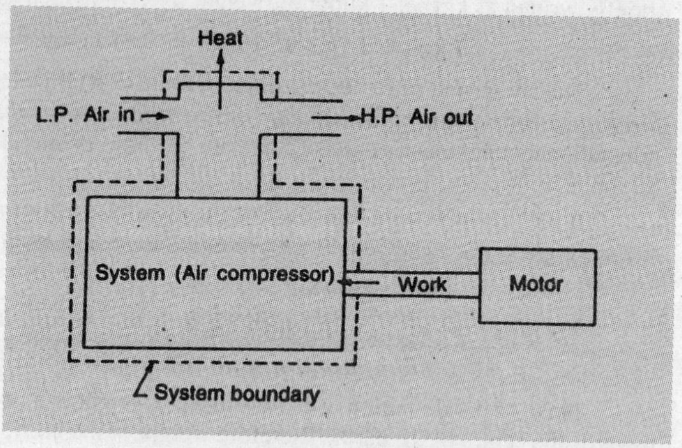

Fig. 1.3. Open thermodynamic system.

The working substance crosses the boundary of the system as the low pressure (L.P.) air enters the compressor and leaves the high pressure (H.P.) air. The work crosses the boundary of the system through the driving shaft and the heat is transferred across the boundary from the cylinder walls.

Thus, an open system permits both mass and energy (heat and work) transfer across the boundaries and the mass within the system may not be constant.

Note. An open system may be referred to as *control volume*. An open system is equivalent in every respect to a control volume, but the term open system is used throughout this text as it specifically implies that the system can have mass and energy crossing the system boundary.

3. *Isolated system.* A system which is completely uninfluenced by the surrounding is called an isolated system. It is a system of fixed mass and no heat or work energy cross its boundary. In other words, an isolated system does not have transfer of either mass or energy (heat or work) with the surroundings. An open system with its surroundings (known as an universe) is an example of an isolated system.

Note. The practical examples of isolated systems are rare. The concept of this system is particularly useful in formulating the principles derived from the Second Law of Thermodynamics.

1.21. Properties of a System

The *state of a system may be identified or described by certain observable quantities such as volume, temperature, pressure and density etc. All the quantities, which identify the state of a system, are called *properties*.

Note. Thermodynamics deals with those quantities also which are not properties of any system. For example, when there is a flow of energy between a system and its surroundings, the energy transferred is not a property of the system or its surroundings.

1.22. Classification of Properties of a System

The thermodynamic properties of a system may be divided into the following two general classes :

1. Extensive properties, and 2. Intensive properties.

---

* Refer Art. 1.23.

1. Extensive properties. A quantity of matter in a given system is divided, notionally into a number of parts. The properties of the system, whose value for the entire system is equal to the sum of their values for the individual parts of the system are called *extensive properties, e.g.* total volume, total mass and total energy of a system are its extensive properties.

2. Intensive properties. It may be noticed that the temperature of the system is not equal to the sum of the temperatures of its individual parts. It is also true for pressure and density of the system. Thus properties like temperature, pressure and density are called *intensive properties.*

Note. The ratio of any extensive property of a system to the mass of the system is called an average specific value of that property (also known as intensive property) *e.g.* specific volume of a system (v_s) is the ratio of the total volume (v) of the system to its total mass (m). Mathematically,

$$v_s = v/m$$

The specific volume is an intensive property.

1.23. State of a System

The state of a system (when the system is in thermodynamic equilibrium) is the condition of the system at any particular moment which can be identified by the statement of its properties, such as pressure, volume, temperature etc. The number of properties which are required to describe a system depends upon the nature of the system.

Consider a system (gas) enclosed in a cylinder and piston arrangement as shown in Fig. 1.4. Let the system is initially in equilibrium when the piston is at position 1, represented by its properties p_1, v_1 and T_1. When the system expands, the piston moves towards right and occupies the final position at 2. At this, the system is finally in the equilibrium state represented by the properties p_2, v_2 and T_2. The initial and final states, on the pressure-volume diagram, are shown in Fig. 1.4.

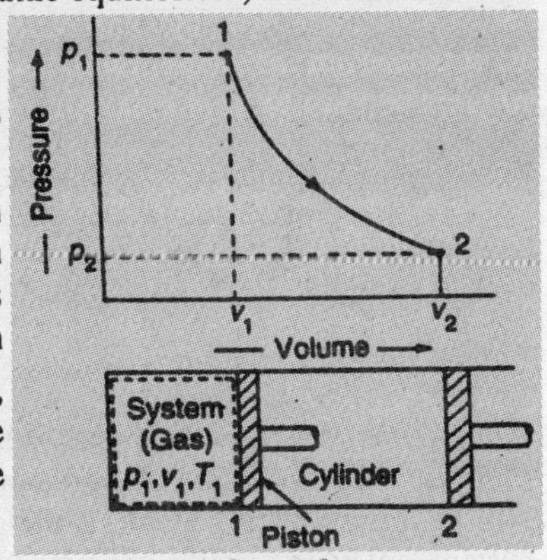

Fig. 1.4. State of a system.

1.24. Path of Change of State

When a system passes through the continuous series of equilibrium states during a change of state (from the initial state to the final state), then it is known as *path of change of state*. When the path is completely specified, it is then known as *path of the process*.

1.25. Thermodynamic Process

When a system changes its state from one equilibrium state to another equilibrium state, then the path of successive states through which the system has passed is known as *thermodynamic process*. In Fig. 1.4, 1-2 represents a thermodynamic process.

1.26. Thermodynamic Cycle or Cyclic Process

When a process or processes are performed on a system in such a way that the final state is identical with the initial state, it is then known as a *thermodynamic cycle* or *cyclic process*. In Fig. 1.5, 1-*A*-2 and 2-*B*-1 are processes whereas 1-*A*-2-*B*-1 is a thermodynamic cycle or cyclic process.

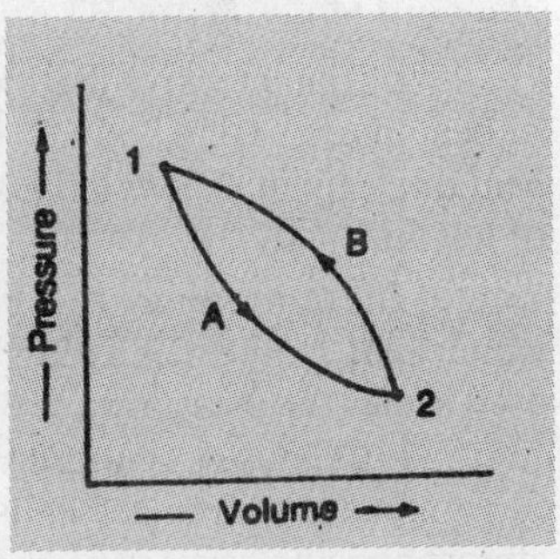

Fig. 1.5. Thermodynamic process or cyclic process.

1.27. Quasi-static or Quasi-equilibrium Process

When the process is carried out in such a way that at every instant, the system deviation from the thermodynamic equilibrium is infinitesimal, then the process is known as *quasi-static* or *quasi-equilibrium process* and each state in the process may be considered as an equilibrium state.

Consider a system (gas) enclosed in a cylinder and piston arrangement as shown in Fig. 1.6 (*a*). Let the system is initially in equilibrium state when the piston is at *A*, where the pressure is p_A, volume v_A and temperature T_A as shown in Fig. 1.6 (*b*). The weight (*W*) on the piston is composed of number of small weights which balances the upward force exerted by the system. If the whole weight is removed from the piston, then there will be unbalanced force between the system and the surroundings and the piston will move upwards till it hits the stops at *B*. At this point *B*, the system again comes to an equilibrium state where the pressure is p_B, volume v_B and temperature T_B. But the intermediate states through which the system has passed, are non-equilibrium states whose properties (pressure, volume and temperature) are not uniform throughout the system and thus the state of the system cannot be well defined. Such a process is called *irreversible* or *non-equilibrium process*, as shown by a broken line in Fig. 1.6 (*b*).

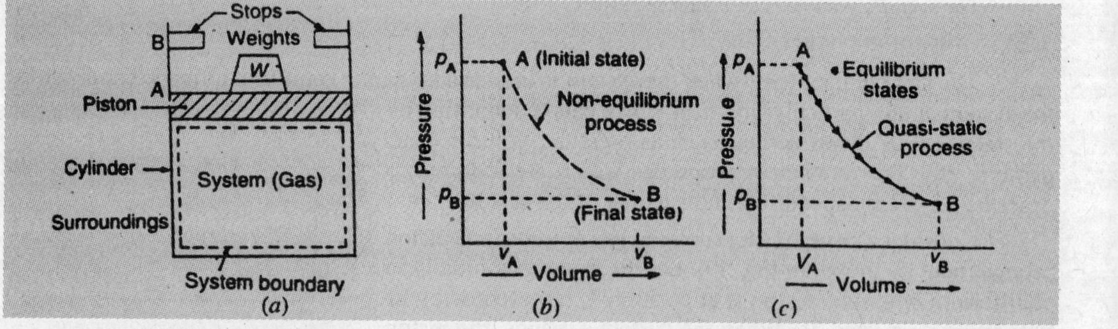

Fig. 1.6. Non-equilibrium and quasi-static (or quasi-equilibrium) process.

Now, if the small weights on the piston are removed one by one very slowly, then at any instant of the upward movement of the piston, the deviation of the state from the thermodynamic equilibrium will be infinitesimally small, if the gas system is isolated. Thus, every state passed through by the system will be in equilibrium state. Such a process, which is the locus of all these equilibrium points passed through the system, is known as *quasi-static* or *quasi-equilibrium process*.

Note : The quasi-static or quasi-equilibrium process is also known as *reversible process*. A process which can be reversed in direction and the system retraces the same equilibrium states is known as reversible process.

1.28. Temperature

It is an intensive thermodynamic property, which determines the degree of hotness or the level of heat intensity of a body. A body is said to be at a *high temperature* or *hot*, if it shows high level of heat intensity in it. Similarly, a body is said to be at a *low temperature* or *cold*, if it shows a low level of heat intensity.

The temperature of a body is measured with the help of an instrument known as *thermometer* which is in the form of a glass tube containing mercury in its stem. Following are the two commonly used scales for measuring the temperature of a body :

1. Celsius or centigrade scale, and 2. Fahrenheit scale.

Each of these scales is based on two fixed points known as freezing point of water under atmospheric pressure or *ice point* and the boiling point of water or *steam point*.

1. *Celsius or centigrade scale*This scale was first used by Celsius in 1742. This scale is mostly used by engineers and scientists. The freezing point of water on this scale is marked as zero, and the boiling point of water as 100. The space between these two points has 100 equal divisions, and each division represents one degree Celsius (written as °C).

2. *Fahrenheit scale*This scale was first used in 1665. In this scale, the freezing point of water is marked as 32 and the boiling point of water as 212. The space between these two points has 180 equal divisions and each division represents one degree Fahrenheit (written as °F).

Note. The relation between Celsius scale and Fahrenheit scale is given by :

$$\frac{C}{100} = \frac{F-32}{180} \quad \text{or} \quad \frac{C}{5} = \frac{F-32}{9}$$

Example 1.1. *Find the temperature which has the same value on both the Celsius and Fahrenheit scales.*

Solution.

Let x = Temperature which has the same value on both the Celsius and Fahrenheit scales.

We know that $\frac{C}{5} = \frac{F-32}{9}$ or $9C = 5(F-32)$

$$\therefore \quad 9x = 5(x-32) = 5x - 160$$

$$9x - 5x = -160 \quad \text{or} \quad 4x = -160$$

$$\therefore \quad x = -160/4 = -40$$

Hence – 40° on the Celsius scale is equal to – 40° on the Fahrenheit scale. **Ans.**

1.29. Absolute Temperature

As a matter of fact, the zero readings of Celsius and Fahrenheit scales are chosen arbitrarily for the purpose of simplicity. It helps us in our calculations, when changes of temperature in a process are known. But, whenever the value of temperature is used in equations relating to fundamental laws, then the value of temperature, whose reference point is true zero or absolute zero, is used. The temperature, below which the temperature of any substance can not fall, is known as absolute zero temperature.

The absolute zero temperature, for all sorts of calculations, is taken as – 273°C in case of Celsius scale and – 460°F in case of Fahrenheit scale. The temperatures measured from this zero are called *absolute temperatures*. The absolute temperature in Celsius scale is called degree Kelvin (briefly written as K)* ; such that K = °C + 273. Similarly, absolute temperature in Fahrenheit scale is called degrees Rankine (briefly written as °R) ; such that °R = °F + 460.

1.30. Thermodynamic Equilibrium

A system is said to be in thermodynamic equilibrium, if it satisfies the following three requirements of equilibrium.

1. *Mechanical equilibrium.* A system is said to be in mechanical equilibrium, when there is no unbalanced forces acting on any part of the system or the system as a whole.

2. *Thermal equilibrium.* A system is said to be in thermal equilibrium, when there is no temperature difference between the parts of the system or between the system and the surroundings.

3. *Chemical equilibrium.* A system is said to be in chemical equilibrium, when there is no chemical reaction within the system and also there is no movement of any chemical constituent from one part of the system to the other.

1.31. Equality of Temperature

Consider two bodies of the same or different materials, one hot and the other cold. When these bodies are brought in contact, the hot body becomes colder, and the cold body becomes warmer. If these bodies remain in contact for some time, a state reaches when there is no further observable change in the properties of the two bodies. This is a state of thermal equilibrium, and at this stage the two bodies have the equal temperatures. It thus follows that when two bodies are in thermal equilibrium with each other, their temperatures are equal.

---

* In S.I. units, degrees Kelvin is not written as °K but only K.

1.32. Pressure

The term 'pressure' may be defined as the normal force per unit area. The unit of pressure depends upon the units of force and area.

In S.I. system of units, the practical unit of pressure is N/mm^2, N/m^2, kN/m^2, MN/m^2 etc. But sometimes a bigger unit of pressure (known as bar) is used, such that,

$$1 \text{ bar} = 1 \times 10^5 \text{ N/m}^2 = 0.1 \times 10^6 \text{ N/m}^2 = 0.1 \text{ MN/m}^2$$

Sometimes the pressure is expressed in another unit, called Pa (named after Pascal) and kPa, such that

$$1 \text{ Pa} = 1 \text{ N/m}^2 \quad \text{and} \quad 1 \text{ kPa} = 1 \text{ kN/m}^2$$

1.33. Gauge Pressure and Absolute Pressure

All the pressure gauges read the difference between the actual pressure in any system and the atmospheric pressure. The reading of the pressure gauge is known as *gauge pressure*, while the actual pressure is called *absolute pressure*. Mathematically,

Absolute pressure = Atmospheric pressure + Gauge pressure

This relation is used for pressures above atmospheric, as shown in Fig. 1.7 (*a*). For pressures below atmospheric, the gauge pressure will be negative. This negative gauge pressure is known as *vacuum pressure*. Therefore

Absolute pressure = Atmospheric pressure − Vacuum pressure

This relation is shown in Fig. 1.7 (*b*).

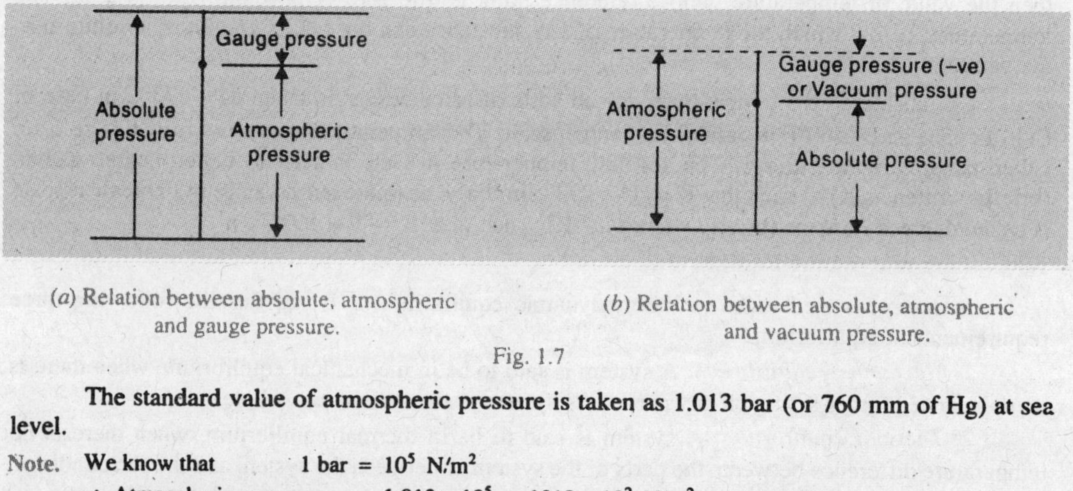

(*a*) Relation between absolute, atmospheric and gauge pressure.

(*b*) Relation between absolute, atmospheric and vacuum pressure.

Fig. 1.7

The standard value of atmospheric pressure is taken as 1.013 bar (or 760 mm of Hg) at sea level.

Note. We know that $1 \text{ bar} = 10^5 \text{ N/m}^2$

$\therefore$ Atmospheric pressure $= 1.013 \times 10^5 = 1013 \times 10^2 \text{ N/m}^2$

We also know that atmospheric pressure

$= 760$ mm of Hg

$\therefore \quad 1 \text{ mm of Hg} = 1013 \times 10^2 / 760 = 133.3 \text{ N/m}^2$

or $\quad 1 \text{ N/m}^2 = 760 / 1013 \times 10^2 = 7.5 \times 10^{-3}$ mm of Hg

1.34. Normal Temperature and Pressure (N.T.P.)

The conditions of temperature and pressure at 0°C (273 K) temperature and 760 mm of Hg pressure are termed as *normal temperature* and *pressure* (briefly written as N.T.P.)

1.35. Standard Temperature and Pressure (S.T.P.)

The temperature and pressure of any gas, under standard atmospheric conditions, is taken as 15°C (288 K) and 760 mm of Hg respectively.

Example 1.2. *The pressure of steam inside a boiler, as measured by pressure gauge, is 1 N/mm^2. The barometric pressure of the atmosphere is 765 mm of mercury. Find the absolute pressure of steam in N/m^2, kPa, bar and N/mm^2.*

Solution. Given : Gauge pressure = 1 N/mm^2 = 1×10^6 N/m^2 ; Atmospheric pressure = 765 mm of Hg

We know that atmospheric pressure

$= 765$ mm of Hg

$= 765 \times 133.3 = 0.102 \times 10^6 \text{ N/m}^2$ ...($\because$ 1 mm of Hg = 133.3 N/m^2)

$\therefore$ Absolute pressure of steam

= Atmospheric pressure + Gauge pressure

$= 0.102 \times 10^6 + 1 \times 10^6 = 1.102 \times 10^6 \text{ N/m}^2$ **Ans.**

$= 1102$ kPa **Ans.** ...($\because$ 1 kPa = 10^3 N/m^2)

$= 11.02$ bar **Ans.** ...($\because$ 1 bar = 10^5 N/m^2)

$= 1.102 \text{ N/mm}^2$ **Ans.** ..($\because$ 1 N/mm^2 = 10^6 N/m^2)

Example 1.3. *In a condenser of a steam power plant, the vacuum is recorded as 700 mm of mercury. If the barometer reading is 760 mm of mercury, find the absolute pressure in the condenser in N/m^2, kPa, bar and N/mm^2.*

Solution. Given : Vacuum pressure = 700 mm of Hg ; Barometer reading = 760 mm of Hg

We know that absolute pressure in the condenser

= Atmospheric pressure – Vacuum pressure

= Barometric pressure – Vacuum pressure

$= 760 - 700 = 60$ mm of Hg

$= 60 \times 133.3 = 7998 \text{ N/m}^2$ **Ans.** ...($\because$ 1 mm of Hg = 133.3 N/m^2)

$= 7.998$ kPa **Ans.** ...($\because$ 1 kPa = 10^3 N/m^2)

$= 0.079\,98$ bar **Ans.** ...($\because$ 1 bar = 10^5 N/m^2)

$= 0.007\,998 \text{ N/mm}^2$ **Ans.** ...($\because$ 1 N/mm^2 = 10^6 N/m^2)

1.36. Energy

The energy is defined as the capacity to do work. In other words, a system is said to possess energy when it is capable of doing work. The energy possessed by a system is of the following two types :

1. Stored energy, and 2. Transit energy (or energy in transition)

The *stored energy* is the energy possessed by a system within its boundaries. The potential energy, kinetic energy and internal energy are the examples of stored energy.

The *transit energy* (or energy in transition) is the energy possessed by a system which is capable of crossing its boundaries. The heat, work and electrical energy are the examples of transit energy.

It may be noted that only the stored energy is a thermodynamic property whereas the transit energy is not a thermodynamic property as it depends upon the path.

1.37. Types of Stored Energy

We have discussed above that the potential energy, kinetic energy and internal energy are the different types of stored energy. These energies are discussed, in detail, as follows :

1. *Potential energy.* It is the energy possessed by a body or a system for doing work, by virtue of its position above the ground level. For example, a body raised to some height above the ground level possesses potential energy because it can do some work by falling on earth's surface.

Let W = Weight of the body,

m = Mass of the body,

z = Distance through which the body falls, and

g = Acceleration due to gravity = 9.81 m/s^2.

∴ Potential energy,

$$PE = W z = m g z$$

It may be noted that

(*a*) When W is in newtons and z in metres, then potential energy will be in N-m.

(*b*) When m is in kg and z in metres, then the potential energy will also be in N-m, as discussed below :

We know that potential energy,

$$PE = m g z = \text{kg} \times \frac{\text{m}}{\text{s}^2} \times \text{m} = \text{N-m} \quad \ldots \left(\because 1 \text{ N} = \frac{1 \text{ kg-m}}{\text{s}^2} \right)$$

2. *Kinetic energy.* It is the energy possessed by a body or a system, for doing work, by virtue of its mass and velocity of motion.

Let m = Mass of the body, and

V = Velocity of the body.

When m is in kg and V is in m/s, then kinetic energy will be in N-m, as discussed below :

We know that kinetic energy,

$$KE = \frac{1}{2} m V^2 = \text{kg} \times \frac{\text{m}^2}{\text{s}^2} = \frac{\text{kg-m}}{\text{s}^2} \times \text{m} = \text{N-m} \ldots \left(\because 1 \text{ N} = \frac{1 \text{ kg-m}}{\text{s}^2} \right)$$

3. *Internal energy.* It is the energy possessed by a body or a system due to its molecular arrangement and motion of the molecules. It is usually represented by U.

In the study of thermodynamics, we are mainly concerned with the change in internal energy (dU) which depends upon the change in temperature of the system.

Notes. 1. The total energy of the system (E) is equal to the sum of the above three types of energies. Mathematically

$$E = PE + KE + U = m g z + \frac{1}{2} \times m V^2 + U$$

Any other form of the energy such as chemical energy, electrical energy etc. is neglected.

For unit mass, the above expression is written as

$$e = pe + ke + u = g z + \frac{V^2}{2} + u$$

2. When the system is stationary and the effect of gravity is neglected, then $PE = 0$, and $KE = 0$. In such a case

$$E = U \quad \text{or} \quad e = u$$

1.38. Law of Conservation of Energy

It states. *"The energy can neither be created nor destroyed, though it can be transformed from one form to any other form, in which the energy can exist."*

1.39. Heat

The heat is defined as the energy transferred, without transfer of mass, across the boundary of a system because of a temperature difference between the system and the surroundings. It is usually represented by Q and is expressed in joule (J) or kilo-joule (kJ).

The heat can be transferred in three distinct ways, *i.e.* conduction, convection and radiation. The transfer of heat through solids takes place by *conduction*, while the transfer of heat through fluids is by *convection*. The *radiation* is an electromagnetic wave phenomenon in which energy can be transported through transparent substances and even through a vacuum. These three modes of heat transfer are quite different, but they have one factor in common. All these modes occur across the surface area of a system because of a temperature difference between the system and the surroundings.

The following points are worth noting about heat :

1. The heat is transferred across a boundary from a system at a higher temperature to a system at a lower temperature by virtue of the temperature difference.
2. The heat is a form of transit energy which can be identified only when it crosses the boundary of a system. It exists only during transfer of energy into or out of a system.
3. The heat flowing *into* a system is considered as *positive* and heat flowing *out* of a system is considered *negative*.

1.40. Specific Heat

The specific heat of a substance may be broadly defined as the amount of heat required to raise the temperature of a unit mass of any substance through one degree. It is generally denoted by c. In S.I. system of units, the unit of specific heat (c) is taken as kJ/kg K. If m kg of a substance of specific heat c is required to raise the temperature from an initial temperature of T_1 to a final temperature of T_2, then

$$\text{Heat required} = m\,c\,(T_2 - T_1)\ \text{kJ}$$

where T_1 and T_2 may be either in °C or in K.

Since the solids and liquids do not change the volume on heating, therefore they have only one specific heat. But the gases have the following two *specific heats depending upon the process adopted for heating the gas.

1. Specific heat at constant pressure (c_p), and
2. Specific heat at constant volume (c_v).

It may be noted that c_p is always greater than c_v. The average values of specific heats for some commonly used substances are given in the following table.

Table 1.4. Values of Specific heat for some commonly used substances.

Solids	*Specific heat (kJ/kg K)*	*Fluids*	*Specific heat (kJ/kg K)*	*Gases*	*Specific heat at constant pressure (kJ/kg K)*
Steel	0.490	Water	4.187	Air	1.000
Copper	0.406	Ice	2.110	Carbon dioxide	0.846
Zinc	0.389	Steam	2.094	Nitrogen	1.043
Mercury	0.138	Petrol	1.817	Oxygen	0.913
Coal	1.010	Alcohol	2.512	Carbon monoxide	1.047
Coke	0.837	Paraffin oil	2.140	Hydrogen	14.257

* *For further details, please refer Art. 2.11.*

1.41. Thermal or Heat Capacity

The thermal or heat capacity of a substance may be defined as the heat required to raise the temperature of whole mass of a substance through one degree. Mathematically,

Thermal or heat capacity = $m\,c$ kJ

where m = Mass of the substance in kg, and

c = Specific heat of the substance in kJ/ kg K.

1.42. Water Equivalent

The water equivalent of a substance may be defined as the quantity of water, which requires the same quantity of heat as the substance to raise its temperature through one degree. Mathematically,

Water equivalent of a substance

$= m\,c$ kg

where m = Mass of the substance in kg, and

c = Specific heat of the substance in kJ/kg K.

Note. The numerical value of the thermal capacity and the water equivalent of the substance are the same but they are expressed in different units.

Example 1.4. *Calculate the quantity of heat required to raise the temperature of a steel forging of mass 180 kg from 300 K to 1265 K. The specific heat of steel = 0.49 kJ/kg K.*

Solution. Given : $m = 180$ kg ; $T_1 = 300$ K ; $T_2 = 1265$ K ; $c = 0.49$ kJ/kg K

We know that the quantity of heat required

$= \text{Mass} \times \text{Sp. heat} \times \text{Rise in temp.} = m\,c\,(T_2 - T_1)$

$= 180 \times 0.49\,(1265 - 300) = 85\,113$ kJ **Ans.**

1.43. Mechanical Equivalent of Heat

It was established by Joule that heat and mechanical energies are mutually convertible. He established, experimentally, that there is a numerical relation between the unit of heat and unit of work. This relation is denoted by J (named after Joule) and is known as Joule's equivalent or mechanical equivalent of heat.

Note. In S.I. system of units, the unit of work done is joule or kilo joule (such that 1 J = 1 N-m or 1 kJ = 1 kN-m). The unit of heat is also joule or kilo joule. So we can straightway convert heat units into mechanical units and vice versa.

1.44. Work

In mechanics, work is defined as the product of the force (F) and the distance moved (x) in the direction of the force. Mathematically, work done,

$$W = F \times x$$

The unit of work depends upon the unit of force and the distance moved. In S.I. system of units, the practical unit of work is newton-metre (briefly written as N-m). The work of 1 N-m is known as joule (briefly written as J) such that 1 N-m = 1 J.

In thermodynamics, work may be defined as follows :

1. According to Obert, work is defined as *the energy transferred (without the transfer of mass) across the boundary of a system because of an intensive property difference other than temperature that exists between the system and surroundings.*

In engineering practice, the intensive property difference is the pressure difference. The pressure difference (between the system and the surrounding) at the surface of the system gives rise to a force and the action of this force over a distance is called mechanical work.

In some cases, the intensive property difference may be the electrical potential difference between the system and the surrounding. In this case, the resulting energy transfer across the system and boundary is known as electrical work.

2. According to Keenan, *work is said to be done by a system during a given operation if the sole effect of the system on things external to the system (surroundings) can be reduced to the raising of a weight.*

The weight may not be actually raised but the net effect external to the system should be the raising of a weight.

For example, consider a system consisting of a storage battery, as shown in Fig. 1.8. The terminals connected to a resistance through a switch constitute external to the system (*i.e.* surroundings). When the switch is closed for a certain period of time, then the current will flow through the battery and the resistance, as a result the resistance becomes warmer. This clearly shows that the system (battery) has interaction with the surroundings. In other words, the energy transfer (electrical energy) has taken place between the system and the surroundings because of potential difference (not the temperature).

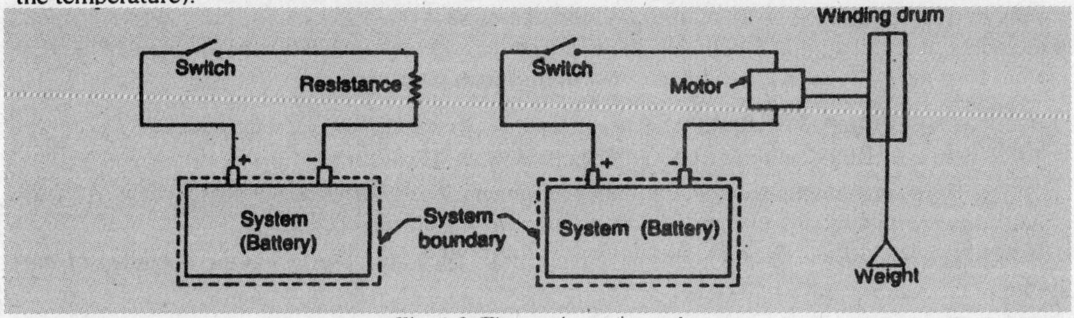

Fig. 1.8. Thermodynamic work.

Now according to the mechanics definition of work, there is no force which moves through a distance. Thus no work is done by the system. However, according to the thermodynamic definition, the work is done by the system because the resistance can be replaced by an ideal motor (100% efficient) driving a winding drum, thereby raising a weight. Thus, the sole effect external to the system (surroundings) has been reduced to the raising of a weight. Hence, thermodynamic work is done by the system.

Note. The work done *by* the system is considered as *positive* work, while the work done *on* the system is considered as *negative* work.

1.45. Heat and Work—A Path Function

Consider that a system from an initial equilibrium state 1 reaches to a final equilibrium state 2 by two different paths 1-*A*-2 and 1-*B*-2, as shown in Fig. 1.9. The processes are quasi-static.

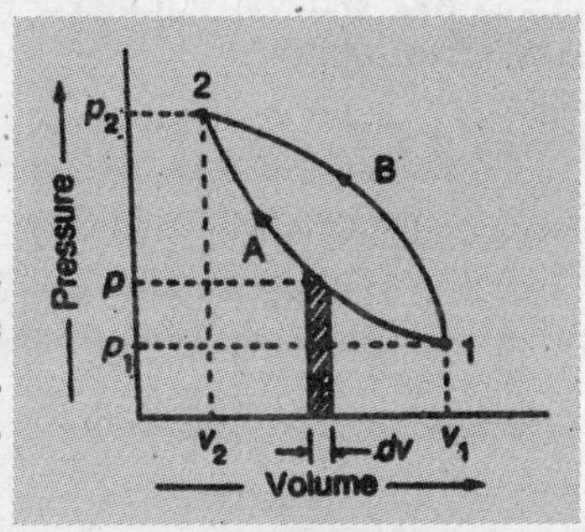

Fig. 1.9. Heat and work-a path function.

When the system changes from its initial state 1 to final state 2, the quantity of heat transfer will depend upon the intermediate stages through which the system passes, *i.e.* its path. In other words, heat is a path function. Thus, heat is an inexact differential and is written as δQ. On integrating, for the path 1-*A*-2,

$$\int_1^2 \delta Q = [Q]_1^2 = Q_{1-2} \quad \text{or} \quad {}_1Q_2$$

It may be noted that $\int_1^2 \delta Q \neq Q_2 - Q_1$, because heat is not a point function. Thus, it is meaningless to say 'heat in a system or heat of a system'. The heat can not be interpreted similar to temperature and pressure.*

The work, like heat, is not a thermodynamic property, therefore it is a path function as its value depends upon the particular path followed during the process. Since the areas under the curves 1-*A*-2 and 1-*B*-2 are different, therefore work done by these two processes will also be different**. Hence, work is an inexact differential and is written as δW. On integration, for the path 1-*A*-2,

$$\int_1^2 \delta W = [W]_1^2 = W_{1-2} \quad \text{or} \quad {}_1W_2$$

As discussed above, $\int_1^2 \delta W \neq W_2 - W_1$, because work is not a point function. Thus, it is meaningless to say 'work in a system or work of a system'. Since the work can not be interpreted similar to temperature and pressure of the system, therefore it is a path function and it depends upon the process. It is not a point function as the temperature and pressure. The work done in taking the system from state 1 to state 2 will be different for different paths.

1.46. Comparison of Heat and Work

There are many similarities between heat and work. These are

1. The heat and work are both transient phenomena. The systems do not possess heat or work. When a system undergoes a change, heat transfer or work done may occur.

2. The heat and work are boundary phenomena. They are observed at the boundary of the system.

3. The heat and work represent the energy crossing the boundary of the system.

4. The heat and work are path functions and hence they are inexact differentials. They are written as δQ and δW.

1.47. Power

It may be defined as the rate of doing work or work done per unit time. Mathematically,

$$\text{Power} = \frac{\text{Work done}}{\text{Time taken}}$$

In S.I. system of units, the unit of power is watt (briefly written as W) which is equal to 1 J/s or 1 N-m/s. Generally, a bigger unit of power called kilowatt (briefly written as kW) is used which is equal to 1000 W.

Notes. 1. If T is the torque transmitted in N-m or J and ω is the angular speed in rad/s, then

$$\text{Power,} \quad P = T\omega = T \times 2\pi N/60 \text{ watts} \qquad \ldots (\because \omega = 2\pi N/60)$$

where N is the speed in r.p.m.

2. The ratio of power output to power input is known as efficiency. It is denoted by a Greek letter eta (η). It is always less than unity and is represented as percentage. Mathematically,

$$\text{Efficiency,} \eta = \frac{\text{Power output}}{\text{Power input}}$$

* Heat is not a thermodynamic property whereas the temperature and pressure are thermodynamic properties.

** The area under the pressure - volume (p-v) diagram represents the work done during the process and is given by $p\,dv$.

1.48. Laws of Thermodynamics

The following three laws of thermoaynamics are important from the subject point of view :

1. Zeroth law of thermodynamics, 2. First law of thermodynamics, and 3. Second law of thermodynamics.

These laws are discussed, in detail, as follows :

1.49. Zeroth Law of Thermodynamics

This law states, "*When two systems are each in thermal equilibrium with a third system, then the two systems are also in thermal equilibrium with one another.*"

This law provides the basis of temperature measurement.

1.50. First Law of Thermodynamics

This law may be stated as follows :

(*a*) "*The heat and mechanical work are mutually convertible*". According to this law, when a closed system undergoes a thermodynamic cycle, the net heat transfer is equal to the net work transfer. In other words, the cyclic integral of heat transfers is equal to the cyclic integral of work transfers. Mathematically,

$$\oint \delta Q = \oint \delta W$$

where symbol $\oint$ stands for cyclic integral (integral around a complete cycle), and δQ and δW represent infinitesimal elements of heat and work transfers respectively. It may be noted that δQ and δW are expressed in same units.

(*b*) *The energy can neither be created nor destroyed though it can be transformed from one form to another.* According to this law, when a system undergoes a change of state (or a thermodynamic process), then both heat transfer and work transfer takes place. The net energy transfer is stored within the system and is known as stored energy or total energy of the system. Mathematically

$$\delta Q - \delta W = dE$$

The symbol δ is used for a quantity which is inexact differential and symbol d is used for a quantity which is an exact differential. The quantity E is an extensive property and represents the total energy of the system at a particular state.

On integrating the above expression for a change of state from 1 to 2, we have

$$Q_{1-2} - W_{1-2} = E_2 - E_1 \quad \text{. . . (}Q, W \text{ and } E \text{ are in same units)}$$

For a unit mass, this expression is written as

$$q_{1-2} - w_{1-2} = e_2 - e_1$$

where Q_{1-2} = Heat transferred to the system during the process from state 1 to state 2,

W_{1-2} = Workdone by the system on the surroundings during the process, and

E_1 = Total energy of the system at state 1

$$= PE_1 + KE_1 + U_1 = {}^*m\,g\,z_1 + \frac{m\,V_1^2}{2} + U_1$$

E_2 = Total energy of the system at state 2

$$= PE_2 + KE_2 + U_2 = m\,g\,z_2 + \frac{m\,V_2^2}{2} + U_2$$

---

* Refer Art. 1.37

Thus the above expression may be written as

$$Q_{1-2} - W_{1-2} = E_2 - E_1 \qquad \ldots (i)$$

$$= (PE_2 + KE_2 + U_2) - (PE_1 + KE_1 + U_1)$$

$$= (PE_2 - PE_1) + (KE_2 - KE_1) + (U_2 - U_1) \qquad \ldots (ii)$$

$$= m\,(g\,z_2 - g\,z_1) + m\left(\frac{V_2^2}{2} - \frac{V_1^2}{2}\right) + (U_2 - U_1)$$

For unit mass, this expression is written as

$$q_{1-2} - w_{1-2} = (g\,z_2 - g\,z_1) + \left(\frac{V_2^2}{2} - \frac{V_1^2}{2}\right) + (u_2 - u_1)$$

Notes. 1. When there is no change in potential energy of the system (*i.e.* when the height of the system from the datum level is same), then $PE_1 = PE_2$. Thus, the above equation (*ii*) is written as

$$Q_{1-2} - W_{1-2} = (KE_2 - KE_1) + (U_2 - U_1) \qquad \ldots (iii)$$

2. When there is no change of PE and also there is no flow of the mass into or out of the system, then $PE_1 = PE_2$ and $KE_1 = KE_2$. Thus, the above equation (*ii*) is written as

$$Q_{1-2} - W_{1-2} = U_2 - U_1 = dU \qquad \ldots (iv)$$

In other words, in a closed or non-flow thermodynamic system,

$$PE = 0 \quad \text{and} \quad KE = 0$$

Thus the equation (*iv*) is known as *Non-flow energy equation.*

3. For an isolated system for which $Q_{1-2} = W_{1-2} = 0$, the above equation (*i*) becomes

$$E_2 = E_1$$

This shows that the first law of thermodynamics is the law of conservation of energy.

1.51. Limitations of First Law of Thermodynamics

We have already discussed that according to first law of thermodynamics that

1. *When a closed system undergoes a thermodynamic cycle, the net heat transfer is equal to the net work transfer.* This statement does not specify the direction of flow of heat and work (*i.e.* whether the heat flows from a hot body to a cold body or from a cold body to a hot body). It also does not give any condition under which these transfers take place.

2. *The heat energy and mechanical work are mutually convertible.* Though the mechanical work can be fully converted into heat energy, but only a part of heat energy can be converted into mechanical work. This means that the heat energy and mechanical work are not fully mutually convertible. In other words, there is a limitation on the conversion of one form of energy into another form.

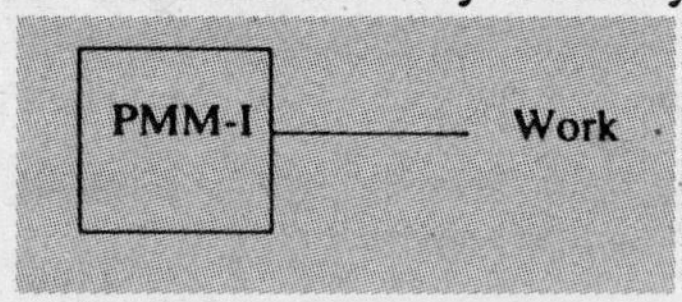

Fig. 1.10. Perpetual motion machine of the first kind.

A machine which violates the first law of thermodynamics (*i.e.* energy can neither be created nor destroyed, but can be transformed from one form to another) is known as *perpetual motion machine of the first kind* (briefly written as PMM-I). It is defined as a machine which produces work energy without consuming an equivalent of energy from other source. Such a machine, as shown in Fig. 1.10, is impossible to obtain in actual practice, because no machine can produce energy of its own without consuming any other form of energy.

1.52. Second Law of Thermodynamics

The second law of thermodynamics may be defined in many ways, but the two common statements according to Kelvin - Planck and Clausius are as follows :

1. *Kelvin - Planck Statement.* According to Kelvin-Planck '*It is impossible to construct an engine working on a cyclic process, whose sole purpose is to convert heat energy from a single thermal reservoir* into an equivalent amount of work*'. In other words, no actual heat engine, working on a cyclic process, can convert whole of the heat supplied to it, into mechanical work. It means that there is a degradation of energy in the process of producing mechanical work from the heat supplied. Thus the Kelvin - Planck statement of the second law of thermodynamics, is sometimes known as *law of degradation of energy*.

A heat engine which violates this statement of the second law of thermodynamics (*i.e.* a heat engine which converts whole of the heat energy into mechanical work) is known as ***perpetual motion machine of the second kind* (briefly written as PMM-II) or 100 percent efficient machine which is impossible to obtain in actual practice, because no machine can convert whole of the heat energy supplied to it, into its equivalent amount of work.

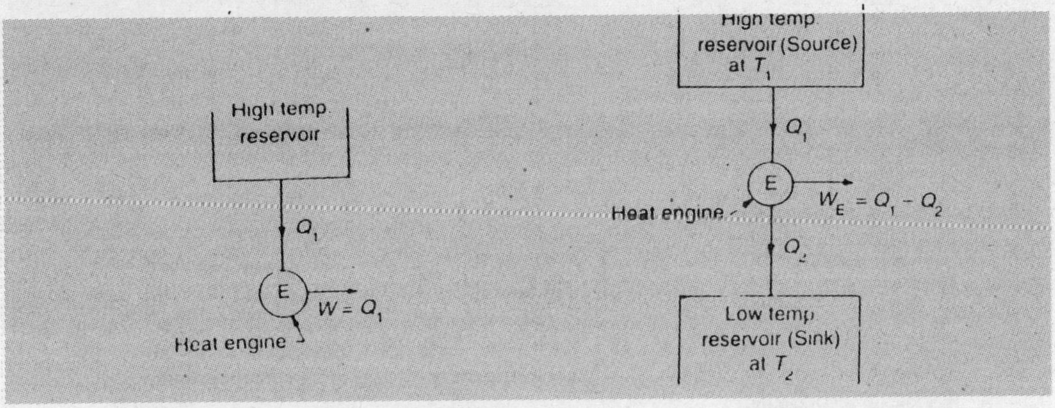

(*a*) Perpetual motion machine of the second kind (impossible).

(*b*) Heat engine.

Fig. 1.11

Thus for the satisfactory operation of a heat engine which is a device used for converting heat energy into mechanical work, there should be at-least two reservoirs of heat, one at a ***higher temperature and the other at a lower temperature, as shown in Fig. 1.11 (*b*). In this case, consider that heat energy (Q_1) from the high temperature reservoir (or source) at temperature T_1 is supplied to the engine. A part of this heat energy is rejected to the low temperature reservoir (or sink) at temperature T_2. If Q_2 is the heat rejected to the sink, then the remaining heat (*i.e.* $Q_1 - Q_2$) is converted into mechanical work. The ratio of the maximum mechanical work obtained to the total heat supplied to the engine is known as maximum thermal efficiency (η_{max}) of the engine. Mathematically,

$$\eta_{max} = \frac{\text{Maximum work obtained}}{\text{Total heat supplied}} = \frac{Q_1 - Q_2}{Q_1} = 1 - \frac{Q_2}{Q_1} = 1 - \frac{T_2}{T_1}$$

Note. For a reversible engine, $Q_1 / T_1 = Q_2 / T_2$.

---

* A thermal reservoir is a body of infinite heat capacity which is capable of absorbing or rejecting an unlimited quantity of heat without affecting its temperature.

** A perpetual motion machine of the second kind (PMM-II) does not violate the first law of thermodynamics as such a machine would not create or destroy energy.

*** In a heat engine, the reservoir (or body) at a higher temperature is known as a *source* and the reservoir at a lower temperature is called a *sink*.

2. *Clausius Statement.* According to Clausius statement "*It is impossible for a self acting machine, working in a cyclic process, to transfer heat from a body at a lower temperature to a body at a higher temperature without the aid of an external agency.*" In other words, heat cannot flow itself from a cold body to a hot body without the help of an external agency (*i.e.* without the expenditure of mechanical work).

The device (such as a refrigerator or a heat pump), as shown in Fig. 1.12 (*a*), violates the Clausius statement because no input work is supplied to the device to transfer heat from a cold body to a hot body. Such a device is called perpetual motion machine of the second kind.

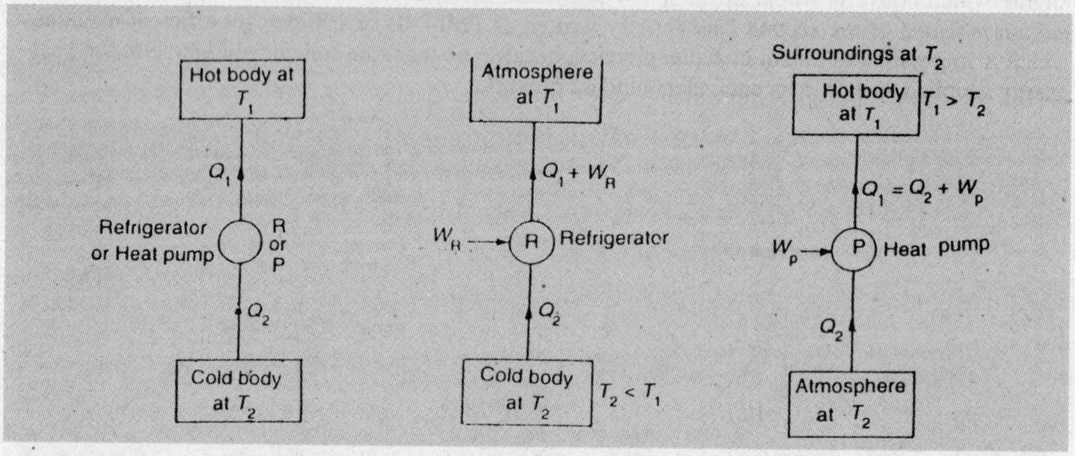

(*a*) Perpetual motion machine of the second kind. (*b*) Refrigerator. (*c*) Heat pump.

Fig. 1.12

In order to achieve the object of transferring heat from a cold body to a hot body, the refrigerator and a heat pump, while operating in a cyclic process, require an input work, as shown in Fig. 1.12 (*b*) and (*c*) respectively. Though there is no difference between the cycle of operations of the refrigerator and a heat pump and achieve the same overall objective, but the basic purpose of each is quite different. A refrigerator is a device which operating in a cyclic process, maintains the temperature of a cold body (refrigerated space) at a temperature lower than the temperature of the surroundings. On the other hand, a heat pump is a device which operating in a cyclic process, maintains the temperature of a hot body (heated space) at a temperature higher than the temperature of surroundings. In other words, a refrigerator works between the cold body temperature and the *atmospheric temperature whereas a heat pump operates between the hot body temperature and the atmospheric temperature.

The **performance of refrigerator and heat pump is measured in terms of coefficient of performance which is defined as the ratio of the maximum heat transferred (*i.e.* heat taken from the cold body) to the amount of work required to produce the desired effect. Mathematically, maximum coefficient of performance for a refrigerator,

$$(\text{C.O.P})_R = \frac{Q_2}{W_R} = \frac{Q_2}{Q_1 - Q_2} = \frac{T_2}{T_1 - T_2}$$

* In case of a refrigerator, the atmosphere acts as a hot body while in case of a heat pump, the atmosphere acts as a cold body.

** The performance of a heat engine is measured in terms of thermal efficiency.

and maximum coefficient of performance for a heat pump,

$$(\text{C.O.P})_P = \frac{Q_1}{W_P} = \frac{Q_1}{Q_1 - Q_2} = \frac{T_1}{T_1 - T_2} = \frac{T_2}{T_1 - T_2} + 1$$

$$= (\text{C.O.P})_R + 1$$

We see that C.O.P of a heat pump is greater than C.O.P of a refrigerator by unity.

1.53. Equivalence of Kelvin-Planck and Clausius Statements

Though Kelvin-Planck and Clausius statements of the second law of thermodynamics appear to be different, from each other, but these two statements are virtually equivalent in all respects. The equivalence of the Kelvin-Planck and Clausius statements can be proved if it can be shown that the violation of Kelvin-Planck statement implies the violation of Clausius statement and vice versa. This is discussed as follows :

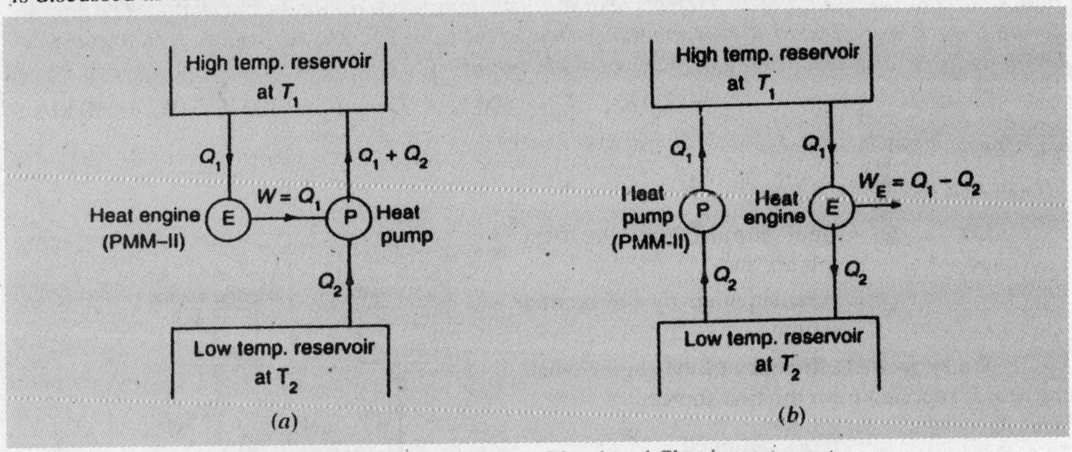

Fig. 1.13. Equivalence of Kelvin-Planck and Clausius statements.

1. Consider a system as shown in Fig. 1.13 (*a*). In this system, a heat engine having 100 percent thermal efficiency (*i.e.* PMM-II) is violating the Kelvin-Planck statement as it converts the heat energy (Q_1) from a single high temperature reservoir at T_1, into an equivalent amount of work (*i.e.* $W = Q_1$). This work output of the heat engine can be used to drive a heat pump (or refrigerator) which receives an amount of heat Q_2 from a low temperature reservoir at T_2 and rejects an amount of heat $(Q_1 + Q_2)$ to a high temperature reservoir at T_1. If the combination of a heat engine and a heat pump (or refrigerator) is considered as a single system, as shown in Fig. 1.13 (*a*), then the result is a device that operates in a cycle and has no effect on the surroundings other than the transfer of heat Q_2 from a low temperature reservoir to a high temperature reservoir, thus violating the Clausius statement. Hence, a violation of Kelvin-Planck statement leads to a violation of Clausius statement.

2. Consider a system as shown in Fig. 1.13 (*b*). In this system, a heat pump or refrigerator (*i.e.* PMM-II) is violating the Clausius statement as it transfers heat from a low temperature reservoir at T_2 to a high temperature reservoir at T_1 without any expenditure of work. Now let a heat engine, operating between the same heat reservoirs, receives an amount of heat Q_1 (as discharged by the heat pump) from the high temperature reservoir at T_1, does work ($W_E = Q_1 - Q_2$) and rejects an amount of heat Q_2 to the low temperature reservoir at T_2. If the combination of the heat pump (or refrigerator) and the heat engine is considered as a single system, as shown in Fig. 1.13 (*b*), then the result is a device that operates in a cycle whose sole effect is to remove heat at the rate of $(Q_1 - Q_2)$ and convert it completely into an equivalent amount of work, thus violating the Kelvin-Planck statement. Hence, a violation of Clausius statement leads to a violation of Kelvin-Planck statement.

From above, we see that the Kelvin-Planck and Clausius statements of the second law of thermodynamics are complimentary to each other. The truth of the first statement implies the truth of the second statement and vice versa.

Example 1.5. *An engine works between the temperature limits of 1775 K and 375 K. What can be the maximum thermal efficiency of this engine ?*

Solution. Given : $T_1 = 1775$ K ; $T_2 = 375$ K

We know that maximum thermal efficiency of the engine,

$$\eta_{max} = \frac{T_1 - T_2}{T_1} = \frac{1775 - 375}{1775} = 0.7887 \text{ or } 78.87\% \textbf{ Ans.}$$

Example 1.6. *A reversible engine is supplied with heat from two constant temperature sources at 900 K and 600 K and rejects heat to a constant temperature sink at 300 K. The engine develops work equivalent to 90 kJ/s and rejects heat at the rate of 56 kJ/s. Estimate : 1. Heat supplied by each source, and 2. Thermal efficiency of the engine.*

Solution. Given : $T_1 = 900$ K ; $T_3 = 600$ K ; $T_2 = T_4 = 300$ K ; $W_E = 90$ kJ/s ; $Q_2 + Q_4 = 56$ kJ/s

1. *Heat supplied by each source*

Let Q_1 = Heat supplied by the first source, and

Q_3 = Heat supplied by the second source.

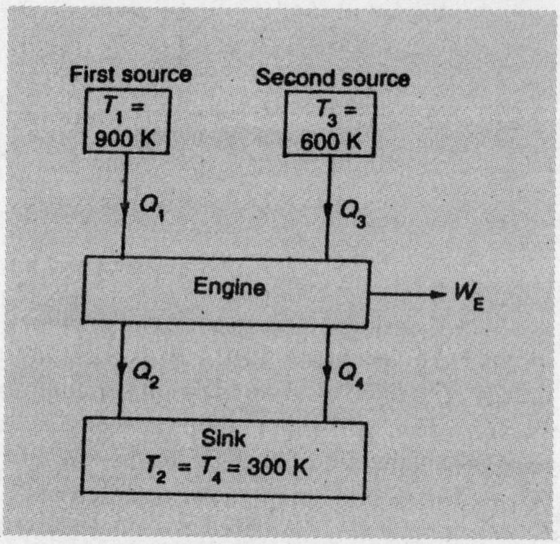

Fig. 1.14

We know that efficiency of the engine when the heat is supplied from the first source,

$$\eta_1 = \frac{\text{Work obtained}}{\text{Heat supplied}} = \frac{W_1}{Q_1}$$

$$= \frac{Q_1 - Q_2}{Q_1} = \frac{T_1 - T_2}{T_1}$$

$$\ldots \left(\because \text{ For a reversible engine, } \frac{Q_1}{T_1} = \frac{Q_2}{T_2}\right)$$

$$= \frac{900 - 300}{900} = 0.67$$

∴ Work obtained by the engine from the first source,

$$W_1 = Q_1 - Q_2 = 0.67\, Q_1$$

and heat rejected to the sink,

$$Q_2 = Q_1 - W_1 = Q_1 - 0.67\, Q_1 = 0.33\, Q_1$$

Similarly, efficiency of the engine when the heat is supplied from the second source

$$\eta_2 = \frac{W_2}{Q_3} = \frac{Q_3 - Q_4}{Q_3} = \frac{T_3 - T_4}{T_3} = \frac{600 - 300}{600} = 0.5$$

∴ Work obtained by the engine from the second source,

$$W_2 = Q_3 - Q_4 = 0.5\, Q_3$$

and heat rejected to the sink,

$$Q_4 = Q_3 - W_2 = Q_3 - 0.5\,Q_3 = 0.5\,Q_3$$

We know that total work obtained from the engine (W_E),

$$90 = W_1 + W_2 = 0.67\,Q_1 + 0.5\,Q_3 \qquad \ldots (i)$$

and total heat rejected to the sink,

$$56 = Q_2 + Q_4 = 0.33\,Q_1 + 0.5\,Q_3 \qquad \ldots (ii)$$

From equations (*i*) and (*ii*),

$$Q_1 = 100 \text{ kJ/s and } Q_3 = 46 \text{ kJ/s} \quad \textbf{Ans.}$$

2. *Thermal efficiency of the engine*

We know that maximum thermal efficiency of the engine,

$$\eta_{max} = \frac{\text{Work obtained}}{\text{Heat supplied}} = \frac{W_E}{Q_1 + Q_3}$$

$$= \frac{90}{100 + 46} = 0.616 \text{ or } 61.6\,\% \quad \textbf{Ans.}$$

Example 1.7. *A cold storage is to be maintained at – 5°C while the surroundings are at 35°C. The heat leakage from the surroundings into the cold storage is estimated to be 29 kW. The actual C.O.P of the refrigeration plant is one -third of an ideal plant working between the same temperatures. Find the power required to drive the plant.*

Solution. Given : T_2 = – 5°C = – 5 + 273 = 268 K ; T_1 = 35°C = 35 + 273 = 308 K

$Q_2 = 29$ kW ; $(\text{C.O.P})_{actual} = \frac{1}{3}(\text{C.O.P})_{ideal}$

The refrigerating plant operating between the temperatures T_1 and T_2 is shown in Fig. 1.15.

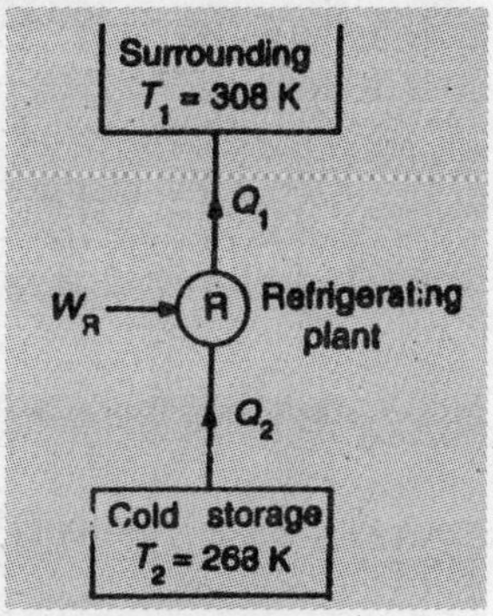

Fig. 1.15

Let W_R = Work or power required to drive the plant.

We know that the coefficient of performance of an ideal refrigeration plant,

$$(\text{C.O.P})_{ideal} = \frac{T_2}{T_1 - T_2}$$

$$= \frac{268}{308 - 268} = 6.7$$

∴ Actual coefficient of performance,

$$(\text{C.O.P})_{actual} = \frac{1}{3} \times (\text{C.O.P})_{ideal} = \frac{1}{3} \times 6.7 = 2.233$$

We also know that $\text{C.O.P.} = \dfrac{Q_2}{W_R}$

$$\therefore \quad W_R = \frac{Q_2}{(\text{C.O.P})_{actual}} = \frac{29}{2.233} = 12.987 \text{ kW} \quad \textbf{Ans.}$$

Example 1.8. *A reversible heat engine operates between two reservoirs at temperatures of 600°C and 40°C. The engine drives a reversible refrigerator which operates between reservoirs at temperatures of 40°C and – 20°C. The heat transfer to the engine is 2 MJ and the net work output of the combined engine and refrigerator plant is 360 kJ. Find the heat transfer to the refrigerant and*

the net heat transfer to the reservoir at 40°C. Also find these values if the efficiency of the heat engine and C.O.P of the refrigerator are each 40% of their maximum possible values.

Solution. Given : $T_1 = 600°C = 600 + 273 = 873$ K ; $T_2 = T_4 = 40°C = 40 + 273 = 313$ K ; $T_3 = -20°C = -20 + 273 = 253$ K ; $Q_1 = 2$ MJ $= 2000$ kJ ; $W = 360$ kJ

The combined heat engine and refrigerator system is shown in Fig. 1.16.

Heat transfer to the refrigerant

Let Q_3 = Heat transfer to the refrigerant.

We know that maximum efficiency of the heat engine,

$$\eta_{max} = 1 - \frac{T_2}{T_1} = 1 - \frac{313}{873} = 0.6415$$

We also know that

$$\eta_{max} = \frac{\text{Heat supplied} - \text{Heat rejected}}{\text{Heat supplied}}$$

$$= \frac{\text{Work done}}{\text{Heat supplied}} = \frac{Q_1 - Q_2}{Q_1}$$

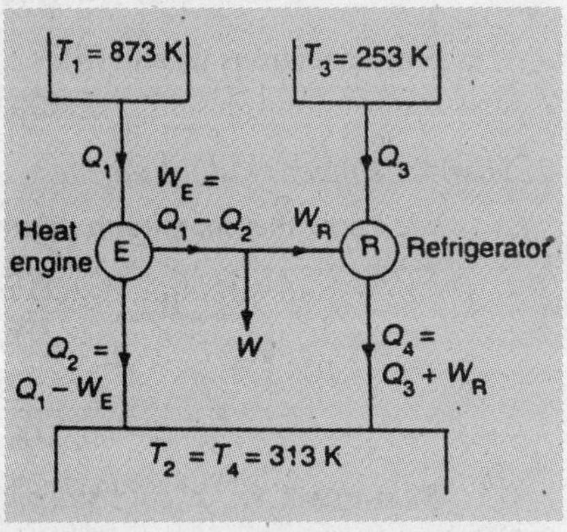

Fig. 1.16

∴ Work done by the heat engine,

$$W_E = Q_1 - Q_2 = \eta_{max} \times Q_1 = 0.6415 \times 2000 = 1283 \text{ kJ}$$

Since the net work output of the combined heat engine and refrigerator plant is $W = W_E - W_R = 360$ kJ, therefore work required for the refrigerator,

$$W_R = W_E - W = 1283 - 360 = 923 \text{ kJ}$$

We know that maximum C.O.P of the refrigerator,

$$(\text{C.O.P})_{max} = \frac{T_3}{T_4 - T_3} = \frac{253}{313 - 253} = 4.217$$

We also know that maximum C.O.P. of the refrigerator,

$$(\text{C.O.P})_{max} = \frac{Q_3}{Q_4 - Q_3} = \frac{Q_3}{W_R}$$

$$\therefore \quad Q_3 = (\text{C.O.P.})_{max} \times W_R = 4.217 \times 923 = 3892.3 \text{ kJ} \quad \textbf{Ans.}$$

Net heat transfer to the reservoir at 40°C

We know that $Q_4 = Q_3 + W_R = 3892.3 + 923 = 4815.3$ kJ

and $Q_2 = Q_1 - W_E = 2000 - 1283 = 717$ kJ

∴ Net heat transfer (*i.e.*, heat rejected) to the reservoir at 40°C

$$= Q_2 + Q_4 = 717 + 4815.3 = 5532.3 \text{ kJ} \quad \textbf{Ans.}$$

When efficiency of the heat engine and C.O.P. of the refrigerator are each 40% of their maximum possible values

We know that the efficiency of the actual heat engine cycle,

$$\eta_{actual} = 40\% \ \eta_{max} = 0.4 \ \eta_{max} = 0.4 \times 0.6415 = 0.2566$$

$$\therefore \quad W_E = \eta_{actual} \times Q_1 = 0.2566 \times 2000 = 513.2 \text{ kJ}$$

$$W_R = W_E - W = 513.2 - 360 = 153.2 \text{ kJ}$$

We know that C.O.P. of the actual refrigerator cycle,

$$(\text{C.O.P})_{actual} = 40\% \, (\text{C.O.P})_{max} = 0.4 \, (\text{C.O.P})_{max}$$

$$= 0.4 \times 4.217 = 1.6868$$

∴ Heat transfer to the refrigerant,

$$Q_3 = (\text{C.O.P})_{actual} \times W_R$$

$$= 1.6868 \times 153.2 = 258.4 \text{ kJ} \quad \textbf{Ans.}$$

We know that $\quad Q_4 = Q_3 + W_R = 258.4 + 153.2 = 411.6 \text{ kJ}$

and $\quad Q_2 = Q_1 - W_E = 2000 - 513.2 = 1486.8 \text{ kJ}$

∴ Net heat transfer (*i.e.*, heat rejected) to the reservoir at 40°C

$$= Q_2 + Q_4 = 1486.8 + 411.6 = 1898.4 \text{ kJ} \quad \textbf{Ans.}$$

EXERCISES

1. The pressure of steam inside a boiler is recorded by a pressure gauge which shows 1.2 N/mm^2. If the barometer reads the atmospheric pressure as 770 mm of mercury, find the absolute pressure of steam inside the boiler in N/m^2, kPa and bar. [**Ans.** 1.3026×10^6 N/m^2 ; 1302.6 kPa ; 13.026 bar]

2. In a condenser, the vacuum is found to be 145 mm of mercury and the barometer reads 735 mm of mercury. Find the absolute pressure in a condenser in N/m^2 ; kPa and N/mm^2. [**Ans.** 78 647 N/m^2 ; 78.647 kPa ; 0.078 647 N/mm^2]

3. A copper vessel of mass 1.35 kg contains 6.75 kg of water at a temperature of 25°C. Find the heat required to warm the vessel and water to 90°C. Take specific heat of copper = 0.406 kJ/kg K and specific heat of water = 4.187 kJ/kg K. [**Ans.** 1872.6 kJ]

4. The net work output of a cyclic process is 45 kN-m. If the heat input is 125 kJ, determine the efficiency of the cycle. [**Ans.** 36%]

5. One kg of air at a temperature of 20°C is heated to a temperature of 60°C. Find the heat supplied to air when heated at constant pressure. The specific heat for air at constant pressure = 1 kJ/kg K. [**Ans.** 40 kJ]

6. A system receives 10×10^6 J in the form of heat energy in a specified process and it produces work of 4×10^6 J. The system velocity changes from 10 m/s to 25 m/s. For 50 kg mass of the system, determine the change in internal energy of the system. [**Ans.** 119.7375 kJ/kg]

[**Hint:** $q_{1-2} = \dfrac{10 \times 10^6}{50}$ J/kg ; $w_{1-2} = \dfrac{4 \times 10^6}{50}$ J/kg

$$ke_1 = \frac{1}{2} \times m \, (V_1)^2 = \frac{1}{2} \times 1 \, (10)^2 = 50 \text{ J/kg}$$

$$ke_2 = \frac{1}{2} \times m \, (V_2)^2 = \frac{1}{2} \times 1 \, (25)^2 = 312.5 \text{ J/kg}$$

We know that

$$q_{1-2} - w_{1-2} = (pe_2 - pe_1) + (ke_2 - ke_1) + (u_2 - u_1)$$

$$\frac{10 \times 10^6}{50} - \frac{4 \times 10^6}{50} = 0 + (312.5 - 50) + (u_2 - u_1)$$

. . . (Taking same datum level, $pe_2 = pe_1$)

$$\therefore \quad u_2 - u_1 = 119\,737.5 \text{ J/kg} = 119.7375 \text{kJ/kg}$$

7. A reversible engine receives heat from a reservoir at 700°C and rejects heat at temperature T_2. A second reversible engine receives the heat rejected by the first engine to a sink at a temperature 37°C. Calculate

the temperature T_2 for 1. equal efficiency of both the engines, and 2. equal output of both the engines. [Ans. 276.2°C ; 368.5°C]

8. A domestic food freezer is to be maintained at temperature of – 15°C. The ambient air temperature is 30°C. If the heat leaks into the freezer at the continuous rate of 1.75 kJ/s, find the power required to pump this heat out continuously. [Ans. 0.305 kJ/s]

9. A heat pump is used for heating the interior of a house in a cold climate. The ambient temperature is – 5°C and the desired interior temperature is 25°C. The compressor of the heat pump is to be driven by a heat engine working between 1000°C and 25°C. Treating both the cycles as reversible, calculate the ratio in which the heat pump and the heat engine share the heating load. [Ans. 7.606]

10. A heat engine is used to drive a heat pump. The heat transfer from the heat engine and from the heat pump are used to heat the water circulating through the radiators of a building. The efficiency of the heat engine is 27% and C.O.P of the heat pump is 4. Show that the ratio of the heat transfer to the circulating water to the heat transfer to the engine is 1.81.

QUESTIONS

1. Define a thermodynamic system. Explain its different types.
2. What do you understand by property of a system ? Distinguish between extensive and intensive properties of a system.
3. Define the following properties :
   (*a*) Specific weight (*b*) Pressure (*c*) Volume
   (*d*) Temperature (*e*) Specific volume (*f*) Density
4. What is a thermodynamic process and a cyclic process ?
5. Explain the non-equilibrium and quasi-static process. Is the quasi-static process a reversible process ?
6. Define temperature. Name the different temperature scales in common use. Establish relation between Celsius and Fahrenheit scales ?
7. What is absolute temperature ? How it is obtained for Celsius and Fahrenheit scales ?
8. Distinguish between gauge pressure and absolute pressure. How the gauge pressure is converted into absolute pressure ?
9. What do you understand by N.T.P. and S.T.P. ? What are their values ?
10. Define energy. What is stored energy and transit energy ? Discuss the types of stored energy.
11. How heat and work is defined ? Are these quantities a path function or point function ?
12. Explain the three laws of thermodynamics.

OBJECTIVE TYPE QUESTIONS

1. A definite area or a space where some thermodynamic process takes place, is known as
   (*a*) thermodynamic cycle (*b*) thermodynamic process
   (*c*) thermodynamic system (*d*) thermodynamic law
2. When neither mass nor energy is allowed to cross the boundary of a system, it is then called
   (*a*) closed system (*b*) open system
   (*c*) isolated system (*d*) none of these
3. Which of the following is the extensive property of a thermodynamic system ?
   (*a*) pressure (*b*) volume
   (*c*) temperature (*d*) density

4. Which of the following is an intensive property of a thermodynamic system ?

(*a*) volume (*b*) temperature

(*c*) mass (*d*) energy

5. Which of the following is not a thermodynamic property ?

(*a*) pressure (*b*) temperature

(*c*) heat (*d*) specific volume

6. When a process or processes are performed on a system in such a way that the final state is identical with the initial state, it is then known as

(*a*) thermodynamic cycle (*b*) thermodynamic property

(*c*) thermodynamic process (*d*) zeroth law of thermodynamics

7. Atmospheric pressure is equal to

(*a*) 1.013 bar (*b*) 101.3 kN/m^2

(*c*) 760 mm of Hg (*d*) all of these

8. First law of thermodynamics deals with

(*a*) conservation of heat (*b*) conservation of momentum

(*c*) conservation of mass (*d*) conservation of energy

9. Second law of thermodynamics defines

(*a*) heat (*b*) work

(*c*) entropy (*d*) internal energy

10. Kelvin-Planck's law deals with

(*a*) conservation of work (*b*) conservation of heat

(*c*) conservation of mass (*d*) conversion of heat into work

ANSWERS

1. (*c*)	2. (*c*)	3. (*b*)	4. (*b*)	5. (*c*)
6. (*a*)	7. (*d*)	8. (*d*)	9. (*c*)	10. (*d*)

2

Properties of Perfect Gases

1. Introduction. 2. Laws of Perfect Gases. 3. Boyle's Law. 4. Charles' Law. 5. Gay-Lussac Law. 6. General Gas Equation. 7. Joule's Law. 8. Characteristic Equation of Gas. 9. Avogadro's Law. 10. Universal Gas Constant or Molar Constant. 11. Specific Heat of a Gas. 12. Specific Heat at Constant Volume. 13. Specific Heat at Constant Pressure. 14. Enthalpy of a Gas. 15. Molar Specific Heats of a Gas. 16. Regnaults' Law. 17. Relation between Specific Heats. 18. Ratio of Specific Heats.

2.1. Introduction

A perfect gas (or an ideal gas) may be defined as a state of a substance, whose evaporation from its liquid state is complete*, and strictly obeys all the gas laws under all conditions of temperature and pressure. In actual practice, there is no real or actual gas which strictly obeys the gas laws over the entire range of temperature and pressure. But, the real gases which are ordinarily difficult to liquify, such as oxygen, nitrogen, hydrogen and air, within certain temperature and pressure limits, may be regarded as perfect gases.

2.2. Laws of Perfect Gases

The physical properties of a gas are controlled by the following three variables :

1. Pressure exerted by the gas, 2. Volume occupied by the gas, and 3. Temperature of the gas.

The behaviour of a perfect gas, undergoing any change in the above mentioned variables, is governed by the following laws which have been established from experimental results.

1. Boyle's law, 2. Charles' law, and 3. Gay-Lussac law.

These laws are discussed, in detail, in the following pages.

2.3. Boyle's Law

This law was formulated by Robert Boyle in 1662. It states, *"The absolute pressure of a given mass of a perfect gas varies inversely as its volume, when the temperature remains constant."* Mathematically,

$$p \propto \frac{1}{v} \quad \text{or} \quad p\,v = \text{Constant}$$

The more useful form of the above equation is :

$$p_1 v_1 = p_2 v_2 = p_3 v_3 = \ldots\ldots = \text{Constant}$$

where suffixes $_1$, $_2$ and $_3$ .... refer to different sets of conditions.

---

* If its evaporation is partial, the substance is called vapour. A vapour, therefore, contains some particles of liquid in suspension, It is thus obvious, that steam, carbon dioxide, sulphur dioxide and ammonia are regarded as vapours. It may be noted that a vapour becomes dry, when it is completely evaporated. If the dry vapour is further heated, the process is called *super heating* and the vapour is called *superheated vapour*. The behaviour of superheated vapour is similar to that of a perfect gas.

2.4. Charles' Law

This law was formulated by a Frenchman Jacques A.C. Charles in about 1787. It may be stated in the following two different forms :

(*i*) *"The volume of a given mass of a perfect gas varies directly as its absolute temperature, when the absolute pressure remains constant."* Mathematically,

$$v \propto T \quad \text{or} \quad \frac{v}{T} = \text{Constant}$$

or
$$\frac{v_1}{T_1} = \frac{v_2}{T_2} = \frac{v_3}{T_3} = ... = \text{Constant}$$

where suffixes $_1$, $_2$ and $_3$ ..... refer to different sets of conditions.

(*ii*) *"All perfect gases change in volume by 1/273th of its original volume at 0° C for every 1° C change in temperature, when the pressure remains constant."*

Let v_0 = Volume of a given mass of gas at 0° C, and

v_t = Volume of the same mass of gas at t° C.

Then, according to the above statement,

$$v_t = v_0 + \frac{1}{273} v_0 t = v_0 \left(\frac{273 + t}{273} \right) = v_0 \times \frac{T}{T_0}$$

or
$$\frac{v_t}{T} = \frac{v_0}{T_0}$$

where T = Absolute temperature corresponding to t° C.

T_0 = Absolute temperature corresponding to 0° C.

A little consideration will show, that the volume of a gas goes on decreasing by 1/273th of its original volume for every 1° C decrease in temperature. It is thus obvious, that at a temperature of – 273° C, the volume of the gas would become *zero. The temperature at which the volume of a gas becomes zero is called *absolute zero temperature*.

Note. In all calculations of a perfect gas, the pressure and temperature values are expressed in absolute units.

2.5. Gay-Lussac Law

This law states, *"The absolute pressure of a given mass of a perfect gas varies directly as its absolute temperature, when the volume remains constant."* Mathematically

$$p \propto T \text{ or } \frac{p}{T} = \text{Constant}$$

or
$$\frac{p_1}{T_1} = \frac{p_2}{T_2} = \frac{p_3}{T_3} = ... = \text{Constant}$$

where suffixes $_1$, $_2$ and $_3$ ... refer to different sets of conditions.

2.6. General Gas Equation

In the previous section we have discussed the gas laws which give us the relation between the two variables when the third variable is constant. But in actual practice, all the three variables *i.e.*, pressure, volume and temperature, change simultaneously. In order to deal with all practical cases, the Boyle's law and Charles' law are combined together, which give us a ***general gas equation***.

---

* It is only theoretical. Its exact value is – 273.16 °C. But for all practical purposes, this value is taken as – 273 °C..

According to Boyle's law

$$p \propto \frac{1}{v} \quad \text{or} \quad v \propto \frac{1}{p} \qquad \text{... (Keeping } T \text{ constant)}$$

and according to Charles' law

$$v \propto T \qquad \text{... (Keeping } p \text{ constant)}$$

It is thus obvious that

$$v \propto \frac{1}{p} \text{ and } T \text{ both} \quad \text{or } v \propto \frac{T}{p}$$

$$\therefore \quad p\,v \propto T \quad \text{or} \quad p\,v = C\,T$$

where C is a constant, whose value depends upon the mass and properties of the gas concerned.

The more useful form of the general gas equation is :

$$\frac{p_1 v_1}{T_1} = \frac{p_2 v_2}{T_2} = \frac{p_3 v_3}{T_3} = \ldots = \text{Constant}$$

where suffixes $_{1}$, $_{2}$ and $_{3}$ refer to different sets of conditions.

Example 2.1. *A gas occupies a volume of 0.1 m³ at a temperature of 20° C and a pressure of 1.5 bar. Find the final temperature of the gas, if it is compressed to a pressure of 7.5 bar and occupies a volume of 0.04 m³.*

Solution. Given : $v_1 = 0.1 \text{ m}^3$; $T_1 = 20° \text{C} = 20 + 273 = 293 \text{ K}$; $p_1 = 1.5 \text{ bar} = {}^*0.15 \times 10^6 \text{ N/m}^2$; $p_2 = 7.5 \text{ bar} = 0.75 \times 10^6 \text{ N/m}^2$; $v_2 = 0.04 \text{ m}^3$

Let T_2 = Final temperature of the gas.

We know that

$$\frac{p_1 v_1}{T_1} = \frac{p_2 v_2}{T_2}$$

$$\therefore \quad T_2 = \frac{p_2 v_2 T_1}{p_1 v_1} = \frac{0.75 \times 10^6 \times 0.04 \times 293}{0.15 \times 10^6 \times 0.1} = 586 \text{ K}$$

$$= 586 - 273 = 313° \text{ C} \quad \textbf{Ans.}$$

2.7. Joule's Law

It states, "*The change of internal energy of a perfect gas is directly proportional to the change of temperature.*" Mathematically

$$dE \propto dT \quad \text{or} \quad dE = m\,c\,dT = m\,c\,(T_2 - T_1)$$

where m = Mass of the gas, and

c = A constant of proportionality, known as specific heat.

An important consequence of this law is that if the temperature of a given mass m of a gas changes from T_1 to T_2, then the internal energy will change from E_1 to E_2 and the change in internal energy $(E_2 - E_1)$ will be same irrespective of the manner how the pressure (p) and volume (v) of the gas have changed.

---

* 1 bar = 0.1×10^6 N/m².

2.8. Characteristic Equation of a Gas

It is a modified form of general gas equation. If the volume (v) in the general gas equation is taken as that of 1 kg of gas (known as its specific volume, and denoted by v_s), then the constant C (in the general gas equation) is represented by another constant R (in the characteristic equation of gas). Thus the general gas equation may be rewritten as :

$$pv_s = RT$$

where R is known as *characteristic gas constant* or simply *gas constant.*

For any mass m kg of a gas, the characteristic gas equation becomes :

$$m\,p\,v_s = m\,R\,T$$

or
$$p\,v = m\,R\,T \qquad \ldots(\because m\,v_s = v)$$

Notes : 1. The units of gas constant (R) may be obtained as discussed below:

$$R = \frac{p\,v}{m\,T} = \frac{\text{N/m}^2 \times \text{m}^3}{\text{kg} \times \text{K}} = \frac{\text{N-m}}{\text{kg} \times \text{K}} = \text{N-m/kg K} = \text{J/kg K}$$

$$\ldots(\because 1 \text{ N-m} = 1 \text{ J})$$

2. The value of gas constant (R) is different for different gases. In S.I. units, its value for atmospheric air is taken 287 J/kg K or 0.287 kJ/kg K.

3. The equation $pv = m\,R\,T$ may also be expressed in another form *i.e.*,

$$p = \frac{m}{v}R\,T = \rho\,R\,T \qquad \ldots\left(\because \frac{m}{v} = \rho\right)$$

where ρ (rho) is the density of the given gas.

Example 2.2. *A vessel of capacity 3 m³ contains air at a pressure of 1.5 bar and a temperature of 25° C. Additional air is now pumped into the system until the pressure rises to 30 bar and temperature rises to 60° C. Determine the mass of air pumped in and express the quantity as a volume at a pressure of 1.02 bar and a temperature of 20° C.*

If the vessel is allowed to cool until the temperature is again 25° C, calculate the pressure in the vessel.

Solution. Given : $v_1 = 3\text{ m}^3$; $p_1 = 1.5\text{ bar} = 0.15 \times 10^6\text{ N/m}^2$; $T_1 = 25°\text{ C} = 25 + 273 = 298\text{ K}$; $p_2 = 30\text{ bar} = 3 \times 10^6\text{ N/m}^2$; $T_2 = 60°\text{ C} = 60 + 273 = 333\text{ K}$; $p_3 = 1.02\text{ bar} = 0.102 \times 10^6\text{ N/m}^2$; $T_3 = 20°\text{ C} = 20 + 273 = 293\text{ K}$

Mass of air pumped in

Let m_1 = Mass of air initially filled in the vessel, and

m_2 = Mass of air in the vessel after pumping.

We know that $p_1\,v_1 = m_1\,R\,T_1$

$$\therefore \quad m_1 = \frac{p_1\,v_1}{R\,T_1} = \frac{0.15 \times 10^6 \times 3}{287 \times 298} = 5.26\text{ kg}$$

$$\ldots(\text{Taking } R \text{ for air} = 287 \text{ J/kg K})$$

Similarly, $p_2\,v_2 = m_2\,R\,T_2$

$$\therefore \quad m_2 = \frac{p_2\,v_2}{R\,T_2} = \frac{3 \times 10^6 \times 3}{287 \times 333} = 94.17\text{ kg} \qquad \ldots(\because v_2 = v_1)$$

∴ Mass of air pumped in,

$$m = m_2 - m_1 = 94.17 - 5.26 = 88.91 \text{ kg } \textbf{Ans.}$$

Volume of air pumped in at a pressure of 1.02 bar and temperature of 20° C

Let v_3 = Volume of air pumped in.

We know that

$$p_3 v_3 = m R T_3$$

$$\therefore \quad v_3 = \frac{m R T_3}{p_3} = \frac{88.91 \times 287 \times 293}{0.102 \times 10^6} = 73.3 \text{ m}^3 \textbf{ Ans.}$$

Pressure in the vessel after cooling

Let p_4 = Pressure in the vessel after cooling.

We know that the temperature after cooling,

$$T_4 = T_1 = 25° \text{C} = 298 \text{ K}$$

Since the cooling is at constant volume, therefore

$$\frac{p_4}{p_2} = \frac{T_4}{T_2}$$

$$\therefore \quad p_4 = \frac{T_4 p_2}{T_2} = \frac{298 \times 3 \times 10^6}{333} = 2.68 \times 10^6 \text{ N/m}^2 = 26.8 \text{ bar } \textbf{Ans.}$$

Example 2.3. *A spherical vessel of 1.5 m diameter, containing air at 40° C is evacuated till the vacuum inside the vessel becomes 735 mm of Hg. Calculate the mass of air pumped out. If the tank is then cooled to 3° C, what will be the final pressure in the tank ? Take atmospheric pressure as 760 mm of Hg.*

Solution. Given : d = 1.5 m ; T_1 = 40° C = 40 + 273 = 313 K ; p_v = 735 mm of Hg ; T_3 = 3° C = 3 + 273 = 276 K; p_1 = 760 mm of Hg

Mass of air pumped out

Let m = Mass of air pumped out.

First of all, let us find out the initial mass of air (m_1) in a vessel. We know that volume of a spherical vessel,

$$v_1 = \frac{\pi d^3}{6} = \frac{\pi (1.5)^3}{6} = 1.767 \text{ m}^3$$

We know that $p_1 v_1 = m_1 R T_1$

$$\therefore \quad m_1 = \frac{p_1 v_1}{R T_1} = \frac{(760 \times 133.3)\, 1.767}{287 \times 313} = 1.993 \text{ kg}$$

...(∵ 1 mm of Hg = 133.3 N/m^2 ; and R for air = 287 J/kg K)

Let m_2 = Mass of air left in the vessel after evacuation.

We know that pressure after evacuation,

$$p_2 = \text{Atmospheric pressure} - \text{Vacuum pressure}$$

$$= 760 - 735 = 25 \text{ mm of Hg} = 25 \times 133.3 = 3332.5 \text{ N/m}^2$$

$$\therefore \quad m_2 = \frac{p_2 v_2}{R T_2} = \frac{3332.5 \times 1.767}{287 \times 313} = 0.066 \text{ kg}$$

$$\ldots (\because v_2 = v_1 \text{ and } T_2 = T_1)$$

∴ Mass of air pumped out,

$$m = m_1 - m_2 = 1.993 - 0.066 = \mathbf{1.927 \text{ kg Ans.}}$$

Final pressure in the tank

Let p_3 = Final pressure in the tank.

Since the cooling is at constant volume, therefore

$$\frac{p_3}{T_3} = \frac{p_2}{T_2}$$

or

$$p_3 = \frac{p_2 T_3}{T_2} = \frac{3332.5 \times 276}{313} = 2938 \text{ N/m}^2 = \mathbf{22 \text{ mm of Hg Ans.}}$$

2.9. Avogadro's Law

It states, "*Equal volumes of all gases, at the same temperature and pressure, contain equal number of molecules.*"

Thus, according to Avogadro's law, 1 m³ of oxygen (O_2) will contain the same number of molecules as 1 m³ of hydrogen (H_2) when the temperature and pressure is the same. Since the molecular mass of hydrogen is 2 and that of oxygen is 16, therefore a molecule of oxygen has a mass which is 32/2 = 16 times the mass of hydrogen molecules. Moreover, as 1 m³ of these two gases contain the same number of molecules, and a molecule of oxygen has a mass 16 times than that of hydrogen molecule, therefore it is evident that density of oxygen is 16 times the density of hydrogen. Hence, the Avogadro's law indicates that the density of any two gases is directly proportional to their molecular masses, if the gases are at the same temperature and pressure.

The density of oxygen at Normal Temperature and Pressure (briefly written as N.T.P.) *i.e.* at 0^0 C and 1.013 bar is 1.429 kg/m³.

∴ Specific volume (of 1 kg) of oxygen at N.T.P.,

$$v_s = \frac{1}{1.429} \text{ m}^3/\text{kg} \qquad \ldots \left[\because \text{Specific volume} = \frac{1}{\text{Density}} \right]$$

and volume of 32 kg (or 1 kg molecule briefly written as 1 kg-mol)

$$= \frac{1}{1.429} \times 32 = 22.4 \text{ m}^3$$

Similarly, it can be proved that the volume of 1 kg mol of any gas at N.T.P. is 22.4 m³.

Note : 1 g-mole (molecular mass expressed in gram) of all gases occupies a volume of 22.4 litres at N.T.P.

The values of molecular mass for some common gases are given in the following table :

Table 2.1. Molecular mass for some common gases.

S.No.	*Gas*	*Molecular mass*	*S.No.*	*Gas*	*Molecular mass*
1.	Hydrogen (H_2)	2	5.	Carbon dioxide (CO_2)	44
2.	Oxygen (O_2)	32	6.	Methane (CH_4)	16
3.	Nitrogen (N_2)	28	7.	Acetylene (C_2H_2)	26
4.	Carbon monoxide (CO)	28	8.	Sulphur dioxide (SO_2)	64

2.10. Universal Gas Constant or Molar Constant

The *universal gas constant* or *molar constant* (generally denoted by R_u) of a gas is the product of the gas constant and the molecular mass of the gas. Mathematically,

$$R_u = MR$$

where M = Molecular mass of the gas expressed in kg-mole, and

R = Gas constant.

In general, if M_1, M_2, M_3, etc. are the molecular masses of different gases and R_1, R_2, R_3, etc. are their gas constants respectively, then

$$M_1 R_1 = M_2 R_2 = M_3 R_3 = ... = R_u$$

Notes : 1. The value of R_u is same for all gases.

2. In S.I. units, the value of R_u is taken as 8314 J/kg-mol K or 8.314 kJ/kg-mol K.

3. The characteristic gas equation (*i.e.* $p\,v = R\,T$) may be written in terms of molecular mass as :

$$p\,v = M\,R\,T$$

Example 2.4. *A mass of 2.25 kg of nitrogen occupying 1.5 m*^{3} *is heated from 25° C to 200° C at a constant volume. Calculate the initial and final pressures of the gas. Take universal gas constant as 8314 J/kg mol K. The molecular mass of nitrogen is 28.*

Solution. Given : m = 2.25 kg ; v_1 = 1.5 m^3 ; T_1 = 25° C = 25 + 273 = 298 K ; T_2 = 200° C = 200 + 273 = 473 K ; R_u = 8314 J/kg mol K ; M = 28

We know that gas constant,

$$R = \frac{R_u}{M} = \frac{8314}{28} = 297 \text{ J/kg K}$$

Initial pressure of the gas

Let p_1 = Initial pressure of the gas.

We know that

$$p_1 v_1 = m\,R\,T_1$$

$$\therefore \quad p_1 = \frac{m\,R\,T_1}{v_1} = \frac{2.25 \times 297 \times 298}{1.5} = 0.133 \times 10^6 \text{ N/m}^2 = 1.33 \text{ bar } \textbf{Ans.}$$

Final pressure of the gas

Let p_2 = Final pressure of the gas.

Since the volume is constant, therefore

$$\frac{p_1}{T_1} = \frac{p_2}{T_2} \quad \text{or} \quad p_2 = \frac{p_1 T_2}{T_1} = \frac{1.33 \times 473}{298} = 2.11 \text{ bar } \textbf{Ans.}$$

Example 2.5. *Nitrogen is to be stored at pressure 140 bar, temperature 27° C in a steel flask of 0.05 m*^{3} *volume. The flask is to be protected against excessive pressure by a fusible plug which will melt and allow the gas to escape if the temperature rises too high. Find :*

1. How many kg of nitrogen will the flask hold at the designed conditions ? Take molecular mass of nitrogen as 28 ; and

2. At what temperature must the fusible plug melt in order to limit the pressure of the full flask to a maximum of 168 bar ?

Solution. Given : p_1 = 140 bar = 14×10^6 N/m^2 ; T_1 = 27° C = 27 + 273 = 300 K ; v_1 = 0.05 m^3 ; $M = 28$

1. *Mass of nitrogen*

Let m = Mass of nitrogen in kg which the flask will hold.

We know that gas constant,

$$R = \frac{\text{Universal gas constant}}{\text{Molecular mass}} = \frac{R_u}{M} = \frac{8314}{28} = 297 \text{ J/kg K}$$

. . . (R_u for all gases = 8314 J/kg K)

We also know that

$$p_1 v_1 = m R T_1$$

$$\therefore \quad m = \frac{p_1 v_1}{R T_1} = \frac{14 \times 10^6 \times 0.05}{297 \times 300} = 7.86 \text{ kg} \quad \textbf{Ans.}$$

2. *Melting temperature of fusible plug*

Let T_2 = Melting temperature of fusible plug, and

p_2 = Maximum pressure = 168 bar = 16.8×10^6 N/m^2 . . . (Given)

Since the gas is heated at constant volume, therefore

$$\frac{p_1}{T_1} = \frac{p_2}{T_2}$$

$$\therefore \quad T_2 = \frac{p_2 T_1}{p_1} = \frac{16.8 \times 10^6 \times 300}{14 \times 10^6} = 360 \text{ K} = 87° \text{ C} \quad \textbf{Ans.}$$

2.11. Specific Heats of a Gas

The specific heat of a substance may be broadly defined as the amount of heat required to raise the temperature of its unit mass through one degree. All the liquids and solids have one specific heat only. But a gas can have any number of specific heats (lying between zero and infinity) depending upon the conditions, under which it is heated. The following two types of specific heats of a gas are important from the subject point of view :

1. Specific heat at constant volume, and 2. Specific heat at constant pressure.

These specific heats are discussed, in detail, as follows :

2.12. Specific Heat at Constant Volume

It is the amount of heat required to raise the temperature of a unit mass of gas through one degree when it is heated at a constant volume. It is generally denoted by c_v.

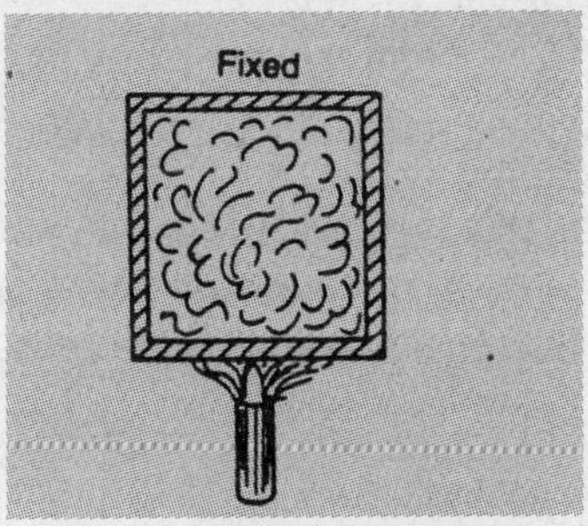

Fig. 2.1. Heat being supplied at constant volume.

Consider a gas contained in a container with a fixed lid as shown in Fig. 2.1. Now, if this gas is heated, it will increase the temperature and pressure of the gas in the container. Since the lid of the container is fixed, therefore the volume of gas remains unchanged.

Let m = Mass of the gas,

T_1 = Initial temperature of the gas, and

T_2 = Final temperature of the gas.

∴ Total heat supplied to the gas at constant volume,

$$Q_{1-2} = \text{Mass} \times \text{Sp. heat at constant volume} \times \text{Rise in temperature}$$

$$= mc_v (T_2 - T_1)$$

It may be noted that whenever a gas is heated at constant volume, no work is done by the gas.* The whole heat energy is utilised in increasing the temperature and pressure of the gas. In other words, all the amount of heat supplied remains within the body of the gas, and represents the *increase in internal energy of the gas.*

2.13. Specific Heat at Constant Pressure

It is the amount of heat required to raise the temperature of a unit mass of a gas through one degree, when it is heated at constant pressure. It is generally denoted by c_p.

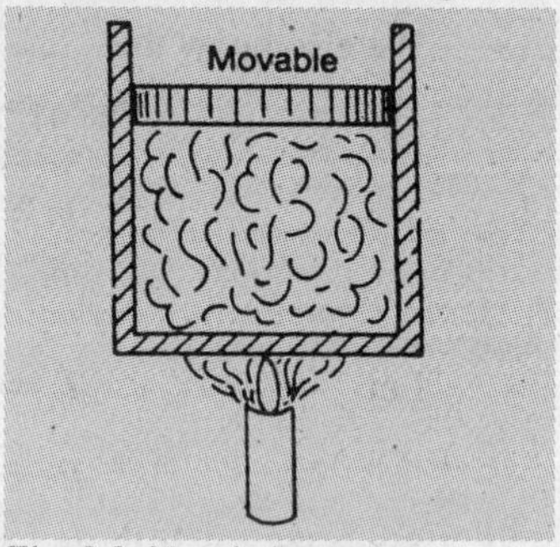

Fig. 2.2. Heat being supplied at constant pressure.

Consider a gas contained in a container with a movable lid as shown in Fig. 2.2. Now if this gas is heated, it will increase the temperature and pressure of the gas in the container. Since the lid of the container is movable, therefore it will move upwards, in order to counterbalance the tendency for pressure to rise.

Let m = Mass of the gas,

T_1 = Initial temperature of the gas,

v_1 = Initial volume of the gas, and

T_2, v_2 = Corresponding values for the final condition of the gas.

∴ Total heat supplied to the gas, at constant pressure,

$$Q_{1-2} = \text{Mass} \times \text{Sp. heat at constant pressure} \times \text{Rise in temperature}$$

$$= mc_p (T_2 - T_1)$$

Whenever a gas is heated at a constant pressure, the heat supplied to the gas is utilised for the following two purposes :

1. To raise the temperature of the gas. This heat remains within the body of the gas, and represents the increase in internal energy Mathematically, increase in internal energy,

$$dU = mc_v (T_2 - T_1)$$

2. To do some external work during expansion. Mathematically, workdone by the gas,

$$W_{1-2} = p(v_2 - v_1) = mR(T_2 - T_1)$$

It is thus obvious, that the specific heat at constant pressure is higher than the specific heat at constant volume.

From above, we may write as

$$Q_{1-2} = dU + W_{1-2} \quad \text{or} \quad {}^{**}Q_{1-2} - W_{1-2} = dU \quad \text{... (First Law of Thermodynamics)}$$

---

* We know that workdone by the gas,

$$W = pdv = p(v_2 - v_1)$$

where p = Pressure of the gas, and

dv = Change in volume = $v_2 - v_1$

When there is no change in volume, then $dv = 0$. Therefore $W = 0$.

** Refer Art. 1.49, note 2 equation (*iv*).

2.14. Enthalpy of a Gas

In thermodynamics, one of the basic quantities most frequently recurring is the *sum* of the internal energy (U) and the product of pressure and volume ($p\,v$). This sum ($U + p\,v$) is termed as *enthalpy* and is written as H. Mathematically,

Enthalpy, $$H = U + p\,v$$

Since ($U + p\,v$) is made up entirely of properties, therefore enthalpy (H) is also a property.

For a unit mass, specific enthalpy,

$$h = u + p\,v_s$$

where u = Specific internal energy, and

v_s = Specific volume.

Note. We know that $Q_{1-2} = dU + W_{1-2} = dU + p\,dv$

When gas is heated at constant pressure from an initial condition 1 to a final condition 2, then change in internal energy,

$$dU = U_2 - U_1$$

and workdone by the gas,

$$W_{1-2} = p\,dv = p\,(v_2 - v_1)$$

$$\therefore \quad Q_{1-2} = (U_2 - U_1) + p\,(v_2 - v_1)$$

$$= (U_2 + p\,v_2) - (U_1 + p\,v_1) = H_2 - H_1$$

and for a unit mass, $$q_{1-2} = h_2 - h_1$$

Thus, for a constant pressure process, the heat supplied to the gas is equal to the change of enthalpy.

2.15. Molar Specific Heats of a Gas

The molar or volumetric specific heat of a gas may be defined as the amount of heat required to raise the temperature of unit mole of gas through one degree. Mathematically, molar specific heat,

$$c_m = Mc$$

where M = Molecular mass of the gas.

In the similar way as discussed in Art. 2.11, the molar specific heat at constant volume,

$$c_{vm} = M\,c_v$$

and molar specific heat at constant pressure,

$$c_{pm} = M\,c_p$$

Example 2.6. *A closed vessel contains 2 kg of carbon dioxide at temperature 20° C and pressure 0.7 bar. Heat is supplied to the vessel till the gas acquires a pressure of 1.4 bar. Calculate : 1. Final temperature ; 2. Work done on or by the gas ; 3. Heat added ; and 4. Change in internal energy. Take specific heat of the gas at constant volume as 0.657 kJ/kg K.*

Solution. Given : m = 2 kg ; T_1 = 20° C = 20 + 273 = 293 K ; p_1 = 0.7 bar ; p_2 = 1.4 bar

1. Final temperature

Let T_2 = Final temperature.

Since the gas is heated in a closed vessel, therefore the volume of gas will remain constant.

We know that $$\frac{p_1}{T_1} = \frac{p_2}{T_2}$$

$$\therefore \quad T_2 = \frac{p_2 T_1}{p_1} = \frac{1.4 \times 293}{0.7} = 586 \text{ K} = 586 - 273 = 313^\circ \text{C} \textbf{ Ans.}$$

2. *Workdone on or by the gas*

Since there is no change in volume, therefore workdone on or by the gas (W_{1-2}) is *zero*. **Ans.**

3. *Heat added*

We know that heat added at constant volume,

$$Q_{1-2} = m\, c_v (T_2 - T_1) = 2 \times 0.657\,(586 - 293) = 385 \text{ kJ} \textbf{ Ans.}$$

4. *Change in internal energy*

Let $\quad dU$ = Change in internal energy.

We know that $\quad Q_{1-2} = W_{1-2} + dU$

$\therefore \quad dU = Q_{1-2} = 385$ kJ **Ans.** $\quad \ldots (\because W_{1-2} = 0)$

Example 2.7. *A mass of 0.25 kg of air in a closed system expands from 2 bar, 60° C to 1 bar and 40° C while receiving 1.005 kJ of heat from a reservoir at 100° C. The surrounding atmosphere is at 0.95 bar and 27° C. Determine the maximum work. How much of this work would be done on the atmosphere ?*

Solution. Given : m = 0.25 kg ; p_1 = 2 bar = 0.2×10^6 N/m^2 ; T_1 = 60° C = 60 + 273 = 333 K ; p_2 = 1 bar = 0.1×10^6 N/m^2 ; T_2 = 40° C = 40 + 273 = 313 K ; Q = 1.005 kJ ; *T_R = 100° C ; p = 0.95 bar = 0.095×10^6 N/m^2 ; *T = 27° C

Maximum Workdone

First of all, let us find the values of initial volume (v_1) and final volume (v_2) of air. We know that

$$v_1 = \frac{m R T_1}{p_1} \quad \ldots (\because pv = mRT)$$

$$= \frac{0.25 \times 287 \times 333}{0.2 \times 10^6} = 0.119 \text{ m}^3 \quad \ldots \text{(Taking } R = 287 \text{ J/kg K)}$$

Similarly

$$v_2 = \frac{m R T_2}{p_2} = \frac{0.25 \times 287 \times 313}{0.1 \times 10^6} = 0.224 \text{ m}^3$$

$\therefore$ Workdone on the atmosphere,

$$W_1 = p\,(v_2 - v_1) = 0.095 \times 10^6\,(0.224 - 0.119) = 9975 \text{ J} = 9.975 \text{ kJ}$$

We know that change of internal energy,

$$dU = m\, c_v (T_2 - T_1) = 0.25 \times 0.712\,(313 - 333) = -3.56 \text{ kJ}$$

$$\ldots \text{(Taking } c_v = 0.712 \text{ kJ/kg K)}$$

The –ve sign shows that there is a *decrease* of internal energy.

Net workdone, $\quad W_2 = Q - dU = 1.005 - (-3.56) = 4.565$ kJ

$\therefore$ Maximum workdone,

$$W = W_1 + W_2 = 9.975 + 4.565 = 14.54 \text{ kJ} \textbf{ Ans.}$$

* Superfluous data

Workdone on the atmosphere

We have calculated above that the workdone on the atmosphere,

$$W_1 = 9.975 \text{ kJ } \textbf{Ans.}$$

Example 2.8. *3 kg of an ideal gas is expanded from a pressure 7 bar and volume 1.5 m³ to a pressure 1.4 bar and volume 4.5 m³. The change in internal energy is 525 kJ. The specific heat at constant volume for the gas is 1.047 kJ/kg K. Calculate : 1. Gas constant ; 2. Change in enthalpy ; and 3. Initial and final temperatures.*

Solution. Given : m = 3 kg ; p_1 = 7 bar = 0.7×10^6 N/m² ; v_1 = 1.5 m³ ; p_2 = 1.4 bar = 0.14×10^6 N/m² ; v_2 = 4.5 m³ ; dU = 525 kJ ; c_v = 1.047 kJ/kg K

1. *Gas constant*

Let R = Gas constant,

T_1 and T_2 = Initial and final temperatures.

We know that $p_1 v_1 = m R T_1$

$$\therefore \quad R T_1 = \frac{p_1 v_1}{m} = \frac{0.7 \times 10^6 \times 1.5}{3} = 0.35 \times 10^6 \quad \dots (i)$$

$$\text{Similarly} \quad R T_2 = \frac{p_2 v_2}{m} = \frac{0.14 \times 10^6 \times 4.5}{3} = 0.21 \times 10^6 \quad \dots (ii)$$

Subtracting equation (*ii*) from equation (*i*),

$$R (T_1 - T_2) = (0.35 - 0.21)\, 10^6 = 0.14 \times 10^6 \quad \dots (iii)$$

We also know that change in internal energy

$$dU = m c_v (T_2 - T_1)$$

Since during expansion, there is a decrease in internal energy, therefore the change in internal energy is *negative.

$$\therefore \quad -525 = 3 \times 1.047 (T_2 - T_1) = -3.141 (T_1 - T_2)$$

$$T_1 - T_2 = 525 / 3.141 = 167.14 \quad \dots (iv)$$

Dividing equation (*iii*) by equation (*iv*), we get

$$R = 0.14 \times 10^6 / 167.14 = 838 \text{ J/kg K} = 0.838 \text{ kJ/ kg K } \textbf{Ans.}$$

2. *Change in enthalpy*

First of all, let us find the value of specific heat at constant pressure (c_p). We know that

$$\therefore \quad c_p - c_v = R \text{ or } c_p = R + c_v = 0.838 + 1.047 = 1.885 \text{ kJ/kg K}$$

∴ Change in enthalpy,

$$dH = m c_p (T_2 - T_1) = 3 \times 1.885 (-167.14) = -945 \text{ kJ } \textbf{Ans.}$$

The –ve sign indicates that there is a decease in enthalpy.

---

* We may also say as follows :
From equation (*iii*), we see that $R (T_1 - T_2)$ is a *positive* equation. This shows that T_1 is greater than T_2, because R is always positive. Thus there is a decrease is internal energy or the change in internal energy is *negative* (*i.e.* $dU = -525$ kJ).

3. *Initial and final temperatures*

From equation (*i*), we find that initial temperature,

$$T_1 = \frac{0.35 \times 10^6}{R} = \frac{0.35 \times 10^6}{838} = 417 \text{ K } \textbf{Ans.}$$

and from equation (*iv*), final temperature,

$$T_2 = T_1 - 167.14 = 417 - 167.14 = 249.86 \text{ K } \textbf{Ans.}$$

2.16. Regnault's Law

This law states, "*The two specific heats of a gas (i.e. the specific heat at constant pressure, c_p and specific heat at constant volume, c_v) do not change with the change in pressure and temperature of the gas.*"

2.17. Relation between Specific Heats

Consider a gas enclosed in a container and being heated, at a constant pressure, from the initial state 1 to the final state 2.

Let m = Mass of the gas,

T_1 = Initial temperature of the gas,

T_2 = Final temperature of the gas,

v_1 = Initial volume of the gas,

v_2 = Final volume of the gas,

c_p = Specific heat at constant pressure,

c_v = Specific heat at constant volume, and

p = Constant pressure.

We know that the heat supplied to the gas at constant pressure,

$$Q_{1-2} = m\, c_p\, (T_2 - T_1)$$

As already discussed, a part of this heat is utilised in doing the external work, and the rest remains within the gas, and is used in increasing the internal energy of the gas.

∴ Heat utilised for external work,

$$W_{1-2} = p\,(v_2 - v_1) \qquad \ldots (i)$$

and increase in internal energy, $dU = m\, c_v\, (T_2 - T_1)$ ... (*ii*)

We know that $Q_{1-2} = W_{1-2} + dU$ ... (*iii*)

$$\therefore \quad m\, c_p\, (T_2 - T_1) = p\,(v_2 - v_1) + m\, c_v\, (T_2 - T_1) \qquad \ldots (iv)$$

Using characteristic gas equation (*i.e.* $p\,v = m\,R\,T$), we have

$$p\,v_1 = m\,R\,T_1 \qquad \ldots \text{(for initial conditions)}$$

and

$$p\,v_2 = m\,R\,T_2 \qquad \ldots \text{(for final conditions)}$$

$$\therefore \quad p\,(v_2 - v_1) = m\,R\,(T_2 - T_1)$$

Now substituting the value of $p\,(v_2 - v_1)$ in equation (*iv*),

$$m\, c_p\, (T_2 - T_1) = m\,R\,(T_2 - T_1) + m\, c_v\, (T_2 - T_1)$$

$$\therefore \quad c_p = R + c_v \quad \text{or} \quad c_p - c_v = R \qquad \ldots (v)$$

The above equation may be rewritten as :

$$c_p - c_v = R \quad \text{or} \quad c_v(\gamma - 1) = R \qquad \ldots \left(\text{where } \gamma = \frac{c_p}{c_v} \right)$$

$$\therefore \quad c_v = \frac{R}{(\gamma - 1)} \qquad \ldots (vi)$$

Notes. 1. The equation (v) gives an important result, as it proves that characteristic constant of a gas (R) is equal to the difference of its two specific heats (*i.e.* $c_p - c_v$).

2. The value of R is taken as 287 J/kg K or 0.287 kJ/kg K.

3. In terms of molar specific heats, the equation (v) may be written as

$$c_{pm} - c_{vm} = R_u$$

where R_u is the universal gas constant and its value is taken 8314 J/kg K or 8.314 kJ/kg K.

2.18. Ratio of Specific Heats

The ratio of two specific heats (*i.e.* c_p/c_v) of a gas is an important constant in the field of Thermodynamics and is represented by a Greek letter gamma (γ). It is also known as *adiabatic index*. Since c_p is always greater than c_v, the value of γ is always greater than unity.

We have seen in Art. 2.17 that

$$c_p - c_v = R \quad \text{or} \quad c_p = c_v + R$$

Dividing both sides by c_v,

$$\frac{c_p}{c_v} = 1 + \frac{R}{c_v} \quad \text{or} \quad \gamma = 1 + \frac{R}{c_v}$$

The values of c_p, c_v and γ for some common gases are given below :

Table 2.2. Values of c_p and c_v for some common gases.

S.No.	*Name of gas*	c_p (kJ/kg K)	c_v (kJ/kg K)	$\gamma = \frac{c_p}{c_v}$
1.	Air	1.000	0.720	1.40
2.	Carbon dioxide (CO_2)	0.846	0.657	1.29
3.	Oxygen (O_2)	0.913	0.653	1.39
4.	Nitrogen (N_2)	1.043	0.745	1.40
5.	Ammonia (NH_3)	2.177	1.692	1.29
6.	Carbon monoxide (CO)	1.047	0.749	1.40
7.	Hydrogen (H_2)	14.257	10.133	1.40
8.	Argon (A)	0.523	0.314	1.67
9.	Helium (He)	5.234	3.153	1.66
10.	Methane (CH_4)	2.169	1.650	1.31

Example 2.9. *One kg of a perfect gas occupies a volume of 0.85 m³ at 15° C and at a constant pressure of 1 bar. The gas is first heated at a constant volume, and then at a constant pressure. Find the specific heat at constant volume and constant pressure of the gas. Take γ = 1.4.*

Solution. Given : m = 1 kg ; v = 0.85 m³ ; T = 15° C = 15 + 273 = 288 K ; p = 1 bar = 0.1×10^6 N/m² ; $\gamma = c_p/c_v = 1.4$

Specific heat of gas dt constant volume

Let c_v = Specific heat of gas at constant volume, and

R = Characteristic gas constant.

We know that

$$p\,v = m\,R\,T$$

$$\therefore \quad R = \frac{p\,v}{m\,T} = \frac{0.1 \times 10^6 \times 0.85}{1 \times 288} = 295 \text{ J/kg K} = 0.295 \text{ kJ/kg K}$$

We also know that

$$c_v = \frac{R}{\gamma - 1} = \frac{0.295}{1.4 - 1} = 0.7375 \text{ kJ/kg K} \quad \textbf{Ans.}$$

Specific heat of gas at constant pressure

We know that specific heat of gas at constant pressure,

$$c_p = 1.4\,c_v = 1.4 \times 0.7375 = 1.0325 \text{ kJ/kg K} \quad \textbf{Ans.} \qquad \ldots (\because\ c_p/c_v = 1.4)$$

Example 2.10. *A gas mixture obeying perfect gas law has a molecular mass of 26.7. Assuming a mean molar specific heat at constant volume of 21.1 kJ/kg K, determine the values of characteristic gas constant, molar specific heat at constant pressure and the ratio of specific heats.*

Solution. Given : $M = 26.7$; $c_{vm} = 21.1$ kJ/kg K

Characteristic gas constant

We know that characteristic gas constant,

$$R = \frac{\text{Universal gas constant}}{\text{Molecular mass}} = \frac{R_u}{M} = \frac{8.314}{26.7} = 0.3114 \text{ kJ/kg K} \quad \textbf{Ans.}$$

$$\ldots (\because\ R_u \text{ for all gases} = 8.314 \text{ kJ / kg K})$$

Molar specific heat at constant pressure

Let c_{pm} = Molar specific heat at constant pressure.

We know that

$$c_{pm} - c_{vm} = R_u \quad \text{or} \quad c_{pm} = R_u + c_{vm} = 8.314 + 21.1 = 29.414 \text{ kJ/kg K} \quad \textbf{Ans.}$$

Ratio of specific heats

We know that ratio of specific heats,

$$\gamma = \frac{c_{pm}}{c_{vm}} = \frac{29.414}{21.1} = 1.394 \quad \textbf{Ans.}$$

Example 2.11. *One kg of ideal gas is heated from 18.3° C to 93.4° C. Assuming R = 0.264 kJ/kg K and γ = 1.18 for the gas, find : 1. Specific heats ; 2. Change in internal energy ; and 3. Change in enthalpy.*

Solution. Given : $m = 1$ kg ; $T_1 = 18.3°\text{ C} = 18.3 + 273 = 291.3$ K ; $T_2 = 93.4°\text{ C} = 93.4 + 273 = 366.4$ K ; $R = 0.264$ kJ/kg K ; $\gamma = c_p/c_v = 1.18$

Specific heats

Let c_p = Specific heat at constant pressure, and

c_v = Specific heat at constant volume.

We know that $c_v = \frac{R}{\gamma - 1} = \frac{0.264}{1.18 - 1} = 1.47$ kJ/kg K **Ans.**

$\therefore \quad c_p = \gamma c_v = 1.18 \times 1.47 = 1.73$ kJ/kg K **Ans.**

Change in internal energy

We know that change in internal energy,

$$dU = m\, c_v (T_2 - T_1) = 1 \times 1.47 \,(366.4 - 291.3) = 110.4 \text{ kJ} \quad \textbf{Ans.}$$

Change in enthalpy

We know that change in enthalpy,

$$dH = m\, c_p (T_2 - T_1) = 1 \times 1.73 \,(366.4 - 291.3) = 130 \text{ kJ} \quad \textbf{Ans.}$$

Example 2.12. *A gas, having initial pressure, volume and temperature as 275 kN/m², 0.09 m³ and 185° C respectively, is compressed at constant pressure until its temperature is 15° C. Calculate the amount of heat transferred and work done during the process. Take R = 290 J/kg K and c_p = 1.005 kJ/kg K.*

Solution. Given : $p_1 = 275$ kN/m² $= 275 \times 10^3$ N/m² ; $v_1 = 0.09$ m³ ; $T_1 = 185°$ C $= 185 + 273 = 458$ K ; $T_2 = 15°$ C $= 15 + 273 = 288$ K ; $R = 290$ J/kg K ; $c_p = 1.005$ kJ/kg K

Amount of heat transferred

First of all, let us find the mass of the gas (m). We know that

$$p_1 v_1 = m R T_1$$

$$\therefore \quad m = \frac{p_1 v_1}{R T_1} = \frac{275 \times 10^3 \times 0.09}{290 \times 458} = 0.186 \text{ kg}$$

We know that the amount of heat transferred,

$$Q_{1-2} = m\, c_p (T_2 - T_1) = 0.186 \times 1.005 \,(288 - 458) \text{ kJ}$$

$$= -31.78 \text{ kJ} \quad \textbf{Ans.}$$

The –ve sign indicates that the heat has been extracted from the gas during the process. In other words, the gas is compressed.

Workdone during the process

First of all, let us find the final volume of the gas (v_2). Since the process takes place at constant pressure, therefore

$$\frac{v_1}{T_1} = \frac{v_2}{T_2} \quad \text{or} \quad v_2 = \frac{v_1 T_2}{T_1} = \frac{0.09 \times 288}{458} = 0.056 \text{ m}^3$$

We know that the workdone during the process,

$$W_{1-2} = p\,(v_2 - v_1) = 275 \times 10^3 \,(0.056 - 0.09) = -9350 \text{ J}$$

$$= -9.35 \text{ kJ} \quad \textbf{Ans.}$$

The –ve sign indicates the work is done on the gas. In other words, the gas is compressed.

Example 2.13. *A certain gas has c_p = 1.96 kJ/kg K and c_v = 1.5 kJ/kg K. Find its molecular mass and gas constant. A constant volume chamber of 0.3 m³ capacity contains 2 kg of this gas at 5° C. The heat is transferred to the gas until the temperature is 100° C. Find the workdone, heat transferred and change in internal energy.*

Solution. Given : c_p = 1.96 kJ/kg K ; c_v = 1.5 kJ/kg K ; $v = 0.3$ m^3 ; m = 2 kg ; T_1 = 5° C = 5 + 273 = 278 K ; T_2 = 100° C = 100 + 273 = 373 K

Molecular mass and gas constant

Let M = Molecular mass, and

R = Gas constant.

We know that gas constant,

$$R = c_p - c_v = 1.96 - 1.5 = 0.46 \text{ kJ/kg K}$$

and molecular mass, $$M = \frac{\text{Universal gas constant}}{\text{Gas constant}} = \frac{R_u}{R} = \frac{8.314}{0.46} = 18 \text{ kg } \textbf{Ans.}$$

. . . ($\because R_u$ for all gases = 8.314 kJ/kg K)

Workdone

Since the volume is constant, therefore workdone ($W_{1\text{-}2}$) is zero **Ans.**

Heat transferred

We know that heat transferred,

$$Q_{1-2} = m\,c_v\,(T_2 - T_1) = 2 \times 1.5\,(373 - 278) = 285 \text{ kJ } \textbf{Ans.}$$

Change in internal energy

Let dU = Change in internal energy.

We know that $Q_{1-2} = W_{1-2} + dU = 0 + dU = dU$

$\therefore$ $dU = Q_{1-2} = 285$ kJ **Ans.**

Example 2.14. *A vessel of 2.5 m^3 capacity contains one kg-mole of nitrogen at 100° C. Evaluate the specific volume and pressure. If the gas is cooled to 30° C, calculate final pressure, change in specific internal energy and specific enthalpy.*

The ratio of specific heats is 1.4 and one kg-mole nitrogen is 28 kg.

Solution. Given : v_1 = 2.5 m^3 ; M = 1 kg-mole = 28 kg ; T_1 = 100° C = 100 + 273 = 373 K ; T_2 = 30° C = 30 + 273 = 303 K ; $\gamma = c_p/c_v = 1.4$

Specific volume and pressure

Let v_s = Specific volume of the gas, and

p_1 = Pressure of the gas.

We know that specific volume of the gas,

$$v_s = \frac{v_1}{M} = \frac{2.5}{28} = 0.089 \text{ m}^3\text{/kg } \textbf{Ans.}$$

Gas constant, $$R = \frac{R_u}{M} = \frac{8314}{28} = 297 \text{ J/kg K}$$ . . . ($\because R_u$ = 8314 J/kg mol K)

We know that $p_1\,v_1 = M\,R\,T_1$

$$\therefore \quad p_1 = \frac{M\,R\,T_1}{v_1} = \frac{28 \times 297 \times 373}{2.5} = 1.24 \times 10^6 \text{ N/m}^2$$

$= 12.4$ bar **Ans.** . . . ($\because$ 1 bar = 0.1×10^6 N/m^2)

Final pressure, change in specific internal energy and specific enthalpy

Let p_2 = Final pressure of the gas.

We know that $\dfrac{p_1 v_1}{T_1} = \dfrac{p_2 v_2}{T_2}$ or $\dfrac{p_1}{T_1} = \dfrac{p_2}{T_2}$ ...($\because v_1 = v_2$)

$$\therefore \quad p_2 = \frac{p_1 T_2}{T_1} = \frac{12.4 \times 303}{373} = 10.07 \text{ bar } \textbf{Ans.}$$

Now, let us find out the values of specific heat at constant pressure (c_p) and specific heat at constant volume (c_v).

We know that $c_p - c_v = R$

$1.4\, c_v - c_v = 297$ ...($\because c_p/c_v = 1.4$)

$$\therefore \quad c_v = 297/0.4 = 742.5 \text{ J/kg} = 0.7425 \text{ kJ/kg K}$$

and $c_p = 1.4\, c_v = 1.4 \times 0.7425 = 1.04$ kJ/kg K

We know that change in specific internal energy,

$$du = c_v (T_2 - T_1) = 0.7425\,(303 - 373) = -52 \text{ kJ/kg } \textbf{Ans.}$$

The –ve sign indicates that the specific internal energy is reduced after the gas is cooled.

We also know that change in specific enthalpy,

$$dh = c_p (T_2 - T_1) = 1.04\,(303 - 373) = -72.8 \text{ kJ/kg } \textbf{Ans.}$$

The –ve sign indicates that the specific enthalpy is reduced after the gas is cooled.

EXERCISES

1. Determine the final pressure of a gas when 2 m^3 of gas at 6 bar is heated by keeping the temperature constant. The final volume is 6 m^3. **[Ans.** 2 bar]

2. A certain quantity of air is cooled at a constant pressure from 300 K to 280 K. If the initial volume of the air is 0.15 m^3, find by how much the volume will diminish ? **[Ans.** 0.01 m^3]

3. A gas at a temperature of 333° C and 20 bar has a volume of 0.06 m^3. It is expanded to a volume of 0.54 m^3. Determine the final pressure of the gas if the temperature of the gas after expansion is 30° C. **[Ans.** 1.33 bar]

4. A gas at a temperature of 20° C and pressure of 1.5 bar occupies a volume of 0.105 m^3. If the gas is compressed to a pressure of 7.5 bar and volume of 0.04 m^3, what will be the final temperature of the gas ? **[Ans.** 285° C]

5. A cylinder contains 3 kg of air at a pressure of 300 bar and a temperature of 27° C. Find the volume of air occupied by the gas. Assume R for air as 287 J/kg K. **[Ans.** 0.0086 m^3]

6. A vessel of capacity 5 m^3 contains 20 kg of an ideal gas having a molecular mass of 25. If the temperature of the gas is 15° C, find its pressure. **[Ans.** 3.83 bar]

$$\left[\textbf{Hint. } R = \frac{R_u}{M} = \frac{8314}{25} = 332.56 \text{ J/kg K} \quad (\because R_u \text{ for an ideal gas} = 8314 \text{ J/kg K})\right]$$

7. A certain gas occupies 0.15 m^3 at a temperature of 20° C and a pressure of 1.2 bar. If the gas has mass of 200 g, calculate (*i*) value of gas constant, and (*ii*) molecular mass of the gas. **[Ans.** 307.2 J/kg K ; 27.06]

8. A certain gas has c_p = 1.96 kJ/kg K and c_v = 1.5 kJ/kg K. Find its molecular mass and the gas constant. The universal gas constant is 8.315 kJ/kg K. **[Ans.** 18 ; 0.46 kJ/kg K]

9. The volume of air at a pressure of 5 bar and 47° C is 0.5 m^3. Calculate the mass of the air, if the specific heats at constant pressure and volume are 1 kJ/kg K and 0.72 kJ/kg K respectively. **[Ans.** 2.8 kg]

10. The heated nitrogen gas expands from 0.2 m^3 to 0.85 m^3 in a quasi-static process at a constant pressure of 1000 kPa. For 1 kg mass of gas, determine the amount of workdone by the gas and the final temperature. $R = 296.8$ J/kg K, for nitrogen. **[Ans.** 650 kJ ; 2863.88 K]

11. The gas constant for atmospheric air is 0.287 kJ/kg K and the specific heat at constant volume is 0.713 kJ/kg K. Find the specific heat at constant pressure and the ratio of specific heats.

[Ans. 1 kJ/kg K ; 1.403]

12. A certain quantity of gas occupies 0.14 m^3 at 12.6 bar and 100° C. Calculate the change in internal energy if the gas is heated to a temperature of 300° C. Take $c_p = 1$ kJ/kg K and $c_v = 0.72$ kJ/kg K.

[Ans. 245 kJ]

13. The temperature of 3.5 kg of gas is raised from 95° C to 225° C at a constant pressure. Find the amount of heat supplied to the gas and the amount of the external workdone. The specific heats at constant pressure and volume are 1 kJ/kg K and 0.72 kJ/kg K respectively. **[Ans.** 455 kJ ; 127.4 kJ]

14. An ideal gas 0.9 kg having gas constant 287 J/kg K is heated at constant pressure of 8 bar from 30° C to 200° C. If the specific heat at constant volume is 0.72 kJ/kg K, find 1. specific heat at constant pressure, 2. total heat supplied to the gas, 3. increase in internal energy, and 4. workdone in expansion.

[Ans. 1.007 kJ/kg K ; 154.1 kJ ; 110.16 kJ ; 43.94 kJ]

15. One kg mole of nitrogen (molecular mass = 28) is contained in a vessel of volume 2.5 m^3 at 100° C. 1. Evaluate the mass, the pressure and the specific volume of the gas ; 2. If the ratio of specific heats is 1.4, evaluate c_p and c_v. 3. If the gas cools to the atmospheric temperature of 30° C, evaluate the final pressure of the gas. 4. Find the increase in specific internal energy and the increase in specific enthalpy.

[Ans. 28 kg, 12.4 bar, 0.089 m^3/kg ; 1.04 kJ/kg K, 0.7425 kJ/kg K ; 10.07 bar ; 52 kJ/kg, 72.8 kJ/kg]

QUESTIONS

1. What is a perfect gas ? Under what conditions does a real gas behave as a perfect gas?
2. Name the variables which control the physical properties of a perfect gas.
3. State Boyle's law and Charles' law and prove that the characteristic gas equation is $p \cdot v = m R T$.
4. What is the difference between universal gas constant and characteristic gas constant ?
5. Define the specific heat at constant volume and at constant pressure.
6. What do you understand by enthalpy ? Show that for a constant pressure process, the heat supplied to the gas is equal to the change of enthalpy.
7. Prove that the difference between two specific heats (c_p and c_v) is equal to characteristic gas constant (R).
8. What is an adiabatic index ? Why its value is always greater than unity ?

OBJECTIVE TYPE QUESTIONS

1. If the temperature remains constant, the volume of a given mass of a gas is inversely proportional to the pressure. This is known as

   (*a*) Charles' law (*b*) Boyle's law (*c*) Joule's law (*d*) Gay-Lussac's law

2. The state of a substance whose evaporation from its liquid state is complete, is known as

   (*a*) steam (*b*) vapour (*c*) air (*d*) perfect gas

3. The characteristic equation of a gas is

   (*a*) pv = constant (*b*) $pv = mR$ (*c*) $pv = mRT$ (*d*) $pv = RT^m$

   where p, v, T and m = Pressure, volume, temperature and mass of the gas respectively, and R = Gas constant.

4. The value of gas constant (R) is

   (*a*) 287 J/kg K (*b*) 28.7 J/kg K (*c*) 2.87 J/kg K (*d*) 0.287 J/kg K

5. The value of universal gas constant (R_u) is

   (*a*) 8.314 J/kg K (*b*) 83.14 J/kg K (*c*) 831.4 J/kg K (*d*) 8314 J/kg K

6. The gas constant (R) is equal to the................of two specific heats.

   (*a*) sum (*b*) difference (*c*) product (*d*) ratio

7. The specific heat at constant pressure is ....that of specific heat at constant volume.

   (*a*) equal to (*b*) less than (*c*) more than

8. The ratio of specific heat at constant pressure (c_p) and specific heat at constant volume (c_v) is

   (*a*) equal to one (*b*) less than one (*c*) more than one (*d*) none of these

9. The value of c_p/c_v for air is

   (*a*) 1 (*b*) 1.4 (*c*) 1.8 (*d*) 2.3

10. When the gas is heated at constant pressure, then the heat supplied

    (*a*) raises the temperature of the gas

    (*b*) increases the internal energy of the gas

    (*c*) does some external work during expansion

    (*d*) both (*a*) and (*b*)

    (*e*) both (*b*) and (*c*)

ANSWERS

1. (*b*)	2. (*d*)	3. (*c*)	4. (*a*)	5. (*d*)
6. (*b*)	7. (*c*)	8. (*c*)	9. (*b*)	10. (*e*)

3

Thermodynamic Processes of Perfect Gases

1. Introduction. 2. Classification of Thermodynamic Processes. 3. Workdone During a Non-flow Process. 4. Application of First Law of Thermodynamics to a Non-flow Process. 5. Heating and Expansion of Gases in Non-flow Processes. 6. Constant Volume Process (or Isochoric Process). 7. Constant Pressure Process (or Isobaric Process). 8. Hyperbolic Process. 9. Constant Temperature Process (or Isothermal Process). 10. Adiabatic Process (or Isentropic Process). 11. Polytropic Process. 12. Rate of Heat Transfer (Absorption or Rejection) per Unit Volume During a Polytropic Process. 13. Determination of Polytropic Index. 14. Free Expansion (or Un-resisted Expansion) Process. 15. General Laws for Expansion and Compression. 16. Summary of Formulae for Heating and Expansion of Perfect Gases in Reversible Non-flow Processes. 17. Flow Processes. 18. Application of First Law of Thermodynamics to a Steady Flow Process. 19. Workdone in a Steady Flow Process. 20. Workdone for Various Steady Flow Processes. 21. Throttling Process. 22. Application of Steady Flow Energy Equation to Engineering Systems.

3.1. Introduction

We have already discussed that when a system changes its state from one equilibrium state to another equilibrium state, then the path of successive states through which the system has passed, is known as a *thermodynamic process*. Strictly speaking, no system is in true equilibrium during the process because the properties (such as pressure, volume, temperature etc.) are changing. However, if the process is assumed to take place sufficiently slowly so that the deviation of the properties at the intermediate states is infinitesimally small, then every state passed through by the system will be in equilibrium. Such a process is called *quasi-static* or *reversible process* and it is represented by a continuous curve on the property diagram (*i.e.* pressure-volume diagram) as shown in Fig. 3.1 (*a*).

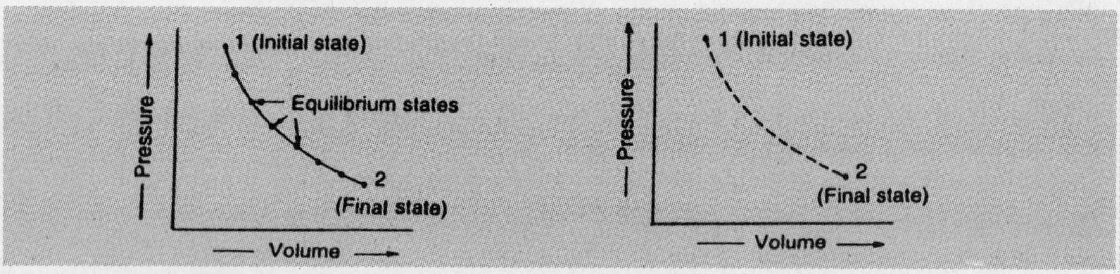

(*a*) Reversible process. (*b*) Irreversible process.

Fig. 3.1. Reversible and irreversible process.

If the process takes place in such a manner that the properties at the intermediate states are not in equilibrium state (except the initial and final state), then the process is said to be *non-equilibrium* or *irreversible process*. This process is represented by the broken lines on the property diagram as shown in Fig. 3.1 (*b*).

3.2. Classification of Thermodynamic Processes

All the thermodynamic processes are classified into the following two groups :

1. Non-flow processes, and 2. Flow processes.

The processes occurring in closed systems which do not permit the transfer of mass across their boundaries, are known as *non-flow processes.* It may be noted that in a non-flow process, the energy crosses the system boundary in the form of heat and work, but there is no mass flow into or out of the system.

The processes occurring in open systems which permit the transfer of mass to and from the system, are known as *flow processes*. It may be noted that in a flow process, the mass enters the system and leaves after enhancing energy. The flow processes may be steady flow and non-steady flow processes. The common examples of steady flow processes are flow through nozzles, turbines compressors etc. and the common examples of non-steady flow processes are filling or evacuation of vessels.

3.3. Workdone During a Non-flow Process

Consider a system contained in a frictionless piston and cylinder arrangement as shown in Fig. 3.2. As the system expands from its original state 1, it overcomes the external resistance (such as rotation of the flywheel) which opposes the motion of the piston by exerting a force through a distance. The variation of the volume and pressure of the system as it expands to final state 2, is drawn on the pressure-volume diagram (briefly called $p\ v$ diagram) as shown in Fig. 3.2.

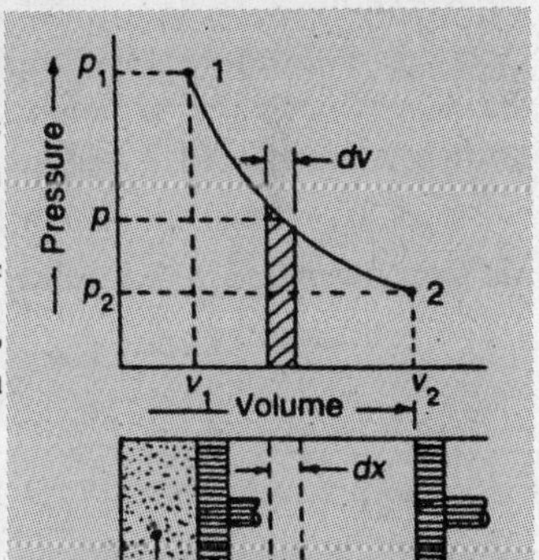

Fig. 3.2. Workdone during a non-flow process.

Let at any small section (shown shaded), the pressure (p) of the system is constant. If A is the cross-sectional area of the piston, then force on the piston ($F = p\,A$) causes the piston to move through a distance dx. Thus, workdone by the system,

$$\delta W = F\,dx = p\,A\,dx = p\,dv$$

$$\ldots (\because\ dv = A\,dx)$$

$\therefore$ Workdone for non-flow process from state 1 to state 2,

$$W_{1-2} = \int_1^2 \delta W = \int_1^2 p\,dv$$

From above, we see that the *workdone is given by the area under the p-v diagram.*

Notes : 1. The workdone by the system is taken as *positive* while the workdone on the system is considered as *negative.*

2. For an irreversible process, $\delta W \neq p\,dv$, because the path of the process is not represented truely on the p-v diagram due to its non-equilibrium states in the process.

3.4. Application of First Law of Thermodynamics to a Non-flow Process

We have already discussed in Chapter 1 (Art. 1.49) that when a system undergoes a change of state or a thermodynamic process, then both the heat transfer and work transfer takes place. The net energy transfer is stored within the system and is known as *stored* or *total energy* of the system. Mathematically

$$Q_{1-2} - W_{1-2} = dE = E_2 - E_1 \quad \ldots (i)$$

where Q_{1-2} = Heat transferred or heat supplied to the system during the process *i.e.* from state 1 to state 2,

W_{1-2} = Workdone by the system on the surrounding during the process *i.e.* from state 1 to state 2,

E_2 = *Total energy or stored energy of the system at the end of the process *i.e.* at state 2, and

E_1 = Total energy or stored energy of the system at the start of the process *i.e.* at state 1

For a non-flow process, the stored energy is the internal energy only. Thus equation (*i*) of the first law of thermodynamics, when applied to a non-flow process or a static system, may be written as

$$Q_{1-2} - W_{1-2} = dU = U_2 - U_1$$

where dU = Change in internal energy $= U_2 - U_1$

It may be noted that heat and work are not a property of the system, but their difference $(Q_{1-2} - W_{1-2})$ during a process is the numerical equivalent of stored energy. Since the stored energy is a property, therefore $(Q_{1-2} - W_{1-2})$ is also a property.

Example 3.1. *When a system is taken from the state A to the state B, in Fig. 3.3, along the path ACB, 80 kJ of heat flows into the system and the system does 30 kJ of work.*

1. How much heat flows into the system along the path ADB, if the workdone is 10 kJ.

2. When the system is returned from the state B to the state A along the curved path, the workdone on the system is 20 kJ. Does the system absorb or liberate heat and how much heat is absorbed or liberated.

3. If the internal energy at A (U_A) = 0 and at D (U_D) = 40 kJ, find the heat absorbed in the processes AD and DB.

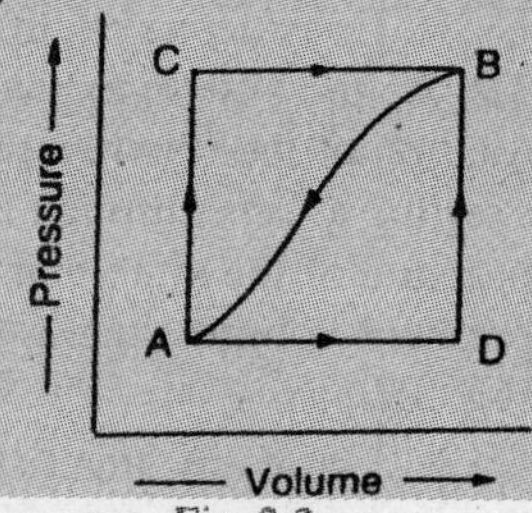

Fig. 3.3

Solution. Given : $Q_{ACB} = 80$ kJ ; $W_{ACB} = 30$ kJ

1. *Heat flowing into the system along the path ADB, if workdone (W_{ADB}) = 10 kJ*

Let Q_{ADB} = Heat flowing into the system along the path *ADB*,

U_A = Internal energy at *A*, and

U_B = Internal energy at *B*.

We know that for path *ACB*,

$$Q_{ACB} = **(U_B - U_A) + W_{ACB}$$

$$\therefore \quad U_B - U_A = Q_{ACB} - W_{ACB} = 80 - 30 = 50 \text{ kJ}$$

We also know that for path *ADB*,

$$Q_{ADB} = (U_B - U_A) + W_{ADB} = 50 + 10 = 60 \text{ kJ} \quad \textbf{Ans.}$$

$$\ldots (\because W_{ADB} = 10 \text{ kJ})$$

2. *Heat absorbed or liberated*

Let Q_{B-A} = Heat absorbed or liberated along the curved path *BA*.

Since the work is done on the system, therefore

$$W_{B-A} = -20 \text{ kJ} \quad \ldots \text{(Given)}$$

* The total energy of a system is the sum of potential energy (*PE*), kinetic energy (*KE*) and internal energy (*U*) plus any other form of the energy such as chemical energy, electrical energy etc.

** Since internal energy is a point function, and depends on the initial and final states, therefore dU is taken as $(U_B - U_A)$ where as heat and work are path functions.

We know that for the curved path *B-A*,

$$Q_{B-A} = (U_A - U_B) + W_{B-A} = -(U_B - U_A) + W_{B-A}$$

$$= -50 - 20 = -70 \text{ kJ } \textbf{Ans.}$$

The –ve sign indicates that heat is liberated by the system.

3. *Heat absorbed in the processes AD and DB*

Let Q_{AD} = Heat absorbed in the process *AD*,

Q_{DB} = Heat absorbed in the process *DB*,

U_A = Internal energy at $A = 0$ ... (Given)

U_D = Internal energy at $D = 40$ kJ ... (Given)

We know that work done along the path *ADB*,

$$W_{ADB} = W_{A-D} + W_{D-B}$$

Since the process *D-B* is a constant volume process, therefore $W_{D-B} = 0$. Thus, workdone during the process *A-D*,

$$W_{A-D} = W_{ADB} = 10 \text{ kJ}$$

We know that $Q_{A-D} = (U_D - U_A) + W_{A-D}$

$$= (40 - 0) + 10 = 50 \text{ kJ } \textbf{Ans.}$$

Similarly $Q_{D-B} = (U_B - U_D) + W_{D-B} = U_B - U_D$ ... ($\because W_{D-B} = 0$)

$$= 50 - 40 = 10 \text{ kJ } \textbf{Ans.}$$

... ($\because U_B - U_A = 50$ kJ and $U_A = 0$, therefore $U_B = 50$ kJ)

Example 3.2. *A fluid is confined in a cylinder by a spring loaded frictionless piston so that the pressure in the fluid is a linear function of the volume ($p = a + b\,v$). The internal energy of the fluid is given by the following equation :*

$$U = 33.5 + 3\,p\,v$$

where U is in kJ, p in kN/m² and v in m³. If the fluid changes from an initial state of 1.7 bar, 0.03 m³ to a final state of 4 bar, 0.06 m³, with no work other than that done on the piston, find the direction and magnitude of the work and heat transfer.

Solution. Given : $p = a + b\,v$; $U = 33.5 + 3\,p\,v$; $p_1 = 1.7$ bar = 170 kN/m² ; $v_1 = 0.03$ m³ ; $p_2 = 4$ bar = 400 kN/m² ; $v_2 = 0.06$ m³

We know that change in internal energy of the fluid,

$$dU = U_2 - U_1 = (33.5 + 3\,p_2\,v_2) - (33.5 + 3\,p_1\,v_1)$$

$$= 3\,(p_2\,v_2 - p_1\,v_1) = 3\,(400 \times 0.06 - 170 \times 0.03) \text{ kJ}$$

$$= 3\,(24 - 5.1) = 56.7 \text{ kJ}$$

We also know that $p = a + b\,v$

$\therefore$ For the initial state of fluid,

$$170 = a + b \times 0.03 \qquad \ldots (i)$$

and for the final state of fluid,

$$400 = a + b \times 0.06 \qquad \ldots (ii)$$

Subtracting equation (*i*) from equation (*ii*),

$$400 - 170 = 0.06\,b - 0.03\,b \quad \text{or} \quad 230 = 0.03\,b$$

$$\therefore \qquad b = 7667$$

Substituting the value of b in equation (*i*),

$$170 = a + 7667 \times 0.03 = a + 230$$

$$\therefore \qquad a = -60$$

We know that work transfer,

$$W_{1-2} = \int_{v_1}^{v_2} p\,dv = \int_{v_1}^{v_2} (a + b\,v)\,dv$$

$$= \left[a\,v + \frac{b\,v^2}{2} \right]_{v_1}^{v_2} = a\,(v_2 - v_1) + b\left(\frac{v_2^2 - v_1^2}{2}\right)$$

$$= -60\,(0.06 - 0.03) + 7667\left[\frac{(0.06)^2 - (0.03)^2}{2}\right]$$

$$= -1.8 + 10.35 = 8.55 \text{ kJ } \textbf{Ans.}$$

$\therefore$ Heat transfer, $\quad Q_{1-2} = W_{1-2} + dU = 8.55 + 56.7 = 65.25$ kJ **Ans.**

Since both work transfer and heat transfer are *positive*, therefore work is done by the fluid and heat is supplied to the fluid.

3.5. Heating and Expansion of Gases in Non-flow Processes

The heating and expanding of a gas may be performed in many ways. But the following are the different non-flow processes (reversible and irreversible) as applied to perfect gas :

1. *Reversible non-flow processes.* These processes are as follows :

(*a*) Constant volume process (or Isochoric process),

(*b*) Constant pressure process (or Isobaric process),

(*c*) Hyperbolic process,

(*d*) Constant temperature process (or Isothermal process)

(*e*) Adiabatic process (or Isentropic process), and

(*f*) Polytropic process

2. *Irreversible non-flow processes.* The free expansion process is an irreversible non-flow process.

These above mentioned processes are discussed, in detail, in the following pages.

Note: The above mentioned processes are also applicable to the cooling and compression of gases. Cooling is regarded as negative heating, and compression as negative expansion.

3.6. Constant Volume Process (or Isochoric Process)

We have already discussed that when a gas is heated at a constant volume, its temperature and pressure will increase. Since there is no change in its volume, therefore no work is done by the gas. All the heat supplied to the gas is stored within the gas in the form of internal energy. It may be noted that this process is governed by Gay-Lussac law. Now consider m kg of a certain gas being heated at constant volume from initial state 1 to a final state 2.

Let $\quad p_1, v_1$ and T_1 = Pressure, volume and temperature at the initial state 1, and

$\quad p_2, v_2$ and T_2 = Pressure, volume and temperature at the final state 2.

The process is shown on the pressure-volume (p-v) and pressure-temperature (p-T) diagrams in Fig. 3.4 (a) and (b) respectively.

(a) p-v diagram. (b) p-T diagram.

Fig. 3.4. Constant volume process.

Now let us derive the following relations for the reversible constant volume process.

1. *Pressure-volume-temperature (p-v-T) relationship*

We know that the general gas equation is

$$\frac{p_1 v_1}{T_1} = \frac{p_2 v_2}{T_2} \qquad \ldots (i)$$

Since the gas is heated at constant volume, therefore $v_1 = v_2$.

$$\therefore \quad \frac{p_1}{T_1} = \frac{p_2}{T_2} \quad \text{or} \quad \frac{p}{T} = \text{Constant} \qquad \ldots \text{[From equation } (i)\text{]}$$

Thus the constant volume process is governed by Gay-Lussac law.

2. *Workdone by the gas*

We know that $\delta W = p\, dv$

On integrating from state 1 to state 2,

$$\int_1^2 \delta W = \int_1^2 p\, dv = p \int_1^2 dv$$

or

$$W_{1-2} = p\,(v_2 - v_1) = 0 \qquad \ldots (\because v_1 = v_2)$$

3. *Change in internal energy*

We know that change in internal energy,

$$dU = m\, c_v\, dT \qquad \ldots \text{(Joule's law)}$$

On integrating from state 1 to state 2,

$$\int_1^2 dU = m\, c_v \int_1^2 dT$$

or

$$U_2 - U_1 = m\, c_v\, (T_2 - T_1)$$

4. *Heat supplied or heat transfer*

We know that $\delta Q = dU + \delta W$

On integrating from state 1 to state 2,

$$\int_1^2 \delta Q = \int_1^2 dU + \int_1^2 \delta W$$

or $$Q_{1-2} = (U_2 - U_1) + W_{1-2}$$

Since $W_{1-2} = 0$, therefore heat supplied or heat transfer,

$$Q_{1-2} = U_2 - U_1 = m\, c_v (T_2 - T_1)$$

This shows that all the heat supplied to the gas is utilised in increasing the internal energy of the gas.

5. *Change in enthalpy*

We know that the change in enthalpy,

$$dH = dU + d(pv)$$

On integrating from state 1 to state 2,

$$\int_1^2 dH = \int_1^2 dU + \int_1^2 d(p\,v)$$

or $$H_2 - H_1 = (U_2 - U_1) + (p_2 v_2 - p_1 v_1)$$

$$= m\, c_v (T_2 - T_1) + m\, R\, (T_2 - T_1)$$

$$\ldots (\because p_1 v_1 = m R T_1 \text{; and } p_2 v_2 = m R T_2)$$

$$= m (T_2 - T_1)(c_v + R) = m\, c_p (T_2 - T_1) \qquad \ldots (\because c_p - c_v = R)$$

Note : **The change in internal energy (dU) and the change in enthalpy (dH) have the same expression for each process.**

Example 3.3. *A certain gas occupies a space of 0.3 m^3 at a pressure of 2 bar and a temperature of 77° C. It is heated at a constant volume, until the pressure is 7 bar. Determine : 1. temperature at the end of the process ; 2. mass of the gas ; 3. change in internal energy ; and 4. change in enthalpy during the process.*

Assume c_p = 1.005 kJ/kg K ; c_v = 0.712 kJ/kg K ; and R = 287 J/kg K.

Solution. Given : $v_1 = 0.3$ m³ ; $p_1 = 2$ bar $= 0.2 \times 10^6$ N/m² ; $T_1 = 77°$ C $= 77 + 273 = 350$ K ; $p_2 = 7$ bar $= 0.7 \times 10^6$ N/m² ; $c_p = 1.005$ kJ/kg K ; $c_v = 0.712$ kJ/kg K ; $R = 287$ J/kg K

1 *Temperature at the end of the process*

Let T_2 = Temperature at the end of the process.

We know that $$\frac{p_1}{T_1} = \frac{p_2}{T_2}$$

$$\therefore \quad T_2 = \frac{p_2 T_1}{p_1} = \frac{0.7 \times 10^6 \times 350}{0.2 \times 10^6} = 1225 \text{ K} = 1225 - 273 = 952° \text{ C} \quad \textbf{Ans.}$$

2. *Mass of the gas*

Let m = Mass of the gas.

We know that $p_1 v_1 = m R T_1$

$$\therefore \qquad m = \frac{p_1 v_1}{R T_1} = \frac{0.2 \times 10^6 \times 0.3}{287 \times 350} = 0.597 \text{ kg } \textbf{Ans.}$$

3. *Change in internal energy*

We know that change in internal energy,

$$dU = U_2 - U_1 = m c_v (T_2 - T_1)$$
$$= 0.597 \times 0.712\,(1225 - 350) = 372 \text{ kJ } \textbf{Ans.}$$

4. *Change in enthalpy*

We know that change in enthalpy,

$$dH = H_2 - H_1 = m c_p (T_2 - T_1)$$
$$= 0.597 \times 1.005\,(1225 - 350) = 525 \text{ kJ } \textbf{Ans.}$$

3.7. Constant Pressure Process (or Isobaric Process)

We have already discussed that when a gas is heated at a constant pressure, its temperature and volume will increase. Since there is a change in its volume, therefore the heat supplied to the gas is utilised to increase the internal energy of the gas and for doing some external work. It may be noted that this process is governed by Charles' law.

Now consider m kg of a certain gas being heated at a constant pressure from an initial state 1 to a final state 2.

Let p_1, v_1 and T_1 = Pressure, volume and temperature at the initial state 1, and

p_2, v_2 and T_2 = Pressure, volume and temperature at the final state 2.

The process is shown on the *p-v* and *p-T* diagrams in Fig. 3.5 (*a*) and (*b*) respectively.

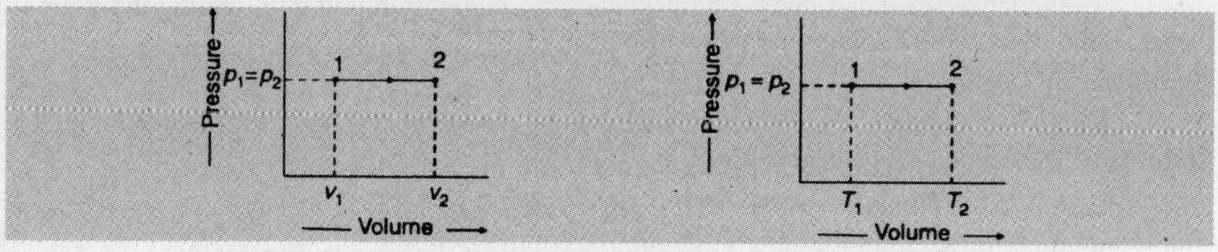

(*a*) *p-v* diagram. (*b*) *p-T* diagram.

Fig. 3.5. Constant pressure process.

Now let us derive the following relations for the reversible constant pressure process.

1. *Pressure -volume-temperature (p-v-T relationship)*

We know that the general gas equation is

$$\frac{p_1 v_1}{T_1} = \frac{p_2 v_2}{T_2} \qquad \ldots (i)$$

Since the gas is heated at constant pressure, therefore $p_1 = p_2$.

$$\therefore \qquad \frac{v_1}{T_1} = \frac{v_2}{T_2} \text{ or } \frac{v}{T} = \text{Constant} \qquad \ldots \text{[From equation } (i)]$$

Thus, the constant pressure process is governed by Charles' law.

2. *Workdone by the gas*

We know that $\delta W = p\,dv$

On integrating from state 1 to state 2,

$$\int_1^2 \delta W = \int_1^2 p\, dv = p \int_1^2 dv$$

or
$$W_{1-2} = p\,(v_2 - v_1) = m\,R\,(T_2 - T_1)$$
$$\ldots (\because p\,v_1 = m\,R\,T_1 \text{ and } p\,v_2 = m\,R\,T_2)$$

3. *Change in internal energy*

We have discussed in the previous article that the change in internal energy (dU) is same for all the processes. Therefore, change in internal energy,

$$dU = U_2 - U_1 = m\,c_v\,(T_2 - T_1)$$

4. *Heat supplied or heat transferred*

We know that
$$\delta Q = dU + \delta W$$

On integrating from state 1 to state 2,

$$\int_1^2 \delta Q = \int_1^2 dU + \int_1^2 \delta W$$

$$\therefore \quad Q_{1-2} = (U_2 - U_1) + W_{1-2} \qquad \ldots (ii)$$
$$= m\,c_v\,(T_2 - T_1) + m\,R\,(T_2 - T_1)$$
$$= m\,(T_2 - T_1)\,(c_v + R)$$
$$= m\,c_p\,(T_2 - T_1) \qquad \ldots (\because c_p - c_v = R)$$

The equation (*ii*) shows that the heat supplied to the gas is utilised in increasing the internal energy of the gas and for doing some external work.

5. *Change in enthalpy*

We have discussed in the previous article that the change in enthalpy (dH) is same for all the processes. Therefore, change in enthalpy,

$$dH = H_2 - H_1 = m\,c_p\,(T_2 - T_1)$$

We see that change in enthalpy is equal to the heat supplied or heat transferred.

Notes. 1. If the gas is cooled at a constant pressure, then there will be a compression. It is thus obvious that, during cooling, the temperature and volume will decrease and work is said to be done on the gas. In this case.

Workdone on the gas, $W_{1-2} = p\,(v_1 - v_2) = m\,R\,(T_1 - T_2)$

Decrease in internal energy, $dU = U_1 - U_2 = m\,c_v\,(T_1 - T_2)$

and heat rejected by the gas, $Q_{1-2} = m\,c_p\,(T_1 - T_2)$

2. During expansion or heating process, work is done by the gas (*i.e.* W_{1-2} is + ve) ; internal energy of the gas increases (*i.e.* dU is +ve) and heat is supplied to the gas (*i.e.* Q_{1-2} is +ve).

3. During compression or cooling process, work is done on the gas (*i.e.* W_{1-2} is −ve) ; internal energy of the gas decreases (*i.e.* dU is −ve) and heat is rejected by the gas (*i.e.* Q_{1-2} is −ve).

Example 3.4. *The values of specific heats at constant pressure and at constant volume for an ideal gas are 0.984 kJ/kg K and 0.728 kJ/kg K. Find the values of characteristic gas constant (R) and ratio of specific heats (γ) for the gas. If one kg of this gas is heated at constant pressure from 25° C to 200° C, estimate the heat added, ideal workdone and change in internal energy. Also calculate the pressure and final volume, if the initial volume was 2 m³.*

Solution. Given : c_p = 0.984 kJ/kg K ; c_v = 0.728 kJ/kg K ; m = 1 kg ; T_1 = 25° C = 25 + 273 = 298 K ; T_2 = 200° C = 200 + 273 = 473 K ; v_1 = 2 m³

The heating of gas at constant pressure is shown in Fig. 3.6.

Characteristic gas constant

We know that characteristic gas constant,

$$R = c_p - c_v = 0.984 - 0.728$$

$$= 0.256 \text{ kJ/kg K } \textbf{Ans.}$$

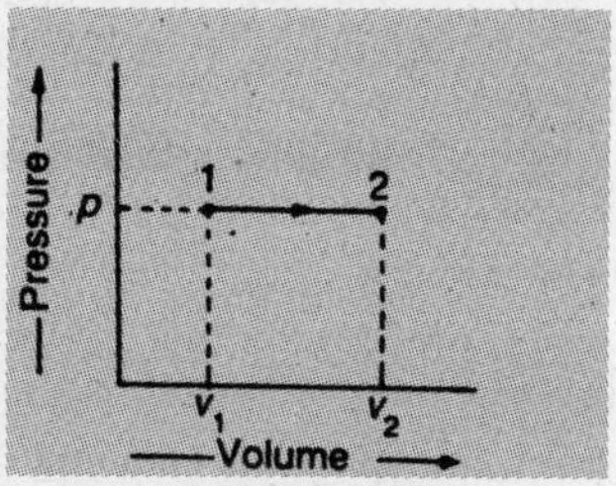

Fig. 3.6

Ratio of specific heats

We know that ratio of specific heats,

$$\gamma = c_p / c_v = 0.984 / 0.728 = 1.35 \textbf{ Ans.}$$

Heat added

We know that heat added during constant pressure operation,

$$Q_{1-2} = m\, c_p (T_2 - T_1) = 1 \times 0.984 (473 - 298) = 172.2 \text{ kJ } \textbf{Ans.}$$

Workdone

We know that workdone during constant pressure operation,

$$W_{1-2} = p (v_2 - v_1) = m\, R (T_2 - T_1) \qquad \ldots (\because p\,v = m\,R\,T)$$

$$= 1 \times 0.256 (473 - 298) = 44.8 \text{ kJ } \textbf{Ans.}$$

Change in internal energy

***We know that change in internal energy,**

$$dU = U_2 - U_1 = m\, c_v (T_2 - T_1) = 1 \times 0.728 (473 - 298) \text{ kJ}$$

$$= 127.4 \text{ kJ } \textbf{Ans.}$$

Pressure and final volume of the gas if the initial volume, $v_1 = 2$ m³

Let $p_1 = p_2$ = **Pressure of the gas, and**

v_2 = **Final volume of the gas.**

We know that $$\frac{p_1 v_1}{T_1} = \frac{p_2 v_2}{T_2} \quad \text{or} \quad \frac{v_1}{T_1} = \frac{v_2}{T_2} \qquad \ldots (\because \text{Pressure is constant})$$

$$\therefore \quad v_2 = \frac{v_1 T_2}{T_1} = \frac{2 \times 473}{298} = 3.17 \text{ m}^3 \textbf{ Ans.}$$

We also know that

$$p_1 v_1 = m\, R\, T_1$$

$$\therefore \quad p_1 = \frac{m\, R\, T_1}{v_1} = \frac{1 \times 256 \times 298}{2} = 38\,140 \text{ N/m}^2 \qquad \ldots (R \text{ is taken in J/kg K})$$

$$= 0.3814 \text{ bar } \textbf{Ans.} \qquad \ldots (\because 1 \text{ bar} = 10^5 \text{ N/m}^2)$$

Example 3.5. *A quantity of gas has a volume of 0.14 m³, pressure 1.5 bar and a temperature 100° C. If the gas is compressed at a constant pressure, until its volume becomes 0.112 m³, determine :*

---

* The change in internal energy may also be obtained from the relation

$$Q_{1-2} = dU + W_{1-2} \quad \text{or} \quad dU = Q_{1-2} - W_{1-2} = 172.2 - 44.8 = 127.4 \text{ kJ}$$

1. the temperature at the end of compression ; 2. work done in compressing the gas ; 3. decrease in internal energy ; and 4. heat given out by the gas.

Assume $c_p = 1.005$ kJ/kg K ; $c_v = 0.712$ kJ/kg K ; and R = 285 J/kg K

Solution. Given : $v_1 = 0.14$ m^3 ; $p = 1.5$ bar $= 0.15 \times 10^6$ N/m^2 ; $T_1 = 100°$ C $= 100 + 273 = 373$ K ; $v_2 = 0.112$ m^3 ; $c_p = 1.005$ kJ/kg K ; $c_v = 0.712$ kJ/kg K ; $R = 285$ J/kg K

The compression of gas at constant pressure is shown in Fig. 3.7.

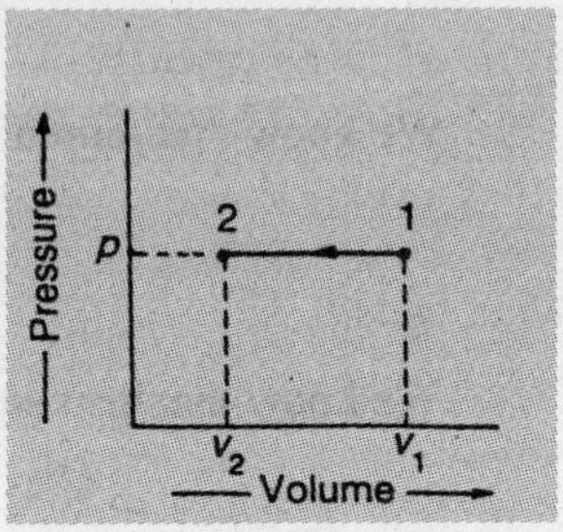

Fig. 3.7

1. Temperature at the end of compression

Let T_2 = Temperature at the end of compression.

We know that $$\frac{v_1}{T_1} = \frac{v_2}{T_2}$$

$$\therefore \quad T_2 = \frac{v_2 T_1}{v_1} = \frac{0.112 \times 373}{0.14} = 298.4 \text{ K}$$

$$= 298.4 - 273 = 25.4° \text{ C } \textbf{Ans.}$$

2. Workdone is compressing the gas

We know that workdone in compressing the gas,

$$W_{1-2} = p(v_1 - v_2) = 0.15 \times 10^6 (0.14 - 0.112) = 4200 \text{ J} = 4.2 \text{ kJ } \textbf{Ans.}$$

3. Decrease in internal energy

First of all, let us find the mass of gas (m) admitted for compression. We know that

$$p_1 v_1 = m R T_1$$

$$\therefore \quad m = \frac{p_1 v_1}{R T_1} = \frac{0.15 \times 10^6 \times 0.14}{285 \times 373} = 0.197 \text{ kg} \quad \ldots (\because p_1 = p)$$

We know that decrease in internal energy,

$$dU = U_1 - U_2 = m c_v (T_1 - T_2)$$

$$= 0.197 \times 0.712 (373 - 298.4) = 10.46 \text{ kJ } \textbf{Ans.}$$

4. Heat given out by the gas

We know that heat given out by the gas,

$$Q_{1-2} = m c_p (T_1 - T_2)$$

$$= 0.197 \times 1.005 (373 - 298.4) = 14.77 \text{ kJ } \textbf{Ans.}$$

3.8. Hyperbolic Process

A process, in which the gas is heated or expanded in such a way that the product of its pressure and volume (*i.e.* $p \times v$) remains constant, is called a *hyperbolic process*.

It may be noted that the hyperbolic process is governed by Boyle's law *i.e.* $p\,v$ = constant. If we plot a graph for pressure and volume, during the process as shown in Fig. 3.8, we shall get a rectangular hyperbola. Hence, this process is termed as hyperbolic process. It is merely a theoretical case, and has a little importance from the subject point of view. Its practical application is isothermal process, which is discussed below.

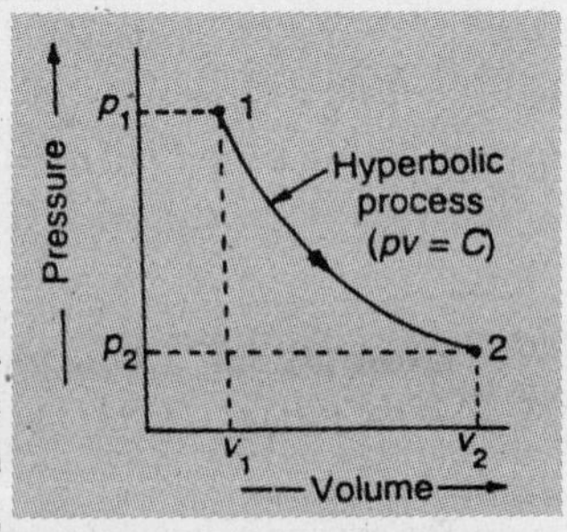

Fig. 3.8. Hyperbolic process.

3.9. Constant Temperature Process (or Isothermal Process)

A process, in which the temperature of the working substance remains constant during its expansion or compression, is called *constant temperature process* or *isothermal process*. This will happen when the working substance remains in a perfect thermal contact with the surroundings, so that the heat 'sucked in' or 'squeezed out' is compensated exactly for the work done by the gas or on the gas respectively. It is thus obvious that in an isothermal process :

1. there is no change in temperature,
2. there is no change in internal energy, and
3. there is no change in enthalpy.

Now consider m kg of a certain gas being heated at constant temperature from an initial state 1 to final state 2.

Let p_1, v_1 and T_1 = Pressure, volume and temperature at the initial state 1, and

p_2, v_2 and T_2 = Pressure, volume and temperature at the final state 2.

The process is shown on the p-v and p-T diagrams in Fig. 3.9 (*a*) and (*b*) respectively.

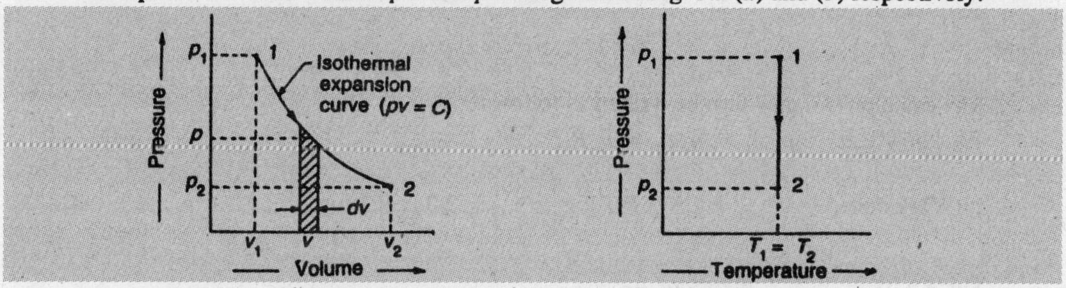

(*a*) p-v diagram. (*b*) p-T diagram.

Fig. 3.9. Constant temperature (Isothermal) process.

Now let us derive the following relations for the reversible constant temperature process or isothermal process.

1. *Pressure-volume-temperature (p-v-T) relationship*

We know that the general gas equation is

$$\frac{p_1 v_1}{T_1} = \frac{p_2 v_2}{T_2} \qquad \ldots (i)$$

Since the gas is heated at constant temperature, therefore $T_1 = T_2$.

$$\therefore \quad p_1 v_1 = p_2 v_2 \quad \text{or} \quad p\,v = \text{Constant} \qquad \ldots \text{[From equation } (i)]$$

Thus, the constant temperature process or isothermal process is governed by Boyle's law.

2. *Workdone by the gas*

We know that $\delta W = p\,dv$

On integrating from state 1 to state 2,

$$\int_1^2 \delta W = \int_1^2 p\,dv$$

or

$$W_{1-2} = \int_1^2 p\,dv \qquad \ldots (ii)$$

Since the expansion of the gas is isothermal, *i.e.* $p\,v = C$, therefore

$$p\,v = p_1\,v_1 \quad \text{or} \quad p = \frac{p_1\,v_1}{v}$$

Substituting this value of p in equation (*ii*), we have

$$W_{1-2} = \int_{v_1}^{v_2} \frac{p_1\,v_1}{v}\,dv = p_1\,v_1 \int_{v_1}^{v_2} \frac{dv}{v}$$

$$= p_1\,v_1 \Big[\log_e v\Big]_{v_1}^{v_2} = p_1\,v_1\,\log_e\left(\frac{v_2}{v_1}\right) \qquad \ldots (iii)$$

The above equation may be expressed in terms of corresponding logarithm to the base 10, *i.e.*

$$W_{1-2} = 2.3\,p_1\,v_1\,\log\left(\frac{v_2}{v_1}\right) = 2.3\,p_1\,v_1\,\log r \qquad \ldots (iv)$$

where $r = \dfrac{v_2}{v_1}$, and is known as *expansion ratio.*

The equation (*iv*) may also be written as follows :

We know that $p_1\,v_1 = p_2\,v_2 = m\,R\,T$

$\therefore$ Work done, $W_{1-2} = 2.3\,m\,R\,T\log\left(\dfrac{v_2}{v_1}\right) = 2.3\,m\,R\,T\log r$

Since $p_1\,v_2 = p_2\,v_2$, therefore $\dfrac{v_2}{v_1} = \dfrac{p_1}{p_2}$

$\therefore$ Work done, $W_{1-2} = 2.3\,p_1\,v_1\,\log\left(\dfrac{p_1}{p_2}\right)$

Notes : (*a*) Expansion ratio, $r = \dfrac{\text{Volume at the end of expansion}}{\text{Volume at the beginning of expansion}}$

(*b*) Compression ratio, $r = \dfrac{\text{Volume at the beginning of compression}}{\text{Volume at the end of compression}}$

3. *Change in internal energy*

We know that change in internal energy,

$$dU = U_2 - U_1 = m\,c_v\,(T_2 - T_1)$$

Since it is a constant temperature process, *i.e.* $T_1 = T_2$, therefore

$$dU = U_2 - U_1 = 0 \quad \text{or} \quad U_1 = U_2$$

4. *Heat supplied or heat transferred*

We know that heat supplied or heat transferred from state 1 to state 2,

$$Q_{1-2} = dU + W_{1-2} = W_{1-2} \qquad \ldots (\because dU = 0)$$

This shows that all the heat supplied to the gas is equal to the workdone by the gas.

5. *Change in enthalpy*

We know that change in enthalpy,

$$dH = H_2 - H_1 = m\,c_p\,(T_2 - T_1)$$

Since it is a constant temperature process, *i.e.* $T_1 = T_2$, therefore

$$dH = H_2 - H_1 = 0 \text{ or } H_1 = H_2$$

Example 3.6. *A quantity of air has a volume of 0.4 m³ at a pressure of 5 bar and a temperature of 80° C. It is expanded in a cylinder at a constant temperature to a pressure of 1 bar. Determine the amount of work done by the air during expansion.*

Solution. Given : $v_1 = 0.4 \text{ m}^3$; $p_1 = 5 \text{ bar} = 0.5 \times 10^6 \text{ N/m}^2$; $^*T = 80^\circ \text{C}$; $p_2 = 1 \text{ bar} = 0.1 \times 10^6 \text{ N/m}^2$

First of all, let us find the volume of air at the end of expansion (*i.e.* v_2). We know that

$$p_1 v_1 = p_2 v_2 \text{ or } v_2 = \frac{p_1 v_1}{p_2} = \frac{0.5 \times 10^6 \times 0.4}{0.1 \times 10^6} = 2 \text{ m}^3$$

and expansion ratio, $r = v_2 / v_1 = 2 / 0.4 = 5$

We know that workdone by the air during expansion,

$$W_{1-2} = 2.3\, p_1 v_1 \log r = 2.3 \times 0.5 \times 10^6 \times 0.4 \log 5 \text{ J}$$

$$= 0.46 \times 10^6 \times 0.699 = 321\,540 \text{ J} = 321.54 \text{ kJ} \textbf{ Ans.}$$

Example 3.7. *0.1 m³ of air at a pressure of 1.5 bar is expanded isothermally to 0.5 m³. Calculate the final pressure of the gas and heat supplied during the process.*

Solution. Given : $v_1 = 0.1 \text{ m}^3$; $p_1 = 1.5 \text{ bar} = 0.15 \times 10^6 \text{ N/m}^2$; $v_2 = 0.5 \text{ m}^3$

Final pressure of the gas

Let p_2 = Final pressure of the gas.

We know that $p_1 v_1 = p_2 v_2$

$$\therefore \quad p_2 = \frac{p_1 v_1}{v_2} = \frac{0.15 \times 10^6 \times 0.1}{0.5} = 0.03 \times 10^6 \text{ N/m}^2 = 0.3 \text{ bar } \textbf{Ans.}$$

Heat supplied during the process

We know that expansion ratio,

$$r = v_2 / v_1 = 0.5 / 0.1 = 5$$

∴ Workdone during the process,

$$W_{1-2} = 2.3\, p_1 v_1 \log r = 2.3 \times 0.15 \times 10^6 \times 0.1 \log 5 \text{ J}$$

$$= 0.0345 \times 10^6 \times 0.699 = 24\,115 \text{ J} = 24.115 \text{ kJ}$$

We know that in an isothermal process, heat supplied (Q_{1-2})is equal to the workdone during the process.

$$\therefore \quad Q_{1-2} = W_{1-2} = 24.115 \text{ kJ } \textbf{Ans.}$$

3.10. Adiabatic Process (or Isentropic Process)

A process, in which the working substance neither receives nor gives out heat to its surroundings, during its expansion or compression, is called an ***adiabatic process*. This will happen when

* Superfluous data

** It may be noted that the adiabatic process may be reversible or irreversible. The reversible adiabatic process or frictionless adiabatic process is known as *isentropic process* (or constant entropy process). But when friction is involved in the process, then the adiabatic process is said to be irreversible, in which case the entropy does not remain constant *i.e.* the entropy increases (See Chapter 4, Art. 4.7).

the working substance remains thermally insulated, so that no heat enters or leaves it during the process. It is thus obvious, that in an adiabatic or isentropic process :

1. No heat leaves or enters the gas,
2. The temperature of the gas changes, as the work is done at the cost of internal energy, and
3. The change in internal energy is equal to the work done.

Now consider m kg of a certain gas being heated adiabatically from an initial state 1 to a final state 2.

Let p_1, v_1 and T_1 = Pressure, volume and temperature at the initial state 1, and

p_2, v_2 and T_2 = Pressure, volume and temperature at the final state 2.

The process is shown on the p-v diagram in Fig. 3.10.

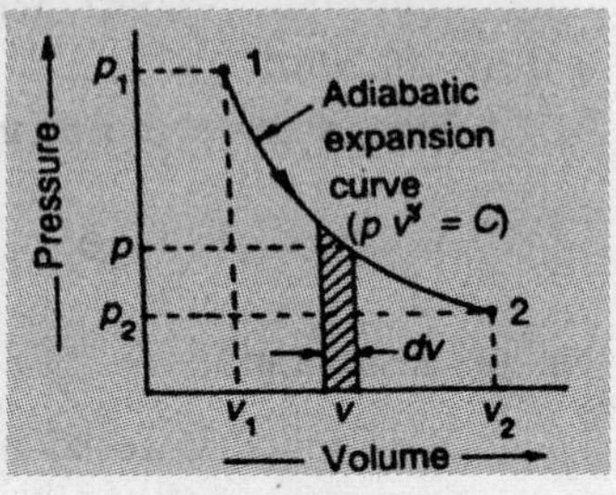

Fig. 3.10. Adiabatic process.

Now let us derive the following relations for a reversible adiabatic process.

1. *Pressure-volume-temperature (p-v-T) relationship*

We know that $$\delta Q = \delta W + dU \quad \text{(First law of thermodynamics)} \quad \ldots (i)$$

Since in an adiabatic process, no heat transfer takes places, therefore

$$\delta Q = 0$$

$$\therefore \quad \delta W + dU = 0$$

or $$p\,dv + m\,c_v\,dT = 0$$

$$\therefore \quad dT = \frac{-p\,dv}{m\,c_v} \quad \ldots (ii)$$

We know that $$p\,v = m\,R\,T$$

Differentiating this expression, we get

$$p\,dv + v\,dp = m\,R\,dT$$

$$\therefore \quad dT = \frac{p\,dv + v\,dp}{m\,R} = \frac{p\,dv + v\,dp}{m\,(c_p - c_v)} \quad \ldots (iii)$$

$$\ldots (\because R = c_p - c_v)$$

Equating equations (*ii*) and (*iii*),

$$\frac{-p\,dv}{m\,c_v} = \frac{p\,dv + v\,dp}{m\,(c_p - c_v)}$$

$$\frac{c_p - c_v}{c_v} = \frac{p\,dv + v\,dp}{-p\,dv} = -1 - \frac{v\,dp}{p\,dv}$$

$$\frac{c_p}{c_v} - 1 = -1 - \left(\frac{v}{dv} \times \frac{dp}{p}\right)$$

$$\gamma = -\left(\frac{v}{dv} \times \frac{dp}{p}\right) \quad \ldots \left(\because \frac{c_p}{c_v} = \gamma\right)$$

$$\therefore \qquad \gamma \times \frac{dv}{v} = -\frac{dp}{p}$$

$$\gamma \times \frac{dv}{v} + \frac{dp}{p} = 0$$

Integrating both sides,

$$\gamma \log_e v + \log_e p = \text{Constant} \quad \text{or} \quad \log_e pv^{\gamma} = \log_e C$$

$$\therefore \qquad pv^{\gamma} = C \quad \text{or} \quad p_1 v_1^{\gamma} = p_2 v_2^{\gamma} = \ldots = C \qquad \ldots (iv)$$

The equation (*iv*) may also be expressed in the following forms ;

$$\frac{p_1}{p_2} = \left(\frac{v_2}{v_1}\right)^{\gamma} \qquad \ldots (v)$$

From the general gas equation, we know that

$$\frac{p_1 v_1}{T_1} = \frac{p_2 v_2}{T_2} \quad \text{or} \quad \frac{p_1}{p_2} = \frac{T_1}{T_2} \times \frac{v_2}{v_1} \qquad \ldots (vi)$$

Equating equations (*v*) and (*vi*),

$$\left(\frac{v_2}{v_1}\right)^{\gamma} = \frac{T_1}{T_2} \times \frac{v_2}{v_1} \quad \text{or} \quad \frac{T_1}{T_2} = \left(\frac{v_2}{v_1}\right)^{\gamma} \times \frac{v_1}{v_2} = \left(\frac{v_2}{v_1}\right)^{\gamma} \left(\frac{v_2}{v_1}\right)^{-1}$$

$$\therefore \qquad \frac{T_1}{T_2} = \left(\frac{v_2}{v_1}\right)^{\gamma - 1} \qquad \ldots (vii)$$

From equation (*iv*), we also know that

$$\frac{v_1}{v_2} = \left(\frac{p_2}{p_1}\right)^{\frac{1}{\gamma}} \qquad \ldots (viii)$$

From the general gas equation, we know that

$$\frac{p_1 v_1}{T_1} = \frac{p_2 v_2}{T_2} \quad \text{or} \quad \frac{v_1}{v_2} = \frac{T_1}{T_2} \times \frac{p_2}{p_1} \qquad \ldots (ix)$$

Equating equations (*viii*) and (*ix*),

$$\left(\frac{p_2}{p_1}\right)^{\frac{1}{\gamma}} = \frac{T_1}{T_2} \times \frac{p_2}{p_1} \quad \text{or} \quad \frac{T_1}{T_2} = \left(\frac{p_2}{p_1}\right)^{\frac{1}{\gamma}} \times \frac{p_1}{p_2} = \left(\frac{p_1}{p_2}\right)^{-\frac{1}{\gamma}+1}$$

$$\therefore \qquad \frac{T_1}{T_2} = \left(\frac{p_1}{p_2}\right)^{\frac{\gamma - 1}{\gamma}} \qquad \ldots (x)$$

2. *Workdone during adiabatic expansion*

We have already discussed that workdone,

$$\delta W = p\, dv$$

On integrating from state 1 to state 2,

$$\int_1^2 \delta W = \int_1^2 p\, dv \quad \text{or} \quad W_{1-2} = \int_1^2 p\, dv \qquad \ldots (xi)$$

Since adiabatic expansion of the gas follows the law,

$$pv^\gamma = p_1 v_1^\gamma \quad \text{or} \quad p = \frac{p_1 v_1^\gamma}{v^\gamma}$$

Substituting this value of p in equation (*xi*),

$$W_{1-2} = \int_1^2 \frac{p_1 v_1^\gamma}{v^\gamma} dv = p_1 v_1^\gamma \int_1^2 v^{-\gamma} dv = p_1 v_1^\gamma \left[\frac{v^{-\gamma+1}}{-\gamma+1} \right]_1^2$$

$$= \frac{p_1 v_1^\gamma}{1-\gamma} \left[v_2^{1-\gamma} - v_1^{1-\gamma} \right]$$

$$= \frac{p_1 v_1^\gamma v_2^{1-\gamma} - p_1 v_1^\gamma v_1^{1-\gamma}}{1-\gamma}$$

$$= \frac{p_2 (v_2^\gamma v_2^{1-\gamma}) - p_1 (v_1^\gamma v_1^{1-\gamma})}{1-\gamma} \qquad \ldots (\because p_1 v_1^\gamma = p_2 v_2^\gamma)$$

$$= \frac{p_2 v_2 - p_1 v_1}{1-\gamma}$$

$$= \frac{p_1 v_1 - p_2 v_2}{\gamma - 1} \qquad \ldots \text{(For expansion)}$$

$$= \frac{p_2 v_2 - p_1 v_1}{\gamma - 1} \qquad \ldots \text{(For compression)}$$

Note : The above equation for work done may also be expressed as :

(*a*) We know that $p_1 v_1 = m R T_1$ and $p_2 v_2 = m R T_2$. Substituting these values in the equation for expansion,

$$W_{1-2} = \frac{m R T_1 - m R T_2}{\gamma - 1} = \frac{m R (T_1 - T_2)}{\gamma - 1} \qquad \ldots \text{(For expansion)}$$

$$= \frac{m R (T_2 - T_1)}{\gamma - 1} \qquad \ldots \text{(For compression)}$$

(*b*) We also know that work done during expansion,

$$W_{1-2} = \frac{p_1 v_1 - p_2 v_2}{\gamma - 1} = \frac{p_1 v_1}{\gamma - 1} \left[1 - \frac{p_2 v_2}{p_1 v_1} \right]$$

$$= \frac{m R T_1}{\gamma - 1} \left[1 - \frac{p_2 v_2}{p_1 v_1} \right] \qquad \ldots (\because p v = m R T)$$

3. *Change in internal energy*

We know that change in internal energy,

$$dU = U_2 - U_1 = m c_v (T_2 - T_1)$$

4. *Heat supplied or heat transferred*

We know that head supplied or heat transferred in case of adiabatic process is zero, therefore

$$Q_{1-2} = 0$$

5. *Change in enthalpy*

We know that change in enthalpy,

$$dH = H_2 - H_1 = m c_p (T_2 - T_1)$$

Example 3.8. *One litre of hydrogen at 0° C is suddenly compressed to one-half of its volume. Find the change in temperature of the gas, if the ratio of two specific heats for hydrogen is 1.4.*

Solution. Given : v_1 = 1 litre ; T_1 = 0° C = 0 + 273 = 273 K ; $v_2 = v_1/2$ = 1/2 = 0.5 litre ; γ = 1.4

Let T_2 = Final temperature of the gas.

We know that
$$\frac{T_1}{T_2} = \left(\frac{v_2}{v_1}\right)^{\gamma-1} = \left(\frac{0.5}{1}\right)^{1.4-1} = (0.5)^{0.4} = 0.758$$

$$\therefore \quad T_2 = T_1 / 0.758 = 273 / 0.758 = 360.16 \text{ K}$$
$$= 360.16 - 273 = 87.16° \text{ C} \quad \textbf{Ans.}$$

Example 3.9. *The initial volume of 0.18 kg of a certain gas was 0.15 m³ at a temperature of 15° C and a pressure of 1 bar. After adiabatic compression to 0.056 m³, the pressure was found to be 4 bar. Find ;*

1. Gas constant ; 2. Molecular mass of the gas ; 3. Ratio of specific heats ; 4. Two specific heats, one at a constant pressure and the other at a constant volume ; and 5. Change of internal energy.

Solution. Given : m = 0.18 kg ; v_1 = 0.15 m³ ; T_1 = 15° C = 15 + 273 = 288 K ; p_1 = 1 bar = 0.1 × 10⁶ N/m² ; v_2 = 0.056 m³ ; p_2 = 4 bar = 0.4 × 10⁶ N/m²

The p-v diagram is shown in Fig. 3.11.

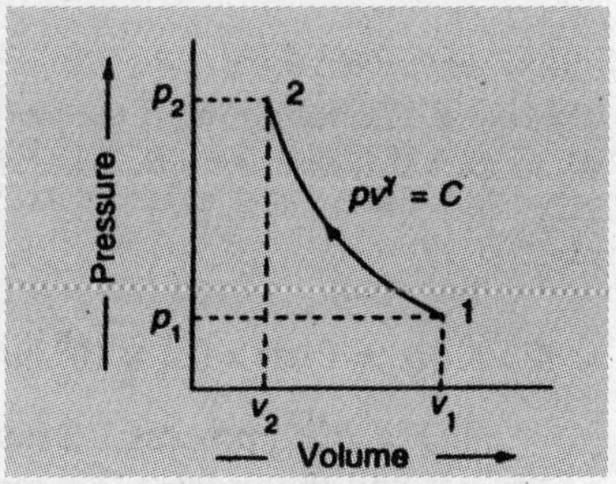

Fig. 3.11

1. *Gas constant*

Let R = Gas constant.

We know that
$$p_1 v_1 = m R T_1$$

$$\therefore \quad R = \frac{p_1 v_1}{m T_1} = \frac{0.1 \times 10^6 \times 0.15}{0.18 \times 288}$$
$$= 289.4 \text{ J/kg K} = 0.2894 \text{ kJ/kg K} \quad \textbf{Ans.}$$

2. *Molecular mass of the gas*

We know that molecular mass of the gas,
$$M = \frac{\text{Universal gas constant } (R_u)}{\text{Characteristic gas constant } (R)} = \frac{8314}{289.4} = 28.73 \text{ kg. } \textbf{Ans.}$$
... ($\because R_u$ = 8314 J/kg K, for all gases)

3. *Ratio of specific heats*

We know that ratio of specific heats,
$$\gamma = \frac{\log\left(\frac{p_2}{p_1}\right)}{\log\left(\frac{v_1}{v_2}\right)} = \frac{\log\left(\frac{0.4 \times 10^6}{0.1 \times 10^6}\right)}{\log\left(\frac{0.15}{0.056}\right)} = \frac{\log 4}{\log 2.678} = \frac{0.6020}{0.4278}$$
$$= 1.407 \quad \textbf{Ans.}$$

4. *Specific heat at a constant volume and constant pressure*

Let c_v = Specific heat at a constant volume, and

c_p = Specific heat at a constant pressure.

We know that $c_p - c_v = R$ or $1.407\, c_v - c_v = 0.2894$ ...($\because \gamma = c_p / c_v = 1.407$)

$\therefore$ $c_v = 0.2894 / 0.407 = 0.711$ kJ/kg K **Ans.**

and $c_p = 1.407\, c_v = 1.407 \times 0.711 = 1$ kJ/kg K **Ans.**

5. *Change in internal energy*

First of all, let us find the final temperature (T_2). We know that

$$\frac{T_1}{T_2} = \left(\frac{p_1}{p_2}\right)^{\frac{\gamma-1}{\gamma}} = \left(\frac{1}{4}\right)^{\frac{1.407-1}{1.407}} = (0.25)^{0.289} = 0.67$$

$$\therefore \quad T_2 = T_1 / 0.67 = 288 / 0.67 = 430 \text{ K}$$

We know that change in internal energy,

$$dU = U_2 - U_1 = m\, c_v (T_2 - T_1) = 0.18 \times 0.711 \,(430 - 288) \text{ kJ}$$

$$= 18.17 \text{ kJ} \quad \textbf{Ans.}$$

Example 3.10. *A system contains 0.15 m³ of a gas at a pressure of 3.8 bar and 150° C. It is expanded adiabatically till the pressure falls to 1 bar. The gas is then heated at a constant pressure till its enthalpy increases by 70 kJ. Determine the total work done. Take c_p = 1 kJ/kg K and c_v = 0.714 kJ/kg K.*

Solution. Given : $v_1 = 0.15$ m³ ; $p_1 = 3.8$ bar $= 0.38 \times 10^6$ N/m² ; $T_1 = 150°$ C $= 150 + 273 = 423$ K ; $p_2 = 1$ bar $= 0.1 \times 10^6$ N/m² ; $dH = 70$ kJ ; $c_p = 1$ kJ/kg K ; $c_v = 0.714$ kJ/kg K

In Fig. 3.12, process 1-2 represents adiabatic expansion of the gas and the process 2-3 represents heating at constant pressure.

Fig. 3.12

First of all, let us find the temperature (T_2) and volume (v_2) after the adiabatic expansion.

We know that adiabatic index,

$$\gamma = c_p / c_v = 1 / 0.714 = 1.4$$

$$\therefore \quad \frac{T_1}{T_2} = \left(\frac{p_1}{p_2}\right)^{\frac{\gamma-1}{\gamma}} = \left(\frac{3.8}{1}\right)^{\frac{1.4-1}{1.4}} = (3.8)^{0.286} = 1.465$$

or $$T_2 = T_1 / 1.465 = 423 / 1.465 = 288.7 \text{ K}$$

and $$\frac{v_1}{v_2} = \left(\frac{p_2}{p_1}\right)^{\frac{1}{\gamma}} = \left(\frac{1}{3.8}\right)^{\frac{1}{1.4}} = (0.263)^{0.714} = 0.385$$

$$\therefore \quad v_2 = v_1 / 0.385 = 0.15 / 0.385 = 0.39 \text{ m}^3$$

Now let us find the temperature (T_3) and volume (v_3) after constant pressure heating.

Let m = Mass of gas contained in the system.

We know that gas constant,

$$R = c_p - c_v = 1 - 0.714 = 0.286 \text{ kJ/kg K} = 286 \text{ J/kg K}$$

and $$p_1 v_1 = m R T_1$$

$$\therefore \quad m = \frac{p_1 v_1}{R T_1} = \frac{0.38 \times 10^6 \times 0.15}{286 \times 423} = 0.47 \text{ kg}$$

We also know that increase in enthalpy (dH),

$$70 = m c_p (T_3 - T_2) = 0.47 \times 1 (T_3 - 288.7) \text{ kJ}$$

$$\therefore \quad T_3 = \frac{70}{0.47} + 288.7 = 437.6 \text{ K}$$

Since the heating is at constant pressure, therefore

$$\frac{v_2}{T_2} = \frac{v_3}{T_3} \text{ or } v_3 = \frac{v_2 T_3}{T_2} = \frac{0.39 \times 437.6}{288.7} = 0.59 \text{ m}^3$$

We know that work done during adiabatic expansion,

$$W_{1-2} = \frac{p_1 v_1 - p_2 v_2}{\gamma - 1} = \frac{0.38 \times 10^6 \times 0.15 - 0.1 \times 10^6 \times 0.39}{1.4 - 1} \text{ J}$$

$$= \frac{57 \times 10^3 - 39 \times 10^3}{0.4} = 45\,000 \text{ J} = 45 \text{ kJ}$$

and workdone during constant pressure heating,

$$W_{2-3} = p_2 (v_3 - v_2) = 0.1 \times 10^6 (0.59 - 0.39) = 20\,000 \text{ J} = 20 \text{ kJ}$$

$\therefore$ Total work done, $W = W_{1-2} + W_{2-3} = 45 + 20 = 65 \text{ kJ}$ **Ans.**

Example 3.11. *0.336 m³ of gas at 10 bar and 150° C expands adiabatically, until its pressure is 4 bar. It is then compressed, isothermally, to its original volume. Find the final temperature and pressure of the gas. Also determine the change in internal energy. Take c_p = 0.996 kJ/kg K ; and c_v = 0.703 kJ/kg K.*

Solution. Given : $v_1 = 0.336 \text{ m}^3$; $p_1 = 10 \text{ bar} = 1 \times 10^6 \text{ N/m}^2$; $T_1 = 150° \text{ C} = 150 + 273 = 423 \text{ K}$; $p_2 = 4 \text{ bar} = 0.4 \times 10^6 \text{ N/m}^2$; $v_3 = v_1 = 0.336 \text{ m}^3$; $c_p = 0.996 \text{ kJ/kg K}$; $c_v = 0.703 \text{ kJ/kg K}$

In Fig. 3.13, process 1-2 represents the adiabatic expansion of the gas and the process 2-3 represents the isothermal compression to its original volume.

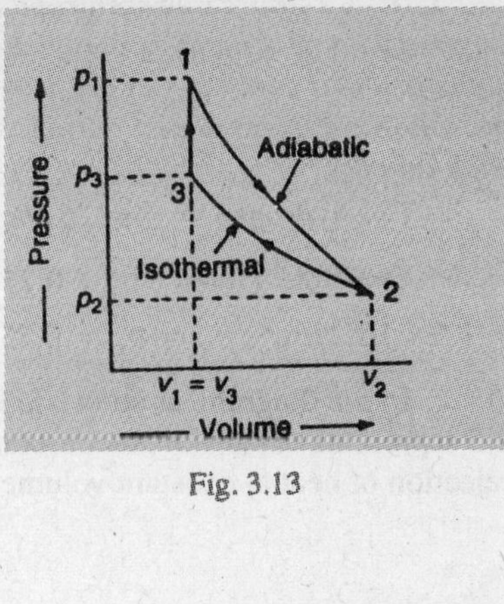

Fig. 3.13

We know that adiabatic index,

$$\gamma = c_p/c_v = 0.996/0.703 = 1.417$$

Final temperature of the gas

Let T_3 = Final temperature of the gas, and

T_2 = Temperature of the gas after adiabatic expansion.

We know that

$$\frac{T_1}{T_2} = \left(\frac{p_1}{p_2}\right)^{\frac{\gamma-1}{\gamma}} = \left(\frac{10}{4}\right)^{\frac{1.417-1}{1.417}} = (2.5)^{0.294} = 1.31$$

$$T_2 = T_1/1.31 = 423/1.31 = 323 \text{ K}$$

Since the compression is isothermal from 2 to 3, therefore

$$T_3 = T_2 = 323 \text{ K} = 50° \text{ C} \textbf{ Ans.}$$

Final pressure of the gas

Let p_3 = Final pressure of the gas.

We know that for a constant volume process 3-1,

$$\frac{p_3}{T_3} = \frac{p_1}{T_1}$$

$$\therefore \quad p_3 = \frac{T_3\, p_1}{T_1} = \frac{323 \times 1 \times 10^6}{423} = 0.76 \times 10^6 \text{ N/m}^2 = 7.6 \text{ bar } \textbf{Ans.}$$

Change in internal energy

First of all, let us find the mass of the gas (m).

We know that gas constant,

$$R = c_p - c_v = 0.996 - 0.703 = 0.293 \text{ kJ/kg K} = 293 \text{ J/kg K}$$

We also know that

$$p_1\, v_1 = m\, R\, T_1$$

or

$$m = \frac{p_1\, v_1}{R\, T_1} = \frac{1 \times 10^6 \times 0.336}{293 \times 423} = 2.7 \text{ kg}$$

$\therefore$ Change in internal energy,

$$dU = U_3 - U_1 = m\, c_v\, (T_3 - T_1) = 2.7 \times 0.703\, (323 - 423) \text{ kJ}$$

$$= -189.8 \text{ kJ } \textbf{Ans.}$$

The *negative* sign indicates that there is a decrease in internal energy.

Example 3.12. *0.75 m³ of hydrogen gas is initially at a pressure of 1 bar and temperature 290 K. It is compressed isentropically to 15 bar. Next it is expanded at constant temperature to original volume. Finally heat rejection takes place at constant volume and the gas pressure is restored to the original condition of pressure. Find : 1. Pressure, volume and temperature at the end of each operation ; 2. Heat added during isothermal expansion ; and 3. Change of internal energy during each process.*

For hydrogen, $R = 4126$ J/kg K ; $c_p = 14.26$ kJ/kg K.

Solution. Given : $v_1 = 0.75$ m³ ; $p_1 = 1$ bar $= 0.1 \times 10^6$ N/m² ; $T_1 = 290$ K ; $p_2 = 15$ bar $= 1.5 \times 10^6$ N/m² ; $v_3 = v_1 = 0.75$ m³ ; $R = 4126$ J/kg K $= 4.126$ kJ/kg K ; $c_p = 14.26$ kJ/kg K

In *p-v* diagram, as shown in Fig. 3.14, process 1-2 represents isentropic compression, process 2-3 represents expansion at constant temperature (isothermal expansion) and process 3-1 represents rejection of heat at constant volume.

1. *Pressure, volume and temperature at the end of each operation*

Let p_2, v_2 and T_2 = Pressure, volume and temperature respectively at the end of isentropic operation 1-2,

p_3, v_3 and T_3 = Pressure, volume and temperature respectively at the end of isothermal operation 2-3.

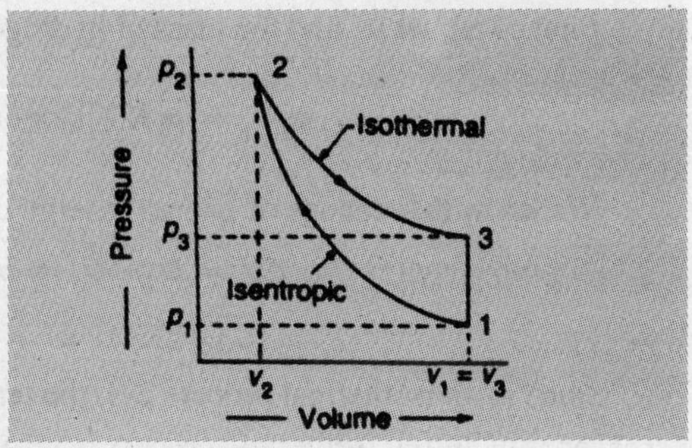

Fig. 3.14

First of all, let us find the value of isentropic index (γ). We know that

$$c_p - c_v = R$$

$$c_v = c_p - R = 14.26 - 4.126 = 10.134 \text{ kJ/kg K}$$

$$\gamma = c_p / c_v = 14.26 / 10.134 = 1.407$$

We know that for isentropic process 1-2,

$$p_1 (v_1)^\gamma = p_2 (v_2)^\gamma \quad \text{or} \quad \left(\frac{v_1}{v_2}\right)^\gamma = \frac{p_2}{p_1}$$

$$\therefore \quad v_2 = v_1 \left(\frac{p_1}{p_2}\right)^{\frac{1}{\gamma}} = 0.75 \left(\frac{1}{15}\right)^{\frac{1}{1.407}} = 0.75 \times 0.146 = 0.11 \text{ m}^3 \textbf{ Ans.}$$

and

$$\frac{p_1 v_1}{T_1} = \frac{p_2 v_2}{T_2}$$

or

$$T_2 = T_1 \times \frac{p_2 v_2}{p_1 v_1} = 290 \times \frac{1.5 \times 10^6 \times 0.11}{0.1 \times 10^6 \times 0.75} = 638 \text{ K } \textbf{Ans.}$$

Now for isothermal process 2-3,

$$p_2 v_2 = p_3 v_3$$

or

$$p_3 = p_2 \times \frac{v_2}{v_3} = 1.5 \times 10^6 \times \frac{0.11}{0.75} = 0.22 \times 10^6 \text{ N/m}^2 = 2.2 \text{ bar } \textbf{Ans.}$$

$$\ldots (\because v_3 = v_1)$$

2. *Heat added during isothermal expansion*

We know that heat added during isothermal expansion 2-3,

$$Q_{2-3} = 2.3\, p_2 v_2 \log \left(\frac{v_3}{v_2}\right)$$

$$= 2.3 \times 1.5 \times 10^6 \times 0.11 \log \left(\frac{0.75}{0.11}\right) = 0.38 \times 10^6 \times 0.834 \text{ J}$$

$$= 0.317 \times 10^6 \text{ J} = 317 \text{ kJ } \textbf{Ans.}$$

3. *Change of internal energy during each process*

First of all, let us find the mass of hydrogen (m). We know that

$$p_1 v_1 = m R T_1 \quad \text{or} \quad m = \frac{p_1 v_1}{R T_1} = \frac{0.1 \times 10^6 \times 0.75}{4126 \times 290} = 0.063 \text{ kg}$$

We know that change of internal energy during isentropic process,

$$dU = U_2 - U_1 = m c_v (T_2 - T_1) = 0.063 \times 10.134 (638 - 290) \text{ kJ}$$

$$= 222.18 \text{ kJ} \quad \textbf{Ans.}$$

Since in an isothermal process 2-3, the temperature is constant ($T_2 = T_3$), therefore change in internal energy during isothermal process is zero. **Ans.**

3.11. Polytropic Process

The polytropic process is also known as the general law for the expansion and compression of gases, and is given by the relation :

$$pv^n = \text{Constant}$$

where n is a polytropic index, which may have any value from zero to infinity, depending upon the manner, in which the expansion or compression has taken place.

The various equations for polytropic process may be expressed by changing the index n for γ in the adiabatic process.

Now consider m kg of a certain gas being heated polytropically from an initial state 1 to a final state 2.

Let p_1, v_1 and T_1 = Pressure, volume and temperature at the initial state 1, and

p_2, v_2 and T_2 = Pressure, volume and temperature at the final state 2.

The process is shown on the p-v diagram in Fig. 3.15. Now let us derive the following relations for the polytropic process.

1. *Pressure-volume-temperature (p-v-T) relationship*

The following relations for the polytropic process are derived in the similar way as discussed for adiabatic process.

(*a*) $p_1 v_1^n = p_2 v_2^n = \ldots = C$

(*b*) $$\frac{T_1}{T_2} = \left(\frac{v_2}{v_1}\right)^{n-1}$$

(*c*) $$\frac{v_1}{v_2} = \left(\frac{p_2}{p_1}\right)^{\frac{1}{n}}$$

(*d*) $$\frac{T_1}{T_2} = \left(\frac{p_1}{p_2}\right)^{\frac{n-1}{n}}$$

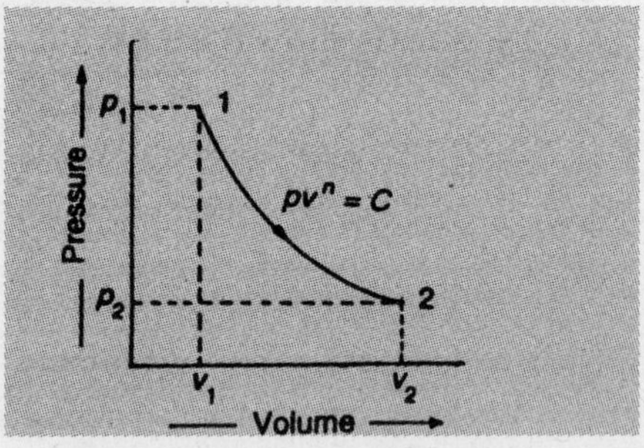

Fig. 3.15. Polytropic process.

2. *Workdone during polytropic expansion*

The equations for the work done during a polytropic process may also be expressed by changing the index n for γ in the adiabatic process.

∴ **Workdone during a polytropic process from state 1 to state 2,**

$$W_{1-2} = \frac{p_1 v_1 - p_2 v_2}{n-1} = \frac{m R (T_1 - T_2)}{n-1} \quad \text{. . . (For expansion)}$$

$$= \frac{p_2 v_2 - p_1 v_1}{n-1} = \frac{m R (T_2 - T_1)}{n-1} \quad \text{. . . (For compression)}$$

3. *Change in internal energy*

We know that change in internal energy,

$$dU = U_2 - U_1 = m c_v (T_2 - T_1)$$

4. *Heat supplied or heat transfer*

We know that the heat supplied or heat transferred,

$$Q_{1-2} = W_{1-2} + dU$$

$$= \frac{p_1 v_1 - p_2 v_2}{n-1} + m c_v (T_2 - T_1)$$

$$= \frac{m R (T_1 - T_2)}{n-1} + m \times \frac{R}{\gamma - 1} (T_2 - T_1) \quad \ldots \left(\because {}^{*}c_v = \frac{R}{\gamma - 1} \right)$$

$$= m R (T_1 - T_2) \left[\frac{1}{n-1} - \frac{1}{\gamma - 1} \right]$$

$$= m R (T_1 - T_2) \left[\frac{(\gamma - 1) - (n-1)}{(n-1)(\gamma - 1)} \right]$$

$$= m R (T_1 - T_2) \left[\frac{\gamma - n}{(n-1)(\gamma - 1)} \right]$$

$$= \frac{\gamma - n}{\gamma - 1} \times \frac{m R (T_1 - T_2)}{n-1}$$

5. *Change in enthalpy*

We know that change in enthalpy,

$$dH = H_2 - H_1 = m c_p (T_2 - T_1)$$

Notes : 1. **The equations for heat transfer may also be expressed as :**

(*a*)

$$Q_{1-2} = \frac{\gamma - n}{\gamma - 1} \times \text{Workdone} = \frac{\gamma - n}{\gamma - 1} \times \frac{p_1 v_1 - p_2 v_2}{(n-1)}$$

(*b*) We know that

$$Q_{1-2} = \frac{\gamma - n}{\gamma - 1} \times \frac{m R (T_1 - T_2)}{n-1} = \frac{\gamma - n}{n-1} \times m c_v (T_1 - T_2) \quad \ldots \left(\because \frac{R}{\gamma - 1} = c_v \right)$$

$$= \frac{\gamma - n}{n-1} \times \text{Change in internal energy}$$

---

* We know that $c_p - c_v = R$

Dividing throughout by c_v, we have

$$\frac{c_p}{c_v} - 1 = \frac{R}{c_v} \quad \text{or} \quad \gamma - 1 = \frac{R}{c_v}, \quad \therefore c_v = \frac{R}{\gamma - 1}$$

2. The above equations give the amount of heat, which has passed into the gas through the cylinder walls as the gas expands. It may be noted that this will happen only when n is less than γ. If however, n is greater than γ, then heat is rejected by the gas.

3. Similarly, during compression, work done will be negative, *i.e.* work will be done on the gas. Moreover, heat will be rejected by the gas. This will happen only when n is less than γ.

4. The relation for work done may also be expressed as :

$$W_{1-2} = \frac{p_1 v_1 - p_2 v_2}{(n-1)} = \frac{p_1 v_1\left(1 - \dfrac{p_2 v_2}{p_1 v_1}\right)}{(n-1)}$$

$$= \frac{p_1 v_1 (1 - r^{n-1})}{(n-1)} \qquad \left[\because \frac{p_2}{p_1} = \left(\frac{v_1}{v_2}\right)^n = r^n \text{ and } \frac{p_2 v_2}{p_1 v_1} = r^n \times \frac{1}{r} = r^{n-1}\right]$$

Example 3.13. *A certain quantity of air has a volume of 0.028 m³ at a pressure of 1.25 bar and 25° C. It is compressed to a volume of 0.0042 m³ according to the law $pv^{1.3}$ = Constant. Find the final temperature and work done during compression. Also determine the reduction in pressure at a constant volume required to bring the air back to its original temperature.*

Solution. Given : $v_1 = 0.028$ m³ ; $p_1 = 1.25$ bar $= 0.125 \times 10^6$ N/m² ; $T_1 = 25°$ C $= 25 + 273 = 298$ K ; $v_2 = 0.0042$ m³ ; $n = 1.3$

The *p-v* diagram is shown in Fig. 3.16.

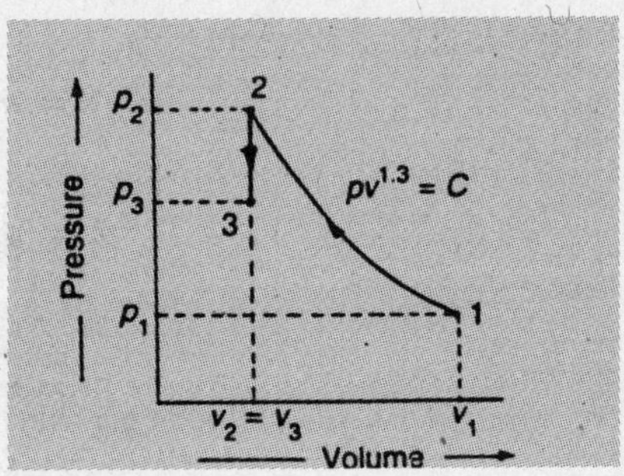

Fig. 3.16

Final temperature

Let T_2 = Final temperature.

We know that

$$\frac{T_1}{T_2} = \left(\frac{v_2}{v_1}\right)^{n-1} = \left(\frac{0.0042}{0.028}\right)^{1.3-1}$$

$$= (0.15)^{0.3} = 0.566$$

$\therefore \quad T_2 = T_1 / 0.566 = 298 / 0.566 = 526.5 \text{ K} = 526.5 - 273 = 253.5° \text{ C}$ **Ans.**

Workdone during compression

First of all, let us find the final pressure (p_2) at the end of compression. We know that

$$p_1 v_1^n = p_2 v_2^n \text{ or } \frac{p_1}{p_2} = \left(\frac{v_2}{v_1}\right)^n = \left(\frac{0.0042}{0.028}\right)^{1.3} = 0.085$$

$\therefore \quad p_2 = p_1 / 0.085 = 1.25 / 0.085 = 14.7 \text{ bar} = 1.47 \times 10^6 \text{ N/m}^2$

We know that workdone during compression,

$$W_{1-2} = \frac{p_2 v_2 - p_1 v_1}{n-1} = \frac{1.47 \times 10^6 \times 0.0042 - 0.125 \times 10^6 \times 0.028}{1.3 - 1}$$

$$= \frac{6174 - 3500}{0.3} = 8913 \text{ J} = 8.913 \text{ kJ}$$ **Ans.**

Pressure at a constant volume

Let p_3 = Pressure at a constant volume required to bring the air back to its initial temperature, $T_1 = 298$ K.

We know that for a constant volume process 2-3,

$$\frac{p_2}{T_2} = \frac{p_3}{T_3} \quad \text{or} \quad p_3 = \frac{p_2 T_3}{T_2} = \frac{14.7 \times 298}{526.5} = 8.32 \text{ bar} \; \textbf{Ans.}$$

$$\ldots (\because v_2 = v_3 \text{ and } T_3 = T_1)$$

Example 3.14. *A gas mixture obeying perfect gas law has a molecular mass of 26.7. The gas mixture is compressed through a compression ratio of 12 according to the law $pv^{1.25}$ = Constant, from initial conditions of 0.9 bar and 333 K. Assuming a mean molar specific heat at constant volume of 21.1 kJ/kg K, find, per kg of mass, the workdone and heat flow across the cylinder walls.*

For the above gas, determine the value of characteristic gas constant, molar specific heat at a constant pressure and ratio of specific heats.

Solution. Given : $M = 26.7$; $r = v_1/v_2 = 12$; $n = 1.25$; $p_1 = 0.9$ bar $= 0.09 \times 10^6$ N/m² ; $T_1 = 333$ K ; $c_{vm} = 21.1$ kJ/kg K ; $m = 1$ kg

The *p-v* diagram is shown in Fig. 3.17.

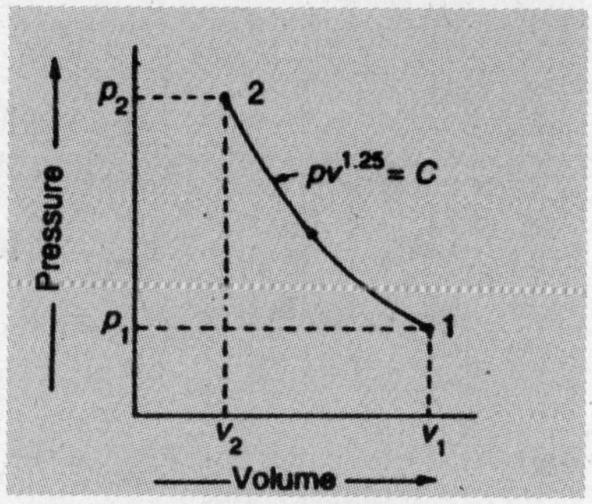

Fig. 3.17

Workdone per kg of gas

First of all, let us find the initial volume (v_1), final volume (v_2) and final pressure (p_2).

We know that $\quad p_1 v_1^n = p_2 v_2^n$

$$\therefore \quad p_2 = p_1 \left(\frac{v_1}{v_2}\right)^n = 0.9\,(12)^{1.25}$$

$$= 20.1 \text{ bar} = 2.01 \times 10^6 \text{ N/m}^2$$

We also know that gas constant,

$$R = \frac{R_u}{M} = \frac{8314}{26.7} = 311.4 \text{ J/kg K}$$

and

$$p_1 v_1 = m R T_1 \quad \text{or} \quad v_1 = \frac{m R T_1}{p_1} = \frac{1 \times 311.4 \times 333}{0.09 \times 10^6} = 1.15 \text{ m}^3$$

$$\therefore \quad v_2 = v_1/12 = 1.15/12 = 0.096 \text{ m}^3 \qquad \ldots (\because v_1/v_2 = 12)$$

We know that workdone

$$W_{1-2} = \frac{p_1 v_1 - p_2 v_2}{n-1} = \frac{0.09 \times 10^6 \times 1.15 - 2.01 \times 10^6 \times 0.096}{1.25 - 1}$$

$$= \frac{103\,500 - 192\,960}{0.25} = -357\,840 \text{ J} = -357.84 \text{ kJ} \; \textbf{Ans.}$$

The *negative* sign indicates that the work is done on the gas.

Heat flow across the cylinder walls

Let $\quad T_2$ = Final temperature.

We know that $$\frac{T_1}{T_2} = \left(\frac{v_2}{v_1}\right)^{n-1} = \left(\frac{1}{12}\right)^{1.25-1} = 0.537$$

$$\therefore \quad T_2 = T_1 / 0.537 = 333 / 0.537 = 620 \text{ K}$$

and specific heat at constant volume,

$$c_v = c_{vm} / M = 21.1 / 26.7 = 0.79 \text{ kJ/kg K}$$

∴ Change in internal energy,

$$dU = U_2 - U_1 = m\,c_v\,(T_2 - T_1) = 1 \times 0.79\,(620 - 333) = 226.7 \text{ kJ}$$

We know that heat flow across the cylinder walls,

$$Q_{1-2} = W_{1-2} + dU = -357.84 + 226.7 = -131.1 \text{ kJ} \quad \textbf{Ans.}$$

The *negative* sign indicates that the heat is rejected through the cylinder walls.

Characteristic gas constant

We know that characteristic gas constant,

$$R = \frac{\text{Universal gas constant}}{\text{Molecular mass}} = \frac{R_u}{M} = \frac{8314}{26.7}$$

... ($\because R_u$ = 8314 J/kg K, for all gases)

$$= 311.4 \text{ J/kg K} = 0.3114 \text{ kJ/kg K} \quad \textbf{Ans.}$$

Molar specific heat at a constant pressure

Let c_{pm} = Molar specific heat at a constant pressure.

We know that $c_p - c_v = R$ or $c_p - 0.79 = 0.3114$

$$\therefore \quad c_p = 0.3114 + 0.79 = 1.1014 \text{ kJ/kg K}$$

and $c_{pm} = M\,c_p = 26.7 \times 1.1014 = 29.4$ kJ/kg K **Ans.**

Ratio of specific heats

We know that ratio of specific heats,

$$\gamma = c_p / c_v = 1.1014 / 0.79 = 1.394 \quad \textbf{Ans.}$$

Example 3.15. *An internal combustion engine has the following dimensions :*

Diameter of cylinder = 550 mm ; Stroke = 750 mm ; Compression ratio = 13.5. At the end of the suction stroke, the pressure is 1 bar and temperature is 316 K. The compression follows the law $pv^{1.37} = C$. Determine :

1. the pressure and temperature at the end of compression ; 2. the mass of the charge ; 3. the work done during compression ; and 4. the heat rejected during compression.

Take c_p = 0.996 kJ/kg K and c_v = 0.707 kJ/kg K.

Solution. Given : D = 550 mm = 0.55 m ; L = 750 mm = 0.75 m ; v_1/v_2 = 13.5 ; p_1 = 1 bar = 0.1×10^6 N/m² ; T_1 = 316 K ; n = 1.37; c_p = 0.996 kJ/kg K ; c_v = 0.707 kJ/kg K

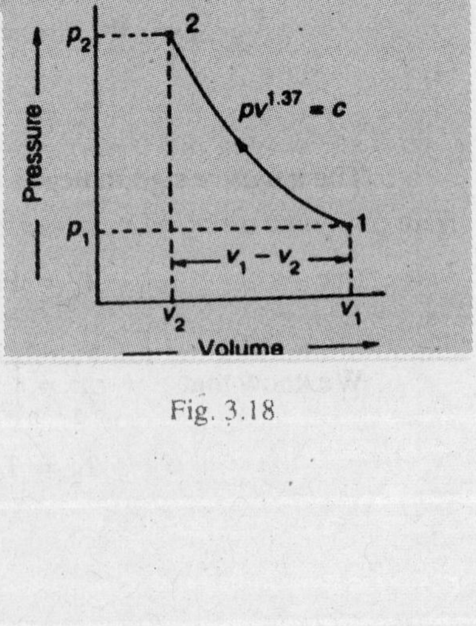

Fig. 3.18

In the *p-v* diagram, as shown in Fig. 3.18, point 1 refers to the end of suction stroke and point 2 refers to the end of compression stroke.

1. Pressure and temperature at the end of compression

Let p_2 and T_2 = Pressure and temperature at the end of compression respectively.

We know that

$$p_1 v_1^n = p_2 v_2^n$$

$$\therefore \qquad p_2 = p_1 \left(\frac{v_1}{v_2}\right)^n = 0.1 \times 10^6 \, (13.5)^{1.37} \text{ N/m}^2$$

$$= 3.536 \times 10^6 \text{ N/m}^2 = 3.536 \text{ bar } \textbf{Ans.}$$

We also know that

$$\frac{T_1}{T_2} = \left(\frac{v_2}{v_1}\right)^{n-1} = \left(\frac{1}{13.5}\right)^{1.37-1} = (0.074)^{0.37} = 0.3816$$

$$\therefore \qquad T_2 = T_1 / 0.3816 = 316 / 0.3816 = 828 \text{ K } \textbf{Ans.}$$

2. *Mass of the charge*

Let $\qquad m$ = **Mass of the charge.**

We know that swept volume,

$$v_1 - v_2 = \frac{\pi}{4} \times D^2 \times L = \frac{\pi}{4} (0.55)^2 \, 0.75 = 0.178 \text{ m}^3$$

or
$$v_1 - \frac{v_1}{13.5} = 0.178 \qquad \ldots \left(\because \frac{v_1}{v_2} = 13.5\right)$$

$$\therefore \qquad v_1 = 0.178 \times 13.5 / 12.5 = 0.192 \text{ m}^3$$

and gas constant,
$$R = c_p - c_v = 0.996 - 0.707 = 0.289 \text{ kJ/kg K} = 289 \text{ J/kg K}$$

We know that $\quad p_1 v_1 = m R T_1$

$$\therefore \qquad m = \frac{p_1 v_1}{R T_1} = \frac{0.1 \times 10^6 \times 0.192}{289 \times 316} = 0.21 \text{ kg } \textbf{Ans.}$$

3. *Workdone during compression*

We know that workdone during compression,

$$W_{1-2} = \frac{m R (T_2 - T_1)}{n - 1} = \frac{0.21 \times 0.289 \, (828 - 316)}{1.37 - 1} = 84 \text{ kJ } \textbf{Ans.}$$

4. *Heat rejected during compression*

We know that adiabatic index,

$$\gamma = c_p / c_v = 0.996 / 0.707 = 1.41$$

∴ Heat rejected during compression,

$$Q_{1-2} = \frac{\gamma - n}{\gamma - 1} \times \text{Workdone during compression}$$

$$= \frac{1.41 - 1.37}{1.41 - 1} \times 84 = 8.195 \text{ kJ } \textbf{Ans.}$$

Example 3.16. *0.2 m³ of mixture of fuel and air at 1.2 bar and 60° C is compressed until its pressure becomes 12 bar and temperature becomes 270° C. Then, it is ignited suddenly at constant volume and its pressure becomes twice the pressure at the end of compression. Calculate the maximum temperature reached and change in internal energy. Also compute the heat transfer during the compression process. Consider mixture as a prefect gas and take* c_p *= 1.072 kJ/kg K ; and R = 294 J/kg K.*

Solution. Given : $v_1 = 0.2\ m^3$; $p_1 = 1.2$ bar $= 0.12 \times 10^6\ N/m^2$; $T_1 = 60°\ C = 60 + 273 = 333$ K ; $p_2 = 12$ bar $= 1.2 \times 10^6\ N/m^2$; $T_2 = 270°\ C = 270 + 273 = 543$ K ; $p_3 = 2\,p_2$; $c_p = 1.072$ kJ/kg K ; $R = 294$ J/kg K $= 0.294$ kJ/kg K.

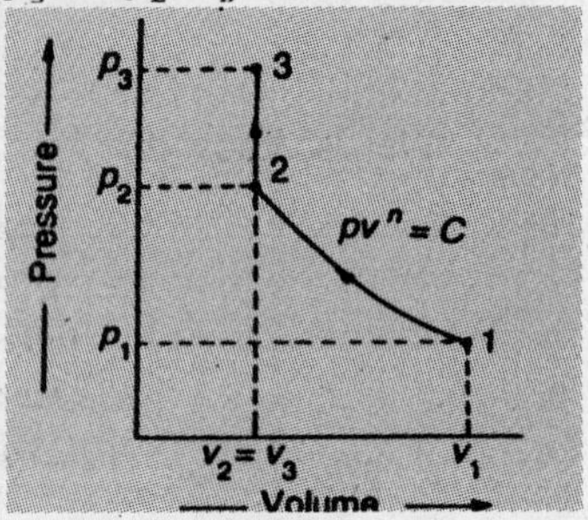

Fig. 3.19

In the *p-v* diagram, as shown in Fig. 3.19, process 1-2 represents polytropic compression (*i.e.* according to the general law $p\,v^n = C$) and the process 2-3 represents the constant volume process.

Maximum temperature

Let T_3 = Maximum temperature.

Since the process 2-3 is a constant volume process, therefore

$$\frac{p_2}{T_2} = \frac{p_3}{T_3}$$

or

$$T_3 = T_2 \times \frac{p_3}{p_2} = 543 \times \frac{2\,p_2}{p_2} = 1086 \text{ K} = 813° \text{ C} \quad \textbf{Ans.}$$

Change in internal energy

First of all, let us find the mass of the mixture compressed (*i.e.* m) and c_v.

We know that $p_1\,v_1 = m\,R\,T_1$ or $m = \dfrac{p_1\,v_1}{R\,T_1} = \dfrac{0.12 \times 10^6 \times 0.2}{294 \times 333} = 0.245$ kg

and $c_v = c_p - R = 1.072 - 0.294 = 0.778$ kJ/kg K $\quad \ldots (\because c_p - c_v = R)$

We know that change in internal energy,

$$dU = U_3 - U_1 = m\,c_v\,(T_3 - T_1) = 0.245 \times 0.778\,(1086 - 333) \text{ kJ}$$

$$= 143.5 \text{ kJ} \quad \textbf{Ans.}$$

Heat transfer during compression process

First of all, let us find the value of polytropic index (n) for the compression process 1-2.

We know that $\dfrac{T_1}{T_2} = \left(\dfrac{p_1}{p_2}\right)^{\frac{n-1}{n}}$ or $\dfrac{333}{543} = \left(\dfrac{1.2}{12}\right)^{\frac{n-1}{n}}$

Taking log on both sides,

$$\log\left(\frac{333}{543}\right) = \frac{n-1}{n}\log\left(\frac{1.2}{12}\right) \quad \text{or} \quad \log 0.613 = \frac{n-1}{n}\log 0.1$$

$$\therefore \quad -0.2125 = \frac{n-1}{n}(-1) \quad \text{or} \quad n = 1.27$$

We know that workdone,

$$W_{1-2} = \frac{m\,R\,(T_1 - T_2)}{n-1} = \frac{0.245 \times 0.294\,(333 - 543)}{1.27 - 1} = -56 \text{ kJ}$$

The *negative* sign shows that the gas is compressed and the work is done on the gas.

We know that change in internal energy,

$$dU = (U_2 - U_1) = m\,c_v\,(T_2 - T_1) = 0.245 \times 0.778\,(543 - 333) = 40 \text{ kJ}$$

∴ Heat transfer during compression process,

$$Q_{1-2} = W_{1-2} + dU = -56 + 40 = -16 \text{ kJ} \textbf{ Ans.}$$

The *negative* sign indicates that the heat is rejected by the gas.

Example 3.17. *An oil engine has a volume of 60 litres and a compression ratio of 14.2 to 1. At the beginning of the compression stroke, the pressure and temperature are 1 bar and 80° C respectively. At the end of compression, the pressure is 30 bar. The charge is now heated at constant pressure until the volume is doubled. Find :*

1. The index of compression ; 2. The temperature at the end of compression, 3. The heat transfer ; and 4. The heat received in constant pressure operation.

Assume c_v = *0.712 kJ/kg K and R = 0.293 kJ/kg K.*

Solution. Given : $v_1 = 60$ litres $= 60 \times 10^{-3}$ m^3 ; $v_1/v_2 = 14.2$; $p_1 = 1$ bar $= 0.1 \times 10^6$ N/m^2 ; $T_1 = 80°$ C $= 80 + 273 = 353$ K ; $p_2 = 30$ bar $= 3 \times 10^6$ N/m^2 ; $v_3 = 2\, v_2$; $c_v = 0.712$ kJ/kg K ; $R = 0.293$ kJ/kg K $= 293$ J/kg K

In Fig. 3.20, 1-2 represents the polytropic compression process and 2-3 represents the constant pressure heating process.

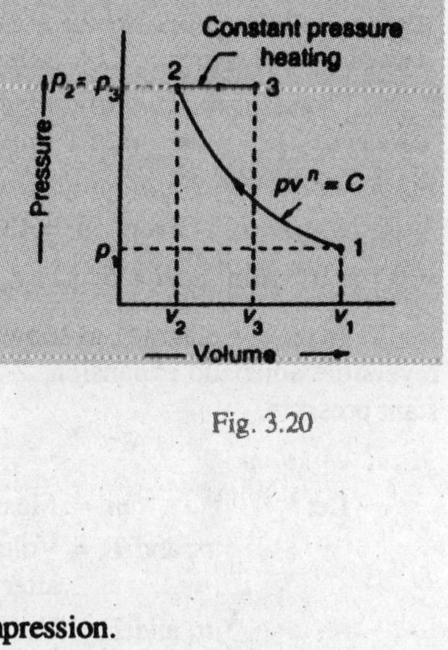

Fig. 3.20

1. *Index of compression*

Let n = Index of compression.

We know that $p_1 v_1^n = p_2 v_2^n$

$$\left(\frac{v_1}{v_2}\right)^n = \frac{p_2}{p_1}$$

$$(14.2)^n = \frac{3 \times 10^6}{0.1 \times 10^6} = 30$$

Taking log on both sides,

$$n \log 14.2 = \log 30$$

$$n \times 1.152 = 1.477$$

$$\therefore \quad n = 1.282 \textbf{ Ans.}$$

2. *Temperature at the end of compression*

Let T_2 = Temperature at the end of compression.

We know that
$$\frac{T_1}{T_2} = \left(\frac{v_2}{v_1}\right)^{n-1} = \left(\frac{1}{14.2}\right)^{1.282-1} = (0.0704)^{0.282} = 0.473$$

$$\therefore \quad T_2 = T_1 / 0.473 = 353 / 0.473 = 746.3 \text{ K} = 473.3° \text{ C} \textbf{ Ans.}$$

3. *Heat transfer*

First of all, let us find the mass of the charge (m) and the ratio of specific heats (γ). We know that

$$p_1 v_1 = m R T_1 \quad \text{or} \quad m = \frac{p_1 v_1}{R T_1} = \frac{0.1 \times 10^6 \times 60 \times 10^{-3}}{293 \times 353} = 0.058 \text{ kg}$$

We also know that

$$c_p - c_v = R \quad \text{or} \quad c_p = c_v + R = 0.712 + 0.293 = 1.005 \text{ kJ/kg K}$$

$$\therefore \quad \gamma = c_p / c_v = 1.005 / 0.712 = 1.41$$

We know that heat transfer,

$$Q_{1-2} = \frac{\gamma - n}{\gamma - 1} \times \frac{m R (T_1 - T_2)}{n - 1} = \frac{1.41 - 1.282}{1.41 - 1} \times \frac{0.058 \times 0.293 (353 - 746.3)}{1.282 - 1}$$

$$= 0.3122 \times - 23.7 = -7.4 \text{ kJ } \textbf{Ans.}$$

The *negative* sign indicates that the heat is rejected.

4. Heat received in constant pressure operation

First of all, let us find the temperature (T_3) at the end of constant pressure operation. We know that, for constant pressure process 2-3,

$$\frac{v_2}{T_2} = \frac{v_3}{T_3} \quad \text{or} \quad T_3 = \frac{v_3 T_2}{v_2} = \frac{2 v_2 \times 746.3}{v_2} = 1492.6 \text{ K} \qquad \ldots (\because v_3 = 2 v_2)$$

We know that heat received,

$$Q_{2-3} = m c_p (T_3 - T_2) = 0.058 \times 1.005 (1492.6 - 746.3) = 43.5 \text{ kJ } \textbf{Ans.}$$

Example 3.18. *A system contains 0.15 m^3 of air at 4 bar and 423 K. A reversible adiabatic expansion takes place till the pressure falls to 1 bar. The air is then heated at constant pressure till enthalpy increases by 67 kJ. Determine the total workdone.*

If these processes are replaced by a single reversible polytropic process giving the same work between the same initial and final states, determine the index of expansion. Take c_p = 1.009 kJ/kg K.

Solution. Given : v_1 = 0.15 m^3 ; p_1 = 4 bar = 0.4 × 10^6 N/m^2 ; T_1 = 423 K ; p_2 = 1 bar = 0.1 × 10^6 N/m^2 ; dH = 67 kJ ; c_p = 1.009 kJ/kg K

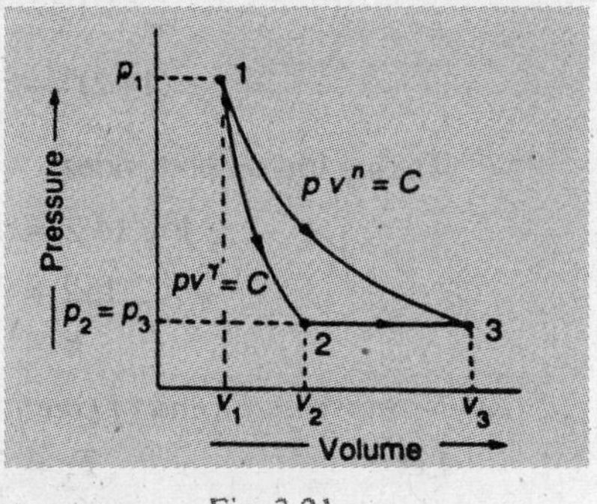

Fig. 3.21

In the *p-v* diagram, as shown in Fig. 3.21, 1-2 represents reversible adiabatic expansion, 2-3 represents heating at constant pressure.

Total workdone

Let m = Mass of air,

v_2 and T_2 = Volume and temperature of air after adiabatic expansion,

v_3 and T_3 = Volume and temperature of air after heating at constant pressure.

We know that

$$p_1 v_1 = m R T_1 \quad \text{or} \quad m = \frac{p_1 v_1}{R T_1} = \frac{0.4 \times 10^6 \times 0.15}{287 \times 423} = 0.494 \text{ kg}$$

...(Taking R = 287 J/kg K)

and $$p_1 (v_1)^\gamma = p_2 (v_2)^\gamma \quad \text{or} \quad \left(\frac{v_2}{v_1}\right)^\gamma = \frac{p_1}{p_2}$$

$$\therefore \quad v_2 = v_1 \left(\frac{p_1}{p_2}\right)^{\frac{1}{\gamma}} = 0.15 \left(\frac{4}{1}\right)^{\frac{1}{1.4}} = 0.4036 \text{ m}^3 \qquad \ldots (\because \gamma \text{ for air} = 1.4)$$

We know that the workdone during adiabatic expansion 1-2,

$$W_{1-2} = \frac{p_1 v_1 - p_2 v_2}{\gamma - 1} = \frac{0.4 \times 10^6 \times 0.15 - 0.1 \times 10^6 \times 0.4036}{1.4 - 1} \text{ J}$$

$$= \frac{60\,000 - 40\,360}{0.4} = 49\,100 \text{ J} = 49.1 \text{ kJ}$$

Now let us find the workdone during heating at constant pressure.

We know that

$$p_2 v_2 = m R T_2 \quad \text{or} \quad T_2 = \frac{p_2 v_2}{m R} = \frac{0.1 \times 10^6 \times 0.4036}{0.494 \times 287} = 284.6 \text{ K}$$

and the increase in enthalpy (dH),

$$67 = m c_p (T_3 - T_2) = 0.494 \times 1.009 (T_3 - 284.6) = 0.5 (T_3 - 284.6)$$

$$\therefore \quad T_3 = \frac{67}{0.5} + 284.6 = 418.6 \text{ K}$$

Since the heating is at constant pressure from 2 to 3, therefore

$$\frac{v_2}{T_2} = \frac{v_3}{T_3} \quad \text{or} \quad v_3 = \frac{v_2 T_3}{T_2} = \frac{0.4036 \times 418.6}{284.6} = 0.594 \text{ m}^3$$

We know that workdone during constant pressure process 2-3,

$$W_{2-3} = p_2 (v_3 - v_2) = 0.1 \times 10^6 (0.594 - 0.4036) = 19\,040 \text{ J} = 19.04 \text{ kJ}$$

$\therefore$ Total workdone,

$$W_{1-2-3} = W_{1-2} + W_{2-3} = 49.1 + 19.04 = 68.14 \text{ kJ} \quad \textbf{Ans.}$$

Index of expansion

Let n = Index of expansion.

The process 1-2 and 2-3 are replaced by a single polytropic process 1-3 giving the same work *i.e.* 68.14 kJ or 68.14×10^3 J.

We know that workdone during a polytropic process 1-3 (W_{1-3}),

$$68.14 \times 10^3 = \frac{p_1 v_1 - p_3 v_3}{n - 1} = \frac{0.4 \times 10^6 \times 0.15 - 0.1 \times 10^6 \times 0.594}{n - 1} = \frac{600}{n - 1}$$

$$\therefore \quad n - 1 = 600 / 68.14 \times 10^3 = 0.0088 \quad \text{or} \quad n = 1.0088 \quad \textbf{Ans.}$$

3.12. Rate of Heat Transfer (Absorption or Rejection) per Unit Volume During a Polytropic Process

In the previous article, we have seen that heat transfer (absorbed or rejected) during a polytropic process,

$$Q_{1-2} = \frac{\gamma - n}{\gamma - 1} \times W_{1-2}$$

where W_{1-2} is the work done during polytropic process.

If dQ is the small quantity of heat transfer during small change of pressure and volume, then

$$dQ = \frac{\gamma - n}{\gamma - 1} \times p \, dv$$

∴ Rate of heat transfer per unit volume,

$$\frac{dQ}{dv} = \frac{\gamma - n}{\gamma - 1} \times p$$

and rate of heat transfer per second,

$$\frac{dQ}{dt} = \frac{dQ}{dv} \times \frac{dv}{dt} = \frac{\gamma - n}{\gamma - 1} \times p \times \frac{dv}{dt}$$

where $\frac{dv}{dt}$ is the swept volume of the piston per second.

Example 3.19. *The law of the expansion curve of a gas engine indicator is found to be $pv^{1.3}$ = Constant, and the ratio of specific heats of the mixture is 1.37. If the piston sweeps out 2 m³/min, when the pressure on this expansion curve is 14 bar, what is the rate of heat transfer per second at this instant ?*

Solution. Given : $n = 1.3$; $\gamma = 1.37$; $dv/dt = 2$ m³/min = 0.0333 m³/s ; $p = 14$ bar $= 1.4 \times 10^6$ N/m²

We know that the rate of heat transfer per second,

$$\frac{dQ}{dt} = \frac{\gamma - n}{\gamma - 1} \times p \times \frac{dv}{dt} = \frac{1.37 - 1.3}{1.37 - 1} \times 1.4 \times 10^6 \times 0.0333 = 8820 \text{ J}$$

$$= 8.82 \text{ kJ} \quad \textbf{Ans.}$$

3.13. Determination of polytropic Index

We know that for a polytropic process,

$$p_1 v_1^n = p_2 v_2^n$$

Taking logarithms on both sides,

$$\log p_1 + n \log v_1 = \log p_2 + n \log v_2$$

or

$$n \log v_1 - n \log v_2 = \log p_2 - \log p_1$$

$$n (\log v_1 - \log v_2) = \log p_2 - \log p_1$$

$$n \log \left(\frac{v_1}{v_2} \right) = \log \left(\frac{p_2}{p_1} \right)$$

$$\therefore \quad n = \frac{\log \left(\frac{p_2}{p_1} \right)}{\log \left(\frac{v_1}{v_2} \right)}$$

Note : Proceeding in the similar way, we can find out that adiabatic index,

$$\gamma = \frac{\log \left(\frac{p_2}{p_1} \right)}{\log \left(\frac{v_1}{v_2} \right)}$$

Example 3.20. *A gas initially at 603 K expands until its volume is 5.2 times the initial volume, according to pv^n = Constant. If the initial and final pressures are observed to be 8.5 bar and 1 bar, determine : 1. the index of expansion, 2. work done per kg of gas, and 3. heat exchange per kg of gas.*

Assume c_v = 0.712 kJ/kg K and γ = 1.4.

Solution. Given : $T_1 = 603$ K ; $v_2 = 5.2\, v_1$; $p_1 = 8.5$ bar $= 0.85 \times 10^6$ N/m^2 ; $p_2 = 1$ bar $= 0.1 \times 10^6$ N/m^2 ; $c_v = 0.712$ kJ/kg K ; $\gamma = c_p / c_v = 1.4$

The p-v diagram is shown in Fig. 3.22.

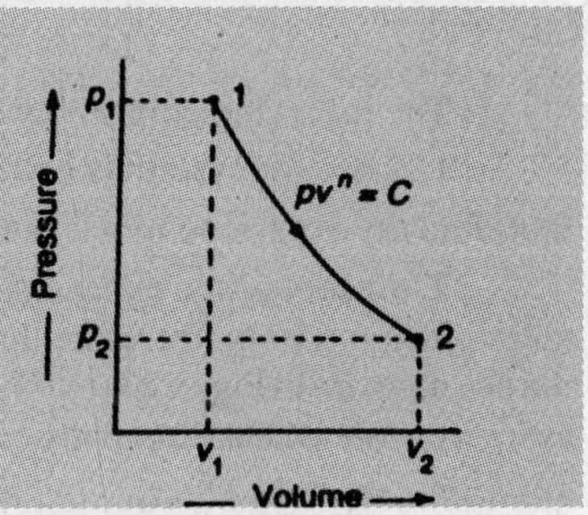

Fig. 3.22

1. *Index of expansion*

We know that index of expansion,

$$n = \frac{\log\left(\dfrac{p_2}{p_1}\right)}{\log\left(\dfrac{v_1}{v_2}\right)} = \frac{\log\left(\dfrac{0.1 \times 10^6}{0.85 \times 10^6}\right)}{\log\left(\dfrac{v_1}{5.2\, v_1}\right)}$$

$$= \frac{\log(0.1176)}{\log(0.1923)} = \frac{-0.9296}{-0.7160} = 1.3 \textbf{ Ans.}$$

2. *Workdone per kg of gas*

First of all, let us find the initial volume (v_1) and final volume (v_2).

We know that gas constant,

$$R = c_p - c_v = 1.4\, c_v - c_v = 0.4\, c_v = 0.4 \times 0.712 = 0.2848 \text{ kJ/kg K}$$

$$= 284.8 \text{ J/kg K} \qquad \ldots(\because c_p / c_v = \gamma = 1.4)$$

and $$p_1 v_1 = m R T_1 \text{ or } v_1 = \frac{m R T_1}{p_1} = \frac{1 \times 284.8 \times 603}{0.85 \times 10^6} = 0.202 \text{ m}^3$$

$$\ldots(\because m = 1 \text{ kg})$$

$$\therefore \quad v_2 = 5.2\, v_1 = 5.2 \times 0.202 = 1.05 \text{ m}^3$$

We know that workdone per kg of gas,

$$W_{1-2} = \frac{p_1 v_1 - p_2 v_2}{n-1} = \frac{0.85 \times 10^6 \times 0.202 - 0.1 \times 10^6 \times 1.05}{1.3 - 1} \text{ J}$$

$$= 0.222 \times 10^6 \text{ J} = 222 \text{ kJ } \textbf{Ans.}$$

3. *Heat exchange per kg of gas*

We know that heat exchange per kg of gas,

$$Q_{1-2} = \frac{\gamma - n}{\gamma - 1} \times \text{Workdone per kg of gas} = \frac{1.4 - 1.3}{1.4 - 1} \times 222 = 55.5 \text{ kJ } \textbf{Ans.}$$

3.14. Free Expansion (or Unresisted Expansion) Process

The free expansion (or unresisted expansion) process is an irreversible non-flow process. A free expansion occurs when a fluid is allowed to expand suddenly into a vacuum chamber through an orifice of large dimensions.

Consider two chambers A and B separated by a partition as shown in Fig. 3.23 (*a*). Let the chamber A contains a perfect gas having volume v_1, pressure p_1, and temperature T_1 and the chamber B is completely evacuated. These chambers are perfectly insulated so that no heat transfer takes place from or to its surroundings. Now, if the partition is removed, the gas will expand freely and occupy the whole space as shown in Fig. 3.23 (*b*). By this, the volume of the gas increases to v_2, pressure decreases to p_2 and the temperature may also decrease to T_2.

Since there is no expansion of the boundary of the system, because it is rigid, therefore no work is done. Thus, for a free expansion process,

$$Q_{1-2} = 0\ ;\ W_{1-2} = 0 \text{ and } dU = 0$$

The following points may be noted regarding the free expansion of a gas :

1. Since the system is perfectly insulated so that no heat transfer takes place (*i.e.* $Q_{1-2} = 0$), therefore the expansion of gas may be called as an adiabatic expansion.

2. Since the free expansion of the gas from the equilibrium state 1 to the equilibrium state 2 takes place suddenly, therefore the intermediate states will not be in equilibrium states, as shown on the *p-v* diagram in Fig. 3.23 (*c*). Thus the process is irreversible and the expansion is, therefore, known as irreversible adiabatic expansion.

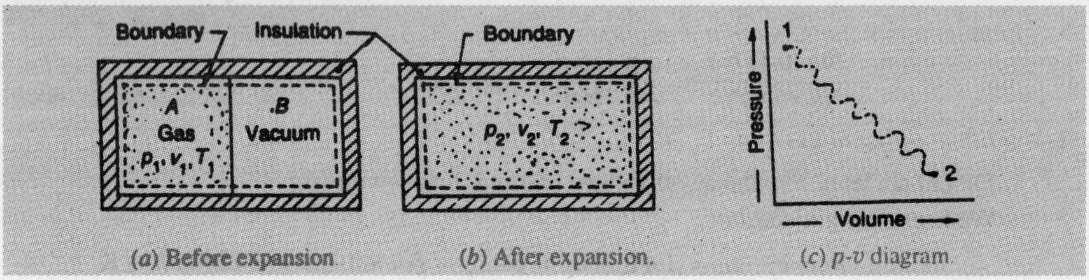

Fig. 3.23. Free expansion.

3. Since there is no resistance to overcome during free expansion process, therefore no work is done by the system (*i.e.* $W_{1-2} = 0$). Thus, the free expansion process is also known as * ***unresisted expansion process.***

4. According to the first law of thermodynamics,

$$Q_{1-2} = W_{1-2} + dU$$

Since for the free expansion, $Q_{1-2} = 0$ and $W_{1-2} = 0$, therefore the change in internal energy,

$$dU = U_2 - U_1 = 0 \text{ or } U_2 = U_1$$

In other words, the internal energy of the system, in a free expansion process remains constant. Thus, the free expansion process is also known as ***constant internal energy process.***

5. We know that change in internal energy,

$$dU = m\, c_v\, dT = m\, c_v\, (T_2 - T_1)$$

Since $dU = 0$, therefore $dT = 0$ or $T_2 = T_1$, *i.e.* there is no change in temperature of the system. In other words, the temperature of the system, in a free expansion process remains constant.

Note : It can not be called an isothermal process because in an actual isothermal process, work is done by the gas during expansion.

6. We know that change in enthalpy,

$$dH = H_2 - H_1 = m\, c_p\, dT$$

Since $dT = 0$, therefore $dH = 0$ or $H_2 = H_1$. In other words, the enthalpy of the system in a free expansion process remains constant. Thus the free expansion process may also be called ***constant enthalpy process.***

---

* In the previous non-flow processes discussed, the expansion was against the resistance offered by the piston. So all the non-flow processes discussed earlier are resisted expansion processes.

3.15. General Laws for Expansion and Compression

The general law of expansion or compression of a perfect gas is pv^n = Constant (Art. 3.11). It gives the relationship between pressure and volume of a given quantity of gas. The value of n depends upon the nature of gas, and condition under which the changes (*i.e.* expansion or compression) take place. The value of n may be between zero and infinity. But the following values of n are important from the subject point of view :

1. When $n = 0$. This means pv^0 = Constant, *i.e.* p = Constant. In other words, for the expansion or compression of a perfect gas at *constant pressure,* $n = 0$.
2. When $n = 1$; then pv = Constant, *i.e.* the expansion or compression is *isothermal* or *hyperbolic.*
3. When n lies between 1 and n, the expansion or compression is *polytropic, i.e.* pv^n = Constant.
4. When $n = \gamma$, the expansion or compression is *adiabatic, i.e.* pv^γ = Constant.
5. When $n = \infty$, the expansion or compression is at constant volume, *i.e.* v = Constant.

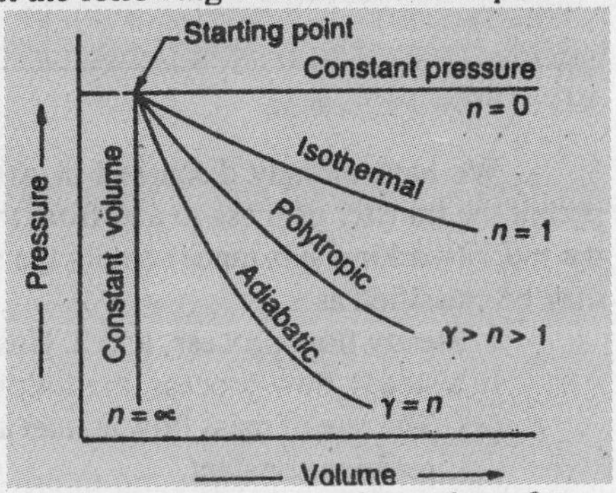

Fig. 3.24. Curves for various values of n.

Fig. 3.24 shows the curves of expansion of a perfect gas for different values of n. It is obvious that greater the value of n, steeper the curve of expansion.

3.16. Summary of Formulae for Heating and Expansion of Perfect Gases in Reversible Non-flow processes

The following table shows the summary of formulae for heating and expansion of perfect gases in reversible non-flow processes :

Table 3.1. Summary of formulae for heating and expansion of perfect gases.

S. No.	*Type of reversible non-flow process*	*p-v-T relation*	*Workdone* (W_{1-2})	*Change of internal energy* ($dU = U_2 - U_1$)	*Heat supplied* (Q_{1-2}) $= W_{1-2} + dU$	*Change in enthalpy* ($dH = H_2 - H_1$)
1.	Constant volume or Isochoric (v = Constant)	$\frac{p_1}{T_1} = \frac{p_2}{T_2}$ (Gay-Lussac Law)	0	$m\,c_v(T_2 - T_1)$	$m\,c_v(T_2 - T_1)$	$m\,c_p(T_2 - T_1)$
2.	Constant pressure or Isobaric (p = Constant)	$\frac{v_1}{T_1} = \frac{v_2}{T_2}$ (Charles' Law)	$p(v_2 - v_1)$ or $mR(T_2 - T_1)$	$m\,c_v(T_2 - T_1)$	$m\,c_p(T_2 - T_1)$	$m\,c_p(T_2 - T_1)$
3.	Hyperbolic or Constant temperature or Isothermal (T = Constant)	$p_1v_1 = p_2v_2$ (Boyle's Law)	$p_1v_1 \log_e\left(\frac{v_2}{v_1}\right)$ or $mRT \log_e\left(\frac{v_2}{v_1}\right)$	0	$p_1v_1 \log_e\left(\frac{v_2}{v_1}\right)$ or $mR\,T \log_e\left(\frac{v_2}{v_1}\right)$	0
4.	Adiabatic or Isentropic (pv^γ = Constant)	$p_1v_1^\gamma = p_2v_2^\gamma$ $\frac{T_1}{T_2} = \left(\frac{v_2}{v_1}\right)^{\gamma-1}$ $\frac{T_1}{T_2} = \left(\frac{p_1}{p_2}\right)^{\frac{\gamma-1}{\gamma}}$	$\frac{p_1v_1 - p_2v_2}{\gamma - 1}$ or $\frac{mR(T_1 - T_2)}{\gamma - 1}$	$mc_v(T_2 - T_1)$	0	$m\,c_p(T_2 - T_1)$

5.	Polytropic or General expansion (pv^n = Constant)	$p_1v_1^n = p_2v_2^n$ $\frac{T_1}{T_2} = \left(\frac{v_2}{v_1}\right)^{n-1}$ $\frac{T_1}{T_2} = \left(\frac{p_1}{p_2}\right)^{\frac{n-1}{n}}$	$\frac{p_1v_1 - p_2v_2}{n-1}$ or $\frac{mR(T_1 - T_2)}{n-1}$	$mc_v(T_2 - T_1)$	$\frac{p_1v_1 - p_2v_2}{n-1}$ $+ mc_v(T_2 - T_1)$ or $\frac{\gamma - n}{\gamma - 1} \times$ $\frac{mR(T_1 - T_2)}{n-1}$	$mc_p(T_2 - T_1)$

3.17. Flow Processes

We have already discussed in Art. 3.2, that the processes occuring in *open system which permit the transfer of mass to and from the system, are known as *flow processes*. In a flow process, the mass (working substance) enters the system and leaves after doing the work. The flow process may be classified as

1. Steady flow process, and 2. Unsteady flow process.

In a steady flow process, the following conditions must be satisfied ;

(*a*) The rate of mass flow at inlet and outlet is same, *i.e.* the mass flow rate through the system remains constant.

(*b*) The rate of heat transfer is constant.

(*c*) The rate of work transfer is constant.

(*d*) The state of working substance at any point within the system is same at all times.

(*e*) There is no change in the chemical composition of the system. Thus no chemical energy is involved.

If any one of these conditions are not satisfied, then the process is said to be non-steady flow process. In engineering, we are mainly concerned with steady flow processes, therefore only these processes are discussed in the following pages.

3.18. Application of First Law of Thermodynamics to a Steady Flow Process

Consider an open system through which the working substance flows at a steady rate, as shown in Fig. 3.25. The working substance enters the system at section 1 and leaves the system at section 2.

Fig. 3.25. Steady flow process.

Let p_1 = Pressure of the working substance entering the system in N/m^2,

v_{s1} = Specific volume of the working substance entering the system in m^3/kg.

V_1 = Velocity of the working substance entering the system in m/s,

u_1 = Specific internal energy of the working substance entering the system in J/kg,

z_1 = Height above datum level for inlet in metres,

p_2, v_{s2}, V_2, u_2 and z_2 = Corresponding values for the working substance leaving the system.

---

* Some authors use the term control volume instead of open system.

q_{1-2} = Heat supplied to the system in J/kg, and

w_{1-2} = Work delivered by the system in J/kg.

Consider 1 kg of mass of the working substance.

We know that total energy entering the system per kg of the working substance,

e_1 = Internal energy + *Flow or displacement energy + Kinetic energy + Potential energy + Heat supplied

$$= u_1 + p_1 v_{s1} + \frac{V_1^2}{2} + g z_1 + q_{1-2} \text{ (in J/kg)}$$

Similarly, total energy leaving the system per kg of the working substance,

$$e_2 = u_2 + p_2 v_{s2} + \frac{V_2^2}{2} + g z_2 + w_{1-2} \text{ (in J/kg)}$$

Assuming no loss of energy during flow, then according to First Law of Thermodynamics (*i.e.* Law of Conservation of Energy), $e_1 = e_2$.

$$\therefore \quad u_1 + p_1 v_{s1} + \frac{V_1^2}{2} + g z_1 + q_{1-2} = u_2 + p_2 v_{s2} + \frac{V_2^2}{2} + g z_2 + w_{1-2}$$

We know that

$u_1 + p_1 v_{s1} = h_1$ = Enthalpy of the working substance entering the system in J/kg, and

$u_2 + p_2 v_{s2} = h_2$ = Enthalpy of the working substance leaving the system in J/kg.

Thus, the above expression may be written as

$$h_1 + \frac{V_1^2}{2} + g z_1 + q_{1-2} = h_2 + \frac{V_2^2}{2} + g z_2 + w_{1-2} \quad \ldots (i)$$

or

$$h_1 + ke_1 + pe_1 + q_{1-2} = h_2 + ke_2 + pe_2 + w_{1-2}$$

It may be noted that all the terms in equaiton (*i*) represent the energy flow per unit mass of the working substance (*i.e.* in J/kg). When the equation (*i*) is multiplied through by the mass of the working substance (m) in kg/s, then all the terms will represent the energy flow per unit time (*i.e.* in J/s).

Thus the equation (*i*) may also be written as

$$m\left(h_1 + \frac{V_1^2}{2} + g z_1 + q_{1-2}\right) = m\left(h_2 + \frac{V_2^2}{2} + g z_2 + w_{1-2}\right) \quad \ldots (ii)$$

---

* The *flow or displacement energy* is the energy required to flow or move the working substance against its pressure. It is also known as *flow work*.

For example, let the working substance with pressure p_1 (in N/m²) flows through area A_1 (in m²) and moves through a distance x_1 (in metres).

∴ Energy or work required to flow the working substance,

$$FE = \text{Force} \times \text{Distance} = (p_1 A_1)\, x_1 = p_1 v_1 \text{ (in joules)} \quad \ldots (\because v_1 = A_1 x_1)$$

where v_1 = Volume of the working substance in m³.

For 1 kg mass of the working substance,

$v_1 = v_{s1}$ = Specific volume of the working substance in m³/kg.

$$\therefore \quad FE = p_1 v_{s1} \text{ (in J/kg)}$$

Both the equations (*i*) and (*ii*) are known as *steady flow energy equations*.

Notes : 1. In a steady flow, the mass flow rate (m) of the working substance entering and leaving the system is given by

$$m = \frac{A_1 V_1}{v_{s1}} = \frac{A_2 V_2}{v_{s2}} \text{ (in kg/s)}$$

This equation is known as *equation of continuity*.

2. The steady flow energy equation (*i*), for unit mass flow may be written as

$$q_{1-2} - w_{1-2} = (h_2 - h_1) + \left(\frac{V_2^2}{2} - \frac{V_1^2}{2}\right) + (g\,z_2 - g\,z_1) \quad \ldots (iii)$$

$$= (h_2 - h_1) + (ke_2 - ke_1) + (pe_2 - pe_1)$$

In differential form, this expression is written as

$$\delta q - \delta w = dh + d\,(ke) + d\,(pe)$$

3. In thermodynamics, the effect of gravity is generally neglected, therefore equation (*iii*) may be written as

$$q_{1-2} - w_{1-2} = (h_2 - h_1) + \left(\frac{V_2^2}{2} - \frac{V_1^2}{2}\right) \quad \ldots (iv)$$

If $V_1 = V_2$, then equation (*iv*) reduces to

$$q_{1-2} - w_{1-2} = h_2 - h_1 \quad \ldots (v)$$

4. In a non-flow process, the flow or displacement energy at inlet and outlet is zero, *i.e.*, $p_1 v_{s1} = 0$ and $p_2 v_{s2} = 0$. Therefore $h_2 = u_2$ and $h_1 = u_1$.

Thus the equation (*v*) may be written as $q_{1-2} - w_{1-2} = u_2 - u_1$; which is same as for non-flow process.

Example 3.21. *A steady flow thermodynamic system receives fluid at the rate of 6 kg/min with an initial pressure of 2 bar, velocity 150 m/s, internal energy 800 kJ/kg and density 27 kg/m³. The fluid leaves the system with a final pressure of 8 bar, velocity 200 m/s, internal energy 800 kJ/kg and density 5 kg/m³. If fluid receives 80 kJ/kg of heat during passing through the system and rises through 60 metres, determine the workdone during the process.*

Solution. Given : $m = 6$ kg/min $= 0.1$ kg/s ; $p_1 = 2$ bar $= 0.2 \times 10^6$ N/m² ; $V_1 = 150$ m/s ; $u_1 = 800$ kJ/kg ; $\rho_1 = 27$ kg/m³ ; $p_2 = 8$ bar $= 0.8 \times 10^6$ N/m² ; $V_2 = 200$ m/s ; $u_2 = 800$ kJ/kg ; $\rho_2 = 5$ kg/m³ ; $q_{1-2} = 80$ kJ/kg ; $z_2 = 60$ m

Consider 1 kg of mass flow.

We know that initial flow energy

$$= p_1 v_{s1} = p_1 / \rho_1 = 0.2 \times 10^6 / 27 = 7407 \text{ J/kg} = 7.407 \text{ kJ/kg}$$

$$\ldots (\because v_{s1} = 1/\rho_1)$$

and final flow energy $= p_2 v_{s2} = p_2 / \rho_2 = 0.8 \times 10^6 / 5 = 160 \times 10^3$ J/kg $= 160$ kJ/kg

∴ Initial enthalpy,

$$h_1 = u_1 + p_1 v_{s1} = 800 + 7.407 = 807.407 \text{ kJ/kg}$$

and final enthalpy, $h_2 = u_2 + p_2 v_{s2} = 800 + 160 = 960$ kJ/kg

Initial kinetic energy,

$$ke_1 = (V_1)^2/2 = (150)^2/2 = 11\,250 \text{ J/kg} = 11.25 \text{ kJ/kg}$$

Final kinetic energy, $ke_2 = (V_2)^2 / 2 = (200)^2 / 2 = 20\,000 \text{ J/kg} = 20 \text{ kJ/kg}$

Initial potential energy, $pe_1 = g\,z_1 = 0$ $\quad \ldots (\because z_1 = 0)$

Final potential energy, $pe_2 = g\,z_2 = 9.81 \times 60 = 588.6 \text{ J/kg} = 0.5886 \text{ kJ/kg}$

We know that the steady flow energy equation for unit mass flow is

$$h_1 + ke_1 + pe_1 + q_{1-2} = h_2 + ke_2 + pe_2 + w_{1-2}$$

or

$$q_{1-2} - w_{1-2} = (h_2 - h_1) + (ke_2 - ke_1) + (pe_2 - pe_1)$$
$$= (960 - 807.407) + (20 - 11.25) + (0.5886 - 0)$$
$$= 161.9316 \text{ kJ/kg}$$

$$\therefore \quad - w_{1-2} = 161.9316 - q_{1-2} = 161.9316 - 80 = 81.9316 \text{ kJ/kg}$$

or

$$w_{1-2} = -81.9316 \times 0.1 = -8.19316 \text{ kJ/s} \quad \textbf{Ans.} \quad (\because m = 0.1 \text{ kg/s})$$

The – *ve* sign indicates that the work is done on the system.

3.19. Workdone in a Steady Flow Process

We know that the steady flow equation for unit mass flow, in the differential form, is

$$\delta q - \delta w = dh + d\,(ke) + d\,(pe) \quad \ldots (i)$$

We also know that $\quad h = u + p\,v_s = u + p\,v \quad \ldots (\because \text{For unit mass, } v_s = v)$

Differentiating this expression,

$$dh = du + d\,(p\,v) = du + p\,dv + v\,dp \quad \ldots (ii)$$

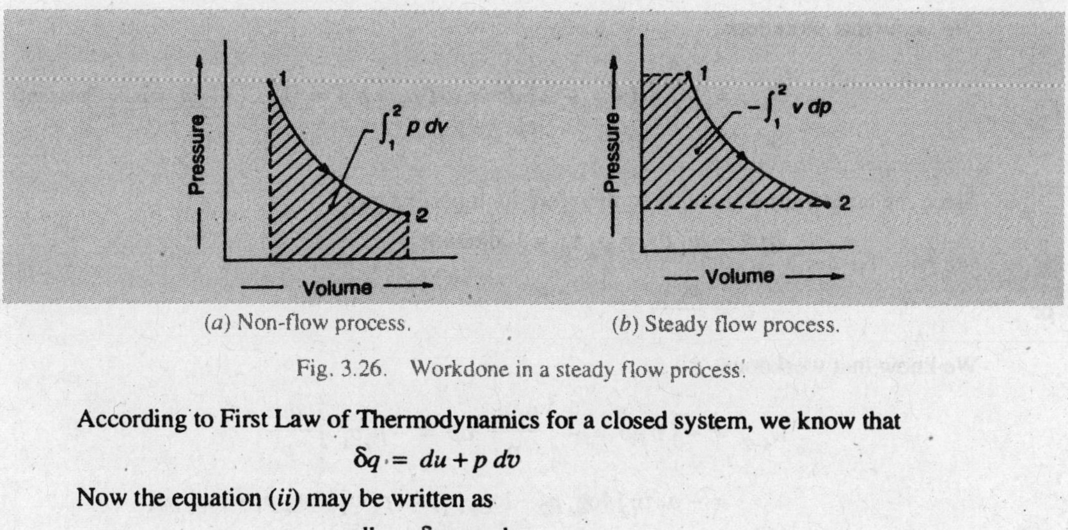

(*a*) Non-flow process. (*b*) Steady flow process.

Fig. 3.26. Workdone in a steady flow process.

According to First Law of Thermodynamics for a closed system, we know that

$$\delta q = du + p\,dv$$

Now the equation (*ii*) may be written as

$$dh = \delta q + v\,dp$$

Substituting this value of *dh* in equation (*i*), we have

$$\delta q - \delta w = (\delta q + v\,dp) + d\,(ke) + d\,(pe)$$

or

$$-\delta w = v\,dp + d\,(ke) + d\,(pe) \quad \ldots (iii)$$

If the changes in kinetic and potential energies are negligible [*i.e.* $d\,(ke) = 0$ and $d\,(pe) = 0$], then equation (*iii*) may be written as

$$-\delta w = v\,dp \quad \text{or} \quad \delta w = -v\,dp$$

On integrating, $\int_1^2 \delta w = -\int_1^2 v\,dp$ or $w_{1-2} = -\int_1^2 v\,dp$ ... (*iv*)

Thus in a steady flow process, the workdone (w_{1-2}) is $-\int_1^2 v\,dp$ instead of $\int_1^2 p\,dv$ in a non-flow process. Fig. 3.26 shows the difference between workdone in a non-flow and steady flow processes.

Note : The –ve sign in equation (*iv*) makes the integral positive during expansion process. In other words, $-\int_1^2 v\,dp$ is a positive quantity and represents the workdone by the system.

3.20. Workdone for Various Steady Flow Processes

The various steady flow processes, like non-flow processes, are as follows :

1. Constant volume process, 2. Constant pressure process, 3. Constant temperature process, 4. Adiabatic process, and 5. Polytropic process.

We shall now derive the expressions for workdone during these processes, as discussed below. The suffixes $_{1,2}$ represents the initial and final conditions respectively.

1. *Constant volume process*

We know that workdone,

$$w_{1-2} = -\int_1^2 v\,dp = -v\int_1^2 dp = -v\,(p_2 - p_1) = v\,(p_1 - p_2)$$

2. *Constant pressure process*

We know that workdone,

$$w_{1-2} = -\int_1^2 v\,dp = -v\int_1^2 dp = v\,(p_1 - p_2) = 0 \quad \ldots (\because p_1 = p_2 = \text{Constant})$$

3. *Constant temperature process*

Since the temperature is constant, therefore for a perfect gas,

$$p\,v = p_1\,v_1 = p_2\,v_2 = \text{Constant}$$

or

$$v = \frac{p_1\,v_1}{p}$$

We know that workdone

$$w_{1-2} = -\int_1^2 v\,dp = -\int_1^2 \frac{p_1\,v_1}{p}\,dp = -p_1 v_1 \int_1^2 \frac{dp}{p}$$

$$= -p_1\,v_1\,[\log_e p_2 - \log_e p_1] = p_1\,v_1\,[\log_e p_1 - \log_e p_2]$$

$$= p_1\,v_1 \log_e\left(\frac{p_1}{p_2}\right) = 2.3\,p_1\,v_1 \log\left(\frac{p_1}{p_2}\right)$$

$$= p_1 v_1 \log_e\left(\frac{v_2}{v_1}\right) = 2.3\,p_1\,v_1 \log\left(\frac{v_2}{v_1}\right)$$

$$\ldots\left(\because p_1\,v_1 = p_2\,v_2 \text{ or } \frac{p_1}{p_2} = \frac{v_2}{v_1}\right)$$

4. *Adiabatic process*

We know that for an adiabatic process,

$$p\,v^{\gamma} = p_1\,v_1^{\gamma} = p_2\,v_2^{\gamma} = \text{Constant}$$

or
$$v = v_1\left(\frac{p_1}{p}\right)^{1/\gamma}$$

We know that workdone,

$$w_{1-2} = -\int_1^2 v\,dp = -\int_1^2 v_1\left(\frac{p_1}{p}\right)^{1/\gamma} dp$$

$$= -v_1\,p_1^{1/\gamma}\int_1^2 p^{-1/\gamma}\,dp = -v_1\,p_1^{1/\gamma}\left[\frac{p^{-1/\gamma+1}}{-\frac{1}{\gamma}+1}\right]_1^2$$

$$= -v_1\,p_1^{1/\gamma}\left[\frac{p^{\frac{\gamma-1}{\gamma}}}{\frac{\gamma-1}{\gamma}}\right]_1^2 = \frac{\gamma}{\gamma-1}\times -v_1\,p_1^{1/\gamma}\left[p_2^{\frac{\gamma-1}{\gamma}} - p_1^{\frac{\gamma-1}{\gamma}}\right]$$

$$= \frac{\gamma}{\gamma-1}(p_1\,v_1 - p_2\,v_2)$$

5. *Polytropic process*

We know that for a polytropic process,

$$p\,v^{n} = p_1\,v_1^{n} = p_2\,v_2^{n} = \text{Constant}$$

$\therefore$ Workdone, $\quad w_{1-2} = \dfrac{n}{n-1}(p_1\,v_1 - p_2\,v_2)$ ... (Substituting $\gamma = n$, for polytropic process)

3.21. Throttling Process

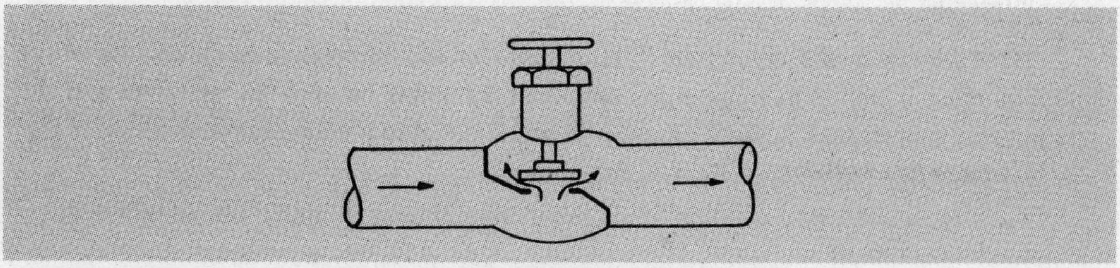

Fig. 3.27. Throttling process.

The throttling process is an irreversible steady flow expansion process in which a perfect gas is expanded through an orifice of minute dimensions such as a narrow throat or a slightly opened valve as shown in Fig. 3.27. Due to the fall in pressure during expansion, the gas should come out with a large velocity, but due to high frictional resistance between the gas and the walls of the aperture, there is no considerable change in velocity. The kinetic energy of the gas is converted into heat which is utilised in warming the gas to its initial temperature. Since no heat is supplied or rejected during the throttling process, and also no work is done, therefore

$$q_{1-2} = 0 \quad \text{and} \quad w_{1-2} = 0$$

We know that steady flow energy equation for unit mass flow is

$$h_1 + \frac{V_1^2}{2} + gz_1 + q_{1-2} = h_2 + \frac{V_2^2}{2} + gz_2 + w_{1-2}$$

Since there is no considerable change in velocity and the inlet and outlet are at the same level, therefore $V_1 = V_2$ and $z_1 = z_2$. Now the steady flow energy equation is written as

$$h_1 = h_2 \qquad \ldots (\because\ q_{1-2} = 0 \text{ and } w_{1-2} = 0)$$

Thus, the throttling process is a *constant enthalpy process*. We have discussed in Art. 3.14, that in a free expansion process, the enthalpy also remains constant. But the difference between the throttling process and the free expansion process is that in the former case, the gas leaves with negligible velocity where as in the latter case, the gas leaves with a large velocity.

The throttling process was investigated by Joule and Thompson during an experiment known as Joule Thompson porous plug experiment, as shown in Fig. 3.28 (*a*).

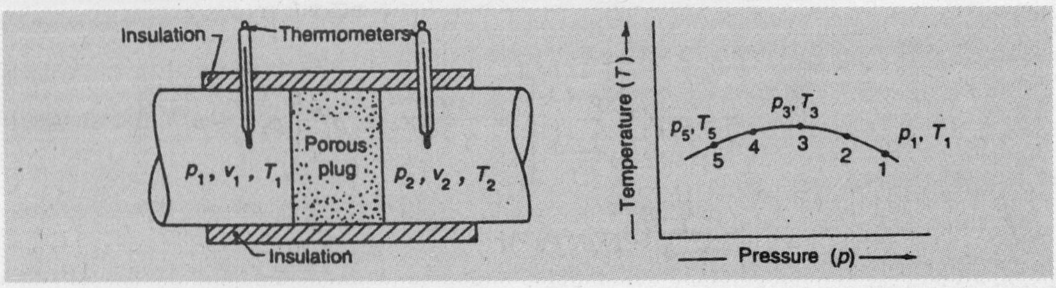

(*a*) Joule Thompson porous plug experiment. (*b*) Constant enthalpy curve.

Fig. 3.28

In this experiment, a stream of gas at pressure p_1 and temperature T_1 is forced continuously through one side of the porous plug, as shown in Fig. 3.28 (*a*). The gas comes out from the other side of the porous plug at a pressure p_2 and temperature T_2. The whole apparatus is completely insulated so that no heat transfer takes place. In the similar way as discussed above, the enthalpy of gas before and after the process, remains constant, *i.e.* $h_1 = h_2$.

If the pressure p_1 and temperature T_1 is kept constant and the pressure on the downstream of the porous plug is varied to p_2, p_3, p_4, p_5 etc., then the graph between the pressures and the corresponding temperatures T_2, T_3, T_4, T_5 etc. will be a curve of constant enthalpy, as shown in Fig. 3.28 (*b*) because $h_1 = h_2 = h_3 = h_4 = h_5$ etc.

The slope of a constant enthalpy line is called *Joule Thompson coefficient* and is denoted by μ whose value is given by

$$\mu = \left(\frac{\partial T}{\partial p}\right)_h$$

The value of Joule Thompson coefficient (μ) at a particular state may be positive, zero or negative. For a perfect gas, the value of μ is zero.

3.22. Application of Steady Flow Energy Equation to Engineering Systems

The application of steady flow energy equation to some of the engineering systems such as boilers, condensers, nozzles, diffusers, compressors and turbines are discussed below :

1. *Boiler*. A boiler, as shown in Fig. 3.29, is a device which supplies heat to water and generates steam. In this system, there is no change in kinetic and potential energies.

Also there is no workdone by the system. In other words,

$$(ke_2 - ke_1) = 0,\ (pe_2 - pe_1) = 0 \text{ and } w_{1-2} = 0$$

We know that the steady flow energy equation for a unit mass flow is

$$q_{1-2} - w_{1-2} = (h_2 - h_1) + (ke_2 - ke_1) + (pe_2 - pe_1)$$

Thus, for a boiler, the steady flow energy equation is written as

$$q_{1-2} = h_2 - h_1$$

This shows that the heat supplied to the system in a boiler increases the enthalpy of the system.

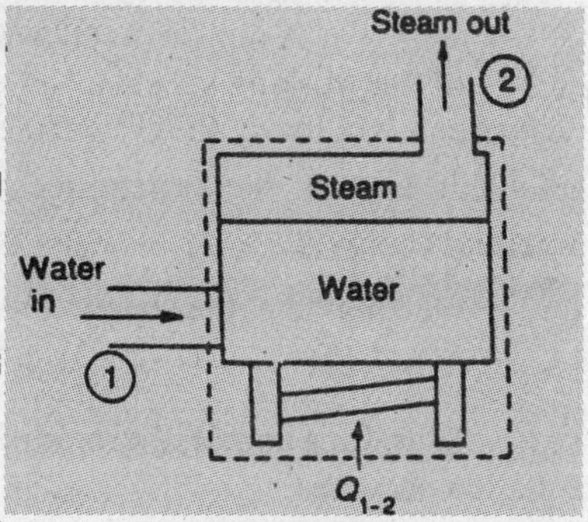

Fig. 3.29. Boiler.

2. *Condenser.* A condenser, as shown in Fig. 3.30. is a device used to condense steam in case of steam power plants using water as the cooling medium, whereas in refrigeration systems, it is used to condense refrigerant vapour using air as the cooling medium. For such a system, there is no change in kinetic and potential energies. Also there is no workdone by the system. In other words,

$$(ke_2 - ke_1) = 0\ ;\ (pe_2 - pe_1) = 0$$

and $$w_{1-2} = 0$$

We know that the steady flow energy equation for a unit mass flow is,

$$q_{1-2} - w_{1-2} = (h_2 - h_1) + (ke_2 - ke_1) + (pe_2 - pe_1)$$

Thus, for a condenser, the steady flow energy equation becomes

$$-q_{1-2} = h_2 - h_1 \quad \text{or} \quad q_{1-2} = h_1 - h_2$$

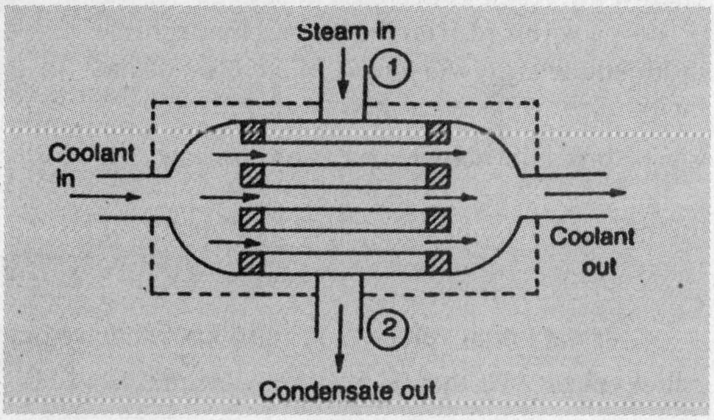

Fig. 3.30. Condenser.

The –ve sign with q_{1-2} is taken because the heat is lost by the coolant while passing through the condenser.

3. *Evaporator.* The evaporator, as shown in Fig. 3.31, is a device used in refrigeration systems in which the liquid refrigerant passes, receives heat and leaves as vapour refrigerant. For such a system, the change in kinetic and potential energies is negligible. Also there is no workdone by the system. In other words,

$$(ke_2 - ke_1) = 0,\ (pe_2 - pe_1) = 0$$

and $$w_{1-2} = 0$$

We know that the steady flow energy equation for a unit mass flow is

$$q_{1-2} - w_{1-2} = (h_2 - h_1) + (ke_2 - ke_1) + (pe_2 - pe_1)$$

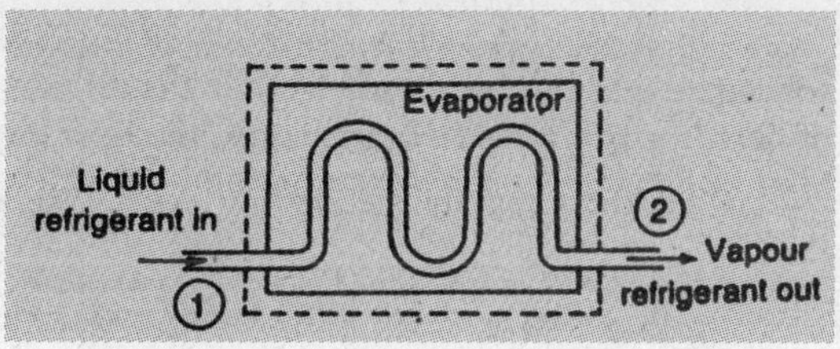

Fig. 3.31. Evaporator.

Thus, for an evaporator, the steady flow energy equation becomes,

$$q_{1-2} = h_2 - h_1$$

Note : The process occuring in an evaporator is the reverse of that of a condenser.

4. *Nozzle.* A nozzle, as shown in Fig. 3.32, is a device which increases the velocity or kinetic energy of the working substance at the expense of its pressure drop. The nozzle is insulated so that no heat enters or leaves the system (*i.e.* $q_{1-2} = 0$). In other words, the flow through nozzles is considered adiabatic. Further, the system does not deliver any work *i.e.* $w_{1-2} = 0$ and there is no change in potential energy, *i.e.* $(pe_2 - pe_1) = 0$.

We know that the steady flow energy equation for a unit mass flow is

$$q_{1-2} - w_{1-2} = (h_2 - h_1) + (ke_2 - ke_1) + (pe_2 - pe_1) = 0$$

Thus, for a nozzle, the steady flow energy equation becomes

$$0 = (h_2 - h_1) + (ke_2 - ke_1)$$

or
$$\frac{V_2^2}{2} - \frac{V_1^2}{2} = h_1 - h_2$$

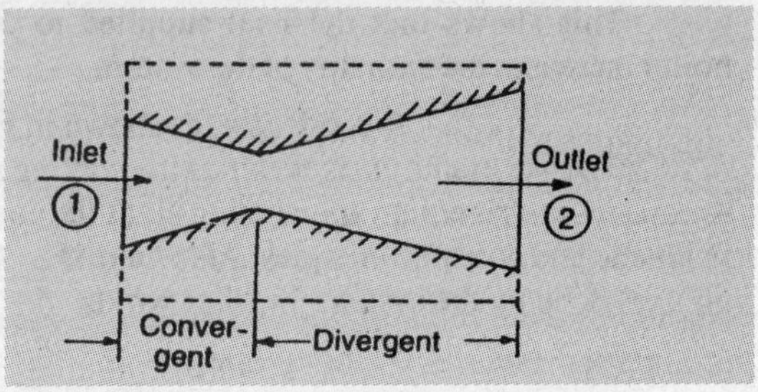

Fig. 3.32. Convergent-divergent nozzle.

This shows that the increase in kinetic energy will result in decrease in enthalpy. If the process is reversed, it is obvious that the decrease in kinetic energy will result in increase of enthalpy. Such a system is known as *diffuser*.

From the above expression, we have

$$V_2^2 - V_1^2 = 2(h_1 - h_2) \quad \text{or} \quad V_2^2 = V_1^2 + 2(h_1 - h_2)$$

$$\therefore \quad V_2 = \sqrt{V_1^2 + 2(h_1 - h_2)}$$

If the initial velocity V_1 (also known as velocity of approach) is very small as compared to outlet velocity V_2, then V_1 may be neglected.

$$\therefore \quad V_2 = \sqrt{2(h_1 - h_2)}$$

Note : If

A_1 = Cross-sectional area of the nozzle at inlet,

V_1 = Velocity of the fluid entering the nozzle,

v_{s1} = Specific volume of the fluid at inlet,

A_2, V_2 and v_{s2} = Corresponding values at outlet.

$\therefore$ For continuous steady flow, mass flow rate,

$$m = \frac{A_1 V_1}{v_{s1}} = \frac{A_2 V_2}{v_{s2}}$$

5. *Turbine.* A turbine, as shown in Fig. 3.33, is a device which converts energy of the working substance (gas or steam) into work in the turbine. The turbine is insulated so that there is no transfer of heat (*i.e.* $q_{1-2} = 0$). In other words, the flow through a turbine is considered adiabatic.

We know that the steady flow energy equation for a unit mass flow is

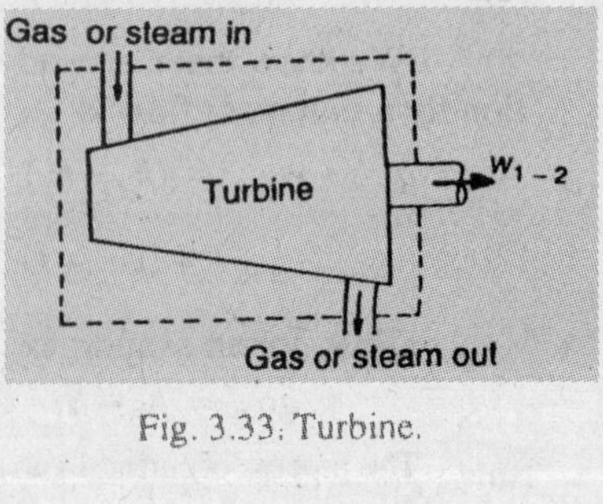

Fig. 3.33. Turbine.

$$q_{1-2} - w_{1-2} = (h_2 - h_1) + (ke_2 - ke_1) + (pe_2 - pe_1)$$

Thus, for a turbine, the steady flow energy equation becomes

$$-w_{1-2} = (h_2 - h_1) + (ke_2 - ke_1) + (pe_2 - pe_1)$$

In case the changes in kinetic and potential energies are negligible [*i.e.* $(ke_2 - ke_1) = 0$ and $(pe_2 - pe_1) = 0$], then the above expression is written as

$$-w_{1-2} = h_2 - h_1 \text{ or } w_{1-2} = h_1 - h_2$$

This shows that the work is done by the system due to decrease in enthalpy of the working substance.

6. Rotary compressor. A rotary compressor, as shown in Fig. 3.34, is a device which compresses air and supplies the same at moderate pressure and in large quantities. The rotary compressor is insulated so that no heat transfer takes place (*i.e.* $q_{1-2} = 0$). In other words, the flow through a rotary compressor is adiabatic.

We know that the steady flow energy equation for a unit mass flow is

$$q_{1-2} - w_{1-2} = (h_2 - h_1) + (ke_2 - ke_1) + (pe_2 - pe_1)$$

Thus, for a rotary compressor, the steady flow energy equation becomes,

$$-(-w_{1-2}) = (h_2 - h_1) + (ke_2 - ke_1) + (pe_2 - pe_1)$$

The $-ve$ sign is used because the work is done on the system. In case the changes in kinetic and potential energies are negligible [*i.e.* $ke_2 - ke_1 = 0$ and $(pe_2 - pe_1) = 0$], then the above expression is written as

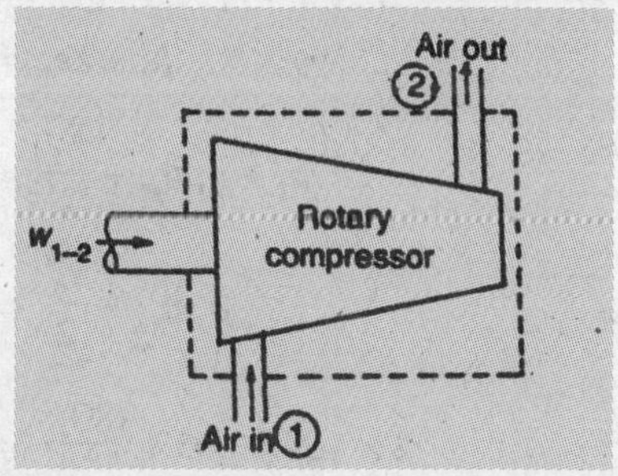

Fig. 3.34. Rotary compressor.

$$w_{1-2} = h_2 - h_1$$

This shows that the work is done due to increase in enthalpy.

7. Reciprocating compressor. A reciprocating compressor, as shown in Fig. 3.35, is a device which compresses air and supplies the same at a considerable higher pressure and in small quantities. The reciprocating compressor is considered as a steady flow system provided it includes the receiver which reduces the fluctuation of flow considerably.

We know that the steady flow energy equation for a unit mass flow is

$$q_{1-2} - w_{1-2} = (h_2 - h_1) + (ke_2 - ke_1) + (pe_2 - pe_1)$$

Since in a reciprocating compressor, the changes in kinetic and potential energies are negligible, therefore $(ke_2 - ke_1) = 0$ and $(pe_2 - pe_1) = 0$

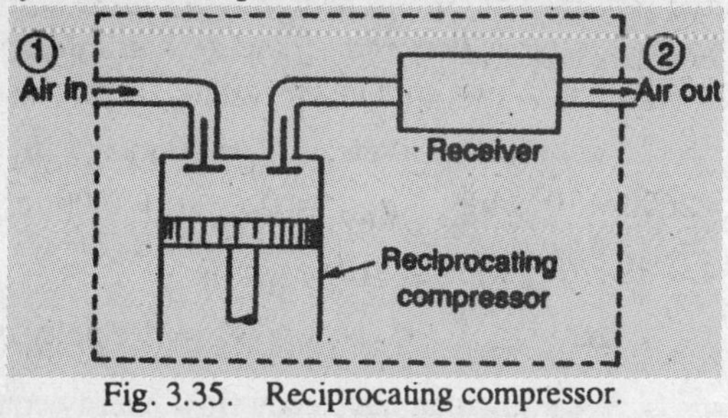

Fig. 3.35. Reciprocating compressor.

Thus, for a reciprocating compressor, the steady flow energy equation becomes

$$-q_{1-2} - (-w_{1-2}) = h_2 - h_1 \text{ or } w_{1-2} = q_{1-2} + (h_2 - h_1)$$

The $-ve$ sign to q_{1-2} and w_{1-2} is used because the heat is rejected and the work is done on the system.

Example 3.22. *A gas expands through an ideally, insulated nozzle following a reversible polytropic law $p\,v^{1.2} = C$. There is no change in potential energy but the pressure drops from 20 bar to 2 bar and specific volume increases from 0.05 m³ to 0.3 m³. If the entrance velocity is 80 m/s, determine the exit velocity.*

Solution. Given : $^*n = 1.2$; $p_1 = 20$ bar $= 2 \times 10^6$ N/m² ; $p_2 = 2$ bar $= 0.2 \times 10^6$ N/m² ; $v_{s1} = 0.05$ m³ ; $v_{s2} = 0.3$ m³ ; $V_1 = 80$ m/s

* Superfluous data

Let V_2 = Exit velocity in m/s.

We know that the steady flow energy equation for a unit mass flow is

$$h_1 + ke_1 + pe_1 + q_{1-2} = h_2 + ke_2 + pe_2 + w_{1-2}$$

Since the nozzle is insulated so that no heat transfer takes place, therefore $q_{1-2} = 0$. Also there is no workdone during expansion of gas through the nozzle (*i.e.* $w_{1-2} = 0$) and there is no change in potential energy (*i.e.* $pe_1 = pe_2$). Thus the steady flow energy equation is written as

$$h_1 + ke_1 = h_2 + ke_2$$

$$u_1 + p_1 v_{s1} + \frac{V_1^2}{2} = u_2 + p_2 v_{s2} + \frac{V_2^2}{2}$$

or $$p_1 v_{s1} + \frac{V_1^2}{2} = p_2 v_{s2} + \frac{V_2^2}{2} \quad \ldots (\because \text{For an insulated nozzle, } u_1 = u_2)$$

$$2 \times 10^6 \times 0.05 + \frac{(80)^2}{2} = 0.2 \times 10^6 \times 0.3 + \frac{V_2^2}{2}$$

$$103\,200 = 60\,000 + \frac{V_2^2}{2}$$

$$\therefore \quad V_2^2 = 2\,(103\,200 - 60\,000) = 86\,400$$

or $$V_2 = 294 \text{ m/s} \quad \textbf{Ans.}$$

Example 3.23. *The velocity and enthalpy of fluid at the inlet of a certain nozzle are 50 m/s and 2800 kJ/kg respectively. The enthalpy at the exit of nozzle is 2600 kJ/kg. The nozzle is horizontal and insulated so that no heat transfer takes place from it. Find : 1. velocity of the fluid at exit of the nozzle ; 2. mass flow rate, if the area at inlet of nozzle is 0.09 m² and the specific volume is 0.185 m³/kg ; and 3. exit area of the nozzle, if the specific volume at the exit of nozzle is 0.495 m³/kg.*

Solution. Given : $V_1 = 50$ m/s ; $h_1 = 2800$ kJ/kg $= 2800 \times 10^3$ J/kg ; $h_2 = 2600$ kJ/kg $= 2600 \times 10^3$ J/kg ; $q_{1-2} = 0$; $A_1 = 0.09$ m² ; $v_{s1} = 0.185$ m³/kg ; $v_{s2} = 0.495$ m³/kg

1. Velocity of fluid at exit of nozzle

Let V_2 = Velocity of fluid at exit of nozzle.

We know that the steady flow energy equation for a unit mass flow is

$$h_1 + ke_1 + pe_1 + q_{1-2} = h_2 + ke_2 + pe_2 + w_{1-2}$$

Since the nozzle is insulated so that no heat transfer takes place, therefore $q_{1-2} = 0$. Also there is no workdone, *i.e.* $w_{1-2} = 0$. Neglecting the potential energy at inlet and outlet of the nozzle because of at same level from datum, the steady flow energy equation is written as

$$h_1 + ke_1 = h_2 + ke_2 \quad \text{or} \quad ke_2 - ke_1 = h_1 - h_2$$

$$\therefore \quad \frac{V_2^2}{2} - \frac{V_1^2}{2} = h_1 - h_2 \quad \text{or} \quad V_2^2 - V_1^2 = 2\,(h_1 - h_2)$$

and $$V_2 = \sqrt{V_1^2 + 2\,(h_1 - h_2)} = \sqrt{(50)^2 + 2\,(2800 \times 10^3 - 2600 \times 10^3)}$$

$$= \sqrt{2500 + 400 \times 10^3} = 634.4 \text{ m/s} \quad \textbf{Ans.}$$

2. Mass flow rate

Let m = Mass flow rate in kg/s.

We know that $m = \frac{A_1 V_1}{v_{s1}} = \frac{0.09 \times 50}{0.185} = 24.3$ kg/s **Ans.**

3. Exit area of nozzle

Let A_2 = Exit area of nozzle in m².

We know that $m = \frac{A_2 V_2}{v_{s2}}$ or $A_2 = \frac{m \times v_{s2}}{V_2} = \frac{24.3 \times 0.495}{634.4} = 0.019 \text{ m}^2$ **Ans.**

Example 3.24. *In an air compressor, air flows steadily at the rate of 15 kg per minute. The air enters the compressor at 5 m/s with a pressure of 1 bar and a specific volume of 0.5 m³/kg. It leaves the compressor at 7.5 m/s with a pressure of 7 bar and a specific volume of 0.15 m³/kg. The internal energy of the air leaving the compressor is 165 kJ/kg greater than that of the air entering. The cooling water in the compressor jackets absorbs heat from the air at the rate of 125 kJ/s. Find : 1. power required to drive the compressor ; and 2. ratio of the inlet pipe diameter to outlet pipe diameter.*

Solution. Given : m = 15 kg/min = 0.25 kg/s ; V_1 = 5 m/s ; p_1 = 1 bar = 0.1×10^6 N/m² ; v_{s1} = 0.5 m³/kg ; V_2 = 7.5 m/s ; p_2 = 7 bar = 0.7×10^6 N/m² ; v_{s2} = 0.15 m³/kg ; $u_2 - u_1$ = 165 kJ/kg = 165×10^3 J/kg ; q_{1-2} = 125 kJ/s = 125/m = 125/0.25 = 500 kJ/kg = 500×10^3 J/kg

1. Power required to drive the compressor

Let w_{1-2} = Work required to drive the compressor in J/kg.

We know that flow energy at inlet

$$= p_1 v_{s1} = 0.1 \times 10^6 \times 0.5 = 50 \times 10^3 \text{ J/kg}$$

Flow energy at outlet $= p_2 v_{s2} = 0.7 \times 10^6 \times 0.15 = 105 \times 10^3$ J/kg

Kinetic energy at inlet,

$$ke_1 = (V_1)^2/2 = 5^2/2 = 12.5 \text{ J/kg}$$

and kinetic energy at outlet,

$$ke_2 = (V_2)^2/2 = (7.5)^2/2 = 28.1 \text{ J/kg}$$

We know that the steady flow energy equation for a unit mass flow is

$$u_1 + p_1 v_{s1} + ke_1 + pe_1 - q_{1-2} = u_2 + p_2 v_{s2} + ke_2 + pe_2 - w_{1-2}$$

In this expression, q_{1-2} and w_{1-2} are taken – *ve*, because heat is *rejected* by the air and work is done *on* the air. Neglecting the potential energy, the steady flow energy equation may be written as

$$w_{1-2} - q_{1-2} = (u_2 - u_1) + (p_2 v_{s2} - p_1 v_{s1}) + (ke_2 - ke_1)$$

$$w_{1-2} - 500 \times 10^3 = 165 \times 10^3 + (105 \times 10^3 - 50 \times 10^3) + (28.1 - 12.5)$$

$$= 220 \times 10^3 \text{ J/kg}$$

... [Neglecting ($ke_2 - ke_1$), as it is very small as compared to other terms]

or $w_{1-2} = 720 \times 10^3$ J/kg

∴ Power required to drive the compressor

$$= m \times w_{1-2} = 0.25 \times 720 \times 10^3 \text{ J/s} = 80 \times 10^3 \text{ J/s}$$

$$= 180 \text{ kJ/s} = 180 \text{ kW} \quad \textbf{Ans.} \qquad \ldots (\because 1 \text{ kJ/s} = 1 \text{kW})$$

2. *Ratio of the inlet pipe diameter to outlet pipe diameter*

Let D_1 = Inlet pipe diameter and

D_2 = Outlet pipe diameter.

We know that $$\frac{A_1 V_1}{v_{s1}} = \frac{A_2 V_2}{v_{s2}} \quad \text{or} \quad \frac{A_1}{A_2} = \frac{V_2 \times v_{s1}}{V_1 \times v_{s2}}$$

$$\therefore \quad \frac{\frac{\pi}{4}(D_1)^2}{\frac{\pi}{4}(D_2)^2} = \frac{V_2 \times v_{s1}}{V_1 \times v_{s2}} \quad \text{or} \quad \frac{D_1}{D_2} = \sqrt{\frac{V_2 \times v_{s1}}{V_1 \times v_{s2}}} = \sqrt{\frac{7.5 \times 0.5}{5 \times 0.15}} = 2.236 \quad \textbf{Ans.}$$

Example 3.25. *In a gas turbine, the gases flow at the rate of 5 kg/s. The gases enter the turbine at a pressure 7 bar with a velocity 120 m/s and leaves at a pressure 2 bar with velocity 250 m/s. The turbine is insulated. If the enthalpy of the gas at inlet is 900 kJ/kg and at outlet 600 kJ/kg, determine the capacity of the turbine.*

Solution. Given : $m = 5$ kg/s ; $p_1 = 7$ bar $= 0.7 \times 10^6$ N/m^2 ; $V_1 = 120$ m/s ; $p_2 = 2$ bar $= 0.2 \times 10^6$ N/m^2 ; $V_2 = 250$ m/s ; $h_1 = 900$ kJ/kg $= 900 \times 10^3$ J/kg ; $h_2 = 600$ kJ/kg $= 600 \times 10^3$ J/kg

We know that the steady flow energy equation for a unit mass flow is

$$h_1 + ke_1 + pe_1 + q_{1-2} = h_2 + ke_2 + pe_2 + w_{1-2}$$

Since the turbine is insulated so that no heat transfer takes place, therefore $q_{1-2} = 0$

Neglecting the potential energy at inlet and outlet, the steady flow energy equation becomes

$$h_1 + ke_1 = h_2 + ke_2 + w_{1-2}$$

or $$w_{1-2} = (h_1 - h_2) + (ke_1 - ke_2) = (h_1 - h_2) + \left(\frac{V_1^2}{2} - \frac{V_2^2}{2}\right)$$

$$= (900 \times 10^3 - 600 \times 10^3) + \left[\frac{(120)^2}{2} - \frac{(250)^2}{2}\right]$$

$$= 300 \times 10^3 - 24.05 \times 10^3 = 275.95 \times 10^3 \text{ J/kg}$$

∴ Capacity of the turbine

$$= m \times w_{1-2} = 5 \times 275.95 \times 10^3 = 1379.75 \times 10^3 \text{ J/s}$$

$$= 1379.75 \text{ kJ/s} = 1379.75 \text{ kW} \quad \textbf{Ans.}$$

EXERCISES

1. One kg of air is heated in a closed vessel, *i.e.*, at a constant volume from a pressure of 2 bar to 5 bar. If the initial temperature of the air is 300 K, determine the change in internal energy. $c_v = 0.712$ kJ/kg K. **[Ans.** 320.4 kJ**]**

2. A volume of 0.5 m^3 of gas at a pressure of 10 bar and 200° C is expanded in a cylinder to 1.2 m^3 at a constant pressure. Calculate the amount of work done by the gas and the increase in internal energy. Assume $c_p = 1.005$ kJ/kg K and $c_v = 0.712$ kJ/kg K. **[Ans.** 700 kJ ; 1697.35 kJ**]**

3. A piston cylinder containing air expands at a constant pressure of 150 kPa from a temperature of 285 K to a temperature of 550 K. The mass of air is 0.05 kg. Find the heat transfer, work transfer and the change

in internal energy during the process. Take $c_p = 1$ kJ/kg K and $R = 0.287$ kJ/kg K.

[Ans. 13.25 kJ ; 3.8 kJ ; 9.45 kJ]

4. One kg of air at a temperature of 40° C is compressed isothermally from a pressure of 1.5 bar to 6 bar. Determine the heat rejected by the air during the process of compression. For air, $c_p = 1.005$ kJ/kg K and $c_v = 0.712$ kJ/kg K. [Ans. 124.2 kJ]

5. An ideal gas at 30° C and 1 bar is compressed adiabatically from 5 m³ to 1 m³. Find the temperature, pressure and the work done. Take $\gamma = 1.4$. [Ans. 304° C ; 9.5 bar ; 1.125 kJ]

6. A 0.568 m³ capacity insulated vessel of oxygen at the pressure of 13.6 bar is stirred by internal paddle until the pressure becomes 21 bar. Find out the heat transferred and the work input.

Take $c_v = 0.658$ kJ/kg K, and $R = 260$ J/kg K. [Ans. zero ; 1063.8 kJ/kg]

7. One kg of a gas expands reversibly and adiabatically. Its temperature during the process falls from 515 K to 390 K, while the volume is doubled. The gas does 92 kJ of work in this process. Find : 1. The value of c_p and c_v ; and 2. The molecular mass of the gas. [Ans. 1.0326 kJ/kg K, 0.736 kJ/kg K ; 28.03]

8. A gas initially at 14.4 bar and 360° C is expanded isothermally to a pressure of 2.24 bar. It is then cooled at constant volume till the pressure falls to 1.02 bar. Finally an adiabatic compression brings the gas back to the initial stage. The mass of the gas is 0.23 kg and $c_p = 1$ kJ/kg K. Draw the *p-v* diagram and determine : 1. the value of the adiabatic index of compression ; and 2. the change of internal energy of the gas during the adiabatic process. [Ans. 1.423 ; 55.8 kJ (decrease)]

9. A cylinder contains 0.084 m³ of hydrogen at 1.05 bar and 18° C. It is compressed adiabatically to 14 bar and then expanded isothermally to the original volume of 0.084 m³. The characteristic constant for hydrogen is 4200 J/kg K and its specific heat at constant pressure is 14.28 kJ/kg K.

Determine the final pressure of the gas and the amount of heat which must be added to the gas during isothermal expansion. Also calculate the heat which must be abstracted from the gas after expansion in order to reduce it to its initial state of pressure. [Ans. 2.25 bar ; 34.5 kJ ; 24.14 kJ]

10. A quantity of gas is compressed according to $pv^{1.25}$ = Constant. The initial temperature and pressure of the gas is 15° C and 1 bar respectively. Find the work done in compressing 1 kg of air at 3 bar and the heat rejected through the walls of the cylinder. $\gamma = 1.4$ for air. [Ans. 82.65 kJ ; 103.3 kJ]

11. A quantity of air has a volume of 56.5 litres and a pressure of 7.03 bar. It is expanded in a cylinder to a pressure of 1.05 bar. Compute the workdone if the expansion is 1. hyperbolic, 2. adiabatic, and 3. $p\,v^{1.2} = C$. Take $\gamma = 1.4$. [Ans. 75.4 kJ ; 42.5 kJ ; 55 kJ]

12. 1.4 m³ of a gas at a pressure of 1.26 bar is compressed to a volume of 0.28 m³. The final pressure is 7 bar. Assuming the compression to be polytropic, calculate the heat transfer and change in internal energy. Assume $\gamma = 1.4$. [Ans. 252.54 kJ (rejected) ; 49 kJ (decrease)]

13. An ideal gas of molecular mass 30 and specific heat ratio 1.38 is compressed according to the law $p\,v^{1.25}$ = constant, from a pressure of 1 bar and 15° C to a pressure of 16 bar. Calculate the temperature at the end of compression, the heat received or rejected and workdone by the gas during the process. Assume 1 kg mass of the gas. Use only calculated values of c_p and c_v. [Ans. 228.7° C ; 81 kJ (rejected) ; 236.78 kJ]

14. A cylinder contains 0.113 m³ of air at 1 bar and 90° C. The air is compressed to a volume of 0.028 m³, the final pressure being 5.8 bar. Determine : 1. mass of the air in the cylinder, 2. value of index (*n*) for the compression process, 3. increase in internal energy, and 4. heat received or rejected by air during compression.

If, after the above process, the air is cooled at a constant pressure to its original temperature of 90° C, find the further work of compression required. $\gamma = 1.4$ and $R = 287$ J/kg K.

[Ans. 0.108 kg ; 1.26 ; 12.3 kJ ; 6.62 kJ (rejected) ; 4.912 kJ]

15. An internal combustion engine cylinder has a diameter 240 mm and length of the stroke 400 mm. The clearance volume is one-fourth of the swept volume. The pressure at the beginning of expansion stroke is 16 bar and the expansion follows the law $pv^{1.3}$ = Constant. Determine : 1. the pressure at the end of expansion stroke, and 2. the work done during the expansion. [Ans. 1.974 bar ; 9.2 kJ]

16. The internal energy and equation of state of a closed gas system, are given by

$$U = (188.4 + 1.256\,t)\text{ kJ/kg} ; \text{ and } p\,v = 600\,T$$

where t is the temperature in °C and T in Kelvin. p is the pressure in bar and v is the specific volume in m³.

If the temperature of 2 kg of gas is raised from 100° C to 200° C at 1. constant pressure, and 2. according to the law $p\,v^{1.2}$ = constant ; find out the heat flow, work flow and change in internal energy of the system. [Ans. 251.2 kJ ; 371.2 kJ ; 600 kJ, 348.8 kJ]

17. Gas at 1.5 bar and 295 K in a closed vessel is compressed to 10 bar. Its temperature then becomes 455 K. If the compression follows the law $pv^n = C$, find the value of n. [Ans. 1.3]

18. One kg of fluid enters the steady flow apparatus at a pressure of 6 bar, velocity 16 m/s and specific volume 0.4 m^3/kg. The inlet is 30 m above the ground level. The fluid leaves the apparatus at a pressure of 1 bar, velocity 275 m/s ; and specific volume 0.6 m^3/kg. The outlet is at the ground level. The total heat loss between the inlet and outlet is 10 kJ/kg of fluid. If 140 kJ/kg of work is done by the system, find the change in specific internal energy and indicate whether this is a increase or decrease. [Ans. 7.4 kJ/kg (decrease)]

19. Air at the rate of 12 kg/min flows steadily through a nozzle. The pressure and temperature of air at the inlet to the nozzle are 20 bar and 390 K respectively. The pressure of air at the exit of nozzle is 5 bar. Assuming adiabatic flow with initial velocity of 100 m/s, determine the exit velocity and the inlet and exit areas.

Take c_p = 1.005 kJ/kg K and γ = 1.4, for air. [Ans. 516.4 m/s ; 112 mm^2, 58 mm^2]

20. An air compressor draws air at 1 bar and 20° C and discharges into a line having an inside diameter of 10 mm. The average air velocity in the line at a point close to the discharge is 7.5 m/s and the discharge pressure is 3 bar. Assuming that compression takes place adiabatically, determine the power required to drive the compressor. The velocity of air entering the compressor has negligible velocity. Take c_p = 1.005 kJ/kg K ; R = 287 J/kg K and γ = 1.4. [Ans. 0.1665 kW]

21. Air is expanded reversibly and adiabatically in a turbine from 3.5 bar and 260° C to 1 bar. The turbine is insulated and the inlet velocity is negligible. The exit velocity is 150 m/s. Find the work output of the turbine per unit mass of air flow. Take for air, c_p = 1.005 kJ/kg K, and γ = 1.4. [Ans. 15 5 kJ/kg]

QUESTIONS

1. What do you understand by a 'thermodynamic process' ? Distinguish between reversible and irreversible process.

2. Explain the difference between non-flow process and a flow process. Derive the equation for workdone during a non-flow process.

3. What is an isothermal process ? Derive an expression for the workdone during an isothermal process.

4. Explain the adiabatic process. Derive an expression for the workdone during the adiabatic expansion of an ideal gas.

5. What is polytropic process ? How does it differ from an adiabatic process.

6. Prove that the heat absorbed or rejected during a polytropic process is $\frac{\gamma - n}{\gamma - 1} \times$ work done, where γ is the ratio of specific heat and n is the polytropic index.

7. Explain what is meant by "polytropic" operation. Starting from the fundamental, show that during a polytropic compression, according to the law pv^n = Constant, the rate of heat rejection per unit change in volume is given by $\left(\frac{\gamma - n}{\gamma - 1}\right) \times p$, where γ is the ratio of specific heat and p is the pressure at a particular point (a mean pressure during the process) at which heat rejection is considered.

8. Explain free expansion process. What is the difference between throttling process and a free expansion process ?

9. What are the assumptions for a steady flow process ? Write the general energy equation for a steady flow system.

10. Write down the simplified steady flow energy equation for a unit mass flow for (*a*) condenser ; (*b*) compressor ; and (*c*) turbine.

11. Apply steady flow energy equation to a nozzle and derive an equation for velocity at exit.

OBJECTIVE TYPE QUESTIONS

1. When a gas is heated at constant volume,

(*a*) its temperature will increase

(*b*) its pressure will increase

(*c*) both temperature and pressure will increase

(*d*) neither temperature nor pressure will increase

2. The heating of a gas at constant pressure is governed by

(*a*) Boyle's law (*b*) Charles' law (*c*) Gay-Lussac law (*d*) Joule's law

3. A process, in which the gas is heated or expanded in such a way that the product of its pressure and volume remains constant, is called

(*a*) isothermal process (*b*) isobaric process

(*c*) adiabatic process (*d*) polytropic process

4. The hyperbolic process is governed by

(*a*) Boyle's law (*b*) Charles' law (*c*) Gay-Lussac law (*d*) Joule's law

5. The heating of gas at constant ....is governed by Boyle's law.

(*a*) volume (*b*) pressure (*c*) temperature

6. In an isothermal process,

(*a*) internal energy increases

(*b*) internal energy decreases

(*c*) there is no change in internal energy

(*d*) internal energy first decreases and then increases

7. The expansion ratio (*r*) is the ratio of

(*a*) $\dfrac{v_1}{v_2}$ (*b*) $\dfrac{v_2}{v_1}$ (*c*) $\dfrac{v_1+v_2}{v_1}$ (*d*) $\dfrac{v_1+v_2}{v_2}$

where v_1 = Volume at the beginning of expansion, and

v_2 = Volume at the end of expansion.

8. When the expansion or compression of the gas takes places according to the law $pv^n = C$, then the process is known as

(*a*) isothermal process (*b*) isobaric process

(*c*) adiabatic process (*d*) polytropic process

9. An adiabatic process is one in which

(*a*) no heat enters or leaves the gas

(*b*) the temperature of the gas changes

(*c*) the change in internal energy is equal to the workdone

(*d*) all of the above

10. The general law of expansion or compression is $pv^n = C$. The process is said to be hyperbolic, if *n* is equal to

(*a*) 0 (*b*) 1 (*c*) γ (*d*) ∞

11. If the value of $n = 0$ in the general law $pv^n = C$, then the process is called

(*a*) isochoric process (*b*) isobaric process

(*c*) isothermal process (*d*) isentropic process

12. The workdone in a free expansion process is

(*a*) zero (*b*) minimum (*c*) maximum (*d*) positive

13. In a steady flow process,

(*a*) the mass flow rate is constant (*b*) the heat transfer rate is constant

(*c*) the work transfer rate is constant (*d*) all of the above

14. The workdone in steady flow process is given by

(*a*) $\int_1^2 p\,dv$ (*b*) $-\int_1^2 p\,dv$ (*c*) $\int_1^2 v\,dp$ (*d*) $-\int_1^2 v\,dp$

15. The throttling process is a

(*a*) non-flow process (*b*) steady flow process

(*c*) non-steady flow process

ANSWERS

1. (*c*)	2. (*b*)	3. (*a*)	4. (*a*)	5. (*c*)
6. (*c*)	7. (*b*)	8. (*d*)	9. (*d*)	10. (*b*)
11. (*b*)	12. (*a*)	13. (*d*)	14. (*d*)	15. (*b*)

4

Entropy of Perfect Gases

1. Introduction. 2. Relation between Heat and Entropy. 3. Importance of Entropy. 4. Available and Unavailable Heat Energy. 5. Units of Entropy. 6. Clausius Inequality. 7. Principle of Increase of Entropy. 8. General Expression for Change of Entropy of a Perfect Gas. 9. Change of Entropy of a Perfect Gas during Various Thermodynamic Processes. 10. Change of Entropy during Constant Volume Process (or Isochoric Process). 11. Change of Entropy during Constant Pressure Process (or Isobaric Process). 12. Change of Entropy during Constant Temperature Process (or Isothermal Process). 13. Change of Entropy during Reversible Adiabatic Process (or Isentropic Process). 14. Change of Entropy during Polytropic Process (pv^n = Constant). 15. Approximate Method for Heat Absorbed.

4.1. Introduction

The term 'entropy' which literally means transformation, was first introduced by Clausius. It is an important thermodynamic property of a working substance, which increases with the addition of heat, and decreases with its removal. As a matter of fact, it is tedious to define the term entropy. But it is comparatively easy to define change of entropy of a working substance. In a reversible process, over a small range of temperature, the increase or decrease of entropy, when multiplied by the absolute temperature, gives the heat absorbed or rejected by the working substance. Mathematically, heat absorbed by the working substance,

$$\delta Q = TdS$$

where T = Absolute temperature, and

dS = Increase in entropy.

Note : The above relation also holds good for heat rejected by the working substance. In that case, dS will be decrease in entropy.

The engineers and scientists use it for providing quick solution, to problems dealing with reversible adiabatic expansion. The entropy is usually represented by S.

4.2. Relation between Heat and Entropy

Consider the heating of a working substance by a reversible process as shown by a curve from 1 to 2, on a graph, whose base represents the entropy and the vertical ordinate represents the absolute temperature as shown in Fig. 4.1. This diagram is known as temperature-entropy (T-S) diagram.

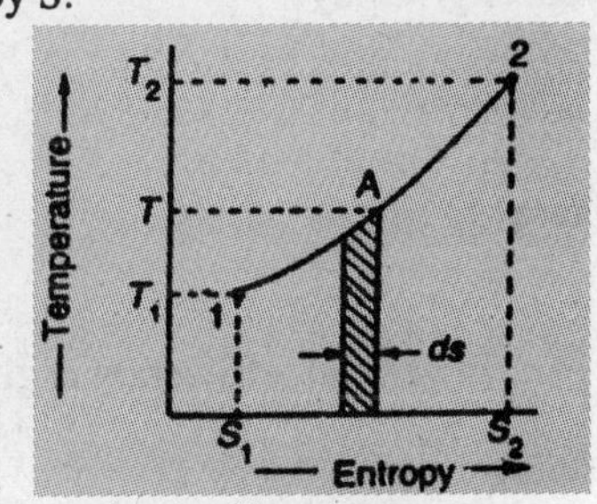

Fig. 4.1. Temperature-entropy diagram.

Now consider any point A on the curve 1-2. At this point, let a small quantity of heat (δQ) be supplied to the working substance, which will increase the entropy by dS. Let the absolute temperature at this instant be T. Then according to the definition of entropy,

$$\delta Q = TdS \quad \ldots (i)$$

From Fig. 4.1, we see that the term TdS represents the area under the curve during this change of entropy. Now the total area under the curve 1-2 may be found out by integrating the equation (*i*), *i.e.*

$$\int \delta Q = \int TdS = \text{Total heat supplied or absorbed}$$

From equation (*i*), we get

$$dS = \frac{\delta Q}{T}$$

The total change in entropy may be obtained by integrating the above expression from state 1 to state 2,

$$\therefore \qquad \int_1^2 dS = \int_1^2 \frac{\delta Q}{T} \qquad \ldots (ii)$$

Notes : 1. The area under the *T-S* diagram of any thermodynamic process represents the heat absorbed or rejected during that process.

2. Since $\int \frac{\delta Q}{T}$ is same for all reversible paths between states 1 and 2, so we conclude that this quantity is independent of a path and is a function of end states only. In other words, the entropy is a point function and thus it is a property of the system. The entropy may be expressed as a function of other thermodynamic properties of the system, such as the pressure and temperature or pressure and volume.

3. We know that according of First Law of Thermodynamics,

$$\delta Q = dU + \delta W = dU + p\,dv \qquad \ldots (\because \delta W = p\,dv) \qquad \ldots (iii)$$

and

$$\delta Q = TdS \qquad \ldots (iv)$$

From equations (*iii*) and (*iv*),

$$TdS = dU + pdv \qquad \ldots (v)$$

It is very interesting to note that in equations (*iii*) and (*iv*), δQ and δW are path functions, therefore these equations are true only for reversible processes. But in equation (*v*), dS, dU and dv are point functions as they depend upon the initial and final equilibrium states, therefore equation (*v*) is true for reversible as well as irreversible processes.

4. The entropy remains constant in a reversible cyclic process and increases in an irreversible cyclic process (see Art. 4.7)

4.3. Importance of Entropy

The maximum possible efficiency obtainable by any engine working on a reversible * Carnot cycle is given by

$$\eta = \frac{T_1 - T_2}{T_1} \qquad \ldots (i)$$

where T_1 = Highest absolute temperature, and

T_2 = Lowest absolute temperature.

In general, efficiency is given by

$$\eta = \frac{\text{Maximum work obtained}}{\text{Heat supplied or absorbed}} = \frac{\delta W}{\delta Q}$$

or

$$\delta W = \delta Q \times \eta = \delta Q \left(\frac{T_1 - T_2}{T_1} \right) \qquad \ldots [\text{From equation } (i)]$$

* For details, please refer to Chapter 6 on Thermodynamic Air Cycles.

For one degree temperature drop, the above expression may be written as

$$\delta W = \frac{\delta Q}{T} = dS = \text{Change in entropy}$$

From this expression, it can be easily understood that

1. The change in entropy represents the maximum amount of work obtainable per degree drop in temperature.
2. The change in entropy may be regarded as a measure of the rate of the availability or *unavailability of heat for transformation into work.
3. The increase in entropy is obtained from a given quantity of heat at a low temperature.

4.4. Available and Unavailable Heat Energy

The heat energy of a system (or heat supplied to the working substance) is considered to have the following two parts :

1. Available heat energy ; and 2. Unavailable heat energy.

The *available heat energy* is that part of the heat energy (or heat supplied) which can be converted into mechanical work by ideal processes which reduces the system in a state of equilibrium.

The *unavailable heat energy* is that part of heat energy (or heat supplied) which can not be converted into mechanical work even by ideal process which reduces the system in a state of equilibrium. The common term used for unavailable heat energy, according to Second Law of Thermodynamics, is the heat rejected by the system to the surroundings.

From above, we have total heat energy or heat supplied to the system,

$$\delta Q = \text{Available heat energy} + \text{Unavailable heat energy}$$

$$= \text{A.H.E} + \text{U.H.E} = \text{Workdone} + \text{Heat rejected}$$

We know that the maximum possible efficiency obtainable by any engine working on a Carnot cycle is given by

$$\eta = \frac{T_1 - T_2}{T_1} = 1 - \frac{T_2}{T_1} \qquad \ldots (i)$$

where T_1 = Maximum absolute temperature, and

T_2 = Lowest absolute temperature.

We also know that efficiency,

$$\eta = \frac{\text{Maximum work obtained}}{\text{Heat supplied or absorbed}} = \frac{\delta W}{\delta Q} \qquad \ldots (ii)$$

From equations (*i*) and (*ii*),

$$\frac{\delta W}{\delta Q} = 1 - \frac{T_2}{T_1} \quad \text{or} \quad \delta W = \delta Q\left(1 - \frac{T_2}{T_1}\right)$$

Since the lowest practical temperature of heat rejection is the temperature of surroundings (T_0), therefore the above expression may be written as

$$\delta W = \delta Q\left(1 - \frac{T_0}{T_1}\right) = \delta Q - \delta Q \times \frac{T_0}{T_1} \qquad \ldots (iii)$$

---

* Refer Art. 4.4 and 4.13

We also know that workdone,

$$\delta W = \text{Heat supplied} - \text{Unavailable heat energy or heat rejected.}$$

$$= \delta Q - \text{U.H.E.} \qquad \ldots (iv)$$

From equations (*iii*) and (*iv*),

$$\delta Q - \delta Q \times \frac{T_0}{T_1} = \delta Q - \text{U.H.E}$$

or

$$\text{U.H.E} = T_0 \left(\frac{\delta Q}{T_1} \right) = T_0 \times dS$$

Thus the unavailable heat energy (U.H.E) or the heat rejected is the product of the lowest temperature of heat rejection and the change of entropy of the system during the process of supplying heat. In other words, the *change in entropy may be regarded as a measure of unavailable form of heat energy* or *irreversibility of the process.*

Note : In the above discussion, the heat rejection takes place from the system to the surrounding (*i.e.* from a higher temperature to a lower temperature). At the end of the process, by virtue of Second Law of Thermodynamics, it is not possible to transfer heat from the system at a lower temperature to the surroundings at a higher temperature. Thus, the above process of heat transfer is *irreversible process.*

4.5. Units of Entropy

The unit of entropy depends upon the unit of heat employed and the absolute temperature. We know that

$$\text{Change in entropy } (dS) = \frac{\text{Heat supplied or rejected } (\delta Q)}{\text{Absolute temperature } (T)}$$

Therefore, if the heat supplied or rejected is in kJ and the temperature is in K, then the unit of entropy is kJ/K. The entropy may be expressed in so many units entropy without assigning any dimensional units. Since the entropy is expressed per unit mass of the working substance, it would be more correct to speak **specific entropy*. The absolute values of entropy cannot be determined, but only the change in entropy may be obtained by using equation (*ii*) in Art. 4.2.

Theoretically, the entropy of a substance is *zero* at *absolute zero temperature*. Hence, in entropy calculations, some convenient datum should be selected from which measurement may be made.

It may be noted that water at 0° C is assumed to have zero entropy, and changes in its entropy are reckoned from this temperature.

4.6. Clausius Inequality

The Clausius inequality states that 'whenever a closed system undergoes a cyclic process, the cyclic integral of $\delta Q/T$ is less than zero (*i.e.* negative) for an irreversible cyclic process and equal to zero for a reversible cyclic process, Mathematically,

$$\oint \frac{\delta Q}{T} < 0, \text{ for an irreversible cyclic process} \qquad \ldots (i)$$

and

$$\oint \frac{\delta Q}{T} = 0, \text{ for a reversible cyclic process} \qquad \ldots (ii)$$

Combining the equations (*i*) and (*ii*), the equation for the Clausius inequality is written as

$$\oint \frac{\delta Q}{T} \leq 0 \qquad \ldots (iii)$$

* The entropy is an extensive property of the system. The ratio of the extensive property of the system to the mass of the system is the specific value of that property as explained in Chapter 1.

The Clausius inequality not only gives mathematical expression to the second law of thermodynamics, but it also gives the quantitative measure of irreversibility of the system. For example, the equation (*i*) for an irreversible cyclic process may be written as

$$\oint \frac{\delta Q}{T} + I = 0$$

where I represents the amount by which the given cyclic process is irreversible. When I is equal to zero, then the given cyclic process will be reversible. Moreover, a cyclic process in which $\oint dQ/T$ is more than zero, is impossible because it voilates the second law of thermodynamics.

4.7. Principle of Increase of Entropy

We have discussed in Art. 4.6 that the equation for Clausius inequality is

$$\oint \frac{\delta Q}{T} \leq 0 \qquad \ldots (i)$$

We know that the change in entropy,

$$dS = \frac{\delta Q}{T}$$

Since the entropy is a thermodynamic property and the cyclic integral of a thermodynamic property is zero, therefore equation (*i*) may be written as

$$\oint \frac{\delta Q}{T} \leq \oint dS$$

or

$$\frac{\delta Q}{T} \leq dS$$

and

$$dS \geq \frac{\delta Q}{T} \qquad \ldots (ii)$$

When the process is reversible, then

$$dS = \frac{\delta Q}{T}$$

and when the process is irreversible, then

$$dS > \frac{\delta Q}{T}$$

If we apply the equation (*ii*) to an isolated system like universe, for which $\delta Q = 0$, then the equation (*ii*) may be written as

$$dS \geq 0$$

For a reversible cyclic process,

$$dS = 0 \text{ or } S = \text{Constant}$$

In other words, the entropy for a reversible cyclic process remains constant. Now for an irreversible cyclic process,

$$dS > 0$$

Since, in practice, all processes are irreversible, therefore the entropy of such a system like universe goes on increasing.

This is known as the *principle of increase of entropy*.

Note : The principle of change of entropy may also be discussed as follows :

Consider a given quantity of heat energy Q rejected by a hot body at temperature T_1 and absorbed by a cold body at temperature T_2.

∴ Loss of entropy by the hot body $= Q / T_1$

and gain of entropy by the cold body $= Q / T_2$

Since T_1 is greater than T_2, therefore the gain of entropy by the cold body is greater than the loss of entropy by the hot body. In other words, we can say that when the temperature *falls in a system (*i.e.* irreversible process) the entropy increases.

This conclusion can be extended to any isolated system (say universe) in which the heat exchange between the system (at a lower temperature) and the surroundings (at a higher temperature) takes place in an irreversible manner. Thus the entropy of an isolated system (universe) increases.

4.8. General Expression for Change of Entropy of a Perfect Gas

Consider a certain quantity of a perfect gas being heated by any thermodynamic process.

Let m = Mass of the gas,

p_1 = Initial pressure of the gas,

v_1 = Initial volume of the gas,

T_1 = Initial temperature of the gas, and

p_2, v_2, T_2 = Corresponding values for the final conditions.

Now the relation for the change of entropy during the process may be expressed in the following three ways :

(*a*) *In terms of volume and absolute temperature*

We know that for a small change in the state of a working substance, the general gas energy equation is,

$$\delta Q = dU + \delta W = mc_v\, dT + p\, dv \quad \ldots (i)$$

where dT = Small change in temperature, and

dv = Small change in volume.

Dividing throughout equation (*i*) by T,

$$\frac{\delta Q}{T} = mc_v \frac{dT}{T} + \frac{p\, dv}{T}$$

Since $pv = mR\,T$ or $\frac{p}{T} = \frac{mR}{v}$ and $\frac{\delta Q}{T} = dS$, therefore

$$dS = mc_v \frac{dT}{T} + \frac{mR}{v}\, dv \quad \ldots (ii)$$

Integrating equation (*ii*) within appropriate limits,

$$\int_{S_1}^{S_2} dS = mc_v \int_{T_1}^{T_2} \frac{dT}{T} + mR \int_{v_1}^{v_2} \frac{dv}{v}$$

$$[\,S\,]_{S_1}^{S_2} = mc_v \Big[\log_e T\Big]_{T_1}^{T_2} + mR\Big[\log_e v\Big]_{v_1}^{v_2}$$

---

* The cold body which is at a lower temperature receives heat from a hot body which is at a higher temperature. The temperature of the hot body falls. It is similar to a case when we supply heat at constant volume or constant pressure. After the process, by virtue of second law of thermodynamics, it not possible to transfer heat from a cold body to a hot body. Such a process is irreversible process.

$$\therefore \qquad S_2 - S_1 = mc_v(\log_e T_2 - \log_e T_1) + mR(\log_e v_2 - \log_e v_1)$$

$$= mc_v \log_e\left(\frac{T_2}{T_1}\right) + mR \log_e\left(\frac{v_2}{v_1}\right) \qquad \ldots (iii)$$

$$= 2.3\, m\left[c_v \log\left(\frac{T_2}{T_1}\right) + R \log\left(\frac{v_2}{v_1}\right)\right]$$

$$= 2.3\, m\left[c_v \log\left(\frac{T_2}{T_1}\right) + (c_p - c_v) \log\left(\frac{v_2}{v_1}\right)\right]$$

(*b*) *In terms of pressure and absolute temperature*

We know from the general gas equation,

$$\frac{p_1 v_1}{T_1} = \frac{p_2 v_2}{T_2} \quad \text{or} \quad \frac{v_2}{v_1} = \frac{p_1}{p_2} \times \frac{T_2}{T_1}$$

Substituting the value of $\left(\frac{v_2}{v_1}\right)$ **in equation (*iii*),**

$$S_2 - S_1 = mc_v \log_e\left(\frac{T_2}{T_1}\right) + mR \log_e\left(\frac{p_1}{p_2} \times \frac{T_2}{T_1}\right)$$

$$= mc_v \log_e\left(\frac{T_2}{T_1}\right) + mR \log_e\left(\frac{p_1}{p_2}\right) + mR \log_e\left(\frac{T_2}{T_1}\right)$$

$$= m \log_e\left(\frac{T_2}{T_1}\right)(c_v + R) + mR \log_e\left(\frac{p_1}{p_2}\right) \qquad \ldots (iv)$$

Now substituting $R = c_p - c_v$ **in the above equation,**

$$S_2 - S_1 = mc_p \log_e\left(\frac{T_2}{T_1}\right) + m(c_p - c_v) \log_e\left(\frac{p_1}{p_2}\right) \qquad \ldots (v)$$

$$= 2.3\, m\left[c_p \log\left(\frac{T_2}{T_1}\right) + (c_p - c_v) \log\left(\frac{p_1}{p_2}\right)\right]$$

(*c*) *In terms of pressure and volume*

We know from the general gas equation,

$$\frac{p_1 v_1}{T_1} = \frac{p_2 v_2}{T_2} \quad \text{or} \quad \frac{T_2}{T_1} = \frac{p_2}{p_1} \times \frac{v_2}{v_1}$$

Substituting the value of $\left(\frac{T_2}{T_1}\right)$ **in equation (*iii*),**

$$S_2 - S_1 = mc_v \log_e\left(\frac{p_2}{p_1} \times \frac{v_2}{v_1}\right) + mR \log_e\left(\frac{v_2}{v_1}\right)$$

Now substituting $R = c_p - c_v$ in the above equation,

$$S_2 - S_1 = mc_v \log_e\left(\frac{p_2}{p_1}\right) + mc_v \log_e\left(\frac{v_2}{v_1}\right) + m\,(c_p - c_v)\log_e\left(\frac{v_2}{v_1}\right)$$

$$= mc_v \log_e\left(\frac{p_2}{p_1}\right) + mc_v \log_e\left(\frac{v_2}{v_1}\right) + mc_p \log_e\left(\frac{v_2}{v_1}\right) - mc_v \log_e\left(\frac{v_2}{v_1}\right)$$

$$= mc_v \log_e\left(\frac{p_2}{p_1}\right) + mc_p \log_e\left(\frac{v_2}{v_1}\right) \qquad \ldots (vi)$$

$$= 2.3\,m\left[c_v \log\left(\frac{p_2}{p_1}\right) + c_p \log\left(\frac{v_2}{v_1}\right)\right]$$

Notes : 1. The expression (*vi*) is valid for both reversible as well as irreversible processes.

2. The change of entropy is *positive* when heat is *absorbed* by the gas and there is *increase of entropy*.

3. The change of entropy is *negative* when heat is *removed* from the gas and there is *decrease of entropy*.

Example 4.1. *0.05 m³ of air at a pressure of 8 bar and temperature 280° C expands to eight times its original volume and the final temperature after expansion is 25° C. Calculate change of entropy of air during the process. Assume c_p = 1.005 kJ/kg K and c_v = 0.712 kJ/kg K.*

Solution. Given : $v_1 = 0.05\ \text{m}^3$; $p_1 = 8\ \text{bar} = 0.8 \times 10^6\ \text{N/m}^2$; $T_1 = 280°\text{C} = 280 + 273 = 553\ \text{K}$; $v_2 = 8\,v_1 = 8 \times 0.05 = 0.4\ \text{m}^3$; $T_2 = 25°\text{C} = 25 + 273 = 298\ \text{K}$; $c_p = 1.005\ \text{kJ/kg K}$; $c_v = 0.712\ \text{kJ/kg K}$

Let m = Mass of air in kg.

We know that gas constant,

$$R = c_p - c_v = 1.005 - 0.712 = 0.293\ \text{kJ/kg K} = 293\ \text{J/kg K}$$

and

$$p_1 v_1 = m R T \text{ or } m = \frac{p_1 v_1}{R T_1} = \frac{0.8 \times 10^6 \times 0.05}{293 \times 553} = 0.247\ \text{kg}$$

∴ Change of entropy,

$$S_2 - S_1 = 2.3\,m\left[c_v \log\left(\frac{T_2}{T_1}\right) + R \log\left(\frac{v_2}{v_1}\right)\right]$$

$$= 2.3 \times 0.247\left[0.712 \log\left(\frac{298}{553}\right) + 0.293 \log\left(\frac{0.4}{0.05}\right)\right]$$

$$= 0.568\,(-0.19 + 0.26) = 0.04\ \text{kJ/K} \quad \textbf{Ans.}$$

4.9. Change of Entropy of a Perfect Gas during Various Thermodynamic Processes

We have already discussed in Chapter 3, the various thermodynamic processes of a perfect gas and have derived the equations for work done, change of internal energy and heat supplied. Now we shall derive expressions for the change of entropy during the following thermodynamic processes :

1. Constant volume process (or isochoric process) ; 2. Constant pressure process (or Isobaric process) ; 3. Constant temperature process (or Isothermal process) ; 4. Adiabatic process (or Isentropic process) ; and 5. Polytropic process.

4.10. Change of Entropy during Constant Volume Process (or Isochoric Process)

Consider a certain quantity of a perfect gas being heated at a constant volume.

Let m = Mass of the gas,

p_1 = Initial pressure of the gas,

T_1 = Initial temperature of the gas,

p_2, T_2 = Corresponding values for the final conditions.

Let this process be represented by the curve 1-2 on *T-S* diagram as shown in Fig. 4.2.

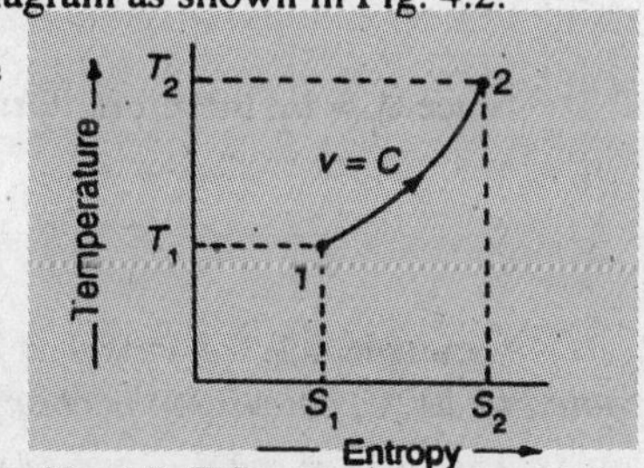

Fig. 4.2. *T-S* curve during constant volume process.

We know that for a small change of temperature (dT), the heat supplied,

$$\delta Q = mc_v \, dT$$

Dividing both sides of the above equation by T,

$$\frac{\delta Q}{T} = mc_v \frac{dT}{T}$$

or

$$^*dS = mc_v \frac{dT}{T} \qquad \ldots \left(\because \frac{\delta Q}{T} = dS\right)$$

Integrating this expression for the total change of entropy,

$$\int_{S_1}^{S_2} dS = mc_v \int_{T_1}^{T_2} \frac{dT}{T} \quad \text{or} \quad [S]_{S_1}^{S_2} = mc_v \left[\log_e T\right]_{T_1}^{T_2}$$

$$\therefore \quad S_2 - S_1 = mc_v \log_e \left(\frac{T_2}{T_1}\right) = 2.3\, mc_v \log \left(\frac{T_2}{T_1}\right) \qquad \ldots (i)$$

The above relation may also be expressed in terms of pressure. We know from the general gas equation,

$$\frac{p_1 v_1}{T_1} = \frac{p_2 v_2}{T_2} \quad \text{or} \quad \frac{T_2}{T_1} = \frac{p_2}{p_1} \qquad \ldots (\because v_1 = v_2)$$

Substituting the value of $\left(\frac{T_2}{T_1}\right)$ in equation (*i*),

$$S_2 - S_1 = 2.3\, mc_v \log \left(\frac{p_2}{p_1}\right) \qquad \ldots (ii)$$

The equations (*i*) and (*ii*) are valid for both reversible as well as irreversible process.

Alternate proofs for change of entropy

We have seen in Art. 4.8 that the general expression for change of entropy in terms of volume and absolute temperature is,

$$S_2 - S_1 = 2.3\, m\left[c_v \log\left(\frac{T_2}{T_1}\right) + R \log\left(\frac{v_2}{v_1}\right)\right]$$

---

* We know that $dS = m\,c_v \frac{dT}{T}$ or $\frac{dT}{dS} = \frac{T}{m\,c_v}$

For 1 kg of a perfect gas, $\frac{dT}{ds} = \frac{T}{c_v}$

The term $\frac{dT}{ds}$ is known as slope of the curve 1-2 on the *T-s* diagram as shown in Fig. 4.2.

Since $v_1 = v_2$, therefore $v_2/v_1 = 1$. Moreover, log 1 = 0

$$\therefore \quad S_2 - S_1 = 2.3\, mc_v \log\left(\frac{T_2}{T_1}\right)$$

Similarly, the general expression for change of entropy in terms of pressure and volume is,

$$S_2 - S_1 = 2.3\, m\left[c_v \log\left(\frac{p_2}{p_1}\right) + c_p \log\left(\frac{v_2}{v_1}\right)\right]$$

Since $v_1 = v_2$, therefore $v_2/v_1 = 1$. Moreover, log 1 = 0.

$$\therefore \quad S_2 - S_1 = 2.3\, mc_v \log\left(\frac{p_2}{p_1}\right)$$

Example 4.2. *A vessel of 2.5 m³ capacity contains 1 kg-mole of N_2 at 100° C. If the gas is cooled to 30° C, calculate the change in specific entropy.*

The ratio of specific heats is 1.4 and one kg-mole nitrogen is 28 kg.

Solution. Given : $*v = 2.5$ m³ ; $M = 1$ kg-mole = 28 kg ; $T_1 = 100°$ C = 100 + 273 = 373 K ; $T_2 = 30°$ C = 30 + 273 = 303 K ; $\gamma = c_p/c_v = 1.4$

Since the universal gas constant (R_u) for all gases is 8.314 kJ/kg K, therefore characteristic gas constant,

$$R = R_u / M = 8.314 / 28 = 0.297 \text{ kJ/kg K}$$

and $$c_p - c_v = R \quad \text{or} \quad 1.4\, c_v - c_v = 0.297 \qquad \ldots (\because c_p/c_v = \gamma = 1.4)$$

$$\therefore \quad c_v = 0.297 / 0.4 = 0.74 \text{ kJ/kg K}$$

We know that change in specific entropy (*i.e.* per kg of gas),

$$s_2 - s_1 = 2.3\, m\, c_v \log\left(\frac{T_2}{T_1}\right) = 2.3 \times 1 \times 0.74 \log\left(\frac{303}{373}\right) \text{ kJ/kg K}$$

$$= -0.1536 \text{ kJ/kg K} \quad \textbf{Ans.}$$

The – *ve* sign indicates that there is a decrease in entropy.

Example 4.3. *A vessel of capacity 3 m³ contains air at a pressure of 1.5 bar and a temperature of 25° C. Additional air is now pumped into the system until the pressure rises to 30 bar and the temperature rises to 60° C. Determine the mass of air pumped in, and express the quantity as a volume at a pressure of 1.02 bar and a temperature of 20° C.*

If the vessel is allowed to cool until the temperature is again 25° C, calculate the pressure in the vessel. Determine the quantity of heat transferred and change of entropy of the gas during the cooling process only. Neglect the effect of heat capacity of the vessel. Assume air as an ideal gas.

Solution. Given : $v_1 = 3$ m³ ; $p_1 = 1.5$ bar $= 0.15 \times 10^6$ N/m² ; $T_1 = 25°$ C = 25 + 273 = 298 K ; $p_2 = 30$ bar $= 3 \times 10^6$ N/m² ; $T_2 = 60°$ C = 60 + 273 = 333 K

Mass of air pumped in

Let m_1 = Mass of air initially filled in the vessel, and

m_2 = Mass of air in the vessel after pumping.

* Superfluous data

We know that $p_1 v_1 = m_1 R T_1$

$$\therefore \quad m_1 = \frac{p_1 v_1}{R T_1} = \frac{0.15 \times 10^6 \times 3}{287 \times 298} = 5.26 \text{ kg} \quad \ldots \text{(Taking } R \text{ for air} = 287 \text{ J/kg K)}$$

Similarly $\quad p_2 v_2 = m_2 R T_2$

$$\therefore \quad m_2 = \frac{p_2 v_2}{R T_2} = \frac{3 \times 10^6 \times 3}{287 \times 333} = 94.17 \text{ kg} \quad \ldots (\because v_2 = v_1)$$

$\therefore$ Mass of air pumped in,

$$m = m_2 - m_1 = 94.17 - 5.26 = 88.91 \text{ kg} \quad \textbf{Ans.}$$

Volume of air pumped in at a pressure of 1.02 bar and temperature of 20° C

Given : $p = 1.02$ bar $= 0.102 \times 10^6$ N/m² ; $T = 20°$ C $= 20 + 273 = 293$ K

Let $\quad v$ = Volume of air pumped in.

We know that $\quad p v = m R T$

$$\therefore \quad v = \frac{m R T}{p} = \frac{88.91 \times 287 \times 293}{0.102 \times 10^6} = 73.3 \text{ m}^3 \quad \textbf{Ans.}$$

Pressure in the vessel after cooling

Let $\quad p_3$ = Pressure in the vessel after cooling.

We know that the temperature after cooling,

$$T_3 = T_1 = 25° \text{ C} = 298 \text{ K}$$

Since the cooling is at constant volume, therefore

$$\frac{p_3}{p_2} = \frac{T_3}{T_2}$$

or

$$p_3 = \frac{T_3 p_2}{T_2} = \frac{298 \times 3 \times 10^6}{333} = 2.68 \times 10^6 \text{ N/m}^2 = 26.8 \text{ bar} \quad \textbf{Ans.}$$

Heat transferred during cooling

Since the vessel is cooled from $T_2 = 333$ K to the initial temperature $T_3 = T_1 = 298$ K, therefore change in internal energy during cooling,

$$dU = m_2 c_v (T_3 - T_2) = 94.17 \times 0.172 (298 - 333) = -2346.7 \text{ kJ} \quad \textbf{Ans.}$$

The – *ve* sign indicates that the internal energy decreases during cooling.

We know that heat transferred is equal to change in internal energy, when the process takes place at constant volume (because work done is zero). Therefore heat transferred,

$$\delta Q = -567 \text{ kJ} \quad \textbf{Ans.}$$

The – *ve* sign indicates that heat is rejected by the gas.

Change of entropy during cooling process

We know that change of entropy during cooling process,

$$S_3 - S_2 = 2.3\, m_2\, c_v \log\left(\frac{T_3}{T_2}\right) = 2.3 \times 94.17 \times 0.712 \log\left(\frac{298}{333}\right) \text{ kJ/K}$$

$$= -7.44 \text{ kJ/K} \quad \textbf{Ans.} \quad \ldots \text{(Taking } c_v = 0.712 \text{ kJ/kg K)}$$

The – *ve* sign indicates that there is a decrease of entropy.

Example 4.4. *An insulated vessel of capacity 0.056 m³ is divided into two compartments A and B by a conducting diaphragm. Each compartment has a capacity of 0.028 m³. The compartment A contains air at a pressure of 1.5 bar and 25° C and the compartment B contains air at a pressure 4.2 bar and 175° C. Find : 1. final equilibrium temperature, 2. final pressure on each side of the diaphragm, and 3. change of entropy of the system.*

Solution. Given : $v_A = v_B = 0.028$ m³ ; $p_A = 1.5$ bar $= 0.15 \times 10^6$ N/m² ; $T_A = 25°$ C $= 25 + 273 = 298$ K ; $p_B = 4.2$ bar $= 0.42 \times 10^6$ N/m² ; $T_B = 175°$ C $= 175 + 273 = 448$ K

1. *Final equilibrium temperature*

Let T_F = Final equilibrium temperature.

First all, let us find the mass of air in compartment A (m_A) and the mass of air in compartment B (m_B). We know that

$$p_A v_A = m_A R T_A \text{ or } m_A = \frac{p_A v_A}{R T_A} = \frac{0.15 \times 10^6 \times 0.028}{287 \times 298} = 0.049 \text{ kg}$$

... ($\because$ R for air = 287 J/kg K)

$$p_B v_B = m_B R T_B \text{ or } m_B = \frac{p_B v_B}{R T_B} = \frac{0.42 \times 10^6 \times 0.028}{287 \times 448} = 0.091 \text{ kg}$$

Since the diaphragm is conducting, therefore

Heat gained by air in compartment A

= Heat rejected by air in compartment B

or $$m_A c_v (T_F - T_A) = m_B c_v (T_B - T_F)$$

$$0.049 \times c_v (T_F - 298) = 0.091 \times c_v (448 - T_F)$$

$$0.049 T_F - 14.6 = 40.77 - 0.091 T_F \quad \text{or} \quad 0.14 T_F = 55.37$$

$\therefore$ $T_F = 395.5$ K **Ans.**

2. *Final pressure on each side of the diaphragm*

Let p_{AF} = Final pressure in compartment A, and

p_{BF} = Final pressure in compartment B.

Since the volume of each compartment is same, therefore

$$\frac{p_A}{T_A} = \frac{p_{AF}}{T_F} \quad \text{or} \quad p_{AF} = \frac{p_A T_F}{T_A} = \frac{1.5 \times 395.5}{298} = 1.99 \text{ bar } \textbf{Ans.}$$

and $$\frac{p_B}{T_B} = \frac{p_{BF}}{T_F} \quad \text{or} \quad p_{BF} = \frac{p_B T_F}{T_B} = \frac{4.2 \times 395.5}{448} = 3.708 \text{ bar } \textbf{Ans.}$$

3. *Change of entropy of the system*

We know that change of entropy for compartment A, whose temperature has been increased from T_A to T_F,

$$(dS)_A = 2.3\, m_A\, c_v \log\left(\frac{T_F}{T_A}\right) = 2.3 \times 0.049 \times 0.712 \log\left(\frac{395.5}{298}\right)$$

... (Taking $c_v = 0.712$ kJ/kg K)

$$= 0.08 \log (1.327) = 0.0098 \text{ kJ/K}$$

and change of entropy for compartment *B*, whose temperature has been decreased from T_B to T_F,

$$(dS)_B = 2.3\, m_B\, c_v \log\left(\frac{T_F}{T_B}\right) = 2.3 \times 0.091 \times 0.712 \log\left(\frac{395.5}{448}\right) \text{kJ/K}$$

$$= 0.149 \log (0.883) = -0.0081 \text{ kJ/K}$$

∴ Change of entropy of the system,

$$dS = (dS)_A + (dS)_B = 0.0098 - 0.0081 = 0.0017 \text{ kJ/K} \quad \textbf{Ans.}$$

4.11. Change of Entropy during Constant Pressure Process (or Isobaric Process)

Consider a certain quantity of a perfect gas being heated at constant pressure.

Let m = Mass of the gas,

v_1 = Initial volume of the gas,

T_1 = Initial temperature of the gas,

v_2, T_2 = Corresponding values for the final conditions.

Let this process be represented by the curve 1-2 on *T-S* diagram as shown in Fig. 4.3. We know that for a small change of temperature (dT), the heat supplied,

$$\delta Q = mc_p\, dT$$

Dividing both sides of the above equation by T,

$$\frac{\delta Q}{T} = mc_p\left(\frac{dT}{T}\right)$$

or

$$^{*}dS = mc_p\left(\frac{dT}{T}\right)$$

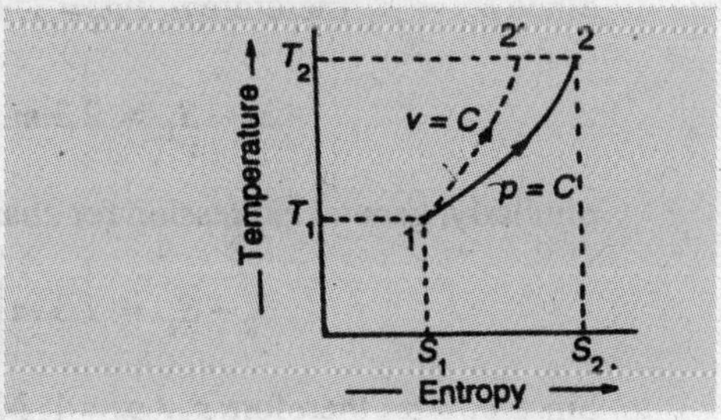

Fig. 4.3. *T-S* curve during constant pressure process.

Integrating this expression for the total change of entropy,

$$\int_{S_1}^{S_2} dS = mc_p \int_{T_1}^{T_2} \frac{dT}{T}$$

---

* We know that $dS = mc_p \dfrac{dT}{T}$ or $\dfrac{dT}{dS} = \dfrac{T}{mc_p}$

For 1 kg of a perfect gas, $\dfrac{dT}{ds} = \dfrac{T}{c_p}$

The term $\dfrac{dT}{ds}$ is known as slope of the curve 1-2 on the *T-S* diagram as shown in Fig. 4.3.

We have already discussed that slope of the curve for constant volume process is

$$\frac{dT}{ds} = \frac{T}{c_v}$$

Since for a perfect gas, $c_v < c_p$, therefore $\dfrac{1}{c_v} > \dfrac{1}{c_p}$ or $\dfrac{T}{c_v} > \dfrac{T}{c_p}$

Thus, the slope of the curve on the *T-S* diagram for constant volume process 1-2′ is higher than that of constant pressure process 1-2, as shown in Fig. 4.3.

or $$S_2 - S_1 = mc_p \log_e\left(\frac{T_2}{T_1}\right) = 2.3\ mc_p \log\left(\frac{T_2}{T_1}\right) \quad \ldots (i)$$

The above relation may also be expressed in terms of volume. We know from the general gas equation,

$$\frac{p_1 v_1}{T_1} = \frac{p_2 v_2}{T_2} \quad \text{or} \quad \frac{T_2}{T_1} = \frac{v_2}{v_1} \quad \ldots (\because p_1 = p_2)$$

Substituting this value of T_2 / T_1 in equation (*i*),

$$S_2 - S_1 = 2.3\ mc_p \log\left(\frac{v_2}{v_1}\right) \quad \ldots (ii)$$

The equations (*i*) and (*ii*) are valid for both reversible as well as irreversible process.

Alternate proof for change of entropy

We have seen in Art. 4.8 that the general expression for change of entropy in terms of pressure and absolute temperature,

$$S_2 - S_1 = 2.3\ m\left[c_p \log\left(\frac{T_2}{T_1}\right) + (c_p - c_v)\log\left(\frac{p_1}{p_2}\right)\right]$$

Since $p_1 = p_2$, therefore $p_1 / p_2 = 1$. Moreover $\log 1 = 0$.

$$\therefore \quad S_2 - S_1 = 2.3\ mc_p \log\left(\frac{T_2}{T_1}\right)$$

Similarly, general expression for change of entropy in terms of pressure and volume,

$$S_2 - S_1 = 2.3\ m\left[c_v \log\left(\frac{p_2}{p_1}\right) + c_p \log\left(\frac{v_2}{v_1}\right)\right]$$

Since $p_1 = p_2$, therefore $p_1 / p_2 = 1$. Moreover $\log 1 = 0$.

$$\therefore \quad S_2 - S_1 = 2.3\ mc_p \log\left(\frac{v_2}{v_1}\right)$$

Example 4.5. *0.5 kg of a perfect gas is heated from 100° C to 300° C at a constant pressure of 2.8 bar. It is then cooled to 100° C at constant volume. Find the overall change in entropy. Take c_p = 1 kJ/kg K and c_v = 0.72 kJ/kg K.*

Solution. Given : m = 0.5 kg ; T_1 = 100° C = 100 + 273 = 373 K ; T_2 = 300° C = 300 + 273 = 573 K ; p = 2.8 bar = 0.28×10^6 N/m^2 ; T_3 = 100° C = 100 + 273 = 373 K ; c_p = 1 kJ/kg K ; c_v = 0.72 kJ/kg K

We know that change of entropy during constant pressure heating,

$$S_2 - S_1 = 2.3\ mc_p \log\left(\frac{T_2}{T_1}\right) = 2.3 \times 0.5 \times 1 \times \log\left(\frac{573}{373}\right) \text{ kJ/K}$$

$$= 0.214 \text{ kJ/K}$$

and change of entropy during constant volume cooling,

$$S_3 - S_2 = 2.3\ mc_v \log\left(\frac{T_3}{T_2}\right) = 2.3 \times 0.5 \times 0.72 \log\left(\frac{373}{573}\right) \text{ kJ/K}$$

$$= -0.154 \text{ kJ/K}$$

The – *ve* sign indicates that there is a decrease of entropy.

∴ Overall change in entropy,

$$S_3 - S_1 = (S_2 - S_1) + (S_3 - S_2) = 0.214 - 0.154 = 0.06 \text{ kJ/K} \quad \textbf{Ans.}$$

Example 4.6. *Cold air from atmosphere is circulated through an air heating system, where the temperature of air is increased from 7.2° C to 21.2° C without any pressure loss. If the hourly consumption of warm air at 21.2° C is 850 m³, calculate how much heat per hour must be imparted to the air ?*

What will be the change in entropy of air circulated per hour ? Assume proper values of the specific heat of air.

Solution. Given : $T_1 = 7.2°\text{ C} = 7.2 + 273 = 280.2\text{ K}$; $T_2 = 21.2°\text{ C} = 21.2 + 273 = 294.2\text{ K}$; $v = 850\text{ m}^3/\text{h}$

Heat imparted to the air per hour

Let $p_1 = p_2$ = Atmospheric pressure (constant).

$= 1.013 \text{ bar} = 0.1013 \times 10^6 \text{ N/m}^2$ ... (Given)

R = Gas constant = 287 J/kg K ... (Assume)

c_p = Specific heat at constant pressure = 1.005 kJ/kg K ... (Assume)

First of all, let us find the mass of air (m). We know that

$$p_2 v_2 = m R T_2 \quad \text{or} \quad m = \frac{p_2 v_2}{R T_2} = \frac{0.1013 \times 10^6 \times 850}{287 \times 294.2} = 1020 \text{ kg/h}$$

We know that heat imparted to the air

$$Q = mc_p (T_2 - T_1) = 1020 \times 1.005 (294.2 - 280.2) \text{ kJ/h}$$

$$= 14\,351.4 \text{ kJ/h} \quad \textbf{Ans.}$$

Change in entropy of air circulated per hour

We know that change in entropy of air,

$$S_2 - S_1 = 2.3\, mc_p \log\left(\frac{T_2}{T_1}\right) = 2.3 \times 1020 \times 1.005 \log\left(\frac{294.2}{280.2}\right) \text{ kJ/K/h}$$

$$= 50 \text{ kJ/K/h} \quad \textbf{Ans.}$$

Example 4.7. *A mass of m_1 kg of a certain gas at a temperature T_1 is mixed at constant pressure with m_2 kg of mass of the same gas at a temperature T_2 ($T_1 > T_2$). The system is thermally insulated. Find the change in entropy of the universe and deduce the same for equal masses of the gas. Show that the change is necessarily positive.*

Solution. First of all, let us find the common temperature (T_c) of the mixture of the gas. This temperature T_c is less than T_1 and greater than T_2. We know that

Heat lost by the gas at temperature T_1

= Heat gained by the gas at temperature T_2

i.e. $m_1 c_p (T_1 - T_c) = m_2 c_p (T_c - T_2)$

... (where c_p is the specific heat of the gas at constant pressure)

$$\therefore \quad T_c = \frac{m_1 T_1 + m_2 T_2}{m_1 + m_2}$$

We know that the change in entropy for the gas whose temperature has been decreased from T_1 to T_c,

$$(dS)_1 = m_1 c_p \log_e \left(\frac{T_c}{T_1} \right) \quad \text{... (This will be negative as } T_1 > T_c\text{)}$$

Similarly, the change in entropy for the gas whose temperature has been increased from T_2 to T_c,

$$(dS)_2 = m_2 c_p \log_e \left(\frac{T_c}{T_2} \right) \quad \text{... (This will be positive as } T_c > T_2\text{)}$$

∴ Change in entropy of the universe (*i.e.* isolated system),

$$(dS)_u = (dS)_1 + (dS)_2$$

$$= m_1 c_p \log_e \left(\frac{T_c}{T_1} \right) + m_2 c_p \log_e \left(\frac{T_c}{T_2} \right)$$

$$= m_1 c_p \log_e \left[\frac{m_1 T_1 + m_2 T_2}{T_1 (m_1 + m_2)} \right] + m_2 c_p \log_e \left[\frac{m_1 T_1 + m_2 T_2}{T_2 (m_1 + m_2)} \right] \quad \textbf{Ans.}$$

If $m_1 = m_2 = m$, then the change in entropy of the universe,

$$(dS)_u = mc_p \log_e \left(\frac{T_1 + T_2}{2\,T_1} \right) + mc_p \log_e \left(\frac{T_1 + T_2}{2\,T_2} \right)$$

$$= mc_p \log_e \left[\left(\frac{T_1 + T_2}{2\,T_1} \right) \left(\frac{T_1 + T_2}{2\,T_2} \right) \right]$$

$$\text{...} \left[\because \log_e x + \log_e y = \log_e (x \times y) \right]$$

$$= mc_p \log_e \left(\frac{T_1 + T_2}{2\sqrt{T_1\,T_2}} \right)^2 = 2\,mc_p \log_e \left(\frac{T_1 + T_2}{2\sqrt{T_1\,T_2}} \right)$$

The term within the bracket is the ratio of arithmetic mean and the geometric mean of the temperatures T_1 and T_2. Since the arithmetic mean is always greater than the geometric mean, therefore $\frac{T_1 + T_2}{2} > \sqrt{T_1\,T_2}$ or $\frac{T_1 + T_2}{2\sqrt{T_1\,T_2}}$ is positive. Thus the change in entropy is always positive.

Example 4.8. *One kg of air at 310 K is heated at constant pressure by bringing it in contact with a hot reservoir at 1150 K. Find the entropy change of air, hot reservoir and of the universe.*

If the air is heated from 310 K to 1150 K by first bringing it in contact with a reservoir at 730 K and then with a reservoir at 1150 K, what will be the change of entropy of the universe ?

Solution. Given : $m = 1$ kg ; $T_A = 310$ K ; $T_R = 1150$ K

Entropy change of air, hot reservoir and of the universe

We know that entropy change of air,

$$(dS)_A = 2.3\,mc_p \log \left(\frac{T_R}{T_A} \right) = 2.3 \times 1 \times 1 \times \log \left(\frac{1150}{310} \right) = 1.309 \text{ kJ/K} \quad \textbf{Ans.}$$

... (Taking c_p for air = 1 kJ/kg K)

We also know that the heat absorbed by air or heat rejected by hot reservoir,

$$Q_A = Q_R = -mc_p(T_R - T_A) = -1 \times 1(1150 - 310) = -840 \text{ kJ}$$

... (–ve sign because of heat rejection)

∴ Change of entropy of the hot reservoir,

$$(dS)_R = \frac{\text{Heat rejected } (Q_R)}{\text{Absolute temperature } (T_R)} = \frac{-840}{1150} = -0.73 \text{ kJ/K} \quad \textbf{Ans.}$$

We know that change of entropy of the universe,

$$(dS)_u = (dS)_A + (dS)_R = 1.309 - 0.73 = 0.579 \text{ kJ/K} \quad \textbf{Ans.}$$

Change of entropy of universe when air is heated in two stages

The air is heated in two stages, first from 310 K to 730 K by bringing it in contact with a first reservoir at 730 K and then from 730 K to 1150 K by bringing it in contact with a second reservoir at 1150 K.

∴ Heat absorbed by air when heated from 310 K to 730 K or heat rejected by first reservoir,

$$Q_{A1} = Q_{R1} = -mc_p(T_{R1} - T_{A1}) = -1 \times 1(730 - 310) = -420 \text{ kJ}$$

... (Here $T_{R1} = 730$ K and $T_{A1} = T_A = 310$ K)

and heat absorbed by air when heated from 730 K to 1150 K or heat rejected by the second reservoir,

$$Q_{A2} = Q_{R2} = -mc_p(T_{R2} - T_{A2}) = -1 \times 1(1150 - 730) = -420 \text{ kJ}$$

... (Here $T_{R2} = 1150$ K and $T_{A2} = 730$ K)

We know that change of entropy of the air,

$$(dS)_A = 2.3\, mc_p \log\left(\frac{T_{R1}}{T_{A1}}\right) + 2.3\, mc_p \log\left(\frac{T_{R2}}{T_{A2}}\right)$$

$$= 2.3 \times 1 \times 1\left[\log\left(\frac{730}{310}\right) + \log\left(\frac{1150}{730}\right)\right]$$

$$= 2.3\,(0.372 + 0.197) = 1.309 \text{ kJ/K}$$

Change of entropy for the first reservoir,

$$(dS)_{R1} = \frac{Q_{R1}}{T_{R1}} = \frac{-420}{730} = -0.575 \text{ kJ/K}$$

Change of entropy for the second reservoir,

$$(dS)_{R2} = \frac{Q_{R2}}{T_{R2}} = \frac{-420}{1150} = -0.365 \text{ kJ/K}$$

∴ Change of entropy of the universe,

$$(dS)_u = (dS)_A + (dS)_{R1} + (dS)_{R2}$$

$$= 1.309 - 0.581 - 0.575 = 0.369 \text{ kJ/K} \quad \textbf{Ans.}$$

4.12. Change of Entropy during Constant Temperature Process (or Isothermal Process)

Consider a certain quantity of a perfect gas being heated at constant temperature.

Let m = Mass of the gas,

p_1 = Initial pressure of gas,

v_1 = Initial volume of gas,

p_2, v_2 = Corresponding values for the final conditions.

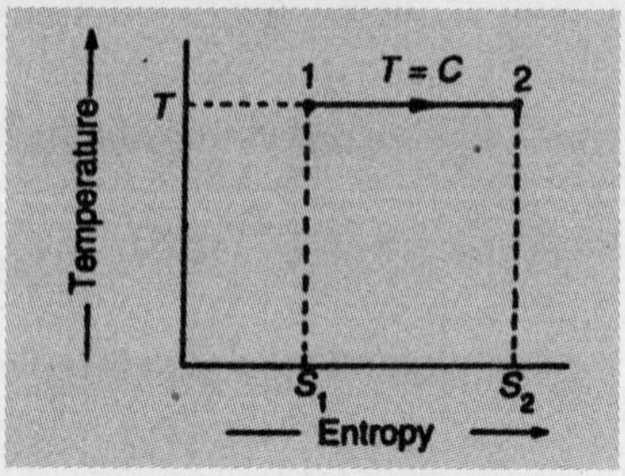

Fig. 4.4. *T-S* curve during constant temperature process

Let this process be represented by the line 1-2 on *T-S* diagram as shown in Fig. 4.4. We know that during constant temperature process (*i.e.* isothermal process), there is no change in internal energy, and the heat supplied is equal to the work done by the gas. We also know that work done during an isothermal process.

$$W_{1-2} = 2.3\, mRT \log\left(\frac{v_2}{v_1}\right)$$

$$\therefore \text{ Heat supplied,} \quad Q_{1-2} = W_{1-2} = 2.3\, mRT \log\left(\frac{v_2}{v_1}\right)$$

We know that change of entropy

$$= \frac{\text{Heat supplied}}{\text{Absolute temperature}}$$

or

$$S_2 - S_1 = 2.3 \frac{mRT}{T} \log\left(\frac{v_2}{v_1}\right)$$

$$= 2.3\, mR \log\left(\frac{v_2}{v_1}\right) = 2.3\, m\,(c_p - c_v) \log\left(\frac{v_2}{v_1}\right) \qquad \ldots (i)$$

The above relation may also be expressed in terms of pressure. We know from the general gas equation,

$$\frac{p_1 v_1}{T_1} = \frac{p_2 v_2}{T_2} \quad \text{or} \quad \frac{v_2}{v_1} = \frac{p_1}{p_2} \qquad \ldots (\because T_1 = T_2)$$

Substituting the value of $\left(\frac{v_2}{v_1}\right)$ in equation (*i*),

$$S_2 - S_1 = 2.3\, mR \log\left(\frac{p_1}{p_2}\right) = 2.3\, m\,(c_p - c_v) \log\left(\frac{p_1}{p_2}\right) \qquad \ldots (ii)$$

The equations (*i*) and (*ii*) are valid for both reversible and irreversible processes.

Alternate proofs for change of entropy

We have seen in Art. 4.8 that the general expression for change of entropy in terms of volume and absolute temperature is,

$$S_2 - S_1 = 2.3\, m\left[c_v \log\left(\frac{T_2}{T_1}\right) + (c_p - c_v) \log\left(\frac{v_2}{v_1}\right)\right]$$

Since $T_1 = T_2$, therefore $T_2 / T_1 = 1$. Moreover $\log 1 = 0$.

$$\therefore \quad S_2 - S_1 = 2.3\, m\, (c_p - c_v) \log \left(\frac{v_2}{v_1} \right)$$

Similarly, general expression for change of entropy in terms of pressure and absolute temperature is,

$$S_2 - S_1 = 2.3\, m \left[c_p \log \left(\frac{T_2}{T_1} \right) + (c_p - c_v) \log \left(\frac{p_1}{p_2} \right) \right]$$

Since $T_1 = T_2$, therefore $T_2 / T_1 = 1$. Moreover $\log 1 = 0$.

$$\therefore \quad S_2 - S_1 = 2.3\, m\, (c_p - c_v) \log \left(\frac{p_1}{p_2} \right)$$

Example 4.9. *A certain quantity of a perfect gas is heated in a reversible isothermal process from 1 bar and 40° C to 10 bar. Find the work done per kg of gas and the change of entropy per kg of gas. Take R = 287 J/kg K.*

Solution. Given : $p_1 = 1$ bar $= 0.1 \times 10^6$ N/m^2 ; $T_1 = 40°$ C $= 40 + 273 = 313$ K ; $p_2 = 10$ bar $= 1 \times 10^6$ N/m^2 ; $R = 287$ J/kg K

Workdone per kg of gas

We know that workdone per kg of gas,

$$w_{1-2} = 2.3\, mRT_1 \log \left(\frac{v_2}{v_1} \right) = 2.3\, mRT_1 \log \left(\frac{p_1}{p_2} \right) \quad \ldots (\because p_1 v_1 = p_2 v_2)$$

$$= 2.3 \times 1 \times 287 \times 313 \log \left(\frac{0.1 \times 10^6}{1 \times 10^6} \right) = -206\,610 \text{ J/kg}$$

$$= -206.61 \text{ kJ/kg} \quad \textbf{Ans.}$$

The $-ve$ sign indicates that the work is done on the gas.

Change in entropy per kg of gas

We know that the change in entropy (*i.e.* change in specific entropy),

$$s_2 - s_1 = 2.3\, mR \log \left(\frac{p_1}{p_2} \right) = 2.3 \times 1 \times 287 \log \left(\frac{0.1 \times 10^6}{1 \times 10^6} \right) \text{ J/kg K}$$

$$= -660.1 \text{ J/kg K} = 0.6601 \text{ kJ/kg K} \quad \textbf{Ans.}$$

The $-ve$ sign indicates that there is a decrease in entropy.

Note : The change in entropy may also be obtained as follows :

We know that the heat supplied (q_{1-2}) in an isothermal process is equal to the workdone (w_{1-2}).

$\therefore$ Heat supplied $= q_{1-2} = w_{1-2} = -206.61$ kJ/kg

$\therefore$ Change in specific entropy,

$$s_2 - s_1 = \frac{\text{Heat supplied}}{\text{Absolute temperature}} = \frac{-206.61}{313} = -0.6601 \text{ kJ/kg K} \quad \textbf{Ans.}$$

Example 4.10. *One kg of air occupies 0.084 m^3 at 12.5 bar and 537° C. It is expanded at a constant temperature to a final volume of 0.336 m^3. Calculate :*

1. the pressure at the end of expansion, 2. work done during expansion, 3. heat absorbed by the air, and 4. change of entropy.

Solution. Given : $m = 1$ kg ; $v_1 = 0.084$ m^3 ; $p_1 = 12.5$ bar $= 1.25 \times 10^6$ N/m^2 ; $T_1 = 537°$ C $= 537 + 273 = 810$ K ; $v_2 = 0.336$ m^3

1. *Pressure at the end of expansion*

Let p_2 = Pressure at the end of expansion.

We know that

$$p_1 v_1 = p_2 v_2$$

$$\therefore \quad p_2 = \frac{p_1 v_1}{v_2} = \frac{1.25 \times 10^6 \times 0.084}{0.336} = 0.3125 \times 10^6 \text{ N/m}^2$$

$$= 3.125 \text{ bar } \textbf{Ans.}$$

2. *Workdone during expansion*

We know that workdone during expansion,

$$W_{1-2} = 2.3\, mR\,T_1 \log\left(\frac{v_2}{v_1}\right) = 2.3\, p_1\, v_1 \log\left(\frac{v_2}{v_1}\right) \quad \ldots (\because p_1 v_1 = m R T_1)$$

$$= 2.3 \times 1.25 \times 10^6 \times 0.084 \log\left(\frac{0.336}{0.084}\right) = 145\,400 \text{ J}$$

$$= 145.4 \text{ kJ } \textbf{Ans.}$$

3. *Heat absorbed by the air*

We know that during constant temperature process, there is no change in internal energy and the heat absorbed is equal to the amount of work done by the air.

$\therefore$ Heat absorbed by the air,

$$Q_{1-2} = \text{Work done by the air} = 145.4 \text{ kJ } \textbf{Ans.}$$

4. *Change of entropy*

We know that change of entropy

$$S_2 - S_1 = \frac{\text{Heat absorbed}}{\text{Absolute temperature}} = \frac{145.4}{810} = 0.18 \text{ kJ/K } \textbf{Ans.}$$

Example 4.11. *One kg of hydrogen (molecular mass 2) is expanded from 1 m^3 to 5 m^3 during a free expansion process. Calculate the change in entropy of the gas and the surroundings. If the expansion between the same two states is carried out by a reversible isothermal process, find the change in entropy of the gas and the surroundings. What will be the net change of entropy of the universe?*

Solution. Given : $m = 1$ kg ; $M = 2$; $v_1 = 1$ m^3 ; $v_2 = 5$ m^3

We know that the characteristic gas constant,

$$R = \frac{\text{Universal gas constant } (R_u)}{\text{Molecular mass } (M)} = \frac{8.314}{2} = 4.157 \text{ kJ/kg K}$$

$$\ldots (\because R_u \text{ for all gases} = 8.314 \text{ kJ/kg K})$$

Change in entropy of the gas and the surroundings

Since in a *free expansion process, the temperature of the gas remains constant, therefore the process is assumed as reversible isothermal process.

---

* Refer Art. 3.14

We know that change in entropy of the gas,

$$(dS)_G = 2.3\, mR \log\left(\frac{v_2}{v_1}\right) = 2.3 \times 1 \times 4.157 \log\left(\frac{5}{1}\right) = 6.67 \text{ kJ/K} \textbf{ Ans.}$$

Also, in a free expansion process, no heat transfer takes place with the surroundings, therefore entropy of surroundings remains constant.

∴ Change of entropy of the surroundings,

$$(dS)_S = 0 \quad \textbf{Ans.}$$

Change in entropy of the gas and the surroundings for reversible isothermal process

We know th hange in entropy of the gas during reversible isothermal expansion,

$$(dS)_G = 2.3\, mR \log\left(\frac{v_2}{v_1}\right) = 2.3 \times 1 \times 4.157 \log\left(\frac{5}{1}\right) = 6.67 \text{ kJ/K} \textbf{ Ans.}$$

In an isothermal expansion, the heat transfer takes place between the gas and the surroundings. The gas absorbs heat and an equal amount of heat is rejected by the surroundings.

∴ Change in entropy of the surroundings,

$$(dS)_S = -6.67 \text{ kJ/K} \textbf{ Ans.}$$

and net change in entropy of the universe,

$$(dS)_U = (dS)_G + (dS)_S = 6.67 - 6.67 = 0 \textbf{ Ans.}$$

Example 4.12. ***2 kg of oxygen at 60° C is mixed with 6 kg of nitrogen at the same temperature. The initial pressure of oxygen and nitrogen is 1.03 bar and remains same after mixing. Find the increase in entropy.***

Solution. Given : $m_O = 2$ kg ; $T_O = T_N = 60°$ C = 60 + 273 = 333 K ; $m_N = 6$ kg ; $p_{O1} = p_{N1} = p_M = 1.03$ bar $= 0.103 \times 10^6$ N/m²

We know that the molecular mass of oxygen (M_O) is 32 and the molecular mass of nitrogen (M_N) is 28. Since the density (*i.e.* mass per unit volume) is directly proportional to the molecular mass, therefore

Initial volume of oxygen,

$$v_O = \frac{m_O}{M_O} = \frac{2}{32} = 0.0625 \text{ m}^3 \qquad \ldots\left(\because v = \frac{m}{\rho} = \frac{m}{M}\right)$$

and initial volume of nitrogen,

$$v_N = \frac{m_N}{M_N} = \frac{6}{28} = 0.2143 \text{ m}^3$$

∴ Volume of the mixture,

$$v_M = v_O + v_N = 0.0625 + 0.2143 = 0.2768 \text{ m}^3$$

We know that the characteristic gas constant for oxygen,

$$R_O = \frac{\text{Universal gas constant } (R_u)}{\text{Molecular mass of oxygen } (M_O)} = \frac{8.314}{32} = 0.26 \text{ kJ/kg K}$$

$$\ldots(\because R_u \text{ for all gases} = 8.314 \text{ kJ/kg K})$$

and characteristic gas constant for nitrogen,

$$R_N = \frac{R_u}{M_N} = \frac{8.314}{28} = 0.297 \text{ kJ/kg K}$$

We know that change in entropy for oxygen

$$(dS)_O = 2.3\, m_O\, R_O \log\left(\frac{v_M}{v_O}\right) = 2.3 \times 2 \times 0.26 \log\left(\frac{0.2768}{0.0625}\right) \text{kJ/K}$$

$$= 1.196 \log(4.429) = 0.773 \text{ kJ/K (increase)}$$

and change in entropy for nitrogen,

$$(dS)_N = 2.3\, m_N\, R_N \log\left(\frac{v_M}{v_N}\right) = 2.3 \times 6 \times 0.297 \log\left(\frac{0.2768}{0.2143}\right) \text{kJ/K}$$

$$= 4.1 \log(1.2916) = 0.456 \text{ kJ/K (increase)}$$

∴ Increase in entropy,

$$dS = (dS)_O + (dS)_N = 0.773 + 0.456 = 1.229 \text{ kJ/K} \quad \textbf{Ans.}$$

4.13. Change of Entropy during Reversible Adiabatic Process (or Isentropic Process)

We have already discussed in Art. 3.10, that in a reversible adiabatic process, no heat enters or leaves the gas. Mathematically,

$$\delta Q = 0$$

$$\therefore \quad dS = 0 \qquad \ldots\left(\because dS = \frac{\delta Q}{T}\right)$$

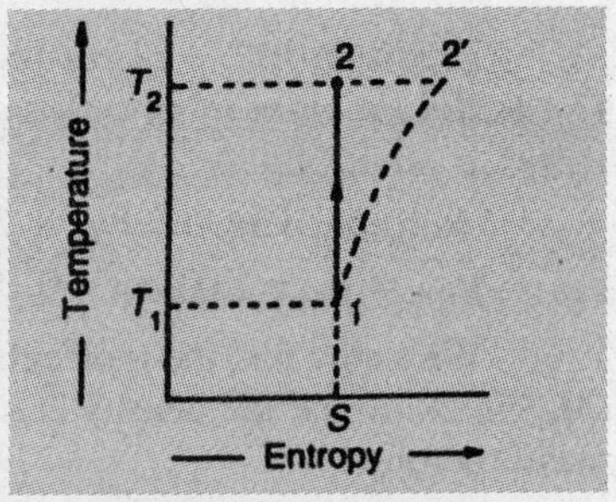

Fig. 4.5. *T-S* curve during adiabatic process.

In other words, change of entropy during a reversible adiabatic process is zero. The reversible adiabatic process on *T-S* graph is shown by a vertical straight line 1-2, as shown in Fig. 4.5.

Since the entropy of the gas remains constant during reversible adiabatic expansion or compression of the gas, this process is said to be isentropic (*i.e.* frictionless adiabatic process). This fact makes the *T-S* diagram quite useful in solving problems on adiabatic expansion.

We have also discussed in Art. 3.10, that in an isentropic process (*i.e.* frictionless adiabatic process), the temperature of the gas changes and the change in internal energy is equal to the work done by the gas during expansion (or work is done on the gas during compression). If the adiabatic process is irreversible (*i.e.* adiabatic process with *friction) as shown by 1-2′ in Fig. 4.5, and the expansion takes places within the same temperature limits T_1 and T_2, then due to internal friction, the internal energy of the gas at the end of the expansion (*i.e.* at point 2′) will be more than that of at point 2 of reversible process. If $\delta Q'$ is the amount of heat absorbed by the gas due to the internal friction, then the ratio $\delta Q'/T$ will be more. Thus the entropy is more at point 2′ (of irreversible process) than at point 2 (of reversible process). Thus, an *irreversible process always results in increase in entropy i.e.* $dS > 0$ (Refer Art. 4.7).

Since the internal energy of the gas at the end of irreversible adiabatic expansion is more than that of isentropic expansion, therefore the amount of work done by the gas will be less than that for

* It may be noted that friction makes the process irreversible as it increases the heat contents of the gas.

reversible expansion. In other words, higher entropy at the end of irreversible process, makes less availability of heat energy for transformation into work. Thus entropy may be regarded as a measure of rate of the availability or non-availability of heat energy for transformation into work.

Example 4.13. *A 0.568 m^3 capacity insulated vessel of oxygen at a pressure of 2 bar is stirred by an internal paddle until the pressure becomes 2.4 bar. Find out 1. Heat transferred, 2. Work output, and 3. Change in entropy per kg.*

Take c_v = 0.657 kJ/kg K, and R = 260 J/kg K.

Solution. Given : $v_1 = 0.568\ m^3$; $p_1 = 2\ bar = 0.2 \times 10^6\ N/m^2$; $p_2 = 2.4\ bar = 0.24 \times 10^6\ N/m^2$; $c_v = 0.657$ kJ/kg K ; $R = 260$ J/kg K

1. *Heat transferred*

Since the vessel is insulated, therefore it is an adiabatic process. We know that in an adiabatic process no heat is transferred. **Ans.**

2. *Work output per kg*

First of all, let us find the initial and final temperature of the gas *i.e.* T_1 and T_2 respectively. Consider 1 kg of mass of the gas.

We know that $p_1 v_1 = m R T_1$ or $T_1 = \dfrac{p_1 v_1}{m R} = \dfrac{0.2 \times 10^6 \times 0.568}{1 \times 260} = 437\ K$

Similarly $p_2 v_2 = m R T_2$ or $T_2 = \dfrac{p_2 v_2}{m R} = \dfrac{0.24 \times 10^6 \times 0.568}{1 \times 260} = 524.3\ K$

$\ldots (\because v_2 = v_1)$

$\therefore$ Change in internal energy,

$$dU = mc_v (T_2 - T_1) = 1 \times 0.657\ (524.3 - 437) = 57.3\ kJ$$

We know that in adiabatic process, the work output is at the cost of change in internal energy.

$\therefore$ Work output $= dU = 57.3$ kJ **Ans.**

3. *Change in entropy per kg*

We know that in an adiabatic process, there is no change in heat, therefore, there is no change in entropy also. **Ans.**

Example 4.14. *An ideal gas of mass 0.25 kg has a pressure of 3 bar, a temperature of 80° C and a volume of 0.07 m^3. The gas undergoes an irreversible adiabatic process to a final pressure of 3 bar and a final volume of 0.10 m^3, during which the workdone on the gas is 25 kJ. Evaluate c_p and c_v of the gas and increase in entropy of the gas.*

Solution. Given : $m = 0.25$ kg ; $p_1 = 3\ bar = 0.3 \times 10^6\ N/m^2$; $T_1 = 80°\ C = 80 + 273 = 353\ K$; $v_1 = 0.07\ m^3$; $p_2 = 3\ bar = 0.3 \times 10^6\ N/m^2$; $v_2 = 0.1\ m^3$; $W_{1-2} = -25$ kJ

Value of c_p and c_v of the gas

First of all, let us find the value of gas constant (R) and the final temperature of the gas (T_2).

We know that $p_1 v_1 = m R T_1$ or $R = \dfrac{p_1 v_1}{m T_1} = \dfrac{0.3 \times 10^6 \times 0.07}{0.25 \times 353} = 238$ J/kg K

and $p_2 v_2 = m R T_2$ or $T_2 = \dfrac{p_2 v_2}{m R} = \dfrac{0.3 \times 10^6 \times 0.1}{0.25 \times 238} = 504\ K$

We know that change in internal energy,

$$dU = mc_v (T_2 - T_1) = 0.25 \times c_v (504 - 353) = 37.75\, c_v$$

We also know that heat transfer,

$$*Q_{1-2} = dU + W_{1-2}$$

$$0 = 37.75\, c_v - 25 \quad \text{or} \quad c_v = 25 / 37.75 = 0.662 \text{ kJ/kg K} \textbf{ Ans.}$$

Now $\quad c_p - c_v = R = 238 \text{ J/kg K} = 0.238 \text{ kJ/kg K}$

$\therefore \quad c_p = R + c_v = 0.238 + 0.662 = 0.9 \text{ kJ/kg K}$ **Ans.**

Increase in entropy of the gas

We know that the change in entropy,

$$S_2 - S_1 = 2.3\, m \left[c_v \log \left(\frac{p_2}{p_1} \right) + c_p \log \left(\frac{v_2}{v_1} \right) \right] \quad \text{. . . (General equation)}$$

$$= 2.3 \times 0.25 \left[0.662 \log \left(\frac{3}{3} \right) + 0.9 \log \left(\frac{0.1}{0.07} \right) \right]$$

$$= 0.575\, [\, 0.662 \log 1 + 0.9 \log 1.428]$$

$$= 0.575 \times 0.9 \times 0.155 = 0.08 \text{ kJ/K} \textbf{ Ans.}$$

The $+ve$ sign indicates that there is an increase in entropy.

4.14. Change of Entropy during Polytropic Process (pv^n = Constant)

Consider a certain quantity of a perfect gas being heated by polytropic process.

Let $\quad m$ = Mass of the gas,

p_1 = Initial pressure of the gas,

v_1 = Initial volume of the gas,

T_1 = Initial temperature of the gas, and

p_2, v_2, T_2 = Corresponding values for the final conditions.

We have already discussed in Art. 3.11 that whenever a gas is heated according to the general law pv^n = constant, the small amount of heat absorbed by the gas during its expansion is given by :

$$\delta Q = \frac{\gamma - n}{\gamma - 1} \times \delta W = \frac{\gamma - n}{\gamma - 1} \times p\, dv \quad \ldots \left(\because \text{ Workdone, } \delta W = pdv \right)$$

Dividing this equation throughout by T,

$$\frac{\delta Q}{T} = \frac{\gamma - n}{\gamma - 1} \times \frac{p\, dv}{T}$$

Substituting $\frac{\delta Q}{T} = dS$, and $\frac{p}{T} = \frac{mR}{v}$, we have $\quad \ldots (\because pv = mRT)$

$$dS = \frac{\gamma - n}{\gamma - 1} \times mR \times \frac{dv}{v}$$

---

* For an adiabatic process, $Q_{1-2} = 0$ and W_{1-2} is $-ve$ as work is done on the gas.

Integrating this expression for the total change of entropy, we have

$$\int_{S_1}^{S_2} dS = \frac{\gamma - n}{\gamma - 1} \times mR \int_{v_1}^{v_2} \frac{dv}{v}$$

or
$$S_2 - S_1 = \frac{\gamma - n}{\gamma - 1} \times mR \log_e \left(\frac{v_2}{v_1} \right) = 2.3\, m \times \frac{\gamma - n}{\gamma - 1} \times R \log \left(\frac{v_2}{v_1} \right) \quad \ldots (i)$$

$$= 2.3\, m \times \frac{\frac{c_p}{c_v} - n}{\gamma - 1} \times c_v (\gamma - 1) \log \left(\frac{v_2}{v_1} \right) \quad \ldots [\because R = c_v (\gamma - 1)]$$

$$= 2.3\, m\, (c_p - n\, c_v) \log \left(\frac{v_2}{v_1} \right) \quad \ldots (ii)$$

The equations (*i*) and (*ii*) are valid for both reversible and irreversible processes.

The above relation may also be expressed in terms of absolute temperature and pressure. We know that in a polytropic process,

$$\frac{T_1}{T_2} = \left(\frac{v_2}{v_1} \right)^{n-1} \quad \text{or} \quad \frac{v_2}{v_1} = \left(\frac{T_1}{T_2} \right)^{\frac{1}{n-1}}$$

Substituting this value of v_2 / v_1 in equation (*i*),

$$S_2 - S_1 = 2.3 \times \frac{\gamma - n}{\gamma - 1} \times mR \log \left(\frac{T_1}{T_2} \right)^{\frac{1}{n-1}}$$

$$= 2.3\, m \times \frac{\gamma - n}{\gamma - 1} \times R \times \frac{1}{n-1} \log \left(\frac{T_1}{T_2} \right)$$

$$= 2.3\, m \times \frac{\gamma - n}{\gamma - 1} \times c_v (\gamma - 1) \times \frac{1}{n-1} \log \left(\frac{T_1}{T_2} \right)$$

$$= 2.3\, m\, c_v \times \frac{\gamma - n}{n-1} \log \left(\frac{T_1}{T_2} \right)$$

We also know that in a polytropic process,

$$\frac{v_1}{v_2} = \left(\frac{p_2}{p_1} \right)^{\frac{1}{n}} \quad \text{or} \quad \frac{v_2}{v_1} = \left(\frac{p_1}{p_2} \right)^{\frac{1}{n}}$$

Substituting this value of v_2 / v_1 in equation (*i*),

$$S_2 - S_1 = 2.3\, m \times \frac{\gamma - n}{\gamma - 1} \times R \log \left(\frac{p_1}{p_2} \right)^{\frac{1}{n}}$$

$$= 2.3\, m \times \frac{\gamma - n}{\gamma - 1} \times R \times \frac{1}{n} \log \left(\frac{p_1}{p_2} \right)$$

$$= 2.3\, m \times \frac{\gamma - n}{\gamma - 1} \times c_v (\gamma - 1) \times \frac{1}{n} \log \left(\frac{p_1}{p_2} \right)$$

$$= 2.3\, m\, c_v \times \frac{\gamma - n}{n} \log \left(\frac{p_1}{p_2} \right)$$

Example 4.15. *A perfect gas is compressed according to the law $pv^{1.25}$ = constant from an initial pressure of 1 bar and volume of $0.9\ m^3$ to a final volume of $0.6\ m^3$. Determine the final pressure and change of entropy per kg of gas during the process. Take $\gamma = 1.4$ and $R = 287$ J/kg K.*

Solution. Given : $n = 1.25$; $p_1 = 1$ bar ; $v_1 = 0.9\ m^3$; $v_2 = 0.6\ m^3$; $\gamma = c_p / c_v = 1.4$; $R = 287$ J/kg K = 0.287 kJ/kg K

Final pressure of the gas

Let p_2 = Final pressure of the gas.

We know that

$$p_1 v_1^{1.25} = p_2 v_2^{1.25}$$

$$\therefore \quad p_2 = p_1 \left(\frac{v_1}{v_2} \right)^{1.25} = 1 \left(\frac{0.9}{0.6} \right)^{1.25} = 1.66 \text{ bar} \quad \textbf{Ans.}$$

Change of entropy per kg of gas

We know that change of entropy per kg of gas,

$$S_2 - S_1 = 2.3\, m \times \frac{\gamma - n}{\gamma - 1} \times R \log \left(\frac{v_2}{v_1} \right)$$

$$= 2.3 \times 1 \times \frac{1.4 - 1.25}{1.4 - 1} \times 0.287 \log \left(\frac{0.6}{0.9} \right) \text{kJ/K}$$

$$= 0.247 \log (0.6667) = -0.0435 \text{ kJ/K} \quad \textbf{Ans.}$$

The –ve sign indicates that there is a decrease of entropy.

Example 4.16. *A certain volume of gas at 320 K and 6.5 bar is expanded to four times its original volume, according to $pv^{1.25}$ = constant. Determine the final temperature of the gas and change of entropy per kg of gas, assuming $c_p = 0.996$ kJ/kg K and $c_v = 0.707$ kJ/kg K.*

Solution. Given : $T_1 = 320$ K ; $^*p_1 = 6.5$ bar ; $v_2 = 4\, v_1$; $n = 1.25$; $c_p = 0.996$ kJ/kg K ; $c_v = 0.707$ kJ/kg K

Final temperature of the gas

Let T_2 = Final temperature of the gas.

We know that

$$\frac{T_1}{T_2} = \left(\frac{v_2}{v_1} \right)^{n-1} = \left(\frac{4v_1}{v_1} \right)^{1.25-1} = (4)^{0.25} = 1.414$$

$$\therefore \quad T_2 = T_1 / 1.414 = 320 / 1.414 = 226.3 \text{ K} \quad \textbf{Ans.}$$

* Superfluous data

Change of entropy per kg of gas

We know that change of entropy per kg of gas,

$$S_2 - S_1 = 2.3\, m\,(c_p - n\,c_v) \log\left(\frac{v_2}{v_1}\right)$$

$$= 2.3 \times 1\,(0.996 - 1.25 \times 0.707) \log\left(\frac{4\,v_1}{v_1}\right) \text{ kJ/K}$$

$$= 0.2576 \log 4 = 0.155 \text{ kJ/K} \quad \textbf{Ans.}$$

Example 4.17. *0.2 kg of air with p_1 = 1.5 bar and T_1 = 300 K is compressed to a pressure of 15 bar, according to the law $pv^{1.25}$ = constant. Determine :*

1. Initial and final parameters of the air, 2. Workdone on or by the air ; 3. Heat flow to or from the air ; and 4. Change of entropy stating whether it is an increase or decrease.

Solution. Given : $m = 0.2$ kg ; $p_1 = 1.5$ bar $= 0.15 \times 10^6$ N/m^2 ; $T_1 = 300$ K ; $p_2 = 15$ bar $= 1.5 \times 10^6$ N/m^2 ; $n = 1.25$

1. *Initial and final parameters of the air*

Let T_2 = Final temperature,

v_1 = Initial volume, and

v_2 = Final volume.

We know that
$$\frac{T_1}{T_2} = \left(\frac{p_1}{p_2}\right)^{\frac{n-1}{n}} = \left(\frac{1.5}{15}\right)^{\frac{1.25-1}{1.25}} = (0.1)^{0.2} = 0.631$$

$$\therefore \quad T_2 = T_1 / 0.631 = 300 / 0.631 = 475.4 \text{ K} \quad \textbf{Ans.}$$

We also know that $p_1\, v_1 = m\,R\,T_1$

$$\therefore \quad v_1 = \frac{m\,R\,T_1}{p_1} = \frac{0.2 \times 287 \times 300}{0.15 \times 10^6} = 0.115 \text{ m}^3 \quad \textbf{Ans.}$$

... (Taking R for air = 287 J/kg K)

and
$$p_1\,v_1^n = p_2\,v_2^n \quad \text{or} \quad v_2 = v_1\left(\frac{p_1}{p_2}\right)^{\frac{1}{n}} = 0.115\left(\frac{1.5}{15}\right)^{\frac{1}{1.25}} \text{ m}^3$$

$$= 0.018 \text{ m}^3 \quad \textbf{Ans.}$$

2. *Workdone on or by the air*

We know that workdone,

$$W_{1-2} = \frac{p_1\,v_1 - p_2\,v_2}{n-1} = \frac{0.15 \times 10^6 \times 0.115 - 1.5 \times 10^6 \times 0.018}{1.25 - 1}$$

$$= -\,39\,000 \text{ J} = -\,39 \text{ kJ} \quad \textbf{Ans.}$$

The –ve sign indicates that work is done on the air.

3. Heat flow to or from the air

We know that change in internal energy,

$$dU = mc_v(T_2 - T_1) = 0.2 \times 0.712 (475.4 - 300) = 25 \text{ kJ}$$

. . . (Taking $c_v = 0.712$ kJ/kg K)

$\therefore$ Heat flow, $$Q_{1-2} = dU + W_{1-2} = 25 - 39 = -14 \text{ kJ} \textbf{ Ans.}$$

The –ve sign indicates that heat flows from the air.

4. Change of entropy

We know that change of entropy

$$S_2 - S_1 = 2.3\, mc_v \times \frac{\gamma - n}{n-1} \log\left(\frac{T_1}{T_2}\right)$$

$$= 2.3 \times 0.2 \times 0.712 \times\times \frac{1.4 - 1.25}{1.25 - 1} \log\left(\frac{300}{475.4}\right) \text{ kJ/K}$$

$$= 0.1965 \log(0.631) = -0.04 \text{ kJ/K} \textbf{ Ans.}$$

The –ve sign indicates that there is a *decrease* in entropy.

Example 4.18. *A mass of 9 kg of air at 1.75 bar and 13° C is compressed to 24.5 bar according to the law $pv^{1.32}$ = constant, and then cooled at constant volume to 15° C. Determine : 1. Volume and temperature at the end of compression, and 2. Change of entropy during compression and during constant volume cooling.*

For air, take c_p = 0.996 kJ/kg K and c_v = 0.712 kJ/kg K.

Solution. Given : $m = 9$ kg ; $p_1 = 1.75$ bar $= 0.175 \times 10^6$ N/m² ; $T_1 = 13°$ C $= 13 + 273 = 286$ K ; $p_2 = 24.5$ bar $= 2.45 \times 10^6$ N/m² ; $n = 1.32$; $T_3 = 15°$ C $= 15 + 273 = 288$ K ; $c_p = 0.996$ kJ/kg K ; $c_v = 0.712$ kJ/kg K

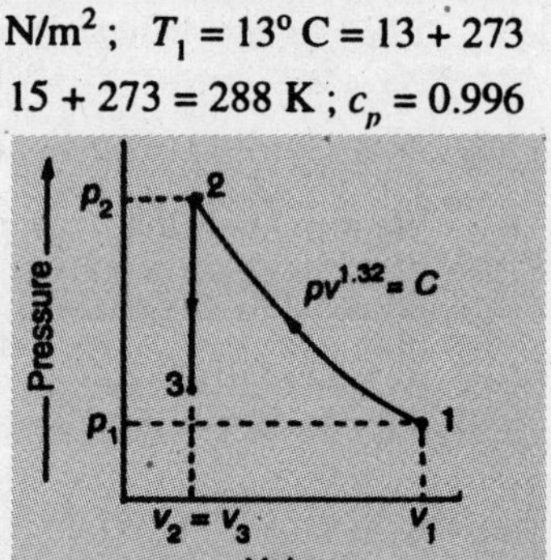

Fig. 4.6

In the *p-v* diagram, as shown in Fig.4.6, the process 1-2 represents the compression of air according to $pv^{1.32} = C$ and the process 2-3 represents cooling of air at constant volume.

1. Volume and temperature at the end of compression

Let v_2 = Volume at the end of compression, and

T_2 = Temperature at the end of compression.

We know that gas constant,

$$R = c_p - c_v = 0.996 - 0.712 = 0.284 \text{ kJ/kg K} = 284 \text{ J/kg K}$$

and $$p_1 v_1 = m R T_1 \quad \text{or} \quad v_1 = \frac{m R T_1}{p_1} = \frac{9 \times 284 \times 286}{0.175 \times 10^6} = 4.18 \text{ m}^3$$

We also know that

$$p_1 v_2^n = p_2 v_2^n$$

$$\therefore \quad v_2 = v_1 \left(\frac{p_1}{p_2}\right)^{\frac{1}{n}} = 4.18 \left(\frac{0.175 \times 10^6}{2.45 \times 10^6}\right)^{\frac{1}{1.32}}$$

$$= 4.18\,(0.0714)^{0.758} = 0.565\text{ m}^3 \quad \textbf{Ans.}$$

We know that

$$\frac{T_1}{T_2} = \left(\frac{p_1}{p_2}\right)^{\frac{n-1}{n}} = \left(\frac{0.175 \times 10^6}{2.45 \times 10^6}\right)^{\frac{1.32-1}{1.32}} = (0.0714)^{0.242} = 0.528$$

$$\therefore \quad T_2 = T_1 / 0.528 = 286 / 0.528 = 542\text{ K} = 269°\text{ C} \quad \textbf{Ans.}$$

2. *Change of entropy*

We know that change of entropy during compression (process 1-2),

$$S_2 - S_1 = 2.3\, m\,(c_p - n\, c_v) \log \left(\frac{v_2}{v_1}\right)$$

$$= 2.3 \times 9\,(0.996 - 1.32 \times 0.712) \log \left(\frac{0.565}{4.18}\right) \text{ kJ/K}$$

$$= 1.16 \log (0.135) = -1.008 \text{ kJ/K} \quad \textbf{Ans.}$$

The –ve sign indicates that there is a decrease of entropy.

We also know that change of entropy during constant volume cooling (process 2-3),

$$S_3 - S_2 = 2.3\, m c_v \log \left(\frac{T_3}{T_2}\right) = 2.3 \times 9 \times 0.712 \log \left(\frac{288}{542}\right) \text{kJ/K}$$

$$= 14.74 \log (0.5314) = -4.04 \text{ kJ/K} \quad \textbf{Ans.}$$

The –ve sign indicates that there is a decrease of entropy.

Example 4.19. *A volume of 0.14 m³ of air at 1 bar and 90° C is compressed to 0.014 m³ according to $pv^{1.3}$ = Constant. Heat is then added at a constant volume until the pressure is 66 bar. Determine : 1. Heat exchange with cylinder walls during compression, and 2. Change of entropy during each portion of the process.*

Assume γ = 1.4 and R = 286 J/kg K.

Solution. Given : $v_1 = 0.14\text{ m}^3$; $p_1 = 1\text{ bar} = 0.1 \times 10^6\text{ N/m}^2$; $T_1 = 90°\text{ C} = 90 + 273 = 363\text{ K}$; $v_2 = 0.014\text{ m}^3$; $n = 1.3$; $p_3 = 66\text{ bar} = 6.6 \times 10^6\text{ N/m}^2$; $\gamma = c_p/c_v = 1.4$; $R = 286\text{ J/kg K} = 0.286\text{ kJ/kg K}$

In the *p-v* diagram, as shown in Fig. 4.7, the process 1-2 represents compression according to $pv^{1.3} = C$ and the process 2-3 represents heating at constant volume.

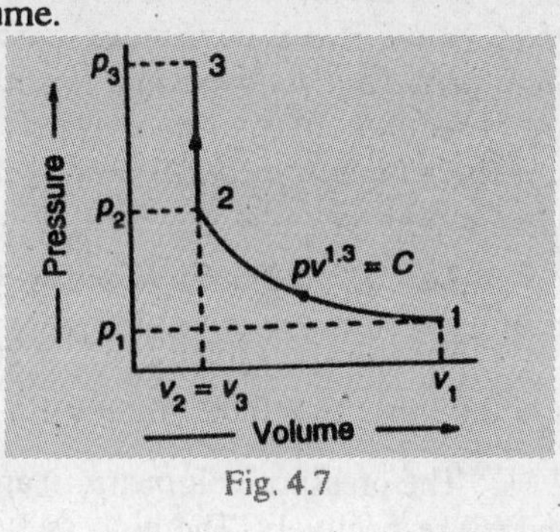

Fig. 4.7

1. *Heat exchange with cylinder walls during compression*

First of all, let us find the pressure at the end of compression (*i.e.* p_2)

We know that

$$\frac{p_2}{p_1} = \left(\frac{v_1}{v_2}\right)^n = \left(\frac{0.14}{0.014}\right)^{1.3}$$

$$= (10)^{1.3} = 19.95$$

$$\therefore \quad p_2 = p_1 \times 19.95 = 1 \times 19.95$$

$$= 19.95\text{ bar} = 1.995 \times 10^6\text{ N/m}^2$$

We know that workdone during compression,

$$W_{1-2} = \frac{p_2 v_2 - p_1 v_1}{n-1} = \frac{1.995 \times 10^6 \times 0.014 - 0.1 \times 10^6 \times 0.14}{1.3 - 1} \text{ J}$$

$$= 44\,635 \text{ J} = 44.635 \text{ kJ}$$

∴ Heat exchange with cylinder walls during compression,

$$Q_{1-2} = \frac{\gamma - n}{\gamma - 1} \times W_{1-2} = \frac{1.4 - 1.3}{1.4 - 1} \times 44.635 = 11.16 \text{ kJ } \textbf{Ans.}$$

2. *Change of entropy during each portion of the process*

First of all let us find out the mass of air (m) and the specific heat at constant volume (c_v).

We know that $\quad p_1 v_1 = m R T_1$

$$\therefore \quad m = \frac{p_1 v_1}{R T_1} = \frac{0.1 \times 10^6 \times 0.14}{286 \times 363} = 0.135 \text{ kg}$$

and $\quad c_p - c_v = R \quad \text{or} \quad 1.4\, c_v - c_v = 0.286$

$$\therefore \quad c_v = 0.286 / 0.4 = 0.715 \text{ kJ/kg K}$$

We know that change of entropy during compression process 1-2,

$$S_2 - S_1 = 2.3\, m \times \frac{\gamma - n}{\gamma - 1} \times R \log \left(\frac{v_2}{v_1}\right)$$

$$= 2.3 \times 0.135 \times \frac{1.4 - 1.3}{1.4 - 1} \times 0.286 \log \left(\frac{0.014}{0.14}\right) \text{ kJ/K}$$

$$= 0.0222 \log (0.1) = -0.0222 \text{ kJ/K } \textbf{Ans.}$$

The – *ve* sign indicates that there is a decrease of entropy.

We also know that change of entropy during constant volume process 2-3,

$$S_3 - S_2 = 2.3\, m c_v \log \left(\frac{p_3}{p_2}\right) = 2.3 \times 0.135 \times 0.715 \log \left(\frac{6.6 \times 10^6}{1.995 \times 10^6}\right)$$

$$= 0.222 \log (3.31) = 0.115 \text{ kJ/K } \textbf{Ans.}$$

Example 4.20. *A mass of air is initially at 206° C at pressure of 7 bar and having volume of 0.03 m³. The air is expanded at constant pressure to 0.09 m³, a polytropic process with n = 1.5 is then carried out, followed by a constant temperature process which completes the cycle. All processes are reversible. Sketch the cycle on pressure-volume and temperature-entropy planes and find : 1. change in entropy during each process ; 2. Heat received and heat rejected in the cycle ; and 3. efficiency of the cycle.*

Take R = 287 J/kg K and c_v = 0.713 kJ/kg K.

Solution. Given : $T_1 = 206°\text{ C} = 206 + 273 = 479 \text{ K}$; $p_1 = p_2 = 7 \text{ bar} = 0.7 \times 10^6 \text{ N/m}^2$; $v_1 = 0.03 \text{ m}^3$; $v_2 = 0.09 \text{ m}^3$; $n = 1.5$; $R = 287 \text{ J/kg K} = 0.287 \text{ kJ/kg K}$; $c_v = 0.713 \text{ kJ/kg K}$

The pressure-volume (p-v) and temperature-entropy (T-S) diagrams are shown in Fig. 4.8 (a) and (b) respectively. The process 1-2 represents the expansion of air at constant pressure ; process

2-3 represents the polytropic expansion *i.e.* according to $pv^{1.5} = C$ and the process 3-1 represents the compression of air at constant temperature (*i.e.* isothermal compression).

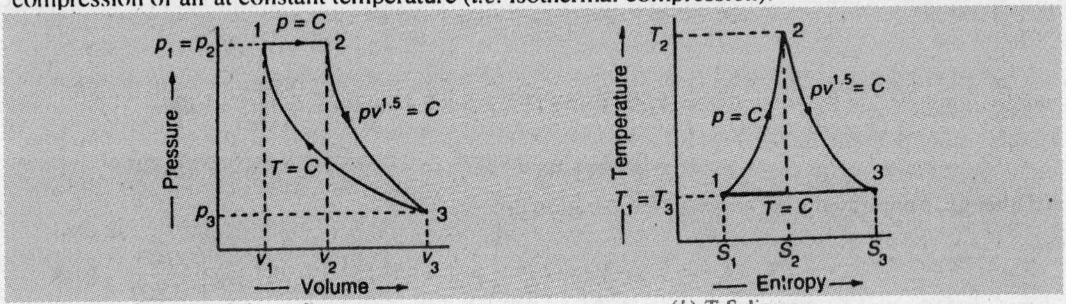

(*a*) *p-v* diagram. (*b*) *T-S* diagram.

Fig. 4.8

1. *Change of entropy during each process*

First of all, let us find the mass of air (*m*). We know that

$$p_1 v_1 = m R T_1 \quad \text{or} \quad m = \frac{p_1 v_1}{R T_1} = \frac{0.7 \times 10^6 \times 0.03}{287 \times 479} = 0.153 \text{ kg}$$

Let T_2 = Temperature at the end of constant pressure process 1-2,

v_3 = Volume at the end of polytropic process 2-3 or at the beginning of constant temperature process 3-1, and

p_3 = Pressure at the end of polytropic process 2-3 or at the beginning of constant temperature process 3-1.

We know that for a constant pressure process 1-2,

$$\frac{v_1}{T_1} = \frac{v_2}{T_2} \quad \text{or} \quad T_2 = \frac{v_2 T_1}{v_1} = \frac{0.09 \times 479}{0.03} = 1437 \text{ K}$$

For a polytropic process 2-3,

$$\frac{T_2}{T_3} = \left(\frac{p_2}{p_3}\right)^{\frac{n-1}{n}} \quad \text{or} \quad \frac{p_2}{p_3} = \left(\frac{T_2}{T_1}\right)^{\frac{n}{n-1}} = \left(\frac{1437}{479}\right)^{\frac{1.5}{1.5-1}} = 3^3 = 27$$

$$\ldots (\because T_3 = T_1)$$

$$\therefore \quad p_3 = p_2 / 27 = 7/27 = 0.26 \text{ bar} = 0.026 \times 10^6 \text{ N/m}^2$$

Now for a constant temperature process 3-1,

$$p_1 v_1 = p_3 v_3 \quad \text{or} \quad v_3 = \frac{p_1 v_1}{p_3} = \frac{7 \times 0.03}{0.26} = 0.807 \text{ m}^3$$

We know that change in entropy during constant pressure process 1-2,

$$S_2 - S_1 = 2.3\, m c_p \log\left(\frac{T_2}{T_1}\right) = 2.3 \times 0.153 \times 1 \log\left(\frac{1437}{479}\right) \text{kJ/K}$$

$$\ldots (\because c_p = R + c_v = 0.287 + 0.713 = 1 \text{ kJ/kg K})$$

$$= 0.352 \log 3 = 0.168 \text{ kJ/K (increase)}$$ **Ans.**

Change in entropy during polytropic process 2-3,

$$S_3 - S_2 = 2.3\ m\,(c_p - n\,c_v)\log\left(\frac{v_3}{v_2}\right)$$

$$= 2.3 \times 0.153\,(1 - 1.5 \times 0.713)\log\left(\frac{0.807}{0.09}\right)\ \text{kJ/K}$$

$$= -0.0246 \log 8.97 = -0.024\ \text{kJ/K (decrease)}\ \textbf{Ans.}$$

and change in entropy during constant temperature process 3-1,

$$S_1 - S_3 = 2.3\ mR\log\left(\frac{v_1}{v_3}\right) = 2.3 \times 0.153 \times 0.287 \log\left(\frac{0.03}{0.807}\right)\ \text{kJ/K}$$

$$= 0.101 \log(0.0372) = -0.144\ \text{kJ/K (decrease)}\ \textbf{Ans.}$$

Note : From above, we find that net change of entropy during a cycle,

$$dS = (S_2 - S_1) + (S_3 - S_2) + (S_1 - S_3)$$

$$= 0.168 - 0.024 - 0.144 = 0 \qquad \text{...(See Art. 4.7)}$$

2. *Heat received and heat rejected in a cycle*

We know that heat transferred during a constant pressure process 1-2,

$$Q_{1-2} = mc_p\,(T_2 - T_1) = 0.153 \times 1\,(1437 - 479) = 146.6\ \text{kJ}\ \textbf{Ans.}$$

The +ve sign indicates that the heat is received by the air.

Workdone during a polytropic process 2-3,

$$W_{2-3} = \frac{p_2 v_2 - p_3 v_3}{n-1} = \frac{0.7 \times 10^6 \times 0.09 - 0.026 \times 10^6 \times 0.807}{1.5 - 1}\ \text{J}$$

$$= \frac{63\,000 - 20\,982}{0.5} = 84\,036\ \text{J} = 84.036\ \text{kJ}$$

Change in internal energy during a polytropic process 2-3,

$$dU = mc_v\,(T_3 - T_2) = 0.153 \times 0.713\,(479 - 1437) = -104.5\ \text{kJ}$$

We know that heat transferred during a polytropic process 2-3,

$$Q_{2-3} = W_{2-3} + dU = 84.036 - 104.5 = -20.464\ \text{kJ}$$

The −ve sign indicates that the heat is rejected by the air.

We also know that heat transferred during a constant temperature process 3-1,

$$Q_{3-1} = \text{Workdone} = 2.3\ mRT_3 \log\left(\frac{v_2}{v_3}\right)$$

$$= 2.3 \times 0.153 \times 0.287 \times 479 \log\left(\frac{0.03}{0.807}\right)\text{kJ}$$

$$= 48.4 \log(0.0372) = -69.2\ \text{kJ}$$

The −ve sign indicates that the heat is rejected by the air.

∴ Total heat rejected in a cycle

$$= 20.464 + 69.2 = 89.664\ \text{kJ}\ \textbf{Ans.}$$

3. *Efficiency of the cycle*

We know that efficiency of the cycle

$$\eta = \frac{\text{Workdone}}{\text{Heat supplied}} = \frac{\text{Heat supplied} - \text{Heat rejected}}{\text{Heat supplied}}$$

$$= \frac{146.6 - 89.664}{146.6} = 0.3884 \text{ or } 38.84\% \textbf{ Ans.}$$

4.15. Approximate Method for Heat Absorbed

Consider 1 kg of a perfect gas being heated by any process. Let this process be represented by the curve 1-2 on *T-S* diagram as shown in Fig. 4.9.

Let T_1 = Initial temperature of the gas,

T_2 = Final temperature of the gas,

S_1 = Initial entropy of the gas, and

S_2 = Final entropy of the gas.

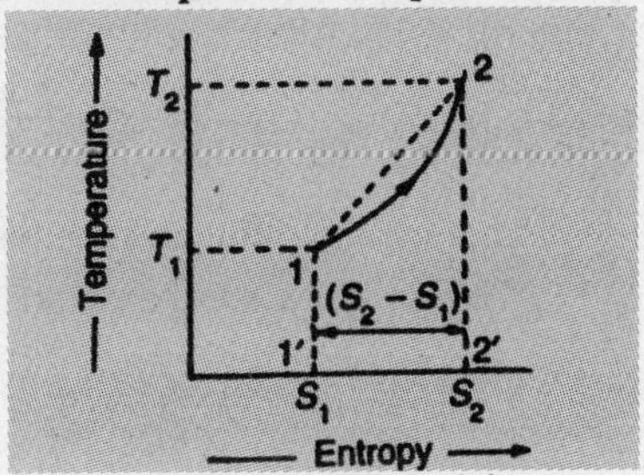

Fig. 4.9. Heat absorbed due to change in entropy.

We know that the heat absorbed during the process is equal to the area of the *T-S* diagram under the curve 1-2 *i.e.* 1-2-2′-1′. Assuming 1-2 to be a straight line (as shown by the dotted line), we find that

Heat absorbed = Area 1-2-2′-1′ = Base × Mean height

or $$Q_{1-2} = (S_2 - S_1)\left(\frac{T_1 + T_2}{2}\right)$$

Thus, the heat absorbed is approximately equal to the change of entropy multiplied by the mean absolute temperature.

Note : This method is called approximate method, as we have taken the curve 1-2 to be a straight line.

Example 4.21. *One kg of air is compressed in a cylinder according to the law $pv^{1.3}$ = Constant. If the initial temperature is 100° C and compression ratio is 15, find the change of entropy of the air.*

Also find the percentage error, if the change in entropy is calculated by the approximate method. Take c_p = 1 kJ/kg K and c_v = 0.714 kJ/kg K.

Solution. Given : m = 1 kg ; n = 1.3 ; T_1 = 100° C = 100 + 273 = 373 K ; v_1/v_2 = 15 ; c_p = 1 kJ/kg K ; c_v = 0.714 kJ/kg K

Change of entropy

First of all, let us find the final temperature of the air (*i.e.* T_2).

We know that $$\frac{T_1}{T_2} = \left(\frac{v_2}{v_1}\right)^{n-1} = \left(\frac{1}{15}\right)^{1.3-1} = 0.4438 \qquad \ldots (\because v_1/v_2 = 15)$$

$$\therefore \quad T_2 = T_1 / 0.4438 = 373 / 0.4438 = 840.5 \text{ K}$$

We know that change of entropy,

$$S_2 - S_1 = 2.3\, m c_v \times \frac{\gamma - n}{n-1} \log\left(\frac{T_1}{T_2}\right)$$

$$= 2.3 \times 1 \times 0.714 \times \frac{1.4 - 1.3}{1.3 - 1} \log\left(\frac{373}{840.5}\right) \text{ kJ/K}$$

$$= 0.547 \log (0.4438) = -0.193 \text{ kJ/K} \quad \textbf{Ans.}$$

The −ve sign indicates that there is a decrease of entropy.

Percentage error

We know that heat transferred during the process,

$$Q_{1-2} = (S_2 - S_1)\left(\frac{T_1 + T_2}{2}\right)$$

or $$S_2 - S_1 = \frac{2 \times Q_{1-1}}{T_1 + T_2} = \frac{2(-111.26)}{373 + 840.5} = -0.1834 \text{ kJ/K}$$

The −ve sign indicates the heat is rejected by the air.

We also know that heat transferred by approximate method,

$$Q_{1-2} = (S_2 - S_1)\left(\frac{T_1 + T_2}{2}\right)$$

or $$S_2 - S_1 = \frac{2 \times Q_{1-2}}{T_1 + T_2} = \frac{2(-111.26)}{373 + 840.5} = -0.1834 \text{ kJ/K}$$

The −ve sign indicates that there is a decrease of entropy.

$$\therefore \text{ Percentage error } = \frac{0.193 - 0.1834}{0.193} = 0.05 \text{ or } 5\% \quad \textbf{Ans.}$$

EXERCISES

1. Calculate the change in entropy per kg when 3 kg of air at a pressure of 2.5 bar and temperature 20° C are expanded to a pressure of 1 bar and temperature of 110° C. Take $R = 0.287$ kJ/kg K ; and $c_v = 0.707$ kJ/kg K. **[Ans.** 0.528 kJ/K]

2. One kg of air is heated at a constant volume from 1 bar and 27° C to a pressure of 5 bar. Calculate the change of entropy. Assume $R = 286$ J/kg K and $c_v = 0.712$ kJ/kg K. **[Ans.** 1.145 kJ/K]

3. A constant volume chamber of 0.3 m³ capacity contains 2 kg of the gas at 5° C. The heat is transferred to the gas until the temperature is 100° C. Find the heat transferred and the change in internal energy, enthalpy and entropy. Take $c_p = 1.985$ kJ/kg K and $c_v = 1.507$ kJ/kg K.

[Ans. 286.3 kJ ; 286.3 kJ ; 377.15 kJ ; 0.885 kJ/K]

4. 0.28 m³ of gas at a pressure of 10.5 bar and temperature of 538° C is expanded at a constant pressure to a volume of 0.34 m³. Determine the change of entropy, assuming $c_v = 0.69$ kJ/kg K and $R = 287$ J/kg K.

[Ans. 0.24 kJ/K]

5. A mixture of ideal gases consists of 3 kg of nitrogen (N_2) and 5 kg of carbon dioxide (CO_2) at a pressure of 3 bar and a temperature of 20° C. If the ratio of specific heats (γ) for the nitrogen is 1.4 kJ/kg K and for carbon dioxide is 1.286 kJ/kg K, find : 1. c_p and c_v for nitrogen, carbon dioxide and of the mixture, and 2. The changes in internal energy, enthalpy and entropy of the mixture, if the mixture is heated to 40° C (*a*) at constant volume, and (*b*) at constant pressure.

The molecular masses for nitrogen and carbon dioxide is 28 and 44 respectively.

[Ans. For nitrogen, $c_p = 1.039$ kJ/kg K, $c_v = 0.742$ kJ/kg K, For carbon dioxide, $c_p = 0.85$ kJ/kg K, $c_v = 0.661$ kJ/kg K, For mixture, $c_p = 0.92$ kJ/kg K, $c_v = 0.69$ kJ/kg K ; 110.4 kJ, 147.2 kJ, 0.368 kJ/kg ; 110.4 kJ, 147.2 kJ, 0.49 kJ/ K]

6. A certain quantity of gas occupies 0.56 m³ at 400° C and 28 bar. Determine the gain in entropy if the gas expands isothermally to a final volume of 2.8 m³. $R = 287$ J/kg K. **[Ans.** 3.746 kJ/K]

7. Calculate the changes of entropy per kg of air in the following cases :

(*a*) Air is heated at constant volume till its final pressure is three times the initial pressure, (*b*) Air is compressed at constant pressure till its final volume is one-half the initial volume ; and

(*c*) Air expands isothermally from 6 bar to 3 bar.

Find the change in entropy if the air undergoes the above three process in sequence.

Take $c_p = 1.01$ kJ/kg K and $c_v = 0.72$ kJ/kg K.

[**Ans.** 0.79 kJ/kg K ; − 0.699 kJ/kg K ; 0.2008 kJ/kg K ; 0.2918 (increase)]

8. A perfect gas is contained in a cylinder and undergoes an isothermal expansion according to the law $p = A + B\,v$, where p is the pressure in bar, v is the volume in m^3 and A and B are constants. The initial and final pressures are 8.4 bar and 2.8 bar and the corresponding volumes are 0.056 m^3 and 0.168 m^3. Find 1. workdone by the gas, 2. heat transferred during the process, and 3. change in entropy per kg of gas during expansion. Take $R = 275$ J/kg K.

[**Ans.** 62.72 kJ ; 62.72 kJ ; 0.3018 kJ/kg K]

9. Calculate the change of entropy when 0.14 kg of gas initially at 170° C expands with a volume ratio of 5.4 according to $pv^{1.24}$ = Constant. Take $\gamma = 1.4$ and $R = 287$ kJ/kg K. [**Ans.** 0.027 kJ/K]

10. One kg of air at a pressure of 7 bar and a temperature of 363 K undergoes a reversible polytropic process which may be represented by $pv^{1.1}$ = Constant. The final pressure is 1.4 bar. Evaluate : 1. The final specific volume, temperature and increase in entropy ; and 2. The workdone and heat transfer during the process. Assume $R = 287$ J/kg K and $\gamma = 1.4$. [**Ans.** 0.643 m^3, 313.3 K, 0.316 kJ/K ; 142.64 kJ, 107.04 kJ]

11. One kg of air at 1 bar and 15° C is compressed according to $pv^{1.25}$ = Constant to a pressure of 16 bar. Calculate the temperature at the end of compression, the heat received or rejected by the air during the process and the change of entropy. Sketch the operation on temperature-entropy diagram.

Take $c_p = 1.005$ kJ/kg K, $c_v = 0.716$ kJ/kg K.

[**Ans.** 228.7° C ; 92.6 kJ (rejected) ; 0.238 kJ/K (decrease)]

12. An ideal gas of molecular mass 30 and specific heat ratio 1.38 is compressed according to the law $pv^{1.25}$ = Constant, from a pressure of 1 bar and 15° C to a pressure of 16 bar. Calculate the temperature at the end of compression, the heat received or rejected and workdone by the gas during the process and the change in entropy. Assume 1 kg mass of the gas. Use only calculated values of c_p and c_v.

[**Ans.** 228.7° C ; 81 kJ (rejected) ; 0.21 kJ/K (decrease)]

13. A gas engine mixture at 95° C and 1 bar is compressed with index of compression 1.3, the volume compression ratio being 6 : 1. The maximum pressure is 25 bar. Assuming the ratio of specific heats as 1.38 and the specific heat at constant volume as 0.754 kJ/kg K, find the change in entropy during compression stroke and during combustion which takes place at constant volume. Represent the process on *p-v* and *T-s* planes.

[**Ans.** 0.108 kJ/K (decrease) ; 0.67 kJ/K]

14. An ideal gas at temperature T_1 is heated at constant pressure to T_2 and then expanded reversibly according to the law pv^n = Constant, until the temperature is again T_1. Find the value of n, if the changes in entropy during the separate processes are equal.

$$\left[\textbf{Ans. } \frac{2\gamma}{\gamma+1}\right]$$

15. The workdone by 0.07 kg of air when it expands according to pv^n = Constant is 7.6 kJ. The temperature of air falls from an initial value of 105° C to a final value of 13° C during the process. Determine : 1. the heat supplied or rejected by the air during the expansion ; 2. the value of index n ; and 3. the change of entropy, stating whether this is an increase or decrease. $\gamma = 1.4$ and $c_v = 0.712$ kJ/kg K.

[**Ans.** 3 kJ ; 1.24 ; 0.009 26 kJ/K (increase)]

16. One kg of air at 1 bar and 27° C is compressed isothermally to one-fifth the original volume. It is then heated at constant volume to a condition such that isentropic expansion from that state will return the system to the original state. Determine the pressure and temperature at the end of constant volume heating. Represent the processes on pressure-volume and temperature-entropy diagrams, and find : 1. the change in entropy during each process ; and 2. net workdone during the cycle.

[**Ans.** 9.518 bar, 571.08 K ; − 0.46 kJ/K, 0.46 kJ/K, zero ; 56.08 kJ]

17. 0.056 m^3 of carbon monoxide is contained in a cylinder at 37° C and 1.4 bar. The gas is compressed to 0.0224 m^3 during the inward stroke of the piston. If the compression process is (*a*) isothermal, and (*b*) adiabatic, find : 1. final temperature and pressure, 2. workdone, and 3. change of entropy. Take $c_p = 1.047$ kJ/kg K and $c_v = 0.749$ kJ/kg K. [**Ans.** 310 K, 3.5 bar, − 7.176 kJ, − 0.0232 kJ/K ; 447.2 K, 5.05 bar, − 8.68 kJ, zero]

18. Calculate the change of entropy when 1 kg of air changes from a temperature of 330 K and a volume of 0.14 m³ to a temperature of 550 K and a volume of 0.56 m³. If the gas expands according to the law pv^n = Constant, determine the value of index n and the heat absorbed or rejected by the air during the expansion. Show that it is approximately equal to the change of entropy multiplied by the mean absolute temperature. R = 286 J/kg K ; and c_v = 0.712 kJ/kg K. **[Ans. 1.365 ; 15.74 kJ]**

QUESTIONS

1. Explain clearly what is meant by 'entropy' of a gas.

2. Show that the specific entropy change for a perfect gas in a process is given by

$$s_2 - s_1 = c_p \log_e \left(\frac{v_2}{v_1}\right) + c_v \log_e \left(\frac{p_2}{p_1}\right)$$

where the subscripts $_1$ and $_2$ relate to the initial and final states and other symbols have their usual meanings.

3. Deduce the expressions in terms of initial and final temperatures and pressures, for the increase in entropy of a perfect gas when heated at a constant volume and at a constant pressure.

4. Show that for an ideal gas, the slope of the constant volume line on a temperature-entropy diagram is higher than that of constant pressure line.

5. Derive an expression for the change of entropy for the isothermal process in terms of volumes.

6. Starting from the fundamental, show that the change of entropy of a gas undergoing a polytropic process according to the law $pv^n = C$ is given by

$$S_2 - S_1 = m\,R \times \frac{\gamma - n}{(\gamma - 1)(n - 1)} \log_e \left(\frac{T_1}{T_2}\right)$$

where
$S_2 - S_1$ = Entropy,
γ = Ratio of specific heats,
n = Index of polytropic operation,
T = Absolute temperature of a gas, and
R = Characteristic gas constant.

7. Establish the equation for the change in entropy of m kg of a perfect gas during a change according to the law pv^n = Constant,

$$S_2 - S_1 = m\,c_v \times \frac{\gamma - n}{n - 1} \log_e \left(\frac{T_1}{T_2}\right)$$

where γ is the ratio of c_p / c_v.

8. Show that when one kg of a perfect gas expands according to pv^n = constant, the change in entropy is given by

$$s_2 - s_1 = (c_p - n\,c_v) \log_e \left(\frac{v_2}{v_1}\right)$$

9. A mass of m kg of a fluid at a temperature T_1 is mixed with an equal mass of the same fluid at a temperature T_2. The system is thermally insulated. Show that the entropy change of the universe is given by

$$dS = 2\,mc_p \log_e \left(\frac{T_1 + T_2}{2\sqrt{T_1 T_2}}\right)$$

and show that it is necessarily positive.

10. Derive an expression for the approximate heat absorbed by a gas when heated from an absolute temperature T_1 to T_2 and an initial entropy of S_1 to S_2.

OBJECTIVE TYPE QUESTIONS

1. The heat absorbed or rejected by the working substance is given by

   (*a*) $\delta Q = TdS$ (*b*) $\delta Q = T/dS$ (*c*) $\delta Q = dS/T$

   where δQ = Heat absorbed or rejected,

   dS = Increase or decrease of entropy, and

   T = Absolute temperature.

2. The property of a working substance which increases or decreases as the heat is supplied or removed in a reversible manner, is known as

   (*a*) enthalpy (*b*) internal energy (*c*) entropy (*d*) external energy

3. The entropy may be expressed as a function of

   (*a*) pressure and temperature (*b*) temperature and volume

   (*c*) heat and work (*d*) all of these

4. The entropy of water at 0° C is assumed to be

   (*a*) 1 (*b*) 0 (*c*) – 1 (*d*) 10

5. The change of entropy when heat is absorbed by the gas is

   (*a*) positive (*b*) negative (*c*) positive or negative

ANSWERS

1. (*a*)	2. (*c*)	3. (*a*)	4. (*b*)	5. (*a*)

5

Kinetic Theory of Gases

1. Introduction. 2. Assumptions in the Kinetic Theory of Gases. 3. Velocity of a Molecule. 4. Pressure Exerted by an Ideal Gas. 5. Kinetic Energy per kg Molecule of a Gas. 6. Kinetic Interpretation of Temperature. 7. Application of Kinetic Theory to Laws of Perfect Gases. 8. Degrees of Freedom. 9. Law of Equipartition of Energy. 10 Ratio of Molar Specific Heats. 11. Van der Waals' Equation of a Real Gas.

5.1. Introduction

A gas consists of an extremely large number of widely separated minute particles, called *molecules*, which are in a state of continuous motion. They rush about in all directions, and in all sorts of manner, and frequently collide with each other. Due to repeated collisions of the molecules, their velocities and directions get changed at random, as a result of which a molecular change prevails within the gas. The kinetic theory of gases covers the behaviour of gases under various physical conditions.

5.2. Assumptions in the Kinetic Theory of Gases

The kinetic theory of gases is based on the following assumptions :

1. The volume of a gas consists of a large number of minute particles called *molecules.* It has been experimentally found that there are about 26.8×10^{18} molecules in *1 ml. of gas at N.T.P.
2. The molecules are mere mass points. In other words, the size of a molecule is assumed to be negligible, as compared to the distance between the molecules.
3. The gas molecules are perfectly elastic spheres and exert negligible force of attraction or repulsion on one another, or on the walls of the containing vessel. Hence in a direct impact, the molecules rebound with the same velocity after each collision.
4. The molecules are continuously colliding against each other, and with the walls of the containing vessel. Between two collisions, a molecule moves in a straight line. This distance is called the free path of the molecule.
5. The time during which a collision takes place is negligible as compared to the time required to transverse the free path, *i.e.* collisions are instantaneous.

---

* Since 1 ml = 10^{-6} m^3, therefore

1 m^3 of any gas at N.T.P. will contain $\dfrac{26.8 \times 10^{18}}{10^{-6}} = 26.8 \times 10^{24}$ molecules.

Since the volume of 1 kg molecule of any gas at N.T.P. is 22.4 m^3, therefore

1 kg molecule of any gas will contain $26.8 \times 10^{24} \times 22.4 = 6.02 \times 10^{26}$ molecules

6. The molecules are continuously in motion. They have velocities in all directions ranging from zero to infinity. The velocity of molecules increases with the increase in temperature. For simplicity, it is assumed that one-third of the molecules move parallel to the X-axis, one-third parallel to Y-axis and one-third parallel to Z-axis.

5.3. Velocity of a Molecule

We have already discussed that molecules of a gas move in an haphazard manner in all directions and have different velocities. The following three velocities are important from the subject point of view ;

1. Mean or average velocity, 2. Root mean square velocity, and 3. Most probable velocity.

These velocities are obtained as discussed below :

1. Mean or average velocity

Consider a gas containing n molecules, and having actual velocities of individual molecules as $C_1, C_2, C_3 \ldots\ldots C_n$.

∴ Mean or average velocity

$$= \frac{C_1 + C_2 + C_3 + \ldots. + C_n}{n} = \sqrt{\frac{8\,kT}{\pi\, m}} = 1.595\sqrt{\frac{kT}{m}}$$

2. Root mean square (r.m.s.) velocity

The root mean square velocity,

$$C = \sqrt{\frac{C_1^2 + C_2^2 + C_3^2 + \ldots. + C_n^2}{n}}$$

The value of r.m.s. velocity may also be obtained from the following relations :

(*i*) $$^{*}C = \sqrt{\frac{3p}{\rho}} = 1.732\sqrt{\frac{p}{\rho}}$$

(*ii*) $$^{**}C = \sqrt{\frac{3\,kT}{m}} = 1.732\sqrt{\frac{kT}{m}}$$

Since the velocity of a molecule changes with temperature, *the root mean square velocity of the molecules remains constant so long as the temperature remains constant.*

3. Most probable velocity

The Maxwell's distribution of molecular velocities and percentage of molecules is shown in Fig. 5.1.

We see that the maximum percentage of molecules (AB) at any temperature move with a velocity of OA. This velocity is known as *most probable velocity*.

The most probable velocity may, therefore, be defined as the *velocity with which the largest percentage of molecules in a gas are found to move at any given temperature.*

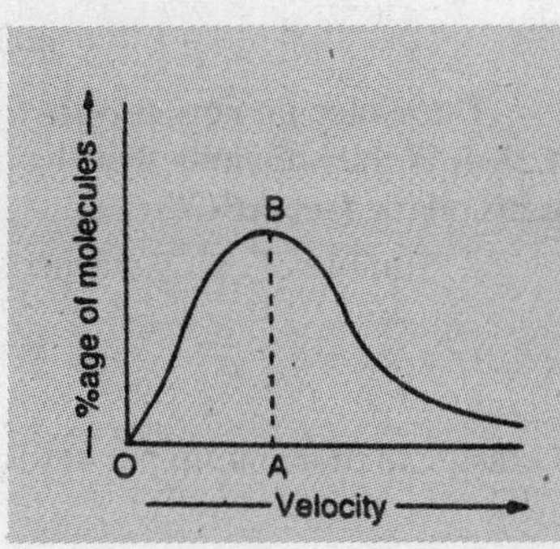

Fig. 5.1. Most probable velocity.

Notes : 1. The value of most probable velocity is given by :

$$\sqrt{\frac{2\,kT}{m}} = 1.414\sqrt{\frac{kT}{m}}$$

* See Art. 5.4.

** See Art. 5.6.

where k = Boltzmann's constant, Its value is same for all gases and may be taken as 1.38×10^{-26} kJ/K,

T = Absolute temperature, and

m = Mass of one molecule.

2. Due to the non-symmetrical shape of the curve, the mean velocity is larger than the most probable velocity, and the r.m.s. velocity is a little larger than either one of them.

3. From above, we see that

Mean or average velocity = $0.9209 \times$ r.m.s. velocity

and most probable velocity = $0.8164 \times$ r.m.s. velocity

5.4. Pressure Exerted by an Ideal Gas

The pressure exerted by a gas is due to the continuous bombardment on the walls of the containing vessel by the rapidly moving gas molecules.

Consider a hollow cubical vessel of each side equal to l, having perfectly elastic walls as shown in Fig. 5.2. Let it contain a perfect gas having a large number of molecules, say, n. Out of these, $n/3$ molecules are travelling back and forth parallel to each of the co-ordinate axis, *i.e.* X-axis, Y-axis and Z-axis with a velocity C (r.m.s. value).

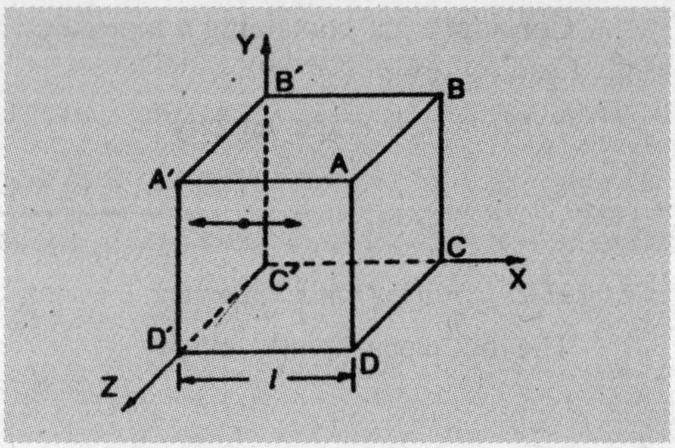

Fig. 5.2. Pressure exerted by an ideal gas.

Let us first consider the pressure exerted by one molecule, travelling back and forth parallel to X-axis. The molecule strikes the face $ABCD$ of the containing vessel and rebounds with the *same velocity* in the reverse direction, since the collision is perfectly elastic.

Let C = Velocity of molecule before collision, and

$-C$ = Velocity after collision.

$\therefore$ Momentum before collision = mC

and momentum after collision = $-mC$

Hence, change in momentum per collision per molecule

$$= mC - (-mC) = 2\,mC$$

The molecule now travels from face $ABCD$ to face $A'B'C'D'$ and travels back again to face $ABCD$. It is thus obvious that the distance travelled by the molecule between the two successive impacts on the face $ABCD$ is $2l$.

$\therefore$ Number of collisions per second

$$= \frac{C}{2l}$$

and change in momentum per collision per second

$$= 2mC \times \frac{C}{2l} = \frac{mC^2}{l}$$

and total change in momentum per second due to the collisions of $n/3$ molecules

$$= \frac{n}{3} \times \frac{mC^2}{l} = \frac{mn}{3l} \times C^2$$

Since the change in momentum per second is equal to the *impressed force, therefore,

Total force on the face *ABCD*

$$= \frac{mn}{3l} \times C^2$$

and** pressure exerted on the face *ABCD*

$$= \frac{\text{Force}}{\text{Area}} = \frac{mn}{3l \times l^2} \times C^2 \qquad \ldots (\because \text{ Area of face} = l^2)$$

$$= \frac{mn}{3\,v} \times C^2 \qquad \ldots (\because \text{ Volume, } v = l^3)$$

Now, $mn = M$, Molecula mass of the gas.

∴ Pressure exerted by an ideal gas,

$$p = \frac{1}{3} \times \frac{M}{v} \times C^2 = \frac{1}{3} \times \rho C^2 \qquad \ldots \left(\because \frac{M}{v} = \rho \right) \qquad \ldots (i)$$

Note : The above equation may also be written as :

$$p = \frac{2}{3} \times \frac{1}{2} \rho C^2 = \frac{2}{3} \times \text{K.E. per molecule} = \frac{2}{3} E$$

Therefore, the pressure exerted by an ideal gas is equal to two-thirds of the kinetic energy of all the molecules contained in a unit volume of the gas.

Example 5.1. *Calculate the root mean square velocity of air at N.T.P. The density of air at N.T.P. is 1.29 kg/m³. Density of mercury = 13 600 kg/m³ and g = 9.81 m/s².*

Solution. Given : p = 760 mm of Hg ; ρ = 1.29 kg/m³ ; ρ_m = 13 600 kg/m³

Let C = Root mean square velocity at N.T.P.

We know that p = 760 mm of Hg = 0.76 m of Hg

$$= 0.76 \times 13\,600 \times 9.81 = 101\,396 \text{ N/m}^2$$

We also know that

$$p = \frac{1}{3} \times \rho C^2 \text{ or } C = \sqrt{\frac{3p}{\rho}} = \sqrt{\frac{3 \times 101\,396}{1.29}} = 485.6 \text{ m/s}$$ **Ans.**

5.5. Kinetic Energy per kg Molecule of a Gas

We know that the pressure exerted by an ideal gas,

$$p = \frac{1}{3} \times \frac{M}{v} \times C^2 \quad \text{or} \quad p\,v = \frac{1}{3} \times MC^2 \qquad \ldots (i)$$

Now, according to perfect gas equation, we have

$$p\,v = R_u\,T \qquad \ldots (ii)$$

where v = Volume of gas per kg molecule, and

R_u = Universal gas constant

* Newton's second law of motion.

** The pressure exerted on any other face is same, *i.e.* $\frac{mn}{3v} \times C^2$

From equations (*i*) and (*ii*), we get

$$\frac{1}{3} \times MC^2 = R_u T \quad \text{or} \quad \frac{1}{2} \times MC^2 = \frac{3}{2} \times R_u T$$

Hence, *the kinetic energy per kg molecule of any gas at absolute temperature T is equal to* $\frac{3}{2} \times R_u T$.

5.6. Kinetic Interpretation of Temperature

We know that the pressure exerted by an ideal gas,

$$p = \frac{1}{3} \times \frac{M}{v} C^2 \quad \text{or} \quad p\,v = \frac{1}{3} \times MC^2$$

We also know that $\quad p\,v = R_u T$

$$\therefore \quad \frac{1}{3} \times MC^2 = R_u T \quad \text{or} \quad \frac{1}{3} \times mNC^2 = R_u T$$

where m = Molecular mass of gas per molecule, and

N = Avogadro's number representing the number of molecules per kg molecule of a gas. Its value is same for all gases and may be taken as 6.02×10^{26}

From, above, we have

$$\frac{1}{3} \times mC^2 = \frac{R_u}{N} \times T \quad \text{or} \quad \frac{1}{2} \times mC^2 = \frac{3}{2} \times \frac{R_u}{N} \times T = \frac{3}{2} \times k\,T$$

where $\quad k = \frac{R_u}{N}$ = Boltzmann's constant.

Hence, kinetic energy of translation of a molecule

$$= \frac{3}{2} \times k\,T$$

In other words, *the kinetic energy of translation of a molecule is proportional to the absolute temperature. This is known as kinetic interpretation of temperature.*

Notes : 1. The K.E. of translation of a molecule at a given absolute temperature is same for all the gases.

2. From equation $\frac{1}{3} \times m\,N\,C^2 = R_u\,T$,

$$C = \sqrt{\frac{3\,R_u\,T}{m\,N}} \quad \ldots (i)$$

$$= \sqrt{\frac{3\,k\,T}{m}} \quad \ldots \left(\because \frac{R_u}{N} = k\right) \ldots (ii)$$

$$= \sqrt{3\,R\,T} \quad \ldots \left(\because m\,N = M \text{ and } \frac{R_u}{M} = R\right) \ldots (iii)$$

where $\quad R$ = Characteristic gas constant.

In other words, $C \propto \sqrt{T}$

3. Since $C \propto \sqrt{T}$ or $C^2 \propto T$, therefore when $T = 0$, the velocity $C = 0$.

Hence, the absolute zero of temperature is that temperature at which the velocities of the molecules of a gas becomes zero.

Example 5.2. *Find the kinetic energy of the molecules of 1 kg of helium at 0° C if R_u = 8.314 kJ/kg K and molecular mass of helium is 4.*

Solution. Given : $T = 0° C = 0 + 273 = 273$ K ; $R_u = 8.314$ kJ/kg K ; $M = 4$

We know that number of molecules in 1 kg of helium

$$= \frac{N}{M} = \frac{N}{4}$$

and K.E. per molecule $= \frac{3}{2} kT$

$\therefore$ K.E. of the molecules in 1 kg of helium

$$= \text{K.E. per molecule} \times \text{No. of molecules}$$

$$= \frac{3}{2} kT \times \frac{N}{4} = \frac{3}{8} R_u T \qquad \ldots (\because R_u = k.N)$$

$$= \frac{3}{8} \times 8.314 \times 273 = 851 \text{ kJ } \textbf{Ans.}$$

Example 5.3. *Avogadro's number is 6.02×10^{26} and Boltzmann's constant k is 1.38×10^{-23} J/K. What is the average velocity of a molecule of oxygen at 27° C ?*

Solution. Given : $N = 6.02 \times 10^{26}$; $k = 1.38 \times 10^{-23}$ J/K ; $T = 27° C = 27 + 273 = 300$ K

We know that molecular mass of oxygen,

$$M = 16 \times 2 = 32$$

$\therefore$ Mass of one molecule,

$$m = \frac{M}{N} = \frac{32}{6.02 \times 10^{26}} = 5.316 \times 10^{-26} \text{ kg}$$

We know that average velocity of a molecule,

$$C = \sqrt{\frac{8\,k\,T}{\pi\, m}} = 1.595 \sqrt{\frac{k\,T}{m}} = 1.595 \sqrt{\frac{1.38 \times 10^{-23} \times 300}{5.316 \times 10^{-26}}} \text{ m/s}$$

$$= 445 \text{ m/s } \textbf{Ans.}$$

Example 5.4. *Calculate the number of molecules in 1 m³ of an ideal gas at 27° C and a pressure of 10 mm of Hg. The mean kinetic energy of a molecule at 27° C is 4×10^{-18} kJ and the density of mercury is 13 600 kg/m³.*

Solution. Given : $T = 27° C$; $p = 10$ mm of Hg ; K.E. $= 4 \times 10^{-18}$ kJ ; $\rho_m = 13\,600$ kg/m³

We know that $\quad p = 10 \text{ mm of Hg} = 0.01 \text{ m of Hg}$

$$= 0.01 \times 13\,600 \times 9.81 = 1334 \text{ N/m}^2$$

and kinetic energy per kg molecule of a gas

$$= \frac{3}{2} \times R_u T = \frac{3}{2} p v \qquad \ldots (\because p v = R_u T)$$

or kinetic energy per kg molecule per unit volume of a gas

$$= \frac{3}{2} p = \frac{3}{2} \times 1334 = 2001 \text{ J} = 2.001 \text{ kJ}$$

$\therefore$ Number of molecules in 1 m³ of gas

$$= \frac{\text{K.E. per unit volume}}{\text{K.E. of a molecule}} = \frac{2.001}{4 \times 10^{-18}} = 5.0025 \times 10^{17} \textbf{ Ans.}$$

5.7. Application of Kinetic Theory to Laws of Perfect Gases

We have already discussed in Chapter 2, the various laws governing the perfect gases. Now we shall attempt to apply the kinetic theory to these laws.

1. *Boyle's law*

We know that pressure exerted by a perfect gas,

$$p = \frac{1}{3} \times \frac{M}{v} \times C^2 \quad \text{or} \quad p\,v = \frac{1}{3} \times MC^2$$

Since C depends on T ($\because\ C \propto \sqrt{T}$), therefore, when T is constant, the right hand side of the above expression is also constant.

$$\therefore \quad p\,v = \text{Constant} \qquad \ldots \text{(when } T \text{ is constant)}$$

2. *Charles' law*

We know that the pressure exerted by a perfect gas,

$$p = \frac{1}{3} \times \frac{M}{v} \times C^2 \quad \text{or} \quad p\,v = \frac{1}{3} \times MC^2$$

Since $C \propto \sqrt{T}$ or $C^2 \propto T$, therefore

$$p\,v \propto T$$

When p is kept constant, then

$$v \propto T \quad \text{or} \quad \frac{v}{T} = \text{Constant}$$

3. *Gay-Lussac law*

We have already seen in the above equation that

$$p\,v \propto T$$

When v is kept constant, then

$$p \propto T \quad \text{or} \quad \frac{p}{T} = \text{Constant}$$

4. *Perfect gas equation*

We have already seen in the above equation that

$$p\,v \propto T$$

or

$$\frac{p\,v}{T} = \text{Constant} \quad \text{or} \quad p\,v = R_u\,T \qquad \ldots \text{(General gas equation)}$$

5. *Avogadro's law*

It states, "*Equal volumes of all gases under the same conditions of temperature and pressure, contain the same number of molecules.*"

Let m_1, n_1 and C_1 represent the mass per molecule, number of molecules per m^3 and root mean square velocity respectively for one gas and m_2, n_2 and C_2 the corresponding values for the other gas. If the two gases exert the same pressure p, then

$$p = \frac{1}{3} m_1\,n_1\,C_1^2 = \frac{1}{3} m_2\,n_2\,C_2^2 \qquad \ldots (i)$$

If the two gases are also at the same temperature, then the mean kinetic energy per molecule in the two gases is equal. In other words

$$\frac{1}{2} m_1 C_1^2 = \frac{1}{2} m_2 C_2^2$$

Dividing equation (*i*) by (*ii*),

$$n_1 = n_2$$

Example 5.5. *At what temperature will the velocity (root mean square velocity) of hydrogen be double of its value at N.T.P., if the pressure remains constant ?*

Solution. Given : $T_0 = 0° C = 0 + 273 = 273$ K

Let C = Root mean square velocity at T_0, and

T = Absolute temperature at which the r.m.s. velocity is $2C$.

We know that $C \propto \sqrt{T}$ or $C^2 = T$

$$\therefore \quad \frac{C^2}{(2C)^2} = \frac{T_0}{T} \quad \text{or} \quad \frac{1}{4} = \frac{273}{T} \quad \text{or} \quad T = 1092 \text{ K} = 819° \text{ C} \textbf{ Ans.}$$

Example 5.6. *Calculate the temperature at which root mean square velocity of a gas molecule is same as that of a molecule of another gas at 47° C. The molecular mass of first and second gases are 64 and 32 respectively.*

Solution. Given : $T_2 = 47° C = 47 + 273 = 320$ K ; $M_1 = 64$; $M_2 = 32$

Let T_1 = Temperature of the first gas, and

C = Root mean square velocity.

We know that $pv = R_u T = \frac{1}{3} MC^2$

or $$\frac{MC^2}{T} = 3 R_u = \text{Constant}$$

$$\therefore \quad \frac{M_1 C_1^2}{T_1} = \frac{M_2 C_2^2}{T_2} \quad \text{or} \quad \frac{M_1}{T_1} = \frac{M_2}{T_2} \qquad \ldots (\because C_1 = C_2)$$

$$\therefore \quad T_1 = T_2 \times \frac{M_1}{M_2} = 320 \times \frac{64}{32} = 640 \text{ K} = 367° \text{ C} \textbf{ Ans.}$$

Example 5.7. *The root mean square velocity of the molecules of a fixed mass of a gas is 895 m/s. Calculate the root mean square velocity of the molecules when the gas is compressed adiabatically to half of its original volume. The ratio of specific heats at constant pressure to that of constant volume is 1.4.*

Solution. Given : $C_1 = 895$ m/s ; $v_2 = 0.5\, v_1$; $\gamma = 1.4$

Let C_2 = Root mean square velocity of the molecules when the gas is compressed,

T_1 = Initial temperature of the gas, and

T_2 = Final temperature of the gas.

We know that for adiabatic compression,

$$\frac{T_1}{T_2} = \left(\frac{v_2}{v_1}\right)^{\gamma - 1} = \left(\frac{0.5\, v_1}{v_1}\right)^{1.4 - 1} = (0.5)^{0.4} = 0.758 \qquad \ldots (i)$$

Since $C \propto \sqrt{T}$ or $C^2 \propto T$, therefore

$$\frac{C_1^2}{C_2^2} = \frac{T_1}{T_2} \quad \text{or} \quad \frac{C_1^2}{C_2^2} = 0.758 \qquad \text{... [From equation (i)]}$$

$$\therefore \quad C_2 = \sqrt{\frac{C_1^2}{0.758}} = \sqrt{\frac{(895)^2}{0.758}} = 1028 \text{ m/s} \quad \textbf{Ans.}$$

5.8. Degrees of Freedom

In order to describe completely the motion of a particle in one plane, only *two* quantities must be known, say its two rectangular components. Similarly, for a particle moving in space, *three* independent quantities must be known to describe its motion. A molecule in a rigid body can have three motions of vibration along any of the three co-ordinate axes in addition to its three motions of translation. It is thus obvious that in order to completely describe the state of motion of a particle, *six* independent quantities must be known.

In general, *the total number of independent quantities, which must be known for describing completely the state of motion of a body, are called its degrees of freedom.*

Now we shall consider the degrees of freedom for the following cases :

1. *Monoatomic gas.* A molecule of a monoatomic gas has only one atom. This molecule can rotate about its polar axis, and can also move bodily in the three perpendicular planes as shown in Fig. 5.3. Hence such a molecule will have *three translational degrees of freedom and one rotational.*

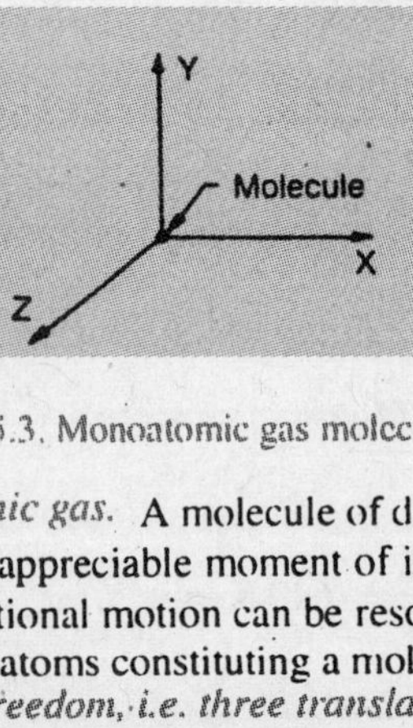

Fig. 5.3. Monoatomic gas molecule.

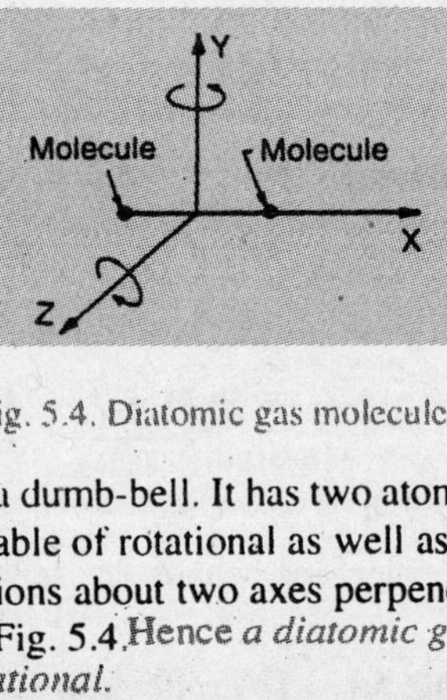

Fig. 5.4. Diatomic gas molecule.

2. *Diatomic gas.* A molecule of diatomic gas is like a dumb-bell. It has two atoms only. Such a molecule has appreciable moment of inertia and it is capable of rotational as well as translational motion. Its rotational motion can be resolved into two motions about two axes perpendicular to the line joining the atoms constituting a molecule as shown in Fig. 5.4. Hence *a diatomic gas possesses five degrees of freedom, i.e. three translational and two rotational.*

It can be observed that more the atoms in a molecule, the greater will be the number of degrees of freedom possessed by it.

5.9. Law of Equipartition of Energy*

This law states, "*The total energy of a molecule is shared equally by the various degrees of freedom possessed by it.*"

In case of monoatomic molecules like argon and helium, the energy possessed by them is only that of translational type, the rotational energy being negligible. We have already discussed in Art. 5.4 that energy of translation per molecule

$$= \frac{3}{2} \times k\,T$$

* This law was devised by Maxwell in 1859 for translational degrees of freedom. But Boltzmann extended it further to rotational and vibrational energies.

This energy is equally distributed among the three translational degrees of freedom.

∴ Energy per molecule per degree of freedom

$$= \frac{1}{3} \times \frac{3}{2} k T = \frac{1}{2} k T$$

Hence, the law of equipartition of energy may be stated as follows :

"The total energy of any dynamical system in thermal equilibrium is divided equally among all its degrees of freedom and the energy associated with each degree of freedom is $\frac{1}{2} k T$*."*

5.10. Ratio of Molar Specific Heats

The ratio of molar specific heats for the following two cases is important from the subject point of view :

1. Monoatomic gas, and 2. Diatomic gas.

1. *Monoatomic gas.* We know that a monoatomic gas has *three* degrees of freedom of translation and energy per molecule of the gas per degree of freedom is $\frac{1}{2} k T$.

∴ Total energy per molecule for 3 degrees of freedom

$$= 3 \times \frac{1}{2} k T = \frac{3}{2} k T$$

and total energy per kg-molecule at temperature T K

$$= \frac{3}{2} k N T \qquad \text{... (}N\text{ being number of molecules)}$$

$$= \frac{3}{2} R_u T \qquad \ldots \left(\because k = \frac{R_u}{N} \right)$$

If the temperature of the gas is raised by 1 K, then total energy per kg-molecule at $(T + 1)$ K

$$= \frac{3}{2} R_u (T + 1)$$

It may be noted that the increase in total energy per kg-molecule per degree rise in temperature is equal to the heat required to raise the temperature of 1 kg molecule through 1°. This is known as *molar specific heat at constant volume, i.e.* c_{vm}.

∴ Molar specific heat at a constant volume,

$$c_{vm} = \frac{3}{2} R_u (T + 1) - \frac{3}{2} R_u T = \frac{3}{2} R_u$$

Since $c_{pm} - c_{vm} = R_u$, therefore

$$c_{pm} = c_{vm} + R_u = \frac{3}{2} R_u + R_u = \frac{5}{2} R_u$$

∴ Ratio of molar specific heats,

$$\gamma = \frac{c_{pm}}{c_{vm}} = \frac{\frac{5}{2} R_u}{\frac{3}{2} R_u} = \frac{5}{3} = 1.67$$

2. *Diatomic gas.* We know that the molecule of a diatomic gas has five degrees of freedom.

∴ Total energy per kg-molecule at temperature T K

$$= \frac{5}{2} R_u T$$

and total energy per kg-molecule at $(T + 1)$ K

$$= \frac{5}{2} R_u (T + 1)$$

Now, molar specific heat at a constant volume,

$$c_{vm} = \frac{5}{2} R_u (T+1) - \frac{5}{2} R_u T = \frac{5}{2} R_u$$

Since $c_{pm} - c_{vm} = R_u$, therefore

$$c_{pm} = c_{vm} + R_u = \frac{5}{2} R_u + R_u = \frac{7}{2} R_u$$

∴ Ratio of molar specific heats,

$$\gamma = \frac{c_{pm}}{c_{vm}} = \frac{\frac{7}{2} R_u}{\frac{5}{2} R_u} = \frac{7}{5} = 1.4$$

5.11. Van der Waals' Equation of a Real Gas

Practically there is no *real gas* which obeys the gas laws perfectly. The deviation is small at ordinary temperatures and pressures, but it is large at high pressures and low temperatures. In deriving the perfect gas laws on the basis of kinetic theory of gases, the following assumptions were made which do not hold true for the real gases :

1. The molecules of a gas are mere mass-points occupying no space ; and
2. There is no attraction or repulsion between the molecules.

In actual practice, the molecules of all actual or real gases do occupy some space and do attract each other. Hence no real gas conforms to the perfect gas equation $p\,v = RT$. The Dutch physicist J.D. Van der Waals was the first scientist to correct this equation by applying corrections for the above two factors.

(i) Correction for the size of molecules. Consider some quantity of a gas contained in a vessel. We know that the molecules of a gas have a finite size, therefore the space available for the molecules to move about is less than the volume of the gas.

Let v = Volume of the gas ; and

b = Volume occupied by all the molecules.

∴ Space actually available or the effective volume of the gas

$$= v - b$$

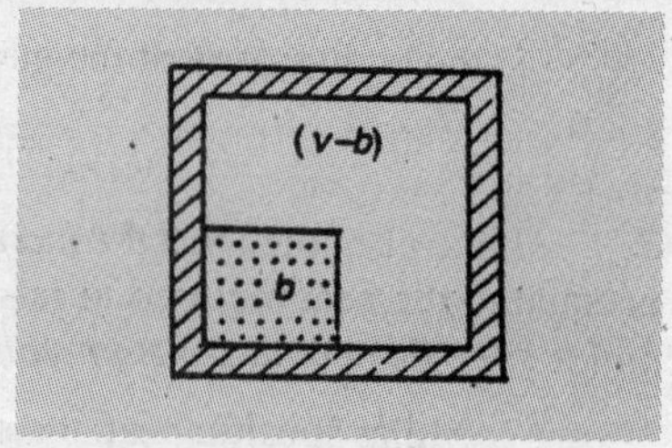

Fig. 5.5. Correction for the size of molecules.

Note : In has been experimentally found that 'b' is four times the total volume of the molecules and not equal to the volume occupied by the molecules.

(ii) Correction for the mutual attraction of molecules. Consider a gas molecule A well within the body of the gas inside the vessel. It is attracted by other molecules in all directions with the same force and the net force on it is zero. But when it strikes the wall of the vessel, it is pulled back by other molecules. Its velocity, and hence the momentum, with which it strikes the wall would be less than the momentum with which it will strike in the absence of the force of attraction. It is obvious that when the pressure of the gas drops, the momentum of the molecules also decreases. It is due to this fact, that the decrease in pressure is proportional to :

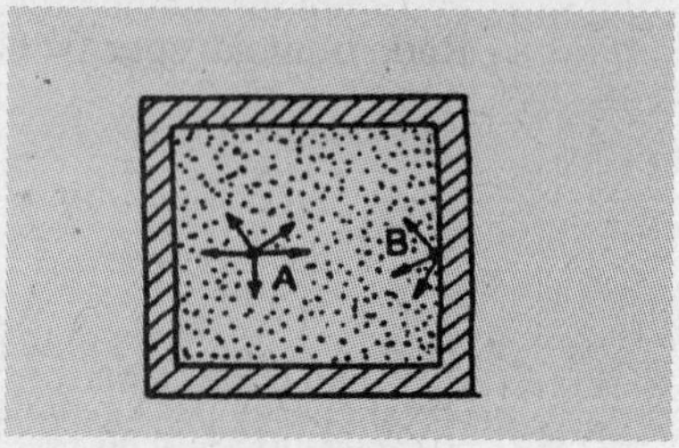

Fig. 5.6. Correction for the mutual attraction of molecules.

1. The number of attracting molecules per unit volume, and
2. The number of attracting molecules striking a unit area of the walls of the containing vessel per unit time.

Each of these factors are proportional to the number of molecules per m^3 or the density of the gas.

∴ Decrease in pressure,

$$p' \propto (\text{density of the gas})^2 \propto \frac{1}{v^2} \quad \text{or} \quad p' = \frac{a}{v^2}$$

where a is constant of proportionality.

$$\therefore \text{ Corrected pressure} = p + p' = p + \frac{a}{v^2}$$

Now with these corrections, the perfect gas equation for the real or actual gas becomes

$$\left(p + \frac{a}{v^2}\right)(v - b) = RT$$

This is known as *Van der Waals' equation for a real gas.*

EXERCISES

1. Calculate the r.m.s. velocity of nitrogen molecules at N.T.P. The density of nitrogen at N.T.P. is 1.25 kg/m^3 and density of mercury is 13 600 kg/m^3. Take g – 9.81 m/s^2 **[Ans. 493.3 m/s]**

2. Calculate the average kinetic energy of a gas molecule at a temperature of 300 K. The Boltzmann's constant is 1.38×10^{-26} kJ/K **[Ans. 621×10^{-26} kJ]**

3. Calculate the root mean square velocity and the mean kinetic energy of 1 kg molecule of hydrogen at N.T.P. The density of hydrogen is 0.09 kg/m^3. If the mass of a molecule of hydrogen is 3.34×10^{-27} kg, find Avogadro's number and Boltzmann's constant. **[Ans. 1838.4 m/s ; 3380 kJ ; 6×10^{26} ; 1.386×10^{-26} kJ/K]**

4. If the Avogadro's number is 6.02×10^{26} per kg mole, find the mass of a molecule of hydrogen. Calculate also the Boltzmann's constant, when R_u = 8.314 kJ/kg K. **[Ans. 0.33×10^{-26} kg ; 1.38×10^{-26} kJ/K]**

5. The density of carbon dioxide gas at 0° C and at a pressure of 1 bar is 1.98 kg/m^3. Find the root mean square velocity of its molecules at 0° C and 30° C. The pressure is constant. **[Ans. 389 m/s ; 410 m/s]**

6. The root mean square velocity of the molecules of hydrogen at N.T.P. is 1840 m/s. Calculate the root mean square velocity of the oxygen molecules at N.T.P, if the molecular mass of hydrogen and oxygen are 2 and 32 respectively. **[Ans. 460 m/s]**

7. The root mean square velocity of oxygen at temperature 1092 K is 920 m/s. If the root mean square velocity of argon at a certain temperature is 450 m/s, find this temperature. The molecular mass of oxygen is 32 and that of argon is 40. **[Ans. 327.6 K]**

QUESTIONS

1. How many molecules are in one m^3 of air ?
2. What are the assumptions made in kinetic theory of gases ?
3. Define the following :
   (*a*) Mean velocity, (*b*) Root mean square velocity, and (*c*) Most probable velocity.
4. Derive an expression for the pressure exerted by a gas.
5. Show that the pressure of a gas in equal to two-thirds of the K.E. of translation per unit volume.
6. Write an expression for the r.m.s. velocity of a molecule in terms of :
   (*a*) Gas pressure and density,
   (*b*) Universal gas constant, absolute temperature and molecular mass of the gas,

(c) Boltzmann's constant, absolute temperature and mass of a gas molecule,

(d) Characteristic gas constant and absolute temperature of the gas.

7. Derive an expression for the average kinetic energy possessed by a gas molecule.

8. Derive the following gas laws as per kinetic theory of gases :

(a) Boyle's law, (b) Charles' law, (c) Gay-Lussac law, (d) Perfect gas equation, and (e) Avogadro's law.

9. What is meant by degrees of freedom possessed by a body ?

10. State the law of equipartition of energy. What is the amount of energy associated with each degree of freedom ?

OBJECTIVE TYPE QUESTIONS

1. The velocity of molecules

(a) increases with the increase of temperature

(b) increases with the decrease of temperature

(c) decreases with the increase of temperature

(d) remain constant at all temperatures

2. The root mean square velocity of the gas molecules (C) is given by

(a) $\sqrt{\frac{3mT}{k}}$ (b) $\sqrt{\frac{3T}{mk}}$ (c) $\sqrt{\frac{3kT}{m}}$ (d) $\sqrt{\frac{mk}{3T}}$

where m = Mass of one molecule of a gas,

k = Boltzmann's constant, and

T = Absolute temperature.

3. The ratio of root mean square velocity to average velocity of gas molecules at a particular temperature is

(a) 0.086 (b) 1.086 (c) 3.086 (d) 4.086

4. The pressure exerted by an ideal gas is .... of the kinetic energy of all the molecules contained in a unit volume of gas.

(a) one-half (b) one-third (c) two-third (d) three-fourth

5. The kinetic energy per kg molecule of any gas at absolute temperature T is equal to

(a) $\frac{1}{2} R_u T$ (b) $\frac{3}{4} R_u T$ (c) $R_u T$ (d) $\frac{3}{2} R_u T$

ANSWERS

1. (a)	2. (c)	3. (b)	4. (c)	5. (d)

6

Thermodynamic Air Cycles

1. Introduction. 2. Assumptions in Thermodynamic Cycles. 3. Classifications of Thermodynamic Cycles. 4. Reversible Cycle. 5. Irreversible Cycle. 6. Reversibility and Irreversibility of Thermodynamic Processes. 7. Relation between Cycle and Engine. 8. Working of an Ideal Engine. 9. Important Terms used in Thermodynamic Cycles. 10. Efficiency of a Cycle. 11. Types of Thermodynamic Cycles. 12. Carnot Cycle. 13. Stirling Cycle. 14. Ericsson Cycle. 15. Joule's Cycle, 16, Otto Cycle. 17. Diesel Cycle. 18. Dual Combustion Cycle.

6.1. Introduction

A thermodynamic cycle or a cyclic process consists of a series of thermodynamic operations (processes), which take place in a certain order, and the initial conditions are restored at the end of the processes. When the operations or processes of cycle are plotted on *p-v* diagram, they form a closed figure, each operation being represented by its own curve. Since the area under each curve gives the work done to some scale, during each operation, it therefore follows that the net work done during one cycle will be given by the enclosed area of the diagram as shown, shaded in Fig. 6.1.

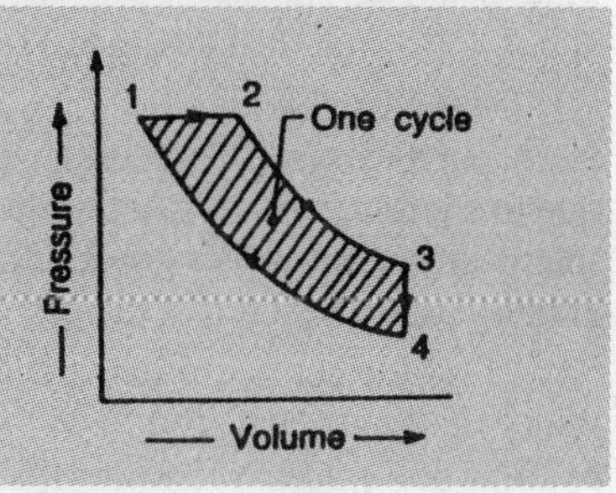

Fig. 6.1. A thermodynamic cycle.

The study of various thermodynamic cycles is very essential for the power developing systems (such as petrol engine diesel engine, gas turbine etc.). These engines use a mixture of fuel and air for their operations. Since the mass of fuel used, as compared to the mass of air is very small, therefore the mixture may be assumed to obey the properties of a perfect gas.

Notes : 1. A cycle, which requires four piston strokes and two complete revolutions of the crank is known as *four stroke cycle*. But a cycle, which requires only two piston strokes and one revolution of the crank, is known as *two stroke cycle*.

2. When air is assumed to be the working substance inside the engine cylinder, the cycle is called as an *air cycle*.

6.2. Assumptions in Thermodynamic Cycles

The analysis of all thermodynamic cycles (or air cycles) is based on the following assumptions :

1. The gas in the engine cylinder is a perfect gas, *i.e.* it obeys the gas laws and constant specific heats.
2. The physical constants of the gas in the engine cylinder are same as those of air at moderate temperatures.
3. All the compression and expansion processes are adiabatic and they take place without any internal friction.

4. Heat is supplied by bringing a hot body in contact with the cylinder at appropriate points during the process. Similarly heat is rejected by bringing a cold body in contact with the cylinder at these points.
5. The cycle is considered to be a closed one and the same air is used again and again to repeat the cycle.
6. No chemical reaction, whatsoever, takes place in the engine cylinder.

6.3. Classifications of Thermodynamic Cycles

The thermodynamic cycles, in general, may be classified into the following two types ;

1. Reversible or ideal cycle, and 2. Irreversible or natural or real cycle.

These cycles are discussed, in detail, in the following pages :

6.4. Reversible Cycle

A thermodynamically reversible cycle consists of reversible processes only. We have already discussed that a reversible process is one which is performed in such a way that at the end of the process, both the system and the surroundings may be restored to their initial states. For example, consider a process in which the system (gas) is expanded from state 1 to state 2 following the path 1-2 as shown in Fig. 6.2. Let during the thermodynamic process 1-2, the workdone by the system is W_{1-2} and the heat absorbed is Q_{1-2}. Now, if by doing the work (W_{1-2}) on the system (*i.e.* by compressing the gas) and extracting heat (Q_{1-2}) from the system, we can bring the system and the surroundings, back from state 2 to state 1 (*i.e.* initial state), following the same path 2-1, then the process is said to be a reversible process.

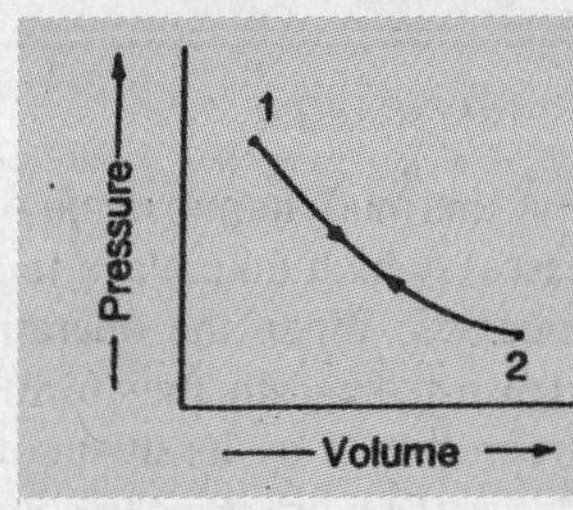

Fig. 6.2. Reversible process.

In a reversible process, there should not be any loss of heat due to friction, radiation or conduction, etc. A cycle will be reversible if all the processes constituting the cycle are reversible. Thus in a reversible cycle, the initial conditions are restored at the end of the cycle.

A little consideration will show, that when the operations are performed in the reversed order, the cycle draws heat from the cold body and rejects it to the hot body. This operation requires an external power to drive the mechanism according to second law of thermodynamics. A machine which operates on a reversed cycle is regarded as a "heat pump", such as a refrigerator, because it pumps heat from the cold body to the hot body. Following are the conditions for reversibility of a cycle :

1. The pressure and temperature of the working substance must not differ, appreciably, from those of the surroundings at any stage in the process.
2. All the processes, taking place in the cycle of operation, must be extremely slow.
3. The working parts of the engine must be friction free.
4. There should be no loss of energy during the cycle of operation.

Note : A reversible cycle should not be confused with a mechanically reversible engine. Steam engine cranks may be made to revolve in a reversed direction by mechanically altering the valve settings. But this does not reverse the cycle, on which it works. A two-stroke petrol engine may be made to revolve in reverse direction by altering the timing of ignition. But this also does not reverse the actual cycle.

6.5. Irreversible cycle

In the previous article, we have discussed that in a reversible process, the heat and work are completely restored back by reversing the process (*i.e.* by compressing the gas). But when the heat and work are not completely restored back by reversing the process, then the process is known as *irreversible process* (also called *natural* or *real process*). In an irreversible process, there is a loss of heat due to friction, radiation or conduction.

In actual practice most of the processes are irreversible to some degree. The main causes for the irreversiblity are:

1. mechanical and fluid friction, 2. unrestricted expansion, and 3. heat transfer with a finite temperature difference. Moreover, friction converts the mechanical work into heat. This heat cannot supply back the same amount of mechanical work, which was consumed for its production. Thus, if there is some friction involved in the process, it becomes irreversible. A cycle will be irreversible if any of the processes, constituting the cycle, is irreversible. Thus in an irreversible cycle, the initial conditions are not restored at the end of the cycle.

6.6. Reversibility and Irreversibility of Thermodynamic Processes

We have already discused the various thermodynamic processes in the last chapter. Now we shall discuss their conditions of reversibility and irreversibility.

1. *Isothermal and adiabatic processes.* It may be noted that a complete process or cycle is only an ideal case. But in actual practice, complete isothermal and adiabatic operations are not achieved. However, they can be approximated. The simple reason for the same is that it is impossible to transfer heat at a constant temperature in case of an isothermal operation. Moreover, it is also impossible to make an absolutely non-conducting cylinder in case of an adiabatic operation. In actual practice, however, an isothermal operation may be approached if the process is so slow that the heat is absorbed or rejected at such a rate that the temperature remains, practically, constant. Similarly, an adiabatic operation may be approached if the process takes place so *quickly* that no time is given to the heat to enter or leave the gas.

In view of the above, the isothermal and adiabatic processes are taken as reversible processes.

2. *Constant volume, constant pressure and constant $p\,v^n$ processes.* We know that when the temperature of the hot body, supplying the heat, remains constant during the process, the temperature of the working substance will vary as the operation proceeds. In view of this, the above three operations are irreversible. But, these can be made to approximate to reversibility by manipulating the temperature of the hot body to vary* so that at any stage the temperature of the working substance remains constant.

In this way, the constant volume, constant pressure and constant pv^n processes are regarded as reversible processes.

3. *Free expansion and throttling processes.* These processes are irreversible, as there is always a loss of heat due to friction when the working substance passes through an orifice.

6.7. Relation between Cycle and Engine

In the study of basic theory of a heat engine, it is assumed that the working fluid is used again and again in the cylinder. We say that the fluid has undergone a cycle, when it passes through different processes and returns back to its original state.

As a matter of fact, the working fluid in an actual engine does not go through a complete cycle, and operates on an open cycle. But for the purpose of simplicity in analysis, we study the closed cycle (*i.e.* ideal cycle), which closely approximates with the open cycle.

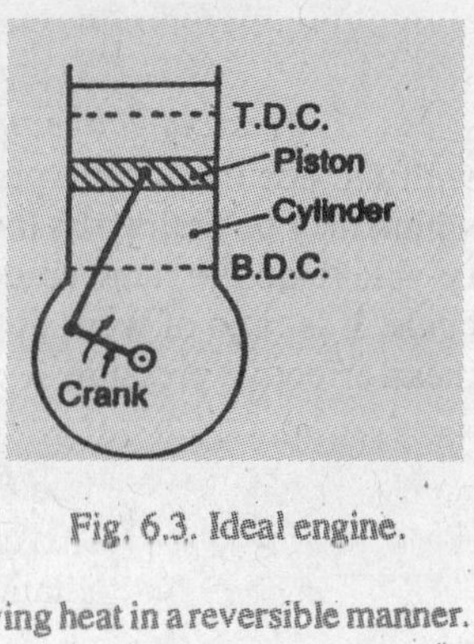

Fig. 6.3. Ideal engine.

6.8. Working of an Ideal Engine

An ideal engine may be defined as a device, which develops work (*i.e.* power) continuously with the help of a working fluid, which undergoes some cyclic process. It is done with the help of a piston and cylinder as shown in Fig. 6.3.

In general, the arrangement of piston and cylinder, of an ideal engine, constitutes the cycle by the following processes :

* An apparatus known as regenerator is used for alternately storing and supplying heat in a reversible manner.

1. Air in the cylinder is heated with the help of some external source. It increases temperature and pressure of the enclosed air.
2. The air expands due to higher pressure and temperature. As a result of this, some work is done by the air.
3. The air then rejects some heat to the external source. Thus the air comes back to the original conditions.
4. The air is then compressed in the cylinder. As a result of this, some work is done on the air.

Note : The scientists, working on the research and development of the engines, have focussed their attention mainly on the process of heating the enclosed air. Thus the various engines are classified according to the process of heat addition.

6.9. Important Terms used in Thermodynamic Cycles

Though there are many terms used in thermodynamic cycles, yet the following are important from the subject point of view :

1. *Cylinder bore.* The inner diameter of the cylinder, in which the piston moves, is known as cylinder bore.

2. *Stroke length.* The piston moves in the cylinder due to rotation of the crank. Its extreme positions are known as top dead centre (TDC) and bottom dead centre (BDC) respectively as shown in Fig. 6.3. The distance between these two extreme positions is known as stroke length or stroke.

3. *Clearance volume.* The volume occupied by the working fluid, when piston reaches the top dead centre, is known as clearance volume. It is generally denoted by (v_c).

4. *Swept volume.* The volume swept by the piston, when it moves between the two extreme positions is known as swept volume or displacement volume or stroke volume. Mathematically, swept volume,

$$v_s = \text{Piston area} \times \text{Stroke length} = \frac{\pi}{4} \times d^2 \times l$$

where d = Cylinder bore or diameter of the piston.

5. *Total cylinder volume.* The volume occupied by the working fluid, when the piston is at the bottom dead centre, is known as total cylinder volume. Mathematically, total cylinder volume is equal to the sum of clearance volume (v_c) and swept volume (v_s).

6. *Compression ratio.* The ratio of total cylinder volume to the clearance volume is known as compression ratio. It is an important term used in an engine. Mathematically, compression ratio,

$$r = \frac{v_c + v_s}{v_c} = 1 + \frac{v_s}{v_c}$$

7. *Mean effective pressure.* As a matter of fact, pressure in the cylinder keeps on changing with the position of the piston. For all sorts of calculations, we need the mean effective pressure, which may be defined as the constant pressure acting on the piston during the working stroke. It will be able to do the same amount of work, as done by the actual varying pressure, produced during the cycle. It is ratio of work done to the stroke volume or piston displacement volume. Mathematically, mean effective pressure,

$$p_m = \frac{\text{Work done}}{\text{Stroke volume}}$$

6.10. Efficiency of a Cycle

It may be defined as the ratio of work done to the heat supplied during a cycle. Mathematically, efficiency of a cycle,

$$\eta = \frac{\text{Work done}}{\text{Heat supplied}}$$

Since the work done during a cycle is equal to heat supplied *minus* the heat rejected, the efficiency of a cycle, therefore, may also be expressed as

$$\eta = \frac{\text{Heat supplied} - \text{Heat rejected}}{\text{Heat supplied}}$$

Notes : 1. The efficiency, as given above, is the theoretical efficiency of the cycle. Therefore it is known as *theoretical thermal efficiency.*

2. It does not take into account the practical losses, which occur in the running of the engine.

3. In order to compare the efficiency of the thermodynamic cycles, air is assumed to be the working substance inside the engine cylinder. Moreover, air is assumed to behave as a perfect gas. The efficiency, thus, obtained is known as *air standard efficiency*. It is also called *ideal efficiency*.

6.11. Types of Thermodynamic Cycles

Though there are many types of thermodynamic cycles, yet the following are important from the subject point of view :

1. Carnot cycle, 2. Stirling cycle, 3. Ericsson cycle, 4. Joule cycle, 5. Otto cycle, 6. Diesel cycle, and 7. Dual combustion cycle.

The above mentioned cycles will be discussed, in detail, in the following pages.

6.12. Carnot Cycle

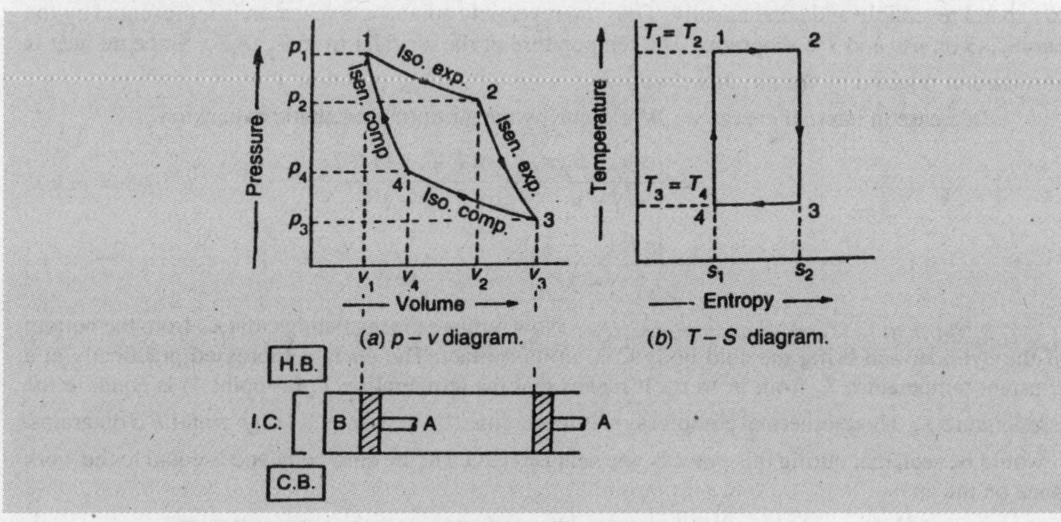

Fig. 6.4. Carnot cycle.

This cycle was devised by *Carnot, who was the first scientist to analyse the problem of the efficiency of a heat engine, disregarding its mechanical details. He focussed his attention on the basic features of a heat engine. In a Carnot cycle, the working substance is subjected to a cyclic operation consisting of two isothermal and two reversible adiabatic or isentropic operations. The *p-v* and *T-S* diagrams of this cycle are shown in Fig. 6.4 (*a*) and (*b*).

* Nicolas Leonard Sadi Carnot was a French engineer. He devised this cycle in his early age.

The engine imagined by Carnot has air (which is supposed to behave like a perfect gas) as its working substance enclosed in a cylinder, in which a frictionless piston A moves. The walls of the cylinder and piston are perfect non-conductor of heat. However, the bottom B of the cylinder can be covered, at will, by an insulating cap (I.C.). The engine is assumed to work between two sources of infinite heat capacity, one at a higher temperature and the other at a lower temperature.

Now, let us consider the four stages of the Carnot's cycle. Let the engine cylinder contain m kg of air at its original condition represented by point 1 on the p-v and T-S diagrams. At this point, let p_1, T_1 and v_1 be the pressure, temperature and volume of the air, respectively.

1. *First stage (Isothermal expansion).* The source (hot body, H.B.) at a higher temperature is brought in contact with the bottom B of the cylinder. The air expands, practically at constant temperature T_1, from v_1 to v_2. It means that the temperature T_2 at point 2 is equal to the temperature T_1. This isothermal expansion is represented by curve 1-2 on p-v and T-S diagrams in Fig. 6.4 (a) and (b). It may be noted that the heat supplied by the hot body is fully absorbed by the air, and is utilised in doing external work.

∴ Heat supplied = *Work done by the air during isothermal expansion

or
$$Q_{1-2} = p_1 v_1 \log_e \left(\frac{v_2}{v_1}\right) = m\,R\,T_1 \log_e \left(\frac{v_2}{v_1}\right) \qquad \ldots (\because p_1 v_1 = mRT_1)$$

$$= 2.3\, m\,R\,T_1 \log r$$

where r = Expansion ratio = v_2/v_1.

2. *Second stage (Reversible adiabatic or isentropic expansion).* The hot body is removed from the bottom of the cylinder B and the insulating cap I.C. is brought in contact. The air is now allowed to expand reversibly and adiabatically. Thus the reversible adiabatic expansion is represented by the curve 2-3 on p-v and T-S diagrams. The temperature of the air falls from T_2 to T_3. Since no heat is absorbed or rejected by the air, therefore

Decrease in internal energy = Workdone by the air during adiabatic expansion

$$= \frac{p_2 v_2 - p_3 v_3}{\gamma - 1} = \frac{m\,R\,T_2 - m\,R\,T_3}{\gamma - 1} \qquad \ldots (\because p\,v = m\,R\,T)$$

$$= \frac{m\,R\,(T_1 - T_3)}{\gamma - 1} \qquad \ldots (\because T_1 = T_2)$$

3. *Third stage (Isothermal compression).* Now remove the insulating cap I.C. from the bottom of the cylinder and bring the cold body C.B. in its contact. The air is compressed practically at a constant temperature T_3 from v_3 to v_4. It means that the temperature T_4 (at point 4) is equal to the temperature T_3. This isothermal compression is represented by the curve 3-4 on p-v and T-S diagrams. It would be seen that during this process, the heat is rejected to the cold body and is equal to the work done on the air.

∴ Heat rejected = Work done on the air during isothermal compression

$$Q_{3-4} = p_3 v_3 \log_e \left(\frac{v_3}{v_4}\right) = m\,R\,T_3 \log_e \left(\frac{v_3}{v_4}\right) \qquad \ldots (\because p\,v = m\,R\,T)$$

$$= 2.3\, m\,R\,T_3 \log r$$

---

* Since the temperature is constant, therefore there is no change in internal energy of the air, *i.e.* $dU = 0$. According to the first law of thermodynamics,

$$Q_{1-2} = dU + W_{1-2} \text{ or } Q_{1-2} = W_{1-2}$$

where r = *Compression ratio $= v_3/v_4$

4. *Fourth stage (Reversible adiabatic or isentropic compression).* Now again the insulated cap I.C. is brought in contact with the bottom of the cylinder B, and the air is allowed to be compressed reversibly and adiabatically. The reversible adiabatic compression is represented by the curve 4-1 on p-v and T-S diagrams. The temperature of the air increases from T_4 to T_1. Since no heat is absorbed or rejected by the air, therefore

Increase in internal energy = Work done on the air during adiabatic compression

$$= \frac{p_1 v_1 - p_4 v_4}{\gamma - 1} = \frac{m R T_1 - m R T_4}{\gamma - 1} \qquad \ldots (\because pv = mRT)$$

$$= \frac{m R (T_1 - T_3)}{\gamma - 1} \qquad \ldots (\because T_3 = T_4)$$

We see from the above discussion that the decrease in internal energy during reversible adiabatic expansion 2-3 is equal to the increase in internal energy during reversible adiabatic compression 4-1. Hence their net effect during the whole cycle is zero. We know that

Work done, W = Heat supplied − Heat rejected

$$= 2.3\, m R T_1 \log r - 2.3\, m R T_3 \log r = 2.3\, m R \log r (T_1 - T_3)$$

and efficiency

$$^{**}\eta = \frac{\text{Work done}}{\text{Heat supplied}} = \frac{2.3\, m R \log r (T_1 - T_3)}{2.3\, m R T_1 \log r}$$

$$= \frac{T_1 - T_3}{T_1} = 1 - \frac{T_3}{T_1}$$

---

* The expansion and compression ratios (r) must be equal, otherwise the cycle would not close.

We know that for reversible adiabatic or isentropic expansion process 2-3,

$$\frac{T_2}{T_3} = \left(\frac{v_3}{v_2}\right)^{\gamma - 1} \quad \text{or} \quad \frac{v_3}{v_2} = \left(\frac{T_2}{T_3}\right)^{\frac{1}{\gamma - 1}} \qquad \ldots (i)$$

Similarly, for reversible adiabatic or isentropic compression process 4-1,

$$\frac{T_1}{T_4} = \left(\frac{v_4}{v_1}\right)^{\gamma - 1} \quad \text{or} \quad \frac{v_4}{v_1} = \left(\frac{T_1}{T_4}\right)^{\frac{1}{\gamma - 1}} \qquad \ldots (ii)$$

Since $T_1 = T_2$ and $T_3 = T_4$, therefore

$$\frac{v_3}{v_2} = \frac{v_4}{v_1} \quad \text{or} \quad r = \frac{v_2}{v_1} = \frac{v_3}{v_4}$$

** **Alternative Proof.**

Heat supplied during isothermal expansion 1-2,

$$Q_{1-2} = T_1 (S_2 - S_1) \qquad \ldots (\because \delta Q = T\,dS)$$

and heat rejected during isothermal compression 3-4,

$$Q_{3-4} = T_4 (S_2 - S_1) = T_3 (S_2 - S_1) \qquad \ldots (\because T_4 = T_3)$$

We know that work done = Heat supplied − Heat rejected

$$= T_1 (S_2 - S_1) - T_3 (S_2 - S_1) = (T_1 - T_3)(S_2 - S_1)$$

$\therefore$ Efficiency,

$$\eta = \frac{\text{Work done}}{\text{Heat supplied}} = \frac{(T_1 - T_3)(S_2 - S_1)}{T_1 (S_2 - S_1)} = \frac{T_1 - T_3}{T_1} = 1 - \frac{T_3}{T_1}$$

The expression for the efficiency of a Carnot cycle may also be written as discussed below :

We know that for reversible adiabatic or isentropic expansion 2-3,

$$\frac{T_2}{T_3} = \left(\frac{v_3}{v_2}\right)^{\gamma-1} \quad \text{or} \quad \frac{T_1}{T_3} = \left(\frac{v_3}{v_2}\right)^{\gamma-1} \qquad \ldots (\because T_2 = T_1) \ldots (i)$$

Similarly, for reversible adiabatic or isentropic compression process 4-1,

$$\frac{T_1}{T_4} = \left(\frac{v_4}{v_1}\right)^{\gamma-1} \quad \text{or} \quad \frac{T_1}{T_3} = \left(\frac{v_4}{v_1}\right)^{\gamma-1} \qquad \ldots (\because T_4 = T_3) \ldots (ii)$$

From equations (*i*) and (*ii*),

$$\frac{v_3}{v_2} = \frac{v_4}{v_1} \quad \text{or} \quad \frac{v_2}{v_1} = \frac{v_3}{v_4} = r$$

where r = Ratio of expansion or compression.

$$\therefore \quad \frac{T_1}{T_3} = (r)^{\gamma-1}$$

We know that efficiency,

$$\eta = 1 - \frac{T_3}{T_1} = 1 - \left(\frac{1}{r}\right)^{\gamma-1} = 1 - \frac{1}{r^{\gamma-1}}$$

Notes : 1. From the above equation, we see that the efficiency of Carnot's cycle increases as T_1 is increased or T_3 is decreased. In other words, the heat should be taken in at as high a temperature as possible, and rejected at as low a temperature as possible. It may be noted that 100% efficiency can be achieved, only, if T_3 reaches absolute zero, though it is impossible to achieve in practice.

2. In the above theory, we have taken temperature at points 1, 2, 3 and 4 as T_1, T_2, T_3 and T_4 respectively in order to keep similarity between Carnot cycle and other cycles. But some authors take it T_1 (for points 1 and 2) and T_2 (for points 3 and 4). In that case, they obtain the relation for efficiency as,

$$\eta = \frac{T_1 - T_2}{T_1} = 1 - \frac{T_2}{T_1}$$

3. It may be noted that it is impossible to make an engine working on Carnot's cycle. The simple reason for the same is that the isothermal expansion 1-2 will have to be carried out extremely slow to ensure that the air is always at temperature T_1. Similarly, the isothermal compression 3-4 will have to be carried out extremely slow. But reversible adiabatic expansion 2-3 and reversible adiabatic compression 4-1 should be carried out as quickly as possible, in order to approach ideal adiabatic conditions. We know that sudden changes in the speed of an engine are not possible in actual practice. Moreover, it is impossible to completely eliminate friction between the various moving parts of the engine, and also heat losses due to conduction, radiation, etc. It is thus obvious, that it is impossible to realise Carnot's engine in actual practice. However, such an imaginary engine is used as the ultimate standard of comparison of all heat engines.

Example 6.1. *A Carnot engine, working between 650 K and 310 K, produces 150 kJ of work. Find thermal efficiency and heat added during the process.*

Solution. $T_1 = 650$ K ; $T_3 = 310$ K ; $W = 150$ kJ

Thermal efficiency

We know that thermal efficiency,

$$\eta = \frac{T_1 - T_3}{T_1} = \frac{650 - 310}{650} = 0.523 \quad \text{or} \quad 52.3\% \ \textbf{Ans.}$$

Heat added during the process

We know that heat added during the process,

$$Q_{1-2} = \frac{W}{\eta} = \frac{150}{0.523} = 286.8 \text{ kJ} \quad \textbf{Ans.}$$

Example 6.2. *A Carnot engine operates between two reservoirs at temperatures T_1 and T_3. The work output of the engine is 0.6 times the heat rejected. The difference in temperatures between the source and the sink is 200° C. Calculate the thermal efficiency, source temperature and the sink temperature.*

Solution. Given : $W = 0.6 \times$ Heat rejected $= 0.6\, Q_{3-4}$; $T_1 - T_3 = 200°$ C

Thermal efficiency

We know that the thermal efficiency,

$$\eta = \frac{\text{Work done}}{\text{Heat supplied}} = \frac{\text{Work done}}{\text{Work done} + \text{Heat rejected}}$$

$$= \frac{0.6\, Q_{3-4}}{0.6\, Q_{3-4} + Q_{3-4}} = \frac{0.6}{1.6} = 0.375 \quad \text{or} \quad 37.5\% \quad \textbf{Ans.}$$

Source and sink temperatures

Let T_1 = Source temperature, and

T_3 = Sink temperature.

We know that thermal efficiency (η),

$$0.375 = \frac{T_1 - T_3}{T_1} = \frac{200}{T_1}$$

$\therefore \quad T_1 = 200/0.375 = 533.3 \text{ K} = 260.3°$ C **Ans.**

and $\quad T_3 = T_1 - 200 = 260.3 - 200 = 60.3°$ C **Ans.**

Example 6.3. *An engineer claims his engine to develop 3.75 kW. On testing, the engine consumes 0.44 kg of fuel per hour having a calorific value of 42 000 kJ/kg. The maximum temperature recorded in the cycle is 1400° C and minimum is 350° C. Find whether the engineer is justified in his claim.*

Solution. Give : $P = 3.75$ kW; Fuel consumed = 0.44 kg/h; Calorific value = 42 000 kJ/kg
$T_1 = 1400°\text{ C} = 1400 + 273 = 1673 \text{ K}$; $T_3 = 350°\text{ C} = 350 + 273 = 623 \text{ K}$

We know that the maximum efficiency, between two specified temperatures, is that of Carnot cycle.

$$\therefore \quad \eta_{carnot} = \frac{T_1 - T_3}{T_1} = \frac{1673 - 623}{1673} = 0.627 \quad \text{or} \quad 62.7\%$$

We also know that the heat supplied to the engine by the fuel

= Fuel consumed × Calorific value of fuel

= 0.44 × 42 000 = 18 480 kJ/h = 5.13 kJ/s

and workdone by the engine = 3.75 kW = 3.75 kJ/s

$$\therefore \text{ Efficiency claimed} = \frac{\text{Workdone}}{\text{Heat supplied}} = \frac{3.75}{5.13} = 0.731 \quad \text{or} \quad 73.1\ \%$$

Since efficiency of the engine claimed (73.1%) is more than the maximum possible efficiency (62.7%), therefore the engineer is not justified in his claim. **Ans.**

Example 6.4. *A Carnot cycle works with isentropic compression ratio of 5 and isothermal expansion ratio of 2. The volume of air at the beginning of the isothermal expansion is 0.3 m³. If the maximum temperature and pressure is limited to 550 K and 21 bar, determine : 1. minimum temperature in the cycle ; 2. thermal efficiency of the cycle ; 3. pressure at all salient points ; 4. change of entropy during the isothermal expansion, 5. work done per cycle, and 6. mean effective pressure. Take ratio of specific heats as 1.4.*

Solution. Given : $v_4 / v_1 = 5$; $v_2 / v_1 = 2$; $v_1 = 0.3\ m^3$; $T_1 = 550$ K ; $p_1 = 21$ bar $= 2.1 \times 10^6$ N/m² ; $\gamma = 1.4$

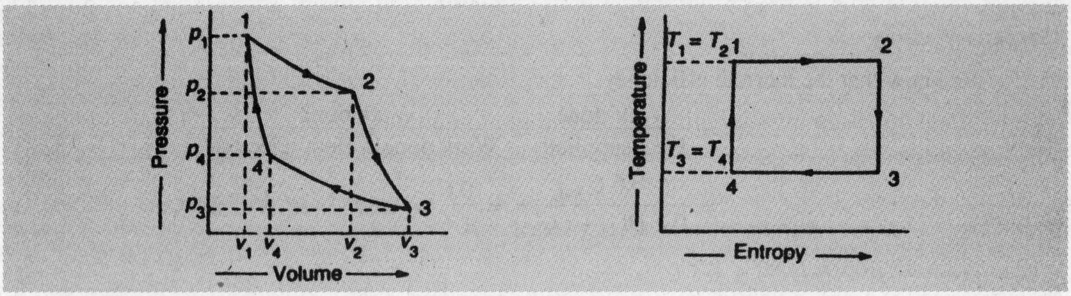

(*a*) *p-v* diagram. (*b*) *T-S* diagram.

Fig. 6.5

1. *Minimum temperature in the cycle*

Let T_4 (or T_3) = Minimum temperature in the cycle.

The cycle on *p-v* and *T-S* diagrams is shown in Fig. 6.5 (*a*) and (*b*) respectively. First of all, let us consider the isentropic compression process 4-1. We know that

$$\frac{T_1}{T_4} = \left(\frac{v_4}{v_1}\right)^{\gamma-1} = (5)^{1.4-1} = (5)^{0.4} = 1.9036$$

$$\therefore \quad T_4 = T_1 / 1.9036 = 550 / 1.9036 = 289 \text{ K} = 16° \text{ C} \quad \textbf{Ans.}$$

2. *Thermal efficiency of the cycle*

We know that thermal efficiency of the cycle,

$$\eta = \frac{T_1 - T_3}{T_1} = \frac{550 - 289}{550} = 0.4745 \text{ or } 47.45\ \% \quad \textbf{Ans.}$$

$$\ldots (\because T_3 = T_4)$$

3. *Pressure at all the salient points*

Let p_2, p_3, p_4 = Pressures at points 2, 3 and 4 respectively.

First of all, let us consider the isothermal expansion process 1-2. We know that in an isothermal expansion,

$$p_1 v_1 = p_2 v_2 \text{ or } p_2 = p_1 \times \frac{v_1}{v_2} = 21 \times \frac{1}{2} = 10.5 \text{ bar} \quad \textbf{Ans.}$$

$$\ldots (\because v_2 / v_1 = 2)$$

Now consider the isentropic expansion process 2-3, we know that

$$p_2 v_2^\gamma = p_3 v_3^\gamma \quad \text{or} \quad p_3 = p_2 \left(\frac{v_2}{v_3}\right)^\gamma = 10.5 \left(\frac{1}{5}\right)^{1.4} = 10.5\,(0.2)^{1.4}$$

$$= 1.103 \text{ bar } \textbf{Ans.} \qquad \ldots \left(\because \frac{v_4}{v_1} = \frac{v_3}{v_2}\right)$$

Now consider the isentropic compression process 4-1. We know that

$$p_4 v_4^\gamma = p_1 v_1^\gamma \quad \text{or} \quad p_4 = p_1 \left(\frac{v_1}{v_4}\right)^\gamma = 21 \left(\frac{1}{5}\right)^{1.4} = 21\,(0.2)^{1.4}$$

$$= 2.206 \text{ bar } \textbf{Ans.}$$

4. *Change of entropy during the isothermal expansion*

We know that change of entropy during the isothermal expansion,

$$S_2 - S_1 = 2.3\, m\, R \log \left(\frac{v_2}{v_1}\right) = 2.3 \times \frac{p_1\, v_1}{T_1} \log \left(\frac{v_2}{v_1}\right) \qquad \ldots (\because p_1\, v_1 = m\, R\, T_1)$$

$$= 2.3 \times \frac{2.1 \times 10^6 \times 0.3}{550} \times \log 2 = 2.636 \times 10^3 \times 0.301 = 793 \text{ J/K}$$

$$= 0.793 \text{ kJ/K } \textbf{Ans.}$$

5. *Workdone per cycle*

We know that heat supplied during the cycle,

$$Q_{1-2} = T_1\,(S_2 - S_1) = 550 \times 0.793 = 436 \text{ kJ}$$

and heat rejected during the cycle,

$$Q_{3-4} = T_3\,(S_2 - S_1) = 289 \times 0.793 = 229 \text{ kJ}$$

∴ Workdone per cycle,

$$W = \text{Heat supplied} - \text{Heat rejected} = 436 - 229 = 207 \text{ kJ } \textbf{Ans.}$$

Note: The heat supplied and heat rejected may also be obtained as discussed below :

We know that heat supplied,

$$Q_{1-2} = 2.3\, p_1\, v_1 \log \left(\frac{v_2}{v_1}\right) = 2.3 \times 2.1 \times 10^6 \times 0.3 \log 2 \text{ J}$$

$$= 1.449 \times 10^6 \times 0.301 = 436 \times 10^3 \text{ J} = 436 \text{ kJ}$$

and heat rejected,

$$Q_{3-4} = 2.3\, p_3\, v_3 \log \left(\frac{v_3}{v_4}\right) = 2.3 \times 0.1103 \times 10^6 \times 3 \log 2$$

$$= 0.761 \times 10^6 \times 0.301 = 229 \times 10^3 \text{ J} = 229 \text{ kJ}$$

$$\ldots \left[\begin{array}{l} \because \dfrac{v_3}{v_4} = \dfrac{v_2}{v_1} = 2 \text{ and } \dfrac{v_4}{v_1} = 5 \\ \text{or } \dfrac{v_3}{v_4} \times \dfrac{v_4}{v_1} = \dfrac{v_3}{v_1} = 2 \times 5 = 10 \\ \therefore v_3 = 10\, v_1 = 10 \times 0.3 = 3 \text{ m}^3 \end{array}\right]$$

6. *Mean effective pressure*

We know that the stroke volume in a Carnot cycle

$$= v_3 - v_1 = 3.0 - 0.3 = 2.7 \text{ m}^3$$

$\therefore$ Mean effective pressure

$$= \frac{\text{Work done}}{\text{Stroke volume}} = \frac{207}{2.7} = 76.7 \text{ kN/m}^2 \quad \textbf{Ans.}$$

6.13. Stirling Cycle

This cycle was devised by *Stirling, which consists of two isothermal processes and two constant volume processes. The last two processes are performed with the help of a regenerator to make this cycle reversible. The *p-v* and *T-S* diagrams of this cycle are shown in Fig. 5.6.

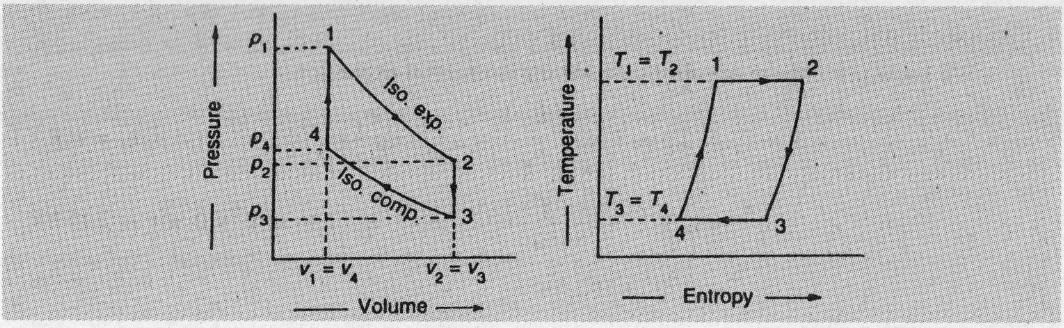

(*a*) *p-v* diagram. (*b*) *T-S* diagram.

Fig. 6.6. Stirling cycle.

Let us now consider the four stages of the Striling cycle. Let the engine cylinder contain *m* kg of air at its original position represented by point 1. At this point, let p_1, T_1 and v_1 be the pressure, temperature and volume of the air respectively.

1. *First stage (Isothermal expansion).* The air expands isothermally, practically, at constant temperature (T_1 or T_2) from v_1 to v_2. It means that the temperature T_2 (at point 2) is equal to temperature T_1. This isothermal expansion is represented by the curve 1-2 in Fig. 6.6 (*a*) and (*b*). The heat supplied by the external source is absorbed during this process.

$\therefore$ Heat supplied = Work done during isothermal expansion

$$Q_{1-2} = p_1 v_1 \log_e \left(\frac{v_2}{v_1}\right) = m R T_1 \log_e \left(\frac{v_2}{v_1}\right) \qquad \ldots (\because p_1 v_1 = m R T_1)$$

$$= 2.3 \, m R T_1 \log r$$

where r = Expansion ratio $= v_2 / v_1$

2. *Second stage (Constant volume cooling).* The air now passes through the regenerator, and is cooled at constant volume to a temperature T_3. This process is represented by a curve 2-3 on *p-v* and *T-S* diagrams. In this process, heat is rejected to the regenerator.

$\therefore$ Heat rejected to the regenerator,

$$Q_{2-3} = m \, c_v (T_2 - T_3)$$

* Robert Stirling devised this cycle in 1845.

3. *Third stage (Isothermal compression).* The air is now compressed isothermally, practically, at a constant temperature (T_3) in the engine cylinder from v_3 to v_4. This process is represented by the curve 3-4 on *p-v* and *T-S* diagrams. Again, heat is rejected by the air during this process.

∴ Heat rejected by the air,

$$Q_{3-4} = p_3 v_3 \log_e \left(\frac{v_3}{v_4}\right) = m R T_3 \log_e \left(\frac{v_3}{v_4}\right) \quad \ldots (\because p_3 v_3 = m R T_3)$$

$$= 2.3 \, m R T_3 \log r$$

where r = Compression ratio = v_3 / v_4

4. *Fourth stage (Constant volume heating).* Lastly, the air is heated at a constant volume to a temperature T_1, by passing it though the regenerator in the reverse direction to that of process 2-3. It may be noted that the air restores initial conditions at the end of this process, and completes the cycle. It is thus obvious, that the heat is absorbed by the air from the regenerator during this process, *i.e.* 4-1.

∴ Heat absorbed by the air,

$$Q_{4-1} = \text{Heat taken in from the regenerator}$$

$$= m \, c_v (T_1 - T_4) = m \, c_v (T_1 - T_3) \quad \ldots (\because T_3 = T_4)$$

We see that the heat rejected to the regenerator during process 2-3 is equal to the heat taken in from the regenerator during process 4-1. Hence there is no interchange of heat to the system during these processes. The only interchange of heat is during the two isothermal processes.

∴ Work done = Heat supplied – Heat rejected

$$= 2.3 \, m R T_1 \log r - 2.3 \, m R T_3 \log r = 2.3 \, m R \log r (T_1 - T_3)$$

and efficiency,

$$\eta = \frac{\text{Work done}}{\text{Heat supplied}} = \frac{2.3 \, m R \log r (T_1 - T_3)}{2.3 \, m R T_1 \log r}$$

$$= \frac{T_1 - T_3}{T_1} = 1 - \frac{T_3}{T_1}$$

Notes : 1. The efficiency of the Stirling cycle is same as that of Carnot cycle. This is due to the fact that the cycle is reversible, and all reversible cycles have the same efficiency.

2. If the efficiency of regenerator is η_R, then heat taken in from the regenerator during process 4-1 will be $m \, c_v (T_1 - T_3)(1 - \eta_R)$. In that case,

$$\eta = \frac{2.3 \, m R \log r (T_1 - T_3)}{2.3 \, m R T_1 \log r + m \, c_v (T_1 - T_3)(1 - n_R)}$$

$$= \frac{2.3 \, R \log r (T_1 - T_3)}{2.3 \, R T_1 \log r + c_v (T_1 - T_3)(1 - \eta_R)}$$

Example 6.5. *An air engine, working on Stirling cycle, has lower limit of temperature as 400° C. The maximum and minimum pressure limits are 12 bar and 2 bar. If the expansion ratio of the cycle is 3, find its ideal efficiency.*

Solution. Given : $T_3 = T_4 = 400°\text{ C} = 400 + 273 = 673\text{ K}$; $p_1 = 12$ bar ; $p_3 = 2$ bar ; $v_2 / v_1 = v_3 / v_4 = 3$

We know that in isothermal expansion process 1-2 (Refer Fig. 6.6),

$$p_1 v_1 = p_2 v_2 \quad \text{or} \quad p_2 = p_1 \times \frac{v_1}{v_2} = 12 \times \frac{1}{3} = 4 \text{ bar}$$

Similarly, in isothermal compression process 3-4,

$$p_3 v_3 = p_4 v_4 \quad \text{or} \quad p_4 = p_3 \times \frac{v_3}{v_4} = 2 \times 3 = 6 \text{ bar}$$

Since the air is heated at constant volume in process 4-1, therefore

$$T_1 = T_4 \times \frac{p_1}{p_4} = 673 \times \frac{12}{6} = 1346 \text{ K} \qquad \ldots \left(\because \frac{p_1}{T_1} = \frac{p_4}{T_4} \right)$$

$\therefore$ Efficiency,

$$\eta = 1 - \frac{T_3}{T_1} = 1 - \frac{673}{1346} = 0.5 \text{ or } 50\% \textbf{ Ans.}$$

Example 6.6. *A Stirling air engine is fitted with a regenerator of efficiency 90%. It operates between the temperature limits of 350° C and 50° C. Determine the efficiency of the engine, assuming isothermal expansion ratio as 2. Take c_p = 1.005 kJ/kg K and c_v = 0.712 kJ/kg K.*

Solution. Given : $\eta_R = 90\% = 0.9$; $T_1 = 350°\text{ C} = 350 + 273 = 623$ K ; $T_3 = 50°\text{ C} = 50 + 273 = 323$ K ; $r = v_2 / v_1 = 2$

We know that gas constant,

$$R = c_p - c_v = 1.005 - 0.712 = 0.293 \text{ kJ/kg K}$$

$\therefore$ Efficiency of the engine,

$$\eta = \frac{2.3\, R \log r\, (T_1 - T_3)}{2.3\, R\, T_1 \log r + c_v (T_1 - T_3)(1 - \eta_R)}$$

$$= \frac{2.3 \times 0.293 \log 2\, (623 - 323)}{2.3 \times 0.293 \times 623 \log 2 + 0.712\, (623 - 323)\, (1 - 0.9)}$$

$$= \frac{60.85}{126.37 + 21.36} = 0.412 \text{ or } 41.2\% \textbf{ Ans.}$$

6.14. Ericsson Cycle

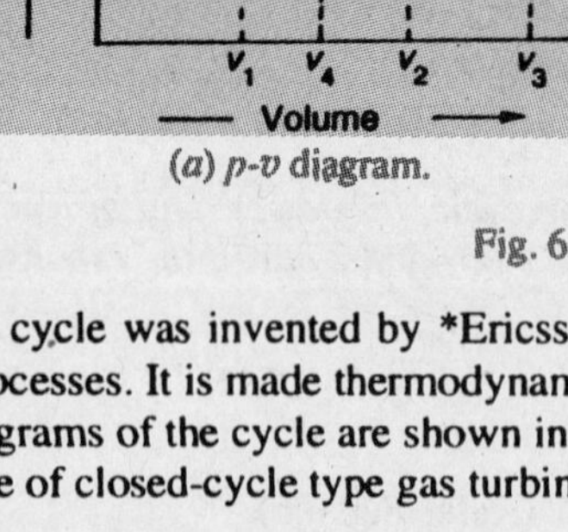

(*a*) *p-v* diagram.

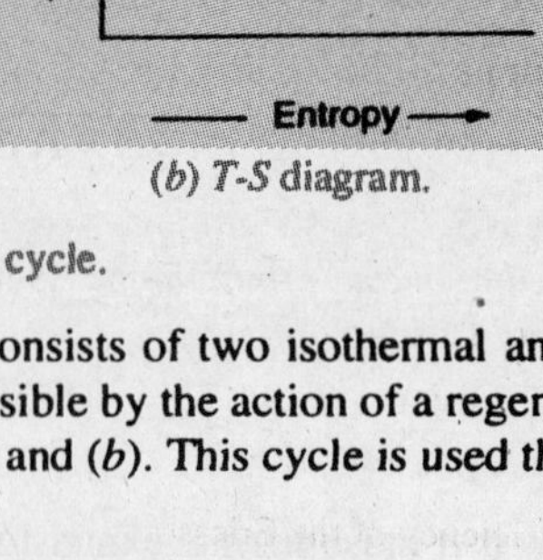

(*b*) *T-S* diagram.

Fig. 6.7. Ericsson cycle.

This cycle was invented by *Ericsson, which consists of two isothermal and two constant pressure processes. It is made thermodynamically reversible by the action of a regenerator. The *p-v* and *T-S* diagrams of the cycle are shown in Fig. 6.7 (*a*) and (*b*). This cycle is used these days in the manufacture of closed-cycle type gas turbines.

---

* J. Ericsson was an American engineer, who invented this engine in 1840. He used a hot air engine, working on this cycle, for running a ship (known as Ericsson) in 1853.

Now, let us consider the four stages of the Ericsson cycle. Let the engine contain m kg of air at its original position represented by point 1 on p-v and T-S diagrams. At this point, let p_1, T_1 and v_1 be the pressure, temperature and volume of the air.

1. *First stage (Constant pressure heating).* The air is heated at constant pressure from initial temperature T_1 to a temperature T_2 represented by the curve 1-2 in Fig. 6.7 (*a*) and (*b*).

∴ Heat supplied to the air,

$$Q_{1-2} = m\,c_p\,(T_2 - T_1)$$

2. *Second stage (Isothermal expansion).* The air is allowed to expand isothermally (*i.e.* at constant temperature $T_2 = T_3$) from initial volume v_2 to v_3 represented by the curve 2-3 in Fig. 6.7 (*a*) and (*b*). We know that a part of the heat supplied in the first stage is utilised for doing work in isothermal expansion,

∴ Heat utilised during isothermal expansion,

$$Q_{2-3} = p_2\,v_2 \log_e\left(\frac{v_3}{v_2}\right) = m\,R\,T_2 \log_e\left(\frac{v_3}{v_2}\right) \qquad \ldots(\because p\,v = m\,R\,T)$$

$$= 2.3\,m\,R\,T_2 \log r$$

where r = Expansion ratio $= v_3 / v_2$

3. *Third stage (Constant pressure cooling).* The air is now cooled at a constant pressure from initial temperature T_3 to a temperature T_4 represented by the curve 3-4 in Fig. 6.7 (*a*) and (*b*).

∴ Heat rejected by the air,

$$Q_{3-4} = m\,c_p\,(T_3 - T_4)$$

4. *Fourth stage (Isothermal compression).* Finally, the air is compressed isothermally (*i.e.* at constant temperature $T_4 = T_1$) from initial volume v_3 to v_4 represented by the curve 4-1 in Fig. 6.7 (*a*) and (*b*). We know that some heat is rejected by the air for doing work on the air.

∴ Heat rejected during isothermal compression,

$$Q_{4-1} = p_4\,v_4 \log_e\left(\frac{v_4}{v_1}\right) = m\,R\,T_4 \log_e\left(\frac{v_4}{v_1}\right) \qquad \ldots(\because p\,v = m\,R\,T)$$

$$= 2.3\,m\,R\,T_4 \log r$$

where r = Compression ratio $= v_4 / v_1$

We see from the above, that the heat supplied during the process 1-2 is equal to the heat rejected during the process 3-4 (because $T_2 - T_1 = T_3 - T_4$).

∴ Work done = Heat supplied – Heat rejected

$$= 2.3\,m\,R\,T_2 \log r - 2.3\,m\,R\,T_4 \log r = 2.3\,m\,R \log r\,(T_2 - T_4)$$

and efficiency,

$$\eta = \frac{\text{Work done}}{\text{Heat supplied}} = \frac{2.3\,m\,R \log r\,(T_2 - T_4)}{2.3\,m\,R\,T_2 \log r}$$

$$= \frac{T_2 - T_4}{T_2} = 1 - \frac{T_4}{T_2} = 1 - \frac{T_1}{T_3}$$

Notes : 1. The efficiency of the Ericsson cycle is same as that of Carnot efficiency *i.e.*

$$\eta = 1 - \frac{\text{Lowest temperature}}{\text{Highest temperature}}$$

2. If the regenerator efficiency is η_R, then heat taken in from the regenerator during process 4-1 will be $mc_p (T_2 - T_4)(1 - \eta_R)$. In that case,

$$\eta = \frac{2.3\, m\, R \log r\,(T_2 - T_4)}{2.3\, m\, R\, T_2 \log r + m\, c_p\,(T_2 - T_4)(1 - \eta_R)}$$

$$= \frac{2.3\, R \log r\,(T_2 - T_4)}{2.3\, R\, T_2 \log r + c_p\,(T_2 - T_4)(1 - \eta_R)}$$

Example 6.7. *An Ericsson regenerative engine works between the temperature limits of 45° C and 230° C. If the ratio of expansion is 2, determine : 1. Work done per kg of air, and 2. Efficiency of the cycle.*

Assume R = 0.287 kJ/kg K.

Solution. Given : $T_1 = T_4 = 45°\text{ C} = 45 + 273 = 318\text{ K}$; $T_2 = T_3 = 230°\text{ C} = 230 + 273 = 503\text{ K}$; $r = 2$; $R = 0.287$ kJ/kg K

1. *Workdone per kg of air*

We know that workdone per kg of air,

$$= 2.3\, m\, R \log r\,(T_2 - T_4)$$

$$= 2.3 \times 1 \times 0.287 \log 2\,(503 - 318) = 36.76 \text{ kJ/kg} \quad \textbf{Ans.}$$

2. *Efficiency of the cycle*

We know that efficiency of the cycle,

$$\eta = 1 - \frac{T_1}{T_3} = 1 - \frac{318}{503} = 0.368 \text{ or } 36.8\ \% \quad \textbf{Ans.}$$

6.15. Joule's Cycle

It consists of two constant pressure and two reversible adiabatic or isentropic processes as shown on *p-v* and *T-S* diagrams in Fig. 6.8 (*a*) and (*b*).

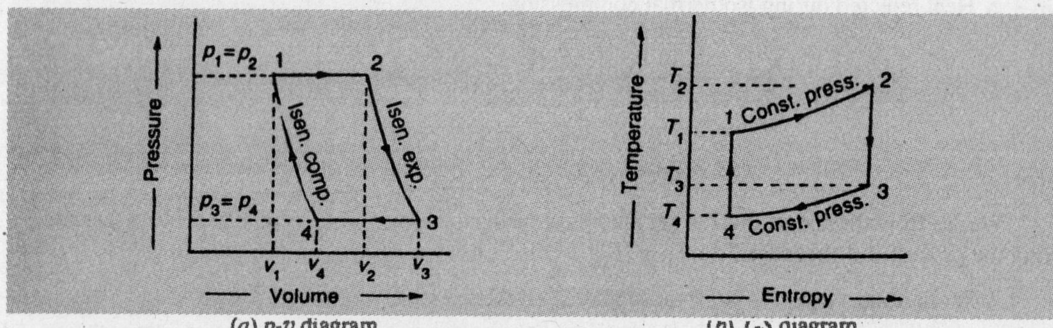

(*a*) *p-v* diagram. (*b*) *T-S* diagram.

Fig. 6.8. Joule's cycle.

Now, let us consider the four stages of the Joule's cycle. Let the engine cylinder contain *m* kg of air at its original condition represented by point 1 on *p-v* and *T-S* diagram. At this point, let p_1, T_1 and v_1 be the pressure, temperature and volume of the air.

1. *First stage (Constant pressure heating).* The air is heated at a constant pressure from initial temperature T_1 to a temperature T_2 represented by the curve 1-2 in Fig. 6.8 (*a*) and (*b*).

∴ Heat supplied to the air,

$$Q_{1-2} = m\, c_p\,(T_2 - T_1)$$

2. *Second stage (Reversible adiabatic or isentropic expansion)* The air is allowed to expand reversibly and adiabatically from v_2 to v_3. The reversible adiabatic expansion is represented by the curve 2-3 in Fig. 6.8 (*a*) and (*b*). The temperature of the air falls from T_2 to T_3. In this process, no heat is absorbed or rejected by the air.

3. *Third stage (Constant pressure cooling).* The air is now cooled at constant pressure from temperature T_3 to a temperature T_4 represented by 3-4 in Fig. 6.8 (*a*) and (*b*).

$\therefore$ Heat rejected by the air,

$$Q_{3-4} = m\,c_p\,(T_3 - T_4)$$

4. *Fourth stage (Reversible adiabatic or isentropic compression).* The air is now compressed reversibly and adiabatically from v_4 to v_1. The reversible adiabatic compression is represented by the curve 4-1 in Fig. 6.8 (*a*) and (*b*). The temperature of the air increases from T_4 to T_1. Again, no heat is absorbed or rejected by the air.

We see, from above, that there is no interchange of heat during the two reversible adiabatic processes. The only interchange of heat takes place during constant pressure processes.

$$\therefore \text{ Work done} = \text{Heat supplied} - \text{Heat rejected} = m\,c_p\,(T_2 - T_1) - m\,c_p\,(T_3 - T_4)$$

and efficiency,

$$\eta = \frac{\text{Work done}}{\text{Heat supplied}} = \frac{m\,c_p\,(T_2 - T_1) - m\,c_p\,(T_3 - T_4)}{m\,c_p\,(T_2 - T_1)}$$

$$= 1 - \frac{T_3 - T_4}{T_2 - T_1} = 1 - \frac{T_3\left(1 - \dfrac{T_4}{T_3}\right)}{T_2\left(1 - \dfrac{T_1}{T_2}\right)} \qquad \ldots (i)$$

We know that for reversible adiabatic or isentropic expansion 2-3,

$$\frac{T_3}{T_2} = \left(\frac{v_2}{v_3}\right)^{\gamma - 1} = \left(\frac{p_3}{p_2}\right)^{\frac{\gamma - 1}{\gamma}} \qquad \ldots (ii)$$

and for reversible adiabatic or isentropic compression 4-1,

$$\frac{T_4}{T_1} = \left(\frac{v_1}{v_4}\right)^{\gamma - 1} = \left(\frac{p_4}{p_1}\right)^{\frac{\gamma - 1}{\gamma}} \qquad \ldots (iii)$$

From equations (*ii*) and (*iii*), we find that

$$\frac{T_3}{T_2} = \frac{T_4}{T_1} \quad \text{or} \quad \frac{T_4}{T_3} = \frac{T_1}{T_2} \qquad \ldots (\because p_1 = p_2 \text{ and } p_3 = p_4)$$

Substituting the value of T_4 / T_3 in equation (*i*),

$$\eta = 1 - \frac{T_3}{T_2} = 1 - \frac{T_4}{T_1} \qquad \ldots \left(\because \frac{T_3}{T_2} = \frac{T_4}{T_1}\right)$$

From equations (*ii*) and (*iii*), we find that

$$\left(\frac{v_2}{v_3}\right)^{\gamma - 1} = \left(\frac{v_1}{v_4}\right)^{\gamma - 1} \qquad \ldots \left(\because \frac{p_3}{p_2} = \frac{p_4}{p_1}\right)$$

or $$\frac{v_2}{v_3} = \frac{v_1}{v_4} = \frac{1}{r} \qquad \ldots \left(\because r = \frac{v_3}{v_2} = \frac{v_4}{v_1} \right)$$

$$\therefore \quad \frac{T_3}{T_2} = \frac{T_4}{T_1} = \left(\frac{1}{r} \right)^{\gamma - 1} = \frac{1}{(r)^{\gamma - 1}}$$

and $$\eta = 1 - \frac{1}{(r)^{\gamma - 1}}$$

Notes : 1. The efficiency of the Joule's cycle is lower than Carnot efficiency. The reason is that all the heat is not taken in at the highest temperature and rejected at the lowest temperature.

2. The cycle is not thermodynamically reversible, because there is no regenerator to provide a constant temperature during heating and cooling at constant pressure.

3. The reversed Joule cycle is known as *Bell-coleman cycle or Brayton cycle and is applied to refrigerators, where air is used as a refrigerant.

Example 6.8. *An air motor works on Joule's cycle between 5 bar and 1 bar. The temperature at the beginning of isentropic expansion is 773 K and at the beginning of isentropic compression is 293 K. Determine the work done per kg of air and ideal efficiency. Assume c_p = 1 kJ/kg K and γ = 1.4.*

Solution. Given : $p_1 = p_2 = 5$ bar ; $p_3 = p_4 = 1$ bar ; $T_2 = 773$ K ; $T_4 = 293$ K ; $c_p = 1$ kJ/kg K ; $\gamma = 1.4$

Work done per kg of air

We know that for isentropic expansion process 2-3 (Refer Fig. 6.8),

$$\frac{T_3}{T_2} = \left(\frac{p_3}{p_2} \right)^{\frac{\gamma - 1}{\gamma}} = \left(\frac{1}{5} \right)^{\frac{1.4 - 1}{1.4}} = (0.2)^{0.286} = 0.6311$$

$$\therefore \quad T_3 = T_2 \times 0.6311 = 773 \times 0.6311 = 488 \text{ K}$$

Similarly, for isentropic compression process 4-1,

$$\frac{T_4}{T_1} = \left(\frac{p_4}{p_1} \right)^{\frac{\gamma - 1}{\gamma}} = \left(\frac{1}{5} \right)^{\frac{1.4 - 1}{1.4}} = (0.2)^{0.286} = 0.6311$$

$$\therefore \quad T_1 = T_4 / 0.6311 = 293 / 0.6311 = 464 \text{ K}$$

We know that heat supplied

$$= m\, c_p (T_2 - T_1) = 1 \times 1\,(773 - 464) = 309 \text{ kJ}$$

and heat rejected $$= m\, c_p (T_3 - T_4) = 1 \times 1\,(488 - 293) = 195 \text{ kJ}$$

$\therefore$ Workdone = Heat supplied – Heat rejected = 309 – 195 = 114 kJ **Ans.**

Ideal efficiency

We know that ideal efficiency,

$$\eta = \frac{\text{Workdone}}{\text{Heat supplied}} = \frac{114}{309} = 0.369 \text{ or } 36.9\%$$ **Ans.**

Example 6.9. *A hot air engine works on Brayton cycle with initial and final pressures of air as 3 bar and 1 bar respectively. If the temperature before isentropic compression and isentropic*

* Please refer Chapter 35 (Art. 35.10).

expansion are 298 K and 923 K, determine : 1. heat supplied per kg of air ; 2. heat rejected per kg of air ; 3. work done per kg of air ; and 4. efficiency of the engine. Take c_p as 1 kJ/kg K, and c_v = 0.715 kJ/kg K.

Solution. Given : $p_1 = p_2 = 3$ bar ; $p_3 = p_4 = 1$ bar ; $T_4 = 298$ K ; $T_2 = 923$ K ; $c_p = 1$ kJ/kg K ; $c_v = 0.715$ kJ/kg K

We know that the ratio of specific heats or isentropic index,

$$\gamma = c_p / c_v = 1 / 0.715 = 1.4$$

Heat supplied per kg of air

We know that in isentropic expansion process 4-1 (Refer Fig. 6.8),

$$\frac{T_1}{T_4} = \left(\frac{p_1}{p_4}\right)^{\frac{\gamma-1}{\gamma}} = \left(\frac{3}{1}\right)^{\frac{1.4-1}{1.4}} = (3)^{0.286} = 1.35$$

$$\therefore \quad T_1 = T_4 \times 1.35 = 298 \times 1.35 = 402 \text{ K}$$

and heat supplied per kg of air during constant pressure process 1-2,

$$Q_{1-2} = m\,c_p\,(T_2 - T_1) = 1 \times 1\,(923 - 402) = 521 \text{ kJ} \quad \textbf{Ans.}$$

Heat rejected per kg of air

We know that in isentropic compression process 2-3,

$$\frac{T_2}{T_3} = \left(\frac{p_2}{p_3}\right)^{\frac{\gamma-1}{\gamma}} = \left(\frac{3}{1}\right)^{\frac{1.4-1}{1.4}} = (3)^{0.286} = 1.35$$

$$\therefore \quad T_3 = T_2 / 1.35 = 923 / 1.35 = 684 \text{ K}$$

and heat rejected per kg of air during constant pressure process 3-4,

$$Q_{3-4} = m\,c_p\,(T_3 - T_4) = 1 \times 1\,(684 - 298) = 386 \text{ kJ} \quad \textbf{Ans.}$$

Work done per kg of air

We know that work done per kg of air,

$$W = Q_{1-2} - Q_{3-4} = 521 - 386 = 135 \text{ kJ} \quad \textbf{Ans.}$$

Efficiency of the engine

We know that efficiency of the engine,

$$\eta = \frac{\text{Work done}}{\text{Heat supplied}} = \frac{135}{521} = 0.259 \text{ or } 25.9\ \% \quad \textbf{Ans.}$$

6.16. Otto Cycle

The first successful engine working on this *cycle was built by A. Otto. These days, many gas, petrol and many of the oil engines run on this cycle. It is the also known as *constant volume cycle*, as the heat is received and rejected at a constant volume.

This cycle is taken as a standard of comparison for internal combustion engines. For the purpose of comparison with other cycles, the air is assumed to be the working substance.

The engine conceived by Otto has air enclosed in a cylinder, whose walls are perfectly non-conductor of heat, but the bottom is a perfect conductor of heat. There is also a hot body and cold

* This cycle was originally devised by a Frenchman Beau-de-Rochas in 1862. The first successful engine, working on this cycle, was built by a German engineer Nicholas A. Otto in 1876.

body and an insulating cap, which are alternately brought in contact with the bottom of the cylinder (*i.e.* a cylinder similar to that of Carnot).

The ideal Otto cycle consists of two constant volume and two reversible adiabatic or isentropic processes as shown on *p-v* and *T-S* diagrams in Fig. 6.9 (*a*) and (*b*).

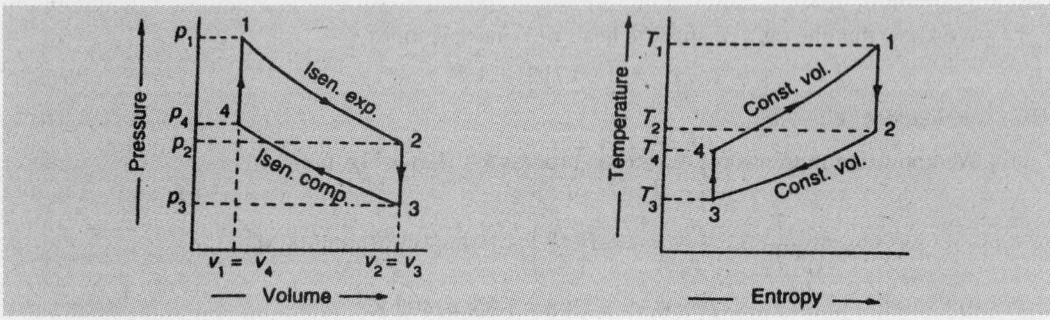

(*a*) *p-v* diagram. (*b*) *T-S* diagram.

Fig. 6.9. Otto cycle.

Let the engine cylinder contain *m* kg of air at point 1. At this point, let p_1, T_1 and v_1 be the pressure, temperature and volume of the air. Following are the four stages of the ideal cycle :

1. *First stage (Reversible adiabatic or isentropic expansion).* The air is expanded reversibly and adiabatically from initial temperature T_1 to a temperature T_2 as shown by the curve 1-2 in Fig. 6.9 (*a*) and (*b*). In this process, no heat is absorbed or rejected by the air.

2. *Second stage (Constant volume cooling).* The air is cooled at constant volume from temperature T_2 to a temperature T_3 as shown by the curve 2-3 in Fig. 6.9 (*a*) and (*b*). We know that heat rejected by the air during this process.

$$Q_{2-3} = m\,c_v\,(T_2 - T_3)$$

3. *Third stage (Reversible adiabatic or isentropic compression).* The air is compressed reversibly and adiabatically from temperature T_3 to a temperature T_4 as shown in by the curve 3-4 in Fig. 6.9 (*a*) and (*b*). In this process, no heat is absorbed or rejected by the air.

4. *Fourth stage (Constant volume heating).* The air is now heated at constant volume from temperature T_4 to a temperature T_1 as shown by the curve 4-1 in Fig. 6.9 (*a*) and (*b*). We know that heat absorbed by the air during this process,

$$Q_{4-1} = m\,c_v\,(T_1 - T_4)$$

We see that the air has been brought back to its original conditions of pressure, volume and temperature, thus completing the cycle.

We know that work done = Heat absorbed – Heat rejected

$$= m\,c_v\,(T_1 - T_4) - m\,c_v\,(T_2 - T_3)$$

∴ Ideal efficiency or air standard efficiency,

$$\eta = \frac{\text{Work done}}{\text{Heat absorbed}}$$

$$= \frac{m\,c_v\,(T_1 - T_4) - m\,c_v\,(T_2 - T_3)}{m\,c_v\,(T_1 - T_4)}$$

$$= 1 - \frac{T_2 - T_3}{T_1 - T_4} = 1 - \frac{T_3\left(\frac{T_2}{T_3} - 1\right)}{T_4\left(\frac{T_1}{T_4} - 1\right)} \quad \ldots (i)$$

We know that for reversible adiabatic or isentropic expansion process 1-2,

$$\frac{T_2}{T_1} = \left(\frac{v_1}{v_2}\right)^{\gamma-1} = \left(\frac{1}{r}\right)^{\gamma-1} \quad \ldots (ii)$$

where r = Expansion ratio = v_2 / v_1

Similarly, for reversible adiabatic or isentropic compression process 3-4,

$$\frac{T_3}{T_4} = \left(\frac{v_4}{v_3}\right)^{\gamma-1} = \left(\frac{1}{r}\right)^{\gamma-1} \quad \ldots (iii)$$

where r = Compression ratio = $v_3 / v_4 = v_2 / v_1$

From equations (*ii*) and (*iii*), we find that

$$\frac{T_2}{T_1} = \frac{T_3}{T_4} = \left(\frac{1}{r}\right)^{\gamma-1} = \frac{1}{(r)^{\gamma-1}} \quad \text{or} \quad \frac{T_1}{T_4} = \frac{T_2}{T_3}$$

Substituting the value of T_1 / T_4 in equation (*i*),

$$\eta = 1 - \frac{T_3}{T_4} = 1 - \frac{T_2}{T_1} = 1 - \frac{1}{(r)^{\gamma-1}} \quad \ldots \left(\because \frac{T_3}{T_4} = \frac{T_2}{T_1}\right) \ldots (iv)$$

Notes : 1. We see from equation (*iv*) that the efficiency of Otto cycle depends on compression ratio (r) only.

2. The efficiency increases with the increase in compression ratio (r). In actual practice, r can not be increased beyond a value of 7 or so.

3. Compression ratio,

$$r = \frac{\text{Total cylinder volume}}{\text{Clearance volume}}$$

$$= \frac{\text{Clearance volume + Stroke volume}}{\text{Clearance volume}} = \frac{v_c + v_s}{v_c}$$

$$\therefore \text{ Clearance volume, } v_c = \frac{\text{Stroke volume}}{r-1} = \frac{v_s}{r-1}$$

4. The relations between pressure and temperature or pressure and volume may be obtained from the usual reversible adiabatic or isentropic processes, *i.e.*

$$\frac{T_1}{T_2} = \left(\frac{p_1}{p_2}\right)^{\frac{\gamma-1}{\gamma}} \quad \text{and} \quad \frac{p_1}{p_2} = \left(\frac{v_2}{v_1}\right)^{\gamma} \quad \text{or} \quad \frac{v_1}{v_2} = \left(\frac{p_2}{p_1}\right)^{\frac{1}{\gamma}}$$

Example 6.10. *In an Otto cycle, the temperature at the beginning and end of the isentropic compression are 316 K and 596 K respectively. Determine the air standard efficiency and the compression ratio. Take γ = 1.4.*

Solution. Given : T_3 = 316 K ; T_4 = 596 K ; γ = 1.4

Compression ratio

Let r = Compression ratio = v_3 / v_4

We know that for isentropic compression 3-4 (Refer Fig. 6.9),

$$\frac{T_3}{T_4} = \left(\frac{v_4}{v_3}\right)^{\gamma-1} = \left(\frac{1}{r}\right)^{1.4-1} = \frac{1}{(r)^{0.4}}$$

$$\therefore \quad (r)^{0.4} = \frac{T_4}{T_3} \text{ or } r = \left(\frac{T_4}{T_3}\right)^{\frac{1}{0.4}} = \left(\frac{596}{316}\right)^{2.5} = 4.885 \textbf{ Ans.}$$

Air standard efficiency

We know that air standard efficiency,

$$\eta = 1 - \frac{1}{(r)^{\gamma-1}} = 1 - \frac{1}{(4.885)^{1.4-1}} = 1 - \frac{1}{1.886}$$

$$= 1 - 0.53 = 0.47 \text{ or } 47\% \textbf{ Ans.}$$

Example 6.11. *An engine, working on the Otto cycle, has a cylinder diameter of 150 mm and a stroke of 225 mm. The clearance volume is 1.25×10^{-3} m³. Find the air standard efficiency of this engine. Take $\gamma = 1.4$.*

Solution. Given : $d = 150$ mm = 0.15 m ; $l = 225$ mm = 0.225 m ; $v_c = 1.25 \times 10^{-3}$ m³ ; $\gamma = 1.4$

We know that swept volume

$$v_s = \frac{\pi}{4} \times d^2 \times l = \frac{\pi}{4}(0.15)^2(0.225) = 3.976 \times 10^{-3} \text{ m}^3$$

$$\therefore \text{ Compression ratio, } r = \frac{v_c + v_s}{v_c} = \frac{1.25 \times 10^{-3} + 3.976 \times 10^{-3}}{1.25 \times 10^{-3}} = 4.18$$

We know that air standard efficiency

$$\eta = 1 - \frac{1}{(r)^{\gamma-1}} - 1 - \frac{1}{(4.18)^{1.4-1}} = 1 - \frac{1}{1.772}$$

$$= 1 - 0.564 = 0.436 \text{ or } 43.6\% \textbf{ Ans.}$$

Example 6.12. *A certain quantity of air at a pressure of 1 bar and temperature 70° C is compressed reversibly and adiabatically until the pressure is 7 bar in an Otto cycle engine. 460 kJ of heat per kg of air is now added at constant volume. Determine : 1. compression ratio of the engine, 2. temperature at the end of compression, and 3. temperature at the end of heat addition.*

Take for air, $c_p = 1$ kJ/kg K and $c_v = 0.707$ kJ/kg K.

Solution. Given : $p_3 = 1$ bar ; $T_3 = 70°$ C = 70 + 273 = 343 K ; $p_4 = 7$ bar ; $Q_{4-1} = 460$ kJ ; $m = 1$ kg ; $c_p = 1$ kJ/kg K ; $c_v = 0.707$ kJ/kg K

We know that ratio of specific heats,

$$\gamma = c_p / c_v = 1/0.707 = 1.41$$

1. *Compression ratio of the engine*

Let r = Compression ratio of the engine = v_3 / v_4.

We know that $p_3 v_3^{\gamma} = p_4 v_4^{\gamma}$

$$\therefore \quad \frac{v_3}{v_4} = \left(\frac{p_4}{p_3}\right)^{\frac{1}{\gamma}} \text{ or } r = \left(\frac{7}{1}\right)^{\frac{1}{1.41}} = (7)^{0.709} = 3.97 \textbf{ Ans.}$$

2. *Temperature at the end of compression*

Let T_4 = Temperature at the end of compression.

We know that $\frac{T_3}{T_4} = \left(\frac{v_4}{v_3}\right)^{\gamma-1} = \left(\frac{1}{r}\right)^{\gamma-1} = \left(\frac{1}{3.97}\right)^{1.41-1} = (0.252)^{0.41} = 0.568$

$\therefore \quad T_4 = T_3 / 0.568 = 343 / 0.568 = 604 \text{ K} = 331° \text{C}$ **Ans.**

3. *Temperature at the end of heat addition*

Let T_1 = Temperature at the end of heat addition.

We know that heat added at constant volume (Q_{4-1}),

$$460 = m\,c_v (T_1 - T_4) = 1 \times 0.707 (T_1 - 604) \text{ kJ}$$

$\therefore \quad T_1 - 604 = 460 / 0.707 = 651$ or $T_1 = 1255 \text{ K} = 982° \text{C}$ **Ans.**

Example 6.13. *In an Otto cycle, air at 1 bar and 290 K is compressed isentropically until the pressure is 15 bar. The heat is added at constant volume until the pressure rises to 40 bar. Calculate the air standard efficiency and the mean effective pressure for the cycle. Take c_v = 0.717 kJ/kg K and R_u = 8.314 kJ/kg mole K.*

Solution. Given : p_3 = 1 bar ; T_3 = 290 K ; p_4 = 15 bar ; p_1 = 40 bar ; c_v = 0.717 kJ/kg K ; R_u = 8.314 kJ/kg mole K

We know that characteristic gas constant,

$$R = \frac{\text{Universal gas constant } (R_u)}{\text{Molecular mass } (M)} = \frac{8.314}{28.97} = 0.287 \text{ kJ/kg K}$$

... ($\because$ *M* for air = 28.97 kg)

and $c_p = R + c_v = 0.287 + 0.717 = 1.004 \text{ kJ/kg K}$

$\therefore \quad \gamma = c_p / c_v = 1.004 / 0.717 = 1.4$

Air standard efficiency

We know that for isentropic compression process 3-4 (Refer Fig. 6.9).

$$p_3 v_3^\gamma = p_4 v_4^\gamma \quad \text{or} \quad \frac{v_3}{v_4} = \left(\frac{p_4}{p_3}\right)^{\frac{1}{\gamma}}$$

$\therefore$ Compression ratio, $r = \frac{v_3}{v_4} = \left(\frac{p_4}{p_3}\right)^{\frac{1}{\gamma}} = \left(\frac{15}{1}\right)^{\frac{1}{1.4}} = (15)^{0.714} = 6.914$

We know that air standard efficiency,

$$\eta = 1 - \frac{1}{(r)^{\gamma-1}} = 1 - \frac{1}{(6.914)^{1.4-1}} = 1 - \frac{1}{2.167}$$

$$= 1 - 0.4615 = 0.5385 \text{ or } 53.85\ \%$$ **Ans.**

Mean effective pressure

First of all, let us find the workdone during the cycle per kg of air.

Let T_4 = Temperature at the end of isentropic compression, and

T_1 = Temperature at the end of constant volume heating.

We know that $\frac{T_3}{T_4} = \left(\frac{v_4}{v_3}\right)^{\gamma-1} = \left(\frac{1}{r}\right)^{\gamma-1} = \left(\frac{1}{6.914}\right)^{1.4-1} = \frac{1}{2.167}$

$\therefore \quad T_4 = T_3 \times 2.167 = 290 \times 2.167 = 628.5 \text{ K}$

Now for constant volume heating process 4-1,

$$\frac{p_4}{T_4} = \frac{p_1}{T_1} \quad \text{or} \quad T_1 = T_4 \times \frac{p_1}{p_4} = 628.5 \times \frac{40}{15} = 1676 \text{ K}$$

We know that heat supplied,

$$Q_{4-1} = m\,c_v\,(T_1 - T_4) = 1 \times 0.717\,(1676 - 628.5) = 751 \text{ kJ}$$

$\therefore$ Workdone during the cycle

$$= \eta \times \text{Heat supplied} = 0.5385 \times 751 = 404.4 \text{ kJ}$$

Now let us find the stroke volume (*i.e.* $v_3 - v_4$ or $v_2 - v_1$). We know that

$$p_3\,v_3 = m\,R\,T_3 \quad \text{or} \quad v_3 = \frac{m\,R\,T_3}{p_3} = \frac{1 \times 287 \times 290}{0.1 \times 10^6} = 0.8323 \text{ m}^3$$

... (R is taken in J/kg K and p_3 in N/m^2)

$\therefore \quad v_4 = v_3 / 6.914 = 0.8323 / 6.914 = 0.1204 \text{ m}^3$

... ($\because r = v_3/v_4 = 6.914$)

and stroke volume $= v_3 - v_4 = 0.8323 - 0.1204 = 0.7119 \text{ m}^3$

We know that mean effective pressure

$$= \frac{\text{Work done}}{\text{Stroke volume}} = \frac{404.4}{0.7119} = 568 \text{ kN/m}^2 = 5.68 \text{ bar } \textbf{Ans.}$$

Example 6.14. *In an air standard Otto cycle, the compression ratio is 7 and the compression begins at 1 bar and 313 K. The heat added is 2510 kJ/kg. Find : 1. maximum temperature and pressure of the cycle ; 2. workdone per kg of air ; 3. cycle efficiency ; and 4. mean effective pressure. Take for air, c_v = 0.713 kJ/kg K and R = 287 J/kg K.*

Solution. Given : $r = v_3 / v_4 = v_2 / v_1 = 7$; p_3 = 1 bar = 0.1×10^6 N/m^2 ; T_3 = 313 K ; Q_{4-1} = 2510 kJ/kg ; c_v = 0.713 kJ/kg K ; R = 287 J/kg K = 0.287 kJ/kg K

We know that $c_p - c_v = R$ or $c_p = R + c_v = 0.287 + 0.713 = 1$ kJ/kg K

and $\gamma = c_p / c_v = 1 / 0.713 = 1.4$

1. *Maximum temperature and pressure of the cycle*

Let T_1 = Maximum temperature, and

p_1 = Maximum pressure.

We know that for isentropic compression process 3-4 (Refer Fig. 6.9),

$$\frac{T_3}{T_4} = \left(\frac{v_4}{v_3}\right)^{\gamma-1} = \left(\frac{1}{r}\right)^{\gamma-1} = \left(\frac{1}{7}\right)^{1.4-1} = \frac{1}{7^{0.4}} = \frac{1}{2.178}$$

$$\therefore \quad T_4 = T_3 \times 2.178 = 313 \times 2.178 = 682 \text{ K}$$

and $$p_3 v_3^\gamma = p_4 v_4^\gamma \text{ or } p_4 = p_3 \left(\frac{v_3}{v_4}\right)^\gamma = 1 \times 7^{1.4} = 15.24 \text{ bar}$$

We know that heat supplied during constant volume process 4-1 (Q_{4-1}),

$$2510 = m\,c_v (T_1 - T_4) = 1 \times 0.713\,(T_1 - 682) \qquad \ldots (\because m = 1 \text{ kg})$$

$$\therefore \quad T_1 - 682 = 2510 / 0.713 = 3520 \text{ or } T_1 = 4202 \text{ K } \textbf{Ans.}$$

We also know that $$\frac{p_4}{T_4} = \frac{p_1}{T_1} \text{ or } p_1 = p_4 \times \frac{T_1}{T_4} = 15.24 \times \frac{4202}{682} = 93.9 \text{ bar } \textbf{Ans.}$$

2. *Workdone per kg of air*

We know that for isentropic expansion process 1-2,

$$\frac{T_2}{T_1} = \left(\frac{v_1}{v_2}\right)^{\gamma - 1} = \left(\frac{1}{r}\right)^{\gamma - 1} = \left(\frac{1}{7}\right)^{1.4 - 1} = \frac{1}{7^{0.4}} = \frac{1}{2.178}$$

$$\therefore \quad T_2 = T_1 / 2.178 = 4202 / 2.178 = 1929 \text{ K}$$

and heat rejected per kg of air,

$$Q_{2-3} = m\,c_v\,(T_2 - T_3) = 1 \times 0.713\,(1929 - 313) = 1152 \text{ kJ}$$

$\therefore$ Workdone per kg of air,

$$w = \text{Heat supplied} - \text{Heat rejected} = Q_{4-1} - Q_{2-3}$$
$$= 2510 - 1152 = 1358 \text{ kJ } \textbf{Ans.}$$

3. *Cycle efficiency*

We know that cycle efficiency,

$$\eta = \frac{\text{Workdone}}{\text{Heat supplied}} = \frac{1358}{2510} = 0.541 \text{ or } 54.1\% \textbf{ Ans.}$$

Note. The cycle efficiency may also be calculated as follows :

We know that $$\eta = 1 - \frac{1}{(r)^{\gamma - 1}} = 1 - \frac{1}{(7)^{1.4-1}} = 1 - \frac{1}{2.178} = 0.541 \text{ or } 54.1\% \textbf{ Ans.}$$

4. *Mean effective pressure*

First of all, let us find the stroke volume (*i.e.* $v_3 - v_4$ or $v_2 - v_1$).

We know that

$$p_3 v_3 = m\,R\,T_3 \text{ or } v_3 = \frac{m\,R\,T_3}{p_3} = \frac{1 \times 287 \times 313}{0.1 \times 10^6} = 0.9 \text{ m}^3$$

and $$v_4 = v_3 / r = 0.9 / 7 = 0.1286 \text{ m}^3 \qquad \ldots (\because v_3 / v_4 = r)$$

We know that stroke volume,

$$v_3 - v_4 = 0.9 - 0.1286 = 0.7714 \text{ m}^3$$

$\therefore$ Mean effective pressure

$$= \frac{\text{Workdone}}{\text{Stroke volume}} = \frac{1358}{0.7714} = 1760 \text{ kN/m}^2 = 17.6 \text{ bar } \textbf{Ans.}$$

6.17. Diesel Cycle

This cycle was devised by Dr. Rudolph Diesel in 1893, with an idea to attain a higher thermal efficiency, with a high compression ratio. This is an important cycle on which all the diesel engines work. It is also known as *constant pressure cycle* as heat is received at a constant pressure.

The engine imagined by Diesel has air enclosed in the cylinder, whose walls are perfectly non-conductor of heat, but bottom is a perfect conductor of *heat. Again, there is a hot body, cold body and an insulating cap, which are alternately brought in contact with the cylinder.

The ideal diesel cycle consists of two reversible adiabatic or isentropic, a constant pressure and a constant volume processes. These processes are represented on *p-v* and *T-S* diagrams as shown in Fig. 6.10 (*a*) and (*b*).

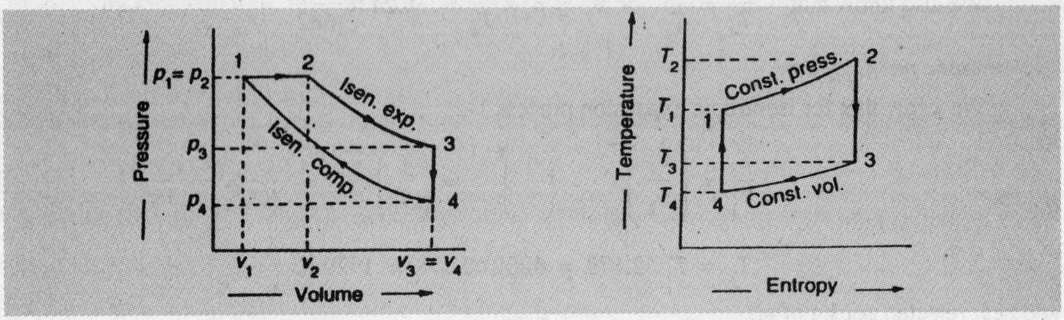

(*a*) *p-v* diagram. (*b*) *T-S* diagram.

Fig. 6.10. Diesel cycle.

Let the engine cylinder contain *m* kg of air at point 1. At this point let, p_1, T_1 and v_1 be the pressure, temperature and volume of the air. Following are four stages of an ideal diesel cycle.

1. *First stage (Constant pressure heating).* The air is heated at constant pressure from initial temperature T_1 to a temperature T_2 represented by the curve 1-2 in Fig. 6.10 (*a*) and (*b*).

∴ Heat supplied to the air,

$$Q_{1-2} = m\, c_p\,(T_2 - T_1)$$

Note : Since the supply of heat is cut off at point 2, therefore it is known as cut-off point.

2. *Second stage (Reversible adiabatic or isentropic expansion).* The air is expanded reversibly and adiabatically from temperature T_2 to a temperature T_3 as shown by the curve 2-3 in Fig. 6.10 (*a*) and (*b*). In this process, no heat is absorbed or rejected by the air.

3. *Third stage (Constant volume cooling).* The air is now cooled at constant volume from temperature T_3 to a temperature T_4 as shown by the curve 3-4 in Fig. 6.10 (*a*) and (*b*).

∴ Heat rejected by the air,

$$Q_{3-4} = m\, c_v\,(T_3 - T_4)$$

4. *Fourth stage (Reversible adiabatic or isentropic compression).* The air is compressed reversibly and adiabatically from temperature T_4 to a temperature T_1 represented by the curve 4-1 in Fig. 6.10 (*a*) and (*b*). In this process, no heat is absorbed or rejected by the air.

We see that the air has been brought back to its original conditions of pressure, volume and temperature, thus completing the cycle. We know that

---

* A similar type of cylinder was designed by Otto.

$$\text{Work done} = \text{Heat absorbed} - \text{Heat rejected}$$

$$= m\,c_p\,(T_2 - T_1) - m\,c_v\,(T_3 - T_4)$$

$\therefore$ Air standard efficiency,

$$\eta = \frac{\text{Work done}}{\text{Heat absorbed}} = \frac{m\,c_p\,(T_2 - T_1) - m\,c_v\,(T_3 - T_4)}{m\,c_p\,(T_2 - T_1)}$$

$$= 1 - \frac{c_v}{c_p}\left(\frac{T_3 - T_4}{T_2 - T_1}\right) = 1 - \frac{1}{\gamma}\left(\frac{T_3 - T_4}{T_2 - T_1}\right) \qquad \ldots (i)$$

Now let compression ratio,

$$r = \frac{v_4}{v_1}$$

Cut-off ratio, $$\rho = \frac{v_2}{v_1}$$

Expansion ratio, $$r_1 = \frac{v_3}{v_2} = \frac{v_4}{v_2} \qquad \ldots (\because v_3 = v_4)$$

$$= \frac{v_4}{v_1} \times \frac{v_1}{v_2} = r \times \frac{1}{\rho} = \frac{r}{\rho}$$

We know for constant pressure heating process 1-2,

$$\frac{v_1}{T_1} = \frac{v_2}{T_2} \qquad \ldots \text{(Charles's law)}$$

$$\therefore \qquad T_2 = T_1 \times \frac{v_2}{v_1} = T_1 \times \rho \qquad \ldots (ii)$$

Similarly, for reversible adiabatic or isentropic expansion process 2-3,

$$\frac{T_3}{T_2} = \left(\frac{v_2}{v_3}\right)^{\gamma-1} = \left(\frac{1}{r_1}\right)^{\gamma-1} = \left(\frac{\rho}{r}\right)^{\gamma-1}$$

$$\therefore \qquad T_3 = T_2\left(\frac{\rho}{r}\right)^{\gamma-1} = T_1 \times \rho\left(\frac{\rho}{r}\right)^{\gamma-1} \qquad \ldots (iii)$$

and for reversible adiabatic or isentropic compression process 4-1,

$$\frac{T_1}{T_4} = \left(\frac{v_4}{v_1}\right)^{\gamma-1} = (r)^{\gamma-1} \quad \text{or} \quad T_1 = T_4\,(r)^{\gamma-1} \qquad \ldots (iv)$$

Substituting the value of T_1 in equations (*ii*) and (*iii*),

$$T_2 = T_4\,(r)^{\gamma-1} \times \rho \qquad \ldots (v)$$

and $$T_3 = T_4\,(r)^{\gamma-1} \times \rho\left(\frac{\rho}{r}\right)^{\gamma-1} = T_4\,\rho^{\gamma} \qquad \ldots (vi)$$

Now substituting the values of T_1, T_2 and T_3 in equation (*i*),

$$\eta = 1 - \frac{1}{\gamma}\left[\frac{(T_4\,\rho^{\gamma}) - T_4}{T_4\,(r)^{\gamma-1}\,\rho - T_4\,(r)^{\gamma-1}}\right]$$

$$= 1 - \frac{1}{(r)^{\gamma-1}}\left[\frac{\rho^{\gamma}-1}{\gamma(\rho-1)}\right] \qquad \ldots (vii)$$

Notes : 1. The efficiency of the ideal Diesel cycle is lower than that of Otto cycle, for the same compression ratio. This is due to the fact that the cut-off ratio (ρ) is always greater than unity and hence the term within the bracket of equation (*vii*) increases with the increase of cut-off ratio. Thus the negative term increases and the efficiency is reduced.

2. The Diesel cycle efficiency increases with decrease in cut-off ratio and approaches maximum (equal to Otto cycle efficiency) when the term within the bracket is unity.

Example 6.15. *In a diesel engine, the compression ratio is 13 : 1 and the fuel is cut-off at 8% of the stroke. Find the air standard efficiency of the engine. Take γ for air as 1.4.*

Solution. Given : $r = v_4 / v_1 = 13$; $\gamma = 1.4$

Since the cut-off takes place at 8% of the stroke, therefore volume at cut-off,

$$v_2 = v_1 + 8\% \text{ of stroke volume} = v_1 + 0.08\,(v_4 - v_1)$$

Let us assume that the clearance volume (v_1) = 1 m^3.

$$\therefore \quad v_4 = 13 \text{ m}^3 \qquad \ldots (\because v_4 / v_1 = 13)$$

and stroke volume, $v_4 - v_1 = 13 - 1 = 12 \text{ m}^3$

$\therefore$ Volume at cut-off

$$v_2 = v_1 + 0.08\,(v_4 - v_1) = 1 + 0.08 \times 12 = 1.96 \text{ m}^3$$

We know that cut-off ratio,

$$\rho = v_2 / v_1 = 1.96 / 1 = 1.96$$

$\therefore$ Air standard efficiency,

$$\eta = 1 - \frac{1}{(r)^{\gamma-1}}\left[\frac{\rho^{\gamma-1}}{\gamma(\rho-1)}\right] = 1 - \frac{1}{(13)^{1.4-1}}\left[\frac{(1.96)^{1.4}-1}{1.4\,(1.96-1)}\right]$$

$$= 1 - 0.417 = 0.583 \text{ or } 58.3\% \textbf{ Ans.}$$

Example 6.16. *In an ideal Diesel cycle, the temperatures at the beginning and end of compression are 57° C and 603° C respectively. The temperatures at the beginning and end of expansion are 1950° C and 870° C respectively. Determine the ideal efficiency of the cycle. $\gamma = 1.4$.*

If the compression ratio is 14 and the pressure at the beginning of the compression is 1 bar, calculate the maximum pressure in the cycle.

Solution. Given : $T_4 = 57°\text{ C} = 57 + 273 = 330\text{ K}$; $T_1 = 603°\text{ C} = 603 + 273 = 876\text{ K}$; $T_2 = 1950°\text{ C} = 1950 + 273 = 2223\text{ K}$; $T_3 = 870°\text{ C} = 870 + 273 = 1143\text{ K}$; $\gamma = 1.4$; $r = v_4/v_1 = 14$; $p_4 = 1$ bar

Ideal efficiency of the cycle

We know that ideal efficiency of the cycle,

$$\eta = 1 - \frac{1}{\gamma}\left(\frac{T_3 - T_4}{T_2 - T_1}\right) = 1 - \frac{1}{1.4}\left(\frac{1143 - 330}{2223 - 876}\right)$$

$$= 1 - 0.431 = 0.569 \text{ or } 56.9\% \textbf{ Ans.}$$

Maximum pressure in the cycle

Let p_1 = Maximum pressure in the cycle.

We know that for reversible adiabatic compression,

$$p_1 v_1^\gamma = p_4 v_4^\gamma \text{ or } p_1 = p_4\left(\frac{v_4}{v_1}\right)^\gamma = 1\,(14)^{1.4} = 40.23 \text{ bar } \textbf{Ans.}$$

Example 6.17. *An ideal Diesel engine has a diameter 150 mm and stroke 200 mm. The clearance volume is 10 per cent of the swept volume. Determine the compression ratio and the air standard efficiency of the engine if the cut-off takes place at 6 per cent of the stroke.*

Solution. Given : d = 150 mm = 0.15 m ; l = 200 mm = 0.2 m ; v_c = 10% of v_s = 0.1 v_s

Compression ratio

We know that stroke volume,

$$v_s = \frac{\pi}{4} \times d^2 \times l = \frac{\pi}{4}(0.15)^2\, 0.2 = 3.53 \times 10^{-3} \text{ m}^3$$

$$\therefore \quad v_c = 0.1\, v_s = 0.1 \times 3.53 \times 10^{-3} = 0.353 \times 10^{-3} \text{ m}^3$$

We know that compression ratio,

$$r = \frac{\text{Total volume}}{\text{Clearance volume}} = \frac{v_c + v_s}{v_c} = \frac{0.353 \times 10^{-3} + 3.53 \times 10^{-3}}{0.353 \times 10^{-3}}$$

$$= 11 \textbf{ Ans.}$$

Air standard efficiency

Since the cut-off takes place at 6% of the stroke, therefore volume at cut-off,

$$v_2 = v_1 + 0.06\, v_s = v_c + 0.06\, v_s \qquad \ldots (\because v_1 = v_c)$$

$$= 0.353 \times 10^{-3} + 0.06 \times 3.53 \times 10^{-3} = 0.565 \times 10^{-3} \text{ m}^3$$

$$\therefore \text{ Cut-off ratio,} \quad \rho = \frac{v_2}{v_1} = \frac{v_2}{v_c} = \frac{0.565 \times 10^{-3}}{0.353 \times 10^{-3}} = 1.6$$

We know that air standard efficiency,

$$\eta = 1 - \frac{1}{(r)^{\gamma-1}}\left[\frac{\rho^\gamma - 1}{\gamma(\rho - 1)}\right] = 1 - \frac{1}{(11)^{1.4-1}}\left[\frac{(1.6)^{1.4} - 1}{1.4\,(1.6 - 1)}\right]$$

$$= 1 - \frac{1}{2.61} \times 1.11 = 1 - 0.4246 = 0.5753 \text{ or } 57.53\% \textbf{ Ans.}$$

Example 6.18. *The compression ratio of an ideal air standard Diesel cycle is 15. The heat transfer is 1465 kJ/kg of air. Find the pressure and temperature at the end of each process and determine the cycle efficiency.*

What is the mean effective pressure of the cycle, if the inlet conditions are 300 K and 1 bar.

Solution. Given : $r = v_4 / v_1 = 15$; Q_{1-2} = 1465 kJ/kg ; T_4 = 300 K ; p_4 = 1 bar = 0.1 × 10^6 N/m^2

Pressure and temperature at the end of each process

The p-v and T-S diagram for the Diesel cycle is shown in Fig. 6.11.

Let p_1, p_2 and p_3 = Pressures at points 1, 2 and 3 respectively.

T_1, T_2 and T_3 = Temperature at points 1, 2 and 3 respectively.

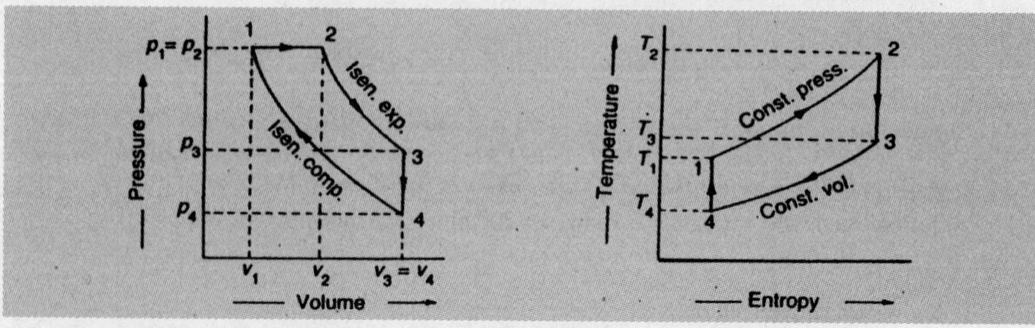

(a) p–v diagram. (b) T–S diagram.

Fig. 6.11

First of all, consider the isentropic compression process 4-1. We know that

$$p_4 v_4^{\gamma} = p_1 v_1^{\gamma} \text{ or } p_1 = p_4 \left(\frac{v_4}{v_1}\right)^{\gamma} = 1\,(15)^{1.4} = 44.3 \text{ bar } \textbf{Ans.}$$

and
$$\frac{T_1}{T_4} = \left(\frac{v_4}{v_1}\right)^{\gamma - 1} = (15)^{1.4-1} = (15)^{0.4} = 2.954$$

$\therefore$
$$T_1 = T_4 \times 2.954 = 300 \times 2.954 = 886.2 \text{ K } \textbf{Ans.}$$

Now consider the constant pressure process 1-2. We know that

$$p_2 = p_1 = 44.3 \text{ bar } \textbf{Ans.}$$

and heat supplied per kg of air during constant pressure process 1-2 (Q_{1-2}),

$$1465 = m\,c_p\,(T_2 - T_1) = 1 \times 1\,(T_2 - 886.2) = T_2 - 886.2$$

... (Taking c_p for air = 1 kJ/kg K)

$\therefore$
$$T_2 = 1465 + 886.2 = 2351.2 \text{ K } \textbf{Ans.}$$

Now consider the isentropic expansion process 2-3. First of all, let us find the volumes at points 2 and 3.

Let v_2 = Volume at point 2, and

v_3 = Volume at point 3.

We know that
$$p_4 v_4 = m\,R\,T_4 \text{ or } v_4 = \frac{m\,R\,T_4}{p_4} = \frac{1 \times 287 \times 300}{0.1 \times 10^6} = 0.861 \text{ m}^3$$

... ($\because$ R for air = 287 J/kg K)

$\therefore$
$$v_3 = 0.861 \text{ m}^3 \qquad \ldots (\because v_3 = v_4)$$

and
$$v_1 = v_4 / 15 = 0.861 / 15 = 0.0574 \text{ m}^3 \qquad \ldots (\because v_4 / v_1 = 15)$$

We also know that for the constant pressure process 1-2,

$$\frac{v_1}{T_1} = \frac{v_2}{T_2} \quad \text{or} \quad v_2 = v_1 \times \frac{T_2}{T_1} = 0.0574 \times \frac{2351.2}{886.2} = 0.1523 \text{ m}^3$$

Now $$\frac{T_2}{T_3} = \left(\frac{v_3}{v_2}\right)^{\gamma-1} = \left(\frac{0.861}{0.1523}\right)^{1.4-1} = (5.65)^{0.4} = 2$$

$\therefore$ $$T_3 = T_2 / 2 = 2351.2 / 2 = 1175.6 \text{ K } \textbf{Ans.}$$

We know that $$p_2 v_2^{\gamma} = p_3 v_3^{\gamma} \quad \text{or} \quad p_3 = p_2 \left(\frac{v_2}{v_3}\right)^{\gamma} = 44.3 \left(\frac{0.1523}{0.861}\right)^{1.4} = 3.92 \text{ bar } \textbf{Ans.}$$

Cycle efficiency

We know that heat rejected per kg of air during the constant volume process 3-4,

$$Q_{3-4} = m\,c_v\,(T_3 - T_4) = 1 \times 0.712\,(1175.6 - 300) = 623.4 \text{ kJ}$$

... (Taking c_v for air = 0.712 kJ/kg K)

$$\therefore \text{Cycle efficiency, } \eta = \frac{\text{Heat supplied} - \text{Heat rejected}}{\text{Heat supplied}} = \frac{1465 - 623.4}{1465}$$

$$= 0.5745 \quad \text{or} \quad 57.45\% \textbf{ Ans.}$$

Mean effective pressure

We know that workdone per kg of air during the cycle

$$= \text{Heat supplied} - \text{Heat rejected} = 1465 - 623.4 = 841.6 \text{ kJ}$$

and stroke volume $$= v_4 - v_1 = 0.861 - 0.0574 = 0.8036 \text{ m}^3$$

$\therefore$ Mean effective pressure

$$= \frac{\text{Workdone}}{\text{Stroke volume}} = \frac{841.6}{0.8036} = 1047.3 \text{ kN/m}^2$$

$$= 10.473 \text{ bar } \textbf{Ans.} \qquad \ldots (\because 1 \text{ bar} = 100 \text{ kN/m}^2)$$

Example 6.19. *Find the air standard efficiencies for the Otto and Diesel cycles on the basis of equal compression ratio of 10 and equal heat rejection of 840 kJ/kg. The suction conditions are 1 bar and 328 K.*

Solution. Given : $r = v_3 / v_4 = 10$; $Q_{2-3} = 840$ kJ/kg ; $p_3 = 1$ bar ; $T_3 = 328$ K

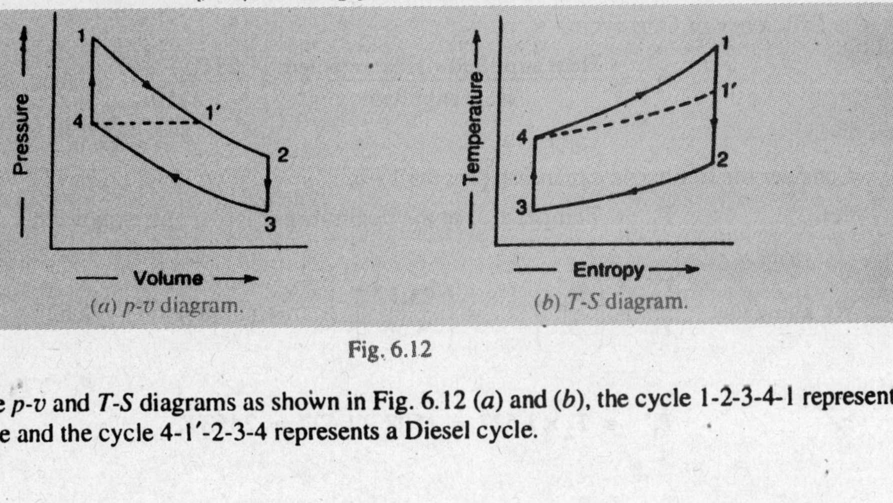

(*a*) *p-v* diagram. (*b*) *T-S* diagram.

Fig. 6.12

In the *p-v* and *T-S* diagrams as shown in Fig. 6.12 (*a*) and (*b*), the cycle 1-2-3-4-1 represents an Otto cycle and the cycle 4-1′-2-3-4 represents a Diesel cycle.

First of all, let us consider isentropic compression process 3-4, which is common for both the cycles.

Let T_4 and p_4 = Temperature and pressure at the end of isentropic compression.

We know that $$\frac{T_4}{T_3} = \left(\frac{v_3}{v_4}\right)^{\gamma-1} = (10)^{1.4-1} = (10)^{0.4} = 2.512$$

$$\therefore \quad T_4 = T_3 \times 2.512 = 328 \times 2.512 = 824 \text{ K}$$

We also know that

$$p_3 v_3^{\gamma} = p_4 v_4^{\gamma} \text{ or } p_4 = p_3 \left(\frac{v_3}{v_4}\right)^{\gamma} = 1\,(10)^{1.4} = 25.12 \text{ bar}$$

Now consider the constant volume process 2-3 which is also common for both the cycles. We know that heat rejected per kg of gas during this process (Q_{2-3}),

$$840 = m\,c_v\,(T_2 - T_3) = 1 \times 0.712\,(T_2 - 328)$$

... (Taking c_v for air = 0.712 kJ/kg K)

$$\therefore \quad T_2 - 328 = 840 / 0.712 = 1180 \text{ or } T_2 = 1180 + 328 = 1508 \text{ K}$$

and pressure at point 2, $p_2 = p_3 \times \dfrac{T_2}{T_3} = 1 \times \dfrac{1508}{328} = 4.6$ bar ... $\left(\because \dfrac{p_2}{T_2} = \dfrac{p_3}{T_3}\right)$

Otto cycle

Consider the isentropic expansion process 1-2.

Let T_1 = Temperature at the beginning of isentropic expansion.

We know that $$\frac{T_1}{T_2} = \left(\frac{v_2}{v_1}\right)^{\gamma-1} = \left(\frac{v_3}{v_4}\right)^{\gamma-1} = (10)^{1.4-1} = 10^{0.4} = 2.512$$

... ($\because v_1 = v_4$ and $v_2 = v_3$)

$$\therefore \quad T_1 = T_2 \times 2.512 = 1508 \times 2.512 = 3788 \text{ K}$$

and heat supplied per kg of air during constant volume process 4-1,

$$Q_{4-1} = m\,c_v\,(T_1 - T_4) = 1 \times 0.712\,(3788 - 824) = 2110 \text{ kJ/kg}$$

∴ Efficiency of Otto cycle,

$$\eta_O = \frac{\text{Heat supplied} - \text{Heat rejected}}{\text{Heat supplied}} = \frac{2110 - 840}{2110} = 0.60 \text{ or } 60\% \text{ Ans.}$$

Diesel cycle

Consider the isentropic expansion process 1′–2.

Let T_1' = Temperature at the beginning of isentropic expansion.

We know that $$\frac{T_1'}{T_2} = \left(\frac{p_1'}{p_2}\right)^{\frac{\gamma-1}{\gamma}} = \left(\frac{25.12}{4.58}\right)^{\frac{1.4-1}{1.4}} = (5.485)^{0.286} = 1.627$$

... ($\because p_1' = p_4 = 25.12$ bar)

$$\therefore \quad T_1' = T_2 \times 1.627 = 1508 \times 1.627 = 2453 \text{ K}$$

and heat supplied per kg of air during constant pressure process 4-1′,

$$Q_{4-1}' = m\,c_p\,(T_1' - T_4) = 1 \times 1\,(2453 - 824) = 1629 \text{ kJ/kg}$$

... (Taking c_p for air = 1 kg/kg K)

∴ Efficiency of the Diesel cycle,

$$\eta_D = \frac{\text{Heat supplied} - \text{Heat rejected}}{\text{Heat supplied}} = \frac{1629 - 840}{1629}$$

$$= 0.484 \text{ or } 48.4\ \% \text{ Ans.}$$

Example 6.20. *Two engines are to operate on Otto and Diesel cycles with the following data :*

Maximum temperature = 1500 K ; Exhaust temperature = 700 K ; Ambient conditions = 1 bar and 300 K.

Compare the compression ratios, maximum pressures, and efficiencies of the two engines.

Solution. Given : $T_1 = 1500$ K ; $T_2 = 700$ K ; $p_3 = 1$ bar ; $T_3 = 300$ K

First of all let us consider an Otto cycle as shown in Fig. 6.13.

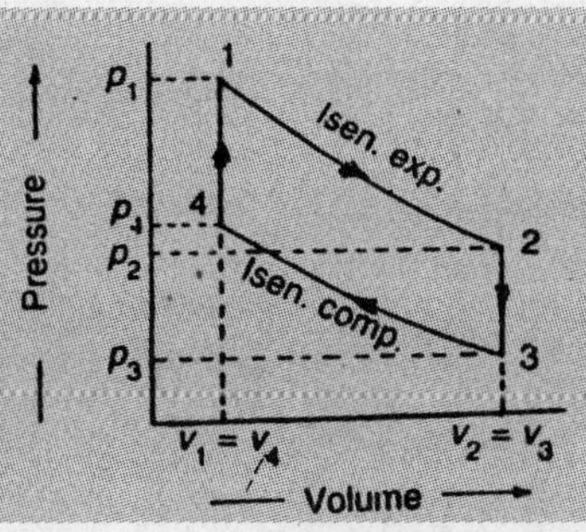

(*a*) *p-v* diagram.

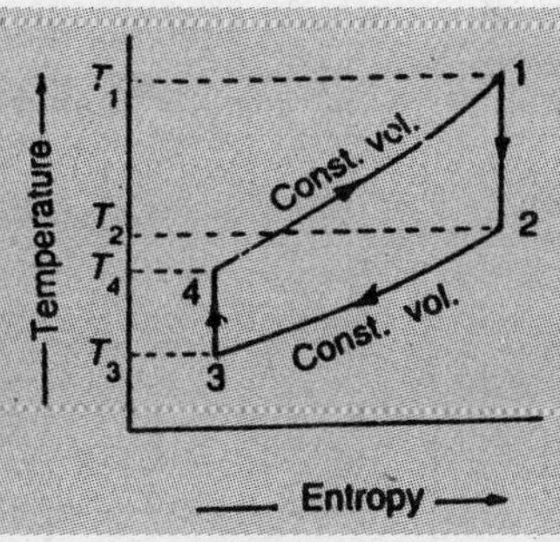

(*b*) *T-S* diagram.

Fig. 6.13. Otto cycle.

Let r = Compression ratio = $v_3 / v_4 = v_2 / v_1$, and

p_1 = Maximum pressure.

We know that for isentropic expansion process 1-2,

$$\frac{T_1}{T_2} = \left(\frac{v_2}{v_1}\right)^{\gamma-1} = r^{\gamma-1}$$

or

$$r = \left(\frac{T_1}{T_2}\right)^{\frac{1}{\gamma-1}} = \left(\frac{1500}{700}\right)^{\frac{1}{1.4-1}} = 6.72$$

Now for isentropic compression process 3-4,

$$\frac{T_4}{T_3} = \left(\frac{v_3}{v_4}\right)^{\gamma-1} = (r)^{\gamma-1} = (6.72)^{1.4-1} = 2.143$$

$$\therefore \quad T_4 = T_3 \times 2.143 = 300 \times 2.143 = 643 \text{ K}$$

and $$p_4 v_4^\gamma = p_3 v_3^\gamma \quad \text{or} \quad p_4 = p_3\left(\frac{v_3}{v_4}\right)^\gamma = p_3 \times r^\gamma = 1\,(6.72)^{1.4} = 14.4 \text{ bar}$$

For constant volume process 4-1,

$$\frac{p_1}{T_1} = \frac{p_4}{T_4} \quad \text{or} \quad p_1 = p_4 \times \frac{T_1}{T_4} = 14.4 \times \frac{1500}{643} = 33.6 \text{ bar}$$

We know that efficiency of Otto cycle,

$$\eta_O = 1 - \frac{1}{(r)^{\gamma-1}} = 1 - \frac{1}{(6.72)^{1.4-1}} = 1 - \frac{1}{2.143} = 1 - 0.467$$

$$= 0.533 \text{ or } 53.3\ \% \textbf{ Ans.}$$

Note. The efficiency may also be calculated as follows :

We know that $$\eta_O = 1 - \frac{T_2 - T_3}{T_1 - T_4} = 1 - \frac{700 - 300}{1500 - 643} = 1 - 0.467 = 0.533 \text{ or } 53.3\%$$

Now let us consider the Diesel cycle as shown in Fig. 6.14. In this case

$T_2 = 1500$ K ; $T_3 = 700$ K ; $T_4 = 300$ K ; $p_4 = 1$ bar

Let r = Compression ratio = v_4 / v_1, and

$p_1 = p_2$ = Maximum pressure.

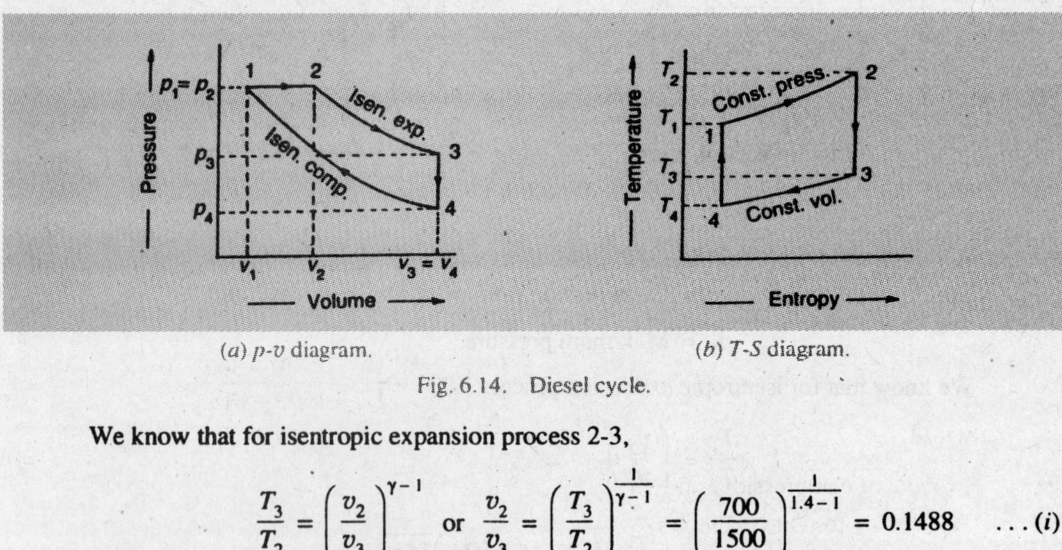

(a) p-v diagram. (b) T-S diagram.

Fig. 6.14. Diesel cycle.

We know that for isentropic expansion process 2-3,

$$\frac{T_3}{T_2} = \left(\frac{v_2}{v_3}\right)^{\gamma-1} \quad \text{or} \quad \frac{v_2}{v_3} = \left(\frac{T_3}{T_2}\right)^{\frac{1}{\gamma-1}} = \left(\frac{700}{1500}\right)^{\frac{1}{1.4-1}} = 0.1488 \quad \ldots (i)$$

Now for isentropic compression process 4-1,

$$\frac{T_1}{T_4} = \left(\frac{v_4}{v_1}\right)^{\gamma-1}$$

or $$T_1 = T_4\left(\frac{v_4}{v_1}\right)^{\gamma-1} = 300\left(\frac{v_4}{v_1}\right)^{1.4-1} = 300\left(\frac{v_4}{v_1}\right)^{0.4} \quad \ldots (ii)$$

and for constant pressure process 1-2,

$$\frac{v_1}{T_1} = \frac{v_2}{T_2} \quad \text{or} \quad T_1 = T_2 \times \frac{v_1}{v_2} = 1500 \times \frac{v_1}{v_2} \qquad \ldots (iii)$$

From equations (*ii*) and (*iii*),

$$300\left(\frac{v_4}{v_1}\right)^{0.4} = 1500 \times \frac{v_1}{v_2} \quad \text{or} \quad \left(\frac{v_4}{v_1}\right)^{0.4} \times \frac{v_2}{v_1} = \frac{1500}{300} \qquad \ldots (iv)$$

We know that
$$\frac{v_2}{v_1} = \frac{v_4}{v_1} \times \frac{v_2}{v_4} = \frac{v_4}{v_1} \times \frac{v_2}{v_3} \qquad \ldots (\because v_4 = v_3)$$

$$= \frac{v_4}{v_1} \times 0.1488 \qquad \ldots \text{[From equation } (i)\text{]}$$

Substituting the value of v_2/v_1 in equation (*iv*), we have

$$\left(\frac{v_4}{v_1}\right)^{0.4} \times \frac{v_4}{v_1} \times 0.1488 = \frac{1500}{300}$$

or
$$\left(\frac{v_4}{v_1}\right)^{1.4} = \frac{1500}{300} \times \frac{1}{0.1488} = 33.6$$

$$\therefore \quad \frac{v_4}{v_1} = (33.6)^{\frac{1}{1.4}} = 12.3 \quad \text{or} \quad r = 12.3$$

Now for isentropic compression process 4-1,

$$p_4 v_4^{\gamma} = p_1 v_1^{\gamma} \quad \text{or} \quad p_1 = p_4\left(\frac{v_4}{v_1}\right)^{\gamma} = 1\,(12.3)^{1.4} = 33.6 \text{ bar}$$

and from equation (*ii*), for isentropic compression process,

$$T_1 = 300\left(\frac{v_4}{v_1}\right)^{0.4} = 300\,(12.3)^{0.4} = 818.6 \text{ K}$$

We know that efficiency of Diesel cycle,

$$\eta_D = 1 - \frac{1}{\gamma}\left(\frac{T_3 - T_4}{T_3 - T_1}\right) = 1 - \frac{1}{1.4}\left(\frac{700 - 300}{1500 - 818.6}\right)$$

$$= 1 - 0.419 = 0.581 \quad \text{or} \quad 58.1\ \%$$

$\therefore$ Ratio of compression ratios,

$$\frac{r \text{ for Otto cycle}}{r \text{ for Diesel cycle}} = \frac{6.72}{12.3} = 0.546 \textbf{ Ans.}$$

Ratio of maximum pressures,

$$\frac{p_1 \text{ for Otto cycle}}{p_1 \text{ for Diesel cycle}} = \frac{33.6}{33.6} = 1 \textbf{ Ans.}$$

Ratio of efficiencies,

$$\frac{\eta \text{ for Otto cycle}}{\eta \text{ for Diesel cycle}} = \frac{0.533}{0.581} = 0.917 \textbf{ Ans.}$$

6.18. Dual Combustion Cycle

This cycle is a combination of Otto and Diesel cycles. It is sometimes called *semi-diesel cycle,* because semi-diesel engines work on this cycle. In this cycle, heat is absorbed partly at a constant volume and partly at a constant pressure.

The ideal dual combustion cycle consists of *two reversible adiabatic or isentropic , two constant volume* and *a constant pressure processes*. These processes are represented on *p-v* and *T-S* diagram as shown in Fig. 6.15 (*a*) and (*b*).

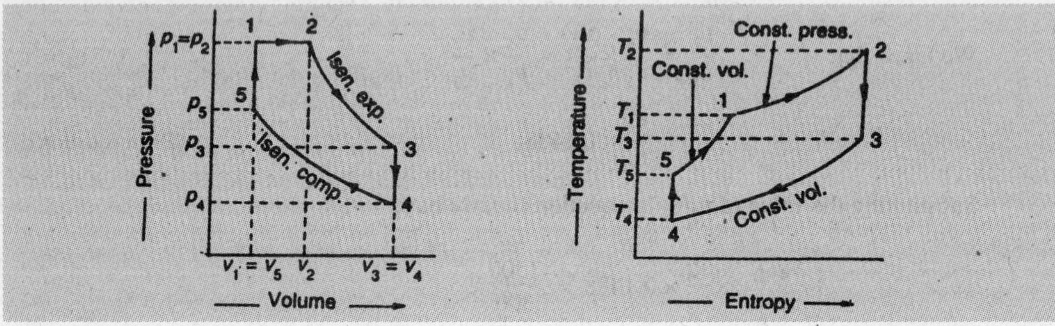

(*a*) *p-v* diagram. (*b*) *T-S* diagram.

Fig. 6.15. Dual combustion cycle.

Let the engine cylinder* contain *m* kg of air at point 1. At this point, let p_1, T_1 and v_1 be the pressure, temperature and volume of the air. Following are the five stages of an ideal dual combustion cycle.

1. *First stage (Constant pressure heating)*The air is heated at constant pressure from initial temperature T_1 to a temperature T_2 represented by the curve 1-2 in Fig. 6.15 (*a*) and (*b*).

$\therefore$ Heat absorbed by the air, $Q_{1-2} = m\, c_p\,(T_2 - T_1)$

2. *Second stage (Reversible adiabatic or isentropic expansion)*The air is expanded reversibly and adiabatically from temperature T_2 to a temperature T_3 as shown by the curve 2-3 in Fig. 6.15 (*a*) and (*b*). In this process, no heat is absorbed or rejected by the air .

3. *Third stage (Constant volume cooling)*The air is now cooled at constant volume from temperature T_3 to temperature T_4 as shown by the curve 3-4 in Fig. 6.15 (*a*) and (*b*).

$\therefore$ Heat rejected by the air, $Q_{3-4} = m\, c_v\,(T_3 - T_4)$

4. *Fourth stage (Reversible adiabatic or isentropic compression)*The air is compressed reversibly and adiabatically from temperature T_4 to a temperature T_5 as shown by the curve 4-5 in Fig. 6.15 (*a*) and (*b*). In this process, no heat is absorbed or rejected by the air.

5. *Fifth stage (Constant volume heating)*The air is finally heated at constant volume from temperature T_5 to a temperature T_1 as shown by the curve 5-1 in Fig. 6.15 (*a*) and (*b*).

$\therefore$ Heat absorbed by the air, $Q_{5-1} = m\, c_v\,(T_1 - T_5)$

We see that the air has been brought back to its original conditions of pressure, volume and temperature, thus completing the cycle. We know that

$$\text{Work done} = \text{Heat absorbed} - \text{Heat rejected}$$

$$= [m\, c_p\,(T_2 - T_1) + m\, c_v\,(T_1 - T_5)] - m\, c_v\,(T_3 - T_4)$$

* The cylinder is of similar type as discussed in Otto and Diesel cycles.

and air standard efficiency, $\eta = \dfrac{\text{Workdone}}{\text{Heat absorbed}}$

$$= \frac{m\,c_p\,(T_2 - T_1) + m\,c_v\,(T_1 - T_5) - m\,c_v\,(T_3 - T_4)}{m\,c_p\,(T_2 - T_1) + m\,c_v\,(T_1 - T_5)}$$

$$= 1 - \frac{c_v\,(T_3 - T_4)}{c_p\,(T_2 - T_1) + c_v\,(T_1 - T_5)}$$

$$= 1 - \frac{T_3 - T_4}{\gamma\,(T_2 - T_1) + (T_1 - T_5)} \qquad \ldots (i)$$

Now, let compression ratio,

$$r = \frac{v_4}{v_5} = \frac{v_3}{v_1} \qquad \ldots (\because v_3 = v_4 \text{ and } v_5 = v_1)$$

Cut-off ratio, $$\rho = \frac{v_2}{v_1} = \frac{v_2}{v_5} \qquad \ldots (\because v_1 = v_5)$$

and pressure ratio,

$$\alpha = \frac{p_1}{p_5}$$

We know that in constant pressure heating process 1-2,

$$\frac{v_1}{T_1} = \frac{v_2}{T_2} \qquad \ldots \text{(Charles' law)}$$

$$\therefore \quad T_2 = T_1 \times \frac{v_2}{v_1} = T_1\,\rho \qquad \ldots (ii)$$

Similarly, in reversible adiabatic or isentropic expansion process 2-3,

$$\frac{T_3}{T_2} = \left(\frac{v_2}{v_3}\right)^{\gamma-1} = \left(\frac{v_2}{v_1} \times \frac{v_1}{v_3}\right)^{\gamma-1} = \left(\frac{\rho}{r}\right)^{\gamma-1}$$

$$\therefore \quad T_3 = T_2\left(\frac{\rho}{r}\right)^{\gamma-1} = T_1\,\rho\left(\frac{\rho}{r}\right)^{\gamma-1} \qquad \ldots (iii)$$

and in reversible adiabatic or isentropic compression process 4-5,

$$\frac{T_5}{T_4} = \left(\frac{v_4}{v_5}\right)^{\gamma-1} = (r)^{\gamma-1}$$

$$\therefore \quad T_5 = T_4\,(r)^{\gamma-1} \qquad \ldots (iv)$$

Now in constant volume heating process 5-1,

$$\frac{p_5}{T_5} = \frac{p_1}{T_1} \qquad \ldots \text{(Gay-Lussac law)}$$

$$\therefore \quad T_1 = T_5 \times \frac{p_1}{p_5} = T_5\,\alpha = T_4\,(r)^{\gamma-1}\,\alpha \qquad \ldots (v)$$

Substituting the value of T_1 in equations (*ii*) and (*iii*),

$$T_2 = T_4 (r)^{\gamma-1} \alpha \rho$$

and

$$T_3 = T_4 (r)^{\gamma-1} \alpha \rho \left(\frac{\rho}{r}\right)^{\gamma-1} = T_4 \alpha \rho^{\gamma}$$

Now substituting the values of T_1, T_2, T_3 and T_5 in equation (*i*),

$$\eta = 1 - \frac{T_4 \alpha \rho^{\gamma} - T_4}{\gamma [T_4 (r)^{\gamma-1} \alpha \rho) - T_4 (r)^{\gamma-1} \alpha] + [T_4 (r)^{\gamma-1} \alpha - T_4 (r)^{\gamma-1}]}$$

$$= 1 - \frac{T_4 (\alpha \rho^{\gamma} - 1)}{T_4 (r)^{\gamma-1} [\gamma (\alpha \rho - \alpha) + (\alpha - 1)]}$$

$$= 1 - \frac{(\alpha \rho^{\gamma} - 1)}{(r)^{\gamma-1} [\gamma \alpha (\rho - 1) + (\alpha - 1)]}$$

$$= 1 - \frac{1}{(r)^{\gamma-1}} \left[\frac{\alpha \rho^{\gamma} - 1}{(\alpha - 1) + \gamma \alpha (\rho - 1)}\right] \quad \ldots (vi)$$

Notes : 1. For Otto cycle, $\rho = 1$. Substituting this value in equation (*vi*),

$$\eta_{otto} = 1 - \frac{1}{(r)^{\gamma-1}} \quad \ldots \text{(Same as before)}$$

2. For Diesel cycle, $\alpha = 1$. Substituting this value in equation (*vi*),

$$\eta_{diesel} = 1 - \frac{1}{(r)^{\gamma-1}} \left(\frac{\rho^{\gamma} - 1}{\gamma (\rho - 1)}\right) \quad \ldots \text{(Same as before)}$$

3. The efficiency of dual combustion cycle is greater than Diesel cycle and less than Otto cycle, for the same compression ratio.

Example 6.21. *An oil engine, working on the dual combustion cycle, has a compression ratio 10 and cut-off takes place at 1/10 of the stroke. If the pressure at the beginning of compression is 1 bar and maximum pressure 40 bar, determine the air standard efficiency of the cycle. Take $\gamma = 1.4$.*

Solution. Given : $r = v_4 / v_5 = 10$; Cut-off = 1/10 of stroke ; $p_4 = 1$ bar ; $p_1 = p_2 = 40$ bar ; $\gamma = 1.4$

Since the cut-off takes place at 1/10 of the stroke, therefore volume at cut-off,

$$v_2 = v_1 + 1/10 \text{ of stroke volume} = v_1 + \frac{1}{10}(v_4 - v_1)$$

$$= v_5 + 0.1 (v_4 - v_5) \quad \ldots (\because v_1 = v_5)$$

$\therefore$ Cut-off ratio,

$$\rho = \frac{v_2}{v_5} = 1 + 0.1 \left(\frac{v_4}{v_5} - 1\right) \quad \ldots \text{(Dividing both sides by } v_5)$$

$$= 1 + 0.1 (10 - 1) = 1.9$$

We know that for isentropic compression process 4-5 (Refer Fig. 6.15),

$$p_4 v_4^{\gamma} = p_5 v_5^{\gamma} \text{ or } p_5 = p_4 \left(\frac{v_4}{v_5}\right)^{\gamma} = 1 (10)^{1.4} = 25.12 \text{ bar}$$

Pressure ratio, $\alpha = p_1 / p_5 = 40 / 25.12 = 1.59$

We know that air standard efficiency,

$$\eta = 1 - \frac{1}{(r)^{\gamma-1}}\left[\frac{\alpha\rho^{\gamma} - 1}{(\alpha - 1) + \gamma\alpha(\rho - 1)}\right]$$

$$= 1 - \frac{1}{(10)^{1.4-1}}\left[\frac{1.59\,(1.9)^{1.4} - 1}{(1.59 - 1) + 1.4 \times 1.59\,(1.9 - 1)}\right]$$

$$= 1 - \frac{1}{2.512}\left[\frac{3.9 - 1}{0.59 + 2}\right] = 1 - 0.446 = 0.554 \text{ or } 55.4\,\% \textbf{ Ans.}$$

Example 6.22. *In a *compression ignition engine, working on a dual combustion cycle, pressure and temperature at the start of compression are 1 bar and 300 K respectively. At the end of compression, pressure reaches a value of 25 bar. 420 kJ of heat is supplied per kg of air during constant volume heating and pressure becomes 2.8 bar at the end of isentropic expansion. Estimate the ideal thermal efficiency. Take c_p = 1.005 kJ/kg K and c_v = 0.712 kJ/kg K.*

Solution. Given : p_4 = 1 bar ; T_4 = 300 K ; p_5 = 25 bar ; Q_{5-1} = 420 kJ/kg ; p_3 = 2.8 bar ; c_p = 1.005 kJ/kg K ; c_v = 0.712 kJ/kg K

We know that ratio of specific heats or isentropic index,

$$\gamma = c_p / c_v = 1.005 / 0.712 = 1.4$$

First of all, consider the isentropic compression process 4-5 (Refer Fig. 6.15).

We know that

$$p_4 v_4^{\gamma} = p_5 v_5^{\gamma}$$

∴ Compression ratio,

$$r = \frac{v_4}{v_5} = \left(\frac{p_5}{p_4}\right)^{\frac{1}{\gamma}} = \left(\frac{25}{1}\right)^{\frac{1}{1.4}} = (25)^{0.714} = 9.96$$

We also know that $$\frac{T_5}{T_4} = \left(\frac{v_4}{v_5}\right)^{\gamma-1} = (9.96)^{1.4-1} = 2.5$$

$$\therefore \quad T_5 = T_4 \times 2.5 = 300 \times 2.5 = 750 \text{ K}$$

Now consider the constant volume heating process 5-1. We know that heat supplied per kg of air (Q_{5-1}),

$$420 = m\,c_v(T_1 - T_5) = 1 \times 0.712\,(T_1 - 750)$$

$$\therefore \quad T_1 - 750 = 420 / 0.712 = 590 \quad \text{or} \quad T_1 = 590 + 750 = 1340 \text{ K}$$

and pressure ratio, $$\alpha = p_1 / p_5 = T_1 / T_5 = 1340 / 750 = 1.787 \quad \ldots (\because p_1 / T_1 = p_5 / T_5)$$

$$\therefore \quad p_1 = p_5 \times 1.787 = 25 \times 1.787 = 44.7 \text{ bar}$$

Now consider the isentropic expansion 2-3. We know that

$$p_2 v_2^{\gamma} = p_3 v_3^{\gamma} \quad \text{or} \quad \frac{v_2}{v_3} = \left(\frac{p_3}{p_2}\right)^{\frac{1}{\gamma}} = \left(\frac{2.8}{44.7}\right)^{\frac{1}{1.4}} = 0.138 \quad \ldots (\because p_2 = p_1)$$

* A Diesel engine is also called compression ignition engine.

and cut-off ratio, $$\rho = \frac{v_2}{v_1} = \frac{v_2}{v_5} = \frac{v_2}{v_3} \times \frac{v_3}{v_5} = \frac{v_2}{v_3} \times \frac{v_4}{v_5} \qquad \ldots (\because v_1 = v_5 \text{ and } v_3 = v_4)$$

$$= 0.138 \times 9.96 = 1.37$$

We know that ideal thermal efficiency,

$$\eta = 1 - \frac{1}{(r)^{\gamma - 1}} \left[\frac{\alpha \rho^{\gamma} - 1}{(\alpha - 1) + \gamma \alpha (\rho - 1)} \right]$$

$$= 1 - \frac{1}{(9.96)^{1.4 - 1}} \left[\frac{1.787 (1.37)^{1.4} - 1}{(1.787 - 1) + 1.4 \times 1.787 (1.37 - 1)} \right]$$

$$= 1 - \frac{1}{2.51} \left[\frac{2.78 - 1}{0.787 + 0.926} \right] = 1 - 0.414 = 0.586 \text{ or } 58.6\ \% \textbf{ Ans.}$$

Example 6.23. *An engine working on a dual combustion cycle has a pressure of 1 bar and 50° C before compression. The air is then compressed isentropically to 1/15th of its original volume. The maximum pressure is twice the pressure at the end of isentropic compression. If the cut-off ratio is 2, determine the temperature at the end of each process and the ideal efficiency of the cycle, γ = 1.4.*

Solution. Given : $p_4 = 1$ bar ; $T_4 = 50° \text{ C} = 50 + 273 = 323$ K ; $v_5 = v_4 / 15$; $p_1 = 2\, p_5$; $\rho = v_2 / v_1 = 2$; $\gamma = 1.4$

We know that $v_5 = v_4 / 15$

∴ Compression ratio,

$$r = v_4 / v_5 = 15$$

Temperature at the end of isentropic compression

Let T_5 = Temperature at the end of isentropic compression.

We know that for isentropic compression process 4-5 (Refer Fig. 6.15),

$$\frac{T_5}{T_4} = \left(\frac{v_4}{v_5} \right)^{\gamma - 1} = (15)^{1.4 - 1} = 2.954$$

$$\therefore \quad T_5 = T_4 \times 2.954 = 323 \times 2.954 = 954 \text{ K} = 681° \text{ C} \textbf{ Ans.}$$

Temperature at the end of constant volume heating

Let T_1 = Temperature at the end of constant volume heating.

We know that for a constant pressure process 5-1,

$$\frac{p_5}{T_5} = \frac{p_1}{T_1} \text{ or } T_1 = T_5 \times \frac{p_1}{p_5} = 954 \times 2 = 1908 \text{ K} = 1635° \text{ C} \textbf{ Ans.}$$

$$\ldots (\because p_1 = 2\, p_5)$$

Temperature at the end of constant pressure heating

Let T_2 = Temperature at the end of constant pressure heating.

We know that for a constant volume process 1-2,

$$\frac{v_1}{T_1} = \frac{v_2}{T_2} \text{ or } T_2 = T_1 \times \frac{v_2}{v_1} = 1908 \times 2 = 3816 \text{ K} = 3543° \text{ C} \textbf{ Ans.}$$

$$\ldots (\because v_2 / v_1 = 2)$$

Temperature at the end of isentropic expansion 2-3.

Let T_3 = Temperature at the end of isentropic expansion.

We know that for isentropic expansion process 2-3,

$$\frac{T_3}{T_2} = \left(\frac{v_2}{v_3}\right)^{\gamma-1} = \left(\frac{v_2}{v_1} \times \frac{v_1}{v_3}\right)^{1.4-1} = \left(2 \times \frac{1}{15}\right)^{0.4} = 0.447$$

$$\ldots \left(\because \frac{v_1}{v_3} = \frac{v_5}{v_4} = \frac{1}{15}\right)$$

$$\therefore \quad T_3 = T_2 \times 0.447 = 3816 \times 0.447 = 1706 \text{ K} = 1433° \text{ C} \textbf{ Ans.}$$

Ideal efficiency of the cycle

We know that ideal efficiency of the cycle,

$$\eta = 1 - \frac{T_3 - T_4}{\gamma(T_2 - T_1) + (T_1 - T_5)}$$

$$= 1 - \frac{1706 - 323}{1.4(3816 - 1908) + (1908 - 954)} = 1 - \frac{1383}{2671 + 954}$$

$$= 1 \quad 0.3815 = 0.6185 \text{ or } 61.85\% \textbf{ Ans.}$$

Example 6.24. *An ideal dual cycle engine works with a stroke volume of 10 litres of air with a compression ratio of 16. The pressure and temperature of air before isentropic compression is 1 bar and 300 K respectively. If the heat is added at a constant pressure of 70 bar and for 5% of the stroke, determine : 1. Pressure ratio ; 2. Cut-off ratio ; 3. Mass of air contained in the cylinder ; 4. Heat added per cycle ; 5. Heat rejected per cycle ; 6. Work done per cycle ; 7. Thermal efficiency of the cycle ; and 8. Mean effective pressure. Take c_p = 1 kJ/kg K and c_v = 0.714 kJ/kg K.*

Solution. Given : $v_s = v_4 - v_5 = v_3 - v_1 = 10$ litres $= 10 \times 10^{-3} = 0.01$ m^3 ; $r = v_4/v_5 = 16$; $p_4 = 1$ bar $= 0.1 \times 10^6$ N/m^2 ; $T_4 = 300$ K ; $p_1 = p_2 = 70$ bar ; Cut-off = 5% of stroke ; $c_p = 1$ kJ/kg K ; $c_v = 0.714$ kJ/kg K

We know that $\gamma = c_p / c_v = 1 / 0.714 = 1.4$

1. *Pressure ratio*

Let p_5 = Pressure at point 5, and

α = Pressure ratio = p_1 / p_5.

We know that in isentropic compression process 4-5 (Refer Fig. 6.15),

$$p_4 v_4^{\gamma} = p_5 v_5^{\gamma} \quad \text{or} \quad p_5 = p_4 \left(\frac{v_4}{v_5}\right)^{\gamma} = 1\,(16)^{1.4} = 48.5 \text{ bar}$$

$\therefore$ Pressure ratio,

$$\alpha = p_1 / p_5 = 70 / 48.5 = 1.44 \textbf{ Ans.}$$

2. *Cut-off ratio*

We know that compression ratio,

$$r = v_4 / v_5 = 16 \quad \text{or} \quad v_4 = 16\, v_5 \qquad \ldots (i)$$

and

$$v_4 - v_5 = 10 \text{ litres} = 0.01 \text{ m}^3 \qquad \ldots (ii)$$

From equations (*i*) and (*ii*),

$$v_5 = 0.67 \times 10^{-3}\ \text{m}^3 \quad \text{and} \quad v_4 = 10.67 \times 10^{-3}\ \text{m}^3$$

Since the cut-off takes place at 5% of the stroke, therefore volume at cut-off,

$$v_2 = v_1 + 5\%\, v_s = v_1 + 0.05\,(v_4 - v_5)$$

$$= 0.67 \times 10^{-3} + 0.05 \times 0.01 = 1.17 \times 10^{-3}\ \text{m}^3 \qquad \ldots (\because v_1 = v_5)$$

$\therefore$ Cut-off ratio, $\rho = v_2 / v_1 = 1.17 \times 10^{-3} / 0.67 \times 10^{-3} =$ 1.746 **Ans.**

3. *Mass of air contained in the cylinder*

Let m = Mass of air contained in the cylinder.

We know that $p_4 v_4 = m R T_4$

$$\therefore \quad m = \frac{p_4 v_4}{R T_4} = \frac{0.1 \times 10^6 \times 10.67 \times 10^{-3}}{287 \times 300} = 0.012 \text{ kg} \quad \textbf{Ans.}$$

... (Taking R = 287 J/kg K)

4. *Heat added per cycle*

First of all, let us find the temperature at points 1 and 2 *i.e.* T_1 and T_2 respectively.

We know that for isentropic compression process 4-5,

$$\frac{T_5}{T_4} = \left(\frac{v_4}{v_5}\right)^{\gamma - 1} = (16)^{1.4 - 1} = 3.03$$

$$\therefore \quad T_5 = T_4 \times 3.03 = 300 \times 3.03 = 909 \text{ K}$$

Similarly for constant volume process 5-1,

$$\frac{p_5}{T_5} = \frac{p_1}{T_1} \quad \text{or} \quad T_1 = T_5 \times \frac{p_1}{p_5} = 909 \times \frac{70}{48.5} = 1312 \text{ K}$$

and for constant pressure process 1-2,

$$\frac{v_1}{T_1} = \frac{v_2}{T_2} \quad \text{or} \quad T_2 = T_1 \times \frac{v_2}{v_1} = 1312 \times \frac{1.17 \times 10^{-3}}{0.67 \times 10^{-3}} = 2291 \text{ K}$$

$$\ldots (\because v_1 = v_5)$$

We know that heat added during constant pressure process 1-2,

$$Q_{1-2} = m\, c_p\, (T_2 - T_1) = 0.012 \times 1\,(2291 - 1312) = 11.75 \text{ kJ}$$

and heat added during constant volume process 5-1,

$$Q_{5-1} = m\, c_v\, (T_1 - T_5) = 0.012 \times 0.714\,(1312 - 909) = 3.45 \text{ kJ}$$

$\therefore$ Total heat added per cycle,

$$Q = Q_{1-2} + Q_{5-1} = 11.75 + 3.45 = 15.20 \text{ kJ} \quad \textbf{Ans.}$$

5. *Heat rejected per cycle*

First of all, let us find the temperature (T_3) at point 3.

We know that for isentropic expansion process 2-3,

$$\frac{T_2}{T_3} = \left(\frac{v_3}{v_2}\right)^{\gamma - 1} = \left(\frac{10.67 \times 10^{-3}}{1.17 \times 10^{-3}}\right)^{1.4 - 1} = 2.42 \qquad \ldots (\because v_3 = v_4)$$

$$\therefore \quad T_3 = T_2 / 2.42 = 2291 / 2.42 = 947 \text{ K}$$

and heat rejected per cycle (during constant volume process 3-4),

$$= m\, c_v (T_3 - T_4) = 0.012 \times 0.714\,(947 - 300) = 5.54 \text{ kJ } \textbf{Ans.}$$

6. *Work done per cycle*

We also know that work done per cycle

$$= \text{Heat added} - \text{Heat rejected} = 15.20 - 5.54 = 9.66 \text{ kJ } \textbf{Ans.}$$

7. *Thermal efficiency of the cycle*

We know that thermal efficiency of the cycle,

$$\eta = \frac{\text{Work done per cycle}}{\text{Heat added per cycle}} = \frac{9.66}{15.20} = 0.636 \text{ or } 63.6\ \% \ \textbf{Ans.}$$

8. *Mean effective pressure.*

We know that mean effective pressure,

$$= \frac{\text{Work done}}{\text{Stroke volume}} = \frac{9.66}{0.01} = 966 \text{ kN/m}^2 = 9.66 \text{ bar } \textbf{Ans.}$$

EXERCISES

1. An ideal engine is imagined to be working on Carnot's cycle. The working fluid receives heat at a temperature of 590 K and rejects at a temperature of 295 K. Find the theoretical efficiency of the cycle.

If the engine working on this cycle absorbs 35 kJ/s from the hot body, calculate the net work done per second. **[Ans.** 50% ; 17.5 kJ/s]

2. In a carnot cycle engine, the temperature of the source and sink are 700° C and 50° C. The heat supply is 84 kJ/s. Find the power developed by the engine.

If the temperature of source is 500° C and the heat supply and workdone are 420 kJ and 200 kJ respectively, find the sink temperature. [**Ans.** 56.112 kW ; 132° C]

3. An inventor claims that his petrol engine operating between temperatures 2000° C and 600° C will produce 0.735 kW consuming 0.12 kg per hour of petrol having 46 000 kJ/kg calorific value. Check the validity of the claim. [**Ans.** Inventor's claim is valid]

4. Determine the efficiency of the regenerative Stirling cycle working between the temperature limits of 700 K and 350 K. The isothermal expansion or compression ratio is 4. [**Ans.** 50%]

5. A regenerative Stirling air engine works between the temperatures of 400° C and 15° C. The ratio of isothermal expansion is 3. Find ideal efficiency of the cycle, if regenerator efficiency is 80%. Assume c_p = 0.996 kJ/kg K, and c_v = 0.707 kJ/kg K. [**Ans.** 45.6%]

6. In a Stirling engine working with a perfect regenerator, the lower limit of temperature is 538° C. The maximum and minimum pressures are 12 bar and 2 bar. Determine the ideal efficiency, if the expansion ratio is 3. [**Ans.** 50%]

7. An Ericsson engine having an expansion ratio 2 works between the temperature ranges of 700 K and 298 K. Calculate the ideal efficiency when :

1. the engine is fitted with a perfect regenerator, and 2. the efficiency of regenerator is 0.85.

Assume c_p = 0.996 kJ/kg K and c_v = 0.712 kJ/kg K [**Ans.** 57.4% ; 40%]

8. A gas turbine working on Joule's cycle receives air at 1 bar and 15° C. It is compressed isentropically to 5 bar and then heated at a constant pressure till its temperature reaches 800° C. The hot air is then expanded isentropically in the turbine back to its original pressure of 1 bar. Determine the work done kg of air supplied and thermal efficiency of the turbine. Assume γ = 1.4 and c_p = 1 kJ/kg K. [**Ans.** 228 kJ ; 37%]

9. The efficiency of an Otto cycle is 50% and γ = 1.5. Find the compression ratio. [**Ans.** 4]

10. An engine working on ideal Otto cycle has temperature and pressure, at the beginning of isentropic compression as 25° C and 1.5 bar respectively. Find the compression ratio, if $\gamma = 1.4$ and thermal efficiency of the engine = 48%. Also find temperature and pressure at the end of compression.

[**Ans.** 5.13 ; 300° C ; 14.8 bar]

11. An engine working on constant volume cycle has a clearance volume of 1 litre and a stroke volume of 6 litres. The suction pressure and temperature are 1 bar and 20° C respectively. The pressure at the end of heat addition is 25 bar. Determine ; 1. pressure and volume at salient points of the cycle ; 2. thermal efficiency, and 3. workdone per cycle.

Take c_v during heat addition = 0.807 kJ/kg K ; c_v during heat rejection = 0.737 kJ/kg K and $\gamma = 1.4$. [**Ans.** p_2 = 1.64 bar = p_4 = 15.24 bar, $v_1 = v_4$ = 1 litre ; $v_2 = v_3$ = 7 litre ; 54.1% ; 192 kJ/kg]

12. A four-cylinder petrol engine has a swept volume of 2×10^{-3} m³ and the clearance volume in each cylinder is 60×10^{-6} m³. If the pressure and temperature at the beginning of compression are 1.02 bar and 24° C and the maximum cycle temperature is 1400° C, calculate 1. the air standard efficiency, and 2. the mean effective pressure. Assume for air, c_p = 1.005 kJ/kg K and c_v = 0.716 kJ/kg K. [**Ans.** 59.1% ; 5.3 bar]

13. An air standard Diesel cycle has a compression ratio of 14. The pressure at the beginning of compression stroke is 1 bar and the temperature is 27° C. The maximum temperature of the cycle is 2500°C. Determine the efficiency of this engine. [**Ans.** 53.6 %]

14. A diesel engine has a bore of 250 mm and a stroke of 400 mm. The cut-off takes place at 5 per cent of the stroke. Estimate : 1. Air standard efficiency, and 2. Mean effective pressure if the clearance volume and the pressure at the end of suction stroke are 1.2 litres and 1 bar respectively. [**Ans.** 64.2% ; 5.11 bar]

15. Estimate the loss in air standard efficiency for an engine of compression ratio of 15 and the cut-off changes from 5 to 15% of the stroke. [**Ans.** 6.6%]

16. An ideal Diesel cycle operates on 1 kg of standard air with initial pressure of 1 bar and a temperature of 35° C. The pressure at the end of compression is 33 bar and the cut-off 6% of the stroke. Determine : 1. the compression ratio, 2. the percentage clearance, 3. the heat supplied, and 4. the heat rejected.

Take $\gamma = 1.4$ and c_p = 1 kJ/kg K. [**Ans.** 12.14 ; 8.98% of stroke ; 559 kJ ; 230 kJ]

17. An ideal Diesel cycle has compression ratio of 16 with lowest temperature and pressure as 27°C and 1 bar respectively. If heat added during the cycle is 500 kJ/kg of air, determine 1. pressure and temperature at each salient point, 2. work done in the cycle/kg, 3. thermal efficiency of the cycle, and 4. mean effective pressure. [**Ans.** $p_1 = p_2$ = 48.5 bar, T_1 = 636° C, T_2 = 1136° C, p_3 = 1.87 bar, T_3 = 175° C ; 394.3 kJ/kg ; 78.9% ; 4.9 bar]

18. Two engines are to operate on Otto and Diesel cycles with the following data :

Maximum temperature = 1500 K ; Exhaust temperature = 700 K ; Ambient condition = 1 bar and 300 K.

Find the compression ratios, maximum pressures and efficiencies.

[**Ans.** Otto cycle : 6.72 ; 33.6 bar ; 53.3 % ; Diesel cycle: 12.3 ; 33.6 bar ; 58.1%]

19. An engine working on the dual combustion cycle has pressure ratio of 1.41. The diameter of engine cylinder is 200 mm and stroke 420 mm. If the compression ratio is 13.5 and cut-off occurs at 5% of the stroke, determine the air standard efficiency of the engine. [**Ans.** 61.8%]

20. An oil engine operates on ideal dual combustion cycle as follows :

Pressure at the beginning of compression = 0.97 bar ; Temperature at the beginning of compression = 55° C ; Pressure ratio = 1.5 ; Cut-off ratio = 1.6 ; Compression ratio = 11 ; Ratio of specific heats, $\gamma = 1.4$.

Determine the temperature at the end of each operation and the efficiency of the engine.

[**Ans.** T_5 = 583° C ; T_1 = 1011° C ; T_2 = 1782° C ; T_3 = 676° C ; 58.8%]

QUESTIONS

1. What is a thermodynamic cycle ? Distinguish between a four-stroke cycle and a two-stroke cycle.
2. Explain reversible and irreversible processes. Is it possible to realise these processes ?

3. What are the conditions of reversibility ? Prove that all reversible engines are equally efficient between the same temperature limits.

4. Do you agree with the statement that the conception of thermodynamic reversibility is purely hypothetical ? Explain fully.

5. Prove that the efficiency of the Carnot engine working between the temperature limits T_1 and T_2 is equal to $\frac{T_1 - T_2}{T_1}$, and show that no engine can be more efficient than this when working over the same temperature range.

6. Derive an expression for the efficiency of a Stirling air engine with the aid of *p-v* and *T-S* diagrams.

7. Deduce an expression for thermal efficiency of Joule's air engine and show that it is less than Carnot efficiency.

8. Explain briefly Otto cycle with the help of *p-v* and *T-S* diagram, and derive an expression for the ideal efficiency of Otto cycle.

9. Show that the efficiency of Otto cycle is a function of compression ratio only.

10. Explain briefly the Diesel cycle with the help of *p-v* and *T-S* diagrams and derive an expression for the ideal efficiency of a Diesel cycle.

11. For a given compression ratio ; the air standard Diesel cycle is less efficient than air standard Otto cycle. Explain.

12. Derive an expression for the ideal efficiency of dual combustion cycle, using ideal air as the working fluid.

OBJECTIVE TYPE QUESTIONS

1. Carnot cycle consists of
   (*a*) two constant volume and two reversible adiabatic processes
   (*b*) two isothermal and two reversible adiabatic processes
   (*c*) two constant pressure and two reversible adiabatic processes
   (*d*) one constant volume, one constant pressure and two reversible adiabatic processes
2. The efficiency of Carnot cycle may be increased by
   (*a*) increasing the highest temperature (*b*) decreasing the highest temperature
   (*c*) increasing the lowest temperature (*d*) decreasing the lowest temperature
3. A cycle consisting of two constant volume and two isothermal processes is known as
   (*a*) Carnot cycle (*b*) Joule cycle (*c*) Diesel cycle (*d*) Stirling cycle
4. The efficiency of Ericsson cycle is ......Carnot cycle.
   (*a*) greater than (*b*) less than (*c*) equal to
5. Otto cycle is also known as
   (*a*) constant pressure cycle (*b*) constant volume cycle
   (*c*) constant temperature cycle (*d*) constant entropy cycle
6. For the same compression ratio, the efficiency of Diesel cycle is ..... Otto cycle.
   (*a*) greater than (*b*) less than (*c*) equal to
7. The efficiency of Diesel cycle approaches to Otto cycle efficiency when
   (*a*) cut-off is increased (*b*) cut-off is decreased
   (*c*) cut-off is zero (*d*) cut-off is constant

8. The air standard efficiency of an Otto cycle is given by

(*a*) $1 - \frac{1}{r^{\gamma - 1}}$ (*b*) $1 + \frac{1}{r^{\gamma - 1}}$ (*c*) $1 - r^{\gamma - 1}$ (*d*) $1 + r^{\gamma - 1}$

9. The efficiency of the dual combustion cycle, for the same compression ratio, is .... Diesel cycle.

(*a*) greater than (*b*) less than (*c*) equal to

10. For the same maximum pressure and temperature of Otto, Diesel and dual combustion air standard cycles,

(*a*) the compression ratios will be same

(*b*) the heat supplied to the cycles will be same

(*c*) the air standard efficiency will have the same value

(*d*) the heat rejected by the engine will be the same

ANSWERS

1. (*b*)	2. (*b*)	3. (*d*)	4. (*c*)	5. (*b*)
6. (*b*)	7. (*b*)	8. (*a*)	9. (*a*)	10. (*d*)

7

Formation and Properties of Steam

1. Introduction. 2. Formation of Steam at a Constant Pressure from Water. 3. Temperature vs. Total Heat Graph during Steam Formation. 4. Important Terms for Steam. 5. Steam Tables and their Uses. 6. Superheated Steam. 7. Advantages of Superheating the Steam. 8. External Work Done During Evaporation. 9. Internal Energy of Steam. 10. Measurement of Dryness Fraction of Steam. 11. Barrel Calorimeter. 12. Separating Calorimeter. 13. Throttling Calorimeter. 14. Combined Separating and Throttling Calorimeter.*

7.1. Introduction

Steam is a vapour of water, and is invisible when pure and dry. It is used as the working substance in the operation of steam engines and steam turbines. Steam does not obey laws of perfect gases, until it is perfectly dry. It has already been discussed that when the dry vapour is heated further, it becomes superheated vapour which behaves, more or less, like a perfect gas.

7.2. Formation of Steam at a Constant Pressure from Water

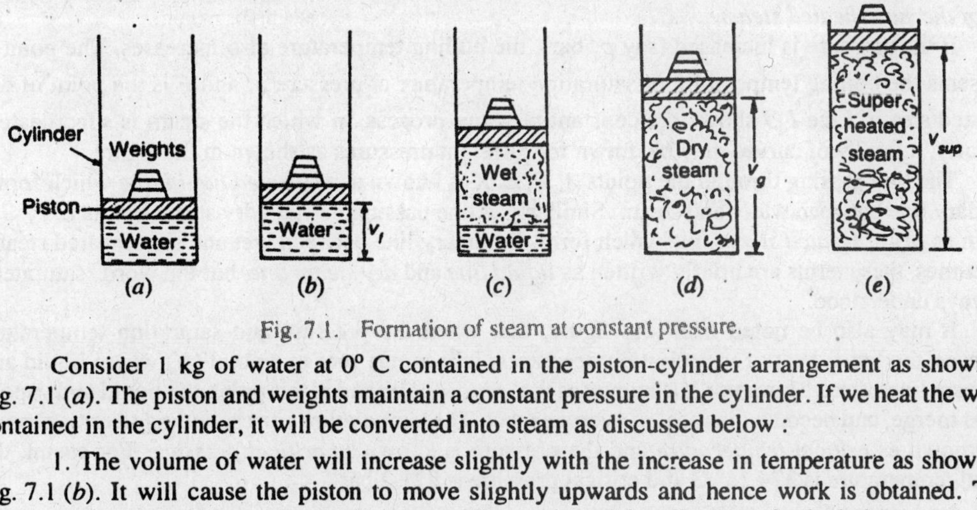

Fig. 7.1. Formation of steam at constant pressure.

Consider 1 kg of water at 0° C contained in the piston-cylinder arrangement as shown in Fig. 7.1 (*a*). The piston and weights maintain a constant pressure in the cylinder. If we heat the water contained in the cylinder, it will be converted into steam as discussed below :

1. The volume of water will increase slightly with the increase in temperature as shown in Fig. 7.1 (*b*). It will cause the piston to move slightly upwards and hence work is obtained. This increase in volume of water (or work) is generally, neglected for all types of calculations.

2. On further heating, temperature reaches boiling point. The boiling point of water, at normal atmospheric pressure of 1.013 bar is 100° C, but it increases with the increase in pressure. When the boiling point is reached, the temperature remains constant and the water evaporates, thus pushing the piston up against the constant pressure. Consequently, the specific volume of steam increases as shown in Fig. 7.1 (*c*). At this stage, the steam will have some particles of water in suspension, and is termed as *wet steam*. This process will continue till the whole water is converted into wet steam.

3. On further heating, the water particles in suspension will be converted into steam. The entire steam, in such a state, is termed as *dry steam* or *saturated steam* as shown in Fig. 7.1 (*d*). Practically, the dry steam behaves like a perfect gas.

4. On further heating, the temperature of the steam starts rising. The steam, in such a state, is termed as *superheated steam* as shown in Fig. 7.1 (*e*).

7.3. Temperature vs. Total Heat Graph during Steam Formation

The process of steam formation, as discussed above, may also be represented on a graph, whose abscissa represents the total heat and the vertical ordinate represents the temperature. The point *A* represents the initial condition of water at 0° C and pressure *p* (in bar) as shown in Fig. 7.2. Line *ABCD* shows the relation between temperature and heat at a specific pressure of *p* (in bar).

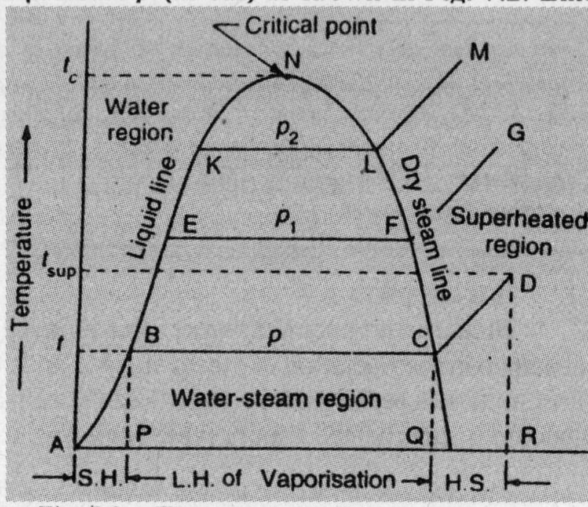

Fig. 7.2. Temperature-total heat graph during steam formation.

During the formation of the superheated steam, from water at freezing point, the heat is absorbed in the following three stages :

1. The heating of water up to boiling temperature or saturation temperature (t) is shown by *AB* in Fig. 7.2. The heat absorbed by the water is *AP*, known as *sensible heat* or *liquid heat* or *total heat of water*.

2. The change of state from liquid to steam is shown by *BC*. The heat absorbed during this stage is *PQ*, known as *latent heat of vaporisation*.

3. The superheating process is shown by *CD*. The heat absorbed during this stage is *QR*, known as *heat of superheat*. Line *AR* represents the *total heat of the superheated steam*.

If the pressure is increased (say p_1 bar), the boiling temperature also increases. The point *E* represents the boiling temperature or saturation temperature at pressure p_1 and *F* is the point of dry saturated steam. Line *FG* shows the constant pressure process, in which the steam is superheated. Similarly, a family of curves may be drawn for different pressures as shown in the figure.

The line passing through the points *A, B, E, K* is known as *saturated liquid line* which forms boundary line between water and steam. Similarly, a line passing through dry steam points *L, F, C* is known as *dry saturated steam line* which forms boundary line between wet and superheated steam. Sometimes, these terms are briefly written as *liquid line* and *dry steam line,* but the word "saturated" is always understood.

It may also be noted from the figure, that when the pressure and saturation temperature increases, the latent heat of vaporisation decreases. it becomes zero at a point (*N*) where liquid and dry steam lines meet. This point *N* is known as the *critical point* and at this point, the liquid and vapour phases merge, and become identical in every respect. The temperature corresponding to critical point *N* is known as *critical temperature* and the pressure is known as *critical pressure*. For steam, the critical temperature is 374.15° C and critical pressure is 221.2 bar.

7.4. Important Terms for Steam

Though there are many terms applied to steam, yet the following are important from the subject point of view :

1. *Wet steam.* When the steam contains moisture or particles of water in suspension, it is said to be *wet steam*. It means that the evaporation of water is not complete, and the whole of the latent heat has not been absorbed.

2. *Dry saturated steam.* When the wet steam is further heated, and it does not contain any suspended particles of water, it is known as *dry saturated steam*. The dry saturated steam has absorbed its full latent heat and behaves practically, in the same way as a perfect gas.

3. *Superheated steam.* When the dry steam is further heated at a constant pressure, thus raising its temperature, it is said to be *superheated steam*. Since the pressure is constant, therefore the volume of superheated steam increases. It may be noted that the volume of one kg of superheated steam is considerably greater than the volume of one kg of dry saturated steam at the same pressure.

In actual practice, the superheated steam is produced in a separate apparatus known as superheater, so that it is out of contact with water from which it was formed.

4. *Dryness fraction or quality of wet steam.* It is the ratio of the mass of actual dry steam, to the mass of same quantity of wet steam, and is generally denoted by 'x'. Mathematically,

$$x = \frac{m_g}{m_g + m_f} = \frac{m_g}{m}$$

where m_g = Mass of actual dry steam,

m_f = Mass of water in suspension, and

m = Mass of wet steam $= m_g + m_f$

Note: The value of dryness fraction, in case of dry steam, is unity. At this stage, the mass of water in suspension (m_f) is zero.

5. *Sensible heat of water.* It is the amount of heat absorbed by 1 kg of water, when heated at a constant pressure, from the freezing point (0° C) to the temperature of formation of steam, *i.e.* saturation temperature (t). The sensible heat is also known as *liquid heat*.

The specific heat of water at constant pressure is usually taken as 4.2 kJ/kg K. Therefore heat absorbed by 1 kg of water from 0° C to t °C or sensible heat

$$= \text{Mass} \times \text{Sp. heat} \times \text{Rise in temperature}$$

$$= 1 \times 4.2\,[(t + 273) - (0 + 273)] = 4.2\,t \text{ kJ/kg}$$

Thus the sensible heat of water in kJ/kg may be obtained directly by multiplying the specific heat of water and the saturation temperature (t) in °C.

It may be noted that the sensible heat of water is taken equal to the * specific enthalpy (*i.e.* total heat) of water. It is, usually, denoted by h_f in steam tables and its value may be determined for any given pressure in kJ/kg.

---

* We know that specific enthalpy (*i.e.* enthalpy per kg) of water = Specific internal energy + Flow work *i.e.* work required in changing the volume against pressure p.

or $h = u + pdv$

Let v_f = Specific volume of water at 0° C (*i.e.* volume of 1 kg of water at 0° C).

When this water is heated from 0° C to t° C, there is a negligible change in volume. In other words, the specific volume of water at t° C is nearly the same *i.e.* v_f.

∴ Specific enthalpy of water at 0° C,

$$h_0 = u_0 + p\,v_f \quad \ldots (i)$$

Similarly, specific enthalpy of water at t° C,

$$h_f = u_f + p\,v_f \quad \ldots (ii)$$

∴ Change in specific enthalpy of water (*i.e.* enthalpy of 1 kg of water when heated from 0°C to t°C),

$$h_f - h_0 = u_f - u_0 \quad \ldots (iii)$$

Since the specific internal energy of water at 0° C is zero and the specific volume (v_f) is negligible, therefore $h_0 = 0$. Now from equation (*iii*),

$h_f = u_f$ *i.e.* specific internal energy at t° C or the heat absorbed by 1 kg of water in raising its temperature from 0° C to t° C

= Sensible heat

6. **Latent heat of vaporisation.* It is the amount of heat absorbed to evaporate 1 kg of water, at its boiling point or saturation temperature without change of temperature. It is denoted by h_{fg} and its value depends upon the pressure. The heat of vaporisation of water or latent heat of steam is 2257 kJ/kg at atmospheric pressure..

It has been experimentally found that the value of h_{fg} decreases as the pressure increases and it is zero at critical pressure. If the steam is wet with a dryness fraction x, then the heat absorbed by it during evaporation is $x\, h_{fg}$.

7. *Enthalpy or total heat of steam.* It is amount of heat absorbed by water from freezing point to saturation temperature *plus* the heat absorbed during evaporation.

∴ Enthalpy or total heat of steam

= Sensible heat + Latent heat

It is denoted by h_g and its value for the dry saturated steam may be read directly from the steam tables. The expressions for the enthalpy of wet steam, dry steam and superheated steam are as follows :

(*i*) *Wet steam.* The enthalpy of wet steam is given by :

$$h = h_f + x\, h_{fg} \qquad \ldots (i)$$

where x is the dryness fraction of steam.

(*ii*) *Dry steam.* We know that in case of dry steam, $x = 1$.

$$\therefore \quad h = h_g = h_f + h_{fg} \qquad \ldots (ii)$$

(*iii*) *Superheated steam.* If we further add heat to the dry steam, its temperature increases while pressure remaining constant. This increase in temperature shows the superheat stage of the steam. Thus, the total heat required for the steam to be superheated is :

$$h_{sup} = \text{Total heat for dry steam} + \text{Heat for superheated steam}$$

$$= h_f + h_{fg} + c_p\,(t_{sup} - t) = h_g + c_p\,(t_{sup} - t) \;**$$

where c_p = Mean specific heat at constant pressure for superheated steam,

t_{sup} = Temperature of the superheated steam, and

t = Saturation temperature at the given constant pressure.

Notes : 1. The difference $(t_{sup} - t)$ is known as *degree of superheat.*

2. The value of c_p for steam lies between 1.67 kJ/kg K to 2.5 kJ/kg K.

8. *Specific volume of steam.* It is the volume occupied by the steam per unit mass at a given temperature and pressure, and is expressed in m^3/kg. It is the reciprocal of density of steam in kg/m^3. It may be noted that the value of specific volume decreases with the increase in pressure. The expressions for the volumes occupied by steam are as follows :

(*i*) *Wet steam.* Consider 1 kg of wet steam of dryness fraction x. We know that this steam will have x kg of dry steam and $((1 - x)$ kg of water. Let v_f be the volume of 1 kg of water, then

Volume of one kg of wet steam

$$= x\, v_g + (1 - x)\, v_f$$

* See also Art. 7.8

** It makes no difference whether the individual temperatures (t_{sup} and t) are in Kelvin or °C because of difference in temperatures.

Since v_f is very small as compared to v_g, therefore the expression $(1-x)\,v_f$ may be neglected.

$\therefore$ Volume of 1 kg of wet steam $= x\,v_g$ m^3

or specific volume of wet steam, $\quad v = x\,v_g$ m^3/kg $\quad \ldots (i)$

(ii) Dry steam. We know that in case of dry steam, the mass of water in suspension is zero and dryness fraction is unity. Therefore specific volume of dry steam

$$= v_g \text{ m}^3\text{/kg} \qquad \ldots (ii)$$

(iii) Superheated steam. We have already discussed that when the dry saturated steam is further heated under a constant pressure, there is an increase in volume with the rise in temperature. The superheated steam behaves more or less like a perfect gas. Therefore, according to Charles' law,

$$\frac{v_{sup}}{T_{sup}} = \frac{v_g}{T} \quad \text{or} \quad v_{sup} = \frac{v_g\,T_{sup}}{T} \qquad \ldots (\because p \text{ is constant}) \ldots (iii)$$

where $\quad v_{sup}$ = Specific volume of superheated steam,

v_g = Specific volume of dry steam at the pressure of steam formation,

T_{sup} = Absolute temperature of superheated steam, and

T = Absolute saturation temperature at the pressure of steam formation.

7.5. Steam Tables and their Uses

The properties of dry saturated steam like its temperature of formation (saturation temperature), sensible heat, latent heat of vaporisation, enthalpy or total heat, specific volume, entropy etc., vary with pressure, and can be found by experiments only. These properties have been carefully determined, and made available in a tabular form known as *steam tables*. It may be noted, that there is a slight difference in figures, quoted by various authors, and even in their various editions of a book. But in this book we shall use the standard properties which are internationally recognised.

There are two important steam tables, one in terms of absolute pressure and other in terms of temperature. An extract of the properties of dry saturated steam is shown in the following two tables :

Table 7.1 (Pressure)

Absolute Pressure (p) in bar	*Temperature (t) in °C*	*Specific volume in m^3/kg*		*Specific enthalpy in kJ/kg*			*Specific entropy in kJ/kg K*		
		Water (v_f)	Steam (v_g)	Water (h_f)	Evaporation (h_{fg})	Steam (h_g)	Water (s_f)	Evaporations (s_{fg})	Steam (s_g)
0.010	6.983	0.001 000	129.21	29.3	2485.1	2514.4	0.106	8.871	8.977
0.015	13.04	0.001 001	87.982	54.7	2470.8	2525.5	0.196	8.634	8.830
0.20	60.09	0.001 017	7.649	251.5	2358.4	2609.9	0.832	7.077	7.909

Table 7.2 (Temperature)

Temperature in (t) in °C	*Absolute pressure (p) in bar*	Specific volume in m^3/kg		*Specific enthalpy in kJ/kg*			*Specific entropy in kg K*		
		Water (v_f)	Steam (v_g)	Water (h_f)	Evaporation (h_{fg})	Steam (h_g)	Water (s_f)	Evaporations (s_{fg})	Steam (s_g)
0	0.006 11	0.001 000	206.31	0.0	2501.6	2501.6	0.000	9.158	9.158
5	0.008 72	0.001 000	147.16	21.0	2489.7	2510.7	0.076	8.951	9.027
10	0.012 27	0.001 000	106.43	42.0	2477.9	2519.9	0.151	8.751	8.902

Example 7.1. *Calculate the enthalpy of 1 kg of steam at a pressure of 8 bar and dryness fraction of 0.8. How much heat would be required to raise 2 kg of this steam from water at 20° C?*

Solution. Given : $p = 8$ bar ; $x = 0.8$

Enthalpy of 1 kg of steam

From steam tables, corresponding to a pressure of 8 bar, we find that

$$h_f = 720.9 \text{ kJ/kg} \text{ and } h_{fg} = 2046.5 \text{ kJ/kg}$$

We know that enthalpy of 1 kg of wet steam,

$$h = h_f + x\,h_{fg} = 720.9 + 0.8 \times 2046.5 = 2358.1 \text{ kJ } \textbf{Ans.}$$

Heat required to raise 2 kg of this steam from water at 20° C

We have calculated above the enthalpy or total heat required to raise 1 kg of steam from water at 0° C. Since the water, in this case, is already at 20° C, therefore

Heat already in water $= 4.2 \times 20 = 84$ kJ

∴ Heat required per kg of steam

$$= 2358.1 - 84 = 2274.1 \text{ kJ}$$

and heat required for 2 kg of steam

$$= 2 \times 2274.1 = 4548.2 \text{ kJ } \textbf{Ans.}$$

Example 7.2. *Determine the quantity of heat required to produce 1 kg of steam at a pressure of 6 bar at a temperature of 25° C, under the following conditions :*

1. when the steam is wet having a dryness fraction 0.9 ;

2. when the steam is dry saturated ; and

3. when it is superheated at a constant pressure at 250° C assuming the mean specific heat of superheated steam to be 2.3 kJ/kg K.

Solution. Given : $p = 6$ bar ; $t_w = 25°$ C ; $x = 0.9$; $t_{sup} = 250°$ C ; $c_p = 2.3$ kJ/kg K

From steam tables, corresponding to a pressure of 6 bar, we find that

$$h_f = 670.4 \text{ kJ/kg} ; \; h_{fg} = 2085 \text{ kJ/kg and } t = 158.8° \text{ C}$$

1. *When the steam is wet*

We know that enthapy or total heat of 1 kg of wet steam,

$$h = h_f + x\,h_{fg} = 670.4 + 0.9 \times 2085 = 2546.9 \text{ kJ}$$

Since the water is at a temperature of 25° C, therefore

Heat already in water $= 4.2 \times 25 = 105$ kJ

∴ Heat actually required $= 2546.9 - 105 = 2441.9$ kJ **Ans.**

2. *When the steam is dry saturated*

We know that enthalpy or total heat of 1 kg of dry saturated steam,

$$h_g = h_f + h_{fg} = 670.4 + 2085 = 2755.4 \text{ kJ}$$

∴ Heat actually required $= 2755.4 - 105 = 2650.4$ kJ **Ans.**

3. *When the steam is superheated*

We know that enthalpy or total heat of 1 kg of superheated steam,

$$h_{sup} = h_g + c_p\,(t_{sup} - t) = 2755.4 + 2.3\,(250 - 158.8) = 2965.16 \text{ kJ}$$

∴ Heat actually required $= 2965.16 - 105 = 2860.16$ kJ **Ans.**

Example 7.3. *Determine the condition of steam in the following cases :*

1. At a pressure of 10 bar and temperature 200° C.

2. At a pressure of 10 bar and volume 0.175 m³/kg.

Solution. Given : $p = 10$ bar ; $t = 200°$ C ; $v = 0.175$ m³/kg

1. Condition of steam at temperature of 200° C

From steam tables, corresponding to a pressure of 10 bar, we find that

$$v_g = 0.194 \text{ m}^3/\text{kg} \; ; \; h_f = 762.6 \text{ kJ/kg} \; ; \text{ and } t = 179.9° \text{C}$$

Since the saturation temperature at 10 bar is (179.9° C) or lower than the given temperature of the steam (200° C), therefore the given steam is superheated. The degree of superheat

$$= 200 - 179.9 = 20.1° \text{C} \textbf{ Ans.}$$

2. Condition of steam at a volume of 0.175 m³/kg

Since the volume of given steam (0.175 m³/kg) is less than the specific volume of the dry saturated steam (0.194 m³/kg), therefore the given steam is wet. The dryness fraction of steam,

$$x = \frac{0.175}{0.194} = 0.902 \textbf{ Ans.}$$

Example 7.4. *Steam enters an engine at a pressure of 12 bar with a 67° C of superheat. It is exhausted at a pressure of 0.15 bar and 0.95 dry. Find the drop in enthalpy of the steam.*

Solution. Given : $p_1 = 12$ bar ; $t_{sup} - t = 67°$ C ; $p_2 = 0.15$ bar ; $x = 0.95$

From steam tables, corresponding to a pressure of 12 bar, we find that

$$h_f = 798.4 \text{ kJ/kg} \; ; \; h_{fg} = 1984.3 \text{ kJ/kg}$$

We know that enthalpy or total heat of 1 kg of superheated steam,

$$h_{sup} = h_f + h_{fg} + c_p(t_{sup} - t) = 798.4 + 1984.3 + 2 \times 67$$

$$= 2916.7 \text{ kJ/kg} \textbf{ Ans.} \qquad \ldots (\text{Taking } c_p = 2 \text{ kJ/kg K})$$

Similarly, from steam tables, corresponding to a pressure of 0.15 bar, we find that

$$h_f = 226 \text{ kJ/kg} \; ; \; h_{fg} = 2373.2 \text{ kJ/kg}$$

We know that enthalpy or total heat of 1 kg of wet steam,

$$h = h_f + x\,h_{fg} = 226 + 0.95 \times 1984.3 = 2111 \text{ kJ/kg}$$

∴ Drop in enthalpy of the steam

$$= h_{sup} - h = 2916.7 - 2111 = 805.7 \text{ kJ/kg} \textbf{ Ans.}$$

Example 7.5. *A steam engine obtains steam from a boiler at a pressure of 15 bar and 0.98 dry. It was observed that the steam loses 21 kJ of heat per kg as it flows through the pipe line, pressure remaining constant. Calculate dryness fraction of the steam, at the engine end of the pipeline.*

Solution. Given : $p = 15$ bar ; $x = 0.98$; Heat loss = 21 kJ/kg

From steam tables, corresponding to a pressure of 15 bar, we find that

$$h_f = 844.6 \text{ kJ/kg} \; ; \; h_{fg} = 1945.3 \text{ kJ/kg}$$

We know that enthalpy of wet steam at the boiler end,

$$h_1 = h_f + x\,h_{fg} = 844.6 + 0.98 \times 1945.3 = 2751 \text{ kJ/kg}$$

Since the steam loses 21 kJ/kg of steam, therefore enthalpy of wet steam at the engine end,

$$h_2 = 2751 - 21 = 2730 \text{ kJ}$$

Let x_2 = Dryness fraction of steam at the engine end.

Since the pressure remains constant, therefore h_f and h_{fg} is same. We know that

$$h_2 = h_f + x_2 h_{fg}$$

$$2730 = 844.6 + x_2 \times 1945.3 \quad \text{or} \quad x_2 = 0.97 \text{ Ans.}$$

7.6. Superheated Steam

We have already discussed that whenever dry steam or saturated steam is further heated, then the steam is termed as superheated steam. The process of superheating is assumed to follow constant pressure process (or Charles' law).

The values of saturation temperature, specific volume, specific enthalpy and specific entropy at a given pressure of the superheated steam are also given in the tabular form known as steam tables of superheated steam.

7.7. Advantages of Superheating the Steam

The following advantages of superheating the steam are important from the subject point of view :

1. The superheated steam contains more heat contents, and hence its capacity to do work is also increased without increasing its pressure.

2. The superheating is done in a superheater, which obtains heat from waste furnace gases. These gases would have otherwise passed, uselessly, through the chimney.

3. The high temperature of the superheated steam results in an increase of thermal efficiency.

4. Since the superheated steam is at a higher temperature than that corresponding to its pressure, therefore it can be considerably cooled during expansion in an engine cylinder. This is done before the temperature of superheated steam falls below that at which it condenses and, thereby, becomes wet. It is thus obvious, that heat losses due to condensation steam on cylinder walls, etc., are avoided to a great extent.

Example 7.6. *Determine the volume of 1 kg of superheated steam at a pressure of 20 bar and a temperature of 300° C.*

Solution. Given : $p = 20$ bar ; $T_{sup} = 300° \text{ C} = 300 + 273 = 573$ K

Let v_{sup} = Volume of 1 kg of superheated steam.

From steam tables, corresponding to a pressure of 20 bar, we find that

$$t = 212.4° \text{ C} \quad \text{and} \quad v_g = 0.1 \text{ m}^3/\text{kg}$$

$$\therefore \quad T = 212.4 + 273 = 485.4 \text{ K}$$

We know that the superheated steam obeys Charles's law.

$$\therefore \quad \frac{v_g}{T} = \frac{v_{sup}}{T_{sup}} \qquad \ldots (\because p \text{ is constant})$$

or

$$v_{sup} = \frac{v_g}{T} \times T_{sup} = \frac{0.1}{485.4} \times 573 = 0.118 \text{ m}^3 \text{ Ans.}$$

Example 7.7. *A boiler is supplied with feed water at a temperature of 45° C. The water is converted into steam at a pressure of 5.5 bar and a temperature of 188° C. Determine the quantity of heat supplied per kg of steam. Assume suitable data.*

Solution. Given : $t_w = 45° C$; $p = 5.5$ bar ; $t_{sup} = 188° C$

From steam tables, corresponding to a pressure of 5.5 bar, we find that

$$h_f = 655.8 \text{ kJ/kg} \; ; \; h_{fg} = 2095.9 \text{ kJ/kg} \; ; \; t = 155.5° C$$

Since the temperature of the given steam (188° C) is higher than the saturation temperature (155.5° C), therefore the steam produced is superheated. Now let us assume the value of c_p for superheated steam as 2.1 kJ/kg K. We know that enthalpy or total heat of superheated steam,

$$h_{sup} = h_f + h_{fg} + c_p (t_{sup} - t)$$

$$= 655.8 + 2095.9 + 2.1 (188 - 155.5) = 28.20 \text{ kJ/kg}$$

Since the water is supplied at 45° C, therefore heat already present in the feed water

$$= 4.2 \times 45 = 189 \text{ kJ}$$

$\therefore$ Heat actually required $= 2820 - 189 = 2631$ kJ/kg **Ans.**

7.8. External Work Done during Evaporation

We have already discussed that whenever water at boiling temperature is heated at a constant pressure, it gets converted into steam after absorbing the latent heat. This latent heat is utilised in the following two ways :

1. In overcoming the internal molecular resistance of water in changing its state from the saturated water to dry saturated steam.
2. In overcoming the external resistance to the movement of the piston due to increase in volume during evaporation.

The first effect is called *internal work* or *internal latent heat*, as the change takes place within the body of the steam, and represents the energy stored in the steam. The second effect is called *external work of evaporation*, and represents the energy which has been taken out of the steam.

Let p = Pressure on the piston in bar $= p \times 10^5$ N/m^2,

v_f = Volume of water in m^3 at pressure p, and

v_g = Volume of steam in m^3 at pressure p.

We know that work done during evaporation,

$$W = \text{Pressure} \times \text{Change in volume}$$

$$= p \times 10^5 (v_g - v_f) \text{ J} = 100 p (v_g - v_f) \text{ kJ}$$

At low pressures, the volume of water (v_f) is very small, as compared to the volume of steam (v_g). Therefore neglecting the value of v_f in the above equation,

$$W = 100 \, p \, v_g \text{ kJ}$$

If the steam is not completely dry, and has dryness fraction of x, then work done,

$$W = 100 \, p \, x \, v_g \text{ kJ}$$

If the steam is superheated, then *work done,

$$W = 100 \, p \, v_{sup} \text{ kJ}$$

---

* The work done by the superheated steam may also be found out as discussed below :

Work done during superheating $= 100 \, p \, v_{sup}$ kJ

$\therefore$ Total work done = Work done during evaporation + Work done during superheating

$$= 100 \, p \, v_g + 100 \, p \, (v_{sup} - v_g) = 100 \, p \, v_{sup} \text{ kJ}$$

Example 7.8. *Find the external work done during evaporation per kg of steam at a pressure of 15 bar when the steam is (a) 90% dry and (b) dry saturated.*

Solution. Given : $p = 15$ bar

From steam tables, corresponding to a pressure of 15 bar, we find that

$$v_g = 0.1317 \text{ m}^3/\text{kg}$$

(a) Workdone when the steam is 90% dry (i.e. $x = 0.9$)

We know that workdone,

$$w = 100\, p\, x\, v_g = 100 \times 15 \times 0.9 \times 0.1317 = 177.8 \text{ kJ/kg} \quad \textbf{Ans.}$$

(b) Workdone when the steam is dry saturated

We know that workdone,

$$w = 100\, p\, v_g = 100 \times 15 \times 0.1317 = 197.5 \text{ kJ/kg} \quad \textbf{Ans.}$$

7.9. Internal Energy of Steam

It is the actual heat energy stored in steam, above the freezing point of water. The internal energy may be calculated by subtracting the external work done during evaporation from the enthalpy or total heat of steam. Mathematically,

Internal energy of steam

= Enthallpy or total heat – External workdone during evaporation

The expressions of internal energy per kg of steam (u) for wet steam, dry saturated steam and superheated steam are given below :

(a) For wet steam

$$u = h - 100\, p\, x\, v_g = h_f + x\, h_{fg} - 100\, p\, x\, v_g \text{ kJ/kg}$$

(b) For dry saturated steam

$$u = h_g - 100\, p\, v_g = h_f + h_{fg} - 100\, p\, v_g \text{ kJ/kg}$$

(c) For superheated steam,

$$u = h_{sup} - 100\, p\, v_{sup} = [h_g + c_p\,(t_{sup} - t)] - 100\, p\, v_{sup} \text{ kJ/kg}$$

Note: If the steam of initial internal energy (u_1) is heated in a cylinder to a final internal energy (u_2), then change in internal energy,

$$u = u_2 - u_1$$

Example 7.9. *Calculate the internal energy of 1 kg of steam at a pressure of 10 bar when the steam is : (a) 0.9 dry ; and (b) dry saturated. The volume of water may be neglected.*

Solution. Given : $p = 10$ bar

From steam tables, corresponding to a pressure of 10 bar, we find that

$$h_f = 762.6 \text{ kJ/kg} \; ; \; h_{fg} = 2013.6 \text{ kJ/kg and } v_g = 0.1943 \text{ m}^3/\text{kg}$$

(a) Internal energy of 1 kg of steam when it is 0.9 dry (i.e. $x = 0.9$)

We know that internal energy for wet steam

$$u = h_f + x\, h_{fg} - 100\, p\, x\, v_g$$

$$= 762.6 + 0.9 \times 2013.6 - 100 \times 10 \times 0.9 \times 0.1943 \text{ kJ}$$

$$= 2574.8 - 174.8 = 2400 \text{ kJ} \quad \textbf{Ans.}$$

(*b*) *Internal energy of 1 kg of steam when it is dry saturated*

We know that internal energy for dry saturated steam,

$$u = h_f + h_{fg} - 100\, p\, v_g$$

$$= 762.6 + 2013.6 - 100 \times 10 \times 0.1943$$

$$= 2776.2 - 194.3 = 2581.9 \text{ kJ} \quad \textbf{Ans.}$$

Example 7.10. *Find the internal energy of 1 kg of superheated steam at a pressure of 10 bar and 280° C. If this steam be expanded to a pressure of 1.6 bar and 0.8 dry, determine the change in internal energy. Assume specific heat of superheated steam as 2.1 kJ/kg K.*

Solution. Given : $p_1 = 10$ bar ; $t_{sup} = 280°$ C ; $p_2 = 1.6$ bar ; $x = 0.8$; $c_p = 2.1$ kJ/kg K

From steam tables, corresponding to a pressure of 10 bar, we find that

$$h_f = 762.6 \text{ kJ/kg} \;;\; h_{fg} = 2013.6 \text{ kJ/kg} \;;\; v_g = 0.1943 \text{ m}^3\text{/kg} \;;\; \text{and } t = 179.9° \text{C}$$

We know that enthalpy of 1 kg of superheated steam,

$$h_{sup} = h_f + h_{fg} + c_p\,(t_{sup} - t)$$

$$= 762.6 + 2013.6 + 2.1\,(280 - 179.9) = 2986.4 \text{ kJ/kg}$$

and
$$v_{sup} = v_g \times \frac{T_{sup}}{T} = 0.1943 \times \frac{(280 + 273)}{(179.9 + 273)} = 0.237 \text{ m}^3\text{/kg}$$

∴ Internal energy of superheated steam,

$$u_{sup} = h_{sup} - 100\, p_1\, v_{sup}$$

$$= 2986.4 - 100 \times 10 \times 0.237 = 2749.4 \text{ kJ/kg}$$

The superheated steam now expands to a pressure of 1.6 bar. From steam tables corresponding to a pressure of 1.6 bar, we find that

$$h_f = 475.4 \text{ kJ/kg} \;;\; h_{fg} = 2220.9 \text{ kJ/kg} \;;\; \text{and } v_g = 1.091 \text{ m}^3\text{/kg}$$

We know that internal energy of the expanded steam,

$$u_e = h_f + x\, h_{fg} - 100\, p_2\, x\, v$$

$$= 475.4 + 0.8 \times 2220.9 - 100 \times 1.6 \times 0.8 \times 1.091$$

$$= 2252.1 - 139.7 = 2112.4 \text{ kJ/kg}$$

∴ Change in internal energy

$$= u_{sup} - u_e = 2749.4 - 2112.4 = 637 \text{ kJ/kg} \quad \textbf{Ans.}$$

Example 7.11. *A vessel contains 2 kg of steam at a pressure of 8 bar. Find the amount of heat, which must be rejected, so as to reduce the quality of steam in the vessel to be 70%.*

Solution. Given : $m = 2$ kg ; $p_1 = 8$ bar ; $x_2 = 70\% = 0.7$

From steam tables, corresponding to a pressure of 8 bar, we find that

$$h_{g1} = 2767.5 \text{ kJ/kg} \;;\; \text{and } v_{g1} = 0.24 \text{ m}^3\text{/kg}$$

∴ Volume of vessel $= 2 \times 0.24 = 0.48 \text{ m}^3$

and initial internal energy per kg of steam,

$$u_1 = h_{g1} - 100\, p_1\, v_{g1} = 2767.5 - 100 \times 8 \times 0.24 = 2575.5 \text{ kJ/kg}$$

Now neglecting the volume of water in the wet steam (with dryness fraction of 0.7), the volume occupied by the steam will be equal to the volume of vessel.

$\therefore \quad 0.48 = 2 \times 0.7 \times v_{g2} = 1.4\, v_{g2} \quad \text{or} \quad v_{g2} = 0.343 \text{ m}^3\text{/kg}$

From steam tables, corresponding to a specific volume of 0.343 m³/kg, we find that

$$p_2 = 5.5 \text{ bar} \; ; \; h_f = 655.8 \text{ kJ/kg} \; ; \text{ and } h_{fg} = 2095.9 \text{ kJ/kg}$$

∴ Final internal energy per kg of steam,

$$u_2 = h_f + x_2\, h_{fg} - 100\, p_2\, x_2\, v_{g2}$$

$$= 655.8 + 0.7 \times 2095.9 - 100 \times 5.5 \times 0.7 \times 0.343$$

$$= 2123 - 132 = 1991 \text{ kJ/kg}$$

and heat rejected $= u_1 - u_2 = 2575.5 - 1991 = 584.5$ kJ/kg

∴ Heat rejected by 2 kg of steam

$$= 2 \times 584.5 = 1169 \text{ kJ} \quad \textbf{Ans.}$$

Example 7.12. *Steam at 18 bar and dryness 0.9 is heated at constant pressure until dry and saturated. Find the increase in volume, heat supplied and work done per kg of steam.*

If the volume is now kept constant, find how much heat must be extracted to reduce the pressure to 14 bar.

Solution. Given : $p = 18$ bar ; $x = 0.9$; $p_1 = 14$ bar

From steam tables, corresponding to a pressure of 18 bar, we find that

$$h_f = 884.5 \text{ kJ/kg} \; ; \; h_{fg} = 1910.3 \text{ kJ/kg} \; ; h_g = 2794.8 \text{ kJ/kg} \; ; \text{ and } v_g = 0.1103 \text{ m}^3\text{/kg}$$

Increase in volume

We know that volume of wet steam,

$$v_f = x\, v_g = 0.9 \times 0.1103 = 0.0993 \text{ m}^3\text{/kg}$$

∴ Increase in volume

$$= v_g - v_f = 0.1103 - 0.0993 = 0.011 \text{ m}^3\text{/kg} \quad \textbf{Ans.}$$

Heat supplied

We know that enthalpy or total heat of wet steam,

$$h = h_f + x\, h_{fg} = 884.5 + 0.9 \times 1910.3 = 2603.8 \text{ kJ/kg}$$

∴ Heat supplied $= h_g - h = 2794.8 - 2603.8 = 191$ kJ/kg **Ans.**

Workdone

We know that workdone,

$$w = 100\, p\, (v_g - v_f) = 100 \times 18 \times 0.011 = 19.8 \text{ kJ/kg} \quad \textbf{Ans.}$$

Heat extracted to reduce the pressure to 14 bar

We know that during constant volume process, no work is done and extraction of heat will result in the reduction of internal energy of the steam. In other words, the steam is cooled, which results in the reduction of dryness fraction.

From steam tables, corresponding to a pressure of 14 bar, we find that

$$h_{f1} = 830.1 \text{ kJ/kg} \; ; \; h_{fg1} = 1957.7 \text{ kJ/kg} \; ; \text{ and } v_{g1} = 0.1407 \text{ m}^3\text{/kg}$$

Let x_1 = Dryness fraction of steam.

Since the volume is constant, therefore

$$v_g = x_1\, v_{g1} \quad \text{or} \quad x_1 = v_g / v_{g1} = 0.1103 / 0.1407 = 0.78$$

Now internal energy of steam at a pressure of 18 bar,

$$u = h_g - 100\, p\, v_g = 2794.8 - 100 \times 18 \times 0.1103 = 2596.3 \text{ kJ/kg}$$

and internal energy of steam at a pressure of 14 bar,

$$u_1 = h_{f1} + x_1 h_{fg1} - 100\, p_1\, x_1\, v_{g1}$$

$$= 830.1 + 0.78 \times 1957.7 - 100 \times 14 \times 0.78 \times 0.1407$$

$$= 2357 - 154 = 2203 \text{ kJ/kg}$$

∴ Heat extracted to reduce the pressure

$$= u - u_1 = 2596.3 - 2203 = 393.3 \text{ kJ/kg} \quad \textbf{Ans.}$$

7.10. Measurement of Dryness Fraction of Steam

Though there are a number of steam calorimeters to determine the dryness fraction of wet steam, yet the following are important from the subject point of view :

1. Barrel calorimeter ; 2. Separating calorimeter, 3. Throttling calorimeter, and 4. Combined separating and throttling calorimeter.

Now we shall discuss these calorimeters, in detail, in the following pages.

7.11. Barrel Calorimeter

In barrel calorimeter, as shown in Fig. 7.3, the known mass of steam sample and of known pressure is condensed by mixing it with a known mass of cold water. The steam condenses into water, thereby increasing the mass and temperature of the water in the calorimeter. The heat supplied by the condensing steam is determined, with the help of the initial and final temperatures of the water and calorimeter.

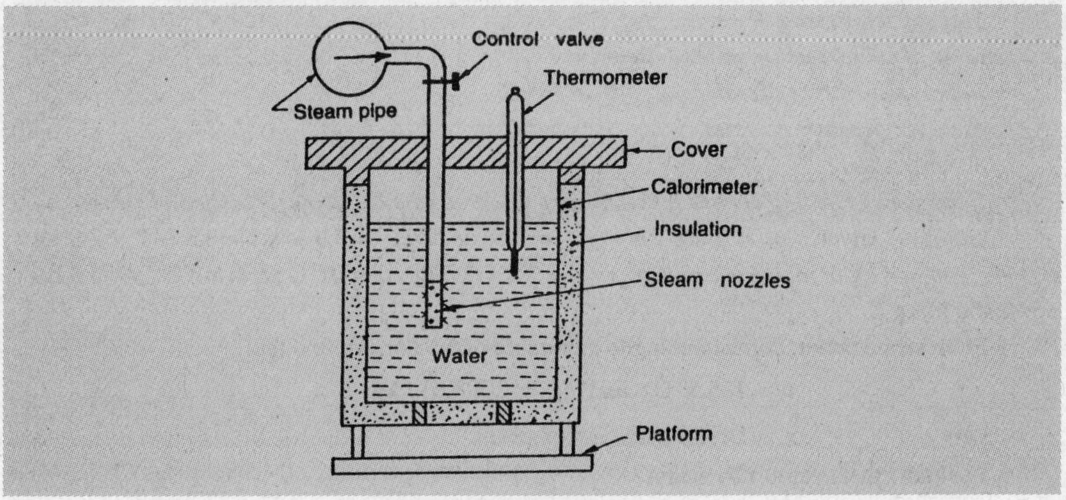

Fig. 7.3. Barrel calorimeter.

Let p = Pressure of the steam in bar,

t = Temperature of steam formation at pressure p (from steam tables),

h_{fg} = Latent heat of steam at pressure p (from steam tables),

m_c = Mass of the calorimeter,

c_c = Specific heat of the calorimeter,

m_e = Water equivalent of the calorimeter $= m_c \times c_c$,

m_s = Mass of the steam condensed,

m_w = Mass of cold water in the calorimeter,

t_1 = Initial temperature of water and calorimeter,

t_2 = Final temperature of water and calorimeter,

c_w = Specific heat of water (4.2 kJ / kg K),

x = Dryness fraction of steam sample.

According to the law of conservation, the heat lost by steam is equal to the heat gained by water and calorimeter, provided there is no addition or loss of heat.

Heat lost by steam $= m_s [(x\, h_{fg}) + c_w (t - t_2)]$ ... (*i*)

Heat gained by water and calorimeter,

$$= (m_w\, c_w + m_c\, c_c)(t_2 - t_1) \quad \ldots (ii)$$

Equating equations (*i*) and (*ii*),

$$m_s [(x\, h_{fg}) + c_w (t - t_2)] = (m_w\, c_w + m_c\, c_c)(t_2 - t_1)$$

From this expression, the dryness fraction of steam (x) may be determined.

Example 7.13. *In a laboratory experiment on wet steam by a barrel calorimeter, the following observations were recorded :*

Mass of copper calorimeter = *1 kg*

Mass of calorimeter + water = *3.8 kg*

Mass of calorimeter + water + steam = *4 kg*

Initial temperature of water = *10° C*

Final temperature of water = *50° C*

Steam pressure = *5.5 bar*

If the specific heat of copper is 0.406 kJ/kg K, determine the dryness fraction of steam.

Solution. Given : $m_c = 1$ kg ; $m_c + m_w = 3.8$ kg or $m_w = 3.8 - m_c = 3.8 - 1 = 2.8$ kg ; $m_c + m_w + m_s = 4$ kg or $m_s = 4 - (m_c + m_w) = 4 - 3.8 = 0.2$ kg ; $t_1 = 10°$ C ; $t_2 = 50°$ C ; $p = 5.5$ bar ; $c_c = 0.406$ kJ/kg K

From steam tables, corresponding to a pressure of 5.5 bar, we find that

$$t = 155.5° \text{ C} ; \text{ and } h_{fg} = 2095.5 \text{ kJ/kg}$$

Let x = Dryness fraction of steam.

We know that heat lost by steam

$$= m_s [(x\, h_{fg}) + c_w (t - t_2)]$$

$$= 0.2 [x \times 2095.5 + 4.2 (155.5 - 50)] \quad \ldots (\because c_w \text{ for water} = 4.2 \text{ kJ/kg K})$$

$$= 419.1\, x + 88.6 \text{ kJ} \quad \ldots (i)$$

and heat gained by water and calorimeter

$$= (m_w\, c_w + m_c\, c_c)(t_2 - t_1)$$

$$= (2.8 \times 4.2 + 1 \times 0.406)(50 - 10) = 486.6 \text{ kJ} \quad \ldots (ii)$$

Equating equations (*i*) and (*ii*),

$$419.1\,x + 88.6 = 486.6 \text{ or } x = 0.95 \textbf{ Ans.}$$

7.12. Separating Calorimeter

The separating calorimeter is used to determine the dryness fraction of steam by mechanically separating the water particles from the wet steam.

The wet steam enters at the top of the calorimeter through a control valve as shown in Fig. 7.4. It strikes the perforated cup and thereby undergoes a quick reversal of directions of motion. The water particles, due to their greater moment of inertia, tend to move on, and consequently get separated from the mixture. The separated water collects at the bottom of the inner chamber, and its amount may be determined by the graduated water gauge. The amount of dry steam leaving the outer chamber of the calorimeter may be measured by condensing it in a weighed quantity of cold water or alternatively by reading the pressure gauge which is calibrated in terms of rate of steam flow for this purpose.

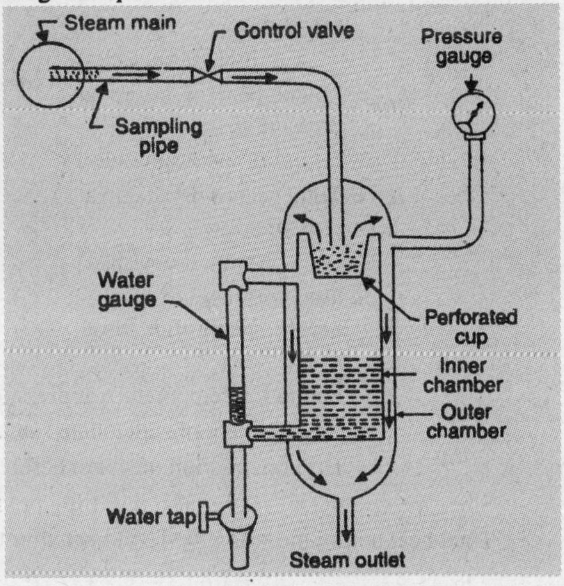

Fig. 7.4. Separating calorimeter.

Let m = Mass of water collected in a certain time,

M = Mass of dry steam passing in the same time, and

x = Dryness fraction of wet steam.

We know that dryness fraction,

$$x = \frac{\text{Mass of dry steam}}{\text{Mass of wet steam}} = \frac{M}{M+m}$$

Example 7.14. *In a laboratory experiment, a sample of wet steam is allowed to pass through a separating calorimeter. At some instant, the water collected in the chamber was 0.1 kg whereas the condensed steam was found to be 1.25 kg. Determine the dryness fraction of the steam entering the calorimeter.*

Solution. Given : $m = 0.1$ kg ; $M = 1.25$ kg

We know that dryness fraction of steam,

$$x = \frac{M}{M+m} = \frac{1.25}{1.25+0.1} = 0.926 \textbf{ Ans.}$$

7.13. Throttling Calorimeter

A throttling calorimeter used to determine the dryness fraction of steam is shown in Fig. 7.5. It consists of a separator *A* into which steam is admitted through a control valve from the steam main. The pressure and temperature are measured by the pressure gauge and the thermometer T_1 provided in this section. It may be noted that temperature recorded by T_1 is same as the saturation temperature corresponding to the pressure of steam in calorimeter *B*. This steam is then throttled through a narrow aperture of restricted valve openings, its total heat remaining constant. The steam is in the superheated state after throttling at a lower pressure than previous. The temperature and pressure of steam leaving the calorimeter B is noted by the thermometer T_2 and manometer respectively.

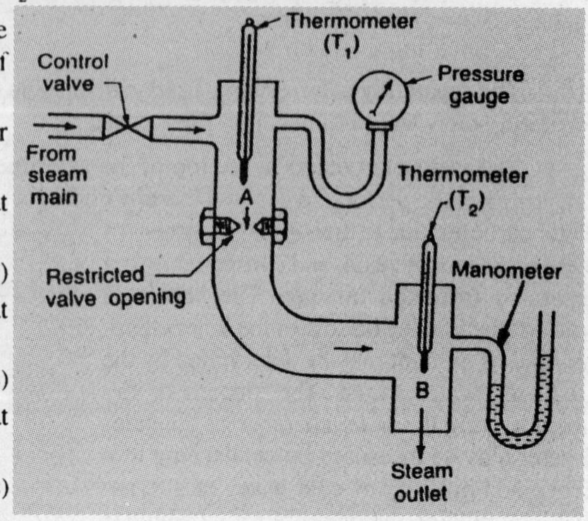

Fig. 7.5. Throttling calorimeter.

Let p_1 = Pressure of steam before throttling, *i.e.* pressure of steam main,

p_2 = Pressure of steam after throttling,

h_{f1} = Sensible heat of water at pressure p_1, ... (From steam tables)

h_{fg1} = Latent heat of steam at pressure p_1, ... (From steam tables)

h_{g2} = Total heat of dry steam at pressure p_2, ... (From steam tables)

t_{sup} = Temperature of superheated steam after throttling,

t_2 = Saturation temperature at pressure p_2, ... (From steam tables)

c_p = Specific heat of superheated steam, and

x = Dryness fraction of steam before throttling.

Since the steam has undergone a throttling process, therefore

Total heat before throttling = Total heat after throttling

$$\therefore \quad h_{f1} + x\, h_{fg1} = h_{g2} + c_p\,(t_{sup} - t_2)$$

Example 7.15. ***In a throttling calorimeter, the steam is admitted at a pressure of 10 bar. If it is discharged at atmospheric pressure and 110° C after throttling, determine the dryness fraction of steam. Assume specific heat of steam as 2.2 kJ/kg K.***

Solution. Given : p_1 = 10 bar ; p_2 = 1.013 bar ; t_{sup} = 110° C ; c_p = 2.2 kJ/kg K

Let x = Dryness fraction of steam.

From steam tables, corresponding to a pressure of 10 bar, we find that

$$h_{f1} = 762.6 \text{ kJ/kg} ; \quad \text{and} \quad h_{fg1} = 2013.6 \text{ kJ/kg}$$

and corresponding to a pressure of 1.013 bar,

$$h_{g2} = 2676 \text{ kJ/kg} ; \text{ and } t_2 = 100° \text{ C}$$

We know that $$h_{f1} + x\, h_{fg1} = h_{g2} + c_p\,(t_{sup} - t_2)$$

$$762.6 + x \times 2013.6 = 2676 + 2.2\,(110 - 100) = 2698$$

$$\therefore \quad x = \frac{2698 - 762.6}{2013.6} = 0.961 \textbf{ Ans.}$$

7.14. Combined Separating and Throttling Calorimeter

A very successful method of measuring the dryness fraction of steam is by a combined separating and throttling calorimeter as shown in Fig. 7.6.

In this calorimeter, the wet steam is first collected in a perforated collecting pipe and then passed through a separating calorimeter. A part of water is removed by the separating calorimeter owing to quick change of direction of flow. The resulting semi-dry steam is throttled into a throttling

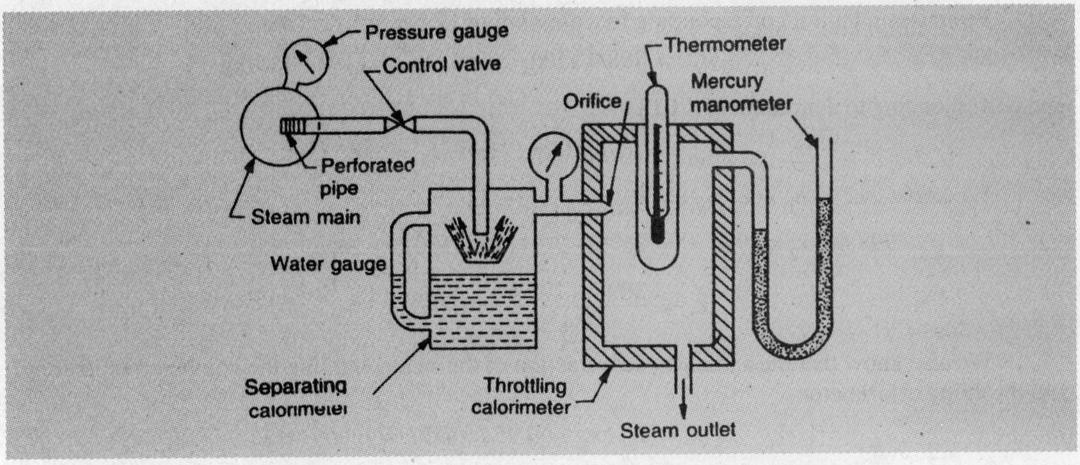

Fig. 7.6. Combined separating and throttling calorimeter.

calorimeter. This method ensures that the steam will be superheated after throttling. This instrument is well insulated to prevent any loss of heat.

Let x_1 = Dryness fraction of steam considering separating calorimeter, and

x_2 = Dryness fraction of steam entering the throttling calorimeter.

Now the actual dryness fraction of steam in the steam main,

$$x = x_1 \times x_2$$

Note : It is not possible to obtain results with this instrument if the final condition of steam is wet. The final condition of steam must be just dry or superheated.

Example 7.16. *In a laboratory experiment, the following observations were recorded to find the dryness fraction of steam by combined separating and throttling calorimeter :*

Total quantity of steam passed = *36 kg*

Water drained from separator = *1.8 kg*

Steam pressure before throttling = *12 bar*

Temperature of steam after throttling = *110° C*

Pressure after throttling = *1.013 bar*

Specific heat of steam = *2.1 kJ/kg K*

Determine the dryness fraction of steam before inlet to the calorimeter.

Solution Given : $m_s + m_w = 36$ kg ; $m_w = 1.8$ kg ; $p_1 = 12$ bar ; $t_{sup} = 110°$ C ; $p_2 = 1.013$ bar ; $c_p = 2.1$ kJ/kg K

Let x_1 = Dryness fraction for separating calorimeter,

x_2 = Dryness fraction for throttling calorimeter,

x = Actual dryness fraction entering the combined separating and throttling calorimeter.

We know that mass of dry steam,

$$m_s = (m_s + m_w) - m_w = 36 - 1.8 = 34.2 \text{ kg}$$

∴ Dryness fraction for separating calorimeter,

$$x_1 = \frac{m_s}{m_s + m_w} = \frac{34.2}{36} = 0.95$$

From steam tables, corresponding to a pressure of 12 bar, we find that

$$h_{f1} = 798.4 \text{ kJ/kg} \; ; \text{ and } h_{fg1} = 1984.3 \text{ kJ/kg}$$

and corresponding to a pressure of 1.013 bar,

$$h_{g2} = 2676 \text{ kJ/kg} \; ; \text{ and } t_2 = 100^\circ \text{ C}$$

We know that $h_{f1} + x_2 h_{fg1} = h_{g2} + c_p (t_{sup} - t_2)$

$$798.4 + x_2 \times 1984.3 = 2676 + 2.1 (110 - 100) = 2697$$

$$\therefore \quad x_2 = \frac{2697 - 798.4}{1984.3} = 0.957$$

We also know that the actual dryness fraction of the steam entering the combined separating and throttling calorimeter,

$$x = x_1 \times x_2 = 0.95 \times 0.957 = 0.909 \text{ Ans.}$$

Note : The value of x may also be obtained as follows :

First of all, the value fo x_2 is obtained as before, *i.e.* $x_2 = 0.957$

Now mass of dry steam entering the throttling calorimeter

= Mass of dry steam leaving the separating calorimeter

$= 34.2 \times 0.957 = 32.73$ kg

∴ Dryness fraction of steam entering the separating calorimeter,

$$x = \frac{32.73}{36} = 0.909 \text{ Ans.}$$

EXERCISES

1. 5 kg of steam at a pressure of 5 bar is produced from water at 20° C. Determine the amount of heat supplied, if the steam is 0.9 dry. **[Ans.** 12 265 kJ]

2. Find the amount of heat required to convert 1 kg of steam at a pressure of 0.5 bar and dryness fraction 0.9 to dry saturated steam. **[Ans.** 230.5 kJ]

3. Find the mass of 0.50 m³ of wet steam at a pressure of 4 bar and dryness fraction 0.8. Also determine the enthalpy of 1 m³ of steam. **[Ans.** 1.35 kg ; 6240 kJ]

4. One kg of water at 47.8° C is heated under a constant pressure of 14 bar until it is converted into steam with 110° C of superheat. Determine the quantity of heat supplied during superheating and the total heat of superheated steam. Assume $c_p = 2.1$ kJ/kg K. **[Ans.** 201 kJ ; 2818 kJ]

5. A certain amount of steam is produced at a pressure of 8 bar and a dryness fraction 0.8. Determine : 1. external workdone during evaporation ; and 2. internal latent heat of steam. **[Ans.** 153.6 kJ ; 1483.6 kJ]

6. Find the internal energy of 1 kg of steam at a pressure of 7 bar under the following conditions :
   1. when it is wet having dryness fraction 0.8 ; and 2. when it is dry saturated. **[Ans.** 2196 kJ/kg ; 2571 kJ/kg]

7. Calculate the internal energy per kg of superheated steam at a pressure of 10 bar and a temperature of 300° C. If the steam is expanded to 1.4 bar and dryness fraction 0.8, find the change of internal energy. **[Ans.** 701 kJ/kg]

8. Find the internal energy of 1 kg of steam at 20 bar in the following cases :
   1. when the stem is wet having dryness fraction 0.9, and 2. when the steam is superheated when its temperature is 400° C.
   Take specific heat for superheated steam as 2.3 kJ/kg K. **[Ans.** 2429.3 kJ/kg ; 2952.7 kJ/kg]

9. Two boilers, one with a superheater and the other without a superheater are delivering equal quantities of steam into a common mains. The pressure in the boiler and the mains is 14.7 bar. The temperature of steam from the boiler with superheater is 325° C and that of steam mains 250° C. Estimate the quality of steam supplied by the other boiler. [Ans. 0.97]

10. Steam for a small boiler is discharged through a pipe into a barrel containing water and after a few minutes, observations were taken of the mass and temperature of the water. From the collected data given below, calculate the dryness fraction of the steam.

Mass of empty barrel = 25 kg ; Mass of barrel + cold water = 135 kg ; Mass of barrel + cold water + steam = 140 kg ; Initial temperature of water = 15° C ; Final temperature of water = 42° C ; Steam pressure = 8 bar ; Specific heat of the material of barrel = 0.4 kJ/kg K [Ans. 0.96]

11. In an experiment, it was found that the steam enters a throttling calorimeter at a pressure of 12.25 bar. After throttling, the pressure and temperature was measured as 1.013 bar and 115° C respectively. Estimate the dryness fraction of steam. [Ans. 0.96]

12. In a combined separating and throttling calorimeter, the following observations were made :

Total quantity of steam = 23.4 kg ; Water drained from separator = 1.2 kg ; Steam pressure before throttling = 8.25 bar ; Temperature of steam on leaving = 111.4° C ; Steam pressure on leaving = 1.013 bar.

Find the dryness fraction of steam on entry. Specific heat of superheated steam is 2 kJ/kg K. [Ans. 0.92]

QUESTIONS

1. What is meant by saturation temperature and saturation pressure ?
2. Explain how the wet steam, dry saturated steam and superheated steam is produced.
3. Define the following terms :
   (*i*) Sensible heat of water ; (*ii*) Latent heat of vaporisation ;
   (*iii*) Total heat of steam ; (*iv*) Dryness fraction of steam ; and
   (*v*) Volume of superheated steam.
4. List out the advantages of superheated steam.
5. Derive an expression for the external work done during evaporation :
   (*i*) when the steam is wet ; (*ii*) when the steam is dry ; and
   (*iii*) when the steam is superheated.
6. What is meant by the internal energy of steam ?

OBJECTIVE TYPE QUESTIONS

1. Water at standard atmospheric conditions
   (*a*) behaves as an ideal gas (*b*) is mostly liquid
   (*c*) is far above its critical state (*d*) is far below its critical state
2. The locus of saturated liquid line and saturated vapour line meets at
   (*a*) boiling point (*b*) critical point (*c*) ice point (*d*) triple point
3. For steam
   (*a*) the critical temperature is 221.2° C and critical pressure is 374.15 bar
   (*b*) the critical temperature is 374.15° C and critical pressure is 221.2 bar
   (*c*) the critical temperature is 221.2° C and critical pressure is 221.2 bar
   (*d*) the critical temperature is 374.15° C and critical pressure is 374.15 bar
4. The latent heat of steam at atmospheric pressure is
   (*a*) 1535 kJ/kg (*b*) 1875 kJ/kg (*c*) 2257 kJ/kg (*d*) 2685 kJ/kg
5. With the increase in pressure
   (*a*) the boiling point of water decreases and enthalpy of evaporation increases

(*b*) the boiling point of water increases and enthalpy of evaporation decreases

(*c*) both the boiling point of water and the enthalpy of evaporation decreases

(*d*) both the boiling point of water and the enthalpy of evaporation increases.

6. The dryness fraction of steam is equal to

(*a*) $\frac{m_g}{m_g + m_f}$ (*b*) $\frac{m_f}{m_g + m_f}$ (*c*) $\frac{m_g}{m_f}$ (*d*) $\frac{m_f}{m_g}$

where m_g = Mass of dry steam, and

m_f = Mass of water in suspension.

7. The specific volume of water when heated from 0° C

(*a*) increases steadily (*b*) decreases steadily

(*c*) first decreases and then increases (*d*) first increases and then decreases

8. The enthalpy of dry saturated steam. . . . . .with the increase in pressure.

(*a*) decreases (*b*) increases (*c*) remains constant

9. The amount of heat absorbed to evaporate 1 kg of water from its saturation temperature, without change of temperature, is called

(*a*) sensible heat of water (*b*) latent heat of vaporisation

(*c*) enthalpy of steam (*d*) entropy of steam

10. The internal energy (u) of a dry steam is given by

(*a*) $u = h_g + 100\, p\, v_g$ (*b*) $u = h_g - 100\, p\, v_g$

(*c*) $u = h_g / 100\, p\, v_g$ (*d*) $u = 100\, p\, v_g / h_g$

where h_g = Enthalpy of 1 kg of dry steam in kJ/kg,

p = Pressure of steam in bar, and

v_g = Specific volume of 1 kg of dry steam in m³/kg.

ANSWERS

1. (*d*)	2. (*b*)	3. (*b*)	4. (*c*)	5. (*b*)
6. (*a*)	7. (*c*)	8. (*a*)	9. (*b*)	10. (*b*)

8

Entropy of Steam

1. Introduction. 2. Entropy of Water. 3. Entropy Increase during Evaporation. 4. Entropy of Wet and Dry Steam. 5. Entropy of Superheated Steam. 6. Temperature-Entropy (T-s) Diagram for Water and Steam. 7. Isothermal Lines on T-s diagram. 8. Isentropic Lines on T-s diagram. 9. Enthalpy-Entropy (h-s) Diagram for Water and Steam or Mollier Chart. 10. Dryness Fraction Lines on h-s Diagram. 11. Constant Volume Lines on h-s Diagram. 12. Constant Pressure Lines on h-s Diagram. 13. Isothermal Lines on h-s Diagram. 14. Isentropic Lines on h-s Diagram. 15. Throttling Lines on h-s Diagram.

8.1. Introduction

We have already discussed, in detail, the entropy of perfect gases in chapter 4. The entropy of steam is also an important property, which increases with the addition of heat, and decreases with its removal.

The entropy of steam consists of :

1. Increase in entropy of water during heating from freezing point to boiling point corresponding to the pressure at which the water is being heated,
2. Increase in entropy during evaporation, and
3. Increase in entropy during superheating.

Notes : (*a*) Since the enthalpy or total heat of steam is reckoned from freezing point of water *i.e.* 0° C, therefore entropy is also reckoned from 0° C.

(*b*) The entropy of water at 0° C is regarded as zero.

(*c*) The absolute temperatures are used in entropy calculations.

8.2. Entropy of Water

Consider 1 kg of water being heated at a constant pressure from freezing temperature (*i.e.* 0° C or 273 K) to the boiling temperature. Now consider an instant, when the absolute temperature of water is T K.

Let, for a small rise in temperature of dT, the heat absorbed by 1 kg of water is δq. Then

$$\delta q = \text{Mass} \times \text{Sp. heat of water} \times \text{Rise in temperature.}$$

$$= 1 \times c_w \times dT = c_w\, dT$$

where c_w = Specific heat of water.

The value of c_w may be taken as 4.2 kJ/kg K.

We know that increase in entropy for rise in temperature dT,

$$ds = \frac{\delta q}{T} = \frac{c_w\, dT}{T} \qquad \ldots (\because \delta q = c_w\, dT)$$

The total increase in entropy of water from freezing point to boiling point, may be obtained by integrating the above expression within the limits 273 K and T K.

$$\int_0^s ds = \int_{273}^T \frac{c_w \, dT}{T}$$

$$\therefore \quad s_f = c_w \log_e \left(\frac{T}{273} \right) = 2.3 \, c_w \log \left(\frac{T}{273} \right)$$

Note : The value of s_f may be directly seen from the steam tables.

8.3. Entropy Increase during Evaporation

When the water is completely evaporated into steam, it absorbs full latent heat (h_{fg}) at constant temperature T, corresponding to the given pressure. We know that

$$\text{Entropy} = \frac{\text{Heat absorbed}}{\text{Absolute temperature}}$$

∴ Increase of entropy during evaporation,

$$s_{fg} = \frac{h_{fg}}{T}$$

If the steam is wet with dryness fraction x, the evaporation will be partial. In such a case, heat absorbed

$$= x \, h_{fg}$$

$$\therefore \text{ Increase in entropy, } \quad s_{fg} = \frac{x \, h_{fg}}{T}$$

8.4. Entropy of Wet and Dry Steam

The entropy of wet and dry steam, above the freezing point of water, is the entropy of water *plus* the entropy during evaporation. Mathematically, increase in entropy

$$= s_f + \frac{x \, h_{fg}}{T} = s_f + x \, s_{fg} \qquad \text{... (For wet steam)}$$

$$= s_f + \frac{h_{fg}}{T} = s_f + s_{fg} = s_g \qquad \text{... (For dry steam)}$$

Note : The entropy of dry steam is denoted by s_g. Its value may also be directly read from the steam tables.

8.5. Entropy of Superheated Steam

We know that during superheating, the heat is supplied at a constant pressure, and the temperature of dry steam (T) increases to the temperature of superheated steam (T_{sup}). For a small rise in temperature dT, the heat absorbed,

$$\delta q = c_p \, dT \text{ per kg of steam}$$

or

$$\frac{\delta q}{T} = \frac{c_p \, dT}{T} \qquad \text{... (Dividing both sides by } T\text{)}$$

$$\therefore \quad ds = \frac{c_p \, dT}{T} \qquad \text{...} \left(\because \frac{\delta q}{T} = ds \right)$$

where c_p = Specific heat of superheated steam at constant pressure. Its value may be taken between 1.67 kJ/kg K to 2.5 kJ/kg K.

The total increase in entropy from T K to T_{sup} K may be obtained by integrating the above expression between the proper limits.

$$\int_{s_g}^{s_{sup}} ds = c_p \int_{T}^{T_{sup}} \frac{dT}{T}$$

or
$$s_{sup} - s_g = c_p \log_e\left(\frac{T_{sup}}{T}\right) = 2.3\, c_p \log\left(\frac{T_{sup}}{T}\right)$$

where $(s_{sup} - s_g)$ is the increase in entropy.

Note : The entropy of 1 kg of superheated steam is given by :

$$s_{sup} = s_g + 2.3\, c_p \log\left(\frac{T_{sup}}{T}\right) = \left[s_f + \frac{h_{fg}}{T}\right] + 2.3\, c_p \log\left(\frac{T_{sup}}{T}\right)$$

Example 8.1. *Find the entropy of 1 kg of dry saturated steam at a pressure of 5.2 bar. The boiling point of water at this pressure is given as 152.6° C and its total heat at this temperature is 2110 kJ/kg.*

Solution. Given : $p = 5.2$ bar ; $T = 152.6°$ C $= 152.6 + 273 = 425.6$ K ; $h_{fg} = 2110$ kJ/kg

We know that entropy of 1 kg of dry saturated steam,

$$s_g = s_f + \frac{h_{fg}}{T} = 2.3\, c_w \log\left(\frac{T}{273}\right) + \frac{h_{fg}}{T}$$

$$= 2.3 \times 4.2 \log\left(\frac{425.6}{273}\right) + \frac{2110}{425.6}$$

$$= 1.86 + 4.96 = 6.82 \text{ kJ/kg K} \quad \textbf{Ans.}$$

Note : The value of entropy of dry saturated steam (s_g) may be directly read from steam tables, corresponding to a pressure of 5.2 bar.

Example 8.2. *Calculate the entropy of 1 kg of wet steam with dryness fraction of 0.9 at a pressure of 8.4 bar.*

Solution. Given ; $x = 0.9$; $p = 8.4$ bar

From steam tables, corresponding to a pressure of 8.4 bar, we find that

$T = 172.4°$ C $= 172.4 + 273 = 445.4$ K ; $h_{fg} = 2039.6$ kJ/kg ; $s_f = 2.066$ kJ/kg K

We know that entropy of 1 kg of wet steam,

$$s = s_f + \frac{x\, h_{fg}}{T} = 2.066 + \frac{0.9 \times 2039.6}{445.4}$$

$$= 2.066 + 4.12 = 6.186 \text{ kJ/kg K} \quad \textbf{Ans.}$$

Note : The entropy of 1 kg of wet steam may also be determined as follows :

From steam tables, corresponding to a pressure of 8.4 bar, we find that

$$s_f = 2.066 \text{ kJ/kg K} \quad \text{and} \quad s_{fg} = 4.577 \text{ kJ/kg K}$$

We know that entropy of 1 kg of wet steam,

$$s = s_f + x\, s_{fg} = 2.066 + 0.9 \times 4.577 = 2.066 + 4.12 = 6.186 \text{ kJ/kg K} \quad \textbf{Ans.}$$

Example 8.3. *Determine the entropy per kg of superheated steam at a pressure of 20 bar and a temperature of 250° C. Assume c_p for superheated steam as 2.2 kJ/kg K.*

Solution. Given : $p = 20$ bar ; $T_{sup} = 250°$C $= 250 + 273 = 523$ K ; $c_p = 2.2$ kJ/kg K

From steam tables, corresponding to a pressure of 20 bar, we find that

$$T = 212.4^\circ\text{C} = 212.4 + 273 = 485.4\text{ K}\,;\; s_g = 6.337 \text{ kJ/kg K}$$

We know that entropy per kg of superheated steam,

$$s_{sup} = s_g + 2.3\,c_p \log\left(\frac{T_{sup}}{T}\right) = 6.337 + 2.3 \times 2.2 \log\left(\frac{523}{485.4}\right)$$

$$= 6.5 \text{ kJ/kg K} \quad \textbf{Ans.}$$

8.6. Temperature-Entropy (*T-s*) Diagram for Water and Steam

The temperature-entropy (briefly written as *T-s*) diagram is very useful in solving the problems on adiabatic expansion and compression of steam. The abscissa of the diagram represents the entropy of 1 kg of water and steam above the freezing point of water. The vertical ordinate shows the values of temperature as shown in Fig. 8.1.

Now consider 1 kg of water being heated at a constant pressure (p). We know that when heat is added to the water, its entropy will increase. If we plot a graph between temperature and entropy, we shall find that the entropy increases along logarithmic curve with the increase in temperature, till the boiling temperature (T) corresponding to the pressure (p) is reached. It is shown by the graph *AB*. The corresponding increase in entropy (s_f) is given by the line *Ab* as shown in Fig. 8.1.

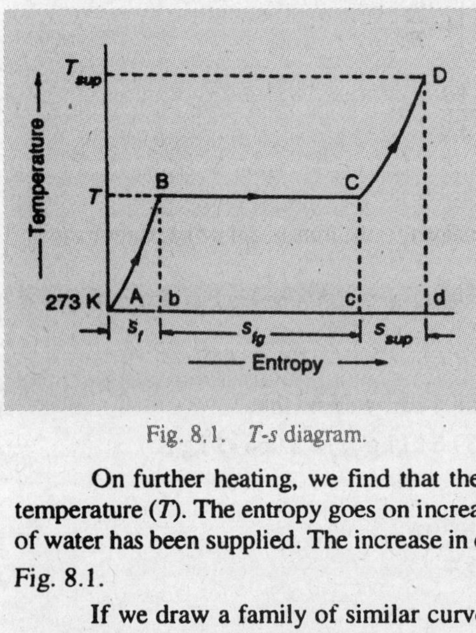

Fig. 8.1. *T-s* diagram.

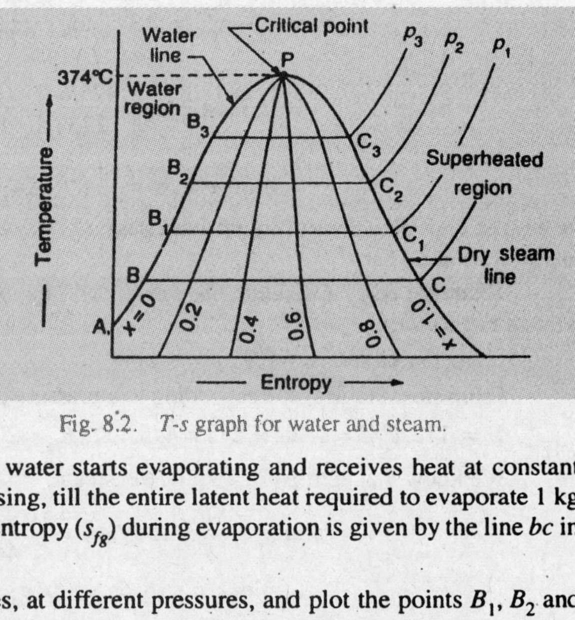

Fig. 8.2. *T-s* graph for water and steam.

On further heating, we find that the water starts evaporating and receives heat at constant temperature (T). The entropy goes on increasing, till the entire latent heat required to evaporate 1 kg of water has been supplied. The increase in entropy (s_{fg}) during evaporation is given by the line *bc* in Fig. 8.1.

If we draw a family of similar curves, at different pressures, and plot the points B_1, B_2 and C_1, C_2 etc., then the line joining the points A, B, B_1, B_2 etc. is called *water line*. Similarly, the line joining the points C, C_1, C_2 etc. is called *dry steam line*. The point, where the water line and dry steam line meets (*i.e. P*) is called *critical point*, and it represents 374.15° C as shown in Fig. 8.2.

Now consider the line *BC*, which represents increase in entropy at constant temperature. We see that dryness fraction increases with entropy from $x = 0$ at B to $x = 1$ at C. The line *BC* is divided into 10 parts, each representing dryness fraction from 0 to 1.0. Now draw similar lines $B_1\,C_1$, $B_2\,C_2$ etc. and mark the values of dryness fraction. The line joining the same values of dryness fraction for different pressures are known as constant dryness fraction lines. In Fig. 8.2, dryness fraction lines at intervals of 0.2 have been drawn.

Notes : 1. In actual practice, the *T-s* diagram in not carried up to the critical point.

2. The advantage of *T-s* diagram over *p-v* diagram is that area under the *T-s* diagram represents heat absorbed or rejected during the process whereas area under the *p-v* diagram represents the workdone.

8.7. Isothermal Lines on *T-s* Diagram

We know that there is no change of temperature during an isothermal process. It is thus obvious, that an isothermal process may be represented by a horizontal line on the *T-s* diagram, as shown in Fig. 8.3.

Consider a sample of 1 kg of wet steam at a temperature of 250° C and dryness fraction of 0.2, which may be shown by the point *A** on the *T-s* diagram. Let this sample be expanded isothermally, till its dryness fraction is 0.8. Now mark this point *B*. This process may be represented by the line *AB* on the *T-s* diagram. The change in entropy during this process may be found out by the difference of readings (s_A and s_B) obtained by drawing vertical lines through *A* and *B*. In this case, the reading at *A* is 3.45 and that at *B* is 5.42. Thus increase in entropy during this process is 5.42 – 3.45 = 1.97 kJ/kg K.

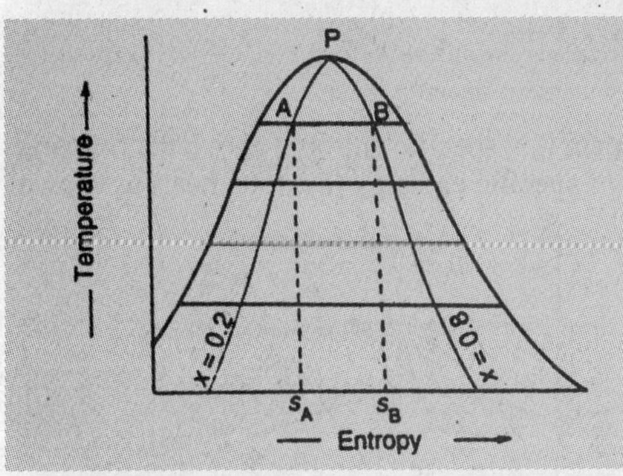

Fig. 8.3. Isothermal lines on *T-s* diagram.

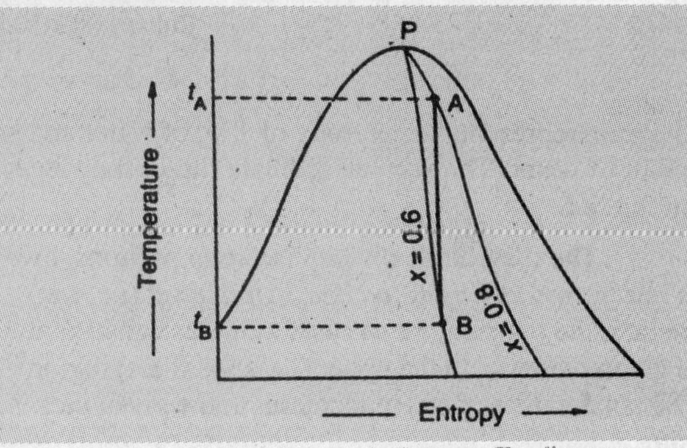

Fig. 8.4. Isentropic lines on *T-s* diagram.

8.8. Isentropic Lines on *T-s* Diagram

We know that there is no increase or decrease in the enthalpy or total heat during an isentropic process. It is thus obvious, that there is no change in entropy also during the process. Therefore, the isentropic process may be represented by a vertical line on the *T-s* diagram, as shown in Fig. 8.4.

Consider a sample of 1 kg of wet steam at a temperature of 310° C and dryness fraction of 0.8, which may be shown by the point *A* on the *T-s* diagram. Let this sample be expanded isentropically till its dryness fraction is 0.6. Now mark this point *B*. This process may be represented by the line *AB* on the *T-s* diagram. The change of temperature during the process may be found out by the difference between the initial temperature of 310° C and the final temperature obtained by drawing horizontal line through *B*. In this case, the final temperature is 40° C. Thus fall in temperature during this process is 310 – 40 = 270° C.

8.9. Enthalpy-Entropy (*h-s*) Diagram for Water and Steam or Mollier Chart

It is a graphical representation of the steam tables, in which the enthalpy (*h*) is plotted along the ordinate and the entropy (*s*) along abscissa. First of all, enthalpy and entropy of water and dry saturated steam, for any particular pressure, are obtained from the steam tables. These values of enthalpies and entropies are plotted and then liquid line and dry saturated line is obtained. Both these lines meet at *C*, *i.e.* the critical point as shown in Fig. 8.5. The critical point corresponds to the enthalpy of liquid and dry saturated steam at 221.2 bar.

---

* The location of point *A* may be marked where the horizontal line through 250° C and logarithmic line of dryness fraction of 0.2 meets. Similarly, the location of point *B* may be marked where the horizontal line through 250° C and logarithmic line of dryness fraction 0.8 meets.

The enthalpy-entropy chart, like temperature-entropy chart, is also very useful in solving the problems on isentropic expansion and compression of steam. In actual diagram, abscissa of the

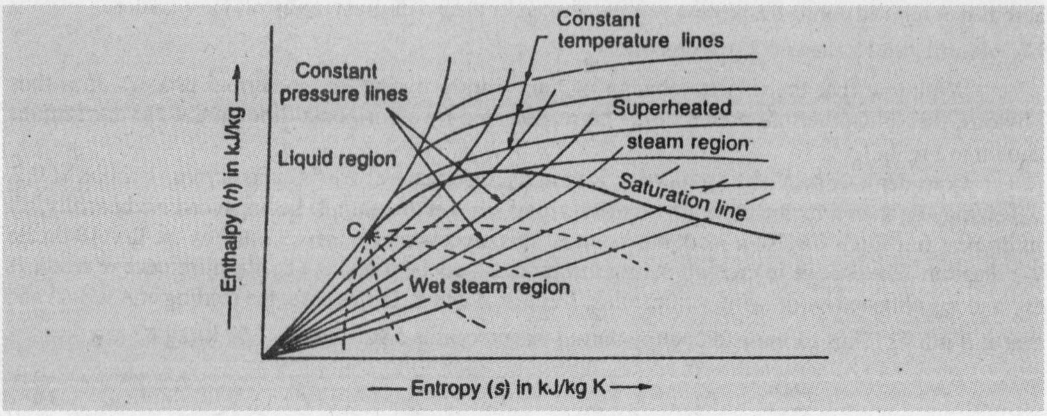

Fig. 8.5. *h-s* diagram for water and steam.

diagram represents the entropy of 1 kg of water and steam (*i.e.* specific entropy) above the freezing point of water. The vertical ordinate shows the values of specific enthalpy (*i.e.* total heat) as shown in Fig. 8.6.

The diagram is divided into two portions, by a line termed as *saturation line*. In the lower (*i.e.* wet) region, the temperature of steam remains constant at a given pressure. In the upper (*i.e.* super heat) region the temperature of steam increases at the given pressure.

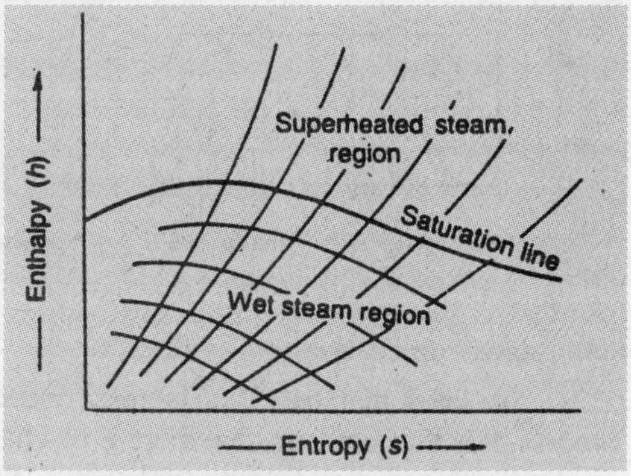

Fig. 8.6. *h-s* diagram or Mollier chart.

The Mollier diagram has the following lines :

1. Dryness fraction lines,
2. Constant volume lines,
3. Constant pressure lines,
4. Isothermal lines,
5. Isentropic lines, and
6. Throttling lines.

These lines are discussed, in detail, in the following pages.

8.10. Dryness Fraction Lines on *h-s* Diagram

The dryness fraction lines are drawn only below the saturation line (which represents dryness fraction equal to 1.0). These lines represent the condition of wet steam between various values of enthalpy and entropy as shown in Fig. 8.7.

Now consider a sample of 1 kg of wet steam of dryness fraction 0.9. In the initial stage, let the sample have enthalpy of 2500 kJ/kg which may be shown by the point *A* on the *h-s* diagram, Let this sample be cooled at constant dryness fraction to an enthalpy of 2400 kJ/kg which may be shown by point *B*. Now the change of entropy during this process may be found out by the difference of initial and final entropies obtained by drawing vertical lines through *A* and *B*. In this case, the entropies at

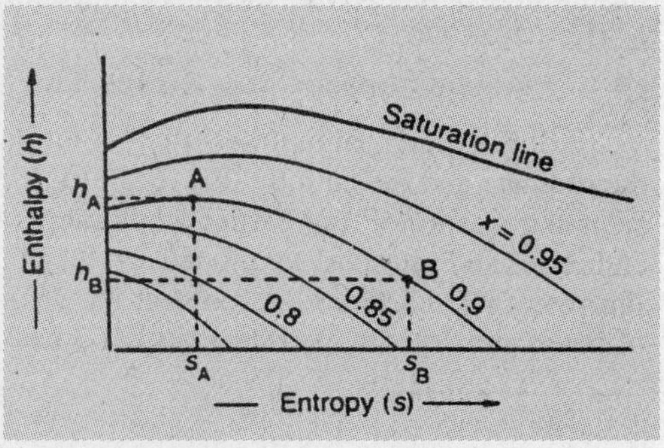

Fig. 8.7. Dryness fraction on *h-s* diagram.

A and *B* are 6.5 and 7.02 respectively. Thus increase in entropy during the process is 7.02 – 6.5 = 0.52 kJ/kg K.

8.11. Constant Volume Lines on *h-s* Diagram

The constant volume lines are drawn in both the wet steam region and superheated steam region. These lines are straight in the wet steam region, but curved upwards above the saturation curve *i.e.* superheated region as shown in Fig. 8.8.

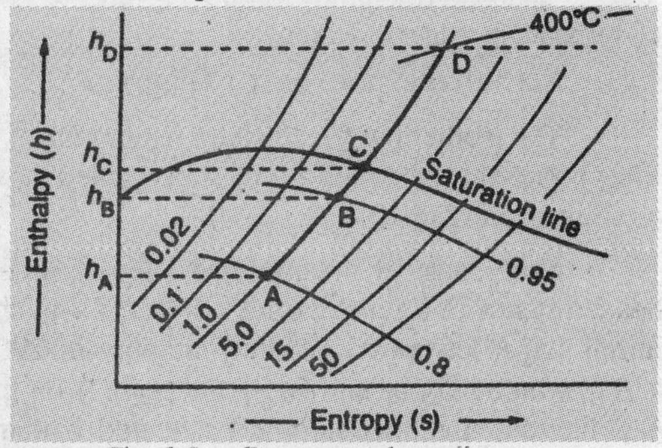

Fig. 8.8. Constant volume lines on *h-s* diagram.

Now consider a sample of wet steam of dryness faction 0.8 having enthalpy (total heat) of 2250 kJ/kg, which may be shown by point *A* on the *h-s* diagram. From the Mollier diagram, we find that the volume of this steam is 1.0 m^3/kg. Let this sample be heated at constant volume of 1.0 m^3/kg till its dryness fraction is 0.95. Now mark this point *B*. This process may be represented by the line *AB* on the Mollier diagram. Now the change of enthalpy (total heat) during this process may be found out by the difference between the initial enthalpy of 2250 kJ/kg and the final enthalpy obtained by drawing a horizontal line through *B*. In this case, the final value of enthalpy is 2590 kJ/kg. Thus increase in enthalpy during this process is 2590-2250 = 340 kJ/kg.

If this sample of steam is further heated at constant volume of 1.0 m^3/kg first into dry saturated steam (shown by point *C*) and then into superheated steam at 400° C (shown by point *D*), we find that enthalpy at *C* and *D* is 2700 kJ/kg and 3275 kJ/kg respectively. Thus increase in enthalpy from *A* to *C* is 2700 – 2250 = 450 kJ/kg. Similarly, increase in enthalpy from *A* to *D* is 3275 – 2250 = 1025 kJ/kg.

Note : Increase in entropy during the process from *A* to *B*, *C* or *D* may be found out by obtaining the corresponding values of entropy.

8.12. Constant Pressure Lines on *h-s* Diagram

The constant pressure lines are drawn in both the wet steam region and superheated steam region. These lines are straight in the wet steam region, because during vaporisation the increase of enthalpy (total heat) is directly proportional to the increase in quality, and hence to the increase in entropy. The constant pressure lines are curved slightly upwards above the saturation curve *i.e.* in the superheated region as shown in Fig. 8.9.

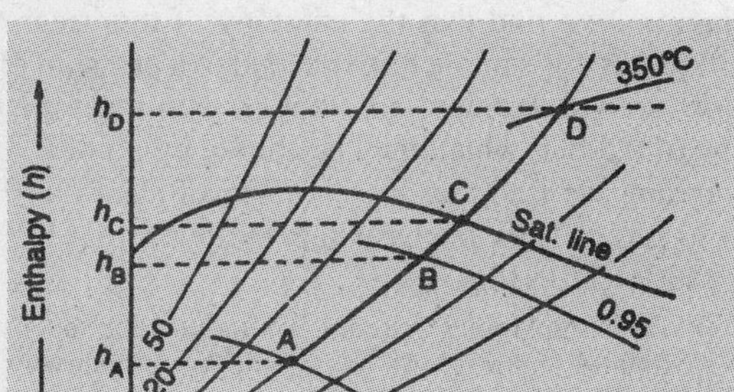

Fig. 8.9. Constant pressure lines on *h-s* diagram.

Now consider a sample of wet steam of dryness fraction 0.75 having enthalpy (total heat) of 2110 kJ/kg which may be shown by the point *A* on the *h-s* diagram. From the *h-s* diagram, we find that the pressure of this steam is 1.0 bar. Let this sample be heated at constant pressure of 1.0 bar till its dryness fraction is 0.95. Now mark this point *B*. This process may be represented by the line *AB* on the Mollier diagram. Now the change of enthalpy during the process may be found out by the difference between the initial enthalpy of 2110 kJ/kg and final enthalpy obtained by drawing a horizontal line through *B*. In this case, the final value of enthalpy is 2570 kJ/kg. Thus increase in enthalpy during this process is 2570 – 2110 = 460 kJ/kg.

If this sample of steam is further heated at constant pressure of 1.0 bar first into dry saturated steam (shown by point *C*) and then into superheated steam at 350° C (shown by point *D*), we find that the enthalpy at *C* and *D* is 2680 kJ/kg and 3170 kJ/kg respectively. Thus increase in enthalpy from *A* to *C* is 2680 – 2110 = 570 kJ/kg. Similarly, increase in enthalpy from *A* to *D* is 3170 – 2110 = 1060 kJ/kg.

Note : Increase of entropy during the process from *A* to *B*, *C* or *D* may be found out by obtaining the corresponding values of entropy.

8.13. Isothermal Lines on *h-s* Diagram

The isothermal (or constant temperature) lines are drawn only above the saturation line. These lines represent the condition of superheated steam between various values of enthalpy and entropy as shown in Fig. 8.10.

Now consider a sample of 1 kg of superheated steam at a temperature of 400° C. In the initial stage, let the sample has enthalpy of 2900 kJ/kg, which may be shown by the point *A* on the *h-s* diagram. Let this sample be heated isothermally (*i.e.* at the constant temperature of 400° C) till the enthalpy is 3270 kJ/kg which may be shown by the point *B*. Now the change of entropy during this process may be found out by the difference of initial and final entropies obtained by drawing vertical lines through *A* and *B*. In this case, the entropies at *A* and *B* are 5.7 and 7.6 respectively. Thus increase in entropy during the process is 7.6 – 5.7 = 1.9 kJ/kg K.

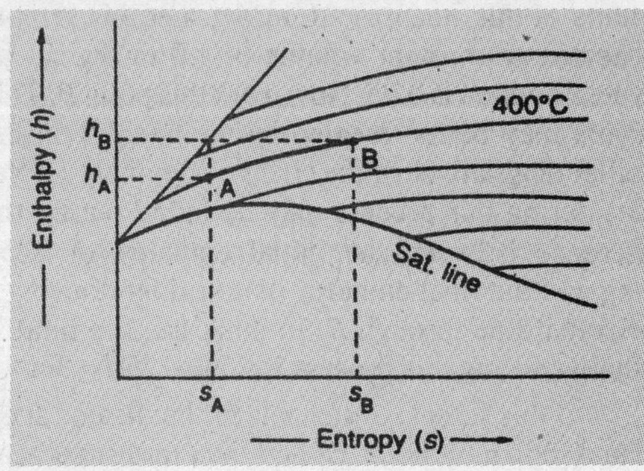

Fig. 8.10. Isothermal lines on *h-s* diagram.

8.14. Isentropic Lines on *h-s* Diagram

We know that during an *isentropic process, there is no change of entropy. It is thus obvious, that the isentropic process is marked by a vertical line *AB* on the *h-s* diagram as shown in Fig. 8.11.

Now consider a sample of superheated steam at a temperature of 300° C and pressure 10 bar which may be shown by the point *A* on the *h-s* diagram. From the Mollier diagram, we find that the enthalpy of this steam is 3050 kJ/kg. Let this sample be expanded to a pressure of 0.5 bar. Now draw a vertical line, through *A* to meet the pressure line of 0.5 bar, which may be shown by point *B*. From the diagram, we find that the enthalpy of steam at 0.5 bar is 2480 kJ/kg, Thus decrease in enthalpy from *A* to *B* is 3050 – 2480 = 570 kJ/kg.

Note: From the Mollier diagram, we find that the specific volume of steam at *A* and *B* is 0.18 and 3.3 m³/kg respectively. Dryness fraction of steam at *B* is 0.928.

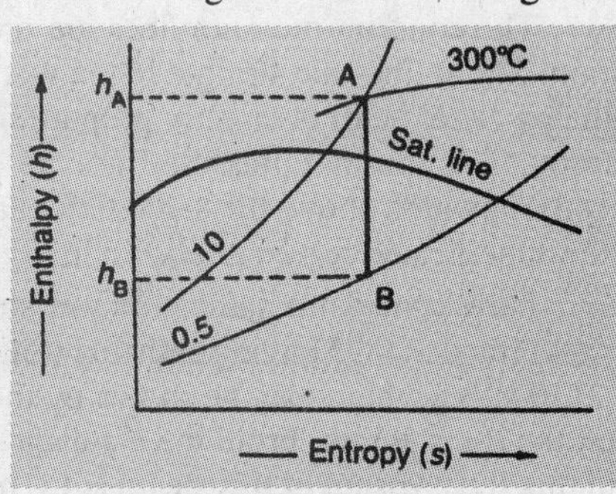

Fig. 8.11. Isentropic lines on *h-s* diagram.

8.15. Throttling Lines on *h-s* Diagram

We know that during a **throttling process, there is no change of enthalpy. It is thus obvious, that the throttling process is marked by a horizontal line *AB* on the *h*-s diagram as shown in Fig. 8.12.

---

* For details, please refer Art 3.10.

** For details, please refer Art. 3.21.

Now consider a sample of wet steam at a pressure of 14 bar and dryness fraction 0.88 which may be shown by the point *A* on the *h-s* diagram. From the Mollier diagram, we find that the entropy of this steam is 5.94 kJ/kg K. Let this sample be expanded by throttling to a pressure of 1.0 bar. Now draw a horizontal line through *A* to meet the pressure line of 1.0 bar, which may be shown by the point *B*. From the Mollier diagram, we find that the entropy at *B* is 7.04 kJ/kg K. Thus increase in entropy from *A* to *B* is 7.04 – 5.94 = 1.1 kJ/kg K.

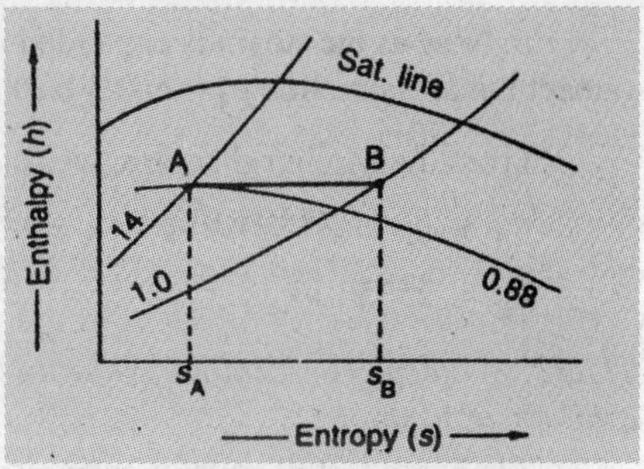

Fig. 8.12. Throttling lines on *h-s* diagram.

Note : From the Mollier diagram, we also find that specific volume of steam at *A* and *B* is 0.14 and 1.7 m³/kg respectively. Dryness fraction of steam at *B* is 0.95.

Example 8.4. *Using Mollier chart, find enthalpy drop and final condition of steam when it is expanded isentropically from an initial pressure of 30 bar and 350° C to a pressure of 1 bar.*

Solution. Given : p_1 = 30 bar ; t_1 = 350° C ; p_2 = 1 bar

1. First of all, on the Mollier chart mark a point *A*, where the pressure line through p_1 (*i.e.* 30 bar) and temperature line through t_1 (*i.e.* 350° C) meets as shown in Fig. 8.13.

2. Since the steam is expanded isentropically (*i.e.* at a constant entropy), therefore draw a vertical line through *A* to meet the pressure line p_2 (*i.e.* 1 bar) at point *B* as shown in Fig. 8.13.

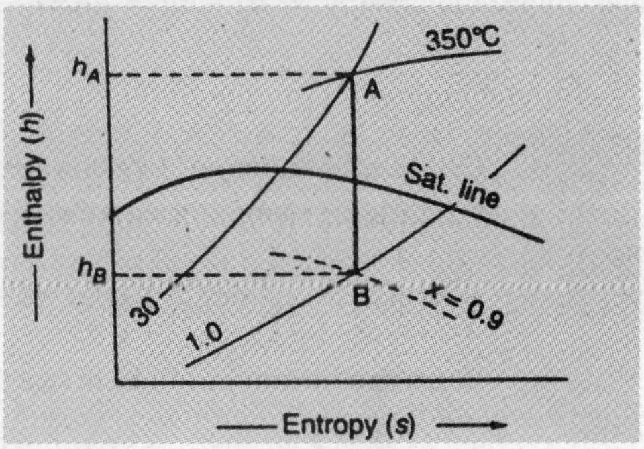

Fig. 8.13

Enthalpy drop during the process

From the Mollier chart, we find that enthalpy at *A*,

$$h_A = 3120 \text{ kJ/kg}$$

and enthalpy at *B*, h_B = 2450 kJ/kg

∴ Enthalpy drop during the process,

$$h = h_A - h_B = 3120 - 2450 = 670 \text{ kJ/kg} \quad \textbf{Ans.}$$

Final condition of the steam

From the Mollier chart, we also find that final condition (*i.e.* dryness fraction) of the steam at *B* is 0.9 **Ans.**

Example 8.5. *A sample of steam at 200° C and a pressure of 7 bar is first expanded at a constant enthalpy to 3 bar, and then at constant entropy to 0.5 bar. With the help of Mollier diagram, determine the change of entropy and enthalpy during the processes.*

Solution. Given : t_1 = 200° C ; p_1 = 7 bar ; p_2 = 3 bar ; p_3 = 0.5 bar

1. First of all, on the Mollier diagram mark point *A* where the pressure line through p_1 (*i.e.* 7 bar) and temperature line through t_1 (*i.e.* 200° C) meets as shown in Fig. 8.14.

2. Since the steam is first expanded at constant enthalpy, therefore draw a horizontal line through *A* to meet the pressure line through p_2 (*i.e.* 3 bar) at point *B* as shown in Fig. 8.14.

3. Now as the steam is expanded at constant entropy, therefore draw a vertical line through B to meet the pressure line p_3 (*i.e.* 0.5 bar) at point C as shown in Fig. 8.14.

From the Mollier diagram, we find that

$$h_A = h_B = 2850 \text{ kJ/kg} \; ; \; s_A = 6.84 \text{ kJ/kg K}$$

$$h_C = 2535 \text{ kJ/kg} \; ; \; s_B = s_C = 7.24 \text{ kJ/kg K}$$

Change of entropy and enthalpy during constant enthalpy process

We know that change of entropy

$$= s_B - s_A = 7.24 - 6.84$$

$$= 0.4 \text{ kJ/kg K} \textbf{ Ans.}$$

and change of enthalpy = 0 **Ans.**

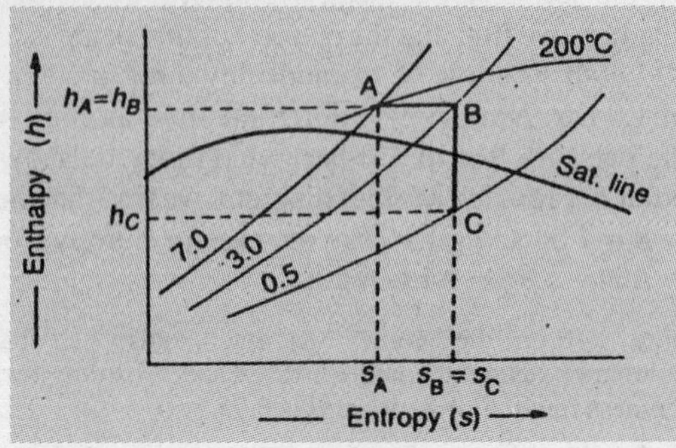

Fig. 8.14

Change of entropy and enthalpy during constant entropy process

We know that change of entropy

$$= 0 \textbf{ Ans.}$$

and change of enthalpy $= h_B - h_C = 2850 - 2535 = 315$ kJ/kg **Ans.**

EXERCISES

1. Calculate the entropy of 1 kg of water at 90° C. [**Ans.** 1.195 kJ/kg K]
2. Calculate the entropy increase during evaporation of 1 kg of dry saturated steam at a pressure of 7 bar. [**Ans.** 4.714 kJ/kg K]
3. Determine the entropy of 1 kg of wet steam at a pressure of 10.5 bar and of dryness fraction 0.9. [**Ans.** 6.127 kJ/kg K]
4. Determine the entropy of 1 kg of superheated steam at 14 bar and 300° C. Assume $c_p = 2.1$ kJ/kg K. [**Ans.** 6.89 kJ/kg K]
5. Calculate the entropy of 1 kg of superheated steam at a pressure of 15 bar and 100° C of superheat. Assume $c_p = 2.3$ kJ/kg K. [**Ans.** 6.883 kJ/kg K]
6. Dry saturated steam at a pressure of 10 bar is expanded isentropically to a pressure of 0.7 bar. Using Mollier diagram, determine the enthalpy drop. [**Ans.** 430 kJ/kg]

QUESTIONS

1. What is the arbitrary temperature of zero entropy chosen for water ?
2. Is the temperature used in calculations for entropy °C or K ?
3. Write the equation for the entropy increase during superheating.
4. Show the following processes on *T-s* diagram :
   (*i*) heating of water from 0° C to boiling temperature ;
   (*ii*) evaporation of water ; and
   (*iii*) superheating of steam.
5. What is meant by constant dryness fraction lines ? How these are plotted on *T-s* diagram ?
6. Write a short note on the Mollier chart and its construction.
7. Name the various uses of Mollier chart.
8. Show the throttling and isentropic expansion of steam on Mollier chart.

OBJECTIVE TYPE QUESTIONS

1. The entropy of water at 0° C is

   (*a*) 0 (*b*) 1 (*c*) 2.3 (*d*) 4.2

2. When water is heated from the freezing point to the boiling point, then the increase in entropy is given by

   (*a*) $c_w \log (273)$ (*b*) $c_w \log (T)$ (*c*) $c_w \log_e \left(\frac{T}{273}\right)$ (*d*) $c_w \log_e \left(\frac{273}{T}\right)$

where c_w = Specific heat of water, and

T = Boiling temperature in K.

3. The entropy of dry steam (s_g) is given by

   (*a*) $s_f + \frac{T}{h_{fg}}$ (*b*) $s_f + \frac{h_{fg}}{T}$ (*c*) $s_f - \frac{T}{h_{fg}}$ (*d*) $s_f - \frac{h_{fg}}{T}$

where s_f = Entropy of water,

T = Absolute temperature, and

h_{fg} = Latent heat of vaporisation.

4. When dry steam at a temperature T is heated upto a temperature of superheated steam (T_{sup}), then the increase in entropy ($s_{sup} - s_g$) is given by

   (*a*) $c_p \log_e \left(\frac{T_{sup}}{T}\right)$ (*b*) $2.3\ c_p \log_e \left(\frac{T_{sup}}{T}\right)$

   (*c*) $c_p \log_e \left(\frac{T}{T_{sup}}\right)$ (*d*) $2.3\ c_p \log_e \left(\frac{T}{T_{sup}}\right)$

where c_p = Specific heat of superheatedsteam.

5. The isentropic process, on the Mollier diagram, is represented by a

   (*a*) horizontal line (*b*) vertical line (*c*) inclined line (*d*) curve

6. The throttling process, on the *h-s* diagram, will be a

   (*a*) horizontal line (*b*) vertical line (*c*) inclined line (*d*) curve

ANSWERS

1. (*a*)	2. (*c*)	3. (*b*)	4. (*a*)	5. (*b*)
6. (*a*)				

9

Thermodynamic Processes of Vapour

1. Introduction. 2. Thermodynamic Processes of Vapour. 3. Constant Volume Process (or Isochoric Process). 4. Constant Pressure Process (or Isobaric Process). 5. Constant Temperature Process (or Isothermal Process). 6. Hyperbolic Process (or $p\,v = C$ Process). 7. Reversible Adiabatic Process (or Isentropic Process or Constant Entropy Process). 8. Polytropic Process (or $p\,v^n = C$ Process). 9. Throttling Process (or Constant Enthalpy or Constant Total Heat Process).

9.1. Introduction

We have already discussed the thermodynamic processes of perfect gases in Chapter 3. These processes are also applicable to vapours, but with different results. In each of these processes, some property of the expanding vapour will remain constant, which enables us to determine the final conditions of vapour after the process. The first and second laws of thermodynamics as applied to gases is also applicable to vapours, for non-flow and flow processes.

We know that for a unit mass,

$$\delta q = du + \delta w = du + p\,dv \qquad \text{... (For non-flow processes)}$$

$$= du - v\,dp \qquad \text{... (For flow processes)}$$

$$= T\,dS \qquad \text{... (For non-flow as well as flow processes)}$$

It may be noted that the general gas equation $p\,v = mRT$ and Joule's law etc. are not applicable to vapours.

9.2. Thermodynamic Processes of Vapour

The following thermodynamic processes for heating or *cooling of vapour are important from the subject point of view :

1. Constant volume process (or Isochoric process),
2. Constant pressure process (or Isobaric process),
3. Constant temperature process (or Isothermal process),
4. Hyperbolic process (or $p\,v = C$ process),
5. Reversible Adiabatic process (or Isentropic process or constant entropy process),
6. Polytropic process (or $p\,v^n = C$ process), and
7. Throttling process (or Constant enthalpy process or constant total heat process).

We shall now discuss these processes, in detail, in the following pages.

* The cooling is regarded as negative heating.

9.3. Constant Volume Process (or Isochoric Process)

The heating or cooling of the steam in a closed vessel is an example of constant volume process. In this process, the volume (or mass) of the steam before and after the process is constant. It may be noted that, in this process, no work is done.

Now consider 1 kg of the wet steam being heated at constant volume from an initial state 1 to the final state 2.

Let p_1 = Initial pressure of the wet steam in bar,

v_{g1} = Specific volume of the dry saturated steam in m³/kg corresponding to initial pressure p_1 (from steam tables),

x_1 = Initial dryness fraction of the wet steam, and

p_2, v_{g2} and x_2 = Corresponding values for the final condition of the steam.

We know that initial volume of 1 kg of wet steam,

$$v_1 = x_1 \, v_{g1}$$

and final volume of 1 kg of wet steam,

$$v_2 = x_2 \, v_{g2}$$

Since the steam is being heated at a constant volume, therefore initial volume of steam is equal to final volume of steam, *i.e.*

$$v_1 = v_2 \quad \text{or} \quad x_1 \, v_{g1} = x_2 \, v_{g2}$$

$$\therefore \quad {}^{*}x_2 = \frac{x_1 \, v_{g1}}{v_{g2}}$$

The following points may be noted :

(*a*) If the value of x_2 is less than x_1, then the steam is being cooled at constant volume.

(*b*) If the final condition of the steam is dry saturated, then $x_2 = 1$. In such a case,

$$v_2 = v_{g2} = x_1 \, v_{g1}$$

(*c*) If the final condition of the steam is superheated, then

$$v_2 = v_{sup}, \quad \textit{i.e. } \text{Specific volume of superheated steam}$$

The final condition of superheated steam is specified by its temperature (*i.e.* T_{sup}). Since the superheated steam behaves like a perfect gas, therefore T_{sup} may be obtained by using Charles' law, *i.e.*

$$\frac{v_{sup}}{T_{sup}} = \frac{v_{g2}}{T_2}$$

---

* The value of x_2 may also be obtained as discussed below :

We know that initial mass of the steam,

$$m_1 = \frac{\text{Total volume of steam in m}^3}{\text{Specific volume of wet steam in m}^3/\text{kg}} = \frac{v_1}{x_1 \, v_{g1}}$$

and final mass of the steam, $$m_2 = \frac{v_2}{x_2 \, v_{g2}}$$

Since in a constant volume process, $m_1 = m_2$ and $v_1 = v_2$, therefore

$$x_1 \, v_{g1} = x_2 \, v_{g2} \quad \text{or} \quad x_2 = \frac{x_1 \, v_{g1}}{v_{g2}}$$

where T_{sup} = Absolute temperature of the superheated steam in K,

v_{g2} = Specific volume of dry steam in m^3/kg, corresponding to pressure p_2 (from steam tables), and

T_2 = Saturation temperature corresponding to pressure p_2 (from steam tables) in K.

Now let us derive the following important relations for the constant volume process.

1. *Workdone during the process*

We know that workdone during the process,

$$w_{1-2} = \text{Pressure} \times \text{Change in volume}$$

Since there is no change in volume, therefore

$$w_{1-2} = 0$$

2. *Change in internal energy of steam*

We know that initial internal energy of steam,

$$u_1 = h_1 - 100\, p_1\, v_1 = h_1 - 100\, p_1\, x_1\, v_{g1} \text{ kJ/kg}$$

and final internal energy of steam,

$$u_2 = h_2 - 100\, p_2\, v_2 \text{ kJ/kg}$$

$$= h_2 - 100\, p_2\, x_2\, v_{g2} \quad \text{... (For wet steam)}$$

$$= h_2 - 100\, p_2\, v_{g2} \quad \text{... (For dry steam)}$$

$$= h_2 - 100\, p_2\, v_{sup} \quad \text{... (For superheated steam)}$$

∴ Change in internal energy of steam,

$$du = u_2 - u_1$$

If *du* is – *ve*, then it indicates that the steam is being cooled and there is a decrease in internal energy.

Note : **The initial and final enthalpy of steam, *i.e.* h_1 and h_2 may be obtained as discussed in Art. 7.4.**

3. *Heat absorbed or heat transferred*

According to the first law of thermodynamics, we know that heat absorbed or heat transferred,

$$q_{1-2} = du + w_{1-2} = (u_2 - u_1) + w_{1-2}$$

Since $w_{1-2} = 0$, therefore

$$q_{1-2} = (u_2 - u_1) \text{ kJ/kg}$$

Thus we find that during a constant volume heating process, the heat absorbed or heat transferred by steam is equal to the change in internal energy.

Note : **If q_{1-2} is –ve, then it indicates that the steam is being cooled and the heat is rejected from the steam.**

Example 9.1. *Steam at a pressure of 4 bar and dryness 0.7 is allowed to expand at a constant volume, until the pressure rises to 5.5 bar. Find the final condition of steam and the heat absorbed by 1 kg of steam.*

Solution. Given : p_1 = 4 bar ; x_1 = 0.7 ; p_2 = 5.5 bar

The expansion of steam at constant volume is shown on p-v, T-s and h-s diagrams in Fig. 9.1.

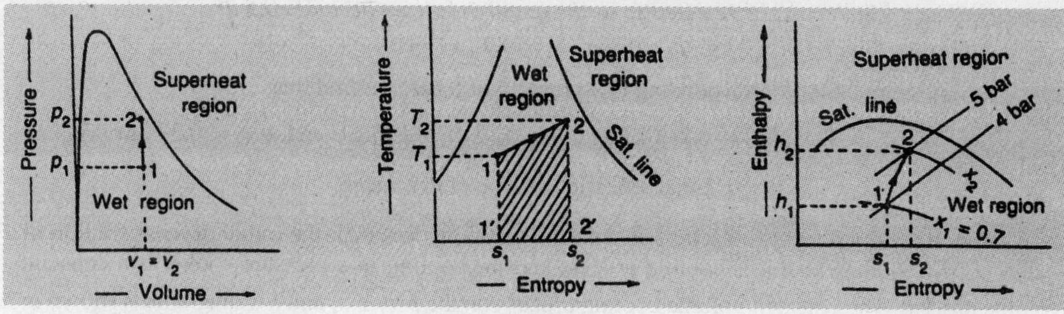

(*a*) p-v diagram. (*b*) T-s diagram. (*c*) h-s diagram.

Fig. 9.1

Final condition of steam

Let x_2 = Final dryness fraction of steam.

From steam tables, corresponding to a pressure of 4 bar, we find that

$$h_{f1} = 604.7 \text{ kJ/kg} \; ; \; h_{fg1} = 2132.9 \text{ kJ/kg} \; ; \text{ and } v_{g1} = 0.462 \text{ m}^3\text{/kg}$$

and corresponding to a pressure of 5.5 bar, we find that

$$h_{f2} = 655.8 \text{ kJ/kg} \; ; h_{fg2} = 2095.9 \text{ kJ/kg} \; ; \text{ and } v_{g2} = 0.342 \text{ m}^3\text{/kg}$$

We know that $$x_2 = \frac{x_1 v_{g1}}{v_{g2}} = \frac{0.7 \times 0.462}{0.342} = 0.945 \textbf{ Ans.}$$

Heat absorbed by 1 kg of steam

We know that initial internal energy of steam,

$$u_1 = h_1 - 100 \, p_1 v_1$$
$$= h_{f1} + x_1 h_{fg1} - 100 \, p_1 x_1 v_{g1}$$
$$= 604.7 + 0.7 \times 2132.9 - 100 \times 4 \times 0.7 \times 0.462$$
$$= 2097.7 - 129.4 = 1968.3 \text{ kJ/kg}$$

and final internal energy of steam,

$$u_2 = h_2 - 100 \, p_2 v_2$$
$$= h_{f2} + x_2 h_{fg2} - 100 \, p_2 x_2 v_{g2}$$
$$= 655.8 + 0.945 \times 2095.9 - 100 \times 5.5 \times 0.945 \times 0.342$$
$$= 2636.4 - 177.7 = 2458.7 \text{ kJ/kg}$$

$\therefore$ Change in internal energy,

$$du = u_2 - u_1 = 2458.7 - 1968.3 = 490.4 \text{ kJ/kg}$$

We know that in constant volume process, heat absorbed by the steam,

$$q_{1-2} = u_2 - u_1 = 490.4 \text{ kJ/kg} \textbf{ Ans.}$$

Notes : 1. The area under the T-s diagram (*i.e.* 1-2-2′-1′) represents the heat absorbed by the steam.

2. The values of h_1 and h_2 may be read directly from the h-s diagram (Mollier chart) corresponding to a pressure $p_1 = 4$ bar, $x_1 = 0.7$ and $p_2 = 5.5$ bar, $x_2 = 0.945$.

Example 9.2. *A pressure cooker contains 2 kg of steam at 5 bar and 0.9 dry. Calculate the quantity of heat which must be rejected so as the quality of steam becomes 0.5 dry.*

Solution. Given : $m = 2$ kg ; $p_1 = 5$ bar ; $x_1 = 0.9$; $x_2 = 0.5$

From steam tables, corresponding to a pressure of 5 bar, we find that

$$h_{f1} = 640.1 \text{ kJ/kg} \; ; \; h_{fg1} = 2107.4 \text{ kJ/kg} \; ; \text{ and } v_{g1} = 0.375 \text{ m}^3\text{/kg}$$

Let v_{g2} = Final specific volume of dry steam.

Since the final dryness fraction of steam ($x_2 = 0.5$) is less than the initial dryness friction of steam ($x_1 = 0.9$), therefore it is cooling process and the cooling in a pressure cooker is a constant volume process. The constant volume cooling process on the *p-v*, *T-s*, and *h-s* diagrams is shown in Fig. 9.2.

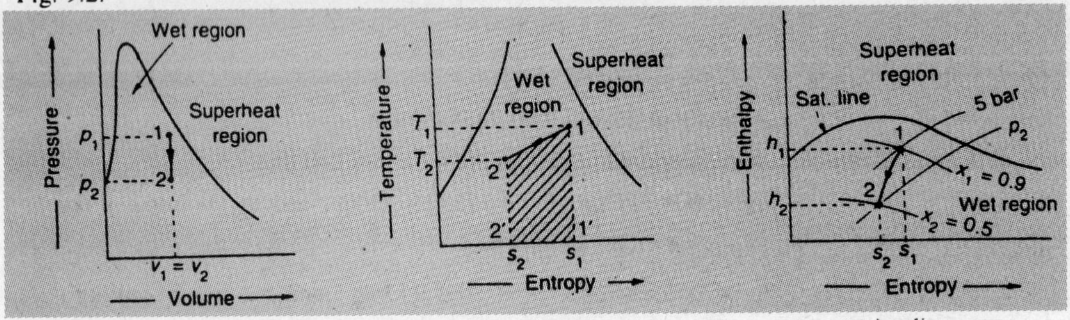

(*a*) *p-v* diagram. (*b*) *T-s* diagram. (*c*) *h-s* diagram.

Fig. 9.2

We know that for constant volume process,

$$x_1 \, v_{g1} = x_2 \, v_{g2} \quad \text{or} \quad v_{g2} = \frac{x_1}{x_2} \times v_{g1} = \frac{0.9}{0.5} \times 0.375 = 0.675 \text{ m}^3\text{/kg}$$

From steam tables, corresponding to $v_{g2} = 0.675$ m³/kg, we find that

$$p_2 = 2.65 \text{ bar} \; ; \; h_{f2} = 543.4 \text{ kJ/kg} \; ; \text{ and } h_{fg2} = 2175.4 \text{ kJ/kg}$$

We know that initial internal energy of steam,

$$u_1 = h_1 - 100 \, p_1 \, x_1 \, v_{g1}$$
$$= h_{f1} + x_1 \, h_{fg1} - 100 \, p_1 \, x_1 \, v_{g1}$$
$$= 640.1 + 0.9 \times 2107.4 - 100 \times 5 \times 0.9 \times 0.375$$
$$= 2536.8 - 168.8 = 2368 \text{ kJ/kg}$$

and final internal energy of steam,

$$u_2 = h_2 - 100 \, p_2 \, x_2 \, v_{g2}$$
$$= h_{f2} + x_2 \, h_{fg2} - 100 \, p_2 \, x_2 \, v_{g2}$$
$$= 543.4 + 0.5 \times 2175.4 - 100 \times 2.65 \times 0.5 \times 0.675$$
$$= 1631 - 89.4 - 1541.6 \text{ kJ/kg}$$

∴ Change in internal energy,

$$du = u_2 - u_1 = 1541.6 - 2368 = -826.4 \text{ kJ/kg}$$

We know that in a constant volume process, heat transferred per kg of steam

$$q_{1-2} = u_2 - u_1 = -826.4 \text{ kJ/kg}$$

$\therefore$ Total heat transferred $= m \times q_{1-2} = 2 \times -826.4 = -1652.8$ kJ **Ans.**

The –ve sign indicates that the heat is rejected from the steam.

Note : The area under the *T-s* diagram (1-2-2′-1) represents the heat rejected from the steam.

Example 9.3. *A closed vessel of 0.2 m³ contains steam at a pressure of 10 bar and a temperature of 250° C. The vessel is cooled till the pressure of steam in the vessel falls to 3.5 bar. Find the final temperature, final dryness fraction, change in internal energy, heat transferred and change in entropy.*

Solution. Given : $v_1 = 0.2 \text{ m}^3$; $p_1 = 10$ bar ; $T_{sup} = 250° \text{ C} = 250 + 273 = 523$ K ; $p_2 = 3.5$ bar

From steam tables, corresponding to a pressure of 10 bar, we find that

$$T_1 = 179.9° \text{ C} = 179.9 + 273 = 452.9 \text{ K} \; ; \; v_{g1} = 0.1943 \text{ m}^3/\text{kg}$$

and corresponding to a pressure of 3.5 bar, we find that

$$v_{g2} = 0.524 \text{ m}^3/\text{kg} \; ; \; h_{f2} = 584.3 \text{ kJ/kg} \; ; \; h_{fg2} = 2147.3 \text{ kJ/kg} \; ;$$

$$s_{f2} = 1.727 \text{ kJ/kg K} \; ; \; s_{fg2} = 5.212 \text{ kJ/kg K}$$

Final temperature of the steam

Let T_2 = Final temperature of the steam.

First of all, let us find the final volume (v_2) of steam at a pressure of 3.5 bar.

Since the initial condition of steam is superheated, therefore specific volume of superheated steam (v_{sup}) is given by

$$\frac{v_{sup}}{T_{sup}} = \frac{v_{g1}}{T_1}$$

$$\therefore \quad v^*_{sup} = v_{g1} \times \frac{T_{sup}}{T_1} = 0.1943 \times \frac{523}{452.9} = 0.2244 \text{ m}^3/\text{kg}$$

We know that mass of steam in the vessel,

$$m = \frac{\text{Total volume of steam in the vessel}}{\text{Specific volume of steam}} = \frac{v_1}{v_{sup}}$$

$$= \frac{0.2}{0.2244} = 0.89 \text{ kg}$$

$\therefore$ Final volume of the steam at a pressure of 3.5 bar,

$$v_2 = m \, v_{g2} = 0.89 \times 0.524 = 0.466 \text{ m}^3$$

Since the final volume of the steam (v_2) is more than the initial volume of the steam (v_1), therefore the final condition of steam at 3.5 bar is **wet. The final temperature must, therefore, be the saturation temperature corresponding to a pressure of 3.5 bar. From steam tables, we find that

$$T_2 = 138.9° \text{ C} \quad \textbf{Ans.}$$

* The value of v_{sup} may be read directly from steam tables for superheated steam corresponding to a pressure of 10 bar and 250° C.

** We may also express in the following way :
Since final specific volume of dry steam (v_{g2}) at a pressure of 3.5 bar is more than initial specific volume of superheated steam (v_{sup}), therefore the final condition of the steam is wet.

Final dryness fraction of steam

Let x_2 = Final dryness fraction of steam.

Since the vessel is a closed one, therefore the cooling takes place at constant volume.

∴ Initial specific volume of superheated steam

= Final specific volume of wet steam

i.e. $$v_{sup} = x_2 v_{g2} \text{ or } x_2 = \frac{v_{sup}}{v_{g2}} = \frac{0.2244}{0.524} = 0.43 \textbf{ Ans.}$$

The constant volume cooling process on the *p-v*, *T-s* and *h-s* diagrams is shown in Fig. 9.3.

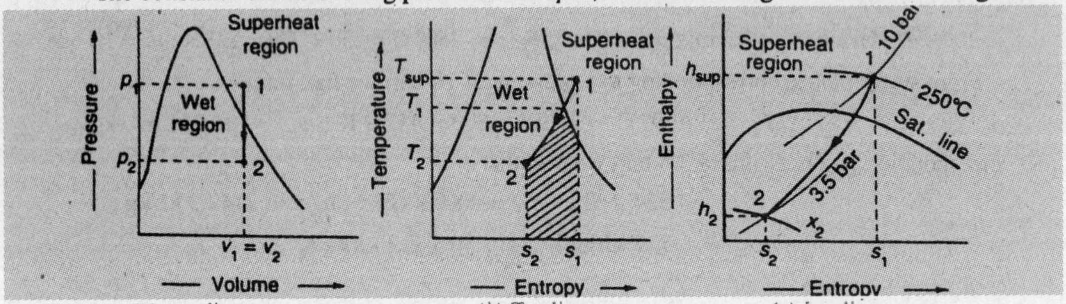

(*a*) *p-v* diagram. (*b*) *T-s* diagram. (*c*) *h-s* diagram.

Fig. 9.3

Change in internal energy

From steam tables for superheated steam, corresponding to a pressure of 10 bar and 250° C, we find that

$$h_{sup} = 2943 \text{ kJ/kg}$$

We know that initial internal energy of superheated steam,

$$u_1 = h_{sup} - 100\, p_1\, v_{sup}$$
$$= 2943 - 100 \times 10 \times 0.2244 = 2718.6 \text{ kJ/kg}$$

and final internal energy of wet steam,

$$u_2 = h_2 - 100\, p_2\, v_2$$
$$= h_{f2} + x_2\, h_{fg2} - 100\, p_2\, x_2\, v_{g2}$$
$$= 584.3 + 0.43 \times 2147.3 - 100 \times 3.5 \times 0.43 \times 0.524$$
$$= 1507.6 - 79 = 1428.6 \text{ kJ/kg}$$

∴ Change in internal energy,

$$du = u_2 - u_1 = 1428.6 - 2718.6 = -1290 \text{ kJ/kg}$$

and total change in internal energy

$$= m\,(u_2 - u_1) = 0.89 \times -1290 = -1148 \text{ kJ } \textbf{Ans.}$$

The –ve sign indicates that there is a decrease in internal energy.

Heat transferred

We know that in a constant volume process, total heat transferred,

$$q_{1-2} = m\,(u_2 - u_1) = -1148 \text{ kJ } \textbf{Ans.}$$

The –ve sign indicates that the heat is rejected from the steam.

Change in entropy

From steam tables for superheated steam, corresponding to a pressure of 10 bar and 250° C, we find that

Initial entropy of superheated steam,

$$s_1 = 6.926 \text{ kJ/kg K}$$

We know that final entropy of wet steam,

$$s_2 = s_{f2} + x_2\, s_{fg2} = 1.727 + 0.43 \times 5.212 = 3.97 \text{ kJ/kg K}$$

$\therefore$ Change in entropy, $ds = s_2 - s_1 = 3.97 - 6.926 = -2.956 \text{ kJ/kg K}$

and total change in entropy $= m\,(s_2 - s_1) = 0.89 \times -2.956 = -2.63 \text{ kJ/K}$ **Ans.**

The –ve sign indicates that there is a decrease in entropy.

Example 9.4. *A closed vessel of 1.2 m³ capacity contains steam at 3 bar and 0.85 dryness. Steam at 10 bar and 0.96 dryness is supplied to the vessel until the pressure in the vessel becomes 5 bar. Calculate the mass of steam supplied to the closed vessel and the final dryness fraction in the vessel. Neglect the volume of the moistures and the thermal capacity of the vessel.*

Solution. Given : $v = 1.2 \text{ m}^3$; $p_1 = 3$ bar ; $x_1 = 0.85$; $p_2 = 10$ bar ; $x_2 = 0.96$; $p_3 = 5$ bar.

From steam tables, corresponding to a pressure of 3 bar, we find that

$$v_{g1} = 0.6055 \text{ m}^3/\text{kg} \;;\; h_{f1} = 561.4 \text{ kJ/kg} \;;\; \text{and } h_{fg1} = 2163.2 \text{ kJ/kg}$$

Corresponding to a pressure of 10 bar, we find that

$$v_{g2} = 0.1943 \text{ m}^3/\text{kg} \;;\; h_{f2} = 762.6 \text{ kJ/kg} \;;\; \text{and } h_{fg2} = 2013.6 \text{ kJ/kg}$$

and corresponding to a pressure of 5 bar, we find that

$$v_{g3} = 0.3746 \text{ m}^3/\text{kg} \;;\; h_{f3} = 640.1 \text{ kJ/kg} \;;\; \text{and } h_{fg3} = 2107.4 \text{ kJ/kg}$$

Mass of steam supplied to the closed vessel

Let m_2 = Mass of steam supplied to the closed vessel at a pressure of 10 bar.

$\therefore$ Enthalpy of steam supplied.

$$h_2 = m_2\,(h_{f2} + x_2\, h_{fg2})$$

$$= m_2\,(762.6 + 0.96 \times 2013.6) = 2696\, m_2 \text{ kJ} \quad \ldots (i)$$

We know that initial mass of steam in the vessel,

$$m_1 = \frac{\text{Total volume of steam in the vessel}}{\text{Specific volume of steam}} = \frac{v}{x_1\, v_{g1}}$$

$$= \frac{1.2}{0.85 \times 0.6055} = 2.33 \text{ kg}$$

$\therefore$ Total initial internal energy of the steam at a pressure of 3 bar,

$$u_1 = m_1\,(h_1 - 100\, p_1\, v_1)$$

$$= m_1\,(h_{f1} + x_1\, h_{fg1} - 100\, p_1\, x_1\, v_{g1})$$

$$= 2.33\,(561.4 + 0.85 \times 2163.2 - 100 \times 3 \times 0.85 \times 0.6055)$$

$$= 2.33\,(2400.1 - 154.4) = 5232.5 \text{ kJ} \quad \ldots (ii)$$

Let m_3 = Final mass of the steam after mixing $= m_1 + m_2 = (2.33 + m_2)$,

x_3 = Final dryness fraction of the steam.

Since it is a closed vessel, therefore volume of steam is constant. We know that mass of steam after mixing,

$$m_3 = \frac{\text{Total volume of steam}}{\text{Specific volume of steam after mixing}}$$

$$(2.33 + m_2) = \frac{1.2}{x_3\, v_{g3}} = \frac{1.2}{x_3 \times 0.3746} = \frac{3.2}{x_3}$$

$$\therefore \quad x_3 = \frac{3.2}{2.33 + m_2} \quad \ldots (iii)$$

We know that final internal energy of steam after mixing,

$$u_3 = m_3\,(h_3 - 100\, p_3\, v_3)$$

$$= m_3\,(h_{f3} + x_3\, h_{fg3} - 100\, p_3\, x_3\, v_{g3})$$

$$= (2.33 + m_2)\,[640.1 + x_3 \times 2107.4 - 100 \times 5 \times x_3 \times 0.3746]$$

$$= (2.33 + m_2)\,(640.1 + 1920\, x_3)$$

$$= (2.33 + m_2)\left(640.1 + \frac{1920 \times 3.2}{2.33 + m_2}\right) \quad \ldots \text{[From equation } (iii)]$$

$$= 640.1\, m_2 + 7632.4$$

Since the enthalpy of steam supplied is equal to the change in internal energy, therefore

$$h_2 = u_3 - u_1$$

or $$2696\, m_2 = 640.1\, m_2 + 7632.4 - 5232.5$$

$$2055.9\, m_2 = 2399.9 \quad \text{or} \quad m_2 = 1.167 \text{ kg} \quad \textbf{Ans.}$$

Final dryness fraction of steam in the vessel

Substituting the value of m_2 in equation (*iii*), we have

$$x_3 = \frac{3.2}{2.33 + 1.167} = 0.917 \quad \textbf{Ans.}$$

9.4. Constant Pressure Process (or Isobaric Process)

The generation of steam in steam boilers is an example of constant pressure process. In this process, the pressure of steam before and after the process is constant. We have already discussed that the heating of wet steam is done at constant pressure in order to convert it into dry saturated steam. Since the wet steam can only exist at a particular temperature corresponding to the saturation temperature (given in the steam tables) at a given pressure, therefore the heating of wet steam to convert it into dry saturated steam at constant pressure is also a constant temperature process (or isothermal process). We also know that the superheating is done at a constant pressure, but its temperature rises according to Charles' law. It, therefore, follows that once the steam is superheated, it will no longer be an isothermal process.

Consider 1 kg of wet steam being heated at constant pressure from an initial state 1 to a final state 2.

Let p = Constant pressure,

x_1 = Initial dryness fraction of steam, and

x_2 = Final dryness fraction of steam.

Since the pressure (p) of steam is constant, therefore the specific volume of dry steam (v_g) corresponding to the pressure (from the steam tables) will also remain constant, *i.e.* $v_{g1} = v_{g2} = v_g$.

We know that initial volume of steam,

$$v_1 = x_1 \, v_g$$

and final volume of steam, $v_2 = x_2 \, v_g$

It may be noted that

(*a*) When the final condition of steam is dry saturated, then

$$x_2 = 1, \text{ and } v_2 = v_g$$

(*b*) When the final condition of steam is superheated, then

$$v_2 = v_{sup}$$

Now let us derive the following relations for the constant pressure process.

1. *Workdone during the process*

We know that workdone during the process,

$$w_{1-2} = \text{Pressure} \times \text{Change in volume}$$

$$= 100\, p\,(v_2 - v_1) \text{ kJ/kg} \quad \text{... (when } p \text{ is in bar)}$$

2. *Change in internal energy*

We know that initial internal energy of steam,

$$u_1 = h_1 - 100\, p\, v_1 \text{ kJ/kg}$$

and final internal energy of steam,

$$u_2 = h_2 - 100\, p\, v_2 \text{ kJ/kg}$$

∴ Change in internal energy,

$$du = (u_2 - u_1) \text{ kJ/kg}$$

3. *Heat absorbed or heat transferred*

According to the first law of thermodynamics, we know that heat absorbed or heat transferred,

$$q_{1-2} = du + w_{1-2} = (u_2 - u_1) + w_{1-2}$$

$$= [(h_2 - 100\, p\, v_2) - (h_1 - 100\, p\, v_1)] + 100\, p\,(v_2 - v_1)$$

$$= (h_2 - h_1) \text{ kJ/kg}$$

Thus we find that during a constant pressure heating process, the heat absorbed or heat transferred by the steam is equal to the change in enthalpy (or total heat) of steam.

Example 9.5. *One kg of steam at a pressure of 17.5 bar and dryness 0.95 is heated at a constant pressure, until it is completely dry. Determine : 1. the increase in volume ; 2. the quantity of heat supplied ; and 3. the change in entropy.*

Solution. Given : p = 17.5 bar ; x_1 = 0.95 ; x_2 = 1

The heating of steam at constant pressure is shown on *p-v*, *T-s* and *h-s* diagrams in Fig. 9.4.

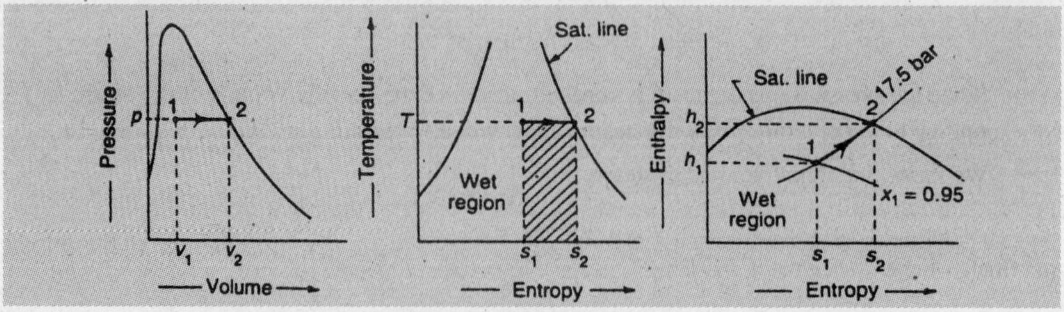

(*a*) *p-v* diagram. (*b*) *T-s* diagram. (*c*) *h-s* diagram.

Fig. 9.4

From steam tables, corresponding to a pressure of 17.5 bar, we find that

$$h_{f1} = h_{f2} = 878.2 \text{ kJ/kg} \,;\; h_{fg1} = h_{fg2} = 1915.9 \text{ kJ/kg} \,;\; v_g = 0.1134 \text{ m}^3\text{/kg} \,;$$

$$s_{f1} = s_{f2} = 2.384 \text{ kJ/kg K} \,;\text{ and } s_{fg1} = s_{fg2} = 4.001 \text{ kJ/kg K}$$

1. *Increase in volume*

We know that initial volume of 1 kg of wet steam,

$$v_1 = x_1\, v_g = 0.95 \times 0.1134 = 0.1077 \text{ m}^3$$

Since the final condition of steam is completely dry, therefore final volume of 1 kg of dry steam,

$$v_2 = v_g = 0.1134 \text{ m}^3$$

$\therefore$ Increase in volume $= v_2 - v_1 = 0.1134 - 0.1077 = 0.0057 \text{ m}^3$ **Ans.**

2. *Quantity of heat supplied*

We know that initial enthalpy or total heat of wet steam,

$$h_1 = h_{f1} + x_1\, h_{fg1} = 878.2 + 0.95 \times 1915.9 = 2698.3 \text{ kJ/kg}$$

and final enthalpy or total heat of dry steam,

$$h_2 = h_{f2} + h_{fg2} = 878.2 + 1915.9 = 2794.1 \text{ kJ/kg}$$

Since the steam is heated at constant pressure, therefore

Heat supplied = Change in enthalpy of steam

$= h_2 - h_1 = 2794.1 - 2698.3 = 95.8$ kJ/kg **Ans.**

3. *Change in entropy*

We know that initial entropy of wet steam,

$$s_1 = s_{f1} + x_1\, s_{fg1} = 2.384 + 0.95 \times 4.001 = 6.185 \text{ kJ/kg K}$$

and final entropy of dry steam,

$$s_2 = s_{f2} + s_{fg2} = 2.384 + 4.001 = 6.385 \text{ kJ/kg K}$$

$\therefore$ Change in entropy,

$$ds = s_2 - s_1 = 6.385 - 6.185 = 0.2 \text{ kJ/kg K} \quad \textbf{Ans.}$$

Example 9.6. *Two boilers discharge equal amounts of steam into the same main. The steam from one is at 18 bar and 380° C, and from the other at 18 bar and 0.95 quality. Determine : 1. the equilibrium condition after mixing, 2. the loss of entropy by the high temperature steam, 3. the gain in entropy by the low temperature steam, and 4. net increase or decrease of entropy.*

Take c_p for superheated steam = 2.3 kJ/kg K.

Solution. Given : $p_1 = 18$ bar ; $T_{sup} = 380°$ C = 380 + 273 = 653 K ; $p_2 = 18$ bar ; $x_2 = 0.95$; $c_p = 2.3$ kJ/kg K

From steam tables, corresponding to a pressure of 18 bar, we find that

$$h_{f1} = h_{f2} = 884.8 \text{ kJ/kg}$$

$$h_{fg1} = h_{fg2} = 1910.3 \text{ kJ/kg}$$

$$h_{g1} = 2795.1 \text{ kJ/kg}$$

$$T_1 = T_2 = 207° \text{ C} = 207 + 273 = 480 \text{ K}$$

$$s_{f2} = 2.398 \text{ kJ/kg K} ; \; s_{fg2} = 3.978 \text{ kJ/kg K}$$

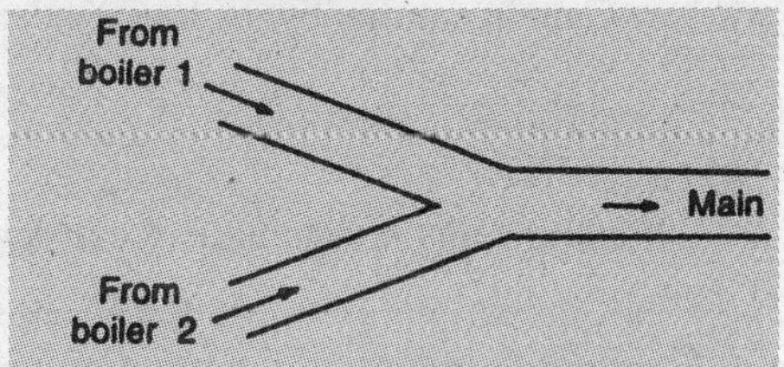

Fig. 9.5

1. *Equilibrium condition after mixing*

Consider that each boiler discharge 1 kg of steam. Therefore

Mass of steam in the main, $m = 2$ kg

We know that enthalpy of steam discharged from boiler 1,

$$h_1 = {}^*h_{sup} = h_{f1} + h_{fg1} + c_p (T_{sup} - T_1)$$

$$= 884.8 + 1910.3 + 2.3\,(653 - 480) = 3193 \text{ kJ/kg}$$

and enthalpy of steam discharged from boiler 2,

$$h_2 = h_{f2} + x_2\, h_{fg2} = 884.8 + 0.95 \times 1910.3 = 2700 \text{ kJ/kg}$$

Let h = Enthalpy of the mixture in kJ/kg.

For equilibrium, $h_1 + h_2 = m\,h$

$\therefore$ $3193 + 2700 = 2 \times h$ or $h = 2946.5$ kJ/kg

Since the enthalpy of the mixture (h = 2946.5 kJ/kg) is more than the enthalpy of the dry saturated steam at a pressure of 18 bar (h_{g1} = 2795.1 kJ/kg) ; therefore the condition of the mixture is superheated.

Let T_m = Temperature of the superheated steam after mixing.

$\therefore$ Enthalpy of the superheated mixture (h),

$$2946.5 = h_{g1} + c_p (T_m - T_1) = 2795.1 + 2.3\,(T_m - 480)$$

or $T_m = 545.8$ K = 272.8° C **Ans.**

2. *Loss of entropy by high temperature steam*

From steam tables for superheated steam, corresponding to a pressure of 18 bar and 380° C, we find that

$$s_1 = 7.113 \text{ kJ/kg K}$$

* The value of h_{sup} may be directly taken from steam tables of superheated steam, corresponding to a pressure of 18 bar and 380° C.

and corresponding to a pressure of 18 bar and 272.8° C, we find that

$$s_m = 6.707 \text{ kJ/kg K}$$

∴ Loss of entropy by high temperature steam

$$= s_1 - s_m = 7.113 - 6.707 = 0.406 \text{ kJ/kg K} \quad \textbf{Ans.}$$

3. *Gain in entropy by the low temperature steam*

We know that entropy of low temperature steam (*i.e.* wet steam from boiler 2),

$$s_2 = s_{f2} + x_2 s_{fg2} = 2.398 + 0.95 \times 3.978 = 6.18 \text{ kJ/kg K}$$

∴ Gain in entropy by the low temperature steam

$$= s_m - s_2 = 6.707 - 6.18 = 0.527 \text{ kJ/kg K}$$

4. *Net increase or decrease of entropy*

We know that net increase of entropy

$$= 0.527 - 0.406 = 0.121 \text{ kJ/kg} \quad \textbf{Ans.}$$

9.5. Constant Temperature Process (or Isothermal Process)

We have already discussed in the previous article that the wet steam has only one temperature (saturation temperature) corresponding to a given pressure. It is thus obvious, that the heating of wet steam at a constant temperature till it becomes dry saturated is similar to that of heating at a constant pressure. Thus for an isothermal process (which is limited to wet steam region only), the same expressions for the workdone, change in internal energy and heat absorbed may be used as for constant pressure process.

Once the steam is superheated, it behaves like a perfect gas. In other words, the superheated steam follows the law

$$\frac{p\,v}{T} = \text{Constant} \quad \text{or} \quad p\,v = \text{Constant} \qquad \ldots (\because T \text{ is constant})$$

The superheating of steam at constant temperature may be regarded as hyperbolic. In other words, it follows Boyle's law.

Example 9.7. *Steam at a pressure of 5.4 bar and dryness fraction 0.8 expands in a cylinder reversibly and isothermally to a pressure of 1 bar. Find : 1. Final condition of steam ; 2. Change in internal energy ; 3. Change of entropy ; 4. Heat transferred ; and 5. Workdone per kg of steam.*

Solution. Given : $p_1 = 5.4$ bar ; $x_1 = 0.8$; $p_2 = 1$ bar

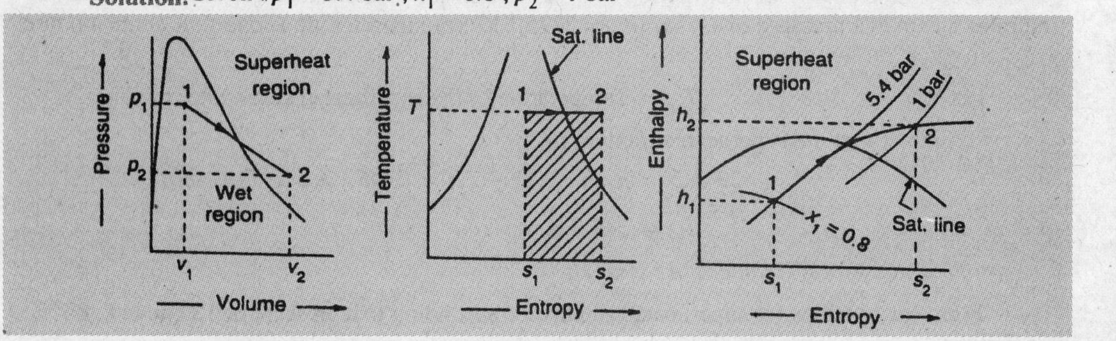

(*a*) *p-v* diagram. (*b*) *T-s* diagram. (*c*) *h-s* diagram.

Fig. 9.6

The isothermal expansion of steam on *p-v*, *T-s* and *h-s* diagram is shown in Fig. 9.6.

From steam tables, corresponding to a pressure of 5.4 bar, we find that

$$T = 154.8°\,C = 154.8 + 273 = 427.8\,K;\; h_{f1} = 652.8\text{ kJ/kg};\; h_{fg1} = 2098\text{ kJ/kg};$$

$$v_{g1} = 0.348\text{ m}^3\text{/kg};\; s_{f1} = 1.89\text{ kJ/kg K};\text{ and } s_{fg1} = 4.903\text{ kJ/kg K}$$

and corresponding to a pressure of 1 bar, we find that

$$h_{f2} = 417.5\text{ kJ/kg};\; h_{fg2} = 2258\text{ kJ/kg}$$

1. *Final condition of steam*

From steam tables, corresponding to a pressure of 5.4 bar, we find that the saturation temperature,

$$T = 154.8°\,C$$

Thus, the final condition of steam at state 2 is superheated having $T_{sup} = T = 154.8°$ C. Ans.

2. *Change in internal energy*

We know that initial internal energy,

$$u_1 = h_1 - 100\,p_1\,v_1 = h_{f1} + x_1\,h_{fg1} - 100\,p_1\,x_1\,v_{g1}$$

$$= 652.8 + 0.8 \times 2098 - 100 \times 5.4 \times 0.8 \times 0.348\text{ kJ/kg}$$

$$= 2331.3 - 150.3 = 2181\text{ kJ/kg}$$

and final internal energy, $u_2 = h_2 - 100\,p_2\,v_2 = {}^*h_{sup} - 100\,p_2\,v_{sup}$

$$= 2785.8 - 100 \times 1 \times 1.96 = 2589.8\text{ kJ/kg}$$

∴ Change in internal energy,

$$du = u_2 - u_1 = 2589.8 - 2181 = 408.8\text{ kJ/kg}\;\textbf{Ans.}$$

3. *Change in entropy*

We know that initial entropy of wet steam,

$$s_1 = s_{f1} + x_1\,s_{fg1} = 1.89 + 0.8 \times 4.903 = 5.81\text{ kJ/kg K}$$

and final entropy of superheated steam from steam tables for superheated steam corresponding to 1 bar and 154.8° C,

$$s_2 = 7.635\text{ kJ/kg K}$$

---

* From steam tables for superheated steam. We shall find the values of h_{sup} and v_{sup} corresponding to a pressure of 1 bar and 154.8° C, as discussed below :

$$h_{sup} = h_{sup}\text{ at }150°\,C + (h_{sup}\text{ at }200°\,C - h_{sup}\text{ at }150°\,C)\;4.8/50$$

$$= 2776.3 + (2875.4 - 2776.3)\;4.8/50 = 2776.3 + 9.5 = 2785.8\text{ kJ/kg}$$

Similarly $\quad v_{sup} = v_{sup}\text{ at }150°\,C + (v_{sup}\text{ at }200°\,C - h_{sup}\text{ at }150°\,C)\;4.8/50$

$$= 1.936 + (2.172 - 1.936)\;4.8/50 = 1.936 + 0.024 = 1.96\text{ m}^3\text{/kg}$$

The value of v_{sup} may also be obtained as follows :

We know that $\quad \dfrac{v_{sup}}{T_{sup}} = \dfrac{v_{g2}}{T_2}$ or $v_{sup} = v_{g2} \times \dfrac{T_{sup}}{T_2}$

From steam tables, corresponding to a pressure of 1 bar, we find that

$$v_{g2} = 1.694\text{ m}^3\text{/kg};\text{ and } T_2 = 99.63°\,C$$

$$\therefore \quad v_{sup} = 1.694 \times \frac{154.8 + 273}{99.63 + 273} = 1.96\text{ m}^3\text{/kg}$$

∴ Change in entropy,

$$ds = s_2 - s_1 = 7.635 - 5.81 = 1.825 \text{ kJ/kg K } \textbf{Ans.}$$

4. *Heat transferred*

We know that heat transferred,

$$q_{1-2} = T\,ds = 427.8 \times 1.825 = 781 \text{ kJ/kg } \textbf{Ans.}$$

5. *Workdone per kg of steam*

We know that workdone,

$$w_{1-2} = q_{1-2} - du \qquad \ldots (\because q_{1-2} = du + w_{1-2})$$

$$= 781 - 408.8 = 372.2 \text{ kJ/kg } \textbf{Ans.}$$

9.6. Hyperbolic Process (or $p\,v = C$ Process)

We have discussed above that superheated steam behaves like a perfect gas and the process of superheating the steam at constant temperature (*i.e.* isothermal process in the superheat region) is regarded as a *hyperbolic process (or $p\,v = C$ process). Now consider 1 kg of wet steam being heated hyperbolically from an initial state 1 to the final state 2.

Let p_1 = Initial pressure of the steam in bar,

v_{g1} = Specific volume of the dry steam in m³/kg corresponding to pressure, p_1 (from steam tables),

x_1 = Initial dryness fraction of the steam, and

p_2, v_{g2}, x_2 = Corresponding values for the final condition of the steam.

We know that initial volume of 1 kg of steam,

$$v_1 = x_1\, v_{g1}$$

and final volume of 1 kg of steam,

$$v_2 = x_2\, v_{g2}$$

Since the steam is heated hyperbolically (*i.e.* $pv = C$), therefore

$$p_1\, v_1 = p_2\, v_2 \quad \text{or} \quad p_1\,(x_1\, v_{g1}) = p_2\,(x_2\, v_{g2})$$

$$\therefore \quad x_2 = \frac{p_1\, x_1\, v_{g1}}{p_2\, v_{g2}} \qquad \ldots (i)$$

The value of x_2 gives us an important information, about the final condition of the steam, *e.g.*

1. If the value of x_2 is less than one, the final condition of steam is *wet.*

2. If the value of x_2 is equal to one, then the final condition of steam is *dry saturated.* In such a case, $v_2 = v_{g2}$.

3. If the value of x_2 is greater than one, then the final condition of steam is *superheated.* In such a case, $v_2 = v_{sup}$.

In case of superheated steam, we shall find out its volume and temperature.

Let v_{sup} = Volume of superheated steam, and

T_{sup} = Temperature of superheated steam in K.

---

* The isothermal process in the wet region is not hyperbolic.

We know that $p_1 x_1 v_{g1} = p_2 v_{sup}$

$$\therefore \quad v_{sup} = \frac{p_1 x_1 v_{g1}}{p_2}$$

Now T_{sup} may be found out by Charles' law, *i.e.*

$$\frac{v_{sup}}{T_{sup}} = \frac{v_{g2}}{T_2}$$

where T_2 = Saturation temperature corresponding to the pressure p_2, in K (from steam tables).

Now let us derive the following important relations for the hyperbolic process.

1. *Workdone during the process*

We know that workdone during the process,

$$w_{1-2} = 100 p_1 v_1 \log_e \left(\frac{v_2}{v_1}\right) = 2.3 \times 100 p_1 v_1 \log \left(\frac{v_2}{v_1}\right) \text{kJ/kg}$$

... (when p is in bar)

$$= 230 p_1 v_1 \log \left(\frac{p_1}{p_2}\right) \text{kJ/kg} \quad \ldots \left(\because p_1 v_1 = p_2 v_2 \text{ or } \frac{v_2}{v_1} = \frac{p_1}{p_2}\right)$$

2. *Change in internal energy of steam*

We know that initial internal energy of steam,

$$u_1 = h_1 - 100 p_1 v_1 \text{ kJ/kg}$$

and final internal energy of steam,

$$u_2 = h_2 - 100 p_2 v_2 \text{ kJ/kg}$$

Since in hyperbolic process, $p_1 v_1 = p_2 v_2$, therefore

$\therefore$ Change in internal energy

$$du = u_2 - u_1 = h_2 - h_1 \text{ kJ/kg}$$

Thus, we find that during hyperbolic process, the change in internal energy is equal to the change in enthalpy.

3. *Heat absorbed or heat transferred*

According to the first law of thermodynamics, we know that heat absorbed or heat transferred,

$$q_{1-2} = du + w_{1-2} = (u_2 - u_1) + w_{1-2} \text{ kJ/kg}$$

Example 9.8. *Steam at a pressure of 10 bar and 0.9 dry expands to atmospheric pressure hyperbolically. Find : 1. Work done, 2. Change in enthalpy, 3. Change in internal energy, and 4. Heat absorbed. Specific heat of steam at constant pressure = 2 kJ/kg K.*

Solution. Given : $p_1 = 10$ bar ; $x_1 = 0.9$; $p_2 = 1.013$ bar ; $c_p = 2$ kJ/kg K

The hyperbolic process on the *p-v*, *T-s* and *h-s* diagram is shown in Fig. 9.7.

From steam tables, corresponding to a pressure of 10 bar, we find that

$$h_{f1} = 762.6 \text{ kJ/kg} ; \; h_{fg1} = 2013.6 \text{ kJ/kg} ; \; v_{g1} = 0.1943 \text{ m}^3/\text{kg} ;$$

$$T_1 = 179.9° \text{ C} = 179.9 + 273 = 452.9 \text{ K}$$

and corresponding to atmospheric pressure of 1.013 bar, we find that

$$h_{f2} = 419 \text{ kJ/kg} \; ; \; h_{fg2} = 2257 \text{ kJ/kg} \; ; \; v_{g2} = 1.673 \text{ m}^3\text{/kg} \; ;$$

$$T_2 = 100^\circ \text{ C} = 100 + 273 = 373 \text{ K}$$

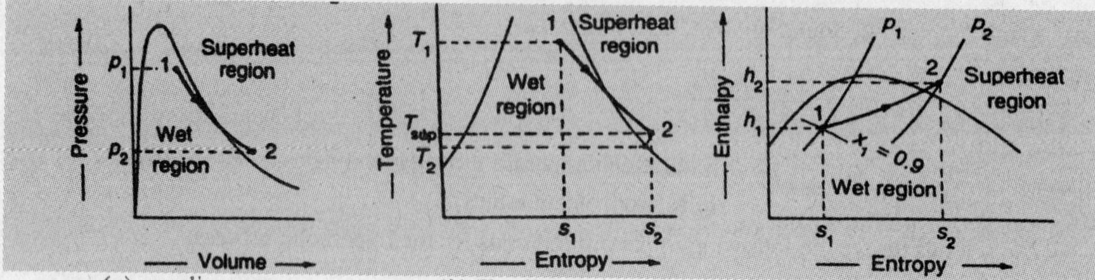

(*a*) *p-v* diagram (*b*) *T-s* diagram (*c*) *h-s* diagram

Fig. 9.7

1. *Workdone*

We know that workdone during hyperbolic expansion,

$$w_{1-2} = 230 \, p_1 \, v_1 \log\left(\frac{v_2}{v_1}\right) = 230 \, p_1 \, x_1 \, v_{g1} \log\left(\frac{p_1}{p_2}\right)$$

$$\ldots (\because v_1 = x_1 \, v_{g1}, \text{ and } v_2 / v_1 = p_1 / p_2)$$

$$= 230 \times 10 \times 0.9 \times 0.1943 \log\left(\frac{10}{1.013}\right) = 400 \text{ kJ/kg} \quad \textbf{Ans.}$$

2. *Change in enthalpy*

First of all, let us find the final condition of steam. We know that for hyperbolic expansion,

$$p_1 \, v_1 = p_2 \, v_2 \quad \text{or} \quad p_1 \, x_1 \, v_{g1} = p_2 \, x_2 \, v_{g2}$$

$$\therefore \quad x_2 = \frac{p_1 \, x_1 \, v_{g1}}{p_2 \, v_{g2}} = \frac{10 \times 0.9 \times 0.1943}{1.013 \times 1.673} = 1.032$$

Since the value of x_2 is more than one, therefore the steam after expansion is *superheated. First of all, let us find out the volume of the superheated steam (v_{sup}) after expansion. We know that

$$p_1 \, x_1 \, v_{g1} = p_2 \, v_{sup}$$

$$\therefore \quad v_{sup} = \frac{p_1 \, x_1 \, v_{g1}}{p_2} = \frac{10 \times 0.9 \times 0.1943}{1.013} = 1.726 \text{ m}^3\text{/kg}$$

Now let us find out the temperature of the superheated steam (T_{sup}) by Charles' law. We know that

$$\frac{v_{sup}}{T_{sup}} = \frac{v_{g2}}{T_2} \quad \text{or} \quad T_{sup} = T_2 \times \frac{v_{sup}}{v_{g2}} = 373 \times \frac{1.726}{1.673} = 385 \text{ K}$$

---

* We may also say as follows :

We know that $p_1 \, v_1 = p_2 \, v_2$ or $v_2 = \dfrac{p_1 \, v_1}{p_2} = \dfrac{p_1 \, x_1 \, v_{g1}}{p_2} = \dfrac{10 \times 0.9 \times 0.1943}{1.013} = 1.726 \text{ m}^3\text{/kg}$

Since $v_2 = 1.726 \text{ m}^3\text{/kg}$ is more than $v_{g2} = 1.673 \text{ m}^3\text{/kg}$, therefore the steam is superheated.

We know that initial enthalpy of wet steam,

$$h_1 = h_{f1} + x_1\, h_{fg1} = 762.6 + 0.9 \times 2013.6 = 2574.8 \text{ kJ/kg}$$

and final enthalpy of superheated steam,

$$h_2 = h_{sup} = h_{f2} + h_{fg2} + c_p\,(T_{sup} - T_2)$$

$$= 419 + 2257 + 2\,(385 - 373) = 2700 \text{ kJ/kg}$$

$\therefore$ Change in enthalpy $dh = h_2 - h_1 = 2700 - 2574.8 = 125.2$ kJ/kg **Ans.**

3. *Change in internal energy*

We know that in hyperbolic process, change in internal energy is equal to the change in enthalpy. Therefore change in internal energy,

$$du = 125.2 \text{ kJ/kg} \quad \textbf{Ans.}$$

4. *Heat absorbed*

We know that heat absorbed,

$$q_{1-2} = du + w_{1-2} = 125.2 + 400 = 525.2 \text{ kJ/kg} \quad \textbf{Ans.}$$

9.7. Reversible Adiabatic Process (or Isentropic Process or Constant Entropy Process)

The adiabatic process may be reversible or irreversible. The reversible adiabatic process (*i.e.* frictionless adiabatic process) is known as *isentropic process* or *constant entropy process*. But when there is some friction involved in the process, it is then known as *irreversible adiabatic process*. We have already discussed that irreversibility leads to increase in entropy. Thus in an irreversible adiabatic process, the entropy *increases.

We know that in a reversible adiabatic or isentropic process :

1. there is no transfer of heat through the cylinder walls.
2. there is no change in entropy, *i.e.* entropy before and after the process is constant.
3. the work is done on the piston by the expanding steam.

Now consider 1 kg of wet steam being heated by reversible adiabatic process *i.e.* isentropically from the initial state 1 to the final state 2.

Let p_1 = Initial pressure of the wet steam in bar,

v_{g1} = Specific volume of the dry saturated steam in m³/kg, corresponding to initial pressure p_1 (from steam tables),

x_1 = Initial dryness fraction of the steam, and

p_2, v_{g2}, x_2 = Corresponding values for the final condition of the steam.

We know that initial entropy of steam before expansion,

$$s_1 = s_{f1} + \frac{x_1\, h_{fg1}}{T_1} = s_{f1} + x_1\, s_{fg1}$$

and final entropy of steam after expansion,

$$s_2 = s_{f2} + \frac{x_2\, h_{fg2}}{T_2} = s_{f2} + x_2\, s_{fg2}$$

* See Example 9.11.

Since the entropy before expansion is equal to the entropy after expansion, therefore

$$s_{f1} + \frac{x_1 h_{fg1}}{T_1} = s_{f2} + \frac{x_2 h_{fg2}}{T_2}$$

or

$$s_{f1} + x_1 s_{fg1} = s_{f2} + x_2 s_{fg2}$$

where s_{f1}, h_{fg1} and T_1 represents entropy of water, latent heat or enthalpy of evaporation and temperature corresponding to pressure p_1,

s_{f2}, h_{fg2} and T_2 represents the corresponding values at pressure p_2, and

s_{fg1} and s_{fg2} represents entropy of evaporation.

From the above equation, the value of *x_2 is obtained. The value of x_2 gives us an important information about the final condition of the steam, *e.g.*

1. If the value of x_2 is less than one, then the final condition of the steam is *wet*.
2. If the value of x_2 is equal to one, then the final condition of the steam is *dry saturated*.
3. If the value of x_2 is more than one, then the final condition of the steam is *superheated*.

In case of superheated steam, the absolute temperature of superheated steam (*i.e.* T_{sup}) may be obtained by using the Charles' law, *i.e.*

$$\frac{v_{sup}}{T_{sup}} = \frac{v_{g2}}{T_2}$$

where v_{sup} = Specific volume of superheated steam in m³/kg, and

T_2 = Saturation temperature corresponding to pressure p_2 (from steam tables) in K.

Now let us derive the following important relations for a reversible adiabatic process or isentropic process.

1. *Change in internal energy*

We know that initial internal energy of the steam,

$$u_1 = h_1 - 100\, p_1 v_1$$

where v_1 = Initial volume of steam $= x_1 v_{g1}$ ... (For wet steam)

and final internal energy of the steam,

$$u_2 = h_2 - 100\, p_2 v_2$$

where v_2 = Final volume of steam

$= x_2 v_{g2}$ ... (For wet steam)

$= v_{g2}$ ... (For dry saturated steam)

$= v_{sup}$ ... (For superheated steam)

∴ Change in internal energy,

$$du = u_2 - u_1$$

* The final condition of steam x_2 may be directly obtained from the Mollier diagram by plotting the initial and final conditions of steam. This has been explained in the previous chapter.

2. Heat absorbed or heat transferred

Since no heat is added or subtracted during an adiabatic process (reversible or irreversible), therefore

$$q_{1-2} = 0$$

3. Workdone during the process

According to first law of thermodynamics, the heat absorbed or heat transferred,

$$q_{1-2} = du + w_{1-2}$$

Since $q_{1-2} = 0$, during a non-flow reversible adiabatic or isentropic process, therefore workdone during the process,

$$w_{1-2} = -du = u_1 - u_2$$

This shows that during a reversible adiabatic process or isentropic process, workdone is equal to the change in internal energy.

If the process is steady flow reversible adiabatic or isentropic process, then according to steady flow energy equation,

$$h_1 + q_{1-2} = h_2 + w_{1-2} \quad \text{... (Neglecting kinetic energy)}$$

Since $q_{1-2} = 0$, therefore

$$w_{1-2} = h_1 - h_2$$

This shows that during a steady flow reversible adiabatic process or isentropic process, workdone is equal to the change in enthalpy.

Notes : 1. The adiabatic expansion of steam, as a rough approximation, is assumed to follow the general law

$$p v^n = \text{Constant}$$

where

$$n = 1.13 \quad \text{... (For wet steam)}$$

$$= 1.135 \quad \text{... (For dry steam)}$$

$$= 1.3 \quad \text{... (For superheated steam)}$$

Then work done on the piston

$$w_{1-2} = \frac{100\,(p_1 v_1 - p_2 v_2)}{(n-1)}$$

This method should not be used unless the value of n for the steam during the expansion is given. The value for the work done on the piston is given by :

$$w_{1-2} = -du = u_1 - u_2$$

2. The dryness fraction of steam decreases during an adiabatic expansion. In other words, the steam becomes wetter during an adiabatic process. But, in some cases such as expansion of steam in the nozzle of a steam turbine, the steam remains in its state of dry saturated or superheated steam because it does not get enough time for expansion.

Example 9.9. *Steam at a pressure of 10 bar and 0.95 dry expands isentropically to a pressure of 4 bar. Determine the final dryness fraction of steam, by 1. Using steam tables, and 2. Using Mollier chart.*

Solution. Given : $p_1 = 10$ bar ; $x_1 = 0.95$; $p_2 = 4$ bar

Let x_2 = Final dryness fraction of steam.

The isentropic process on *p-v*, *T-s* and *h-s* diagrams is shown in Fig 9.8.

1. *Using steam tables*

From steam tables, corresponding to a pressure of 10 bar, we find that

$$s_{f1} = 2.138 \text{ kJ/kg K} ; \quad s_{fg1} = 4.445 \text{ kJ/kg K}$$

and corresponding to a pressure of 4 bar, we find that

$$s_{f2} = 1.776 \text{ kJ/kg K} ; \quad s_{fg2} = 5.118 \text{ kJ/kg K}$$

We know that for an isentropic process,

$$s_{f1} + x_1 s_{fg1} = s_{f2} + x_2 s_{fg2}$$

$$2.138 + 0.95 \times 4.445 = 1.776 + x_2 \times 5.118$$

$$2.138 + 4.223 = 1.776 + 5.118\, x_2 \quad \text{or} \quad x_2 = 0.896 \textbf{ Ans.}$$

2. *Using Mollier chart*

First of all, mark point 1 where the pressure line through p_1 (*i.e.* 10 bar) and dryness fraction x_1 (*i.e.* 0.95) meets, as shown in Fig. 9.8 (*c*). Since the steam expands isentropically, therefore draw a vertical line through 1 to meet the pressure line through p_2 (*i.e.* 4 bar) at point 2.

The final dryness fraction of the steam, as read from the chart at point 2, is $x_2 = 0.896$. **Ans.**

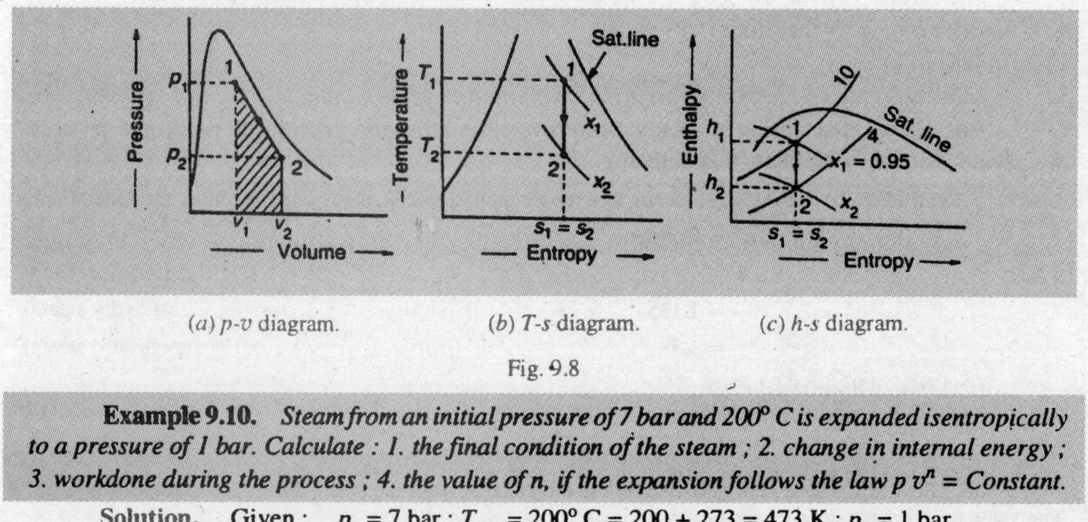

(*a*) *p-v* diagram. (*b*) *T-s* diagram. (*c*) *h-s* diagram.

Fig. 9.8

Example 9.10. *Steam from an initial pressure of 7 bar and 200° C is expanded isentropically to a pressure of 1 bar. Calculate : 1. the final condition of the steam ; 2. change in internal energy ; 3. workdone during the process ; 4. the value of n, if the expansion follows the law $p\,v^n$ = Constant.*

Solution. Given : $p_1 = 7$ bar ; $T_{sup} = 200°\text{ C} = 200 + 273 = 473$ K ; $p_2 = 1$ bar

The isentropic process on the *p-v*, *T-s* and *h-s* diagrams is shown in Fig. 9.9.

From steam tables of superheated steam, corresponding to a pressure of 7 bar and 200° C, we find that

$$s_{sup} = 6.886 \text{ kJ/kg K} ; \; h_{sup} = 2844 \text{ kJ/kg} ; \text{ and } \; v_{sup} = 0.3 \text{ m}^3\text{/kg}$$

and from steam tables for saturated steam, corresponding to a pressure of 1 bar, we find that

$$s_{f2} = 1.303 \text{ kJ/kg K} ; \; s_{fg2} = 6.057 \text{ kJ/kg K} ; \; h_{f2} = 417.5 \text{ kJ/kg} ;$$

$$h_{fg2} = 2258 \text{ kJ/kg} ; \text{ and } \; v_{g2} = 1.694 \text{ m}^3\text{/kg}$$

1. *Final condition of steam*

Let x_2 = Final dryness fraction of steam.

Since the initial condition of steam is superheated, therefore for isentropic process,

$$s_{sup} = s_2 = s_{f2} + x_2 \, s_{fg2}$$

$$6.886 = 1.303 + x_2 \times 6.057 \quad \text{or} \quad x_2 = 0.92 \text{ Ans.}$$

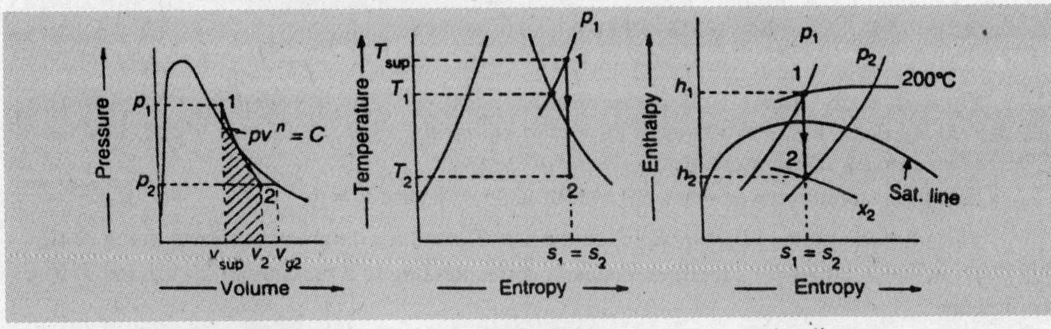

(*a*) *p-v* diagram. (*d*) *T-s* diagram. (*c*) *h-s* diagram.

Fig. 9.9

Note : The final condition of steam (*i.e.* x_2) may also be obtained by using Mollier diagram as shown in Fig. 9.9 (*c*). The vertical line drawn from point 1 of an initial condition of steam, *i.e.* pressure 7 bar and temperature 200° C, upto the pressure line of 1 bar. The dryness fraction of steam is read at the point of intersection (*i.e.* point 2) from which $x_2 = 0.92$. **Ans.**

2. *Change in internal energy*

We know that initial internal energy of the steam,

$$u_1 = h_1 - 100\, p_1\, v_1 = h_{sup} - 100\, p_1\, v_{sup}$$

$$= 2844 - 100 \times 7 \times 0.3 = 2634 \text{ kJ/kg}$$

and final internal energy of the steam,

$$u_2 = h_2 - 100\, p_2\, v_2 = h_{f2} + x_2\, h_{fg2} - 100\, p_2\, x_2\, v_{g2}$$

$$= 417.5 + 0.92 \times 2258 - 100 \times 1 \times 0.92 \times 1.694$$

$$= 2495 - 156 = 2339 \text{ kJ/kg}$$

$\therefore$ Change in internal energy,

$$du = u_2 - u_1 = 2339 - 2634 = -295 \text{ kJ/kg} \text{ Ans.}$$

The –ve sign indicates that there is a decrease in internal energy.

3. *Workdone during the process*

Since the heat transferred (q_{1-2}) during an isentropic process is zero, therefore workdone during the process,

$$w_{1-2} = -du = -(-295) = 295 \text{ kJ/kg Ans.} \qquad \ldots (\because q_{1-2} = du + w_{1-2})$$

4. *Value of n*

Since the expansion follows the law $p\,v^n$ = Constant, therefore

$$p_1\, v_1^n = p_2\, v_2^n \quad \text{or} \quad p_1\,(v_{sup})^n = p_2\,(x_2\, v_{g2})^n$$

Taking logarithm on both sides,

$$\log p_1 + n \log (v_{sup}) = \log p_2 + n \log (x_2\, v_{g2})$$

or $$n\,[\log (v_{sup}) - \log (x_2\, v_{g2})] = \log p_2 - \log p_1$$

$$\therefore \quad n = \frac{\log p_2 - \log p_1}{\log (v_{sup}) - \log (x_2\, v_{g2})} = \frac{\log (1 \times 10^5) - \log (7 \times 10^5)}{\log (0.3) - \log (0.92 \times 1.694)}$$

. . . (p_1 and p_2 are taken in N/m²)

$$= \frac{5 - 5.845}{-0.523 - 0.1927} = \frac{-0.845}{-0.7157} = 1.18 \textbf{ Ans.}$$

Example 9.11. *Steam at a pressure of 20 bar and 350° C expands adiabatically to a pressure of 1 bar till the steam is dry saturated at the end of expansion. Find : 1. Change of internal energy, 2. Workdone ; and 3. Change of entropy during the process.*

Solution. Given : $p_1 = 20$ bar ; $T_{sup} = 350°$ C ; $p_2 = 1$ bar ; $x_2 = 1$

The adiabatic process (irreversible) on the *p-v*, *T-s* and *h-s* diagrams is shown in Fig. 9.10.

From steam tables for superheated steam, corresponding to a pressure of 20 bar and 350° C, we find that

$$h_{sup} = 3138.6 \text{ kJ/kg} \; ; \; v_{sup} = 0.1386 \text{ m}^3\text{/kg} \; ; \; s_{sup} = 6.96 \text{ kJ/kg K}$$

and from steam tables for saturated steam, corresponding to a pressure of 1 bar, we find that

$$h_{f2} = 417.5 \text{ kJ/kg} \; ; \; h_{fg2} = 2258 \text{ kJ/kg} \; ; \; v_{g2} = 1.694 \text{ m}^3\text{/kg} \; ;$$

$$s_{f2} = 1.303 \text{ kJ/kg K} \; ; \; s_{fg2} = 6.057 \text{ kJ/kg K}$$

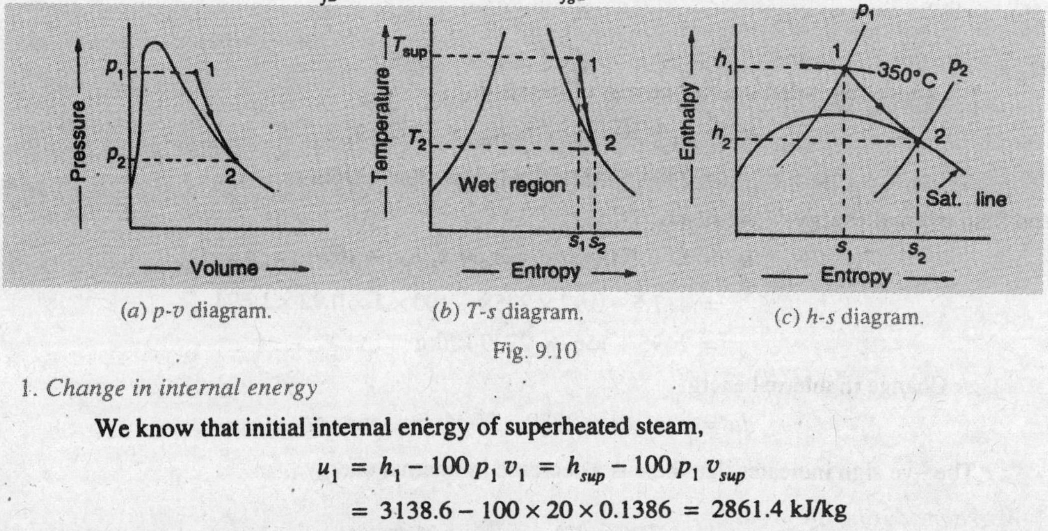

(*a*) *p-v* diagram. (*b*) *T-s* diagram. (*c*) *h-s* diagram.

Fig. 9.10

1. *Change in internal energy*

We know that initial internal energy of superheated steam,

$$u_1 = h_1 - 100\, p_1\, v_1 = h_{sup} - 100\, p_1\, v_{sup}$$

$$= 3138.6 - 100 \times 20 \times 0.1386 = 2861.4 \text{ kJ/kg}$$

and final internal energy of dry saturated steam,

$$u_2 = h_2 - 100\, p_2\, v_2 = h_{f2} + h_{fg2} - 100\, p_2\, v_{g2}$$

$$= 417.5 + 2258 - 100 \times 1 \times 1.694 = 2506.1 \text{ kJ/kg}$$

$\therefore$ Change of internal energy,

$$du = u_2 - u_1 = 2506.1 - 2861.4 = -355.3 \text{ kJ/kg } \textbf{Ans.}$$

The –ve sign indicates that there is a decrease in internal energy.

2. *Workdone*

Since the heat transferred (q_{1-2}) is zero during an adiabatic process, therefore workdone

$$w_{1-2} = -du = -(-355.3) = 355.3 \text{ kJ/kg } \textbf{Ans.}$$

3. *Change of entropy*

We know that initial entropy of superheated steam,

$$s_1 = s_{sup} = 6.96 \text{ kJ/kg K}$$

and final entropy of dry saturated steam,

$$s_2 = s_{f2} + s_{fg2} = 1.303 + 6.057 = 7.36 \text{ kJ/kg K}$$

$\therefore$ Change in entropy $= s_2 - s_1 = 7.36 - 6.96 = 0.4$ kJ/kg K **Ans.**

Note : From above, we see that the entropy has increased. Thus the adiabatic process is irreversible.

9.8. Polytropic Process ($pv^n = C$ Process)

Consider 1 kg of wet steam being heated polytropically from an initial state 1 to a final state 2.

Let p_1 = Initial pressure of the steam in bar,

v_{g1} = Specific volume of dry steam, corresponding to a pressure p_1 (from steam tables),

x_1 = Initial dryness fraction of the steam,

p_2, v_{g2}, x_2 = Corresponding values for the final condition of the steam, and

n = Polytropic index.

We know that initial volume of 1 kg of steam,

$$v_1 = x_1 v_{g1}$$

and final volume of 1 kg of steam,

$$v_2 = x_2 v_{g2}$$

Since the steam is heated polytropically (*i.e.* $pv^n = C$), therefore

$$p_1 v_1^n = p_2 v_2^n \quad \text{or} \quad p_1 (x_1 v_{g1})^n = p_2 (x_2 v_{g2})^n$$

$$\therefore \quad x_2 = \frac{x_1 v_{g1}}{v_{g1}} \times \left(\frac{p_1}{p_2}\right)^{1/n}$$

The value of x_2 gives us an important information about the final condition of the steam, e.g.

1. If the value of x_2 is less than one, then the final condition of the steam is *wet*.
2. If the value of x_2 is equal to one, then the final condition of the steam is *dry saturated*. In such a case, $v_2 = v_{g2}$.
3. If the value of x_2 is more than one, then the final condition of the steam is *superheated*. In such a case, $v_2 = v_{sup}$.

In case of superheated steam, we shall find out its volume and temperature as follows :

Let v_{sup} = Volume of the superheated steam, and

T_{sup} = Absolute temperature of the superheated steam.

We know that $p_1 (x_1 v_{g1})^n = p_2 (v_{sup})^n$

$$\therefore \quad v_{sup} = x\, v_{g1} \left(\frac{p_1}{p_2}\right)^{1/n}$$

Now T_{sup} may be found out by Charles' law, *i.e.*

$$\frac{v_{sup}}{T_{sup}} = \frac{v_{g2}}{T_2} \quad \text{or} \quad T_{sup} = T_2 \times \frac{v_{sup}}{v_{g2}}$$

where T_2 = Saturation temperature corresponding to pressure p_2 (from steam tables) in K.

Now let us derive the following important relations for the polytropic process :

1. *Workdone during the process*

We know that workdone during the polytropic process,

$$w_{1-2} = \frac{100\,(p_1 v_1 - p_2 v_2)}{n-1} \text{ kJ/kg} \quad \text{... (when } p \text{ is in bar)}$$

2. *Change in internal energy*

We know that initial internal energy of steam,

$$u_1 = h_1 - 100\,p_1 v_1 \text{ kJ/kg}$$

and final internal energy of steam,

$$u_2 = h_2 - 100\,p_2 v_2 \text{ kJ/kg}$$

3. *Heat absorbed or heat transferred*

According to the first law of thermodynamics, we know that heat absorbed or heat transferred,

$$q_{1-2} = du + w_{1-2} = (u_2 - u_1) + w_{1-2} \text{ kJ/kg}$$

Example 9.12. *One kg of steam at a pressure of 1 bar and 0.8 dry is compressed in a cylinder to a pressure of 2 bar. The law of compression is $pv^{1.2}$ = Constant. Find : 1. the final condition of the steam ; 2. change in internal energy ; and 3. the heat that passes through the cylinder walls.*

Solution. Given : p_1 = 1 bar ; x_1 = 0.8 ; p_2 = 2 bar ; n = 1.2

From steam tables, corresponding to a pressure of 1 bar, we find that

$$h_{f1} = 417.5 \text{ kJ/kg} \; ; \; h_{fg1} = 2258 \text{ kJ/kg} \; ; \text{ and } v_{g1} = 1.694 \text{ m}^3\text{/kg}$$

and corresponding to a pressure of 2 bar, we find that

$$h_{f2} = 504.7 \text{ kJ/kg} \; ; \; h_{fg2} = 2201.6 \text{ kJ/kg} \; ; \text{ and } v_{g2} = 0.885 \text{ m}^3\text{/kg}$$

1. *Final condition of the steam*

Let x_2 = Final dryness fraction of the steam.

We know that $p_1 v_1^n = p_2 v_2^n$ or $p_1 (x_1 v_{g1})^n = p_2 (x_2 v_{g2})^n$

$$\therefore \quad x_2 = \frac{x_1 v_{g1}}{v_{g2}} \left(\frac{p_1}{p_2}\right)^{\frac{1}{n}} = \frac{0.8 \times 1.694}{0.885} \left(\frac{1}{2}\right)^{\frac{1}{1.2}} = 0.86$$

2. *Change in internal energy*

We know that initial internal energy of the steam,

$$u_1 = h_1 - 100\,p_1 v_1 = h_{f1} + x_1 h_{fg1} - 100\,p_1 x_1 v_{g1}$$

$$= 417.5 + 0.8 \times 2258 - 100 \times 1 \times 0.8 \times 1.694 \text{ kJ/kg}$$

$$= 2224 - 135.5 = 2088.5 \text{ kJ/kg}$$

and final internal energy of the steam,

$$u_2 = h_2 - 100\, p_2 v_2 = h_{f2} + x_2 h_{fg2} - 100\, p_2 x_2 v_{g2}$$

$$= 504.7 + 0.86 \times 2201.6 - 100 \times 2 \times 0.86 \times 0.885 \text{ kJ/kg}$$

$$= 2398 - 152.2 = 2245.8 \text{ kJ/kg}$$

$\therefore$ Change in internal energy,

$$du = u_2 - u_1 = 2245.8 - 2088.5 = 157.3 \text{ kJ/kg} \quad \textbf{Ans.}$$

3. *Heat passing through the cylinder walls*

We know that workdone during the compression process,

$$w_{1-2} = \frac{100\,(p_1 v_1 - p_2 v_2)}{n-1} = \frac{100\,(p_1 x_1 v_{g1} - p_2 x_2 v_{g2})}{n-1}$$

$$= \frac{100\,(1 \times 0.8 \times 1.694 - 2 \times 0.86 \times 0.885)}{1.2 - 1} = -83.5 \text{ kJ/kg}$$

The –ve sign indicates that the work is done on the steam.

We know that heat passing through the cylinder walls or heat transferred,

$$q_{1-2} = du + w_{1-2} = 157.3 - 83.5 = 73.84 \text{ kJ/kg} \quad \textbf{Ans.}$$

Example 9.13. *Steam of dryness fraction 0.96 is expanded in a cylinder according to $pv^{1.02}$ = Constant. The pressure at the beginning of expansion is 20 bar, and is continued till the pressure reduces to 1 bar. Determine : 1. the final condition of the steam ; 2. the workdone during expansion ; 3. the change in internal energy, and 4. the heat exchange that occurs between the steam and cylinder walls per kg.*

Solution. Given : $x_1 = 0.96$; $n = 1.02$; $p_1 = 20$ bar ; $p_2 = 1$ bar

The polytropic process on *p-v*, *T-s* and *h-s* diagrams is shown in Fig. 9.11.

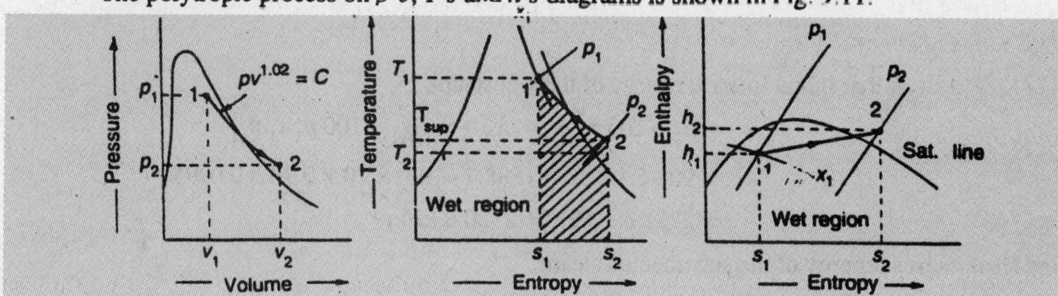

(*a*) *p-v* diagram. (*b*) *T-s* diagram. (*c*) *h-s* diagram.

Fig. 9.11

From steam tables, corresponding to a pressure of 20 bar, we find that

$$h_{f1} = 908.6 \text{ kJ/kg} \; ; \; h_{fg1} = 1888.7 \text{ kJ/kg} \; ; \text{ and } v_{g1} = 0.0995 \text{ m}^3\text{/kg}$$

and corresponding to a pressure of 1 bar, we find that

$$h_{f2} = 417.5 \text{ kJ/kg} \; ; \; h_{fg2} = 2258 \text{ kJ/kg} \; ; v_{g2} = 1.694 \text{ m}^3\text{/kg} \; ; \text{ and}$$

$$T_2 = 99.6^\circ \text{C} = 99.6 + 273 = 372.6 \text{ K}$$

1. *Final condition of the steam*

Let x_2 = Final dryness fraction of the steam.

We know that $p_1 v_1^n = p_2 v_2^n$ or $p_1 (x_1 v_{g1})^n = p_2 (x_2 v_{g2})^n$

$$\therefore \quad x_2 = \frac{x_1 v_{g1}}{v_{g2}}\left(\frac{p_1}{p_2}\right)^{\frac{1}{n}} = \frac{0.96 \times 0.0995}{1.694}\left(\frac{20}{1}\right)^{\frac{1}{1.02}} = 1.062$$

Since the final dryness fraction of the steam in greater than one, therefore the final condition of steam at the end of expansion is superheated.

Let v_{sup} = Volume of the superheated steam in m³/kg, and

T_{sup} = Temperature of the superheated steam in K.

We know that $p_1 v_1^n = p_2 v_2^n$ or $p_1 (x_1 v_{g1})^n = p_2 (v_{sup})^n$

$$\therefore \quad v_{sup} = x_1 v_{g1}\left(\frac{p_1}{p_2}\right)^{\frac{1}{n}} = 0.96 \times 0.0995\left(\frac{20}{1}\right)^{\frac{1}{1.02}} = 1.8 \text{ m}^3/\text{kg}$$

Now let us find out the final condition of the superheated steam (*i.e.* T_{sup}) by Charles' law. We know that

$$\frac{v_{sup}}{T_{sup}} = \frac{v_{g2}}{T_2} \quad \text{or} \quad T_{sup} = T_2 \times \frac{v_{sup}}{v_{g2}} = 372.6 \times \frac{1.8}{1.694} = 396 \text{ K} \quad \textbf{Ans.}$$

2. *Workdone during expansion*

We know that workdone during expansion,

$$w_{1-2} = \frac{100\,(p_1 v_1 - p_2 v_2)}{n-1} = \frac{100\,(p_1 x_1 v_{g1} - p_2 v_{sup})}{n-1}$$

$$= \frac{100\,(20 \times 0.96 \times 0.0995 - 1 \times 1.8)}{1.02 - 1} = 552 \text{ kJ/kg} \quad \textbf{Ans.}$$

3. *Change in internal energy*

We know that initial internal energy of the wet steam,

$$u_1 = h_1 - 100\,p_1 v_1 = h_{f1} + x_1 h_{fg1} - 100\,p_1 x_1 v_{g1}$$

$$= 908.6 + 0.96 \times 1888.7 - 100 \times 20 \times 0.96 \times 0.0995$$

$$= 2721.8 - 191 = 2530.8 \text{ kJ/kg}$$

and final internal energy of the superheated steam,

$$u_2 = h_2 - 100\,p_2 v_2 = {}^*h_{sup} - 100\,p_2 v_{sup}$$

$$= [h_{f2} + h_{fg2} + c_p (T_{sup} - T_2)] - 100\,p_2 v_{sup}$$

$$= [417.5 + 2258 + 2\,(396 - 372.6)] - 100 \times 1 \times 1.8$$

$$= 2722.3 - 180 = 2542.3 \text{ kJ/kg}$$

... (Taking c_p for superheated steam = 2 kJ/kg K)

∴ Change in internal energy,

$$du = u_2 - u_1 = 2542.3 - 2530.8 = 11.5 \text{ kJ/kg} \quad \textbf{Ans.}$$

* The value of h_{sup} may also be read directly from steam tables for superheated steam, corresponding to a pressure of 1 bar and 123° C.

4. *Heat exchange between the steam and cylinder walls*

We know that heat exchange between the steam and cylinder walls or heat transferred,

$$q_{1-2} = du + w_{1-2} = 11.5 + 552 = 563.5 \text{ kJ/kg} \quad \textbf{Ans.}$$

9.9. Throttling Process (Constant Enthalpy or Constant Total Heat Process)

It is also known as wire drawing when the working substance is vapour. Steam is said to be throttled when it passes through a restricted opening such as a narrow aperture or a slightly opened valve. The leakage of a fluid through a crack in the vessel is an example of throttling.

During the throttling process, the fluid is forced out by its pressure through the restricted opening or aperture. The aperture is so narrow that due to frictional resistance between the fluid and the sides of the aperture, the velocity of the outcoming fluid is almost reduced to zero. The kinetic energy is converted into heat by friction. Hence, if the steam is initially wet, it will start drying up as it expands. If throttled, the dry steam will become superheated.

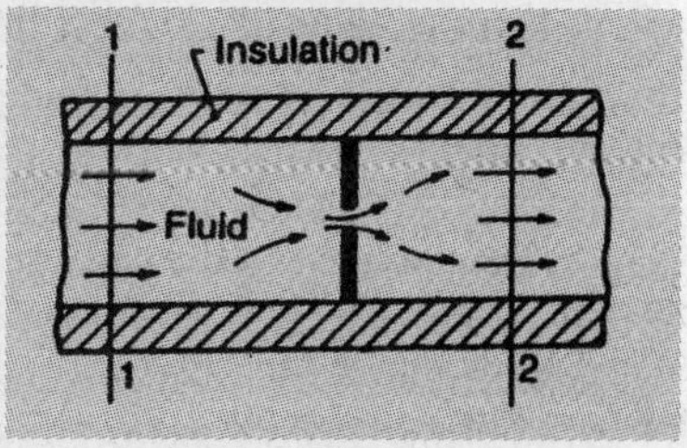

Fig. 9.12. Throttling process.

It may be noted that during a throttling process :

1. No heat is supplied or rejected (*i.e.* $q_{1-2} = 0$).
2. No work is done by the expanding fluid (*i.e.* $w_{1-2} = 0$).
3. No change in the internal energy of fluid (*i.e.* $du = 0$).
4. The enthalpy or total heat of the fluid remains constant (*i.e.* $h_1 = h_2$).

Consider 1 kg of wet steam being throttled through a narrow aperture, from an initial state 1 to the final state 2, as shown in Fig. 9.12. The throttling process is irreversible adiabatic process.

Applying *steady flow equation between sections 1-1 and 2-2, we have for a unit mass of fluid,

$$h_1 + \frac{V_1^2}{2} + g\,z_1 + q_{1-2} = h_2 + \frac{V_2^2}{2} + g\,z_2 + w_{1-2}$$

Since the datum level is same, therefore $z_1 = z_2$. Also the velocities V_1 and V_2 are small, therefore $V_1^2/2$ and $V_2^2/2$ are negligible. Hence the above equation reduces to

$$h_1 = h_2 \qquad \ldots(\because q_{1-2} = 0 \text{ and } w_{1-2} = 0)$$

This shows that during throttling process, enthalpy remains constant.

Let p_1 = Initial pressure of the steam in bar,

v_{g1} = Specific volume of dry steam in m³/kg corresponding to an initial pressure p_1 (from steam tables).

x_1 = Initial dryness fraction of the steam, and

p_1, v_{g2}, x_2 = Corresponding values for the final condition of the steam.

We know that initial enthalpy or total heat of steam before expansion,

$$h_1 = h_{f1} + x_1\,h_{fg1}$$

* For details, please refer chapter 3, Art 3.18.

and final enthalpy or total heat of steam after expansion,

$$h_2 = h_{f2} + x_2 h_{fg2}$$

Since enthalpy remains constant during throttling process, therefore

$$h_1 = h_2 \quad \text{or} \quad h_{f1} + x_1 h_{fg1} = h_{f2} + x_2 h_{fg2}$$

$$\therefore \quad x_2 = \frac{h_{f1} + x_1 h_{fg1} - h_{f2}}{h_{fg2}}$$

The value of x_2 gives us an important information, *e.g.*

1. If the value of x_2 is less than one, then the final condition of the steam is *wet*.
2. If the value of x_2 is equal to one, then the final condition of the steam is *dry saturated.*
3. If the value of x_2 is greater than one, then the final condition of the steam is *superheated.*

Note : The throttling process may be represented on Mollier chart by a horizontal line as discussed in the previous chapter. The final condition of steam may also be read directly from the chart.

Example 9.14. *One kg of steam at 14 bar pressure and dryness fraction 0.9 is throttled to 1.4 bar. Determine the final condition of the steam by using : 1. Steam tables, and 2. Mollier chart.*

Solution. Given : $p_1 = 14$ bar ; $x_1 = 0.9$; $p_2 = 1.4$ bar

Let x_2 = Final dryness fraction of the steam.

1. *Using steam tables*

From steam tables, corresponding to a pressure of 14 bar, we find that

$$h_{f1} = 830.1 \text{ kJ/kg} ; \; h_{fg1} = 1957.7 \text{ kJ/kg}$$

and corresponding to a pressure of 1.4 bar, we find that

$$h_{f2} = 458.4 \text{ kJ/kg} ; \; h_{fg2} = 2232 \text{ kJ/kg}$$

We know that for a throttling process,

$$h_{f1} + x_1 h_{fg1} = h_{f2} + x_2 h_{fg2}$$

$$830.1 + 0.9 \times 1957.7 = 458.4 + x_2 \times 2232$$

$$\therefore \quad x_2 = 0.96 \textbf{ Ans.}$$

2. *Using Mollier chart*

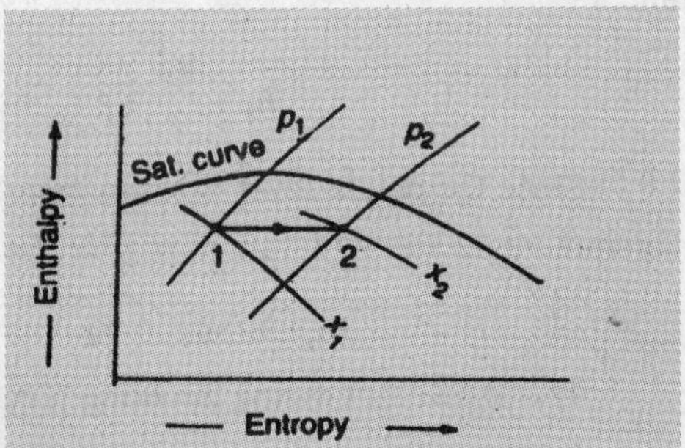

Fig. 9.13

First of all, mark point 1 where the pressure p_1 (*i.e.* 14 bar) and dryness fraction x_1 (*i.e.* 0.9) meet as shown in Fig. 9.13. Since the steam is throttled, therefore draw a horizontal line through 1 to meet the pressure line p_2 (*i.e.* 1.4 bar) at point 2.

The final dryness fraction of the steam, as read from the point 2 is $x_2 = 0.96$. **Ans.**

Example 9.15. *Steam at 10 bar and 0.9 dryness is throttled to a pressure of 2 bar. Using steam tables only, evaluate the final dryness fraction or degree of superheat. Estimate the change of entropy during this process.*

Solution. Given : $p_1 = 10$ bar ; $x_1 = 0.9$; $p_2 = 2$ bar

From steam tables, corresponding to a pressure of 10 bar, we find that

$$h_{f1} = 762.6 \text{ kJ/kg} ; \; h_{fg1} = 2013.6 \text{ kJ/kg} ; \; s_{f1} = 2.138 \text{ kJ/kg K} ; \text{ and}$$

$$s_{fg1} = 4.445 \text{ kJ/kg K}$$

and corresponding to a pressure of 2 bar, we find that

$$h_{f2} = 504.7 \text{ kJ/kg} \; ; \; h_{fg2} = 2201.6 \text{ kJ/kg} \; ; \; s_{f2} = 1.53 \text{ kJ/kg K} \; ; \text{ and}$$

$$s_{fg2} = 5.597 \text{ kJ/kg K}$$

Let x_2 = Final dryness fraction of steam.

We know that for a throttling process,

$$h_{f1} + x_1 h_{fg1} = h_{f2} + x_2 h_{fg2}$$

$$762.6 + 0.9 \times 2013.6 = 504.7 + x_2 \times 2201.6$$

$$\therefore \quad x_2 = 0.94 \text{ **Ans.**}$$

Change of entropy

We know that entropy before expansion,

$$s_1 = s_{f1} + x_1 s_{fg1} = 2.138 + 0.9 \times 4.445 = 6.14 \text{ kJ/kg K}$$

and entropy after expansion, $s_2 = s_{f2} + x_2 s_{fg2} = 1.53 + 0.94 \times 5.597 = 6.79$ kJ/kg K

$\therefore$ Change of entropy, $s = s_2 - s_1 = 6.79 - 6.14 = 0.65$ kJ/kg K **Ans.**

Note : From above we see that the entropy increases during the throttling process. Hence, the throttling process is irreversible.

Example 9.16. *Steam at a pressure of 14 bar with 80 K of superheat is throttled to a pressure of 8.5 bar. Determine the temperature of steam after throttling. The specific heat of superheated steam may be assumed as 2.2 kJ/kg K.*

Solution. Given : $p_1 = 14$ bar ; $T_{sup} - T_1 = 80$ K ; $p_2 = 8.5$ bar ; $c_p = 2.2$ kJ/kg K

Let T = Temperature of steam after throttling.

From steam tables, corresponding to a pressure of 14 bar, we find that

$$h_{g1} = 2787.8 \text{ kJ/kg}$$

and corresponding to a pressure of 8.5 bar, we find that

$$h_{g2} = 2769 \text{ kJ/kg and } T_2 = 172.9^\circ\text{ C} = 445.9 \text{ K}$$

We know that initial enthalpy of superheated steam,

$$h_1 = h_{g1} + c_p (T_{sup} - T_1) = 2787.8 + 2.2 \times 80 = 2963.8 \text{ kJ/kg} \quad \ldots (i)$$

and final enthalpy of superheated steam,

$$h_2 = h_{g2} + c_p (T - T_2)$$

$$= 2769 + 2.2 (T - 445.9) = 1788 + 2.2\, T \quad \ldots (ii)$$

Equating equations (*i*) and (*ii*),

$$2963.8 = 1788 + 2.2\, T \quad \text{or} \quad T = 534.4 \text{ K} = 261.4^\circ\text{ C} \text{ **Ans.**}$$

Example 9.17. *Steam initially at a pressure of 15 bar and 0.95 dryness expands isentropically to 7.5 bar and then throttled until it is dry. Using steam tables only, Calculate : 1. change in entropy, and 2. change in internal energy per kg of steam during the entire process. Show the process in a h-s plane. Is the entire process reversible. Justify your statement.*

Solution. Given : $p_1 = 15$ bar ; $x_1 = 0.95$; $p_2 = 7.5$; $x_3 = 1$

First of all, let us find the dryness fraction of steam (x_2) after isentropic expansion and pressure (p_3) at the end of throttling process.

From steam tables, corresponding to a pressure of 15 bar, we find that

$$h_{f1} = 844.6 \text{ kJ/kg} \; ; \; h_{fg1} = 1945.3 \text{ kJ/kg} \; ;$$

$$v_{g1} = 0.1317 \text{ m}^3\text{/kg} \; ; \; s_{f1} = 2.314 \text{ kJ/kg K} \; ;$$

$$s_{fg1} = 4.126 \text{ kJ/kg K}$$

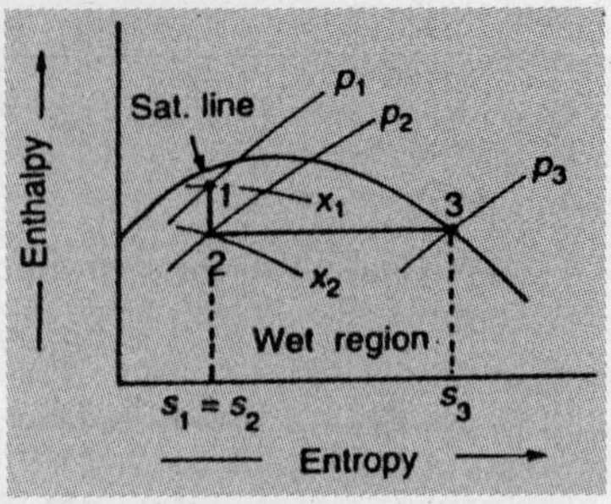

Fig. 9.14

and corresponding to a pressure of 7.5 bar, we find that

$$h_{f2} = 709.3 \text{ kJ/kg} \; ; \; h_{fg2} = 2055.5 \text{ kJ/kg} \; ;$$

$$v_{g2} = 0.2553 \text{ m}^3\text{/kg} \; ; \; s_{f2} = 2.019 \text{ kJ/kg K} \; ;$$

$$s_{fg2} = 4.662 \text{ kJ/kg K}$$

We know that entropy of wet steam before isentropic expansion,

$$s_1 = s_{f1} + x_1 \, s_{fg1} = 2.314 + 0.95 \times 4.126 = 6.324 \text{ kJ/kg K}$$

and entropy of wet steam after isentropic expansion,

$$s_2 = s_{f2} + x_2 \, s_{fg2} = 2.019 + x_2 \times 4.662 = 2.019 + 4.662 \, x_2$$

Since the entropy remains constant during isentropic process, therefore

$$s_1 = s_2$$

$$6.324 = 2.019 + 4.662 \, x_2 \quad \text{or} \quad x_2 = 0.923$$

We know that enthalpy of wet steam before isentropic expansion,

$$h_1 = h_{f1} + x_1 \, h_{fg1} = 844.6 + 0.95 \times 1945.3 = 2693 \text{ kJ/kg}$$

and enthalpy of wet steam after isentropic expansion or before throttling,

$$h_2 = h_{f2} + x_2 \, h_{fg2} = 709.3 + 0.923 \times 2055.5 = 2606 \text{ kJ/kg}$$

We know that enthalpy of steam after throttling,

$$h_3 = \text{Enthalpy of steam before throttling } (h_2) = 2606 \text{ kJ/kg}$$

From steam tables, corresponding to the enthalpy of dry steam equal to 2606 kJ/kg, we find that pressure after throttling,

$$p_3 = 0.18 \text{ bar} \; ; \; v_{g3} = 8.445 \text{ m}^3\text{/kg} \; ; \text{ and } s_3 = 7.946 \text{ kJ/kg K}$$

1. *Change of entropy*

We know that change of entropy during the entire process,

$$= s_3 - s_1 = 7.946 - 6.324 = 1.622 \text{ kJ/kg K} \; \textbf{Ans.}$$

2. *Change in internal energy*

We know that initial internal energy of the steam,

$$u_1 = h_1 - 100 \, p_1 \, v_1 = h_1 - 100 \, p_1 \, x_1 \, v_{g1}$$

$$= 2693 - 100 \times 15 \times 0.95 \times 0.1317 = 2505 \text{ kJ/kg}$$

and final internal energy of the steam,

$$u_3 = h_3 - 100 \, p_3 \, v_3 = h_3 - 100 \, p_3 \, x_3 \, v_{g3}$$

$$= 2606 - 100 \times 0.18 \times 1 \times 8.445 = 2454 \text{ kJ/kg}$$

$$\ldots (\because h_3 = h_2 \text{ and } x_3 = 1)$$

∴ Change in internal energy during the entire process

$$= u_3 - u_1 = 2454 - 2505 = -51 \text{ kJ/kg} \quad \textbf{Ans.}$$

The –ve sign indicates that the internal energy has decreased.

The process on the *h-s* diagram is shown in Fig. 9.14. We know that for a reversible process, the change in entropy is zero whereas for an irreversible process, the entropy increases. Thus from above, we see that the isentropic process 1-2 is reversible because $s_1 = s_2$ or $(s_2 - s_1) = 0$ and throttling process 2-3 is irreversible because s_3 is greater than s_2 or entropy increases.

EXERCISES

1. Steam at a pressure of 6 bar and dryness 0.8 is heated at a constant volume to a pressure of 7 bar. Determine the final dryness fraction and heat absorbed by 1 kg of steam. [**Ans.** 0.92 ; 234 kJ/kg]

2. A pressure cooker contains 1.5 kg of saturated steam at a pressure of 5 bar. Find the quantity of heat which must be rejected so as to reduce the quality of steam to 60% dry. Determine the pressure and temperature of the steam at the new state. [**Ans.** 1218.5 kJ ; 2.9 bar ; 132.4° C]

3. A closed vessel of 0.255 m³ contains steam at a pressure of 18 bar and 0.75 dry. The vessel is cooled till the pressure in the vessel falls to 12 bar. Find 1. the final condition of steam ; 2. the heat transferred, and 3. the change in entropy. [**Ans.** 0.507 ; 1426 kJ (rejected); 3 kJ/K (decrease)]

4. A quantity of steam at 10 bar and 0.85 dryness occupies 0.15 m³. The steam is heated at constant pressure to raise its temperature upto 300° C. Determine : 1. the workdone, 2. the change in internal energy, and 3. the heat supplied. [**Ans.** 84.4 kJ ; 440.4 kJ ; 524.8 kJ]

5. One kg of steam at 18 bar and 280° C undergoes a constant pressure process until the quality of steam becomes 0.5 dry. Find the workdone, the heat transferred and the change in entropy. [**Ans.** – 142 kJ/kg ; – 1142.8 kJ/kg ; – 2.351 kJ/kg K]

6. A cylinder contains steam at a pressure of 7 bar and 0.95 dry. The steam is expanded isothermally and reversibly to a pressure of 1.5 bar. Find : 1. the final condition of steam ; 2. the change in enthalpy ; 3. the change in internal energy ; 4. the change in entropy ; 5. the heat transferred ; and 6. the workdone per kg of steam. [**Ans.** 165° C ; 143.5 kJ/kg ; 127 kJ/kg ; 1.015 kJ/kg K ; 445 kJ/kg ; 318 kJ/kg]

7. One kg of dry saturated steam at 230° C expands isothermally to a pressure of 4 bar. Find : 1. change in enthalpy ; 2. change in internal energy ; 3. change in entropy ; 4. heat transferred, and 5. workdone during the process. [**Ans.** 121 kJ/kg ; 93 kJ/kg ; 1.0854 kJ/kg K ; 546 kJ/kg ; 453 kJ/kg]

8. One kg of steam at a pressure of 8 bar and dryness 0.8 is expanded hyperbolically in a cylinder to a pressure of 0.5 bar. Determine the final condition of steam and the heat that passes through the cylinder walls into the steam. [**Ans.** 0.95 ; 598.6 kJ/kg]

9. Dry saturated steam at a pressure of 7 bar is expanded hyperbolically to 0.36 bar. Find: 1. the final condition of steam ; 2. change of entropy ; and 3. heat exchange through the cylinder walls per kg of steam. [**Ans.** Superheated (143.5° C) ; 1.354 kJ/kg K ; 564 kJ/kg]

10. 1 m³ of steam at a pressure of 25 bar and dryness fraction 0.9 expands to 2.5 bar. If the steam expands hyperbolically, find : 1. final condition of the steam ; 2. workdone ; and 3. heat transferred. [**Ans.** Dry ; 5755 kJ ; 7142.2 kJ]

11. 0.05 m³ of steam at a pressure of 1.5 bar and dryness 0.8 is contained in a cylinder. The steam is compressed till its volume becomes one-fifth of the original volume. If the compression follows the law $p\,v$ = constant, find : 1. the final pressure of steam ; 2. the final dryness fraction of steam ; 3. the change in internal energy ; 4. the change in entropy ; and 5. the heat flow to or from the steam during the compression process. [**Ans.** 7.5 bar ; 0.726 ; – 0.46 kJ ; – 0.0354 kJ/K ; – 20.57 kJ]

12. Steam at a pressure of 5 bar and a temperature of 200° C expands isentropically to a pressure 0.7 bar. Find the final dryness of steam by using 1. Steam tables, and 2. Mollier chart. [**Ans.** 0.93]

13. Steam at a pressure of 22 bar and 380° C expands isentropically in a perfectly insulated cylinder till the temperature falls to 90° C. Determine the quality of steam at the final state and the workdone if the process is 1. non-flow, and 2. steady flow. [**Ans.** 0.925 ; 574 kJ/kg ; 712 kJ/kg]

14. Steam at a pressure of 120 bar and 400° C expands adiabatically to a pressure of 38 bar. If the final condition of steam is dry saturated, find the workdone by the steam. State whether the process is reversible or irreversible. [**Ans.** 200 kJ/kg ; Reversible]

15. Dry saturated steam at 7 bar is expanded isentropically to 4 bar after which it is condensed at a constant volume to 1 bar. Determine the final dryness fraction of the steam. [**Ans.** 0.26]

16. Steam having dryness fraction 0.9 expands isentropically in a cylinder from a pressure of 4 bar to 1 bar. Determine the final condition of the steam and the work done on the piston.

Assuming the expansion to be according to pv^n = constant, find the value of n.

[**Ans.** 0.84 ; 184 kJ/kg ; 1.127]

17. Dry saturated steam at a pressure of 11 bar expands polytropically according to $pv^{1.13}$ = constant, to a pressure of 1 bar. Determine 1. the final condition of steam ; 2. the amount of heat transferred per kg of steam ; and 3. the change of entropy. [**Ans.** 0.874 ; 325.7 kJ/kg ; 0.075 kJ/kg K]

18. Steam at a pressure of 15 bar and 250° C expands according to $p\,v^{1.25}$ = constant, to a pressure of 1.5 bar. Find 1. the final condition of steam ; 2. the workdone ; 3. the heat transferred ; and 4. change in entropy. [**Ans.** 0.827 ; 336 kJ/kg ; – 195 kJ/kg ; – 0.487 kJ/kg K]

19. Steam is throttled from a pressure of 11.5 bar to a pressure 1.4 bar. If the steam is dry saturated at the end of expansion, what is its dryness fraction at the beginning ? By how much is the entropy of the steam increased by throttling ? [**Ans.** 0.954 ; 0.912 kJ/kg K]

20. Steam at a pressure of 10 bar and 200° C is throttled to a pressure of 3 bar and then expanded isentropically to a pressure of 0.5 bar. By using Mollier chart, find out the change in entropy and enthalpy during these two processes. Also find the quality of steam at the end of each process.

[**Ans.** 0.54 kJ/kg K, 0 ; 0, – 300 kJ/kg ; 184° C, 0.94]

QUESTIONS

1. Show that the heat supplied is equal to the change of internal energy, when steam is expanded at a constant volume.
2. Is the expansion of superheated steam isothermal ? If not, state the reason.
3. Prove that during expansion of steam according to $pv = C$, the change in internal energy is equal to the change in total heat of steam.
4. What are the characteristics of adiabatic expansion ?
5. Find the amount of heat absorbed or rejected through the cylinder walls when the steam expands polytropically.
6. Explain throttling process of steam.
7. Show the throttling expansion of steam on a *h-s* plane.

OBJECTIVE TYPE QUESTIONS

1. In a constant volume process, heat transferred is equal to

(*a*) workdone (*b*) change in internal energy

(*c*) change in enthalpy (*d*) none of these

2. The process in which heat transferred is equal to the change of enthalpy, is known as

(*a*) constant pressure process (*b*) constant volume process

(*c*) constant temperature process (*d*) constant entropy process

3. The heating of wet steam at a constant temperature till it becomes dry saturated is similar to that of heating at a

(*a*) constant volume (*b*) constant pressure

(*c*) constant entropy (*d*) none of these

4. The superheating of steam at ...... may be regarded as hyperbolic.
(a) constant volume (b) constant pressure
(c) constant entropy (d) constant temperature

5. In an isentropic process,
(a) workdone is zero (b) change in internal energy is zero
(c) change in entropy is zero (d) change in enthalpy is zero

6. The non-flow process, in which the workdone is equal to the change in internal energy, is known as
(a) isentropic process (b) isothermal process
(c) isobaric process (d) isochroic process

7. In a steady flow reversible adiabatic process, workdone is equal to
(a) change in internal energy (b) change in entropy
(c) change in enthalpy (d) heat transferred

8. The isentropic process on *h-s* diagram will be a
(a) vertical line (b) horizontal line (c) curve

9. In a throttling process .... remains constant.
(a) enthalpy (b) temperature (c) entropy

10. The throttling process on *h-s* diagram will be a
(a) vertical line (b) horizontal line (c) curve

ANSWERS

1. (b)	2. (a)	3. (b)	4. (d)	5. (c)
6. (a)	7. (c)	8. (a)	9. (a)	10. (b)

10

Thermodynamic Vapour Cycles

1. Introduction. 2. Carnot Cycle with Steam as Working Substance. 3. Performance Criteria for Thermodynamic Cycles. 4. Rankine Cycle. 5. Rankine Cycle with Incomplete Evaporation. 6. Rankine Cycle with Superheated Steam. 7. Modified Rankine Cycle. 8. Efficiency of Modified Rankine Cycle.

10.1. Introduction

We have already discussed the thermodynamic cycles of air in chapter 6. But in this chapter, we shall discuss the thermodynamic cycles of vapour. In a vapour cycle, all the theory remains the same, except the working substance, which is steam. The steam may be in any form *i.e.* wet, dry saturated or superheated. Though there are many vapour cycles, yet the following are important from the subject point of view :

10.2. Carnot Cycle with Steam as Working Substance

The schematic diagram of a Carnot engine is shown in Fig. 10.1 and the Carnot cycle using steam as the working substance is represented on *p-v* and *T-s* diagrams as shown in Fig. 10.2 (*a*) and (*b*) respectively.

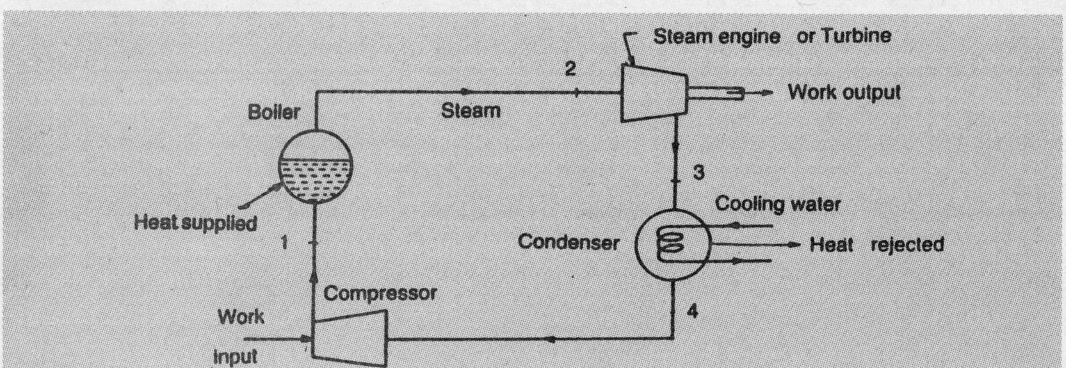

Fig. 10.1. Schematic diagram of a Carnot engine.

Consider 1 kg of saturated water at pressure p_1 and absolute temperature T_1, as represented by point 1 in Fig. 10.2 (*a*) and (*b*). The cycle is completed by the following four processes :

1. *Process 1-2.* The saturated water at point 1 is isothermally converted into dry saturated steam, in a boiler, and the heat is absorbed at a constant temperature T_1 and pressure p_1. The dry state of steam is represented by point 2. It means that the temperature T_2 (*i.e.* at point 2) and pressure p_2 (*i.e.* at point 2) is equal to temperature T_1 and pressure p_1 respectively. This isothermal process is represented by curve 1-2 on *p-v* and *T-s* diagrams in Fig. 10.2 (*a*) and (*b*).

We know that the heat absorbed by the saturated water during its conversion into dry steam is its latent heat of evaporation (*i.e.* $h_{fg1} = h_{fg2}$) corresponding to a pressure p_1 or p_2 ($\because p_1 = p_2$).

We also know that the area 1-2-*b*-*a* in the *T*-*s* diagram represents the heat absorbed to some scale, during the isothermal process.

$\therefore$ Heat absorbed during isothermal process (area 1-2-*b*-*a*),

$$q_{1-2} = \text{Change in entropy} \times \text{Absolute temperature}$$

$$= (s_2 - s_1)\,T_1 = (s_2 - s_1)\,T_2 \qquad \ldots (\because T_1 = T_2) \qquad \ldots (i)$$

2. *Process 2-3.* The dry steam at point 2 now expands isentropically in a steam engine or turbine. The pressure and temperature falls from p_2 to p_3 and T_2 to T_3 respectively. Since no heat is supplied or rejected during this process, therefore there is no change of entropy. The isentropic expansion is represented by the curve 2-3 as shown in Fig. 10.2 (*a*) and (*b*).

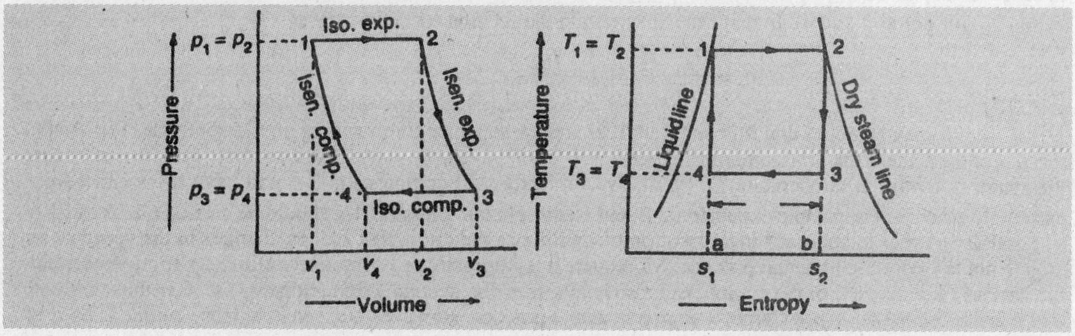

(*a*) *p*-*v* diagram. (*b*) *T*-*s* diagram.

Fig. 10.2. Carnot cycle.

3. *Process 3-4.* The wet steam at point 3 is now isothermally condensed in a condenser and the heat is rejected at a constant temperature T_3 and pressure p_3. It means that the temperature T_4 (*i.e.* at point 4) and pressure p_4 (*i.e.* at point 4) is equal to the temperature T_3 and pressure p_3 respectively. This isothermal process is represented by the curve 3-4 on *p*-*v* and *T*-*s* diagrams as shown in Fig. 10.2 (*a*) and (*b*).

We know that area 3-4-*b*-*a* in the *T*-*s* diagram represents the heat rejected to some scale during the isothermal process.

$\therefore$ Heat rejected during isothermal compression (area 3-4-*a*-*b*),

$$q_{3-4} = (s_2 - s_1)\,T_3 = (s_2 - s_1)\,T_4 \qquad \ldots (\because T_3 = T_4) \qquad \ldots (ii)$$

4. *Process 4-1.* The wet steam at point 4 is finally compressed isentropically in a compressor, till it returns back to its original state (point 1). The pressure and temperature rises from p_4 to p_1 and T_4 to T_1 respectively. The isentropic compression is represented by the curve 4-1 as shown in Fig. 10.2 (*a*) and (*b*). Since no heat is absorbed or rejected during this process, therefore entropy remains constant. This completes the cycle.

We know that work done during the cycle

$$= \text{Heat absorbed} - \text{Heat rejected}$$

$$= (s_2 - s_1)\,T_1 - (s_2 - s_1)\,T_3 = (s_2 - s_1)(T_1 - T_3)$$

and efficiency of the Carnot cycle,

$$\eta = \frac{\text{Work done}}{\text{Heat absorbed}}$$

$$= \frac{(s_2 - s_1)(T_1 - T_3)}{(s_2 - s_1)\, T_1} = \frac{T_1 - T_3}{T_1} = 1 - \frac{T_3}{T_1}$$

where T_1 = Highest temperature corresponding to the boiler pressure $p_1 = p_2$, and

T_3 = Lowest temperature corresponding to the condenser pressure $p_3 = p_4$.

Notes : 1. Since the heat absorbed is at the highest temperature and rejected at the lowest temperature, the Carnot cycle would give a maximum possible efficiency.

2. In the above theory, we have taken temperature at points 1, 2, 3 and 4 as T_1, T_2, T_3 and T_4 respectively in order to keep similarity between Carnot cycle and other cycles. But some authors take it T_1 (for points 1 and 2) and T_2 (for points 3 and 4). In that case, they obtain the relation for efficiency as,

$$\eta = \frac{T_1 - T_2}{T_1} = 1 - \frac{T_2}{T_1}$$

3. It may be noted that it is impossible to make a steam engine working on Carnot cycle. The simple reason for the same is that the isothermal expansion 1-2 will have to be carried out extremely slow to ensure that the steam is always at temperature T_1. Similarly, the isothermal compression 3-4 will have to be carried out extremely slow. But isentropic expansion 2-3 and isentropic compression 4-1 should be carried out as quickly as possible in order to approach ideal isentropic conditions. We know that sudden changes in the speed of an engine are not possible in actual practice. Moreover, it is impossible to completely eliminate friction between the various moving parts of the engine, and also heat losses due to conduction, radiation etc. It is thus obvious, that it is impossible to realise Carnot's engine in actual practice. However, such an imaginary engine is used as the ultimate standard of comparison of all steam engines.

Example 10.1. *A power plant is supplied with dry saturated steam at a pressure of 16 bar and exhausts at 0.2 bar. Using steam tables, find the efficiency of the Carnot cycle.*

Solution. Given : $p_1 = 16$ bar ; $p_2 = 0.2$ bar

From steam tables, corresponding to a pressure of 16 bar, we find that

$$T_1 = 201.4^\circ\text{C} = 201.4 + 273 = 474.4\,\text{K}$$

and corresponding to a pressure of 0.2 bar, we find that

$$T_3 = 60.1^\circ\text{C} = 60.1 + 273 = 333.1\,\text{K}$$

We know that efficiency of the Carnot cycle,

$$\eta = \frac{T_1 - T_3}{T_1} = \frac{474.4 - 333.1}{474.4} = 0.298 \text{ or } 29.8\% \textbf{ Ans.}$$

Example 10.2. *In a Carnot cycle, heat is supplied at 350° C and is rejected at 25° C. The working fluid is water, which while receiving heat, evaporates from liquid at 350° C to steam at 350° C. From the steam tables, the entropy change for this process is 1.438 kJ/kg K.*

If the cycle operates on a stationary mass of 1 kg of water, find the heat supplied, work done and heat rejected per cycle. What is the pressure of water during heat reception ?

Solution. Given : $T_1 = 350^\circ\text{C} = 350 + 273 = 623$ K ; $T_3 = 25^\circ\text{C} = 25 + 273 = 298$ K ; $s_2 - s_1 = 1.438$ kJ/kg K

Heat supplied per cycle

We know that the heat supplied per cycle,

$$= (s_2 - s_1)\, T_1 = 1.438 \times 623 = 895.87 \text{ kJ/kg } \textbf{Ans.}$$

Work done per cycle

We know that the work done per cycle,

$$= (s_2 - s_1)(T_1 - T_3) = 1.438\,(623 - 298) = 467.35 \text{ kJ/kg} \quad \textbf{Ans.}$$

Heat rejected per cycle

We know that the heat rejected per cycle

$$= (s_2 - s_1)\,T_3 = 1.438 \times 298 = 428.52 \text{ kJ/kg} \quad \textbf{Ans.}$$

Pressure of water during heat reception

The pressure of water during heat reception is the formation pressure of steam corresponding to 350° C. From steam tables, corresponding to 350° C, the pressure is 165.35 bar. **Ans.**

10.3. Performance Criteria for Thermodynamic Vapour Cycles

Though, theoretically, the Carnot cycle is the most efficient cycle, yet it is not considered as a standard of reference for the comparison of performance of thermodynamic vapour cycles.

The following terms, in addition to the efficiency, are commonly used for the comparison of performance of thermodynamic vapour cycles.

1. *Efficiency ratio.* It is also known as relative efficiency. It is defined as the ratio of thermal efficiency (or actual cycle efficiency) to Rankine efficiency (or ideal cycle efficiency). Mathematically,

$$\text{Efficiency ratio} = \frac{\text{Thermal efficiency}}{{}^{*}\text{ Rankine efficiency}}$$

Note : Thermal efficiency

$$= \frac{\text{Heat equivalent to one kilowatt hour (kWh)}}{\text{Total heat supplied to the steam per kWh}} = \frac{3600 \times P}{m_s\,(h_2 - h_{f3})}$$

where m_s = Mass of steam supplied in kg/h, and

P = Power developed in kW.

2. *Work ratio.* It is defined as the ratio of net work output to the gross (engine or turbine) output. Mathematically,

$$\text{Work ratio} = \frac{\text{Net work output}}{\text{Gross output}} = \frac{\text{Turbine work} - \text{Compressor work}}{\text{Turbine work}}$$

It may be noted that the Carnot cycle, despite of its high ideal thermal efficiency, has low work ratio. It is one of the reason that Carnot cycle is not attempted. In order to have better performance of the plant, both efficiency ratio and work ratio are the important criteria. It is desirable to have the value of work ratio almost unity. The higher value of work ratio also means a smaller size of the plant.

3. *Specific steam consumption.* It is also known as steam rate or specific rate of flow of steam. It is defined as the mass of steam that must be supplied to a steam engine or turbine in order to develop a unit amount of work or power output. The amount of work or power output is usually expressed in kilowatt hour (kWh). Mathematically

Specific steam consumption

$$= \frac{1 \text{ kWh}}{w} = \frac{3600}{w} = \frac{3600}{h_2 - h_3} \text{ kg / kWh} \qquad \ldots (\because 1 \text{ kWh} = 3600 \text{ kJ})$$

where w = Net workdone or power output = $(h_2 - h_3)$ kJ/kg

Note : In case of steam plants, the specific steam consumption is an indicator of the relative size of the plant.

---

* Refer Art. 10.4.

10.4. Rankine Cycle

The Rankine cycle is an ideal cycle for comparing the performance of steam plants. It is modified form of Carnot cycle, in which the condensation process (3-4) is continued until the steam is condensed into water. The schematic diagram of a steam engine or a turbine plant is shown in Fig. 10.3.

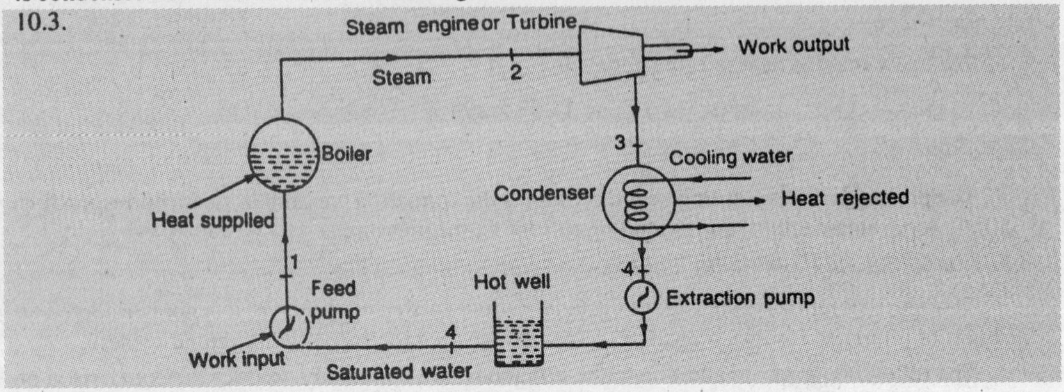

Fig. 10.3. Schematic diagram of a steam engine or turbine plant.

A Rankine cycle, using steam as a working substance, is represented on *p-v* and *T-s* diagrams as shown in Fig. 10.4 (*a*) and (*b*).

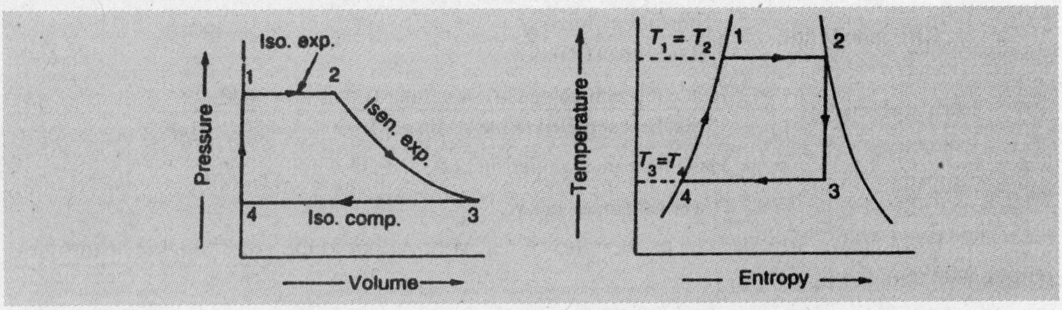

(*a*) *p-v* diagram. (*b*) *T-s* diagram.

Fig. 10.4. Rankine cycle.

Consider 1 kg of saturated water at pressure p_1 and absolute temperature T_1 as represented by point 1 in Fig. 10.4 (*a*) and (*b*). The cycle is completed by the following four processes :

1. *Process 1-2.* The saturated water at point 1 is isothermally converted into dry saturated steam in a boiler, and the heat is absorbed at a constant temperature T_1 and pressure p_1. The dry state of steam is represented by point 2. It means that the temperature T_2 (*i.e.* at point 2) and pressure p_2 (*i.e.* at point 2) is equal to temperature T_1 and pressure p_1 respectively. This isothermal process is represented by curve 1-2 on *p-v* and *T-s* diagrams in Fig. 10.4 (*a*) and (*b*).

We know that the heat absorbed during isothermal process by water during its conversion into dry steam is its latent heat of vaporisation (*i.e.* $h_{fg1} = h_{fg2}$), corresponding to a pressure p_1 or p_2 ($\because p_1 = p_2$).

2. *Process 2-3.* The dry saturated steam at point 2, now expands isentropically in an engine or turbine. The pressure and temperature falls from p_2 to p_3 and T_2 to T_3 respectively with a dryness fraction x_3. Since no heat is supplied or rejected during this process, therefore there is no change of entropy. The isentropic expansion is represented by the curve 2-3 as shown in Fig. 10.4 (*a*) and (*b*).

3. *Process 3-4.* The wet steam at point 3 is now isothermally condensed in a condenser and the heat is rejected at constant temperature T_3 and pressure p_3 until the whole steam is condensed into water. It means that the temperature T_4 and pressure p_4 (*i.e.* at point 4) is equal to the temperature T_3 and pressure p_3 respectively. The isothermal compression is represented by curve 3-4 on *p-v* and *T-s* diagrams in Fig. 10.4 (*a*) and (*b*). The heat rejected by steam is its latent heat (equal to $x_3\, h_{fg3}$).

4. *Process 4-1.* The water at point 4 is now warmed in a boiler at constant volume from temperature T_4 to T_1. Its pressure also rises from p_4 to p_1. This warming operation is represented by the curve 4-1 on *p-v* and *T-s* diagrams in Fig. 10.4 (*a*) and (*b*). The heat absorbed by water during this operation is equal to the sensible heat or liquid heat corresponding to the pressure p_1 *i.e.* equal to sensible heat at point 1 *minus* sensible heat at point 4.

Let $h_{f1} = h_{f2} =$ Sensible heat or enthalpy of water at point 1 corresponding to a pressure of p_1 or p_2 ($\because p_1 = p_2$), and

$h_{f4} = h_{f3} =$ Sensible heat or enthalpy of water at point 4 orresponding to a pressure of p_4 or p_3 ($\because p_4 = p_3$).

$\therefore$ Heat absorbed during warming operation 4-1

$$= h_{f1} - h_{f4} = h_{f2} - h_{f3}$$

and heat absorbed during the complete cycle

= Heat absorbed during isothermal operation 1-2
+ Heat absorbed during warming operation 4-1

$$= h_{fg2} + (h_{f2} - h_{f3}) = h_{f2} + h_{fg2} - h_{f3} = h_2 - h_{f3}$$

$$\ldots (\because \text{For dry steam, } h_2 = h_{f2} + h_{fg2}) \quad \ldots (i)$$

We know that heat rejected during the cycle

$$= h_3 - h_{f4} = h_{f3} + x_3\, h_{fg3} - h_{f4} = x_3\, h_{fg3} \qquad \ldots (\because h_{f3} = h_{f4})$$

$\therefore$ *Workdone during the cycle

= Heat absorbed − Heat rejected

$$= (h_2 - h_{f3}) - x_3\, h_{fg3}$$

$$= h_2 - (h_{f3} + x_3\, h_{fg3}) = h_2 - h_3 \qquad \ldots (\because h_3 = h_{f3} + h_{fg3}) \quad \ldots (ii)$$

and efficiency (also called Rankine efficiency),

$$\eta_R = \frac{\text{Work done}}{\text{Heat absorbed}} = \frac{h_2 - h_3}{h_2 - h_{f3}} \qquad \ldots (iii)$$

Notes : 1. The difference of enthalpies $(h_2 - h_3)$ is known as isentropic heat drop.

2. If the expansion of steam (2-3) is not isentropic and follows the general law pv^n = constant, then work done during the process will not be $(h_2 - h_3)$. The work done in this case will be given by the relation :

$$W = p_2 v_2 + \frac{p_2 v_2 - p_3 v_3}{(n-1)} - p_3 v_3 = \frac{n\,(p_2 v_2 - p_3 v_3)}{n-1}$$

---

* Since the workdone by the extraction and boiler feed pumps in increasing the pressure of water from the condenser pressure ($p_3 = p_4$) to the boiler pressure ($p_1 = p_2$) is very small, therefore it is neglected.

Example 10.3. *A steam power plant is supplied with dry saturated steam at a pressure of 12 bar and exhausts into a condenser at 0.1 bar. Calculate the Rankine efficiency by using 1. Steam tables, and 2. Mollier chart.*

Solution. Given : $x_2 = 1$; $p_2 = 12$ bar ; $p_3 = 0.1$ bar

1. *Using steam tables*

From steam tables, corresponding to a pressure of 12 bar, we find that

$$h_{f2} = 798.4 \text{ kJ/kg} ; \; h_{fg2} = 1984.3 \text{ kJ/kg} ; \; s_2 = s_{g2} = 6.519 \text{ kJ/kg K}$$

and corresponding to a pressure of 0.1 bar, we find that

$$h_{f3} = 191.8 \text{ kJ/kg} ; \; h_{fg3} = 2393 \text{ kJ/kg} ; \; s_{f3} = 0.649 \text{ kJ/kg K} ; \; s_{fg3} = 7.502 \text{ kJ/kg K}$$

First of all, let us find the dryness fraction (x_3) of steam after isentropic expansion. We know that in isentropic expansion,

Entropy before expansion (s_2) = Entropy after expansion (s_3)

$$6.519 = s_{f3} + x_3 \, s_{fg3} = 0.649 + x_3 \times 7.502$$

$$\therefore \quad x_3 = 0.783$$

Now enthalpy or total heat of dry saturated steam before expansion,

$$h_2 = h_{g2} = h_{f2} + x_2 \, h_{fg2} = 798.4 + 1 \times 1984.3 = 2782.7 \text{ kJ/kg}$$

and enthalpy or total heat of wet steam after expansion,

$$h_3 = h_{f3} + x_3 \, h_{fg3} = 191.8 + 0.783 \times 2393 = 2065.5 \text{ kJ/kg}$$

We know that Rankine efficiency,

$$\eta_R = \frac{h_2 - h_3}{h_2 - h_{f3}} = \frac{2782.7 - 2065.5}{2782.7 - 191.8} = 0.277 \text{ or } 27.7\% \textbf{ Ans.}$$

2. *Using Mollier chart*

In Fig. 10.5, the initial condition of steam is marked at point 2 and final condition at point 3. From Mollier chart, we find that

$$h_2 = 2780 \text{ kJ/kg} ; \; h_3 = 2060 \text{ kJ/kg} ; \text{ and}$$

$$h_{f3} = 191.8 \text{ kJ/kg} \qquad \text{... (From steam tables)}$$

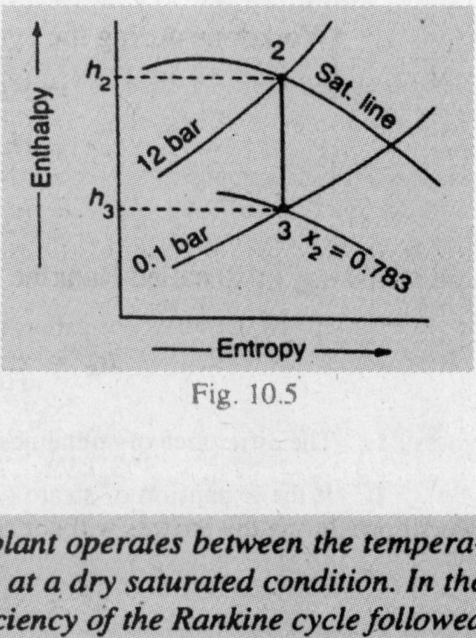

Fig. 10.5

We know that the Rankine efficiency,

$$\eta_R = \frac{h_2 - h_3}{h_2 - h_{f3}} = \frac{2780 - 2060}{2780 - 191.8}$$

$$= 0.278 \text{ or } 27.8\% \textbf{ Ans.}$$

Example 10.4. *A simple Rankine cycle steam power plant operates between the temperatures of 260° C and 95° C. The steam is supplied to the turbine at a dry saturated condition. In the turbine, it expands in an isentropic manner. Determine the efficiency of the Rankine cycle followed by the turbine, and the efficiency of the Carnot cycle, operating between these two temperature limits. Draw the turbine cycle on a h-s or on a T-s diagram.*

Solution. Given : $T_1 = T_2 = 260° \text{ C} = 260 + 273 = 533$ K ; $T_3 = T_4 = 95° \text{ C} = 95 + 273 = 368$ K

From steam tables, we find that the pressure corresponding to the highest temperature of 260° C,

$$p_2 = 46.94 \text{ bar}$$

and the pressure corresponding to the lowest temperature of 95° C,

$$p_3 = 0.845 \text{ bar}$$

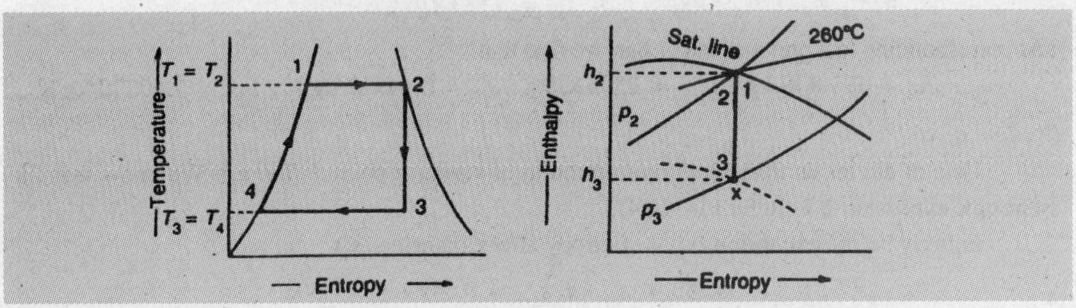

(*a*) *T-s* diagram. (*b*) *h-s* diagram.

Fig. 10.6

The initial and final conditions of steam are shown in Fig. 10.6 (*a*) and (*b*). From Mollier chart we find that

$$h_2 = 2800 \text{ kJ/kg and } h_3 = 2150 \text{ kJ/kg}$$

From steam tables, we find that the enthalpy of water at 95° C,

$$h_{f3} = 398 \text{ kJ/kg}$$

Efficiency of the Rankine cycle

We know that efficiency of the Rankine cycle,

$$\eta_R = \frac{h_2 - h_3}{h_2 - h_{f3}} = \frac{2800 - 2150}{2800 - 398} = 0.27 \text{ or } 27\% \textbf{ Ans.}$$

Efficiency of the Carnot cycle

We know that efficiency of the Carnot cycle,

$$\eta_c = \frac{T_1 - T_3}{T_1} = \frac{533 - 368}{533} = 0.3096 \text{ or } 30.96\% \textbf{ Ans.}$$

Note : The value of h_2 and h_3 may be determined by using steam tables, as discussed below :

From steam tables, corresponding to a temperature of 260° C, we find that

$$h_2 = h_{g2} = 2796.4 \text{ kJ/kg} ; \; s_2 = s_{g2} = 6.001 \text{ kJ/kg K}$$

and corresponding to a temperature of 95° C, we find that

$$h_{f3} = 398 \text{ kJ/kg} ; h_{fg3} = 2270.2 \text{ kJ/kg} ; s_{f3} = 1.25 \text{ kJ/kg} ; s_{fg3} = 6.167 \text{ kJ/kg K}$$

First of all, let us find the dryness fraction of steam at point 3 (*i.e.* x_3). We know that for isentropic expansion 2-3,

$$s_2 = s_3$$

or $$6.001 = s_{f3} + x_3 s_{fg3} = 1.25 + x_3 \times 6.167$$

$\therefore$ $$x_3 = 0.77$$

We know that $$h_3 = h_{f3} + x_3 h_{fg3} = 398 + 0.77 \times 2270.2 = 2146 \text{ kJ/kg}$$

Example 10.5. *The steam consumption of a steam engine is 20 tonnes per shift of 8 hours when developing 220 kW. Dry and saturated steam enters the engine at 10 bar pressure and leaves it at 0.1 bar pressure. Estimate the Rankine efficiency and the thermal efficiency of the engine.*

Solution. Given : $m_s = 20/8 = 2.5$ t/h $= 2500$ kg/h ; $P = 220$ kW ; $x_2 = 1$; $p_1 = p_2 = 10$ bar ; $p_3 = p_4 = 0.1$ bar

From steam tables, corresponding to a pressure of 10 bar, we find that

$$h_2 = h_{g2} = 2776.2 \text{ kJ/kg} \; ; \; s_2 = s_{g2} = 6.583 \text{ kJ/kg K}$$

and corresponding to a pressure of 0.1 bar, we find that

$$h_{f3} = 191.8 \text{ kJ/kg} \; ; \; h_{fg3} = 2393 \text{ kJ/kg} \; ; \; s_{f3} = 0.649 \text{ kJ/kg K} \; ; \; s_{fg3} = 7.502 \text{ kJ/kg K}$$

Rankine efficiency

First of all, let us find the dryness fraction of steam at point 3 (*i.e.* x_3). We know that for isentropic expansion 2-3 (Refer Fig. 10.4),

Entropy before expansion (s_2) = Entropy after expansion (s_3)

$$6.583 = s_{f3} + x_3 \, s_{fg3} = 0.649 + x_3 \times 7.502$$

$$\therefore \quad x_3 = 0.791$$

We know that enthalpy or total heat of steam at point 3,

$$h_3 = h_{f3} + x_3 \, h_{fg3} = 191.8 + 0.791 \times 2393 = 2084.7 \text{ kJ/kg}$$

$$\therefore \text{ Rankine efficiency,} \quad \eta_R = \frac{h_2 - h_3}{h_2 - h_{f3}} = \frac{2776.2 - 2084.7}{2776.2 - 191.8} = 0.267 \text{ or } 26.7\% \text{ Ans.}$$

Thermal efficiency of the engine

We know that thermal efficiency of the engine

$$\eta_{th} = \frac{3600 \times P}{m_s (h_2 - h_{f3})} = \frac{3600 \times 220}{2500\,(2776.2 - 191.8)}$$

$$= 0.1226 \text{ or } 12.26\% \text{ Ans.}$$

10.5. Rankine Cycle with Incomplete Evaporation

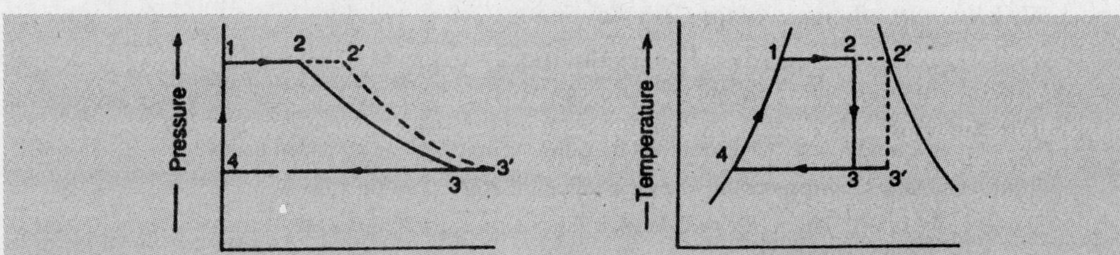

(*a*) *p-v* diagram. (*b*) *T-s* diagram.

Fig. 10.7. Rankine cycle with incomplete evaporation.

We have already discussed in the last article that in isothermal expansion of a Rankine cycle, the water is converted into dry saturated steam at a constant temperature T_1 and pressure p_1. Sometimes, the steam produced is not completely dry, but it is wet with dryness fraction equal to x_2. In such a case, the Rankine cycle may be represented on *p-v* and *T-s* diagrams as shown in Fig. 10.7 (*a*) and (*b*).

It may be noted from the above figure, that 1-2-3-4 represents the Rankine cycle with incomplete evaporation, whereas 1-2′-3′-4 represents the cycle with complete evaporation. In such a case, heat absorbed during the isothermal expansion 1-2 is equal to $x_2\, h_{fg2}$.

∴ Heat absorbed during the complete cycle

$$= x_2\, h_{fg2} + h_{f2} - h_{f3} = h_2 - h_{f3}$$

$$\ldots (\because \text{ For wet steam, } h_2 = h_{f2} + x_2\, h_{fg2})$$

Example 10.6. *Dry saturated steam at 10 bar is supplied to a prime mover and the exhaust takes place at 0.2 bar. Determine the Rankine efficiency, efficiency ratio and specific steam consumption of the prime mover, if the indicated thermal efficiency is 20%.*

Also find the percentage change in the Rankine efficiency, if the steam is initially 90% dry.

Solution. Given : $p_2 = 10$ bar ; $p_3 = 0.2$ bar ; $n_{th} = 20\% = 0.2$

Rankine efficiency

The initial and final conditions of steam are shown in Fig. 10.8 (*a*) and (*b*). From Mollier chart, we find that

$$h_2 = 2775 \text{ kJ/kg}\ ; \quad \text{and} \quad h_3 = 2150 \text{ kJ/kg}$$

From steam tables, we find that enthalpy of water at 0.2 bar,

$$h_{f3} = 251.5 \text{ kJ/kg}$$

We know that Rankine efficiency,

$$\eta_R = \frac{h_2 - h_3}{h_2 - h_{f3}} = \frac{2775 - 2150}{2775 - 251.5} = 0.247 \text{ or } 24.7\% \textbf{ Ans.}$$

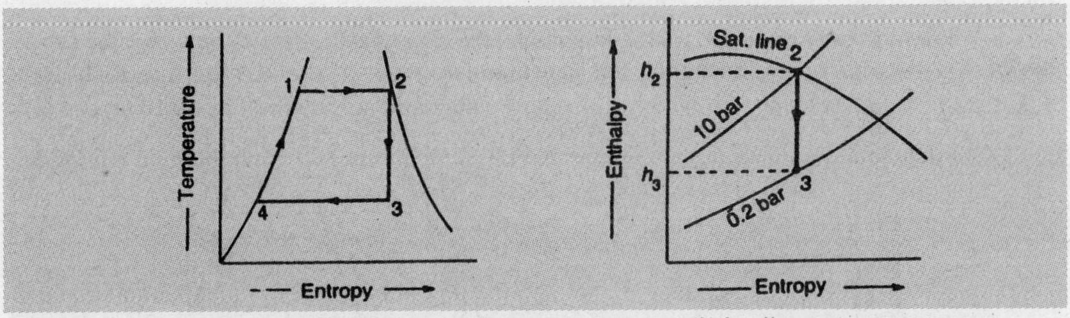

(*a*) *T-s* diagram. (*b*) *h-s* diagram.

Fig. 10.8

Efficiency ratio

We know that efficiency ratio

$$= \frac{\text{Indicated thermal efficiency}}{\text{Rankine efficiency}} = \frac{0.2}{0.247} = 0.81 \text{ or } 81\ \% \textbf{ Ans.}$$

Specific steam consumption

We know that specific steam consumption

$$= \frac{3600}{h_2 - h_3} = \frac{3600}{2775 - 2150} = 5.76 \text{ kg/kWh} \textbf{ Ans.}$$

Percentage change in the Rankine efficiency if the steam is initially 90% dry

The initial and final conditions of steam are shown in Fig. 10.9 (*a*) and (*b*). From Mollier chart, we find that

$$h_2 = 2580 \text{ kJ/kg} \; ; \; h_3 = 2030 \text{ kJ/kg}$$

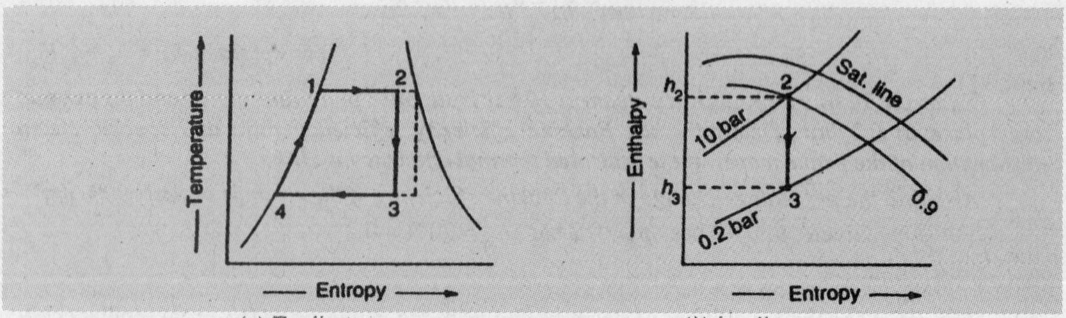

(*a*) *T-s* diagram. (*b*) *h-s* diagram.

Fig. 10.9

We know that Rankine efficiency,

$$\eta_R = \frac{h_2 - h_3}{h_2 - h_{f3}} = \frac{2580 - 2030}{2580 - 251.5} = 0.236 \quad \text{or} \quad 23.6 \%$$

$\therefore$ Percentage change in Rankine efficiency

$$= \frac{24.7 - 23.6}{24.7} \times 100 = 4.45 \% \quad \textbf{Ans.}$$

10.6. Rankine Cycle with Superheated Steam

We have already discussed, in the last article, the case of a Rankine cycle where the steam produced is wet with dryness fraction x_2. But sometimes, the steam produced is superheated. In such a case, the Rankine cycle may be shown on *p-v* and *T-s* diagrams as shown in Fig. 10.10 (*a*) and (*b*).

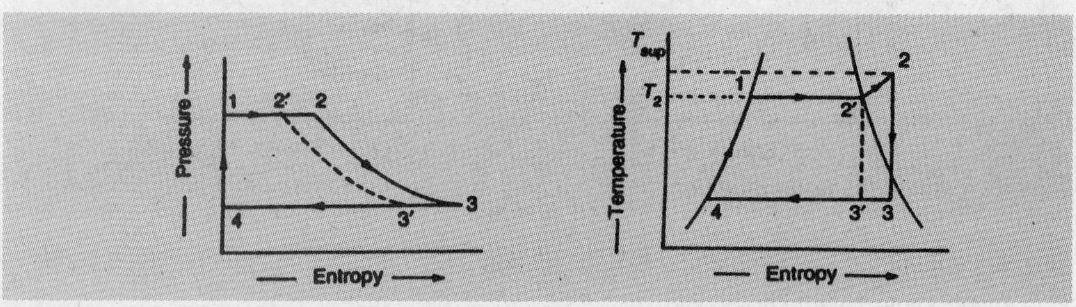

(*a*) *p-v* diagram. (*b*) *T-s* diagram.

Fig. 10.10. Rankine cycle with superheated steam.

In may be noted from the above figure, that 1-2-3-4 represents the Ranking cycle with superheated steam, where as 1-2′-3′-4 represents the cycle with complete evaporation. In such a case, heat absorbed during isothermal expansion,

$$h_2 = h_{sup} = h_{g2} + c_p (T_{sup} - T_2)$$

where h_{g2} = Enthalpy of dry saturated steam corresponding to a pressure of $p_1 = p_2$ (from steam tables),

T_{sup} = Temperature of superheated steam in K,

T_2 = Saturation temperature in K. corresponding to a pressure of $p_1 = p_2$ (from steam tables), and

c_p = Specific heat of superheated steam.

Note : The value of h_{sup} can be read directly from steam tables for superheated steam or Mollier chart corresponding to a pressure of p_2 and temperature T_{sup} in °C.

The superheating of steam before entering for isentropic expansion has the following advantages :

1. The work done (equal to the area 2′-2-3-3′ as shown in Fig. 10.10) increases.
2. The dryness fraction of steam at the end of isentropic expansion (*i.e.* leaving the steam engine or turbine) increases.
3. The specific steam consumption decreases.
4. The net efficiency of the cycle increases with the increase in degree of superheat (*i.e* $T_{sup} - T_2$). In other words, as the degree of superheat increases, the average temperature of heat addition of the cycle also increases and thus the efficiency increases.

Example 10.7. *A steam turbine receives steam at 15 bar and 350° C, and exhausts to the condenser at 0.06 bar. Determine the thermal efficiency of the ideal Rankine cycle operating between these two limits.*

Solution. Given : $p_1 = p_2 = 15$ bar ; $T_{sup} = 350°$ C ; $p_3 = p_4 = 0.06$ bar

From steam tables of superheated steam, corresponding to a pressure of 15 bar and 350° C, we find that

$$h_2 = h_{sup} = 3148.7 \text{ kJ/kg} \; ; \text{ and } s_2 = s_{sup} = 7.104 \text{ kJ/kg K}$$

and corresponding to a pressure of 0.06 bar, we find that

$$h_{f3} = 151.5 \text{ kJ/kg} \; ; h_{fg3} = 2416 \text{ kJ/kg} \; ; s_{f3} = 0.521 \text{ kJ/kg K} \; ; \text{and } s_{fg3} = 7.81 \text{ kJ/kg K}$$

First of all, let us find out the dryness fraction of steam (x_3) after expansion. We know that

Entropy before expansion (s_2) = Entropy after expansion (s_3)

or

$$s_{sup} = s_{f3} + x_3 \, s_{fg3}$$

$$7.104 = 0.521 + x_3 \times 7.81 \quad \text{or} \quad x_3 = 0.843$$

∴ Enthalpy of steam at point 3,

$$h_3 = h_{f3} + x_3 \, h_{fg3} = 151.5 + 0.843 \times 2416 = 2188 \text{ kJ/kg}$$

We know that thermal efficiency of the ideal Rankine cycle,

$$\eta_R = \frac{h_2 - h_3}{h_2 - h_{f3}} = \frac{3148.7 - 2188}{3148.7 - 151.5} = 0.32 \text{ or } 32\% \text{ **Ans.**}$$

Note : The value of isentropic heat drop ($h_2 - h_3$) may be obtained directly from the Mollier chart.

Example 10.8. *A steam turbine receives superheated steam at a pressure of 17 bar and having a degree of superheat of 110° C. The exhaust pressure is 0.07 bar and the expansion of steam takes place isentropically. Calculate 1. the heat supplied, 2. the heat rejected, 3. net workdone, and 4. the thermal efficiency.*

Solution. Given : $p_1 = p_2 = 17$ bar ; $T_{sup} - T_2 = 110°$ C ; $p_3 = p_4 = 0.07$ bar

The Rankine cycle with superheated steam is shown in Fig. 10.11.

From steam tables, corresponding to a pressure of 17 bar, we find that saturation temperature,

$$T_2 = 212.4° \text{C}$$

∴ Temperature of superheated steam,

$$T_{sup} = 110 + 212.4 = 322.4° \text{C}$$

Now from steam tables for superheated steam, corresponding to a pressure of 17 bar and 322.4° C, we find that

$$h_2 = h_{sup} = 3083.4 \text{ kJ/kg ; and}$$

$$s_2 = s_{sup} = 6.939 \text{ kJ/kg K}$$

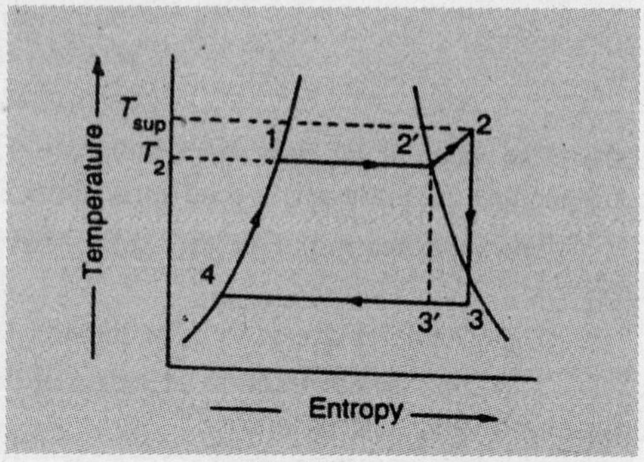

Fig. 10.11

and from steam tables for dry saturated steam,corresponding to a pressure of 0.07 bar, we find that

$$h_{f3} = h_{f4} = 163.4 \text{ kJ/kg} ; h_{fg3} = 2409.2 \text{ kJ/kg} ; s_{f3} = 0.559 \text{ kJ/kg K} ; \text{and } s_{fg3} = 7.718 \text{ kJ/kg K}$$

First of all, let us find the dryness fraction of steam at point 3 (*i.e.* x_3). We know that for isentropic expansion 2-3,

Entropy before expansion (s_2) = Entropy after expansion (s_3)

or
$$s_{sup} = s_{f3} + x_3 \, s_{fg3}$$

$$6.939 = 0.559 + x_3 \times 7.718 \quad \text{or} \quad x_3 = 0.827$$

We know that enthalpy of steam at point 3,

$$h_3 = h_{f3} + x_3 \, h_{fg3} = 163.4 + 0.827 \times 2409.2 = 2155.8 \text{ kJ/kg}$$

1. *Heat supplied*

We know that heat supplied

$$= h_2 - h_{f3} = 3083.4 - 163.4 = 2920 \text{ kJ/kg} \quad \textbf{Ans.}$$

2. *Heat rejected*

We know that heat rejected

$$= h_3 - h_{f4} = h_{f3} + x_3 \, h_{fg3} - h_{f3} = x_3 \, h_{fg3} \qquad \ldots (\because h_{f4} = h_{f3})$$

$$= 0.827 \times 2409.2 = 1992.4 \text{ kJ/kg} \quad \textbf{Ans.}$$

3. *Net workdone*

We know that net workdone = Heat supplied – Heat rejected

$$= 2920 - 1992.4 = 927.6 \text{ kJ/kg} \quad \textbf{Ans.}$$

4. *Thermal efficiency*

We know that thermal efficiency

$$= \frac{\text{Net workdone}}{\text{Heat supplied}} = \frac{927.6}{2920} = 0.317 \text{ or } 31.7\% \quad \textbf{Ans.}$$

Example 10.9. ***Steam at 50 bar, 400° C expands in a Rankine cycle to 0.34 bar. For a mass flow rate of 150 kg/s of steam, determine 1. the power developed, 2. the thermal efficiency, and 3 specific steam consumption***

Solution. Given ; $p_1 = p_2 = 50$ bar ; $T_{sup} = 400°$ C ; $p_3 = p_4 = 0.34$ bar ; $m_s = 150$ kg/s

From steam tables for superheated steam, corresponding to a pressure of 50 bar and 400° C, we find that

$$h_2 = h_{sup} = 3198.3 \text{ kJ/kg} \; ; \text{ and } \; s_2 = s_{sup} = 6.651 \text{ kJ/kg K}$$

and from steam tables for dry saturated steam, corresponding to a pressure of 0.34 bar, we find that

$h_{f3} = 301.5$ kJ/kg ; $h_{fg3} = 2329$ kJ/kg ; $s_{f3} = 0.98$ kJ/kg K ; and $s_{fg3} = 6.747$ kJ/kg K

First of all, let us find the dryness fraction of steam at point 3 (*i.e.* x_3). We know that for isentropic expansion 2-3 (Refer Fig. 10.10),

Entropy before expansion (s_2)

= Entropy after expansion (s_3)

or $$6.651 = s_{f3} + x_3 \, s_{fg3} = 0.98 + x_3 \times 6.747 \quad \text{or} \quad x_3 = 0.84$$

We know that enthalpy of steam at point 3,

$$h_3 = h_{f3} + x_3 \, h_{fg3} = 301.5 + 0.84 \times 2329 = 2258 \text{ kJ/kg}$$

1. *Power developed*

We know that workdone,

$$w = h_2 - h_3 = 3198.3 - 2258 = 940.3 \text{ kJ/kg}$$

and total workdone $= m_s \times w = 150 \times 940.3 = 141\,045$ kJ/s

Power developed = **141 045 kW Ans.** ...($\because$ 1 kJ/s = 1 kW)

2. *Thermal efficiency*

We know that thermal efficiency,

$$\eta_R = \frac{h_2 - h_3}{h_2 - h_{f3}} = \frac{3198.3 - 2258}{3198.3 - 301.5} = 0.3246 \quad \text{or} \quad \mathbf{32.46\%} \textbf{ Ans.}$$

3. *Specific steam consumption*

We know that specific steam consumption

$$= \frac{3600}{h_2 - h_3} = \frac{3600}{w} = \frac{3600}{940.3} = \mathbf{3.83 \text{ kg/kWh}} \textbf{ Ans.}$$

10.7. Modified Rankine Cycle

We have seen in the Rankine cycle, that the steam is expanded to the extreme toe of the *p-v* diagram (at point 3) as shown in Fig. 10.12 (*a*). But in actual reciprocating steam engines, it is found to be too uneconomical (due to larger size of the cylinder) to expand steam to the full limit (*i.e.* upto the point 3).

It may be noted that the diagram is very narrow at the toe, and the amount of work done (represented by area 5-3-6) during this final portion of the expansion stroke is extremely small. In fact, it is too small to overcome even the friction of the moving parts in the steam engine. The expansion of steam, therefore, is carried on in the engine cylinder at a pressure higher than that of the condenser pressure or exhaust pressure or back pressure. This higher pressure is known as release pressure (p_5).

In order to overcome the above mentioned difficulty, the Rankine cycle is slightly modified. In a modified Rankine cycle, the expansion stroke of the piston is stopped at point 5 by cutting the toe of Rankine cycle, and the steam is exhausted from the cylinder at a constant volume. This causes a sudden drop of pressure from p_5 to p_6. The expansion of steam is, therefore, completed by a constant volume line 5-6 as shown on *p-v* and *T-s* diagram in Fig. 10.12 (*a*) and (*b*). By doing so, the size of the cylinder and stroke length is considerably reduced.

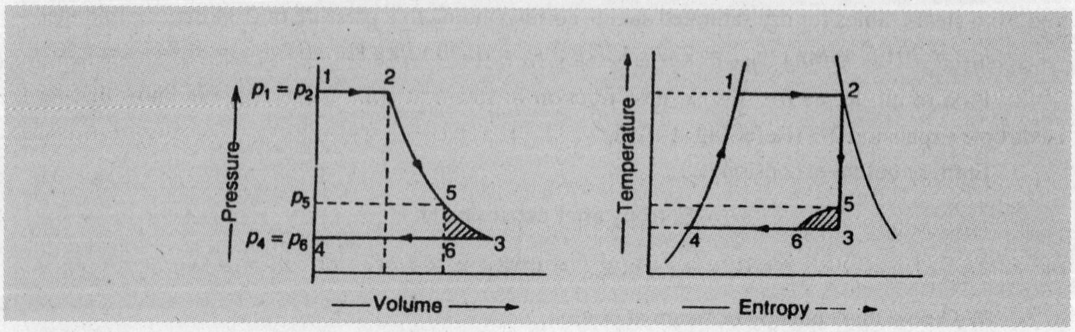

(*a*) *p-v* diagram. (*b*) *T-s* diagram.

Fig. 10.12. Modified Rankine cycle.

10.8. Efficiency of Modified Rankine Cycle

Consider a modified Rankine cycle whose processes are shown in Fig. 10.13 (*a*) and (*b*).

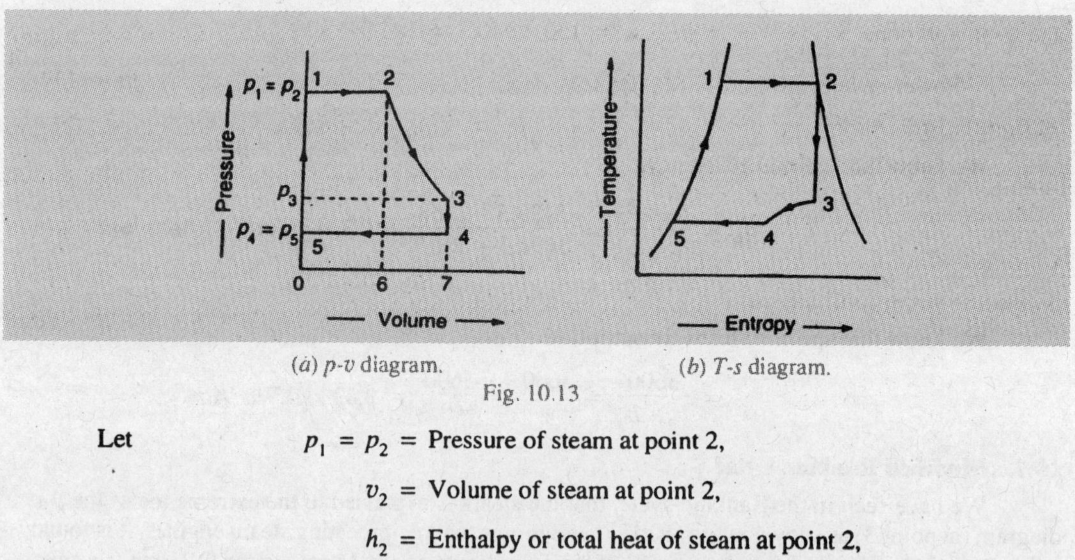

(*a*) *p-v* diagram. (*b*) *T-s* diagram.

Fig. 10.13

Let

$p_1 = p_2$ = Pressure of steam at point 2,

v_2 = Volume of steam at point 2,

h_2 = Enthalpy or total heat of steam at point 2,

u_2 = Internal energy of steam at point 2,

p_3, v_3, h_3, u_3 = Corresponding values of steam at point 3.

p_4 = Back pressure of steam at point 4, and

h_{f4} = Sensible heat or enthalpy of water at point 4.

We know that work done during constant pressure process 1-2,

$$= \text{Area 1-2-6-0} = 100\, p_2\, v_2 \quad \ldots (i)$$

We also know that work done during isentropic expansion 2-3

$$= \text{Area 2-3-7-6} = \text{Change in internal energy}$$

$$= u_2 - u_3 \qquad \ldots (ii)$$

and work done during constant pressure process 4-5

$$= \text{Area 0-5-4-7} = 100\, p_4\, v_4 \qquad \ldots (iii)$$

∴ Work done during the cycle per kg of steam,

$$W = \text{Area 1-2-3-4-5}$$

$$= \text{Area 1-2-6-0} + \text{Area 2-3-7-6} - \text{Area 0-5-4-7}$$

$$= 100\, p_2\, v_2 + (u_2 - u_3) - 100\, p_4\, v_4$$

$$= 100\, p_2\, v_2 + [(h_2 - 100\, p_2\, v_2) - (h_3 - 100\, p_3\, v_3)] - 100\, p_4\, v_4$$

$$= h_2 - h_3 + 100\,(p_3 - p_4)\, v_3 \qquad \ldots (\because v_3 = v_4)$$

We know that heat supplied per cycle

$$= h_2 - h_{f5} = h_2 - h_{f4} \qquad \ldots (\because h_{f5} = h_{f4})$$

∴ Efficiency of the modified Rankine cycle,

$$\eta_{MR} = \frac{\text{Work done}}{\text{Heat supplied}} = \frac{(h_2 - h_3) + 100\,(p_3 - p_4)\, v_3}{h_2 - h_{f4}}$$

Example 10.10. *A steam engine admits steam at a pressure of 5.6 bar and 0.9 dry. The cut off occurs at one-half stroke, and pressure at release is 2.6 bar. The back pressure is 1 bar. Find the modified Rankine efficiency of the engine, assuming the pressure drop at release to take place at a constant volume. Neglect clearance volume.*

Solution. Given : $p_2 = 5.6$ bar ; $x_2 = 0.9$; $v_2 = v_3 / 2$; $p_3 = 2.6$ bar ; $p_4 = 1$ bar

From steam tables, corresponding to a pressure of 5.6 bar, we find that

$$h_{f2} = 658.8 \text{ kJ/kg} \,;\; h_{fg2} = 2093.7 \text{ kJ/kg} \,;\; v_{g2} = 0.3367 \text{ m}^3\text{/kg} \,;$$

$$s_{f2} = 1.904 \text{ kJ/kg K} \,;\; s_{fg2} = 4.877 \text{ kJ/kg K}$$

Similarly, corresponding to a pressure of 2.6 bar, we find that

$$h_{f3} = 540.9 \text{ kJ/kg} \,;\; h_{fg3} = 2177.3 \text{ kJ/kg} \,;\; v_{g3} = 0.6925 \text{ m}^3\text{/kg} \,;$$

$$s_{f3} = 1.621 \text{ kJ/kg K} \,;\; s_{fg3} = 5.418 \text{ kJ/kg K}$$

and corresponding to a pressure of 1 bar, we find that

$$h_{f4} = 417.5 \text{ kJ/kg}$$

We know that enthalpy or total heat of steam at point 2,

$$h_2 = h_{f2} + x_2\, h_{fg2} = 658.8 + 0.9 \times 2093.7 = 2543 \text{ kJ/kg}$$

and volume at point 2, $v_2 = x_2\, v_{g2} = 0.9 \times 0.3367 = 0.303 \text{ m}^3\text{/kg}$

∴ Volume of stroke at point 3 (*i.e.* stroke volume),

$$v_3 = 2 \times v_2 = 2 \times 0.303 = 0.606 \text{ m}^3\text{/kg} \qquad \ldots (\because v_2 = v_3 / 2)$$

Now first of all, let us find out the dryness fraction of steam at point 3 (*i.e.* x_3). We know that for isentropic expansion 2-3 (Refer Fig. 10.13),

Entropy before expansion (s_2)

$$= \text{Entropy after expansion } (s_3)$$

or

$$s_{f2} + x_2 s_{fg2} = s_{f3} + x_3 s_{fg3}$$

$$1.904 + 0.9 \times 4.877 = 1.621 + x_3 \times 5.418 \quad \text{or} \quad x_3 = 0.86$$

Since the dryness fraction of steam at point 3 (*i.e.* x_3) is less than one, therefore the condition of steam at point 3 is wet.

∴ Enthalpy or total heat of steam at point 3,

$$h_3 = h_{f3} + x_3 h_{fg3} = 540.9 + 0.86 \times 2177.3 = 2413.4 \text{ kJ/kg}$$

We know that modified Rankine efficiency,

$$\eta_{MR} = \frac{(h_2 - h_3) + 100\,(p_3 - p_4)\,v_3}{h_2 - h_{f4}}$$

$$= \frac{(2543 - 2413.4) + 100\,(2.6 - 1)\,0.606}{2543 - 417.5}$$

$$= 0.1066 \text{ or } 10.66\ \% \textbf{ Ans.}$$

Example 10.11. *A steam engine is supplied dry saturated steam at 15 bar. The pressure at release is 3 bar and the back pressure is 1 bar. Using steam tables or Mollier chart, determine the efficiency of the modified Rankine cycle.*

Solution. Given : $p_2 = 15$ bar ; $p_3 = 3$ bar ; $p_4 = 1$ bar

First of all, mark a point 2, where pressure line through p_2 (*i.e.* 15 bar) meets the saturation line. Since the steam expands isentropically, therefore draw a vertical line through the point 2 to meet the pressure line through p_3 (*i.e.* 3 bar) at point 3, as shown in Fig. 10.14.

Now from the Mollier diagram, we find that the enthalpy of steam at point 2,

$$h_2 = 2790 \text{ kJ/kg}$$

Similarly, enthalpy of steam at point 3,

$$h_3 = 2510 \text{ kJ/kg}$$

and dryness fraction of steam at point 3,

$$x_3 = 0.9$$

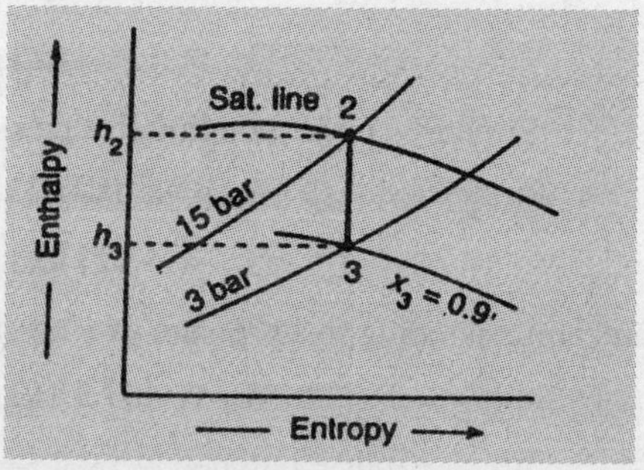

Fig. 10.14

From steam tables, corresponding to a pressure of 3 bar, we find that specific volume of steam,

$$v_{g3} = 0.6055 \text{ m}^3\text{/kg}$$

∴ Volume of steam at point 3,

$$v_3 = x_3\,v_{g3} = 0.9 \times 0.6055 = 0.545 \text{ m}^3\text{/kg}$$

Similarly, from steam tables, corresponding to a pressure of 1 bar, we find that sensible heat of water,

$$h_{f4} = 417.5 \text{ kJ/kg}$$

We know that modified Rankine efficiency,

$$\eta_{MR} = \frac{(h_2 - h_3) + 100\,(p_3 - p_4)\,v_3}{h_2 - h_{f4}}$$

$$= \frac{(2790 - 2510) + 100\,(3 - 1)\,0.545}{2790 - 417.5}$$

$$= 0.164 \text{ or } 16.4\% \textbf{ Ans.}$$

Example 10.12. *A steam engine takes dry steam at 20 bar and exhausts at 1.2 bar. The pressure at the release is 3 bar. Find ; 1. the theoretical loss of work per kg of steam due to incomplete expansion ; and 2. the loss in Rankine efficiency due to restricted expansion of steam.*

Solution. Given : $x_2 = 1$; $p_1 = p_2 = 20$ bar ; $p_4 = p_5 = 1.2$ bar ; $p_3 = 3$ bar

In the *p-v* and *T-s* diagram, as shown in Fig. 10.15 (*a*) and (*b*), 1-2-3-4-5-1 represents the modified Rankine cycle and 1-2-3′-5-1 represents the Rankine cycle.

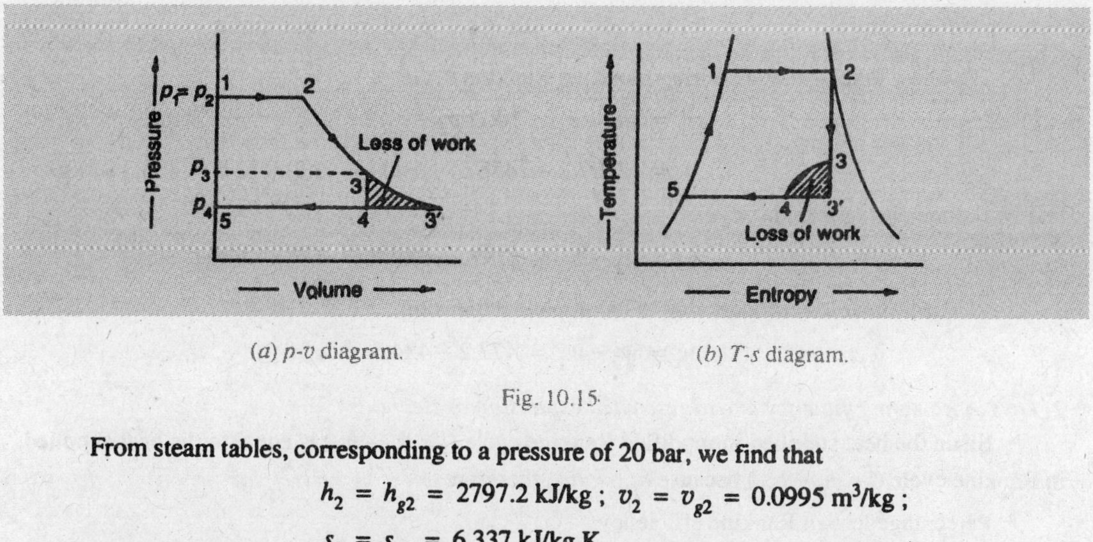

(*a*) *p-v* diagram. (*b*) *T-s* diagram.

Fig. 10.15

From steam tables, corresponding to a pressure of 20 bar, we find that

$$h_2 = h_{g2} = 2797.2 \text{ kJ/kg} \; ; \; v_2 = v_{g2} = 0.0995 \text{ m}^3/\text{kg} \; ;$$

$$s_2 = s_{g2} = 6.337 \text{ kJ/kg K}$$

Similarly, corresponding to a pressure of 3 bar, we find that

$$h_{f3} = 561.4 \text{ kJ/kg} \; ; \; h_{fg3} = 2163.2 \text{ kJ/kg} \; ; \; v_{g3} = 0.6055 \text{ m}^3/\text{kg} \; ;$$

$$s_{f3} = 1.672 \text{ kJ/kg K} \; ; \; s_{fg3} = 5.319 \text{ kJ/kg K}$$

and corresponding to a pressure of 1.2 bar, we find that

$$h_{f4} = h_{f3'} = 439.4 \text{ kJ/kg} \; ; \; h_{fg3'} = 2244.1 \text{ kJ/kg} \; ;$$

$$s_{f3'} = 1.361 \text{ kJ/kg K} \; ; \; s_{fg3'} = 5.937 \text{ kJ/kg K}$$

1. Theoretical loss of work per kg of steam due to incomplete expansion

First of all, let us find the dryness fraction of steam at points 3 and 3′ (*i.e.* x_3 and $x_{3'}$). We know that for isentropic expansion 2-3 (for modified Rankine cycle),

Entropy before expansion (s_2) = Entropy after expansion (s_3)

$$6.337 = s_{f3} + x_3\, s_{fg3} = 1.672 + x_3 \times 5.319$$

$$\therefore \quad x_3 = 0.877$$

Similarly, for isentropic expansion 2-3′ (for Rankine cycle),

$$s_2 = s_{3'}$$

$$6.337 = s_{f3'} + x_{3'}\, s_{fg3'} = 1.361 + x_{3'} \times 5.937$$

$$\therefore \quad x_{3'} = 0.838$$

We know that enthalpy or total heat of steam at point 3,

$$h_3 = h_{f3} + x_3\, h_{fg3} = 561.4 + 0.877 \times 2163.2 = 2458.5 \text{ kJ/kg}$$

and enthalpy or total heat of steam at point 3′,

$$h_{3'} = h_{f3'} + x_{3'}\, h_{fg3'}$$

$$= 439.4 + 0.838 \times 2244.1 = 2320 \text{ kJ/kg}$$

We also know that volume of steam at point 3,

$$v_3 = x_3\, v_{g3} = 0.877 \times 0.6055 = 0.531 \text{ m}^3\text{/kg}$$

We know that workdone during modified Rankine cycle,

$$w_1 = (h_2 - h_3) + 100\,(p_3 - p_4)\, v_3$$

$$= (2797.2 - 2458.5) + 100\,(3 - 1.2)\, 0.531 = 434.3 \text{ kJ/kg}$$

and workdone during Rankine cycle

$$w_2 = h_2 - h_{3'} = 2797.2 - 2320 = 477.2 \text{ kJ/kg}$$

$\therefore$ Theoretical loss of work due to incomplete expansion,

$$w = w_2 - w_1 = 477.2 - 434.3 = 42.9 \text{ kJ/kg} \quad \textbf{Ans.}$$

2. Loss in Rankine efficiency due to restricted expansion of steam

Since the heat supplied in modified Rankine cycle (*i.e.* $h_2 - h_{f4}$) is equal to the heat supplied in Rankine cycle (*i.e.* $h_2 - h_{f3'}$) because $h_{f4} = h_{f3'}$, therefore

Percentage loss in Rankine efficiency

$$= \text{Percentage loss of work}$$

$$= \frac{w}{w_2} \times 100 = \frac{42.9}{477.2} \times 100 = 9\% \quad \textbf{Ans.}$$

Example 10.13. *A steam engine receives steam at 8 bar superheated at 200° C, pressure at release being 2.8 bar and at exhaust 1 bar. Assuming isentropic expansion and constant volume conditions between release and commencement of exhaust, determine :*

1. Workdone in N-m per kg of steam, and 2. Efficiency of the unit.

Also compare these values with those for Rankine cycle working between the same pressure and temperature limits.

Solution. Given : $p_2 = 8$ bar ; $T_{sup} = 200°$ C ; $p_3 = 2.8$ bar ; $p_4 = 1$ bar

The modified Rankine cycle with superheated steam is shown in Fig. 10.16.

From steam tables of superheated steam, corresponding to a pressure of 8 bar and 200° C, we find that

$$h_2 = h_{sup} = 2838.6 \text{ kJ/kg} \; ; \; v_2 = v_{sup} = 0.3 \text{ m}^3\text{/kg} \; ; \text{ and } s_2 = s_{sup} = 6.886 \text{ kJ/kg K}$$

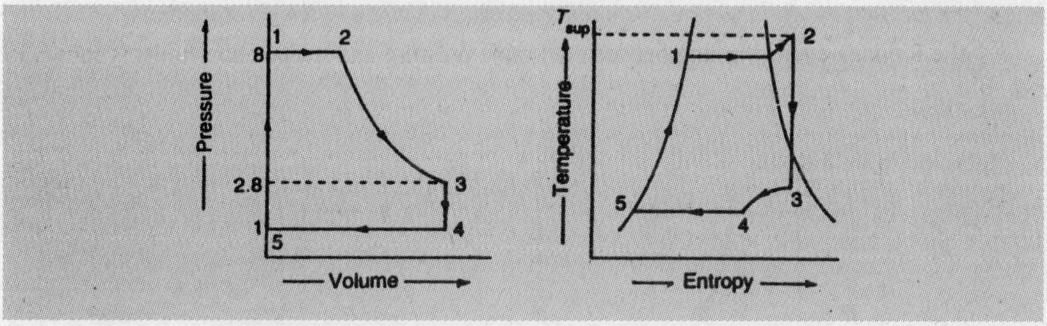

(*a*) *p-v* diagram. (*b*) *T-s* diagram.

Fig. 10.16

From steam tables of dry saturated steam, corresponding to a pressure of 2.8 bar, we find that

$$h_{f3} = 551.5 \text{ kJ/kg} \; ; h_{fg3} = 2170 \text{ kJ/kg} \; ; v_{g3} = 0.646 \text{ m}^3\text{/kg} \; ; s_{f3} = 1.647 \text{ kJ/kg K} \; ; \text{and}$$

$$s_{fg3} = 5.367 \text{ kJ/kg K}$$

and corresponding to a pressure of 1 bar, we find that

$$h_{f4} = 417.5 \text{ kJ/kg}$$

First of all, let us find the dryness fraction of steam at point 3 (*i.e.* x_3). We know that for isentropic expansion 2-3,

Entropy before expansion (s_2)

$$= \text{Entropy after expansion } (s_3)$$

or $$6.886 = s_{f3} + x_3 \, s_{fg3} = 1.647 + x_3 \times 5.367$$

$\therefore$ $$x_3 = 0.976$$

Since the dryness fraction of steam at point 3 (*i.e.* x_3) is less than one, therefore the condition of steam at point 3 is wet.

$\therefore$ Enthalpy or total heat of steam at point 3,

$$h_3 = h_{f3} + x_3 \, h_{fg3} = 551.5 + 0.976 \times 2170 = 2669.4 \text{ kJ/kg}$$

and volume at point 3, $$v_3 = x_3 \, v_{g3} = 0.976 \times 0.646 = 0.63 \text{ m}^3\text{/kg}$$

1. *Workdone per kg of steam*

We know that workdone per kg of steam

$$= (h_2 - h_3) + 100\,(p_3 - p_4)\,v_3$$

$$= (2838.6 - 2669.4) + 100\,(2.8 - 1)\,0.63 \qquad \ldots (\because h_2 = h_{sup})$$

$$= 282.6 \text{ kJ/kg} = 282.6 \text{ kN-m/kg} = 282\text{-}600 \text{ N-m/kg}$$ **Ans.**

2. *Efficiency of the unit*

We know that heat supplied

$$= h_2 - h_{f4} = 2838.6 - 417.5 = 2421.1 \text{ kJ/kg}$$

∴ Efficiency of the unit,

$$\eta_{MR} = \frac{\text{Work done}}{\text{Heat supplied}} = \frac{282.6}{2421.1} = 0.117 \text{ or } 11.7\% \textbf{ Ans.}$$

Workdone for Rankine cycle working between the same pressure and temperature limits

The Rankine cycle working between the same pressure and temperature limits is shown in Fig. 10.17.

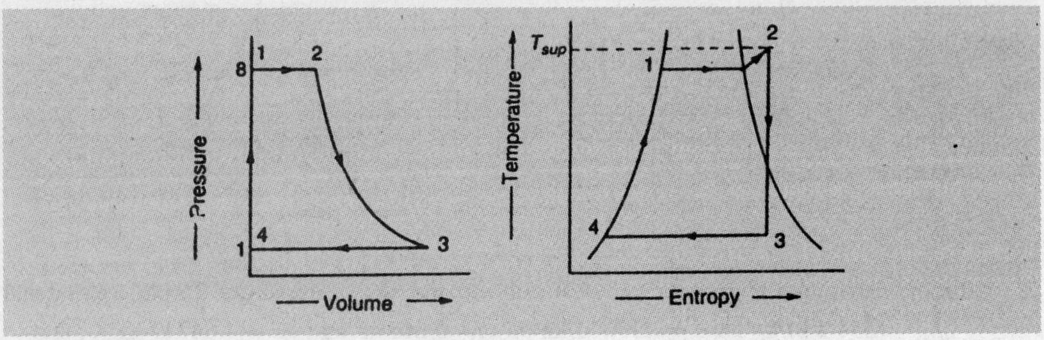

(*a*) *p-v* diagram. (*b*) *T-s* diagram.

Fig. 10.17

From steam tables, corresponding to a pressure of 1 bar, we find that

h_{f3} = 417.5 kJ/kg ; h_{fg3} = 2258 kJ/kg ; s_{f3} = 1.303 kJ/kg K ; and s_{fg3} = 6.057 kJ/kg K

First of all, let us find the dryness fraction of steam at point 3 (*i.e.* x_3). We know that for isentropic expansion 2-3,

Entropy before expansion (s_2)

$$= \text{Entropy after expansion } (s_3)$$

or
$$6.886 = s_{f3} + x_3\, s_{fg3} = 1.303 + x_3 \times 6.057$$

∴
$$x_3 = 0.92$$

We know that enthalpy or total heat of steam at point 3,

$$h_3 = h_{f3} + x_3\, h_{fg3} = 417.5 + 0.92 \times 2258 = 2495 \text{ kJ/kg}$$

∴ Workdone during the Rankine cycle

$$= h_2 - h_3 = 2838.6 - 2495 = 343.6 \text{ kJ/kg} \qquad \ldots (\because h_2 = h_{sup})$$

$$= 343.6 \text{ kN-m/kg} = 343\,600 \text{ N-m/kg } \textbf{Ans.}$$

Efficiency of the Rankine cycle

We know that heat supplied

$$= h_2 - h_{f3} = 2838.6 - 417.5 = 2421.1 \text{ kJ/kg}$$

∴ Efficiency of the Rankine cycle,

$$\eta_R = \frac{\text{Workdone}}{\text{Heat supplied}} = \frac{343.6}{2421.1} = 0.142 \text{ or } 14.2\% \textbf{ Ans.}$$

Example 10.14. *The cylinder of a steam engine is 300 mm in diameter and piston stroke is 580 mm. The steam at admission is at 10 bar and 300° C. It expands isentropically to 0.7 bar and then reduced at constant volume to a condenser at 0.28 bar. Determine : 1. the modified Rankine efficiency ; 2. the new stroke if the same amount of steam from the original condition is expanded isentropically to condenser pressure ; 3 the new Rankine efficiency ; and 4. the workdone by the extraction and boiler feed pumps per kg of wate. returned to the boiler.*

Solution. Given : $D = 300$ mm = 0.3 m ; $L = 580$ mm = 0.5 m ; $p_1 = p_2 = 10$ bar ; $T_{sup} = 300°$ C ; $p_3 = 0.7$ bar ; $p_4 = p_{3'} = p_5 = 0.28$ bar

From steam tables for superheated steam, corresponding to a pressure of 10 bar and 300° C, we find that

$$h_2 = h_{sup} = 3052.1 \text{ kJ/kg} ; \; s_2 = s_{sup} = 7.125 \text{ kJ/kg K}$$

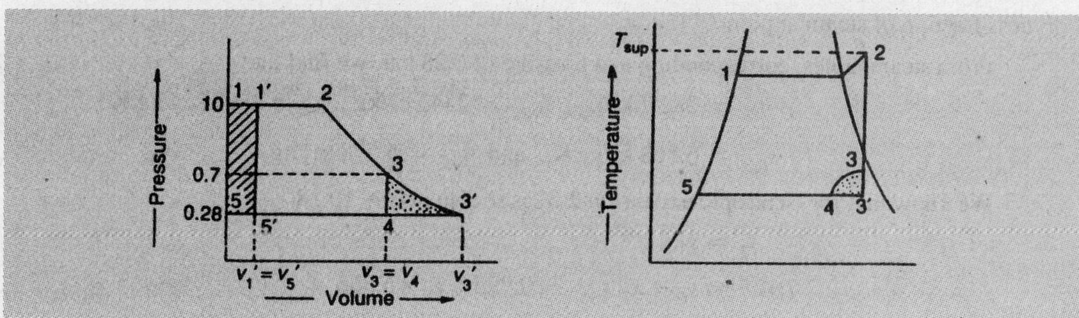

(a) p-v diagram. (b) T-s diagram.

Fig. 10.18

Similarly, from steam tables for dry saturated steam, corresponding to a pressure of 0.7 bar, we find that

$$h_{f3} = 376.8 \text{ kJ/kg} ; \quad h_{fg3} = 2283.3 \text{ kJ/kg} ; \quad s_{f3} = 1.192 \text{ kJ/kg K} ;$$

$$s_{fg3} = 6.288 \text{ kJ/kg} ; \text{ and } v_{g3} = 2.365 \text{ m}^3\text{/kg}$$

and corresponding to a pressure of 0.28 bar, we find that

$$h_{f4} = 282.7 \text{ kJ/kg}$$

1. Modified Rankine efficiency

First of all, let us find the dryness fraction of steam at point 3 (*i.e.* x_3). We know that for isentropic expansion 2-3, as shown in Fig. 10.18,

Entropy before expansion (s_2)

$$= \text{Entropy after expansion } (s_3)$$

$$7.125 = s_{f3} + x_3 \, s_{fg3} = 1.192 + x_3 \times 6.288$$

$$\therefore \quad x_3 = 0.943$$

We know that enthalpy or total heat of steam at point 3,

$$h_3 = h_{f3} + x_3 \, h_{fg3} = 376.8 + 0.943 \times 2283.3 = 2530 \text{ kJ/kg}$$

and volume of steam at point 3,

$$v_3 = x_3 \, v_{g3} = 0.943 \times 2.365 = 2.23 \text{ m}^3\text{/kg}$$

We know that modified Rankine efficiency,

$$\eta_{MR} = \frac{(h_2 - h_3) + 100\,(p_3 - p_4)\,v_3}{h_3 - h_{f4}}$$

$$= \frac{(3052.1 - 2530) + 100\,(0.7 - 0.28)\,2.23}{3052.1 - 282.7}$$

$$= 0.2223 \text{ or } 22.23\% \textbf{ Ans.}$$

2. *New stroke*

Let L' = New stroke length.

Since the same amount of steam as at point 3 is expanded isentropically from the original condition (*i.e.* at 10 bar and 300° C) to the condenser pressure $p_{3'} = 0.28$ bar, therefore let us find the dryness fraction of steam at point 3′ (*i.e.* $x_{3'}$).

From steam tables, corresponding to a pressure of 0.28 bar, we find that

$$h_{f3'} = 282.7 \text{ kJ/kg}\ ; \ h_{fg3'} = 2340 \text{ kJ/kg}\ ; \ s_{f3'} = 0.925 \text{ kJ/kg K}\ ;$$

$$s_{fg3'} = 6.868 \text{ kJ/kg K}\ ; \text{ and } v_{g3'} = 5.578 \text{ m}^3\text{/kg}$$

We know that for isentropic expansion 2-3′ as shown in Fig. 10.18,

$$s_2 = s_{3'}$$

$$7.125 = s_{f3'} + x_{3'}\,s_{fg3'} = 0.925 + x_{3'} \times 6.868$$

$$\therefore \quad x_{3'} = 0.903$$

We know that cylinder volume,

$$v = \frac{\pi}{4} \times D^2\,L = \frac{\pi}{4}\,(0.3)^2\,0.58 = 0.041 \text{ m}^3$$

∴ Mass of steam at point 3,

$$m_3 = \frac{v}{v_3} = \frac{0.041}{2.23} = 0.0184 \text{ kg/stroke}$$

and volume of steam at point 3′,

$$v_{3'} = m_3\,x_{3'}\,v_{g3'} = 0.0184 \times 0.903 \times 5.578 = 0.0927 \text{ m}^3 \quad \ldots (i)$$

We also know that new stroke volume at point 3′,

$$v_{3'} = \frac{\pi}{4} \times D^2 \times L' = \frac{\pi}{4}\,(0.3)^2\,L' = 0.0707\,L' \text{ m}^3 \quad \ldots (ii)$$

From equations (*i*) and (*ii*),

$$L' = 0.0927 / 0.0707 = 1.31 \text{ m } \textbf{Ans.}$$

3. *New Rankine efficiency*

We know that enthalpy or total heat of steam at point 3′,

$$h_{3'} = h_{f3'} + x_{3'}\,h_{fg3'} = 282.7 + 0.903 \times 2340 = 2395.7 \text{ kJ/kg}$$

∴ New Rankine efficiency,

$$\eta_R = \frac{h_2 - h_{3'}}{h_2 - h_{f3'}} = \frac{3052.1 - 2395.7}{3052.1 - 282.7} = 0.237 \text{ or } 23.7\% \textbf{ Ans.}$$

4. Workdone by the extraction and boiler feed pumps

The work is done by the extraction and boiler feed pumps in raising the pressure of water from condenser pressure (0.28 bar) to boiler pressure (10 bar). In Fig. 10.18, area 1-1′-5′-5 represents the required work done.

From steam tables, corresponding to a pressure of 0.28 bar, we find that

Volume of water, $v_5 = 0.001\,021\ m^3/kg$

∴ Workdone by the extraction and boiler feed pumps

$$= 100\,(p_1 - p_5)\,v_5 = 100\,(10 - 0.28)\,0.001\,021\ \text{kJ/kg}$$

$$= 0.9924\ \text{kJ/kg}\ \textbf{Ans.}$$

EXERCISES

1. A Carnot engine works between pressure limits of 35 bar and 0.7 bar. Find the amount of work done per kg of steam and thermal efficiency of the engine. [**Ans.** 518.2 kJ/kg ; 29.6%]

2. A simple Rankine cycle uses steam as the working medium and operates between 0.5 bar and 20 bar. If the steam entering the turbine is dry at 20 bar, determine the quality of steam as it leaves the turbine.

Entropy of water at 0.5 bar = 1.091 kJ/kg K ; Entropy of saturated steam at 20 bar = 6.341 kJ/kg K ; and Entropy of saturated steam at 0.5 bar = 7.524 kJ/kg K. [**Ans.** 0.816]

3. Dry and saturated steam at a pressure of 11 bar is supplied to a turbine and is expanded isentropically to a pressure of 1 bar. Calculate 1. Heat supplied, 2. Heat rejected, and 3. Theoretical thermal efficiency. [**Ans.** 2382.2 kJ/kg ; 1955.5 kJ/kg ; 21.35%]

4. A boiler supplies dry saturated steam to a steam power plant at a pressure of 12.5 bar. The back pressure is 0.2 bar. Find the efficiency of the plant, if it operates on Rankine cycle by using : 1. Steam tables only, and 2. Mollier chart. [**Ans.** 25.3%]

5. Dry and saturated steam at 15 bar is supplied to a steam engine. The exhaust takes place at 1.1 bar. Calculate 1. Rankine efficiency ; 2. Mean effective pressure ; 3. Steam consumption per I.P. hour, if the efficiency ratio is 0.65 ; and 4. Carnot cycle efficiency for the given pressure limits using steam as the working fluid. [**Ans.** 18.8% ; 3.36 bar ; 12.5 kg/kWh ; 20.4%]

6. In a Rankine cycle, the maximum pressure of steam supplied is 6 bar. The dryness fraction is 0.9. The exhaust pressure is 0.7 bar. Find the theoretical workdone and Rankine efficiency. [**Ans.** 325.3 kJ/kg ; 15%]

7. Compare the efficiencies of Carnot cycle and Rankine cycle if the maximum and minimum temperatures are 400° C and 40° C. The steam in the case of Rankine cycle is supplied at 20 bar. [**Ans.** 53.5% ; 33.35%]

8. A steam engine is supplied with 90% dry steam at a pressure of 10 bar. The exhaust takes place at 1.1 bar. Determine : 1. Rankine efficiency ; 2. Percentage increase in efficiency if the steam has a temperature of 250° C before entering the cylinder. [**Ans.** 16.1% ; 2.4%]

9. Steam is supplied to a Rankine engine at 21 bar with 110° C of superheat. The back pressure is 0.20 bar. Find : 1. Rankine efficiency, and 2. Volume of steam entering the engine per hour if the steam supplied is 100 kg per hour. [**Ans.** 29.1% ; 11.64 m³/h]

10. Steam is supplied to a steam turbine at a pressure of 20 bar and 230° C. It is then expanded isentropically to a pressure of 1 bar. Determine : 1. Rankine efficiency, 2. Specific volume of steam at the end of expansion, and 3. Carnot efficiency between the same temperature limits. [**Ans.** 21% ; 1.44 m³/kg]

11. In an ideal Rankine cycle, the steam condition at turbine inlet is 20 bar and 350° C. The condenser pressure is 0.08 bar. Determine the cycle efficiency. If the steam flow rate is 2000 kg/h, what is the power out-put in kW ? [**Ans.** 32.4% ; 533.5 kW]

12. Steam at a pressure of 15 bar and at a temperature of 300° C is supplied to a steam turbine working on the Rankine cycle. If the exhaust takes place at 0.15 bar, evaluate the Rankine efficiency. Calculate the steam consumption in kg/h to develop 750 kW, if efficiency ratio is 0.6. [**Ans.** 28.3% ; 5653 kg/h]

13. A steam engine uses dry saturated steam at a pressure of 10 bar and the back pressure is 0.7 bar. The pressure at release is 3.8 bar. Assuming the pressure drop to take place at a constant volume, find the efficiency of the modified Rankine cycle. Neglect clearance. [**Ans.** 13%]

14. A steam engine receives steam at a pressure of 5.6 bar and 190° C. The pressure at release is 2.8 bar, and back pressure 1.0 bar. Determine : 1. Work done in kJ/kg of steam ; and 2. Efficiency of the cycle. [**Ans.** 246 kJ/kg ; 10.15%]

15. Steam at 17 bar and 250° C is supplied to a steam engine where it is expanded isentropically to a release pressure of 3.5 bar. The steam is released from the engine at constant volume into the condenser where it is condensed at a pressure of 0.5 bar and the condensate is pumped back to the boiler. The steam flow rate is 1200 kg/h. Neglecting pump work, determine with the help of steam tables : 1. the power output of the engine in kW ; and 2. the efficiency of the cycle. [**Ans.** 153.3 kW ; 17.87 %]

QUESTIONS

1. Describe the Carnot vapour cycle and derive the expression for its efficiency.
2. Why the Carnot cycle can not be considered as the theoretical cycle for steam power plants even though its efficiency is maximum ?
3. Show Rankine cycle on *p-v* and *T-s* diagrams and explain the processes involved. Also draw the mechanical system to show different processes of the Rankine cycle.
4. Draw the Rankine cycle on *T-s* diagram using dry saturated steam and obtain an expression for the Rankine cycle efficiency.
5. Describe briefly the Rankine cycle using superheated steam and show in what respect this cycle differs from Carnot cycle between the same temperatures.
6. Prove that the efficiency of a Rankine cycle using superheated steam is greater than the efficiency of a corresponding Rankine cycle using steam without superheat. Both the cycles operate between the same boiler and condenser pressure limits.
7. State the reasons for modifying the Rankine cycle for operation of steam engines.
8. Explain with the help of *p-v* and *T-s* diagrams, the sequence of operations in the modified Rankine cycle.

OBJECTIVE TYPE QUESTIONS

1. The ideal cycle on which a steam engine works, is

   (*a*) Carnot cycle (*b*) Rankine cycle (*c*) Otto cycle (*d*) Joule cycle

2. Rankine cycle comprises

   (*a*) two isothermal and two isentropics (*b*) two isobarics and two isothermals

   (*c*) two isobarics and two isentropics (*d*) two isothermals and two isochorics

3. The Rankine cycle, as compared to Carnot cycle, has .... work ratio,

   (*a*) high (*b*) low

4. In a Rankine cycle with superheated steam,

   (*a*) the workdone increases

   (*b*) the dryness fraction of steam after isentropic expansion increases

   (*c*) the specific steam consumption decreases

   (*d*) all of the above

5. In the operation of steam engines, the vapour cycle adopted is

   (*a*) Carnot cycle (*b*) Rankine cycle

   (*c*) modified Rankine cycle (*d*) Regenerative cycle

ANSWERS

1. (*b*)	2. (*c*)	3. (*a*)	4. (*d*)	5. (*c*)

11

Fuels

1. Introduction. 2. Classification of Fuels. 3. Solid Fuels. 4. Liquid Fuels. 5. Merits and Demerits of Liquid Fuels over Solid Fuels. 6. Gaseous Fuels. 7. Merits and Demerits of Gaseous Fuels. 8. Requirements of a Good Fuel. 9. Calorific Value of Fuels. 10. Gross or Higher Calorific Value. 11. Net or Lower Calorific Value. 12. Experimental Determination of Higher Calorific Value. 13. Bomb Calorimeter. 14. Boy's Gas Calorimeter.

11.1. Introduction

A fuel, in general terms, may be defined as a substance (containing mostly carbon and hydrogen) which, on burning with oxygen in the atmospheric air, produces a large amount of heat. The amount of heat generated is known as calorific value of the fuel.

As the principal constituents of a fuel are carbon and hydrogen, therefore, it is also known as hydrocarbon fuel. Sometimes, a few traces of sulphur are also present in it.

11.2. Classification of Fuels

The fuels may be classified into the following three general forms :

1. Solid fuels,
2. Liquid fuels, and
3. Gaseous fuels.

Each of these fuels may be further subdivided into the following two types :

(*a*) Natural fuels, and

(*b*) Prepared fuels.

11.3. Solid Fuels

The natural solid fuels are wood, peat, lignite or brown coal, bituminous coal and anthracite coal. The prepared solid fuels are wood charcoal, coke, briquetted coal and pulverised coal.

The following solid fuels are important from the subject point of view :

1. *Wood.* At one time it was extensively used as a fuel. It consists of mainly carbon and hydrogen. The wood is converted into coal when burnt in the absence of air. It is not considered as a commercial fuel, except in industries, where a large amount of waste wood is available. The calorific value of wood varies with its kind and moisture content. The average calorific value of the wood is 19 700 kJ/kg.

2. *Peat.* It is a spongy humid substance found in boggy land. It may be regarded as the first stage in the formation of coal. It has a large amount of water contents (upto 30%) and therefore has to be dried before use. It has a characteristic odour at the time of burning, and has a smoky flame. Its average calorific valuc is 23 000 kJ/kg.

3. *Lignite or brown coal.* It represents the next stage of peat in the coal formation, and is an intermediate variety between bituminous coal and peat. It contains nearly 40% moisture and 60% of carbon. When dried, it crumbles and hence does not store well. Due to its brittleness, it is converted into briquettes, which can be handled easily. Its average calorific value is 25 000 kJ/kg.

4. *Bituminous coal.* It represents the next stage of lignite in the coal formation and contains very little moisture (4 to 6%) and 75 to 90% of carbon. It is weather-resistant and burns with a yellow flame. The average calorific value of bituminous coal is 33 500 kJ/kg. The bituminous coal is of the following two types :

(*a*) Caking bituminous coal, and (*b*) Non-caking bituminous coal.

(*a*) *Caking bituminous coal* : It softens and swells on heating and its pieces adhere together forming a pasty mass which makes firing difficult. It burns with a fairly long flame. Its specific gravity is 1.26 to 1.36. The caking variety is very useful for manufacturing gas. It is also known as soft coal. Its average calorific value is 35 000 kJ/kg.

(*b*) *Non-caking bituminous coal* : It burns with a shorter flame than the caking coal, and gives off little or no smoke. Its specific gravity is 1.22 to 1.42. The non-caking variety is mostly used as fuel for steam boilers, hence it is known as steam coal. Its average calorific value is 33 000 kJ/kg.

Note : The bituminous coal is non-caking, if its carbon content is 78 to 81%. If the percentage of carbon is 81 to 82.5%, it is slightly caking. In medium caking bituminous coal, the carbon content is 82.5 to 84%. But if the carbon content is 84 to 89%, it makes the coal strongly caking.

5. *Anthracite coal.* It represents the final stage in the coal formation, and contains 90% or more carbon with a very little volatile matter. It is thus obvious, that the anthracite coal is comparatively smokeless, and has very little flame. It possesses a high calorific value of about 36 000 kJ/kg and is therefore, very valuable for steam raising and general power purposes.

6. *Wood charcoal.* It is made by heating wood with a limited supply of air to a temperature not less than 280° C. It is a good prepared solid fuel, and is used for various metallurgical processes.

7. *Coke.* It is produced when coal is strongly heated continuously for 42 to 48 hours in the absence of air in a closed vessel. This process is known as *carbonisation of coal.* Coke is dull black in colour, porous and smokeless. It has a high carbon content (85 to 90%) and has a higher calorific value than coal.

If the carbonisation of coal is carried out at 500 to 700° C, the resulting coke is called *lower temperature coke* or *soft coke*. It is used as a domestic fuel. The coke produced by carbonisation of coal at 900 to 1100° C, is known as *hard coke.* The hard coke is mostly used as a blast furnace fuel for extracting pig iron from iron ores, and to some extent as a fuel in cupola furnace for producing cast iron.

8. *Briquetted coal.* It is produced from the finely ground coal by moulding under pressure with or without a binding material. The binding materials usually used are pitch, coal tar, crude oil and clay etc. The briquetted coal has the advantage of having, practically, no loss of fuel through grate openings and thus it increases the heating value of the fuel.

9. *Pulverised coal.* The low grade coal with a high ash content, is powdered to produce pulverised coal. The coal is first dried and then crushed into a fine powder by pulverising machines. The pulverised coal is widely used in the cement industry and also in metallurgical processes.

Note : Out of all the above mentioned types of solid fuels, anthracite coal is commonly used in all types of heat engines.

11.4. Liquid Fuels

Almost all the commercial liquid fuels are derived from natural petroleum (or crude oil). The crude oil is obtained from bore-holes in the earth's crust in certain parts of the world. The liquid fuels consist of hydrocarbons. The natural petroleum may be separated into petrol or gasoline, paraffin oil or kerosene, fuel oils and lubricating oils by boiling the crude oil at different temperatures and subsequent fractional distillation* or by a process such as cracking.** The solid products like vaseline and paraffin wax are recovered from the residue in the still.

The following liquid fuels are important from the subject point of view :

1. *Petrol or gasoline.* It is the lightest and most volatile liquid fuel, mainly used for light petrol engines. It is distilled at a temperature from 65° to 220° C.

2. *Kerosene or paraffin oil.* It is heavier and less volatile fuel than the petrol, and is used as heating and lighting fuel. It is distilled at a temperature from 220° to 345° C.

3. *Heavy fuel oils.* The liquid fuels distilled after petrol and kerosene are known as heavy fuel oils. These oils are used in diesel engines and in oil-fired boilers. These are distilled at a temperature from 345° to 470° C.

11.5. Merits and Demerits of Liquid Fuels over Solid Fuels

Following are the merits and demerits of liquid fuels over solid fuels :

Merits

1. Higher calorific value.
2. Lower storage capacity required.
3. Better economy in handling.
4. Better control of consumption by using valves.
5. Better cleanliness and freedom from dust.
6. Practically no ashes.
7. Non-deterioration in storage.
8. Non-corrosion of boiler plates.
9. Higher efficiency.

Demerits

1. Higher cost.
2. Greater risk of fire.
3. Costly containers are required for storage and transport.

11.6. Gaseous Fuels

The natural gas is, usually, found in or near the petroleum fields, under the earth's surface. It, essentially, consists of marsh gas or methane (CH_4) together with small amounts of other gases such as ethane (C_2H_6), carbon dioxide (CO_2) and carbon monoxide (CO).

The following prepared gases, which are used as fuels, are important from the subject point of view :

1. *Coal gas.* It is also known as a *town gas*. It is obtained by the carbonisation of coal and consists mainly of hydrogen, carbon monoxide and various hydrocarbons. The quality of coal gas depends upon the quality of the coal used, temperature of the carbonisaiton and the type of plant. It is very rich among combustible gases, and is largely used in towns for street and domestic lighting and heating. It is also used in furnaces and for running gas engines. Its calorific value is about 21 000 to 25 000 kJ/m^3.

---

* It is the distillation by stages, *i.e.* distillation carried out in such a way so that the liquid with the lowest boiling point is first evaporated and recondensed. The liquid with the next higher boiling point is then evaporated and recondensed, and so on until all the available liquid fuels are separately recovered in the sequence of their boiling points.

** Cracking is a special process of heating crude oil to a high temperature under a very high pressure (exceeding 50 atmospheres) to increase the yield of lighter distillates, particularly petrol. The residue left after distillation by cracking is called cracked residue pressure tar and is used in road construction.

2. *Producer gas.* It is obtained by the partial combustion of coal, coke, anthracite coal or charcoal in a mixed air-steam blast. It is, mostly, used for furnaces particularly for glass melting and also for power generation. Its manufacturing cost is low, and has a calorific value of about 5000 to 6700 kJ/m^3.

3. *Water gas.* It is a mixture of hydrogen and carbon monoxide and is made by passing steam over incandescent coke. As it burns with a blue flame, it is also known as *blue water gas.*

The water gas is usually converted into carburetted (enriched) water gas by passing it through a carburetter into which a gas oil is sprayed. It is, usually, mixed with coal gas to form town gas. The water gas is used in furnaces and for welding.

4. *Mond gas.* It is produced by passing air and a large amount of steam over waste coal at about 650° C. It is used for power generation and heating. It is also suitable for use in gas engines. Its calorific value is about 5850 kJ/m^3.

5. *Blast furnace gas.* It is a by-product in the production of pig iron in the blast furnace. This gas serves as a fuel in steel works, for power generation in gas engines, for steam raising in boilers and for preheating the blast for furnace. It is extensively used as fuel for metallurgical furnaces. The gas, leaving the blast furnace, has a high dust content the proportion of which varies with the operation of the furnace. It has a low heating value of about 3750 kJ/m^3.

6. *Coke oven gas.* It is a by-product from coke oven, and is obtained by the carbonisation of bituminous coal. Its calorific value varies from 14 500 to 18 500 kJ/m^3. It is used for industrial heating and power generation.

11.7. Merits and Demerits of Gaseous Fuels

Following are the merits and demerits of the gaseous fuels :

Merits

1. The supply of fuel gas, and hence the temperature of furnace is easily and accurately controlled.
2. The high temperature is obtained at a moderate cost by pre-heating gas and air with heat of waste gases of combustion.
3. They are directly used in internal combustion engines.
4. They are free from solid and liquid impurities.
5. They do not produce ash or smoke.
6. They undergo complete combustion with minimum air supply.

Demerits

1. They are readily inflammable.
2. They require large storage capacity.

11.8. Requirements of a Good Fuel

Though there are many requirements of a good fuel, yet the following are important from the subject point of view :

1. A good fuel should have a low ignition point.
2. It should have a high calorific value.
3. It should freely burn with a high efficiency, once it is ignited.
4. It should not produce harmful gases.
5. It should produce least quantity of smoke and gases.
6. It should be economical, easy to store and convenient for transportation.

11.9. Calorific Value of Fuels

The calorific value (briefly written as C.V.) or heat value of a solid or liquid fuel may be defined as the amount of heat given out by the complete combustion of 1 kg of fuel. It is expressed in terms of kJ/kg of fuel. The calorific value of gaseous fuels is, however, expressed in terms of kJ/m^3 at a specified temperature and pressure.

Following are the two types of the calorific value of fuels :

1. Gross or higher calorific value, and 2. Net or lower calorific value.

These calorific values are discussed, in detail, in the following articles.

11.10. Gross or Higher Calorific Value

All fuels, usually, contain some percentage of hydrogen. When a given quantity of a fuel is burnt, some heat is produced. Moreover, some hot flue gases are also produced. The water, which takes up some of the heat evolved, is converted into steam. If the heat, taken away by the hot flue gases and the steam is taken into consideration, *i.e.* if the heat is recovered from flue gases and steam is condensed back to water at room temperature (15° C), then the amount of total heat produced per kg is known as *gross* or *higher calorific value of fuel*. In other words, the amount of heat obtained by the complete combustion of 1 kg of a fuel, when the products of its combustion are cooled down to the temperature of supplied air (usually taken as 15° C), is called the gross or higher calorific value of fuel. It is briefly written as H.C.V.

If the chemical analysis of a fuel is available, then the higher calorific value of the fuel is determined by the following formula, known as Dulong's formula :

$$\text{H.C.V.} = 33\,800\,C + 144\,000\,H_2 + 9270\,S \text{ kJ/kg} \qquad \ldots (i)$$

where C, H_2 and S represent the mass of carbon, hydrogen and sulphur in 1 kg of fuel, and the numerical values indicate their respective calorific values.

If the fuel contains oxygen (O_2), then it is assumed that the whole amount is combined with hydrogen having mass equal to 1/8th of that of oxygen. Therefore, while finding the calorific value of fuel, this amount of hydrogen should be subtracted.

$$\therefore \quad \text{H.C.V.} = 33\,800\,C + 144\,000\left(H_2 - \frac{O_2}{8}\right) + 9270\,S \text{ kJ/kg} \qquad \ldots (ii)$$

11.11. Net or Lower Calorific Value

When the heat absorbed or carried away by the products of combustion is not recovered (which is the case in actual practice), and the steam formed during combustion is not condensed, then the amount of heat obtained per kg of the fuel is known as *net* or *lower calorific value*. It is briefly written as L.C.V.

If the higher calorific value is known, then the lower calorific value may be obtained by subtracting the amount of heat carried away by products of combustion (especially steam) from H.C.V.

$$\therefore \quad \text{L.C.V.} = \text{H.C.V.} - \text{Heat of steam formed during combustion}$$

Let m_s = Mass of steam formed in kg per kg of fuel = $9\,H_2$

Since the amount of heat per kg of steam is the latent heat of vaporisation of water corresponding to a standard temperature of 15° C, is 2466 kJ/kg, therefore

$$\text{L.C.V} = \text{H.C.V.} - m_s \times 2466 \text{ kJ/kg}$$

$$= \text{H.C.V.} - 9\,H_2 \times 2466 \text{ kJ/kg} \qquad \ldots (\because m_s = 9\,H_2)$$

Example 11.1. *A fuel consists of 85% carbon ; 12.5% hydrogen ; 2.5% residual matter by mass. Working from first principles, find the higher and lower calorific values per kg of the fuel.*

Solution. Given : $C = 85\% = 0.85$ kg ; $H_2 = 12.5\% = 0.125$ kg ; * Residual matter = 2.5% = 0.025 kg

Higher calorific value per kg of fuel

We know that higher calorific value per kg of fuel,

$$\text{H.C.V.} = 33\,800\,C + 144\,000\,H_2$$

$$= 33\,800 \times 0.85 + 144\,000 \times 0.125 = 46\,730 \text{ kJ/kg} \quad \textbf{Ans.}$$

Lower calorific value per kg of fuel

We know that lower calorific value per kg of fuel,

$$\text{L.C.V.} = \text{H.C.V.} - (9H_2 \times 2466)$$

$$= 46\,730 - (9 \times 0.125 \times 2466) = 43\,956 \text{ kJ/kg} \quad \textbf{Ans.}$$

Example 11.2. *A sample of coal has the following composition by mass :*

Carbon 75% ; hydrogen 6% ; oxygen 8% ; nitrogen 2.5% ; sulphur 1.5% ; and ash 7%. Calculate its higher and lower calorific values per kg of coal.

Solution. Given : $C = 75\% = 0.75$ kg ; $H_2 = 6\% = 0.06$ kg ; $O_2 = 8\% = 0.08$ kg ; $*N_2 = 2.5\% = 0.025$ kg ; $S = 1.5\ \% = 0.015$ kg ; *Ash = 7% = 0.07 kg

Higher calorific value per kg of coal

We know that higher calorific value per kg of coal,

$$\text{H.C.V.} = 33\,800\,\text{C} + 144\,000\left(H_2 - \frac{O_2}{8}\right) + 9270\,S$$

$$= 33\,800 \times 0.75 + 144\,000\left(0.06 - \frac{0.08}{8}\right) + 9270 \times 0.015$$

$$= 25\,350 + 7200 + 139 = 32\,689 \text{ kJ/kg} \quad \textbf{Ans.}$$

Lower calorific value per kg of coal

We know that lower calorific value per kg of coal,

$$\text{L.C.V.} = \text{H.C.V.} - (9\,H_2 \times 2466)$$

$$= 32\,689 - 9 \times 0.06 \times 2466 = 31\,358.3 \text{ kJ/kg} \quad \textbf{Ans.}$$

11.13. Experimental Determination of Higher Calorific Value

The method of determining higher calorific value, as explained in Art. 11.10, gives approximate results only. The most satisfactory method of obtaining the calorific value of a fuel is by actual experiment. In all these experimental methods, a known mass of fuel is burnt in a suitable calorimeter, and the heat so evolved is found by measuring the rise in temperature of the surrounding water. The calorimeters used for finding the calorific value of fuels are known as fuel calorimeters.

The following two fuel calorimeters are important from the subject point of view :

1. Bomb calorimeter, and 2. Boy's gas calorimeter.

These calorimeters are discussed, in detail, in the following pages.

---

* Superfluous data.

11.14. Bomb Calorimeter

It is used for finding the higher calorific value of solid and liquid fuels. In this calorimeter, as shown in Fig. 11.1, the fuel is burnt at a constant volume and under a high pressure in a closed vessel called bomb.

The bomb is made mainly of acid-resisting stainless steel, machined from the solid metal, which is capable of withstanding high pressure (upto 100 bar), heat and corrosion. The cover or head of the bomb carries the oxygen valve for admitting oxygen and a release valve for exhaust gases. A cradle or carrier ring, carried by the ignition rods, supports the silica crucible, which in turn holds the sample of fuel under test. There is an ignition wire of platinum or nichrome which dips into the crucible. It is connected to a battery, kept outside, and can be sufficiently heated by passing current through it so as to ignite the fuel.

The bomb is completely immersed in a measured quantity of water. The heat, liberated by the combustion of fuel, is absorbed by this water, the bomb and copper vessel. The rise in the temperature of water is measured by a precise thermometer, known as Beckmann thermometer which reads upto 0.01° C.

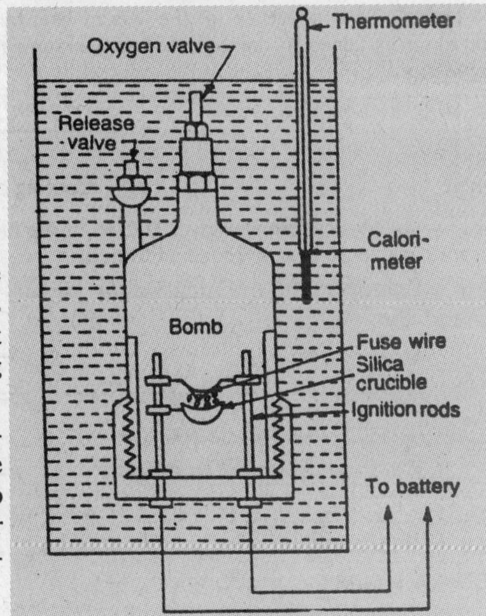

Fig. 11.1. Bomb calorimeter.

Procedure

A carefully weighed sample of the fuel (usually one gram or so) is placed in the crucible. Pure oxygen is then admitted through the oxygen valve, till pressure inside the bomb rises to 30 atmosphere. The bomb is then completely submerged in a known quantity of water contained in a large copper vessel. This vessel is placed within a large insulated copper vessel (not shown in the figure) to reduce loss of heat by radiation. When the bomb and its contents have reached steady temperature (this temperature being noted), fuse wire is heated up electrically. The fuel ignites, and continues to burn till whole of it is burnt. The heat released during combustion is absorbed by the surrounding water and the apparatus itself. The rise in temperature of water is noted.

Let m_f = Mass of fuel sample bunt in the bomb in kg,

H.C.V. = Higher calorific value of the fuel sample in kJ/kg,

m_w = Mass of water filled in the calorimeter in kg,

m_e = Water equivalent of apparatus in kg,

t_1 = Initial temperature of water and apparatus in °C, and

t_2 = Final temperature of water and apparatus in °C.

We know that heat liberated by fuel

$$= m_f \times \text{H.C.V.} \qquad \ldots (i)$$

and heat absorbed by water and apparatus

$$= (m_w + m_e)\, c_w\, (t_2 - t_1) \qquad \ldots (ii)$$

Since the heat liberated is equal to the heat absorbed (neglecting losses), therefore equating equations (*i*) and (*ii*),

$$m_f \times \text{H.C.V.} = (m_w + m_e)\, c_w\, (t_2 - t_1)$$

$$\therefore \qquad \text{H.C.V.} = \frac{(m_w + m_e)\, c_w\, (t_2 - t_1)}{m_f} \text{ kJ/kg}$$

Notes : 1. To compensate for the loss of heat by radiation (which cannot be totally eliminated), a cooling correction is added to the observed temperature rise. This corrected temperature rise is used in the above expression.

$$\therefore \qquad \text{H.C.V.} = \frac{(m_w + m_e)\, c_w\, [(t_2 - t_1) + t_c]}{m_f}$$

where $\qquad t_c$ = Cooling correction.

2. This calorimeter gives H.C.V. of the fuel because any steam formed is condensed (since it cannot escape) and hence heat is recovered from it.

Example 11.3. *Calculate the higher calorific value of a coal specimen from the following data :*

Mass of coal burnt	*= 1 g*
Quantity of water in calorimeter	*= 2.5 kg*
Increase in temperature of water	*= 2.6° C*
Water equivalent of apparatus	*= 390 g*

If the fuel used contains 6% of hydrogen, calculate its lower calorific value.

Solution. Given : m_f = 1 g = 0.001 kg ; m_w = 2.5 kg ; $t_2 - t_1$ = 2.6° C ; m_e = 390 g = 0.39 kg ; H_2 = 6% = 0.06

Higher calorific value

We know that higher calorific value,

$$\text{H.C.V.} = \frac{(m_w + m_e)\, c_w\, (t_2 - t_1)}{m_f} = \frac{(2.5 + 0.39)\, 4.2 \times 2.6}{0.001} \text{ kJ/kg}$$

$$= 31\,560 \text{ kJ/kg } \textbf{Ans.} \qquad \ldots (\because c_w = 4.2 \text{ kJ/kg K})$$

Lower calorific value

We know that lower calorific value

$$= \text{H.C.V.} - (9H_2 \times 2466)$$

$$= 31\,560 - 9 \times 0.06 \times 2466 = 30\,228.3 \text{ kJ/kg } \textbf{Ans.}$$

11.14. Boy's Gas Calorimeter

It is primarily used for gaseous fuels, though it can be modified for liquid fuels also. It gives the higher calorific value only. But the lower calorific value may be calculated, since the amount of water produced can be collected and measured.

It consists of a suitable gas burner at *B*, in which a known volume of gas at a known pressure is burnt. The hot gases, produced by combustion, rise up in the copper chimney or combustion chamber, which is surrounded by a double metal tubing through which a continuous flow of water under a constant head is maintained. From the top of inner chamber, hot gases are deflected downwards through the space containing the inner water tubes *M*. From here the gases are deflected upwards through the space containing the outer water tubes *N*. Then the gases escape into the atmosphere from the top and their temperature is recorded just before their exit. During this process of playing up and down the water tubes, the gases give out, practically, whole of their heat so that any steam formed during combustion is condensed back into water.

The temperature of the circulating water is measured at inlet and outlet by thermometers T_1 and T_2 respectively, as shown in Fig. 11.2. After an initial warming up period during which conditions are established, simultaneous readings are taken of :

1. The volume of gas burnt in a certain time.
2. The quantity of water passing through the tube during the same time, and
3. The rise in temperature of water.

The volume of the gas consumed is reduced to some standard conditions of pressure and temperature (15° C and 760 mm of Hg), the normal temperature and pressure being 0° C and 760 mm of Hg.

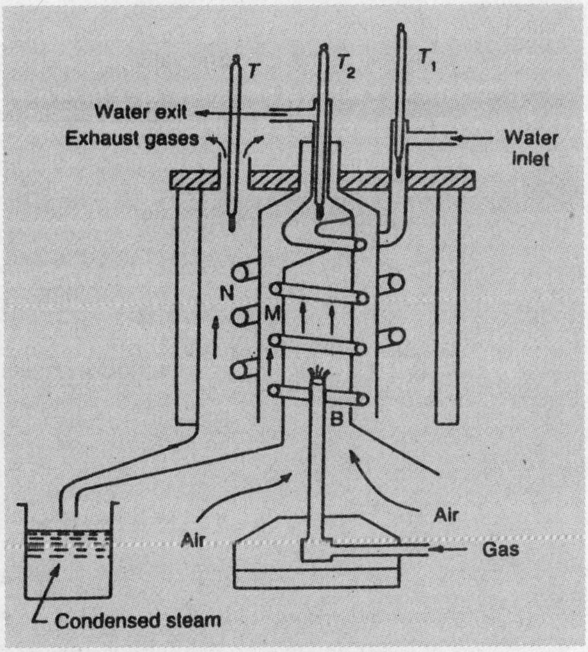

Fig. 11.2. Boy's gas calorimeter.

Let v = Volume of gas burnt at standard temperature and pressure (S.T.P.) in m^3,

m_w = Mass of cooling water used in kg,

H.C.V. = Higher calorific value of the fuel in kJ/m^3,

t_1 = Temperature of water at inlet, in °C, and

t_2 = Temperature of water at outlet in °C.

We know that heat produced by the combustion of fuel

$$= v \times \text{H.C.V.} \quad \ldots (i)$$

and heat absorbed by circulating water

$$= m\, c_w (t_2 - t_1) \quad \ldots (ii)$$

Since the heat produced by the combustion of fuel is equal to the heat absorbed by the circulating water (neglecting losses), therefore equating equations (*i*) and (*ii*),

$$v \times \text{H.C.V.} = m_w\, c_w (t_2 - t_1)$$

$$\therefore \quad \text{H.C.V.} = \frac{m_w\, c_w (t_2 - t_1)}{v}$$

Notes : 1. The lower calorific value may be calculated if the water produced during the combustion is drained off from the bottom of the calorimeter, collected and weighed. If this mass is m, then

$$\text{L.C.V.} = \text{H.C.V.} - m \times 2466 \text{ kJ/kg}$$

2. The modification of this gas calorimeter is known as Junker's calorimeter.

Example 11.4. *The following data refers to a calorific value test of a fuel by means of a gas calorimeter.*

Volume of gas used = 0.7 m^3 (reckoned at S.T.P.) ; mass of water heated = 25 kg ; rise in temperature of water at inlet and outlet = 14° C ; mass of steam condensed = 0.028 kg. Find the higher and lower calorific values per m^3 at S.T.P. Take the heat liberated in condensing water vapour and cooling the condensate as 2475 kJ/kg.

Solution. Given : $v = 0.7\ m^3$; $m_w = 25$ kg ; $t_2 - t_1 = 14°$ C ; $m_s = 0.028$ kg ; Heat liberated = 2475 kJ/kg

Higher calorific value

We know that higher calorific value,

$$\text{H.C.V.} = \frac{m_w\, c_w\,(t_2 - t_1)}{v} = \frac{25 \times 4.2 \times 14}{0.7} = 2100 \text{ kJ/m}^3 \textbf{ Ans.}$$

Lower calorific value

We know that lower calorific value,

$$\text{L.C.V.} = \text{H.C.V.} - \text{Mass of steam condensed in kg/m}^3 \times \text{Heat liberated}$$

$$= 2100 - \frac{0.028}{0.7} \times 2475 = 2001 \text{ kJ/m}^3 \textbf{ Ans.}$$

Example 11.5. *The following results were obtained when a sample of gas was tested in a Junker's gas calorimeter :*

Gas burnt	$= 0.03\ m^3$
Pressure of the gas	*= 54.4 mm of water*
Barometer reading	*= 750 mm of Hg*
Temperature of gas	*= 27° C*
Temperature of water at inlet	*= 28° C*
Temperature of water at outlet	*= 40° C*
Mass of water passing through the calorimeter	*= 10 kg*
Steam condensed during test	*= 0.025 kg*

Determine the higher and lower calorific values of gas at 15° C and a standard barometer of 760 mm of Hg.

Solution. Given : $v_1 = 0.03\ m^3$; $p_1 = 54.4$ mm of water $= \frac{54.4}{13.6} + 750 = 754$ mm of Hg ; $T_1 = 27°\text{ C} = 27 + 273 = 300$ K ; $t_1 = 28°$ C ; $t_2 = 40°$ C ; $m_w = 10$ kg ; $m_s = 0.025$ kg ; $T_2 = 15°$ C $= 15 + 273 = 288$ K ; $p_2 = 760$ mm of Hg

First of all, let us find the volume of gas (v_2) at temperature T_2 and pressure p_2.

We know that for a perfect gas,

$$\frac{p_1\, v_1}{T_1} = \frac{p_2\, v_2}{T_2} \quad \text{or} \quad v_2 = \frac{p_1\, v_1\, T_2}{p_2\, T_1} = \frac{754 \times 0.03 \times 288}{760 \times 300} = 0.029 \text{ m}^3$$

$$\therefore \text{Mass of steam condensed} = \frac{0.025}{0.029} = 0.86 \text{ kg/m}^3$$

Higher calorific value of the gas

We know that higher calorific value of the gas,

$$\text{H.C.V.} = \frac{m_w\, c_w\,(t_2 - t_1)}{v_2} = \frac{10 \times 4.2\,(40 - 28)}{0.029} = 17\,380 \text{ kJ/m}^3 \textbf{ Ans.}$$

Lower calorific value of the gas

We know that lower calorific value of the gas,

$$\text{L.C.V} = \text{H.C.V.} - \text{Mass of steam condensed in kg/m}^3 \times \text{Heat liberated}$$

$$= 17\,380 - 0.86 \times 2466 = 15\,259.3 \text{ kJ/m}^3 \textbf{ Ans.}$$

EXERCISES

1. A sample of fuel on analysis is found to contain carbon 85% ; hydrogen 10% ; sulphur 2% and ash 3%. Find the higher calorific value of the fuel. **[Ans.** 43 315.4 kJ/kg**]**

2. The composition of sample of anthracite coal was found to be ; C 91%, H_2 3%, O_2 2%, N_2 0.8%, S 0.8%, and the remainder is ash.

   Calculate the higher and lower calorific values of the fuel. **[Ans.** 34 792 kJ/kg ; 34 126 kJ/kg**]**

3. The following data were recorded during an experiment to find the calorific value of a sample of coal.

   Mass of coal burnt = 1 g
   Mass of water in the calorimeter = 1020 g
   Water equivalent of the calorimeter = 170 g
   Initial temperature of water = 23.3° C
   Final temperature of water = 26.2° C

   Determine the calorific value of the sample of the coal. **[Ans.** 14 494 kJ/kg**]**

4. A bomb calorimeter is used to determine the calorific value of a sample of coal and the following results are obtained :

   Mass of coal burnt = 1 g
   Mass of water in the calorimeter = 2.5 kg
   Water equivalent of the apparatus = 0.75 kg
   Initial temperature of water = 17.5° C
   Maximum observed temperature of water = 20° C
   Cooling correction = + 0.015° C

   If the fuel contains 4% of hydrogen, find the lower calorific value of the fuel. **[Ans.** 33 442 kJ/kg**]**

5. The following observations were made during a test on coal gas :

   Volume of gas used = 0.06 m^3
   Mass of cooling water circulated = 9.8 kg
   Mass of condensed steam collected = 0.009 kg
   Rise in temperature of cooling water = 6.3° C
   Pressure of gas tested above atmosphere = 45 mm of water
   Temperature of gas tested = 14° C
   Barometric pressure = 750 mm of Hg

   Calculate the higher and lower calorific values at N.T.P. **[Ans.** 4581 kJ/m^3 ; 4189 kJ/m^3**]**

QUESTIONS

1. What is meant by the term fuel ? What are its constituents ?
2. List out the merits and demerits of liquid fuels over solid fuels.
3. What are the advantages of gaseous fuels ?
4. Define the calorific value of a solid fuel and also that of a gaseous fuel.
5. Distinguish between higher and lower calorific value of a fuel.
6. Explain, briefly, the method used to determine the higher calorific value of the liquid fuel.

OBJECTIVE TYPE QUESTIONS

1. The principal constituents of a fuel are

   (*a*) carbon and hydrogen (*b*) oxygen and hydrogen
   (*c*) sulphur and oxygen (*d*) sulphur and hydrogen

2. The fuel mostly used in boilers is

(*a*) brown coal (*b*) peat

(*c*) caking bituminous coal (*d*) non-caking bituminous coal

3. Which of the following fuel has the highest calorific value ?

(*a*) Peat (*b*) Coke

(*c*) Bituminous coal (*d*) Anthracite coal

4. The fuel mostly used in blast furnace for extracting pig iron from iron ores is

(*a*) hard coke (*b*) soft coke

(*c*) pulverised coal (*d*) bituminous coal

5. Steam coal is a

(*a*) pulverised coal (*b*) brown coal

(*c*) caking bituminous coal (*d*) non-caking bituminous coal

6. A process of heating crude oil to a high temperature under a very high pressure to increase the yield of lighter distillates, is known as

(*a*) cracking (*b*) carbonisation

(*c*) fractional distillation (*d*) full distillation

7. Petrol is distilled at

(*a*) 65° to 220° C (*b*) 220° to 345° C (*c*) 345° to 470° C (*d*) 470° to 550° C

8. Which of the following gas has the highest calorific value ?

(*a*) Producer gas (*b*) Coal gas (*c*) Mond gas (*d*) Blast furnace gas

9. A bomb calorimeter is used for finding the . . . calorific value of solid and liquid fuels.

(*a*) higher (*b*) lower

10. Which of the following statement is incorrect ?

(*a*) The liquid fuels consist of hydrocarbons.

(*b*) The liquid fuels have higher calorific value than solid fuels.

(*c*) The solid fuels have higher efficiency than liquid fuels.

(*d*) A good fuel should have low ignition point.

ANSWERS

1. (*a*)	2. (*d*)	3. (*c*)	4. (*a*)	5. (*d*)
6. (*a*)	7. (*a*)	8. (*b*)	9. (*a*)	10. (*c*)

12

Combustion of Fuels

1. Introduction. 2. Elements and Compounds. 3. Atoms and Molecules. 4. Atomic Mass 5. Molecular Mass. 6. Symbols for Elements and Compounds. 7. Combustion Equations of Solid Fuels. 8. Combustion Equations of Gaseous Fuels. 9. Theoretical or Minimum Mass of Air Required for Complete Combustion. 10. Theoretical or Minimum Volume of Air Required for Complete Combustion. 11. Conversion of Volumetric Analysis into Mass Analysis or Gravimetric Analysis. 12. Conversion of Mass Analysis into Volumetric Analysis. 13. Mass of Carbon in Flue Gases. 14. Mass of Flue Gases per kg of Fuel Burnt. 15. Excess Air Supplied. 16. Mass of Excess Air Supplied. 17. Flue Gas Analysis by Orsat Apparatus.

12.1. Introduction

The combustion of fuels may be defined as a chemical combination of oxygen, in the atmospheric air, and hydro-carbons. It is, usually, expressed both qualitatively and quantitatively by equations known as *chemical equations*. A chemical equation shows, in a concise form, the complete nature of the chemical action or reaction taking place.

12.2. Elements and Compounds

The elements are those substances which have so far not been resolved by any means into other substances of simpler form. About 109 such elements are known so far. The examples are hydrogen, oxygen, nitrogen, helium, iron, carbon, etc.

The compounds are formed by the combination of different elements in simple proportion. The number of compounds, that can be formed, is almost infinite. The examples are water (combination of hydrogen and oxygen), carbon dioxide (combination of carbon and oxygen), methane gas (combination of carbon and hydrogen), etc.

12.3. Atoms and Molecules

The elements are made up of minute and chemically indivisible particles known as *atoms*.

The smallest quantity of a substance, which can exist by itself, in a chemically recognizable form, is known as *molecule*. A molecule may consist of one atom, two atoms, three atoms or even more. Such a molecule is known as monoatomic, diatomic, triatomic or polyatomic molecule respectively.

12.4. Atomic Mass

Hydrogen is the lightest known substance. By taking the mass of hydrogen atom as unity, it is possible to obtain relative masses of other atoms and molecules. It may be noted that actual masses of these atoms are extremely small. The atomic mass of an element is the number of times, the atom of that element is heavier than the hydrogen atom. For example, the atomic mass of oxygen is 16. It means that the oxygen atom is 16 times heavier than the hydrogen atom.

12.5. Molecular Mass

The molecular mass of a substance is the number of times a molecule of that substance is heavier than the hydrogen atom. For example, one molecule of oxygen consists of two atoms of oxygen, each of which is 16 times heavier than hydrogen atom. It is thus obvious, that a molecule of oxygen is $2 \times 16 = 32$ times heavier than hydrogen atom. In other words, the molecular mass of oxygen is 32.

12.6. Symbols for Elements and Compounds

The elements and compounds are, generally, represented by symbols. The symbol of an atom is, usually, the initial letter of the element in capital. The symbol for carbon atom, for example, is C ; for hydrogen atom H ; for oxygen atom O ; and so on. A molecule is represented by a single expression. For example, molecule of oxygen is written as O_2, where the suffix 2 represents the number of atoms in one molecule of oxygen.

Similarly CO_2 stands for one molecule of carbon dioxide gas, which consists of one atom of carbon and two atoms of oxygen. The coefficients or prefix of an expression indicates the number of molecules of a substance, *e.g.* 7 CO_2 means 7 molecules of carbon dioxide, each of which consists of one atom of carbon and two atoms of oxygen.

The symbols, atomic mass and molecular mass of the following substances are important from the subject point of view :

Table 12.1. Symbols with atomic mass and molecular mass.

Substance	*Symbol*	*Atomic Mass*	*Molecular Mass*
Hydrogen	H_2	1	2
Oxygen	O_2	16	32
Nitrogen	N_2	14	28
Carbon	C	12	-
Sulphur	S	32	-
Carbon monoxide	CO	-	28
Methane or Marsh gas	CH_4	-	16
Acetylene	C_2H_2	-	26
Ethylene	C_2H_4	-	28
Ethane	C_2H_6	-	30
Carbon dioxide	CO_2	-	44
Sulphur dioxide	SO_2	-	64
Steam or water	H_2O	-	18

Note : If the atomic mass of a substance is known, then its molecular mass may be easily obtained, For example, molecular mass of $CO_2 = 12 + (2 \times 16) = 44$

12.7. Combustion Equations of Solid Fuels

We have already discussed in Art. 12.1 that the combustion of a fuel is the chemical combination of oxygen. Though there are many chemical equations for the chemical combination of oxygen (representing combustion), yet the following equations are important from the subject point of view :

1. When carbon burns in *sufficient quantity* of oxygen, carbon dioxide is produced along with a release of large amount of heat. This is represented by the following chemical equation :

$$C + O_2 = CO_2$$

$$1 \text{ mol.} + 1 \text{ mol.} = 1 \text{ mol.} \qquad \text{...(By volume)}$$

or $$12 \text{ kg} + 32 \text{ kg} = 44 \text{ kg} \qquad \text{...(By mass)}$$

i.e. $$1 \text{ kg} + \frac{8}{3} \text{ kg} = \frac{11}{3} \text{ kg}$$

It means that 1 kg of carbon requires 8/3 kg of oxygen for its complete combustion, and produces 11/3 kg of carbon dioxide gas.

2. If sufficient oxygen is not available, then combustion of carbon is incomplete. It then produces carbon monoxide instead of carbon dioxide. It is represented by the following chemical equation :

$$2C + O_2 = 2CO$$

$$2 \text{ mol.} + 1 \text{ mol.} = 2 \text{ mol.} \qquad \text{...(By volume)}$$

or $$2 \times 12 \text{ kg} + 2 \times 16 \text{ kg} = 2 \times 28 \text{ kg} \qquad \text{...(By mass)}$$

i.e. $$1 \text{ kg} + \frac{4}{3} \text{ kg} = \frac{7}{3} \text{ kg}$$

It means that 1 kg of carbon requires 4/3 kg of oxygen, and produces 7/3 kg of carbon monoxide.

3. If carbon monoxide is burnt further, it is converted into carbon dioxide. Thus

$$2CO + O_2 = 2CO_2$$

$$2 \text{ mol.} + 1 \text{ mol.} = 2 \text{ mol.} \qquad \text{...(By volume)}$$

or $$2 \times 28 \text{ kg} + 2 \times 16 \text{ kg} = 2 \times 44 \text{ kg} \qquad \text{...(By mass)}$$

i.e. $$1 \text{ kg} + \frac{4}{7} \text{ kg} = \frac{11}{7} \text{ kg}$$

It means that 1 kg of carbon monoxide requires 4/7 kg of oxygen, and produces 11/7 kg of carbon dioxide.

4. When sulphur burns with oxygen, it produces sulphur dioxide. This is represented by the following chemical equation :

$$S + O_2 = SO_2$$

$$1 \text{ mol.} + 1 \text{ mol.} = 1 \text{ mol.} \qquad \text{...(By volume)}$$

or $$32 \text{ kg} + 2 \times 16 \text{ kg} = 64 \text{ kg} \qquad \text{...(By mass)}$$

i.e. $$1 \text{ kg} + 1 \text{ kg} = 2 \text{ kg}$$

It means that 1 kg of sulphur requires 1 kg of oxygen for complete combustion and produces 2 kg of sulphur dioxide.

Note :We see from the above chemical equations that there is no loss of mass although changes in volume do occur. Hence the total mass on both sides of a chemical equation is the same.

12.8. Combustion Equations of Gaseous Fuels

The gaseous fuels are, generally measured by volume (in m^3) than by mass (in kg). It is thus obvious, that their combustion equations are usually stated quantitatively in volume form, though we may use masses also. Following are the important equations for the chemical combination of oxygen (representing combustion) from the subject point of view :

1. $$2CO + O_2 = 2CO_2$$

or $$\left[\begin{array}{c} 2 \text{ Volumes} + 1 \text{ Volume} = 2 \text{ Volumes} \\ 2 \text{ m}^3 + 1 \text{ m}^3 = 2 \text{ m}^3 \end{array}\right] \qquad \text{...(By volume)}$$

i.e.
$$\begin{bmatrix} 2\times 28\text{ kg} + 1\times 32\text{ kg} = 2\times 44\text{ kg} \\ 1\text{ kg} + \frac{4}{7}\text{ kg} = \frac{11}{7}\text{ kg} \end{bmatrix} \quad \text{. . . (By mass)}$$

It means that, 2 volumes of carbons monoxide require 1 volume of oxygen and produces 2 volumes of carbon dioxide.

Or in other words, 1 kg of carbon monoxide requires 4/7 kg of oxygen and produce 11/7 kg of carbon dioxide.

2.
$$2H_2 + O_2 = 2H_2O$$

or
$$\begin{bmatrix} 2\text{ Vol.} + 1\text{ Vol.} = 2\text{ Vol.} \\ 2\text{ m}^3 + 1\text{ m}^3 = 2\text{ m}^3 \end{bmatrix} \quad \text{. . . (By volume)}$$

i.e.
$$\begin{bmatrix} 2\times 2\text{ kg} + 1\times 32\text{ kg} = 2\times 18\text{ kg} \\ 1\text{ kg} + 8\text{ kg} = 9\text{ kg} \end{bmatrix} \quad \text{. . . (By mass)}$$

It means that, 2 volumes of hydrogen require 1 volume of oxygen and produce 2 volumes of water or steam.

Or in other words, 1 kg of hydrogen requires 8 kg of oxygen and produces 9 kg of water or steam.

3.
$$CH_4 + 2O_2 = CO_2 + 2H_2O$$

or
$$\begin{bmatrix} 1\text{ Vol.} + 2\text{ Vol.} = 1\text{ Vol.} + 2\text{ Vol.} \\ 1\text{ m}^3 + 2\text{ m}^3 = 1\text{ m}^3 + 2\text{ m}^3 \end{bmatrix} \quad \text{. . . (By volume)}$$

i.e.
$$\begin{bmatrix} 16\text{ kg} + 2\times 32\text{ kg} = 44\text{ kg} + 2\times 18\text{ kg} \\ 1\text{ kg} + 4\text{ kg} = \frac{11}{4}\text{ kg} + \frac{9}{4}\text{ kg} \end{bmatrix} \quad \text{. . . (By mass)}$$

It means that, 1 volume of methane requires 2 volumes of oxygen and produces 1 volume of carbon dioxide and 2 volumes of water or steam.

Or in other words, 1 kg of methane requires 4 kg of oxygen and produces 11/4 kg of carbon dioxide and 9/4 kg of water or steam.

4.
$$C_2H_4 + 3O_2 = 2CO_2 + 2H_2O$$

or
$$\begin{bmatrix} 1\text{ Vol.} + 3\text{ Vol.} = 2\text{ Vol.} + 2\text{ Vol.} \\ 1\text{ m}^3 + 3\text{ m}^3 = 2\text{ m}^3 + 2\text{ m}^3 \end{bmatrix} \quad \text{. . . (By volume)}$$

i.e.
$$\begin{bmatrix} 28\text{ kg} + 3\times 32\text{ kg} = 2\times 44\text{ kg} + 2\times 18\text{ kg} \\ 1\text{ kg} + \frac{24}{7}\text{ kg} = \frac{22}{7}\text{ kg} + \frac{9}{7}\text{ kg} \end{bmatrix} \quad \text{. . . (By mass)}$$

It means that, 1 volume of ethyene requires 3 volumes of oxygen and produces 2 volumes of carbon dioxide and 2 volumes of water or steam.

Or in other words, 1 kg of ethyene requires 24/7 kg of oxygen and produces 22/7 kg of carbon dioxide and 9/7 kg of water or steam.

12.9. Theoretical or Minimum Mass of Air Required for Complete Combustion

We have seen in the previous articles that adequate supply of oxygen is very essential for the complete combustion of a fuel, and hence for obtaining maximum amount of heat from a fuel.

The theoretical or minimum mass (or volume) of oxygen required for complete combustion of 1 kg of fuel may be calculated from the chemical analysis of the fuel. The mass of oxygen, required by each of the constituents of the fuel, may be calculated from the chemical equations in Art. 12.8.

Now consider 1 kg of a fuel.

Let Mass of carbon = C kg

Mass of hydrogen = H_2 kg

Mass of sulphur = S kg

We know that 1 kg of carbon requires 8/3 kg of oxygen for its complete combustion. Similarly, 1 kg of hydrogen requires 8 kg of oxygen and 1 kg of sulphur requires 1 kg of oxygen for its complete combustion.

∴ Total oxygen required for complete combustion of 1 kg of fuel

$$= \frac{8}{3} C + 8 H_2 + S \text{ kg} \quad \ldots (i)$$

If some oxygen (say O_2 kg) is already present in the fuel, then total oxygen for the complete ombustion of 1 kg of fuel

$$= \left[\frac{8}{3} C + 8 H_2 + S\right] - O_2 \text{ kg} \quad \ldots (ii)$$

It may be noted that the oxygen has to be obtained from atmospheric air, which mainly consists of nitrogen and oxygen along with a small amount of carbon dioxide and negligible amounts of rare gases like argon, neon, and krypton etc. But for all calculations, the composition of air is taken as :

Nitrogen (N_2) = 77% ; Oxygen (O_2) = 23% ... (By mass)

and Nitrogen (N_2) = 79% ; Oxygen (O_2) = 21% ... (By volume)

It is thus obvious, that for obtaining 1 kg of oxygen, amount of air required

$$= \frac{100}{23} = 4.35 \text{ kg} \quad \ldots \text{(By mass)}$$

∴ Theoretical or minimum air required for complete combustion of 1 kg of fuel

$$= \frac{100}{23}\left[\left(\frac{8}{3} C + 8 H_2 + S\right) - O_2\right] \text{ kg}$$

Example 12.1. *A fuel has the following composition by mass :*

Carbon 86%, Hydrogen 11.75%, Oxygen 2.25%.

Calculate the theoretical air supply per kg of fuel, and the mass of products of combustion per kg of fuel.

Solution. Given : C = 86% = 0.86 kg ; H_2 = 11.75 % = 0.1175 kg ; O_2 = 2.25% = 0.0225 kg

Theoretical air supply per kg of fuel

We know that theoretical air supply per kg of fuel

$$= \frac{100}{23}\left[\left(\frac{8}{3} C + 8 H_2 + S\right) - O_2\right] \text{ kg}$$

$$= \frac{100}{23}\left[\left(\frac{8}{3} \times 0.86 + 8 \times 0.1175 + 0\right) - 0.0225\right] \text{ kg}$$

$$= 13.96 \text{ kg} \quad \textbf{Ans.}$$

Mass of products of combustion

The chemical equations of carbon and hydrogen with oxygen are

(*i*) $C + O_2 = CO_2$

(*ii*) $2H_2 + O_2 = 2H_2O$

We know that 1 kg of carbon produces 11/3 kg of carbon dioxide and 1 kg of hydrogen produces 9 kg of water.

∴ Total mass of the products of combustion

$$= \frac{11}{3} \times C + 9H_2 \text{ kg}$$

$$= \frac{11}{3} \times 0.86 + 9 \times 0.1175 = 4.21 \text{ kg } \textbf{Ans.}$$

12.10. Theoretical or Minimum Volume of Air Required for Complete Combustion

We have discussed in the last article, the minimum mass of air required for complete combustion of a fuel. Similarly, the volume of oxygen required for complete combustion of 1 m^3 of gaseous fuel may be calculated from the chemical analysis of the fuel. The volume of oxygen, required by each of the constituents of the fuel, may be calculated from the chemical equations as discussed in Art. 12.8. Now consider 1 m^3 of a gaseous fuel.

Let volume of carbon monoxide $= CO \text{ m}^3$

Volume of hydrogen $= H_2 \text{ m}^3$

Volume of methane $= CH_4 \text{ m}^3$

Volume of ethylene $= C_2H_4 \text{ m}^3$

We know that 2 volumes of carbon monoxide require 1 volume of oxygen. Or in other words, 1 volume of carbon monoxide requires 0.5 volume of oxygen for its complete combustion . Similarly, 2 volumes of hydrogen require 1 volume of oxygen. Or in other words, 1 volume of hydrogen requires 0.5 volume of oxygen for its complete combustion. Similarly, 1 volume of methane requires 2 volumes of oxygen and 1 volume of ethylene requires 3 volumes of oxygen for its complete combustion.

∴ Total oxygen required for complete combustion of 1 m^3 of fuel

$$= 0.5\, CO + 0.5\, H_2 + 2CH_4 + 3\, C_2\, H_4 \text{ m}^3$$

If some oxygen (say O_2 m^3) is already present in the fuel, then total oxygen required for complete combustion of 1 m^3 of fuel

$$= [0.5\, CO + 0.5\, H_2 + 2CH_4 + 3C_2H_2] - O_2 \text{ m}^3$$

Since the oxygen present in the air is 21% by volume, therefore theoretical or minimum volume of air required for complete combustion of 1 m^3 of fuel

$$= \frac{100}{21}\left[\left(0.5\, CO + 0.5\, H_2 + 2CH_4 + 3C_2H_4\right) - O_2\right] \text{m}^3$$

Example 12.2. *A producer gas, used as a fuel, has the following volumetric composition : H_2 28% ; CO 12% ; CH_4 2% ; CO_2 16% and N_2 42%.*

Find the volume of air required for complete combustion of 1 m^3 of this gas. Air contains 21% by volume of oxygen.

Solution. Given : $H_2 = 28\% = 0.28 \text{ m}^3$; $CO = 12\% = 0.12 \text{ m}^3$; $CH_4 = 2\% = 0.02 \text{ m}^3$; $CO_2 = 16\% = 0.16 \text{ m}^3$; $N_2 = 42\% = 0.42 \text{ m}^3$

We know that theoretical air required

$$= \frac{100}{21}\left[\left(0.5\, CO + 0.5\, H_2 + 2CH_4 + 3C_2H_4\right) - O_2\right] \text{m}^3$$

$$= \frac{100}{21}[\,(0.5 \times 0.12 + 0.5 \times 0.28 + 2 \times 0.02 + 0) - 0\,] \text{ m}^3$$

$$= 1.143 \text{ m}^3 \textbf{ Ans.}$$

12.11. Conversion of Volumetric Analysis into Mass Analysis or Gravimetric Analysis

When the volumetric composition of any fuel gas is known, it can be converted to gravimetric composition by applying Avogadro's Law (Art. 2.9). Thus

1. Multiply the volume of each constituent by its own molecular mass. This gives the proportional mass of the constituents.
2. Add up these masses and divide each mass by this total mass, and express it as a percentage.
3. This gives percentage analysis by mass.

The conversion of volumetric analysis into mass analysis may be clearly understood by the following example.

Example 12.3. *The volumetric analysis of a gas is CO_2 14%, CO 1%, O_2 5% and N_2 80%. Calculate the fuel gas composition by mass.*

Solution. Given : CO_2 = 14% = 0.14 m^3 ; CO = 1% = 0.01 m^3 ; O_2 = 5% = 0.05 m^3 ; N_2 = 80% = 0.8 m^3

The volumetric analysis may be converted into mass analysis by completing the table as follows :

Constituent	*Volume in 1 m^3 of flue gas* (a)	*Molecular mass* (b)	*Proportional mass* $(c) = (a \times b)$	*Mass in kg per kg of flue gas* $(d) = \frac{(c)}{\Sigma (c)}$	*% by mass* $= (d) \times 100$
CO_2	0.14	44	6.16	$\frac{6.16}{30.44} = 0.202$	20.2%
CO	0.01	28	0.28	$\frac{0.28}{30.44} = 0.009$	0.9%
O_2	0.05	32	1.60	$\frac{1.60}{30.44} = 0.053$	5.3%
N_2	0.80	28	22.40	$\frac{22.40}{30.44} = 0.736$	73.6%
Total	1.00		$\Sigma (c) = 30.44$	1.0000	100.0

The fuel gas composition, by mass is given in the last column, *i.e.*

CO_2 = 20.2 % ; CO = 0.9% ; O_2 = 5.3% and N_2 = 73.6% **Ans.**

12.12. Conversion of Mass Analysis into Volumetric Analysis

The conversion of mass analysis of a fuel gas into the volumetric analysis may be done by the following steps :

1. Divide the percentage mass of each constituent by its own molecular mass. This gives the proportional volumes of the constituents.
2. Add these volumes and divide each volume by this total volume, and express it as a percentage.
3. This gives percentage analysis by volume.

Example 12.4. *A fuel gas has the following percentage composition by mass :*

CO_2 13.3 % ; CO 0.95 % ; O_2 8.35% and N_2 77.4%. Convert this into volumetric analysis.

Solution. Given : CO_2 = 13.3 % ; CO = 0.95 % ; O_2 = 8.35% ; N_2 = 77.4%

The mass analysis may be converted into volumetric analysis by completing the table as follows :

Constituent	% Mass analysis (a)	Molecular mass (b)	Proportional volume $(c) = \frac{(a)}{(b)}$	Volume in 1 m^3 of flue gas $(d) = \frac{(c)}{\Sigma(c)}$	% Volumetric analysis $= (d) \times 100$
CO_2	13.3	44	$\frac{13.3}{44} = 0.302$	$\frac{0.302}{3.357} = 0.090$	9.0
CO	0.95	28	$\frac{0.95}{28} = 0.034$	$\frac{0.034}{3.357} = 0.010$	1.0
O_2	8.35	32	$\frac{8.35}{32} = 0.261$	$\frac{0.261}{3.357} = 0.078$	7.8
N_2	77.4	28	$\frac{77.4}{28} = 2.76$	$\frac{2.76}{3.357} = 0.822$	82.2
Total	100.0		$\Sigma(c) = 3.357$	1.000	100.0

The fuel gas composition by volume, is given in the last column, *i.e.*

$CO_2 = 9\ \%$; $CO = 1\ \%$; $O_2 = 7.8\ \%$ and $N_2 = 82.2\ \%$ **Ans.**

12.13. Mass of Carbon in Flue Gases

The mass of carbon, contained in 1 kg of flue or exhaust gases, may be calculated from the mass of carbon dioxide and carbon monoxide present in them.

We know that 1 kg of carbon produces 11/3 kg of carbon dioxide. Hence 1 kg of carbon dioxide will contain 3/11 kg of carbon. Also 1 kg of carbon produces 7/3 kg of carbon monoxide, hence 1 kg of carbon monoxide will contain 3/7 kg of carbon.

∴ Mass of carbon per kg of flue gas

$$= \frac{3}{11} CO_2 + \frac{3}{7} CO$$

where CO_2 and CO represent the quantities of carbon dioxide and carbon monoxide present in 1 kg of flue gases.

12.14. Mass of Flue Gases per kg of Fuel Burnt

The mass of dry flue gases may be obtained by comparing the mass of carbon present in the flue gases with the mass of carbon in the fuel, since there is no loss of carbon during combustion.

Mathematically, mass of flue gas per kg of fuel

$$= \frac{\text{Mass of carbon in 1 kg of fuel}}{\text{Mass of carbon in 1 kg of flue gas}}$$

Example 12.5. *The ultimate analysis of dry coal burnt in a boiler is C 84%, H_2 9% and incombustibles 7% by mass Determine the mass of dry flue gases per kg of coal burnt, if the volumetric composition of the flue gas is : CO_2 8.75 %, CO 2.25 %, O_2 8% and N_2 81 %.*

Solution. Given : C = 84% = 0.84 kg ; H_2 = 9% = 0.09 kg ; Incombustibles = 7% = 0.07 kg ; CO_2 = 8.75 % = 0.0875 m^3 ; CO_2 = 2.25 % = 0.0225 m^3 ; O_2 = 8% = 0.08 m^3 ; N_2 = 81 % = 0.81 m^3

First of all, let us convert the volumetric analysis of flue gas to mass analysis for the mass of carbon present in 1 kg of flue gas as given in the following table :

Constituent	*Volume in 1 m³ of flue gas* (a)	*Molecular mass* (b)	*Proportional mass* (c) = a × b	*Mass of constituent in kg per kg of flue gas* $d = \frac{(c)}{\Sigma(c)}$
CO_2	0.0875	44	3.85	$\frac{3.85}{29.72} = 0.130$
CO	0.0225	28	0.63	$\frac{0.63}{29.72} = 0.021$
O_2	0.08	32	2.56	$\frac{2.56}{29.72} = 0.086$
N_2	0.81	28	22.68	$\frac{22.68}{29.72} = 0.763$
Total	1.0000		Σ(c) = 29.72	1.000

We know that mass of carbon in 1 kg of flue gas

$$= \frac{3}{11}CO_2 + \frac{3}{7}CO = \frac{3}{11} \times 0.130 + \frac{3}{7} \times 0.021 = 0.044 \text{ kg}$$

∴ Mass of dry flue gas per kg of coal burnt

$$= \frac{\text{Mass of carbon in 1 kg of coal}}{\text{Mass of carbon in 1 kg flue gas}} = \frac{0.84}{0.044} = 19.1 \text{ kg} \quad \textbf{Ans.}$$

Example 12.6. *A blast furnace gas has the following volumetric composition :*
CO_2 = 11% ; CO = 27% ; H_2 = 2% and N_2 = 60%.

Find the theoretical volume of air required for the complete combustion of 1 m³ of the gas. Find the percentage composition of dry flue gases by volume. Assume that air contains 21 % of O_2 and 79% of N_2 by volume.

Solution. Given : CO_2 = 11% = 0.11 m³ ; CO = 27% = 0.27 m³ ; H_2 = 2% = 0.02 m³ ; N_2 = 60% = 0.6 m³

Theoretical volume of air required

We know that theoretical volume of air required

$$= \frac{100}{21}\left[(0.5\,CO + 0.5\,H_2 + 2\,CH_4 + 3\,C_2H_4) - O_2\right] m^3$$

$$= \frac{100}{21}[0.5 \times 0.27 + 0.5 \times 0.02] = 0.69 \text{ m}^3 \quad \textbf{Ans.}$$

. . . (∵ CH_4, C_2H_4 and O_2 are equal to zero)

Percentage composition of dry flue gases

We know that 1 m³ of CO produces 1 m³ of CO_2, therefore

Volume of CO_2 obtained from the given 0.27 m³ of CO

$$= 0.27 \text{ m}^3$$

and volume of CO_2 already present in the fuel,

$$= 0.11 \text{ m}^3$$ . . . (Given)

∴ Total volume of CO_2 in the flue gas

$$= 0.11 + 0.27 = 0.38 \text{ m}^3$$

Now volume of N_2 from the theoretical air supplied

$$= \frac{79}{100} \times 0.69 = 0.545 \text{ m}^3$$

and volume of N_2 already present in the fuel

$$= 0.6 \text{ m}^3 \quad \text{. (Given)}$$

∴ Total volume of N_2 in the flue gas

$$= 0.6 + 0.545 = 1.145 \text{ m}^3$$

and total volume of the dry flue gas

$$= \text{Total volume of } CO_2 + \text{Total volume of } N_2$$

$$= 0.38 + 1.145 = 1.525 \text{ m}^3$$

∴ Percentage of CO_2 in the dry flue gas

$$= \frac{0.38}{1.525} \times 100 = 24.92\% \text{ Ans.}$$

and percentage of N_2 in the dry flue gas

$$= \frac{1.145}{1.525} \times 100 = 75.08\ \% \text{ Ans.}$$

12.15. Excess Air Supplied

In the last two articles, we have discussed the theoretical or minimum air required for complete combustion. But in actual practice, to ensure complete and rapid combustion of a fuel, some quantity of air, in excess of the theoretical or minimum air, is supplied. It is due to the fact, that the excess air does not come in contact with the fuel particles. If just minimum amount of air is supplied, a part of the fuel may not burn properly.

The amount of excess air supplied varies with the type of fuel and firing conditions. It may approach to a value of 100 per cent, but the modern tendency is to use 25 to 50 per cent excess air.

12.16. Mass of Excess Air Supplied

The mass of excess air supplied may be determined by the mass of unused oxygen, found in the flue gases. We know that in order to supply one kg of oxygen, we need 100/23 kg of air.

Similarly, mass of excess air supplied

$$= \frac{100}{23} \times \text{Mass of excess oxygen}$$

∴ Total mass of air supplied

$$= \text{Mass of necessary air} + \text{Mass of excess air}$$

Notes : 1. The mass of excess air supplied per kg of fuel may approximately be obtained from the following formula :

*Excess air supplied per kg of fuel

$$= \frac{79 \times O_2 \times C}{21 \times 33\ (CO_2 + CO)}$$

where O_2, CO_2 and CO represents the percentages of oxygen, carbon dioxide and carbon monoxide respectively in flue gases (by volume) in the percentage of carbon in fuel by mass.

---

* The excess air supplied per kg of fuel may also be obtained as discussed in Example 12.9 (Note 2).

2. The amount of air actually required per kg of fuel may approximately be obtained from the following formula :

Actual air required per kg of fuel

$$= \frac{N_2 \times C}{33\ (CO_2 + CO)}$$

where N_2 is percentage of nitrogen in the flue gases by volume.

3. The richness of air-fuel mixture supplied to an internal combustion engine is measured by the ratio of the air to the fuel by mass.

$\therefore$ Air-fuel ratio = Mass of air supplied per kg of fuel.

Example 12.7. *A steam boiler uses pulverised coal in the furnace. The ultimate analysis of coal (by mass) as received is :*

C 78% ; H_2 3% ; O_2 3% ; S 1% ; ash 10% ; and moisture 5%.

Excess air supplied is 30%. Calculate the mass of air to be supplied and mass of gaseous product formed per kg of coal burnt.

Solution. Given : C = 78% = 0.78 kg ; O_2 = 3% = 0.03 kg ; S = 1% = 0.01 kg ; Ash = 10% = 0.1 kg ; Moisture = 5% = 0.05 kg ; Excess air supplied = 30%

Mass of air to be supplied per kg of coal burnt

We know that theoretical air required to burn 1 kg of coal

$$= \frac{100}{23}\left[\left(\frac{8}{3}C + 8H_2 + S\right) - O_2\right] \text{kg}$$

$$= \frac{100}{23}\left[\left(\frac{8}{3} \times 0.78 + 8 \times 0.03 + 0.01\right) - 0.03\right] = 10.1 \text{ kg}$$

$\therefore$ Excess air supplied per kg of coal

$$= \frac{30 \times 10.1}{100} = 3.03 \text{ kg}$$

and mass of air to be supplied per kg of coal burnt

$$= \text{Theoretical air} + \text{Excess air}$$

$$= 10.1 + 3.03 = 13.13 \text{ kg} \quad \textbf{Ans.}$$

Mass of gaseous products formed per kg of coal

The gaseous products formed are carbon dioxide (CO_2), water (H_2O), sulphur dioxide (SO_2), excess oxygen (O_2) and nitrogen (N_2). We know that 1 kg of carbon produces 11/3 kg of carbon dioxide, 1 kg of hydrogen produces 9 kg of water and 1 kg of sulphur produces 2 kg of sulphur dioxide.

$\therefore$ Mass of CO_2 produced per kg of coal

$$= \frac{11}{3} \times 0.78 = 2.86 \text{ kg} \qquad \ldots (i)$$

Mass of H_2O produced per kg of coal

$$= 9 \times 0.03 = 0.27 \text{ kg} \qquad \ldots (ii)$$

Mass of SO_2 produced per kg of coal

$$= 2 \times 0.01 = 0.02 \text{ kg} \qquad \ldots (iii)$$

Mass of excess O_2 produced per kg of coal

$$= \frac{23}{100} \times \text{Excess air supplied} = \frac{23}{100} \times 3.03 = 0.70 \text{ kg} \qquad \ldots (iv)$$

Mass of N_2 per kg of coal

$$= \frac{77}{100} \times \text{Actual mass of air supplied} = \frac{77}{100} \times 13.13 = 10.1 \text{ kg} \quad \ldots (v)$$

Hence the mass of the gaseous products of combustion per kg of coal are : CO_2 = 2.86 kg, H_2O = 0.27 kg, SO_2 = 0.02 kg, excess O_2 = 0.70 kg and N_2 = 10.1 kg. **Ans.**

Example 12.8. *A gas fuel has the following percentage composition by volume :*

CO 10% ; H_2 50% ; CH_4 26% ; O_2 3% ; CO_2 2% ; and N_2 9%.

Estimate the minimum volume of air required for complete combustion of 1 m³ of the gas. If 50% excess air is supplied, give the volumes of each of the dry constituents of the flue gas. Air contains 21% by volume of oxygen.

Solution. Given : CO = 10% = 0.1 m³ ; H_2 = 50% = 0.5 m³ ; CH_4 = 26% = 0.26 m³ ; O_2 = 3% = 0.03 m³ ; CO_2 = 2% = 0.02 m³ ; N_2 = 9% = 0.09 m³ ; Excess air supplied = 50%

Minimum volume of air required for complete combustion

We know that minimum volume of air required

$$= \frac{100}{21} [(0.5 \, CO + 0.5 \, H_2 + 2CH_4) - O_2] \text{ m}^3$$

$$= \frac{100}{21} [(0.5 \times 0.1 + 0.5 \times 0.5 + 2 \times 0.26) - 0.03] = 3.76 \text{ m}^3 \quad \textbf{Ans.}$$

Volume of dry constituents of the flue gas

Since 50% excess air is supplied, therefore actual amount of air supplied

$$= 3.76 \times 1.5 = 5.64 \text{ m}^3$$

∴ Excess air supplied

$$= 5.64 - 3.76 = 1.88 \text{ m}^3$$

and excess oxygen supplied

$$= 1.88 \times \frac{21}{100} = 0.395 \text{ m}^3$$

The dry constituents of the flue gas contain CO_2 from the combustion of fuel as well as from the combustion of CO and CH_4, excess O_2 and N_2 from the combustion of the fuel as well as from actual air supplied.

Now volume of CO_2 contained in the fuel (as given)

$$= 0.02 \text{ m}^3 \quad \ldots (i)$$

We know that 1 m³ of CO produces 1 m³ of CO_2. Therefore volume of CO_2 obtained from 0.1 m³ of CO

$$= 0.10 \text{ m}^3 \quad \ldots (ii)$$

We also know that 1 m³ of CH_4 produces 1 m³ of CO_2. Therefore volume of CO_2 obtained from 0.26 m³ of CH_4

$$= 0.26 \text{ m}^3 \quad \ldots (iii)$$

∴ Total volume of CO_2 obtained from the flue gas

$$= 0.02 + 0.10 + 0.26 = 0.38 \text{ m}^3$$

Volume of excess O_2 in flue gas

$$= 0.395 \text{ m}^3 \quad \ldots \text{(As above)}$$

Volume of N_2 in the fuel $= 0.09 \text{ m}^3$

Volume of N_2 from the actual air supplied

$$= \frac{79}{100} \times \text{Volume of actual air} = \frac{79}{100} \times 5.64 = 4.46 \text{ m}^3$$

$\therefore$ Total volume of N_2 in the flue gas

$$= 0.09 + 4.46 = 4.55 \text{ m}^3$$

Hence the volume of dry constituents of the flue gas are :

*$CO_2 = 0.38 \text{ m}^3$, Excess $O_2 = 0.395 \text{ m}^3$, and $N_2 = 4.55 \text{ m}^3$ **Ans.**

Example 12.9. *During a boiler trial, the dry flue gas analysis by volume was reported as*

CO_2 13% : CO 0.3% : O_2 6% ; N_2 80.7%

The coal analysis by mass was reported as

C 62.4% ; H_2 4.2% ; O_2 4.5% ; moisture 15% ; ash 13.9%.

Calculate : 1. Minimum air required to burn 1 kg of coal, 2. Mass of air actually supplied per kg of coal, 3. The amount of excess air supplied per kg of coal burnt.

Solution. Given : For dry flue gas (by volume) : CO_2 =13% = 0.13 m³ ; CO = 0.3% = 0.003 m³ ; O_2 = 6% = 0.06 m³ ; N_2 = 80.7% = 0.807 m³ ; For coal (by mass) : C = 62.4% = 0.624 kg ; H_2 = 4.2% = 0.042 kg ; O_2 = 4.5% = 0.045 kg ; Moisture 15% = 0.15 kg ; Ash 13.9% = 0.139 kg

1. Minimum air required to burn 1 kg of coal

We know that minimum air required to burn 1 kg of coal

$$= \frac{100}{23}\left[\left(\frac{8}{3}C + 8H_2 + S\right) - O_2\right] \text{kg}$$

$$= \frac{100}{23}\left[\left(\frac{8}{3} \times 0.624 + 8 \times 0.042\right) - 0.045\right] = 8.5 \text{ kg} \quad \textbf{Ans.}$$

2. Mass of air actually supplied per kg of coal

First of all, let us convert the volumetric analysis of flue gas to mass analysis as given in the following table :

Constituent	*Volume in 1 m³ of flue gas* (a)	*Molecular mass* (b)	*Proportional mass in kg* (c) = a × b	*Mass of constituent in kg per kg of flue gas* (d) = $\frac{(c)}{\Sigma(c)}$
CO_2	0.13	44	5.720	$\frac{5.720}{30.319} = 0.189$
CO	0.003	28	0.084	$\frac{0.084}{30.319} = 0.003$
O_2	0.06	32	1.920	$\frac{1.920}{30.319} = 0.063$
N_2	0.807	28	22.595	$\frac{22.595}{30.319} = 0.745$
Total	1.000		$\Sigma(c) = 30.319$	1.000

* It may be noted that carbon dioxide and nitrogen do not require any oxygen.

We know that mass of carbon in 1 kg of flue gas

$$= \frac{3}{11} CO_2 + \frac{3}{7} CO = \frac{3}{11} \times 0.189 + \frac{3}{7} \times 0.003 = 0.053 \text{ kg}$$

∴ Total mass of flue gas per kg of fuel burnt

$$= \frac{\text{Mass of carbon in 1 kg of coal}}{\text{Mass of carbon in 1 kg of flue gas}} = \frac{0.624}{0.053} = 11.77 \text{ kg}$$

and total mass of N_2 present in 11.77 kg of flue gas

$$= 0.745 \times 11.77 = 8.77 \text{ kg / kg of coal}$$

Since the given coal does not contain nitrogen, therefore the nitrogen is supplied by air only.

∴ Mass of air actually supplied per kg of coal

$$= \frac{100}{77} \times 8.77 = 11.4 \text{ kg} \quad \textbf{Ans.}$$

3. *Amount of excess air supplied per kg of fuel*

We know that the amount of excess air supplied per kg of fuel

= Mass of air actually supplied – Minimum air

$= 11.4 - 8.5 = 2.9$ kg **Ans.**

Notes : 1. The mass of air actually supplied and the amount of excess air supplied may also be obtained as discussed below :

We know that mass of air actually supplied per kg of coal

$$= \frac{N_2 \times C}{33\,(CO_2 + CO)} = \frac{80.7 \times 62.4}{33\,(13 + 0.3)} = 11.47 \text{ kg} \quad \textbf{Ans.}$$

and the amount of excess air supplied per kg of coal

$$= \frac{79 \times O_2 \times C}{21 \times 33\,(CO_2 + CO)} = \frac{79 \times 6 \times 62.4}{21 \times 33\,(13 + 0.3)} = 3.2 \text{ kg} \quad \textbf{Ans.}$$

2. The amount of excess air supplied per kg of coal may also be obtained by the following two methods :

(*a*) The oxygen in the flue gas is unused oxygen.

∴ Mass of unused O_2 per kg of coal

$$= 0.063 \times \text{Total mass of flue gas} = 0.063 \times 11.77 = 0.7415 \text{ kg}$$

and mass of CO per kg of coal $= 0.003 \times 11.77 = 0.0353$ kg

We know that 1 kg of CO requires 4/7 kg of oxygen and produces 1/7 kg of CO_2. Therefore Oxygen (O_2) required to burn CO

$$= 0.0353 \times \frac{4}{7} = 0.0202 \text{ kg}$$

∴ Excess $O_2 = 0.7415 - 0.0202 = 0.7213$ kg

and excess air $= 0.7213 \times \frac{100}{23} = 3.136$ kg **Ans.**

(*b*) We know that oxygen in the flue gas is 6% by volume.

∴ Amount of N_2 associated with this oxygen

$$= 6 \times \frac{79}{21} = 22.57\%$$

N_2 already present in the flue gas

= 80.7%

$$\therefore \text{Excess air} = \frac{22.57}{80.7} \times \text{Mass of actual air supplied}$$

$$= \frac{22.57}{80.7} \times 11.4 = 3.2 \text{ kg} \quad \textbf{Ans.}$$

Example 12.10. *The percentage composition by mass of a sample of coal as found by analysis is given as :*

C 90, H_2 3.3, O_2 3.0, N_2 0.8, S 0.9 and ash 2.0.

Calculate the minimum mass of air required for the complete combustion of 1 kg of this fuel.

If 50% excess air is supplied, find the total mass of dry flue gases per kg of fuel and the percentage composition of the dry flue gases by volume.

Solution. Given : C = 90% = 0.9 kg ; H_2 = 3.3% = 0.033 kg ; O_2 = 3% = 0.03 kg ; N = 0.8% = 0.008 kg ; S = 0.9% = 0.009 kg ; Ash = 2% = 0.02 kg ; Excess air supplied = 50 %

Minimum mass of air required for complete combustion

We know that minimum mass of air required for complete combustion of 1 kg of fuel

$$= \frac{100}{23}\left[\left(\frac{8}{3}C + 8\,H_2 + S\right) - O_2\right] \text{kg}$$

$$= \frac{100}{23}\left[\left(\frac{8}{3} \times 0.9 + 8 \times 0.033 + 0.009\right) - 0.03\right] = 11.5 \text{ kg} \quad \textbf{Ans.}$$

Total mass of dry flue gases per kg of fuel

Since 50% excess air is supplied, therefore actual amount of air supplied per kg of coal

$$= 11.5 \times 1.5 = 17.25 \text{ kg}$$

$$\therefore \text{Excess air supplied} = 17.25 - 11.5 = 5.75 \text{ kg}$$

The products of combustion are represented by the following chemical equations :

(*i*) $C + O_2 = CO_2$

(*ii*) $2H_2 + O_2 = 2H_2O$

(*iii*) $S + O_2 = SO_2$

In addition to carbon dioxide, water and sulphur dioxide, the excess oxygen and nitrogen will be available in the products of combustion. It may be noted that H_2O (water vapour) is a wet gas, therefore the dry flue gases are only carbon dioxide, sulphur dioxide, excess oxygen and nitrogen. Let us now find the mass of each of these flue gases per kg of fuel.

We know that 1 kg of carbon produces 11/3 kg of carbon dioxide and 1 kg of sulphur produces 2 kg of sulphur dioxide.

$\therefore$ Mass of CO_2 contained in 0.9 kg of carbon per kg of fuel

$$= \frac{11}{3} \times 0.9 = 3.3 \text{ kg} \quad \ldots(i)$$

and mass of SO_2 in 0.009 kg of sulphur per kg of fuel

$$= 2 \times 0.009 = 0.018 \text{ kg} \quad \ldots(ii)$$

We also know that the mass of excess O_2 per kg of fuel

$$= \frac{23}{100} \times \text{Excess air supplied} = \frac{23}{100} \times 5.75 = 1.323 \text{ kg} \quad \ldots(iii)$$

and mass of nitrogen in the products of combustion per kg of fuel

$$= \frac{77}{100} \times \text{Actual air supplied}$$

$$= \frac{77}{100} \times 17.25 = 13.283 \text{ kg} \quad \ldots (iv)$$

$\therefore$ Total mass of dry flue gases per kg of fuel

$$= 3.3 + 0.018 + 1.323 + 13.283 \text{ kg} = 17.924 \text{ kg} \textbf{ Ans.}$$

Percentage composition of dry flue gases by volume

First of all, let us find out the percentage composition of the dry flue gases from the above data by mass. We know that the composition of :

$$CO_2 = \frac{3.3}{17.924} = 0.184 = 18.4\%$$

$$SO_2 = \frac{0.018}{17.924} = 0.001 = 0.1\%$$

$$\text{Excess } O_2 = \frac{1.323}{17.924} = 0.074 = 7.4\%$$

$$N_2 = \frac{13.283}{17.924} = 0.741 = 74.1\%$$

Now let us convert this mass analysis of dry flue gases into volumetric analysis as given below :

Constituent	*% Mass analysis (a)*	*Molecular mass (b)*	*Proportional volume* $(c) = \frac{(a)}{(b)}$	*Volume in 1 m³ of flue gas* $(d) = \frac{(c)}{\Sigma(c)}$	*% Volumetric analysis* $(d) \times 100$
CO_2	18.4	44	$\frac{18.4}{44} = 0.418$	$\frac{0.418}{3.2966} = 0.1268$	12.68
SO_2	0.1	64	$\frac{0.1}{64} = 0.0016$	$\frac{0.0016}{3.2966} = 0.0005$	0.05
Excess O_2	7.4	32	$\frac{7.4}{32} = 0.231$	$\frac{0.231}{3.2966} = 0.0701$	7.01
N_2	74.1	28	$\frac{74.1}{28} = 2.646$	$\frac{2.646}{3.2966} = 0.8026$	80.26
Total	100.0		$\Sigma(c) = 3.2966$		100.00

The percentage composition of dry flue gases (by volume) is given in last column, *i.e.*

$CO_2 = 12.68\%$, $SO_2 = 0.06\%$; Excess $O_2 = 7.01\%$; and $N_2 = 80.25\%$. **Ans.**

Example 12.11. *A fuel oil has following analysis by mass :*

C 85%, H_2 12.5%, O_2 2% and the residue 0.5%.

The dry flue has the following composition by volume :

CO_2 9%, CO 1 %, O_2 7.77% and N_2 82.23 %.

Determine the air fuel ratio.

Solution. Given : For fuel oil (by mass) C = 85% = 0.85 kg ; H_2 = 12.5% = 0.125 kg ; O_2 = 2% = 0.02 kg ; Residue = 0.5% = 0.005 kg

For dry flue (by volume) : $CO_2 = 9\% = 0.09$ m^3 ; CO = 1 % = 0.01 m^3 ; $O_2 = 7.77\% = 0.0777$ m^3 ; $N_2 = 82.23\ \% = 0.8223$ m^3

We know that minimum air required per kg of fuel

$$= \frac{100}{23}\left[\left(\frac{8}{3}C + 8H_2 + S\right) - O_2\right] \text{kg}$$

$$= \frac{100}{23}\left[\left(\frac{8}{3} \times 0.85 + 8 \times 0.125\right) - 0.02\right] = 14.1 \text{ kg} \qquad \ldots (\because S = 0)$$

First of all, let us convert the volumetric analysis of dry flue gas into mass analysis as given in the following table.

Constituent	*Volume in 1 m^3 of the flue gas* (a)	*Molecular mass* (b)	*Proportional mass* (c) = (a) × (b)	*Mass in 1 kg per kg of flue gas* $(d) = \frac{(c)}{\Sigma(c)}$	*% Mass analysis* = (d) × 100
CO_2	0.09	44	3.96	$\frac{3.96}{29.75} = 0.133$	13.3
CO	0.01	28	0.28	$\frac{0.28}{29.75} = 0.009$	0.9
O_2	0.0777	32	2.49	$\frac{2.49}{29.75} = 0.084$	8.4
N_2	0.8223	28	23.02	$\frac{23.02}{29.75} = 0.774$	77.4
Total	1.0000		$\Sigma(c) = 29.75$	1.000	100.00

We know that 1 kg of CO_2 contains 3/11 kg of carbon and 1 kg of CO contains 3/7 kg of carbon. Therefore, mass of carbon per kg of flue gas

$$= \frac{3}{11}CO_2 + \frac{3}{7}CO = \frac{3}{11} \times 0.133 + \frac{3}{7} \times 0.009 = 0.04 \text{ kg}$$

We also know that the mass of flue gas per kg of fuel

$$= \frac{\text{Mass of carbon in 1 kg of fuel}}{\text{Mass of carbon in 1 kg of flue gas}} = \frac{0.85}{0.04} = 21.25 \text{ kg}$$

Since 1 kg of CO requires 4/7 kg of oxygen, therefore mass of excess oxygen per kg of flue gas

$$= \text{Oxygen already present in flue gas} - \frac{4}{7}CO$$

$$= 0.084 - \frac{4}{7} \times 0.009 = 0.079 \text{ kg}$$

and mass of excess oxygen per kg of fuel

$$= \text{Mass of excess } O_2 \text{ per kg of the gas} \times \text{Mass of flue gas per kg of fuel}$$

$$= 0.079 \times 21.25 = 1.679 \text{ kg}$$

∴ Mass of excess air per kg of fuel

$$= \frac{100}{23} \times \text{Mass of excess oxygen} = \frac{100}{23} \times 1.679 = 7.3 \text{ kg}$$

Total air required per kg of fuel

$$= \text{Minimum air} + \text{Excess air} = 14.1 + 7.3 = 21.4 \text{ kg}$$

∴ Air-fuel ratio $= 21.4 : 1$ **Ans.**

Example 12.12. *A sample of coal with C = 0.78 ; H_2 = 0.05 ; O_2 = 0.08 ; S = 0.02 ; N_2 = 0.02 and ash = 0.05 is burnt in a furnace with 50% excess air. The flue gases enter the chimney at 325° C and the atmospheric temperature is 15° C. Take c_p for O_2, N_2 and air = 1.008 kJ/kg K and c_p for CO_2 and SO_2 from the flue gas = 1.05 kJ/kg K. Assume that the heat carried away per kg of moisture in flue gases is 2940 kJ. Calculate the quantity of heat carried away by the flue gases in kJ/kg of coal.*

Solution. Given : *C = 0.78 kg ; H_2 = 0.05 kg ; O_2 = 0.08 kg ; S = 0.02 kg ; N_2 = 0.02 kg , Ash = 0.05 kg ; Excess air = 50% ; t_g = 325° C ; t_a = 15° C ; c_p for O_2, N_2 and air = 1.008 kJ/kg K ; c_p for CO_2 and SO_2 = 1.05 kJ/kg K ; Heat carried away per kg of moisture = 2940 kJ

First of all, let us find the mass of the flue gases produced per kg of coal.

We know that minimum air required to burn 1 kg of coal

$$= \frac{100}{23}\left[\left(\frac{8}{3}C + 8H_2 + S\right) - O_2\right] \text{kg}$$

$$= \frac{100}{23}\left[\left(\frac{8}{3} \times 0.78 + 8 \times 0.05 + 0.02\right) - 0.08\right] = 10.52 \text{ kg}$$

∴ Excess air supplied per kg of coal

$$= \frac{50}{100} \times 10.52 = 5.26 \text{ kg}$$

and mass of air to be supplied per kg of coal burnt

$$= \text{Minimum air} + \text{Excess air} = 10.52 + 5.26 = 15.78 \text{ kg}$$

The flue gases produced by the combustion of coal are CO_2, H_2O, SO_2, excess O_2 and N_2.

We know that 1 kg of carbon produces 11/3 kg of carbon dioxide (CO_2) ; 1 kg of hydrogen produces 9 kg of water (H_2O) and 1 kg of sulphur produces 2 kg of sulphur dioxide (SO_2).

∴ Mass of CO_2 produced by 0.78 kg of carbon per kg of coal

$$= \frac{11}{3} \times 0.78 = 2.86 \text{ kg}$$

Mass of H_2O produced by 0.05 kg of hydrogen per kg of coal

$$= 9 \times 0.05 = 0.45 \text{ kg}$$

Mass of SO_2 produced by 0.02 kg of sulphur per kg of coal

$$= 2 \times 0.02 = 0.04 \text{ kg}$$

Mass of excess O_2 produced per kg of coal

$$= \frac{23}{100} \times \text{Excess air supplied} = \frac{23}{100} \times 5.26 = 1.21 \text{ kg}$$

and mass of N_2 produced per kg of coal

$$= \frac{77}{100} \times \text{Actual mass of air supplied} = \frac{77}{100} \times 15.78 = 12.15 \text{ kg}$$

---

* A sample of 1 kg of coal is considered.

We know that heat carried away by CO_2

$$= \text{Mass} \times \text{Specific heat} \times \text{Rise in temperature}$$

$$= 2.86 \times 1.05\,(325 - 15) = 931 \text{ kJ/kg}$$

Heat carried away by SO_2

$$= 0.04 \times 1.05 \times (325 - 15) = 13.02 \text{ kJ/kg}$$

Heat carried away by excess O_2

$$= 1.21 \times 1.008\,(325 - 15) = 378 \text{ kJ/kg}$$

Heat carried away by N_2 $= 12.15 \times 1.008\,(325 - 15) = 3797$ kJ/kg

Since the heat carried away by moisture is given as 2940 kJ/kg, therefore

Heat carried away by H_2O

$$= 0.45 \times 2940 = 1323 \text{ kJ/kg}$$

$\therefore$ Total heat carried away by flue gases

$$= 931 + 13.02 + 378 + 3797 + 1323 = 6442.02 \text{ kJ/kg of coal}$$ **Ans.**

12.17. Flue Gas Analysis by Orsat Apparatus

To check the combustion efficiency of boilers, it is considered essential to determine the constituents of the flue gases. Such an analysis is carried out with the help of Orsat apparatus as shown in Fig. 12.1.

It consists of a graduated measuring glass tube (known as eudiometer tube) and three flasks *A*, *B* and *C*, each containing different chemicals for absorbing carbon dioxide, carbon monoxide and oxygen. An aspirator bottle containing water is connected to the bottom of the eudiometer tube by means of a rubber tube. It can be moved up and down, at will, for producing a suction or pressure effect on the sample of the flue gas.

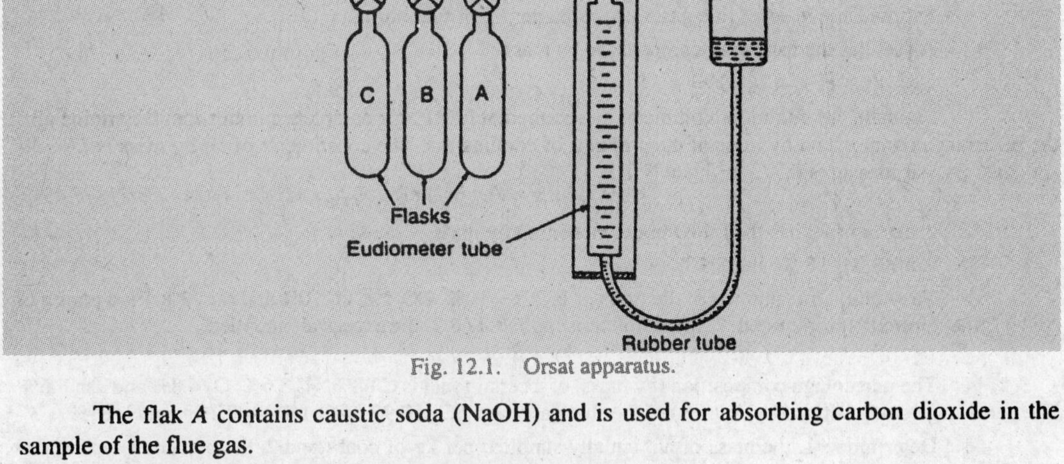

Fig. 12.1. Orsat apparatus.

The flak *A* contains caustic soda (NaOH) and is used for absorbing carbon dioxide in the sample of the flue gas.

The flask *B* contains caustic soda (NaOH) and pyrogallic acid, which absorbs oxygen from the sample of the flue gas.

The flak *C* contains a solution of cuprous chloride (Cu_2Cl_2) in hydrochloric acid (HCl). It absorbs carbon monoxide (CO) from the sample of the flue gas.

Each of the three flasks has stop cock '*a*', '*b*' and '*c*' respectively and a three-way cock '*d*' which can be opened to either atmosphere or flue gas.

The sample of the flue gas to be analysed is first sucked in the eudiometer tube, and its volume is noted. It can take usually 100 cm^3 of gas. By manipulating the level of aspirator bottle, the flue gas can, in turn, be forced into either of the flasks *A*, *B* and *C* by opening the respective cocks *a*, *b* and *c*. The flue gas is left there for sometime and then sucked back into eudiometer tube. The chemicals, in the three flasks absorb carbon dioxide, and carbon monoxide and the resulting contraction in volume enables the percentage of each gas present in the sample to be read on the eudiometer tube.

Since the flue is collected over water in the tube, therefore any steam present in it will be condensed. Similarly, the sulphur dioxide present in it will also be absorbed. Hence the percentages of dry flue gases are only obtained by Orsat apparatus.

EXERCISES

1. The percentage composition of a sample of coal is found by analysis as : C 91%, H_2 3%, O_2 = 2%, N_2 0.8%, S 0.8% and the remaining is ash.

   Calculate the minimum mass of air for complete combustion of 1 kg of coal. [Ans. 11.5 kg]

2. A sample of coal was found on analysis to have following composition by mass :

   Carbon 72.2%, hydrogen 3.2%, oxygen 18.5%, sulphur 2.4%, and the remainder being incombustible matter.

   Calculate the theoretical mass of air required for the complete combustion of 1 kg of this coal. [Ans. 8.51 kg]

3. A coal contains by mass 81% carbon, 6% hydrogen and the remainder ash. Find : 1. minimum mass of air required to burn 1 kg of coal, and 2. the mass of produces of combustion. [Ans. 11.48 kg ; CO = 2.97 kg, H_2O = 0.54 kg]

4. The volumetric analysis of a flue gas is CO_2 15% ; CO 2.2% ; O_2 1.6% and N_2 81.2%. Convert this volumetric analysis into percentage analysis by mass. [Ans. CO_2 21.6% ; CO 2.2 % ; O_2 1.7% ; N_2 74.5%]

5. The percentage composition of a certain fuel by mass is

   C = 87.1 ; H = 4.4 ; O = 1.2 and ash 7.3.

   The percentage volumetric composition of dry flue gases is

   CO_2 = 15 ; CO = 2.2 ; O = 1.6 ; N_2 = 81.2.

   Estimate the mass of flue gases produced per kg of fuel burnt. [Ans. 12.85 kg]

6. A fuel has the following composition by mass :

   C = 86 ; H = 12 ; O = 1 ; S = 1.

   Estimate the minimum volume of air required at N.T.P. for complete combustion. Determine also the percentage composition by mass of the products of combustion. The constituents of air by mass is 77% N_2 and 23% O_2. Air measures 0.773 m^3/kg at N.T.P.
   [Ans. 10.93 m^3/kg of fuel ; CO_2 22.42%, SO_2 0.142%, N_2 77.43%]

7. A certain fuel has the following composition by mass :

   C 80%, H_2 10 %, and S 10%.

   The volumetric analysis of the fuel gas is : CO_2 10%, CO 1%, O_2 10% and N_2 79%. Find per kg of coal : 1. the minimum air required, 2. the actual air supplied, and 3. the excess air supplied.
   [Ans. 13.2 kg ; 17.4 kg ; 4.2 kg]

8. The percentage composition (by mass) of a certain fuel is C 88%, H_2 3.6%, O_2 4.8% and ash 3.6%. The percentage composition (by volume) of the flue gases are CO_2 10.9%, CO 1%, O_2 7.1% and N_2 81%.

   Determine : 1. the mass of air actually supplied per kg of coal ; and 2. the percentage excess air supplied. [Ans. 18.15 kg, 38.3%]

9. The mass analysis of a fuel is carbon 75%, hydrogen 8%, oxygen 6% and remainder incombustible. If 16 kg of air were supplied per kg of coal ; find ;

   1. The percentage of excess air ; and

   2. The percentage (by mass) of CO_2 in the dry products of combustion. Take the percentage of oxygen (by mass) in air as 23.1. [Ans. 43.6% ; 12.35%]

10. A producer gas has the following percentage analysis by volume :

H_2 15, CH_4 2, CO 20, CO_2 6, O_2 3 and N_2 54. If 50% of excess air is supplied for the combustion, determine :

1. the volume of air supplied per m^3 of the gas, and
2. the volumetric analysis of the dry products of combustion.

[**Ans.** 1.322 m^3 ; CO_2 14.3%, N_2 80.8%, O_2 4.7%]

11. The ultimate analysis (by mass) of a fuel used in a boiler is : carbon = 75% ; hydrogen = 14% and the remaining being incombustible. The air supplied is 52% in excess of that required for complete combustion. If 2.4% of carbon in the fuel is burnt only to CO and the rest to CO_2, estimate the volumetric analysis of dry flue gases. Air contains 23% by mass and 21% by volume of O_2.

[**Ans.** CO_2 8.93% ; CO 0.3% ; O_2 7.44% ; N_2 83.33%]

12. Calculate the stoichiometric air-fuel ratio for the combustion of a sample of dry coal of the following by mass:

C = 88% ; H_2 = 4% ; O_2 = 2.5% ; S = 0.5% and rest ash.

Also determine the volumetric composition of the dry products of combustion if 20% excess air is supplied. Assume that air contains 23.3% O_2 and rest N_2 by mass.

[**Ans.** 11.36 ; CO_2 15.84%, SO_2 0.03% ; O_2 3.56% ; N_2 80.57%]

13. A water gas at 21°C and 760 mm of Hg is burnt with dry air supplied at 21°C and 760 mm of Hg. The fuel gas composition is

CO_2 = 6% ; N_2 = 5.5% ; H_2 = 48 % ; O_2 = 0.5% ; CH_4 = 2% and CO = 38%.

The Orsat analysis of the flue gas showed CO_2 = 15.5% ; O_2 = 4.76% and CO = 0.2%. Find the percentage of excess air supplied for combustion and the volume of the flue gas at 232° C and 1.013 bar formed per m^3 of the fuel. [**Ans.** 29.52% ; 5.025 m^3]

14. The coal supplied to boiler furnace has the following composition by mass :

Carbon = 82% ; Hydrogen = 5% ; Oxygen = 7% ; Nitrogen = 1% and rest ash. The volumetric analysis of dry flue gas is found to be as

CO_2 = 10% ; CO = 1.3% ; O_2 = 7.5% and N_2 = 81.2.

If the flue gas temperature is 200° C and the boiler room temperature is 50°C, calculate 1. percentage of excess air supplied to the boiler furnace, and 2. heat carried away by the dry flue gas per kg of coal.

The specific heats at constant pressure for the gases are :

CO_2 = 0.882 kJ/kg K ; CO = 1.05 kJ/kg K ; O_2 = 1.05 kJ/kg K and N_2 = 1.025 kJ/kg K.

Assume that air contains 23% O_2 and 77% N_2 by mass. [**Ans.** 53% ; 2750 kJ]

QUESTIONS

1. Define the following terms :
   (*a*) element, (*b*) compound, (*c*) atom, (*d*) molecule,
   (*e*) atomic mass, and (*f*) molecular mass
2. What do you understand by 'minimum air' and 'excess air' in context of combustion ?
3. Fill in the following blanks :
   (*i*) 1 kg of carbon requires ...... kg of oxygen and produces ...... kg of carbon dioxide.
   (*ii*) 1 kg of carbon requires ...... kg of oxygen and produces ...... kg of carbon monoxide.
   (*iii*) 1 kg of carbon monoxide requires ...... kg of oxygen and produces ...... kg of carbon dioxide.
   (*iv*) 1 kg of sulphur requires ...... kg of oxygen and produces ...... kg of sulphur dioxide.
   (*v*) 1 kg of hydrogen requires .... kg of oxygen and produce .... kg of water.
   (*vi*) .... m^3 of hydrogen requires .... m^3 of oxygen and produces ...... m^3 of water.
   (*vii*) 1 m^3 of .... requires .... m^3 of oxygen and produces .... m^3 of CO_2 and .... m^3 of H_2O.

4. Give chemical reactions and numerical values for estimating the air requirement for complete combustion of coal.

5. Lay down the procedure for determination of minimum air required for complete combustion of coal.

6. Sketch and explain the use of Orsat apparatus used in determining the percentage of flue or exhaust gases. Does this help in controlling combustion ?

OBJECTIVE TYPE QUESTIONS

1. The smallest quantity of a substance, which can exist by itself in a chemically recongnizable form is known as

   (a) element (b) compound (c) atom (d) molecule

2. The molecular mass of oxygen is

   (a) 12 (b) 14 (c) 16 (d) 32

3. The molecular mass of nitrogen is ...... oxygen.

   (a) equal to (b) less than (c) more than

4. Which of the following has minimum molecular mass ?

   (a) Oxygen (b) Nitrogen (c) Hydrogen (d) Water

5. One kg of carbon monoxide (CO) requires 4/7 kg of oxygen and produces

   (a) 11/3 kg of CO_2 (b) 7/3 kg of CO (c) 11/7 kg of CO_2 (d) 8/3 kg of CO

6. One kg of carbon requires ...... of oxygen and produces 7/3 kg of carbon monoxide.

   (a) 4/3 (b) 7/3 (c) 8/3 (d) 11/3

7. One kg of ethylene (C_2H_4) requires 2 kg of oxygen and produces 22/7 kg of carbon dioxide and

   (a) 9/7 kg of water (b) 11/7 kg of water
   (c) 7/4 kg of water (d) 11/4 kg of water

8. The mass of carbon per kg of flue gas is given by

   (a) $\frac{3}{11} CO_2 + \frac{3}{7} CO$ (b) $\frac{3}{7} CO_2 + \frac{11}{3} CO$
   (c) $\frac{7}{3} CO_2 + \frac{3}{11} CO$ (d) $\frac{3}{11} CO_2 + \frac{7}{3} CO$

9. The mass of flue gas per kg of fuel is the ratio of the

   (a) mass of oxygen in 1 kg of flue gas to the mass of oxygen in 1 kg of fuel
   (b) mass of oxygen in 1 kg of fuel to the mass of oxygen in 1 kg of flue gas
   (c) mass of carbon in 1 kg of flue gas to the mass of carbon in 1 kg of fuel
   (d) mass of carbon in 1 kg of fuel to the mass of carbon in 1 kg of flue gas

10. The mass of excess air supplied is equal to

   (a) $\frac{23}{100} \times$ Mass of excess carbon (b) $\frac{23}{100} \times$ Mass of excess oxygen
   (c) $\frac{100}{23} \times$ Mass of excess carbon (d) $\frac{100}{23} \times$ Mass of excess oxygen

ANSWERS

1. (a)	2. (d)	3. (b)	4. (c)	5. (c)
6. (a)	7. (a)	8. (a)	9. (c)	10. (d)

13

Steam Boilers

1. Introduction. 2. Important Terms for Steam Boilers. 3. Essentials of a Good Steam Boiler. 4. Selection of a Steam Boiler. 5. Classifications of Steam Boilers. 6. Simple Vertical Boiler. 7. Cochran Boiler or Vertical Multi-tubular Boiler. 8. Scotch Marine Boiler. 9. Lancashire Boiler. 10. Cornish Boiler. 11. Locomotive Boiler. 12. Babcock and Wilcox Boiler. 13. La-Mont Boiler. 14. Loeffler Boiler. 15. Benson Boiler. 16. Comparison Between Water Tube and Fire Tube Boiler.

13.1. Introduction

A steam generator or boiler is, usually, a closed vessel made of steel. Its function is to transfer the heat produced by the combustion of fuel (solid, liquid or gaseous) to water, and ultimately to generate steam. The steam produced may be supplied :

1. to an external combustion engine, *i.e.* steam engines and turbines,
2. at low pressures for industrial process work in cotton mills, sugar factories, breweries, etc., and
3. for producing hot water, which can be used for heating installations at much lower pressures.

13.2. Important Terms for Steam Boilers

Though there are many terms used in steam boilers, yet the following are important from the subject point of view :

1. *Boiler shell.* It is made up of steel plates bent into cylindrical form and riveted or welded together. The ends of the shell are closed by means of end plates. A boiler shell should have sufficient capacity to contain water and steam.

2. *Combustion chamber.* It is the space, generally below the boiler shell, meant for burning fuel in order to produce steam from the water contained in the shell.

3. *Grate.* It is a platform, in the combustion chamber, upon which fuel (coal or wood) is burnt. The grate, generally, consists of cast iron bars which are spaced apart so that air (required for combustion) can pass through them. The surface area of the grate, over which the fire takes place, is called *grate surface*.

4. *Furnace.* It is the space, above the grate and below the boiler shell, in which the fuel is actually burnt. The furnace is also called *fire box.*

5. *Heating surface.* It is that part of boiler surface, which is exposed to the fire (or hot gases from the fire).

6. *Mountings.* These are the fittings which are mounted on the boiler for its proper functioning. They include water level indicator, pressure gauge, safety valve etc. It may be noted that a boiler cannot function safely without the mountings.

7. *Accessories.* These are the devices, which form an integral part of a boiler, but are not mounted on it. They include superheater, economiser, feed pump etc. It may be noted that the accessories help in controlling and running the boiler efficiently.

13.3. Essentials of a Good Steam Boiler

Following are the important essentials of a good steam boiler :

1. It should produce maximum quantity of steam with the minimum fuel consumption.
2. It should be economical to instal, and should require little attention during operation.
3. It should rapidly meet the fluctuation of load.
4. It should be capable of quick starting.
5. It should be light in weight.
6. It should occupy a small space.
7. The joints should be few and accessible for inspection.
8. The mud and other deposits should not collect on the heating plates.
9. The refractory material should be reduced to a minimum. But it should be sufficient to secure easy ignition, and smokeless combustion of the fuel on reduced load.
10. The tubes should not accumulate soot or water deposits, and should have a reasonable margin of strength to allow for wear or corrosion.
11. The water and flue gas circuits should be designed to allow a maximum fluid velocity without incurring heavy frictional losses.
12. It should comply with safety regulations as laid down in the Boilers Act.

13.4. Selection of a Steam Boiler

The selection of type and size of a steam boiler depends upon the following factors :

1. The power required and the working pressure.
2. The rate at which steam is to be generated.
3. The geographical position of the power house.
4. The fuel and water available.
5. The type of fuel to be used.
6. The probable permanency of the station.
7. The probable load factor.

13.5. Classifications of Steam Boilers

Though there are many classifications of steam boilers, yet the following are important from the subject point of view :

1. *According to the contents in the tube.* The steam boilers, according to the contents in the tube may be classified as :

(*a*) Fire tube or smoke tube boiler, and (*b*) Water tube boiler.

In *fire tube steam boilers*, the flames and hot gases, produced by the combustion of fuel, pass through the tubes (called multi-tubes) which are surrounded by water. The heat is conducted through the walls of the tubes from the hot gases to the surrounding water. Examples of fire tube boilers are : Simple vertical boiler, Cochran boiler, Lancashire boiler, Cornish boiler, Scotch marine boiler, Locomotive boiler, and Velcon boiler.

In *water tube steam boilers,* the water is contained inside the tubes (called water tubes) which are surrounded by flames and hot gases from outside. Examples of water tube boilers are : Babcock and Wilcox boiler, Stirling boiler, La-Mont boiler, Benson boiler, Yarrow boiler and Loeffler boiler.

2. *According to the position of the furnace.* The steam boilers, according to the position of the furnace are classified as :

(*a*) Internally fired boilers, and (*b*) Externally fired boilers.

In *internally fired steam boilers*, the furnace is located inside the boiler shell. Most of the fire tube steam boilers are internally fired.

In *externally fired steam boilers*, the furnace is arranged underneath in a brick-work setting. Water tube steam boilers are always externally fired.

3. *According to the axis of the shell.* The steam boilers, according to the axis of the shell, may be classified as :

(*a*) Vertical boilers, and (*b*) Horizontal boilers.

In *vertical steam boilers*, the axis of the shell is vertical. Simple vertical boiler and Cochran boiler are vertical boilers.

In *horizontal steam boilers*, the axis of the shell is horizontal. Lancashire boiler, Locomotive boiler and Babcock and Wilcox boiler are horizontal boilers.

4. *According to the number of tubes.* The steam boilers, according to the number of tubes, may be classified as :

(*a*) Single tube boilers, and (*b*) Multitubular boilers.

In *single tube steam boilers*, there is only one fire tube or water tube. Simple vertical boiler and Cornish boiler are single tube boilers.

In *multitubular steam boilers*, there are two or more fire tubes or water tubes. Lancashire boiler, Locomotive boiler, Cochran boiler, Babcock and Wilcox boiler are multitubular boilers.

5. *According to the method of circulation of water and steam.* The steam boilers, according to the method of circulation of water and steam, may be classified as :

(*a*) Natural circulation boilers, and (*b*) Forced circulation boilers.

In *natural circulation steam boilers*, the circulation of water is by natural convection currents, which are set up during the heating of water. In most of the steam boilers, there is a natural circulation of water.

In *forced circulation steam boilers*, there is a forced circulation of water by a centrifugal pump driven by some external power. Use of forced circulation is made in high pressure boilers such as La-Mont boiler, Benson boiler, Loeffler boiler and Velcon boiler.

6. *According to the use.* The steam boilers, according to their use, may be classified as

(*a*) Stationary boilers, and (*b*) Mobile boilers.

The *stationary steam boilers* are used in power plants, and in industrial process work. These are called stationary because they do not move from one place to another.

The *mobile steam boilers* are those which move from one place to another. These boilers are locomotive and marine boilers.

7. *According to the source of heat.* The steam boilers may also be classified according to the source of heat supplied for producing steam. These sources may be the combustion of solid, liquid or gaseous fuel, hot waste gases as by-products of other chemical processes, electrical energy or nuclear energy, etc.

13.6. Simple Vertical Boiler

A simple vertical boiler produces steam at a low pressure and in small quantities. It is, therefore, used for low power generation or at places where the space is limited. The construction of this type of boiler is shown in Fig. 13.1.

It consists of a cylindrical shell surrounding a nearly cylindrical fire box. The fire box is slightly tapered towards the top to allow the ready passage of the steam to the

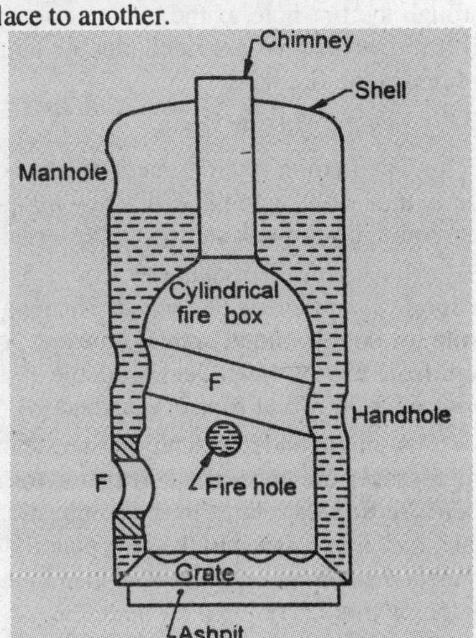

Fig. 13.1. Simple vertical boiler.

surface. At the bottom of the fire box, is a grate. The fire box is fitted with two or more inclined cross tubes *F*, *F*. The inclination is provided to increase the heating surface as well as to improve the circulation of water. An uptake tube passes from the top of the fire box to the chimney. The handholes are provided opposite to the end of each water tube for cleaning deposits. A manhole is provided at the top for a man to enter and clean the boiler. A mudhole is provided at the bottom of the shell to remove the mud, that settles down. The space between the boiler shell and fire box is filled with water to be heated.

13.7. Cochran Boiler or Vertical Multitubular Boiler

There are various designs of vertical multitubular boilers. A Cochran boiler is considered to be one of the most efficient type of such boilers. It is an improved type of simple vertical boiler.

This boiler consists of an external cylindrical shell and a fire box as shown in Fig. 13.2. The shell and fire box are both hemispherical, The hemispherical crown of the boiler shell gives maximum space and strength to withstand the pressure of steam inside the boiler. The hemispherical crown of the fire box is also advantageous for resisting intense heat. The fire box and the combustion chamber is connected through a short pipe. The flue gases from the combustion chamber flow to the smoke box through a number of smoke tubes. These tubes generally have 62.5 mm external diameter and are 165 in number. The gases from the smokebox pass to the atmosphere through a chimney. The combustion chamber is lined with fire bricks on the shell side. A manhole near the top of the crown on the shell is provided for cleaning.

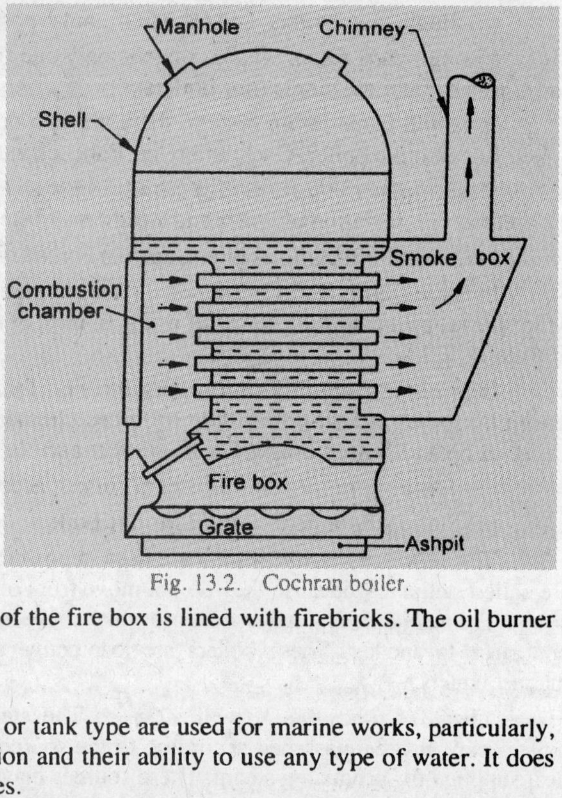

Fig. 13.2. Cochran boiler.

At the bottom of the fire box, there is a grate (in case of coal firing) and the coal is fed through the fire hole. If the boiler is used for oil firing, no grate is provided, but the bottom of the fire box is lined with firebricks. The oil burner is fitted at the fire hole.

13.8. Scotch Marine Boiler

The marine steam boilers of the scotch or tank type are used for marine works, particularly, due to their compactness, efficiency in operation and their ability to use any type of water. It does not require brick work setting and external flues.

It has a drum of diameter from 2.5 to 3.5 metres placed horizontally. These steam boilers may be *single ended* or *double ended.* The length of a single ended steam boiler may be upto 3.5 meters while for double ended upto 6.5 meters. A single ended boiler has one to four furnaces which enter from front end of the boiler. A double ended boiler has furnaces on both of its ends, and may have furnaces from two to four in each end.

A single ended scotch marine steam boiler is fired by four furnaces, as shown in Fig. 13.3. The furnaces are generally corrugated for strength. Each furnace has its own combustion chamber. There are fine flat plates in the combustion chamber, which require staying, *i.e.* the top plate, back plate, two side plates and the tube plate. There are a number of smoke tubes placed horizontally and connect the combustion chamber to chimney. The front and back plates of the shell are strengthened by longitudinal stays.

The combustion chamber walls form the best heating surface. The furnace tubes, smoke tubes and the combustion chamber, all being surrounded by water, give a very large heating surface area in proportion to the cubical size of boiler.. The water circulates around the smoke tubes. The level of water is maintained a little above the combustion chamber. The flue gases, from the combustion chamber, are forwarded by draught through the smoke tubes, and finally up the chimney. The smoke box is provided with a door for cleaning the tubes and smoke box.

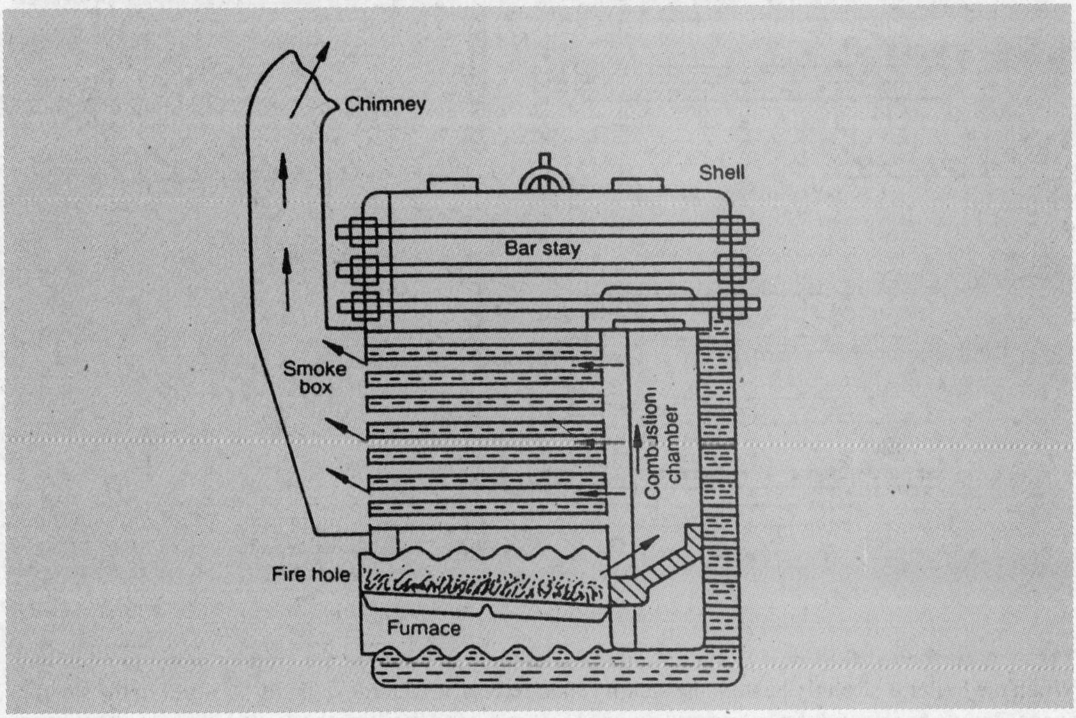

Fig. 13.3. Scotch marine boiler.

13.9. Lancashire Boiler

It is a stationary, fire tube, internally fired, horizontal and natural circulation boiler. It is used where working pressure and power required are moderate. These boilers have a cylindrical shell of 1.75 m to 2.75 m diameter. Its length varies from 7.25 m to 9 m. It has two internal flue tubes having diameter about 0.4 times that of shell. This type of boiler is set in brick work forming external flue so that part of the heating surface is on the external shell.

A Lancashire boiler with brick work setting is shown in Fig. 13.4. This boiler consists of a long cylindrical external shell (1) built of steel plates, in sections riveted together. It has two large internal flue tubes (2). These are reduced in diameter at the back end to provide access to the lower part of the boiler. A fire grate (3) also called furnace, is provided at one end of the flue tubes on which solid fuel is burnt. At the end of the fire grate, there is a brick arch (5) to deflect the flue gases upwards. The hot flue gases, after leaving the internal flue tubes pass down to the bottom tube (6). These flue gases move to the front of the boiler where they divide and flow into the side flue (7). The flue gases then enter the main flue (9), which leads them to chimney.

The damper (8) is fitted at the end of side flues to control the draught (*i.e.* rate of flow of air) and thus regulate the rate of generation of steam. These dampers are operated by chain passing over a pulley on the front of the boiler.

A spring loaded safety valve (10) and a stop valve (11) is mounted as shown in Fig. 13.4. The stop valve supplies steam to the engine as required. A high steam and low water safety valve (12) is also provided.

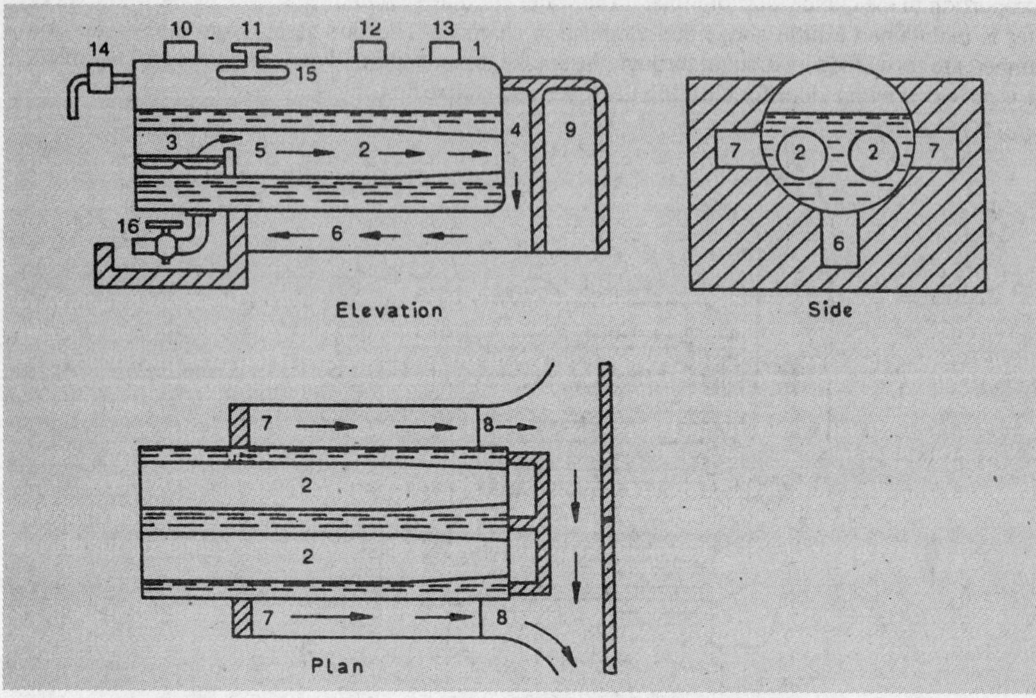

Fig. 13.4. Elevation, side and plan of lancashire boiler.

A perforated feed pipe (14) controlled by a feed valve is used for feeding water uniformly. When the boiler is strongly heated, the steam generated carries a large quantity of water in the steam space, known as *priming*. An antipriming pipe (15) is provided to separate out water as far as possible. The stop valve thus receives dry steam.

A blow-off cock (16) removes mud, etc., that settles down at the bottom of the boiler, by forcing out some of the water. It is also used to empty water in the boiler, whenever required for inspection. Manholes are provided at the top and bottom of the boiler for cleaning and repair purposes.

13.10. Cornish Boiler

It is similar to a Lancashire boiler in all respects, except there is only one flue tube in Cornish boiler instead of two in Lancashire boiler, as shown in Fig. 13.5. The diameter of Cornish boiler is generally 1 m to 2 m and its length varies from 5 m to 7.5 m. The diameter of flue tube may be about 0.6 times that of shell. The capacity and working pressure of a Cornish boiler is low as compared to Lancashire boiler.

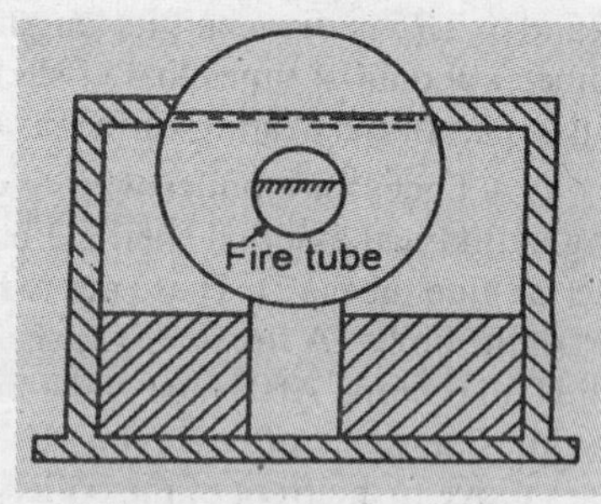

Fig. 13.5. Cornish boiler.

13.11. Locomotive Boiler

It is a multi-tubular, horizontal, internally fired and mobile boiler. The principal feature of this boiler is to produce steam at a very high rate. A modern type of a locomotive boiler is shown in Fig. 13.6.

It consists of a shell or barrel having 1.5 metres diameter and 4 metres in length. The coal is fed into the fire box through the fire door and burns on grate. The flue gases from the grate are deflected by a brick arch, and thus whole of the fire box is properly heated. There are about 157 thin tubes or fire tubes *F* (47.5 mm diameter) and 24 thick or superheated tubes *G* (130 mm diameter). The flue gases after passing through these tubes enter a smoke box. The gases are then lead to atmosphere through a chimney. The barrel contains water around the tubes, which is heated up by the flue gases and gets converted into steam.

A stop valve as regulator is provided inside a cylindrical steam dome. This is operated by a regulator shaft from the engine room by a driver. The header is divided into two portions, one is the superheated steam chamber and the other is the saturated steam chamber. The steam pipe leads the steam from the regulator to the saturated steam chamber. It then leads the steam to the superheated tubes, and after passing through these tubes, the steam returns back to the superheated steam chamber. The superheated steam now flows through the steam pipe to the cylinder, one on each side. The draught is due to the exhaust steam from the cylinders, which is discharged through the exhaust pipe. The front door can be opened for cleaning or repairing the smoke box.

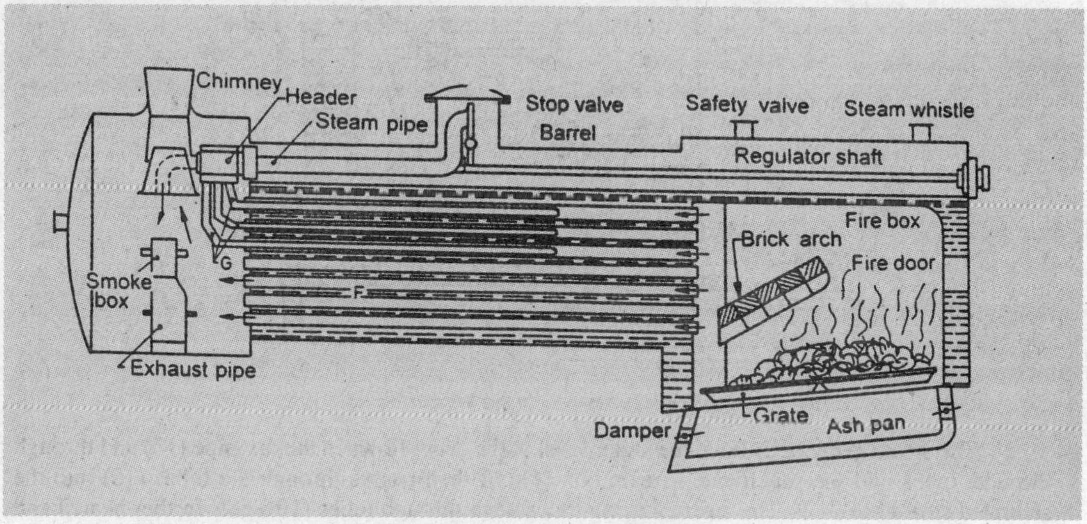

Fig. 13.6. Locomotive boiler.

The safety valves and a steam whistle are provided as shown in Fig. 13.6. The ash from the grate is collected in ash pan and is discharged out from time to time by opening it with the help of dampers operated by rods and levers.

13.12. Babcock and Wilcox Boiler

It is a straight tube, stationary type water tube boiler, as shown in Fig. 13.7. It consists of a steam and water drum (1). It is connected by a short tube with uptake header or riser (2) at the back end.

The water tubes (5) (100 mm diameter) are inclined to the horizontal and connects the uptake header to the downtake header. Each row of the tubes is connected with two headers, and there are plenty of such rows. The headers are curved when viewed in the direction of tubes so that one tube is not in the space of other, and hot gases can pass properly after heating all the tubes. The headers are provided with hand holes in the front of the tubes and are covered with caps (18).

A mud box (6) is provided with each downtake header and the mud, that settles down is removed. There is a slow moving automatic chain grate on which the coal is fed from the hopper (21). A fire bricks baffle causes hot gases to move upwards and downwards and again upwards before leaving the chimney. The dampers (17) are operated by a chain (22) which passes over a pulley to the front of a boiler to regulate the draught.

The boiler is suspended on steel girders, and surrounded on all the four sides by fire brick walls. The doors (4) are provided for a man to enter the boiler for repairing and cleaning. Water circulates from the drum (1) into the header (2) and through the tubes (5) to header (3) and again to the drum. Water continues to circulate like this till it is evaporated. A steam superheater consists of a large number of steel tubes (10) and contains two boxes ; one is superheated steam box (11) and other is saturated steam box (12).

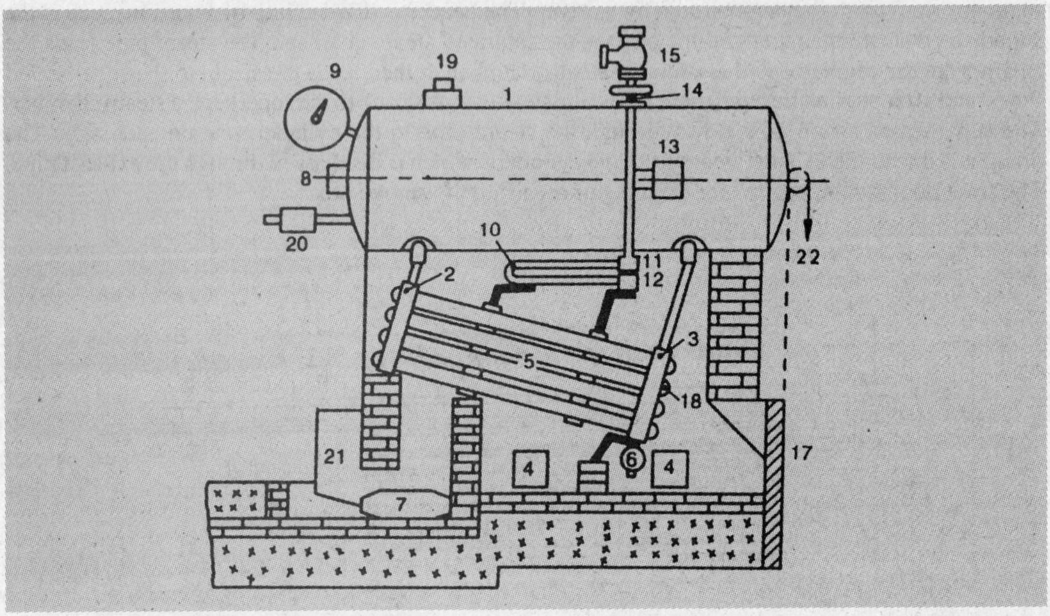

Fig. 13.7. Babcock and Wilcox boiler.

The steam generated above the water level in the drum flows in the dry pipe (13) and through the inlet tubes into the superheated steam box (11). It then passes through the tubes (10) into the saturated steam box (12). The steam, during its passage through tubes (10), gets further heated and becomes superheated. The steam is now taken through the outlet pipe (14) to the stop valve (15).

The boiler is fitted with usual mountings, such as safety valve (19), feed valve (20), water level indicator (8) and pressure gauge (9).

13.13. La-Mont Boiler

This is a modern high pressure water tube steam boiler working on a forced circulation. The circulation is maintained by a centrifugal pump, driven by a steam turbine, using steam from the boiler. The forced circulation causes the feed water to circulate through the water walls and drums equal to ten times the mass of steam evaporated. This prevents the tubes from being overheated.

A diagrammatic sketch of La-Mont steam boiler is shown in Fig. 13.8. The feed water passes through the economiser to an evaporating drum. It is then drawn to the circulating pump through

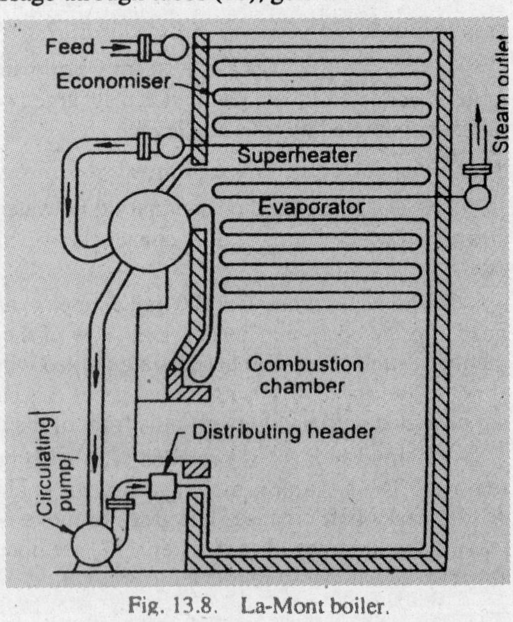

Fig. 13.8. La-Mont boiler.

the tube. The pump delivers the feed to the headers, at a pressure above the drum pressure. The header distributes water through nozzles into the generating tubes acting in parallel. The water and steam from these tubes passes into the drum. The steam in the drum is then drawn through the superheater.

13.14. Loeffler Boiler

This is a water tube boiler using a forced circulation. Its main principle of working is to evaporate the feed water by means of superheated steam from the superheater. The hot gases from the furnace are used for superheating.

A diagrammatic sketch of a Loeffler steam boiler is shown in Fig. 13.9. The feed water from the economiser tubes is forced to mix with the superheated steam in the evaporating drum. The saturated steam, thus formed, is drawn from the evaporating drum by a steam circulating pump. This steam passes through the tubes of the combustion chamber walls and then enters the superheater. From the superheater, about one-third of the superheated steam passes to the turbine and the remaining two-third is used to evaporate the feed water in the evaporating drum.

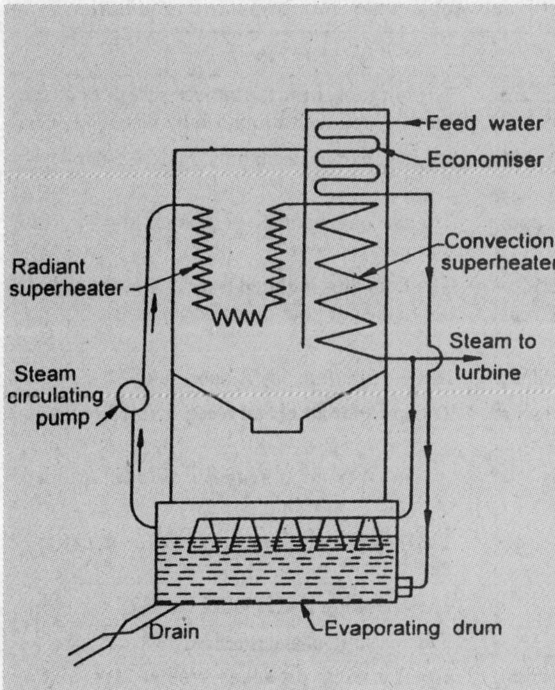

Fig. 13.9. Loeffler boiler.

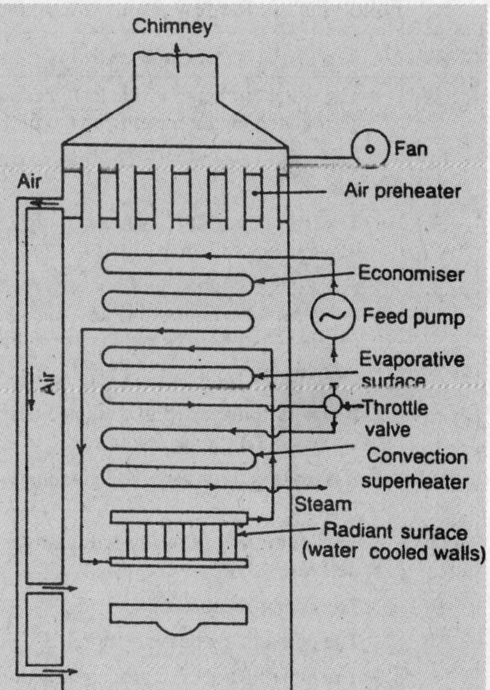

Fig. 13.10. Benson boiler.

13.15. Benson Boiler

It is a high pressure, drum less, water tube steam boiler using forced circulation. In this boiler, the feed water enters at one end and discharges superheated steam at the other end. The feed pump increases the pressure of water to supercritical pressure (*i.e.* above the critical pressure of 225 bar) and thus the water directly transforms into *steam without boiling.

The diagrammatic sketch of a Benson boiler is shown in Fig. 13.10. The feed water passes through the economiser to the water cooled walls of the furnace. The water receives heat by radiation

* We know that at critical pressure, the latent heat of vaporisation is zero. Thus the water transforms into steam without boiling.

and the temperature rises to almost critical temperature. It then enters the evaporator and may get superheated to some degree. Finally, it is passed through the superheater to obtain desired superheated steam.

The Benson boiler is also known as light-weight boiler as there is no large water and steam drum. The thermal efficiency upto 90 percent may be achieved by this boiler. The average operating pressure and capacity of such boilers are 250 bar and 135 tonnes/h. It can be started within 15 minutes.

Following are the advantages of Benson boiler :

1. The initial cost of boiler is low because there is no water and steam drum.
2. Since there is no pressure limit, therefore supercritical pressure may be employed.
3. The high pressure avoids the bubble formation in the tubes which increases heat transfer rate.
4. It is a light-weight boiler.
5. The boiler can be started within 15 minutes.

13.16. Comparison Between Water Tube and Fire Tube Boilers

Following are the few points of comparison between a water tube and a fire tube boiler.

S.No.	*Water tube boiler*	*Fire tube boiler*
1.	The water circulates inside the tubes which are surrounded by hot gases from the furnace.	The hot gases from the furnace pass through the tubes which are surrounded by water.
2.	It generates steam at a higher pressure upto 165 bar.	It can generate steam only upto 24.5 bar.
3.	The rate of generation of steam is high, *i.e.* upto 450 tonnes per hour.	The rate of generation of steam is low, *i.e.* upto 9 tonnes per hour.
4.	For a given power, the floor area required for the generation of steam is less, *i.e.* about 5 m^2 per tonne per hour of steam generation.	The floor area required is more, *i.e.* about 8 m^2 per tonne per hour of steam generation.
5.	Overall efficiency with economiser is upto 90%.	Its overall efficiency is only 75%.
6.	It can be transported and erected easily as its various parts can be separated.	The transportation and erection is difficult.
7.	It is preferred for widely fluctuating loads.	It can also cope reasonably with sudden increase in load but for a shorter period.
8.	The direction of water circulation is well defined.	The water does not circulate in a definite direction.
9.	The operating cost is high.	The operating cost is less.
10.	The bursting chances are more.	The bursting chances are less.
11.	The bursting does not produce any destruction to the whole boiler.	The bursting produces greater risk to the damage of the property.
12.	It is used for large power plants.	It is not suitable for large plants.

QUESTIONS

1. What is a steam boiler ? How they are classified ?

2. Explain the construction and working of a Lancashire boiler with the help of suitable sketches.

3. What is the difference between a Cornish boiler and a Lancashire boiler ?

4. Draw a neat sketch of a Locomotive boiler and label the parts. Explain its working also.

5. Describe with a neat diagram, the construction and working of a Babcock and Wilcox water tube boiler.

6. Explain with a neat sketch the working of a La-Mont boiler.

7. Describe with a neat line sketch of a Benson boiler mentioning its distinguishing features. State the advantages for this type of boilers.

8. What are the differentiating features between a water tube and a fire tube boiler ?

OBJECTIVE TYPE QUESTIONS

1. The water tubes in a simple vertical boiler are
   (*a*) horizontal (*b*) vertical (*c*) inclined
2. Lancashire boiler is a
   (*a*) stationary fire tube boiler (*b*) internally fired boiler
   (*c*) horizontal boiler (*d*) natural circulation boiler
   (*e*) all of the above (*f*) none of the above
3. The diameter of internal flue tubes of a Lancashire boiler is about .... that of its shell.
   (*a*) one-fourth (*b*) one-third (*c*) two-fifth (*d*) one-half
4. Locomotive boiler is a
   (*a*) single tube, horizontal, internally fired and stationary boiler
   (*b*) single tube, vertical, externally fired and stationary boiler
   (*c*) multi-tubular, horizontal, internally fired and mobile boiler
   (*d*) multi-tubular, horizontal, externally fired and stationary boiler
5. Which of the following is a water tube boiler ?
   (*a*) Lancashire boiler (*b*) Babcock and Wilcox boiler
   (*c*) Locomotive boiler (*d*) Cochran boiler
6. In fire tube boilers
   (*a*) water passes through the tubes which are surrounded by flames and hot gases
   (*b*) the flames and hot gases pass through the tubes which are surrounded by water
   (*c*) forced circulation takes place
   (*d*) none of the above
7. Which of the following boiler is best suited to meet the fluctuating demand of steam ?
   (*a*) Locomotive boiler (*b*) Lancashire boiler
   (*c*) Cornish boiler (*d*) Babcock and Wilcox boiler
8. Water tube boilers produce steam at a ...... pressure than that of fire tube boilers.
   (*a*) lower (*b*) higher
9. The locomotive boiler has
   (*a*) 137 fire tubes and 44 superheated tubes
   (*b*) 147 fire tubes and 34 superheated tubes
   (*c*) 157 fire tubes and 24 superheated tubes
   (*d*) 167 fire tubes and 14 superheated tubes
10. La-Mont boiler, is a ..... pressure water tube steam boiler working on forced circulation.
    (*a*) low (*b*) high

ANSWERS

1. (*c*)	2. (*e*)	3. (*c*)	4. (*c*)	5. (*b*)
6. (*b*)	7. (*a*)	8. (*b*)	9. (*c*)	10. (*b*)

14

Boiler Mountings and Accessories

1. Introduction. 2. Boiler Mountings. 3. Water Level Indicator. 4. Pressure Gauge. 5. Safety Valves. 6. Lever Safety Valve. 7. Dead Weight Safety Valve. 8. High Steam and Low Water Safety Valve. 9. Spring Loaded Safety Valve. 10. Steam Stop Valve. 11. Blow off Cock. 12. Feed Check Valve. 13. Fusible Plug. 14. Boiler Accessories. 15. Feed Pump. 16. Superheater. 17. Economiser. 18. Air Preheater.

14.1. Introduction

We have already discussed in Art. 13.2 that boiler mountings and accessories are required for the proper and satisfactory functioning of the steam boilers. Now in this chapter, we shall discuss these fittings and appliances which are commonly used these days.

14.2. Boiler Mountings

These are the fittings, which are mounted on the boiler for its proper and safe functioning. Though there are many types of boiler mountings, yet the following are important from the subject point of view :

1. Water level indicator ; 2. Pressure gauge ; 3. Safety valves ; 4. Stop valve ; 5. Blow off cock ; 6. Feed check valve ; and 7. Fusible plug.

14.3. Water Level Indicator

It is an important fitting, which indicates the water level inside the boiler to an observer. It is a safety device, upon which the correct working of the boiler depends. This fitting may be seen in front of the boiler, and are generally two in number.

A water level indicator, mostly employed in the steam boiler is shown in Fig. 14.1. It consists of three cocks and a glass tube. Steam cock C_1 keeps the glass tube in connection with the steam space. Water cock C_2 puts the glass tube in connection with the water in the boiler. Drain cock C_3 is used at frequent intervals to ascertain that the steam and water cocks are clear.

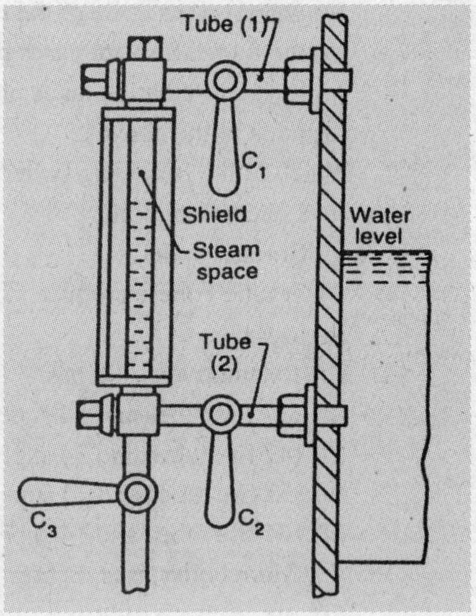

Fig. 14.1. Water level indicator.

In the working of a steam boiler and for the proper functioning of the water level indicator, the steam and water cocks are opened and the drain cock is closed. In this case, the handles are placed in a vertical position as shown in Fig. 14.1. The rectangular passage at the ends of the glass tube contains two balls.

In case the glass tube is broken, the two balls are carried along its passages to the ends of the glass tube. It is thus obvious, that water and steam will not escape out. The glass tube can be easily replaced by closing the steam and water cocks and opening the drain cock.

When the steam boiler is not working, the bolts may be removed for cleaning. The glass tube is kept free from leaking by means of conical ring and the gland nut.

14.4. Pressure Gauge

A pressure gauge is used to measure the pressure of the steam inside the steam boiler. It is fixed in front of the steam boiler. The pressure gauges generally used are of Bourden type.

A Bourden pressure gauge, in its simplest form, consists of an elliptical elastic tube *ABC* bent into an arc of a circle, as shown in Fig. 14.2. This bent up tube is called Bourden's tube.

One end of the tube gauge is fixed and connected to the steam space in the boiler. The other end is connected to a sector through a link. The steam, under pressure, flows into the tube. As a result of this increased pressure, the Bourden's tube tends to straighten itself. Since the tube is encased in a circular curve, therefore it tends to become circular instead of straight. With the help of a simple pinion and sector arrangement, the elastic deformation of the Bourden's tube rotates the pointer. This pointer moves over a calibrated scale, which directly gives the gauge pressure.

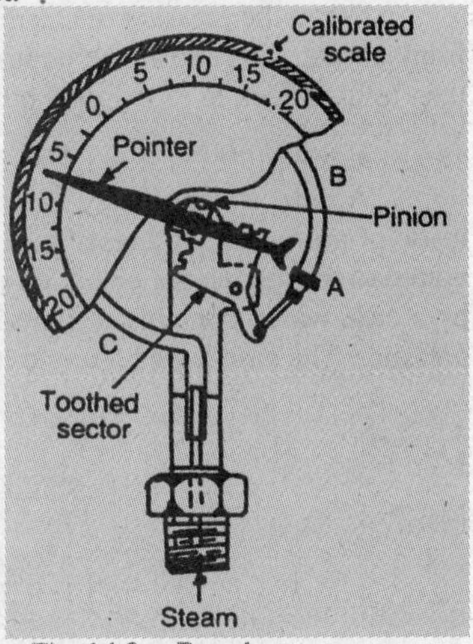

Fig. 14.2. Bourden type pressure gauge.

14.5. Safety Valves

These are the devices attached to the steam chest for preventing explosions due to excessive internal pressure of steam. A steam boiler is, usually, provided with two safety valves. These are directly placed on the boiler. In brief, the function of a safety valve is to blow off the steam when the pressure of steam inside the boiler exceeds the working pressure. The following are the four types of safety valves :

1. Lever safety valve, 2. Dead weight safety valve, 3. High steam and low water safety valve, and 4. Spring loaded safety valve.

It may be noted that the first three types of the safety valves are usually employed with stationary boilers, but the fourth type is mainly used for locomotive and marine boilers.

14.6. Lever Safety Valve

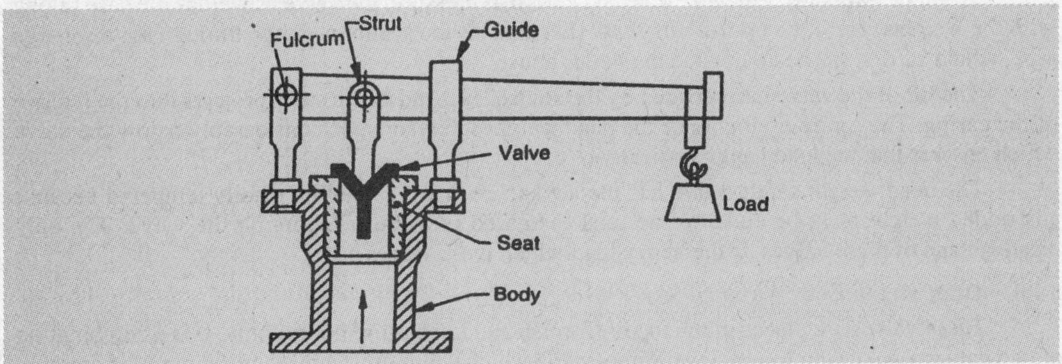

Fig. 14.3. Lever safety valve.

A lever safety valve used on steam boilers is shown in Fig. 14.3. It serves the purpose of maintaining constant safe pressure inside the steam boiler. If the pressure inside the boiler exceeds the designed limit, the valve lifts from its seat and blows off the steam pressure automatically.

A lever safety valve consists of a valve body with a flange fixed to the steam boiler. The bronze valve seat is screwed to the body, and the valve is also made of bronze. It may be noted that by using the valve and seat of the same material, rusting is considerably reduced. The thrust on the valve is transmitted by the strut. The guide keeps the lever in a vertical plane. The load is properly adjusted at the other end of the lever.

When the pressure of steam exceeds the safe limit, the upward thrust of steam raises the valve from its seat. This allows the steam to escape till the pressure falls back to its normal value. The valve then returns back to its original position.

14.7. Dead Weight Safety Valve

A dead weight safety valve, used for stationary boilers, is shown in Fig. 14.4. The valve is made of gun metal, and rests on its gun metal seat. It is fixed to the top of a steel pipe. This pipe is bolted to the mountings block, riveted to the top of the shell. Both the valve and the pipe are covered by a case which contains weights. These weights keep the valve on its seat under normal working pressure. The case hangs freely over the valve to which it is secured by means of a nut.

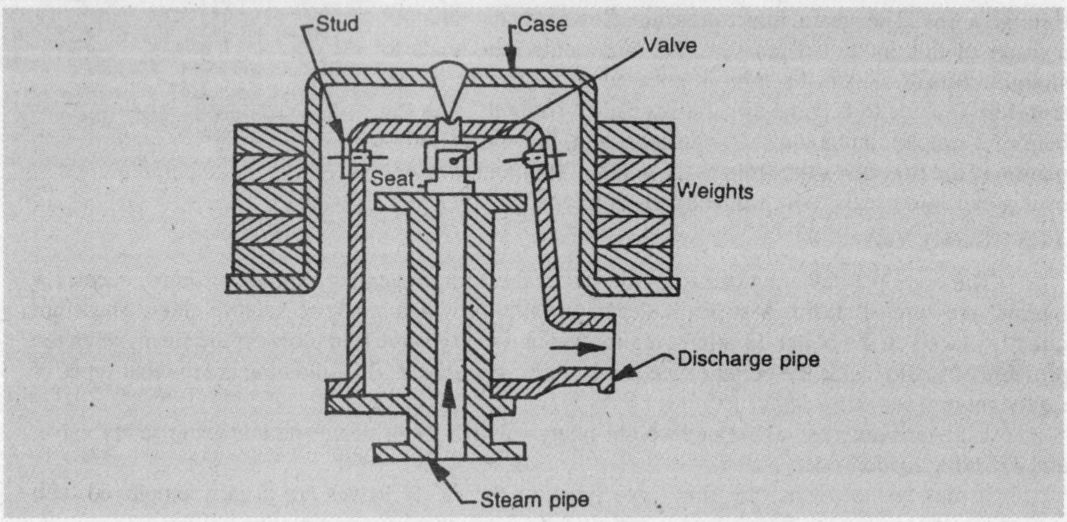

Fig. 14.4. Dead weight safety valve.

When the pressure of steam exceeds the normal pressure, the valve as well as the case (along with the weights) are lifted up from its seat. This enables the steam to escape through the discharge pipe, which carries the steam outside the boiler house.

The lift of the valve is controlled by the studs. The head of the studs projects into the interior of the casing. The centre of gravity of the dead weight safety valve is considerably below the valve which ensures that the load hangs vertically.

The dead weight safety valve has the advantage that it cannot be readily tempered because any added weight must be equal to the total increased pressure of steam on the valve. The only disadvantage of these valves, is the heavy load which these valves carry.

14.8. High Steam Low Water Safety Valve

These valves are placed at the top of Cornish and Lancashire boilers only. It is a combination of two valves, one of which is the lever safety valve which blows off steam when the working pressure of steam exceeds. The second valve operates by blowing off the steam when the water level becomes too low.

A best known combination of high steam low water safety valve is shown in Fig. 14.5. It consists of a main valve (known as lever safety valve) and rests on its seat. In the centre of the main valve, a seat for a hemispherical valve is formed for low water operation. This valve is loaded directly by the dead weights attached to the valve by a long rod. There is a lever *J-K*, which has its fulcrum at *K*. The lever has a weight *E* suspended at the end *K*. When it is fully immersed in water, it is balanced by a weight *F* at the other end *J* of the lever.

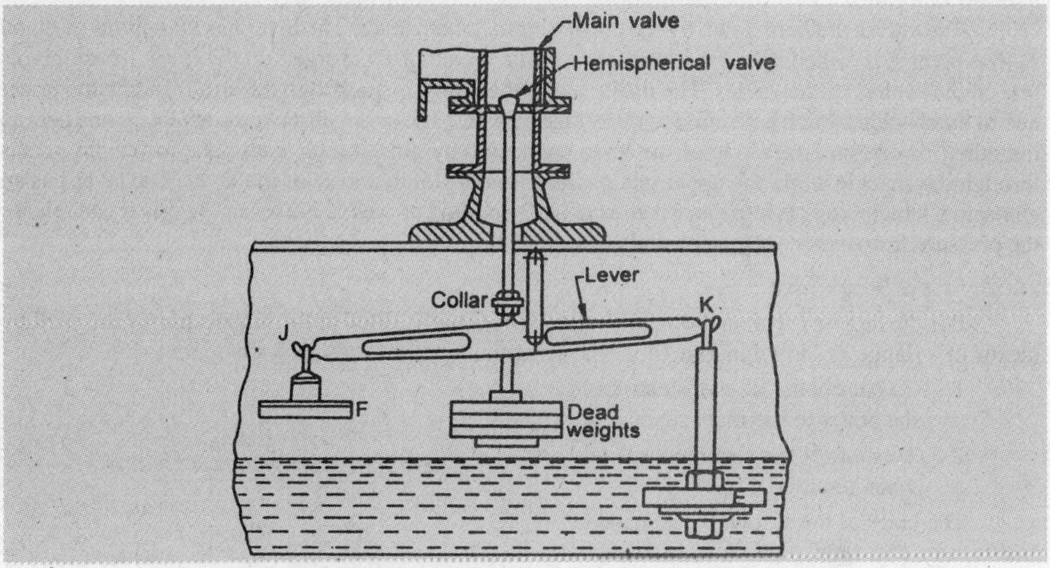

Fig. 14.5. High steam low water safety valve.

When the water level falls, the weight *E* comes out of water and the weight *F* will not be sufficient to balance weight *E*. Therefore weight *E* comes down. There are two projections on the lever to the left of the fulcrum which comes in contact with a collar attached to the rod. When weight *E* comes down, the hemispherical valve is lifted up and the steam escapes with a loud noise, which warns the operator. A drain pipe is provided to carry water, which is deposited in the valve casing.

14.9. Spring Loaded Safety Valve

A spring loaded safety valve is mainly used for locomotives and marine boilers. It is loaded with spring instead of weights. The spring is made of round or square spring steel rod in helical form. The spring may be in tension or compression, as the steam pressure acts along the axis of the spring. In actual practice, the spring is placed in compression.

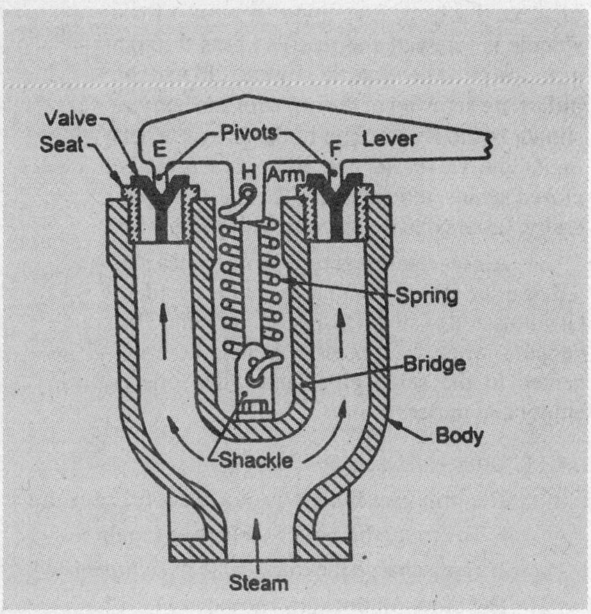

Fig. 14.6. Spring loaded safety valve.

A *Ramsbottom spring loaded safety valve is shown in Fig. 14.6. It is, usually, fitted to locomotives. It consists of a cast iron body connected to the top of a boiler. It has two separate valves of the

* It was introduced by John Ramsbottom of London.

same size. These valves have their seatings in the upper ends of two hollow valve chests. These valve chests are united by a bridge and a base. The base is bolted to a mounting block on the top of a boiler over the fire box.

The valves are held down by means of a spring and a lever. The lever has two pivots at E and F. The pivot E is joined by a pin to the lever, while the pivot F is forged on the lever. These pivots rest on the centres of the valves. The upper end of the spring is hooked to the arm H, while the lower end to the shackle, which is secured to the bridge by a nut. The spring has two safety links, one behind the other, or one on either side of the lever connected by pins at their ends. The lower pin passes through the shackle while the upper one passes through slot in arm H of the lever. The lever has an extension, which projects into the driver's cabin. By pulling or raising the lever, the driver can release the pressure from either valve separately.

14.10. Steam Stop Valve

It is the largest valve on the steam boiler. It is, usually, fitted to the highest part of the shell by means of a flange as shown in Fig. 14.7. The principal functions of a stop valve are :

1. To control the flow of steam from the boiler to the main steam pipe.
2. To shut off the steam completely when required.

The body of the stop valve is made of cast iron or cast steel. The valve, valve seat and the nut through which the valve spindle works, are made of brass or gun metal.

The spindle passes through a gland and stuffing box. The spindle is rotated by means of a hand wheel. The upper portion of the spindle is screwed and made to pass through a nut in a cross head carried by two pillars. The pillars are screwed in the cover of the body as shown in the figure. The boiler pressure acts under the valve, so that the valve must be closed against the pressure. The valve is, generally, fastened to the spindle which lifts it up.

A non-return valve is, sometimes, fitted near the stop valve to prevent the accidental admission of steam from other boilers. This happens when a number of boilers are connected to the same pipe, and when one is empty and under repair.

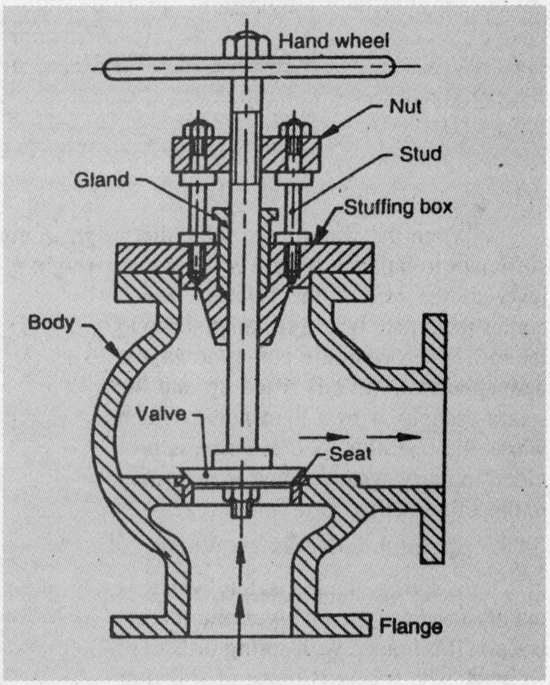

Fig. 14.7. Steam stop valve.

14.11. Blow off Cock

The principal functions of a blow-off cock are :

1. To empty thê boiler whenever required.
2. To discharge the mud, scale or sediments which are accumulated at the bottom of the boiler.

The blow-off cock, as shown in Fig. 14.8, is fitted to the bottom of a boiler drum and consists of a conical plug fitted to the body or casing. The casing is packed, with asbestos packing, in grooves round the top and bottom of the plug. The asbestos packing is made tight and plug bears on the packing. It may be noted that the cocks packed in this way keep the grip better under high pressure and easily operated than unpacked.

The shank of plug passes through a gland and stuffing box in the cover. The plug is held down by a yoke and two stud bolts (not shown in the figure). The yoke forms a guard on it. There are two vertical slots on the inside of a guard for the box spanner to be used for operating the cock.

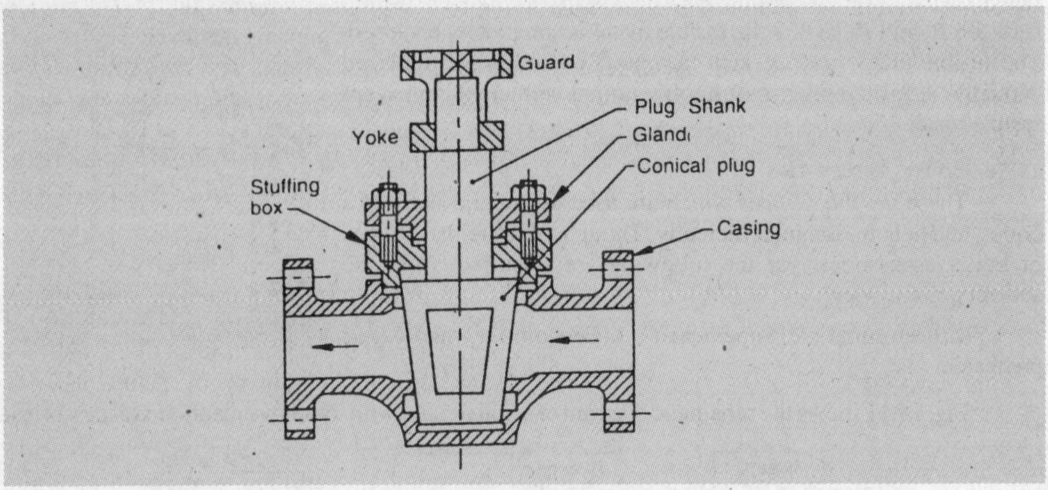

Fig. 14.8. Blow off cock.

14.12. Feed Check Valve

It is a non-return valve, fitted to a screwed spindle to regulate the lift. Its function is to regulate the supply of water, which is pumped into the boiler, by the feed pump. This valve must have its spindle lifted before the pump is started. It is fitted to the shell slightly below the normal water level of the boiler.

A feed check valve for marine boilers is shown in Fig. 14.9. It consists of a valve whose lift is controlled by a spindle and hand wheel. The body of the valve is made of brass casting and except spindle, its every part is made of brass. The spindle is made of muntz metal. A flange is bolted to the end of boiler at a point from which perforated pipe leads the feed water. This pipe distributes the water in the boiler uniformly.

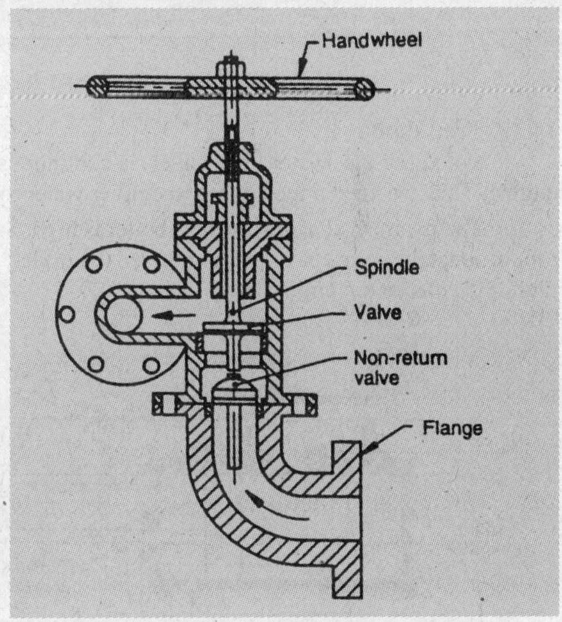

Fig. 14.9. Feed check valve.

14.13. Fusible Plug

It is fitted to the crown plate of the furnace or the fire box. Its object is to put off the fire in the furnace of the boiler when the level of water in the boiler falls to an unsafe limit, and thus avoids the explosion which may take place due to overheating of the furnace plate.

A fusible plug consists of a hollow gun metal plug P_1 as shown in Fig. 14.10. It is screwed to the furnace crown. A second hollow gun metal plug P_2 is screwed to the first plug. There is also a third hollow gun metal plug P_3 separated from P_1 by a ring of fusible metal. The inner surface of P_2 and outer surface of P_3 are grooved so that when the fusible metal is poured into the plug, P_2

and P_3 are locked together. A hexagonal flange is provided on plug P_1 to take a spanner for fixing or removing the plug P_1. There is a hexagonal flange on plug P_2 for fixing or removing it. The fusible metal is protected from fire by the flange on the lower end of plug P_2. There is also a contact at the top between P_2 and P_3 so that the fusible metal is completely enclosed. The fusible plugs must be kept in a good condition and replaced annually. A fusible plug must not be refilled with anything except fusible metal.

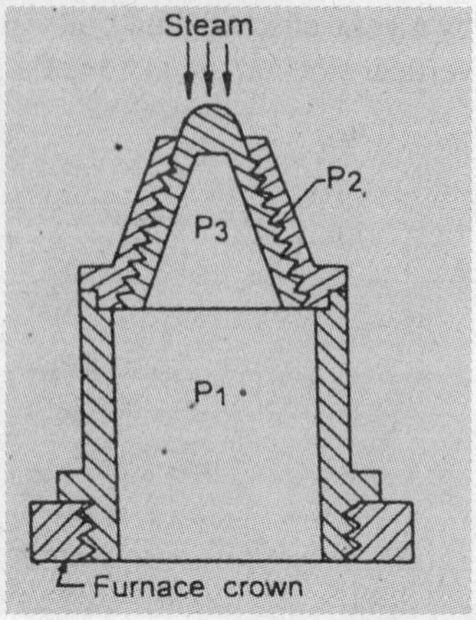

Fig. 14.10. Fusible plug.

14.14. Boiler Accessories

These are the devices which are used as integral parts of a boiler, and help in running efficiently. Though there are many types of boiler accessories, yet the following are important from the subject point of view :

1. Feet pump ; 2. Superheater ; 3. Economiser ; and 4. Air preheater.

Fig. 14.11 shows the schematic diagram of a boiler plant with the above mentioned accessories.

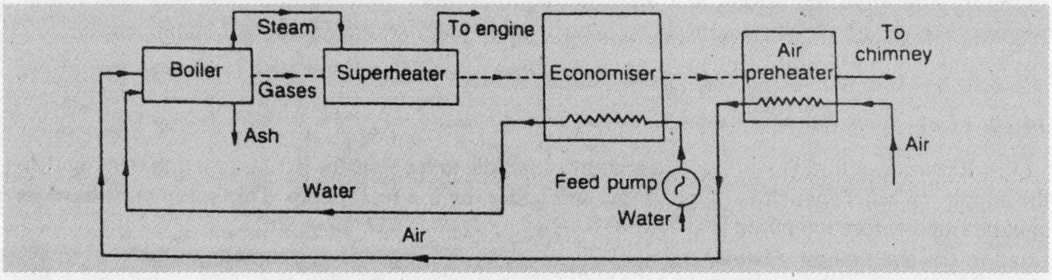

Fig. 14.11. Schematic diagram of a boiler plant.

14.15. Feed Pump

We know that water, in a boiler, is continuously converted into steam, which is used by the engine. Thus we need a feed pump to deliver water to the boiler.

The pressure of steam inside a boiler is high. So the pressure of feed water has to be increased proportionately before it is made to enter the boiler. Generally, the pressure of feed water is 20% more than that in the boiler.

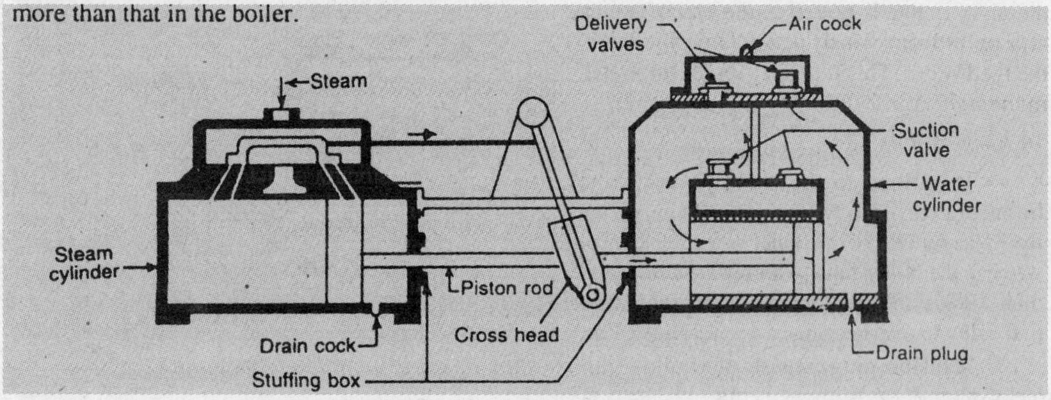

Fig. 14.12. Duplex feed pump.

A feed pump may be of centrifugal type or reciprocating type. But a doub'e acting reciprocating pump is commonly used as a feed pump these days. The reciprocating pumps are run by the steam from the same boiler in which water is to be fed. These pumps may be classified as simplex, duplex and triplex pumps according to the number of pump cylinders. The common type of pump used is a duplex feed pump, as shown in Fig. 14.12. This pump has two sets of suction and delivery valves for forward and backward stroke. The two pumps work alternately so as to ensure continuous supply of feed water.

14.16. Superheater

A superheater is an important device of a steam generating unit. Its purpose is to increase the temperature of saturated steam without raising its pressure. It is generally an integral part of a boiler, and is placed in the path of hot flue gases from the furnace. The heat, given up by these flue gases, is used in superheating the steam. Such superheaters, which are installed within the boiler, are known as integral superheaters.

A Sudgen's superheater commonly employed with Lancashire boilers is shown in Fig. 14.13. It consists of two mild steel boxes or heaters from which hangs a group of solid drawn tubes bent to U-form. The ends of these tubes are expanded into the headers. The tubes are arranged in groups of four and one pair of headers generally carries ten of these groups or forty tubes in all. The outside of the tubes can be cleaned through the space between the headers. This space is closed by covers.

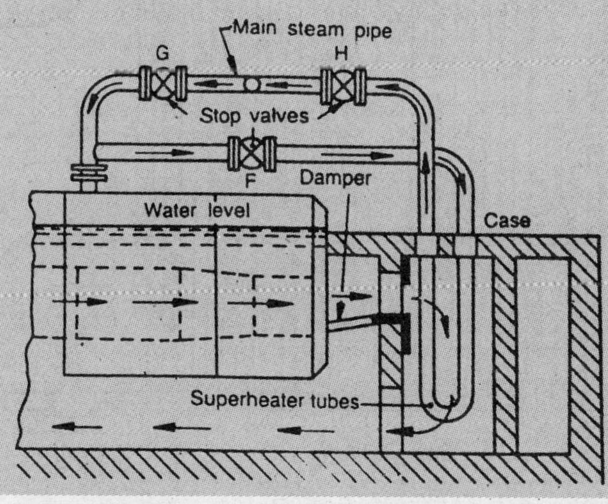

Fig. 14.13. Superheater.

The steam enters at one end of the rear header and leaves at the opposite end of the front header. The overheating of superheater tubes is prevented by the use of a balanced damper which is operated by the handle. The superheater is in action when the damper is in a position as shown in the figure. If the damper is in vertical position, the gases pass directly into the bottom flue without passing over the superheater tubes. In this way, the superheater is out of action. By placing the damper in intermediate position, some of the gases will pass over the superheater tubes and the remainder will pass directly to the bottom flue. It is thus obvious, that required degree of heat for superheating may be obtained by altering the position of the damper.

It may be noted that when the superheater is in action, the stop valves *G* and *H* are opened and *F* is closed. When the steam is taken directly from the boiler, the valves *G* and *H* are closed and *F* is open.

14.17. Economiser

An economiser is a device used to heat feed water by utilising the heat in the exhaust flue gases before leaving through the chimney. As the name indicates, the economiser improves the economy of the steam boiler.

A well known type of economiser is Greens economiser. It is extensively used for stationary boilers, especially those of Lancashire type. It consists of a large number of vertical pipes or tubes placed in an enlargement of the flue gases between the boiler and chimney as shown in Fig. 14.14. These tubes are 2.75 metres long, 114 mm in external diameter and 11.5 mm thick and are made of cast iron.

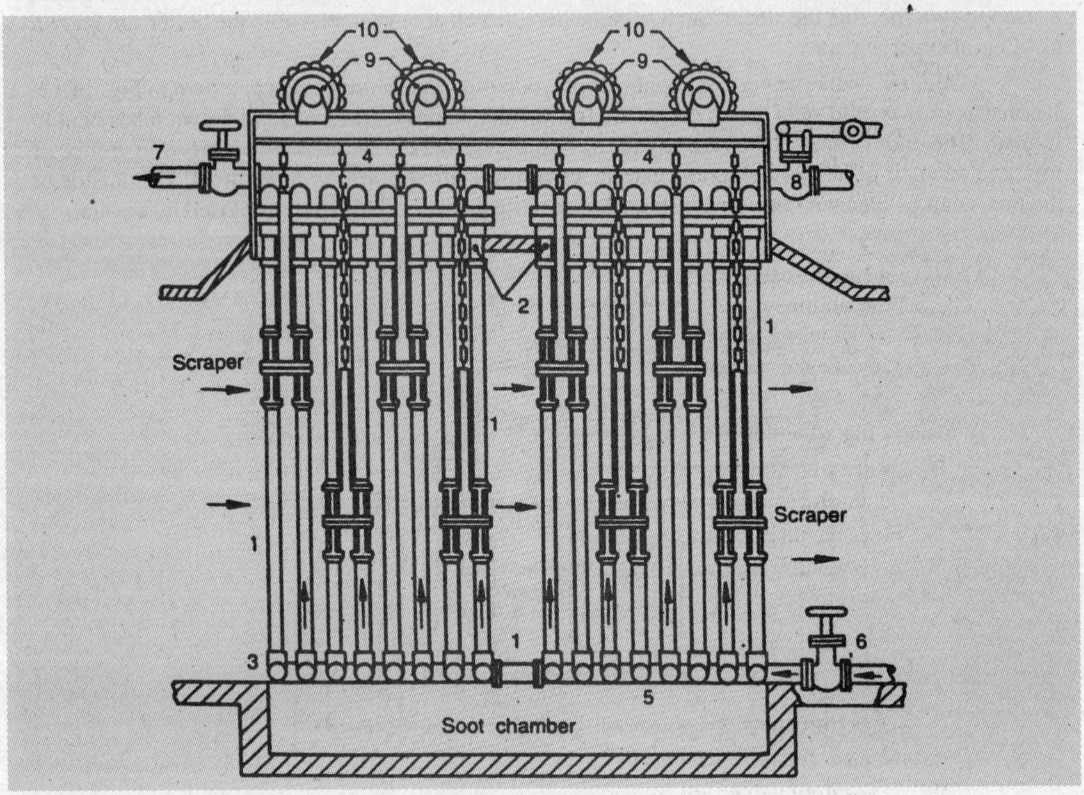

Fig. 14.14. Economiser.

The economiser is built-up of transverse section. Each section consists of generally six or eight vertical tubes (1). These tubes are joined to horizontal pipes or boxes (2) and (3) at the top and bottom respectively. The top boxes (2) of the different sections are connected to the pipe (4), while the bottom boxes are connected to pipe (5). The pipes (4) and (5) are on opposite sides, which are outside the brickwork enclosing the economiser.

The feed water is pumped into the economiser at (6) and enters the pipe (5). It then passes into the bottom boxes (3) and then into the top boxes (2) through the tubes (1). It is now led by the pipe (4) to the pipe (7) and then to the boiler. There is a blow-off cock at the end of the pipe (5) opposite to the feed inlet (6). The purpose of this valve is to remove mud or sediment deposited in the bottom boxes. At the end of pipe (4) (opposite to the feed outlet) there is a safety valve.

It is essential that the vertical tubes may be kept free from deposits of soot, which greatly reduce the efficiency of the economicser. Each tube is provided with scraper for this purpose. The scrapers of two adjoining sections of tubes are grouped together, and coupled by rods and chains to the adjacent group of scrapers. The chain passes over a pulley (9) so that one group of scrapers balance the adjacent group. The pulley (9) of each chain is connected to a worm wheel (10) which is driven by a worm on a longitudinal shaft (not shown in the figure). The scrapers automatically reverse when they reach the top or bottom end of the tubes. These are kept in motion continuously when the economiser is in use. The speed of scraper is about 46 m/h.

It may be noted that the temperature of feed should not be less than about 35° C, otherwise there is a danger of corrosion due to the moisture in the flue gases being deposited in cold tubes. Following are the advantages of using an economiser :-

1. There is about 15 to 20% of coal saving.
2. It increases the steam raising capacity of a boiler because it shortens the time required to convert water into steam.
3. It prevents formation of scale in boiler water tubes, because the scale formed in the economiser tubes, can be cleaned easily.
4. Since the feed water entering the boiler is hot, therefore strains due to unequal expansion are minimised.

14.18. Air Preheater

An air preheater is used to recover heat from the exhaust flue gases. It is installed between the economiser and the chimney. The air required for the purpose of combustion is drawn through the air preheater where its temperature is raised. It is then passed through ducts to the furnace. The air is passed through the tubes of the heater internally while the hot flue gases are passed over the outside of the tubes.

The following advantages are obtained by using an air preheater :

1. The preheated air gives higher furnace temperature which results in more heat transfer to the water and thus increases the evaporative capacity per kg of fuel.
2. There is an increase of about 2% in the boiler efficiency for each 35-40° C rise in temperature of air.
3. It results in better combustion with less soot, smoke and ash.
4. It enables a low grade fuel to be burnt with less excess air.

QUESTIONS

1. Describe with a neat sketch, water level indicator for a boiler.
2. Explain how the flow of steam of water is automatically stopped when the glass tube breaks.
3. Why the safety valves are needed in a boiler ? Sketch and describe a Ramsbottom spring loaded safety valve.
4. Differentiate between lever safety valve and dead weight safety valve.
5. What is the purpose of a steam stop valve ? Explain its working.
6. Explain the functions of blow off cock and feed check valve.
7. What is the function of a superheater ? Describe Sugden's superheater.
8. Discuss, briefly, the working of an economiser in a boiler plant giving a neat sketch.
9. Explain why air preheaters are used in a high pressure boiler.

OBJECTIVE TYPE QUESTIONS

1. A device attached to the steam chest for preventing explosions due to excessive internal pressure of steam is called

(*a*) safety valve (*b*) water level indicator
(*c*) pressure gauge (*d*) fusible plug

2. **A safety valve mainly used with locomotive and marine boilers is**

   (*a*) lever safety valve **(*b*) high pressure and low water safety valve**

   (*c*) dead weight safety valve **(*d*) spring loaded safety valve**

3. **A device used in a boiler to control the flow of steam from the boiler to the main pipe and to shut off the steam completely when required, is known as**

   (*a*) blow off cock **(*b*) fusible plug** **(*c*) stop valve** **(*d*) economiser**

4. **A device used to put off fire in the furnace of the boiler when the level of water in the boiler falls to an unsafe limit, is called**

   (*a*) blow off cock **(*b*) fusible plug** **(*c*) superheater** **(*d*) economiser**

5. **A device used to increase the temperature of saturated steam without raising its pressure. is called**

   (*a*) blow off cock **(*b*) fusible plug** **(*c*) superheater** **(*d*) economiser**

6. **A device used to heat feed water by utilising the heat in the exhaust flue gases before leaving through the chimney, is known as**

   (*a*) superheater **(*b*) economiser** **(*c*) blow off cock** **(*d*) stop valve**

7. **Which of the following are boiler accessories ?**

   (*a*) safety valve **(*b*) pressure gauge** **(*c*) superheater** **(*d*) economiser**

   (*e*) both (*a*) and (*b*) **(*f*) both (*c*) and (*d*)**

8. **An economiser ...... the steam raising capacity of a boiler.**

   (*a*) increases **(*b*) decreases** **(*c*) has no effect on**

9. **The pressure of feed water has to be raised before its entry into the boiler. The pressure is raised by a device known as**

   (*a*) feed check valve **(*b*) feed pump** **(*c*) pressure gauge** **(*d*) injector**

10. **An air preheater**

    (*a*) increases evaporative capacity of the boiler

    (*b*) increases the efficiency of the boiler

    (*c*) enables low grade fuel to be burnt

    (*d*) all of the above

ANSWERS

1. (*a*)	**2. (*d*)**	**3. (*c*)**	**4. (*b*)**	**5. (*c*)**
6. (*b*)	**7. (*f*)**	**8. (*a*)**	**9. (*b*)**	**10. (*d*)**

15

Performance of Steam Boilers

1. Introduction. 2. Equivalent Evaporation. 3. Boiler Efficiency. 4. Boiler Trial. 5. Heat Losses in a Boiler. 6. Heat Balance Sheet.

15.1. Introduction

The performance of a steam boiler is measured in terms of its *evaporative capacity. However, the evaporative capacities of two boilers cannot be compared unless both the boilers have the same feed water temperature, working pressure, fuel and the final condition of steam. In actual practice, the feed water temperature and working pressure varies considerably. It is thus obvious, that the comparison of two boilers becomes difficult unless some standard feed temperature and working pressure is adopted.

The feed temperature usually adopted is 100° C and the working pressure as normal atmospheric pressure, *i.e.* 1.013 bar. It is assumed that the boiler is supplied with water at the boiling temperature (100° C) corresponding to the atmospheric pressure.

15.2. Equivalent Evaporation

It is the amount of water evaporated from feed water at 100° C and formed into dry and saturated steam at 100° C at normal atmospheric pressure. It is, usually, written as *"from and at 100°C"*.

As the water is already at the boiling temperature, it requires only latent heat at 1.013 bar to convert it into steam at the temperature (100° C). The value of this latent heat is taken as 2257 kJ/kg. Mathematically,

Equivalent evaporation "from and at 100° C",

$$E = \frac{\text{Total heat required to evaporate feed water}}{2257}$$

Let t_1 = Temperature of feed water in °C,

h_{f1} = Enthalpy or sensible heat of feed water in kJ/kg of steam corresponding to t_1 °C (from steam tables),

h = Enthalpy or total heat of steam in kJ/kg of steam corresponding to a given working pressure (from steam tables),

$= h_f + x\,h_{fg}$ . . . (For wet steam)

$= h_f + h_{fg} = h_g$ . . . (For dry wet steam)

---

* The evaporative capacity or power of a boiler is the amount of water evaporated or steam produced in kg/h. It may also be expressed in kg/kg of fuel burnt or kg/h/m² of heating surface.

$$= h_g + c_p (t_{sup} - t) \quad \text{... (For superheated steam)}$$

m_e = Mass of water actually evaporated or steam produced in kg/h or kg/kg of fuel burnt.

We know that heat required to evaporate 1 kg of water

$$= h - h_{f1}$$

∴ Total heat required to evaporate m_e kg of water

$$= m_e (h - h_{f1})$$

and equivalent evaporation 'from and at 100° C'

$$E = \frac{m_e (h - h_{f1})}{2257}$$

Note : The factor $\frac{h - h_{f1}}{2257}$ is known as factor of evaporation, and is usually denoted by F_e. Its value is always greater than unity for all boilers.

15.3. Boiler Efficiency

It may be defined as the ratio of heat actually used in producing the steam to the heat liberated in the furnace. It is also known as thermal efficiency of the boiler. Mathematically,

Boiler efficiency or thermal efficiency,

$$\eta = \frac{\text{Heat actually used in producing steam}}{\text{Heat liberated in the furnace}} = \frac{m_e (h - h_{f1})}{C}$$

where m_e = Mass of water actually evaporated or actual evaporation in kg / kg of fuel, and

C = Calorific value of fuel in kJ/kg of fuel.

Notes : 1. If m_s = Total mass of water evaporated into steam in kg.

and m_f = Mass of fuel used in kg.

Then $m_e = \frac{m_s}{m_f}$ kg/kg of fuel

and $\eta = \frac{m_s (h - h_{f1})}{m_f \times C}$

2. If a boiler consisting of an economiser and superheater is considered to be a single unit, then the efficiency is termed as overall efficiency of the boiler.

Example 15.1. *A boiler evaporates 3.6 kg of water per kg of coal into dry saturated steam at 10 bar. The temperature of feed water is 32° C. Find the equivalent evaporation "from and at 100° C" as well as the factor of evaporation.*

Solution. Given : m_e = 3.6 kg / kg of coal ; p = 10 bar ; t_1 = 32° C

Equivalent evaporation 'from and at 100° C'

From steam tables, corresponding to a feed water temperature of 32° C, we find that

$$h_{f1} = 134 \text{ kJ/kg}$$

and corresponding to a steam pressure of 10 bar, we find that

$$h = h_g = 2776.2 \text{ kJ/kg} \quad \text{... (For dry saturated steam)}$$

We know that equivalent evaporation 'from and at 100° C',

$$E = \frac{m_e (h - h_{f1})}{2257} = \frac{3.6 (2776.2 - 134)}{2257} = 4.2 \text{ kg / kg of coal } \textbf{Ans.}$$

Factor of evaporation

We know that factor of evaporation

$$= \frac{h - h_{f1}}{2257} = \frac{2776.2 - 134}{2257} = 1.17 \textbf{ Ans.}$$

Example 15.2. *The following observations were made in a boiler trial :*

Coal used 250 kg of calorific value 29 800 kJ/kg, water evaporated 2000 kg, steam pressure 11.5 bar, dryness fraction of steam 0.95 and feed water temperature 34° C.

Calculate the equivalent evaporation "from and at 100° C" per kg of coal and the efficiency of the boiler.

Solution. Given : $m_f = 250$ kg ; $C = 29\,800$ kJ/kg ; $m_s = 2000$ kg ; $p = 11.5$ bar ; $x = 0.95$; $t_1 = 34°$ C

Equivalent evaporation 'from and at 100° C'

From steam tables, corresponding to a feed water temperature of 34° C, we find that

$$h_{f1} = 142.4 \text{ kJ/kg}$$

and corresponding to a steam pressure of 11.5 bar, we find that

$$h_f = 790 \text{ kJ/kg} ; \quad h_{fg} = 1991.4 \text{ kJ/kg}$$

We know that enthalpy or total heat of steam,

$$h = h_f + x\, h_{fg} = 790 + 0.95 \times 1991.4 = 2681.8 \text{ kJ/kg}$$

and mass of water evaporated per kg of coal

$$m_e = m_s / m_f = 2000 / 250 = 8 \text{ kg / kg of coal}$$

$\therefore$ Equivalent evaporation 'from and at 100° C'

$$E = \frac{m_e (h - h_{f1})}{2257} = \frac{8 (2681.8 - 142.4)}{2257} = 0.682 \text{ kg/kg of coal } \textbf{Ans.}$$

Efficiency of the boiler

We know that efficiency of the boiler,

$$\eta = \frac{m_e (h - h_{f1})}{C} = \frac{8 (2681.8 - 142.4)}{29\,800} = 0.682 \text{ or } 68.2\% \textbf{ Ans.}$$

Example 15.3. *A Lancashire boiler generates 2400 kg of dry steam per hour at a pressure of 11 bar. The grate area is 3 m² and 90 kg of coal is burnt per m² of grate area per hour. The calorific value of the coal is 33 180 kJ/kg and the temperature of feed water is 17.5° C. Determine : 1. Actual evaporation per kg of coal, 2. Equivalent evaporation 'from and at 100° C', and 3. Efficiency of the boiler.*

Solution. Given : $m_s = 2400$ kg/h ; $p = 11$ bar ; Grate area = 3 m² ; Coal burnt = 90 kg/m² /h ; $C = 33\,180$ kJ/kg ; $t_1 = 17.5°$ C

1. Actual evaporation per kg of coal

We know that mass of coal burnt per hour,

$$m_f = 90 \times 3 = 270 \text{ kg/h}$$

∴ Actual evaporation per kg of coal,

$$m_e = m_s / m_f = 2400 / 270 = 8.89 \text{ kg / h} \quad \textbf{Ans.}$$

2. *Equivalent evaporation 'from and at 100° C'*

From steam tables, corresponding to a feed water temperature of 17.5° C, we find that

$$h_{f1} = 73.4 \text{ kJ/kg}$$

and corresponding to a steam pressure of 11 bar, we find that

$$h = h_g = 2779.7 \text{ kJ/kg} \qquad \text{. . . (For dry steam)}$$

We know that equivalent evaporation 'from and at 100° C'

$$E = \frac{m_e (h - h_{f1})}{2257} = \frac{8.89 (2779.7 - 73.4)}{2257} = 10.66 \text{ kg/h} \quad \textbf{Ans.}$$

3. *Efficiency of the boiler*

We know that efficiency of the boiler,

$$\eta = \frac{m_e (h - h_{f1})}{C} = \frac{8.89 (2779.7 - 73.4)}{33\,180} = 0.725 \text{ or } 72.5\ \% \quad \textbf{Ans.}$$

Example 15.4. *A coal fired boiler plant consumes 400 kg of coal per hour. The boiler evaporates 3200 kg of water at 44.5° C into superheated steam at a pressure of 12 bar and 274.5° C. If the calorific value of fuel is 32 760 kJ/kg of coal, determine : 1. Equivalent evaporation "from and at 100° C," and 2. Thermal efficiency of the boiler.*

Assume specific heat of superheated steam as 2.1 kJ/kg K.

Solution. Given : $m_f = 400$ kg/h ; $m_s = 3200$ kg ; $t_1 = 44.5°$ C ; $p = 12$ bar ; $t_{sup} = 274.5°$ C ; $C = 32\,760$ kJ/kg of coal ; $c_p = 2.1$ kJ/kg K

1. *Equivalent evaporation 'from and at 100° C'*

We know that mass of water evaporated per kg of coal

$$m_e = m_s / m_f = 3200 / 400 = 8 \text{ kg}$$

From steam tables, corresponding to a feed water temperature of 44.5° C, we find that

$$h_{f1} = 186.3 \text{ kJ/kg}$$

and corresponding to a steam pressure of 12 bar, we find that

$$h_g = 2782.7 \text{ kJ/kg} \;\; ; \text{ and } t = 188° \text{ C}$$

We know that enthalpy or total heat required for 1 kg of superheated steam,

$$h_{sup} = h_g + c_p (t_{sup} - t)$$

$$= 2782.7 + 2.1 (274.5 - 188) = 2964.4 \text{ kJ/kg}$$

∴ Equivalent evaporation 'from and at 100° C',

$$E = \frac{m_e (h_{sup} - h_{f1})}{2257} = \frac{8 (2964.4 - 186.3)}{2257} \text{ kg/kg of coal}$$

$$= 9.85 \text{ kg/kg of coal} \quad \textbf{Ans.}$$

2. *Thermal efficiency of the boiler*

We know that thermal efficiency of the boiler,

$$\eta = \frac{m_e (h_{sup} - h_{f1})}{C} = \frac{8 (2964.4 - 186.3)}{32\,760} = 0.678 \text{ or } 67.8\ \% \quad \textbf{Ans.}$$

Example 15.5. *The following observations were made on a boiler plant during one hour test :*

Steam pressure = 20 bar ; Steam temperature = 260° C ; Steam generated = 37 500 kg , Temperature of water entering the economiser = 15° C ; Temperature of water leaving the economiser = 90° C ; Fuel used = 4400 kg ; Energy of combustion of fuel = 30 000 kJ/kg.

Calculate : 1. The equivalent evaporation per kg of fuel ; 2. The thermal efficiency of the plant ; and 3. The percentage heat energy of the fuel energy utilised by the economiser.

Solution. Given : $p = 20$ bar ; $t_{sup} = 260°$ C ; $m_s = 37\,500$ kg/h ; $t_1 = 15°$ C ; $t_2 = 90°$ C ; $m_f = 4400$ kg/h ; $C = 30\,000$ kJ/kg

1. *Equivalent evaporation per kg of fuel*

We know that mass of water actually evaporated,

$$m_e = m_s / m_f = 37\,500 / 4400 = 8.52 \text{ kg/kg of fuel}$$

From steam tables, corresponding to a feed water temperature of 15° C, we find that

$$h_{f1} = 62.9 \text{ kJ/kg}$$

and corresponding to a steam pressure of 20 bar, we find that

$$h_g = 2797.2 \text{ kJ/kg} \; ; \text{ and } t = 212.4° \text{C}$$

We know that enthalpy or total heat of 1 kg of superheated steam,

$$h_{sup} = h_g + c_p (t_{sup} - t)$$

$$= 2797.2 + 2.1 (260 - 212.4) = 2897 \text{ kJ/kg}$$

... (Taking $c_p = 2.1$ kJ/kg K)

∴ Equivalent evaporation,

$$E = \frac{m_e (h_{sup} - h_{f1})}{2257} = \frac{8.52 (2897 - 62.9)}{2257} = 10.7 \text{ kg/kg of fuel } \textbf{Ans.}$$

2. *Thermal efficiency of the plant*

We know that thermal efficiency of the plant,

$$\eta = \frac{m_e (h_{sup} - h_{f1})}{C} = \frac{8.52 (2897 - 62.9)}{30\,000} = 0.805 \text{ or } 80.5\,\% \textbf{ Ans.}$$

3. *Percentage heat energy of the fuel energy utilised by the economiser*

From steam tables, corresponding to a temperature of 90° C, we find that

$$h_{f2} = 376.9 \text{ kJ/kg}$$

We know that heat utilised by the economiser per kg of fuel

$$= m_e (h_{f2} - h_{f1}) = 8.52 (376.9 - 62.9) = 2675 \text{ kJ}$$

∴ Percentage of heat utilised by the economiser

$$= \frac{2675}{30\,000} = 0.089 \text{ or } 8.9\,\% \textbf{ Ans.}$$

Example 15.6. *The following particulars refer to a steam plant consisting of a boiler, economiser and a superheater :*

Steam pressure = 14 bar ; Mass of steam generated = 5000 kg/h ; Mass of coal used = 675 kg/h ; Calorific value of coal = 29 800 kJ/kg of coal ; Temperature of feed water entering the

economiser = 30° C ; Temperature of feed water leaving the economiser = 130° C ; Dryness fraction of steam leaving the boiler = 0.97 ; Temperature of steam leaving the superheater = 320° C.

Determine : 1. Overall efficiency of the plant, and 2. The percentage of the available heat utilised in the boiler, economiser and superheater respectively.

Solution. Given : p = 14 bar ; m_s = 5000 kg/h ; m_f = 675 kg/h ; C = 29 800 kJ/kg of coal ; t_1 = 30° C ; t_2 = 130° C ; x = 0.97 ; t_{sup} = 320° C

1. *Overall efficiency of the plant*

We know that mass of water actually evaporated per kg of coal,

$$m_e = m_s / m_f = 5000 / 675 = 7.41 \text{ kg}$$

From steam tables, corresponding to a feed water temperature of 30° C, we find that

$$h_{f1} = 125.7 \text{ kJ/kg}$$

and corresponding to a steam pressure of 14 bar, we find that

$$h_f = 830.1 \text{ kJ/kg} ;\ h_{fg} = 1957.7 \text{ kJ/kg} ;\ h_g = 2787.8 \text{ kJ/kg} ; \text{ and } t = 195° \text{ C}$$

We know that enthalpy or total heat of superheated steam,

$$h_{sup} = h_g + c_p (t_{sup} - t)$$

$$= 2787.8 + 2.1 (320 - 195) = 3050.3 \text{ kJ/kg} \quad \ldots (\text{Taking } c_p = 2.1 \text{ kJ/kg K})$$

∴ Overall efficiency of the plant,

$$\eta = \frac{m_e (h_{sup} - h_{f1})}{C} = \frac{7.41 (3050.3 - 125.7)}{29\ 800} = 0.727 \text{ or } 72.7\% \text{ Ans.}$$

2. *Percentage of available heat utilised*

Here we shall consider the following three cases :

(*a*) *Considering the boiler*

We know that enthalpy or sensible heat of feed water leaving the economiser or entering the boiler at 130° C (from steam tables),

$$h_{f2} = 546.3 \text{ kJ/kg}$$

Heat utilised in the boiler for 7.41 kg of steam at 14 bar and 0.97 dryness from water at 130° C

$$= m_e [(h_f + x\, h_{fg}) - h_{f2}]$$

$$= 7.41 [(830.1 + 0.97 \times 1957.7) - 546.3] = 16\ 174 \text{ kJ}$$

We know that the heat available per kg of coal

$$= 29\ 800 \text{ kJ/kg of coal}$$

∴ Percentage of available heat used in boiler

$$= \frac{16\ 174}{29\ 800} = 0.543 \text{ or } 54.3\% \text{ Ans.}$$

(*b*) *Considering the economiser*

We know that heat utilised in the economiser per kg of coal

$$= m_e\, c_p (t_2 - t_1) = 7.41 \times 4.2 (130 - 30) = 3112.2 \text{ kJ}$$

$$\ldots (\because c_p \text{ for water} = 4.2 \text{ kJ/kg K})$$

∴ Percentage of available heat used in the economiser

$$= \frac{3112.2}{29\,800} = 0.104 \text{ or } 10.4\ \% \textbf{ Ans.}$$

(c) Considering the superheater

Heat utilised in the superheater per kg of steam

$$= \text{Total heat of superheated steam} - \text{Total heat of wet steam}$$

$$= [h_g + c_p\,(t_{sup} - t)] - [(h_f + x\,h_{fg})]$$

$$= [2787.8 + 2.1\,(320 - 195)] - [(830.1 + 0.97 \times 1957.7)]$$

$$= 3050.3 - 2729 = 321.3 \text{ kJ}$$

∴ Total heat utilised in the superheater

$$= 7.41 \times 321.3 = 2381 \text{ kJ}$$

and percentage of heat utilised in the superheater

$$= \frac{2381}{29\,800} = 0.08 \text{ or } 8\% \textbf{ Ans.}$$

Note : As a check, the percentage of heat utilised in boiler

$$= 54.3 + 10.4 + 8 = 72.7\ \%$$

which is equal to the overall efficiency of the plant *i.e.* 72.7%.

15.4. Boiler Trial

The main objects of a boiler trial are :

1. To determine the generating capacity of the boiler.
2. To determine the thermal efficiency of the boiler when working at a definite pressure.
3. To prepare heat balance sheet for the boiler.

We have already discussed the first two objects in the previous articles. Now we shall discuss the third object, *i.e.* to prepare heat balance sheet.

15.5. Heat Losses in a Boiler

We know that the efficiency of a boiler is the ratio of heat utilised in producing steam to the heat liberated in the furnace. Also the heat utilised is always less than the heat liberated in the furnace. The difference of heat liberated in the furnace and heat utilised in producing steam is known as *heat lost in the boiler*. The loss of heat may be divided into various heads, but the following are important from the subject point of view :

1. Heat lost in dry flue gases

Heat lost to dry flue gases per kg of fuel

$$= m_g \times c_{pg}\,(t_g - t_b) \quad \ldots (i)$$

where m_g = Mass of dry flue gases per kg of fuel,

c_{pg} = Mean specific heat of dry flue gases,

t_g = Temperature of flue gases leaving chimney, and

t_b = Temperature of boiler room.

This loss is maximum in a boiler.

2. Heat lost in moisture present in the fuel

It is assumed that the moisture is converted into superheated steam at atmospheric pressure (1.013 bar).

Heat lost in moisture present in the fuel

$$= m_m (h_{sup} - h_b) = m_m [h_g + c_p (t_g - t) - h_b]$$

$$= m_m [2676 + c_p (t_g - 100) - h_b]$$

... [From steam tables, corresponding to 1.013 bar, h_g = 2676 kJ/kg and t = 100° C]

where m_m = Mass of moisture per kg of fuel,

c_p = Mean specific heat of superheated steam in flue gases,

t_g = Temperature of flue gases leaving chimney,

t_b = Temperature of boiler room, and

h_b = Enthalpy or sensible heat of water at boiler room temperature.

3. *Heat lost to steam formed by combustion of hydrogen per kg of fuel*

Let H_2 = Mass of hydrogen present per kg of fuel.

∴ Mass of steam formed

$$= 9 H_2$$

Then the heat lost to steam per kg of fuel

$$= 9H_2 [2676 + c_p (t_g - 100) - h_b]$$

Note : Heat lost to steam and moisture per kg of fuel

$$= (9H_2 + m_m) [2676 + c_p (t_g - 100) - h_b]$$

where m_m is the mass of moisture per kg of fuel.

4. *Heat lost due to unburnt carbon in ash pit*

The heat lost due to unburnt carbon per kg of fuel

$$= m_1 \times C_1$$

where m_1 = Mass of carbon in ash pit per kg of fuel.

C_1 = Calorific value of carbon.

5. *Heat lost due to incomplete combustion of carbon to carbon monoxide (CO)*

This loss, generally, occurs in a boiler due to insufficient air supply.

Heat lost due to incomplete combustion

$$= m_2 \times C_2$$

where m_2 = Mass of carbon monoxide in flue gas per kg of fuel, and

C_2 = Calorific value of carbon monoxide.

6. *Heat lost due to radiation*

There is no direct method for finding the heat lost due to radiation. This loss is calculated by subtracting the heat utilised in raising steam and heat losses from the heat supplied.

15.6. Heat Balance Sheet

A heat balance sheet shows the complete account of heat supplied by 1 kg of dry fuel* and heat consumed. The heat supplied is mainly utilised for raising the steam and the remaining heat is

* It is equal to the calorific value of the fuel.

lost. We know that heat utilised in raising steam per kg of fuel

$$= m_e (h - h_{f1})$$

The heat balance sheet for a boiler trial per kg of fuel is drawn as below :

Heat supplied	*kJ*		*Heat consumed*	*kJ*	*%*
Heat supplied by 1 kg of dry fuel	X	1.	Heat utilisied in raising steam	x_1	...
		2.	Heat lost in dry flue gases	x_2	...
		3.	Heat lost in moisture in fuel	x_3	...
		4.	Heat lost to steam by combustion of hydrogen	x_4	...
		5.	Heat lost due to unburnt carbon in ash pit	x_5	..
		6.	Heat lost due to incomplete combustion	x_6	...
		7.	Heat lost due to radiation, etc. (by difference)	$X - (x_1 + x_2 + x_3 + x_4 + x_5 + x_6)$	...
Total	X		Total	X	100%

Example 15.7. *In a boiler, the following observations were made :*

Pressure of steam	*= 10 bar*
Steam condensed	*= 540 kg/h*
Fuel used	*= 65 kg/h*
Moisture in fuel	*= 2% by mass*
Mass of dry flue gases	*= 9 kg/kg of fuel*
Lower calorific value of fuel	*= 32 000 kJ/kg*
Temperature of the flue gases	*= 325° C*
Temperature of boiler house	*= 28° C*
Feed water temperature	*= 50° C*
Mean specific heat of flue gases	*= 1 kJ/kg K*
Dryness fraction of steam	*= 0.95*

Draw up a heat balance sheet for the boiler.

Solution. Given : $p = 10$ bar ; $m_s = 540$ kg/h ; $m_f = 65$ kg/h ; $m_m = 0.02$ kg/kg of fuel ; $m_g = 9$ kg/kg of fuel ; $C = 32\,000$ kJ/kg ; $t_g = 325°$ C ; $t_b = 28°$ C ; $t_1 = 50°$ C ; $c_{pg} = 1$ kJ/kg K ; $x = 0.95$

First of all, let us find the heat supplied by 1 kg of fuel. Since the moisture in fuel is 0.02 kg, therefore heat supplied by 1 kg of fuel

$$= (1 - 0.02)\,32\,000 = 31\,360 \text{ kJ} \quad \ldots (i)$$

1. *Heat utilised in raising steam per kg of fuel*

We know that the mass of water actually evaporated per kg of fuel,

$$m_e = m_s / m_f = 540 / 65 = 8.31 \text{ kg}$$

From steam tables, corresponding to a feed water temperature of 50° C, we find that

$$h_{f1} = 209.3 \text{ kJ/kg}$$

and corresponding to a steam pressure of 10 bar, we find that

$$h_f = 762.6 \text{ kJ/kg} ; \; h_{fg} = 2013.6 \text{ kJ/kg}$$

∴ Heat utilised in raising steam per kg of fuel

$$= m_e (h - h_{f1}) = m_e (h_f + x\, h_{fg} - h_{f1})$$

$$= 8.31 (762.6 + 0.95 \times 2013.6 - 209.3) = 20\,495 \text{ kJ} \qquad \ldots (ii)$$

2. *Heat carried away by dry flue gas*

We know that heat carried away by dry flue gas

$$= m_g\, c_{pg} (t_g - t_b) = 9 \times 1 (325 - 28) = 2673 \text{ kJ} \qquad \ldots (iii)$$

3. *Heat carried away by moisture in fuel per kg of fuel*

From steam tables, corresponding to a temperature of 28° C, we find that

$$h_b = 117.3 \text{ kJ/kg}$$

We know that heat carried away by moisture in fuel

$$= m_m [2676 + c_p (t_g - 100) - h_b]$$

$$= 0.02 [2676 + 2.1 (325 - 100) - 117.3] = 60.6 \text{ kJ} \qquad \ldots (iv)$$

. . . (Taking c_p for superheated steam = 2.1 kJ/kg K)

4. *Heat lost by radiation etc.*

We know that heat lost by radiation etc. (by difference)

$$= 31\,360 - (20\,495 + 2673 + 60.6) = 8131.4 \text{ kJ} \qquad \ldots (v)$$

Now complete heat balance sheet per kg of fuel is given below :

Heat supplied	*kJ*	*Heat expenditure*	*kJ*	*%*
Heat supplied by 1 kg of fuel	31 360	1. Heat utilised in raising steam	20 495	65.35
		2. Heat carried away by dry flue gases	2673	8.53
		3. Heat carried away by moisture in fuel	60.6	0.19
		4. Heat lost by radiation etc. (by difference).	8131.4	25.93
Total	31 360	Total	31 360	100

Example 15.8. *The following observations were made during a boiler trial :*

Mass of feed water per hour = 635 kg ; Temperature of feed water = 65° C ; Steam pressure = 10.5 bar ; Oil fired per hour = 52 kg ; Higher calorific value = 44 900 kJ/kg.

Percentage composition of oil by mass, C = 84.75 ; H_2 = 13 ; S = 1.25.

Analysis of dry flue gases by volume, CO_2 = 12.4 ; O_2 = 4.3 ; N_2 = 83.3.

Temperature of gases leaving the boiler = 362° C ; Specific heat of dry flue gases = 1.005 kJ/kg K ; Boiler room temperature = 21° C ; Throttling calorimeter temperature at outlet = 125° C ; Pressure of steam after throttling = 101 mm of mercury ; Barometer reading = 760 mm of Hg ; Heating surface of boiler = 20 m² ; Specific heat of superheated steam = 2.1 kJ/kg K ; Partial pressure of steam in flue gases = 0.07 bar.

Draw up a complete heat balance sheet and calculate the boiler efficiency and equivalent evaporation per kg of fuel and per m² of heating surface per hour.

Solution. Given : $m_s = 635$ kg/h ; $t_1 = 65°$ C ; $p = 10.5$ bar ; $m_f = 52$ kg/h ; $C = 44\,900$ kJ/kg ; $t_g = 362°$ C ; $c_{pg} = 1.005$ kJ/kg K ; $t_b = 21°$ C ; $t_{sup} = 125°$ C ; c_p for superheated steam = 2.1 kJ/kg K ; $p_2 = 0.07$ bar

First of all, let us find the dryness fraction of steam before throttling (*i.e.* x).

We know that absolute pressure of steam after throttling

$$= 101 + 760 = 861 \text{ mm of Hg}$$

$$= 861 \times 133.3 = 115\,000 \text{ N/m}^2 \quad \ldots (\because 1 \text{ mm of Hg} = 133.3 \text{ N/m}^2)$$

$$= 1.15 \text{ bar} \quad \ldots (\because 1 \text{ bar} = 10^5 \text{ N/m}^2)$$

From steam tables, corresponding to an initial pressure of 10.5 bar, we find that

$$h_f = 772 \text{ kJ/kg} \;;\; h_{fg} = 2006 \text{ kJ/kg}$$

and corresponding to a final pressure of 1.15 bar, we find that

$$h_{g2} = 2681.6 \text{ kJ/kg} \;;\; t_2 = 103.6° \text{ C}$$

We know that enthalpy or total heat of steam before throttling

= Enthalpy or total heat of steam after throttling

i.e. $$h_f + x\,h_{fg} = h_{g2} + c_p\,(t_{sup} - t_2)$$

$$772 + x \times 2006 = 2681.6 + 2.1\,(125 - 103.6)$$

$$\therefore \quad x = 0.974$$

In order to draw the heat balance sheet, we shall determine the values of the following heats :

1. *Heat utilised in raising steam per kg of oil*

We know that the mass of water actually evaporated per kg of oil

$$m_e = m_s / m_f = 635/52 = 12.21 \text{ kg}$$

From steam tables, corresponding to a feed water temperature of 65° C, we find that

$$h_{f1} = 272 \text{ kJ/kg}$$

∴ Heat utilised in raising steam per kg of oil

$$= m_e\,(h - h_{f1})$$

$$= m_e\,[(h_f + x\,h_{fg}) - h_{f1}]$$

$$= 12.21\,[(772 + 0.974 \times 2006) - 272] = 29\,960 \text{ kJ} \quad \ldots (i)$$

2. *Heat carried away by dry flue gases*

First of all, let us find the mass of dry flue gases per kg of fuel (*i.e.* m_g).

The percentage composition of oil by mass is given as :

$$C = 84.75\% = 0.8475 \text{ kg} \;;\; H_2 = 13\% = 0.13 \text{ kg} \;;\; \text{and } N_2 = 1.25\% = 0.0125 \text{ kg}$$

and analysis of dry flue gases by volume is given as :

$$CO_2 = 12.4\% = 0.124 \text{ m}^3 \;;\; O_2 = 4.3\% = 0.043 \text{ m}^3 \;;\; \text{and } N_2 = 83.3\% = 0.833 \text{ m}^3$$

The volumetric analysis of the dry flue gases is converted into mass analysis as given in the following table :

Constituent	Volume in 1 m^3 of the flue gas (a)	Molecular mass (b)	Proportional mass (c) = (a) × (b)	Mass of constituent in kg per kg of flue gas (d) = $\frac{(c)}{\Sigma (c)}$
CO_2	0.124	44	5.456	$\frac{5.456}{30.156} = 0.181$
O_2	0.043	32	1.376	$\frac{1.376}{30.156} = 0.046$
N_2	0.833	28	23.324	$\frac{23.324}{30.156} = 0.773$
Total	1.000		$\Sigma(c) = 30.156$	1.000

We know that mass of carbon in 1 kg of flue gases

$$= \frac{3}{11} CO_2 + \frac{3}{7} CO = \frac{3}{11} CO_2 \qquad \ldots (\because CO = 0)$$

$$= \frac{3}{11} \times 0.181 = 0.0494 \text{ kg}$$

∴ Mass of dry flue gases per kg of oil burnt,

$$m_g = \frac{\text{Mass of carbon in 1 kg of fuel}}{\text{Mass of carbon in 1 kg of flue gas}}$$

$$= \frac{0.8475}{0.0494} = 17.16 \text{ kg/kg of oil}$$

We know that heat carried away by dry flue gases per kg of oil

$$= m_g \, c_{pg} \, (t_g - t_b)$$

$$= 17.16 \times 1.005 \, (362 - 21) = 5880 \text{ kJ} \qquad \ldots (ii)$$

3. *Heat carried away by steam*

Since H_2 is present in the oil, therefore mass of steam formed in the flue gases due to combustion of H_2 is given by

$$m = 9\, H_2 = 9 \times 0.13 = 1.17 \text{ kg}$$

From steam tables, corresponding to a partial pressure of steam in the flue gases, *i.e.* 0.07 bar, we find that

$$h_g = 2572.6 \text{ kJ/kg}; \text{ and } t = 39^\circ \text{C}$$

and corresponding to a boiler room temperature of 21° C, we find that

$$h_b = 88 \text{ kJ/kg}$$

∴ Heat carried away by steam in the flue gases per kg of oil

$$= m\, [h_g + c_p \, (t_g - t) - h_b]$$

$$= 1.17\, [2572.6 + 2.1\, (362 - 39) - 88] = 3700 \text{ kJ} \qquad \ldots (iii)$$

4. *Heat lost by radiation etc.*

We know that heat lost by radiation etc. (by difference)

$$= 44\,900 - (29\,960 + 5880 + 3700) = 5360 \text{ kJ}$$

Now complete heat balance sheet per kg of oil is given below :

Heat supplied	*kJ*		*Heat expenditure*	*kJ*	*%*
Heat supplied by 1 kg of oil	44 900	1.	Heat utilised in raising steam	29 960	66.73
		2.	Heat carried away by dry flue gases	5880	13.10
		3.	Heat carried away by steam	3700	8.24
		4.	Heat lost by radiation etc. (by difference)	5360	11.93
Total	44 900		Total	44 900	100

Boiler efficiency

We know that boiler efficiency,

$$\eta = \frac{m_e (h - h_{f1})}{C} = \frac{29\,960}{44\,900} = 0.6673 \text{ or } 66.73\ \% \textbf{ Ans.}$$

... [From equation (*i*)]

Equivalent evaporation per kg of fuel

We know that equivalent evaporation per kg of fuel

$$= \frac{m_e (h - h_{f1})}{2257} = \frac{29\,960}{2257} = 13.27 \text{ kg } \textbf{Ans.}$$

Equivalent evaporation per m² per hour

We know that equivalent evaporation per m^2 per hour

$$= \frac{\text{Equivalent evaporation per kg of fuel}}{\text{Heating surface area}} = \frac{13.27}{20} \text{ or } 0.6635 \text{ kg } \textbf{Ans.}$$

EXERCISES

1. A boiler produces 4 kg of steam per kg of coal from feed water at 45° C. The steam pressure is 10.5 bar. If the dryness fraction of steam is 0.98, determine the equivalent evaporation from and at 100° C. **[Ans. 4.52 kg]**

2. A boiler raises 3.7 kg of water per kg of coal from feed water at 54.5° C, to steam at the pressure of 34 bar and temperature of 370° C. Assuming specific heat of superheated steam as 2.6, calculate equivalent evaporation/kg of coal. **[Ans. 4.77 kg]**

3. In a boiler trial, the following observations were recorded :

Boiler pressure = 10 bar ; Dryness fraction of steam = 0.95 ; Coal consumption = 500 kg/h ; Calorific value of coal = 30 500 kJ/kg ; Feed water temperature = 50° C ; Feed water supplied = 4 tonnes/h.

Find the evaporation factor and the equivalent evaporation from and at 100° C in kg per kg of coal fired. Take specific heat of feed water as 4.187 kJ/kg K. **[Ans. 1.09 ; 8.72 kg/kg of fuel]**

4. A boiler produces 9000 kg of steam while 1 tonne of coal is burnt. The steam is produced at 10 bar from water at 15° C. The dryness fraction of steam is 0.9. Determine the efficiency of the boiler when the calorific value of the coal is 32 000 kJ/kg. **[Ans. 70.65%]**

5. A boiler delivers steam at 100 bar and 500° C. The feed water inlet temperature is 160° C. The steam is produced at the rate of 100 tonnes/h and the boiler efficiency is 88%. Estimate the fuel burning rate in kg/h. if the calorific value of the fuel is 21 MJ/kg. **[Ans. 14.6 tonnes/h]**

6. In a boiler test, the following observations were made :

Feed water temperature = 12° C ; Pressure of steam = 11 bar ; Dryness fraction of steam = 0.95 ; Mass of coal burnt = 300 kg/h ; Calorific value of coal = 32 000 kJ/kg of coal ; Mass of water supplied to boiler in 7 hrs 14 min = 14 625 kg.

The mass of water in the boiler at the end of the test was less than that at the commencement by 900 kg. Calculate : 1. Actual evaporation per kg of coal ; 2. Equivalent evaporation from and at 100° C per kg of coal ; and 3. Thermal efficiency of the boiler. [**Ans.** 7.15 kg ; 8.33 kg ; 58.75 %]

7. A steam plant consisting of a boiler, superheater and economiser has the following particulars :

Steam pressure = 12.6 bar ; Temperature of steam leaving superheater = 245° C ; Fuel used per hour = 1000 kg ; Feed water per hour = 9000 kg ; Temperature of feed water entering the economiser = 40° C ; Temperature of feed water leaving the economiser = 115° C ; Dryness fraction of steam leaving the boiler = 0.9 ; Calorific value of fuel used = 30 240 kJ/kg.

Calculate : 1. Overall efficiency of the plant, and 2. Percentage of heat in fuel used in the boiler, economiser and superheater. [**Ans.** 81.3% ; 62.6% ; 9.37% ; 9.33%]

8. The following observations were made during the trial of a boiler plant consisting of a battery of 6 lancashire boilers and an economiser :

Calorific value of coal per kg = 30 MJ/kg ; Mass of feed water per kg of dry coal = 9.1 kg ; Equivalent evaporation from and at 100° C per kg of dry coal = 9.6 kg ; Temperature of feed water to economiser = 12° C ; Temperature of feed water to boiler = 105° C ; Air temperature = 13° C ; Temperature of flue gases entering economiser = 370° C ; Mass of flue gases entering economiser = 18.2 kg/kg of coal ; Mean specific heat of flue gases = 1.005 kJ/kg K.

Find : 1. the efficiency of the boiler alone ; 2. the efficiency of the economiser alone ; and 3. the efficiency of the whole boiler plant. [**Ans.** 72.22% ; 54.43% ; 84.1%]

9. The following particulars were recorded during a steam boiler trial :

Pressure of steam = 11 bar ; Mass of feed water = 4600 kg/h ; Temperature of feed water = 75° C ; Dryness fraction of steam = 0.96 ; Coal used = 490 kg/h ; Calorific value of coal = 35 700 kJ/kg ; Moisture in coal = 4% by mass ; Mass of dry flue gases = 18.57 kg/kg of coal ; Temperature of flue gases = 300° C ; Boiler house temperature = 16° C ; Specific heat of flue gases = 0.97 kJ/kg K.

Draw the heat balance sheet of the boiler per kg of coal.

Ans. Heat balance sheet

Heat supplied	*kJ*		*Heat consumed*	*kJ*	*%*
Heat supplied in 1 kg of coal	34 272	1.	Heat utilised in raising steam.	22 428	65.44
		2.	Heat carried away by dry flue gases.	5116	14.93
		3.	Heat lost in moisture.	121	0.35
		4.	Heat lost in radiation, etc. (by difference)	6607	19.28
Total	34 272		Total	34 272	100

10. In a boiler trial, the following observations were obtained :

Mass of feed water = 1520 kg/h ; Temperature of feed water = 30° C ; Dryness fraction of steam = 0.95 ; Pressure of steam = 8.5 bar ; Coal burnt/hour = 200 kg ; Calorific value of coal = 27 300 kJ/kg of coal ; Ash and unburnt coal collected = 16 kg/h ; Calorific value of ash and unburnt coal = 3780 kJ/kg ; Mass of flue gases = 17.3 kg/kg of coal ; Temperature of flue gases = 330° C ; Boiler room temperature = 17° C ; Mean specific heat of flue gases = 1 kJ/kg K.

Estimate the thermal efficiency of the boiler and draw the heat balance sheet. [**Ans.** 70.8%]

Ans. Heat balance sheet

Heat supplied	*kJ*	*Heat consumed*	*kJ*	*%*
Heat supplied by 1 kg of coal	27 300	1. Heat utilised in raising steam	19 329	70.80
		2. Heat carried away by the flue gases	5415	19.83
		3. Heat lost in ash and unburnt coal	302	1.11
		4. Heat lost in radiation, etc. (by difference)	2254	8.26
Total	27 300	Total	27 300	100

QUESTIONS

1. What do you understand by the evaporative capacity of a boiler ?
2. Explain clearly the equivalent evaporation from and at 100° C.
3. Discuss briefly the term boiler efficiency.
4. Enlist the various heat losses in a boiler. Which is the biggest loss ?
5. Draw the heat balance sheet of a boiler.

OBJECTIVE TYPE QUESTIONS

1. The equivalent evaporation is defined as the
   (*a*) ratio of heat actually used in producing the steam to the heat liberated in the furnace
   (*b*) amount of water evaporated or steam produced in kg per kg of fuel burnt.
   (*c*) amount of water evaporated 'from and at 100° C' into dry and saturated steam
   (*d*) none of the above

2. When the enthalpy or total heat of steam is h kJ/kg and the enthalpy or sensible heat of feed water is h_{f1} kJ/kg, then the factor of evaporation is given by

   (*a*) $\dfrac{h - h_{f1}}{2257}$ (*b*) $\dfrac{h + h_{f1}}{2257}$ (*c*) $\dfrac{h \times h_{f1}}{2257}$ (*d*) $\dfrac{h}{h_{f1} \times 2257}$

3. The amount of water evaporated in kg per kg of fuel burnt is called
   (*a*) equivalent evaporation 'from and at 100° C'
   (*b*) evaporative capacity of a boiler
   (*c*) boiler efficiency
   (*d*) none of the above

4. The ratio of heat actually used in producing the steam to the heat liberated in the furnace is called
   (*a*) equivalent evaporation 'from and at 100° C'
   (*b*) evaporative capacity of a boiler
   (*c*) boiler efficiency
   (*d*) none of the above

5. In a boiler, various heat losses take place. The biggest loss is due to
   (*a*) moisture in fuel (*b*) dry flue gases (*c*) steam formation (*d*) unburnt carbon

ANSWERS

1. (*c*)	2. (*a*)	3. (*b*)	4. (*c*)	5. (*b*)

16

Boiler Draught

1. Introduction. 2. Classification of Draughts. 3. Types of Draughts. 4. Advantages and Disadvantages of Mechanical Draught. 5. Comparison between Forced Draught and Induced Draught. 6. Balanced Draught. 7. Height of Chimney. 8. Condition for Maximum Discharge through the Chimney. 9. Power Required to Drive a Fan. 10. Efficiency of Chimney.

16.1. Introduction

In the last chapters, we have already discussed the formation of steam and combustion of fuels. It may be noted that the rate of steam generation, in a boiler, depends upon the rate at which the fuel is burnt. The rate of fuel burning depends upon the availability of oxygen or in other words availability of fresh air. The fresh air will enter the fuel bed, if the gases of combustion are exhausted from the combustion chamber of the boiler. This is possible only if a difference of pressure is maintained above and below the fire grate. This difference of pressure is known as *draught*.

The main objects of producing draught in a boiler are :

1. To provide an adequate supply of air for the fuel combustion.
2. To exhaust the gases of combustion from the combustion chamber.
3. To discharge these gases to the atmosphere through the chimney.

16.2. Classification of Draughts

In general, the draughts may be classified into the following two types :

1. *Natural draught.* It is the draught produced by a chimney due to the difference of densities between the hot gases inside the chimney and cold atmospheric air outside it.

2. *Artificial draught.* The artificial draught may be a mechanical draught or a steam jet draught. The draught produced by a fan or blower is known as *mechanical* or *fan draught* where as the draught produced by a steam jet is called *steam jet draught.* The artificial draught is provided, when natural draught is not sufficient. It may be induced or forced.

16.3. Types of Draughts

In general, the draughts are of the following three types :

1. *Chimney draught.* The draught produced by means of a chimney alone is known as *chimney draught.* It is a natural draught and has induced effect. Since the atmospheric air (outside the chimney) is heavier than the hot gases (inside the chimney), the outside air will flow through the furnace into the chimney. It will push the hot gases to pass through the chimney. The chimney draught varies with climatic conditions, temperature of furnace gases and height of chimney.

2. *Mechanical or fan draught.* The draught, produced by means of a fan or blower, is known as *mechanical draught* or *fan draught.* The fan used is, generally, of centrifugal type and is driven by an electric motor.

In an *induced fan draught,* a centrifugal fan is placed in the path of the flue gases before they enter the chimney. It draws the flue gases from the surface and forces them up through the chimney. The action of this type of draught is similar to that of the natural draught.

In case of *forced fan draught*, the fan is placed before the grate, and air is forced into the grate through the closed ash pit.

3. *Steam jet draught.* It is a simple and cheap method of producing artificial draught. In a steam jet draught, the exhaust steam, from a non-condensing steam engine, is used for producing draught. It is mostly used in locomotive boilers, where the exhaust steam from the engine cylinder is discharged through a blast pipe placed at the smoke box and below the chimney.

In an *induced steam jet draught*, the steam jet issuing from a nozzle, is placed in the chimney. But in a *forced steam jet draught*, the steam jet issuing from a nozzle is placed in the ash pit under the fire grate of the furnace.

16.4. Advantages and Disadvantages of Mechanical Draught

These days the mechanical draught is widely used in large boiler plants. It has the following advantages and disadvantages over the natural draught.

Advantages

1. It is more economical.
2. It is better in control.
3. The flow of air through the grate and furnace is uniform.
4. It produces more draught.
5. Its rate of combustion is very high.
6. Low grade fuel can be used.
7. The air flow can be regulated according to the changing requirements.
8. It is not affected by the atmospheric temperature.
9. It reduces the amount of smoke.
10. It reduces height of chimney.
11. It increases efficiency of the plant.
12. It reduces the fuel consumption. About 15% of fuel is saved for the same amount of work.

Disadvantages

1. Its initial cost is high.
2. Its running cost is also high.
3. It has increased maintenance cost.

16.5. Comparison between Forced Draught and Induced Draught

The following table gives the comparison between forced draught and induced draught.

S. No.	*Forced draught*	*Induced draught*
1.	The fan is placed before the fire grate.	The fan is placed after the fire grate.
2.	The pressure inside the furnace is above the atmospheric pressure.	The pressure inside the furnace is below the atmospheric pressure.
3.	It forces fresh air into the combustion chamber.	It sucks hot gases from the combustion chamber, and forces them into the chimney.

4.	It requires less power as the fan has to handle cold air only. Moreover, volume of air handled is less because of low temperature of the cold air.	It requires more power as the fan has to handle hot air and flue gases. Moreover, volume of air and gases is more because of high temperature of the air and gases.
5.	The flow of air through grate and furnace is more uniform.	The flow of air through grate and furnace is less uniform.
6.	As the leakages are outward, therefore there is a serious danger of blow out when the fire doors are opened and the fan is working.	As the leakages are inward, therefore there is no danger of blow out. But if the fire doors are opened and the fan is working there will be a heavy air infiltration.

16.6. Balanced Draught

It is an improved type of draught, and is a combination of induced and forced draught. It is produced by running both induced and forced draught fans simultaneously.

16.7. Height of Chimney

We have already discussed that natural draught is produced by means of a chimney. Since the amount of draught depends upon the height of chimney, therefore its height should be such that it can produce a sufficient draught.

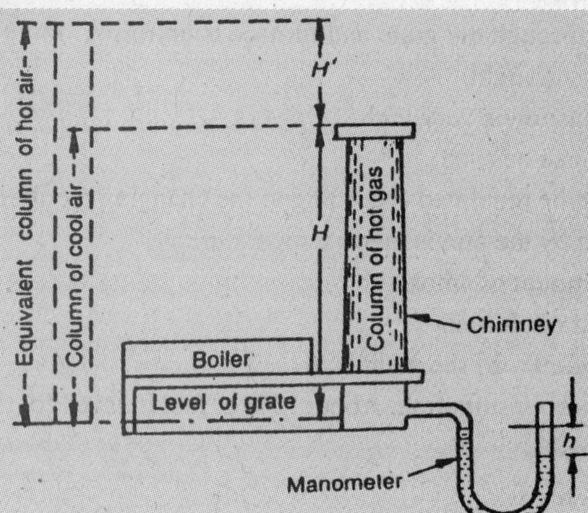

Fig. 16.1. Height of chimney for a given draught.

Let H = Height of chimney above the fire grate in metres.

h = Draught required in terms of mm of water.

T_1 = Absolute temperature of air outside the chimney in K.

T_2 = Absolute temperature of the flue gas inside the chimney in K.

v_1 = Volume of outside air at temperature T_1 in m^3/ kg of fuel.

v_2 = Volume of flue gases inside the chimney at temperature T_2 in m^3/ kg of fuel

m = Mass of air actually used in kg / kg of fuel.

$\therefore$ $m + 1$ = Mass of flue gases in kg per kg of fuel.

First of all, let us find the volume of outside air per kg of fuel at N.T.P. (*i.e.* at 0° C temperature and 1.013 bar pressure).

Let v_0 = Volume of air at 0° C.

∴ Absolute temperature,

$$T_0 = 0° + 273 = 273\text{ K}$$

Atmospheric pressure,

$$p_0 = 1.013 \text{ bar} = 1.013 \times 10^5 \text{ N/m}^2 \qquad \ldots (\because 1 \text{ bar} = 10^5 \text{ N/m}^2)$$

We know that $p \times v = m\,R\,T$

or
$$v_0 = \frac{m\,R\,T_0}{p_0} = \frac{m \times 287 \times 273}{1.013 \times 10^5} = 0.773\,m \text{ m}^3\text{/ kg of fuel}$$

$$\ldots (\because \text{ For air, } R = 287 \text{ J/kg K})$$

Volume of outside air at T_1 K,

$$v_1 = \frac{v_0 \times T_1}{T_0} \qquad \ldots \left(\because \frac{v_0}{T_0} = \frac{v_1}{T_1}\right)$$

$$= \frac{0.773\,m \times T_1}{273} = \frac{m\,T_1}{353} \text{ m}^3\text{/kg of fuel}$$

Density of outside air at T_1 K,

$$\rho_1 = \frac{m}{\dfrac{m\,T_1}{353}} = \frac{353}{T_1} \text{ kg/m}^3 \qquad \ldots \left(\because \text{Density} = \frac{\text{Mass}}{\text{Volume}}\right)$$

∴ Pressure due to a similar column of outside (cold) air,

$$p_1 = \text{Density} \times \text{Height} \times g = \rho_1\,H\,g$$

$$= \frac{353}{T_1} \times H \times 9.81 = \frac{3463\,H}{T_1} \text{ N/m}^2$$

According to Avogadro's law, *the flue gas at N.T.P. occupies the same volume as that of air used at N.T.P.*

∴ Volume of flue gases at 0° C

$$= 0.773\,m \text{ m}^3\text{/ kg of fuel}$$

and volume of flue gases at T_2 K,

$$v_2 = \frac{m\,T_2}{353} \text{ m}^3\text{/ kg of fuel}$$

Density of flue gases at T_2 K,

$$\rho_2 = \frac{m+1}{\dfrac{m\,T_2}{353}} = \frac{353\,(m+1)}{m\,T_2} \text{ kg/m}^3$$

∴ Pressure due to column of hot gases at the base of chimney,

$$p_2 = \rho_2\,H\,g = \frac{353\,(m+1)\,H \times 9.81}{m\,T_2} = \frac{3463\,(m+1)\,H}{m\,T_2} \text{ N/m}^2$$

We know that the draught pressure is due to the pressure difference between the hot column of gas in the chimney and a similar column of cold air outside the chimney. Therefore draught pressure,

$$p = p_1 - p_2 = \frac{3463\,H}{T_1} - \frac{3463\,(m+1)\,H}{m\,T_2}\ \text{N/m}^2$$

$$= 3463\,H\left(\frac{1}{T_1} - \frac{m+1}{m\,T_2}\right)\text{N/m}^2 \quad \ldots (i)$$

In actual practice, the draught pressure is expressed in mm of water as indicated by a manometer. Since* 1 N/m^2 = 0.101 937 mm of water, therefore

$$h = 353\,H\left(\frac{1}{T_1} - \frac{m+1}{m\,T_2}\right)\ \text{mm of water} \quad \ldots (ii)$$

Notes : 1. The equations (*i*) and (*ii*) give only the theoretical value of the draught and is known as *static draught*. The actual value of the draught is less than the theoretical value due to the following reasons :

(*a*) The effect of frictional resistance offered to the passage of air through the fire bars, fire flues and chimney is to reduce the draught *h*.

(*b*) The temperature of flue gases inside the chimney diminishes for every metre of its height.

2. The draught may also be expressed in terms of column of hot gases. If H' is the height in metres of the hot gas column which would produce the draught pressure *p*, then

$$p = \text{Density} \times H' \times g$$

$$= \frac{353\,(m+1)}{m\,T_2} \times H' \times 9.81 = \frac{3463\,(m+1)}{m\,T_2} \times H'\ \text{N/m}^2$$

Substituting this value in equation (*i*) above,

$$\frac{3463\,(m+1)}{m\,T_2} \times H' = 3463\,H\left(\frac{1}{T_1} - \frac{m+1}{m\,T_2}\right)$$

$$\therefore \quad H' = H\left[\left(\frac{m}{m+1} \times \frac{T_2}{T_1}\right) - 1\right] \text{metres}$$

3. The velocity of flue gases through the chimney under a static draught of H' metres is given by

$$V = \sqrt{2gH'} = 4.43\sqrt{H'} \quad \ldots \text{(Neglecting friction)}$$

Example 16.1. *A chimney is 28 m high and the temperature of the hot gases in the chimney is 320° C. The temperature of outside air is 23° C and the furnace is supplied with 15 kg of air per kg of coal burnt. Calculate draught in mm of water.*

Solution. Given : $H = 28$ m ; $T_2 = 320°\text{ C} = 320 + 273 = 593$ K ; $T_1 = 23°\text{ C} = 23 + 273 = 296$ K ; $m = 15$ kg / kg of coal

We know that draught,

$$h = 353\,H\left(\frac{1}{T_1} - \frac{m+1}{m\,T_2}\right) = 353 \times 28\left(\frac{1}{296} - \frac{15+1}{15 \times 593}\right)$$

$$= 15.6 \text{ mm of water } \textbf{Ans.}$$

---

* We know that $1\ \text{kg/m}^2 = 1$ mm of water

or $9.81\ \text{N/m}^2 = 1$ mm of water

$\therefore \quad 1\ \text{N/m}^2 = \frac{1}{9.81} = 0.101\,937$ mm of water

Example 16.2. *A boiler uses 18 kg air per kg of fuel. Determine the minimum height of chimney required to produce a draught of 25 mm of water. The mean temperature of chimney gases is 315° C and that of outside air 27° C.*

Solution. Given : m = 18 kg/kg of fuel ; h = 25 mm of water ; T_2 = 315° C = 315 + 273 = 588 K ; T_1 = 27° C = 27 + 273 = 300 K

Let H = Minimum height of chimney required.

We know that draught (h),

$$25 = 353\,H\left(\frac{1}{T_1} - \frac{m+1}{m\,T_2}\right) = 353\,H\left(\frac{1}{300} - \frac{18+1}{18 \times 588}\right) = 0.53\,H$$

$\therefore$ H = 47.2 m **Ans.**

Example 16.3. *The following data pertain to a steam power plant :*

Height of chimney = 30 m ; Draught produced = 16.5 mm of water gauge ; Temperature of flue gas = 360° C ; Temperature of boiler house = 28° C ; Atmospheric pressure = 1.013 bar.

Determine the quantity of air used per kg of fuel burnt in the boiler.

Solution. Given : H = 30 m ; h = 16.5 mm of water ; T_2 = 360° C = 360 + 273 = 633 K ; T_1 = 28° C = 28 + 273 = 301 K ; p_0 = 1.013 bar

Let m = Quantity of air used per kg of fuel.

We know that draught (h),

$$16.5 = 353\,H\left(\frac{1}{T_1} - \frac{m+1}{m\,T_2}\right) = 353 \times 30\left(\frac{1}{301} - \frac{m+1}{m \times 633}\right)$$

$$\frac{m+1}{m \times 633} = \frac{1}{301} - \frac{16.5}{353 \times 30} = 0.001\,76$$

$$m + 1 = 0.001\,76\,(m \times 633) - 1.114\,m$$

$\therefore$ m = 8.772 kg / kg of fuel **Ans.**

Example 16.4. *A 30 m high chimney is used to discharge hot gases at 297° C to the atmosphere which is at 27° C. Find the mass of air actually used per kg of fuel, if the draught produced is 15 mm of water. If the coal burnt in the combustion chamber contains 80% carbon, 6% moisture and remaining ash, determine the percentage of excess air supplied.*

Solution. Given : H = 30 m ; T_2 = 297° C = 297 + 273 = 570 K ; T_1 = 27° C = 27 + 273 = 300 K ; h = 15 mm of water

Mass of air used per kg of fuel

Let m = Mass of air used per kg of fuel.

We know that draught,

$$h = 353\,H\left(\frac{1}{T_1} - \frac{m+1}{mT_2}\right)$$

$$\therefore \quad 15 = 353 \times 30\left(\frac{1}{300} - \frac{m+1}{m \times 570}\right) = 35.3 - 18.58\left(\frac{m+1}{m}\right)$$

or

$$\frac{m+1}{m} = \frac{35.3 - 15}{18.58} = 1.09$$

$\therefore$ m = 11.11 kg / kg of fuel **Ans.**

Percentage of excess air supplied

We know that 1 kg of carbon requires 8/3 kg of oxygen or in other words 1 kg of carbon requires

$$\frac{8}{3} \times \frac{100}{23} = 11.6 \text{ kg of air}$$

Since 1 kg of coal contains 0.8 kg of carbon, therefore air required for complete combustion of 0.8 kg of carbon

$$= 0.8 \times 11.6 = 9.28 \text{ kg/kg of fuel}$$

∴ Percentage of excess air supplied

$$= \frac{11.11 - 9.28}{9.28} = 0.197 \text{ or } 19.7\ \% \textbf{ Ans.}$$

Example 16.5. *A boiler is equipped with a chimney of 30 metres height. The flue gases, which pass through the chimney are at temperature of 288° C, whereas the atmospheric temperature is 21° C. If the air flow through the combustion chamber is 18 kg/kg of fuel burnt, find : 1. the theoretical draught produced in mm of water and in height of hot gases column, and 2. velocity of the flue gases passing through the chimney, if 50% of the theoretical draught is lost in friction at the grate and passage.*

Solution. Given : $H = 30$ m ; $T_2 = 288°$ C $= 288 + 273 = 561$ K ; $T_1 = 21°$ C $= 21 + 273 = 294$ K ; $m = 18$ kg/kg of fuel

1. *Theoretical draught produced in mm of water*

We know that theoretical draught produced in mm of water,

$$h = 353\,H\left(\frac{1}{T_1} - \frac{m+1}{m\,T_2}\right) = 353 \times 30\left(\frac{1}{294} - \frac{18+1}{18 \times 561}\right)$$

$$= 16.1 \text{ mm of water } \textbf{Ans.}$$

Theoretical draught produced in height of hot gases column

We know that theoretical draught produced in height of hot gases column,

$$H' = H\left[\left(\frac{m}{m+1} \times \frac{T_2}{T_1}\right) - 1\right] = 30\left[\left(\frac{18}{18+1} \times \frac{561}{294}\right) - 1\right]$$

$$= 24.2 \text{ m } \textbf{Ans.}$$

2. *Velocity of flue gases passing through the chimney*

Since 50% of the theoretical draught is lost in friction, therefore net draught available,

$$H' = 24.2 \times 0.5 = 12.1 \text{ m}$$

∴ Velocity of flue gases passing through the chimney,

$$V = 4.43\sqrt{H'} = 4.43\sqrt{12.1} = 15.4 \text{ m/s } \textbf{Ans.}$$

16.8. Condition for Maximum Discharge through the Chimney

We have already discussed in the last article that the height of hot gas column producing the draught,

$$H' = H\left[\left(\frac{m}{m+1} \times \frac{T_2}{T_1}\right) - 1\right] \text{ metres} \qquad \ldots (i)$$

and the velocity of flue gases through the chimney,

$$V = \sqrt{2gH'} = \sqrt{2gH\left[\left(\frac{m}{m+1}\times\frac{T_2}{T_1}\right)-1\right]} \text{ m/s}$$

where H = Height of chimney,

m = Mass of air actually used in kg/kg of fuel,

T_1 = Absolute temperature of the air outside the chimney or atmospheric temperature, and

T_2 = Absolute temperature of the flue gas inside the chimney.

Now consider a chimney discharging hot gases to the atmosphere under the action of the natural draught.

Let A = Area of cross-section of chimney in m^2, and

ρ = Density of hot gases in kg/m^3.

$\therefore$ Mass of hot gases discharged per second,

$$M = \text{Volume of hot gases} \times \text{Density of hot gases} = A\,V\,\rho \qquad \ldots (ii)$$

We know that density of the hot gas is inversely proportional to its temperature, *i.e.*

$$\rho \propto \frac{1}{T_2} \quad \text{or} \quad \rho = \frac{K}{T_2}$$

where K is a constant of proportionality.

Now substituting the value of V and ρ in equation (*ii*),

$$M = A\sqrt{2gH\left[\left(\frac{m}{m+1}\times\frac{T_2}{T_1}\right)-1\right]}\times\frac{K}{T_2}$$

Substituting $A.\,K = K_1$, another constant, we have

$$M = \frac{K_1}{T_2}\sqrt{2gH\left[\left(\frac{m}{m+1}\times\frac{T_2}{T_1}\right)-1\right]}$$

Again substituting $K_1\sqrt{2gH} = K_2$, another constant,

$$M = \frac{K_2}{T_2}\sqrt{\left(\frac{m}{m+1}\times\frac{T_2}{T_1}\right)-1}$$

$$= K_2\sqrt{\left(\frac{m}{m+1}\times\frac{1}{T_1\,T_2}\right)-\frac{1}{T_2^2}} = K_2\left(\frac{m}{m+1}\times\frac{1}{T_1T_2}-\frac{1}{T_2^2}\right)^{1/2}$$

Differentiating M respect to T_2, for maximum discharge and equating to zero.

i.e. $$\frac{dM}{dT_2} = 0$$

$$K_2\times\frac{1}{2}\left(\frac{m}{m+1}\times\frac{1}{T_1T_2}-\frac{1}{T_2^2}\right)^{-1/2}\times\left(-\frac{m}{m+1}\times\frac{1}{T_1}\times\frac{1}{T_2^2}+\frac{2}{T_2^3}\right) = 0$$

or $$K_2\times\frac{1}{2}\times\frac{-\dfrac{m}{m+1}\times\dfrac{1}{T_1}\times\dfrac{1}{T_2^2}+\dfrac{2}{T_2^3}}{\sqrt{\dfrac{m}{m+1}\times\dfrac{1}{T_1T_2}-\dfrac{1}{T_2^2}}} = 0$$

$$\therefore \quad -\frac{m}{m+1} \times \frac{1}{T_1 T_2^2} + \frac{2}{T_2^3} = 0$$

$$\frac{m}{m+1} \times \frac{1}{T_1 T_2^2} = \frac{2}{T_2^3} \quad \text{or} \quad \frac{m}{m+1} \times \frac{1}{T_1} = \frac{2}{T_2}$$

$$\therefore \quad T_2 = 2\left(\frac{m+1}{m}\right) T_1 \qquad \ldots (ii)$$

Thus we see that for maximum discharge, temperature of the flue gases (T_2) should be slightly more than the atmospheric temperature (T_1).

Notes : 1. The height of hot gas column (H')producing the draught for maximum mass of hot gases to be discharged is obtained by substituting the value of T_2 in equation (*i*).

$$H' = H\left[\left(\frac{m}{m+1} \times \frac{2(m+1)T_1}{m \times T_1}\right) - 1\right] = H \text{ metres}$$

It shows that for maximum discharge, the height of hot gas column producing the draught is equal to the height of the chimney.

2. We know that draught pressure,

$$h = 353\, H\left(\frac{1}{T_1} - \frac{m+1}{m\, T_2}\right) \text{mm of water}$$

$\therefore$ The draught pressure for maximum discharge,

$$h = 353\, H\left[\frac{1}{T_1} - \frac{m+1}{m} \times \frac{1}{2\left(\frac{m+1}{m}\right) T_1}\right] \qquad \ldots \left[\because T_2 = 2\left(\frac{m+1}{m}\right) T_1\right]$$

$$= \frac{353\, H}{2\, T_1} = \frac{176.5\, H}{T_1} \text{ mm of water}$$

Example 16.6. *A chimney is 30 m high and the temperature of atmosphere is 12° C. Calculate the draught produced in mm of water under the conditions of maximum discharge.*

Solution. Given : $H = 30$ m ; $T_1 = 12°$ C $= 12 + 273 = 285$ K

We know that draught produced for maximum discharge,

$$h = \frac{176.5\, H}{T_1} = \frac{176.5 \times 30}{285} = 18.6 \text{ mm of water } \textbf{Ans.}$$

Example 16.7. *A chimney is 60 metres high and the temperature of atmospheric air is 27° C. If 15 kg of air / kg of fuel is used ; find for maximum discharge of hot gases : 1. the temperature of hot gases, and 2. the draught pressure in mm of water.*

Solution. Given : $H = 60$ m ; $T_1 = 27°$ C $= 27 + 273 = 300$ K ; $m = 15$ kg / kg of fuel

1. *Temperature of hot gases for maximum discharge*

We know that temperature of hot gases for maximum discharge,

$$T_2 = 2\left(\frac{m+1}{m}\right) T_1 = 2\left(\frac{15+1}{15}\right) 300 = 640 \text{ K} = 367° \text{C } \textbf{Ans.}$$

2. *Draught pressure in mm of water*

We know that draught pressure for maximum discharge,

$$h = \frac{176.5\, H}{T_1} = \frac{176.5 \times 60}{300} = 35.3 \text{ mm of water } \textbf{Ans.}$$

16.9. Power Required to Drive a Fan

The power of a fan (or air power) is the power required to do internal work on the air or gas to deliver its certain volume.

Let p = Draught pressure in N/m^2 = $9.81 \times h$ mm of water,

v = Volume of air or gas flowing through the fan in m^3/min, and

η_f = Efficiency of the fan.

We know that work done by a fan

$= \text{Pressure} \times \text{Volume} = p\,v$ N-m/min

and power required to drive the fan or power of the motor driving the fan,

$$P = \frac{p\,v}{60 \times \eta_f} \text{ watts} \quad \ldots (i)$$

Now let us find the value of P for forced draught fan and induced draught fan.

Let m = Mass of air actually used in kg / kg of fuel,

M = Mass of fuel in kg/min,

T_1 = Absolute temperature of outside air (cold air) in K, and

T_2 = Absolute temperature of hot gases in K.

∴ Mass of air used/min

$= m\,M$ kg

We know that the volume of air at N.T.P.,

$v_0 = 0.773 \times \text{Mass of air used} = 0.773\, m\, M$ kg/min ... (Refer Art. 16.7)

∴ Volume of air handled by a forced draught fan at temperature T_1 K,

$$v_1 = \frac{v_0\,T_1}{T_0} = \frac{0.773\, m\, M\, T_1}{273} = \frac{1}{353}\, m\, M\, T_1 \text{ m}^3 \quad \ldots \left(\because \frac{v}{T_1} = \frac{v_0}{T_0} \right)$$

∴ Power required to drive the forced draught fan,

$$P = \frac{p\, m\, M\, T_1}{60 \times 353 \times \eta_f} = \frac{9.81\, h\, m\, M\, T_1}{60 \times 353 \times \eta_f} = \frac{h\, m\, M\, T_1}{60 \times 36 \times \eta_f} \text{ watts}$$

Now mass of flue gases drawn by the induced draught fan,

$= (m + 1)$ kg / kg of fuel

Mass of fuel per min

$= M$ kg

∴ Mass of flue gases handled by induced draught fan

$= M\,(m + 1)$ kg/min

We know that the volume of flue gases at N.T.P. is equal to the volume of air used at N.T.P.

∴ Volume of flue gases at 0° C,

$v_0 = 0.773\, m\, M$ kg/min

and volume of flue gases handled by induced draught fan at T_2 K,

$$v_2 = \frac{v_0 \times T_2}{T_0} = \frac{0.773\, m\, M\, T_2}{273} = \frac{m\, M\, T_2}{353} \text{ m}^3$$

∴ Power required to drive the induced draught fan,

$$P = \frac{p\,m\,M\,T_2}{60 \times 353 \times \eta_f} = \frac{h\,m\,M\,T_2}{60 \times 36 \times \eta_f} \text{ watts}$$

Note : The comparison of the power required to drive the induced draught and forced draught fan can be made by assuming that :

(*i*) both the fans have the same efficiency, and

(*ii*) they produce the equal draught.

$$\therefore \quad \frac{\text{Power required for I.D. fan}}{\text{Power required for F.D. fan}} = \frac{T_2}{T_1}$$

Example 16.8. *A boiler fitted with a forced draught fan has the following particulars :*

Mass of air required = 20 kg/kg of fuel

Mass of fuel used = 1500 kg/h

Temperature of outside air = 42° C

Temperature of chimney gas = 168° C

Draught pressure = 40 mm of water

Efficiency of fan = 70%

Determine the power required to drive the fan. If the boiler is equipped with induced draught fan, instead of forced draught fan, what will be the power required to drive it ?

Solution. Given : $m = 20$ kg / kg of fuel ; $M = 1500$ kg/h $= 25$ kg/min ; $T_1 = 42° \text{C} = 42 + 273 = 315$ K ; $T_2 = 168° \text{C} = 168 + 273 = 441$ K ; $h = 40$ mm of water ; $\eta_f = 70\% = 0.7$

Power required to drive the forced draught fan

We know that power required to drive the forced draught fan,

$$P_1 = \frac{h\,m\,M\,T_1}{60 \times 36 \times \eta_f} = \frac{40 \times 20 \times 25 \times 315}{60 \times 36 \times 0.7} = 4167 \text{ W}$$

$$= 4.167 \text{ kW} \quad \textbf{Ans.}$$

Power required to drive the induced draught fan

We know that power required to drive the induced draught fan,

$$P_2 = \frac{h\,m\,M\,T_2}{60 \times 36 \times \eta_f} = \frac{40 \times 20 \times 25 \times 441}{60 \times 36 \times 0.7} = 5833 \text{ W}$$

$$= 5.883 \text{ kW} \quad \textbf{Ans.}$$

16.10. Efficiency of Chimney

It may be defined as the ratio of the energy required to produce the artificial draught (expressed in metres head or J/kg of flue gas) to the mechanical equivalent of extra heat carried away per kg of flue gases due to the natural draught.

Let H' = Height of the flue gas column or the artificial draught produced in metres.

T_2 = Temperature of flue gases in chimney with natural draught in K,

T = Temperature of flue gases in chimney with artificial draught in K,

c_p = Specific heat of flue gases in kJ/kg K. Its value may be taken as 1.005 kJ/kg K.

We know that energy required to produce the artificial draught, per kg of flue gas

$$= H' \, g \ \text{J/kg of flue gas}$$

and extra heat carried away per kg of flue gas due to natural draught

$$= 1 \times c_p \, (T_2 - T) \ \text{kJ/kg}$$

∴ Mechanical equivalent of extra heat carried away

$$= 1000 \, c_p \, (T_2 - T) \ \text{J/kg of flue gas}$$

and efficiency,

$$\eta_c = \frac{H' g}{1000 \, c_p \, (T_2 - T)}$$

Notes : 1. In the above expression the value of H' may be substituted as :

$$H' = H\left[\left(\frac{m}{m+1} \times \frac{T_2}{T_1}\right) - 1\right] \text{ metres}$$

2. The efficiency of chimney is less than 1 percent.

Example. 16.9. *In a chimney of height 50 metres, temperature of flue gases with natural draught is 367° C. The temperature of waste gases by using artificial draught is 127° C. The temperature of outside air is 27° C. If air supplied is 19 kg/kg of fuel burnt, determine the efficiency of chimney. Assume c_p = 1.005 kJ/kg K for flue gases.*

Solution. Given : $H = 50$ m ; $T_2 = 367°$ C $= 367 + 273 = 640$ K ; $T = 127°$ C $= 127 + 273 = 400$ K ; $T_1 = 27°$ C $= 27 + 273 = 300$ K ; $m = 19$ kg/kg of fuel ; $c_p = 1.005$ kJ/kg K

We know that height of the artificial draught,

$$H' = H\left[\left(\frac{m}{m+1} \times \frac{T_2}{T_1}\right) - 1\right] = 50\left[\left(\frac{19}{19+1} \times \frac{640}{300}\right) - 1\right] = 51.33 \text{ m}$$

∴ Efficiency of chimney,

$$\eta_c = \frac{H' g}{1000 \, c_p \, (T_2 - T)} = \frac{51.33 \times 9.81}{1000 \times 1.005 \, (640 - 400)}$$

$$= 0.0021 \text{ or } 0.21\ \% \ \textbf{Ans.}$$

EXERCISES

1. Find the draught in mm of water column produced by a chimney 36 m high when the mean temperature of hot gases is 300° C, the temperature of outside air is 27° C and 19 kg of air is supplied per kg of fuel burnt in the furnace. [**Ans.** 19 mm of water]

2. A boiler uses 14 kg of air per kg of fuel. The temperature of the hot gases inside the chimney is 597° C and that of outside air 17° C. If the draught produced is 26 mm of water, determine the minimum height of the chimney required. [**Ans.** 33.2 m]

3. A 30 m high chimney is used to produce a natural draught of 15 mm of water. The temperature of hot gases in the chimney is 287° C. If the temperature of outside air is 27° C ; find the mass of air used per kg of fuel. [**Ans.** 15.6 kg/kg of fuel]

4. Calculate the height of chimney required to generate a pressure difference of 100 mm of water if the mean temperature of gases in the chimney is 150° C and the ambient temperature is 30° C. Neglect flow friction in the chimney. Assume usual value of R for air. [**Ans.** 302 m]

5. A chimney is 30 m high. The temperature of outside air is 21° C. The air is supplied at the rate of 18 kg/kg of fuel for the complete combustion of fuel. Find the temperature of the hot gases inside the chimney if the draught produced is 16 mm of water. [**Ans.** 286° C]

6. A boiler is provided with a chimney of 25 m height. The boiler house temperature is 30° C and the temperature of flue gases leaving the chimney is 300°C. If the air supplied to the boiler is 19 kg per kg of fuel,

estimate : 1. Draught in mm of water ; and 2. Velocity of flue gases passing through the chimney with 50% loss of draught in friction. [**Ans.** 13.24 mm of water ; 14 m/s]

7. A chimney 30 m high is full of hot gases at a temperature of 307° C. The air required for the complete combustion of 1 kg of fuel is 18 kg. If the temperature of atmospheric air is 27° C, find the draught :

1. in terms of water column, and 2. in terms of column of hot gases.

[**Ans.** 16 mm of water ; 25 m]

8. A chimney has a height of 60 metres. The temperature of air is 27° C. Find the draught in mm of water when the temperature of chimney gases is such as to cause the mass of these gases discharged in a given time to be maximum. [**Ans.** 35.3 mm of water]

9. Calculate the efficiency of chimney from the following data :

Temperature of flue gases with natural draught = 350° C ; Temperature of waste gases with artificial draught = 150° C ; Temperature of atmospheric air = 38° C ; Amount of air supplied/kg of fuel = 20 kg ; Height of chimney = 40 m ; Specific heat of flue gases = 1.005 kJ/kg K. [**Ans.** 0.177%]

QUESTIONS

1. What is the significance of draught in boiler practice ?
2. What are the functions of a boiler chimney ? Why is no chimney provided on a locomotive boiler ?
3. Describe briefly various types of artificial draught system used in steam boilers indicating their main advantages.
4. Explain the terms forced draught, induced draught and balanced draught.
5. What are the advantages of artificial draught system over natural draught system ?
6. Deduce a relation for the calculation of natural draught in a boiler plant and state the parameters on which the draught depends.
7. Prove that for maximum mass of hot gases discharged through the chimney,

$$\frac{T_2}{T_1} = \frac{2(m+1)}{m}$$

where T_1 = Absolute temperature of cold air,

T_2 = Absolute temperature of hot air, and

m = Mass of air supplied in kg/kg of fuel.

8. Show that under maximum discharge conditions, the draught produced in terms of hot gas column is equal to the height of chimney.
9. What is the relation between draught pressure (h), height of chimney (H) and absolute temperature of outside air (T_1), when the discharge of hot gases is maximum.
10. Deduce an expression for the power required to drive (*a*) a forced draught fan, and (*b*) an induced draught fan.
11. Define efficiency of chimney and write the expression to calculate the same.

OBJECTIVE TYPE QUESTIONS

1. The air pressure at the fuel bed is reduced below that of atmosphere by means of a fan placed at or near the bottom of the chimney to produce a draught. Such a draught is called

(*a*) natural draught (*b*) induced draught
(*c*) forced draught (*d*) balanced draught

2. The draught may be produced by a

(*a*) mechanical fan (*b*) chimney (*c*) steam jet (*d*) all of these

3. The draught in locomotive boilers is produced by a
   (*a*) chimney (*b*) centrifugal fan (*c*) steam jet (*d*) none of these
4. The draught produced by a steam jet issuing from a nozzle placed in the chimney, is called
   (*a*) induced steam jet draught (*b*) forced steam jet draught
   (*c*) chimney draught (*d*) none of these
5. The chimney draught varies with
   (*a*) climatic conditions (*b*) temperature of furnace gases
   (*c*) height of chimney (*d*) all of these
6. The mechanical draught produces .... draught than natural draught.
   (*a*) more (*b*) less
7. The mechanical draught ...... the amount of smoke.
   (*a*) increases (*b*) decreases (*c*) does not effect
8. The efficiency of the plant ...... with the mechanical draught.
   (*a*) increases (*b*) decreases (*c*) remains constant
9. The velocity of flue gases (V) through the chimney under a static draught of H' metres is given by
   (*a*) $4.43\,H'$ (*b*) $4.43\sqrt{H'}$ (*c*) $(4.43\,H')^2$ (*d*) $4.43\,(H')^2$
10. Which of the following statement is wrong ?
    (*a*) The mechanical draught reduces the height of chimney.
    (*b*) The natural draught reduces the fuel consumption.
    (*c*) A balanced draught is a combination of induced and forced draught.
    (*d*) all of the above
    (*e*) none of the above

ANSWERS

1. (*b*)	2. (*d*)	3. (*c*)	4. (*a*)	5. (*d*)
6. (*a*)	7. (*b*)	8. (*a*)	9. (*b*)	10. (*b*)

17

Simple Steam Engines

1. Introduction. 2. Classification of Steam Engines. 3. Important Parts of a Steam Engine. 4. Working of a Single Cylinder Double Acting Horizontal Reciprocating Steam Engine. 5. Important Terms used in Steam Engines. 6. Indicator Diagram of a Simple Steam Engine. 7. Theoretical or Hypothetical Indicator Diagram. 8. Theoretical or Hypothetical Mean Effective Pressure. 9. Actual Indicator Diagram. 10. Diagram Factor. 11. Power Developed by a Simple Steam Engine. 12. Indicated Power. 13. Effect of Piston Rod in Double Acting Steam Engine. 14. Brake Power. 15. Measurement of Brake Power. 16. Prony Brake Dynamometer. 17. Rope Brake Dynamometer.

17.1. Introduction

In all steam engines, the steam is used as the working substance. These engines operate on the principle of first law of thermodynamics, *i.e.* heat and work are mutually convertible. In a reciprocating steam engine, as the heat energy in the steam is converted into mechanical work by the reciprocating (to and fro) motion of the piston, it is also called *reciprocating steam engine*. Moreover, as the combustion of the fuel takes place outside the engine cylinder, it is also called an *external combustion engine*.

17.2. Classification of Steam Engines

The steam engines have been classified by various scientists on different basis. But the following classifications are important from the subject point of view.

1. *According to number of working strokes*

(*a*) Single acting steam engine, and (*b*) Double acting steam engine.

When steam is admitted on one side of the piston, and one working stroke is produced during each revolution of the crankshaft, it is said to be a *single acting steam engine*. But when the steam is admitted, in turn, on both sides of the piston and two working strokes are produced during each revolution of the crankshaft, it is said to be a *double acting steam engine*. A double acting steam engine produces double the power than that produced by a single acting steam engine.

2. *According to the position of the cylinder*

(*a*) Horizontal steam engine, and (*b*) Vertical steam engine.

When the axis of the cylinder is horizontal, it is said to be a *horizontal steam engine*. But when the axis of the cylinder is vertical, it is called a *vertical steam engine*. A vertical steam engine requires less floor area than the horizontal steam engine.

3. *According to the speed of the crankshaft*

(*a*) Slow speed steam engine, (*b*) Medium speed steam engine, and (*c*) High speed steam engine.

When the speed of the crankshaft is less than 100 revolutions per minute (r.p.m.), it is called a *slow speed steam engine*. But when the speed of the crankshaft is between 100 r.p.m. and 250 r.p.m., it is called a *medium speed steam engine*. Similarly, when the speed of the crankshaft is above 250 r.p.m., it is known as a *high speed steam engine.*

4. *According to the type of exhaust*

(*a*) Condensing steam engine, and (*b*) Non-condensing steam engine.

When steam after doing work in the cylinder passes into a condenser, which condenses the steam into water at a pressure less than the atmospheric pressure, it is said to be a *condensing steam engine*. But when the steam after doing work in the cylinder is exhausted into the atmosphere, it is said to be a *non-condensing steam engine*. The steam pressure in the cylinder is, therefore, not allowed to fall below the atmospheric pressure.

5. *According to the expansion of the steam in the engine cylinder*

(*a*) Simple steam engine, and (*b*) Compound steam engine.

When the expansion of the steam is carried out in a single cylinder and then exhausted into the atmosphere or a condenser, it is said to be a *simple steam engine*. But when the expansion of the steam is completed in two or more cylinders, the engine is called a *compound steam engine*. The compound steam engines are generally condensing engines. But some of them may be non-condensing also.

6. *According to the method of governing employed*

(*a*) Throttling steam engine, and (*b*) Automatic cut-off steam engine.

When the engine speed is controlled by means of a throttle valve in the steam pipe, which regulates the pressure of steam to the engine, it is called a *throttling steam engine*. But when the speed is controlled by controlling the steam pressure with an automatic cut-off governor, it is called an *automatic cut-off steam engine*.

17.3. Important Parts of a Steam Engine

All the parts of a steam engine may be broadly divided into two groups *i.e.* stationary parts and moving parts. Though a steam engine consists of innumerable parts, both stationary and moving, yet the following are important from the subject point of view :

1. *Frame.* It is a heavy cast iron part, which supports all the stationary as well as moving parts and holds them in proper position. It generally, rests on engine foundations.

2. *Cylinder.* It is also a cast iron cylindrical hollow vessel, in which the piston moves to and fro under the steam pressure. Both ends of the cylinder are closed and made steam tight. In small steam engines, the cylinder is made an integral part of the frame.

3. *Steam chest.* It is casted as an integral part of the cylinder. It supplies steam to the cylinder with the movement of D-slide valve.

4. *D-slide valve.* It moves in the steam chest with simple harmonic motion. Its function is to exhaust steam from the cylinder at proper movement.

5. *Inlet and exhaust ports.* These are holes provided in the body of the cylinder for the movement of steam. The steam is admitted from the steam chest alternately to either sides of the cylinder through the inlet ports. The steam, after doing its work in the cylinder, is exhausted through the exhaust port.

6. *Piston.* It is a cylindrical disc, moving to and fro, in the cylinder because of the steam pressure. Its function is to convert heat energy of the steam into mechanical work. Piston rings, made from cast iron, are fitted in the grooves in the piston. Their purpose is to prevent the leakage of steam.

7. *Piston rod.* It is a circular rod, which is connected to the piston on one side and cross head to the other. Its main function is to transfer motion from the piston to the cross-head.

8. *Cross-head.* It is a link between the piston rod and connecting rod. Its function is to guide motion of the piston rod and to prevent it from bending.

9. *Connecting rod.* It is made of forged steel, whose one end is connected to the cross head and the other to the crank. Its function is to convert reciprocating motion of the piston (or cross head) into rotary motion of the crank.

10. *Crank shaft.* It is the main shaft of the engine having a crank. The crank works on the lever principle and produces rotary motion of the shaft. The crank shaft is supported on main bearing of the engine.

11. *Eccentric.* It is generally made of cast iron, and is fitted to the crank shaft. Its function is to provide reciprocating motion to the slide valve.

12. *Eccentric rod and valve rod.* The eccentric rod is made of forged steel, whose one end is fixed to the eccentric and other to the valve rod. Its function is to convert rotary motion of the crankshaft into to and fro motion of the valve rod. The valve rod connects the eccentric and the D-slide valve. Its function is to provide simple harmonic motion to the D-slide valve.

13. *Flywheel.* It is a heavy cast iron wheel, mounted on the crank shaft. Its function is to prevent the fluctuation of engine. It also prevents the jerks to the crankshaft.

14. *Governor.* It is a device to keep the engine speed, more or less, uniform at all load conditions. It is done either by controlling the quantity or pressure of the steam supplied to the engine.

17.4. Working of a Single Cylinder Double Acting Horizontal Reciprocating Steam Engine

The principal parts of a single cylinder, double acting horizontal reciprocating steam engine are shown in Fig. 17.1.

The superheated steam at a high pressure (about 20 atmospheres) from the boiler is led into the steam chest. After that the steam makes its way into the cylinder through any of the ports '*a*' or '*b*' depending upon the position of the D-slide valve. When port '*a*' is open, the steam rushes to the left side of the piston and forces it to the right. At this stage, the slide valve covers the exhaust port and the other steam port '*b*' as shown in Fig. 17.1. Since the pressure of steam is greater on the left side than that on right side, the piston moves to the right.

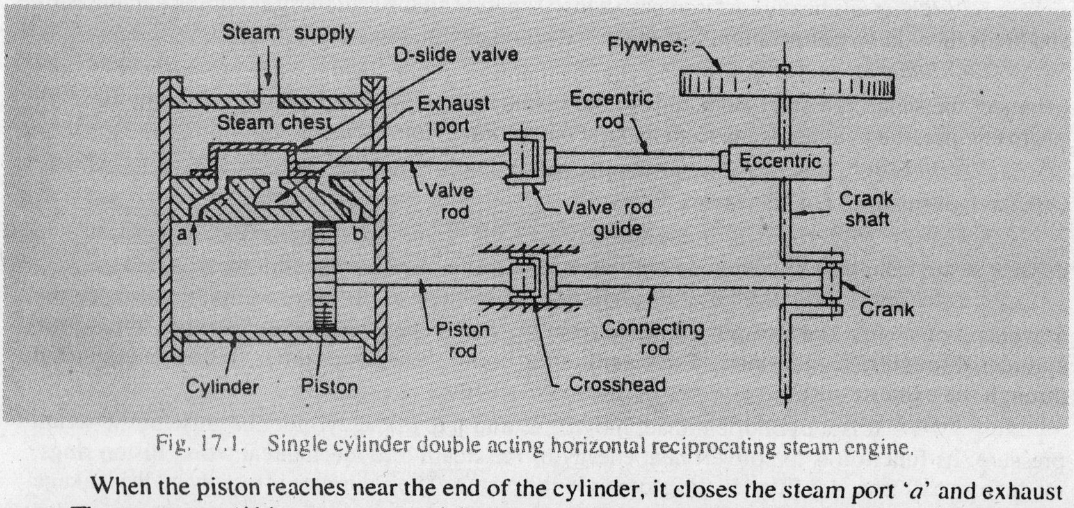

Fig. 17.1. Single cylinder double acting horizontal reciprocating steam engine.

When the piston reaches near the end of the cylinder, it closes the steam port '*a*' and exhaust port. The steam port '*b*' is now open, and the steam rushes to the right side of the piston. This forces the piston to the left and at the same time the exhaust steam goes out through the exhaust pipe, and thus completes the cycle of operation. The same process is repeated in other cycles of operation, and as such the engine works.

Note : At the end of each stroke, the piston changes its direction of motion and is momentarily stopped. The crank comes in line with the piston rod. The extreme left and right positions of the crank, where the piston rod exerts no turning tendency on the main shaft, are called dead centres of the crank.

17.5. Important Terms used in Steam Engines

The *theoretical indicator diagram for a simple steam engine is shown in Fig. 17.2. The following are some important terms used in steam engines.

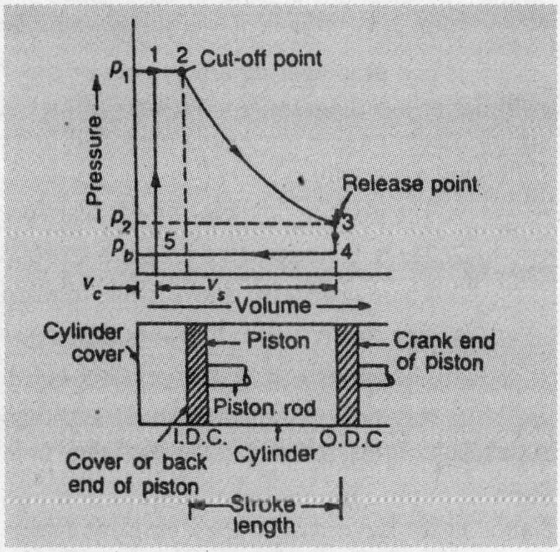

Fig. 17.2. Important terms used in steam engines.

1. *Bore.* The internal diameter of the cylinder of the engine is known as *bore*.

2. *Dead centres.* The extreme positions of the piston inside the cylinder during its motion are known as dead centres. There are two dead centres, *i.e.*

**(*a*) Inner dead centre (I.D.C.), and (*b*) Outer dead centre (O.D.C.).

In a horizontal engine, the inner most position of the piston (towards the cylinder cover end) is known as *inner dead centre*, whereas the outer most position of the piston towards the crank end is called *outer dead centre*, as shown in Fig. 17.2.

3. *Clearance volume.* The volume of space between the cylinder cover and the piston, when the piston is at I.D.C. position is called clearance volume (v_c). It is usually represented as a percentage of stroke volume.

4. *Stroke volume or swept volume.* The volume swept by the piston when it moves from I.D.C. to O.D.C., is known as stroke volume or swept volume (v_s). It is also known as *piston displacement*. Mathematically, stroke volume or swept volume,

$$v_s = \frac{\pi}{4} \times D^2 \times L$$

where D = Bore or internal diameter of the cylinder, and

L = Length of the stroke.

5. *Cut-off volume.* Theoretically, the steam from the boiler enters the clearance space and pushes the piston outward doing external work. At some point during outward movement of the piston, the supply of steam is stopped. The point or the volume where the cut-off of steam takes place is called the point of cut-off or cut-off volume.

6. *Average piston speed.* The distance travelled by the piston per unit time is known as average piston speed. Mathematically,

Average piston speed $= LN$ m/min, for single acting steam engine

$= 2\,LN$ m/min, for double acting steam engine

where L = Length of the stroke in metres, and

N = Speed in R.P.M.

---

* For further details, see Art. 17.7.

** In a vertical engine, these centres are known as bottom dead centre (B.D.C.) and top dead centre (T.D.C.).

7. *Mean effective pressure.* The average pressure on the piston during the working stroke is called mean effective pressure. It is given by the mean depth of the *p-v* diagram. Mathematically, mean effective pressure,

$$p_m = \frac{\text{Workdone per cycle}}{\text{Stroke volume}}$$

17.6. Indicator Diagram of a Simple Steam Engine

It is a graphical representation of the variation in pressure and volume of steam inside the cylinder or *p-v* diagram. As a matter of fact, the theoretical or hypothetical indicator diagram of a simple steam engine has been developed from that of a modified Rankine cycle. It is based on the following assumptions :

1. The opening and closing of steam ports is instantaneous.
2. There is no pressure drop due to condensation.
3. There is no wire drawing due to restricted valve opening.
4. The steam is admitted at boiler pressure and exhausted at condenser pressure.
5. The expansion (or compression) of the steam is hyperbolic (*i.e.* $p\,v = C$)

It may be noted that the above assumptions are not correct from the practical point of view. As a result of this, it has lead to the change in the indicator diagram from the basic modified Rankine cycle.

Note : In the succeeding articles, we shall discuss work done and mean effective pressure (also known as theoretical work done and theoretical mean effective pressure) from the theoretical indicator diagrams.

17.7. Theoretical or Hypothetical Indicator Diagram

The theoretical or hypothetical indicator diagram without clearance and with clearance is shown in Fig. 17.3. In other words, if there is no steam in the cylinder (or there is zero volume of steam at point 1), the indicator diagram will be as shown in Fig. 17.3 (*a*). Similarly, if there is some steam in the cylinder at point 1, the indicator diagram will be as shown in Fig. 17.3 (*b*).

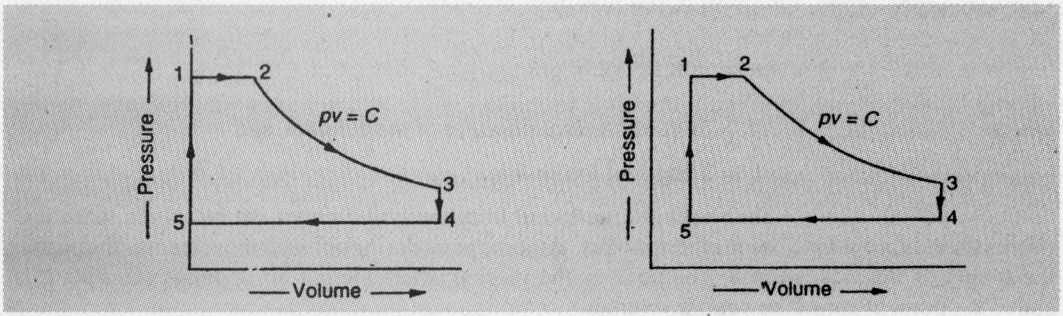

(*a*) Without clearance. (*b*) With clearance.

Fig. 17.3. Theoretical or hypothetical indicator diagram.

The sequence of processes is given below :

1. *Process 1-2.* At point 1, the steam is admitted into the cylinder through the inlet port. As the piston moves towards right, therefore the steam is admitted at constant pressure. Since the supply of steam is cut off at point 2, therefore this point is known as *cut-off point.*

2. *Process 2-3.* At point 2, expansion of steam, in the cylinder, starts with movement of the piston till it reaches the dead end. This expansion takes place hyperbolically (*i.e.* $pv = C$) and pressure falls considerably as shown in Fig. 17.3.

3. *Process 3-4.* At point 3, the exhaust port opens and steam is released from the cylinder to

the exhaust. As a result of steam exhaust, pressure in the cylinder falls suddenly (without change in volume) as shown in Fig. 17.3. The point 3 is known as *release point.*

4. *Process 4-5.* At point 4, return journey of the piston starts. Now the used steam is exhausted at constant pressure, till the exhaust port is closed, and the inlet port is open. The steam pressure at point 4 is called ***back pressure.***

5. *Process 5-1.* At point 5, the inlet port is opened and some steam suddenly enters into the cylinder, which increases the pressure of steam (without change in volume). This process continues till the original position is restored.

17.8. Theoretical or Hypothetical Mean Effective Pressure

The theoretical or hypothetical mean effective pressure may be determined as discussed below, by considering the theoretical indicator diagram without clearance and with clearance as shown in Fig. 17.4 (*a*) and (*b*) respectively.

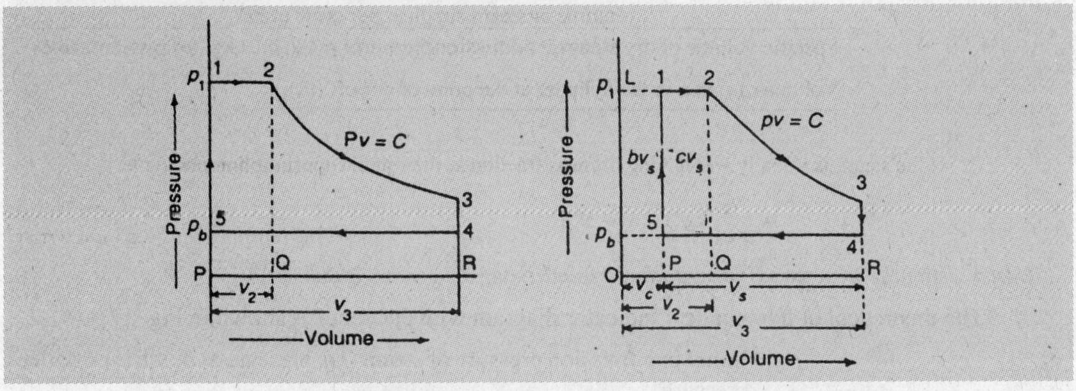

(*a*) Without clearance. (*b*) With clearance.

Fig. 17.4. Theoretical or hypothetical indicator diagram.

1. Considering the theoretical or hypothetical indicator diagram without clearance

The theoretical or hypothetical indicator diagram without clearance is shown in Fig. 17.4 (*a*).

Let p_1 = Initial or admission pressure of steam (*i.e.* pressure at point 1) or boiler pressure,

p_b = Back pressure (*i.e.* pressure at point 4 or 5) or condenser pressure,

v_2 = Volume of steam in the cylinder at the point of cut-off (*i.e.* volume at point 2), and

v_3 = Stroke volume or swept volume or piston displacement volume. It is the volume of steam in the cylinder at the end of stroke (*i.e.* volume at point 3 or 4).

We know that theoretical or hypothetical workdone per cycle

$$= \text{Area of figure } 123451$$

$$= \text{Area } 12\,QP + \text{Area } 23\,RQ - \text{Area } 45PR$$

$$= p_1 v_2 + 2.3\, p_1 v_2 \log (v_3 / v_2) - p_b v_3$$

and theoretical or hypothetical mean effective pressure,

$$p_m = \frac{\text{Workdone per cycle}}{\text{Stroke volume}} = \frac{p_1 v_2 + 2.3\, p_1 v_2 \log (v_3 / v_2) - p_b v_3}{v_3}$$

$$= p_1 \times \frac{v_2}{v_3} + 2.3\, p_1 \times \frac{v_2}{v_3} \log (v_3 / v_2) - p_b = \frac{p_1}{r} (1 + 2.3 \log r) - p_b$$

where $r = v_3 / v_2$ = Expansion ratio

$$= \frac{\text{Volume of steam in the cylinder at the end of stroke}}{\text{Volume of steam at the point of cut–off}}$$

Notes : 1. The volume of steam in the cylinder at the end of stroke, neglecting clearance, is equal to stroke volur.

2. The ratio v_2 / v_3 (*i.e.* reciprocal of expansion ratio)is termed as cut-off ratio. It is defined as the ratio of volume between the points of admission and cut-off of steam and the stroke volume.

3. The steam consumption in kg per cycle may be obtained as follows :

Steam consumption per cycle

$$= \frac{\text{Volume of steam supplied per cycle in m}^3}{\text{Specific volume of dry steam at admission pressure} (p_1) \text{ in m}^3/\text{kg, from steam tables}}$$

$$= \frac{\text{Volume of steam in the cylinder at the point of cut-off } (v_2)}{v_g}$$

If the steam is initially wet having dryness fraction x, then steam consumption per cycle

$$= \frac{v_2}{x\, v_g} \text{ kg}$$

2. *Considering the theoretical or hypothetical indicator diagram with clearance*

The theoretical or hypothetical indicator diagram with clearance is shown in Fig. 17.4 (*b*).

Let p_1 = Initial or admission pressure of steam (*i.e.* pressure at point 1) or boiler pressure,

p_b = Back pressure (*i.e.* pressure at point 4 or 5) or condenser pressure,

v_c = Clearance volume of the cylinder,

v_2 = Volume of steam at the point of cut-off (*i.e.* volume at point 2)

v_s = Stroke volume or swept volume or piston displacement volume.

v_3 = Total volume of steam in the cylinder = $v_c + v_s$,

b = Ratio of clearance volume to stroke volume = v_c / v_s,

c = Ratio of volume between the points of admission and cut-off of steam to the stroke volume = $(v_2 - v_c) / v_s$

We know that theoretical or hypothetical workdone per cycle.

$$= \text{Area of figure 123451} = \text{Area 12 } QP + \text{Area 23 } RQ - \text{Area 45 } PR$$

$$= p_1 (v_2 - v_c) + 2.3\, p_1\, v_2 \log (v_3 / v_2) - p_b\, v_s \quad \ldots (i)$$

and theoretical or hypothetical mean effective pressure,

$$p_m = \frac{\text{Workdone per cycle}}{\text{Stroke volume}} = \frac{p_1 (v_2 - v_c) + 2.3\, p_1\, v_2 \log (v_3 / v_2) - p_b\, v_s}{v_s}$$

$$= p_1 \left(\frac{v_2 - v_c}{v_s} \right) + 2.3 \frac{p_1\, v_2}{v_s} \log \left(\frac{v_3}{v_2} \right) - p_b$$

$$= p_1\,c + 2.3\,p_1 \left(\frac{b\,v_s + c\,v_s}{v_s}\right) \log\left(\frac{v_c + v_s}{b\,v_s + c\,v_s}\right) - p_b$$

$$= p_1\,c + 2.3\,p_1\,(b + c) \log\left(\frac{b+1}{b+c}\right) - p_b \qquad \ldots (ii)$$

Notes : 1. We know that cut-off ratio

$$= \frac{\text{Volume of steam in the cylinder at the point of cut–off}}{\text{Volume of steam in the cylinder at the end of stroke}}$$

$$= \frac{v_2}{v_3} = \frac{b\,v_s + c\,v_s}{v_c + v_s} = \frac{b+c}{b+1}$$

and expansion ratio, $\quad r = \dfrac{v_3}{v_2} = \dfrac{b+1}{b+c}$

2. When clearance is neglected, then $b = 0$ or $r = 1/c$. Now from equation (*ii*),

$$p_m = \frac{p_1}{r}\,(1 + 2.3 \log r) - p_b \qquad \ldots \text{(same as before)}$$

Example 17.1. *A steam engine cylinder receives steam at a pressure of 11.5 bar and cut-off takes place at half of the stroke. Find the theoretical mean effective pressure, if the back pressure of the steam is 0.15 bar. Neglect clearance.*

Solution. Given : $p_1 = 11.5$ bar ; $v_2 = 0.5\,v_3$; $p_b = 0.15$ bar

We know that expansion ratio,

$$r = v_3 / v_2 = v_3 / 0.5\,v_3 = 2$$

We know that theoretical mean effective pressure,

$$p_m = \frac{p_1}{r}\,(1 + 2.3 \log r) - p_b$$

$$= \frac{11.5}{2}\,(1 + 2.3 \log 2) - 0.15 = 9.58 \text{ bar} \quad \textbf{Ans.}$$

Example 17.2. *The cylinder of a non-condensing steam engine is supplied with steam at a pressure of 12 bar. The clearance volume is 1/10 of the stroke volume and the cut-off takes place at 0.25 of the stroke. The back pressure is 1.1 bar. Find the mean effective pressure of the steam on the piston. Assume hyperbolic expansion.*

Solution. Given : $p_1 = 12$ bar ; $b = v_c / v_s = 1/10 = 0.1$; $c = (v_2 - v_c)/v_s = 0.25$; $p_b = 1.1$ bar

We know that mean effective pressure,

$$p_m = p_1\,c + 2.3\,p_1\,(b + c) \log\left(\frac{b+1}{b+c}\right) - p_b$$

$$= 12 \times 0.25 + 2.3 \times 12\,(0.1 + 0.25) \log\left(\frac{0.1+1}{0.1+0.25}\right) - 1.1$$

$$= 3 + 9.66 \log 3.143 - 1.1 = 6.7 \text{ bar} \quad \textbf{Ans.}$$

17.9. Actual Indicator Diagram

The comparison of actual indicator diagram (drawn with firm line) and theoretical indicator diagram (drawn with dotted line) is shown in Fig. 17.5. The following points, regarding actual indicator diagram, are important from the subject point of view :

1. The pressure of steam in the engine cylinder at the beginning of the stroke is less than the boiler pressure. This happens because of the fact that a certain pressure drop is necessary to produce a flow of steam from boiler to the engine cylinder.

2. During the forward stroke of the piston, there is always a slight fall in pressure (shown by line *AB*) due to wire drawing through the steam ports.

3. As the inlet port can not close instantaneously, the point of cut-off will not be sharp as 2, but rounded off as at *B*. The rounding of the cut-off point depends upon the type of valve and valve-mechanism employed.

4. The exhaust port opens before the end of the forward stroke (as shown by point *C*) due to wire drawing through exhaust ports. This causes the rounding off the toe of the diagram.

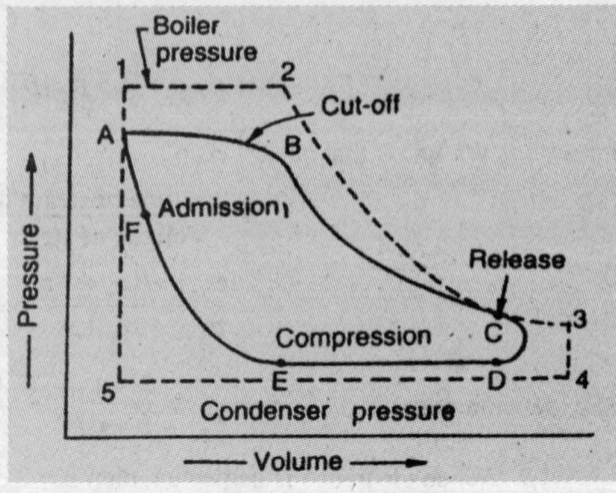

Fig. 17.5. Actual indicator diagram.

5. During the exhaust stroke, the pressure in the cylinder is higher than that of condenser pressure in case of condensing steam engines and higher than atmospheric pressure in case of non-condensing steam engines.

6. The exhaust valve closes at some point *E*, and the remaining steam in the cylinder is compressed along the curve *EF* before the end of the exhaust stroke. This reduces the wire drawing when the inlet valve opens at *F* and also reduces initial condensation. This also serves the purpose of cushioning, which gradually brings the piston, to rest, and thus prevents the shock on the connecting rod bearings, which would otherwise be produced.

7. Due to wire drawing effects, the steam is admitted just before the end of exhaust stroke at *F*. The pressure produced by compression upto this point is raised to admission pressure by the time piston has reached at the end of exhaust stroke.

Notes : 1. The effect of wire drawing is to decrease the area of indicator diagram. In other words, work done by the engine is reduced. This is, however, compensated by the fact that the wire drawing or throttling dries the steam slightly.

2. The effect of clearance, at the first sight, appears to increase the steam consumption from *E* to *F*. But the clearance increases the mean effective pressure and thus increases the work done. However, the net effect of the clearance is to decrease the efficiency. The increase in steam consumption may be reduced by making the point of compression earlier, and thus increasing the pressure obtained at *F* when the fresh steam is admitted to the cylinder. Earlier compression, however, decreases the area of indicator diagram, *i.e.* work done is reduced.

17.10. Diagram Factor

The diagram factor (usually denoted by *K*) is the ratio of the area of actual indicator diagram to the area of theoretical or hypothetical indicator diagram. Mathematically, diagram factor,

$$K = \frac{\text{Area of actual indicator diagram}}{\text{Area of theoretical indicator diagram}}$$

We know that the area of the indicator diagram represents the work done per stroke. Therefore, diagram factor may also be expressed mathematically,

$$K = \frac{\text{Actual work done per stroke}}{\text{Theoretical work done per stroke}}$$

We also know that work done per stroke

$$= \text{Mean effective pressure} \times \text{Swept volume}$$

$\therefore$ Actual work done per stroke

$$= \text{Actual m.e.p} \times \text{Swept volume}$$

and theoretical work done per stroke

$$= \text{Theoretical m.e.p} \times \text{Swept volume}$$

$$\therefore \quad K = \frac{\text{Actual mean effective pressure} (p_a)}{\text{Theoretical mean effective pressure} (p_m)}$$

The diagram factor may, therefore, be defined as the ratio of actual mean effective pressure to the theoretical mean effective pressure.

Notes : 1. The value of the diagram factor, to be used in any particular case, depends upon a number of factors such as initial condition of steam, initial pressure of steam, back pressure, speed of the engine, type of the engine, type of the valves, etc.

2. An average value of K lies between 0.65 and 0.9.

3. Actual mean effective pressure,

$$p_a = \text{Theoretical m.e.p} \times \text{Diagram factor} = p_m \times K$$

Example 17.3. *The steam is supplied at a pressure of 8.4 bar and cut-off occurs at 0.35 of the stroke. The back pressure is 1.25 bar. If the diagram factor is 0.75, determine the actual mean effective pressure. Neglect clearance.*

Solution. Given : $p_1 = 8.4$ bar ; $v_2 = 0.35\, v_3$; $p_b = 1.25$ bar ; $K = 0.75$

We know that expansion ratio,

$$r = v_3 / v_2 = v_3 / 0.35\, v_3 = 2.86$$

We know that theoretical mean effective pressure,

$$p_m = \frac{p_1}{r}(1 + 2.3 \log r) - p_b$$

$$= \frac{8.4}{2.86}(1 + 2.3 \log 2.86) - 1.25 = 4.77 \text{ bar}$$

$\therefore$ Actual mean effective pressure,

$$p_a = p_m \times K = 4.77 \times 0.75 = 3.58 \text{ bar} \quad \textbf{Ans.}$$

17.11. Power Developed by a Simple Steam Engine

The term 'power' may be defined as the rate of doing work. It is thus the measure of performance of a steam engine, *e.g.* an engine doing a certain amount of work in one second will be twice as powerful as an engine doing the same amount of work in two seconds. Mathematically, power developed by an engine,

$$P = \frac{\text{Work done}}{\text{Time taken}}$$

In S.I. system of units, the unit of power is watt (briefly written as W) which is equal to 1 N-m/s or 1 J/s. Generally, a bigger unit of power known as kilowatt (briefly written as kW) is used which is equal to 1000 W.

In case of steam engines, the following two terms are commonly used for the power developed :

1. Indicated power, and 2. Brake power.

The indicated power and brake power are discussed, in detail, in the following pages.

17.12. Indicated Power

The actual power generated in the engine cylinder is called *power input* or *indicated power* (briefly written as I.P.). Since the instrument used to draw the *p-v* diagram (from which work done during the stroke is obtained), is known as indicator, that is why this power is called indicated power. Now consider a simple steam engine, whose indicated power is required to be found out.

Let p_a = Actual mean effective pressure in N/m^2,

A = Area of the cylinder or piston in m^2,

L = Length of the stroke in metres, and

N = Speed of the crankshaft in revolution per minute (r.p.m.).

We know that force on the piston

$$= \text{Pressure} \times \text{Area} = p_a \times A \text{ (in N)}$$

and work done per stroke $= p_a \times A \times L$ N-m

$\therefore$ Work done per minute

$$= p_a \times A \times L \times N \quad \text{... (For single acting)}$$

$$= p_a \times A \times L \times 2N \quad \text{... (For double acting)}$$

and indicated power,

$$\text{I.P.} = \frac{p_a L A N}{60} \text{ W} \quad \text{... (For single acting)}$$

$$[\because 1 \text{ N-m/s or } 1 \text{ J/s} = 1 \text{ W}]$$

$$= \frac{2 p_a L A N}{60} \text{ W} \quad \text{... (For double acting)}$$

When the actual mean effective pressure is given in bar, then

$$\text{I.P.} = \frac{p_a \times 10^5 L A N}{60} \text{ W} \quad \text{... } (\because 1 \text{ bar} = 10^5 \text{ N/m}^2$$

$$= \frac{p_a \times 10^5 \times L A N}{60 \times 1000} \text{ kW} = \frac{100\, p_a L A N}{60} \text{ kW} \quad \text{... (For single acting)}$$

$$= \frac{200\, p_a L A N}{60} \text{ kW} \quad \text{... (For double acting)}$$

Note : The value of actual mean effective pressure (p_a) may also be obtained from the following expression :

$$p_a = \frac{\text{Area of actual indicator diagram in m}^2 \times \text{Spring strength in bar/m}}{\text{Length of actual indicator diagram in m}} \text{ (in bar)}$$

Example 17.4. *A double acting single cylinder has 200 mm stroke, 160 mm diameter. It runs at 250 r.p.m. and the cut-off is 25% of the stroke. The pressure at cut-off is 15 bar and exhaust is at 0.3 bar for a diagram factor of 0.75. Estimate the indicated power in kW.*

Given : L = 200 mm = 0.2 m ; D = 160 mm = 0.16 m ; N = 250 r.p.m. ; v_2 = 25% of stroke = 0.25 v_3 ; p_1 = 15 bar ; p_b = 0.3 bar ; K = 0.75

We know that expansion ratio,

$$r = \frac{v_3}{v_2} = \frac{v_3}{0.25\, v_3} = 4$$

$\therefore$ Theoretical mean effective pressure,

$$p_m = \frac{p_1}{r}(1 + 2.3 \log r) - p_b = \frac{15}{4}(1 + 2.3 \log 4) - 0.3 = 8.64 \text{ bar}$$

and actual mean effective pressure,

$$p_a = p_m \times K = 8.64 \times 0.75 = 6.48 \text{ bar}$$

Area of the cylinder,

$$A = \frac{\pi}{4} \times D^2 = \frac{\pi}{4}(0.16)^2 = 0.02 \text{ m}^2$$

We know that indicated power,

$$\text{I.P.} = \frac{200\, p_a\, L\, A\, N}{60} = \frac{200 \times 6.48 \times 0.2 \times 0.02 \times 250}{60} \text{ kW}$$

$$= 21.6 \text{ kW} \quad \textbf{Ans.}$$

Example 17.5. *Calculate the indicated power and steam consumption in kg/h of a double acting steam engine from the following data :*

Diameter of cylinder = 300 mm ; Stroke = 450 mm ; R.P.M. = 120 ; Steam pressure = 7 bar, and 0.9 dry ; Back pressure = 1.2 bar ; Cut-off takes place at 32 % of stroke for both ends.

Solution. Given : $D = 300$ mm $= 0.3$ m ; $L = 450$ mm $= 0.45$ m ; $N = 120$ r.p.m. ; $p_1 = 7$ bar ; $x = 0.9$; $p_b = 1.2$ bar

Since the cut-off takes place at 32% of the stroke for both ends, therefore

Volume at cut-off $= 0.32 \times$ Stroke volume

$$= 0.32 \times \frac{\pi}{4} \times D^2 \times L = 0.32 \times \frac{\pi}{4}(0.3)^2\, 0.45 = 0.01 \text{ m}^3$$

and expansion ratio,

$$r = \frac{1}{0.32} = 3.125$$

Indicated power

We know that actual mean effective pressure,

$$p_a = K\left[\frac{p_1}{r}(1 + 2.3 \log r) - p_b\right]$$

$$= 1\left[\frac{7}{3.125}(1 + 2.3 \log 3.125) - 1.2\right] = 3.6 \text{ bar} \quad \ldots (\text{Taking } K = 1)$$

Area of cylinder,

$$A = \frac{\pi}{4} \times D^2 = \frac{\pi}{4}(0.3)^2 = 0.07 \text{ m}^2$$

∴ Indicated power,

$$\text{I.P.} = \frac{200\, p_a\, L\, A\, N}{60} = \frac{200 \times 3.6 \times 0.45 \times 0.07 \times 120}{60} = 45.4 \text{ kW} \quad \textbf{Ans.}$$

Steam consumption

From steam tables, corresponding to a pressure of 7 bar, we find that specific volume of dry steam,

$$v_g = 0.2727 \text{ m}^3/\text{kg}$$

We know that mass of steam used per stroke

$$= \frac{\text{Volume of steam at cut-off}}{x\, v_g} = \frac{0.01}{0.9 \times 0.2727} = 0.0407 \text{ kg}$$

and mass of steam used per minute $= 0.0407 \times 2N = 0.0407 \times 2 \times 120 = 9.77$ kg/min

$\therefore$ Steam consumption per hour

$$= 9.77 \times 60 = 586.2 \text{ kg/h} \quad \textbf{Ans.}$$

Example 17.6. *Determine the stroke and diameter of a double acting steam engine cylinder developing 180 kW under the following conditions :*

Initial steam pressure 7 bar ; back pressure 1.12 bar ; crank speed 100 r.p.m. ; average piston speed 135 m/min ; diagram factor 0.8 ; cut-off at 0.4 of the stroke.

Solution. Given : I.P. = 180 kW ; $p_1 = 7$ bar ; $p_b = 1.12$ bar ; $N = 100$ r.p.m. ; Average piston speed = 135 m/min ; $K = 0.8$; $v_2 = 0.4\, v_3$

Stroke of the cylinder

Let L = Length of the stroke in metres.

We know that average piston speed,

$$2LN = 135 \text{ or } L = 135/2N = 135/2 \times 100 = 0.675 \text{ m} \quad \textbf{Ans.}$$

Diameter of the cylinder

Let D = Diameter of the cylinder in metres.

$$\therefore \quad \text{Area, } A = \frac{\pi}{4} \times D^2 = 0.7854\, D^2 \text{ m}^2$$

First of all, let us find the actual mean effective pressure (p_a).

We know that expansion ratio,

$$r = v_3 / v_2 = v_3 / 0.4 v_3 = 2.5$$

$\therefore$ Theoretical mean effective pressure,

$$p_m = \frac{p_1}{r}(1 + 2.3 \log r) - p_b$$

$$= \frac{7}{2.5}(1 + 2.3 \log 2.5) - 1.12 = 4 \text{ bar}$$

and actual mean effective pressure,

$$p_a = p_m \times K = 4 \times 0.8 = 3.2 \text{ bar}$$

We know that indicated power (I.P.),

$$180 = \frac{200\, p_a\, L\, A\, N}{60} = \frac{200 \times 3.2 \times 0.675 \times 0.7854\, D^2 \times 100}{60}$$

$$= 565.5\, D^2$$

$$\therefore \quad D^2 = 0.318 \text{ or } D = 0.564 \text{ m} \quad \textbf{Ans.}$$

Example 17.7. *A single cylinder double acting steam engine is supplied with steam at 11.5 bar and exhaust occurs at 1.1 bar. The cut-off occurs at 40% of the stroke. If the stroke is equal to 1.25 times the cylinder bore and the engine develops an indicated power of 60 kW at 90 r.p.m., determine the bore and stroke of the engine.*

Assume hyperbolic expansion and a diagram factor of 0.79. Also estimate the theoretical steam consumption in m³/min.

Solution. Given : $p_1 = 11.5$ bar ; $p_b = 1.1$ bar ; $v_2 = 40\%\, v_3 = 0.4\, v_3$; $L = 1.25\, D$; I.P. = 60 kW ; $N = 90$ r.p.m. ; $K = 0.79$

Bore and stroke of the engine

Let D = Bore of the cylinder in metres,

L = Length of the stroke in metres = 1.25 D . . . (Given)

We know that expansion ratio,

$$r = v_3 / v_2 = v_3 / 0.4\, v_3 = 2.5$$

and actual mean effective pressure,

$$p_a = K\left[\frac{p_1}{r}(1 + 2.3 \log r) - p_b\right]$$

$$= 0.79\left[\frac{11.5}{2.5}(1 + 2.3 \log 2.5) - 1.1\right] = 6.1 \text{ bar}$$

Area of the cylinder,

$$A = \frac{\pi}{4} \times D^2 = 0.7854 D^2 \text{ m}^2$$

We know that indicated power (I.P.)

$$60 = \frac{200\, p_a\, L\, A\, N}{60} = \frac{200 \times 6.1 \times 1.25\, D \times 0.7854\, D^2 \times 90}{60}$$

$$= 1797\, D^3$$

$\therefore$ $D^3 = 0.0334$ or $D = 0.322$ m **Ans.**

and $L = 1.25\, D = 1.25 \times 0.322 = 0.403$ m **Ans.**

Theoretical steam consumption

We know that stroke volume,

$$v_3 = A\, L = 0.7854\, D^2\, L = 0.7854\, (0.322)^2\, 0.403 = 0.0328 \text{ m}^3$$

Since the cut-off occurs at 40% of the stroke, therefore volume of steam at the point of cut-off per stroke,

$$v_2 = 0.4 \times 0.0328 = 0.013\,12 \text{ m}^3$$

$\therefore$ Theoretical steam consumption per min.

$$= v_2 \times 2\, N = 0.013\,12 \times 2 \times 90 = 2.36 \text{ m}^3\text{/min} \textbf{ Ans.}$$

17.13. Effect of Piston Rod in Double Acting Steam Engines

In a double acting steam engine, the piston rod reduces the effective area of the piston during return stroke as shown in Fig. 17.6.

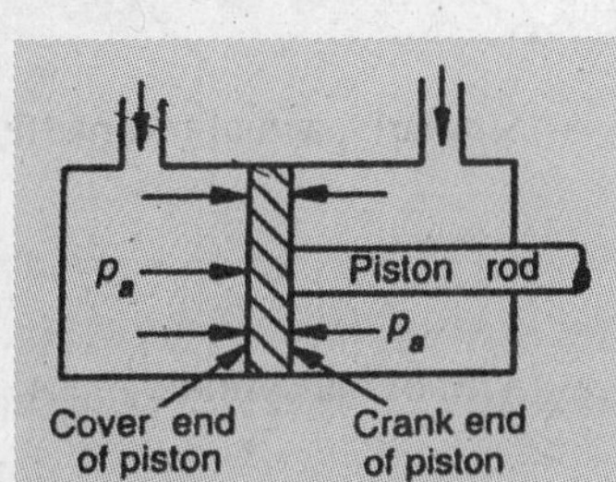

Fig. 17.6. Double acting steam engine.

Let p_a = Actual mean effective pressure on each side of piston in bar,

A = Area of the piston on the cover end or back end side in m², and

a = Area of the piston rod in m².

$\therefore$ Effective area of the piston on the crank end side,

$$A_1 = (A - a) \text{ m}^2$$

We know that indicated power for the cover end or back end side of the piston,

$$(\text{I.P.})_1 = \frac{100\, p_a\, L\, A\, N}{60} \text{ kW}$$

and indicated power for the crank end side of the piston,

$$(\text{I.P.})_2 = \frac{100\, p_a\, L\, A_1\, N}{60} = \frac{100\, p_a\, L\,(A-a)\, N}{60} \text{ kW}$$

∴ Total indicated power of a double acting steam engine,

$$\text{I.P.} = (\text{I.P.})_1 + (\text{I.P.})_2$$

$$= \frac{100\, p_a\, L\, A\, N}{60} + \frac{100\, p_a\, L\,(A-a)\, N}{60} \text{ kW}$$

$$= \frac{100\, p_a\, L\, N\,(A+A-a)}{60} = \frac{100\, p_a\, L\, N\,(2A-a)}{60} \text{ kW}$$

Note : If the actual mean effective pressure on both sides of the piston is different, then total indicated power,

$$\text{I.P.} = \frac{100\, p_{a1}\, L\, A\, N}{60} + \frac{100\, p_{a2}\, L\,(A-a)\, N}{60} \text{ kW}$$

$$= \frac{100\, L\, N}{60} [\, p_{a1}\, A + p_{a2}\,(A-a)] \text{ kW}$$

Example 17.8. *Following data refer to a double acting steam engine :*

Bore = 300 mm ; Stroke = 550 mm ; Piston rod diameter = 30 mm ; Speed = 97 r.p.m. ; Base of both indicator diagrams = 100 mm ; Area of cover and crank side diagrams = 765 and 741 mm² respectively ; Spring number = 500 bar/m ; Steam pressure at inlet = 7.5 bar ; Steam pressure at exhaust = 0.3 bar ; Cut-off = 2/5 of stroke.

Neglect clearance volume of the cylinder and calculate the indicated power of the engine.

Solution : Given : $D = 300$ mm = 0.3 m ; $L = 550$ mm = 0.55 m ; $d = 30$ mm = 0.03 m ; $N = 97$ r.p.m. ; $b = 100$ mm = 0.1 m ; $a_1 = 765\text{ mm}^2 = 765 \times 10^{-6}\text{ m}^2$; $a_2 = 741\text{ mm}^2 = 741 \times 10^{-6}\text{ m}^2$; $s = 500$ bar/m ; *$p_1 = 7.5$ bar ; $p_b = 0.3$ bar ; Cut-off = 2/5 of stroke

We know that actual mean effective pressure for the cover side,

$$p_{a1} = \frac{\text{Area of indicator diagram for cover side} \times \text{Spring number}}{\text{Length or base of indicator diagram}}$$

$$= \frac{a_1 \times s}{b} = \frac{765 \times 10^{-6} \times 500}{0.1} = 3.825 \text{ bar}$$

Similarly, actual mean effective pressure for the crank side,

$$p_{a2} = \frac{a_2\, s}{b} = \frac{741 \times 10^{-6} \times 500}{0.1} = 3.705 \text{ bar}$$

Area of the cylinder, $A = \frac{\pi}{4} \times D^2 = \frac{\pi}{4}(0.3)^2 = 0.0707 \text{ m}^2$

and area of the piston rod, $a = \frac{\pi}{4} \times d^2 = \frac{\pi}{4}(0.03)^2 = 0.707 \times 10^{-3} \text{ m}^2$

* Superfluous data

We know that indicated power for the cover side,

$$(\text{I.P.})_1 = \frac{100\, p_{a1}\, L A N}{60} = \frac{100 \times 3.825 \times 0.55 \times 0.0707 \times 97}{60} = 24.04 \text{ kW}$$

and indicated power for the crank side,

$$(\text{I.P.})_2 = \frac{100\, p_{a2}\, L\,(A-a)\, N}{60} = \frac{100 \times 3.705 \times 0.55\,(0.0707 - 0.707 \times 10^{-3})\, 97}{60}$$

$$= 23.06 \text{ kW}$$

$\therefore$ Total indicated power,

$$\text{I.P.} = (\text{I.P.})_1 + (\text{I.P.})_2 = 24.04 + 23.06 = 47.1 \text{ kW } \textbf{Ans.}$$

17.14. Brake Power

The power available at the crankshaft of an engine is called *power output* or *brake power* (briefly written as B.P.). It has been observed that all the power generated by the engine cylinder is not available at the crankshaft for doing useful work. This happens because some of the power is utilised in overcoming the internal friction of the moving parts of the engine. This power lost in friction is known as *frictional power* (briefly written as F.P.). Thus

$$\text{B.P.} = \text{I.P} - \text{F.P.}$$

17.15. Measurement of Brake Power

The brake power of an engine is measured by an apparatus known as dynamometer. Though there are many types of dynamometers for measuring the brake power of an engine, yet the absorption type dynamometers are important from the subject point of view. These dynamometers are of the following two types :

1. Prony brake dynamometer, and 2. Rope brake dynamometer.

These dynamometers are discussed, in detail, in the following pages.

17.16. Prony Brake Dynamometer

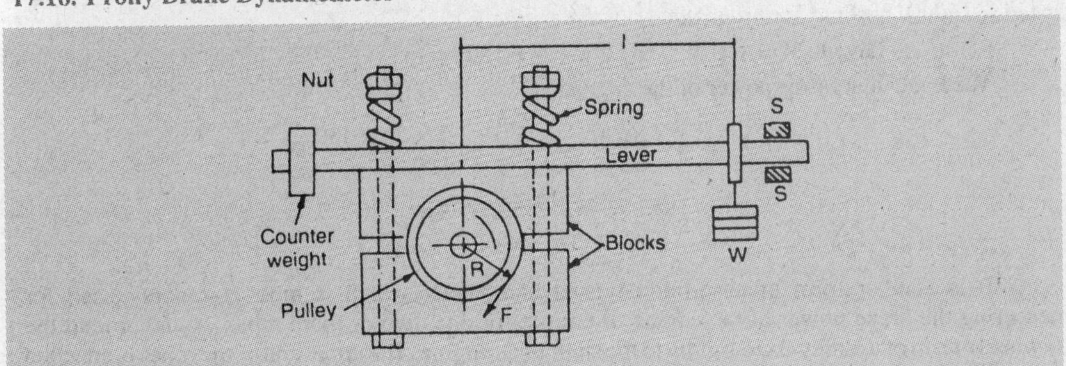

Fig. 17.7. Prony brake dynamometer.

A simplest form of an absorption type dynamometer is a prony brake dynamometer, as shown in Fig. 17.7. It consists of two wooden blocks placed around a pulley fixed to the shaft of an engine whose power is required to be measured. The blocks are clamped by means of two bolts and nuts, as shown in Fig. 17.7. A helical spring is provided between the nut and the upper block to adjust the pressure on the pulley to control its speed. The upper block has a long lever attached to it and carries a weight W at its outer end. A counter weight is placed at the other end of the lever which balances the brake when unloaded. Two stops S, S are provided to limit the motion of the lever.

When the brake is to be put in operation, the long end of the lever is loaded with suitable weights W and the nuts are tightened until the engine shaft runs at a constant speed and the lever is in horizontal position. Under these conditions, the moment due to the weight W must balance the moment of the frictional resistance between the blocks and the pulley.

Let W = Weight at the outer end of the lever in newtons.

l = Horizontal distance of the weight W from the centre of the pulley in metres,

F = Frictional resistance between the blocks and the pulley in newtons,

R = Radius of the pulley in metres, and

N = Speed of the shaft in r.p.m.

We know that the moment of the frictional resistance or torque on the shaft,

$$T = W\,l = F\,R \text{ N-m}$$

Workdone in one revolution

$$= \text{Torque} \times \text{Angle turned in radians} = T \times 2\pi \text{ N-m}$$

$\therefore$ Workdone per minute $= T \times 2\pi N$ N-m

We know that brake power of the engine,

$$\text{B.P.} = \frac{\text{Workdone per min}}{60} = \frac{T \times 2\pi N}{60} = \frac{W\,l \times 2\pi N}{60} \text{ watts}$$

Notes : 1. From the above expression, we see that while determining the brake power of an engine with the help of a prony brake dynamometer, it is not necessary to know the radius of the pulley, the coefficient of friction between the wooden blocks and the pulley and the pressure exerted by tightening of the nuts.

2. When the driving torque on the shaft is not uniform, this dynamometer is subjected to severe oscillations.

Example 17.9. *Following observations were recorded during the trial of a prony brake dynamometer.*

Weight hung from the lever = 100 N ; Distance between weight and pulley = 1.2 m ; Shaft speed = 150 r.p.m. Find the brake power of the engine.

Solution. Given : $W = 100$ N ; $l = 1.2$ m ; $N = 150$ r.p.m.

We know that brake power of the engine,

$$B.P. = \frac{W.l \times 2\pi N}{60} = \frac{100 \times 1.2 \times 2\pi \times 150}{60} \text{ W}$$

$$= 1885 \text{ W} = 1.885 \text{ kW} \quad \textbf{Ans.}$$

17.17. Rope Brake Dynamometer

It is another form of absorption type dynamometer which is most commonly used for measuring the brake power of the engine. It consists of one, two or more ropes wound around the flywheel or rim of a pulley fixed rigidly to the shaft of an engine. The upper end of the ropes is attached to a spring balance while the lower end of the ropes is kept in position by applying a dead weight, as shown in Fig. 17.8. In order to prevent the slipping of the rope over the flywheel, wooden blocks are placed at intervals around the circumference of the flywheel.

In the operation of the brake, the engine is made to run at a constant speed. The frictional torque, due to the rope, must be equal to the torque being transmitted by the engine.

Let W = Dead load in newtons,

S = Spring balance reading in newtons,

D = Diameter of the wheel in metres,

d = Diameter of rope in metres, and

N = Speed of the engine shaft in r.p.m.

$\therefore$ Net load on the brake

$$= (W - S)\, N$$

We know that distance moved in one revolution

$$= \pi (D + d) \text{ m}$$

$\therefore$ Workdone per revolution

$$= (W - S)\, \pi (D + d) \text{ N-m}$$

and workdone per minute

$$= (W - S)\, \pi (D + d)\, N \text{ N-m}$$

$\therefore$ Brake power of the engine,

$$\text{B.P.} = \frac{\text{Workdone per min.}}{60}$$

$$= \frac{(W - S)\, \pi (D + d)\, N}{60} \text{ watts}$$

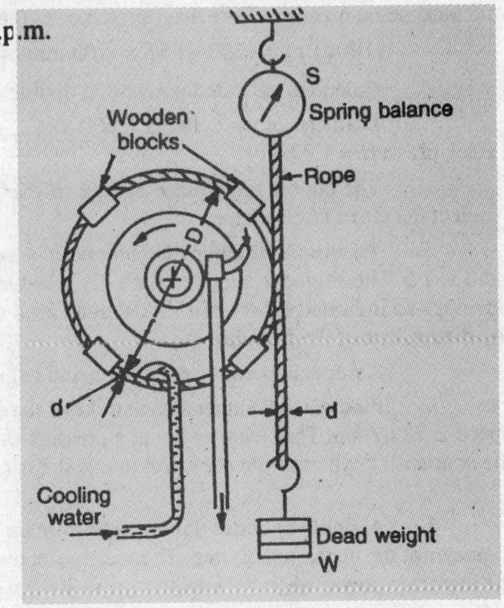

Fig. 17.8. Rope brake dynamometer.

If the diameter of the rope (d) is neglected, then brake power of the engine,

$$\text{B.P.} = \frac{(W - S)\, \pi DN}{60} \text{ watts}$$

Note : Since the energy produced by the engine is absorbed by the frictional resistances of the brake and is transformed into heat, therefore it is necessary to keep the flywheel of the engine cool with soapy water. The flywheels have their rims made of a channel section so as to receive a stream of water which is being whirled round by the wheel. The water is kept continually flowing into the rim and is drained away by a sharp edged scoop on the other side, as shown in Fig. 17.8.

Example 17.10. *The following data were recorded in laboratory experiment with the rope brake :*

Diameter of the flywheel = 1.2 m ; Diameter of the rope = 12.5 mm ; Engine speed = 200 r.p.m. ; Dead load on the brake = 600 N ; Spring balance reading = 150 N.

Calculate brake power of the engine.

Solution. Given : $D = 1.2$ m ; $d = 12.5$ mm $= 0.0125$ m ; $N = 200$ r.p.m. ; $W = 600$ N ; $S = 150$ N

We know that brake power of the engine,

$$\text{B.P.} = \frac{(W - S)\, \pi (D + d)\, N}{60} = \frac{(600 - 150)\, \pi (1.2 + 0.0125)\, 200}{60} \text{ W}$$

$$= 5715 \text{ W} = 5.715 \text{ kW} \quad \textbf{Ans.}$$

EXERCISES

1. In a reciprocating simple steam engine, the steam is supplied at 7 bar and cut-off occurs at 1/4 of the stroke. Determine the mean effective pressure, if the back pressure of the steam is 1 bar **[Ans.** 3.17 bar**]**

2. Steam at a pressure of 10 bar is admitted into a steam engine with expansion ratio of 4.5. If the mean effective pressure required is 3.5 bar, determine the back pressure. Neglect compression and clearance of the engine. **[Ans.** 2.06 bar**]**

3. Find the actual mean effective pressure of a double acting single cylinder steam engine with admission pressure of 10 bar and cut-off at 1/3 of the stroke. The exhaust pressure is 657 mm vacuum with 760 mm barometric pressure. Take diagram factor as 0.68. [Ans. 4.66 bar]

[Hint : p_b = 760 – 657 = 103 mm of mercury = 103 × 133.3 = 13 730 N/m^2 = 0.1373 bar]

4. Calculate indicated power of a double acting steam engine from the following data :

Diameter of the cylinder = 600 mm ; Stroke = 900 mm ; R.P.M. = 88 ; Admission pressure = 8 bar ; Back pressure = 1.8 bar.

The cut-off takes place at 20% of the stroke for both ends. Assume the diagram factor to be 0.8. Neglect the effect of clearance. [Ans. 141.6 kW]

5. The stroke length to diameter ratio in a double acting steam engine cylinder is 1.3 and its expansion ratio is 2.5. The engine is supplied with dry saturated steam at 9.8 bar which exhausts at 1.05 bar. The engine develops an indicated power of 185 kW at a speed of 200 r.p.m. Assuming a diagram factor of 0.8; determine the dimensions of the cylinder.

Assume hyperbolic expansion and neglect clearance. [Ans. 374.5 mm ; 486.8 mm]

6. Find out the diameter and stroke of the engine to develop 35 kW at a speed of 120 r.p.m. The piston speed is 72 m/min. The steam enters at a pressure of 11 bar for 3 / 8 of the stroke. The steam engine is to be of the condensing type and the back pressure is 0.3 bar. Assume the diagram factor to be 0.82 and no clearance. [Ans. 240 mm ; 300 mm]

7. A single cylinder double acting steam engine has piston diameter 250 mm, stroke 400 mm and diameter of the piston rod 50 mm. The mean effective pressure on both sides of the piston is 2.5 bar. Determine the indicated power when the engine runs at 200 r.p.m. [Ans. 32.08 kW]

8. A double acting steam engine with a bore of 300 mm and stroke of 400 mm runs at 300 r.p.m. The inlet is at 8 bar and the back pressure is 1.2 bar. The cut-off occurs at 30% of the stroke. Determine the power developed, taking the diagram factor as 0.8. If the steam is dry saturated at 8 bar at the point of cut-off, determine the steam consumption in kg/h. [Ans. 91.6 kW ; 1260 kg/h]

9. Dry saturated steam is supplied to a single cylinder double acting steam engine at a pressure of 9 bar and is exhausted at 1.4 bar. The cut-off takes place at 0.4 stroke and the engine develops 25 kW. The stroke-bore ratio of the engine is 1.25 and the speed 250 r.p.m. Assume diagram factor of 0.75, Calculate the cylinder bore and piston stroke of the engine. Neglect clearance and assume hyperbolic expansion.

If the actual steam consumption is 1.3 times the theoretical quantity, find specific steam consumption in kg/kWh. [Ans. 0.195 m, 0.244 m ; 21.17 kg/kW h]

10. The following observations were recorded during a test on a single acting, non condensing, single cylinder steam engine : Effective brake diameter = 2.75 m ; Net load on the brake = 1650 N ; Speed = 100 r.p.m.

Find the brake power of the engine. [Ans 23.76 kW]

11. In a single cylinder double acting steam engine, steam is admitted at a pressure of 12 bar and is exhausted at 1.3 bar. The diameter of the cylinder is 250 mm and the stroke length is 450 mm. The cut-off of steam occurs when the piston has moved 150 mm from its I.D.C. position. The r.p.m. of the engine is 260. Neglecting clearance and assuming a diagram factor of 0.9, determine the indicated power of the engine. If the power lost in friction amounts to 7.5 kW, what will be the brake power of the engine. [Ans. 114.86 kW]

QUESTIONS

1. State the classifications of steam engine.
2. Describe, with a neat sketch, the working of a single cylinder, double acting reciprocating steam engine.
3. Explain the following terms as applied to steam engines : (*a*) clearance volume ; (*b*) stroke volume ; (*c*) cut-off volume ; (*d*) release and back pressure ; (*e*) mean effective pressure.
4. What do you understand by hypothetical indicator diagram ? Derive an expression to determine hypothetical mean effective pressure of a steam engine having clearance.
5. How and why does the hypothetical indicator diagram differ from actual indicator diagram ?

6. Discuss, with the help of pressure-volume diagram, the effect of clearance and compression on the workdone per stroke in a steam engine.

7. What is diagram factor ? State the reasons why its value is less than unity.

8. Differentiate between indicated power and brake power of a steam engine.

9. Describe the effect of piston rod in a double acting steam engine.

10. Explain the method of measuring brake power of a steam engine.

OBJECTIVE TYPE QUESTIONS

1. All steam engines work on
   (*a*) Zeroth law of thermodynamics (*b*) first law of thermodynamics
   (*c*) second law of thermodynamics (*d*) none of these
2. A single acting steam engine produces ..... power than that of double acting steam engine.
   (*a*) half (*b*) double (*c*) four times
3. The function of a D-slide valve in a steam engine is
   (*a*) to guide motion of the piston rod and prevent it from bending
   (*b*) to transfer motion from the piston to the crosshead
   (*c*) to convert heat energy of the steam into mechanical work
   (*d*) to exhaust steam from the cylinder at proper moment
4. The ratio of clearance volume to the swept volume is called
   (*a*) cut-off ratio (*b*) expansion ratio
   (*c*) clearance ratio (*d*) none of these
5. In case of condensing steam engines, the pressure of steam in the cylinder during exhaust stroke is ..... condenser pressure.
   (*a*) equal to (*b*) lower than (*c*) higher than
6. The clearance in the engine cylinder
   (*a*) increases the mean effective pressure (*b*) increases the workdone
   (*c*) decreases the efficiency of the engine (*d*) all of these
7. The diagram factor is the ratio of the
   (*a*) area of the actual indicator diagram to the area of theoretical indicator diagram
   (*b*) actual workdone per stroke to the theoretical workdone per stroke
   (*c*) actual mean effective pressure to the theoretical mean effective pressure
   (*d*) all of the above
8. The average value of diagram factor lies between
   (*a*) 0.2 to 0.5 (*b*) 0.6 to 0.65 (*c*) 0.65 to 0.9 (*d*) 0.9 to 1.2
9. For the same length of stroke and speed of crankshaft, the piston speed for a double acting steam engine is ..... the piston speed of single acting steam engine.
   (*a*) equal to (*b*) twice (*c*) four times
10. The actual power supplied by the engine crankshaft is called
   (*a*) indicated power (*b*) brake power (*c*) frictional power

ANSWERS

1. (*b*)	2. (*a*)	3. (*d*)	4. (*b*)	5. (*c*)
6. (*d*)	7. (*d*)	8. (*c*)	9. (*b*)	10. (*b*)

18

Compound Steam Engines

1. Introduction. 2. Arrangement of Cylinders in Compound Steam Engines. 3. Advantages of Compounding of Steam Engines. 4. Classification of Compound Steam Engines. 5. Tandem Type Compound Steam Engine. 6. Woolf Type Compound Steam Engine. 7. Receiver Type Compound Steam Engine. 8. Combined Indicator Diagram of a Compound Steam Engine. 9. Work Done and Power Developed by a Compound Steam Engine. 10. Determination of Cylinder Dimensions for a Compound Steam Engine. 11. Two Cylinder Compound Steam Engine with Complete Continuous Expansion in both the H.P. and L.P. Cylinders. 12. Two Cylinder Compound Steam Engine with Complete Expansion in H.P. Cylinder and Incomplete Expansion in L.P. Cylinder. 13. Two Cylinder Compound Steam Engine with Incomplete Expansion in both the H.P. and L.P. Cylinders. 14. Three Cylinder Compound Steam Engines.

18.1. Introduction

In the previous chapter, we have discussed the working of Simple Steam Engines. The scientists and engineers, working for the improvement of simple steam engines, thought of various methods. One of the method was to use a high pressure steam. They faced the following difficulties in expanding a high pressure steam in a single cylinder :

1. The steam, when admitted into a cylinder, comes in contact with a relatively cold cylinder walls which cause initial condensation.
2. When the steam is expanded down to the condenser pressure, there is a greater range of pressure difference. This causes a larger temperature range in the cylinder.
3. Due to the greater range of pressure difference, the ratio of expansion is large.
4. The stroke of the piston is large due to the large ratio of expansion.

To overcome these difficulties, the expansion of steam is divided into stages, each stage taking place in a different cylinder. In this method, the high pressure steam is first partly expanded in the high pressure cylinder and then exhausted into the low pressure cylinder, where the expansion is completed. This method reduces the ratio of expansion in the cylinder, as a result of which the length of stroke is also reduced. Moreover, it reduces the temperature range in each cylinder. This lead to the development of compound steam engines.

18.2. Arrangement of Cylinders in Compound Steam Engines

We have already discussed that a steam engine, in which the expansion of steam takes place, in more than one cylinder, is known as a compound steam engine. The cylinder, which receives the high pressure steam, is known as high pressure (H.P.) cylinder. The steam after expanding in the high pressure cylinder, exhausts into a larger cylinder known as low pressure (L.P.) cylinder. In this cylinder, the last stage of expansion is performed. The L.P. cylinder generally exhausts into a condenser. That is why, the compound steam engines are generally condensing type, but they may be non-condensing also. If the expansion of steam takes place in three cylinders, the engine is called triple expansion engine. Similarly, if the expansion is carried out in four cylinders, it is known quadruple expansion engine.

In case of triple expansion engines, the first stage of expansion is performed in H.P. cylinder, intermediate expansion takes place in intermediate pressure (I.P.) cylinder, and the last expansion is completed in L.P. cylinder. In quadruple expansion engine, the intermediate expansion is carried out in two I.P. cylinders.

Note : The high pressure cylinder is, generally, of smaller size than the low pressure cylinder.

18.3. Advantages of Compounding of Steam Engines

Following are the advantages of compounding the expansion of steam in two or more cylinders :

1. There is a considerable economy in steam for high pressure operations.
2. The temperature range per cylinder is reduced, with a corresponding reduction in the condensation.
3. The ratio of expansion is reduced, thus reducing the length of stroke.
4. The leakage past the valves and piston is reduced, because of the reduced pressure difference across these parts.
5. The steam can be reheated after expansion in one cylinder, and before entering the next.
6. The mechanical balance can be made more nearly perfect, and therefore high speeds are possible.
7. In case of a breakdown, the engine can be modified to continue working on reduced load.
8. More uniform turning moment is exerted on the crank shaft, by spacing the cranks at 90° in the case of a two cylinder engines or at 120° in triple expansion engines. Thus a lighter flywheel is required.
9. The forces in the working parts are reduced, as the forces are distributed over more parts.
10. The cost of the engine, for the same power and economy, is less than that of a simple steam engine.

18.4. Classification of Compound Steam Engines

The compound steam engines may be classified according to the arrangement of cranks, and the angles between them. Two-cylinder compound engines are generally classified as:

1. Tandem type compound steam engines,
2. Woolf type compound steam engines, and
3. Receiver type compound steam engines.

These compound engines are discussed, in detail, in the following pages :

18.5. Tandem Type Compound Steam Engine

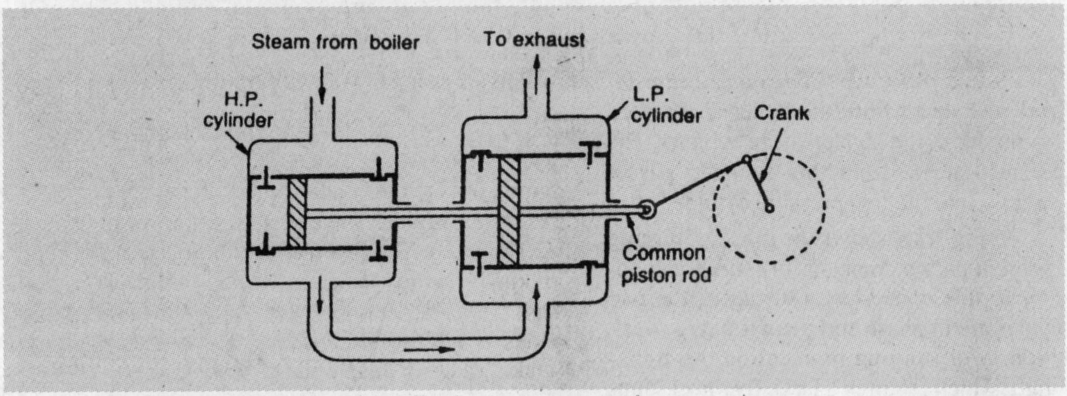

Fig. 18.1. Tandem type compound steam engine.

In this engine, the cylinders (H.P. and L.P.) have a common piston rod working on the same crank as shown in Fig. 18.1.

In a *Tandem type compound steam engine*, the steam from the boiler is admitted to one side of the high pressure cylinder. The exhaust steam, from this cylinder, passes directly into the low pressure cylinder. Since both the pistons are at the end of their strokes, these cylinders may therefore be regarded as having cranks at 0° to each other.

Fig. 18.2 shows the graph of turning moment of the crankshaft *versus* crank angle. It may be noticed from the graph that their cycles are in phase, therefore maximum and minimum turning moments on the crankshaft due to each cylinder will act at the same time (*i.e.* at the same crank angle). This is the disadvantage of this type of compound engine, as a larger flywheel is required to overcome these fluctuations in turning moment.

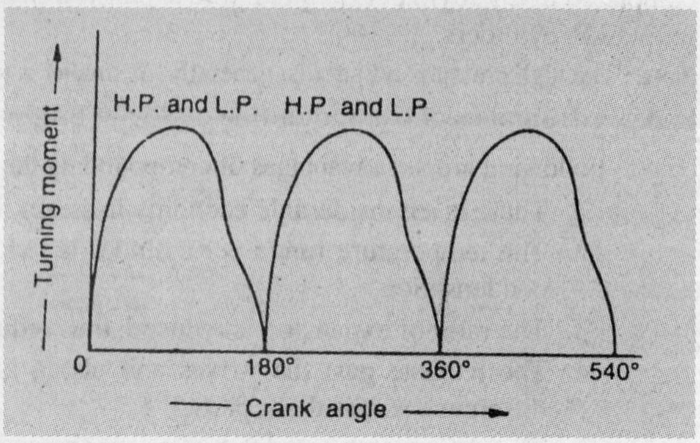

Fig. 18.2. Turning moment-crank angle graph for Tandem type compound steam engine.

18.6. Woolf Type Compound Steam Engine

In this engine, the two cylinders (H.P. and L.P.) have different piston rods attached to two different cranks set of 180° to each other. These cranks are cast in the same crank shaft as shown in Fig. 18.3.

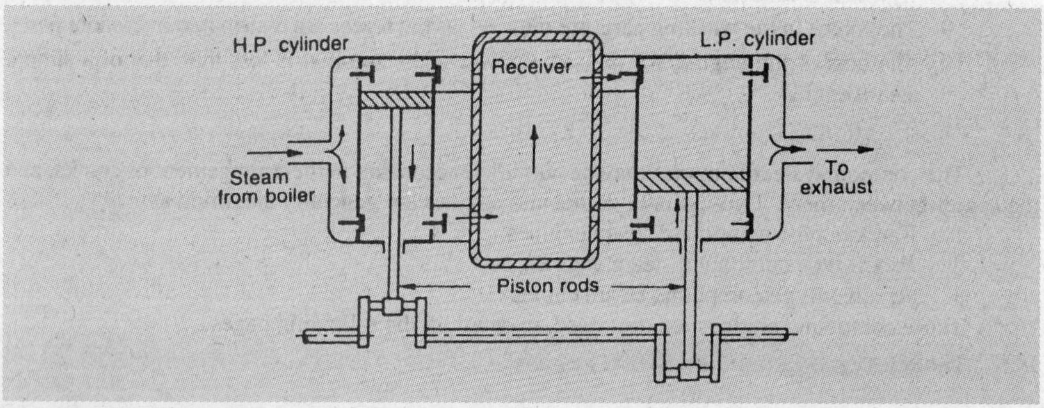

Fig. 18.3. Woolf type compound steam engine.

In a *Woolf type compound steam engine*, the steam from the boiler is admitted to one side of the high pressure cylinder. The exhaust steam, from this cylinder, passes directly into the low pressure cylinder.

Fig. 18.4 shows the graph of turning moment on the crankshaft *versus* crank angle. As the cranks are 180° apart, the two cycles are in phase and there is a large variation in the turning moment on the crankshaft, which requires a large flywheel. Thus the Woolf type compound steam engine has the same disadvantage as the Tandem type compound steam engine.

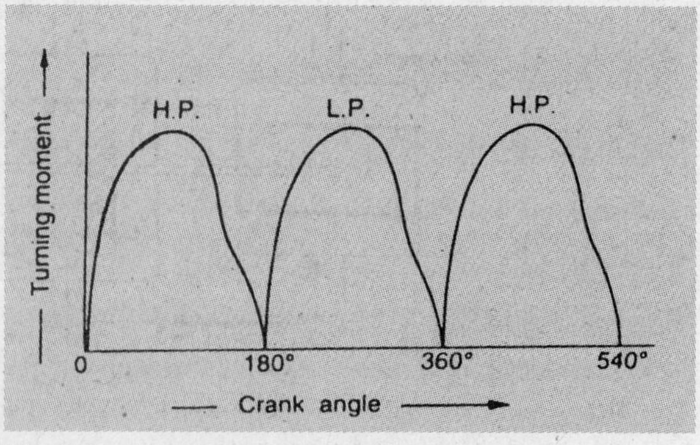

Fig. 18.4. Turning moment-crank angle graph for Woolf type compound steam engine.

18.7. Receiver Type Compound Steam Engine

In this engine, the two cylinders (H.P. and L.P.) have different piston rods attached to two different cranks set at 90° to each other. These cranks are cast in the same crank shaft as shown in Fig. 18.5.

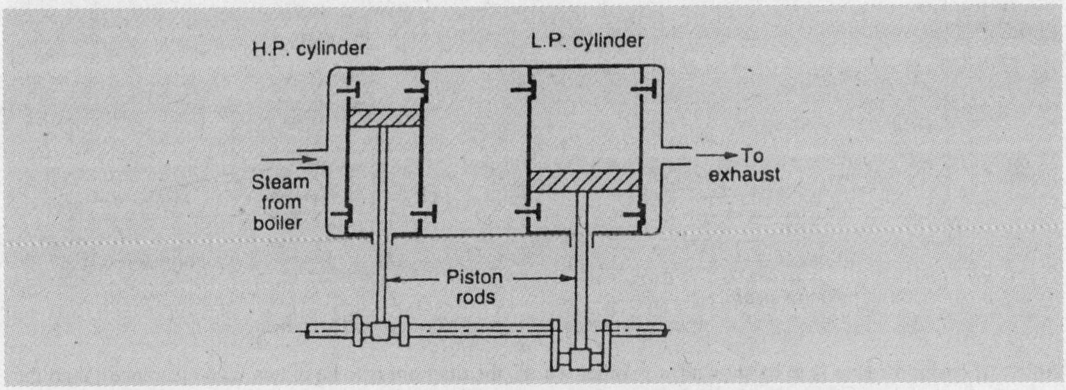

Fig. 18.5. Receiver type compound steam engine.

In a *receiver type compound steam engine*, the steam from the boiler is admitted to one side of the high pressure cylinder. Since the two cranks are set at 90°, therefore the two cylinders are out of phase. As a result of this, the steam can not directly pass from the high pressure cylinder to the low pressure cylinder. It is, therefore, essential to introduce an intermediate vessel known as receiver between the high pressure cylinder and low pressure cylinder, as shown in Fig. 18.5. The steam from the high pressure cylinder enters the receiver, from which it enters the low pressure cylinder.

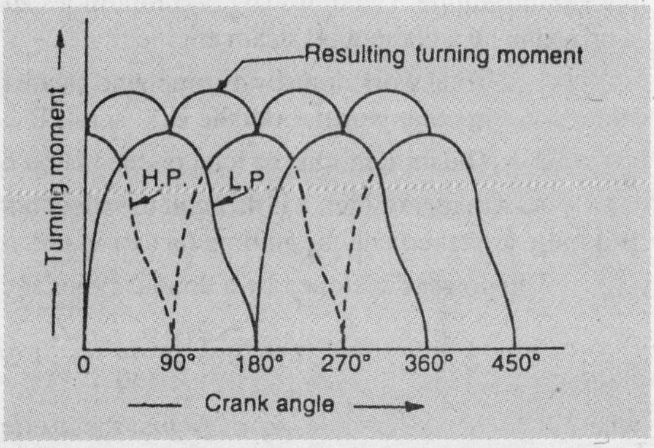

Fig. 18.6. Turning moment-crank angle graph for receiver type compound steam engine.

Fig. 18.6 shows the graph of turning moment *versus* crank angle of a receiver type compound engine. As the cranks are 90° apart, the two cycles are out of phase by 90°. The resulting turning moment diagram is also shown in Fig. 18.6. It will be noticed that variation of turning moment is considerably reduced by placing the cranks at 90°, hence a lighter flywheel is required. This is the chief advantage of the receiver type compound engine.

Note : Woolf type and Receiver type compound steam engines are *cross compound steam engines*. In cross compounding, the cylinders are arranged side by side and each cylinder has separate piston, connecting rod and crank.

18.8. Combined Indicator Diagram of a Compound Steam Engine

We have already discussed in the last chapter, that an indicator diagram is a graphical representation of the variation in pressure and volume of steam inside the cylinder on *p-v* diagram. In a compound steam engine, since there are two (or more) cylinders, therefore separate indicator diagrams are first drawn for H.P. and L.P. cylinders. These two diagrams are then combined together into one diagram, as shown in Fig. 18.7, as discussed below :

1. First of all draw the average indicator diagrams for both sides of the H.P. and L.P. cylinders.

2. Replot both the diagrams to the same *scale of pressure and volume.
3. Now plot the two diagrams together to give a combined indicator diagram, as shown in Fig. 18.7.

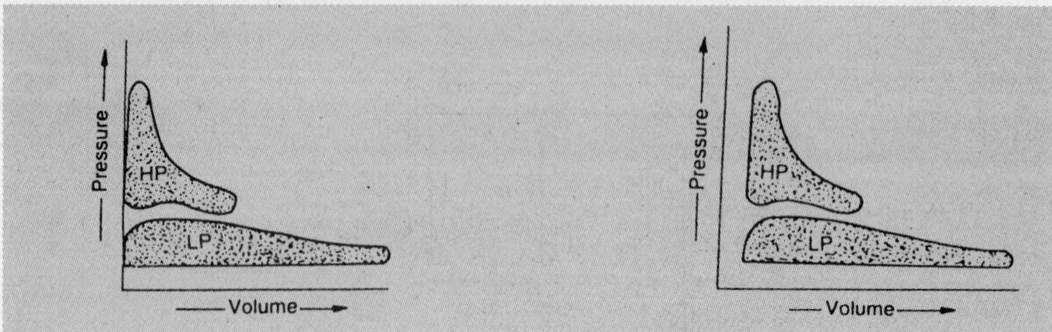

(a) Without cushion steam. (b) With cushion steam.

Fig. 18.7. Combined indicator diagram of a compound steam engine.

Note : If cushion steam is to be shown (or in other words, the diagram is to be drawn with clearance), then the same may be done as shown in Fig. 18.7 (*b*).

18.9. Work Done and Power Developed by a Compound Steam Engine

We have already discussed, in the last chapter, work done and power developed by a simple steam engine. The same equation, in a slightly modified form, is used for the work done by a compound engine. The following two conditions are generally regarded as the guiding factors for the work done by a compound steam engine :

1. Total work done by a compound steam engine is shared equally by both the cylinders. Or in other words, both the H.P. and L.P. cylinders do the same amount of work.
2. The initial thrust or load on the piston of both the H.P. and L.P. cylinders is the same.

As a matter of fact, it is difficult to satisfy both the conditions from the practical point of view. But these are taken only as guiding factors in order to derive the equations for the work done.

The power developed by a double cylinder compound steam engine is given by the relation :

$$\text{I.P.} = \frac{200\, p_a\, L\, A\, N}{60}\ \text{kW}$$

where p_a = Actual mean effective pressure referred to L.P. cylinder in bar,

L = Stroke length in metres,

A = Area of the L.P. cylinder or piston in m^2, and

N = Engine speed in r.p.m.

Note : The actual mean effective pressure referred to L.P. cylinder may be obtained as discussed in Art. 18.11.

18.10. Determination of Cylinder Dimensions for a Compound Steam Engine

The estimation of cylinder dimensions is one of the most important step in the calculation and design of the compound steam engines. The common cylinder dimensions are diameter of H.P. cylinder, diameter of L.P. cylinder, stroke length etc.

In the following pages, we shall discuss following types of compound steam engines :

1. Two cylinder compound steam engine with complete continuous expansion in both the cylinders.

---

* It is a common practice to use different scales of pressure and volume for the indicator diagrams of H.P. and L.P. cylinders.

2. Two cylinder compound steam engine with complete expansion in H.P. cylinder but incomplete expansion in L.P. cylinder.
3. Two cylinder compound steam engine with incomplete expansion in both the cylinders.

18.11. Two Cylinder Compound Steam Engine with Complete Continuous Expansion in both the H.P. and L.P. Cylinders

Consider a two cylinder compound steam engine having complete continuous expansion of steam in both the H.P. and L.P. cylinders. Let the theoretical indicator diagram (neglecting cushion steam or clearance volume) of such an engine be drawn, as shown in Fig. 18.8.

Let p_1 = Pressure of steam admitted into the H.P. cylinder,

p_2 = Pressure of steam at the release point of H.P. cylinder (or admission point of L.P. cylinder),

p_b = Back pressure of the L.P. cylinder,

v_1 = Volume of steam admitted into the H.P. cylinder,

v_2 = Volume of steam at the release point of H.P. cylinder (or admission point of L.P. cylinder),

v_3 = Volume of steam at the release point of L.P. cylinder,

A_H = Area of H.P. cylinder,

A_L = Area of L.P. cylinder,

L = Length of stroke,

N = Speed of the engine in r.p.m, and

K = Overall diagram factor for the combined indicator diagram.

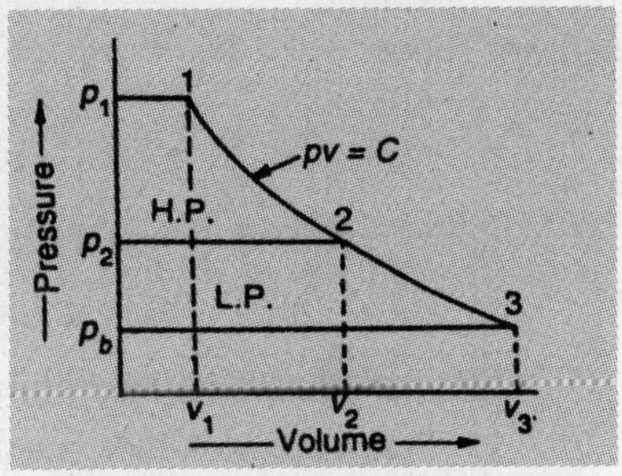

Fig. 18.8. Compound steam engine with complete continuous expansion.

We know that volume of H.P. cylinder (neglecting clearance volume),

$$v_2 = A_H \times L$$

and volume of L.P. cylinder (neglecting clearance volume),

$$v_3 = A_L \times L$$

We also know that expansion ratio in H.P. cylinder,

$$r_H = \frac{v_2}{v_1}$$

Similarly, expansion ratio in L.P. cylinder,

$$r_L = \frac{v_3}{v_2}$$

We know that the actual mean effective pressure in H.P. cylinder,

$$p_{aH} = K\left[\frac{p_1}{r_H}(1 + 2.3 \log r_H) - p_2\right]$$

and work done,

$$W_H = p_{aH} \times v_2 = K\left[\frac{p_1}{r_H}(1 + 2.3 \log r_H) - p_2\right] v_2$$

$$= K\left[\frac{p_1}{v_2/v_1}(1 + 2.3 \log r_H) - p_2\right] v_2 \qquad \ldots\left(\because r_H = \frac{v_2}{v_1}\right)$$

$$= K\,[p_1 v_1 (1 + 2.3 \log r_H) - p_2 v_2]$$

$$= K\,[p_1 v_1 + 2.3\, p_1 v_1 \log r_H - p_2 v_2]$$

Since expansion of steam follows the law $p\,v = C$, therefore

$$p_1 v_1 = p_2 v_2 = p_b v_3$$

$$\therefore \quad W_H = K \times 2.3\, p_1 v_1 \log r_H \qquad \ldots (i)$$

Similarly, mean effective pressure in L.P. cylinder,

$$p_{aL} = K\left[\frac{p_2}{r_L}(1 + 2.3 \log r_L) - p_b\right]$$

and work done,

$$W_L = p_{aL} \times v_3 = K\left[\frac{p_2}{v_3/v_2}(1 + 2.3 \log r_L) - p_b\right] v_3$$

$$= K\,[p_2 v_2 (1 + 2.3 \log r_L) - p_b v_3]$$

$$= K\,[p_2 v_2 + 2.3\, p_2 v_2 \log r_L - p_b v_3]$$

$$= K \times 2.3\, p_2 v_2 \log r_L \qquad \ldots (ii)$$

Now let us consider the following two conditions :

1. When the workdone is both the cylinders is equal, and
2. When the initial load or thrust on both the pistons is equal.

1. *When the workdone in both the cylinders is equal*

Considering that the workdone in both the cylinders (*i.e.* in H.P. and L.P. cylinders) is equal. Therefore equating equations (*i*) and (*ii*),

$$K \times 2.3\, p_1 v_1 \log r_H = K \times 2.3\, p_2 v_2 \log r_L$$

$$\therefore \quad r_H = r_L \qquad \ldots (\because p_1 v_1 = p_2 v_2)$$

or

$$\frac{v_2}{v_1} = \frac{v_3}{v_2} \qquad \ldots \left(\because r_H = \frac{v_2}{v_1} \text{ and } r_L = \frac{v_3}{v_2}\right)$$

and

$$\frac{p_1}{p_2} = \frac{p_2}{p_b} \qquad \ldots \left(\because \frac{v_2}{v_1} = \frac{p_1}{p_2} \text{ and } \frac{v_3}{v_2} = \frac{p_2}{p_b}\right)$$

or

$$p_2 = \sqrt{p_1\, p_b}$$

2. *When the initial load or thrust on both the pistons is equal*

Considering that the initial load or thrust on both the pistons (*i.e.* the pistons of both the H.P. and L.P. cylinders) is equal. Therefore

$$(p_1 - p_2)\, A_H = (p_2 - p_b)\, A_L$$

or

$$\frac{(p_1 - p_2)}{(p_2 - p_b)} = \frac{A_L}{A_H} = \frac{v_3}{v_2} = \frac{p_2}{p_b} \qquad \ldots \left[\because v_3 = A_L \times L;\ v_2 = A_H \times L \text{ and } \frac{v_3}{v_2} = \frac{p_2}{p_b}\right]$$

$$\therefore \quad p_1 p_b - p_2 p_b = p_2^2 - p_2 p_b$$

or

$$p_2 = \sqrt{p_1\, p_b} \qquad \ldots \text{(same as above)}$$

Notes : 1. It is only a theoretical case, for double cylinder compound steam engine. However, its relations are used in a triple cylinder compound steam engine.

2. The actual mean effective presssure for the combined indicator diagram is generally known as actual mean effective pressure referred to L.P. cylinder and is given by

$$p_a = K\left[\frac{p_1}{R}(1 + 2.3 \log R) - p_b\right]$$

where $R = \text{Total expansion ratio} = \dfrac{v_3}{v_1} = \dfrac{v_2}{v_1} \times \dfrac{v_3}{v_2} = r_H \times r_L$

and total workdone, $W = p_a \times v_3 = K\left[\dfrac{p_1}{R}(1 + 2.3 \log R) - p_b\right]v_3$

18.12. Two Cylinder Compound Steam Engine with Complete Expansion in H.P. Cylinder and Incomplete Expansion in L.P. Cylinder

Consider a two cylinder compound steam engine having complete expansion in H.P. cylinder and incomplete expansion in L.P. cylinder. Let the theoretical indicator diagram (neglecting cushion steam or clearance volume) of such an engine be drawn as shown in Fig. 18.9.

Let p_1 = Pressure of steam admitted into the H.P. cylinder,

p_2 = Pressure of steam at the release point of H.P. cylinder (or admission point of L.P. cylinder),

p_3 = Pressure of steam at the release point of L.P. cylinder,

p_b = Back pressure of steam in L.P. cylinder,

v_1 = Volume of steam admitted into the H.P. cylinder,

v_2 = Volume of steam at the release point of H.P. cylinder (or admission point of L.P. cylinder),

v_3 = Volume of steam at the release point of L.P. cylinder,

A_H = Area of H.P. cylinder,

A_L = Area of L.P. cylinder,

L = Length of stroke,

N = Speed of the engine in r.p.m., and

K = Overall diagram factor for the combined indicator diagram.

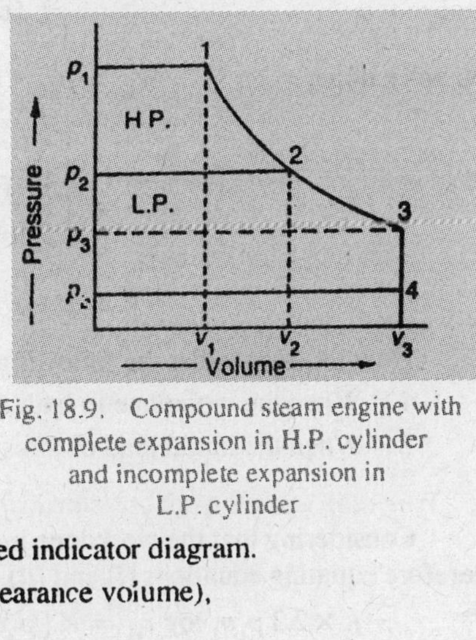

Fig. 18.9. Compound steam engine with complete expansion in H.P. cylinder and incomplete expansion in L.P. cylinder

We know that volume of H.P. cylinder (neglecting clearance volume),

$$v_2 = A_H \times L$$

and volume of L.P. cylinder (neglecting clearance volume),

$$v_3 = A_L \times L$$

We also know that expansion ratio in H.P. cylinder,

$$r_H = \frac{v_2}{v_1}$$

Similarly, expansion ratio in L.P. cylinder,

$$r_L = \frac{v_3}{v_2}$$

We know that actual mean effective pressure in H.P. cylinder,

$$p_{aH} = K\left[\frac{p_1}{r_H}(1+2.3\log r_H) - p_2\right]$$

and work-done.

$$W_H = p_{aH}\times v_2 = K\left[\frac{p_1}{r_H}(1+2.3\log r_H) - p_2\right]v_2$$

$$= K\left[\frac{p_1}{v_2/v_1}(1+2.3\log r_H) - p_2\right]v_2$$

$$= K[p_1v_1(1+2.3\log r_H) - p_2v_2]$$

$$= K[p_1v_1 + 2.3\,p_1v_1\log r_H - p_2v_2]$$

Since expansion of steam follows the law $p\,v = C$, therefore

$$p_1v_1 = p_2v_2 = p_3v_3$$

$$\therefore\quad W_H = K\times 2.3\,p_1v_1\log r_H \qquad \ldots(i)$$

Similarly, mean effective pressure in L.P. cylinder,

$$p_{aL} = K\left[\frac{p_2}{r_L}(1+2.3\log r_L) - p_b\right]$$

and work done,

$$W_L = p_{aL}\times v_3 = K\left[\frac{p_2}{r_L}(1+2.3\log r_L) - p_b\right]v_3$$

$$= K\left[\frac{p_2}{v_3/v_2}(1+2.3\log r_L) - p_b\right]v_3$$

$$= K[p_2v_2(1+2.3\log r_L) - p_bv_3] \qquad \ldots(ii)$$

Now let us consider the following two conditions :

1. When the workdone in both the cylinders is equal, and
2. When the initial load or thrust on both the pistons is equal.

1. *When the workdone in both the cylinders is equal*

Considering that the workdone in both the cylinders (*i.e.* in H.P. and L.P. cylinders) is equal. Therefore equating equations (*i*) and (*ii*),

$$K\times 2.3\,p_1v_1\log r_H = K[p_2v_2(1+2.3\log r_L) - p_bv_3]$$

$$2.3\,p_1v_1\log r_H = p_2v_2(1+2.3\log r_L) - p_bv_3$$

$$2.3\log r_H = \frac{p_2v_2}{p_1v_1}(1+2.3\log r_L) - \frac{p_bv_3}{p_1v_1}$$

$$= 1 + 2.3\log r_L - \frac{p_bv_3}{p_1v_1} \qquad \ldots(\because p_1v_1 = p_2v_2)$$

$$2.3\log\left(\frac{r_H}{r_L}\right) = 1 - \frac{p_bv_3}{p_1v_1}$$

$$2.3 \log\left(\frac{v_2}{v_1}\times\frac{v_2}{v_3}\right) = 1 - \frac{p_b v_3}{p_1 v_1}$$

$$2.3 \log\left(\frac{v_2^2}{v_1 v_3}\right) = 1 - \frac{p_b v_3}{p_1 v_1}$$

2. *When the initial load or thrust on both the pistons is equal*

Considering that the initial load or thrust on both the pistons (*i.e.* the piston of both the H.P. and L.P. cylinders) is equal. Therefore

$$(p_1 - p_2)\, A_H = (p_2 - p_b)\, A_L$$

or
$$\frac{p_1 - p_2}{p_2 - p_b} = \frac{A_L}{A_H} = \frac{v_3}{v_2} \qquad \ldots (\because v_3 = A_L \times L \text{ and } v_2 = A_H \times L)$$

Example 18.1. *A two cylinder compound steam engine is to develop 90 kW at 110 r.p.m. The steam is supplied at 7.35 bar and the condenser pressure is 0.21 bar. The stroke of each piston is equal to L.P. cylinder diameter. The total expansion ratio is 15. Allow a diagram factor of 0.7. Assume hyperbolic expansion and neglect clearance and receiver loss. Determine the diameter of the cylinders so that they may develop equal power.*

Solution. Given : I.P. = 90 kW ; N = 110 r.p.m. ; p_1 = 7.35 bar ; p_b = 0.21 bar ; $L = D_L$; $R = v_3/v_1 = 15$; $K = 0.7$

Diameter of L.P. cylinder

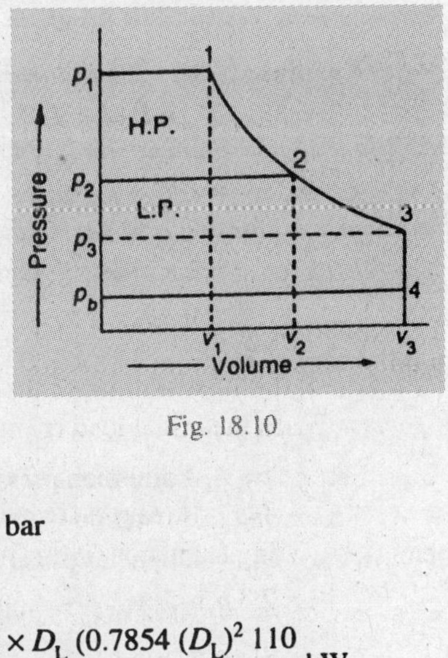

Fig. 18.10

Let D_L = Diameter of L.P. cylinder is metres,

$$A = \text{Area of L.P. cylinder in m}^2 = \frac{\pi}{4}(D_L)^2$$

$$= 0.7854\,(D_L)^2$$

L = Length of stroke in metres = D_L ... (Given)

We know that actual mean effective pressure referred to L.P. cylinder,

$$p_a = K\left[\frac{p_1}{R}(1 + 2.3 \log R) - p_b\right]$$

$$= 0.7\left[\frac{7.35}{15}(1 + 2.3 \log 15) - 0.21\right] = 1.124 \text{ bar}$$

We also know that indicated power (I.P.),

$$90 = \frac{200\, p_a\, L\, A\, N}{60} = \frac{200 \times 1.124 \times D_L\,(0.7854\,(D_L)^2\, 110}{60} \text{ kW}$$

$$= 323.7\,(D_L)^3$$

$\therefore$ $(D_L)^3 = 0.278$ or $D_L = 0.527$ m **Ans.**

Diameter of H.P. cylinder

Let D_H = Diameter of H.P. cylinder in metres, and

$$A_H = \text{Area of H.P. cylinder in m}^2 = \frac{\pi}{4}(D_H)^2 = 0.7854\,(D_H)^2 \text{ m}^2$$

For equal powers in H.P. and L.P. cylinders, we know that

$$2.3 \log\left(\frac{v_2^2}{v_1\, v_3}\right) = 1 - \frac{p_b\, v_3}{p_1\, v_1}$$

$$2.3 \log\left(\frac{v_2^2}{v_1 \times 15\, v_1}\right) = 1 - \frac{0.21}{7.35} \times 15 = 0.5714 \qquad \ldots (\because v_3/v_1 = 15)$$

$$\log\left(\frac{v_2^2}{15\, v_1^2}\right) = \frac{0.5714}{2.3} = 0.2484$$

$$\frac{v_2^2}{15\, v_1^2} = 1.772 \qquad \ldots \text{(Taking antilog of 0.2484)}$$

$$\therefore \qquad \frac{v_2}{v_1} = \sqrt{15 \times 1.772} = 5.155$$

We know that

$$\frac{v_2}{v_3} = \frac{v_2}{v_1} \times \frac{v_1}{v_3} = \frac{5.155}{15} = 0.3437 \qquad \ldots (\because v_3 = 15\, v_1)$$

or

$$\frac{0.7854\,(D_H)^2\, L}{0.7854\,(D_L)^2\, L} = 0.3437$$

$$\therefore \qquad D_H = \sqrt{0.3437\,(D_L)^2} = \sqrt{0.3437\,(0.527)^2} = 0.309 \text{ m} \quad \textbf{Ans.}$$

Example 18.2. *The following data refer to a double acting compound steam engine :*

I.P. = 375 kW ; R.P.M = 420 ; Stroke = 600 mm ; Admission pressure = 10 bar ; Back pressure = 0.3 bar ; Expansion ratio = 10 ; Diagram factor = 0.8.

Assuming complete expansion in H.P. cylinder and equal initial load and expansion follows the law p v = constant and neglecting clearance, determine

1. The admission pressure for the low pressure cylinder, and 2. The diameter of each cylinder.

Solution : Given : I.P. = 375 kW ; $N = 420$ r.p.m. ; $L = 600$ mm = 0.6 m ; $p_1 = 10$ bar ; $p_b = 0.3$ bar ; $R = v_3/v_1 = 10$; $K = 0.8$

1. Admission pressure for the low pressure cylinder

Let p_2 = Admission pressure for the low pressure cylinder,

D_L = Diameter of L.P. cylinder,

A_L = Area of L.P. cylinder,

$= \frac{\pi}{4}(D_L)^2 = 0.7854\,(D_L)^2,$

D_H = Diameter of H.P. cylinder, and

A_H = Area of H.P. cylinder,

$= \frac{\pi}{4}(D_H)^2 = 0.7854\,(D_H)^2$

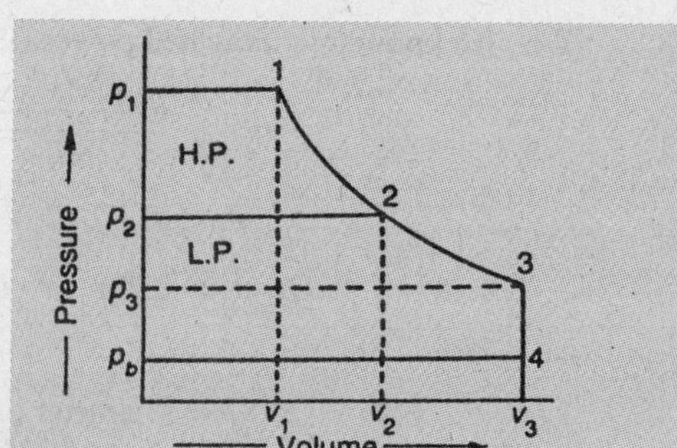

Fig. 18.11

Since the expansion follows the law $p\,v$ = constant, therefore

$$p_1 v_1 = p_2 v_2 = p_3 v_3$$

$$\therefore \quad p_3 = \frac{p_1 v_1}{v_3} = \frac{10}{10} = 1 \text{ bar} \qquad \ldots (\because v_3/v_1 = 10)$$

and
$$p_2 = \frac{p_3 v_3}{v_2} = \frac{p_3 A_L L}{A_H L} = \frac{1 \times 0.7854 (D_L)^2 L}{0.7854 (D_H)^2 L} = \frac{(D_L)^2}{(D_H)^2} \qquad \ldots (i)$$

For equal initial load on the pistons of H.P. and L.P. cylinders, we know that

$$(p_1 - p_2) A_H = (p_2 - p_b) A_L$$

$$\frac{p_1 - p_2}{p_2 - p_b} = \frac{A_L}{A_H} = \frac{0.7854 (D_L)^2}{0.7854 (D_H)^2} = \frac{(D_L)^2}{(D_H)^2} \qquad \ldots (ii)$$

From equations (*i*) and (*ii*),

$$\frac{p_1 - p_2}{p_2 - p_b} = p_2 \quad \text{or} \quad p_1 - p_2 = p_2^2 - p_2 p_b$$

$$\therefore \quad 10 - p_2 = p_2^2 - p_2 \times 0.3 \quad \text{or} \quad p_2^2 + 0.7 p_2 - 10 = 0$$

$$\therefore \quad p_2 = \frac{-0.7 \pm \sqrt{(0.7)^2 + 4 \times 10}}{2} = \frac{-0.7 \pm 6.36}{2}$$

$$= 2.83 \text{ bar } \textbf{Ans.} \qquad \ldots \text{(Taking +ve sign)}$$

2. Diameter of each cylinder

We know that actual mean effective pressure referred to L.P. cylinder,

$$p_a = K\left[\frac{p_1}{R}(1 + 2.3 \log R) - p_b\right]$$

$$= 0.8\left[\frac{10}{10}(1 + 2.3 \log 10) - 0.3\right] = 2.4 \text{ bar}$$

and indicated power (I.P.),
$$375 = \frac{200 \times p_a L A N}{60} = \frac{200 \times 2.4 \times 0.6 \times 0.7854 (D_L)^2\, 420}{60} \text{ kW}$$

$$= 1583 (D_L)^2$$

$$\therefore \quad (D_L)^2 = 0.237 \quad \text{or} \quad D_L = 0.4867 \text{ m } \textbf{Ans.}$$

From equation (*i*),
$$(D_H)^2 = \frac{(D_L)^2}{p_2} = \frac{0.237}{2.83} = 0.0837$$

$$\therefore \quad D_H = 0.2894 \text{ m } \textbf{Ans.}$$

Example 18.3. *A double acting compound steam engine with two cylinders is supplied with steam at 14 bar and 0.9 dry. The steam is exhausted into the condenser at 0.35 bar. Both the cylinders have stroke length of 350 mm and have equal loads on their pistons initially. The diameters of H.P. and L.P. cylinders are 200 mm and 300 mm respectively. If the engine runs at 300 r.p.m., find : 1. intermediate pressure ; 2. indicated power ; and 3. steam consumption of the engine in kg/hour. Assume diagram factor as 0.8 and complete expansion of steam in H.P. cylinder.*

Solution. Given : $p_1 = 14$ bar ; $x = 0.9$; $p_b = 0.35$ bar ; $L = 350$ mm $= 0.35$ m ; $D_H = 200$ mm $= 0.2$ m ; $D_L = 300$ mm $= 0.3$ m ; $N = 300$ r.p.m. ; $K = 0.8$

We know that volume of H.P. cylinder,

$$v_2 = \frac{\pi}{4}(D_H)^2 L = \frac{\pi}{4}(0.2)^2\, 0.35 = 0.011 \text{ m}^3$$

and volume of L.P. cylinder,

$$v_3 = \frac{\pi}{4}(D_L)^2 L = \frac{\pi}{4}(0.3)^2\, 0.35 = 0.0247 \text{ m}^3$$

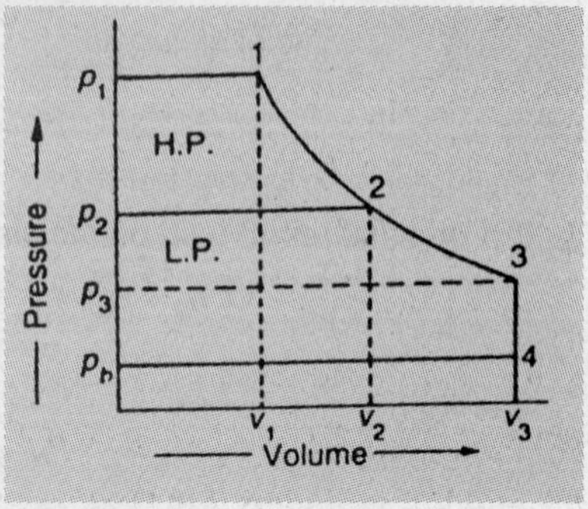

Fig. 18.12

1. *Intermediate pressure*

Let p_2 = Intermediate pressure.

We know that for equal initial load on the pistons,

$$\frac{(p_1 - p_2)}{(p_2 - p_b)} = \frac{v_3}{v_2}$$

$$\frac{(14 - p_2)}{(p_2 - 0.35)} = \frac{0.0247}{0.011} = 2.25$$

$$14 - p_2 = 2.25\, p_2 - 2.25 \times 0.35 \quad \text{or} \quad p_2 = 4.55 \text{ bar} \;\; \textbf{Ans.}$$

2. *Indicated power*

Since expansion of the steam is hyperbolic, therefore

$$p_1 v_1 = p_2 v_2$$

or

$$v_1 = \frac{p_2 v_2}{p_1} = \frac{4.55 \times 0.011}{14} = 0.0036 \text{ m}^3$$

$\therefore$ Total expansion ratio,

$$R = \frac{v_3}{v_1} = \frac{0.0247}{0.0036} = 6.86$$

We know that actual mean effective pressure referred to L.P. cylinder,

$$p_a = K\left[\frac{p_1}{R}(1 + 2.3 \log R) - p_b\right]$$

$$= 0.8\left[\frac{14}{6.86}(1 + 2.3 \log 6.86) - 0.35\right] = 4.5 \text{ bar}$$

Area of L.P. cylinder,

$$A = \frac{\pi}{4}(D_L)^2 = \frac{\pi}{4}(0.3)^2 = 0.0707 \text{ m}^2$$

We know that indicated power,

$$\text{I.P.} = \frac{200\, p_a\, L A N}{60} = \frac{200 \times 4.5 \times 0.35 \times 0.0707 \times 300}{60} \text{ kW}$$

$$= 111.3 \text{ kW} \;\; \textbf{Ans.}$$

3. *Steam consumption of the engine in kg/hour*

From the steam tables, we find that the specific volume of steam at 14 bar (*i.e.* admission pressure).

$$v_s = 0.1407 \text{ m}^3/\text{kg}$$

Since the steam admitted into the H.P. cylinder is 0.9 dry, therefore actual volume of steam (this is equal to the steam admitted in one stroke).

$$= x\,v_s = 0.9 \times 0.1407 = 0.1266 \text{ m}^3/\text{kg}$$

Total volume of steam admitted (or consumed) in one hour

$$= v_1 \times 2N \times 60 = 0.0036 \times 2 \times 300 \times 60 = 129.6 \text{ m}^3/\text{h}$$

∴ Steam consumption of the engine

$$= \frac{129.6}{0.1266} = 1023.7 \text{ kg/h} \textbf{ Ans.}$$

18.13. Two Cylinder Compound Steam Engine with Incomplete Expansion in both the H.P. and L.P. Cylinders

Consider a two cylinder compound steam engine having incomplete expansion in both the H.P. and L.P. cylinders. Let the theoretical indicator diagram (neglecting cushion steam or clearance volume) of such an engine be drawn as shown in Fig. 18.13.

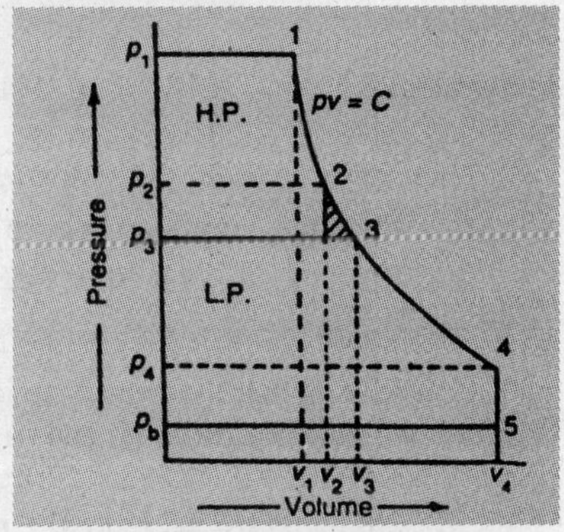

Fig. 18.13. Compound steam engine with incomplete expansion.

Let p_1 = Pressure of steam admitted into the H.P. cylinder,

p_2 = Pressure of steam at release point of H.P. cylinder,

p_3 = Pressure of steam admitted into the L.P. cylinder,

p_4 = Pressure of steam at the release point of L.P. cylinder,

p_b = Back pressure of steam in L.P. cylinder,

v_1 = Volume of steam admitted into the H.P. cylinder,

v_2 = Volume of steam at the release point of H.P. cylinder,

v_3 = Volume of steam admitted into the L.P. cylinder,

v_4 = Volume of steam at the release point of L.P. cylinder,

A_H = Area of H.P. cylinder,

A_L = Area of L.P. cylinder,

L = Length of stroke,

N = Speed of the engine in r.p.m, and

K = Overall diagram factor for the combined indicator diagram.

We know that volume of H.P. cylinder (neglecting clearance volume),

$$v_2 = A_H \times L$$

and volume of L.P. cylinder (neglecting clearance volume),

$$v_4 = A_L \times L$$

We also know that expansion ratio in H.P. cylinder,

$$r_H = \frac{v_2}{v_1}$$

Similarly, expansion ratio in L.P. cylinder,

$$r_L = \frac{v_4}{v_3}$$

We know that actual mean effective pressure in H.P. cylinder,

$$p_{aH} = K\left[\frac{p_1}{r_H}(1 + 2.3 \log r_H) - p_3\right]$$

and work done,

$$W_H = p_{aH} \times v_2 = K\left[\frac{p_1}{r_H}(1 + 2.3 \log r_H) - p_3\right] v_2 \qquad \ldots (i)$$

Similarly, actual mean effective pressure in L.P. cylinder,

$$p_{aL} = K\left[\frac{p_3}{r_L}(1 + 2.3 \log r_L) - p_b\right]$$

and work done,

$$W_L = p_{aL} \times v_4 = K\left[\frac{p_3}{r_L}(1 + 2.3 \log r_L) - p_b\right] v_4 \qquad \ldots (ii)$$

Now let us consider the following two conditions :

1. When the workdone in both the cylinders is equal, and
2. When the initial load or thrust on both the pistons is equal.

1. *When the workdone in both the cylinders is equal*

Considering that the workdone in both the cylinders (*i.e.* in H.P. and L.P. cylinders) is equal. Therefore equating equations (*i*) and (*ii*),

$$K\left[\frac{p_1}{r_H}(1 + 2.3 \log r_H) - p_3\right] v_2 = K\left[\frac{p_3}{r_L}(1 + 2.3 \log r_L) - p_b\right] v_4$$

or

$$\left[\frac{p_1}{r_H}(1 + 2.3 \log r_H) - p_3\right] = \frac{v_4}{v_2}\left[\frac{p_3}{r_L}(1 + 2.3 \log r_L) - p_b\right] = \frac{R}{r_H}\left[\frac{p_3}{r_L}(1 + 2.3 \log r_L) - p_b\right]$$

$$\ldots \left(\because \frac{v_4}{v_2} = \frac{v_4}{v_1} \times \frac{v_1}{v_2} = R \times \frac{1}{r_H}\right)$$

2. *When the initial load or thrust on both the pistons is equal*

Considering that the initial load or thrust on both the pistons (*i.e.* pistons of both the H.P. and L.P. cylinders) is equal. Therefore

$$(p_1 - p_3) A_H = (p_3 - p_b) A_L$$

or

$$\frac{p_1 - p_3}{p_3 - p_b} = \frac{A_L}{A_H} = \frac{v_4}{v_2} \qquad \ldots (\because v_4 = A_L \times L \text{ and } v_2 = A_H \times L)$$

Example 18.4. *The following data refer to a double acting two cylinder compound steam engine :*

	H.P. cylinder	*L.P. cylinder*
Piston diameter	*250 mm*	*450 mm*
Stroke length	*600 mm*	*600 mm*
Cut-off (percentage of stroke length)	*25%*	*35%*
Expansion follows the law	*hyperbolic*	*hyperbolic*
Diagram factor	*0.75*	*0.65*
R.P.M.	*120*	*120*

If the steam is supplied to H.P. cylinder at 10 bar and exhaust takes place at 0.11 bar, determine : 1. Mean effective pressure of H.P. and L.P. cylinders ; and 2. Ratio of workdone in two cylinders. Assume no clearance and compression.

Solution. Given : $D_H = 250$ mm $= 0.25$ m ; $D_L = 450$ mm $= 0.45$ m ; $L = 600$ mm $= 0.6$ m ; $v_1 = 25\%\ v_2 = 0.25\ v_2$; $v_3 = 35\%\ v_4 = 0.35\ v_4$; $K_H = 0.75$; $K_L = = 0.65$; $N = 120$ r.p.m. ; $p_1 = 10$ bar ; $p_b = 0.11$ bar

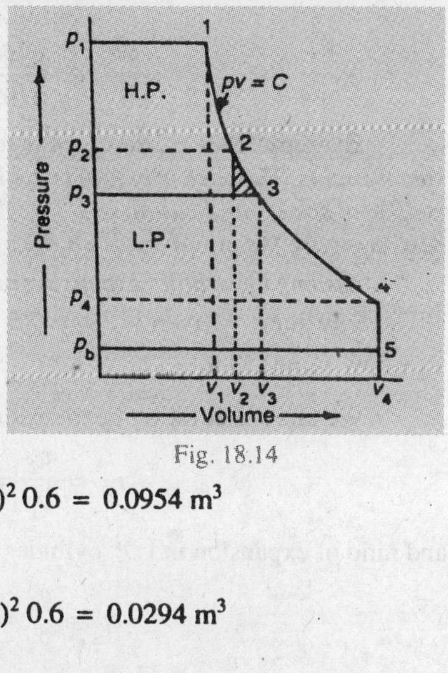

Fig. 18.14

1. *Mean effective pressure of H.P. and L.P. cylinders*

We know that expansion ratio in H.P. cylinder,

$$r_H = \frac{v_2}{v_1} = \frac{v_2}{0.25\ v_2} = 4$$

and expansion ratio in L.P. cylinder,

$$r_L = \frac{v_4}{v_3} = \frac{v_4}{0.35\ v_4} = 2.857$$

We also know that volume of L.P. cylinder,

$$v_4 = A_L \times L = \frac{\pi}{4}(D_L)^2 L = \frac{\pi}{4}(0.45)^2\, 0.6 = 0.0954 \text{ m}^3$$

and volume of H.P. cylinder,

$$v_2 = A_H \times L = \frac{\pi}{4}(D_H)^2 L = \frac{\pi}{4}(0.25)^2\, 0.6 = 0.0294 \text{ m}^3$$

$$\therefore \qquad \frac{v_4}{v_2} = \frac{0.0954}{0.0294} = 3.245$$

Since the expansion is hyperbolic (*i.e.* $p\,v$ = constant), therefore

$$p_1 v_1 = p_2 v_2 = p_3 v_3 = p_4 v_4$$

$$\therefore \qquad p_3 = \frac{p_1 v_1}{v_3} = p_1 \times \frac{v_1}{v_2} \times \frac{v_2}{v_4} \times \frac{v_4}{v_3} = 10 \times \frac{1}{4} \times \frac{1}{3.245} \times 2.857 = 2.2 \text{ bar}$$

We know that actual mean effective pressure of H.P. cylinder,

$$p_{aH} = K_H \left[\frac{p_1}{r_H}(1 + 2.3 \log r_H) - p_3 \right]$$

$$= 0.75 \left[\frac{10}{4}(1 + 2.3 \log 4) - 2.2 \right] = 2.82 \text{ bar} \quad \textbf{Ans.}$$

and actual mean effective pressure of L.P. cylinder,

$$p_{aL} = K_L\left[\frac{p_3}{r_L}(1 + 2.3 \log r_L) - p_b\right]$$

$$= 0.65\left[\frac{2.2}{2.857}(1 + 2.3 \log 2.857) - 0.11\right] = 0.954 \text{ bar} \quad \textbf{Ans.}$$

2. *Ratio of workdone in H.P. and L.P. cylinders*

We know that workdone in H.P. cylinder per minute,

$$W_H = p_{aH} \times v_2 \times N = 2.82 \times 10^5 \times 0.0294 \times 120 = 9.92 \times 10^5 \text{ N-m}$$

... (p_{aH} is taken in N/m²)

and workdone in L.P. cylinder per minute

$$W_L = p_{aL} \times v_4 \times N = 0.954 \times 10^5 \times 0.0954 \times 120 = 10.92 \times 10^5 \text{ N-m}$$

... (p_{aL} is taken in N/m²)

$$\therefore \quad \frac{W_H}{W_L} = \frac{9.95 \times 10^5}{10.92 \times 10^5} = 0.911 \quad \textbf{Ans.}$$

Example 18.5. *The steam is supplied at 7.5 bar to a double acting, two cylinder compound steam engine. The back pressure is 0.2 bar. If the cylinder volume ratio is 3.5, cut-off in H.P. cylinder is 40% of stroke, and cut-off in L.P. cylinder is 53% of stroke, determine the L.P. receiver pressure. Also compare the initial loads on the piston.*

Assume hyperbolic expansion and neglect clearance.

Solution. Given : $p_1 = 7.5$ bar ; $p_b = 0.2$ bar ; $v_4/v_2 = 3.5$; $v_1 = 40\% \; v_2 = 0.4\, v_2$; $v_3 = 53\%\, v_4 = 0.53\, v_4$

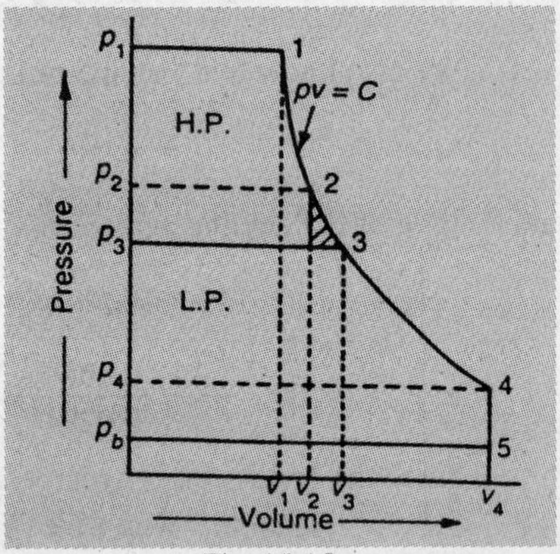

Fig. 18.15

We know that ratio of expansion in H.P. cylinder,

$$r_H = \frac{v_2}{v_1} = \frac{v_2}{0.4\, v_2} = 2.5$$

and ratio of expansion in L.P. cylinder,

$$r_L = \frac{v_4}{v_3} = \frac{v_4}{0.53\, v_4} = 1.89$$

L.P. receiver pressure

Let $\quad p_3$ = L.P. receiver pressure.

Since the expansion is assumed hyperbolic, therefore

$$p_1 v_1 = p_3 v_3$$

or

$$p_1 \times \frac{v_1}{v_2} \times v_2 = p_3 \times \frac{v_3}{v_4} \times v_4$$

$$p_1 \times \frac{1}{r_H} = p_3 \times \frac{1}{r_L} \times \frac{v_4}{v_2}$$

$$p_3 = p_1 \times \frac{r_L}{r_H} \times \frac{v_2}{v_4} = 7.5 \times \frac{1.89}{2.5} \times \frac{1}{3.5} = 1.62 \text{ bar} \quad \textbf{Ans.}$$

Comparison of initial loads on the piston

We know that

$$\frac{\text{Initial load on H.P. piston}}{\text{Initial load on L.P. piston}} = \frac{(p_1 - p_3)\,A_H}{(p_3 - p_b)\,A_L} = \frac{(p_1 - p_3)\,v_2}{(p_3 - p_b)\,v_4} \quad \ldots (\because A_H = v_2/L \text{ and } A_L = v_4/L)$$

$$= \frac{7.5 - 1.62}{1.62 - 0.2} \times \frac{1}{3.5} = 1.183 \textbf{ Ans.}$$

Example 18.6. *A compound steam engine is to develop 260 kW when taking steam at 8.75 bar and exhausting at 0.15 bar. The engine speed is 140 r.p.m. and the piston speed is 150 m/min. The cut-off in the H.P. cylinder is to be 0.4 and the cylinder volume ratio is 3.7. Allow a diagram factor of 0.83 for the combined cards and determine suitable dimensions of the cylinders. If the diagram factor for the H.P. cylinder alone is 0.85, determine the separate powers developed in the two cylinders when the L.P. cut-off is arranged to give equal initial loads on the pistons. Assume hyperbolic expansion and neglect clearance effects.*

Solution. Given : I.P. = 260 kW ; $p_1 = 8.75$ bar ; $p_b = 0.15$ bar ; $N = 140$ r.p.m. ; Piston speed = 150 m/min ; $v_1 = 0.4\,v_2$; $v_4/v_2 = 3.7$; $K = 0.83$; $K_H = 0.85$

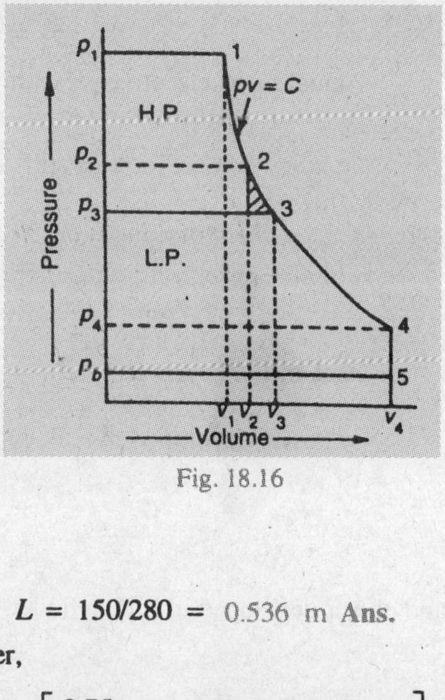

Fig. 18.16

We know that expansion ratio in H.P. cylinder,

$$r_H = \frac{v_2}{v_1} = \frac{v_2}{0.4\,v_2} = 2.5$$

and total expansion ratio,

$$R = \frac{v_4}{v_1} = \frac{v_4}{v_2} \times \frac{v_2}{v_1} = 3.7 \times 2.5 = 9.25$$

Suitable dimensions of the cylinders

Let L = Length of stroke in metres,

D_L = Diameter of L.P. cylinder in metres, and

D_H = Diameter of H.P. cylinder in metres.

We know that piston speed,

$$150 = 2\,L\,N = 2\,L \times 140 = 280\,L \quad \text{or} \quad L = 150/280 = 0.536 \text{ m } \textbf{Ans.}$$

Actual mean effective pressure referred to L.P. cylinder,

$$p_a = K\left[\frac{p_1}{R}(1 + 2.3 \log R) - p_b\right] = 0.83\left[\frac{8.75}{9.25}(1 + 2.3 \log 9.25) - 0.15\right]$$

$$= 2.4 \text{ bar}$$

Area of L.P. cylinder,

$$A = \frac{\pi}{4}(D_L)^2 = 0.7854\,(D_L)^2 \text{ m}^2$$

We know that indicated power (I.P.),

$$260 = \frac{200\,p_a\,L\,A\,N}{60} = \frac{200 \times 2.4 \times 0.536 \times 0.7854\,(D_L)^2\,140}{60} = 470\,(D_L)^2$$

$$\therefore \quad (D_L)^2 = 0.553 \text{ or } D_L = 0.744 \text{ m } \textbf{Ans.}$$

We know that ratio of L.P. cylinder volume to H.P. cylinder volume,

$$3.7 = \frac{v_4}{v_2} = \frac{0.7854\,(D_L)^2\,L}{0.7854\,(D_H)^2\,L} = \frac{(D_L)^2}{(D_H)^2}$$

$$\therefore \quad (D_H)^2 = \frac{(D_L)^2}{3.7} = \frac{0.553}{3.7} = 0.1495 \text{ or } D_H = 0.386 \text{ m} \quad \textbf{Ans.}$$

Power developed in H.P. cylinder

First of all, let us find the pressure of steam at release point of H.P. cylinder (*i.e.* p_3).

We know that for equal initial loads on both the pistons,

$$\frac{p_1 - p_3}{p_3 - p_b} = \frac{v_4}{v_2} = 3.7$$

$$\frac{8.75 - p_3}{p_3 - 0.15} = 3.7 \text{ or } 8.75 - p_3 = 3.7\,p_3 - 3.7 \times 0.15$$

$$\therefore \quad p_3 = 1.98 \text{ bar}$$

Actual mean effective pressure in H.P. cylinder,

$$p_{aH} = K_H \left[\frac{p_1}{r_H}(1 + 2.3 \log r_H) - p_3 \right]$$

$$= 0.85 \left[\frac{8.75}{2.5}(1 + 2.3 \log 2.5) - 1.98 \right] = 4 \text{ bar}$$

Area of H.P. cylinder, $A_H = \frac{\pi}{4}(D_H)^2 = \frac{\pi}{4}(0.386)^2 = 0.1174 \text{ m}^2$

$\therefore$ Power developed in H.P. cylinder

$$= \frac{200\,p_{aH}\,L\,A_H\,N}{60} = \frac{200 \times 4 \times 0.536 \times 0.1174 \times 140}{60} \text{ kW}$$

$$= 117.4 \text{ kW} \quad \textbf{Ans.}$$

and power developed in L.P. cylinder

$$= 260 - 117.4 = 142.6 \text{ kW} \quad \textbf{Ans.}$$

Example 18.7. *A compound steam engine develops 200 kW at 150 r.p.m. when the steam enters the H.P. cylinder at 10 bar and leaves the L.P. cylinder at 0.2 bar. Cut-off takes place at 0.45 stroke in H.P. cylinder and ratio of L.P. cylinder volume to H.P. cylinder volume is 3.2. Calculate the cylinder diameters and stroke length assuming a diagram factor of 0.75 and mean piston speed as 180 m/min.*

Also find the fraction of stroke at which cut-off occurs in L.P. cylinder for approximately equal initial force on both the pistons.

Solution. Given : I.P. = 200 kW ; N = 150 r.p.m. ; p_1 = 10 bar ; p_b = 0.2 bar ; $v_1 = 0.45\,v_2$; $v_4/v_2 = 3.2$; K = 0.75 ; Mean piston speed = 180 m/min

We know that expansion ratio in H.P. cylinder,

$$r_H = \frac{v_2}{v_1} = \frac{v_2}{0.45\,v_2} = 2.22$$

and total expansion ratio, $R = \dfrac{v_4}{v_1} = \dfrac{v_4}{v_2} \times \dfrac{v_2}{v_1} = 3.2 \times 2.22 = 7.1$

Stroke length

Let L = Stroke length in metres.

We know that mean piston speed,

$$180 = 2LN = 2L \times 150 = 300\,L$$

or $L = 180/300 = 0.6$ m **Ans.**

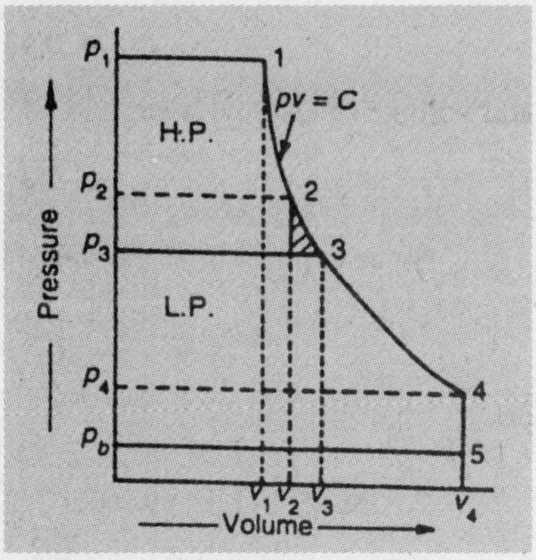

18.17

Diameter of L.P. cylinder

Let D_L = Diameter of L. P. cylinder in metres.

$\therefore$ Area, $A = \dfrac{\pi}{4}(D_L)^2 = 0.7854\,(D_L)^2$

We know that actual mean effective pressure referred to L.P. cylinder,

$$p_a = K\left[\frac{p_1}{R}(1 + 2.3 \log R) - p_b\right]$$

$$= 0.75\left[\frac{10}{7.1}(1 + 2.3 \log 7.1) - 0.2\right] = 2.97 \text{ bar}$$

and indicated power (I.P.),

$$200 = \frac{200\,p_a\,L\,A\,N}{60} = \frac{200 \times 2.97 \times 0.6 \times 0.7854\,(D_L)^2\,150}{60} = 700\,(D_L)^2$$

$\therefore$ $(D_L)^2 = 200/700 = 0.286$ or $D_L = 0.534$ m **Ans.**

Diameter of H.P. cylinder

Let D_H = Diameter of H.P. cylinder in metres.

We know that ratio of L.P. cylinder volume to H.P. cylinder volume,

$$3.2 = \frac{v_4}{v_2} = \frac{0.7854\,(D_L)^2\,L}{0.7854\,(D_H)^2\,L} = \frac{(D_L)^2}{(D_H)^2}$$

$$\therefore \quad D_H = \sqrt{\frac{(D_L)^2}{3.2}} = \sqrt{\frac{(0.534)^2}{3.2}} = 0.298 \text{ m} \quad \textbf{Ans.}$$

Fraction of stroke at which cut-off occurs in L.P. cylinder

First of all, let us find the pressure of steam at release point of H.P. cylinder (*i.e.* p_3). We know that for equal initial loads on both the pistons,

$$\frac{p_1 - p_3}{p_3 - p_b} = \frac{v_4}{v_2} = 3.2$$

$$\frac{10 - p_3}{p_3 - 0.2} = 3.2 \quad \text{or} \quad 10 - p_3 = 3.2\,p_3 - 3.2 \times 0.2$$

$$\therefore \quad p_3 = 2.53 \text{ bar}$$

We know that volume of H.P. cylinder,

$$v_2 = \frac{\pi}{4}(D_H)^2 L = \frac{\pi}{4}(0.298)^2\, 0.6 = 0.042 \text{ m}^3$$

$$\therefore \quad v_1 = 0.45 \times 0.042 = 0.019 \text{ m}^3 \qquad \ldots (\because v_1 = 0.45\, v_2)$$

and volume of L.P. cylinder,

$$v_4 = \frac{\pi}{4}(D_L)^2 L = \frac{\pi}{4}(0.534)^2\, 0.6 = 0.1344 \text{ m}^3$$

Since expansion of the steam follows the law $pv = C$, therefore

$$p_1 v_1 = p_3 v_3$$

or

$$v_3 = \frac{p_1 v_1}{p_3} = \frac{10 \times 0.019}{2.53} = 0.075 \text{ m}^3$$

∴ Cut-off in L.P. cylinder

$$= \frac{v_3}{v_4} = \frac{0.075}{0.1344} = 0.56 \textbf{ Ans.}$$

Note : The cut-off in L.P. cylinder may also be obtained as follows :

We know that $p_1 v_1 = p_3 v_3$

$$\therefore \quad \frac{v_3}{v_1} = \frac{p_1}{p_3} = \frac{10}{2.53} = 3.95$$

Now

$$\frac{v_3}{v_4} = \frac{v_3}{v_1} \times \frac{v_1}{v_2} \times \frac{v_2}{v_4} = 3.95 \times \frac{1}{2.22} \times \frac{1}{3.2} = 0.56 \textbf{ Ans.}$$

18.14. Three Cylinder Compound Steam Engines

In the previous articles, we have been discussing double cylinder, (or in other words double expansion) compound steam engines. In such engines, we have always been referring to H.P. and L.P. cylinders.

But sometimes, we use three cylinder compound steam engines instead of two cylinder compound steam engines. As the name indicates, a three cylinder (or triple expansion) compound steam engine consists of three cylinders namely H.P. (high pressure) cylinder, I.P. (intermediate pressure) cylinder and L.P. (low pressure) cylinder. In a triple expansion compound steam engine, the steam is first admitted into the H.P. cylinder. After expansion in the H.P. cylinder, it is admitted into the I.P. cylinder. Similarly, after further expansion in the I.P. cylinder, the steam is admitted into the L.P. cylinder. Again after further expansion in the L.P. cylinder, the steam is exhausted.

All the relations of double cylinder compound steam engines also hold good for triple cylinder compound steam engines. The following examples will illustrate the theory of triple cylinder compound steam engines.

Example 18.8. *A triple expansion engine is supplied with steam at 13 bar and the condenser pressure is 0.2 bar. The overall expansion ratio is 13. Neglecting clearance effects, assuming no pressure drop at release in the high pressure and intermediate pressure cylinders, and assuming hyperbolic expansion, determine the ratio of cylinder volumes. Take the high pressure cylinder volume as unity, in order that equal powers may be developed in the three cylinders.*

Solution. Given : $p_1 = 13$ bar ; $p_b = 0.2$ bar ; $R = v_4/v_1 = 13$; $v_2 = 1$

Let v_3 = Volume of I.P. cylinder, and

v_4 = Volume of L.P. cylinder.

We have discussed in Art 18.11 that work done in H.P. cylinder,

$$W_H = K \times 2.3\, p_1 v_1 \log r_H \quad \ldots (i)$$

Similarly, work done in I.P. cylinder,

$$W_I = K \times 2.3\, p_2 v_2 \log r_I \quad \ldots (ii)$$

and work done in L.P. cylinder,

$$W_L = K [\, p_3 v_3 (1 + 2.3 \log r_L) - p_b v_4] \quad \ldots (iii)$$

Since power developed in each cylinder is the same, therefore

$$W_H = W_I = W_L$$

$$K \times 2.3\, p_1 v_1 \log r_H = K \times 2.3\, p_2 v_2 \log r_I \quad \ldots \text{(Equating } W_H \text{ and } W_I)$$

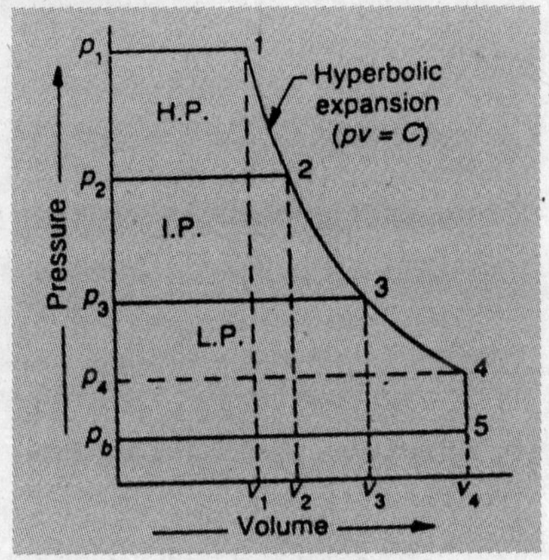

Fig. 18.18

$$\therefore \qquad r_H = r_I \qquad \ldots (\because p_1 v_1 = p_2 v_2)$$

or
$$\frac{v_2}{v_1} = \frac{v_3}{v_2} \qquad \ldots \left(\because r_H = \frac{v_2}{v_1} \text{ and } r_I = \frac{v_3}{v_2}\right)$$

$$\therefore \qquad v_3 = \frac{v_2 \times v_2}{v_1} = \frac{1}{v_1} \qquad \ldots \text{(Given } v_2 = 1)$$

Again $$K \times 2.3\, p_1 v_1 \log r_H = K [\, p_3 v_3 (1 + 2.3 \log r_L) - p_b v_4] \qquad \ldots \text{(Equating } W_H = W_L)$$

$$\therefore \qquad 2.3 \log r_H = \frac{p_3 v_3}{p_1 v_1}\left[1 + 2.3 \log\left(\frac{v_4}{v_3}\right)\right] - \frac{p_b v_4}{p_1 v_1} \qquad \ldots \left(\because r_L = \frac{v_4}{V_3}\right)$$

$$2.3 \log\left(\frac{v_2}{v_1}\right) = 1 + 2.3 \log\left(\frac{v_4}{v_3}\right) - \frac{p_b v_4}{p_1 v_1} \qquad \ldots (\because p_1 v_1 = p_2 v_2 = p_3 v_3)$$

$$2.3 \log\left(\frac{v_2}{v_1}\right) - 2.3 \log\left(\frac{v_4}{v_3}\right) = 1 - \frac{p_b}{p_1} \times \frac{v_4}{v_1} = 1 - \frac{0.2}{13} \times 13 = 0.8 \qquad \ldots (\because v_4/v_1 = 13)$$

$$\log\left(\frac{v_2}{v_1} \times \frac{v_3}{v_4}\right) = \frac{0.8}{2.3} = 0.3478$$

or
$$\frac{v_2}{v_1} \times \frac{v_3}{v_4} = 2.227 \qquad \ldots \text{(Taking antilog of 0.3478)}$$

Substituting the values of $v_2 = 1$, $v_3 = 1/v_1$ and $v_4 = 13\, v_1$, in the above equation,

$$\frac{1}{v_1} \times \frac{1}{v_1 \times 13\, v_1} = 2.227 \text{ or } \frac{1}{(v_1)^3} = 13 \times 2.227 = 28.95$$

$$\therefore \qquad \frac{1}{v_1} = 3.07 \text{ or } v_1 = 0.326 \text{ m}^3$$

We know that $v_3 = \frac{1}{v_1} = 3.07$

and $v_4 = 13\,v_1 = 13 \times 0.326 = 4.24$

$\therefore$ Ratio of cylinder volumes,

$$v_2 : v_3 : v_4 = 1 : 3.07 : 4.24 \textbf{ Ans.}$$

Example 18.9. *A triple expansion engine is required to develop 3000 kW I.P. at 100 r.p.m. The steam is supplied at 10.5 bar and back pressure is 0.07 bar. The piston speed is 180 m/min, and diagram factor is 0.6. If the ratio of cylinder volumes is 1 : 3 : 7.5 and cut-off in H.P. cylinder is 0.7, determine the size of cylinders.*

Solution. Given : I.P. = 3000 kW ; N = 100 r.p.m. ; p_1 = 10.5 bar ; p_b = 0.07 bar ; Piston speed = 180 m/min ; $K = 0.6$; $v_4 / v_2 = 3$; $v_6 / v_2 = 7.5$; $v_1 / v_2 = 0.7$

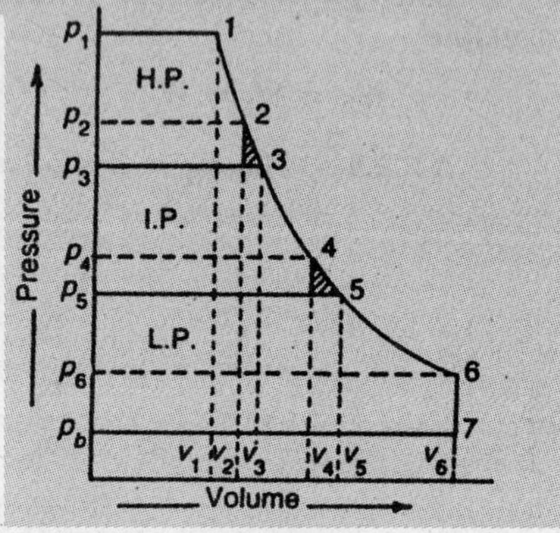

Fig. 18.19

We know that total expansion ratio,

$$R = \frac{v_6}{v_1} = \frac{v_6}{v_2} \times \frac{v_2}{v_1}$$

$$= \frac{7.5}{0.7} = 10.7$$

and actual m.e.p. referred to L.P. cylinder,

$$p_a = K\left[\frac{p_1}{R}(1 + 2.3 \log R) - p_b\right]$$

$$= 0.6\left[\frac{10.5}{10.7}(1 + 2.3 \log 10.7) - 0.07\right] = 1.94 \text{ bar}$$

Diameter of L.P. cylinder

Let D_L = Diameter of L.P. cylinder in metres.

$\therefore$ Area, $A = \frac{\pi}{4}(D_L)^2 = 0.7854\,(D_L)^2 \text{ m}^2$

We know that indicated power (I.P.),

$$3000 = \frac{200\,p_a\,L\,A\,N}{60} = \frac{100\,p_a\,A \times 2L\,N}{60}$$

$$= \frac{100 \times 1.94 \times 0.7854\,(D_L)^2\,180}{60} = 457\,(D_L)^2 \quad \ldots (\because\ 2\,L\,N = 180 \text{ m/min})$$

$\therefore$ $(D_L)^2 = 5.56$ or $D_L = 2.56$ m **Ans.**

Diameter of H.P. cylinder

Let D_H = Diameter of H.P. cylinder in metres.

We know that volume of L.P. cylinder,

$$v_6 = \frac{\pi}{4}(D_L)^2 L = 0.7854\,(2.56)^2 L = 5.147\,L$$

and volume of H.P. cylinder,

$$v_2 = \frac{\pi}{4}(D_H)^2 L = 0.7854\,(D_H)^2 L$$

We are given that $\quad \dfrac{v_6}{v_2} = 7.5 \quad$ or $\quad \dfrac{5.147\,L}{0.7854\,(D_H)^2\,L} = 7.5$

$$\therefore \qquad (D_H)^2 = 0.8738 \quad \text{or} \quad D_H = 0.935 \text{ m \textbf{Ans.}}$$

Diameter of I.P. cylinder

Let $\qquad D_I$ = Diameter of I.P. cylinder in metres,

We know that volume of I.P. cylinder,

$$v_4 = \frac{\pi}{4}(D_I)^2\,L = 0.7854(D_I)^2\,L$$

We are given that $\quad \dfrac{v_4}{v_2} = 3 \quad$ or $\quad \dfrac{0.7854\,(D_I)^2\,L}{0.7854\,(0.935)^2\,L} = 3$

$$\therefore \qquad (D_I)^2 = 2.623 \quad \text{or} \quad D_I = 1.62 \text{ m \textbf{Ans.}}$$

Example 18.10. *In a triple cylinder compound steam engine, the steam is supplied at 12.5 bar and the back pressure is 0.2 bar. All the cylinders have restricted expansions with volumes in the ratio of 1 : 2.5 : 6 and overall expansion ratio is 15. The cut-off in I.P. cylinder is 50% of the stroke and in L.P. cylinder is 65% of the stroke. Assuming hyperbolic expansion and neglecting clearance, find : 1. the theoretical mean effective pressure of the three cylinders referred to L.P. cylinder ; 2. the theoretical mean effective pressure referred to L.P. cylinder when there is a complete expansion in H.P. and I.P. cylinders. The overall expansion ratio is 12.5 ; and 3. the percentage loss of power due to restricted expansion in H.P. and I.P. cylinders.*

Solution. Given : p_1 = 12.5 bar ; $p_b = 0.2$ bar ; $v_4 / v_2 = 2.5$; $v_6 / v_2 = 6$; $R = v_6 / v_2 = 15$; $v_3 / v_4 = 50\%$ of stroke or $v_3 = 0.5\,v_4$; $v_5 / v_6 = 65\%$ of stroke or $v_5 = 0.65\,v_6$

We know that cut-off in H.P. cylinder,

$$\frac{v_1}{v_2} = \frac{v_1}{v_6} \times \frac{v_6}{v_2} = \frac{1}{15} \times 6 = 0.4$$

1. *Theoretical mean effective pressure of the three cylinders referred to L.P. cylinder*

Let $\qquad p_m$ = Theoretical mean effective pressure.

From the geometry of the cylinders, we find that expansion ratio in H.P. cylinder,

$$r_H = \frac{v_2}{v_1} = \frac{1}{0.4} = 2.5$$

Similarly, $\qquad r_I = \dfrac{v_4}{v_3} = \dfrac{1}{0.5} = 2$

and $\qquad r_L = \dfrac{v_6}{v_5} = \dfrac{1}{0.65} = 1.54$

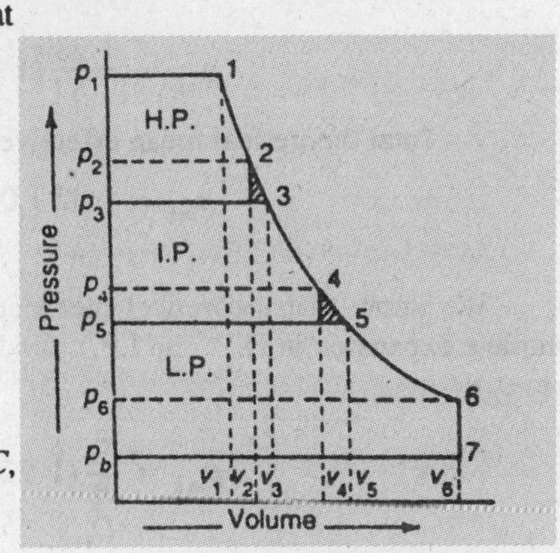

Fig. 18.20

Since expansion of steam follows the law $p\,v = C$, therefore

$$p_1 v_1 = p_3 v_3 = p_5 v_5$$

$$\therefore \qquad p_3 = p_1 \times \frac{v_1}{v_3} = p_1 \times \frac{v_1}{v_2} \times \frac{v_2}{v_4} \times \frac{v_4}{v_3}$$

$$= 12.5 \times 0.4 \times \frac{1}{2.5} \times 2 = 4 \text{ bar}$$

Similarly

$$p_5 = p_3 \times \frac{v_3}{v_5} = p_3 \times \frac{v_3}{v_4} \times \frac{v_4}{v_6} \times \frac{v_6}{v_5}$$

$$= 4 \times 0.5 \times \frac{2.5}{6} \times 1.54 = 1.28 \text{ bar} \qquad \ldots \left(\frac{v_4}{v_6} = \frac{v_4}{v_2} \times \frac{v_2}{v_6} = 2.5 \times \frac{1}{6} \right)$$

We know that theoretical mean effective pressure of H.P. cylinder,

$$p_{mH} = \frac{p_1}{r_H}(1 + 2.3 \log r_H) - p_3$$

$$= \frac{12.5}{2.5}(1 + 2.3 \log 2.5) - 4 = 5.58 \text{ bar}$$

and theoretical mean effective pressure of H.P. cylinder referred to L.P. cylinder

$$= p_{mH} \times \frac{v_2}{v_6} = 5.58 \times \frac{1}{6} = 0.93 \text{ bar} \qquad \ldots \left(\because \frac{v_2}{v_6} = \frac{1}{6} \right)$$

Similarly, theoretical mean effective pressure of I.P. cylinder,

$$p_{mI} = \frac{p_3}{r_I}(1 + 2.3 \log r_I) - p_5$$

$$= \frac{4}{2}(1 + 2.3 \log 2) - 1.28 = 2.1 \text{ bar}$$

and theoretical mean effective pressure of I.P. cylinder referred to L.P. cylinder

$$= p_{mI} \times \frac{v_4}{v_6} = 2.1 \times \frac{2.5}{6} = 0.88 \text{ bar} \qquad \ldots \left(\because \frac{v_4}{v_6} = \frac{2.5}{6} \right)$$

Similarly, theoretical mean effective pressure of L.P. cylinder,

$$p_{mL} = \frac{p_5}{r_L}(1 + 2.3 \log r_L) - p_b$$

$$p_{mL} = \frac{1.28}{1.54}(1 + 2.3 \log 1.54) - 0.2 = 0.99 \text{ bar}$$

$\therefore$ Total theoretical mean effective pressure referred to L.P. cylinder,

$$p_{m1} = 0.93 + 0.88 + 0.99 = 2.80 \text{ bar } \textbf{Ans.} \qquad \ldots (i)$$

2. *Theoretical mean effective pressure referred to L.P. cylinder for complete expansion*

We know that theoretical mean effective pressure referred to L.P. cylinder when there is complete expansion in H.P. and I.P. cylinders (or in other words there is a complete expansion from v_1 to v_6),

$$p_{m2} = \frac{p_1}{R}(1 + 2.3 \log R) - p_b$$

$$= \frac{12.5}{12.5}(1 + 2.3 \log 12.5) - 0.2 = 3.3 \text{ bar } \textbf{Ans.} \qquad \ldots (\text{Given } R = 12.5)$$

3. *Percentage loss of power due to restricted expansion in H.P. and I.P. cylinders*

We know that percentage loss of power due to restricted expansion in H.P. and I.P. cylinders

$$= \frac{p_{m2} - p_{m1}}{p_{m2}} \times 100 = \frac{3.3 - 2.8}{3.3} = 0.1515 \text{ or } 15.15\ \% \textbf{ Ans.}$$

EXERCISES

1. A horizontal, double acting compound steam engine is to develop 360 kW under the following conditions :

Pressure at steam chest = 12 bar ; Back pressure = 0.2 bar ; Average piston speed = 180 m/min ; R.P.M. = 180 ; Ratio of L.P. to H.P. cylinder areas = 4 ; Total expansion ratio = 15 ; Overall diagram factor = 0.8.

Determine the main dimensions of the engine. Assume complete expansion in H.P. cylinder and hyperbolic expansion. Neglect clearance. [**Ans.** L = 0.5 m ; D_H = 0.415 m ; D_L = 0.831 m]

2. A compound steam engine in required to develop 110 kW at 150 r.p.m. The steam is supplied at 7.5 bar and back pressure is 0.2 bar. Assuming complete hyperbolic expansion ratio of 15, calculate the cylinder diameters, so that power developed by each cylinder is equal. Take stroke length equal to diameter of L.P. cylinder and diagram factor as 0.7. Neglect clearance. [**Ans.** D_H = 0.367 m ; D_L = 0.623 m]

3. The following data refer to a double acting compound steam engine :

Indicated power = 300 kW ; Speed = 120 r.p.m ; Stroke = 900 mm ; Initial pressure = 12 bar ; Back pressure = 0.14 bar ; Expansion ratio = 15 ; Diagram factor = 0.85.

Assuming complete expansion in H.P. cylinder, equal initial load, hyperbolic expansion and neglecting clearance, determine : 1. the receiver pressure for L.P. cylinder ; and 2. the diameters of H.P. and L.P. cylinders. [**Ans.** 2.7 bar ; D_L = 0.669 m ; D_H = 0.364 m]

4. A double acting, two cylinder compound steam engine develops 185 kW at 120 r.p.m., when supplied with steam at a pressure of 7.5 bar. The condenser pressure is 0.2 bar ; allowable piston speed is 150 m/s; the ratio of cylinder volumes is 3.5; diagram factor is 0.8; and cut-off in H.P. cylinder is at 0.4 stroke.

Assuming hyperbolic expansion and neglecting clearance, determine the cylinder dimensions. [**Ans.** D_L = 0.684 m ; D_H = 0.366 m]

5. A double acting compound engine working between pressures 10.5 bar and 0.3 bar has a total ratio of expansion as 10. The cut-off in the H.P. cylinder takes place at 0.4 of stroke. Assuming hyperbolic expansion, find : 1. the cut-off in the L.P. cylinder if the initial loads on two pistons are equal, and 2. the ratio of workdone in the two cylinders. [**Ans.** 0.45 of stroke ; 1.117]

6. A double acting compound steam engine is required to develop 400 kW at 125 r.p.m. The steam is supplied at 13 bar with cut-off at 0.5 stroke in the H.P. cylinder. The condenser pressure is 0.14 bar, the mean piston speed 235 metres/min; overall expansion ratio 14; overall diagram factor 0.7; the stroke of H.P. and L.P. cylinders are same. Neglecting clearance and assuming hyperbolic expansion, calculate the L.P. and H.P. cylinder diameters and the stroke. Also find the cut-off in L.P. cylinder for equal load on both the pistons. [**Ans.** L = 0.94 m ; D_L = 0.758 m ; D_H = 0.286 m ; 0.53]

7. A double acting compound steam engine has cylinder diameter H.P. = 300 mm ; L.P. = 600 mm and stroke of both cylinders is 400 mm. The engine develops 32 kW while running at 160 r.p.m. The steam is supplied at 13.8 bar and back pressure is 0.28 bar. If cut-off in the H.P. cylinder is at one-third stroke, find the actual and theoretical mean effective pressure referred to L.P. cylinder and hence find the overall diagram factor. Neglect clearance. [**Ans.** 2.64 bar ; 3.73 bar ; 0.82]

8. A two cylinder compound steam engine receives steam at a pressure of 8 bar and discharges at 0.2 bar. The cylinder volume ratio is 4. The cut-off in H.P. and L.P. cylinder is 0.4 and 0.5 of the stroke respectively. Determine : 1. the actual mean effective pressure in H.P. and L.P. cylinders, if the diagram factor for the H.P. and L.P. cylinders is 0.75 and 0.65 respectively, and 2. the ratio of workdone in H.P. and L.P. cylinders. [**Ans.** 3.4 bar, 0.75 bar ; 0.985]

9. A compound steam engine develops 330 kW indicated power when taking steam at 10.5 bar and exhausting at 0.21 bar. The rotational speed is 180 r.p.m. and the piston speed is 183 m/min. The cut-off in H.P. cylinder is to be 0.4 of the stroke; cut-off in L.P. cylinder is 0.5 of stroke, and the cylinder volume ratio is

4. Assuming hyperbolic expansion and neglecting clearance, determine suitable cylinder dimensions. Take diagram factor 0.84. Find also the L.P. receiver pressure and compare the initial loads on the piston.

[**Ans.** L = 0.508 m ; D_L = 0.71 m, D_H = 0.355 m ; 2.1 bar ; 1.11]

10. A double acting compound steam engine working between pressures of 10 bar and 0.21 bar is to develop 150 kW indicated power at 90 r.p.m. The H.P. cylinder cut-off is at 0.35 of stroke and ratio of L.P. cylinder to H.P. cylinder volume is 3.2. Determine the cylinder diameters and stroke assuming a diagram factor of 0.7 and mean piston speed of 135 metres/min.

Find also the fraction of stroke at which the cut-off should occur in L.P. cylinder for equal initial load on both the pistons. [**Ans.** L = 0.75 m ; D_L = 0.603 m ; D_H = 0.337 m ; 0.43 of stroke]

11. A double acting compound steam engine is developing 45 kW at 200 r.p.m. The engine is supplied with steam at 10 bar. The back pressure is 650 mm of Hg and the barometer reads 750 mm of Hg. The cut-off in H.P. cylinder takes place at 50% of the stroke and the overall expansion ratio is 12. The ratio of stroke to L.P. cylinder diameter is 1.2 and the overall diagram factor is 0.7. Find the cylinder diameters, stroke and L.P. cut-off for equal initial piston loads. [**Ans.** D_L = 0.333 m, D_H = 0.136 m, L = 0.4 m ; 0.54 of stroke]

12. A double acting compound steam engine is supplied with steam at 8 bar and exhausts into a condenser at a pressure of 0.3 bar. The ratio of cylinder volumes is 4 and cut-off in H.P. cylinder takes place at 50% of the stroke. The length of stroke of both the cylinders is same. Assuming hyperbolic expansion and neglecting clearance, find :

1. the cut-off in L.P. cylinder for equal initial load on the pistons,
2. the mean effective pressure of each cylinder referred to L.P. cylinder, and
3. the percentage loss of work due to incomplete expansion in H.P. cylinder.

[**Ans.** 0.543 ; 1.23 bar, 1.31 bar ; 8.63 %]

13. A triple expansion engine is to develop 2200 kW indicated power at a piston speed of 400 m/min. The volumes of H.P., I.P. and L.P. cylinders are in the ratio of 1 : 2.5 : 7.5. The initial pressure of steam is 17 bar and the exhaust pressure is 0.2 bar. The cut-off in H.P. cylinder is at 0.5 of the stroke and overall diagram factor is 0.65.

Assuming complete expansion in H.P. and I.P. cylinders and hyperbolic expansion, determine the diameters of the cylinders. Neglect clearance. [**Ans.** D_L = 1.272 m ; D_I = 0.734 m ; D_H = 0.464 m]

14. In a triple expansion steam engine, the steam is supplied at 15 bar and back pressure is 0.14 bar. All the cylinders have restricted expansion. The cylinder volumes are in the ratio of 1 : 2.5 : 7.5 and overall expansion ratio is 15. If the cut-off in I.P. cylinder occurs at 50% of stroke and cut-off in L.P. cylinder is 60% of stroke, calculate the total mean effective pressure of all the cylinders referred to L.P. cylinder. Assume hyperbolic expansion and neglect clearance [**Ans.** 3.402 bar]

15. A triple expansion engine is required to develop 2950 kW at a piston speed of 210 m/min, under the following conditions :

Pressure in the steam chest = 15 bar ; Back pressure = 0.15 bar ; Cylinder volume ratios = 1 : 2.4 : 7.2; Total expansion ratio = 18; Overall diagram factor = 0.62.

Assuming equal initial load on each piston, find : 1. cylinder diameters ; 2. receiver pressures ; and 3. cut-off points in each cylinder.

Neglect clearance and assume hyperbolic expansion.

[**Ans.** D_L = 2.367 m, D_I = 1.367 m, D_H = 0.882 m ; 5.46 bar, 1.475 bar ; 0.4, 0.458, 0.565]

QUESTIONS

1. Explain what is meant by compounding of steam engines.
2. State the advantages of compounding a steam engine.
3. Explain, with the help of line diagrams, the various methods of compounding a steam engine.
4. What are the main factors in deciding the sizes of the cylinder in a compound steam engine ?

OBJECTIVE TYPE QUESTIONS

1. In a compound steam engine, the last stage of expansion is carried out in a
(*a*) low pressure cylinder (*b*) high pressure cylinder
(*c*) Intermediate pressure cylinder (*d*) none of these

2. By compounding the expansion of steam in two or more cylinders, the ratio of expansion
(*a*) increases (*b*) decreases (*c*) does not change

3. By compounding the expansion of steam in two or more cylinders, the length of stroke
(*a*) increases (*b*) decreases (*c*) does not change

4. A compound steam engine requires ..... flywheel than simple steam engine.
(*a*) lighter (*b*) heavier

5. In a compound steam engine, the diameter of high pressure cylinder is ..... the low pressure cylinder.
(*a*) equal to (*b*) less than (*c*) greater than

6. In a Tandem type compound steam engine, the high pressure and low pressure cylinders
(*a*) have common piston rod (*b*) are set at 90°
(*c*) have separate piston rods (*d*) are set in V-arrangement

7. The high pressure and low pressure cylinders in a Woolf type compound steam engine are regarded as having cranks
(*a*) 180° to each other (*b*) 90° to each other
(*c*) 0° to each other (*d*) none of these

8. In a receiver type compound steam engine, the high pressure and low pressure cylinders
(*a*) have common piston rod (*b*) are set at 90°
(*c*) have separate piston rods (*d*) are set in V-arrangement

9. A compound steam engine in which the piston rods of high and low pressure cylinders are attached to two different cranks set at 180° to each other, is called
(*a*) receiver type compound engine (*b*) Woolf type compound engine
(*c*) Tandem type compound engine (*d*) both (*a*) and (*b*)

10. The high pressure and low pressure cylinders in a receiver type compound steam engine are regarded as having cranks
(*a*) 180° to each other (*b*) 90° to each other
(*c*) 0° to each other (*d*) none of these

ANSWERS

1. (*a*)	2. (*b*)	3. (*b*)	4. (*a*)	5. (*b*)
6. (*a*)	7. (*a*)	8. (*c*)	9. (*b*)	10. (*b*)

19

Performance of Steam Engines

1. Introduction. 2. Efficiencies of a Steam Engine. 3. Mass of Steam in Engine Cylinder. 4. Missing Quantity. 5. Methods of Reducing Missing Quantity or Cylinder Condensation. 6. Heat Balance Sheet. 7. Governing of Steam Engines. 8. Governing of Simple Steam Engines. 9. Throttle Governing of Simple Steam Engines. 10. Cut-off Governing of Simple Steam Engines. 11. Governing of Compound Steam Engines. 12. Throttle Governing of Compound Steam Engines. 13. Cut-off Governing of Compound Steam Engines. 14. Steam Consumption (Willian's Law).

19.1. Introduction

In the last two chapters, we have discussed simple and compound steam engines. In these chapters, we have discussed power generated and cylinder dimensions of the steam engines. But in this chapter, we shall discuss their performance *i.e.* efficiencies, governing etc.

19.2. Efficiencies of a Steam Engine

The efficiency of an engine is defined as the ratio of work done to the energy supplied to an engine. The following efficiencies of a steam engine are important from the subject point of view :

1. *Mechanical efficiency.* It is the ratio of the *brake power (B.P.) to the indicated power (I.P.). Mathematically, mechanical efficiency,

$$\eta_m = \frac{\text{B.P.}}{\text{I.P.}}$$

It may be observed that the mechanical efficiency is always less than unity (*i.e.* 100%) because some power is lost in overcoming the engine friction. In other words, the indicated power is always greater than brake power. This power which is lost in overcoming the engine friction is known as *frictional power*. Therefore, frictional power,

$$\text{F.P.} = \text{I.P.} - \text{B.P.}$$

2. *Overall efficiency.* It is the ratio of the work obtained at the crank shaft in a given time to the energy supplied by fuel during the same time.

Let m_f = Mass of fuel burnt in kg per hour, and

C = Calorific value of fuel in kJ/kg of fuel.

∴ Energy supplied by fuel/min

$$= \frac{m_f \times C}{60} \text{ kJ}$$

and work obtained at the crank shaft/min

$$= \text{B.P.} \times 60 \text{ kJ} \qquad \ldots (\because \text{B.P. is in kW and 1 kW = 1 kJ/s})$$

* The indicated power (I.P.) and brake power (B.P.) has already been discussed in chapter 17.

∴ Overall efficiency,

$$\eta_0 = \frac{\text{B.P.} \times 60 \times 60}{m_f \times C} = \frac{\text{B.P.} \times 3600}{m_f \times C}$$

3. *Indicated thermal efficiency.* It is the ratio of heat equivalent of indicated power to the energy in the steam supplied per minute.

Let m_s = Mass of steam used in kg/min,

h_1 = Enthalpy or total heat of steam supplied at admission pressure p_1 in kJ/kg (from steam tables), and

h_{fb} = Enthalpy or sensible heat of feed water at back pressure p_b in kJ/kg (from steam tables).

∴ Energy in steam supplied/min.

$$= m_s (h_1 - h_{fb}) \text{ kJ/min}$$

and heat equivalent to I.P. $= \text{I.P.} \times 60$ kJ/min

∴ Indicated thermal efficiency

$$= \frac{\text{I.P.} \times 60}{m_s (h_1 - h_{fb})}$$

Note : The mass of steam used in kg per indicated power or brake power per hour (*i.e.* in kg/kWh) is known as *specific steam consumption.*

We know that mass of steam used per hour

$$= m_s \times 60 \text{ kg/h}$$

∴ Specific steam consumption

$$= \frac{m_s \times 60}{\text{I.P. or B.P.}} \text{ kg/kWh}$$

4. *Brake thermal efficiency.* It is the ratio of the heat equivalent of brake power to the energy in the steam supplied per minute. Mathematically,

Brake thermal efficiency

$$= \frac{\text{B.P.} \times 60}{m_s (h_1 - h_{fb})}$$

Note : Whenever thermal efficiency is mentioned without qualifying the name, *i.e.* "indicated" or "brake", the indicated thermal efficiency should be calculated.

5. *Relative efficiency.* The relative efficiency is also known as *efficiency ratio.* It is the ratio of thermal efficiency to the Rankine efficiency. Mathematically, relative efficiency,

$$\eta = \frac{\text{Thermal efficiency}}{\text{Rankine efficiency}}$$

Example 19.1. *During a test on a single acting non-condensing, single cylinder steam engine, the following observations were recorded :*

Bore = 225 mm ; Stroke = 600 mm ; Speed = 100 r.p.m. ; Effective brake diameter = 2.75 m ; Net load on the brake = 1650 N ; Area of indicator diagram = 2500 mm² ; Length of indicator diagram = 100 mm ; Spring strength = 530 bar/m.

Determine: 1. Indicated power ; 2. Brake power ; and 3. Mechanical efficiency.

Solution. Given : $D = 225$ mm $= 0.225$ m ; $L = 600$ mm $= 0.6$ m ; $N = 100$ r.p.m. ; $D_1 = 2.75$ m ; $(W - S) = 1650$ N ; $a_1 = 2500 \text{ mm}^2 = 2500 \times 10^{-6} \text{ m}^2$; $b = 100$ mm $= 0.1$ m ; $s = 530$ bar/m

We know that actual mean effective pressure,

$$p_a = \frac{\text{Area of indicator diagram} \times \text{Spring strength}}{\text{Length of indicator diagram}}$$

$$= \frac{a_1 \times s}{b} = \frac{2500 \times 10^{-6} \times 530}{0.1} = 13.25 \text{ bar}$$

Area of cylinder, $A = \frac{\pi}{4} \times D^2 = \frac{\pi}{4}(0.225)^2 = 0.04 \text{ m}^2$

1. *Indicated power*

We know that indicated power,

$$\text{I.P.} = \frac{100\, p_a L A N}{60} = \frac{100 \times 13.25 \times 0.6 \times 0.04 \times 100}{60} = 53 \text{ kW} \textbf{ Ans.}$$

... ($\because$ Engine is single acting)

2. *Brake Power*

We know that brake power,

$$\text{B.P.} = \frac{(W - S)\, \pi D_1 N}{60} = \frac{1650 \times \pi \times 2.75 \times 100}{60} = 23\,760 \text{ W}$$

$$= 23.76 \text{ kW} \textbf{ Ans.}$$

3. *Mechanical efficiency,*

We know that mechanical efficiency,

$$\eta_m = \frac{\text{B.P.}}{\text{I.P.}} = \frac{23.76}{53} = 0.448 \text{ or } 44.8\% \textbf{ Ans.}$$

Example 19.2. *Estimate the brake power of simple steam engine having 250 mm piston diameter, and 40 mm piston rod diameter with 250 mm stroke length operating at 300 r.p.m. The initial and back pressure of steam is 8.5 bar and 1.2 bar respectively. Assume 90% mechanical efficiency, cut-off at 25% of the forward stroke and 0.73 diagram factor. Neglect clearance and compression.*

Solution. Given : D = 250 mm = 0.25 m ; d = 40 mm = 0.04 m ; L = 250 mm = 0.25 m ; N = 300 r.p.m. ; p_1 = 8.5 bar ; p_b = 1.2 bar ; η_m = 90 % = 0.9 ; $v_1 = 0.25\, v_2$; K = 0.73

We know that expansion ratio,

$$r = \frac{v_2}{v_1} = \frac{v_2}{0.25\, v_2} = 4$$

and actual mean effective pressure,

$$p_a = K\left[\frac{p_1}{r}(1 + 2.3 \log r) - p_b\right] = 0.73\left[\frac{8.5}{4}(1 + 2.3 \log 4) - 1.2\right]$$

$$= 2.82 \text{ bar}$$

Area of piston, $A = \frac{\pi}{4} \times D^2 = \frac{\pi}{4}(0.25)^2 = 0.0491 \text{ m}^2$

and area of piston rod, $a = \frac{\pi}{4} \times d^2 = \frac{\pi}{4}(0.04)^2 = 0.001\,26 \text{ m}^2$

∴ Indicated power,

$$\text{I.P.} = \frac{100\, p_a\, L\, A\, N}{60} + \frac{100\, p_a\, L\,(A-a)\, N}{60} = \frac{100\, p_a\, L\,(2A-a)\, N}{60}$$

$$= \frac{100 \times 2.82 \times 0.25\,(2 \times 0.0491 - 0.001\ 26)\ 300}{60} = 34.2 \text{ kW}$$

We know that mechanical efficiency (η_m),

$$0.9 = \frac{\text{B.P}}{\text{I.P.}} \quad \text{or B.P.} = 0.9 \times \text{I.P.} = 0.9 \times 34.2 = 30.78 \text{ kW} \quad \textbf{Ans.}$$

Example 19.3. *A double acting steam engine with cylinder 150 mm diameter and 200 mm stroke is to develop 18 kW at 300 r.p.m. with cut-off at 20% of the stroke. The back pressure is 0.3 bar. Determine the admission pressure if diagram factor is 0.7. Also calculate the indicated thermal efficiency of the engine if it receives 220 kg of dry steam per hour. Neglect clearance.*

Solution. Given : $D = 150$ mm $= 0.15$ m ; $L = 200$ mm $= 0.2$ m ; I.P. $= 18$ kW ; $N = 300$ r.p.m. ; $v_2 = 0.2\, v_3$; $p_b = 0.3$ bar ; $K = 0.7$; $m_s = 220$ kg/h $= 3.67$ kg/min

Admission pressure

Let p_1 = Admission pressure in bar, and

p_a = Actual mean effective pressure in bar.

We know that area of the cylinder,

$$A = \frac{\pi}{4} \times D^2 = \frac{\pi}{4}(0.15)^2 = 0.0177 \text{ m}^2$$

and indicated power (I.P.),

$$18 = \frac{200\, p_a\, L\, A\, N}{60} = \frac{200 \times p_a \times 0.2 \times 0.0177 \times 300}{60} = 3.54\, p_a$$

$$\therefore \quad p_a = 5.085 \text{ bar}$$

Expansion ratio, $r = v_3 / v_2 = v_3 / 0.2\, v_3 = 5$

We also know that actual mean effective pressure (p_a),

$$5.085 = K\left[\frac{p_1}{r}(1 + 2.3 \log r) - p_b\right] = 0.7\left[\frac{p_1}{5}(1 + 2.3 \log 5) - 0.3\right]$$

$$= 0.7\,(0.52\, p_1 - 0.3) = 0.364\, p_1 - 0.21$$

$$\therefore \quad p_1 = \frac{5.085 + 0.21}{0.364} = 14.5 \text{ bar} \quad \textbf{Ans.}$$

Indicated thermal efficiency

From steam tables, corresponding to a pressure of 14.5 bar, we find that

$$h_1 = h_g = 2789 \text{ kJ/kg} \qquad \text{... (For dry steam)}$$

and corresponding to a pressure of 0.3 bar,

$$h_{fb} = 289.3 \text{ kJ/kg}$$

We know that indicated thermal efficiency.

$$= \frac{\text{I.P.} \times 60}{m_s\,(h_1 - h_{fb})} = \frac{18 \times 60}{3.67\,(2789 - 289.3)} = 0.118 \quad \text{or} \quad 11.8\,\% \quad \textbf{Ans.}$$

Example 19.4. *A double acting single cylinder steam engine runs at 250 r.p.m. and develops 30 kW. The pressure limits of operation are 10 bar and 1 bar. The cut-off is at 40% of the stroke. The stroke/bore ratio is 1.25 and the diagram factor is 0.75. Assume dry saturated steam at inlet, hyperbolic expansion and negligible effect of piston rod.*

Find : 1. the mean effective pressure ; 2. the cylinder dimensions, and 3. the indicated thermal efficiency.

Solution. Given : $N = 250$ r.p.m. ; I.P.= 30 kW ; $p_1 = 10$ bar ; $p_b = 1$ bar ; $v_2 = 0.4\, v_3$; $L/D = 1.25$; $K = 0.75$

1. *Mean effective pressure*

We know that expansion ratio,

$$r = v_3 / v_2 = v_3 / 0.4\, v_3 = 2.5$$

and actual mean effective pressure,

$$p_a = K\left[\frac{p_1}{r}(1 + 2.3 \log r) - p_b\right]$$

$$= 0.75\left[\frac{10}{2.5}(1 + 2.3 \log 2.5) - 1\right] = 5 \text{ bar} \quad \textbf{Ans.}$$

2. *Cylinder dimensions*

Let D = Diameter of the cylinder in metres, and

L = Length of the stroke in metres = $1.25\, D$ ... (Given)

Area of the cylinder, $A = \frac{\pi}{4} \times D^2 = 0.7854\, D^2$ m²

We know that indicated power (I.P.),

$$30 = \frac{200\, p_a\, L\, A\, N}{60} = \frac{200 \times 5 \times 1.25\, D \times 0.7854\, D^2 \times 250}{60}$$

$$= 4090\, D^3$$

$$\therefore \quad D^3 = 0.0733 \quad \text{or} \quad D = 0.194 \text{ m} \quad \textbf{Ans.}$$

and $$L = 1.25\, D = 1.25 \times 0.194 = 0.2425 \text{ m} \quad \textbf{Ans.}$$

3. *Indicated thermal efficiency*

First of all, let us find the mass of steam (m_s) used per minute.

From steam tables, corresponding to a pressure of 10 bar, we find that for dry saturated steam,

$$v_g = 0.1943 \text{ m}^3/\text{kg}, \text{ and } h_1 = 2776.2 \text{ kJ/kg}$$

and corresponding to a pressure of 1 bar,

$$h_{fb} = 417.5 \text{ kJ/kg}$$

We know that stroke volume,

$$v_3 = \frac{\pi}{4} \times D^2 \times L = \frac{\pi}{4}(0.194)^2\, 0.2425 = 0.007\,17 \text{ m}^3$$

and volume of steam in the cylinder at the point of cut-off,

$$v_2 = 0.4\, v_3 = 0.4 \times 0.007\,17 = 0.002\,87 \text{ m}^3$$

We know that mass of steam used per stroke

$$= \frac{v_2}{v_g} = \frac{0.002\,87}{0.1943} = 0.0147 \text{ kg}$$

and mass of steam used per minute,

$$m_s = 0.0147 \times 2\,N = 0.0147 \times 2 \times 250 = 7.35 \text{ kg/min}$$

∴ Indicated thermal efficiency

$$= \frac{\text{I.P.} \times 60}{m_s\,(h_1 - h_{fb})} = \frac{30 \times 60}{7.35\,(2776.2 - 417.5)} = 0.104 \text{ or } 10.4\,\% \text{ Ans.}$$

Example 19.5. *The following data were obtained during test on double acting steam engine :*

Indicated mean effective pressure = 2.5 bar ; R.P.M. = 104 ; Bore = 250 mm ; Stroke = 300 mm ; Net brake load = 1150 N ; Effective brake drum diameter = 1.65 m.

The steam is supplied at 7 bar and is dry and saturated. The condenser pressure = 0.07 bar ; condenser temperature = 22° C and condensate quantity = 3.3 kg/min.

Determine : 1. indicated power ; 2. brake power ; 3. mechanical efficiency ; and 4. brake thermal efficiency.

Solution. Given : $p_a = 2.5$ bar ; $N = 104$ r.p.m. ; $D = 250$ mm $= 0.25$ m ; $L = 300$ mm $= 0.3$ m ; $(W - S) = 1150$ N ; $D_1 = 1.65$ m ; $p_1 = 7$ bar ; $p_b = 0.07$ bar ; $t = 22°$ C ; $m_s = 3.3$ kg/min

1. *Indicated power*

We know that area of piston,

$$A = \frac{\pi}{4} \times D^2 = \frac{\pi}{4}(0.25)^2 = 0.0491 \text{ m}^2$$

∴ Indicated power,

$$\text{I.P.} = \frac{200\,p_a\,L\,A\,N}{60} = \frac{200 \times 2.5 \times 0.3 \times 0.0491 \times 104}{60} = 12.8 \text{ kW Ans.}$$

2. *Brake power*

We know that brake power,

$$\text{B.P.} = \frac{(W - S)\,\pi\,D_1\,N}{60} = \frac{1150 \times \pi \times 1.65 \times 104}{60} = 10.33 \times 10^3 \text{ W}$$

$$= 10.33 \text{ kW Ans.}$$

3. *Mechanical efficiency*

We know that mechanical efficiency,

$$\eta_m = \frac{\text{B.P.}}{\text{I.P.}} = \frac{10.33}{12.8} = 0.807 \text{ or } 80.7\,\% \text{ Ans.}$$

4. *Brake thermal efficiency*

From steam tables, corresponding to a pressure of 7 bar, we find that

$$h_1 = h_g = 2762 \text{ kJ/kg} \qquad \text{. . . (For dry saturated steam)}$$

and corresponding to a condenser pressure of 0.07 bar,

$$h_{fb} = 163.4 \text{ kJ/kg}$$

We know that brake thermal efficiency

$$= \frac{\text{B.P.} \times 60}{m_s (h_1 - h_{fb})} = \frac{10.33 \times 60}{3.3\,(2762 - 163.4)} = 0.0723 \text{ or } 7.23\,\% \textbf{ Ans.}$$

Example 19.6. *Steam is supplied at a pressure of 12 bar and 0.95 dry to a simple double acting non-condensing steam engine working with the following data :*

B.P. = 40 kW ; Cut-off = 0.5 of stroke ; Back pressure = 1.1 bar ; Clearance = 5% of stroke; Mean piston speed = 125 m/min; R.P.M. = 300 ; Diagram factor = 0.8; Mechanical efficiency = 90%.

Assume hyperbolic expansion and neglect compression effect of piston rod. If the steam consumption is 700 kg/h, calculate 1. stroke length ; 2. cylinder diameter ; and 3. brake thermal efficiency of the engine.

Solution. Given : $p_1 = 12$ bar ; $x = 0.95$; B.P. = 40 kW ; $c = (v_2 - v_c)/v_s = 0.5$; $p_b = 1.1$ bar ; $b = v_c/v_s = 0.05$; Mean piston speed = 125 m/min ; $N = 300$ r.p.m. ; $K = 0.8$; $\eta_m = 90\% = 0.9$; $m_s = 700$ kg/h = 11.67 kg/min

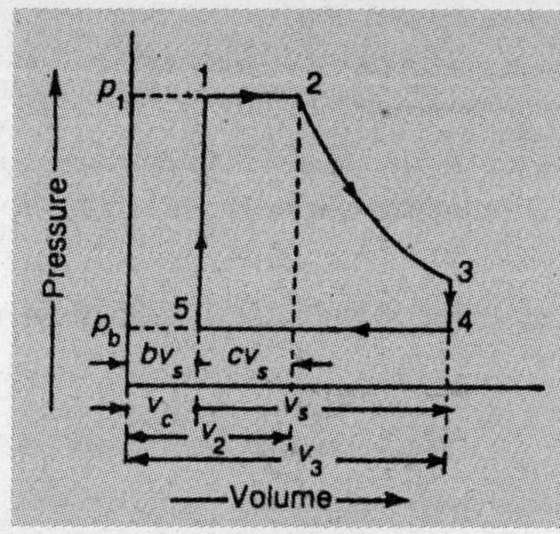

Fig. 19.1

1. *Stroke length*

Let L = Length of stroke in metres.

We know that mean piston speed,

$$125 = 2\,L\,N = 2\,L \times 300 = 600\,L$$

$$\therefore \quad L = 125/600 = 0.208 \text{ m } \textbf{Ans.}$$

2. *Cylinder diameter*

Let D = Cylinder diameter is metres.

$$\therefore \text{ Area,} \quad A = \frac{\pi}{4} \times D^2 = 0.7854\, D^2 \text{ m}^2$$

We know that theoretical mean effective pressure,

$$^* p_m = p_1\, c + 2.3\, p_1\, (b + c) \log \left(\frac{b+1}{b+c} \right) - p_b$$

$$= 12 \times 0.5 + 2.3 \times 12\,(0.05 + 0.5) \log \left(\frac{0.05 + 1}{0.05 + 0.5} \right) - 1.1$$

$$= 6 + 15.18 \log 1.91 - 1.1 = 9.17 \text{ bar}$$

and actual mean effective pressure,

$$p_a = K \times p_m = 0.8 \times 9.17 = 7.34 \text{ bar}$$

We know that indicated power,

$$\text{I.P.} = \text{B.P.} / \eta_m = 40/0.9 = 44.44 \text{ kW}$$

We also know that indicated power (I.P.),

$$44.44 = \frac{200\, p_a\, L\, A\, N}{60} = \frac{200 \times 7.34 \times 0.208 \times 0.7854\, D^2 \times 300}{60} = 1200\, D^2$$

$$\therefore \quad D^2 = 0.037 \text{ or } D = 0.192 \text{ m } \textbf{Ans.}$$

---

* Refer Chapter 17 Art. 17.8.

3. *Brake thermal efficiency*

From steam tables, corresponding to a pressure of 12 bar, we find that

$$h_{f1} = 798.4 \text{ kJ/kg, and } h_{fg1} = 1984.3 \text{ kJ/kg}$$

and corresponding to a pressure of 1.1 bar,

$$h_{fb} = 428.8 \text{ kJ/kg}$$

We know that enthalpy or total heat of steam supplied,

$$h_1 = h_{f1} + x \cdot h_{fg1} = 798.4 + 0.95 \times 1984.3 = 2683.5 \text{ kJ/kg}$$

∴ Brake thermal efficiency

$$= \frac{\text{B.P.} \times 60}{11.67\,(h_1 - h_{fb})} = \frac{40 \times 60}{11.67\,(2683.5 - 428.8)} = 0.0912$$

$$= 9.12\ \% \textbf{ Ans.}$$

Example 19.7. *A steam engine is supplied with dry saturated steam at 7 bar and exhausts at 1.4 bar. The steam consumption was found to be 2 kg/min, when the engine output was 4.4 kW. Using steam tables or chart, find the relative efficiency of the engine.*

Solution. Given : $p_1 = 7$ bar ; $p_2 = 1.4$ bar ; $m_s = 2$ kg/min ; I.P. = 4.4 kW

From steam tables, corresponding to a pressure of 7 bar, we find that

$$h_1 = h_g = 2762 \text{ kJ/kg} \qquad \text{... (For dry saturated steam)}$$

and corresponding to a pressure of 1.4 bar,

$$h_{fb} = 458.4 \text{ kJ/kg}$$

We know that indicated thermal efficiency

$$= \frac{\text{I.P.} \times 60}{m_s\,(h_1 - h_{fb})} = \frac{4.4 \times 60}{2\,(2762 - 458.4)} = 0.0573 \text{ or } 5.73\%$$

Now the enthalpy or total heat of steam at 1.4 bar (h_2) may be obtained from Mollier chart by plotting the initial condition of steam at 14 bar on the saturation curve and then drawing a vertical line up to the pressure line of 1.4 bar. The corresponding value of h_2 is 2470 kJ/kg.

$$\therefore \text{ Rankine efficiency } = \frac{h_1 - h_2}{h_1 - h_{fb}} = \frac{2762 - 2470}{2762 - 458.4} = 0.127 \text{ or } 12.7\%$$

We know that relative efficiency

$$= \frac{\text{Thermal efficiency}}{\text{Rankine efficiency}} = \frac{0.0573}{0.127} = 0.451 \text{ or } 45.1\% \textbf{ Ans.}$$

19.3. Mass of Steam in the Engine Cylinder

A steam engine takes a certain mass of steam, from its boiler, during each stroke of the piston. This is known as *cylinder feed*. In addition to the cylinder feed, there is some steam left behind in the clearance space from the previous stroke. The steam left in the clearance space is known as *cushion steam*. Mathematically, total mass of steam in the engine cylinder,

$$m = m_s + m_c$$

m_s = Mass of steam used per hour, and

m_c = Mass of cushion steam.

The mass of steam used per hour may be calculated by measuring the mass of steam condensed in the condenser over a specified period. The mass of cushion steam is given by the relation ;

$$m_c = \frac{v_h}{v_{gh}}$$

where v_h = Volume of steam at any point (H) in the compression curve as shown in Fig. 19.2.

v_{gh} = Specific volume of steam at the given pressure at point H (from steam tables).

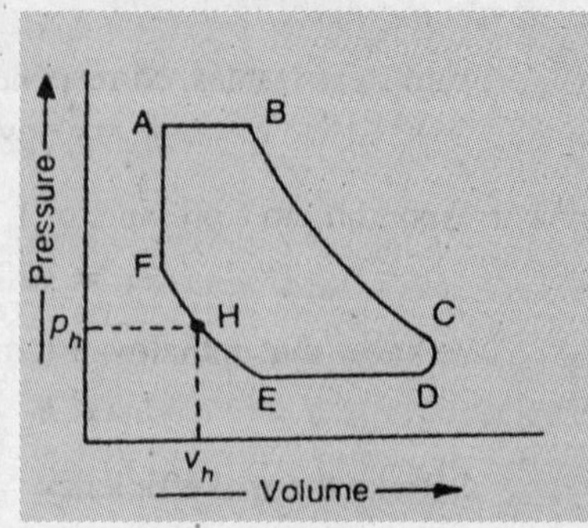

Fig. 19.2. *p-v* diagram of steam in the engine cylinder.

Example 19.8. ***The volume and pressure at a certain point on the compression curve of an actual indicator diagram is 0.002 m³ and 2.4 bar respectively. Find the mass of cushion steam and the total mass of steam in the cylinder using expansion stroke. The steam consumption per minute is 6.5 kg and speed is 100 r.p.m. Assume the engine to be double acting.***

Solution. Given : $v_h = 0.002$ m^3 ; $p_h = 2.4$ bar ; $m = 6.5$ kg/min ; $N = 100$ r.p.m.

Mass of cushion steam

From steam tables, corresponding to a pressure of 2.4 bar, we find that specific volume of steam at point H,

$$v_{gh} = 0.7464 \text{ m}^3/\text{kg}$$

We know that mass of cushion steam,

$$m_c = \frac{v_h}{v_{gh}} = \frac{0.002}{0.7464} = 0.0027 \text{ kg } \textbf{Ans.}$$

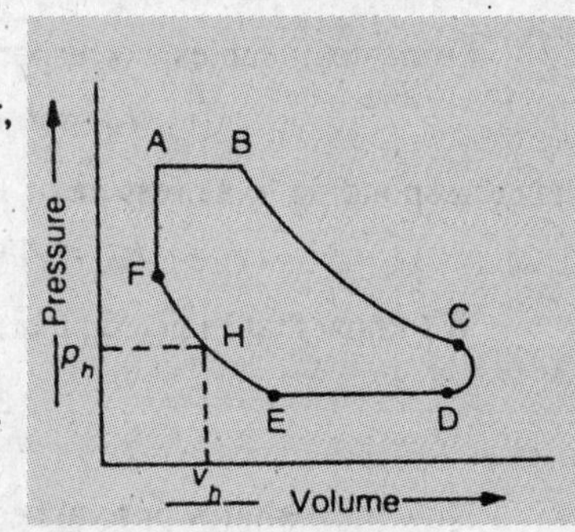

Fig. 19.3

Total mass of steam

We know that mass of steam consumed per stroke,

$$m_s = \frac{\text{Mass of steam in kg/min}}{\text{Number of strokes /min}} = \frac{m}{2N} = \frac{6.5}{2 \times 100} = 0.0325 \text{ kg}$$

$\therefore$ Total mass of steam in the cylinder for expansion stroke

$$= m_s + m_c = 0.0325 + 0.0027 = 0.0352 \text{ kg } \textbf{Ans.}$$

19.4. Missing Quantity

When a *p-v* curve for the mass of dry saturated steam is plotted on the actual indicator diagram, a curve (such as MN) is obtained. This curve is known as *saturation curve*. With the help of the saturation curve, the dryness fraction of steam, at all points of the expansion curve BC, is obtained. Thus, at any point K,

Dryness fraction of steam

$$= \frac{JK}{JL} = \frac{\text{Volume of steam in cylinder at } K}{\text{Volume of steam at } K, \text{ if dry}}$$

Now the volume of steam represented by the line KL is called the volume of *missing quantity* at point K. The missing quantity per stroke at point K may also be regarded as the difference of the total mass of steam in the cylinder during expansion stroke and the mass of steam in the cylinder assuming dry saturated at point K. Therefore,

Missing quantity per stroke at K

= Total mass of steam in the cylinder during expansion stroke – Mass of steam in the cylinder assuming dry saturated at K

$$= (m_s + m_c) - \frac{v_k}{v_{gk}}$$

where v_k = Total volume of steam at point K, and

v_{gk} = Specific volume of dry saturated steam at pressure p_k (from steam tables).

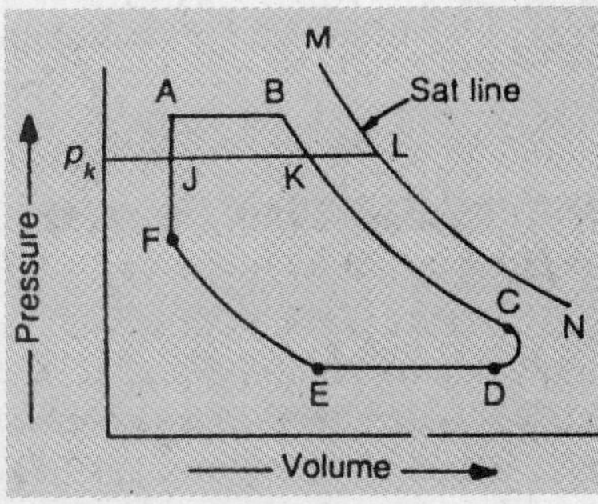

Fig. 19.4. Saturation curve in the p-v diagram of steam in the engine cylinder.

Notes : 1. The ratio of v_k / v_{gk} is known as indicated mass of steam (m_i) at any point K on the expansion curve. Therefore missing quantity per stroke

= Cylinder feed – Indicated mass of steam

2. The missing quantity is mainly due to cylinder condensation and a small amount of steam leakage past the valves and piston.

19.5. Methods of Reducing Missing Quantity or Cylinder Condensation

As a matter of fact, missing quantity is the loss of work in each stroke in an engine cylinder. It is, therefore, desirable to reduce the missing quantity. The cylinder condensation or missing quantity may be reduced by the following methods :

1. By the efficient steam jacketing of the cylinder walls.
2. By superheating the steam supplied to the engine cylinder.
3. By lagging in pipe from the boiler to the engine cylinder with a non-conducting material.
4. By compounding the expansion of steam in two cylinders, instead of one cylinder. Or, in other words, by keeping the expansion ratio small in each cylinder.
5. By increasing the speed of the engine.

Example 19.9. *The total mass of steam in an engine cylinder is 0.032 kg. It is found from the indicator diagram that at a pressure of 3.5 bar, the total cylinder volume occupied by steam is 0.0134 m³. Find the dryness fraction of steam at this pressure and also the missing quantity.*

Solution. Given : $m = 0.032$ kg ; $p_k = 3.5$ bar ; $v_k = 0.0134$ m^3

From steam tables, we find that at a pressure of 3.5 bar, specific volume of steam at point K,

$$v_{gk} = 0.524 \text{ m}^3/\text{kg}$$

$\therefore$ Total volume of steam at point K, if dry

$$= 0.524 \times 0.032 = 0.0167 \text{ m}^3$$

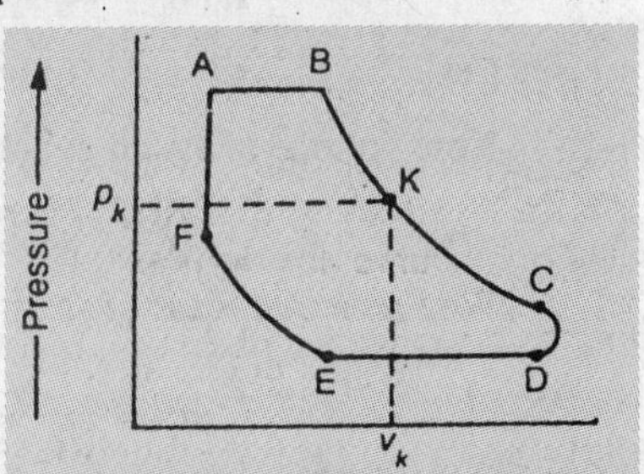

Fig. 19.5

Dryness fraction of steam

We know that dryness fraction of steam,

$$x = \frac{\text{Volume of steam in the cylinder at point } K}{\text{Volume of steam at point } K, \text{ if dry}} = \frac{0.0134}{0.0167}$$

$= 0.802$ **Ans.**

Missing quantity

We also know that volume of missing quantity

$$= \text{Volume of steam at } K\text{, if dry} - \text{Volume of steam in the cylinder at point } K$$

$$= 0.0167 - 0.0134 = 0.0033 \text{ m}^3 \quad \textbf{Ans.}$$

Example 19.10. *Estimate the dryness fraction of steam in a cylinder at 0.7 of the stroke from the following data :*

R.P.M. = 100 ; Cut-off = 0.5 of stroke ; Steam condensed per minute = 45 kg/min ; Clearance = 8 % ; Swept volume = 0.1062 m³ ; Pressure of steam at 0.7 stroke = 4.2 bar ; Pressure of steam at 0.8 of return stroke on compression curve = 1.33 bar ; Volume of 1 kg of steam at 4.2 bar = 0.438 m³ ; Volume of 1 kg of steam at 1.33 bar = 1.296 m³.

Solution. Given : $N = 100$ r.p.m ; $v_B - v_A = 0.5\, v_s$; $m = 45$ kg/min ; $v_c = 0.08\, v_s$; $v_s = 0.1062 \text{ m}^3$; $p_k = 4.2$ bar ; $p_h = 1.33$ bar ; $v_{gk} = 0.438 \text{ m}^3/\text{kg}$; $v_{gh} = 1.296 \text{ m}^3/\text{kg}$

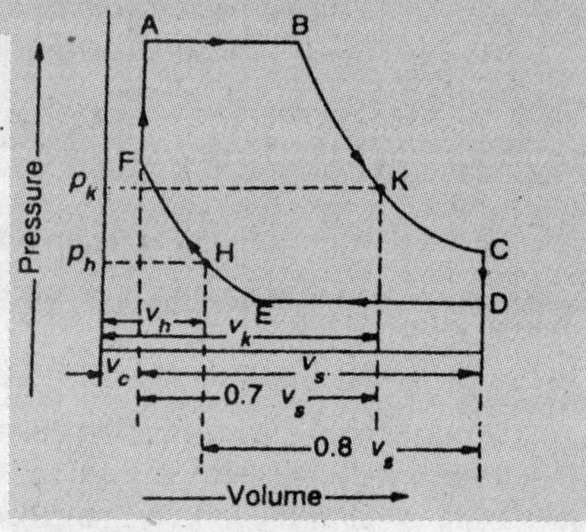

Fig. 19.6

We know that mass of steam used per stroke or cylinder feed,

$$m_s = \frac{\text{Mass of steam in kg/min}}{\text{No. of strokes/min}}$$

$$= \frac{m}{N} = \frac{45}{100} = 0.45 \text{ kg}$$

... (Assuming single acting engine)

Clearance volume,

$$v_c = 0.08\, v_s = 0.08 \times 0.1062 = 0.0085 \text{ m}^3$$

and volume of steam at point H on the compression curve (*i.e.* at 0.8 of the return stroke).

$$v_h = v_c + (1 - 0.8)\, v_s = 0.0085 + 0.2 \times 0.1062 = 0.0297 \text{ m}^3$$

We know that mass of cushion steam,

$$m_c = \frac{v_h}{v_{gh}} = \frac{0.0297}{1.296} = 0.023 \text{ kg}$$

∴ Total mass of steam used during expansion stroke,

$$m_k = m_s + m_c = 0.45 + 0.023 = 0.473 \text{ kg}$$

Let x = Dryness fraction of steam at 0.7 of the stroke *i.e.* at point K,

Now volume of steam at point K (*i.e.* at 0.7 of the stroke),

$$v_k = v_c + 0.7\, v_s = 0.0085 + 0.7 \times 0.1062 = 0.083 \text{ m}^3$$

We know that mass of steam used during expansion stroke,

$$m_k = \frac{\text{Volume of steam at } K}{\text{Volume of steam at } K\text{, if dry}} = \frac{v_k}{x\, v_{gk}}$$

$$\therefore \quad x = \frac{v_k}{m_k\, v_{gk}} = \frac{0.083}{0.473 \times 0.438} = 0.4 \quad \textbf{Ans.}$$

Example 19.11. *The following results were obtained by measurements taken on an indicator diagram of a double acting steam engine :*

(i) Immediately after cut-off : Volume = 0.1 m³ ; and pressure = 12.5 bar.

(ii) Immediately after compression has begun : Volume = 0.03 m³ ; and pressure = 4 bar.

The speed of the engine was 105 r.p.m. and the steam supply per minute was 180 kg. Find the missing quantity in kg/min.

Solution. Given : $v_k = 0.1$ m³ ; $p_k = 12.5$ bar ; $v_h = 0.03$ m³ ; $p_h = 4$ bar ; $N = 105$ r.p.m. ; $m = 180$ kg/min

We know that mass of steam used per stroke, *i.e.* cylinder feed,

$$m_s = \frac{\text{Mass of steam used/min}}{\text{No. of strokes/min}} = \frac{m}{2N} = \frac{180}{2 \times 105} = 0.857 \text{ kg}$$

From steam tables, corresponding to a pressure of 12.5 bar, we find that specific volume of steam,

$$v_{gk} = 0.157 \text{ m}^3/\text{kg}$$

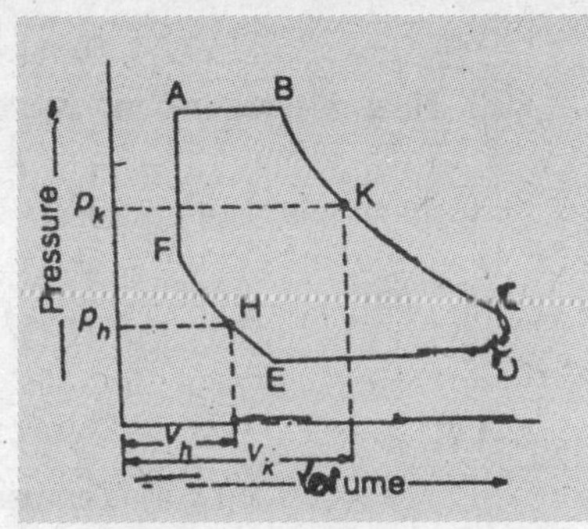

Fig. 19.7

and corresponding to a pressure of 4 bar, specific volume of steam,

$$v_{gh} = 0.462 \text{ m}^3/\text{kg}$$

∴ Mass of cushion steam,

$$m_c = \frac{v_h}{v_{gh}} = \frac{0.03}{0.462} = 0.065 \text{ kg}$$

and indicated mass of steam,

$$m_i = \frac{v_k}{v_{gk}} = \frac{0.1}{0.157} = 0.637 \text{ kg}$$

We know that missing quantity per stroke

$$= (m_s + m_c) - m_i = (0.857 + 0.065) - 0.637 = 0.285 \text{ kg}$$

∴ Missing quantity/min

$$= 0.285 \times 2 \times 105 = 59.85 \text{ kg} \quad \textbf{Ans.}$$

Note : This is the missing quantity at the beginning of the expansion stroke.

19.6. Heat Balance Sheet

The complete record of heat supplied and heat rejected during a certain time (say one minute) by a steam engine in a tabulation form is known as *heat balance sheet*. The following values are generally required to complete the heat balance sheet of a steam engine.

1. *Heat supplied to cylinder per minute*

Let m_s = Mass of steam supplied to cylinder in kg/min,

m_j = Mass of steam supplied to jackets in kg/min, and

h = Total heat of steam supplied in kJ/kg.

∴ Heat supplied to cylinder/min.

$$= h\,(m_s + m_j) \text{ kJ/min} \qquad \ldots (i)$$

Note : If some water is received from receiver drain (m_r), then heat supplied to the cylinder/min.

$$= h\,(m_s + m_j + m_r) \text{ kJ/min}$$

2. *Heat absorbed in indicated power*

We know that heat absorbed in indicated power (I.P.) per minute

$$= \text{I.P.} \times 60 \text{ kJ/min} \qquad \ldots (ii)$$

3. *Heat rejected to the cooling water*

Let m_c = Mass of the cooling water/min,

t_o = Outlet temperature of cooling water, and

t_i = Inlet temperature of cooling water.

∴ Heat rejected to the cooling water

$$= m_c \times c_w (t_0 - t_i) = 4.2\, m_c (t_0 - t_i) \text{ kJ/min} \qquad \ldots (iii)$$

...(∵ Sp. heat of water, c_w = 4.2 kJ/kg K)

4. *Heat rejected in condensate*

Let t_c = Temperature of the condensate.

∴ Heat rejected in condensate

$$= m_s\, c_c\, t_c = 4.2\, m_s\, t_c \text{ kJ/min} \qquad \ldots (iv)$$

...(∵ Sp. heat of condensate, c_c = 4.2 kJ/kg K)

Note : The sum of heat rejected to the cooling water and condensate is known as heat rejected to exhaust steam.

5. *Heat rejected in jacket drain*

Let m_j = Mass of water drained for jackets/min, and

t_j = Temperature of jacket water.

∴ Heat rejected in jacket drain

$$= m_j\, c_w\, t_j = 4.2\, m_j\, t_j \text{ kJ/min} \qquad \ldots (v)$$

6. *Unaccounted heat*

There is always some loss of heat due to friction, leakage, radiation etc., which can not be determined experimentally. In order to complete the heat balance sheet, this loss is obtained by the difference of heat supplied to cylinder per minute and heat rejected in I.P., exhaust steam and jacket drain.

Finally, the heat balance sheet is prepared as given below :

S.No.	*Particulars*	*Heat in*	
		kJ/min	*%*
	Total heat supplied	...	*100*
1.	Heat absorbed in I.P.	...	...
2.	Heat in cooling water	...	...
3.	Heat in condensate	...	...
4.	Heat in jacket drain	...	...
5.	Unaccounted heat	...	...

Example 19.12. *The following observations were made during a trial of a jacketed simple steam engine :*

Pressure of steam supplied = *10 bar*

Cylinder feed = *13.5 kg/min*
Jacket feed = *1.5 kg/min*
Condition of cylinder and jacket feed = *95% dry*
Mass of circulating water = *220 kg/min*
Outlet temperature = *35° C*
Inlet temperature = *15° C*
Condensate temperature = *50° C*
Temperature of jacket drain = *150° C*
Indicated power = *80 kW*
Draw a heat balance sheet for the engine and also find indicated thermal efficiency.

Solution. Given : $p = 10$ bar ; $m_s = 13.5$ kg/min ; $m_j = 1.5$ kg/min ; $x = 95\% = 0.95$; $m_c = 220$ kg/min ; $t_o = 35°$ C ; $t_i = 15°$ C ; $t_c = 50°$ C ; $t_j = 150°$ C ; I.P. = 80 kW

Heat balance sheet

From steam tables, corresponding to a pressure of 10 bar, we find that

$$h_f = 762.6 \text{ kJ/kg} ; \quad \text{and} \quad h_{fg} = 2013.6 \text{ kJ/kg}$$

We know that total heat in 1 kg of steam,

$$h = h_f + x\,h_{fg} = 762.6 + 0.95 \times 2013.6 = 2675.5 \text{ kJ/kg}$$

∴ Total heat supplied to the cylinder per minute

$$= h\,(m_s + m_j) = 2675.5\,(13.5 + 1.5) = 40\,132 \text{ kJ/min}$$

Heat absorbed in I.P. per minute

$$= \text{I.P.} \times 60 = 80 \times 60 = 4800 \text{ kJ/min} \qquad \ldots (i)$$

Heat rejected to cooling water per minute

$$= m_c\,c_w\,(t_o - t_i) = 220 \times 4.2\,(35 - 15) = 18\,480 \text{ kJ/min} \qquad \ldots (ii)$$

Heat in condensate per min

$$= m_s\,c_c\,t_c = 13.5 \times 4.2 \times 50 = 2835 \text{ kJ/min} \qquad \ldots (iii)$$

Heat rejected in jacket drain

$$= m_j\,c_w\,t_j = 1.5 \times 4.2 \times 150 = 945 \text{ kJ/min} \qquad \ldots (iv)$$

Unaccounted heat per min (by difference)

$$= 40\,132 - (4800 + 18\,480 + 2835 + 945) = 13\,072 \text{ kJ/min} \qquad \ldots (v)$$

Now prepare the heat balance sheet as given below :

S.No.	*Particulars*	*Heat in*	
		kJ/min	*%*
	Total heat supplied	40 132	100
1.	Heat absorbed in I.P.	4800	11.96
2.	Heat in cooling water	18 480	46.05
3.	Heat in condensate	2835	7.06
4.	Heat in jacket drain	945	2.35
5.	Unaccounted heat	13 072	32.58

Indicated thermal efficiency

We know that heat equivalent to I.P./min

$$= 4800 \text{ kJ/min}$$

and net heat supplied/min = Total heat supplied – Heat in condensate

$$= 40\,132 - 2835 = 37\,297 \text{ kJ/min}$$

∴ Indicated thermal efficiency

$$= \frac{\text{Heat equivalent to I.P./min}}{\text{Net heat supplied / min}} = \frac{4800}{37\,297} = 0.1287 \text{ or } 12.87\ \% \quad \textbf{Ans.}$$

Example 19.13. *The following observations were recorded during trial on a jacketed double acting compound steam engine supplied with dry saturated steam :*

I.P. of the steam engine	*= 180 kW*
Pressure of steam supplied	*= 6 bar*
Receiver pressure	*= 2.5 bar*
Cylinder feed	*= 17.8 kg/min*
Discharge from jacket drain	*= 2.2 kg/min*
Discharge from receiver drain	*= 1.4 kg/min*
Mass of circulating water	*= 350 kg/min*
Rise of temperature in cooling water	*= 25° C*
Condensate temperature	*= 50° C*
Temperature of jacket drain	*= 154.5° C*
Average brake torque	*= 4500 N-m*
Engine speed	*= 250 r.p.m.*

Find mechanical and brake thermal efficiency of the steam engine. Also prepare heat balance sheet on minute as well as percentage basis.

Solution. Given : I.P. = 180 kW ; $p = 6$ bar ; $p_r = 2.5$ bar ; $m_s = 17.8$ kg/min ; $m_j = 2.2$ kg/min ; $m_r = 1.4$ kg/min ; $m_w = 350$ kg/min ; $t_o - t_i = 25°$ C ; $t_c = 50°$ C ; $t_j = 154.5°$ C ; $T = 4500$ N-m; $N = 250$ r.p.m.

Mechanical efficiency

We know that brake power of the engine,

$$\text{B.P.} = \frac{2\pi N T}{60} = \frac{2\pi \times 250 \times 4500}{60} = 117\,800 \text{ W} = 117.8 \text{ kW}$$

∴ Mechanical efficiency,

$$\eta_m = \frac{\text{B.P.}}{\text{I.P.}} = \frac{117.8}{180} = 0.654 \text{ or } 65.4\ \% \quad \textbf{Ans.}$$

Brake thermal efficiency

We know that heat in B.P. per minute

$$= \text{B.P.} \times 60 = 117.8 \times 60 = 7068 \text{ kJ/min}$$

Heat in condensate per minute

$$= m_s\, c_c\, t_c = 17.8 \times 4.2 \times 50 = 3738 \text{ kJ/min}$$

From steam tables, corresponding to a pressure of 6 bar, we find that heat in 1 kg of dry saturated steam,

$$h = 2755.5 \text{ kJ/kg}$$

∴ Heat supplied to the cylinder per minute

$$= h(m_s + m_j + m_r) = 2755.5\,(17.8 + 2.2 + 1.4) = 58\,970 \text{ kJ/min}$$

and net heat supplied per minute

= Total heat supplied – Heat in condensate

$$= 58\,970 - 3738 = 55\,232 \text{ kJ/min}$$

∴ Brake thermal efficiency

$$= \frac{\text{Heat in B.P. per minute}}{\text{Net heat supplied per minute}} = \frac{7068}{55\,232}$$

$$= 0.128 \text{ or } 12.8\ \% \textbf{ Ans.}$$

Heat balance sheet

Heat in I.P. per minute

$$= \text{I.P} \times 60 = 180 \times 60 = 10\,800 \text{ kJ/min}$$

We know that heat rejected to the cooling water per minute

$$= m_w\, c_w\,(t_o - t_i) = 350 \times 4.2 \times 25 = 36\,750 \text{ kJ/min}$$

and heat rejected in jacket drain per minute

$$= m_j\, c_w\, t_j = 2.2 \times 4.2 \times 154.5 = 1428 \text{ kJ/min}$$

From steam tables, corresponding to a receiver pressure of 2.5 bar, we find that heat in 1 kg of saturated water,

$$h_f = 535.4 \text{ kJ/kg}$$

∴ Heat rejected in receiver drain per min

$$= 1.4 \times 535.4 = 750 \text{ kJ/min}$$

Unaccounted heat per minute (by difference)

$$= 58\,970 - (10\,800 + 36\,750 + 3738 + 1428 + 750) = 5504 \text{ kJ/min}$$

Now prepare the heat balance sheet as given below :

S.No.	*Particulars*	*Heat in*	
		kJ/min	*%*
	Total heat supplied	58 970	100
1.	Heat absorbed in I.P.	10 800	18.31
2.	Heat in cooling water	36 750	62.32
3.	Heat in condensate	3738	6.34
4.	Heat in jacket drain	1428	2.42
5.	Heat in receiver drain	750	1.27
6.	Unaccounted heat	5504	9.34

19.7. Governing of Steam Engines

As a matter of fact, simple and compound steam engines are always designed to run at a particular speed. But in actual practice, load on the engine keeps on fluctuating from time to time. The change of load, on the steam engine, is sure to change its speed and rate of steam flow. It has been observed that if load on the steam engine is decreased, without changing the quantity of steam, the engine will run at a higher speed. Similarly, if load on the steam engine is increased, without changing the quantity of steam, the engine will run at a lower speed.

Now, in order to have a high efficiency of a steam engine, at different load conditions, its speed must be kept constant as far as possible. The process of providing any arrangement, which will keep the speed constant (according to the changing load conditions) is known as *governing of steam engines*.

We know that work done in a cylinder of a steam engine, is equal to the area of the indicator diagram to some scale. It is thus obvious, that if the engine is subjected to an increased load, more work must be done in the cylinder. Similarly, if the engine is subjected to a reduced load, less work must be done in the cylinder. Now in order to do a greater amount of work in the cylinder, area of the indicator diagram must be increased. Similarly, in order to do a less amount of work in the cylinder, area of the indicator diagram must be decreased. A little consideration will show, that in order to increase the area of indicator diagram either steam pressure or volume of the intake steam should be increased (or decreased for decrease in area of the indicator diagram). These days, both the above mentioned methods are used for the governing of a steam engine. In the following pages, we shall discuss the governing of simple and compound steam engines.

19.8. Governing of Simple Steam Engines

The governing of simple steam engines is done by a number of methods. But the following two methods are important from the subject point of view :

1. Throttle governing, and 2. Cut-off governing.

These two methods of governing of simple steam engines are discussed, in detail, in the following pages.

19.9. Throttle Governing of Simple Steam Engines

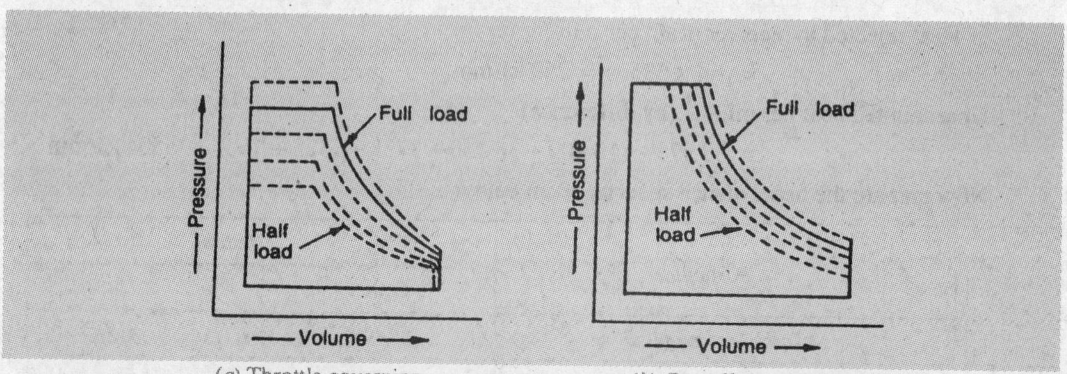

(*a*) Throttle governing. (*b*) Cut-off governing.

Fig. 19.8. Governing of simple steam engines.

The throttle governing, of simple steam engines, is a method of controlling the engine output by varying pressure of the intake steam. The pressure of intake steam is varied by opening or closing the throttle valve under the control of a centrifugal governor. This is illustrated in Fig. 19.8 (*a*), In this figure, actual indicator diagram of the simple steam engine at full load on *p-v* diagram is shown. Whenever load on the engine is decreased, it tends to run the engine at a higher speed. Now, in order to run the engine at the designed speed, less work is required to be done in the engine cylinder. Or in other words, area of the indicator diagram is required to be reduced. In throttle governing, area of the indicator diagram is reduced by reducing pressure of the admission steam as shown in Fig. 19.8 (*a*). Similarly, if the load further decreases, pressure of the admission steam is also further reduced.

Notes : 1. In throttle governing, cut-off of the engine cylinder remains the same.

2. If load on the engine is increased, then pressure of the admission steam is also increased to suit the increased load.

3. The throttle governing of a simple steam engine results in the reduction of its thermal efficiency, and as such is somewhat wasteful with regard to the steam. The reason for the same is that at any load (below the full load), the full pressure of the steam is not used. Moreover, due to constant cut-off, a large quantity of steam is used in every stroke. This also tends to lower the thermal efficiency of the engine.

19.10. Cut-off Governing of Simple Steam Engines

The cut-off governing, of a simple steam engine, is a method of controlling the engine output by varying volume of intake steam. This is done by varying the cut-off point by a slide valve under the control of a centrifugal governor. This is illustrated in Fig. 19.8 (*b*). In this figure, actual indicator diagram of the simple steam engine at full load on *p-v* diagram is shown. Whenever load on the engine is decreased, it tends to run the engine at a higher speed. Now, in order to run the engine at the designed speed, less work is required to be done in the engine cylinder. Or in other words, area of the indicator diagram is required to be reduced. In cut-off governing, area of the indicator diagram is reduced by reducing volume of the intake steam, as shown in Fig. 19.8 (*b*). Similarly, if the load further decreases, volume of the intake steam is also further reduced.

Notes : 1. In a cut-off governing, pressure of the intake steam remains the same.

2. If load on the engine is increased, then pressure of the admission steam is also increased to suit the increased load.

3. This method is more economical and efficient. That is why, these days, cut-off governing is mostly used. But it requires a special valve gear.

19.11 Governing of Compound Steam Engines

The governing of compound steam engines is done by a number of methods. But, like the governing of simple steam engines, the following two methods are important from the subject point of view :

1. Throttle governing, and 2. Cut-off governing.

These two methods of governing compound steam engines are discussed, in detail, in the following pages.

19.12. Throttle Governing of Compound Steam Engines

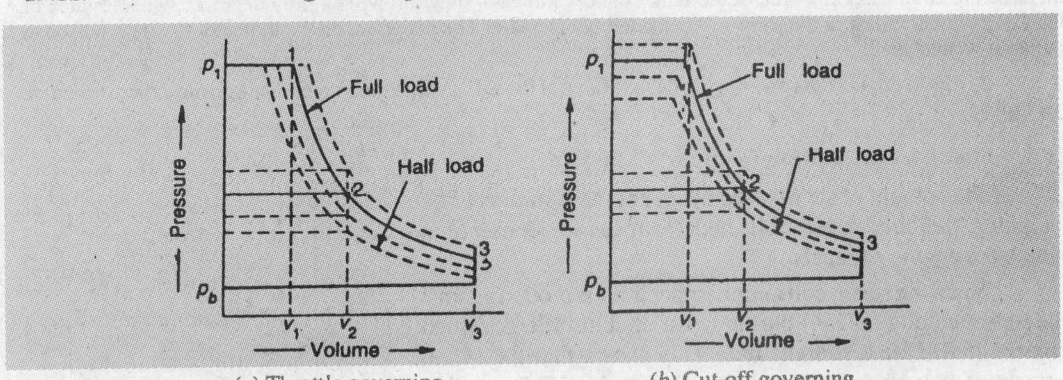

(*a*) Throttle governing. (*b*) Cut-off governing.

Fig. 19.9. Governing of compound steam engines.

The throttle governing, of a compound steam engine, is a method of controlling the engine output by varying pressure of the intake steam. The pressure of intake steam is varied by opening or closing the throttle valve under the control of a centrifugal governor. This is illustrated in Fig. 19.9 (*a*). In this figure, indicator diagram of a compound steam engine is shown with complete expansion in H.P. cylinder and incomplete expansion in L.P. cylinder at full load. Whenever load on the engine is decreased, it tends to run the engine at a higher speed. Now in order to run the engine at the designed speed, less work is required to be done in the engine cylinders. Or in other words, area of the indicator diagram is required to be reduced. In throttle governing, area of the indicator diagram is reduced by reducing pressure of the admission steam as shown in Fig. 19.9 (*a*). Similarly, if the load further decreases, pressure of the admission steam is also further reduced.

Notes : 1. In throttle governing, cut-off point of the engine cylinder remains the same.

2. If load on the engine is increased, then pressure of the admission steam is also increased to suit the increased load.

3. In throttle governing, work done in H.P. cylinder is greatly reduced, whereas work done in L.P. cylinder is slightly reduced.

19.13. Cut-off Governing of Compound Steam Engines

The cut-off governing of a compound steam engine, is a method of controlling the engine output by varying volume of intake steam in the H.P. cylinder. This is done by varying the cut-off point under the control of a centrifugal governor. This is illustrated in Fig. 19.9 (*b*). In this figure, actual indicator diagram of a compound steam engine at full load on *p-v* diagram is shown. Whenever load on the engine is decreased, it tends to run the engine at a higher speed. Now, in order to run the engine at the designed speed, less work is required to be done in the engine cylinder. Or in other words, area of the indicator diagram is required to be reduced. In this method, area of the indicator diagram is reduced by reducing the volume of the intake steam as shown in Fig. 19.9 (*b*). Similarly, if the load further decreases, volume of the intake steam is also further reduced.

From the geometry of the indicator diagram, we see that when volume of the intake steam is reduced, it will reduce the pressure of steam at the release point of H.P. cylinder (or admission into the L.P. cylinder). This will happen as the volume of H.P. cylinder (v_2) will remain constant. In such a case, area of H.P. cylinder indicator diagram will slightly increase ; whereas area of L.P. cylinder indicator diagram will considerably decrease. The net effect on the indicator diagram will be to reduce its total area. Or in other words to reduce the total work done in both the cylinders.

Notes : 1. In this method, pressure of the intake steam remains the same.

2. If load on the engine is increased, then volume of the admission steam is also increased to suit the increased load. In this case, area of the H.P. cylinder indicator diagram will slightly decrease; whereas area of the L.P. cylinder indicator diagram will considerably increase. The net effect on the indicator diagram will be to increase its total area.

3. If the engine runs on a very little load, area of the L.P. cylinder indicator diagram may almost become negligible.

19.14. Steam Consumption (Willian's Law)

The amount of steam used by an engine is measured by weighing the condensate collected from the condenser into which the engine exhausts.

When the steam consumption per hour is plotted against the indicated power (I.P.) during a test on a throttle governed engine, it will be a straight line. This shows that the steam consumption per hour is directly proportional to I.P. It is called as *Willian's law* and the straight line is called *Willian's line* as shown in Fig. 19.10.

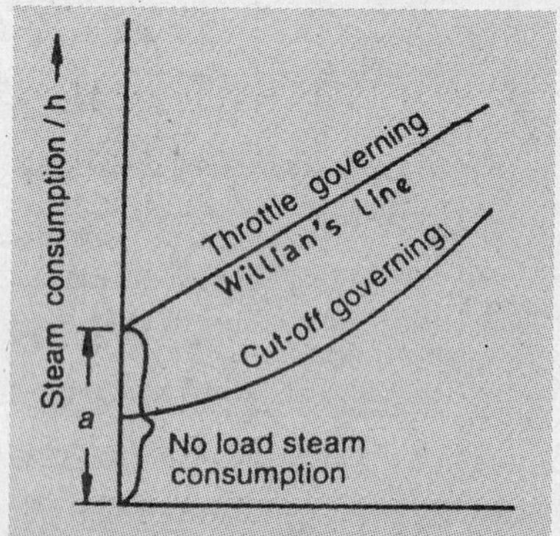

Fig. 19.10. Willian's law for steam consumption.

Willian's law holds good only for a throttle-governed engine, because the ratio of expansion remains constant. This condition is not fulfilled in a cut-off governed engine. The Willian's line follows the law :

$$m = a + b \times I.P.$$

where m = Steam consumption per hour,

a = A constant, *i.e.* no load consumption per hour,

b = Another constant representing the shape of the Willian's line, and

I.P. = Indicated power.

Example 19.14. *A throttle-governed steam engine develops 15 kW with 280 kg of steam used per hour and 37.5 kW with 520 kg of steam per hour. Determine the steam in kg/h when steam engine develops 20 kW.*

Solution. Given : m_1 = 280 kg/h when I.P. = 15 kW ; m_2 = 520 kg/h when I.P. = 37.5 kW

We know that Willian's law is

$$m = a + b \times \text{I.P.}$$

$$\therefore \quad 280 = a + b \times 15 \quad \ldots (i)$$

and

$$520 = a + b \times 37.5 \quad \ldots (ii)$$

Subtracting equation (*i*) from equation (*ii*),

$$240 = 22.5\,b \quad \text{or} \quad b = 10.67$$

Substituting the value of b = 10.67 in equation (i), we get a = 120.

Now the equation of Willian's line becomes

$$m = 120 + 10.67 \times \text{I.P.}$$

∴ Steam cosumption at 20 kW

$$= 120 + 10.67 \times 20 = 333.4 \text{ kg/h}$$ **Ans.**

Example 19.15. *A steam engine of 400 kW I.P., governed by throttle control consumes 10 kg of steam per hour per I.P. at full load and 12.5 kg at half load. Find the steam consumption, if it runs at two-third the load for 4 hours.*

Solution. Given : I.P. = 400 kW

We know that steam consumption at full load

$$= 10 \text{ kg/I.P./h}$$

∴ Total steam consumption per hour at full load *i.e.* at 400 kW,

$$m_1 = 10 \times 400 = 4000 \text{ kg/h}$$

We also know that steam consumption at half load

$$= 12.5 \text{ kg/I.P./h}$$

∴ Total steam consumption per hour at half load *i.e.* at 200 kW,

$$m_2 = 12.5 \times 200 = 2500 \text{ kg/h}$$

According to Willian's law :

$$4000 = a + b \times 400 \quad \ldots (i)$$

and

$$2500 = a + b \times 200 \quad \ldots (ii)$$

Subtracting equation (*ii*) from (*i*),

$$1500 = 200\,b \quad \text{or} \quad b = 7.5$$

Substituting the value of b = 7.5 in equation (*i*), we get a = 1000.

Now the equation of Willian's line becomes

$$m = 1000 + 7.5 \times \text{I.P.}$$

$\therefore$ Steam consumption at 2/3 load *i.e.* at $2 \times 400/3$ kW,

$$= 1000 + 7.5 \times \frac{2}{3} \times 400 = 3000 \text{ kg/h}$$

and steam consumption for 4 hours

$$= 4 \times 3000 = 12\,000 \text{ kg} \quad \textbf{Ans.}$$

Example 19.16. *A steam engine with throttle governing and running at 140 r.p.m. develops 4.2 kW and 9.5 kW indicated power while steam consumption is 104 kg/h and 179 kg/h respectively. The steam is supplied to the engine at a pressure of 6 bar and 0.9 dryness fraction. The condensate temperature is 40° C. Find the indicated thermal efficiency of the engine when it runs at 140 r.p.m. and develops 7.5 kW.*

Solution. Given : $^*N = 140$ r.p.m. ; $(\text{I.P.})_1 = 4.2$ kW ; $(\text{I.P.})_2 = 9.5$ kW ; $m_1 = 104$ kg/h ; $m_2 = 179$ kg/h ; $p = 6$ bar ; $x = 0.9$; $t_c = 40°$ C

We know that Willian's law is

$$m = a + b \times \text{I.P.}$$

$$\therefore \quad 104 = a + b \times 4.2 \quad \ldots (i)$$

and

$$179 = a + b \times 9.5 \quad \ldots (ii)$$

Subtracting equation (*i*) from (*ii*),

$$75 = 5.3\, b \quad \text{or} \quad b = 14.15$$

Substituting the value of $b = 14.15$ in equation (*i*), we get $a = 44.57$.

Now the equation of Willian's line becomes

$$m = 44.57 + 14.15 \times \text{I.P.}$$

$\therefore$ Steam consumption when engine develops 7.5 kW,

$$m_s = 44.57 + 14.15 \times 7.5 = 150.7 \text{ kg/h} = 2.51 \text{ kg/min}$$

From steam tables corresponding to a pressure of 6 bar, we find that

$$h_{f1} = 670.4 \text{ kJ/kg} ; \text{ and } h_{fg1} = 2085 \text{ kJ/kg}$$

and corresponding to a condensate temperature of 40° C,

$$h_{fb} = 167.5 \text{ kJ/kg}$$

We know that total heat of steam supplied,

$$h_1 = h_{f1} + x\, h_{fg1} = 670.4 + 0.9 \times 2085 = 2547 \text{ kJ/kg}$$

$\therefore$ Indicated thermal efficiency

$$= \frac{\text{I.P.} \times 60}{m_s (h_1 - h_{fb})} = \frac{7.5 \times 60}{2.51\,(2547 - 167.5)} = 0.075 \text{ or } 7.5\% \quad \textbf{Ans.}$$

EXERCISES

1. A double acting steam engine with piston diameter 275 mm, stroke length 650 mm and cut-off 50% of stroke length is supplied steam at a pressure of 7 bar. The back pressure is 1.2 bar. Assuming a diagram factor of 0.75, find the indicated power of the engine when it runs at 250 r.p.m. Also find mechanical efficiency of the engine if its brake power is 100 kW. Neglect clearance. [**Ans.** 114 kW ; 87.7 %]

---

* Superfluous data.

2. The following data pertains to a single cylinder double acting steam engine :

Admission pressure = 14 bar ; Back pressure = 0.35 bar ; Cut-off = 0.4 of stroke ; Diagram factor = 0.7 ; Cylinder diameter = 300 mm ; Stroke = 1.5 Cylinder bore ; Mechanical efficiency = 80%.

Neglecting clearance, estimate brake power developed by the engine running at 200 r.p.m.

[**Ans.** 123.2 kW]

3. A single cylinder double acting steam engine of 280 mm bore and 450 mm stroke works on a supply pressure of 10 bar and back pressure of 0.15 bar. The cut-off takes place at 25% of stroke. Assuming a diagram factor of 0.7 and engine speed 3 revolutions per second, find the I.P. of the engine. If the engine consumes 15 kg of dry saturated steam per minute, find the indicated thermal efficiency of the engine. Neglect clearance.

[**Ans.** 10.62%]

4. A single cylinder double acting steam engine with cylinder diameter 150 mm ; stroke length 200 mm is required to develop 20 kW of indicated power at 300 r.p.m. with cut-off at 20% of the stroke. Find the admission pressure if the diagram factor is 0.72 and back pressure is 0.28 bar. Also calculate the indicated thermal efficiency of the engine, when it receives 240 kg of dry steam per hour. [**Ans.** 15.6 bar ; 11.96 %]

5. The following observations were made during the trial of a steam engine :

Radius of the brake wheel = 500 mm ; Load on the brake = 1350 N ; Spring balance reading = 50 N ; Steam supplied per minute = 7 kg ; Admission pressure of steam = 11 bar ; Quality of steam = 0.98 dry ; Back pressure = 1.1 bar ; Engine speed = 240 r.p.m.

Find the brake thermal efficiency of the engine. [**Ans.** 8.5%]

6. The following data relates to a single cylinder double acting steam engine using dry saturated steam :

Bore = 300 mm ; Stroke = 450 mm ; Hypothetical mean effective pressure = 3.5 bar ; Cut-off = 25% of stroke ; Exhaust pressure = 0.5 bar ; Steam consumption = 560 kg/h ; Speed = 160 r.p.m. ; Diagram factor = 0.8 ; Mechanical efficiency = 80%.

Calculate brake thermal efficiency. Assume hyperbolic expansion and neglect clearance and compression. [**Ans.** 10.1%]

7. A single cylinder double acting steam engine is supplied with dry saturated steam at 11 bar. The bore is 300 mm and the stroke 375 mm. The engine runs at 250 r.p.m. with a cut-off ratio of 1/3 and against a back pressure of 0.36 bar. If the diagram factor is 0.8, calculate the indicated power if mechanical efficiency is 85% and brake thermal efficiency 15%. Calculate also the brake power and the specific steam consumption in kg/B.P./h. Assume that the condensate is used as the feed water to the boiler and the expansion of steam is hyperbolic. Neglect clearance [**Ans.** 129.5 kW ; 110 kW ; 9.7 kg/B.P./h]

8. The following observations were recorded during a trial of a single cylinder double acting non-condensing steam engine :

Cylinder diameter = 250 mm ; Piston rod diameter = 50 mm ; Stroke = 350 mm ; Speed = 240 r.p.m.; Cut-off = 0.3 of stroke ; Length of indicator diagram = 53 mm ; Area of indicator diagram for cover end = 1570 mm^2 ; Area of indicator diagram for crank end = 1440 mm^2 ; Spring number = 120 bar/m ; Diameter of brake wheel = 1.67 m ; Diameter of brake rope = 20 mm ; Dead load on the brake = 2000 N ; Reading of spring balance = 200 N ; Pressure of steam supplied = 10.5 bar ; Dryness fraction of steam supplied = 0.9.

Find : 1. Indicated power ; 2. Brake power ; 3. Mechanical efficiency ; 4. Specific steam consumption on I.P. and B.P. basis ; 5. Indicated thermal efficiency ; and 6. Brake thermal efficiency.

[**Ans.** 45.9 kW ; 38.23 kW ; 83.3% ; 19.35 kg/kWh, 23.23 kg/kWh ; 7.2% ; 6%]

9. A steam engine uses 500 kg of dry saturated steam per hour at a pressure of 20 bar and exhaust takes place at a pressure of 0.2 bar with dryness fraction of 0.78. Find Rankine efficiency and relative efficiency of the engine, if it develops 40 kW at full load. [**Ans.** 26.8% ; 50.3%]

10. The volume and pressure at a certain point on the compression curve of an actual indicator diagram is 0.028 m^3 and 4 bar respectively. Estimate the total mass of steam in the cylinder per stroke when the steam supplied is 210 kg/min at 105 r.p.m. [**Ans.** 1.06 kg]

11. In a double acting engine running at 100 r.p.m., the following values were noted :

(*i*) Immediately after cut-off :

Volume = 0.0588 m^3 ; pressure = 5 bar

(*ii*) Immediately after compression has begun :

Volume = 0.0211 m^3 ; pressure = 1.4 bar

Find the missing quantity in kg/min. The steam supplied is 40 kg/min. [**Ans.** 18 kg/min]

12. The following values were recorded during a test on a single cylinder double acting jacketed steam engine :

Pressure of steam supplied = 5.5 bar ; Mass of steam used in engine cylinder = 5.4 kg/min ; Mass of steam used in jacket = 0.5 kg/min ; Dryness fraction of steam = 0.95 ; Mass of circulating water = 126.5 kg/min ; Rise in temperature of circulating water = 20° C ; Condensate temperature = 40° C ; Temperature of jacket drain = 140° C ; Indicated power = 25 kW.

Draw a heat balance sheet for the engine and find its indicated thermal efficiency. [**Ans.** 8.16%]

13. The following observations were recorded during a trial on a single cylinder double acting condensing type steam engine :

Pressure of steam supply = 8 bar ; Mass of condensate available from condenser per hour = 95 kg ; Quality of steam supplied to the engine = dry and saturated ; Condenser vacuum = 300 mm of Hg ; Engine speed = 150 r.p.m. ; Mean effective pressure on cover end side = 1.52 bar ; Mean effective pressure on crank end side = 1.22 bar ; Diameter of cylinder = 200 mm ; Length of stroke = 300 mm ; Piston rod diameter = 38 mm ; Net load on the brake = 950 N ; Effective diameter of brake wheel = 0.75 m ; Mass of cooling water per minute = 45 kg ; Rise in temperature of cooling water = 18° C ; Condensate temperature = 55° C.

Find the mechanical efficiency and brake thermal efficiency. Also draw a heat balance sheet.

[**Ans.** 88.14%; 8.36%]

14. A reciprocating steam engine, governed by throttling, uses 530 kg of steam per hour, when developing 55 kW I.P. It uses 2160 kg per hour when developing 280 kW I.P. Find the approximate power of this engine when the steam consumption is 1580 kg/h, assuming the Willian's relation holds good.

[**Ans.** 200 kW]

15. A throttle governed steam engine requires 500 kg of steam per hour while developing 37.5 kW and 2000 kg per hour when developing 187.5 kW. Find the thermal efficiency of the engine when it develops 115 kW assuming the steam supplied to be dry saturated at a pressure of 15 bar and exhaust pressure to be 0.3 bar.

[**Ans.** 13%]

QUESTIONS

1. What is meant by 'efficiency' of a steam engine ?
2. Differentiate between :

   (*a*) Mechanical efficiency and overall efficiency.

   (*b*) Indicated thermal efficiency and brake thermal efficiency.

   (*c*) Relative efficiency and Rankine efficiency.
3. Define missing quantity. Discuss the method to find it.
4. What are the methods used to reduce cylinder condensation in steam engine ?
5. Explain the term 'heat balance sheet'. What important light does it throw on the working of a steam engine.
6. What do you understand by the term 'governing of steam engines'? Explain its necessity.
7. Discuss the methods of governing a simple steam engine.
8. Explain clearly, the effect on the distribution of work between the cylinders of a two cylinder compound steam engine, when governing is done by (*a*) throttling, and (*b*) the alterations of cut-off in the H.P. cylinder.
9. State and explain Willian's law.

OBJECTIVE TYPE QUESTIONS

1. There is always some steam left in the clearance space from the previous stroke. This steam left in the clearance space is known as

    (*a*) wet steam (*b*) saturated steam (*c*) superheated steam (*d*) cushion steam

2. The missing equality per stroke is equal to

    (*a*) cylinder feed – indicated mass of steam

    (*b*) cylinder feed + indicated mass of steam

    (*c*) mass of cushion steam + indicated mass of steam

    (*d*) mass of cushion steam – cylinder feed

3. The throttle governing of steam engines is a method of controlling the engine output by varying

    (*a*) volume of intake steam (*b*) pressure of intake steam

    (*c*) temperature of intake steam (*d*) all of these

4. Willian's line for a steam engine is a straight line relationship between the steam consumption per hour and

    (*a*) indicated power (*b*) brake power (*c*) efficiency (*d*) pressure of steam

5. Wilian's law holds good for .... governed engine.

    (*a*) cut-off (*b*) throttle

ANSWERS

1. (*d*)	2. (*a*)	3. (*b*)	4. (*a*)	5. (*b*)

20

Steam Condensers

1. Introduction. 2. Advantages of a Condenser in a Steam Power Plant. 3. Requirements of a Steam Condensing Plant. 4. Classification of Condensers. 5. Jet Condensers. 6. Types of Jet Condensers. 7. Parallel Flow Jet Condensers. 8. Counterflow or Low Level Jet Condensers. 9. Barometric or High Level Jet Condensers. 10. Ejector Condensers. 11. Surface Condensers. 12. Types of Surface Condensers. 13. Down Flow Surface Condensers. 14. Central Flow Condensers. 15. Regenerative Surface Condensers. 16. Evaporative Condensers. 17. Comparison of Jet and Surface Condensers. 18. Mixture of Air and Steam (Dalton's Law of Partial Pressures). 19. Measurement of Vacuum in a Condenser. 20. Vacuum Efficiency. 21. Condenser Efficiency. 22. Mass of Cooling Water Required or Condensation of Steam. 23. Sources of Air into the Condenser. 24. Effects of Air Leakage. 25. Air Pump. 26. Edwards Air Pump. 27. Cooling Towers. 28. Types of Cooling Towers.

20.1. Introduction

A steam condenser is a closed vessel into which the steam is exhausted, and condensed after doing work in an engine cylinder or turbine. A steam condenser has the following two objects :

1. The primary object is to maintain a low pressure (below atmospheric pressure) so as to obtain the maximum possible energy from steam and thus to secure a high efficiency.
2. The secondary object is to supply pure feed water to the hot well, from where it is pumped back to the boiler.

Note : The low pressure is accompanied by low temperature and thus all condensers maintain a vacuum under normal conditions. The condensed steam is called condensate. The temperature of condensate is higher on leaving the condenser than that of circulating water at inlet. It is thus obvious, that the condensate will have a considerable liquid heat.

20.2. Advantages of a Condenser in a Steam Power Plant

Following are the main advantages of incorporating a condenser in a steam power plant :

1. It increases expansion ratio of steam, and thus increases efficiency of the plant.
2. It reduces back pressure of the steam, and thus more work can be obtained.
3. It reduces temperature of the exhaust steam, and thus more work can be obtained.
4. The reuse of condensate (*i.e.* condensed steam) as feed water for boilers reduces the cost of power generation.
5. The temperature of condensate is higher than that of fresh water. Therefore the amount of heat supplied per kg of steam is reduced.

20.3. Requirements of a Steam Condensing Plant

The principle requirements of a condensing plant, as shown in Fig. 20.1, are as follows :

1. *Condenser.* It is a closed vessel is which steam is condensed. The steam gives up heat energy to coolant (which is water) during the process of condensation.

2. *Condensate pump.* It is a pump, which removes condensate (*i.e.* condensed steam) from the condenser to the hot well.
3. *Hot well.* It is a sump between the condenser and boiler, which receives condensate pumped by the condensate pump.

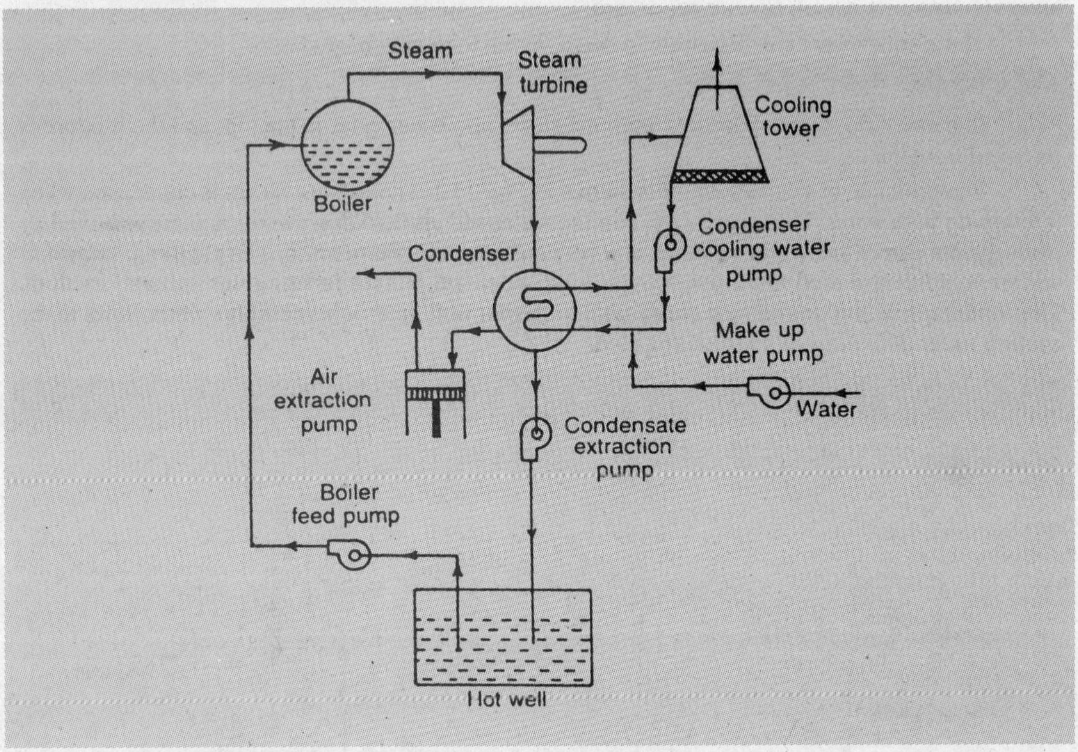

Fig. 20.1. Steam condensing plant.

4. *Boiler feed pump.* It is a pump, which pumps the condensate from the hot well to the boiler. This is done by increasing the pressure of condensate above the boiler pressure.
5. *Air extraction pump.* It is a pump which extracts (*i.e.* removes) air from the condenser.
6. *Cooling tower.* It is a tower used for cooling the water which is discharged from the condenser.
7. *Cooling water pump.* It is a pump, which circulates the cooling water through the condenser.

20.4. Classification of Condensers

The steam condensers may be broadly classified into the following two types, depending upon the way in which the steam is condensed :

1. Jet condensers or mixing type condensers, and
2. Surface condensers or non-mixing type condensers.

20.5. Jet Condensers

These days, the jet condensers are seldom used because there is some loss of condensate during the process of condensation and high power requirements for the pumps used. Moreover, the condensate can not be used as feed water to the boiler as it is not free from salt. However, jet condensers may be used at places where water of good quality is easily available in sufficient quantity

20.6. Types of Jet Condensers

The jet condensers may be further classified, on the basis of the direction of flow of the condensate and the arrangement of the tubing system, into the following four types :

1. Parallel flow jet condenser, 2. Counterflow or low level jet condenser, 3. Barometric or high level jet condenser, and 4. Ejector condenser.

These condensers are discussed, in detail, in the following pages.

20.7. Parallel Flow Jet Condensers

In parallel flow jet condensers, both the steam and water enter at the top, and the mixture is removed from the bottom.

The principle of this condenser is shown in Fig. 20.2. The exhaust steam is condensed when it mixes up with water. The condensate, cooling water and air flow downwards and are removed by two separate pumps known as air pump and condensate pump. Sometimes, a single pump known as wet air pump, is also used to remove both air and condensate. But the former gives a greater vacuum. The condensate pump delivers the condensate to the hot well, from where surplus water flows to the cooling water tank through an overflow pipe.

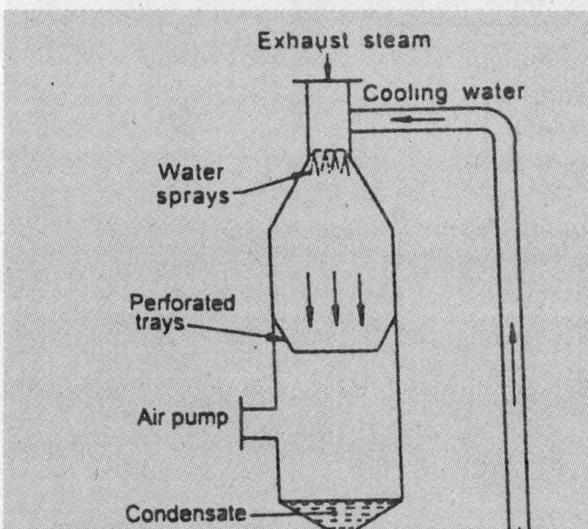

Fig. 20.2. Parallel flow jet condenser.

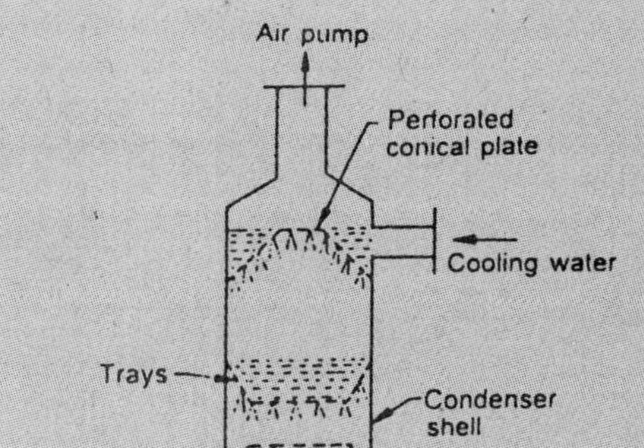

Fig. 20.3. Counterflow jet condenser.

20.8. Counterflow or Low Level Jet Condensers

In counterflow or low level jet condensers, the exhaust steam enters at the bottom, flows upwards and meets the downcoming cooling water.

The vacuum is created by the air pump, placed at the top of the condenser shell. This draws the supply of cooling water, which falls in a large number of jets, through perforated conical plate as shown in Fig. 20.3. The falling water is caught in the trays, from which it escapes in a second series of jets and meets the exhaust steam entering at the bottom. The rapid condensation occurs, and the condensate and cooling water descends through a vertical pipe to the condensate pump, which delivers it to the hot well.

20.9. Barometric or High Level Jet Condensers

These condensers are provided at a high level with a long vertical discharge pipe as shown in Fig. 20.4. In high level jet condensers, exhaust steam enters at the bottom, flows upwards and meets the downcoming cooling water in the same way as that of low level jet condenser. The vacuum is created by the air pump, placed at the top of the condenser shell. The condensate and cooling water descends through a vertical pipe to the hot well without the aid of any pump. The surplus water from the hot well flows to the cooling water tank through an overflow pipe.

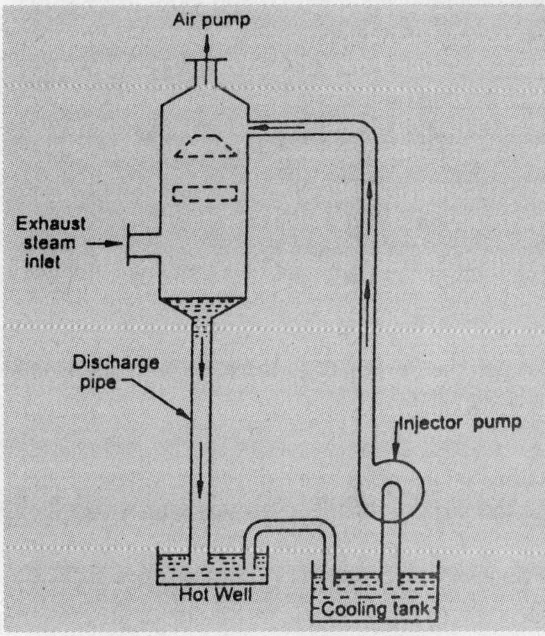

Fig. 20.4. High level jet condenser.

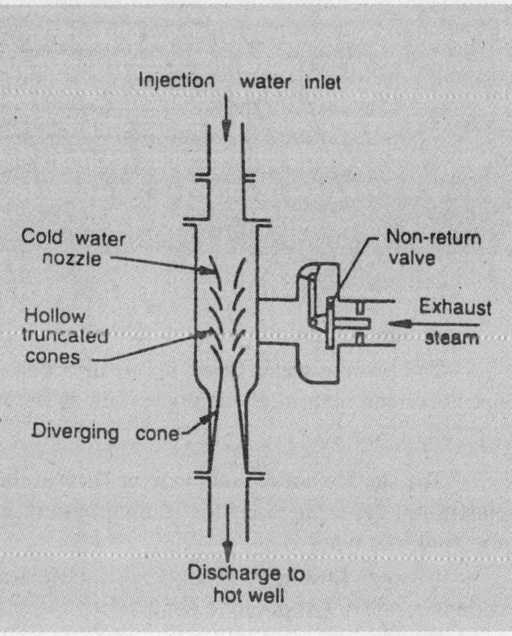

Fig. 20.5. Ejector condenser.

20.10. Ejector Condensers

In ejector condensers, the steam and water mix up while passing through a series of metal cones. Water enters at the top through a number of guide cones. The exhaust steam enters the condenser through non-return valve arrangement. The steam and air then passes through the hollow truncated cones. After that it is dragged into the diverging cones where its kinetic energy is partly transformed to pressure energy. The condensate and cooling water is then discharged to the hot well as shown in Fig. 20.5.

20.11. Surface Condensers

A surface condenser has a great advantage over the jet condensers, as the condensate does not mix up with the cooling water. As a result of this, whole condensate can be reused in the boiler. This type of condenser is essential in ships which can carry only a limited quantity of fresh water for the boilers. It is also widely used in land installations, where inferior water is available or the better quality of water for feed is to be used economically.

Fig. 20.6 shows a longitudinal section of a two pass surface condenser. It consists of a horizontal cast iron cylindrical vessel packed with tubes, through which the cooling water flows. The ends of the condenser are cut off by vertical perforated type plates into which water tubes are fixed. This is done in such a manner that the leakage of water into the centre condensing space is prevented.

The water tubes pass horizontally through the main condensing space for the steam. The steam enters at the top and is forced to flow downwards over the tubes due to the suction of the extraction pump at the bottom.

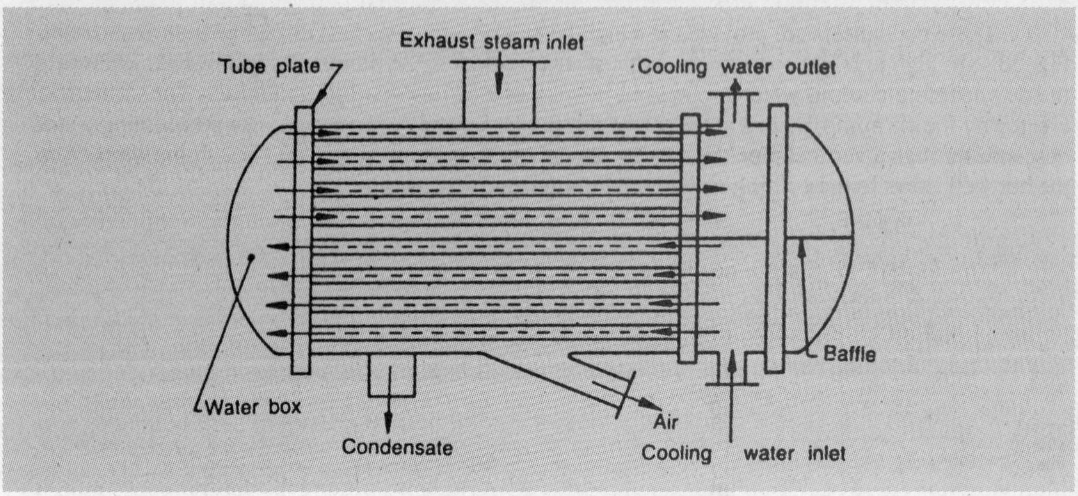

Fig. 20.6. Surface condenser.

The cooling water flows in one direction through the lower half of the tubes and returns in opposite direction through the upper half, as shown in Fig. 20.6.

20.12. Types of Surface Condensers

The surface condensers may be further classified on the basis of the direction of flow of the condensate, the arrangement of tubing system and the position of the extraction pump, into the following four types :

1. Down flow surface condenser, 2. Central flow surface condenser, 3. Regenerative surface condenser, and 4. Evaporative condenser.

These condensers are discussed, in detail, in the following pages.

20.13. Down Flow Surface Condensers

In down flow surface condensers, the exhaust steam enters at the top and flow downwards over the tubes due to force of gravity as well as suction of the extraction pump fitted at the bottom. The condensate is collected at the bottom and then pumped by the extraction pump. The dry air pump suction pipe, which is provided near the bottom, is covered by a baffle so as to prevent the entry of condensed steam into it, as shown in Fig. 20.7.

As the steam flows perpendicular to the direction of flow of cooling water (inside the tubes), this is also called a cross-surface condenser.

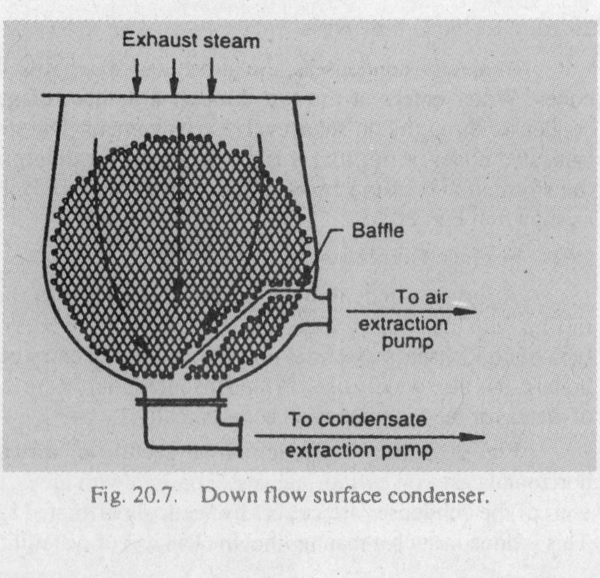

Fig. 20.7. Down flow surface condenser.

20.14. Central Flow Surface Condensers

In central flow surface condensers, the exhaust steam enters at the top and flow downwards. The suction pipe of the air extraction pump is placed in the centre of the tube nest as shown in Fig. 20.8. This causes the steam to flow radially inwards over the tubes towards the suction pipe. The condensate is collected at the bottom and then pumped by the extraction pump as shown in Fig. 20.8.

The central flow surface condenser is an improvement over the down flow type as the steam is directed radially inwards by a volute casing around the tube nest. It, thus, gives an access to the whole periphery of the tubes.

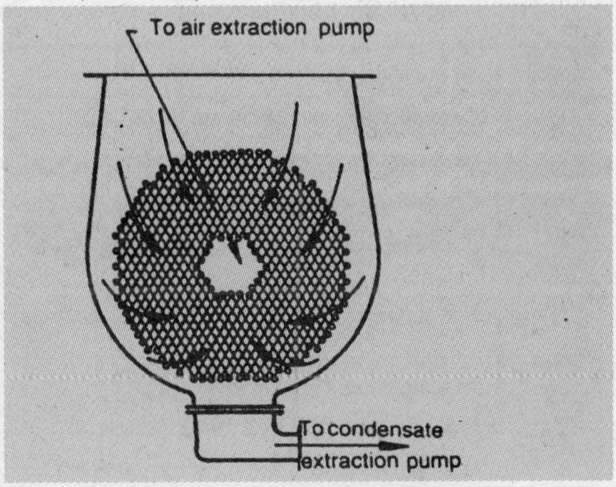

Fig. 20.8. Central flow surface condenser.

20.15. Regenerative Surface Condensers

In regenerative surface condensers, the condensate is heated by a regenerative method. The condensate after leaving the tubes is passed through the exhaust steam from the engine or turbine. It thus, raises its temperature for use as feed water for the boiler.

20.16. Evaporative Condenser

The steam to be condensed enters at the top of a series of pipes outside of which a film of cold water is falling. At the same time, a current of air circulates over the water film, causing rapid evaporation of some of the cooling water. As a result of this, the steam circulating inside the pipe is condensed. The remaining cooling water is collected at an increased temperature and is reused. Its original temperature is restored by the addition of the requisite quantity of cold water.

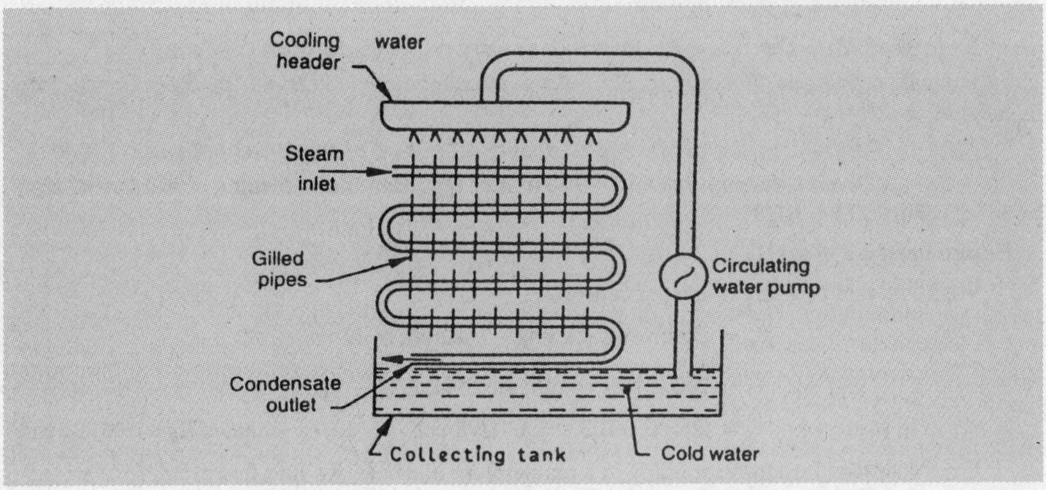

Fig. 20.9. Evaporative condenser.

The evaporative condensers are provided when the circulating water is to be used again and again. These condensers consist of sheets of gilled piping, which is bent backwards and forwards and placed in a vertical plane, as shown in Fig. 20.9.

20.17. Comparison of Jet and Surface Condensers

Following are the important points of comparison between jet and surface condensers :

S. No.	*Jet condensers*	*Surface condensers*
1.	Cooling water and steam are mixed up.	Cooling water and steam are not mixed up.
2.	Less suitable for high capacity plants.	More suitable for high capacity plants.
3.	Condensate is wasted.	Condensate is reused.
4.	It requires less quantity of circulating water.	It requires a large quantity of circulating water.
5.	The condensing plant is economical and simple.	The condensing plant is costly and complicated.
6.	Its maintenance cost is low.	Its maintenance cost is high.
7.	More power is required for air pump.	Less power is required for air pump.
8.	High power is required for water pumping.	Less power is required for water pumping.

20.18. Mixture of Air and Steam (Dalton's Law of Partial Pressures)

It states "*The pressure of the mixture of air and steam is equal to the sum of the pressures, which each constitutent would exert, if it occupied the same space by itself.*" Mathematically, pressure in the condenser containing mixture of air and steam,

$$p_c = p_a + p_s$$

where p_a = Partial pressure of air, and

p_s = Partial pressure of steam.

Note : In most of the cases, we are required to find partial pressure of air, therefore Dalton's law may also be used as :

$$p_a = p_c - p_s$$

Example 20.1. *The following observations were recorded during a condenser test ;*

Vacuum reading = 700 mm of Hg ; Barometer reading = 760 mm of Hg ; Condensate temperature = 34° C

Find : 1. Partial pressure of air, and 2. Mass of air per m³ of condenser volume.

Solution. Given : Vacuum reading = 700 mm of Hg ; Barometer reading = 760 mm of Hg ; $T = 34° \text{ C} = 34 + 273 = 307 \text{ K}$

1. *Partial pressure of air*

We know that pressure in the condenser

$$p_c = \text{Barometer reading} - \text{Vacuum reading}$$

$$= 760 - 700 = 60 \text{ mm of Hg}$$

$$= 60 \times 0.001\,33 = 0.0798 \text{ bar} \quad \ldots (\because 1 \text{ mm of Hg} = 0.001\,33 \text{ bar})$$

From steam tables, corresponding to a temperature of 34° C, we find that pressure of steam,

$$p_s = 0.0532 \text{ bar}$$

∴ Partial pressure of air,

$$p_a = p_c - p_s = 0.0798 - 0.0532 = 0.0266 \text{ bar} \quad \textbf{Ans.}$$

2. *Mass of air per m³ of condenser volume*

We know that mass of air per m³ of condenser volume,

$$m_a = \frac{p_a v}{R\,T} = \frac{0.0266 \times 10^5 \times 1}{287 \times 307} = 0.03 \text{ kg} \quad \textbf{Ans.}$$

. . (p_a is taken in N/m² and R for air = 287 J/kg K)

20.19. Measurement of Vacuum in a Condenser

The vacuum, in general may be defined as the difference between the atmospheric pressure and the absolute pressure. In the study of condensers, the vacuum is generally converted to correspond with a standard atmospheric pressure, which is taken as the barometric pressure of 760 mm of mercury. Mathematically, vacuum gauge reading corrected to standard barometer or in other words

Corrected vacuum = 760 – (Barometer reading – Vacuum gauge reading)

Note : We know that

Atmospheric pressure = 760 mm of Hg = 1.013 bar

$$\therefore \text{ 1 mm of Hg} = \frac{1.013}{760} = 0.00133 \text{ bar} = 133 \text{ N/m}^2 \qquad \ldots (\because 1 \text{ bar} = 10^5 \text{ N/m}^2)$$

Example 20.2. *A vacuum gauge fitted to a condenser reads 680 mm of Hg, when the barometer reads 750 mm of Hg. Determine the corrected vacuum in terms of mm of Hg and bar.*

Solution. Given : Vacuum gauge reading = 680 mm of Hg ; Barometer reading = 750 mm of Hg

We know that pressure in the condenser

= 750 – 680 = 70 mm of Hg

and corrected vacuum = 760 – 70 = 690 mm of Hg **Ans.**

= 690 × 0.001 33 = 0.918 bar **Ans.**

20.20. Vacuum Efficiency

The minimum absolute pressure (also called ideal pressure) at the steam inlet of a condenser is the pressure corresponding to the temperature of the condensed steam. The corresponding vacuum (called ideal vacuum) is the maximum vacuum that can be obtained in a condensing plant, with no air present at that temperature. The pressure in the actual condenser is greater than the ideal pressure by an amount equal to the pressure of air present in the condenser. The ratio of the actual vacuum to the ideal vacuum is known as *vacuum efficiency.* Mathematically, vacuum efficiency,

$$\eta_v = \frac{\text{Actual vacuum}}{\text{Ideal vacuum}}$$

where Actual vacuum = Barometric pressure – Actual pressure

and Ideal vacuum = Barometric pressure – Ideal pressure

Example 20.3. *Calculate the vacuum efficiency from the following data :*

Vacuum at steam inlet to condenser = 700 mm of Hg ; Barometer reading = 760 mm of Hg ; Hot well temperature = 30° C.

Solution. Given : Vacuum reading or actual vacuum = 700 mm of Hg ; Barometer reading = 760 mm of Hg ; t = 30° C

We know that pressure in the condenser

= 760 – 700 = 60 mm of Hg

From steam tables, corresponding to a temperature of 30° C, we find that ideal pressure of steam,

$$= 0.0424 \text{ bar} = \frac{0.0424}{0.001\,33} = 31.88 \text{ mm of Hg}$$

We know that ideal vacuum = Barometer reading – Ideal pressure

$= 760 - 31.88 = 728.12$ mm of Hg

$\therefore$ Vacuum efficiency, $\eta_v = \dfrac{\text{Actual vacuum}}{\text{Ideal vacuum}} = \dfrac{700}{728.12} = 0.9614$ or 96.14 % **Ans.**

Example 20.4. *The vacuum efficiency of a condenser is 96%. The temperature of condensate is 40° C. If the barometer reads 752 mm of Hg, find the vacuum gauge reading of the condenser.*

Solution. Given : $\eta_v = 96\% = 0.96$; $t = 40°$ C ; Barometer reading = 752 mm of Hg

From steam tables, corresponding to a temperature of 40° C, we find that ideal pressure of steam,

$$= 0.0737 \text{ bar} = \frac{0.0737}{0.001\,33} = 55.4 \text{ mm of Hg}$$

$\therefore$ Ideal vacuum = Barometer reading – Ideal pressure

$= 752 - 55.4 = 696.6$ mm of Hg

We know that vacuum efficiency (η_v),

$$0.96 = \frac{\text{Actual vacuum}}{\text{Ideal vacuum}} = \frac{\text{Actual vacuum}}{696.6}$$

$\therefore$ Actual vacuum or vacuum gauge reading of the condenser

$= 0.96 \times 696.6 =$ 668.74 mm of Hg **Ans.**

Example 20.5. *In a surface condenser, the vacuum maintained is 700 mm of Hg. The barometer reads 754 mm. If the temperature of condensate is 18° C, determine : 1. mass of air per kg of steam ; and 2. vacuum efficiency.*

Solution. Given : Actual vacuum = 700 mm of Hg ; Barometer reading = 754 mm of Hg ; $T = 18°\text{ C} = 18 + 273 = 291$ K

We know that pressure in the condenser,

$$p_c = 754 - 700 = 54 \text{ mm of Hg}$$

From steam tables, corresponding to 18° C, we find that absolute or ideal pressure of steam,

$$p_s = 0.0206 \text{ bar} = \frac{0.0206}{0.001\,33} = 15.5 \text{ mm of Hg}$$

and specific volume of steam, $v_s = 65.09$ m³/kg

Mass of air per kg of steam

We know that pressure of air (as per Dalton's law),

$$p_a = p_c - p_s = 54 - 15.5 = 38.5 \text{ mm of Hg}$$

$$= 38.5 \times 0.001\,33 = 0.0512 \text{ bar} = 0.0512 \times 10^5 \text{ N/m}^2$$

and mass of air per kg of steam,

$$m_a = \frac{p_a\, v}{R\, T} = \frac{0.0512 \times 10^5 \times 65.09}{287 \times 291} = 4 \text{ kg } \textbf{Ans.} \quad \ldots (\because pv = mRT)$$

Vacuum efficiency

We know that ideal vacuum = Barometer reading – Ideal pressure

$= 754 - 15.5 = 738.5$ mm of Hg

and vacuum efficiency, $\eta_v = \dfrac{\text{Actual vacuum}}{\text{Ideal vacuum}} = \dfrac{700}{738.5} = 0.948$ or 94.8% **Ans.**

Example 20.6. *A surface condenser fitted with separate air and water extraction pumps, has a portion of the tubes near the air pump suction screened off from the steam so that the air is cooled below the condensate temperature. The steam enters the condenser at 38° C and the condensate is removed at 37° C. The air removed has a temperature of 36° C. If the total air infiltration from all sources together is 5 kg/h, determine the volume of air handled by the air pump per hour. What would be the corresponding value of the air handled if a combined air and condensate pump was employed ? Assume uniform pressure in the condenser.*

Solution. Given : $T_s = 38°\text{ C} = 38 + 273 = 311\text{ K}$; $T_c = 37°\text{ C} = 37 + 273 = 310\text{ K}$; $T_a = 36°\text{ C} = 36 + 273 = 309\text{ K}$; $m_a = 5\text{ kg/h}$

1. *Volume of air handled by the air pump per hour*

Since the pressure at entry to the condenser (p_c) is equal to the pressure of steam corresponding to 38° C, therefore from steam tables,

$$p_c = 0.0662\text{ bar}$$

and pressure of steam at the air pump suction, corresponding to 36° C (from steam tables),

$$p_s = 0.0594\text{ bar}$$

∴ Pressure of air at the air pump suction (as per Dalton's law),

$$p_a = p_c - p_s = 0.0662 - 0.0594 \doteq 0.0068\text{ bar}$$

$$= 0.0068 \times 10^5 = 680\text{ N/m}^2$$

We know that volume of air handled by the air pump,

$$v_a = \frac{m_a R T_a}{p_a} = \frac{5 \times 287 \times 309}{680} = 652\text{ m}^3\text{/h} \quad \textbf{Ans.}$$

2. *Volume of air handled when a combined air and condensate pump is employed*

From steam tables, corresponding to a condensate temperature of 37° C, we find that pressure of steam,

$$p_s = 0.0627\text{ bar}$$

∴ Pressure of air (as per Dalton's law),

$$p_a = p_c - p_s = 0.0662 - 0.0627 = 0.0035\text{ bar}$$

$$= 0.0035 \times 10^5 = 350\text{ N/m}^2$$

We know that volume of air handled,

$$v_a = \frac{m_a R T_c}{p_a} = \frac{5 \times 287 \times 310}{350} = 1271\text{ m}^3\text{/h} \quad \textbf{Ans.}$$

Example 20.7. *The air leakage into a surface condenser operating with a steam turbine is estimated as 84 kg/h. The vacuum near the inlet of air pump is 700 mm of Hg when barometer reads 760 mm of Hg. The temperature at inlet of vacuum pump is 20° C. Calculate :*

1. The minimum capacity of the air pump is m³/h ; 2. The dimensions of the reciprocating air pump to remove the air if it runs at 200 r.p.m. Take L / D ratio = 1.5 and volumetric efficiency = 100 percent ; and 3. The mass of vapour extracted per minute.

Solution. Given : $m_a = 84\text{ kg/h}$; Vacuum = 700 mm of Hg ; Barometer reading = 760 mm of Hg ; $T = 20°\text{ C} = 20 + 273 = 293\text{ K}$

1. *Minimum capacity of the air pump*

We know that pressure in the condenser,

$$p_c = \text{Barometer reading} - \text{Condenser Vacuum}$$

$$= 760 - 700 = 60 \text{ mm of Hg} = 60 \times 0.001\,33 = 0.0798 \text{ bar}$$

From steam tables, corresponding to a temperature of 20° C, we find that pressure of steam,

$$p_s = 0.0234 \text{ bar}$$

∴ Pressure of air (as per Dalton's law),

$$p_a = p_c - p_s = 0.0798 - 0.0234 = 0.0564 \text{ bar}$$

$$= 0.0564 \times 10^5 = 5640 \text{ N/m}^2$$

We know that minimum capacity of the air pump,

$$v_a = \frac{m_a R T}{p_a} = \frac{84 \times 287 \times 293}{5640} = 1252.4 \text{ m}^3\text{/h} \quad \textbf{Ans.}$$

2 *Dimensions of the reciprocating pump*

Let D = Diameter of the cylinder is metres,

L = Length of the stroke in metres = $1.5\,D$ ... (Given)

η_v = Volumetric efficiency = 100% = 1

N = Speed of the pump = 200 r.p.m. ... (Given)

We know that minimum capacity of the air pump (v_a),

$$\frac{1252.4}{60} = \frac{\pi}{4} \times D^2 \times L \times N = \frac{\pi}{4} \times D^2 \times 1.5\,D \times 200 = 235.6\,D^3$$

... (v_a is taken in m³/min)

∴ $D^3 = 0.0886$ or $D = 0.446$ m **Ans.**

and $L = 1.5\,D = 1.5 \times 0.446 = 0.669$ m **Ans.**

3. *Mass of vapour extracted per minute*

From steam tables, corresponding to a temperature of 20° C, we find that specific volume of steam,

$$v_g = 57.84 \text{ m}^3\text{/kg}$$

∴ Mass of vapour extracted per minute

$$= \frac{v_a}{v_g} = \frac{1252.4}{60 \times 57.84} = 0.361 \text{ kg/min} \quad \textbf{Ans.}$$

Example 20.8. *The vacuum at the extraction pipe in a condenser is 710 mm of mercury and the temperature is 35.82° C. The barometer reads 760 mm of mercury. The air leakage into the condenser is 4 kg per 10 000 kg of steam. Determine : 1. the volume of air to be dealt with by the dry air pump per kg of steam entering the condenser, and 2. the mass of water vapour associated with this air.*

Take R = 287 J/kg K for air.

Solution. Given : Vacuum = 710 mm of Hg ; T = 35.82° C = 35.82 + 273 = 308.82 K ; Barometer reading = 760 mm of Hg ; m_a = 4 kg per 10 000 kg of steam = 0.0004 kg / kg of steam

1. *Volume of air per kg of steam entering the condenser*

Let v_a = Volume of air per kg of steam entering the condenser.

We know that pressure in the condenser

$$p_c = \text{Barometer reading} - \text{Condenser vacuum}$$

$$= 760 - 710 = 50 \text{ mm of Hg} = 50 \times 0.001\,33 = 0.0665 \text{ bar}$$

From steam tables, corresponding to the temperature of 35.82° C, we find that the pressure of steam,

$$p_s = 0.0588 \text{ bar}$$

∴ Pressure of air (as per Dalton's law),

$$p_a = p_c - p_s = 0.0665 - 0.0588 = 0.0077 \text{ bar}$$

$$= 0.0077 \times 10^5 = 770 \text{ N/m}^2$$

We know that $p_a v_a = m_a R T$

$$\therefore \quad v_a = \frac{m_a R T}{p_a} = \frac{0.0004 \times 287 \times 308.82}{770} = 0.046 \text{ m}^3\text{/kg of steam} \quad \textbf{Ans.}$$

2. *Mass of water vapour associated with this air*

From steam tables, corresponding to a temperature of 35.82° C, we find that specific volume of steam,

$$v_g = 24.2 \text{ m}^3\text{/kg}$$

∴ Mass of water vapour associated with the air

$$= \frac{v_a}{v_g} = \frac{0.046}{24.2} = 0.0019 \text{ kg} \quad \textbf{Ans.}$$

Example 20.9. *The air entering a steam condenser with steam is estimated at 6 kg per hour. The temperature at inlet to air cooler section is 30° C and at the outlet 26° C. The vacuum in the shell is essentially constant throughout and is 721 mm of Hg, while the barometer reads 758 mm of Hg. Calculate :*

1. *The volume of air entering the cooling section per hour ;*
2. *The mass of moisture contained in the air ; and*
3. *The mass of steam condensed per hour in the cooling section.*

Solution. Given : $m_a = 6$ kg/h ; $T_1 = 30°\text{ C} = 30 + 273 = 303$ K ; $T_2 = 26°\text{ C} = 26 + 273 = 299$ K ; Condenser vacuum = 721 mm of Hg ; Barometer reading = 758 mm of Hg

We know that pressure in the condenser,

$$p_c = \text{Barometer reading} - \text{Condenser vacuum}$$

$$= 758 - 721 = 37 \text{ mm of Hg} = 37 \times 0.001\,33 = 0.0492 \text{ bar}$$

Considering the inlet to the cooling section. From steam tables, corresponding to a temperature of 30° C, we find that pressure of steam,

$$p_s = 0.0424 \text{ bar}$$

∴ Pressure of air (as per Dalton's law),

$$p_a = p_c - p_s = 0.0492 - 0.0424 = 0.0068 \text{ bar}$$

$$= 0.0068 \times 10^5 = 680 \text{ N/m}^2$$

1. *Volume of air entering the cooling section per hour*

We know that volume of air entering the cooling section,

$$v_a = \frac{m_a R T_1}{p_a} = \frac{6 \times 287 \times 303}{680} = 767 \text{ m}^3\text{/h} \textbf{ Ans.}$$

2. *Mass of moisture in the air*

From steam tables, corresponding to a temperature of 30° C (or pressure 0.0424 bar), we find that specific volume of steam

$$v_g = 32.93 \text{ m}^3\text{/kg}$$

∴ Mass of moisture or steam associated with the air,

$$m_1 = \frac{v_a}{v_g} = \frac{767}{32.93} = 23.3 \text{ kg/h} \textbf{ Ans.}$$

3. *Mass of steam condensed per hour in the cooling section*

Now considering the outlet of the cooling section. From steam tables, corresponding to a temperature of 26° C, we find that pressure of steam,

$$p_s = 0.0336 \text{ bar}$$

∴ Pressure of air (as per Dalton's law),

$$p_a = p_c - p_s = 0.0492 - 0.0336 = 0.0156 \text{ bar}$$

$$= 0.0156 \times 10^5 = 1560 \text{ N/m}^2$$

We know that volume of air at outlet to the cooling section,

$$v_a = \frac{m_a R T_2}{p_a} = \frac{6 \times 287 \times 299}{1560} = 330 \text{ m}^3\text{/h}$$

From steam tables, corresponding to a temperature of 26° C (or pressure 0.0336 bar), we find that specific volume of steam,

$$v_g = 41.034 \text{ m}^3\text{/kg}$$

and mass of steam associated with the air,

$$m_2 = \frac{v_a}{v_g} = \frac{330}{41.034} = 8.04 \text{ kg/h}$$

∴ Mass of steam condensed

$$= m_1 - m_2 = 23.3 - 8.04 = 15.26 \text{ kg/h} \textbf{ Ans.}$$

20.21. Condenser Efficiency

The condenser efficiency may be *defined as the ratio of temperature rise of cooling water to the vacuum temperature *minus* inlet cooling water temperature. Mathematically, condenser efficiency,

$$\eta_c = \frac{\text{Temperature rise of cooling water}}{\text{Vacuum temperature} - \text{Inlet cooling water temperature}}$$

$$= \frac{t_o - t_i}{t_v - t_i}$$

* This definition was proposed by M/s C.A. Parsons & Co., well known manufacturers of steam turbines. Today this definition is widely used.

where t_o = Outlet temperature of cooling water,

t_i = Inlet temperature of cooling water,

t_v = Vacuum temperature. It is the saturation temperature corresponding to the condenser pressure.

Example 20.10. ***The inlet and outlet temperatures of cooling water in a condenser are 27° C and 35° C respectively. If the vacuum in the condenser is 700 mm of Hg against barometric pressure of 760 mm of Hg, calculate the efficiency of the condenser.***

Solution. Given : $t_i = 27°$ C ; $t_0 = 35°$ C ; Condenser vacuum = 700 mm of Hg ; Barometric pressure = 760 mm of Hg

We know that pressure in the condenser,

$$= 760 - 700 = 60 \text{ mm of Hg}$$

$$= 60 \times 0.001\,33 = 0.0798 \text{ bar}$$

From steam tables, corresponding to a pressure of 0.0798 bar, we find that vacuum temperature,

$$t_v = 41.5° \text{C}$$

∴ Condenser efficiency,

$$\eta_c = \frac{t_o - t_i}{t_v - t_i} = \frac{35 - 27}{41.5 - 27} = 0.552 \text{ or } 55.2\% \text{ Ans.}$$

Example 20.11. ***The following data were obtained from the test of a surface condenser :***

Condenser vacuum = 711 mm of Hg ; Hot well temperature = 32° C ; Inlet temperature of circulated water = 12° C ; Outlet temperature of circulated water = 28° C ; Barometer reading = 760 mm of Hg.

Compute the vacuum efficiency and the efficiency of the condenser.

Solution. Given : Condenser vacuum or actual vacuum = 711 mm of Hg ; $t_v = 32°$ C ; $t_i = 12°$ C ; $t_o = 28°$ C ; Barometer reading = 760 mm of Hg

Vacuum efficiency

We know that pressure in the condenser,

$$p_c = 760 - 711 = 49 \text{ mm of Hg}$$

From steam tables, corresponding to a temperature of 32° C, we find that ideal pressure of steam

$$= 0.0475 \text{ bar} = \frac{0.0475}{0.001\,33} = 35.7 \text{ mm of Hg}$$

We know that ideal vacuum

$$= \text{Barometer reading} - \text{Ideal pressure}$$

$$= 760 - 35.7 = 724.3 \text{ mm of Hg}$$

∴ Vacuum efficiency,

$$\eta_v = \frac{\text{Actual vacuum}}{\text{Ideal vacuum}} = \frac{711}{724.3} = 0.9816 \text{ or } 98.16\% \text{ Ans.}$$

Condenser efficiency

We know that condenser efficiency,

$$\eta_c = \frac{t_o - t_i}{t_v - t_i} = \frac{28 - 12}{32 - 12} = 0.8 \text{ or } 80\% \text{ Ans.}$$

20.22. Mass of Cooling Water Required for Condensation of Steam

In the previous articles, we have discussed various types of condensers and their working. Now we shall discuss the amount of cooling water required for the condensation of steam.

Let m_w = Mass of cooling water,

m_s = Mass of steam condensed (*i.e.* condensate),

h = Total heat of steam entering the condenser,

h_{f1} = Total heat in condensate.

t_i = Inlet temperature of circulating water, and

t_o = Outlet temperature of circulating water.

We know that heat lost by steam

$$= m_s (h - h_{f1})$$

Heat gained by cooling water

$$= m_w c_w (t_o - t_i)$$

We also know that heat gained by cooling water

= Heat lost by steam

$$m_w c_w (t_o - t_i) = m_s (h - h_{f1})$$

$$\therefore \quad m_w = \frac{m_s (h - h_{f1})}{c_w (t_o - t_i)}$$

Note : The above equation is applicable to both jet and surface condensers.

Example 20.12. *A surface condenser is designed to handle 10 000 kg of steam per hour. The steam enters at 0.08 bar and 0.9 dryness and the condensate leaves at the corresponding saturation temperature. The pressure is constant throughout the condenser. Estimate the cooling water flow rate per hour, if the cooling water temperature rise is limited to 10° C.*

Solution. Given : m_s = 10 000 kg/h ; p = 0.08 bar ; x = 0.9 ; $t_0 - t_i$ = 10° C

From steam tables, corresponding to a pressure of 0.08 bar, we find that

$$h_f = 173.9 \text{ kJ/kg} ; h_{fg} = 2403.2 \text{ kJ/kg} ; \text{and } t = 41.5° \text{ C}$$

∴ Total heat of the entering steam,

$$h = h_f + x\, h_{fg} = 173.9 + 0.9 \times 2403.2 = 2336.8 \text{ kJ/kg}$$

Since the condensate temperature is equal to the saturation temperature of 41.5° C, therefore heat in condensate corresponding to 41.5° C,

$$h_{f1} = 173.9 \text{ kJ/kg}$$

We know that the cooling water flow rate per hour,

$$m_w = \frac{m_s (h - h_{f1})}{c_w (t_0 - t_i)} = \frac{10\,000\,(2336.8 - 173.9)}{4.2 \times 10} = 514\,980 \text{ kg/h} \text{ Ans.}$$

Example 20.13. *In a condenser test, the following observations were made :*

Vacuum = 690 mm of Hg ; Barometer reading = 750 mm of Hg ; Mean temperature of condensation = 35° C ; Hot well temperature = 28° C ; Mass of cooling water = 50 000 kg/h ; Inlet temperature = 17° C ; Outlet temperature = 30° C ; Mass of condensate per hour = 1250 kg.

Find : 1. The mass of air present per m³ of condenser volume ; 2. The state of steam entering the condenser ; and 3. The vacuum efficiency.

Take R for air = 287 J/kg K.

Solution. Given : Vacuum = 690 mm of Hg ; Barometer reading = 750 mm of Hg ; $t_c = 35°$ C ; $t_h = 28°$ C ; $m_w = 50\,000$ kg/h ; $t_i = 17°$ C ; $t_o = 30°$ C ; $m_s = 1250$ kg/h ; $R = 287$ J/kg K

1. *Mass of air present per m³ of condenser volume*

We know that pressure in the condenser,

$$p_c = 750 - 690 = 60 \text{ mm of Hg} = 60 \times 0.001\,33 = 0.08 \text{ bar}$$

From steam tables, corresponding to a condensation temperature of 35° C, we find that the pressure of steam,

$$p_s = 0.0562 \text{ bar}$$

∴ Pressure of air (as per Dalton's law),

$$p_a = p_c - p_s = 0.08 - 0.0562 = 0.0238 \text{ bar}$$

$$= 0.0238 \times 10^5 = 2380 \text{ N/m}^2$$

We know that mass of air per m³ of condenser volume,

$$m_a = \frac{p_a v}{R\,T} = \frac{2380 \times 1}{287\,(35 + 273)} = 0.027 \text{ kg} \quad \textbf{Ans.}$$

2. *State of steam entering the condenser*

Let x = Dryness fraction (*i.e.* state)of steam entering the condenser.

From steam tables, corresponding to a pressure of 0.0562 bar (or 35° C), we find that

$$h_f = 146.6 \text{ kJ/kg}, \text{ and } h_{fg} = 2418.8 \text{ kJ/kg}$$

and corresponding to a hot well temperature of 28° C,

$$h_{f1} = 117.3 \text{ kJ/kg}$$

We know that total heat of entering steam,

$$h = h_f + x\,h_{fg} = 146.6 + x \times 2418.8 \text{ kJ/kg}$$

We also know that mass of cooling water (m_w),

$$50\,000 = \frac{m_s\,(h - h_{f1})}{c_w\,(t_o - t_i)} = \frac{1250\,[146.6 + x \times 2418.8 - 117.3]}{4.2\,(30 - 17)}$$

$$= 22.894\,(29.3 + x \times 2418.8)$$

or $$29.3 + x \times 2418.8 = 50\,000 / 22.894 = 2184$$

∴ $$x = \frac{2184 - 29.3}{2418.8} = 0.89 \quad \textbf{Ans.}$$

3. *Vacuum efficiency*

We know that corresponding to a condensation temperature of 35° C, ideal pressure of steam

$$= 0.0562 \text{ bar} = \frac{0.0562}{0.001\ 33} = 42.25 \text{ mm of Hg}$$

$\therefore$ Ideal vacuum = Barometer reading – Ideal pressure

= 750 – 42.25 = 707.75 mm of Hg

We know that vacuum efficiency,

$$\eta_v = \frac{\text{Actual vacuum}}{\text{Ideal vacuum}} = \frac{690}{707.75} = 0.975 \text{ or } 97.5\ \% \text{ Ans.}$$

Example 20.14. *The following observations were recorded during a test on a steam condenser :*

Barometer reading	*= 765 mm of Hg*
Condenser vacuum	*= 710 mm of Hg*
Mean condenser temperature	*= 35° C*
Condensate temperature	*= 28° C*
Condensate collected per hour	*= 2 tonnes*
Quantity of cooling water per hour	*= 60 tonnes*
Temperature of cooling water at inlet	*= 10° C*
Temperature of cooling water at outlet	*= 25° C*

Find : 1. vacuum corrected to the standard barometer reading ; 2. vacuum efficiency of the condenser ; 3. undercooling of the condensate ; 4. condenser efficiency ; 5. quality of the steam entering the condenser ; 6. mass of air per m^3 *of condenser volume ; and 7. mass of air per kg of uncondensed steam.*

Solution. Given : Barometer reading = 765 mm of Hg ; Condenser vacuum = 710 mm of Hg ; $T = 35°\text{ C} = 35 + 273 = 308$ K ; $t_c = 28°$ C ; $m_s = 2$ t/h = 2000 kg/h ; $m_w = 60$ t/h = 60 000 kg/h ; $t_i = 10°$ C ; $t_o = 25°$ C

1. *Vacuum corrected to the standard barometer reading*

We know that absolute pressure in the condenser

= Barometer reading – Condenser vacuum

= 765 – 710 = 55 mm of Hg

and vacuum corrected to the standard barometer reading (assuming 760 mm of Hg)

= 760 – 55 = 705 mm of Hg **Ans.**

2. *Vacuum efficiency of the condenser*

From steam tables, corresponding to the mean condenser temperature of 35° C, we find that ideal pressure of steam,

$$p_s = 0.0562 \text{ bar} = \frac{0.0562}{0.001\ 33} = 42.2 \text{ mm of Hg}$$

We know that ideal vacuum

= Barometer pressure – Ideal pressure

= 765 – 42.2 = 722.8 mm of Hg

and vacuum efficiency, $\eta_v = \frac{\text{Actual vacuum}}{\text{Ideal vacuum}} = \frac{710}{722.8} = 0.982$ or 98.2 % **Ans.**

3. *Undercooling of the condensate*

We know that undercooling of the condensate

$$= \text{Mean condenser temp.} - \text{Condensate temp.}$$

$$= 35 - 28 = 7° \text{ C } \textbf{Ans.}$$

4. *Condenser efficiency*

We have already found that pressure in the condenser,

$$p_c = 765 - 710 = 55 \text{ mm of Hg}$$

$$= 55 \times 0.001\,33 = 0.073 \text{ bar}$$

From steam tables, corresponding to a pressure of 0.073 bar, we find that vacuum temperature

$$t_v = 39.83° \text{ C}$$

∴ Condenser efficiency,

$$\eta_c = \frac{\text{Temperature rise of cooling water}}{\text{Vacuum temperature} - \text{Inlet cooling temperature}} = \frac{t_o - t_i}{t_v - t_i}$$

$$= \frac{25 - 10}{39.83 - 10} = 0.503 \text{ or } 50.3\ \% \ \textbf{Ans.}$$

5. *Quality of steam entering the condenser*

Let x = Quality of steam entering the condenser.

From steam tables, corresponding to a pressure of 0.073 bar, we find that

$$h_f = 166.7 \text{ kJ/kg}\ ; \text{ and } h_{fg} = 2407.4 \text{ kJ/kg}$$

and corresponding to a condensate temperature of 28° C, heat in condensate,

$$h_{f1} = 117.3 \text{ kJ/kg}$$

We know that total heat of entering steam,

$$h = h_f + x\, h_{fg} = 166.7 + x \times 2407.4$$

We also know that mass of cooling water (m_w),

$$60\,000 = \frac{m_s\,(h - h_{f1})}{c_w\,(t_o - t_i)} = \frac{2000\,(166.7 + x \times 2407.4 - 117.3)}{4.2\,(25 - 10)}$$

$$= 31.7\,(49.4 + x \times 2407.4)$$

or

$$x \times 2407.4 = \frac{60\,000}{31.7} - 49.4 = 1843.3$$

$$\therefore \quad x = 0.76 \ \textbf{Ans.}$$

6. *Mass of air per m^3 of condenser volume*

We know that absolute pressure of air (as per Dalton's law),

$$p_a = p_c - p_s = 0.073 - 0.0562 = 0.0168 \text{ bar}$$

$$= 0.0168 \times 10^5 = 1680 \text{ N/m}^2$$

∴ Mass of air per m^3 of condenser volume,

$$m_a = \frac{p_a\, v}{R\,T} = \frac{1680 \times 1}{287 \times 308} = 0.019 \text{ kg } \textbf{Ans.} \quad \ldots (\because pv = mRT \text{ and } v = 1 \text{ m}^3)$$

7. *Mass of air per kg of uncondensed steam*

From steam tables, corresponding to 35° C (*i.e.* mean condenser temperature), specific volume of steam,

$$v_g = 25.245 \text{ m}^3/\text{kg}$$

Thus air associated with 1 kg of steam at 35° C will occupy the same volume *i.e.* 25.245 m³.

∴ Mass of air per kg of uncondensed steam,

$$m_a = \frac{p_a v_g}{R\,T} = \frac{1680 \times 25.245}{287 \times 308} = 0.48 \text{ kg} \quad \textbf{Ans.}$$

20.23. Sources of Air into the Condenser

The following are the main sources through which the air may enter into the condenser :

1. The dissolved air in the feed water enters into the boiler, which in turn enters into the condenser with the exhaust steam.
2. The air leaks into the condenser, through various joints, due to high vacuum pressure in the condenser.
3. In case of jet condensers, dissolved air with the injection water enters into the condenser.

20.24. Effects of Air Leakage

Following are the effects of air leakage on the performance of condensing plants :

1. It reduces the vacuum pressure in the condenser.
2. Since air is a poor heat conductor, particularly at low densities, it reduces the rate of heat transmission.
3. It requires a larger air pump. Moreover, an increased power is required to drive the pump.

20.25. Air Pump

The main function of an air pump is to maintain a vacuum in the condenser as nearly as possible, corresponding to exhaust steam temperature. This is done by removing uncondensable air from the condenser. Another common, but not the essential function of the pump, is to remove both air and condensate from the condenser.

The air pump, which extracts both the condensate and air, is called a *wet air pump*. But a pump which extracts only moist air is known as *dry air pump*. The air pumps may be of reciprocating type or rotary type. But here we shall discuss only a reciprocating type air pump (or Edward's air pump) which is commonly used.

20.26. Edward's Air Pump

It is a wet air pump of the reciprocating type. The special features of Edward's air pump is the absence of suction valve and bucket valve, which are necessary in the ordinary reciprocating type air pumps.

The Edward's air pump consists of delivery or head valves, as shown in Fig. 20.10. These valves are placed in the cover which is on the top of the pump barrel lever. The reciprocating piston of the pump is flat on its upper surface and conical at the bottom as shown. The pump lever has a ring of ports around its lower end for the whole circumference. This communicates with the condenser.

When the piston is at the top of the barrel, the condensate and air from the condenser is collected in the conical portion of the lower part of the barrel, through the ports. On the downward stroke of the reciprocating piston, the vacuum is produced above it, since the head valves are closed and sealed by water. The piston uncovers the ports. When it moves downwards, the mixture of condensate, vapour and air rushes into the space above the piston. This mixture is compressed, when the piston goes to the top and raises the pressure slightly above the atmospheric pressure. The head valves are now open, which allow the mixture to pass on the top of the cover. The condensate flows over the weir to the hot well, which is at atmospheric pressure. A relief valve is placed in the base of the cylinder to release the pressure.

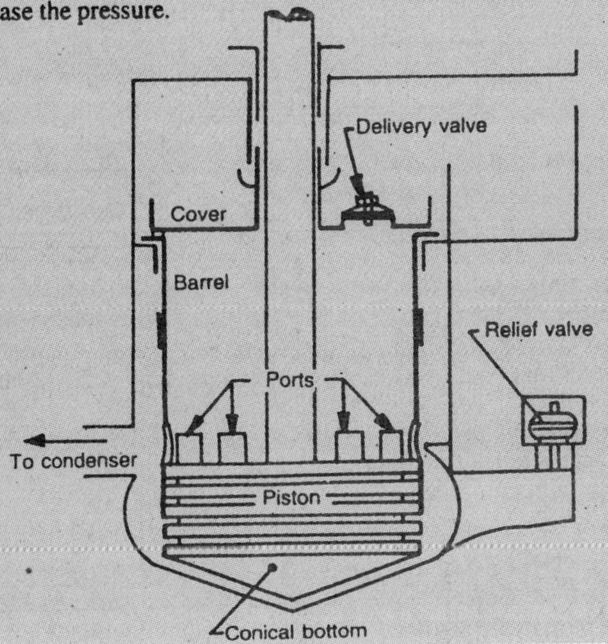

Fig. 20.10. Edward's air pump.

20.27. Cooling Towers

The cooling towers are used in many applications in engineering. The main applications are in power plants and refrigeration plants. Its function is to cool the hot water from the condenser by exposing it to the atmospheric air, so that the cold water may be used again for circulation. The cooling towers are used in steam power plants where there is a limited supply of cooling water. It is placed at a certain height (at about 9 metres from the ground level). The hot water falls down in radial sprays from a height and the atmospheric air enters from the base of the tower. The partial evaporation of water takes place which reduces the temperature of circulating water. This cooled water is collected in the pond at the base of the tower and pumped into the condenser.

Following are some factors which affect the cooling of water in a cooling tower :

1. Size and height of cooling tower,
2. Arrangement of plates in cooling tower,
3. Velocity of air entering the cooling tower,
4. Temperature of air,
5. Humidity of air, and
6. Accessibility of air to all parts of the cooling tower.

20.28. Type of Cooling Towers

The cooling towers may be classified as follows :

1. *According to the type of draught.* The cooling towers, according to the type of draught are (*a*) Natural draught cooling towers, (*b*) Forced draught cooling towers, and (*c*) Induced draught cooling towers.

In a *natural draught cooling tower,* as shown in Fig. 20.11, the circulation of air is produced by the pressure difference of air inside and outside the cooling tower.

In a *forced draught cooling tower*, as shown in Fig. 20.12, the circulation of air is produced by means of fans placed at the base of the tower.

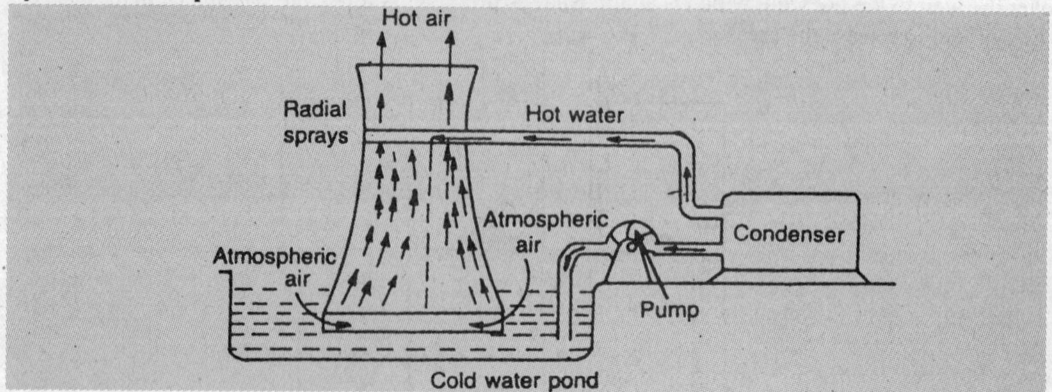

Fig. 20.11. Natural draught cooling tower.

In an *induced draught cooling tower*, as shown in Fig. 20.13, the circulation of air is provided by means of fans placed at the top of the tower.

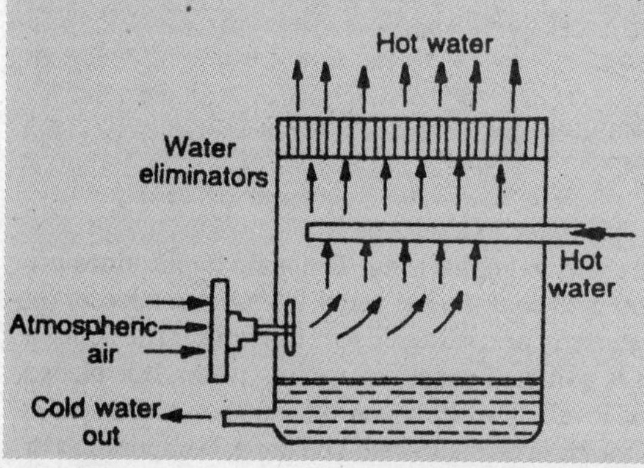

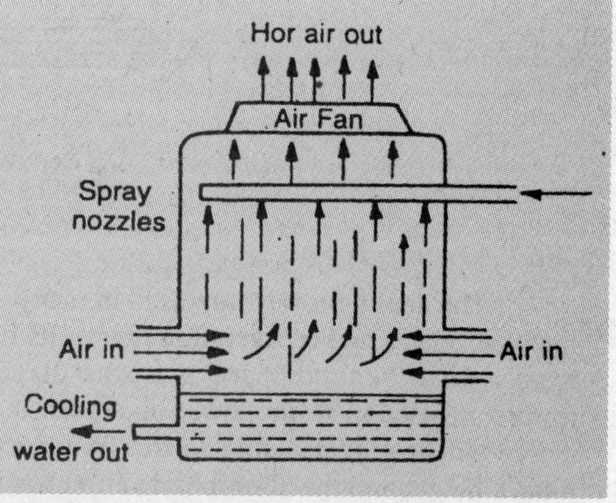

Fig. 20.12. Forced draught cooling tower. 20.13. Induced draught cooling tower.

2. *According to the material used.* The cooling towers, according to the material used are (*a*) Timber cooling towers, (*b*) Concrete (ferro-concrete, multideck concrete hyperbolic) cooling towers, and (*c*) Steel duct type cooling towers.

The timber cooling towers are rarely used due to the following disadvantages :

(*i*) Short life, (*ii*) High maintenance charges, (*iii*) Limited cooling capacity, (*iv*) Rots easily due to exposure to sun, wind, water etc., (*v*) Design generally does not facilitate proper circulation of air.

The concrete cooling towers has the following advantages :

(*i*) Large capacity, (*ii*) Improved draught and air circulation, (*iii*) Increased stability under pressure, (*iv*) Low maintenanee.

The duct type cooling towers are rarely used in case of modern power plants due to their small capacity.

EXERCISES

1. The vacuum in a condenser is 665 mm of Hg and the barometer reading is 762 mm of Hg. The temperature of steam at inlet to the condenser is 35° C. Calculate the mass of air entering the condenser per unit volume of the condenser. [Ans. 0.082 kg/m^3]

2. The barometer stands at 760 mm and condenser vacuum is at 710 mm and temperature is 30° C. Calculate the mass of air per kg of uncondensed steam. [Ans. 0.913 kg]

3. A vacuum gauge fitted to a condenser reads 660 mm of Hg, when the barometer reads 750 mm of Hg. Find the corrected vacuum referred to standard barometer of 760 mm of Hg. [Ans. 670 mm of Hg]

4. Calculate the vacuum efficiency of a condenser from the following data : Vacuum at steam inlet to condenser = 710 mm of Hg ; Barometer reading = 760 mm of Hg ; Hot well temperature = 32° C. [Ans. 98%]

5. The vacuum gauge on a condenser reads 655 mm of Hg at a barometric pressure of 760 mm of Hg. Steam condenses at 48° C. Find the ratio of mass of water vapour (steam) to air to be dealt by dry air pump. If the air leakage into the condenser were reduced by 50%, find the resulting alteration in vacuum in mm of Hg. [Ans. 2.48; 665.6 mm of Hg]

6. A steam condenser has separate air and condensate pumps. The entry to the air pump section is screened. Steam enters the condenser at 38° C and the condensate is removed at 37° C. The air removed has a temperature of 36° C. If the quantity of air infiltration from various sources is 5 kg/h, determine the volume of air handled by the air pump per hour. Compare this with the quantity that would have to be dealt with by using a combined air and condensate pump. Neglect the pressure due to air at the entry of steam and assume uniform pressure in the condenser. [Ans. 652 m^3/h ; 1271 m^3/h]

7. The vacuum in the shell of a condenser is 710 mm of Hg and atmospheric pressure is 760 mm of Hg. The temperature at inlet and outlet of the air cooling section are 35° C and 30° C respectively. Calculate for a leakage of 0.5 kg of air per hour : 1. The volume of air entering the cooling section per hour, and 2. The mass of steam condensed per hour in the section. Take R = 294 J/kg K. [Ans. 42.9 m^3/h ; 1.152 kg/h]

8. A 110 kW steam engine has a steam consumption of 9.5 kg per kWh. The back pressure of the engine which is approximately the same as the condenser pressure is 0.15 bar. The temperature of condensate is 35° C. The cooling water temperature at inlet and outlet are 18° C and 34° C respectively. Estimate the quantity of cooling water required per hour if the steam exhausted to the condenser is dry [Ans. 38140 kg/h]

9. A steam turbine uses 45 000 kg of steam per hour which it exhausts at a dryness fraction 0.9 into a condenser fitted with water extraction and air pumps. With the barometer at 760 mm of mercury, the vacuum at the air pump suction is 716.8 mm and the temperature 32° C. The air leakage is estimated at 1 kg per 1000 kg of steam condensed. Estimate the capacity of the air pump in m^3/min and the quantity of circulating water required in kg/min if the temperature rise is 15° C. [Ans. 66.3 m^3/min ; 25 990 kg/min]

10. A surface condenser receives exhaust steam at 0.14 bar from an engine developing 130 kW. The circulating water enters the condenser at 15° C and leaves at 40° C. The final temperature of the condensed steam is 50° C. If the engine consumes 12.25 kg of steam per kW hour, determine the quality of steam entering the condenser, if the mass of circulating water per hour is 31 600 kg. [Ans. 0.87]

11. The vacuum in a condenser dealing with 8100 kg of steam per hour is found to be 710 mm of Hg when the barometer reads 750 mm of Hg. The temperature in the condenser is 20° C. The air leakage amounts to 8.1 kg per 1000 kg of steam. Determine the capacity of a suitable dry air pump in m^3 per minute required for the condenser. Take volumetric efficiency of pump as 0.85. [Ans. 36.3 m^3/min]

12. The exhaust steam having a dryness fraction of 0.84 enters a surface condenser where the vacuum is 695 mm of Hg and is condensed to water at 35.8° C. The temperature of the hot well is 32.6° C. The circulating water enters the condenser at 15° C and leaves at 35° C. The barometric pressure is 756 mm of Hg. Calculate : 1. The mass of circulating water required per kg of steam, and 2. The mass of air extracted per m^3 of condenser volume. [Ans. 24.3 kg/kg of steam ; 0.0364 kg/m^3]

13. A surface condenser receives 15 150 kg of steam per hour after the steam does work in the turbine. Steam at 10 bar and 250° C enters the turbine. The vacuum in the condenser is maintained at 650 mm of Hg and the barometer reads 752 mm of Hg. The rise in temperature of cooling water is limited to 15° C. The temperature of condensate leaving the condenser is 35° C. What is the amount of cooling water required per hour ?

[Ans. 485 470 kg/h]

[**Hint** : First of all, find the dryness fraction of steam entering the condenser by equating the entropy of steam entering the turbine to the entropy of steam leaving the turbine or entering the condenser]

14. A turbine consumes 14 000 kg of steam per hour while developing 2500 kW. Steam is supplied at 30 bar and 300° C. The exhaust from the turbine is condensed in a condenser at a vacuum of 725 mm of Hg and the barometer reads 758 mm of Hg. The condensate is removed from the condenser at a temperature of 28° C. The temperature of cooling water increases from 7° C to 27° C. Assuming no radiation losses, find : 1. The

dryness fraction of steam entering the condenser, and 2. The mass of circulating water per hour.

[**Ans.** 0.965 ; 391 870 kg/h]

15. The following observations refer to a surface condenser :

Mean temperature of condensation = 34.9° C ; Temperature of hot well = 29.7° C ; Condenser vacuum = 701 mm of Hg ; Barometer = 763 mm of Hg ; Mass of cooling water = 45 500 kg/h ; Inlet temperature of cooling water = 16.5° C ; Outlet temperature of cooling water = 30.6° C ; Mass of condensate = 1180 kg/h.

Find : 1. the mass of air present per m^3 of condenser volume ; 2. the state of steam entering the condenser ; and 3. the vacuum efficiency. [**Ans.** 0.03 kg/m^3 ; 0.935 ; 97.24%]

QUESTIONS

1. What are functions of the condenser in a steam plant ?
2. Describe the principle requirements of a steam condensing plant.
3. Explain the principles of operation of different types of jet condensers. Describe with a sketch a low level jet condenser of the counter flow type.
4. Describe with a neat sketch the working of a surface condenser.
5. Compare the merits and demerits of surface condenser over jet condenser.
6. State Dalton's law of partial pressures.
7. What do you understand by the term vacuum efficiency of a condensing plant ? On what factors does this efficiency depend ?
8. Prove with the help of an example that the vacuum efficiency decreases with the increase in barometric pressure.
9. What are the various sources of air leakage into a steam condenser ? How does it affect the performance of the condensing plant ?
10. Explain the construction and working of Edward's air pump.
11. What part is played by a cooling tower ? What are the different types of cooling towers ? Mention advantage and disadvantage of each type.

OBJECTIVE TYPE QUESTIONS

1. A condenser in a steam power plant

   (*a*) increases expansion ratio of steam (*b*) reduces back pressure of steam

   (*c*) reduces temperature of exhaust steam (*d*) all of these

2. The temperature of condensate is ...... on leaving the condenser than that of circulating water at inlet.

   (*a*) higher (*b*) lower

3. A condenser where circulating water flows through tubes which are surrounded by steam, is known as

   (*a*) surface condenser (*b*) jet condenser

   (*c*) barometric condenser (*d*) evaporative condenser

4. The ratio of actual vacuum to the ideal vacuum in a condenser is called

   (*a*) condenser efficiency (*b*) vacuum efficiency

   (*c*) boiler efficiency (*d*) nozzle efficiency

5. The actual vacuum in a condenser is equal to

   (*a*) barometric pressure + actual pressure (*b*) barometric pressure – actual pressure

   (*c*) gauge pressure + atmospheric pressure (*d*) gauge pressure – atmospheric pressure

ANSWERS

1. (*d*)	2. (*a*)	3. (*a*)	4. (*b*)	5. (*b*)

21

Steam Nozzles

1. Introduction. 2. Types of Steam Nozzles. 3. Flow of Steam through Convergent-divergent Nozzle. 4. Friction in a Nozzle or Nozzle Efficiency. 5. Velocity of Steam Flowing through a Nozzle. 6. Mass of Steam Discharged through a Nozzle. 7. Condition for Maximum Discharge through a Nozzle (Critical Pressure Ratio). 8. Values for Maximum Discharge through a Nozzle. 9. Values for Critical Pressure Ratio. 10. Physical Significance of Critical Pressure Ratio. 11. Diameters of Throat and Exit for Maximum Discharge. 12. Supersaturated Flow or Metastable Flow through Nozzles. 13. Effect of Supersaturation. 14. Steam Injector. 15. Steam Injector Calculations.

21.1. Introduction

A steam nozzle is a passage of varying cross-section, which converts heat energy of steam into kinetic energy. During the first part of the nozzle, the steam increases its velocity. But in its later part, the steam gains more in volume than in velocity. Since the mass of steam, passing through any section of the nozzle remains constant, the variation of steam pressure in the nozzle depends upon the velocity, specific volume and dryness fraction of steam. A well designed nozzle converts the heat energy of steam into kinetic energy with a minimum loss.

The main use of steam nozzle in steam turbines, is to produce a jet of steam with a high velocity. The smallest section of the nozzle is called throat.

21.2. Types of Steam Nozzles

Following three types of nozzles are important from the subject point of view :

1. *Convergent nozzle.* When the cross-section of a nozzle decreases continuously from entrance to exit, it is called a convergent nozzle as shown in Fig. 21.1 (*a*).

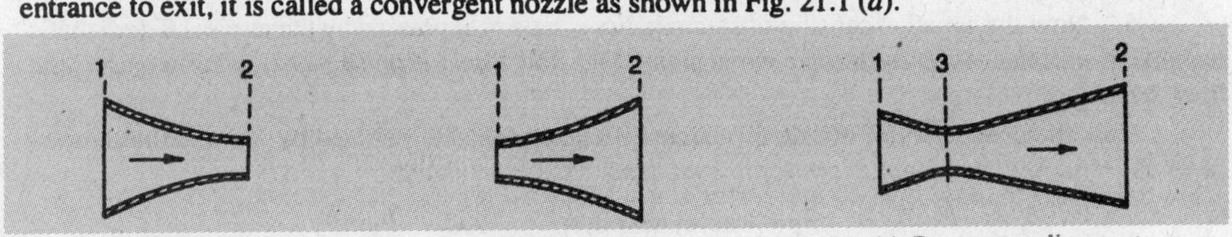

(*a*) Convergent. (*b*) Divergent. (*c*) Convergent-divergent.

Fig. 21.1. Types of steam nozzles.

2. *Divergent nozzle.* When the cross-section of a nozzle increases continuously from entrance to exit, it is called a divergent nozzle, as shown in Fig. 21.1 (*b*).

3. *Convergent-divergent nozzle.* When the cross-section of a nozzle first decreases from its entrance to throat, and then increases from its throat to exit, it is called a convergent-divergent nozzle as shown in Fig. 21.1 (*c*). This type of nozzle is widely used these days in various types of steam turbines.

21.3. Flow of Steam through Convergent-divergent Nozzle

The steam enters the nozzle with a high pressure, but with a negligible velocity. In the converging portion (*i.e.* from the inlet to the throat), there is a drop in the steam pressure with a rise in its velocity. There is also a drop in the enthalpy or total heat of the steam. This drop of heat is not utilised in doing some external work, but is converted into kinetic energy. In the divergent portion (*i.e.* from the throat to outlet), there is further drop of steam pressure with a further rise in its velocity. Again, there is a drop in the enthalpy or total heat of steam, which is converted into kinetic energy.

It will be interesting to know that the steam enters the nozzle with a high pressure and negligible velocity. But leaves the nozzle with a high velocity and small pressure. The pressure, at which the steam leaves the nozzle, is known as back pressure. Moreover, no heat is supplied or rejected by the steam during flow through a nozzle. Therefore, it is considered as isentropic flow, and the corresponding expansion is considered as an isentropic expansion.

21.4. Friction in a Nozzle or Nozzle Efficiency

As a matter fact, when the steam flows through a nozzle, some loss in its enthalpy or total heat takes place due to friction between the nozzle surface and the flowing steam. This can be best understood with the help of *h-s* diagram or Mollier chart, as shown in Fig. 21.2., which can be completed as discussed below :

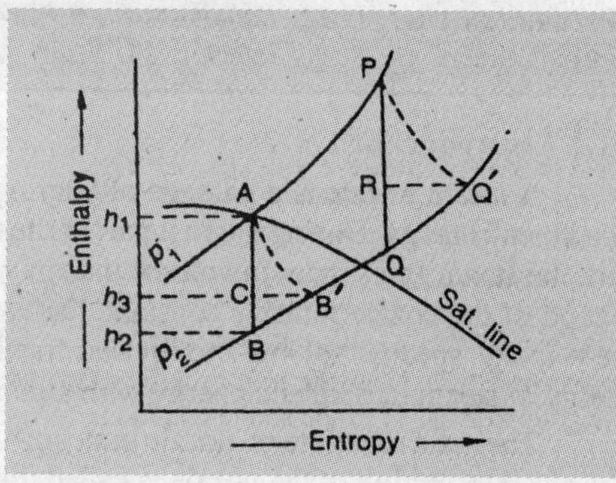

Fig. 21.2. *h-s* graph for nozzle efficiency.

1. First of all, locate the point *A* for the initial conditions of the steam. It is a point, where the saturation line meets the initial pressure (p_1) line.

2. Now draw a vertical line through *A* to meet the final pressure (p_2) line. This is done as the flow through the nozzle is isentropic, which is expressed by a vertical line *AB*. The heat drop ($h_1 - h_2$) is known as *isentropic heat drop*.

3. Due to friction in the nozzle the actual heat drop in the steam will be less than ($h_1 - h_2$). Let this heat drop be shown as *AC* instead of *AB*.

4. As the expansion of steam ends at the pressure p_2, therefore final condition of steam is obtained by drawing a horizontal line through *C* to meet the final pressure (p_2) line at *B'*.

5. Now the actual expansion of steam in the nozzle is expressed by the curve *AB'* (adiabatic expansion) instead of *AB* (isentropic expansion). The actual heat drop ($h_1 - h_3$) is known as *useful heat drop*.

Now the coefficient of nozzle or nozzle efficiency (usually denoted by *K*) is defined *as the ratio of useful heat drop to the isentropic heat drop*. Mathematically,

$$K = \frac{\text{Useful heat drop}}{\text{Isentropicheat drop}} = \frac{AC}{AB} = \frac{h_1 - h_3}{h_1 - h_2}$$

Notes : 1. We see from Fig. 21.2, that the dryness fraction of steam at *B'* is greater than that at *B*. It is thus obvious, that the effect of friction is to increase the dryness fraction of steam. This is due to the fact that the energy lost in friction is transferred into heat, which tends to dry or superheat the steam.

2. A similar effect is produced when the steam is superheated at the entrance of the nozzle.

Let PR = Useful heat drop, and PQ = Isentropicheat drop.

$\therefore$ Nozzle efficiency, $K = \frac{PR}{PQ}$

3. In general, if 15% of the heat drop is lost in friction, then efficiency of the nozzle is equal to $100 - 15 = 85\% = 0.85$.

21.5. Velocity of Steam Flowing through a Nozzle

Consider a unit mass flow of steam through a nozzle.

Let V_1 = Velocity of steam at the entrance of nozzle in m/s

V_2 = Velocity of steam at any section considered in m/s,

h_1 = Enthalpy or total heat of steam entering the nozzle in kJ/kg, and

h_2 = Enthalpy or total heat of steam at the section considered in kJ/kg.

We know that for a steady flow process in a nozzle,

$$h_1 + \frac{*1}{1000}\left(\frac{V_1^2}{2}\right) = h_2 + \frac{1}{1000}\left(\frac{V_2^2}{2}\right) + \text{Losses}$$

Neglecting losses in a nozzle,

$$\frac{1}{1000}\left(\frac{V_2^2}{2} - \frac{V_1^2}{2}\right) = h_1 - h_2$$

$$\therefore \quad V_2 = \sqrt{V_1^2 + 2000(h_1 - h_2)} = \sqrt{V_1^2 + 2000\, h_d} \quad \ldots (i)$$

where h_d = Enthalpy or heat drop during expansion of steam in a nozzle

$= h_1 - h_2$

Since the entrance velocity or velocity of approach (V_1) is negligible as compared to V_2, therefore from equation (*i*),

$$V_2 = \sqrt{2000\, h_d} = 44.72\sqrt{h_d} \quad \ldots (ii)$$

Note : In actual practice, there is always a certain amount of friction present between the steam and nozzle surfaces. This reduces the heat drop by 10 to 15 percent and thus the exit velocity of steam is also reduced correspondingly. Thus the above relation may be written as :

$$V_2 = 44.72\sqrt{K\, h_d}$$

where K is the nozzle coefficient or nozzle efficiency.

Example 21.1. *Dry saturated steam at 5 bar with negligible velocity expands isentropically in a convergent nozzle to 1 bar and dryness fraction 0.94. Determine the velocity of steam leaving the nozzle.*

Solution. Given : $p_1 = 5$ bar ; $p_2 = 1$ bar ; $x_2 = 0.94$

From steam tables, corresponding to a pressure of 5 bar, we find that enthalpy or total heat of dry saturated steam,

$$h_1 = h_{g1} = 2747.5 \text{ kJ/kg}$$

and corresponding to a pressure of 1 bar, we find that

$$h_{f2} = 417.5 \text{ kJ/kg, and } h_{fg2} = 2257.9 \text{ kJ/kg}$$

$\therefore$ Enthalpy or total heat of final steam,

$$h_2 = h_{f2} + x_2\, h_{fg2} = 417.5 + 0.94 \times 2257.9 = 2540 \text{ kJ/kg}$$

and enthalpy or heat drop, $h_d = h_1 - h_2 = 2747.5 - 2540 = 207.5$ kJ/kg

We know that velocity of steam leaving the nozzle,

$$V_2 = 44.72\sqrt{h_d} = 44.72\sqrt{207.5} = 644 \text{ m/s} \quad \textbf{Ans.}$$

---

* We know that $\text{K.E.} = \frac{1}{2} mV^2 = \frac{1}{2} \times 1 \times V^2 = \frac{V^2}{2} \text{ J} = \frac{1}{1000}\left(\frac{V^2}{2}\right) \text{kJ} \quad \ldots (\because m = 1 \text{ kg})$

Example 21.2. *Dry saturated steam at a pressure of 15 bar enters in a nozzle and is discharged at a pressure of 1.5 bar. Find the final velocity of the steam, when the initial velocity of the steam is negligible.*

If 10% of the heat drop is lost in friction, find the percentage reduction in the final velocity.

Solution. Given : $p_1 = 15$ bar ; $p_2 = 1.5$ bar

Final velocity of the steam

From steam tables, corresponding to a pressure of 15 bar, we find that enthalpy of dry saturated steam,

$$h_1 = 2789.9 \text{ kJ/kg}$$

and corresponding to a pressure of 1.5 bar, enthalpy of dry saturated steam,

$$h_2 = 2693.4 \text{ kJ/kg}$$

∴ Heat drop, $h_d = h_1 - h_2 = 2789.9 - 2693.4 = 96.5$ kJ/kg

We know that final velocity of the steam,

$$V_2 = 44.72\sqrt{h_d} = 44.72\sqrt{96.5} = 439.3 \text{ m/s } \textbf{Ans.}$$

Percentage reduction in the final velocity

We know that heat drop lost in friction

$$= 10\% = 0.1 \quad \text{...(Given)}$$

∴ Nozzle coefficient or nozzle efficiency

$$K = 1 - 0.1 = 0.9$$

We know that final velocity of the steam,

$$V_2 = 44.72\sqrt{K\,h_d} = 44.72\sqrt{0.9 \times 96.5} = 416.8 \text{ m/s}$$

∴ Percentage reduction in final velocity

$$= \frac{439.3 - 416.8}{439.3} = 0.051 \text{ or } 5.1\% \textbf{ Ans.}$$

Example 21.3. *Dry saturated steam at 10 bar is expanded isentropically in a nozzle to 0.1 bar. Using steam tables only, find the dryness fraction of the steam at exit. Also find the velocity of steam leaving the nozzle when 1. initial velocity is negligible, and 2. initial velocity of the steam is 135 m/s.*

Solution. Given : $p_1 = 10$ bar ; $p_2 = 0.1$ bar

Dryness fraction of the steam at exit

Let x_2 = Dryness fraction of the steam at exit.

From steam tables, corresponding to a pressure of 10 bar, we find that entropy of dry saturated steam,

$$s_1 = s_{g1} = 6.583 \text{ kJ/kg K}$$

and corresponding to a pressure of 0.1 bar, we find that

$$s_{f2} = 0.649 \text{ kJ/kg K}, \quad \text{and} \quad s_{fg2} = 7.502 \text{ kJ/kg K}$$

Since the expansion of steam is isentropic, therefore

Entropy of steam at inlet (s_1) = Entropy of steam at exit (s_2)

$$6.583 = s_{f2} + x_2\, s_{fg2} = 0.649 + x_2 \times 7.502$$

$$\therefore \quad x_2 = 0.791 \textbf{ Ans.}$$

1. *Velocity of steam leaving the nozzle when initial velocity is negligible*

From steam tables, corresponding to a pressure of 10 bar, we find that enthalpy or total heat of dry saturated steam,

$$h_1 = h_{g1} = 2776.2 \text{ kJ/kg}$$

and corresponding to a pressure of 0.1 bar,

$$h_{f2} = 191.8 \text{ kJ/kg, and } h_{fg2} = 2392.9 \text{ kJ/kg}$$

∴ Enthalpy or total heat of steam of exit,

$$h_2 = h_{f2} + x_2 h_{fg2}$$

$$= 191.8 + 0.791 \times 2392.9 = 2084.6 \text{ kJ/kg}$$

and heat drop, $h_d = h_1 - h_2 = 2776.2 - 2084.6 = 691.6 \text{ kJ/kg}$

We know that velocity of steam leaving the nozzle,

$$V_2 = 44.72\sqrt{h_d} = 44.72\sqrt{691.6} = 1176 \text{ m/s} \quad \textbf{Ans.}$$

2. *Velocity of steam leaving the nozzle when initial velocity, V_1 = 135 m/s*

We know that velocity of steam leaving the nozzle,

$$V_2 = \sqrt{V_1^2 + 2000\, h_d} = \sqrt{(135)^2 + 2000 \times 691.6} = 1184 \text{ m/s} \quad \textbf{Ans.}$$

Example 21.4. *Dry saturated steam at a pressure of 10 bar is expanded in a nozzle to a pressure of 0.7 bar. With the help of Mollier diagram find the velocity and dryness fraction of steam issuing from the nozzle, if the friction is neglected.*

Also find the velocity and dryness fraction of the steam, if 15% of the heat drop is lost in friction.

Solution. Given : p_1 = 10 bar ; p_2 = 0.7 bar

Velocity and dryness fraction of steam issuing from the nozzle, if friction is neglected

The process on the Mollier diagram, as shown in Fig. 21.3, is drawn as discussed below :

1. First of all, locate the point *A* on the saturation line (because the steam is initially dry saturated) where the initial pressure line (10 bar) meets it.

2. Since the expansion in the nozzle is isentropic, therefore draw a vertical line through *A* to meet the final pressure line (0.7 bar) at point *B*,

Now from the Mollier diagram, we find that

$$h_1 = 2772 \text{ kJ/kg,}$$

and $$h_2 = 2310 \text{ kJ/kg}$$

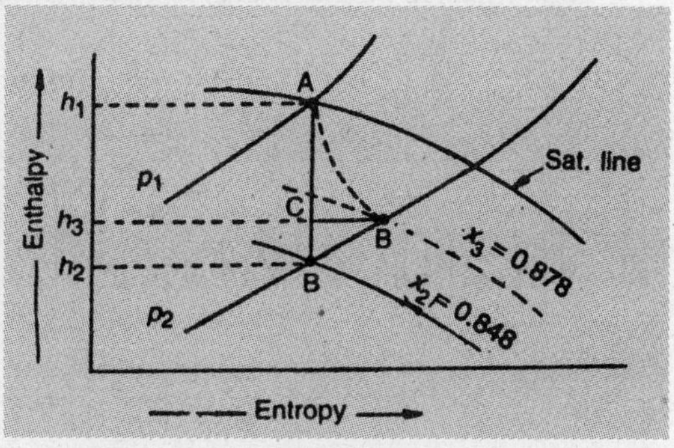

Fig 21.3

∴ Heat drop,

$$h_d = h_1 - h_2 = 2772 - 2310 = 462 \text{ kJ/kg}$$

We know that velocity of steam issuing from the nozzle,

$$V_2 = 44.72\sqrt{h_d} = 44.72\sqrt{462} = 961 \text{ m/s} \quad \textbf{Ans.}$$

From Mollier diagram, we also find that the dryness fraction of steam issuing from the nozzle (*i.e.* at point *B*) is x_2 = 0.848. **Ans.**

Velocity and dryness fraction of steam issuing from the nozzle if 15% of the heat drop is lost in friction.

Since 15% heat drop is lost in friction, therefore nozzle coefficient or nozzle efficiency,

$$K = 100 - 15 = 85\% = 0.85$$

and heat drop due to friction $= 462 \times 0.15 = 69.3$ kJ/kg

We know that velocity of steam issuing from the nozzle,

$$V_2 = 44.72\sqrt{K\,h_d} = 44.72\sqrt{0.85 \times 462} = 886 \text{ m/s} \quad \textbf{Ans.}$$

Now let us complete the Mollier diagram as discussed below :

1. Locate point *C* on the vertical line *AB*, such that $BC = 69.3$ kJ/kg.
2. Now through *C*, draw a horizontal line CB' to meet the final pressure line (0.7 bar) at B'.

From the Mollier diagram, we find that the dryness fraction of steam issuing from the nozzle, (*i.e.* at point B') is $x_3 = 0.878$. **Ans.**

21.6. Mass of Steam Discharged through Nozzle

We have already discussed that the flow of steam, through the nozzle is isentropic, which is approximately represented by the general law:

$$pv^n = \text{Constant}$$

We know that gain in kinetic energy

$$= \frac{V_2^2}{2} \quad \text{. . . (Neglecting initial velocity of steam)}$$

and

$$\text{Heat drop} = \text{Work done during Rankine cycle}$$

$$= \frac{n}{n-1}(p_1 v_1 - p_2 v_2)$$

Since gain in kinetic energy is equal to heat drop, therefore

$$\frac{V_2^2}{2} = \frac{n}{n-1}(p_1 v_1 - p_2 v_2)$$

$$= \frac{n}{n-1} \times p_1 v_1 \left(1 - \frac{p_2 v_2}{p_1 v_1}\right) \quad \text{. . . (i)}$$

We know that $p_1 v_1^n = p_2 v_2^n$

$$\frac{v_2}{v_1} = \left(\frac{p_1}{p_2}\right)^{\frac{1}{n}} = \left(\frac{p_2}{p_1}\right)^{-\frac{1}{n}}$$

$$\therefore \quad v_2 = v_1\left(\frac{p_2}{p_1}\right)^{-\frac{1}{n}} \quad \text{. . . (ii)}$$

Substituting, the value of v_2 / v_1 in equation (*i*),

$$\frac{V_2^2}{2} = \frac{n}{n-1} \times p_1 v_1 \left[1 - \frac{p_2}{p_1}\left(\frac{p_2}{p_1}\right)^{-\frac{1}{n}}\right]$$

$$= \frac{n}{n-1} \times p_1 v_1 \left[1 - \left(\frac{p_2}{p_1} \right)^{\frac{n-1}{n}} \right]$$

or
$$V_2 = \sqrt{2 \times \frac{n}{n-1} \times p_1 v_1 \left[1 - \left(\frac{p_2}{p_1} \right)^{\frac{n-1}{n}} \right]} \quad \ldots (iii)$$

Now the volume of steam flowing per second

$$= \text{Cross-sectional area of nozzle} \times \text{Velocity of steam} = A\,V_2$$

and volume of 1 kg of steam *i.e.* specific volume of steam at pressure p_2

$$= v_2 \text{ m}^3/\text{kg}$$

∴ Mass of steam discharged through nozzle per second,

$$m = \frac{\text{Volume of steam flowing per second}}{\text{Volume of 1 kg of steam at pressure } p_2}$$

$$= \frac{A V_2}{v_2} = \frac{A}{v_2} \sqrt{2 \times \frac{n}{n-1} \times p_1 v_1 \left[1 - \left(\frac{p_2}{p_1} \right)^{\frac{n-1}{n}} \right]}$$

Substituting the value of v_2 from equation (*ii*),

$$m = \frac{A}{v_1} \left(\frac{p_1}{p_2} \right)^{-\frac{1}{n}} \sqrt{\frac{2n}{n-1} \times p_1 v_1 \left[1 - \left(\frac{p_2}{p_1} \right)^{\frac{n-1}{n}} \right]}$$

$$= \frac{A}{v_1} \left(\frac{p_2}{p_1} \right)^{\frac{1}{n}} \sqrt{\frac{2n}{n-1} \times p_1 v_1 \left[1 - \left(\frac{p_2}{p_1} \right)^{\frac{n-1}{n}} \right]}$$

$$= A \sqrt{\left(\frac{p_2}{p_1} \right)^{\frac{2}{n}} \times \frac{2n}{n-1} \times \frac{p_1}{v_1} \left[1 - \left(\frac{p_2}{p_1} \right)^{\frac{n-1}{n}} \right]}$$

$$= A \sqrt{\frac{2n}{n-1} \times \frac{p_1}{v_1} \left[\left(\frac{p_2}{p_1} \right)^{\frac{2}{n}} - \left(\frac{p_2}{p_1} \right)^{\frac{n+1}{n}} \right]} \quad \ldots (iv)$$

Example 21.5. *Dry air at a temperature of 27° C and pressure of 20 bar enters a nozzle and leaves at a pressure of 4 bar. Find the mass of air discharged, if the area of the nozzle is 200 mm².*

Solution. Given : $T_1 = 27° \text{ C} = 27 + 273 = 300 \text{ K}$; $p_1 = 20 \text{ bar} = 20 \times 10^5 \text{ N/m}^2$; $p_2 = 4 \text{ bar} = 4 \times 10^5 \text{ N/m}^2$; $A = 200 \text{ mm}^2 = 200 \times 10^{-6} \text{ m}^2$

Let v_1 = Specific volume of air in m³/kg.

We know that $p_1 v_1 = m R T_1$

$$\therefore \quad v_1 = \frac{m R T_1}{p_1} = \frac{1 \times 287 \times 300}{20 \times 10^5} = 0.043 \text{ m}^3/\text{kg} \quad \ldots (\because R \text{ for air} = 287 \text{ J/kg K})$$

We know that mass of steam discharged through the nozzle,

$$m = A\sqrt{\frac{2n}{n-1}\times\frac{p_1}{v_1}\left[\left(\frac{p_2}{p_1}\right)^{\frac{2}{n}}-\left(\frac{p_2}{p_1}\right)^{\frac{n+1}{n}}\right]}$$

$$= 200\times10^{-6}\sqrt{\frac{2\times1.4}{1.4-1}\times\frac{20\times10^5}{0.043}\left[\left(\frac{4}{20}\right)^{\frac{2}{1.4}}-\left(\frac{4}{20}\right)^{\frac{1.4+1}{1.4}}\right]}$$

$$= 200\times10^{-6}\sqrt{3256\times10^5\,[\,0.1-0.06\,]} = 0.72\text{ kg/s} \quad \textbf{Ans.}$$

21.7. Condition for Maximum Discharge through a Nozzle (Critical Pressure Ratio)

A nozzle is, normally, designed for maximum discharge by designing a certain throat pressure which produces this condition.

Let p_1 = Initial pressure of steam in N/m²,

p_2 = Pressure of steam at throat in N/m²,

v_1 = Volume of 1 kg of steam at pressure (p_1) in m³,

v_2 = Volume of 1 kg of steam at pressure (p_2) in m³, and

A = Cross-sectional area of nozzle at throat, in m².

We have derived an equation in the previous article that the mass of steam discharged through nozzle,

$$m = A\sqrt{\frac{2n}{n-1}\times\frac{p_1}{v_1}\left[\left(\frac{p_2}{p_1}\right)^{\frac{2}{n}}-\left(\frac{p_2}{p_1}\right)^{\frac{n+1}{n}}\right]} \qquad \ldots (i)$$

There is only one value of the ratio p_2/p_1, which produces maximum discharge from the nozzle. This ratio p_2/p_1, is obtained by differentiating the right hand side of the equation. We see from this equation that except p_2/p_1, all other values are constant. Therefore, only that portion of the equation which contains p_2/p_1, is differentiated and equated to zero for maximum discharge.

$$\therefore \quad \frac{d}{d\left(\frac{p_2}{p_1}\right)}\left[\left(\frac{p_2}{p_1}\right)^{\frac{2}{n}}-\left(\frac{p_2}{p_1}\right)^{\frac{n+1}{n}}\right] = 0$$

or

$$\frac{2}{n}\left(\frac{p_2}{p_1}\right)^{\frac{2}{n}-1}-\frac{n+1}{n}\left(\frac{p_2}{p_1}\right)^{\frac{n+1}{n}-1} = 0$$

$$\therefore \quad \frac{2}{n}\left(\frac{p_2}{p_1}\right)^{\frac{2-n}{n}} = \frac{n+1}{n}\left(\frac{p_2}{p_1}\right)^{\frac{1}{n}}$$

$$\left(\frac{p_2}{p_1}\right)^{\frac{2-n}{n}}\times\left(\frac{p_2}{p_1}\right)^{-\frac{1}{n}} = \frac{n+1}{n}\times\frac{n}{2}$$

$$\left(\frac{p_2}{p_1}\right)^{\frac{2-n}{n}-\frac{1}{n}} = \frac{n+1}{2}$$

$$\left(\frac{p_2}{p_1}\right)^{\frac{1-n}{n}} = \frac{n+1}{2}$$

$$\frac{p_2}{p_1} = \left(\frac{n+1}{2}\right)^{\frac{n}{1-n}} = \left(\frac{n+1}{2}\right)^{\frac{-n}{-(1-n)}}$$

$$= \left(\frac{2}{n+1}\right)^{\frac{n}{n-1}} \qquad \ldots (ii)$$

Notes : 1. The ratio p_2/p_1 is known as *critical pressure ratio,* and the pressure p_2 at the throat is known as *critical pressure.*

2. The maximum value of the discharge per second is obtained by substituting the value of p_2/p_1 in equation (*i*).

$$\therefore \text{ Maximum discharge, } m_{max} = A\sqrt{\frac{2n}{n-1}\times\frac{p_1}{v_1}\left[\left(\frac{2}{n+1}\right)^{\frac{2}{n-1}}-\left(\frac{2}{n+1}\right)^{\frac{n+1}{n-1}}\right]}$$

$$= A\sqrt{\frac{2n}{n-1}\times\frac{p_1}{v_1}\left(\frac{2}{n+1}\right)^{\frac{2}{n-1}}\left[1-\left(\frac{2}{n+1}\right)^{\frac{n+1}{n-1}-\frac{2}{n-1}}\right]}$$

$$= A\sqrt{\frac{2n}{n-1}\times\frac{p_1}{v_1}\left(\frac{2}{n+1}\right)^{\frac{2}{n-1}}\left[1-\left(\frac{2}{n+1}\right)\right]}$$

$$= A\sqrt{\frac{2n}{n-1}\times\frac{p_1}{v_1}\left(\frac{2}{n+1}\right)^{\frac{2}{n-1}}\left[\frac{n-1}{n+1}\right]}$$

$$= A\sqrt{\frac{2n}{n+1}\times\frac{p_1}{v_1}\left(\frac{2}{n+1}\right)^{\frac{2}{n-1}}} \qquad \ldots (iii)$$

3. We see from the above equation that in a convergent-divergent nozzle, the discharge depends upon the area of nozzle at throat and the initial conditions of the steam (*i.e.* pressure p_1 and volume v_1). It is independent of the exit conditions of the steam. It is thus obvious, that the discharge remains constant after the throat (*i.e.* in the divergent portion of the nozzle).

4. The equations derived above are true for gases also.

21.8. Values for Maximum Discharge through a Nozzle

In the last article we have derived a relation for the maximum discharge through a nozzle, *i.e.*

$$m_{max} = A\sqrt{\frac{2n}{n+1}\times\frac{p_1}{v_1}\left(\frac{2}{n+1}\right)^{\frac{2}{n-1}}}$$

Now we shall discuss the values of maximum discharge for the following three conditions :

1. *When the steam is initially dry saturated*

We know that for dry saturated steam, $n = 1.135$. Therefore substituting the value of n in the relation for maximum discharge, we have

$$m_{max} = 0.637\, A \sqrt{\frac{p_1}{v_1}}$$

2. *When the steam is initially superheated*

We know that for superheated steam, $n = 1.3$. Therefore substituting the value of n in the relation for maximum discharge, we have

$$m_{max} = 0.666\, A \sqrt{\frac{p_1}{v_1}}$$

3. *For gases*

We know that for gases, $n = 1.4$. Therefore substituting the value of n in the relation for maximum discharge, we have

$$m_{max} = 0.685\, A \sqrt{\frac{p_1}{v_1}}$$

Example 21.6. *Dry air at a pressure of 12 bar and 300°C is expanded isentropically through a nozzle at a pressure of 2 bar. Determine the maximum discharge through the nozzle of 150 mm² area.*

Solution. Given : $p_1 = 12$ bar $= 12 \times 10^5$ N/m² ; $T_1 = 300°C = 300 + 273 = 573$ K ; $p_2 = 2$ bar ; $A = 150$ mm² $= 150 \times 10^{-6}$ m²

Let v_1 = Specific volume of air in m³/kg.

We know that $p_1 v_1 = m R T_1$

$$\therefore \quad v_1 = \frac{m R T_1}{p_1} = \frac{1 \times 287 \times 573}{12 \times 10^5} = 0.137 \text{ m}^3/\text{kg}$$

We know that maximum discharge through the nozzle,

$$m_{max} = 0.685\, A \sqrt{\frac{p_1}{v_1}} = 0.685 \times 150 \times 10^{-6} \sqrt{\frac{12 \times 10^5}{0.137}} \text{ kg/s}$$

$$= 0.304 \text{ kg/s } \textbf{Ans.}$$

Example 21.7. *Steam at a pressure of 10 bar and 210°C is supplied to a convergent divergent nozzle with a throat area of 1500 mm². The exit is below critical pressure. Find the coefficient of discharge, if the flow is 7200 kg of steam per hour.*

Solution. Given : $p_1 = 10$ bar $= 10 \times 10^5$ N/m² ; $T_1 = 210°$ C ; $A = 1500$ mm² $= 1500 \times 10^{-6}$ m² ; $m = 7200$ kg/h $= 2$ kg/s

From steam tables, for superheated steam, corresponding to a pressure of 10 bar and 210° C, we find that specific volume of steam,

$$v_1 = 0.2113 \text{ m}^3/\text{kg}$$

We know that for superheated steam, $n = 1.3$.

$$\therefore \text{ Maximum discharge, } m_{max} = A \sqrt{\frac{2n}{n+1} \times \frac{p_1}{v_1} \left(\frac{2}{n+1}\right)^{\frac{2}{n-1}}}$$

$$= 1500 \times 10^{-6} \sqrt{\frac{2 \times 1.3}{1.3+1} \times \frac{10 \times 10^5}{0.2113} \left(\frac{2}{1.3+1} \right)^{\frac{2}{1.3-1}}} \text{ kg/s}$$

$$= 2.17 \text{ kg/s}$$

We know that coefficient of discharge

$$= \frac{\text{Actual discharge}}{\text{Maximum discharge}} = \frac{2}{2.17} = 0.922 \textbf{ Ans.}$$

Note : The maximum discharge for superheated steam may also be calculated by using the relation,

$$m_{max} = 0.666\, A \sqrt{\frac{p_1}{v_1}}$$

21.9. Values for Critical Pressure Ratio

We have also discussed in Art. 21.7 that the critical pressure ratio,

$$\frac{p_2}{p_1} = \left(\frac{2}{n+1} \right)^{\frac{n}{n-1}}$$

where p_2 / p_1 = Critical pressure ratio.

We shall now discuss the values of critical pressure ratio for the following conditions :

1. *When the steam is initially saturated*

We know that for dry saturated steam, $n = 1.135$.

$$\therefore \quad \frac{p_2}{p_1} = \left(\frac{2}{1.135+1} \right)^{\frac{1.135}{1.135-1}} = 0.577 \quad \text{or} \quad p_2 = 0.577\, p_1$$

2. *When the steam is initially superheated*

We know that for superheated steam, $n = 1.3$.

$$\therefore \quad \frac{p_2}{p_1} = \left(\frac{2}{1.3+1} \right)^{\frac{1.3}{1.3-1}} = 0.546 \quad \text{or} \quad p_2 = 0.546\, p_1$$

3. *When the steam is initially wet*

It has been experimentally found that the critical pressure ratio for wet steam,

$$\frac{p_2}{p_1} = 0.582 \quad \text{or} \quad p_2 = 0.582\, p_1$$

4. *For gases*

We know that for gases, $n = 1.4$.

$$\therefore \quad \frac{p_2}{p_1} = \left(\frac{2}{1.4+1} \right)^{\frac{1.4}{1.4-1}} = 0.528 \quad \text{or} \quad p_2 = 0.528\, p_1$$

21.10. Physical Significance of Critical Pressure Ratio

In the previous article, we discussed the values of critical pressure ratio for various forms of steam. But now we shall discuss the physical significance of the critical pressure ratio.

Now consider two vessels *A* and *B* connected by a convergent nozzle as shown in Fig. 21.4 (*a*). Let the vessel *A* contains steam at a high and steady pressure (p_1), and the vessel *B* contains steam at another pressure (p_2) which may be varied at will.

First of all, let the pressure (p_2) in the vessel *B* be made equal to the pressure (p_1) in the vessel *A*. In this case, there will be no flow of steam through the nozzle. Now if the pressure (p_2) in the vessel *B* is gradually reduced, the discharge through the nozzle will increase accordingly as shown in Fig. 21.4 (*b*). As the pressure (p_2) in the vessel *B* approaches the critical value, the rate of discharge will also approach its maximum value. If the pressure (p_2) in the vessel *B* is further reduced, it will not increase the rate of discharge. But the discharge will remain the same as that at critical pressure as shown in Fig. 21.4 (*b*). The ratio of exit pressure to the inlet pressure is called ***critical pressure ratio.***

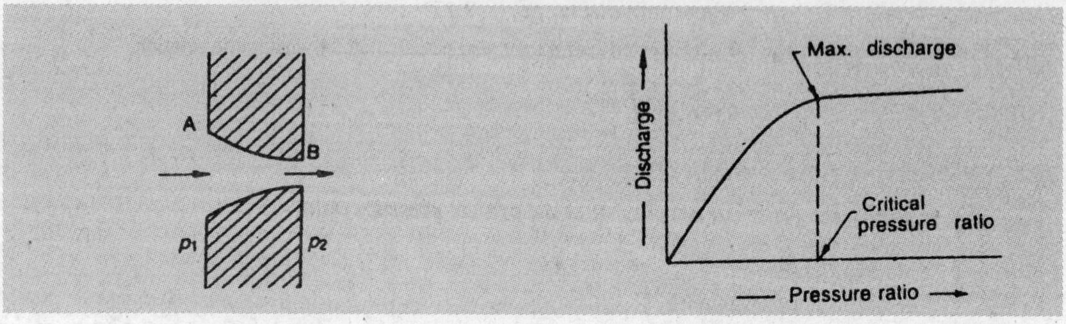

(*a*) Nozzle section. (*b*) Discharge-pressure ratio curve.

Fig. 21.4

We know that the velocity of steam at any section in the nozzle [Refer Art. 21.6, equation (*iii*)],

$$V_2 = \sqrt{\frac{2n}{n-1} \times p_1 v_1 \left[1 - \left(\frac{p_2}{p_1} \right)^{\frac{n-1}{n}} \right]} \qquad \ldots (i)$$

and the critical pressure ratio for maximum discharge,

$$\frac{p_2}{p_1} = \left(\frac{2}{n+1} \right)^{\frac{n}{n-1}} \quad \text{or} \quad \left(\frac{p_2}{p_1} \right)^{\frac{n-1}{n}} = \frac{2}{n+1}$$

Substituting this value in equation (*i*),

$$V_2 = \sqrt{\frac{2n}{n-1} \times p_1 v_1 \left[1 - \frac{2}{n+1} \right]} = \sqrt{\frac{2n}{n-1} \times p_1 v_1 \left[\frac{n+1-2}{n+1} \right]}$$

$$= \sqrt{\frac{2n}{n+1} \times p_1 v_1} = \sqrt{\frac{2n}{n+1} \times \frac{p_1}{\rho_1}} \qquad \ldots (ii)$$

$$\ldots \left[\because \text{Volume } (v) = \frac{1}{\text{Density } (\rho)} \right]$$

We also know that for isentropic expansion,

$$p_1 v_1^n = p_2 v_2^n$$

or

$$\frac{p_1}{\rho_1^n} = \frac{p_2}{\rho_2^n} \qquad \ldots \left(\because v = \frac{1}{\rho} \right)$$

or

$$\frac{1}{\rho_1} = \frac{1}{\rho_2} \left(\frac{p_2}{p_1} \right)^{\frac{1}{n}}$$

$$\therefore \quad \frac{p_1}{\rho_1} = \frac{p_1}{\rho_2}\left(\frac{p_2}{p_1}\right)^{\frac{1}{n}} \qquad \text{(Multiplying by } p_1\text{)}$$

$$= \frac{p_2}{\rho_2} \times \frac{p_1}{p_2}\left(\frac{p_2}{p_1}\right)^{\frac{1}{n}} = \frac{p_2}{\rho_2}\left(\frac{p_2}{p_1}\right)^{\frac{1-n}{n}}$$

$$= \frac{p_2}{\rho_2}\left(\frac{2}{n+1}\right)^{\frac{n}{n-1} \times \frac{1-n}{n}} = \frac{p_2}{\rho_2}\left(\frac{n+1}{2}\right) \ldots \left[\because \frac{p_2}{p_1} = \left(\frac{2}{n+1}\right)^{\frac{n}{n-1}}\right]$$

Substituting the value of p_1/ρ_1, in equation (*ii*),

$$V_2 = \sqrt{2\left(\frac{n}{n+1}\right)\frac{p_2}{\rho_2}\left(\frac{n+1}{2}\right)} = \sqrt{\frac{n\,p_2}{\rho_2}} \qquad \ldots (iii)$$

This is the value of velocity of sound in the medium at pressure p_2 and is known as *sonic velocity.*

Notes : 1. The critical pressure gives the velocity of steam at the throat equal to the velocity of sound.

2. The flow in the convergent portion of the nozzle is sub-sonic and in the divergent portion it is supersonic.

3. To increase the velocity of steam above sonic velocity (supersonic) by expanding steam below the critical pressure, the divergent portion for the nozzle is necessary.

21.11. Diameters of Throat and Exit for Maximum Discharge

Consider a convergent-divergent nozzle discharging steam, as shown in Fig. 21.5 (*a*),

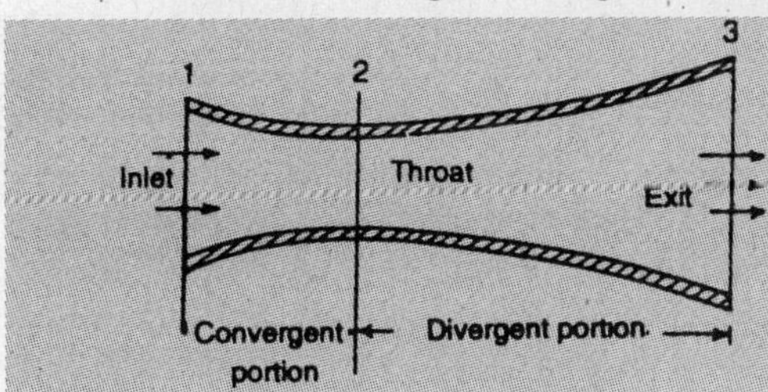

(*a*) Convergent-divergent nozzle.

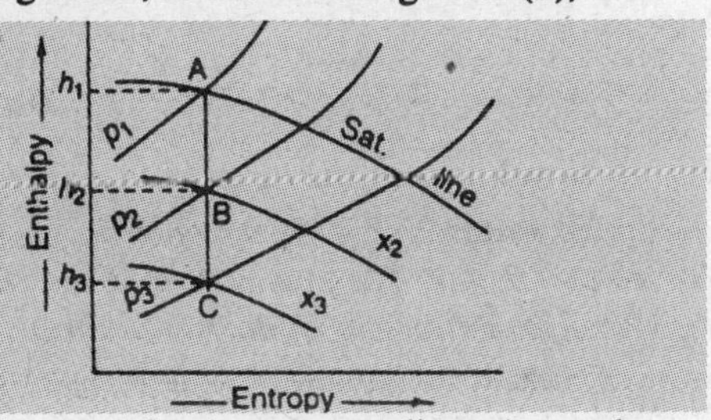

(*b*) *h-s* graph for a convergent-divergent nozzle.

Fig. 21.5

Let

p_1 = Initial pressure of steam,

h_1 = Enthalpy or total heat of steam at inlet,

p_2, p_3, h_2, h_3 = Corresponding values at throat and outlet,

x_2 = Dryness fraction of steam at throat,

V_2 = Velocity of steam at throat,

v_{g2} = Specific volume of steam at throat corresponding to pressure p_2 (from steam tables),

A_2 = Area of throat,

x_3, V_3, v_{g3}, A_3 = Corresponding values at exit, and

m = Mass of steam discharged.

First of all, find the value of critical pressure (p_2) as discussed in Art. 21.7.

Now complete the *h-s* diagram, as shown in Fig. 21.5 (*b*), for the expansion of steam through the convergent-divergent nozzle as discussed below :

1. First of all, locate the point *A* for the initial conditions of steam. It is a point, where the saturation line meets the initial pressure (p_1) line.
2. Now draw a vertical line through *A* to meet the critical pressure (p_2) line at *B*. This represents the throat of the nozzle.
3. Now extend the vertical line *AB* to meet the outlet pressure (p_3) line at *C*. This represents the outlet of the nozzle.
4. Now find the values of h_1, h_2, h_3, x_2 and x_3 from the *h-s* graph.

First of all, consider the flow of steam from the inlet to the throat. We know that

Enthalpy or heat drop, $h_{d2} = h_1 - h_2$

∴ Velocity of steam at throat,

$$V_2 = 44.72\sqrt{h_{d2}} \quad \text{... (Neglecting friction)}$$

We know that mass of steam discharged per second,

$$m = \frac{\text{Volume of steam flowing at throat}}{\text{Volume of 1 kg of steam at pressure } p_2}$$

$$= \frac{A_2 V_2}{v_2} = \frac{A_2 V_2}{x_2 v_{g2}} \quad \text{... } (\because v_2 = x_2 v_{g2})$$

Similarly, for exit conditions,

$$m = \frac{A_3 V_3}{x_3 v_{g3}} = \frac{A_2 V_2}{x_2 v_{g2}}$$

Now knowing the value of *m*, we can determine the area or diameter of throat and exit.

Example 21.8. *Steam enters a group of nozzles of a steam turbine at 12 bar and 220°C and leaves at 1.2 bar. The steam turbine develops 220 kW with a specific steam consumption of 13.5 kg/kWh. If the diameter of nozzles at throat is 7 mm, calculate the number of nozzles.*

Solution. Given : p_1 = 12 bar ; T_1 = 220°C ; p_3 = 1.2 bar ; Power developed = 220 kW ; m_s = 13.5 kg/kWh ; d_2 = 7 mm

We know that for superheated steam, pressure of steam at throat,

$$p_2 = 0.546\, p_1 = 0.546 \times 12 = 6.552 \text{ bar}$$

The Mollier diagram for the expansion of steam through the nozzle is shown in Fig. 21.6.

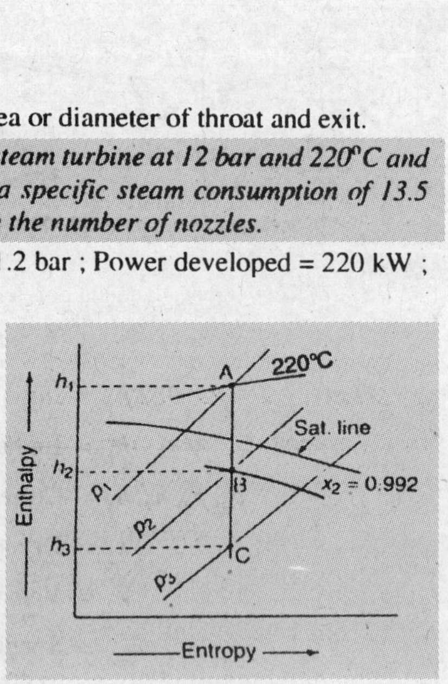

Fig. 21.6

From the Mollier diagram, we find that enthalpy of steam at entrance (*i.e.* at 12 bar and 220° C),

$$h_1 = 2860 \text{ kJ/kg}$$

Enthalpy of steam at throat (*i.e.* at pressure 6.552 bar),

$$h_2 = 2750 \text{ kJ/kg}$$

and dryness fraction of steam at throat,

$$x_2 = 0.992$$

From steam tables, we find that specific volume of dry saturated steam at throat (*i.e.* at pressure 6.552 bar),

$$v_{g2} = 0.29 \text{ m}^3/\text{kg}$$

We know that heat drop from entrance to throat,

$$h_{d2} = h_1 - h_2 = 2860 - 2750 = 110 \text{ kJ/kg}$$

$\therefore$ Velocity of steam at throat,

$$V_2 = 44.72\sqrt{h_{d2}} = 44.72\sqrt{110} = 470 \text{ m/s}$$

Area of nozzle at throat,

$$A_2 = \frac{\pi}{4}(d_2)^2 = \frac{\pi}{4} \times 7^2 = 38.5 \text{ mm}^2 = 38.5 \times 10^{-6} \text{ m}^2$$

$\therefore$ Mass flow rate per nozzle,

$$m = \frac{A_2 V_2}{v_2} = \frac{A_2 V_2}{x_2 v_{g2}} = \frac{38.5 \times 10^{-6} \times 470}{0.992 \times 0.29} = 0.063 \text{ kg/s}$$

We know that total mass flow rate

$$= 13.5 \times 220 = 2970 \text{ kg/h} = 0.825 \text{ kg/s}$$

$$\therefore \text{ Number of nozzles } = \frac{\text{Total mass flow rate}}{\text{Mass flow rate per nozzle}} = \frac{0.825}{0.063} = 13.1 \text{ say } 14 \text{ Ans.}$$

Example 21.9. *Estimate the mass flow rate of steam in a nozzle with the following data : Inlet pressure and temperature = 10 bar and 200° C ; Back pressure = 0.5 bar ; Throat diameter = 12 mm.*

Solution. Given : $p_1 = 10$ bar ; $T_1 = 200°$ C ; $p_3 = 0.5$ bar ; $d_2 = 12$ mm

We know that for superheated steam, pressure of steam at throat,

$$p_2 = 0.546\, p_1 = 0.546 \times 10 = 5.46 \text{ bar}$$

The Mollier diagram for the expansion of steam through the nozzle is shown in Fig. 21.7. From the Mollier diagram, we find that

$$h_1 = 2825 \text{ kJ/kg}$$

$$h_2 = 2710 \text{ kJ/kg}$$

and $$x_2 = 0.982$$

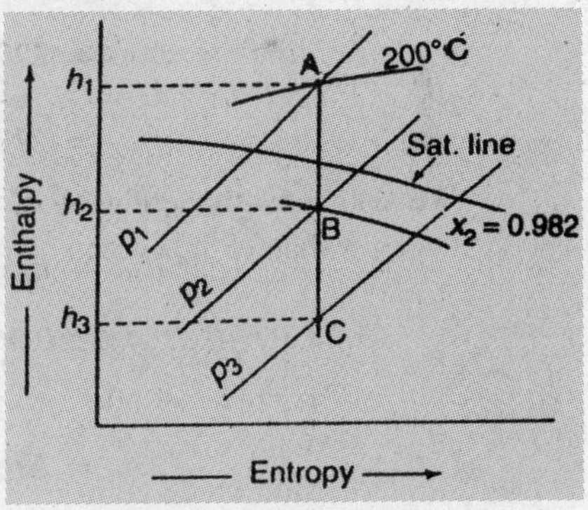

Fig. 21.7

We know that heat drop,

$$h_d = h_1 - h_2 = 2825 - 2710 = 115 \text{ kJ/kg}$$

$\therefore$ Velocity of steam at throat,

$$V_2 = 44.72\sqrt{h_d} = 44.72\sqrt{115} = 480 \text{ m/s}$$

From steam tables, corresponding to a pressure of 5.46 bar, we find that specific volume of steam at throat,

$$v_{g2} = 0.345 \text{ m}^3/\text{kg}$$

Area of nozzle at throat,

$$A_2 = \frac{\pi}{4}(d_2)^2 = \frac{\pi}{4}(12)^2 = 113 \text{ mm}^2 = 113 \times 10^{-6} \text{ m}^2$$

∴ Mass flow rate of steam,

$$m = \frac{A_2 V_2}{x_2 v_{g2}} = \frac{113 \times 10^{-6} \times 480}{0.982 \times 0.345} = 0.16 \text{ kg/s}$$

$$= 0.16 \times 3600 = 576 \text{ kg/h} \quad \textbf{Ans.}$$

Example 21.10. *Dry saturated steam enters a nozzle at a pressure of 10 bar and with an initial velocity of 90 m/s. The outlet pressure is 6 bar and the outlet velocity is 435 m/s. The heat loss from the nozzle is 9 kJ/kg of steam flow.*

Calculate the dryness fraction and the area at the exit, if the area at the inlet is 1256 mm².

Solution. Given : $p_1 = 10$ bar ; $V_1 = 90$ m/s ; $p_3 = 6$ bar ; $V_3 = 435$ m/s ; Loses = 9 kJ/kg ; $A_1 = 1256 \text{ mm}^2 = 1256 \times 10^{-6} \text{ m}^2$

Dryness fraction of steam

Let x_3 = Dryness fraction of steam at the exit.

From steam tables, corresponding to a pressure of 10 bar, we find that enthalpy of dry saturated steam,

$$h_1 = 2776.2 \text{ kJ/kg} \; ; \text{ and } v_{g1} = 0.1943 \text{ m}^3\text{/kg}$$

and corresponding to a pressure of 6 bar, we find that

$$h_{f3} = 670.4 \text{ kJ/kg} \; ; \; h_{fg3} = 2085 \text{ kJ/kg} \; ; \text{ and } v_{g3} = 0.3155 \text{ m}^3\text{/kg}$$

We know that for a steady flow through the nozzle,

$$h_1 + \frac{1}{1000}\left(\frac{V_1^2}{2}\right) = h_3 + \frac{1}{1000}\left(\frac{V_3^2}{2}\right) + \text{Losses}$$

$$\therefore \quad h_3 = h_1 + \frac{1}{2000}(V_1^2 - V_3^2) - \text{Losses}$$

$$= 2776.2 + \frac{1}{2000}\left[(90)^2 - (435)^2\right] - 9$$

$$= 2776.2 - 99.6 = 2676\ 6 \text{ kJ/kg}$$

We also know that enthalpy of wet steam (h_3),

$$2676.6 = h_{f3} + x_3 h_{fg3} = 670.4 + x_3 \times 2085$$

$$\therefore \quad x_3 = 0.962 \quad \textbf{Ans.}$$

Area at exit

Let A_3 = Area at exit in m².

We know that $\dfrac{A_1 V_1}{x_1 v_{g1}} = \dfrac{A_3 V_3}{x_3 v_{g3}}$ or $\dfrac{1256 \times 10^{-6} \times 90}{1 \times 0.1943} = \dfrac{A_3 \times 435}{0.962 \times 0.3155}$

... (For dry saturated steam, $x_1 = 1$)

$$\therefore \quad A_3 = 406 \times 10^{-6} \text{ m}^2 = 406 \text{ mm}^2 \quad \textbf{Ans.}$$

Example 21.11. *Dry saturated steam at a pressure of 8 bar enters a convergent-divergent nozzle and leaves it at a pressure of 1.5 bar, If the flow is isentropic, and the corresponding expansion index is 1.135 ; find the ratio of cross-sectional area at exit and throat for maximum discharge.*

Solution. Given : $p_1 = 8$ bar ; $p_3 = 1.5$ bar ; $n = 1.135$

Let A_2 = Cross-sectional area at throat,

A_3 = Cross-sectional area at exit, and

m = Mass of steam discharged per second.

We know that for dry saturated steam (or when n = 1.135), critical pressure ratio,

$$\frac{p_2}{p_1} = 0.577$$

$$\therefore \quad p_2 = 0.577\, p_1 = 0.577 \times 8 = 4.616 \text{ bar}$$

Now complete the Mollier diagram for the expansion of steam through the nozzle, as shown in Fig. 21.8.

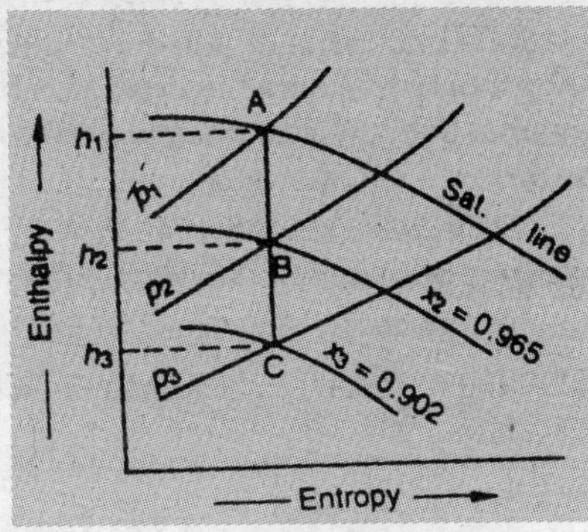

Fig. 21.8

From Mollier diagram, we find that

$h_1 = 2775$ kJ/kg ; $h_2 = 2650$ kJ/kg ; $h_3 = 2465$ kJ/kg ; $x_2 = 0.965$; and $x_3 = 0.902$

From steam tables, we also find that the specific volume of steam at throat corresponding to 4.616 bar,

$$v_{g2} = 0.405 \text{ m}^3/\text{kg}$$

and specific volume of steam at exit corresponding to 1.5 bar,

$$v_{g3} = 1.159 \text{ m}^3/\text{kg}$$

Heat drop between entrance and throat,

$$h_{d2} = h_1 - h_2 = 2775 - 2650 = 125 \text{ kJ/kg}$$

$\therefore$ Velocity of steam at throat,

$$V_2 = 44.72\sqrt{h_{d2}} = 44.72\sqrt{125} = 500 \text{ m/s}$$

and
$$m = \frac{A_2\, V_2}{x_2\, v_{g2}}$$

or
$$A_2 = \frac{m\, x_2\, v_{g2}}{V_2} = \frac{m \times 0.965 \times 0.405}{500} = 0.000\,786\, m \qquad \ldots (i)$$

Heat drop between entrance and exit,

$$h_{d3} = h_1 - h_3 = 2775 - 2465 = 310 \text{ kJ/kg}$$

$\therefore$ Velocity of steam at exit,

$$V_3 = 44.72\sqrt{h_{d3}} = 44.72\sqrt{310} = 787.4 \text{ m/s}$$

and
$$m = \frac{A_3\, V_3}{x_3\, v_{g3}}$$

or
$$A_3 = \frac{m\, x_3\, v_{g3}}{V_3} = \frac{m \times 0.902 \times 1.159}{787.4} = 0.001\,33\, m \qquad \ldots (ii)$$

$\therefore$ Ratio of cross-sectional area at exit and throat,

$$\frac{A_3}{A_2} = \frac{0.001\,33\, m}{0.000\,786\, m} = 1.7 \text{ **Ans.**}$$

Example 21.12. *A convergent-divergent nozzle is required to discharge 2 kg of steam per second. The nozzle is supplied with steam at 7 bar and 180° C and discharge takes place against a back pressure of 1 bar. The expansion upto throat is isentropic and the frictional resistance between the throat and exit is equivalent to 63 kJ/kg of steam. Taking approach velocity of 75 m/s and throat pressure of 4 bar, estimate :*

1. Suitable areas for the throat and exit ; and 2. Overall efficiency of the nozzle based on the enthalpy drop between the actual inlet pressure and temperature and the exit pressure.

Solution. Given : $m = 2$ kg/s ; $p_1 = 7$ bar ; $T_1 = 180°$ C ; $p_3 = 1$ bar ; Frictional resistance = 63 kJ/kg of steam ; $V_1 = 75$ m/s ; $p_2 = 4$ bar

1. *Suitable areas for the throat and exit*

Let A_2 = Area at the throat, and

A_3 = Area at the exit.

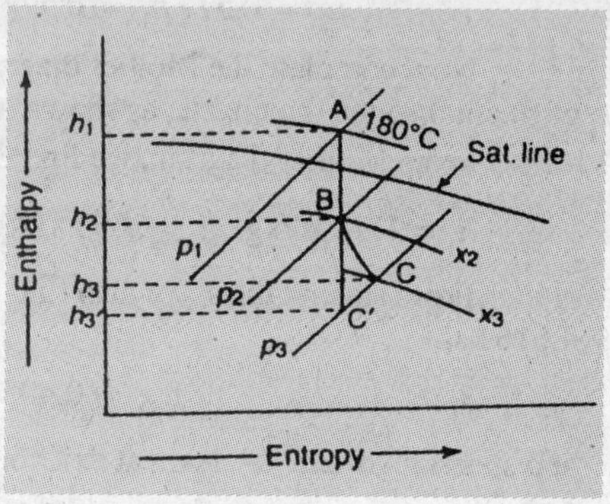

Fig. 21.9

The expansion of steam through the nozzle on the Mollier diagram is shown in Fig. 21.9. From the Mollier diagram, we find that

$h_1 = 2810$ kJ/kg ; $h_2 = 2680$ kJ/kg ; $h_{3'} = 2470$ kJ/kg ; $x_2 = 0.97$; $x_3 = 0.934$

From steam tables, we also find that the specific volume of steam at throat corresponding to 4 bar,

$$v_{g2} = 0.462 \text{ m}^3/\text{kg}$$

and specific volume of steam corresponding to 1 bar,

$$v_{g3} = 1.694 \text{ m}^3/\text{kg}$$

We know that heat drop between entrance and throat,

$$h_{d2} = h_1 - h_2 = 2810 - 2680 = 130 \text{ kJ/kg}$$

∴ Velocity of steam at throat,

$$V_2 = \sqrt{V_1^2 + 2000\, h_{d2}} = \sqrt{(75)^2 + 2000 \times 130} = 515 \text{ m/s}$$

and

$$m = \frac{A_2 V_2}{x_2 v_{g2}}$$

or

$$A_2 = \frac{m x_2 v_{g2}}{V_2} = \frac{2 \times 0.97 \times 0.462}{515} = 1.74 \times 10^{-3} \text{ m}^2$$

$$= 1740 \text{ mm}^2 \quad \textbf{Ans.}$$

Since there is a frictional resistance of 63 kJ/kg of steam between the throat and exit, therefore

$$h_3 - h_{3'} = 63 \text{ or } h_3 = h_{3'} + 63 = 2470 + 63 = 2533 \text{ kJ/kg}$$

and heat drop between entrance and exit,

$$h_{d3} = h_1 - h_3 = 2810 - 2533 = 277 \text{ kJ/kg}$$

∴ Velocity of steam at exit,

$$V_3 = \sqrt{V_1^2 + 2000\, h_{d3}} = \sqrt{(75)^2 + 2000 \times 277} = 748 \text{ m/s}$$

and $$m = \frac{A_3 V_3}{x_3 v_{g3}}$$

or $$A_3 = \frac{m x_3 v_{g3}}{V_3} = \frac{2 \times 0.934 \times 1.694}{748} = 4.23 \times 10^{-3} \text{ m}^2$$

$$= 4230 \text{ mm}^2 \textbf{ Ans.}$$

2. Overall efficiency of the nozzle

We know that overall efficiency of the nozzle,

$$\eta = \frac{\text{Useful heat drop}}{\text{Isentropic heat drop}} = \frac{h_1 - h_3}{h_1 - h_{3'}}$$

$$= \frac{2810 - 2533}{2810 - 2470} = 0.815 \text{ or } 81.5\% \textbf{ Ans.}$$

Example 21.13. *Steam at a pressure of 10 bar and 0.9 dry discharges through a nozzle having throat area of 450 mm². If the back pressure is 1 bar, find 1. final velocity of the steam, and 2. cross-sectional area of the nozzle at exit for maximum discharge.*

Solution. Given : $p_1 = 10$ bar ; $x_1 = 0.9$; $A_2 = 450 \text{ mm}^2 = 450 \times 10^{-6} \text{ m}^2$; $p_3 = 1$ bar

Final velocity of steam

Let V_3 = Final velocity of steam.

We know that for maximum discharge, pressure of steam at throat (for wet steam),

$$p_2 = 0.582\, p_1 = 0.582 \times 10 = 5.82 \text{ bar}$$

Now complete the Mollier diagram for the expansion of steam through the nozzle as shown in Fig. 21.10.

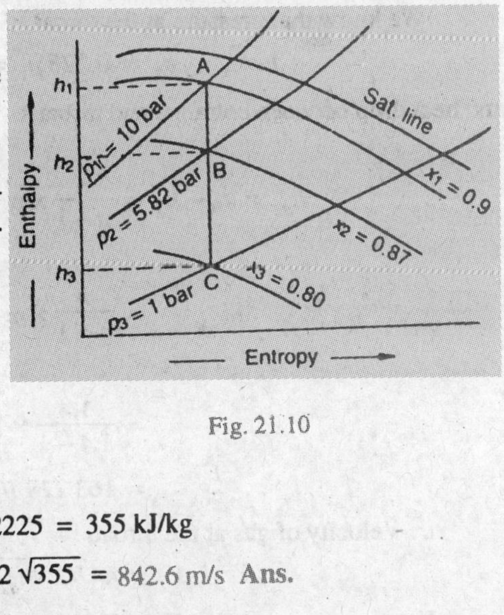

Fig. 21.10

From the Mollier diagram, we find that

$$h_1 = 2580 \text{ kJ/kg} ;\ h_2 = 2485 \text{ kJ/kg} ;$$

$$h_3 = 2225 \text{ kJ/kg} ;\ x_2 = 0.87 ;$$

and $$x_3 = 0.8$$

∴ Heat drop from entrance to exit,

$$h_{d3} = h_1 - h_3 = 2580 - 2225 = 355 \text{ kJ/kg}$$

and velocity of steam, $$V_3 = 44.72\sqrt{h_{d3}} = 44.72\sqrt{355} = 842.6 \text{ m/s} \textbf{ Ans.}$$

Cress-sectional area of the nozzle at exit

Let A_3 = Cross-sectional area of the nozzle at exit.

From steam tables, we find that specific volume of steam at throat corresponding to a pressure of 5.82 bar,

$$v_{g2} = 0.3254 \text{ m}^3\text{/kg}$$

and specific volume of steam at exit corresponding to a pressure of 1 bar,

$$v_{g3} = 1.694 \text{ m}^3\text{/kg}$$

We know that heat drop from entrance to throat,

$$h_{d2} = h_1 - h_2 = 2580 - 2485 = 95 \text{ kJ/kg}$$

∴ Velocity of steam at throat,

$$V_2 = 44.72\sqrt{h_{d2}} = 44.72\sqrt{95} = 436 \text{ m/s}$$

*Since the mass flow rate is same at throat and exit, therefore

$$\frac{A_2 V_2}{x_2 v_{g2}} = \frac{A_3 V_3}{x_3 v_{g3}}$$

$$\frac{450 \times 10^{-6} \times 436}{0.87 \times 0.3254} = \frac{A_3 \times 842.6}{0.8 \times 1.694}$$

$$\therefore \quad A_3 = 1114 \times 10^{-6} \text{ m}^2 = 1114 \text{ mm}^2 \textbf{ Ans.}$$

Example 21.14. *A gas expands in a convergent-divergent nozzle from 5 bar to 1.5 bar, the initial temperature being 700° C and the nozzle efficiency is 90%. All the losses take place after the throat. For 1 kg/s mass flow rate of the gas, find throat and exit areas.*

Take n = 1.4 and R = 287 J/kg K.

Solution. Given : $p_1 = 5$ bar ; $p_3 = 1.5$ bar ; $T_1 = 700°\text{ C} = 700 + 273 = 973$ K ; $K = 90\% = 0.9$; $m = 1$ kg/s ; $n = 1.4$; $R = 287$ J/kg K

Throat area

We know that pressure at the throat,

$$p_2 = 0.528\,p_1 = 0.528 \times 5 = 2.64 \text{ bar}$$

and heat drop between entrance and throat,

$$h_{d2} = h_1 - h_2 = \frac{n}{n-1} \times p_1 v_1 \left[1 - \left(\frac{p_2}{p_1}\right)^{\frac{n-1}{n}}\right] \quad \text{... (Refer Art. 21.6)}$$

$$= \frac{n}{n-1} \times m R T_1 \left[1 - \left(\frac{p_2}{p_1}\right)^{\frac{n-1}{n}}\right]$$

$$= \frac{1.4}{1.4-1} \times 1 \times 287 \times 973 \left[1 - \left(\frac{2.64}{5}\right)^{\frac{1.4-1}{1.4}}\right] \text{ J/kg}$$

$$= 163\,220 \text{ J/kg} = 163.22 \text{ kJ/kg}$$

∴ Velocity of gas at the throat,

$$V_2 = 44.72\sqrt{h_{d2}} = 44.72\sqrt{163.22} = 571.3 \text{ m/s}$$

Let T_2 = Temperature of gas at the throat,

v_2 = Volume of gas at the throat, and

---

* This may also be found as discussed below :

We know that mass of steam discharged per second,

$$m = \frac{A_2 V_2}{x_2 v_{g2}} = \frac{450 \times 10^{-6} \times 436}{0.87 \times 0.3254} = 0.693 \text{ kg}$$

Similarly, $$0.693 = \frac{A_3 V_3}{x_3 v_{g3}} = \frac{A_3 \times 842.6}{0.8 \times 1.694} = 621.5\,A_3$$

$$\therefore \quad A_3 = 1114 \times 10^{-6} \text{ m}^2 = 1114 \text{ mm}^2 \textbf{ Ans.}$$

$$A_2 = \text{Area at the throat.}$$

We know that $\dfrac{T_1}{T_2} = \left(\dfrac{p_1}{p_2}\right)^{\frac{n-1}{n}} = \left(\dfrac{5}{2.64}\right)^{\frac{1.4-1}{1.4}} = 1.2$

$\therefore$ $T_2 = T_1 / 1.2 = 973 / 1.2 = 810.8\text{ K}$

and $p_2 v_2 = m R T_2$

or $v_2 = \dfrac{m R T_2}{p_2} = \dfrac{1 \times 287 \times 810.8}{2.64 \times 10^5} = 0.88\text{ m}^3/\text{kg}$

We know that $m = \dfrac{A_2 V_2}{v_2}$

or $A_2 = \dfrac{m v_2}{V_2} = \dfrac{1 \times 0.88}{571.3} = 1.54 \times 10^{-3}\text{ m}^2 = 1540\text{ mm}^2$ **Ans.**

Exit area

We know that heat drop between entrance and exit,

$$h_{d3} = h_1 - h_3 = \frac{n}{n-1} \times m R T_1 \left[1 - \left(\frac{p_3}{p_1}\right)^{\frac{n-1}{n}}\right]$$

$$= \frac{1.4}{1.4-1} \times 1 \times 287 \times 973 \left[1 - \left(\frac{1.5}{5}\right)^{\frac{1.4-1}{1.4}}\right] \text{J/kg}$$

$$= 284\,420\text{ J/kg} = 284.42\text{ kJ/kg}$$

$\therefore$ Velocity of gas at exit,

$$V_3 = 44.72\sqrt{K\,h_{d3}} = 44.72\sqrt{0.9 \times 284.42} = 715.5\text{ m/s}$$

... ($\because$ losses takes place after throat)

Let $T_{3'}$ = Temperature of gas at exit when friction is neglected,

T_3 = Temperature of gas at exit when friction is considered,

v_3 = Volume of gas at exit, and

A_3 = Area at exit.

We know that $\dfrac{T_1}{T_{3'}} = \left(\dfrac{p_1}{p_3}\right)^{\frac{n-1}{n}} = \left(\dfrac{5}{1.5}\right)^{\frac{1.4-1}{1.4}} = 1.41$

$\therefore$ $T_{3'} = T_1 / 1.41 = 973 / 1.41 = 690\text{ K}$

Since the nozzle efficiency is 90% (*i.e.* $K = 0.9$), therefore heat drop lost in friction is

$$= (1 - K)\,h_{d3} = (1 - 0.9)\,284.42 = 28.442\text{ kJ/kg}$$

∴ Increase in temperature due to friction

$$= * \ 28.442 \text{ K}$$

$$\therefore \quad T_3 = 690 + 28.442 = 718.442 \text{ K}$$

and $$p_3 v_3 = m R T_3$$

or $$v_3 = \frac{m R T_3}{p_3} = \frac{1 \times 287 \times 718.442}{1.5 \times 10^5} = 1.375 \text{ m}^3/\text{kg}$$

We know that $$m = \frac{A_3 V_3}{v_3}$$

or $$A_3 = \frac{m v_3}{V_3} = \frac{1 \times 1.375}{715.5} = 1.92 \times 10^{-3} \text{ m}^2 = 1920 \text{ mm}^2 \quad \textbf{Ans.}$$

21.12. Supersaturated Flow or Metastable Flow through Nozzle

When dry saturated steam is expanded adiabatically or isentropically, it becomes wet and is shown by a vertical line on Mollier diagram.

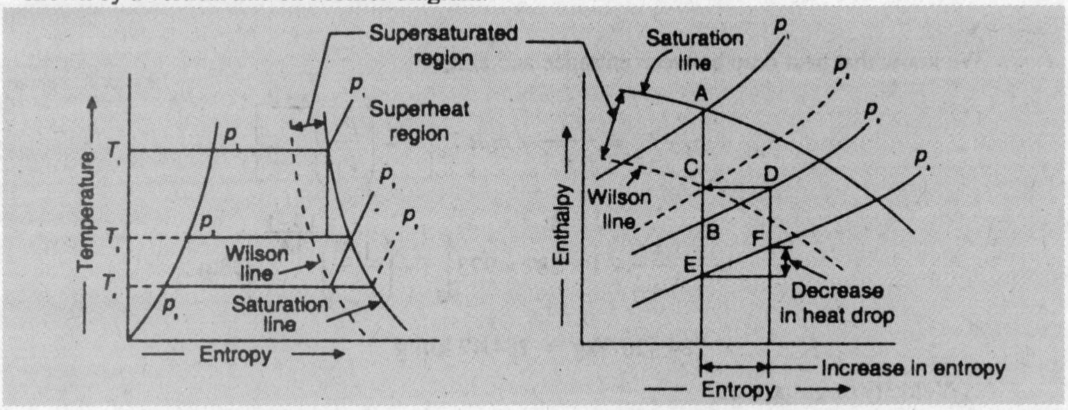

(a) T-s diagram (b) h-s diagram

Fig. 21.11 Supersaturated flow on T-s and h-s diagram

We have already discussed that expansion of steam in an ideal nozzle is isentropic, which is accompanied by condensation process. If the steam is initially superheated, the condensation should start after it has become dry saturated. This is possible when the steam has proceeded through some distance in the nozzle and in a short interval of time. But from practical point of view, the steam has a great velocity (sometimes sonic and even supersonic). Thus the phenomenon of condensation does not take place at the expected rate. As a result of this, equilibrium between the liquid and vapour phase is delayed and the steam continues to expand in a dry state. The steam in such a set of conditions, is said to be ***supersaturated*** or in ***metastable state.*** It is also called ***supercooled steam***, as its temperature at any pressure is less than the saturation temperature corresponding to the pressure. The flow of supersaturated steam, through the nozzle is called ***supersaturated flow*** or ***metastable flow***.

---

* We know that heat drop lost in friction

$$= \text{Mass} \times \text{Sp. heat} \times \text{Increase in temp.}$$

$$\therefore \quad \text{Increase in temp} = \frac{\text{Heat drop lost in friction}}{\text{Mass} \times \text{Sp. heat}} = \frac{28.442}{1 \times 1} = 28.442 \text{ K}$$

(∵ Sp. heat is taken as 1 kJ/kg K)

Experiments of supersaturated flow of steam have shown that there is a limit to which the supersaturated flow is possible. This limit is represented by *Wilson line on $T-s$ and h-s diagram as shown in Fig. 21.11 (*a*) and (*b*) respectively. It may be noted that the Willson line closely follows the 0.97 dryness fraction line. Beyond this Wilson line, there is no supersaturation. The steam suddenly condenses and restores its normal equilibrium state.

In Fig. 21.11 (*b*) is shown the isentropic expansion of steam in a nozzle. The point A represents the position of initial dry saturated steam at pressure p_1. The line AC represents the isentropic expansion of steam in the supersaturated region. The metastable state (point C) is obtained by drawing a vertical line through A to meet the Wilson line. At C, the steam condenses suddenly. The line CD represents the condensation of steam at constant enthalpy. The point D is obtained by drawing a horizontal line through C to meet the throat pressure (p_2) of the nozzle. The line DF represents the isentropic expansion of steam in the divergent portion.

Notes : 1. The same theory is applicable, if the steam is initially superheated.

2. The difference of supersaturated temperature and saturation temperature at that pressure is known as *degree of undercooling*. Mathematically, degree of undercooling

$$= T_2 - T_2'$$

3. The ratio of pressures corresponding to temperatures T_2 and T_2' is known as *degree of supersaturation*. Mathematically, degree of supersaturation

$$= \frac{\text{Pressure corresponding to } T_2}{\text{Pressure corresponding to } T_2'} = \frac{p_2}{p_2'}$$

4. The following relations may be used in solving problem on supersaturated flow.

(*i*) $v = \dfrac{0.0023\,(h-1943)}{p}$ (*ii*) $pv^{1.3}$ = Constant ; and (*iii*) $\dfrac{p}{T^{13/3}}$ = Constant

where v = Volume of steam in m^3/kg,

p = Pressure of steam in bar,

h = Enthalpy or total heat of steam in kJ/kg, and

T = Absolute temperature of supersaturated steam in K.

21.13. Effects of Supersaturation

The following effects in a nozzle, in which supersaturation occurs, are important from the subject point of view ;

1. Since the condensation does not take place during supersaturated expansion , so the temperature at which the supersaturation occurs will be *less* than the saturation temperature corresponding to the pressure. Therefore, the density of supersaturated steam will be more** than for the equilibrium conditions, which gives the increase in the mass of steam discharged.

2. The supersaturation increases the entropy and specific volume of the steam.

3. The supersaturation reduces the heat drop (for the same pressure limits) below that for thermal equilibrium. Hence the exit velocity of the steam is reduced.

4. The supersaturation increases dryness fraction of steam.

Example 21.15. *The dry saturated steam is expanded in a nozzle from pressure of 10 bar to a pressure of 5 bar. If the expansion is supersaturated, find : 1. the degree of undercooling ; and 2. the degree of supersaturation.*

---

* The limit of supersaturated expansion was first shown by the experiments done by C.T.R. Wilson in 1857. The subsequent work by H.M. Martin has enabled a curve which was termed by him as the Wilson line.

** It has been found that the density of supersaturated steam is about *eight times* that of the ordinary saturated vapour at the corresponding pressure.

Solution. Given : $p_1 = 10$ bar ; $p_2 = 5$ bar

1. *Degree of undercooling*

From steam tables, corresponding to a pressure of 10 bar, we find that the initial temperature of steam,

$$T_1 = 179.9° \text{C} = 179.9 + 273 = 452.9 \text{ K}$$

Let $T_2' =$ Temperatureat which supersaturation occurs.

We know that for supersaturated expansion,

$$\frac{p_1}{(T_1)^{13/3}} = \frac{p_2}{(T_2')^{13/3}} \quad \text{or} \quad \frac{T_2'}{T_1} = \left(\frac{p_2}{p_1}\right)^{3/13} = \left(\frac{5}{10}\right)^{3/13} = 0.852$$

$$\therefore \quad T_2' = T_1 \times 0.852 = 452.9 \times 0.852 = 385.9 \text{ K}$$

$$= 385.9 - 273 = 112.9° \text{C}$$

From steam tables, corresponding to a pressure of 5 bar, we find that the saturation temperature,

$$T_2 = 151.9° \text{C}$$

$\therefore$ Degree of undercooling $= T_2 - T_2' = 151.9 - 112.9 = 39°$ C **Ans.**

2. *Degree of supersaturation*

From steam tables, corresponding to a temperature of 112.9° C, we find that

$$p_2' = 1.584 \text{ bar}$$

$\therefore$ Degree of supersaturation

$$= p_2 / p_2' = 5 / 1.584 = 3.16 \textbf{ Ans.}$$

Example 21.16. *Find the percentage increase in discharge from a convergent-divergent nozzle expanding steam from 8.75 bar dry to 2 bar, when ; 1. the expansion is taking place under thermal equilibrium, and 2. the steam is in metastable state during part of its expansion.*

Take area of nozzle as 2500 mm².

Solution. Given : $p_1 = 8.75$ bar ; $p_2 = 2$ bar ; $A_2 = 2500 \text{ mm}^2 = 2500 \times 10^{-6} \text{ m}^2$

1. *Mass of steam discharged when the expansion is under thermal equilibrium*

Let $m_1 =$ Mass of steam discharged.

The expansion of steam under conditions of *thermal equilibrium is shown on Mollier diagram as in Fig. 21.12.

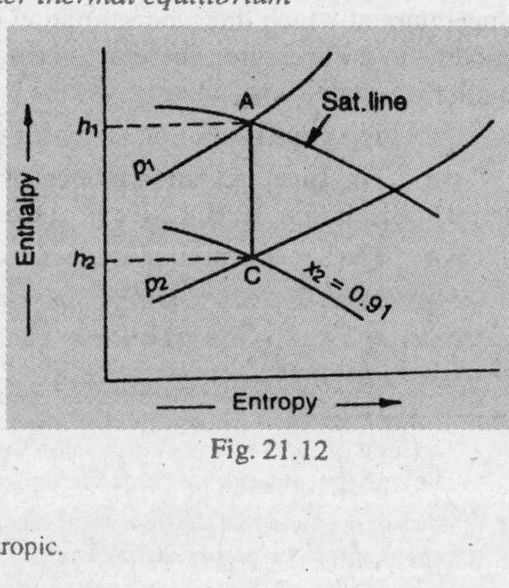

Fig. 21.12

From Mollier diagram, we find that

$h_1 = 2770$ kJ/kg ; $h_2 = 2515$ kJ/kg ; and $x_2 = 0.91$

From steam tables, at a pressure of 2 bar, we find that the specific volume of steam at exit,

$$v_{g2} = 0.885 \text{ m}^3/\text{kg}$$

* Thermal equilibrium means that the flow of steam of isentropic.

We know that heat drop from inlet to exit,

$$h_{d2} = h_1 - h_2 = 2770 - 2515 = 255 \text{ kJ/kg}$$

∴ Velocity of steam at exit,

$$V_2 = 44.72\sqrt{h_{d2}} = 44.72\sqrt{255} = 714 \text{ m/s}$$

and $$m_1 = \frac{A_2 V_2}{v_2} = \frac{A_2 V_2}{x_2 v_{g2}} = \frac{2500 \times 10^{-6} \times 714}{0.91 \times 0.885} = 2.21 \text{ kg/s} \quad \textbf{Ans.}$$

2. *Mass of steam discharged when it is in *metastable state*

Let m_2 = Mass of steam discharged.

We know that volume of steam at inlet,

$$v_1 = \frac{0.0023\,(h_1 - 1943)}{p_1} = \frac{0.0023\,(2770 - 1943)}{8.75} = 0.217 \text{ m}^3/\text{kg}$$

and volume of steam at exit,

$$v_2 = v_1 \left(\frac{p_1}{p_2}\right)^{1/1.3} = 0.217 \left(\frac{8.75}{2}\right)^{1/1.3} = 0.675 \text{ m}^3/\text{kg}$$

$$\ldots (\because p_1 v_1^{1.3} = p_2 v_2^{1.3})$$

We know that volume of steam at exit (v_2),

$$0.675 = \frac{0.0023\,(h_2 - 1943)}{p_2} = \frac{0.0023\,(h_2 - 1943)}{2}$$

$$\therefore \quad h_2 = 2530 \text{ kJ/kg}$$

We know that heat drop from inlet to exit,

$$h_{d2} = h_1 - h_2 = 2770 - 2530 = 240 \text{ kJ/kg}$$

∴ Velocity of steam at exit,

$$V_2 = 44.72\sqrt{h_{d2}} = 44.72\sqrt{240} = 693 \text{ m/s}$$

and $$m_2 = \frac{A_2 V_2}{v_2} = \frac{2500 \times 10^{-6} \times 693}{0.675} = 2.57 \text{ kg/s} \quad \textbf{Ans.}$$

∴ Percentage increase in discharge

$$= \frac{m_2 - m_1}{m_1} = \frac{2.57 - 2.21}{2.21} = 0.163 \text{ or } 16.3\,\% \quad \textbf{Ans.}$$

21.14. Steam Injector

The principle of a steam nozzle may also be applied to a steam injector. It utilises the kinetic energy of a steam jet for increasing the pressure and velocity of water. It is mostly used for forcing the feed water into steam boilers under pressure. The action of a steam injector is shown in Fig. 21.13.

The steam from the boiler is expanded to a high velocity by passing it through a convergent nozzle A. The steam jet enters the mixing cone and imparts its momentum to the incoming water supply from the feed tank** The cold water causes the steam to condense. The resulting jet at B,

---

* The problems on metastable flow cannot be solved by Mollier diagram unless Wilson line is drawn.

** The feed tank may be above or below the level of the steam injector.

formed by the steam and water is at atmospheric pressure, and has a large velocity. The mixture then enters delivery pipe at C through a diverging cone or diffuser, in which the kinetic energy is reduced and converted into pressure energy. This pressure energy is sufficient to overcome the boiler pressure

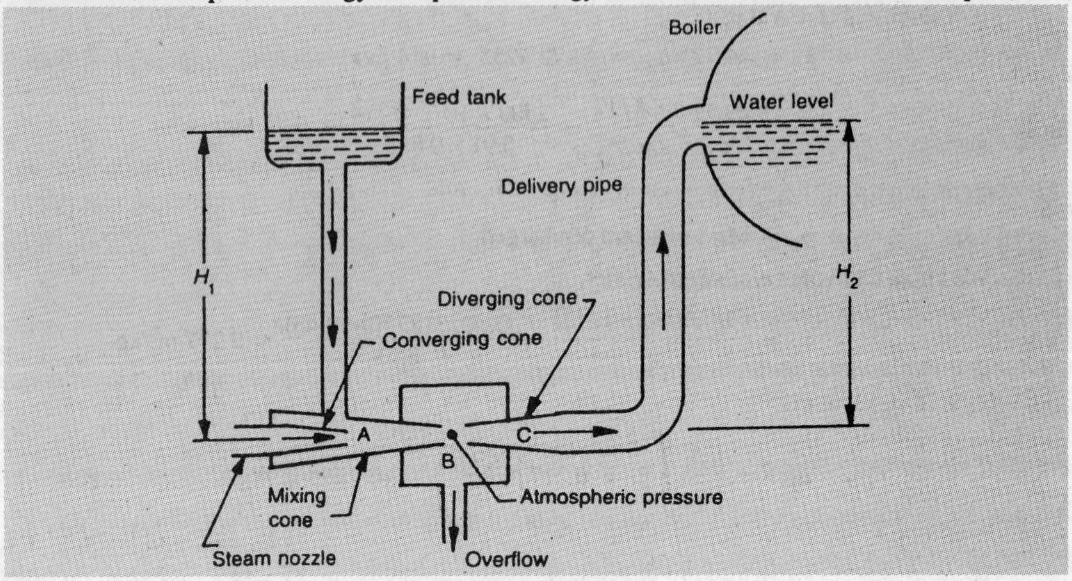

Fig. 21.13 Steam injector

and to lift the water through a height H_2. The pressure of water on leaving the delivery pipe must be about 20% higher than the boiler pressure in order to overcome all resistances. The gap between the mixing cone and diverging cone is provided with an outlet through which any excess water may overflow during the starting of the injector.

21.15. Steam Injector Calculations

The following calculations of a steam injector are important from the subject point of view :

1. *Amount of water injected*

Let m_w = Mass of water entering the mixing cone in kg/kg of steam,

V_s = Velocity of steam leaving the converging cone or nozzle,

V_w = Velocity of water entering the mixing cone, and

V_m = Velocity of the mixture leaving the mixing cone.

According to the principle of conservation of momentum,

Momentum of steam + Momentum of water = Momentum of mixture

$$1 \times V_s + m_w V_w = (1 + m_w) V_m = V_m + m_w V_m \quad \ldots (i)$$

or

$$V_s - V_m = m_w (V_m - V_w)$$

$$\therefore \quad m_w = \frac{(V_s - V_m)}{(V_m - V_w)} \text{ kg/kg of steam}$$

Note : In case the feed tank is below the level of the steam injector, the equation (*i*) may be written as :

$$1 \times V_s - m_w V_w = (1 + m_w) V_m$$

or

$$m_w = \frac{(V_s - V_m)}{(V_m + V_w)} \text{ kg/kg ofsteam}$$

2. *Velocity of steam leaving the nozzle*

Let p_1 = Initial pressure of steam,

p_2 = Pressure of steam leaving the nozzle at *A*, and

h_d = Isentropic heat drop.

∴ Velocity of steam leaving the nozzle,

$$V_s = 44.72\sqrt{h_d}$$

3. *Velocity of water entering the mixing cone*

Let H_1 = Height of water in the feed tank from the level of steam injector.

∴ Velocity of water entering the mixing cone,

$$V_w = \sqrt{2gH_1}$$

4. *Velocity of mixture leaving the mixing cone*

Let p_1 = Pressure of steam in the boiler in bar,

p_m = Pressure of the mixture leaving the mixing cone at *B* in bar,

ρ = Density of the mixture at *B* in kg/m³, and

V_m = Velocity of mixture at *B*.

∴ Total energy per kg of water at *B*

$$= \frac{10^5 p_m}{\rho} + \frac{V_m^2}{2} \text{ (in joules)} \quad \ldots (i)$$

This energy must be sufficient to lift the water through a height H_2 metres and inject it into the boiler. The final pressure head on leaving at *C*, must be somewhat greater than the height H_2 *plus* the boiler pressure. If *H* is the necessary excess head in metres, then

Total energy per kg of water at *B*

$$= \frac{10^5 p_1}{\rho} + g\,(H_2 + H) \text{ (in joules)} \quad \ldots (ii)$$

Equating equations (*i*) and (*ii*),

$$\frac{10^5 p_m}{\rho} + \frac{V_m^2}{2} = \frac{10^5 p_1}{\rho} + g\,(H_2 + H)$$

or

$$V_m = \sqrt{\frac{2 \times 10^5\,(p_1 - p_m)}{\rho} + 2g\,(H_2 + H)} \quad \ldots (iii)$$

We have already discussed that the pressure of the mixture leaving the mixing cone is atmospheric. Therefore, taking the value of p_m as 1.013 bar and assuming the density of the mixture as 1000 kg/m³ (equal to density of water), we get

$$V_m = \sqrt{\frac{2 \times 10^5\,(p_1 - 1.013)}{1000} + 2g\,(H_2 + H)}$$

$$= 4.43\sqrt{10.2\,(p_1 - 1.013) + H_2 + H}$$

Note : If V_d is the velocity in the delivery pipe, then

$$H = \frac{V_d^2}{2g}$$

5. *Nozzle areas*

Let A_a = Area of steam nozzle at A,

A_b = Area of combining nozzle (or mixing cone) at B,

V_s = Velocity of steam leaving the converging cone,

v_a = Specific volume of steam after expansion in nozzle at A,

m = Mass of water required to be delivered in kg/s, and

m_w = Mass of water entering the mixing cone in kg/kg of steam.

∴ Mass of steam supplied,

$$m_s = \frac{A_a V_s}{v_a} = \frac{m}{m_w}$$

or

$$A_a = \frac{m}{m_w} \times \frac{v_a}{V_s}$$

Also

$$m_s + m = A_b V_m \rho$$

or

$$A_b = \frac{m_s + m}{1000 V_m} = \frac{m\left(1 + \frac{1}{m_w}\right)}{1000 V_m} \qquad \ldots (\because \rho = 1000 \text{ kg/m}^3)$$

6. *Heat balance per kg of steam*

Let h = Enthalpy or total heat of steam entering the injector in kJ,

t_w = Temperature of water in feed tank in °C,

h_{fw} = Sensible heat of water supplied to the injector, corresponding to a temperature of t_w in kJ/kg,

t_w = Temperature of water leaving the mixing cone at B in °C, and

h_{fm} = Sensible heat of water leaving the mixing cone at B, correspondir to a temperature of t_m in kJ/kg.

Then heat supplied in steam + Heat supplied in water + Kinetic energy of water

= Heat in mixture + Kinetic energy of mixture

$$h + m_w h_{fw} \pm \frac{m_w V_w^2}{2000} = (1 + m_w) h_{fm} + \frac{(1 + m_w) V_m^2}{2000}$$

From this equation, the value of h_{fm} may be determined and hence the temperature of the mixture t_m is known.

Note : In the above equation *plus* sign is used when feed tank is above the level of the steam injector, while *negative* sign is used when it is below the level of injector.

Example 21.17. *An injector is required to deliver 120 kg of water per minute from a tank, whose constant water level is 3 m below the level of injector, into a boiler in which the steam pressure is 15 bar. The water level in the boiler is 0.7 metre above the level of the injector. The steam for the injector is taken from the same boiler and it is assumed to be dry and saturated. The pressure of steam leaving steam nozzle is 0.6 times that of the supply pressure. The temperature of the water in the feed tank is 25° C. If the velocity in the delivery pipe is 15 m/s, find :*

1. Mass of water injected per kg of steam, 2. Area of mixing cone, 3. Area of steam nozzle, and 4. Temperature of water leaving the injector.

Solution. Given : $m = 120$ kg/min $= 2$ kg/s ; $H_1 = 3$ m below the level of injector ; $p_1 = 15$ bar ; $H_2 = 0.7$ m above the level of injector ; $p_2 = 0.6\,p_1 = 0.6 \times 15 = 9$ bar ; $t_w = 25°$ C ; $V_d = 15$ m/s

1. *Mass of water injected per kg of steam*

From Mollier chart, the isentropic heat drop between pressure 15 bar dry and 9 bar,

$$h_d = h_1 - h_2 = 2795 - 2700 = 95 \text{ kJ/kg}$$

and dryness fraction of steam after expansion,

$$x_2 = 0.965$$

We know that velocity of steam leaving the nozzle,

$$V_s = 44.72\sqrt{h_d} = 44.72\sqrt{95} = 436 \text{ m/s}$$

We know that velocity of water entering the mixing cone,

$$V_w = \sqrt{2gH_1} = \sqrt{2 \times 9.81 \times 3} = 7.67 \text{ m/s}$$

and velocity of mixture leaving the mixing cone,

$$V_m = 4.43\sqrt{10.2\,(p_1 - 1.013) + H_2 + H}$$

$$= 4.43\sqrt{10.2\,(15 - 1.013) + 0.7 + 11.47} = 55 \text{ m/s}$$

$$\ldots \left[\because H = \frac{V_d^2}{2g} = \frac{15^2}{2 \times 9.81} = 11.47 \text{ m}\right]$$

$\therefore$ Mass of water injected per kg of steam,

$$m_w = \frac{V_s - V_m}{V_m + V_w} = \frac{436 - 55}{55 + 7.67} = 6.08 \text{ kg} \quad \textbf{Ans.}$$

2. *Area of mixing cone*

We know that area of the mixing cone

$$A_b = \frac{m\left(1 + \dfrac{1}{m_w}\right)}{1000\,V_m} = \frac{2\left(1 + \dfrac{1}{6.08}\right)}{1000 \times 55} = 42.3 \times 10^{-6} \text{ m}^2$$

$$= 42.3 \text{ mm}^2 \quad \textbf{Ans.}$$

3. *Area of steam nozzle*

From steam tables, corresponding to a pressure of 9 bar, we find that specific volume of steam,

$$v_{ga} = 0.2148 \text{ m}^3\text{/kg}$$

We know that area of steam nozzle,

$$A_a = \frac{m}{m_w} \times \frac{v_a}{V_s} = \frac{m}{m_w} \times \frac{x_2\,v_{ga}}{V_s} = \frac{2}{6.08} \times \frac{0.965 \times 0.2148}{436}$$

$$= 156 \times 10^{-6} \text{ m}^2 = 156 \text{ mm}^2 \quad \textbf{Ans.}$$

4. *Temperature of water leaving the injector*

From steam tables, corresponding to a temperature of $t_w = 25°$ C, we find that

$$h_{fw} = 104.8 \text{ kJ/kg}$$

We know that $\quad h_1 + m_w\,h_{fw} - \dfrac{m_w\,V_w^2}{2000} = (1 + m_w)\,h_{fm} + \dfrac{(1 + m_w)\,V_m^2}{2000}$

$$2795 + 6.08 \times 104.8 - \frac{6.08\,(7.67)^2}{2000} = (1 + 6.08)\, h_{fm} + \frac{(1 + 6.08)\,(55)^2}{2000}$$

$$3432 = 7.08\, h_{fm} + 10.71$$

or $$h_{fm} = 483.2 \text{ kJ/kg}$$

∴ Temperature of water leaving the injector (from steam tables corresponding to 483.2 kJ/kg),

$$t_m = 115^\circ \text{ C} \quad \textbf{Ans.}$$

EXERCISES

1. The dry and saturated steam at a pressure of 5 bar is expanded isentropically in a nozzle to a pressure of 0.2 bar. Find the velocity of steam leaving the nozzle. **[Ans.** 1000 m/s]

2. The dry and saturated steam at a pressure of 10.5 bar is expanded isentropically in a nozzle to a pressure of 0.7 bar. Determine the final velocity of the steam issuing from the nozzle, when (*a*) friction is neglected, and (*b*) 10% of the heat drop is lost in friction.

   The initial velocity of steam may be neglected. **[Ans.** (*a*) 905 m/s; (*b*) 859 m/s]

3. Steam at a pressure of 6.3 bar and 200° C is expanded in a nozzle to a pressure of 0.2 bar. Find the final velocity and dryness fraction of steam, if

   (*a*) friction is neglected ; and (*b*) 10% of the heat drop is lost in friction.

   [Ans. (*a*) 1039 m/s, 0.83 ; (*b*) 996 m/s, 0.852]

4. Steam is supplied to a nozzle at 3.5 bar and 0.96 dry. The steam enters the nozzle at 240 m/s. The pressure drops to 0.8 bar. Determine the velocity and dryness fraction of the steam when it leaves the nozzle.

   [Ans. 545.5 m/s ; 0.92]

5. Steam expands through an ideally, insulated nozzle following a reversible polytropic law $p\,v^{1.2} = C$. There is no change in potential energy but the pressure drops from 20 bar to 2 bar and the specific volume increases from 0.05 m^3 to 0.3 m^3. If the entrance velocity is 80 m/s, determine the exit velocity.

   [Ans. 697.5 m/s]

6. Calculate the throat area of nozzle supplied with steam at 10 bar and 200°C. The rate of flow of steam is 1.2 kg/s. Neglect friction and assume the velocity at inlet to be small. **[Ans.** 837 mm^2]

7. Steam expands isentropically from the state of 8 bar and 250° C to 1.5 bar in a convergent-divergent nozzle. The steam flow rate is 0.75 kg/s. Find : 1. the velocity of steam at exit from the nozzle ; and 2. the exit area of nozzle. Neglect the inlet velocity of steam. **[Ans.** 800 m/s ; 1054 mm^2]

8. Steam enters a group of convergent-divergent nozzles at a pressure of 22 bar and with a temperature of 240° C. The exit pressure is 4 bar and 9% of the total heat drop is lost in friction. The mass flow rate is 10 kg/s and the flow upto the throat may be assumed friction less. Calculate

   1. the throat and exit velocities, and 2. the throat and exit areas.

   [Ans. 529 m/s, 775 m/s ; 3000 mm^2, 5500 mm^2]

9. The throat diameter of a nozzle is 5 mm. If dry and saturated steam at 10 bar is supplied to the nozzle, calculate the mass flow per second. The exhaust pressure is 1.5 bar. Assume friction less adiabatic flow and index of expansion, $n = 1.135$.

   If 10 percent of the isentropic heat drop is lost in friction, what should be the correct diameter at outlet for steam to issue at the same exhaust pressure ? **[Ans.** 103.3 kg/h ; 7.12 mm]

10. Calculate the throat and exit diameters of a convergent divergent nozzle which will discharge 820 kg of steam per hour from a pressure of 8 bar superheated to 220° C into a chamber having a pressure of 1.5 bar. The friction loss in the divergent part of the nozzle may be taken as 0.15 of the total enthalpy drop.

    [Ans. 15.2 mm ; 20.6 mm]

11. A steam turbine develops 185 kW with a consumption of 16.5 kg/kW/h. The pressure and temperature of the steam entering the nozzle are 12 bar and 220° C. The steam leaves the nozzle at 1.2 bar. The diameter of the nozzle at throat is 7 mm. Find the number of nozzles.

If 8% of the total enthalpy drop is lost in friction in the diverging part of the nozzle, determine the diameter at the exit of the nozzle and the exit velocity of the leaving steam.

Sketch the skeleton Mollier diagram and show on it the values of pressure, temperature or dryness fraction, enthalpy and specific volume at inlet, throat and exit. [Ans. 14 ; 11.1 mm ; 847 m/s]

12. Steam expands in a nozzle under the following conditions : Inlet pressure = 15 bar ; Inlet temperature = 250° C; Final pressure = 4 bar ; Mass flow = 1 kg/s.

Calculate the required throat and exit areas, using Mollier diagram, when 1. the expansion is frictionless, and 2. the friction loss at any pressure amounts to 10 percent of the total heat drop down to that pressure. [Ans. 480 mm^2, 606 mm^2 ; 508 mm^2, 650 mm^2]

13. Gases expand in a convergent divergent nozzle from 3.6 bar and 425° C to a back pressure of 1 bar, at the rate of 18 kg/s. If the nozzle efficiency is 0.92, calculate the required throat and exit areas of the nozzle. Neglect inlet velocity and friction in the convergent part. For the gases, take c_p = 1.113 kJ/kg K and γ = 1.33. [Ans. 0.0325 m^2 ; 0.04 m^2]

14. The dry saturated steam expands in a nozzle from a pressure of 2 bar to 1 bar. If the expansion is supersaturated, determine the degree of undercooling and the degree of supersaturation. [Ans. 37.63°C ; 4.58]

15. Steam at 42 bar and 260° C enters a nozzle and leaves at 28 bar. Neglecting initial kinetic energy and considering super-saturation, determine the discharge area for a flow of 10 350 kg/h and a nozzle velocity coefficient of 96%. [Ans. 484 mm^2]

16. Compare the mass of discharge from a convergent-divergent nozzle expanding from 8 bar and 210° C to 2 bar, when

1. the expansion takes place under thermal equilibrium, and 2. the steam is in super-saturated condition during a part of its expansion.

Take area of nozzle as 2400 mm^2. [Ans. 8.3%]

17. An injector is to deliver 100 kg of water per minute from a tank, whose constant water level is 1.2 m below the level of the injector into a boiler in which the steam pressure is 14 bar. The water level in the boiler is 1.5 metre above the level of the injector. The steam for the injector is taken from the same boiler and it is assumed to be dry and saturated. The pressure of steam leaving steam nozzle is 0.5 times that of the supply pressure. If the velocity in the delivery pipe is 13.5 m/s, find :

1. Mass of water pumped per kg of steam ; 2. Area of mixing cone ; 3. Area of steam nozzle ; and 4. Temperature of water leaving the injector, if the temperature of water in the feed tank is 15° C.

[Ans. 7.9 kg ; 35.5 mm^2 ; 107.4 mm^2 ; 87.8° C]

QUESTIONS

1. Explain the function of nozzles used with steam turbines.
2. Discuss the functions of the convergent portion, the throat and the divergent portion of a convergent-divergent nozzle with reference to flow of steam.
3. What is steady flow energy equation as applied to steam nozzles ? Explain its use in the calculation of steam velocity at the exit of a nozzle.
4. Discuss the effect of friction during the expansion of steam through a convergent-divergent nozzle when

   (*i*) the steam at entry to the nozzle is saturated, and

   (*ii*) the steam at entry is superheated.

   Assume the pressure of steam to be initially same in both the cases. Mark the processes on a sketch of enthalpy-entropy diagram.
5. Explain what is meant by critical pressure ratio of a nozzle.
6. Starting from fundamentals, show that for maximum discharge through a nozzle, the ratio of throat pressure to inlet pressure is given by $\left(\frac{2}{n+1}\right)^{\frac{n}{n-1}}$ where n is the index for isentropic expansion through the nozzle.

7. Derive an expression for maximum discharge through convergent divergent nozzle for steam.

8. Draw the 'discharge' versus 'ratio of pressures at outlet to inlet' curve for a convergent steam nozzle. Discuss the physical significance of critical pressure ratio.

9. Explain the supersaturated or metastable flow of steam through a nozzle and the significance of Wilson's line.

10. What are the effects of supersaturation on discharge and heat drop ?

OBJECTIVE TYPE QUESTIONS

1. The steam leaves the nozzle at a

   (*a*) high pressure and low velocity (*b*) high pressure and high velocity
   (*c*) low pressure and low velocity (*d*) low pressure and high velocity

2. The effect of friction in a nozzle ...... dryness fraction of steam.

   (*a*) increases (*b*) decreases

3. The velocity of steam leaving the nozzle (*V*) is given by

   (*a*) $V = 44.72\,K\,h_d$ (*b*) $V = 44.72\,K\sqrt{h_d}$
   (*c*) $V = 44.72\sqrt{K\,h_d}$ (*d*) $V = 44.72\,h_d\sqrt{K}$

where K = Nozzle coefficient, and
h_d = Enthalpy drop during expansion.

4. The critical pressure ratio for initially dry saturated steam is

   (*a*) 0.528 (*b*) 0.546 (*c*) 0.577 (*d*) 0.582

5. The critical pressure ratio for initially superheated steam is ...... as compared to initially dry saturated steam.

   (*a*) more (*b*) less

6. The flow of steam is super sonic

   (*a*) at the entrance to the nozzle (*b*) at the throat of the nozzle
   (*c*) in the convergent portion of the nozzle (*d*) in the divergent portion of the nozzle

7. The difference of supersaturated temperature and saturation temperature at that pressure is known as

   (*a*) degree of supersaturation (*b*) degree of superheat
   (*c*) degree of undercooling (*d*) none of these

8. In a nozzle, the effect of supersaturation is to

   (*a*) decrease the dryness fraction of steam (*b*) decrease the specific volume of steam
   (*c*) increase the entropy (*d*) increase the enthalpy drop

9. The density of supersaturated steam is about .... that of the ordinary saturated vapour at the corresponding pressure.

   (*a*) same as (*b*) 2 times (*c*) 4 times (*d*) 8 times

10. When the back pressure of a nozzle is below the designed value of pressure at exit of nozzle, the nozzle is said to be

   (*a*) choked (*b*) under damping (*c*) over damping

ANSWERS

1. (*d*)	2. (*a*)	3. (*c*)	4. (*c*)	5. (*b*)
6. (*d*)	7. (*c*)	8. (*c*)	9. (*d*)	10. (*b*)

22

Impulse Turbines

1. Introduction. 2. Advantages of Steam Turbines over Reciprocating Steam Engines. 3. Classification of Steam Turbines. 4. Impulse Turbine. 5. De-laval Impulse Turbine. 6. Pressure and Velocity of Steam in an Impulse Turbine. 7. Velocity Triangles for Moving Blade of an Impulse Turbine. 8. Combined Velocity Triangle for Moving Blade. 9. Power Produced by an Impulse Turbine. 10. Effect of Friction on the Combined Velocity Triangle. 11. Combined Velocity Diagram for Axial Discharge. 12. Velocity Diagram for Two Stage Impulse Turbine.

22.1. Introduction

A steam turbine is a prime mover in which rotary motion is obtained by the gradual change of momentum of the steam. We have already discussed that in a reciprocating steam engines, the steam acts on the piston, as a load or weight, *i.e.*, the action of steam is *static*. The statical pressure of steam gives to and fro motion to the piston, and conversion of energy takes place through crank and connecting rod.

In a steam turbine, the force exerted on the blades is due to the velocity of steam. This is due to the fact that the curved blades by changing the direction of steam receive a force or impulse. The action of steam in this case is said to be *dynamic*. Thus, the dynamical pressure of steam rotates the vanes, buckets or blades directly. The turbine blades are curved in such a way that the steam directed upon them enters without shock, though there is always some loss of energy by the friction upon the surface of blades. In general, a steam turbine, essentially, consists of the following two parts :

1. The nozzle in which the heat energy of high pressure steam is converted into kinetic energy, so that the steam issues from the nozzle with a very high velocity.

2. The blades which change the direction of steam issuing from the nozzle, so that a force acts on the blades due to change of momentum and propel them.

Thus, the basic principle of operation of a steam turbine is the generation of high velocity steam jet by the expansion of high pressure steam and then conversion of kinetic energy, so obtained into mechanical work on rotor blades.

22.2. Advantages of Steam Turbines over Reciprocating Steam Engines

Following are the important advantages of steam turbines over reciprocating steam engines :

1. A steam turbine may develop higher speeds and a greater steam range is possible.
2. The efficiency of a steam turbine is higher.
3. The steam consumption is less.
4. Since all the moving parts are enclosed in a casing, the steam turbine is comparatively safe.
5. A steam turbine requires less space and lighter foundations, as there are little vibrations.
6. There is less frictional loss due to fewer sliding parts.

7. The applied torque is more uniform to the driven shaft.
8. A steam turbine requires less attention during running. Moreover, the repair costs are generally less.

22.3. Classification of Steam Turbines

The steam turbines may be classified into the following types :

1. *According to the mode of steam action*

   (*i*) Impulse turbine, and (*ii*) Reaction turbine.
2. *According to the direction of steam flow*

   (*i*) Axial flow turbine, and (*ii*) Radial flow turbine.
3. *According to the exhaust condition of steam*

   (*i*) Condensing turbine, and (*ii*) Non-condensing turbine.
4. *According to the pressure of steam*

   (*i*) High pressure turbine, (*ii*) Medium pressure turbine, and (*iii*) Low pressure turbine.
5. *According to the number of stages*

   (*i*) Single stage turbine, and (*ii*) Multi-stage turbine.

In this chapter, we shall discuss impulse turbines only. All the above mentioned other steam turbines will be discussed at the appropriate places in the book.

22.4. Impulse Turbine

An *impulse turbine, as the name indicates, is a turbine which runs by the impulse of steam jet In this turbine, the steam is first made to flow through a nozzle. Then the steam jet impinges on the turbine blades (which are curved like buckets) and are mounted on the circumference of the wheel. The steam jet after impinging glides over the concave surface of the blades and finally leave the turbine.

Note : The action of the jet of steam, impinging on the blades, is said to be an *impulse* and the rotation of the rotor is due to the impulsive forces of the steam jets.

22.5. De-Level Impulse Turbine

A **De-Level turbine is the simplest type of impulse steam turbine, and is commonly used. It has the following main components :

1. *Nozzle.* It is a circular guide mechanism, which guides the steam to flow at the designed direction and velocity. It also regulates the flow of steam. The nozzle is kept very close to the blades, in order to minimise the losses due to windage.

2. *Runner and blades.* The runner of a De-Laval impulse turbine essentially consists of a circular disc fixed to a horizontal shaft. On the periphery of the runner, a number of blades are fixed uniformly. The steam jet impinges on the buckets, which move in the direction of the jet. This movement of the blades makes the runner to rotate.

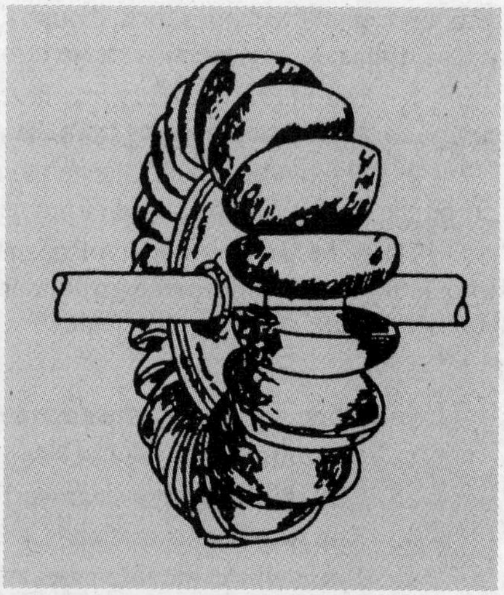

Fig. 22.1 Runner and bucket of impulse turbine.

The surface of the blades is made very smooth to minimise the frictional losses. The blades are generally

* The first impulse turbine was devised by Giovanni Branca in 1629.

** Named after the Swedish engineer De-Lavel, who devised this turbine in 1881.

made of special steel alloys. In most of the cases, the blades are bolted to the runner disc. But sometimes the blades and disc are cast as a single unit.

It has been experienced that all the blades do not wear out equally with the time. A few of them get worn out and damaged early and need replacement. This can be done only if the blades are bolted to the disc.

3. *Casing.* It is an air-tight metallic case, which contains the turbine runner and blades. It controls the movement of steam from the blades to the condenser, and does not permit it to move into the space. Moreover, it is essential to safeguard the runner against any accident.

22.6. Pressure and Velocity of Steam in an Impulse Turbine

The pressure of steam jet is reduced in the nozzle and remains constant while passing through the moving blade. The velocity of steam is increased in the nozzle, and is reduced while passing through the moving blades.

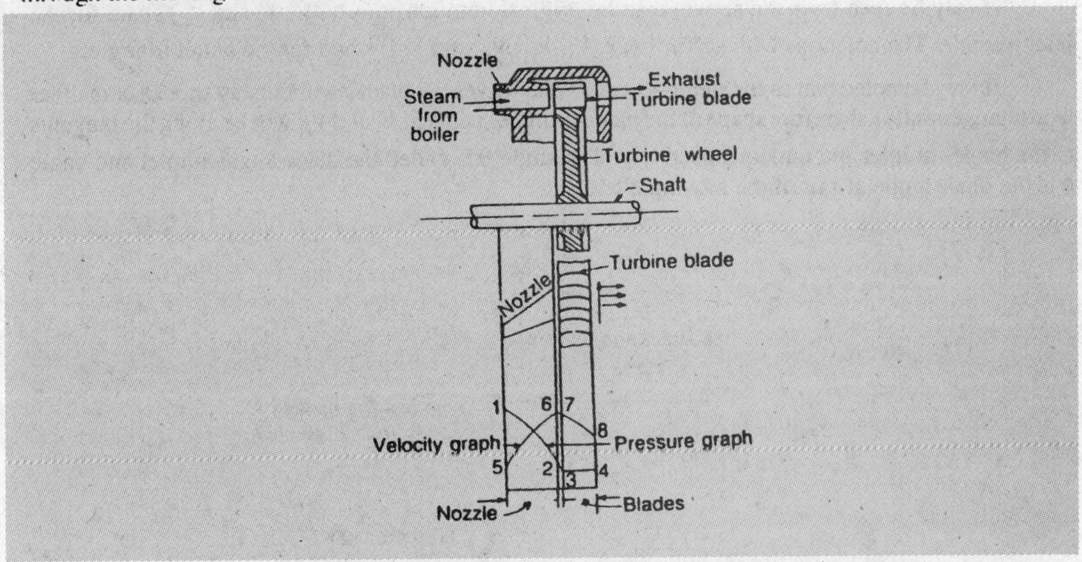

Fig. 22.2 Pressure and velocity graph of steam in a simple impulse turbine.

Fig. 22.2 shows the pressure and velocity graphs of the steam in a simple impulse turbine while it flows in the nozzle and blades. The pressure graph 1-2-3-4 represents steam pressure at entrance of the nozzle, exit of the nozzle, entrance of the blades and exit of the blades respectively. Similarly, velocity graph 5-6-7-8 represents the velocity of steam at entrance of the nozzle, exit of the nozzle, entrance of the blades and exit of the blades respectively.

22.7. Velocity Triangles for Moving Blade of an Impulse Turbine

We have already discussed that in an impulse turbine, the steam jet after leaving the nozzle impinges on one end of the blade. The jet then glides over the inside surface of the blade and finally leaves from the other edge, as shown in Fig. 22.3. It may be noted that the jet enters and leaves the blades tangentially for shockless entry and exit.

Consider a steam jet entering a curved blade after leaving the nozzle at *C*. Now let the jet glides over the inside surface and leaves the blade at *D*, as shown in Fig. 22.3. Now let us draw the velocity triangles at inlet and outlet tips of the moving blade, as shown in Fig. 22.3.

Let V_b = Linear velocity of the moving blade (*AB*),

V = Absolute velocity of steam entering the moving blade (*AC*),

V_r = Relative velocity of jet to the moving blade (*BC*). It is the vectorial difference between V_b and V.

V_f = Velocity of flow at entrance of the moving blade. It is the vertical component of V.

V_w = Velocity of whirl at entrance of the moving blade. It is the horizontal component of V.

θ = Angle which the relative velocity of jet to the moving blade (V_r) makes with the direction of motion of the blade.

α = Angle with the direction of motion of the blade at which the jet enters the blade.

$V_1, V_{r1}, V_{f1}, V_{w1}, \beta, \phi$ = Corresponding values at exit of the moving blade.

It may be seen from the above, that the original notations (*i.e.* V, V_r, V_f and V_w) stand for the inlet triangle. The notations with suffix 1 (*i.e.* V_1, V_{r1}, V_{f1} and V_{w1}) stand for the outlet triangle.

It may be noted that as the steam jet enters and leaves the blades without any shock (or in other words tangentially), therefore shape of the blades will be such that V_r and V_{r1} will be along the tangents to the blades at inlet and outlet respectively. The angle θ is called the blade angle at inlet and angle ϕ is the blade angle at exit of the moving blade.

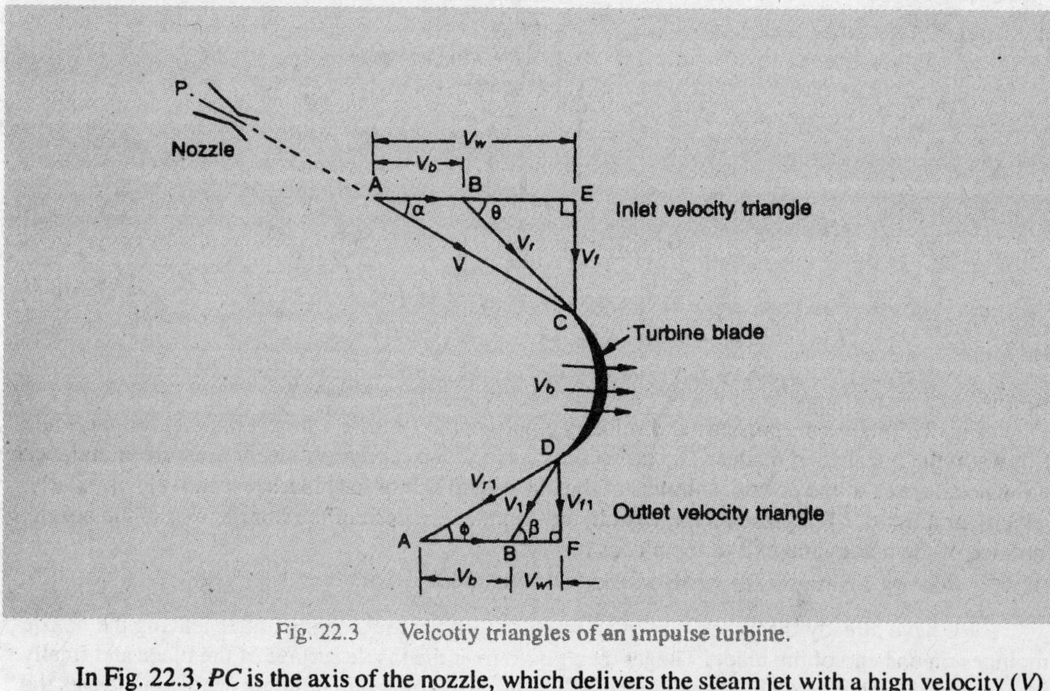

Fig. 22.3 Velcotiy triangles of an impulse turbine.

In Fig. 22.3, *PC* is the axis of the nozzle, which delivers the steam jet with a high velocity (V) at an angle α with the direction of motion of the blade. The jet impinges on a series of turbine blades mounted on the runner disc.

The axial component of V (*i.e. EC*) which does no work on the blade, is known as *velocity of flow* (V_f). It causes the steam to flow through the turbine and also an axial thrust on the rotor. The tangential component of V (represented by *BE*) is known as velocity of whirl at inlet (V_w). The linear velocity or mean velocity of the blade (*i.e.* V_b) is represented by *AB* in magnitude and direction. The length *AC* represents the relative velocity (V_r) of the steam jet with respect to the blade.

The jet now glides over and leaves the blade with a relative velocity V_{r1}, which is represented by *DA*. The absolute velocity of jet (V_1) as it leaves the blade, is represented by *DB* inclined at an angle β with the direction of the blade motion. The tangential component of V_1 (represented by *BF*) is known as velocity of whirl at exit (V_{w1}). The axial component of V_1 (represented by *DF*) is known as velocity of flow at exit (V_{f1}).

Notes : 1. The inlet triangle of velocities is represented by *BEC*, whereas the outlet triangle by *AFD*.

2. The relations between inlet and outlet velocity triangle (until and unless given) is :

$$V_r = V_{r1}$$

22.8. Combined Velocity Triangle for Moving Blades

In the last article, we have discussed the inlet and outlet velocity triangles separately. For the sake of simplification, a combined velocity triangle for the moving blade is drawn, for solving problems on steam turbines, as shown in Fig. 22.4, and as discussed below :

1. First of all, draw a horizontal line, and cut off *AB* equal to velocity of blade (V_b) to some suitable scale.

2. Now at *B*, draw a line *BC* at an angle α with *AB*. Cut off *BC* equal to *V* (*i.e.* velocity of steam jet at inlet of the blade) to the scale.

3. Join *AC*, which represents the relative velocity at inlet (V_r). Now at *A*, draw a line *AD* at an angle ϕ with *AB*.

4. Now with *A* as centre and radius equal to *AC*, draw an arc meeting the line through *A* at *D*, such that *AC* = *AD* or $V_r = V_{r1}$.

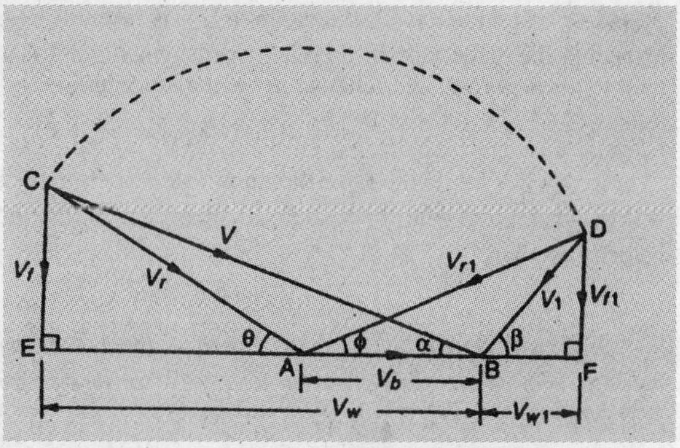

Fig. 22.4 Combined velocity triangle for an impulse turbine.

5. Join *BD*, which represents velocity of jet at exit (V_1) to the scale.

6. From *C* and *D*, draw perpendiculars meeting the line *AB* produced at *E* and *F* respectively.

7. Now *EB* and *CE* represents the velocity of whirl and velocity of flow at inlet (V_w and V_f) to the scale. Similarly, *BF* and *DF* represents the velocity of whirl and velocity of flow at outlet (V_{w1} and V_{f1}) to the scale.

22.9. Power Produced by an Impulse Turbine

Consider an impulse turbine working under the action of a steam jet. Let us draw a combined velocity triangle, for the impulse turbine as shown in Fig. 22.4.

Let m = Mass of the steam flowing through the turbine in kg/s,

$(V_w + V_{w1})$ = Change in the velocity of whirl in m/s.

We know that according to the Newton's second law of motion, force in the direction of motion of the blades,

$$F_X = \text{Mass of steam flowing per second} \times \text{Change in the velocity of whirl}$$

$$= m\,[V_w - (-V_{w1})]$$

$$= m\,[V_w + V_{w1}] = m \times EF \text{ N} \qquad \ldots (i)$$

and work done in the direction of motion of the blades

$$= \text{Force} \times \text{Distance}$$

$$= m\,[V_w + V_{w1}]\,V_b \text{ N-m/s} \quad \dots (ii)$$

$$= m \times EF \times AB \text{ N-m/s}$$

∴ Power produced by the turbine,

$$P = m \times EF \times AB \text{ watts} \quad \dots (\because 1 \text{ N-m/s} = 1 \text{ watt})$$

$$= m\,(V_w + V_{w1})\,V_b \text{ watts}$$

Similarly, we can find out the axial thrust on the wheel which is due to the difference of velocities of flow at inlet and outlet. Mathematically, axial thrust on the wheel,

$$F_Y = \text{Mass of steam flowing per second} \times \text{Change in the velocity of flow}$$

$$= m\,(V_f - V_{f1}) = m\,(CE - DF) \text{ N} \quad \dots (iii)$$

Notes : 1. In equation (*i*), the value of V_{w1} is taken as *negative* because of the opposite direction of V_w with respect to the blade motion. In other words, when point *F* in the velocity diagram lies on the right of point *B*, then V_{w1} is *negative*. Thus change in velocity of whirl,

$$= V_w - (-V_{w1}) = V_w + V_{w1}$$

2. If V_{w1} is in the same direction with respect to the blade motion, then V_{w1} is taken as positive. In other words, when point *F* in the velocity diagram lies on the left of point *B*, then V_{w1} is positive. Thus change in velocity of whirl

$$= V_w - (+V_{w1}) = V_w - V_{w1}$$ (See Example 22.3)

Example 22.1. *In a De-lavel turbine, the steam enters the wheel through a nozzle with a velocity of 500 m/s and at an angle of 20° to the direction of motion of the blade. The blade speed is 200 m/s and the exit angle of the moving blade is 25°. Find the inlet angle of the moving blade, exit velocity of steam and its direction and work done per kg of steam.*

Solution. Given : $V = 500$ m/s ; $\alpha = 20°$; $V_b = 200$ m/s ; $\phi = 25°$

Now let us draw the combined velocity triangle, as shown in Fig. 22.5, as discussed below :

1. First of all, draw a horizontal line and cut off *AB* equal to 200 m/s, to some suitable scale, representing the blade speed (V_b).

2. Now at *B*, draw a line *BC* at an angle of 20° (nozzle angle, α) and cut off *BC* equal to 500 m/s to the scale to represent the velocity of steam jet entering the blade (*V*).

3. Join *AC*, which represents the relative velocity at inlet (V_r).

4. At *A*, draw a line *AD* at an angle of 25° (exit angle of the moving blade, ϕ). Now with *A* as centre, and radius equal to *AC*, draw an arc meeting the line through *A* at *D*.

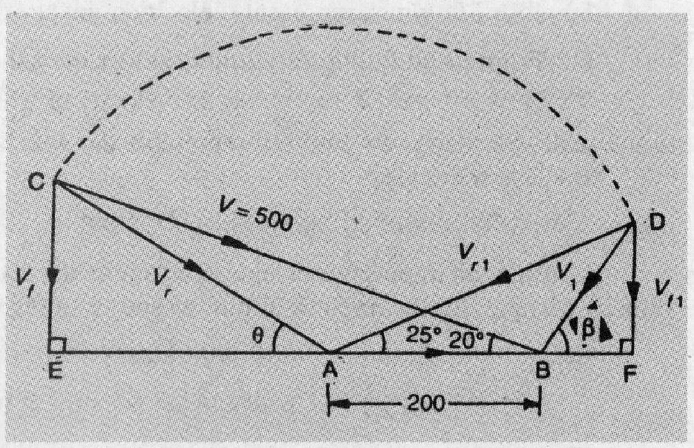

Fig. 22.5

5. Join *BD*, which represents the velocity of steam jet at outlet (V_1).

6. From *C* and *D*, draw perpendiculars meeting the line *AB* produced at *E* and *F* respectively. *CE* and *DF* represents the velocity of flow at inlet (V_f) and outlet (V_{f1}) respectively.

The following values are *measured from the velocity diagram :

$$\theta = 32^\circ ; \beta = 59^\circ ; V_1 = BD = 165 \text{ m/s}$$

$$V_w = BE = 470 \text{ m/s and } V_{w1} = BF = 90 \text{ m/s}$$

Inlet angle of moving blade

By measurement from the velocity diagram, we find that the inlet angle of the moving blade,

$$\theta = 32^\circ \textbf{ Ans.}$$

Exit velocity of steam

By measurement from the velocity diagram, we also find that the exit velocity of steam,

$$V_1 = 165 \text{ m/s}$$

Direction of the exit steam

By measurement from the velocity diagram, we also find that the direction of the exit steam,

$$\beta = 59^\circ \textbf{ Ans.}$$

Workdone per kg of steam

We know that workdone per kg of steam

$$= m\,(V_w + V_{w1})$$

$$= 1\,(470 + 90) = 560 \text{ N-m } \textbf{Ans.} \qquad \ldots (\because m = 1 \text{ kg})$$

22.10. Effect of Friction on the Combined Velocity Triangle

In the last article, we have discussed that the relative velocity of steam jet is the same at the inlet and outlet tips of the blade. In other words, we have assumed that the inner side of the curved blade offers no resistance to the steam jet. But in actual practice, some resistance is always offered by the blade surface to the gliding steam jet, whose effect is to reduce the relative velocity of the jet. *i.e.* to make V_{r1} less than V_r. The ratio of V_{r1} to V_r is known as *blade velocity coefficient* or coefficient of velocity or friction factor, (usually denoted by *K*). Mathematically, blade velocity coefficient,

$$K = \frac{V_{r1}}{V_r}$$

---

* These values may also be found out from the geometry of the velocity diagram as discussed below :

$$V_w = V \cos 20^\circ = 500 \times 0.9397 = 469.9 \text{ m/s}$$

$$V_f = V \sin 20^\circ = 500 \times 0.342 = 171 \text{ m/s}$$

$$\tan\theta = \frac{V_f}{V_w - V_b} = \frac{171}{469.9 - 200} = 0.6335 \text{ or } \theta = 32.35^\circ$$

$$V_r = V_{r1} = \frac{V_f}{\sin\theta} = \frac{171}{\sin 32.35^\circ} = \frac{171}{0.5351} = 319.4 \text{ m/s}$$

$$V_{f1} = V_{r1} \sin 25^\circ = 319.4 \times 0.4226 = 135 \text{ m/s}$$

$$V_{w1} = (V_{r1} \cos 25^\circ) - 200 = (319.4 \times 0.9063) - 200 = 89.5 \text{ m/s}$$

$$\tan\beta = \frac{V_{f1}}{V_{w1}} = \frac{135}{89.5} = 1.51 \text{ or } \beta = 56.47^\circ$$

$$V_1 = \frac{V_{f1}}{\sin\beta} = \frac{135}{\sin 56.47^\circ} = \frac{135}{0.8336} = 162 \text{ m/s}$$

It may be noted that the effect of friction on the combined velocity triangle will be to reduce the relative velocity at outlet (V_{r1}) as shown in Fig. 22.6.

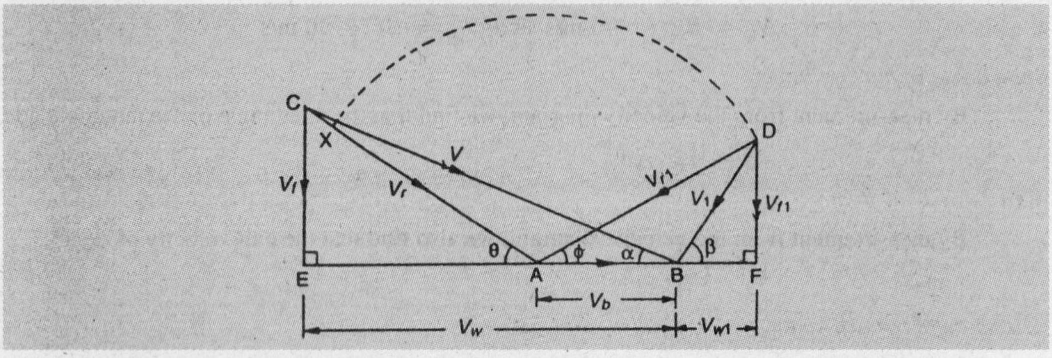

Fig. 22.6. Effect of friction on the combined velocity triangle.

Notes : 1. Since V_{r1} is decreased due to friction, therefore work done per kg of steam is also reduced.

2. The value of K varies from 0.75 to 0.85 depending upon the shape of the blades.

Example 22.2. *The velocity of steam, leaving the nozzles of an impulse turbine, is 1200 m/s and the nozzle angle is 20°. The blade velocity is 375 m/s and the blade velocity coefficient is 0.75. Assuming no loss due to shock at inlet, calculate for a mass flow of 0.5 kg/s and symmetrical blading : (a) blade inlet angle ; (b) driving force on the wheel ; (c) axial thrust on the wheel ; and (d) power developed by the turbine.*

Solution. Given : $V = 1200$ m/s ; $\alpha = 20°$; $V_b = 375$ m/s ; $K = V_{r1}/V_r = 0.75$; $m = 0.5$ kg/s ; $\theta = \phi$, for symmetrical blading.

Now draw the combined velocity triangle, as shown in Fig. 22.7, as discussed below :

1. First of all, draw a horizontal line, and cut off AB equal to 375 m/s to some suitable scale representing the velocity of blade (V_b).

2. Now at B, draw a line BC at an angle of 20° (Nozzle angle, α) and cut off BC equal to 1200 m/s to the scale to represent the velocity of steam jet entering the blade (V).

3. Join CA, which represents the relative velocity at inlet (V_r). By measurement, we find that $CA = V_r$ = 860 m/s. Now cut off AX equal to $860 \times 0.75 = 645$ m/s to the scale to represent the relative velocity at exit (V_{r1}).

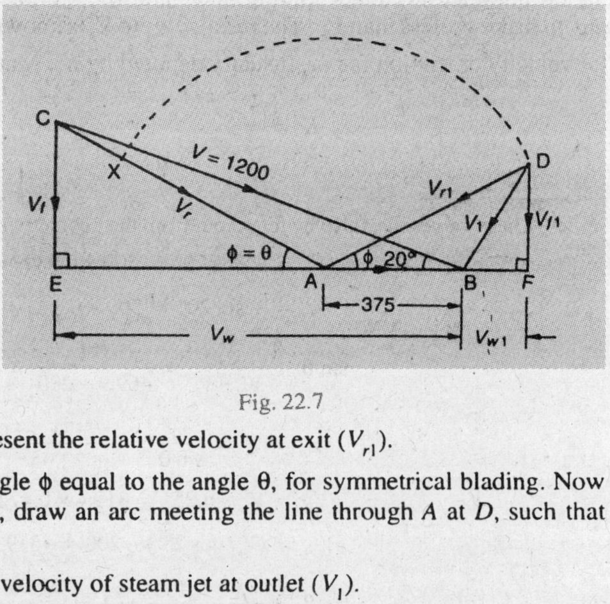

Fig. 22.7

4. At A, draw a line AD at an angle φ equal to the angle θ, for symmetrical blading. Now with A as centre, and radius equal to AX, draw an arc meeting the line through A at D, such that $AD = V_{r1}$.

5. Join BD, which represents the velocity of steam jet at outlet (V_1).

6. From C and D, draw perpendiculars meeting the line AB produced at E and F respectively. CE and DF represents the velocity of flow at inlet (V_f) and outlet (V_{f1}) respectively.

The following values are *measured from the velocity diagram :

$$\theta = 29° \; ; \; V_w = BE = 1130 \text{ m/s} \; ; \; V_{w1} = BF = 190 \text{ m/s}$$

$$V_f = CE = 410 \text{ m/s and } V_{f1} = DF = 310 \text{ m/s}$$

(*a*) *Blade inlet angle*

By measurement from the velocity diagram, we find that the blade angle at inlet,

$$\theta = 29° \text{ Ans.}$$

(*b*) *During force on the wheel*

We know that driving force on the wheel,

$$F_X = m(V_w + V_{w1}) = 0.5(1130 + 190) = 660 \text{ N Ans.}$$

(*c*) *Arial thrust on the wheel*

We know that axial thrust on the wheel,

$$F_Y = m(V_f - V_{f1}) = 0.5(410 - 310) = 50 \text{ N Ans.}$$

(*d*) *Power developed by the turbine*

We know that power developed by the turbine,

$$P = m(V_w + V_{w1})V_b = 0.5(1130 + 190)\,375 = 247\,500 \text{ W}$$

$$= 247.5 \text{ kW Ans.}$$

Example 22.3. *A steam jet entres the row of blades with a velocity of 375 m/s at an angle of 20° with the direction of motion of the moving blades. If the blade speed is 165 m/s, find the suitable inlet and outlet blade angles assuming that there is no thrust on the blades. The velocity of steam passing over the blades is reduced by 15%. Also determine power developed by the turbine per kg of steam flowing over the blades per second.*

Solution. Given : $V = 375$ m/s ; $\alpha = 20°$; $V_b = 165$ m/s ; $m = 1$ kg/s

Since there is no thrust on the blades, therefore

$$V_f = V_{f1}$$

Also, the velocity of steam passing over the blades is reduced by 15%, therefore

$$V_{r1} = 0.85\, V_r$$

---

* These values may also be found out from the geometry of the velocity diagram as discussed below :

$$V_w = V\cos 20° = 1200 \times 0.9397 = 1128 \text{ m/s}$$

$$V_f = V\sin 20° = 1200 \times 0.3420 = 410.4 \text{ m/s}$$

$$\tan\theta = \frac{V_f}{V_w - V_b} = \frac{410.4}{1128 - 375} = 0.5450 \quad \text{or} \quad \theta = 28.6°$$

$$V_r = \frac{V_f}{\sin\theta} = \frac{410.4}{\sin 28.6°} = \frac{410.4}{0.4787} = 856.5 \text{ m/s}$$

$$\therefore \quad V_{r1} = 0.75 \times 856.5 = 642.4 \text{ m/s}$$

$$V_{f1} = V_{r1}\sin 28.6° = 642.4 \times 0.4787 = 307.5 \text{ m/s}$$

$$V_{w1} = (V_{r1}\cos 28.6°) - 375 = (642.4 \times 0.8780) - 375 = 189 \text{ m/s}$$

Now draw the combined velocity triangle, as shown in Fig. 22.8, as discussed below :

1. First of all, draw a horizontal line and cut off AB equal to 165 m/s to some suitable scale representing the blade speed (V_b).

2. Now draw inlet velocity triangle ABC on the base AB with $\alpha = 20^\circ$ and $V = 375$ m/s to the scale. From the velocity triangle, we find that $V_f = 30$ m/s and $V_r = 230$ m/s.

3. Similarly, draw outlet velocity triangle ABD on the same base AB with $V_{r1} = 0.85\ V_r = 0.85 \times 230 = 195.5$ m/s to the scale and $V_{f1} = V_f = 130$ m/s.

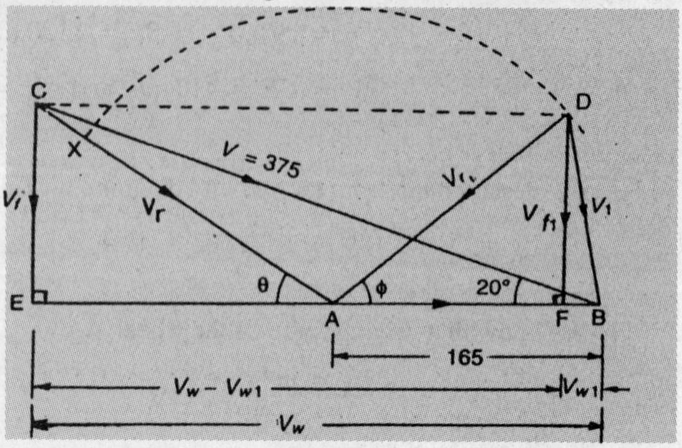

Fig. 22.8

4. From C and D draw perpendiculars to meet the line AB produced at E and F. From the geometry of the figure, we find that V_{w1} is in the opposite direction of V_w.

The following values are measured from the combined velocity triangle :

$$\theta = 34^\circ\ ;\ \phi = 41^\circ \text{ and } (V_w - V_{w1}) = 320 \text{ m/s}$$

Inlet and outlet blade angle

By measurement from the velocity diagram, we find that inlet blade angle,

$$\theta = 34^\circ \quad \textbf{Ans.}$$

and outlet blade angle,

$$\phi = 41^\circ \quad \textbf{Ans.}$$

Power developed by the turbine

We know that power developed by the turbine,

$$P = m\,(V_w - V_{w1})\,V_b$$

$$= 1 \times 320 \times 165 = 52\,800 \text{ W} = 52.8 \text{ kW} \quad \textbf{Ans.}$$

Example 22.4. *The blade speed of a single ring impulse blading is 250 m/s and nozzle angle is 20°. The heat drop is 550 kJ/kg and nozzle efficiency is 0.85. The blade discharge angle is 30° and the machine develops 30 kW, when consuming 360 kg of steam per hour. Draw the velocity diagram and calculate : 1. axial thrust on the blading, and 2. the heat equivalent per kg of steam friction of the blading.*

Solution. Given : $V_b = 250$ m/s ; $\alpha = 20^\circ$; $h_d = 550$ kJ/kg ; $K = 0.85$; $\phi = 30^\circ$; $P = 30$ kW $= 30 \times 10^3$ W ; $m = 360$ kg/h $=$ 0.1 kg/s

We know that the velocity of steam entering the blades,

$$V = 44.72\sqrt{K\,h_d}$$

$$= 44.72\sqrt{0.85 \times 550} = 967 \text{ m/s}$$

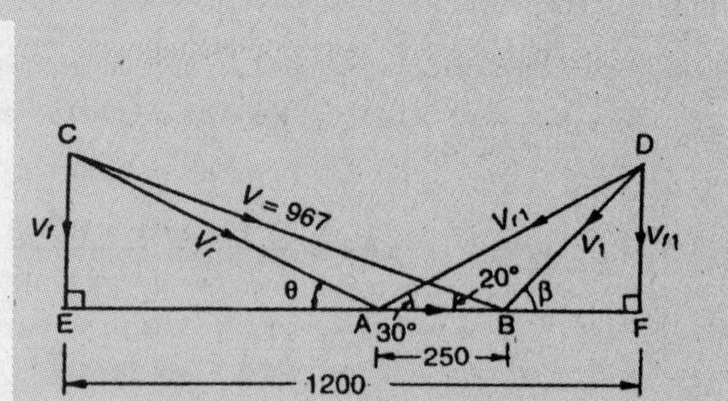

Fig. 22.9

and power developed (P),

$$30 \times 10^3 = m\,(V_w + V_{w1})\,V_b = 0.1\,(V_w + V_{w1})\,250$$

$$\therefore \quad V_w + V_{w1} = 1200 \text{ m/s}$$

Now draw the combined velocity triangle, as shown in Fig. 22.9, as discussed below :

1. First of all, draw a horizontal line and cut off *AB* equal to 250 m/s to some suitable scale representing the blade speed (V_b).

2. Now draw inlet velocity triangle *ABC* on the base *AB* with $\alpha = 20°$ and $V = 967$ m/s to the scale.

3. From point *C*, draw *CE* perpendicular to *AB* produced and take

$$EF = V_w + V_{w1} = 1200 \text{ m/s}$$

4. Draw a line at point *A* at an angle 30° intersecting the perpendicular drawn through *F*, at *D*. Now complete the outlet velocity triangle *ABD* as shown in the figure.

The following values are measured from the combined velocity triangle :

$$V_f = 400 \text{ m/s} \; ; \; V_{f1} = 325 \text{ m/s} \; ; \; V_r = 895 \text{ m/s and } V_{r1} = 650 \text{ m/s}$$

1. *Axial thrust on the blading*

We know that axial thrust on the blading,

$$F_Y = m(V_f - V_{f1}) = 0.1\,(400 - 325) = 7.5 \text{ N } \textbf{Ans.}$$

2. *Heat equivalent per kg of steam of friction of the blading*

We know that work lost in friction of the blading

$$= \frac{(V_r)^2 - (V_{r1})^2}{2000} = \frac{(895)^2 - (650)^2}{2000} = 189.26 \text{ kJ}$$

$\therefore$ Heat equivalent per kg of steam friction of the blading

$$= 189.26 \text{ kJ } \textbf{Ans.}$$

Example 22.5. *Steam at 5 bar and 200° C is first made to pass through nozzles. It is then supplied to an impulse turbine at the rate of 30 kg/minute. The steam is finally exhausted to a condenser at 0.2 bar. The blade speed is 300 m/s. The nozzles are inclined at 25° with the direction of motion of the blades and the outlet blade angle is 35°. Neglecting friction, find the theoretical power developed by the turbine.*

Solution. Given : $p_1 = 5$ bar ; $T_1 = 200°$ C ; $m = 30$ kg/min $= 0.5$ kg/s ; $p_2 = 0.2$ bar ; $V_b = 300$ m/s ; $\alpha = 25°$; $\phi = 35°$

First of all, let us draw Mollier diagram for the flow of steam through the nozzle, as shown in Fig. 22.10. From this diagram, we find that heat drop during the flow,

$$h_d = h_1 - h_2$$
$$= 2850 - 2340 = 510 \text{ kJ/kg}$$

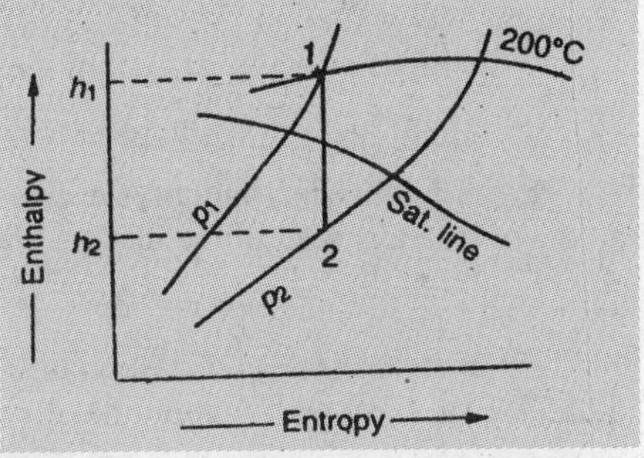

Fig. 22.10

$\therefore$ Velocity of steam at inlet of the blade,

$$V = 44.72\sqrt{510} = 1010 \text{ m/s}$$

Now draw the combined velocity triangle, as shown in Fig. 22.11, as discussed below :

1. First of all, draw a horizontal line and cut off *AB* equal to 300 m/s, to some suitable scale, to represent the blade speed (V_b).

2. Now draw inlet velocity triangle *ABC* on the base *AB* with $\alpha = 25°$ and $V = 1010$ m/s. From the velocity triangle, we find that $V_r = 850$ m/s to the scale.

3. Similarly, draw the outlet velocity triangle ABD on the same base AB with $\phi = 35°$ and $V_{r1} = V_r = 850$ m/s to the scale.

4. From C and D draw perpendiculars to meet the line AB produced at E and F.

By measurement from the velocity diagram, we find that

$$EF = V_w + V_{w1} = 1400 \text{ m/s}$$

We know theoretical power developed by the turbine,

$$P = m\,(V_w + V_{w1})\,V_b$$
$$= 0.5 \times 1400 \times 300$$
$$= 210\,000 \text{ W}$$
$$= 210 \text{ kW} \quad \textbf{Ans.}$$

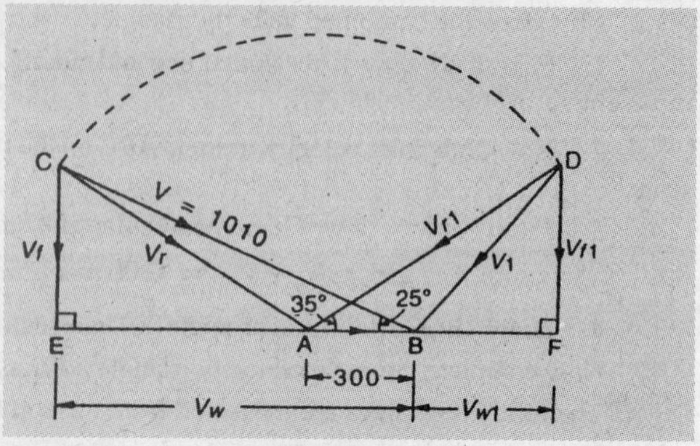

Fig. 22.11

Example 22.6. *In a certain stage of an impulse turbine, the nozzle angle is 20° with the plane of the wheel. The mean diameter of the blade ring is 2.8 metres. It develops 55 kW at 2400 r.p.m. Four nozzles, each of 10 mm diameter expand steam isentropically from 15 bar and 250° C to 0.5 bar. The axial thrust is 3.5 N. Calculate :*

1. blade angles at entrance and exit, and 2. power lost in blade friction.

Solution. Given : $\alpha = 20°$; $D = 2.8$ m ; $P = 55$ kW $= 55 \times 10^3$ W ; $N = 2400$ r.p.m. ; $n = 4$; $d = 10$ mm ; $p_1 = 15$ bar ; $T_1 = 250°$ C ; $p_2 = 0.5$ bar ; $F_Y = 3.5$ N

We know that blade speed,

$$V_b = \pi DN/60 = \pi \times 2.8 \times 2400/60 = 352 \text{ m/s}$$

1. *Blade angles at entrance and exit*

First of all, let us draw Mollier diagram for the flow of steam through the nozzle, as shown in Fig. 22.12. From this diagram, we find that heat drop during the flow,

$$h_d = h_1 - h_2 = 2920 - 2330 = 590 \text{ kJ/kg}$$

∴ Velocity of steam at inlet of the blade,

$$V = 44.72\sqrt{590} = 1086 \text{ m/s}$$

From the Mollier diagram, we also find that dryness fraction of steam at inlet of the blade,

$$x = 0.86$$

From steam tables, corresponding to a pressure of 0.5 bar, we find that specific volume of steam,

$$v_g = 3.24 \text{ m}^3/\text{kg}$$

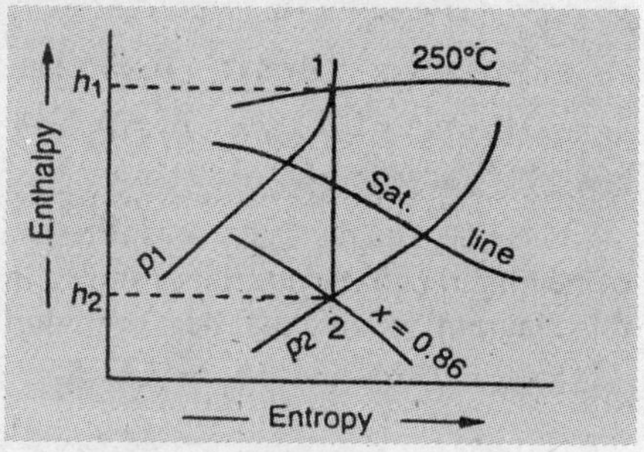

Fig. 22.12

We know that area of each nozzle

$$A = \frac{\pi}{4} \times d^2 = \frac{\pi}{4} \times 10^2 = 78.55 \text{ mm}^2 = 78.55 \times 10^{-6} \text{ m}^2$$

∴ Mass of steam discharged through the nozzle,

$$m = \frac{n\,A\,V}{x\,v_g} = \frac{4 \times 78.55 \times 10^{-6} \times 1086}{0.86 \times 3.24} = 0.12 \text{ kg/s}$$

We know that power developed (P),

$$55 \times 10^3 = m(V_w + V_{w1}) V_b = 0.12(V_w + V_{w1}) 352 = 42.24(V_w + V_{w1})$$

$$\therefore \quad V_w + V_{w1} = 1302 \text{ m/s}$$

We also know that axial thrust (F_Y),

$$3.5 = m(V_f - V_{f1}) = 0.12(V_f - V_{f1})$$

$$\therefore \quad V_f - V_{f1} = 29.2 \text{ m/s} \quad \text{or} \quad V_{f1} = (V_f - 29.2) \text{ m/s}$$

Now draw the combined velocity triangle, as shown in Fig. 22.13, as discussed below :

1. First of all, draw a horizontal line and cut off AB equal to 352 m/s, to some suitable scale, representing the blade speed (V_b).

2. Now draw the inlet velocity triangle ABC on the base AB with $\alpha = 20°$ and $V = 1086$ m/s to the scale. By measurement, we find that $V_f = 370$ m/s.

3. Similarly, draw outlet velocity triangle ABD on the same base AB with EF (*i.e.* $V_w + V_{w1}$) = 1302 m/s to the scale and $V_{f1} = V_f - 29.2 = 370 - 29.2 = 340.8$ m/s .

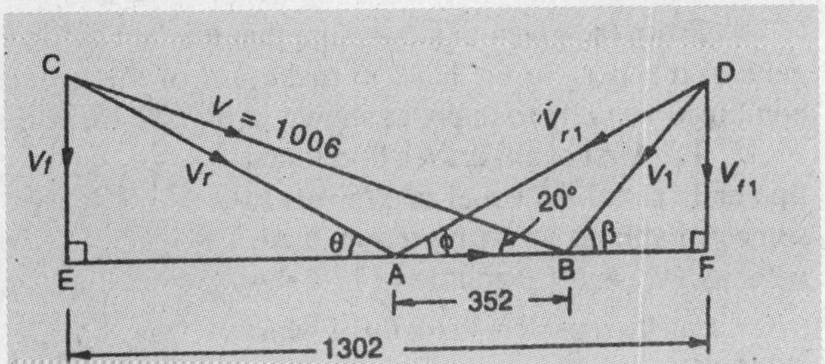

Fig. 22.13

4. From C and D, draw perpendiculars to meet the line AB produced at E and F.

By measurement from the velocity diagram, we find that

$$\theta = 29° \text{ ; and } \phi = 28° \textbf{ Ans.}$$

2. *Power lost in friction*

By measurement from the velocity diagram, we also find that

$$V_r = 765 \text{ m/s ; and } V_{r1} = 730 \text{ m/s}$$

We know that power lost in blade friction

$$= \frac{(V_r)^2 - (V_{r1})^2}{2000} = \frac{(765)^2 - (730)^2}{2000} = 26.16 \text{ kJ/s}$$

$$= 26.16 \text{ kW } \textbf{Ans.} \qquad \ldots (\because 1 \text{ kJ/s} = 1 \text{kW})$$

22.11. Combined Velocity Diagram for Axial Discharge

Sometimes, the steam leaves the blade at its exit tip at 90° to the direction of the blade motion. In such a case, the turbine is said to have an axial discharge. The combined velocity diagram for axial discharge is drawn as shown in Fig. 22.14. It may be noted that in such a turbine, velocity of whirl at outlet (V_{w1}) is equal to zero. Therefore power developed by the turbine,

$$P = m \times V_w \times V_b \text{ watts} \qquad \ldots (\because V_{w1} = 0)$$

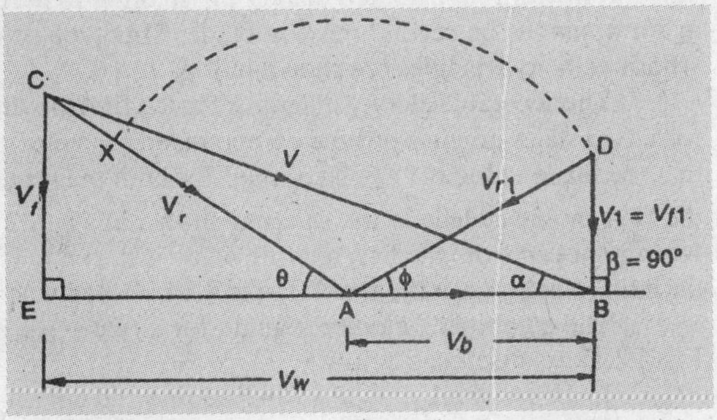

Fig. 22.14. Combined velocity diagram for axial discharge.

Example 22.7. *The steam supply to an impulse turbine with a single row of moving blades is 2 kg/s. The turbine develops 130 kW, the blade velocity being 175 m/s. The steam flows from a nozzle with a velocity of 400 m/s and the velocity coefficient of blades is 0.9. Find the nozzle angle, blade angle at entry and exit, if the steam flows axially after passing over the blades.*

Solution. Given : $m = 2$ kg/s ; $P = 130$ kW $= 130 \times 10^3$ W ; $V_b = 175$ m/s ; $V = 400$ m/s ; $K = 0.9$

Let V_w = Velocity of whirl at inlet.

We know that power developed (P),

$$130 \times 10^3 = m \times V_w \times V_b = 2 \times V_w \times 175 = 350\,V_w$$

$$\therefore \quad V_w = 371.4 \text{ m/s}$$

From the given data, we find that it is not sufficient to draw the inlet velocity triangle in the general manner. So we have to make use of the geometry of the axial discharge. Now draw the combined velocity triangle, as shown in Fig. 22.15, as discussed below :

1. First of all, draw a horizontal line and cut off AB equal to 175 m/s to some suitable scale to represent blade velocity (V_b) and EB equal to 371.4 m/s to the scale to represent velocity of whirl at inlet (V_w).

2. Now draw velocity triangle EBC on the base EB with $\angle CEB = 90°$ and $V = 400$ m/s to the scale. Join CA, which represents relative velocity at inlet (V_r). By measurement, we find that $V_r = 240$ m/s.

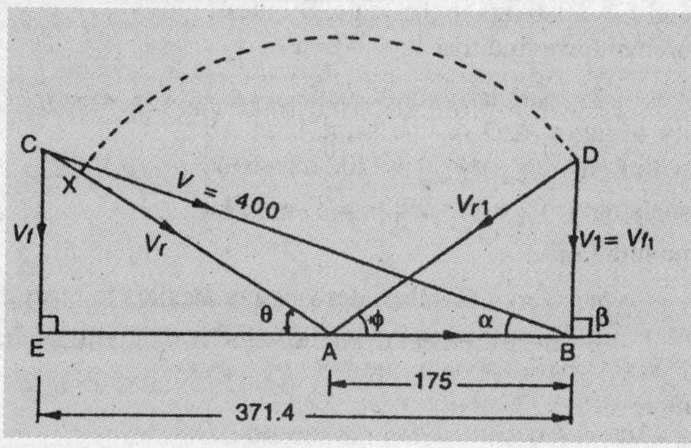

Fig. 22.15

3. Now draw outlet velocity triangle ABD on the base AB with $\beta = 90°$ and $V_{r1} = 0.9\,V_r = 0.9 \times 240 = 216$ m/s to the scale.

By measurement, we find that $\alpha = 19°$; $\theta = 33°$ and $\phi = 36°$ **Ans.**

22.12. Velocity Diagram for Two Stage Impulse Turbine

In the previous articles, we have been discussing the impulse turbine in which the steam after leaving the nozzle impinges on one end of the blade, glides over the inner surface, leaves the blade and then exhausts into the condenser. But sometimes, the steam after leaving the moving blade is made to flow through a fixed blade ring (in order to make the steam to flow at a designed angle) and again impinges on second moving blade. This type of turbine is called *two-stage impulse turbine*, whose velocity triangles are shown in Fig. 22.16.

The separate velocity diagrams for the first moving, first fixed and second moving blade rings for a two stage impulse turbine compounded for velocity are shown in Fig. 22.16. It may be noted, that the blade velocity (V_b) is constant for both the stages. The inlet and outlet velocity diagram for the first moving blade is the same as shown in Fig. 22.16. It may also be noted, that the absolute velocity at exit from the first moving blade is the entrance velocity to the fixed blade ring. Similarly, the exit velocity from the fixed ring is the entrance velocity to the second moving blade ring.

The combined velocity triangle for a two stage impulse turbine may be drawn, as shown in Fig. 22.17, as discussed below :

1. First of all, draw a horizontal line and cut off AB equal to the given blade velocity (V_b) to some suitable scale.

2. Now draw the inlet velocity triangle *ABC* for the first moving ring on the base *AB* with the help of nozzle angle of the first moving ring (α) and velocity of steam entering the turbine (*V*).

3. Now cut off *CX* equal to the friction of the blades on the first moving ring. The length *AX* will give the value of relative velocity at exit of the first moving ring (V_{r1}).

4. Now draw the outlet velocity triangle *ABD* for the first moving ring on the same base *AB* with the help of exit blade angle for the first moving ring (ϕ) and relative velocity at exit for the first moving ring (V_{r1}).

5. Now cut off *DY* equal to the friction of the blades of the fixed ring. The length *BY* will give the exit velocity of steam from the fixed ring. It will also be equal to the velocity of steam entering the second moving ring (V_1).

6. Now draw the inlet velocity triangle *ABC'* for the second moving ring on the same base *AB* with the help of nozzle angle of the second moving ring (α′) and velocity of steam entering the second moving ring (*V'*)

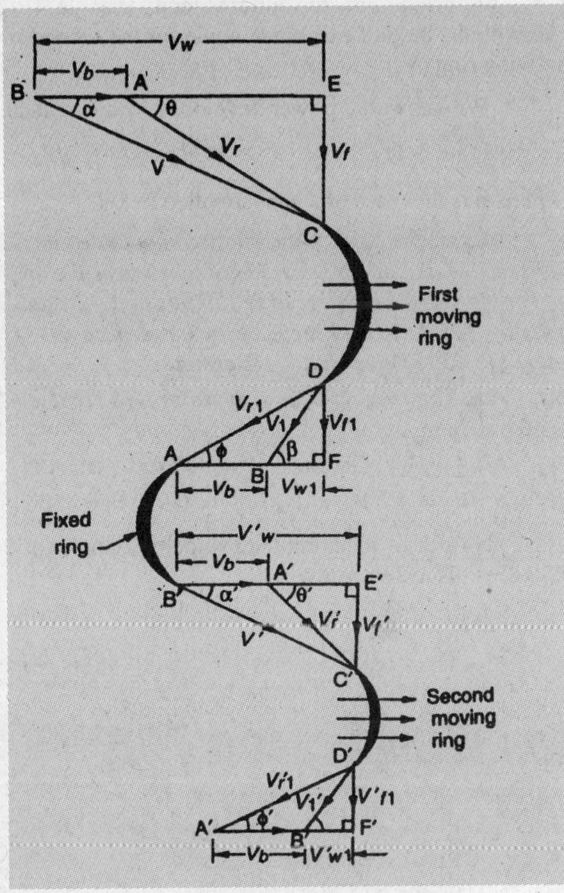

Fig. 22.16. Two-stage impulse turbine.

7. Now cut off *C'Z* equal to the friction of blades on the second moving ring. The length *AZ* will give the value of relative velocity at exit of the second moving ring (V_r').

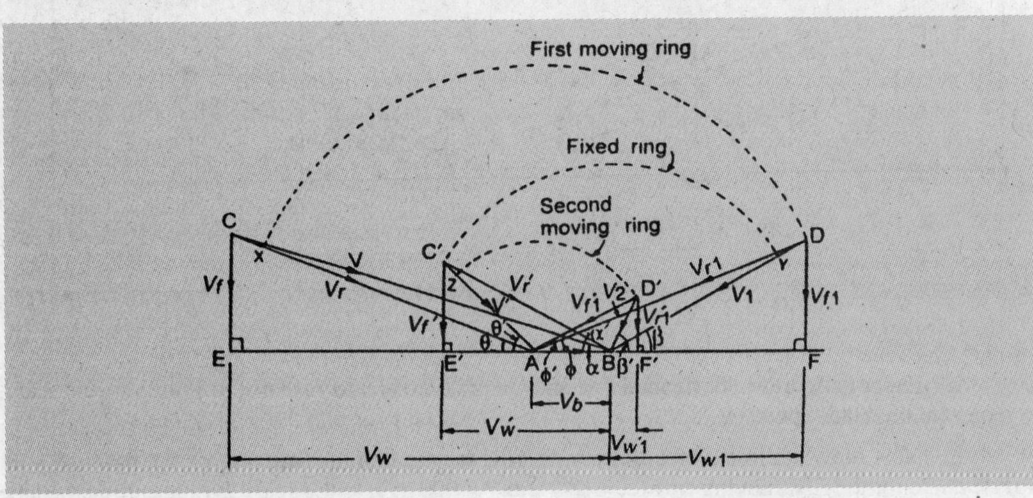

Fig. 22.17. Combined velocity triangle for two stage impulse turbine.

8. Now draw the outlet velocity triangle ABD' for the second moving ring on the same base AB with the help of exit blade angle for the second moving ring (ϕ') and exit velocity of the second moving ring (V_2).

We know that power developed by a two stage impulse turbine,

$$P = m\,(EF + E'F')\,V_b \text{ watts}$$

where m is the mass of steam supplied in kg/s.

Example 22.8. *A velocity compounded impulse turbine has two rows of moving blades with a fixed row of guide blades. The steam leaves the nozzle at 900 m/s in a direction at 18° to the plane of rotation. The blade speed is 150 m/s and the blade outlet angles are 24°, 26° and 30° for the first moving, first fixed and second moving respectively. The friction factor is 0.9 for all rows. The steam supply is 4500 kg per hour. Determine :*

(a) Tangential force on the rotor ; (b) Total work done on the blades ; and (c) Power developed by the turbine.

Solution. Given : $V = 900$ m/s ; $\alpha = 18°$; $V_b = 150$ m/s ; $\phi = 24°$; $\alpha' = 26°$; $\phi' = 30°$; $K = V_{r1}/V_r = V'/V_1 = V_{r1}'/V_r' = 0.9$; $m = 4500$ kg/h $= 1.25$ kg/s

Now draw the combined velocity triangle for the two stage impulse turbine, as shown in Fig. 22.18, as discussed below :

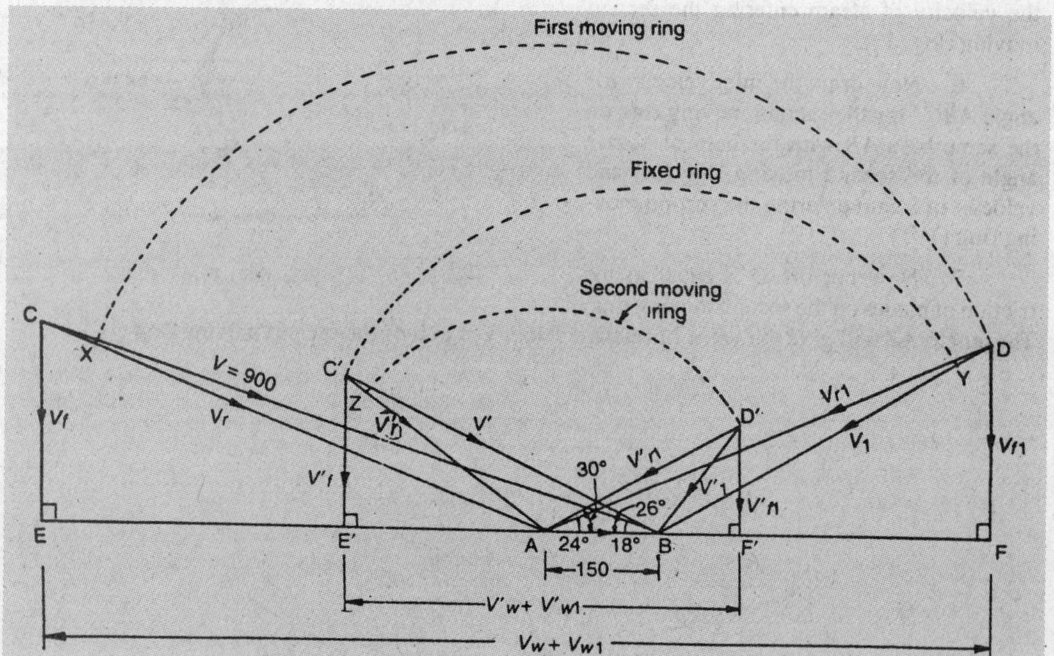

Fig. 22.18

1. First of all, draw a horizontal line and cut off AB equal to 150 m/s, to some suitable scale, to represent the blade speed (V_b).

2. Now draw the inlet velocity triangle ABC for the first moving ring on the base AB with nozzle angle for the first moving ring ($\alpha = 18°$) and $V = 900$ m/s, to the scale. By measuring the side AC, we find that $V_r = 750$ m/s.

3. Now draw the outlet velocity triangle *ABD* for the first moving ring on the same base *AB* with $\phi = 24°$ and $V_{r1} = 0.9\ V_r = 0.9 \times 750 = 675$ m/s, to the scale. By measuring the side *BD*, we find that $V_1 = 540$ m/s.

4. Now draw the inlet velocity triangle *ABC'* for the second moving ring on the same base *AB* with $\alpha' = 26°$ and $V' = 0.9\ V_1 = 0.9 \times 540 = 486$ m/s. By measuring the side *AC'*, we find that $V_r' = 350$ m/s.

5. Now draw the outlet velocity triangle *ABD'* for the second moving ring on the same base *AB* with $\phi' = 30°$ and $V_{r1}' = 0.9\ V_r' = 0.9 \times 350 = 315$ m/s.

The following values are measured from the combined velocity diagram

$$EF = (V_w + V_{w1}) = 1400 \text{ m/s}$$

and

$$E'F' = (V_w' + V_{w1}') = 690 \text{ m/s}$$

(*a*) *Tangential force on the rotor*

We know that tangential force on the rotor,

$$F_x = m\,(EF + E'F') = 1.25\,(1400 + 690) = 2612.5 \text{ N} \quad \textbf{Ans.}$$

(*b*) *Total workdone on the blades*

We know that total workdone on the blades,

$$W.D. = m\,(EF + E'F')\,V_b = 1.25\,(1400 + 690)\,150 = 391\,880 \text{ N-m/s}$$

$$= 391.88 \text{ kN-m/s} \quad \textbf{Ans.}$$

(*c*) *Power developed by the turbine*

We know that power developed by the turbine,

$$P = m\,(EF + E'F')\,V_b = 391.88 \text{ kW} \quad \textbf{Ans.} \qquad \ldots(\because 1 \text{ kN-m/s} = 1 \text{ kW})$$

Example 22.9. *Steam issuing from a nozzle at 600 m/s enters the first set of blades of a two row wheel impulse turbine. The tips of both the set of moving blades are inclined at 30° to the plane of motion. Find the speed of the blades, so that the steam is finally discharged axially. Neglect friction.*

Also find the power developed by the turbine, if the mass of steam supplied to the turbine is 3 kg/s.

Solution. Given : $V = 600$ m/s ; $\theta = \theta' = \phi = 30°$; $\beta = 90°$ (for axial discharge) ; $m = 3$ kg/s

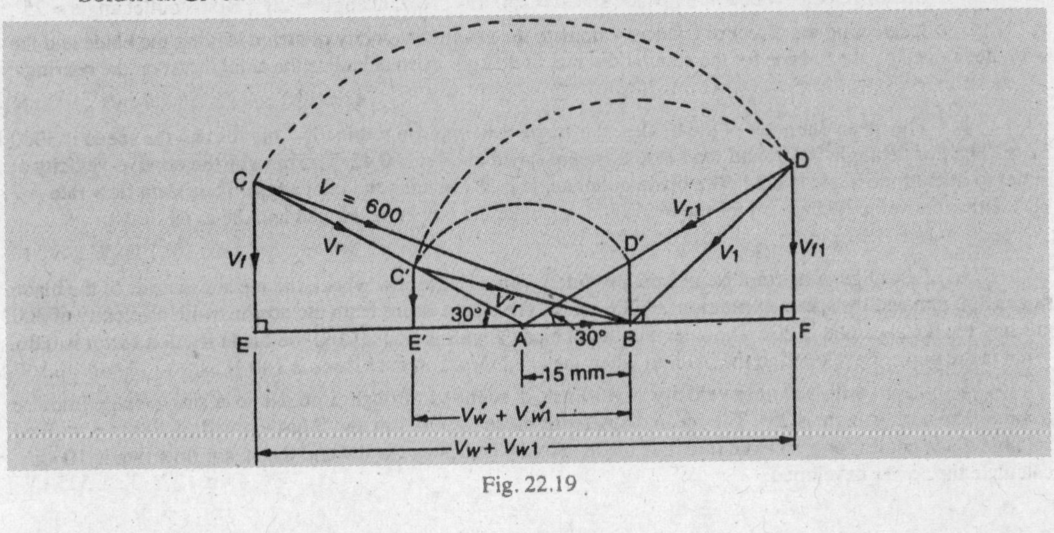

Fig. 22.19

From the given data, we find that it is not sufficient to draw inlet velocity triangle for the first moving ring. Therefore, we have to start the diagram from the outlet velocity triangle for the second moving ring. Now draw the combined velocity diagram for the two stages of the turbine, as shown in Fig. 22.19, as discussed below :

1. First of all, draw a horizontal line and mark *AB* equal to 15 mm to represent the blade velocity (which is required to be found out).

2. Now draw the outlet velocity triangle *ABD'* for the second moving ring, on the base *AB* with $\phi = 30°$ and $\beta = 90°$. From the geometry of the figure, we find that *D'B* represents V_{f1}' and V_1'. Similarly, *D'A* represents V_{r1}'. Since there is no friction in the turbine, therefore $V_{r1}' = V_r'$.

3. Now draw the inlet velocity triangle *ABC'* for the second moving ring on the same base *AB* with $\theta' = 30°$ and $V_r' = V_{r1}'$ (*i.e.* length $AC' = AD'$). From the geometry of the figure, we find that *C'A* represents V_{r1}'. Since there is no friction in the turbine, therefore $V_1 = V_r'$.

4. Now draw the outlet velocity triangle *ABD* for the first moving ring on the same base *AB* with $\phi = 30°$ and $V_r' = V_1$ (*i.e.* length $BC' = BD$). From the geometry of the figure, we find that *AD* represents V_{r1}. Since there is no friction in the turbine, therefore $V_r = V_{r1}$.

5. Now draw the inlet velocity triangle *ABC* for the first moving ring on the same base *AB* with $\theta = 30°$ and $V_{r1} = V_r$ (*i.e.* length $AD = AC$).

From the combined velocity diagram, we find that the length $BC = 56$ mm.

$\therefore$ Scale, $\quad 1 \text{ mm} = 600/56 = 10.7$ m/s

and blade speed, $\quad V_b = 15 \times 10.7 = 160.5$ m/s **Ans.**

From the figure, we find that length $(EF + E'B)$

$$= 103 \text{ mm} = 103 \times 10.7 = 1102 \text{ m/s}$$

We know that power developed by the turbine,

$$P = m\,(EF + E'B)\,V_b = 3 \times 1102 \times 160.5 = 530\,610 \text{ W}$$

$$= 530.61 \text{ kW} \quad \textbf{Ans.}$$

EXERCISES

1. The following data relates to a single stage impulse turbine :

Steam velocity = 600 m/s ; Blade speed = 250 m/s ; Nozzle angle = 20° ; Blade outlet angle = 25°.

Neglecting the effect of friction, calculate the absolute velocity of steam leaving the blade and the work developed by the turbine for the steam flow rate of 20 kg/s. Also calculate the axial thrust on the bearings.

[**Ans.** 180 m/s ; 3250 kN-m/s ; 1000 N]

2. The mean diameter of the blades of a single row impulse turbine is 2 metres and the speed is 3000 r.p.m. The nozzle angle is 18° and the blade to steam speed at inlet is 0.42. The ratio of the relative velocity at outlet to inlet of the blade is 0.84. The blade outlet angle is 3° less than the inlet angle. The steam flow rate is 7 kg/s. Draw the velocity diagram for the blade and calculate the power developed and the axial thrust.

[**Ans.** 1671.5 kW ; 490 N]

3. In a stage of an impulse turbine provided with a single row wheel, the mean diameter of the blade ring is 800 mm and the speed of rotation is 3000 r.p.m. The steam issues from the nozzles with a velocity of 300 m/s and the nozzle angle is 20°. The inlet and outlet blade angles are equal and the blade friction factor is 0.86. What is the power developed in the blading when the axial thrust on the blades is *140 N*. [**Ans.** 358 kW]

4. Steam with absolute velocity of 400 m/s is supplied through a nozzle to a single stage impulse turbine. The nozzle angle is 25°. The mean diameter of blade rotor is 1 m and it has a speed of 2000 r.p.m. Find suitable blade angles for zero axial thrust. If blade velocity coefficient is 0.9 and the steam flow rate is 10 kg/s, calculate the power developed.

[**Ans.** 27°, 32° ; 525 kW]

5. The steam enters an impulse wheel having a nozzle angle of 20° at a velocity of 450 m/s. The exit angle of the moving blade is 20° and the relative velocity of the steam may be assumed to remain constant over the moving blades. If the blade speed is 180 m/s, determine :

1. Blade angle at inlet, 2. Work done per kg of steam, and 3. Power of the wheel, when the turbine is supplied with 1.8 kg of steam per second. [**Ans.** 33° ; 92.7 kN-m ; 167 kW]

6. A De-Lavel turbine it supplied with dry steam and works on a pressure range from 10.5 bar to 0.3 bar. The nozzle angle is 20° and the blade exit angle is 30°. The mean blade speed is 270 m/s. If there is a 10% loss due to friction in the nozzle and blade velocity coefficient 0.82, find the thrust on the shaft per kW power developed. [**Ans.** 0.154 N/kW]

7. Steam issues from the nozzles of single stage impulse turbine at 1000 m/s and the nozzles are inclined at 24° to the direction of motion of the blades, which have a speed of 400 m/s. The blade angles at inlet and outlet are equal. If the steam enters and leaves the blades without shock and the flow over the blades is frictionless, find the inlet blade angle. Also determine the force exerted on the blades in the direction of their motion and power developed when the steam flows at thc rate of 4000 kg/h. [**Ans.** 39° ; 1.135 kN ; 454 kW]

8. The following particulars refer to a velocity compounded impulse turbine having two rows of moving blades with a fixed row of guide blades between them :

The velocity of steam leaving the nozzle is 1250 m/s, nozzle angle is 20° and blade speed is 300 m/s. The blade angles of the first moving blade are symmetrical and the blade output angle of the second moving blade is 30°. The friction factor for all rows is 0.9.

Draw the velocity diagram and determine the power developed and the axial thrust on the rotor for a steam rate of 5000 kg/h. [**Ans.** 871 kW ; 173.6 N]

QUESTIONS

1. What is a turbine ? How does it differ from a steam engine ?
2. Give the classification of steam turbines.
3. Explain the principle of impulse turbine.
4. Show by graphical method, variation in the pressure and velocity of steam in an impulse turbine.
5. Describe the use of combined velocity triangle of an impulse turbine.
6. Describe a relation for the power developed by an impulse turbine.
7. What do you understand by the term 'friction' in an impulse turbine. How does it effect the combined velocity triangle.
8. Define two-stage impulse tu.. ıne. How will you draw the combined velocity triangle for such a turbine ?

OBJECTIVE TYPE QUESTIONS

1. The action of steam in a steam turbine is

(*a*) static (*b*) dynamic

(*c*) static and dynamic (*d*) neither static nor dynamic

2. In an impulse turbine

(*a*) the steam is expanded in nozzles only and there is a pressure drop and heat drop

(*b*) the steam is expanded both in fixed and moving blades continuously

(*c*) the steam is expanded in moving blades only

(*d*) the pressure and temperature of steam remains constant.

3. De-Laval turbines are mostly used

(*a*) where low speeds are required

(*b*) for small power purposes and low speeds

(c) for small power purposes and high speeds

(d) for large power purposes

4. In impulse turbines, when friction is neglected, the relative velocity of steam at outlet tip of the blade is ..... the relative velocity of steam at inlet tip of the blade.

(a) equal to (b) less than (c) greater than

5. The blade friction in the impulse turbine reduces the velocity of steam by ..... while it passes over the blades.

(a) 10 to 15% (b) 15 to 20% (c) 20 to 30% (d) 30 to 40%

ANSWERS

1. (b)	2. (a)	3. (c)	4. (a)	5 (a)

23

Reaction Turbines

1. Introduction. 2. Parson's Reaction Turbine. 3. Pressure and Velocity in a Reaction Turbine. 4. Comparison between Impulse Turbine and Reaction Turbine. 5. Velocity Triangles for Moving Blades of a Reaction Turbine. 6. Combined Velocity Triangle for Moving Blades. 7. Power Produced by a Reaction Turbine. 8. Degree of Reaction. 9. Height of Blades of a Reaction Turbine.

23.1. Introduction

In a reaction turbine, the steam enters the wheel under pressure and flows over the blades. The steam, while gliding, propels the blades and make them to move. As a matter of fact, the turbine runner is rotated by the reactive forces of steam jets. The backward motion of the blades is similar to the recoil of a gun. It may be noted that an absolute reaction turbine is rarely used in actual practice.

23.2. Parson's Reaction Turbine

A *Parson's turbine is the simplest type of reaction steam turbine, and is commonly used. It has the following main components :

1. *Casing.* It is an air-tight metallic case, in which the steam from the boiler, under a high pressure and temperature, is distributed around the fixed blades (guide mechanism) in the casing. The casing is designed in such a way that the steam enters the fixed blades with a uniform velocity.

2. *Guide mechanism.* It is a mechanism, made up with the help of guide blades, in the form of a wheel. This wheel is, generally, fixed to the casing; that is why these guide blades are also called fixed blades. The guide blades are properly designed in order to :

(*a*) allow the steam to enter the runner without shock. This is done by keeping the relative velocity at inlet of the runner tangential to the blade angle.

(*b*) allow the required quantity of steam to enter the turbine. This is done by adjusting the openings of the blades.

The guide blades may be opened or closed by rotating the regulating shaft, thus allowing the steam to flow according to the need. The regulating shaft is operated by means of a governor whose function is to govern the turbine (*i.e.* to keep to speed constant at varying loads).

Fig. 23.1. Turbine runner.

* A Parson's turbine is also known as 50% reaction turbine. (See Art. 23.8)

3. *Turbine runner.* The turbine runner of a Parson's reaction turbine essentially consists of runner blades fixed to a shaft or rings, depending upon the type of turbine. The blades, fixed to the runner, are properly designed in order to allow the steam to enter and leave the runner without shock, as shown in Fig. 23.1.

The surface of the turbine runner is made very smooth to minimise the frictional losses. The turbine runner is, generally, cast in one piece. But sometimes, it is made up of separate steel plates welded together.

4. *Draft tube.* The steam, after passing through the runner, flows into the condenser through a tube called draft tube. It may be noted that if this tube is not provided in the turbine, then the steam will move freely and will cause steam eddies.

23.3. Pressure and Velocity of Steam in a Reaction Turbine

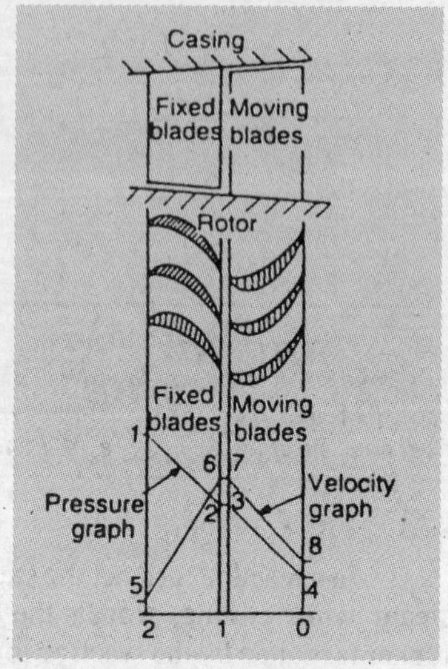

Fig. 23.2. Pressure and velocity graphs of a reaction turbine.

It will be interesting to know that the pressure in a reaction turbine is reduced in the fixed blades as well as in moving blades. The velocity of steam is increased in the fixed blades, and is reduced while passing through the moving blades.

Fig. 23.2 shows the pressure and velocity graphs of the steam while it flows in the fixed and moving blades of a reaction turbine. The pressure graph 1-2-3-4 represents steam pressure at entrance of the fixed blades, exit of the fixed blades, entrance of the moving blades and exit of the moving blades respectively. Similarly, velocity graph 5-6-7-8 represents the velocity of steam at entrance of the fixed blades, exit of the fixed blades, entrance of the moving blades and exit of the moving blades respectively.

23.4. Comparison between Impulse Turbine and Reaction Turbine

Following are the few points of comparison between an impulse turbine and a reaction turbine :

S. No.	*Impulse turbine*	*Reaction turbine*
1.	The steam flows through the nozzles and impinges on the moving blades.	The steam flows first through guide mechanism and then through the moving blades.
2.	The steam impinges on the buckets with kinetic energy.	The steam glides over the moving vanes with pressure and kinetic energy.
3.	The steam may or may not be admitted over the whole circumference.	The steam must be admitted over the whole circumference.
4.	The steam pressure remains constant during its flow through the moving blades.	The steam pressure is reduced during its flow through the moving blades.
5.	The relative velocity of steam while gliding over the blades remains constant (assuming no friction).	The relative velocity of steam while gliding over the moving blades increases (assuming no friction).
6.	The blades are symmetrical.	The blades are not symmetrical.
7.	The number of stages required are less for the same power developed.	The number of stages required are more for the same power developed.

23.5. Velocity Triangles for Moving Blades of a Reaction Turbine

We have already discussed that in a reaction turbine, the steam enters at one end (say *C*) of the moving blades from the guide mechanism. The jet then glides over the inside surface of the blades and finally leaves from the other edge (say *D*), as shown in Fig. 23.3. It may be noted that the jet enters and leaves the blades tangentially for shockless entry and exit.

Consider steam, in the form of a jet, entering the curved blade (after leaving the fixed blade) at *C*. Let the jet glides over the inside surface and leaves the blade at *D*, as shown in Fig. 23.3. Now let us draw the velocity triangles at inlet and outlet tips of the moving blade, as shown in Fig. 23.3.

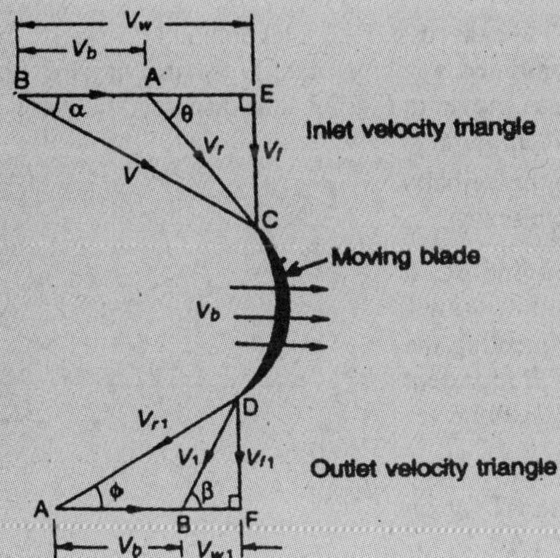

Fig. 23.3. Velocity triangles for a reaction turbine.

Let V_b = Linear velocity of the moving blade (*AB*).

V = Absolute velocity of steam entering the moving blade (*BC*).

V_r = Relative velocity of jet to the moving blade (*AC*). It is the vectorial difference of V_b and V.

V_f = Velocity of flow at entrance (*EC*). It is the vertical component of V.

V_w = Velocity of whirl at entrance (*BE*). It is the horizontal component of V.

α = Angle with the direction of motion of the blade at which the steam enters the blade.

θ = Angle which the relative velocity of jet (V_r) makes with the direction of motion of the blade.

$V_1, V_{r1}, V_{f1}, V_{w1}, \beta, \phi$ = Corresponding values at exit of the moving blade.

It may be seen, from the above, that original notations (*i.e.* V, V_r, V_f, V_w) stand for inlet triangle. The notations with suffix 1 (*i.e.* $V_1, V_{r1}, V_{f1}, V_{w1}$) stand for outlet triangle. It may be noted that as the steam enters and leaves the blades without any shock (or in other words tangentially), therefore shape of the blades will be such that V_r and V_{r1} will be along the tangents to the blades at inlet and outlet respectively.

The steam jet enters the blades along *BC* with a velocity (V) at an angle α with the direction of motion of the blade. The axial component of V (*i.e. CE*) which does no work on the blade, is known as *velocity of flow* (V_f). It causes the steam to flow through the blade and also an axial thrust on the rotor. The linear velocity or mean velocity of the blades (V_b) is represented by *AB* in magnitude and direction. The length *AC* represents the relative velocity (V_r) of the steam with respect to the blade. The jet now glides over and leaves the blade with a relative velocity (V_{r1}) which is represented by *DA*. The absolute velocity of jet (V_1) as it leaves the blade is represented by *DB* inclined at an angle

β with the direction of blade motion. The tangential component of V_1 (represented by BF) is known as *velocity of whirl* at exit (V_{w1}). The axial component of V_1 (represented by DF) is known as *velocity of flow* at exit (V_{f1}).

Note : The inlet triangle of velocities is represented by BEC whereas the outlet triangle by AFD.

23.6. Combined Velocity Triangle for Moving Blades

In the last article, we have discussed the inlet and outlet velocity triangles separately. For the sake of simplification, a combined velocity triangle for the moving blade is drawn, for solving problems on steam turbines, as shown in Fig. 23.4, as discussed below :

1. First of all, draw a horizontal line and cut off AB equal to the velocity of blade (V_b), to some suitable scale.

2. Now at B, draw a line BC at angle α with AB. Similarly at A, draw a line AC at angle θ with EA meeting the first line at C. Now CA and CB represent the relative velocity (V_r) and absolute velocity (V) of steam at inlet, to the scale.

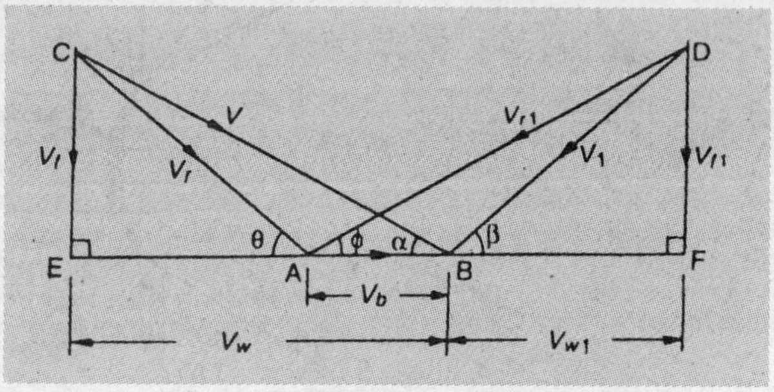

Fig. 23.4. Combined velocity triangle for a Parson's reaction turbine.

3. At A, draw a line AD at an angle ϕ (such that $\phi = \alpha$) with AB. Similarly at B draw a line BD at an angle β (such that $\beta = \theta$) with AB meeting the first line at D. Now DA and DB represent the relative velocity (V_{r1}) and absolute velocity (V_1) of steam at outlet, to the scale.

4. From C and D draw perpendiculars meeting the line AB produced at E and F.

5. Now EB and CE represent the velocity of whirl and velocity of flow at inlet (V_w and V_f) to the scale. Similarly BF and DF represent the velocity of whirl and velocity of flow at outlet (V_{w1} and V_{f1}), to the scale.

Note : A careful study of the combined velocity diagram of Parson's reaction turbine will reveal that it is symmetrical about the central line. Therefore following relations exist in the combined velocity diagram :

$$V_f = V_{f1} \,;\, V = V_{r1} \,;\, V_r = V_1 \,;\, EA = BF$$

23.7. Power Produced by a Reaction Turbine

Consider a reaction turbine working under the action of steam pressure. Let us draw a combined velocity triangle for the reaction turbine, as shown in Fig. 23.4.

Let m = Mass of the steam flowing through the turbine in kg/s, and

$(V_w + V_{w1})$ = Change in the velocity of whirl in m/s.

We know that according to the Newton's second law of motion, force in the direction of motion of the blades,

$$F_X = \text{Mass of steam flowing/second} \times \text{Change in the velocity of whirl}$$

$$m\,[V_w - (-V_{w1})] = m\,[V_w + V_{w1}] = m \times EF \text{ N} \quad \ldots (i)$$

and work done in the direction of motion of the blades

$$= \text{Force} \times \text{Distance} = {}^*m\,(V_w + V_{w1})\,V_b = m \times EF \times AB \text{ N-m/s} \quad \ldots (ii)$$

---

* Workdone per kg of steam = $V_b\,(2\,V\cos\alpha - V_b)$ ... Refer Chapter 24 Art. 24.4, page 548.

∴ Power produced by the turbine,

$$P = m\,(V_w + V_{w1})\,V_b \text{ watts} \qquad \ldots (\because 1 \text{ N-m/s} = 1 \text{ watt})$$

Similarly, we can find out the axial thrust on the wheel, which is due to difference of velocities of flow at inlet and outlet. Mathematically, axial thrust,

$$F_Y = \text{Mass of steam flowing /second} \times \text{Change in the velocity of flow}$$

$$= m\,(V_f - V_{f1}) = m\,(CE - DF) \text{ N} \qquad \ldots (iii)$$

Note : In equation (*i*), the value of V_{w1} is taken as *negative* because of the opposite direction of V_w with respect to the blade motion.

23.8. Degree of Reaction

We have already discussed in Art. 23.3 that in a reaction turbine, the pressure drop takes place in both the fixed and moving blades. In other words, there is an enthalpy drop in both the fixed and moving blades as shown on *h-s* diagram in Fig. 23.5. The ratio of the enthalpy or heat drop in the moving blades to the total enthalpy or heat drop in the stage is known as *degree of reaction.* Mathematically,

Degree of reaction

$$= \frac{\text{Enthalpy or heat drop in the moving blades}}{\text{Total enthalpy or heat drop in the stage}} = \frac{h_2 - h_3}{h_1 - h_3}$$

The enthalpy drop in the fixed blades per kg of steam is given by

$$h_1 - h_2 = \frac{V^2 - V_1^2}{2000} \text{ kJ/kg}$$

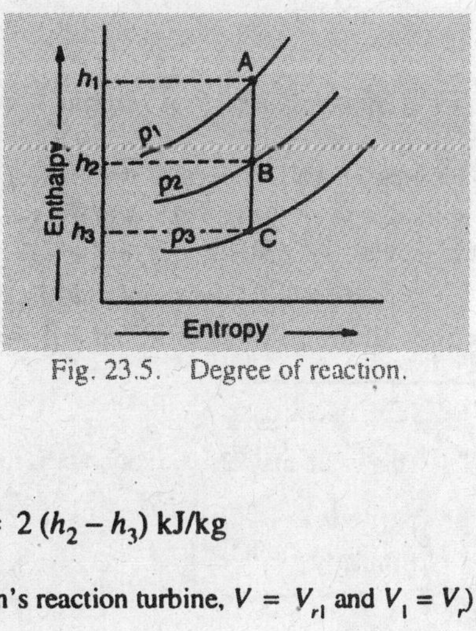

Fig. 23.5. Degree of reaction.

and enthalpy drop in the moving blades,

$$h_2 - h_3 = \frac{V_{r1}^2 - V_r^2}{2000} \text{ kJ/kg}$$

∴ Total enthalpy drop in the stage,

$$h_1 - h_3 = (h_1 - h_2) + (h_2 - h_3)$$

$$= \frac{V^2 - V_1^2}{2000} + \frac{V_{r1}^2 - V_r^2}{2000} = \frac{2\,(V_{r1}^2 - V_r^2)}{2000} = 2\,(h_2 - h_3) \text{ kJ/kg}$$

$$\ldots (\because \text{ For Parson's reaction turbine, } V = V_{r1} \text{ and } V_1 = V_r)$$

We know that degree of reaction

$$= \frac{h_2 - h_3}{h_1 - h_3} = \frac{h_2 - h_3}{2\,(h_2 - h_3)} = \frac{1}{2} = 0.5 \text{ or } 50\%$$

Thus we see that a Parson's reaction turbine is a 50 percent reaction turbine.

Example 23.1. *In one stage of a reaction steam turbine, both the fixed and moving blades have inlet and outlet blade tip angles of 35° and 20° respectively. The mean blade speed is 80 m/s and the steam consumption is 22 500 kg per hour. Determine the power developed in the pair, if the isentropic heat drop for the pair is 23.5 kJ per kg.*

Solution. Given : $\theta = \beta = 35^\circ$; $\phi = \alpha = 20^\circ$; $V_b = 80$ m/s ; $m = 22\,500$ kg/h = 6.25 kg/s ; $*h_d = 23.5$ kJ/kg

Now let us draw the combined velocity triangle, as shown in Fig. 23.6, as discussed below :

1. First of all, draw a horizontal line and cut off AB equal to 80 m/s (V_b) to some suitable scale.

2. Now at B, draw a line BC at an angle $\alpha = 20^\circ$, with AB. Similarly, at A draw a line AC at an angle $\theta = 35^\circ$ with BA meeting the first line at C.

3. At A, draw a line AD at angle $\phi = 20^\circ$ (because $\phi = \alpha$) with AB. Similarly, at B draw a line BD at an angle $\beta = 35^\circ$ (because $\beta = \theta$) with AB meeting the first line at D.

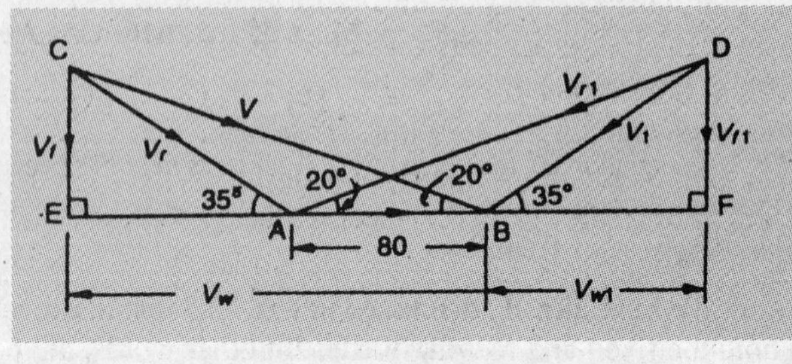

Fig. 23.6

4. From C and D draw perpendiculars meeting the line AB produced at E and F.

By **measurement, we find that the change in the velocity of whirl,

$$(V_w + V_{w1}) = 235 \text{ m/s}$$

We know that power developed in the pair,

$$P = m\,(V_w + V_{w1})\,V_b = 6.25 \times 235 \times 80 = 117\,500 \text{ W}$$
$$= 117.5 \text{ kW } \textbf{Ans.}$$

Example 23.2. *A Parson's reaction turbine, while running at 400 r.p.m. consumes 30 tonnes of steam per hour. The steam at a certain stage is at 1.6 bar with dryness fraction of 0.9 and the stage develops 10 kW. The axial velocity of flow is constant and equal to 0.75 of the blade velocity. Find mean diameter of the drum and the volume of steam flowing per second. Take blade tip angles at inlet and exit as 35° and 20° respectively.*

Solution. Given : $N = 400$ r.p.m. ; $m = 30$ t/h = 8.33 kg/s ; $p = 1.6$ bar ; $x = 0.9$; $P = 10$ kW $= 10 \times 10^3$ W ; $V_f = 0.75\,V_b$; $\theta = \beta = 35^\circ$; $\phi = \alpha = 20^\circ$

---

* Superfluous data

** This value may also be found out from the geometry of the velocity diagram as discussed below :

$$V_f = V \sin 20^\circ = 0.3240\,V$$

Similarly, $$V_f = V_r \sin 35^\circ = 0.5736\,V_r$$

$\therefore$ $$0.3240\,V = 0.5736\,V_r \quad \text{or} \quad V = 1.77\,V_r$$

$$V_w = V \cos 20^\circ = 0.9397\,V$$

Similarly, $$V_w = V_r \cos 35^\circ + 80 = 0.8192\,V_r + 80$$

$\therefore$ $$0.9397\,V = 0.8192\,V_r + 80$$

or $$0.9397\,(1.77\,V_r) = 0.8192\,V_r + 80$$

$\therefore$ $$1.663\,V_r = 0.8192\,V_r + 80$$

$$0.8438\,V_r = 80 \quad \text{or} \quad V_r = 94.8 \text{ m/s}$$

and $$EA = V_r \cos 35^\circ = 94.8 \times 0.8192 = 77.7$$

$\therefore$ $$(V_w + V_{w1}) = EA + AB + BF = 77.7 + 80 + 77.7 = 235.4 \text{ m/s}$$

Mean diameter of the drum

Let D = Mean diameter of the durm.

Now let us draw the combined velocity triangle, as shown in Fig. 23.7, as discussed below :

1. First of all, draw a horizontal line and cut off AB equal to 25 mm to represent the blade velocity (which is required to be found out).

2. Now at B, draw a line BC at an angle $\alpha = 20°$ with AB. Similarly at A, draw a line AC at an angle $\theta = 35°$ meeting the first line at C.

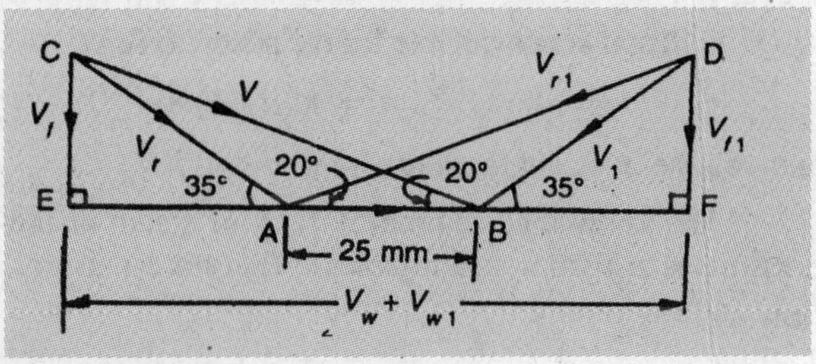

Fig. 23.7

3. At A, draw a line AD at an angle $\phi = 20°$ with AB. Similarly at B, draw a line BD at an angle $\beta = 35°$ with AB meeting the first line at D.

4. From C and D draw perpendiculars meeting the line AB produced at E and F.

By *measurement, we find that the change in the velocity of whirl,

$$(V_w + V_{w1}) = 73.5 \text{ mm}$$

$$\therefore \quad \frac{(V_w + V_{w1})}{V_b} = \frac{73.5}{25} = 2.94$$

or

$$(V_w + V_{w1}) = 2.94\, V_b$$

We know that the power developed (P),

$$10 \times 10^3 = m\,(V_w + V_{w1})\, V_b = 8.33 \times 2.94\, V_b \times V_b = 24.49 (V_b)^2$$

$$\therefore \quad (V_b)^2 = 408.3 \text{ or } V_b = 20.2 \text{ m/s}$$

We know that the blade velocity (V_b),

$$20.2 = \frac{\pi D N}{60} = \frac{\pi D \times 400}{60} = 20.94\, D$$

$$\therefore \quad D = 0.965 \text{ m} = 965 \text{ mm} \textbf{ Ans.}$$

Volume of steam flowing per second

From steam tables, corresponding to a pressure of 1.6 bar, we find that specific volume of steam,

$$v_g = 1.091 \text{ m}^3/\text{kg}$$

∴ Volume of steam flowing per second

$$= m \times v_g = 8.33 \times 0.9 \times 1.091 = 8.18 \text{ m}^3/\text{s} \textbf{ Ans.}$$

23.9. Height of Blades of a Reaction Turbine

We have already discussed that in a reaction turbine, the steam enters the moving blades over the whole circumference. As a result of this, the area through which the steam flows is always full of steam. Now consider a reaction turbine whose end view of the blade ring in shown in Fig. 23.8.

---

* It may also be found out analytically in the same way as in the last example. Since the blade angles in this example as well as in the last example are same, therefore

$$(V_w + V_{w1}) = \frac{235.4}{80} \times 25 = 73.56 \text{ mm}$$

Let d = Diameter of rotor drum,

*h = Height of blades, and

V_{f1} = Velocity of flow at exit.

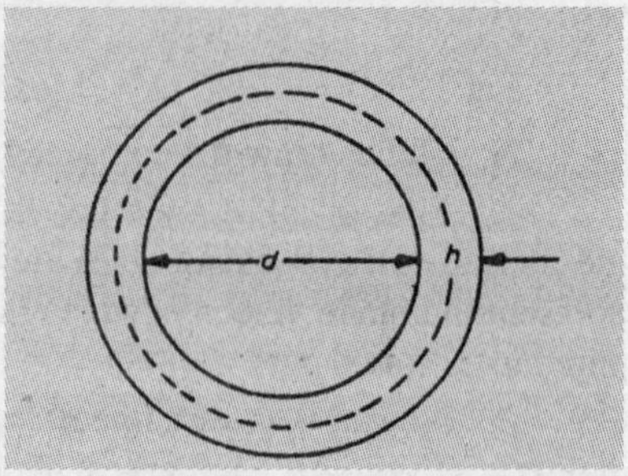

Fig. 23.8. Height of blades for reaction turbines.

∴ Total area available for the steam to flow,

$$A = \pi (d + h)\, h$$

and volume of steam flowing $= \pi (d + h)\, h\, V_{f1}$

We know that volume of 1 kg of steam at the given pressure is v_g (from steam tables). Therefore mass of steam flowing,

$$m = \frac{\pi (d + h)\, h\, V_{f1}}{v_g} \text{ kg/s}$$

If the steam has a dryness fraction of x, then mass of steam flowing,

$$m = \frac{\pi (d + h)\, h\, V_{f1}}{x\, v_g} = \frac{\pi\, d_m\, h\, V_{f1}}{x\, v_g} \text{ kg/s}$$

where d_m is the mean blade diameter and is equal to $(d + h)$.

Note : In most of the reaction turbines, the velocity of flow is constant at inlet and outlet (*i.e.* $V_f = V_{f1}$). Therefore, we can use the value of V_f instead of V_{f1} in the above relation.

Example 23.3. *In a reaction turbine, the blade tips are inclined at 35° and 20° in direction of motion. The guide blades are of the same shape as the moving blades, but reversed in direction. At a certain place in the turbine, the drum diameter is 1 metre and the blades are 100 mm high. At this place, steam has a pressure of 1.7 bar and dryness 0.935. If the speed of the turbine is 250 r.p.m. and the steam passes through the blades without shock, find the mass of steam flow and the power developed in the ring of the moving blades.*

Solution. Given : $\theta = \beta = 35°$; $\phi = \alpha = 20°$; $d = 1$ m ; $h = 100$ mm $= 0.1$ m ; $p = 1.7$ bar ; $x = 0.935$; $N = 250$ r.p.m.

We know that blade speed,

$$V_b = \frac{\pi (d + h) N}{60} = \frac{\pi (1 + 0.1)\, 250}{60} = 14.4 \text{ m/s}$$

Now let us draw the combined velocity triangle, as shown in Fig. 23.9, as discussed below :

1. First of all, draw a horizontal line, and cut off AB equal to 14.4 m/s to some suitable scale to represent the velocity of blade (V_b).

2. Now draw inlet velocity triangle ABC on the base AB with $\alpha = 20°$ and $\theta = 35°$.

3. Similarly draw outlet velocity triangle ABD on the same base AB with $\phi = 20°$ and $\beta = 35°$.

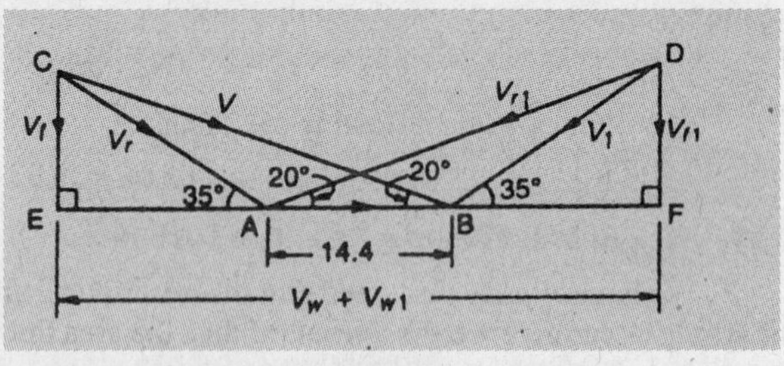

Fig. 23.9

4. From C and D draw perpendiculars to meet the line AB produced at E and F.

---

* Generally the height of blades is taken as $k\, d$, where k is a design constant, whose value is usually taken as 1/12 and d is the diameter of the rotor drum.

By measurement from velocity triangle, we find that

Change in the velocity of whirl, $(V_w + V_{w1}) = EF = 42.5$ m/s

and velocity of flow at outlet, $V_{f1} = DF = 10$ m/s

Mass of steam flow

From steam tables, corresponding to a pressure of 1.7 bar, we find that the specific volume of steam, $v_g = 1.031$ m³/kg.

We know that mass of steam flow,

$$m = \frac{\pi (d+h)\, h\, V_{f1}}{x\, v_g} = \frac{\pi (1 + 0.1)\, 0.1 \times 10}{0.935 \times 1.031} = 3.58 \text{ kg/s} \quad \textbf{Ans.}$$

Power developed in the ring of the moving blades

We know that power developed in the ring of the moving blades,

$$P = m\,(V_w + V_{w1})\, V_b = 3.58 \times 42.5 \times 14.4 = 2191 \text{ W}$$

$$= 2.191 \text{ kW} \quad \textbf{Ans.}$$

Example 23.4. *A reaction turbine runs at 300 r.p.m. and its steam consumption is 15 400 kg/h. The pressure of steam at a certain pair is 1.9 bar ; its dryness 0.93 and power developed by the pair is 3.5 kW. The discharging blade tip angle is 20° for both fixed and moving blades and the axial velocity of flow is 0.72 of the blade velocity. Find the drum diameter and blade height. Take the tip leakage steam as 8%, but neglect blade thickness.*

Solution. Given : $N = 300$ r.p.m. ; $m_1 = 15\,400$ kg/h $= 4.28$ kg/s ; $p = 1.9$ bar ; $x = 0.93$; $P = 3.5 \text{ kW} = 3.5 \times 10^3$ W ; $\alpha = \phi = 20°$; $V_f = 0.72\, V_b$

Since the tip leakage steam is 8%, therefore actual mass of steam flowing over the blades,

$$m = 4.28 - (4.28 \times 0.08) = 3.94 \text{ kg/s}$$

Blade height

Let h = Blade height, and

d_m = Mean diameter of the blades.

We know that blade velocity,

$$V_b = \frac{\pi\, d_m N}{60} = \frac{\pi\, d_m \times 300}{60} = 15.71\, d_m \text{ m/s}$$

$$\therefore \quad V_f = 0.72 \times 15.71\, d_m = 11.3\, d_m \text{ m/s}$$

Now let us draw the combined velocity triangle, as shown in Fig. 23.10, as discussed below :

1. First of all, draw a horizontal line, and cut off AB equal to $15.71\, d_m$, to some suitable scale representing the blade velocity (V_b).

2. Now draw inlet velocity triangle ABC on the base AB with $\alpha = 20°$ and $BC = V_f / \sin 20° = 11.3\, d_m / 0.342 = 33\, d_m$, to the scale.

3. Similarly, draw outlet velocity triangle on the same base AB with $\phi = 20°$ and $V_{r1} = V_{f1} / \sin 20° = 11.3\, d_m / 0.342 = 33\, d_m$, to the scale.

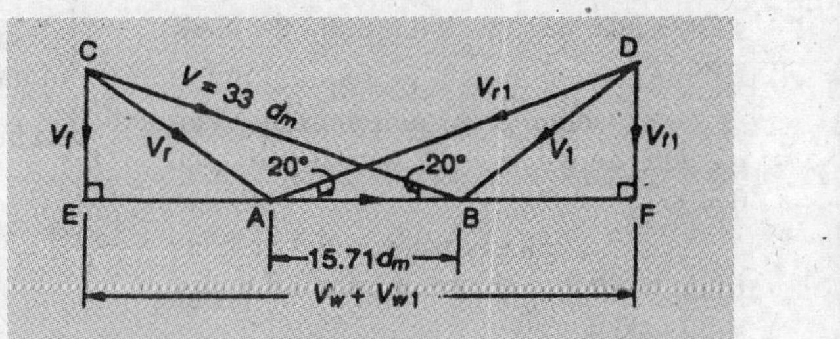

Fig. 23.10

4. From C and D draw perpendiculars to meet the line AB produced at E and F.

By measurement from velocity triangle, we find that change in the velocity of whirl,

$$(V_w + V_{w1}) = 46\, d_m \text{ m/s}$$

We know that power developed (P),

$$3.5 \times 10^3 = m\,(V_w + V_{w1})\, V_b = 3.94 \times 46\, d_m \times 15.71\, d_m = 2845\, d_m^2$$

$$\therefore \quad d_m^2 = 1.23 \quad \text{or} \quad d_m = 1.11 \text{ m}$$

and $$V_{f1} = V_f = 11.3\, d_m = 11.3 \times 1.11 = 12.54 \text{ m/s}$$

From steam tables, corresponding to a pressure of 1.9 bar, we find that specific volume of steam,

$$v_g = 0.929 \text{ m}^3\text{/kg}$$

We know that mass of steam flow (m),

$$3.94 = \frac{\pi\, d_m\, h\, V_{f1}}{x\, v_g} = \frac{\pi \times 1.11 \times h \times 12.54}{0.93 \times 0.929} = 50.6\, h$$

$$\therefore \quad h = 0.078 \text{ m} = 78 \text{ mm} \quad \textbf{Ans.}$$

Drum diameter

We know that drum diameter,

$$d = d_m - h = 1.11 - 0.078 = 1.032 \text{ m} \quad \textbf{Ans.}$$

Example 23.5. *At a particular stage of a reaction steam turbine, the mean blade speed is 60 m/s. Steam is at a pressure of 3 bar with a temperature of 200° C. If the fixed and moving blades, at this stage, have inlet angle 30° and exit angle 20°, determine : (a) blade height at this stage, if the blade height is 1/10 of the mean blade ring diameter and the steam flow is 10 kg/s, (b) power developed by a pair of fixed and moving blade rings at this stage, and (c) the heat drop required by the pair if the steam expand with an efficiency of 85%.*

Solution. Given : $V_b = 60$ m/s ; $p = 3$ bar ; $T = 200°$ C ; $\theta = \beta = 30°$; $\phi = \alpha = 20°$

(*a*) *Blade height*

Let d = Mean diameter of blade ring,

h = Blade height = $d/10$, and ... (Given)

m = Mass of steam flow = 10 kg/s ... (Given)

Now let us draw the combined velocity triangle, as shown in Fig. 23.11, as discussed below :

1. First of all, draw a horizontal line and cut off AB equal to 60 m/s, to some suitable scale, to represent the blade speed (V_b).

2. Now draw inlet velocity triangle ABC on the base AB with $\alpha = 20°$ and $\theta = 30°$.

3. Similarly draw outlet velocity triangle ABD on the same base AB with $\phi = 20°$ and $\beta = 30°$.

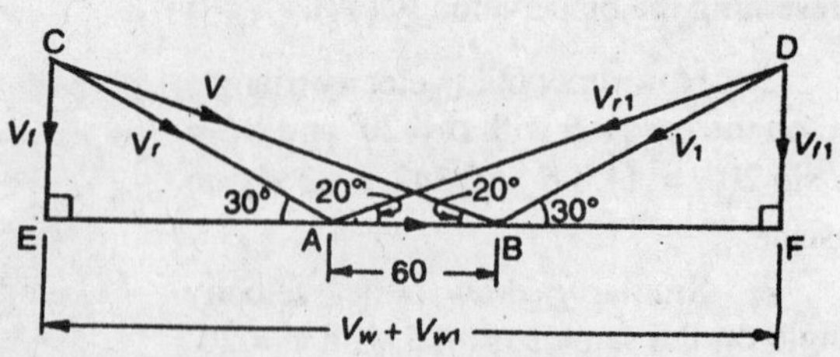

Fig. 23.11

4. From C and D draw perpendiculars to meet the line AB produced at E and F.

By measurement* from the velocity diagram, we find that

Change in the velocity of whirl, $(V_w + V_{w1}) = EF = 265$ m/s

and velocity of flow at exit, $V_{f1} = DF = 60$ m/s

From steam tables of superheated steam, corresponding to a pressure of 3 bar and 200° C, we find that the specific volume of steam,

$$v_{sup} = 0.7164 \text{ m}^3/\text{kg}$$

We know that mass of steam flow (m),

$$10 = \frac{\pi (d+h)\, h\, V_{f1}}{v_{sup}} = \frac{\pi (10\,h + h)\, h \times 60}{0.7164} = 2894\, h^2$$

$$\therefore \quad h = 0.059 \text{ m} = 59 \text{ say } 60 \text{ mm} \quad \textbf{Ans.}$$

(b) Power developed

We know that power developed by a pair of fixed and moving blade rings,

$$P = m\,(V_w + V_{w1})\, V_b = 10 \times 265 \times 60 = 159\,000 \text{ W}$$

$$= 159 \text{ kW} \quad \textbf{Ans.}$$

(c) Heat drop required by the pair

Since the steam expands with an efficiency of 85%, therefore heat drop required by the pair

$$= \frac{159}{0.85} = 187 \text{ kJ/s} \quad \textbf{Ans.} \qquad \ldots (\because 1 \text{ kW} = 1 \text{ kJ/s})$$

Example 23.6. *At a certain pair in a reaction turbine, the steam leaves the fixed blade at a pressure of 3 bar with a dryness fraction of 0.98 and a velocity of 130 m/s. The blades are 20 mm high and dicharge angle for both the rings is 20°. The ratio of axial velocity of flow to the blade velocity is 0.7 at inlet and 0.76 at exit from the moving blade. If the turbine uses 4 kg of steam per second with 5% tip leakage, find the mean blade diameter and the power developed in the ring.*

Solution. Given : $p = 3$ bar ; $x = 0.98$; $V = 130$ m/s ; $h = 20$ mm $= 0.02$ m ; $\alpha = \phi = 20°$; $V_f = 0.7\, V_b$; $V_{f1} = 0.76\, V_b$; $m_1 = 4$ kg/s

Since the tip leakage is 5%, therefore actual mass of steam flow in the turbine,

$$m = 4 - (4 \times 0.05) = 3.8 \text{ kg/s}$$

Mean blade diameter

Let d_m = Mean blade diameter.

---

* These values may also be found out analytically by the geometry of the velocity diagram by method discussed on page 526 or by the sine rule as discussed below :

$$\frac{V}{\sin 150°} = \frac{V_r}{\sin 20°} = \frac{60}{\sin 10°}$$

$$\therefore \quad V_r = \frac{60}{\sin 10°} \times \sin 20° = \frac{60}{0.1736} \times 0.3420 = 118.2 \text{ m/s}$$

and $\quad V_f = V_r \sin 30° = 118.2 \times 0.5 = 59.1$ m/s

$$\therefore \quad EA = V_r \cos 30° = 118.2 \times 0.866 = 102.4 \text{ m/s}$$

and $\quad (V_w + V_{w1}) = EA + AB + BF = 102.4 + 60 + 102.4 = 264.8$ m/s

From the combined velocity of triangle, we find that the velocity of flow at inlet,

$$V_f = V \sin 20° = 130 \times 0.342 = 44.46 \text{ m/s}$$

and blade velocity, $$V_b = \frac{V_f}{0.7} = \frac{44.46}{0.7} = 63.5 \text{ m/s}$$

∴ Velocity of flow at exit,

$$V_{f1} = 0.76\, V_b = 0.76 \times 63.5 = 48.3 \text{ m/s}$$

From steam tables, corresponding to a pressure of 3 bar, we find that specific volume of steam,

$$v_g = 0.6055 \text{ m}^3\text{/kg}$$

We know that mass of steam flow (m),

$$3.8 = \frac{\pi\, d_m \times h\, V_{f1}}{x\, v_g} = \frac{\pi\, d_m \times 0.02 \times 48.3}{0.98 \times 0.6055} = 5.1\, d_m$$

$$\therefore \quad d_m = 0.745 \text{ m} = 745 \text{ mm} \textbf{ Ans.}$$

Power developed in the ring

Now let us draw the combined velocity triangle, as shown in Fig. 23.12, as discussed below :

1. First of all, draw a horizontal line and cut off AB equal to 63.5 m/s, to some suitable scale, representing the blade velocity (V_b).

2. Now draw inlet velocity triangle ABC on the base AB with $\alpha = 20°$ and $V = 130$ m/s to the scale.

3. Similarly, draw outlet velocity triangle ABD on the same base AB with $\phi = 20°$ and V_{r1} equal to $V_{f1}/\sin 20°$ $= 48.3 / 0.342 = 141.2$ m/s, to the scale.

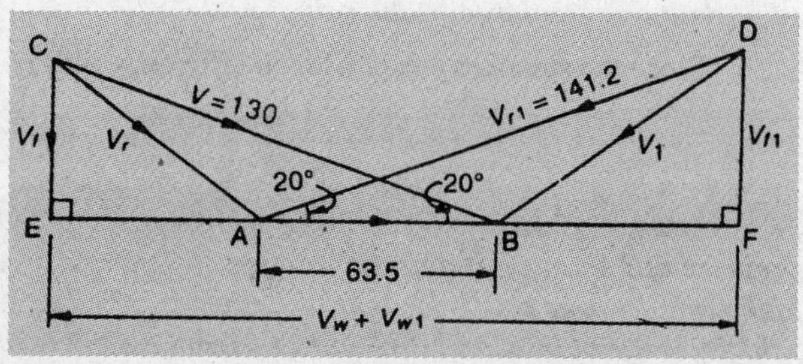

Fig. 23.12

4. From C and D draw perpendiculars to meet the line AB produced at E and F.

By measurement, we find that change in the velocity of whirl,

$$(V_w + V_{w1}) = EF = 190 \text{ m/s}$$

We know that power developed in the ring,

$$P = m\,(V_w + V_{w1})\, V_b = 3.8 \times 190 \times 63.5 = 45\,850 \text{ W}$$

$$= 45.85 \text{ kW} \textbf{ Ans.}$$

EXERCISES

1. The following particulars refer to a stage of a Parson's steam turbine, comprising one ring of fixed blades and one ring of moving blades :

Mean diameter of blade ring = 700 mm ; R.P.M. = 3000 ; Steam velocity at exit of blades = 160 m/s ; Blade outlet angle = 20° ; Steam flow through blades = 7 kg/s.

Draw a neat velocity diagram and find (*a*) Blade inlet angle ; (*b*) Tangential force on the ring of moving blades ; and (*c*) Power developed in the stage. **[Ans.** 33.5° ; 2625 N ; 288.75 kW]

2. A reaction turbine running at 360 r.p.m. consumes 5 kg of steam per second. The leakage is 10%. The discharge blade tip angle for both moving and fixed blades is 20°. The axial velocity of flow is 0.75 times blade velocity. The power developed by a certain pair is 4.8 kW where the pressure is 2 bar and dryness fraction is 0.95. Find the drum diameter and blades height. [Ans. 0.931 m ; 83 mm]

3. A 50% reaction turbine (with symmetrical velocity triangles) running at 400 r.p.m. has the exit angle of the blades as 20° and the velocity of steam relative to the blades at exit is 1.35 times the mean blade speed. The steam flow rate is 8.33 kg/s and at a particular stage, the specific volume is 1.381 m³/kg. Calculate for this stage 1. a suitable blade height, assuming the rotor mean diameter 12 times the blade height, and 2. the diagram work. [Ans. 138 mm ; 153.14 N-m/s]

4. The outlet angle of a blade of Parson's turbine is 20° and the axial velocity of flow of steam is 0.5 times the mean blade velocity. Draw the velocity diagram for a stage consisting of one fixed and one moving row of blades.

It is given that mean diameter = 710 mm and speed of rotation = 3000 r.p.m. Find the inlet angle of blades if the steam is to enter the blade channels without shock.

If the blade height is 64 mm, the mean steam pressure 5.6 bar, the steam dry saturation (v_g = 0.3434 m³/kg) ; find the power developed in the stage. [Ans. 50° ; 516.9 kW]

5. At a particular ring of a reaction turbine, the blade speed is 66 m/s and the flow of steam is 4 kg/s dry saturated at 1.4 bar. Both fixed and moving blades have inlet and exit angles of 35° and 20° respectively. Calculate : 1. the required blade height which is to be one-tenth of the mean blade ring diameter, 2. the power developed by the pair of rings, and 3. the heat drop required by the pair if the steam expand with an efficiency of 80 percent.

[Ans. 55 mm ; 57 kW ; 71.3 kJ/s]

6. The blade angles of both fixed and moving blades of a reaction steam turbine are 35° at the receiving tips and 20° at the discharging tips. At a certain point in the turbine, the drum diameter is 1.37 m and the blade height is 127 mm. The pressure of steam supply to a ring of fixed blades at this point is 1.25 bar and the dryness fraction is 0.925. Find the workdone in next row of moving blades for 1 kg of steam at 600 r.p.m., the steam passing through the blades without shock.

Assuming an efficiency of 85% for the pair of rings of fixed and moving blades, find the heat drop in the pair and the state of steam at entrance to the next row of fixed blades. [Ans. 6.77 kN-m/s ; 7.962 kJ/s]

QUESTIONS

1. Distinguish between impulse and reaction turbine.
2. Explain the functions of the blading of a reaction turbine.
3. Draw the combined velocity triangle for a single stage reaction turbine and derive an expression for workdone per stage.
4. Define the term 'degree of reaction' as applied to a reaction turbine. Show that for a Parson's reaction turbine, the degree of reaction is 50 percent.
5. What do you understand by the term 'height of blades' as applied to a reaction turbine.

OBJECTIVE TYPE QUESTIONS

1. In a reaction turbine
   (*a*) the steam is allowed to expand in the nozzle, where it gives a high velocity before it enters the moving blades
   (*b*) the expansion of steam takes place partly in the fixed blades and partly in the moving blades
   (*c*) the steam is expanded from a high pressure to a condenser pressure in one or more nozzles
   (*d*) the pressure and temperature of steam remains constant.

2. The Parson's reaction turbine has
   (*a*) only moving blades (*b*) only fixed blades
   (*c*) identical fixed and moving blades (*d*) fixed and moving blades of different shape
3. The degree of reaction is defined as the ratio of
   (*a*) heat drop in the fixed blades to the heat drop in the moving blades
   (*b*) heat drop in the moving blades to the heat drop in the fixed blades
   (*c*) heat drop in the moving blades to the total heat drop in the fixed and moving blades
   (*d*) total heat drop in the fixed and moving blades to the heat drop in the moving blades
4. For a Parson's reaction turbine, the degree of reaction is
   (*a*) 20% (*b*) 30% (*c*) 40% (*d*) 50%
5. In a reaction turbine, when the degree of reaction is zero, then there is
   (*a*) no heat drop in the moving blades
   (*b*) no heat drop in the fixed blades
   (*c*) maximum heat drop in the moving blades
   (*d*) maximum heat drop in the fixed blades

ANSWERS

1. (*b*)	2. (*c*)	3. (*c*)	4. (*d*)	5. (*a*)

24

Performance of Steam Turbines

1. Introduction. 2. Efficiencies of Steam Turbine. 3. Condition for Maximum Efficiency of an Impulse Turbine. 4. Condition for Maximum Efficiency of a Reaction Turbine. 5. Compounding of Impulse Steam Turbines (Methods of Reducing Rotor Speeds). 6. Velocity Compounding of an Impulse Turbine. 7. Pressure Compounding of an Impulse Turbine. 8. Pressure-velocity Compounding of an Impulse Turbine. 9. Internal Losses in Turbines. 10. Governing of Steam Turbines. 11. Throttle Governing of Steam Turbines.

24.1. Introduction

In the last two chapters, we have discussed impulse and reaction steam turbines. In these chapters, we have discussed power generated in these turbines. But in this chapter, we shall discuss their performance *i.e.* efficiencies and governing.

24.2. Efficiencies of Steam Turbine

The following efficiencies of impulse as well as reaction steam turbines are important from the subject point of view :

1. *Diagram or blading efficiency.* It is the ratio of the work done on the blades to the energy supplied to the blades.

Let V = Absolute velocity of inlet steam in m/s, and

m = Mass of steam supplied in kg/s.

∴ Energy supplied to the blade per second,

$$= \frac{mV^2}{2} \text{ J/s}$$

We know that work done on the blades per second

$$= m(V_w + V_{w1})\, V_b \text{ J/s}$$

∴ Diagram* or blading efficiency,

$$\eta_b = \frac{m(V_w + V_{w1})\, V_b}{mV^2/2} = \frac{2(V_w + V_{w1})\, V_b}{V^2}$$

The work done on the turbine blades may also be obtained from the kinetic energy at inlet and exit as discussed below :

Let V_1 = Absolute velocity of exit steam in m/s.

We know that kinetic energy at inlet per second

$$= \frac{mV^2}{2} \text{ J/s}$$

* It is called diagram efficiency because the quantities involved are obtained from velocity diagram.

and kinetic energy at exit per second

$$= \frac{mV_1^2}{2} \text{ J/s}$$

$\therefore$ Work done on the blades per second,

$$= \text{Loss of kinetic energy}$$

$$= \frac{mV^2}{2} - \frac{mV_1^2}{2} = \frac{m}{2}(V^2 - V_1^2) \text{ J/s}$$

and power developed, $$P = \frac{m(V^2 - V_1^2)}{2} \text{ watts} \quad \ldots (\because 1 \text{ J/s} = 1 \text{ watt})$$

$\therefore$ Blading efficiency, $$\eta_b = \frac{\frac{m}{2}(V^2 - V_1^2)}{\frac{mV^2}{2}} = \frac{V^2 - V_1^2}{V^2}$$

2. *Gross or stage efficiency.* It is the ratio of the work done on the blades per kg of steam to the total energy supplied per stage per kg of steam.

Let h_1 = Enthalpy or total heat of steam before expansion through the nozzle in kJ/kg of steam, and

h_2 = Enthalpy or total heat of steam after expansion through the nozzle in kJ/kg of steam.

$\therefore$ Enthalpy or heat drop in the nozzle ring of an impulse wheel,

$$h_d = h_1 - h_2 \text{ (in kJ/kg)}$$

and total energy supplied per stage $= 1000\, h_d$ J/kg of steam

We know that work done on the blade per kg of steam

$$= 1\,(V_w + V_{w1})\, V_b \text{ J/kg of steam}$$

$\therefore$ Gross or stage efficiency,

$$\eta_s = \frac{(V_w + V_{w1})\, V_b}{1000\, h_d} = \frac{(V_w + V_{w1})\, V_b}{1000\,(h_1 - h_2)}$$

3. *Nozzle efficiency.* It is the ratio of energy supplied to the blades per kg of steam to the total energy supplied per stage per kg of steam.

We know that energy supplied to the blades per kg of steam

$$= V^2/2 \text{ (in joules)}$$

$\therefore$ Nozzle efficiency, $$\eta_n = \frac{V^2/2}{1000\, h_d} = \frac{V^2}{2000\, h_d}$$

Note : We know that stage efficiency,

$$\eta_s = \frac{(V_w + V_{w1})\, V_b}{1000\, h_d} = \frac{2\,(V_w + V_{w1})\, V_b}{V^2} \times \frac{V^2}{2000\, h_d}$$

$$= \eta_b \times \eta_n$$

i.e. Stage efficiency = Blading efficiency × Nozzle efficiency

Example 24.1. *The velocity of steam at inlet to a simple impulse turbine is 1000 m/s and the nozzle angle is 20°. The mean blade speed is 400 m/s and the blades are symmetrical. The mass flow*

rate of steam is 0.75 kg/s. The friction effects on the blades are negligible. Estimate : (a) the blade angles ; (b) the tangential force on the blades; (c) the axial thrust; (d) the diagram power; and (e) the diagram efficiency.

Solution. Given : $V = 1000$ m/s ; $\alpha = 20°$; $V_b = 400$ m/s ; $\theta = \phi$ for symmetrical blades, $m = 0.75$ kg/s

(*a*) *Blade angles*

Now draw the combined velocity triangle, as shown in Fig. 24.1, as discussed below :

1. First of all, draw a horizontal line and cut off *AB* equal to 400 m/s, to some suitable scale, to represent the blade speed (V_b).

2. Now draw inlet velocity triangle *ABC* on the base *AB* with $\alpha = 20°$ and $BC = V = 1000$ m/s. By measurement, we find that blade angle at inlet,

$$\theta = 33° \text{ Ans.}$$

3. Similarly, draw outlet velocity triangle *ABD* on the same base *AB* with $\phi = \theta = 33°$ and $V_{r1} = V_r$.

4. From *C* and *D*, draw perpendiculars to meet the line *AB* produced at *E* and *F*.

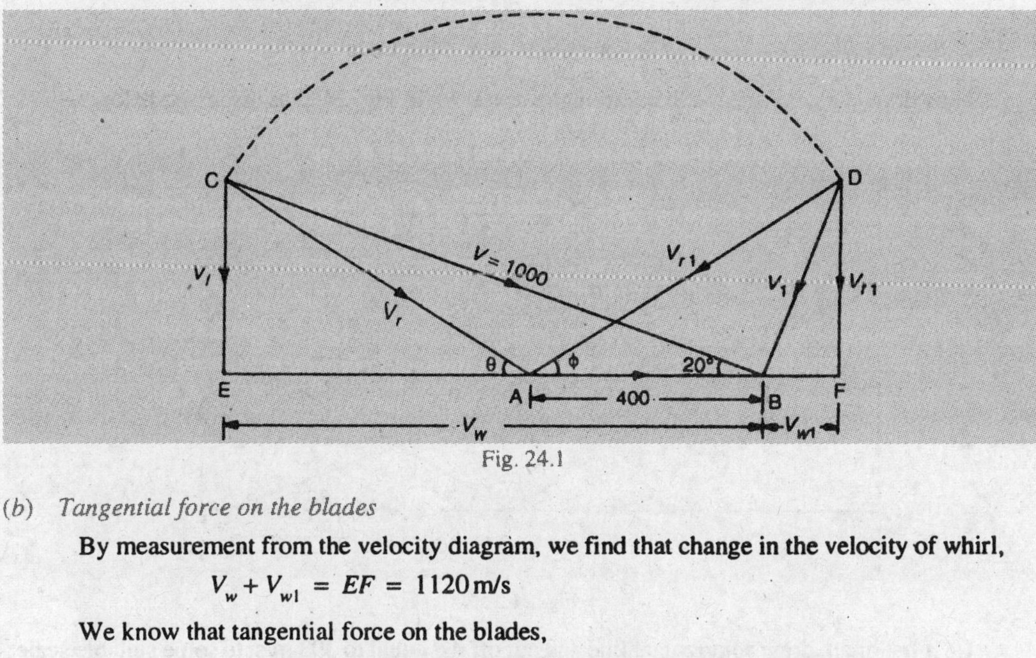

Fig. 24.1

(*b*) *Tangential force on the blades*

By measurement from the velocity diagram, we find that change in the velocity of whirl,

$$V_w + V_{w1} = EF = 1120 \text{ m/s}$$

We know that tangential force on the blades,

$$F_X = m\,(V_w + V_{w1}) = 0.75 \times 1120 = 840 \text{ N Ans.}$$

(*c*) *Axial thrust*

We know that axial thrust,

$$F_Y = m\,(V_f - V_{f1}) = 0 \text{ Ans.}$$

Note : Since the blades are symmetrical and friction effects on the blades are neglected, therefore $V_f = V_{f1}$.

(*d*) *Diagram power*

We know that diagram power,

$$P = m\,(V_w + V_{w1})\,V_b = 0.75 \times 1120 \times 400 = 336\,000 \text{ W}$$

$$= 336 \text{ kW Ans.}$$

(*e*) *Diagram efficiency*

We know that diagram efficiency,

$$\eta_b = \frac{2(V_w + V_{w1})V_b}{V^2} = \frac{2 \times 1120 \times 400}{(1000)^2} = 0.896 \text{ or } 89.6\% \text{ Ans.}$$

Example 24.2. *The following particulars refer to a single row impulse turbine :*

Mean diameter of blade ring = 2.5 m ; Speed = 3000 r.p.m. ; Nozzle angle = 20° ; Ratio of blade velocity to steam velocity = 0.4 ; Blade friction factor = 0.8 ; Blade angle at exit = 3° less than that at inlet ; Steam flow = 36 000 kg/h.

Draw velocity diagram for moving blade and estimate (a) Power developed ; (b) Blade efficiency ; and (c) Steam consumption is kg/kW h.

Solution. Given : $d_m = 2.5$ m ; $N = 3000$ r.p.m. ; $\alpha = 20°$; $V_b / V = 0.4$; $K = V_{r1} / V_r = 0.8$; $\phi = \theta - 3°$; $m = 36\,000$ kg/h = 10 kg/s

We know that blade velocity,

$$V_b = \frac{\pi d_m N}{60} = \frac{\pi \times 2.5 \times 3000}{60} = 393 \text{ m/s}$$

$\therefore$ Steam velocity, $V = V_b / 0.4 = 393 / 0.4 = 982.5$ m/s

Now draw the combined velocity triangle, as shown in Fig. 24.2, as discussed below :

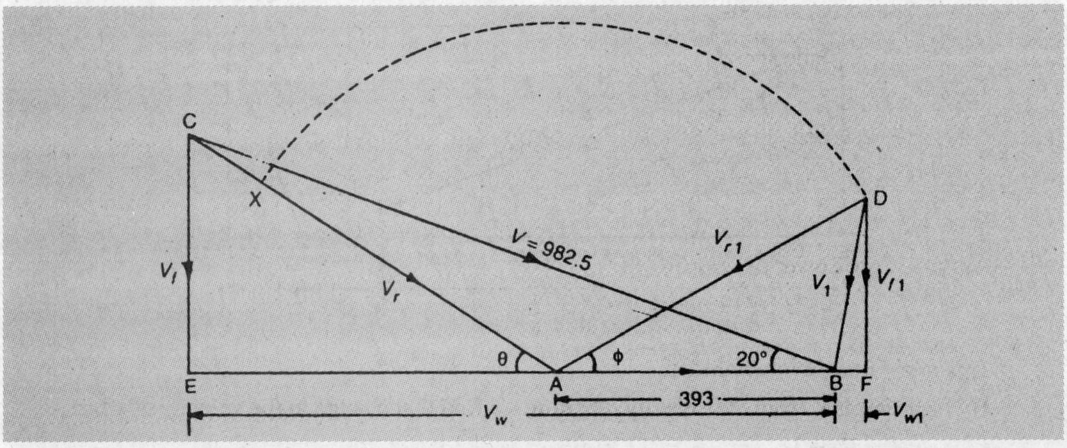

Fig. 24.2

1. First of all, draw a horizontal line and cut off AB equal to 393 m/s, to some suitable scale, to represent the blade velocity (V_b).

2 Now draw inlet velocity triangle ABC on the base AB with $\alpha = 20°$ and $BC = V = 982.5$ m/s. By measurement, we find that blade angle at inlet (θ) = 32.5°, and relative velocity of steam at inlet (V_r) = 626.7 m/s.

3. Similarly, draw outlet velocity triangle ABD on the same base AB with $\phi = \theta - 3° = 32.5 - 3 = 29.5°$ and $V_{r1} = AX = 0.8\,V_r = 0.8 \times 626.7 = 501.4$ m/s.

4. From C and D, draw perpendiculars to meet the line AB produced at E and F.

By measurement from velocity triangle, we find that change in the velocity of whirl,

$$V_w + V_{w1} = EF = 967 \text{ m/s}$$

(*a*) *Power developed*

We know that power developed,

$$P = m(V_w + V_{w1})V_b = 10 \times 967 \times 393 = 3800 \times 10^3 \text{ W}$$

$$= 3800 \text{ kW } \textbf{Ans.}$$

(*b*) *Blade efficiency*

We know that blade efficiency,

$$\eta_b = \frac{2(V_w + V_{w1})V_b}{V^2} = \frac{2 \times 967 \times 393}{(982.5)^2} = 0.787 \text{ or } 78.7\% \textbf{ Ans.}$$

(*c*) *Steam consumption in kg/kWh*

We know that steam consumption

$$= \frac{m}{P} = \frac{36\,000}{3800} = 9.47 \text{ kg/kWh } \textbf{Ans.}$$

Example 24.3. *A single row impulse turbine receives 3 kg/s steam with a velocity of 425 m/s. The ratio of blade speed to jet speed is 0.4 and the stage output is 170 kW. If the internal losses due to disc friction etc. amount to 15 kW, determine the blading efficiency and the blade velocity coefficient. The nozzle angle is 16° and the blade exit angle is 17°.*

Solution. Given : $m = 3$ kg/s ; $V = 425$ m/s ; $V_b / V = 0.4$; Stage output = 170 kW ; Internal losses = 15 kW ; $\alpha = 16°$; $\phi = 17°$

Blading efficiency

We know that blade speed,

$$V_b = V \times 0.4 = 425 \times 0.4 = 170 \text{ m/s}$$

and total power developed, P = Stage output + Internal losses

$$= 170 + 15 = 185 \text{ kW} = 185 \times 10^3 \text{ W}$$

Let $V_w + V_{w1}$ = Change in the velocity of whirl.

We know that power developed (P),

$$185 \times 10^3 = m(V_w + V_{w1})V_b = 3(V_w + V_{w1})\,170$$

$$\therefore \quad V_w + V_{w1} = 185 \times 10^3 / 3 \times 170 = 363 \text{ m/s}$$

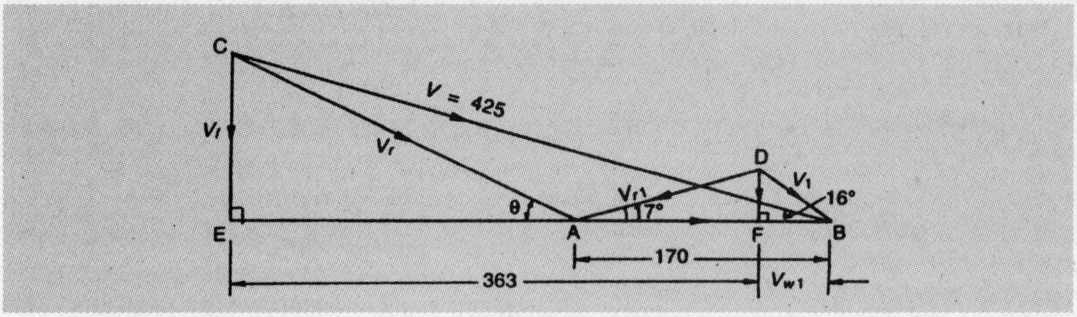

Fig. 24.3

Now draw the combined velocity triangle, as shown in Fig. 24.3, as discussed below :

1. First of all, draw a horizontal line and cut off AB equal to 170 m/s, to some suitable scale, to represent the blade speed (V_b).

2. Now draw the inlet velocity triangle *ABC* on the base *AB* with $\alpha = 16°$ and $V = 425$ m/s, to the scale.

3. Similarly, draw the outlet velocity triangle *ABD* on the same base *AB* with $\phi = 17°$ and $(V_w + V_{w1}) = 363$ m/s to the scale.

4. From *C* and *D* draw perpendiculars to meet the line *AB* at *E* and *F*. From the geometry of the figure, we find that V_{w1} is in the opposite direction of V_w. Therefore $(V_w - V_{w1}) = 363$ m/s.

By *measurement from the velocity diagram, we find that

Relative velocity at inlet, $V_r = 265$ m/s

and relative velocity at outlet, $V_{r1} = 130$ m/s

We know that blading efficiency,

$$\eta_b = \frac{2\,(V_w - V_{w1})\,V_b}{V^2} = \frac{2 \times 363 \times 170}{(425)^2} = 0.683 \quad \text{or} \quad 68.3\ \%\ \textbf{Ans.}$$

Blade velocity coefficient

We know that blade velocity coefficient,

$$K = \frac{V_{r1}}{V_r} = \frac{130}{265} = 0.49\ \textbf{Ans.}$$

Example 24.4. *A two row curtis wheel operates at a blade speed of 150 m/s, when receiving 3 kg of steam per second at 10.5 bar dry and saturated. The ratio of blade speed to the steam speed at exit from the nozzle is 0.21 and nozzle efficiency is 90%. The nozzles are inclined at 16° to the plane of the wheel. The outlet angles of the first row moving, fixed and second row moving blades are respectively 20°, 24° and 32° with respective blade velocity coefficients 0.79, 0.83 and 0.88. Determine : (a) The pressure of steam at exhaust; (b) Diagram efficiency ; and (c) Stage efficiency.*

Solution. Given : $V_b = 150$ m/s; $m = 3$ kg/s ; $p = 10.5$ bar ; $V_b/V = 0.21$; $k = 90\% = 0.9$; $\alpha = 16°$; $\phi = 20°$; $\alpha' = 24°$; $\phi' = 32°$; $V_{r1}/V_r = 0.79$ for first moving blades ; $V'/V_1 = 0.83$ for fixed blades ; $V_{r1}'/V_r = 0.88$ for second moving blades

We know that steam speed at exit from the nozzle,

$$V = V_b / 0.21 = 150 / 0.21 = 714.3 \text{ m/s}$$

---

* These values may also be obtained from the geometry of the combined velocity triangle, as discussed below :

$$V_f = 425 \sin 16° = 425 \times 0.2756 = 117.1 \text{ m/s}$$

Now $$\tan\theta = \frac{117.1}{425 \cos 16° - 170} = \frac{117.1}{(425 \times 0.9613) - 170} = 0.4909$$

$\therefore$ $$\theta = 26.15° \text{ and } \angle ACB = 26.15° - 16° = 10.15°$$

Now in triangle *ABC*,

$$\frac{V_r}{\sin 16°} = \frac{170}{\sin 10.15°}$$

$\therefore$ $$V_r = \frac{170}{\sin 10.15°} \times \sin 16° = \frac{170}{0.1763} \times 0.2756 = 265.8 \text{ m/s}$$

We know that $$AF = 361 - V_r \cos 26.15° = 361 - (265.8 \times 0.8976) = 122.4 \text{ m/s}$$

$$V_{r1} = \frac{122.4}{\cos 17°} = \frac{122.4}{0.9563} = 128 \text{ m/s}$$

(*a*) *Pressure of steam at exhaust*

Let p_2 = Pressure of steam at exhaust, and

h_d = Enthalpy or heat drop from pressure p_1 to p_2.

We know that the velocity of steam at exit from nozzle (*V*),

$$714.3 = 44.72\sqrt{k\,h_d} = 44.72\sqrt{0.9\,h_d}$$

$$\therefore \quad h_d = \left(\frac{714.3}{44.72}\right)^2 \times \frac{1}{0.9} = 283.5 \text{ kJ/kg}$$

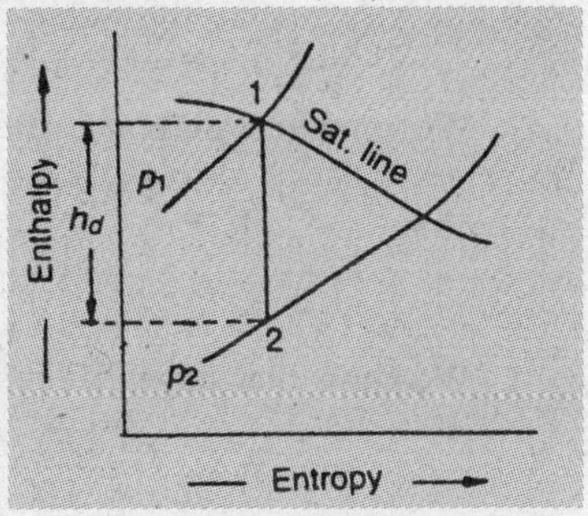

Fig. 24.4

Now let us complete the Mollier diagram for the steam flow through the nozzle, as shown in Fig. 24.4. From the Mollier diagram, we find that

p_2 = 2 bar **Ans.**

(*b*) *Diagram efficiency*

Now draw the combined velocity triangle for the two stage impulse turbine, as shown in Fig. 24.5, as discussed below :

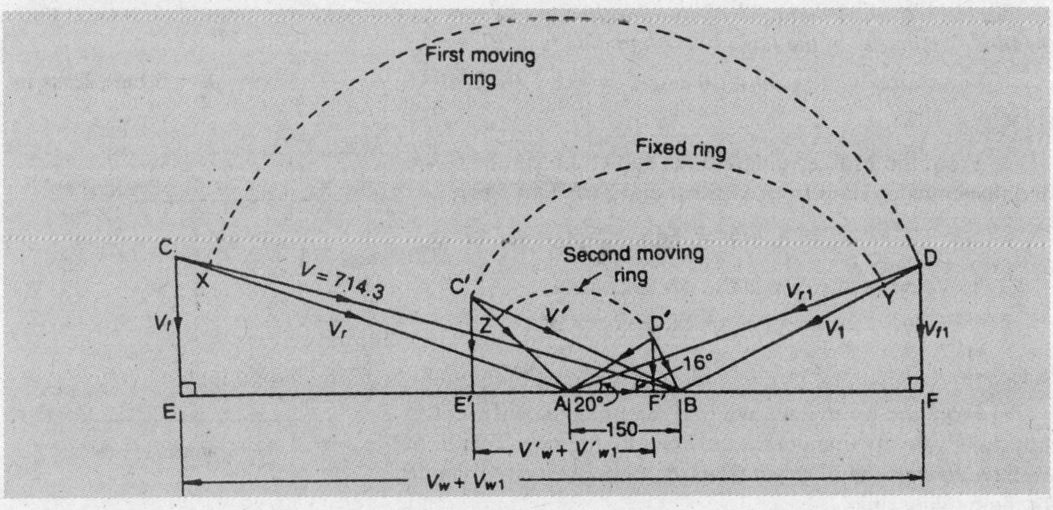

Fig. 24.5

1. First of all, draw a horizontal line, and cut off *AB* equal to 150 m/s to a suitable scale to represent the blade velocity (V_b).

2. Now draw the inlet velocity triangle *ABC* for the first moving ring on the base *AB* with nozzle angle for the first moving ring (α = 16°) and *V* = 714.3 m/s. By measuring the side *AC*, we find that V_r = 570 m/s.

3 Now draw the outlet velocity triangle *ABD* for the first moving ring on the same base *AB* with ϕ = 20° and $V_{r1} = 0.79\ V_r = 0.79 \times 570 = 450.3$ m/s, to the scale. By measuring the side *BD*, we find that V_1 = 375 m/s.

4. Now draw the inlet velocity triangle *ABC'* for the second moving ring on the same base *AB* with α' = 24° and $V' = 0.83 \times 375 = 311.2$ m/s, to the scale. By measuring the side *AC'*, we find that V_r' = 185 m/s.

5. Now draw the outlet velocity triangle ABD' for the second moving ring on the same base AB with $\phi' = 32^\circ$ and $V_{r1}' = 0.88 \times V_r' = 0.88 \times 185 = 162.8$ m/s to the scale.

The following values are measured from the combined velocity diagram :

$$EF = (V_w + V_{w1}) = 1020 \text{ m/s}$$

$$E'F' = (V_w' + V_{w1}') = 275 \text{ m/s}$$

We know that diagram efficiency,

$$\eta_b = \frac{2\,AB\,(EF + E'F')}{V^2} = \frac{2 \times 150\,(1020 + 275)}{(714.3)^2} = 0.761 \text{ or } 76.1\% \textbf{ Ans.}$$

(c) *Stage efficiency*

We know that stage efficiency,

$$\eta_s = \frac{AB\,(EF + E'F')}{1000\,h_d} = \frac{150\,(1020 + 275)}{1000 \times 283.5} = 0.685 \text{ or } 68.5\% \textbf{ Ans.}$$

Example 24.5. *In an impulse turbine, there are two rings of moving blades separated by fixed blades. The tips of the moving blades are inclined at 30° to the plane of motion. The steam enters the wheel chamber through a nozzle to which the steam is supplied at 240° C and 11.5 bar. The pressure in the wheel chamber is 5 bar. Find the speed of the blades, so that the steam finally discharged is axial. Assume a 10% loss in velocity, while passing through a blade ring. Also determine the blade tip angles of the fixed blade and diagram efficiency.*

Solution. Given : ϕ, ϕ', θ and $\theta' = 30^\circ$; $T = 240^\circ$ C ; $p_1 = 11.5$ bar ; $p_2 = 5$ bar ; Loss in velocity = 10%

From the Mollier diagram, as shown in Fig. 24.6, we find that enthalpy drop from 11.5 bar and 240° C to 5 bar,

$$h_d = h_1 - h_2 = 2915 - 2760 = 155 \text{ kJ/kg}$$

∴ Velocity of steam at the wheel chamber,

$$V = 44.72\sqrt{155} = 556.6 \text{ m/s}$$

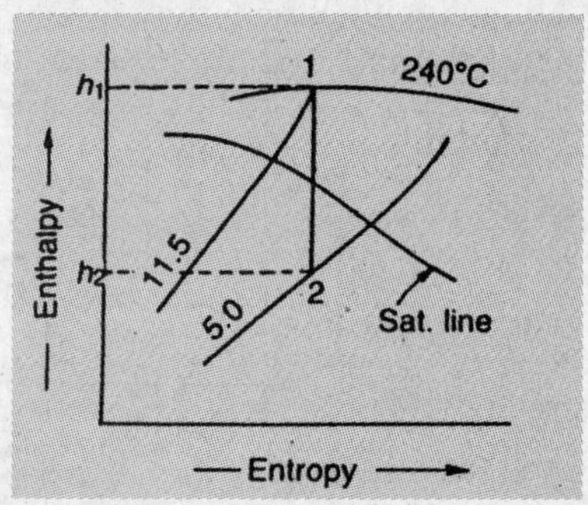

Fig. 24.6

Blade speed

From the given data, we find that it is not sufficient to draw inlet velocity triangle for the first moving ring. Therefore we have to start the diagram from the outlet velocity triangle for the moving ring.

Now draw the combined velocity diagram for the two stages of the turbine, as shown in Fig. 24.7, as discussed below :

1. First of all, draw a horizontal line and mark AB equal to 15 mm to represent the blade velocity (which is required to be found out).

2. Now draw the outlet velocity triangle ABD' for the second moving ring on the base AB with $\phi' = 30^\circ$ and $\beta' = 90^\circ$. From the geometry of the figure, we find that $D'B$ represents V_f' and $D'A$ represents V_{r1}'. By measurement, we find that $V_{r1}' = 17.3$ mm. Since there is a 10% loss in velocity due to friction, therefore, $V_r' = V_{r1}' / 0.9 = 17.3 / 0.9 = 19.2$ mm.

3. Now draw the inlet velocity triangle ABC' for the second moving ring on the same base AB with $\theta' = 30^\circ$ and $V_r' = 19.2$ mm. From the geometry of the figure, we find that $C'A$ and $C'B$ represents V_r' and V' respectively to the scale. By measurement, we find $C'B = 33$ mm. Since there is a 10% loss in velocity due to friction, therefore $V_1 = V'/0.9 = 33/0.9 = 36.7$ mm.

4. Now draw the outlet velocity triangle *ABD* for the first moving ring on the same base *AB* with $\phi = 30°$ and $V_1 = 36.7$ mm. From the geometry of the figure, we find that *DF* and *DA* represent V_{f1} and V_{r1} respectively to the scale. By measurement, we find that $V_{r1} = 49$ mm. Since there is 10% loss in velocity due to friction, therefore $V_r = V_{r1} / 0.9 = 49 / 0.9 = 54.4$ mm.

5. Now draw the inlet velocity triangle *ABC* for the first moving ring on the same base *AB* with $\theta = 30°$ and $V_r = 54.4$ mm.

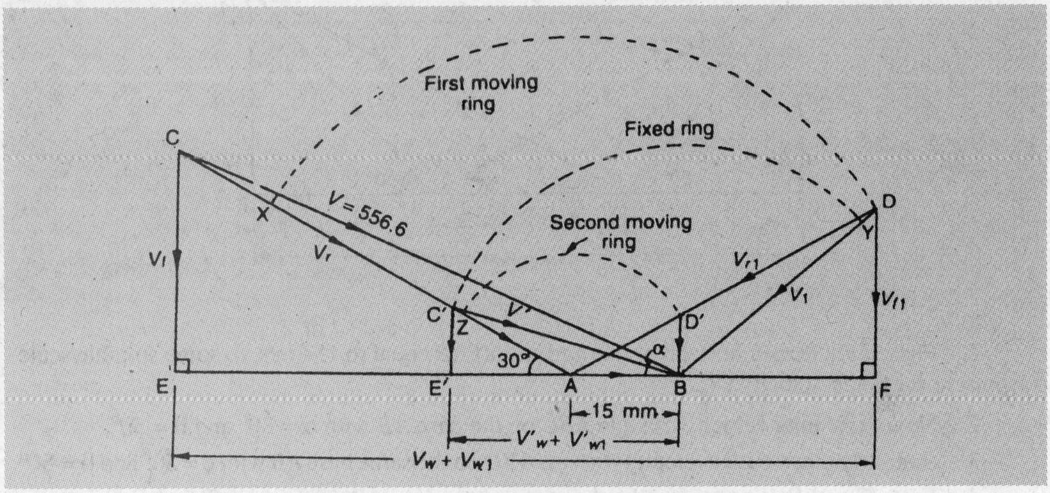

Fig. 24.7

From the combined velocity diagram, we find that the length of $BC = 67$ mm.

$$\therefore \quad \text{Scale, 1 mm} = \frac{556.6}{67} = 8.31 \text{ m/s}$$

and blade speed, $\quad V_b = 15 \times 8.31 = 124.6$ m/s

Blade tip angles

By measurement from the velocity diagram, we find that

$$\alpha' = 17° \text{ and } \beta = 43° \textbf{ Ans.}$$

Diagram efficiency

By measurement from the velocity diagram, we find that $EF = V_w + V_{w1} = 88$ mm, and $E'F' = V_w' + V_{w1}' = 31$ mm.

We know that diagram efficiency,

$$\eta_b = \frac{2\,AB\,(EF + E'F')}{V^2} = \frac{2 \times 15\,(88 + 31)}{67^2} = 0.795 \text{ or } 79.5\% \textbf{ Ans.}$$

Example 24.6. *A reaction turbine with a mean blade diameter of 1 m runs at a speed of 50 rev/second. The blades are designed with exit angles of 50° and inlet angles of 30°. If the turbine is supplied with steam at the rate of 20 kg/s and gross efficiency is 85%, determine : (a) power output of the stage; (b) specific enthalpy drop in the stage; and (c) percentage increase in relative velocity in the moving blades due to steam expansion.*

Solution. Given : $d_m = 1$ m ; $N = 50$ r.p.s. ; $\theta = \beta = 50°$; $\phi = \alpha = 30°$; $m = 20$ kg/s ; $\eta_s = 85\% = 0.85$

Power output of the stage

We know that blade velocity,

$$V_b = \pi d_m N = \pi \times 1 \times 50 = 157 \text{ m/s}$$

Now let us draw the combined velocity triangle, as shown in Fig. 24.8, as discussed below :

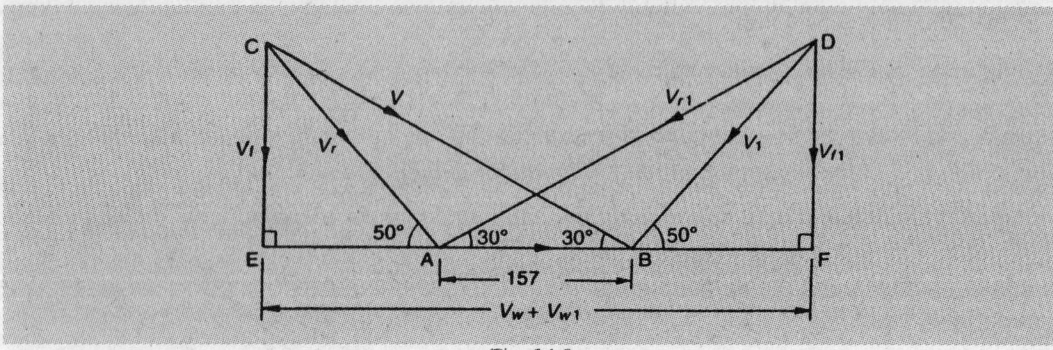

Fig. 24.8

1. First of all, draw a horizontal line and cut off *AB* equal to 157 m/s, to some suitable scale, to represent the blade velocity (V_b).

2. Now draw inlet velocity triangle *ABC* on the base *AB* with $\alpha = 30°$ and $\theta = 50°$,

3. Similarly, draw outlet velocity triangle *ABD* on the same base *AB* with $\phi = 30°$ and $\beta = 50°$.

4. From *C* and *D*, draw perpendiculars to meet the line *AB* produced at *E* and *F*.

By *measurement from the velocity triangle, we find that change in the velocity of whirl,

$$V_w + V_{w1} = EF = 450 \text{ m/s}$$

Relative velocity at inlet, $V_r = CA = 230$ m/s

and relative velocity at outlet, $V_{r1} = DA = 350$ m/s

We know that power output of the stage,

$$P = m\,(V_w + V_{w1})\,V_b = 20 \times 450 \times 157 = 1\,413\,000 \text{ W}$$

$$= 1413 \text{ kW} \quad \textbf{Ans.}$$

---

* These values may also be obtained from the geometry of the combined velocity triangle as discussed below :

In the inlet velocity triangle *ABC*,

$$\frac{V_r}{\sin 30°} = \frac{V}{\sin 130°} = \frac{157}{\sin 20°}$$

$$\therefore \quad V_r = \frac{157}{\sin 20°} \times \sin 30° = \frac{157}{0.342} \times 0.5 = 229.5 \text{ m/s}$$

and $$V \text{ (or } V_{r1}) = \frac{157}{\sin 20°} \times \sin 130° = \frac{157}{\sin 20°} \times \sin 50°$$

$$= \frac{157}{0.342} \times 0.766 = 351.6 \text{ m/s}$$

$$\therefore \quad EA = V_r \cos 50° = 229.5 \times 0.6428 = 147.5 \text{ m/s}$$

and $$(V_w + V_{w1}) = EA + AB + BF = 147.5 + 157 + 147.5 \doteq 452 \text{ m/s}$$

(b) Specific enthalpy drop in the stage

Let U = Specific enthalpy drop in the stage.

We know that stage efficiency (η_s),

$$0.85 = \frac{(V_w + V_{w1})\, V_b}{1000\, h_d} = \frac{450 \times 157}{1000\, h_d} = \frac{70.65}{h_d}$$

$$\therefore \quad h_d = 70.65 / 0.85 = 83.1 \text{ kJ } \textbf{Ans.}$$

(c) Percentage increase in the relative velocity

We know that increase in the relative velocity in the moving blades due to steam expansion

$$= \frac{V_{r1} - V_r}{V_r} = \frac{351.6 - 229.5}{229.5} = 0.532 \text{ or } 53.2\,\% \textbf{ Ans.}$$

24.3. Condition for Maximum Efficiency of an Impulse Turbine

We have already discussed in Art. 24.2 that the blading efficiency (or diagram efficiency) of an impulse turbine,

$$\eta_b = \frac{V^2 - V_1^2}{V^2} = \frac{2\,(V_w + V_{w1}) V_b}{V^2}$$

It may be noted that the blading efficiency will be maximum when V_1 is minimum. From the combined velocity triangle, we see that the value of V_1 will be minimum, when β is equal to 90°. In other words, for maximum efficiency, the steam should leave the turbine blades at right angles to their motion. The modified combined velocity triangle for maximum efficiency is shown in Fig. 24.9. It may also be noted that for maximum efficiency, V_{w1} is zero.

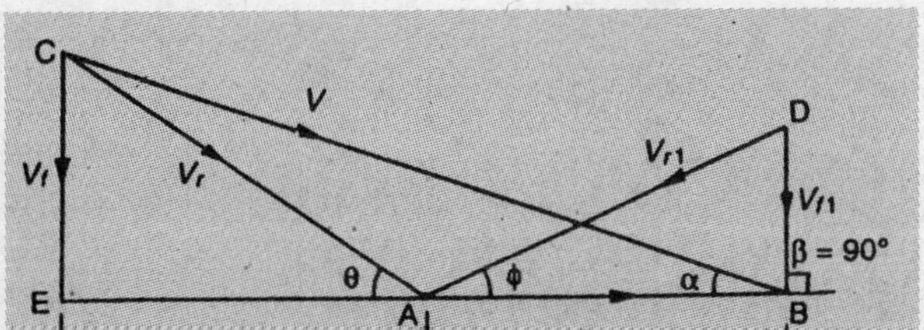

Fig. 24.9. Combined velocity triangle for maximum efficiency.

$\therefore$ For maximum efficiency, substituting $V_{w1} = 0$ in the general expression for efficiency of an impulse turbine, we have

$$\eta_{max} = \frac{2 \times V_w \times V_b}{V^2}$$

We know that in De-lavel turbine, $\theta = \phi$, considering $V_r = V_{r1}$ (or in other words, neglecting blade friction),

$$V_b = \frac{1}{2} V_w = 0.5\, V_w \qquad \ldots (\because \Delta EAC = \Delta ADB)$$

$$= 0.5\, V \cos \alpha \qquad \ldots (\because V_w = V \cos \alpha)$$

$$\therefore \quad \eta_{max} = \frac{2 \times V_w \times 0.5\, V_w}{V^2} = \frac{V_w^2}{V^2} = \cos^2 \alpha \qquad \ldots (\because V_w / V = \cos \alpha)$$

Example 24.7. *The nozzles of the impulse stage of a turbine receive steam at 15 bar and 300° C and discharge it at 1 bar. The nozzle efficiency is 95% and the nozzle angle is 20°. The blade speed is that required for maximum efficiency and the entry of steam to the blades is without shock.*

The blade exit angle is 5° less than the blade inlet angle. The blade velocity coefficient is 0.9 and the steam flow rate is 1350 kg/h.

Calculate (a) the diagram power, and (b) the stage efficiency.

Solution. Given : p_1 = 15 bar ; T_1 = 300° C ; p_2 = 1 bar ; K = 95% = 0.95 ; α = 20° ; $\phi = \theta - 5°$; $K = V_{r1}/V_r = 0.9$; m = 1350 kg/h = 0.375 kg/s

From Mollier diagram, as shown in Fig. 24.10, we find that enthalpy drop from 15 bar and 300° C to 1 bar,

$$h_d = h_1 - h_2$$

$$= 3040 - 2520 = 520 \text{ kJ/kg}$$

∴ Velocity of steam at inlet to the blade,

$$V = 44.72\sqrt{K\,h_d}$$

$$= 44.72\sqrt{0.95 \times 520} = 994 \text{ m/s}$$

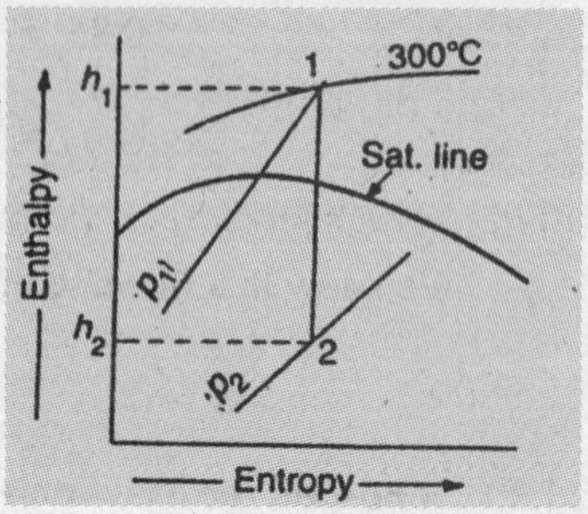

Fig. 24.10

We know that for maximum efficiency, blade velocity,

$$V_b = 0.5\,V\cos\alpha = 0.5 \times 994\cos 20° = 467 \text{ m/s}$$

Now draw the combined velocity triangle, as shown in Fig. 24.11, as discussed below :

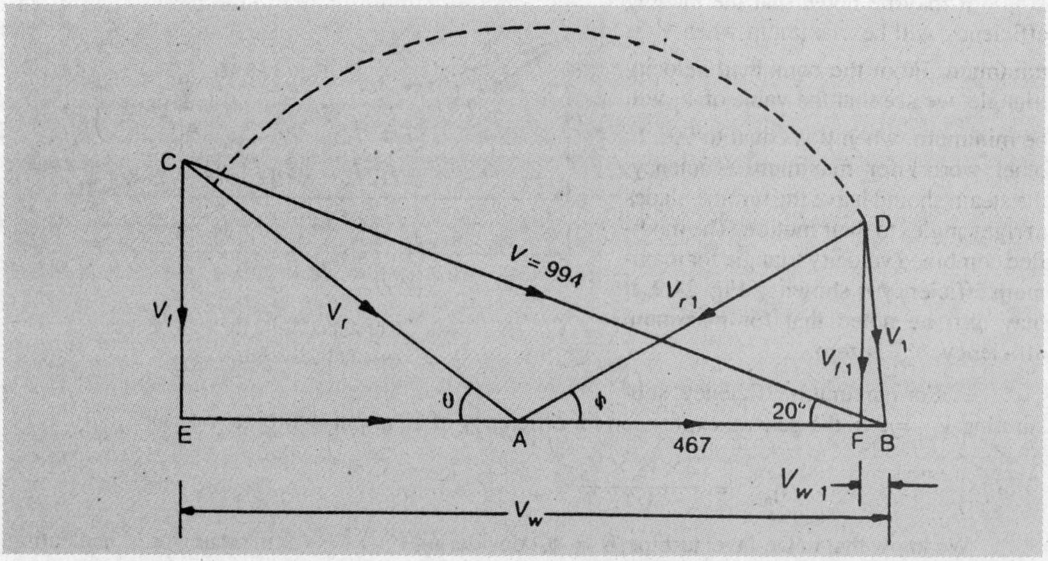

Fig. 24.11

1. First of all, draw a horizontal line and cut off AB equal to 467 m/s, to some suitable scale, to represent the blade velocity (V_b).

2. Now draw the velocity triangle ABC on the base AB with α = 20° and V = 994 m/s, to the scale. By measurement, we find that blade inlet angle (θ) = 36° and relative velocity of steam at inlet (V_r) = 555 m/s.

3. Similarly, draw outlet triangle ABD on the same base AB with $\phi = \theta - 5° = 36 - 5 = 31°$ and $V_{r1} = 0.9\,V_r = 0.9 \times 555 = 499.5$ m/s, to the scale.

4. From C and D, draw perpendiculars to meet the line AB produced at E and F.

By measurement from the velocity diagram, we find that change in the velocity of whirl,

$$* \; V_w - V_{w1} = EF = 880 \text{ m/s}$$

(*a*) *Diagram power*

We know that diagram power,

$$P = m(V_w - V_{w1}) V_b = 0.375 \times 880 \times 467 = 154\,110 \text{ W}$$

$$= 154.11 \text{ kW } \textbf{Ans.}$$

(*b*) *Stage efficiency*

We know that stage efficiency,

$$\eta_s = \frac{(V_w - V_{w1}) V_b}{1000 \, h_d} = \frac{880 \times 467}{1000 \times 520} = 0.79 \text{ or } 79\% \textbf{ Ans.}$$

Example 24.8. *A De-Laval steam turbine receives dry and saturated steam at a pressure of 15 bar and discharges at a pressure of 2 bar. The nozzle is inclined at an angle of 20° to the direction of motion of the blades and 10% of the available enthalpy drop is lost in friction of the nozzle. The mean diameter of the blade ring is 1 m and the rotational speed is 6000 r.p.m.*

Assuming a blade friction of 0.8 and taking the blade to be symmetrical, determine (a) absolute velocity of the steam as it leaves the blade, (b) diagram (blade) efficiency, and (c) maximum efficiency of the turbine.

Solution. Given : $p_1 = 15$ bar ; $p_2 = 2$ bar ; $\alpha = 20°$; Enthalpy drop lost in friction = 10% ; $d_m = 1$ m ; $N = 6000$ r.p.m. ; $V_{r1}/V_r = 0.8$; $\theta = \phi$ for symmetrical blades

(*a*) *Absolute velocity of steam as it leaves the blade*

We know that blade velocity,

$$V_b = \frac{\pi \, d_m \, N}{60} = \frac{\pi \times 1 \times 6000}{60} = 314.2 \text{ m/s}$$

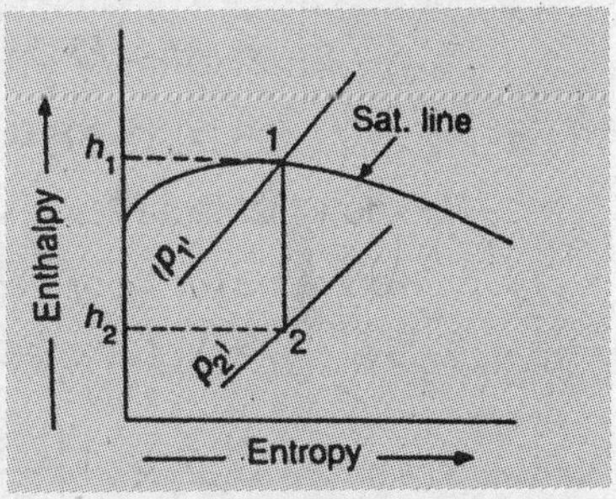

Fig. 24.12

From the Mollier diagram, as shown in Fig. 24.12, we find that enthalpy drop from 15 bar and dry saturated to 2 bar,

$$h_d = h_1 - h_2$$

$$= 2790 - 2440 = 350 \text{ kJ/kg}$$

Since the enthalpy drop lost in friction is 10 percent, therefore nozzle coefficient,

$$K = 1 - 0.1 = 0.9$$

We know that velocity of steam at inlet of the blade,

$$V = 44.72 \sqrt{K \, h_d} = 44.72 \sqrt{0.9 \times 350} = 793.7 \text{ m/s}$$

Now draw the combined velocity triangle, as shown in Fig. 24.13, as discussed below :

1. First of all, draw a horizontal line and cut off *AB* equal to 314.2 m/s, to some suitable scale, to represent the blade velocity (V_b).

2. Now draw the inlet velocity triangle *ABC* on the base *AB* with $\alpha = 20°$ and $V = 793.7$ m/s to the scale. By measurement, we find that $V_r = 510$ m/s and $\theta = 28°$.

---

* Since V_w and V_{w1} are in the same direction with respect to blade motion, therefore change in the velocity of whirl

$$= V_w - (+ V_{w1}) = V_w - V_{w1} \qquad \text{. . . (See chapter 22, Art. 22.8 Note 2)}$$

3. Similarly draw the outlet velocity triangle *ABD* on the same base *AB* with $\phi = \theta = 28°$ and $V_{r1} = 0.8\, V_r = 0.8 \times 510 = 408$ m/s to the scale.

4. From *C* and *D* draw perpendiculars to meet the line *AB* produced at *E* and *F*.

By measurement from the velocity diagram, we find that the absolute velocity of steam as it leaves the blades,

$$V_1 = BD = 200 \text{ m/s} \quad \textbf{Ans.}$$

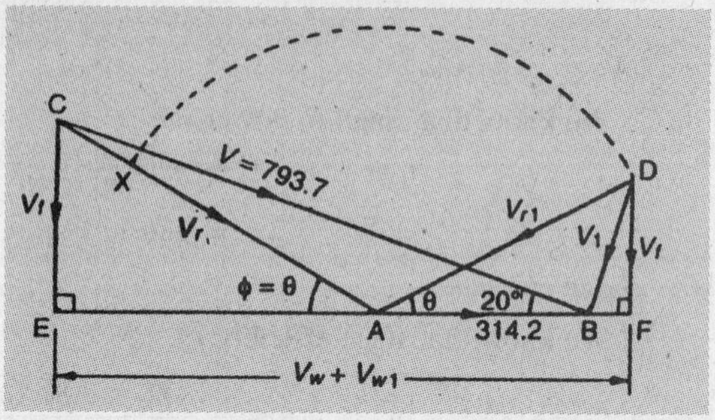

Fig. 24.13

(*b*) *Diagram (blade) efficiency*

By measurement, from the velocity diagram, we find that

$$V_w + V_{w1} = 810 \text{ m/s}$$

We know that diagram (blade) efficiency,

$$\eta_b = \frac{2\,(V_w + V_{w1})\,V_b}{V^2} = \frac{2 \times 810 \times 314.2}{(793.7)^2} = 0.808 \quad \text{or} \quad 80.8\% \quad \textbf{Ans.}$$

(*c*) *Maximum efficiency of the turbine*

We know that maximum efficiency of the turbine,

$$\eta_{max} = \cos^2\alpha = \cos^2 20° = (0.9397)^2 = 0.883 \quad \text{or} \quad 88.3\,\% \quad \textbf{Ans.}$$

24.4. Condition for Maximum Efficiency of a Reaction Turbine

We have already discussed in Art. 23.7 that work done by a reaction turbine per kg of steam

$$= AB \times EF = V_b\,(EB + AF - AB)$$

$$= V_b\,(V \cos\alpha + V_{r1} \cos\phi - V_b)$$

We know that in a Parson's reaction turbine, $\alpha = \phi$; $V = V_{r1}$; and $V_1 = V_r$.

∴ Work done per kg of steam

$$= V_b\,(2V \cos\alpha - V_b)$$

$$= V_b \times V^2 \left(\frac{2\cos\alpha}{V} - \frac{V_b}{V^2} \right) \qquad \text{. . . (Multiplying and dividing by } V^2)$$

$$= V^2 \left(\frac{2\,V_b \cos\alpha}{V} - \frac{V_b^2}{V^2} \right)$$

$$= V^2\,(2\,\rho \cos\alpha - \rho^2) \qquad \text{. . .} \left(\text{Substituting } \frac{V_b}{V} = \rho \right)$$

We know that kinetic energy supplied to the fixed blade per kg of steam

$$= \frac{V^2}{2}$$

and kinetic energy supplied to the moving blade per kg of steam

$$= \frac{(V_{r1})^2 - (V_r)^2}{2} = \frac{V^2 - V_1^2}{2} \qquad \text{. . . } (\because V_{r1} = V; \text{ and } V_r = V_1)$$

$\therefore$ Total energy supplied to the turbine

$$= \frac{V^2}{2} + \frac{V^2 - V_1^2}{2} = \frac{2V^2 - V_1^2}{2}$$

From the combined velocity triangle, we find that

$$V_1 = V_r = \sqrt{V^2 + V_b^2 - 2V V_b \cos\alpha}$$

$\therefore$ Total energy supplied to the turbine,

$$= \frac{2V^2 - (V^2 + V_b^2 - 2V V_b \cos\alpha)}{2} = \frac{V^2 - V_b^2 + 2V V_b \cos\alpha}{2}$$

$$= \frac{V^2}{2}\left(1 - \frac{V_b^2}{V^2} + \frac{2V_b}{V}\cos\alpha\right)$$

$$= \frac{V^2}{2}(1 - \rho^2 + 2\rho\cos\alpha) \qquad \ldots\left(\because \frac{V_b}{V} = \rho\right)$$

We know that diagram or blading efficiency,

$$\eta_b = \frac{\text{Work done}}{\text{Energy supplied}} = \frac{V^2(2\rho\cos\alpha - \rho^2)}{\frac{V^2}{2}(1 - \rho^2 + 2\rho\cos\alpha)}$$

$$= \frac{2(2\rho\cos\alpha - \rho^2)}{(1 - \rho^2 + 2\rho\cos\alpha)} \qquad \ldots(i)$$

It may be noted that the efficiency of the turbine will be maximum when $(1 - \rho^2 + 2\rho\cos\alpha)$ is minimum. Now for $(1 - \rho^2 + 2\rho\cos\alpha)$ to be minimum, differentiate this expression with respect to ρ and equate the same to zero, *i.e.*

$$\frac{d}{d\rho}(1 - \rho^2 + 2\rho\cos\alpha) = 0$$

or $$2\rho - 2\cos\alpha = 0$$

$$\therefore \quad \rho = \cos\alpha \text{ or } \frac{V_b}{V} = \cos\alpha \qquad \ldots\left(\because \rho = \frac{V_b}{V}\right)$$

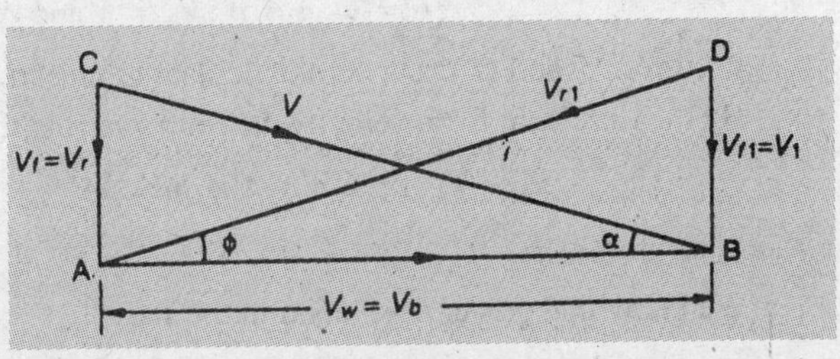

Fig. 24.14

$$\therefore \quad V_b = V\cos\alpha$$

It may be noted that when $V_b = V\cos\alpha$, the value of V_{w1} will be zero (a condition for maximum efficiency of impulse turbine also) as shown in Fig. 24.14. In such a case $V_b = V_w$.

Now substituting $\rho = \cos\alpha$ in equation (*i*) for maximum efficiency,

$$\eta_{max} = \frac{2(2\cos^2\alpha - \cos^2\alpha)}{(1 - \cos^2\alpha + 2\cos^2\alpha)} = \frac{2\cos^2\alpha}{1 + \cos^2\alpha}$$

Example 24.9. *A certain stage of a Parson's turbine consists of one row of fixed blades and one row of moving blades. The details of the turbine are as follows :*

Mean diameter of the blades = 680 mm ; R.P.M. of the turbine = 3000 ; Mass of steam passing per second = 13.5 kg ; Steam velocity at exit from fixed blades = 143.7 m/s ; Blade outlet angle = 20°.

Calculate the power developed in the stage and the gross efficiency assuming carry over coefficient as 0.74 and the efficiency of conversion of heat energy into kinetic energy in the blade channels as 0.92.

Solution. Given : $D = 680$ mm $= 0.68$ m ; $N = 3000$ r.p.m. ; $m = 13.5$ kg/s ; $V = 143.7$ m/s ; $\alpha = 20°$; $K = 0.74$; $\eta = 0.92$

We know that blade velocity,

$$V_b = \frac{\pi DN}{60} = \frac{\pi \times 0.68 \times 3000}{60} = 106.8 \text{ m/s}$$

Now draw the combined velocity triangle, as shown in Fig. 24.15, as discussed below :

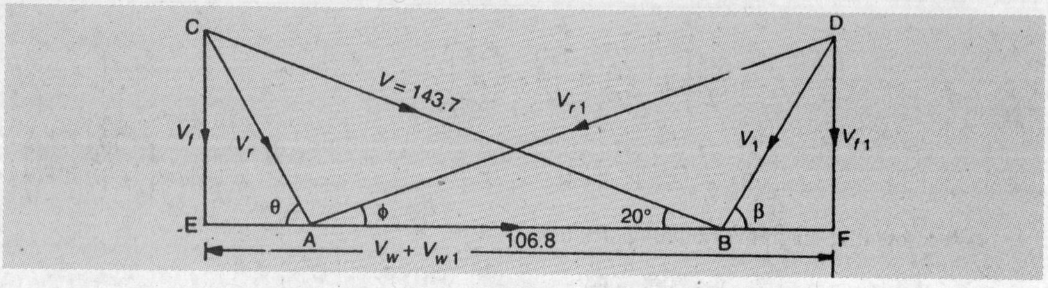

Fig. 24.15

1. First of all, draw a horizontal line and cut off AB equal to 106.8 m/s, to some suitable scale, to represent the blade velocity (V_b).

2. Now draw inlet velocity triangle ABC on the base AB with $\alpha = 20°$ and $V = 143.7$ m/s, to the scale.

3. Now draw outlet velocity triangle ABD on the same base AB with $\phi = \alpha = 20°$ and $V_{r1} = V = 143.7$ m/s, to the scale.

Note : In a Parson's turbine, the blades are symmetrical. Therefore

$$\phi = \alpha \,;\, \theta = \beta \,;\, V_{r1} = V \text{ and } V_1 = V_r.$$

4. From C and D, draw perpendiculars to meet the line AB produced at E and F.

By measurement from velocity diagram, we find that the change in the velocity of whirl,

$$V_w + V_{w1} = EF = 164 \text{ m/s}$$

Power developed in the stage

We know that power developed in the stage,

$$P = m\,(V_w + V_{w1})\,V_b = 13.5 \times 164 \times 106.8 = 236\,460 \text{ W} = 236.46 \text{ kW} \textbf{ Ans.}$$

Gross efficiency

By measurement from velocity diagram, we find that

$$V_1 = V_r = 58.7 \text{ m/s}$$

We know that heat drop,

$$h_d = 2\left(\frac{V^2 - KV_1^2}{2\,\eta}\right) = 2\left[\frac{(143.7)^2 - 0.74\,(58.7)^2}{2 \times 0.92}\right] = 19\,674 \text{ J} = 19.674 \text{ kJ}$$

∴ Gross efficiency,

$$\eta_s = \frac{(V_w + V_{w1})\,V_b}{1000\,h_d} = \frac{164 \times 106.8}{1000 \times 19.674} = 0.89 \text{ or } 89\% \textbf{ Ans.}$$

Example 24.10. *At a stage of a reaction turbine, the rotor diameter is 1.4 m and speed ratio 0.7. If the blade outlet angle is 20° and the rotor speed 3000 r.p.m., find the blade inlet angle and diagram efficiency.*

Also find the percentage increase in diagram efficiency and rotor speed, if the turbine is designed to run at the best theoretical speed.

Solution. Given : $D = 1.4$ m ; $\rho = V_b / V = 0.7$; $\alpha = \phi = 20°$; $N = 3000$ r.p.m.

We know that blade velocity,

$$V_b = \frac{\pi DN}{60} = \frac{\pi \times 1.4 \times 3000}{60} = 220 \text{ m/s}$$

and velocity of steam at inlet to the blade,

$$V = V_b / 0.7 = 220 / 0.7 = 314.3 \text{ m/s}$$

Blade inlet angle

Now draw the combined velocity triangle, as shown in Fig. 24.16, as discussed below :

1. First of all, draw a horizontal line and cut off AB equal to 220 m/s, to some suitable scale, to represent the blade velocity (V_b).

2. Now draw inlet velocity triangle ABC on the base AB with $\alpha = 20°$ and $V = 314.3$ m/s, to the scale.

3. Similarly, draw outlet velocity triangle ABD on the same base AB with $\phi = 20°$ and $V_{r1} = 314.3$ m/s, to the scale.

4. From C and D, draw perpendiculars to meet the line AB produced at E and F.

By measurement from velocity diagram, we find that the blade inlet angle,

$$\theta = 55° \textbf{ Ans.}$$

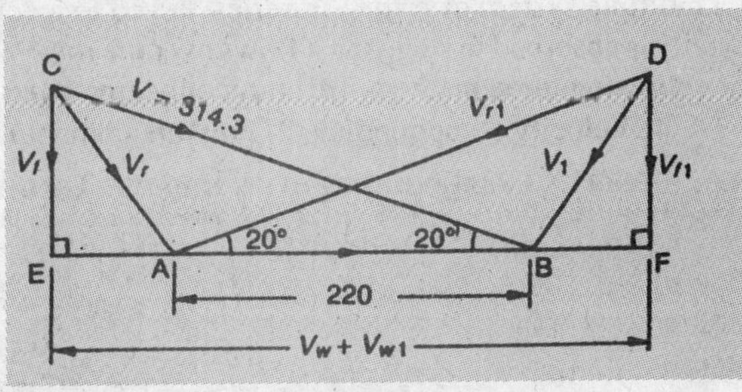

Fig. 24.16

Diagram efficiency

By measurement from velocity diagram, we find that velocity of steam at outlet,

$$V_1 = 130 \text{ m/s}$$

We know that diagram efficiency,

$$\eta_b = \frac{V^2 - V_1^2}{V^2} = \frac{(314.3)^3 - (130)^2}{(314.3)^2} = 0.829 \quad \text{or} \quad 82.9\ \% \textbf{ Ans.}$$

Percentage increase in diagram efficiency for best theoretical speed

We know that maximum efficiency of the turbine,

$$\eta_{max} = \frac{2\cos^2\alpha}{1+\cos^2\alpha} = \frac{2 \times \cos^2 20°}{1+\cos^2 20°} = \frac{2\,(0.9397)^2}{1+(0.9397)^2} = 0.938$$

$\therefore$ Percentage increase in diagram efficiency

$$= \frac{0.938 - 0.829}{0.829} = 0.131 \quad \text{or} \quad 13.1\% \textbf{ Ans.}$$

Percentage increase in rotor speed for best theoretical speed

Let N_1 = Maximum rotor speed.

We know that for best theoretical speed (or in other words, for maximum efficiency),the blade velocity,

$$V_b = V\cos\alpha = 314.3\cos 20° = 314.3 \times 0.9397 = 295.3 \text{ m/s}$$

We also know that blade velocity (V_b),

$$\therefore \quad 295.3 = \frac{\pi D N_1}{60} = \frac{\pi \times 1.4 N_1}{60} = 0.073 N_1$$

or $$N_1 = 4044 \text{ r.p.m.}$$

∴ Percentage increase in rotor speed

$$= \frac{4044 - 3000}{3000} = 0.348 \quad \text{or} \quad 34.8\ \% \ \textbf{Ans.}$$

24.5. Compounding of Impulse Steam Turbines (Methods of Reducing Rotor Speeds)

In the recent years, high pressure (100 to 140 bar) and high temperature steam is used in the power plants to increase their thermal efficiency. If the entire pressure drop (from boiler pressure to condenser pressure (say from 125 bar to 1 bar) is carried out in one stage only, then the velocity of steam entering into the turbine will be extremely high. It will make the turbine rotor to run at a very high speed (even up to 30 000 r.p.m.). From practical point of view, such a high speed of the turbine rotor is bound to have a number of disadvantages.

In order to reduce the rotor speed, various methods are employed. All of these methods consist of a multiple system of rotors, in series, keyed to a common shaft and the steam pressure or the jet velocity is absorbed in stages as it flows over the rotor blades. This process is known as *compounding*. The following three methods are commonly employed for reducing the rotor speed :

1. Velocity compounding, 2. Pressure compounding, and 3. Pressure-velocity compounding.

24.6. Velocity Compounding of an Impulse Turbine

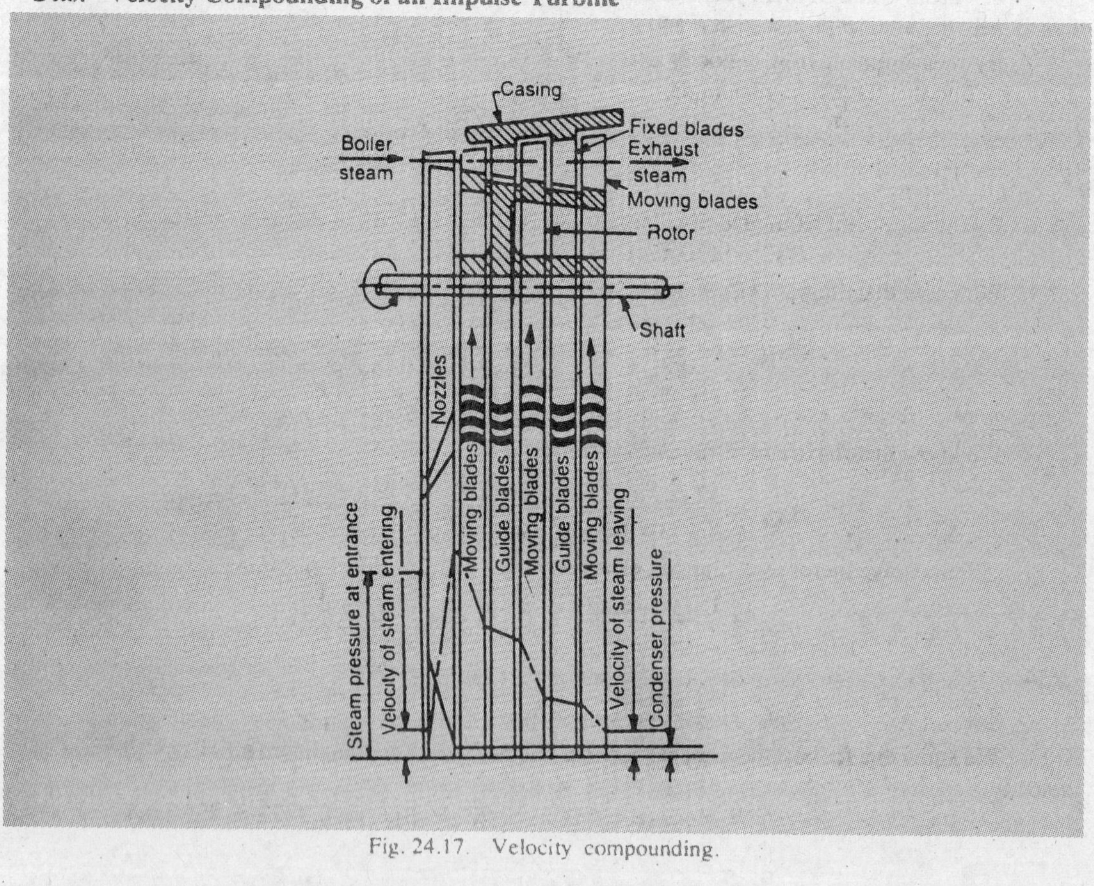

Fig. 24.17. Velocity compounding.

In velocity compounding of an impulse turbine, the expansion of steam takes place in a nozzle or a set of nozzles from the boiler pressure to the condenser pressure. The impulse wheel carries two or three rows of moving blades. Fig. 24.17 shows the three rings of moving blades, separated by two rings of fixed or guide blades in the reverse manner.

The steam, after expanding through nozzles, enters the first ring of moving blades at a high velocity. A portion of this high velocity is absorbed by this blade ring and the remaining is passed on to the next ring of fixed blades. The fixed blades change the direction of steam and direct it to the second ring of moving blades, without altering the velocity appreciably. After passing through this second ring of moving blades, a further portion of velocity is absorbed. The steam is now directed by the second ring of fixed blades to the third ring of moving blades and then enters into the condenser.

In Fig. 24.17, the curves of velocity and pressure on a base representing the axis of the turbine are shown. It may be noted, from the figure, that no pressure drop occurs either in the fixed or moving blades. All the pressure drop occures in the nozzles. This turbine can run at about one-third of the speed of De-Laval turbine, for the same pressure drop and diameter of the wheel.

24.7. Pressure Compounding of an Impulse Turbine

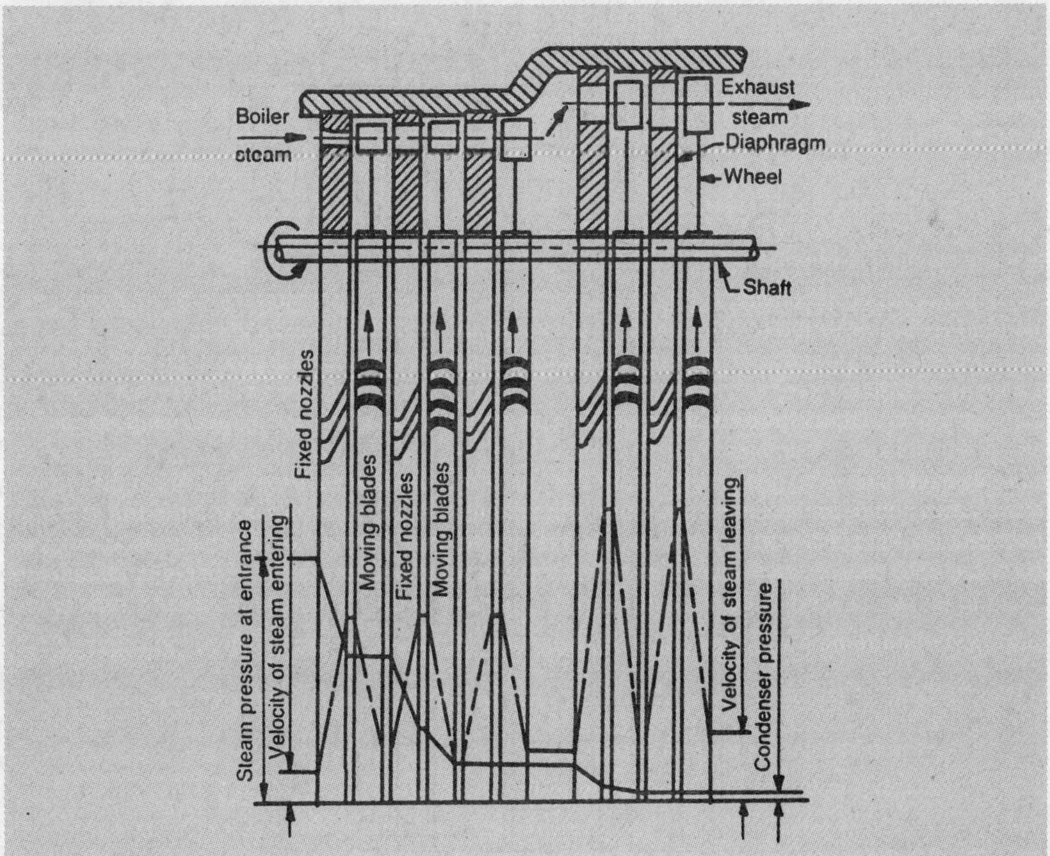

Fig. 24.18. Pressure compounding.

In a pressure compounding of an impulse turbine, the rings of the moving blades, each having a ring of fixed nozzles, are keyed to the turbine shaft in series, as shown in Fig. 24.18. The total pressure drop, of the steam, does not take place in the first nozzle ring, but is divided equally among all the nozzle rings.

The steam from the boiler is passed through the first nozzle ring, where only a small pressure drop occurs with an increase in velocity of steam. The steam is now directed on the first moving blade ring, where the pressure of steam does not alter, but the velocity decreases. This constitutes one stage. It may be noted that a stage consists of a fixed nozzle ring and a moving blade ring. The steam from the first moving blade ring enters the second nozzle ring, where its pressure is further reduced. A little consideration will show, that the pressure drop per stage in the nozzle rings is not the same. but the number of heat units, converted into velocity energy in each stage, is the same. The process is repeated in the remaining rings, until the condenser pressure is reached.

In Fig. 24.18, the curves of velocity and pressure on a base representing the axis of the turbine are also shown. It may be noted, from the figure, that by arranging a small pressure drop per stage, the velocity of steam entering the moving blades, and hence the speed of rotor is reduced.

The Rateau and Zoelly turbines are the examples of pressure compounded turbines.

24.8. Pressure-velocity Compounding of an Impulse Turbine

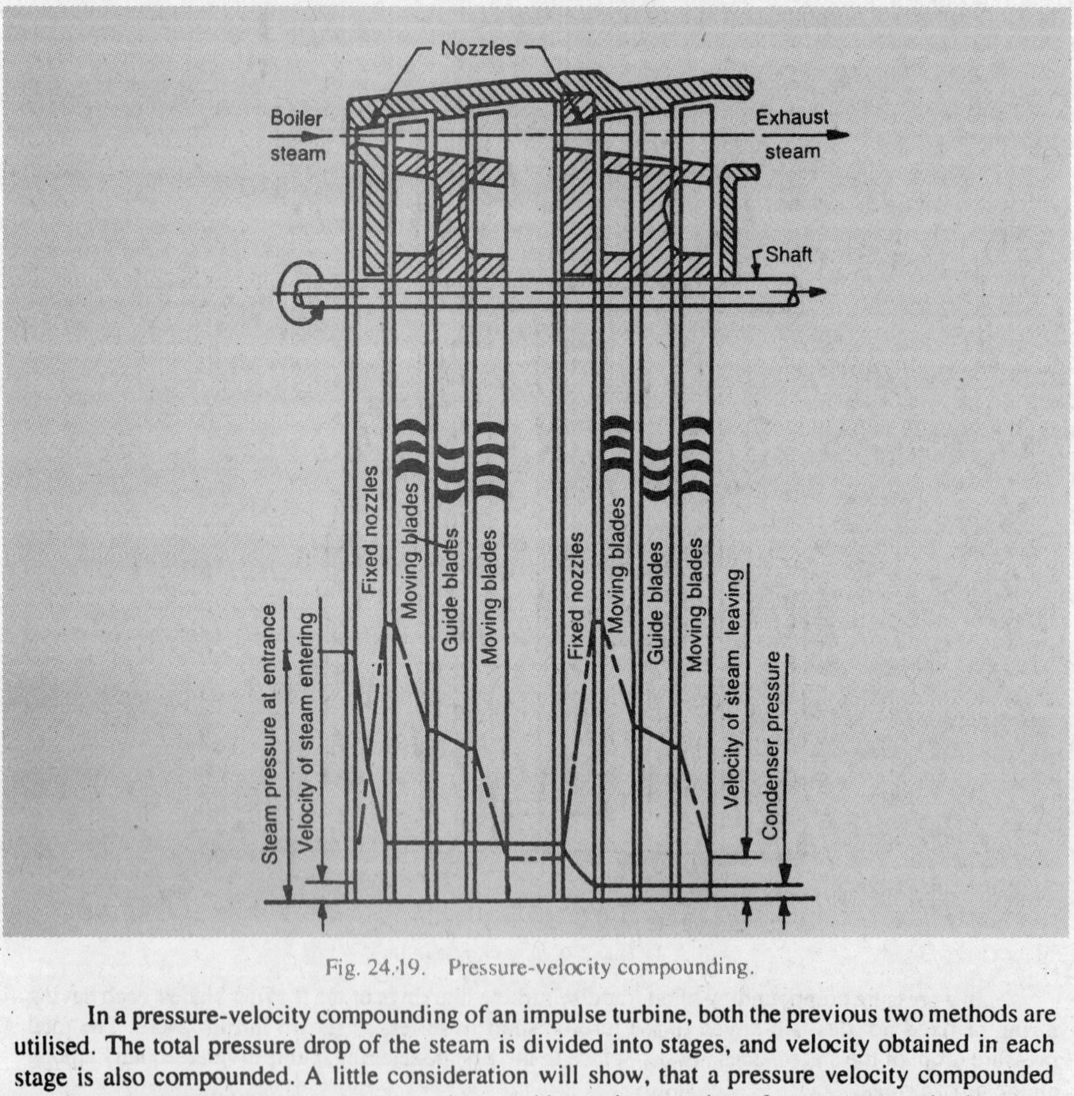

Fig. 24.19. Pressure-velocity compounding.

In a pressure-velocity compounding of an impulse turbine, both the previous two methods are utilised. The total pressure drop of the steam is divided into stages, and velocity obtained in each stage is also compounded. A little consideration will show, that a pressure velocity compounded impulse turbine allows a bigger pressure drop, and hence less number of stages are required.

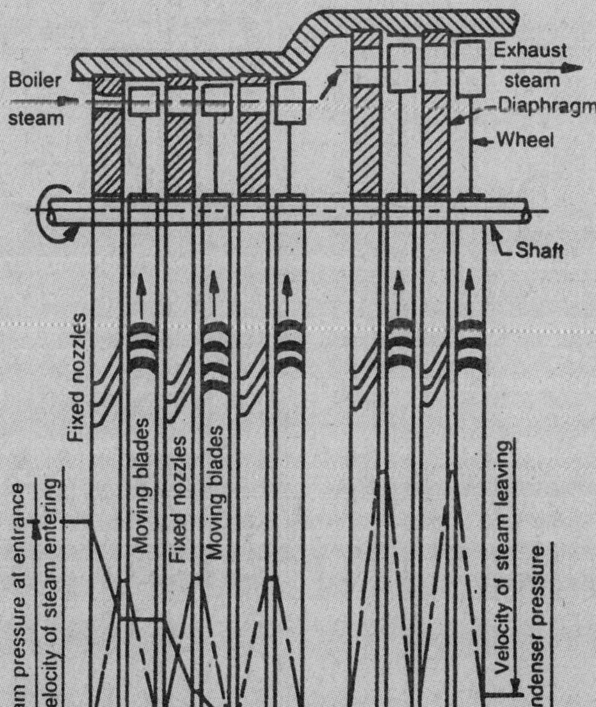

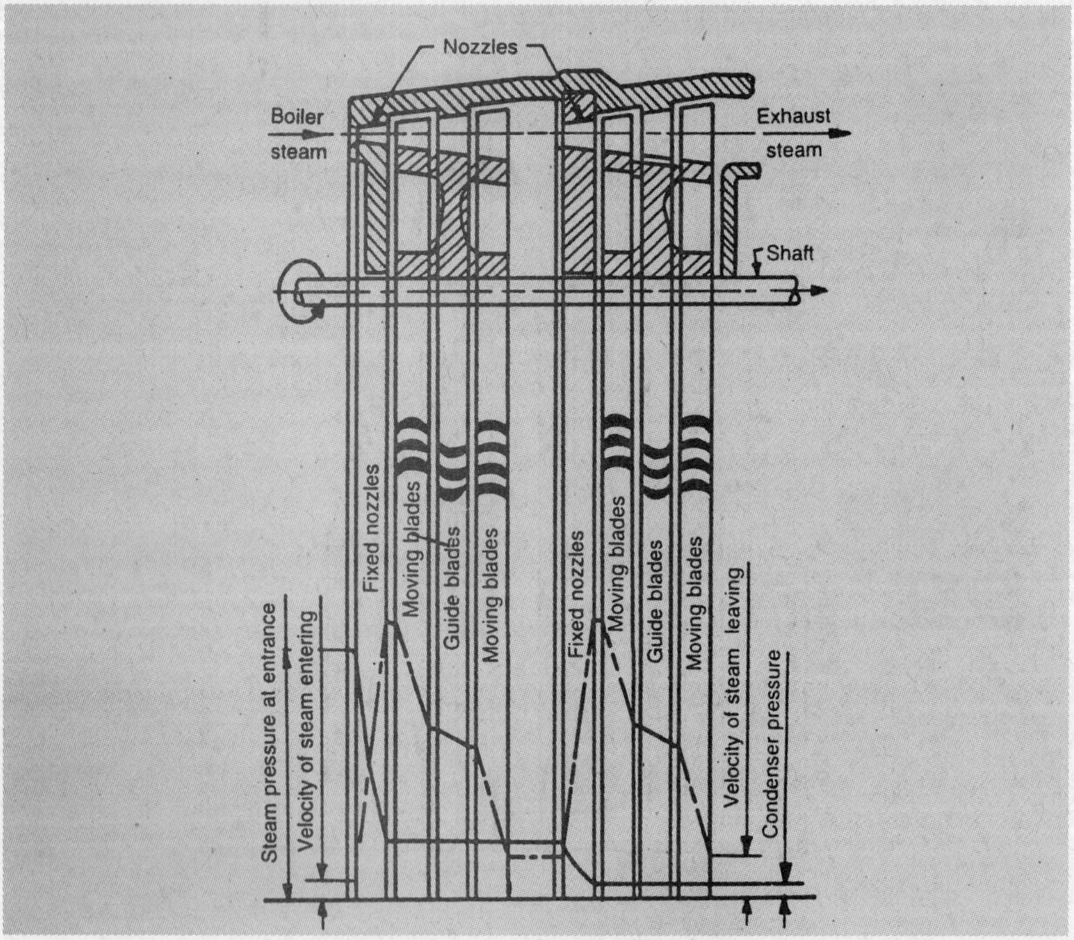

Fig. 24.19 shows the curves of pressure and velocity for this type of turbine. It may be noted that the diameter of the turbine is increased at each stage, to allow the increasing volume of steam at the lower pressures. A ring of nozzles is fixed at the commencement of each stage as shown in Fig. 24.19. A curtis turbine is an example of pressure-velocity compounded impulse turbine.

24.9. Internal Losses in Turbines

Strictly speaking, an ideal turbine (having 100 per cent gross efficiency) will do the work equivalent to the isentropic enthalpy or heat drop of the steam used in the turbine. But in actual practice, the work done by a turbine is much less than isentropic heat drop of the steam used. There are several factors, which effect the performance of a steam turbine. All these factors, which reduce the output of the turbine, are known as *internal losses*. Though there are many internal losses in a steam turbine, yet the following are important from the subject point of view :

1. *Nozzle loss.* It is an important loss in impulse turbines, which occurs when the steam flows through the nozzle. This loss takes place due to friction in the nozzle and the formation of eddies.

2. *Blade friction loss.* It is an important loss in both the impulse and reaction turbines, which occurs when the steam glides over the blades. This loss takes place due to friction of the surface of blades. As a result of the blade friction, the relative velocity of the steam is reduced while gliding over the blade.

3. *Wheel friction loss.* It is another important loss in both the impulse and reaction turbines, which occurs when the turbine wheel rotates in the steam. This loss takes place due to the resistance offered by the steam to the moving turbine wheel or disc. As a result of this loss, the turbine wheel rotates at a lower speed.

4. *Mechanical friction loss.* It is a loss in both the turbines, which occurs due to friction between the shaft and wheel bearing as well as regulating the valves. This loss can be reduced by lubricating the moving parts of the turbine.

5. *Leakage loss.* It is a loss in both the turbines, which occurs due to leakage of the steam at each stage of the turbine, blade tips and glands.

6. *Residual velocity loss.* It is a loss in both the turbines, which occurs due to the kinetic energy of the steam, as it leaves the turbine wheel. This loss is reduced by using multi-stage wheels.

7. *Moisture loss.* It is a loss in both the turbines, which takes place due to moisture present in the steam. This loss occurs when the steam, passing through lower stages, becomes wet. The velocity of water particles is less than that of steam. As a result of this, the steam has to drag the water particles, which reduce the kinetic energy of the steam.

8. *Radiation losses.* It is a loss in both the turbines, which takes place due to difference of the temperatures between the turbine casing and the surrounding atmosphere. This is reduced by properly insulating the turbine.

9. *Governing loss.* It is a loss in both the turbines, which occur due to throttling of the steam at the main stop valve of the governor.

24.10. Governing of Steam Turbines

We have already discussed in Art. 19.7, the necessity and processes of governing of steam engines. In the same way, we provide an arrangement for the governing of steam turbines also. Though there are many methods of governing steam turbines, yet the throttle governing is important from the subject point of view :

24.11. Throttle Governing of Steam Turbines

The throttle governing of a steam turbine is a method of controlling the turbine output by varying the quantity of steam entering into the turbine. This method is also known as ***servometer method***, whose operation is given below :

The centrifugal governor is driven from the main shaft of turbine by belt or gear arrangement. The control valve controls the direction of flow of the oil (which is pumped by gear pump) either in the pipe *AA* or *BB*. The servometer or relay valve has a piston whose motion (towards left or right depends upon the pressure of the oil flowing through the pipes *AA* or *BB*) is connected to a spear or needle which moves inside the nozzle, as shown in Fig. 24.20.

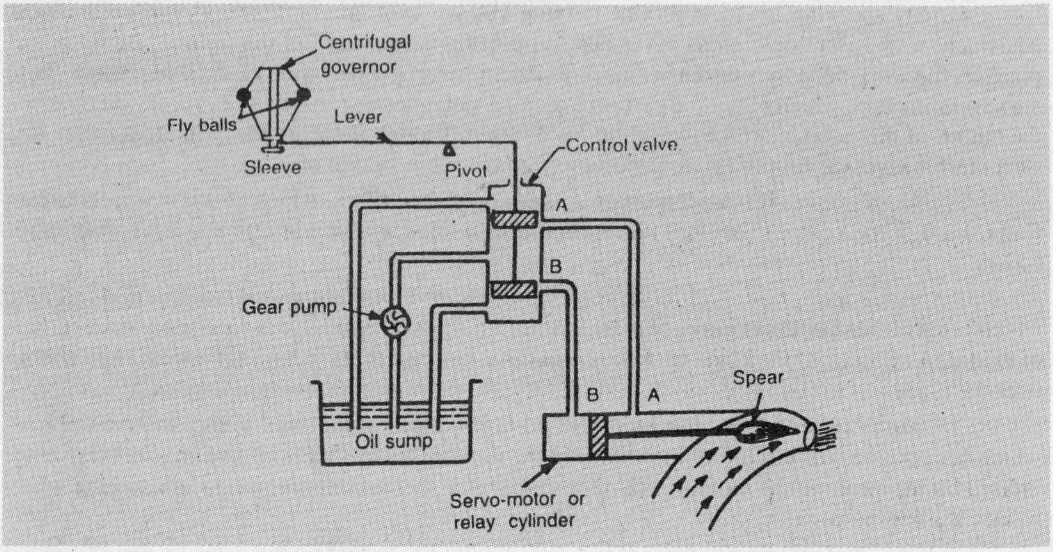

Fig. 24.20. Throttle governing of steam turbines.

We know that when the turbine is running at its normal speed, the positions of piston in the servometer, control valve, fly balls of centrifugal governor will be in their normal positions as shown in the figure. The oil pumped by the gear pump into the control valve, will come back into the oil sump as the mouths of both pipes *AA* or *BB* are closed by the two wings of control valve.

Now let us consider an instant, when load on the turbine increases. As a result of load increase, the turbine speed will be decreased. This decrease in the speed of the turbine wheel will also decrease the speed of centrifugal governor. As a result of this, the fly balls will come down (due to decrease in centrifugal force) thus decreasing their amplitude. As the sleeve is connected to the central vertical bar of the centrifugal governor, therefore coming down of the flyballs will also bring down the sleeve. This downward movement of the sleeve will raise the control valve rod, as the sleeve is connected to the control valve rod through a lever pivoted on the fulcrum. Now, a slight upward movement of the control valve rod will open the mouth of the pipe *AA* (still keeping the mouth of the pipe *BB* closed). Now, the oil under pressure will rush from the control valve to the right-side of the piston in the servometer through the pipe *AA*. This oil, under pressure, will move the piston as well as spear towards left, which will open more area of the nozzle. This increase in the area of flow will increase the rate of steam flow into the turbine. As a result of increase in the steam flow, there will be an increase in the turbine output as well as its speed. When speed of the turbine wheel will come up to its normal range, fly balls will move up. Now the sleeve as well as control valve rod will occupy its normal position, and the turbine will run at its normal speed.

It may be noted that when load on the turbine decreases, it will increase its speed. As a result of this, the fly balls will go up (due to increase in centrifugal force) and the sleeve will also go up. This will push the control valve rod downwards. This downward movement of the control valve rod will open the mouth of the pipe *BB* (still keeping the mouth of the pipe *AA* closed). Now, the oil under pressure, will rush from the control valve to the left side of the piston in the servometer through the pipe *BB*. This oil under pressure, will move the piston and spear towards right, which will decrease the area of the nozzle. This decrease in the area of the flow will decrease the rate of steam flow into the turbine. As a result of decrease in the steam flow, there will be a decrease in the turbine output as well as its speed. When the speed of the turbine is reduced to its normal range, the flyballs will come down. Now the sleeve as well as control valve rod will occupy its normal positions, and the turbine will run at its normal speed.

EXERCISES

1. The following data refer to one stage of an impulse turbine :

Steam velocity = 500 m/s ; Blade speed = 200 m/s ; Nozzle angle = 20° ; Exit angle of moving blades = 25° ; Steam flow = 5 kg/s.

Neglecting the effect of friction when passing through passages, determine : 1. the inlet angle of moving blade ; 2. the power developed ; 3. the axial thrust, and 4. the diagram efficiency.

[**Ans.** 33° ; 560 kW ; 200 N ; 89.6%]

2. In a De-Lavel turbine, the blade angles are equal. The steam enters at a velocity of 300 m/s with a nozzle angle of 16°. The blade friction factor is 0.84. If the blade speed and steam speed ratio is 0.47, estimate : 1. blade efficiency ; 2. power developed for a mass flow rate of 2 kg/s. [**Ans.** 84.6% ; 76.14 kW]

3. Following is the data pertaining to a single-row impulse turbine :

Mean diameter of blade ring = 2 m ; Speed = 3000 r.p.m. ; Nozzle angle = 18° ; Ratio of blade velocity to steam velocity = 0.5 ; Blade friction factor = 0.9 ; Blade angle at exit = 3° more than at inlet ; Steam supply = 30 000 kg/h.

Draw velocity diagram for the moving blade and estimate : 1. power developed ; 2. diagram efficiency ; and 3. steam consumption in kg/kWh. [**Ans.** 1387.16 kW ; 84.3% ; 21.6 kg/kWh]

4. A steam turbine running at 3600 r.p.m. takes 4.5 kg of steam per second. The nozzle angle is 16° and the mean diameter of the blade ring 1.2 metre. The blade outlet angle is 18° and the isentropic heat drop in the nozzle is 165 kJ/kg of steam. The shaft power is 485 kW. Assuming a nozzle efficiency as 92% and blade velocity coefficient 0.85, find 1. Blading efficiency, 2. Stage efficiency, and 3. Power lost in friction.

[**Ans.** 85.7% ; 78.8% ; 100.3 kW]

5. Steam is supplied to a simple impulse turbine at 10 bar, 250° C. The pressure in the wheel casing is 1.2 bar and the nozzle efficiency is 91%. Determine the exit area required for a steam flow rate of 5000 kg/h.

The nozzles are inclined to the plane of rotation at 20°. The blades are equiangular and have a velocity coefficient of 0.8. Assuming that the ratio of mean blade speed to steam speed at nozzle exit to be 0.45, calculate 1. the blade angles ; 2. the power developed ; and 3. the stage efficiency.

[**Ans.** 35° ; 442.26 kW ; 74%]

6. The velocity of steam as it flows out of a nozzle is 440 m/s which is compounded in an impulse turbine by passing it successively through moving, fixed and finally through a second ring of moving blades. The tip angles of the moving blades throughout the turbine are 30°. Assume a loss of 10% in velocity due to friction when the steam passes over a blade ring. Find the velocity of the moving blades in order to have final discharge of steam axial. Also determine the diagram efficiency. [**Ans.** 97 m/s ; 76%]

7. The first stage of an impulse turbine is compounded for velocity and has two rings of moving blades and one ring of fixed blades. The nozzle angle is 20° and the leaving angles of the blades are respectively as follows :

First moving 20°, fixed 25° and second moving 30°.

The velocity of steam leaving the nozzle is 600 m/s and the steam velocity relative to the blade is reduced by 10 percent during the passage through each ring. Find the diagram efficiency and the power developed for a steam flow of 4 kg/s. The blade speed may be taken as 125 m/s. [**Ans.** 78.5% ; 565 kW]

8. The following particulars relate to a two row velocity compounded impulse wheel :

Steam velocity at nozzle outlet = 650 m/s ; Mean blade velocity = 125 m/s ; Nozzle outlet angle = 16° ; Outlet angle – first row of moving blades = 18° ; Outlet angle – fixed guide blades = 22° ; Outlet angle-second row of moving blades = 36° ; Steam flow = 2.5 kg/s.

If the ratio of relative velocity at outlet to that inlet is 0.84 for all blades, find the following :

1. the axial thrust on the blades ; 2. the power developed ; and 3. the efficiency of the wheel.

[**Ans.** 142.5 N ; 390 kW ; 73.8%]

9. An impulse stage of a turbine has two rows of moving blades separated by fixed blades. The steam leaves the nozzles at an angle of 20° with the direction of the motion of blades. The blade exit angles are : first moving 30°, fixed 22° and second moving 30°.

If the isentropic heat drop for the nozzle is 186 kJ/kg and the nozzle efficiency 90%, find the blade speed necessary if the final velocity of steam is to be axial. Assume a loss of 15 percent in relative velocity for all blade passages. Find also the blade efficiency and the stage efficiency. [**Ans.** 117.2 m/s ; 71.9% ; 64.7%]

10. In a reaction turbine pair, the fixed and moving blades are of the same shape but reversed in direction. The angles of the receiving tips are 35° and of discharging tip, 20°. The mean velocity of the blades is 37.5 m/s and steam flows at the rate of 64 kg/s. If the isentropic heat drop for this turbine pair is 6 kJ/kg, calculate the diagram power and efficiency of the pair.

[**Ans.** 285.6 kW ; 74.3%]

QUESTIONS

1. Define the following terms :

   (*a*) Diagram efficiency, and (*b*) Stage efficiency.

2. Derive the condition for maximum efficiency of an impulse turbine and show that the maximum efficiency is $\cos^2 \alpha$, where α is the angle at which the steam enters the blade.

3. Determine the condition for maximum efficiency of a 50 percent reaction turbine and show that the maximum efficiency for such a turbine is $2\cos^2 \alpha / (1 + \cos^2 \alpha)$, where α is the angle at which the steam enters the blades.

4. Explain the term 'Compounding of steam turbine'. What are the different methods of reducing rotor speed ?

5. Discuss the method of velocity compounding of an impulse turbine for achieving rotor speed reduction.

6. Enumerate the different losses in a steam turbine.

7. What are the methods of governing a steam turbine ? Describe any one method of governing steam turbines.

OBJECTIVE TYPE QUESTIONS

1. The ratio of workdone on the blades per kg of steam to the energy supplied to the blades is called

   (*a*) diagram or blading efficiency (*b*) nozzle efficiency
   (*c*) gross or stage efficiency (*d*) mechanical efficiency

2. The stage efficiency (η_s) is given by

   (*a*) η_b / η_n (*b*) η_n / η_b (*c*) $\eta_b \eta_n$

   where η_b = Blading efficiency, and

   η_n = Nozzle efficiency.

3. The maximum efficiency of a De-Laval turbine is

   (*a*) $\sin^2 \alpha$ (*b*) $\cos^2 \alpha$ (*c*) $\tan^2 \alpha$ (*d*) $\cot^2 \alpha$

   where α = Nozzle angle.

4. The maximum efficiency of a reaction turbine is

   (*a*) $\dfrac{2\sin^2 \alpha}{1+\sin^2 \alpha}$ (*b*) $\dfrac{1+\sin^2 \alpha}{2\sin^2 \alpha}$ (*c*) $\dfrac{2\cos^2 \alpha}{1+\cos^2 \alpha}$ (*d*) $\dfrac{1+\cos^2 \alpha}{2\cos^2 \alpha}$

5. The pressure-velocity compounded impulse turbine allows a bigger pressure drop and hence ..... number of stages are required.

(*a*) more (*b*) less

6. Rateu turbine is a

(*a*) simple impulse turbine (*b*) simple reaction turbine

(*c*) velocity compounded turbine (*d*) pressure compounded turbine

7. Curtis turbine is a

(*a*) velocity compounded turbine (*b*) pressure compounded turbine

(*c*) pressure-velocity compounded turbine (*d*) none of these

8. De-Laval turbine is a

(*a*) impulse reaction turbine (*b*) single rotor impulse turbine

(*c*) multi-rotor impulse turbine (*d*) none of these

9. The compounding of turbines is done is order to

(*a*) reduce speed of rotor (*b*) improve efficiency

(*c*) reduce exit losses (*d*) all of these

10. The purpose of governing in steam turbines is to

(*a*) reduce the effective heat drop

(*b*) reheat the steam and improve its quality

(*c*) completely balance against end thrust

(*d*) maintain the speed of the turbine

ANSWERS

1. (*a*)	2. (*c*)	3. (*b*)	4. (*c*)	5. (*b*)
6. (*d*)	7. (*c*)	8. (*b*)	9. (*a*)	10. (*d*)

25

Modern Steam Turbines

1. Introduction. 2. Reheating of Steam. 3. Advantages of Reheating of Steam. 4. Reheat Cycle. 5. Multi-stage Turbines. 6. Reheat Factor. 7. Efficiencies of a Multi-stage Turbine. 8. Regenerative Cycle. 9. Bleeding. 10. Regenerative Cycle with Single Feed Water Heater. 11. Regenerative Cycle with Two Feed Water Heaters. 12. Binary Vapour Plants. 13. Binary Vapour Cycle. 14. Some Special Turbines. 15. Passout or Extraction Turbine. 16. Back Pressure Turbine. 17. Exhaust or Low Pressure Turbine. 18. Future Power Plants.

25.1. Introduction

In the last three chapters, we have discussed impulse turbines, reaction turbines and their performance. In these chapters, we discussed, apart from other things, power developed by the turbines and their efficiencies. The scientists and engineers, working in research centres all over the world, concentrated their attention to produce more power and to improve efficiencies of these turbines. They have listed a number of methods for this purpose. but the following are important from the subject point of view :

1. Reheating of steam, 2. Regenerative feed heating, and 3. Binary vapour plants.

All the above mentioned methods will be discussed, in detail, in this chapter.

25.2. Reheating of Steam

We have already discussed that efficiency of the ordinary Rankine cycle can be improved by increasing the pressure and temperature of the steam entering into the turbine, A little consideration will show, that the increase in the initial steam pressure will increase the expansion ratio, and steam will become quite wet at the end of expansion. As a matter of fact, it is not desirable that the steam may become wet at the end of expansion. The wet steam causes erosion of the turbine blades and increases internal losses. This will ultimately reduce the blade efficiency of the turbine.

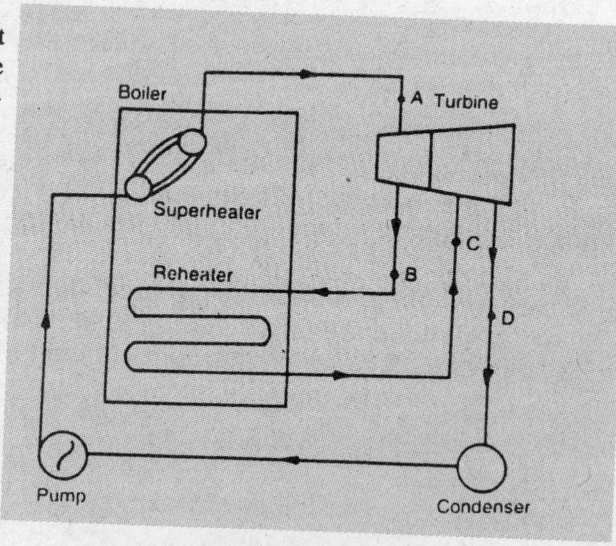

Fig. 25.1. Reheating of steam.

The above mentioned difficulty may be overcome by reheating of the steam. In this system, the steam is removed from the turbine when it becomes wet. It is then reheated at a constant pressure by the flue gases, until it is again in the superheated state. It is then returned to the next stage in the turbine. The schematic diagram of the process is shown in Fig. 25.1.

25.3. Advantages of Reheating of Steam

The reheating of steam in a turbine has the following advantages :

1. It increases the work done through the turbine.
2. It increases the efficiency of the turbine.
3. It reduces the erosion of the blades, because of increase in dryness fraction of steam at exhaust.
4. The amount of water required in the condenser of the turbine is reduced, due to reduction in the specific steam consumption.

25.4. Reheat Cycle

In a reheat cycle, the steam enters the turbine in a superheated state at point A. The steam then expands isentropically while flowing through the turbine, as shown by the vertical line AB in Fig. 25.2.

After expansion, the steam becomes wet, which is reheated at a constant pressure generally up to the same temperature as that at A shown by the point C, where it is again in superheated state. The steam again expands isentropically while flowing through the next stage of the turbine as shown by the vertical line CD in Fig. 25.2.

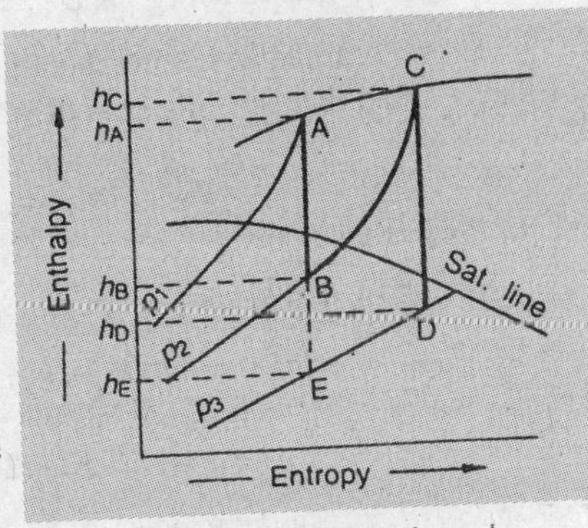

Fig. 25.2. Reheat cycle on h-s diagram.

Now consider a steam turbine with a reheating system as shown in Fig. 25.2.

Let h_A = Enthalpy or total heat of steam at A,

h_B, h_C, h_D = Corresponding values at B, C and D,

h_{fD} = Enthalpy or sensible heat of water at D.

We know that the total heat supplied to steam is the sum of the total heat at A and the heat supplied during reheating between B and C.

$\therefore$ Total heat supplied = Total heat at A + Heat supplied between B and C

$$= h_A + [(h_C - h_B) - h_{fD}]$$

We also know that work done

$$= \text{Total heat drop} = (h_A - h_B) + (h_C - h_D)$$

and efficiency,

$$\eta = \frac{\text{Work done}}{\text{Total heat supplied}} = \frac{(h_A - h_B) + (h_C - h_D)}{h_A + [(h_C - h_B) - h_{fD}]}$$

Notes : 1. If there had been no reheating of steam, the expansion through the turbine would have been along the line AE. In that case,

Efficiency, $$\eta = \frac{h_A - h_E}{h_A - h_{fE}} \quad \ldots \text{(Rankine efficiency)}$$

where h_E = Total heat of steam at E, and

h_{fE} = Total heat of water at E.

2. For simplicity, we have taken only one stage of reheating. But in actual practice, there may be more than one stage of reheating.

Example 25.1. *Steam at a pressure of 15 bar and 250° C is first expanded through a turbine to a pressure of 4 bar. It is then reheated at a constant pressure to the initial temperature of 250° C and is finally expanded to 0.1 bar. Using Mollier chart, estimate the work done per kg of steam flowing through the turbine and the amount of heat supplied during the process of reheat.*

Also find the work output when the expansion is direct from 15 bar to 0.1 bar without any reheat. Assume all expansion processes to be isentropic.

Solution. Given : $p_1 = 15$ bar ; $T_1 = 250°$ C ; $p_2 = 4$ bar ; $T_2 = 250°$ C ; $p_3 = 0.1$ bar

The reheating of steam is represented on the Mollier chart as shown in Fig. 25.3. From the chart, we find that

$h_A = 2930$ kJ/kg ; $h_B = 2660$ kJ/kg ; $h_C = 2965$ kJ/kg ; $h_D = 2345$ kJ/kg ; and $h_E = 2130$ kJ/kg

From steam tables, corresponding to a pressure of 0.1 bar, we find that sensible heat of water at *D*,

$$h_{fD} = h_{fE} = 191.8 \text{ kJ/kg}$$

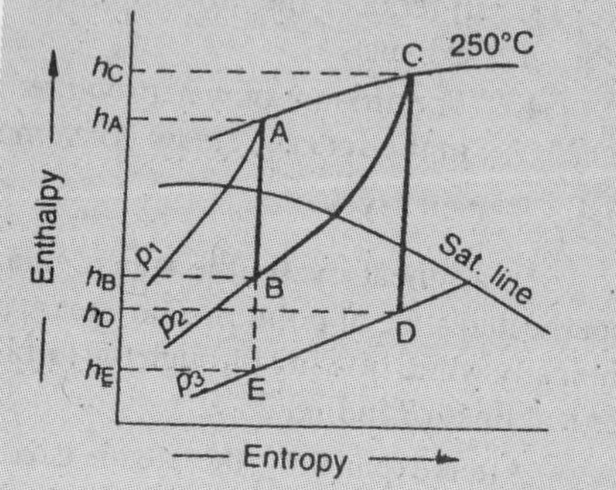

Fig. 25.3

Workdone per kg of steam

We know that workdone per kg of steam,

$$w = (h_A - h_B) + (h_C - h_D)$$

$$= (2930 - 2660) + (2965 - 2345) = 890 \text{ kJ/kg} \quad \textbf{Ans.}$$

Heat supplied during the process of reheat

We know that the heat supplied during the process of reheat,

$$h = \text{Heat supplied between } B \text{ and } C$$

$$= (h_C - h_B) - h_{fD} = (2965 - 2660) - 191.8 = 113.2 \text{ kJ/kg} \quad \textbf{Ans.}$$

Work output when the expansion is direct

The direct expansion from 15 bar to 0.1 bar is shown by the line *AE* in Fig. 25.3.

We know that work output

$$= \text{Total heat drop} = h_A - h_E = 2930 - 2130 = 800 \text{ kJ/kg} \quad \textbf{Ans.}$$

Example 25.2. *In a thermal plant, the steam is supplied at a pressure of 30 bar and temperature of 300° C to the high pressure side of steam turbine where it is expanded to 5 bar. The steam is then removed and reheated to 300° C at a constant pressure. It is then expanded to the low pressure side of the turbine to 0.5 bar. Find the efficiency of the cycle with and without reheating.*

Solution. Given : $p_1 = 30$ bar ; $T_1 = 300°$ C ; $p_2 = 5$ bar ; $T_2 = 300°$ C ; $p_3 = 0.5$ bar

Efficiency of the cycle with reheating

The reheating of steam is represented on the Mollier chart as shown in Fig. 25.4. From the chart, we find that

$h_A = 2990$ kJ/kg ; $h_B = 2625$ kJ/kg ; $h_C = 3075$ kJ/kg ; $h_D = 2595$ kJ/kg ; and $h_E = 2280$ kJ/kg

From steam tables, corresponding to a pressure of 0.5 bar, we find that sensible heat of water at *D*,

$$h_{fD} = h_{fE} = 340.6 \text{ kJ/kg}$$

We know that efficiency of the cycle with reheating,

$$\eta_1 = \frac{(h_A - h_B) + (h_C - h_D)}{h_A + [(h_C - h_B) - h_{fD}]} = \frac{(2990 - 2625) + (3075 - 2595)}{2990 + [(3075 - 2625) - 340.6]}$$

$$= \frac{845}{3099.4} = 0.273 \text{ or } 27.3\% \textbf{ Ans.}$$

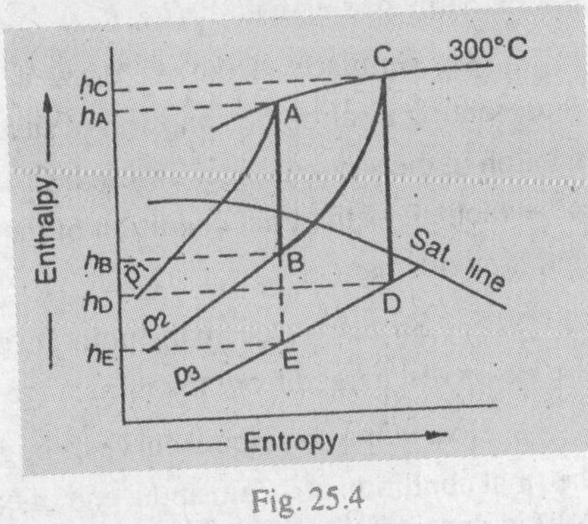

Fig. 25.4

Efficiency of the cycle without reheating

We know that efficiency of the cycle without reheating,

$$\eta_2 = \frac{h_A - h_E}{h_A - h_{fE}} = \frac{2990 - 2280}{2990 - 340.6}$$

$$= 0.268 \text{ or } 26.8\% \textbf{ Ans.}$$

25.5. Multi-stage Turbines

We have already discussed a two-stage impulse turbine in Art. 22.12. In this turbine, we have seen that the steam after leaving the moving blade is made to flow through a fixed ring and again it impinges on the blades fixed to the second moving ring. If the steam, from the second moving ring, is made to flow into the condenser, it is known as two-stage turbine. But sometimes, we make the steam to pass through a number of stages in order to get more work (or precisely to develop more power). Such a turbine is known as *multi-stage turbine.*

25.6. Reheat Factor

It is an important term used for the multi-stage turbines, which may be broadly defined as the ratio of cumulative heat drop to the isentropic heat drop. Now consider a multi-stage turbine (say three-stage turbine) whose reheat factor is required to be found out.

Let p_1 = Initial pressure of the steam,

p_2 = Pressure of steam leaving the first stage,

p_3 = Pressure of steam leaving the second stage,

p_4 = Final pressure of the steam.

t_1 = Initial temperature of the steam, and

η = Stage efficiency for each stage of the turbine.

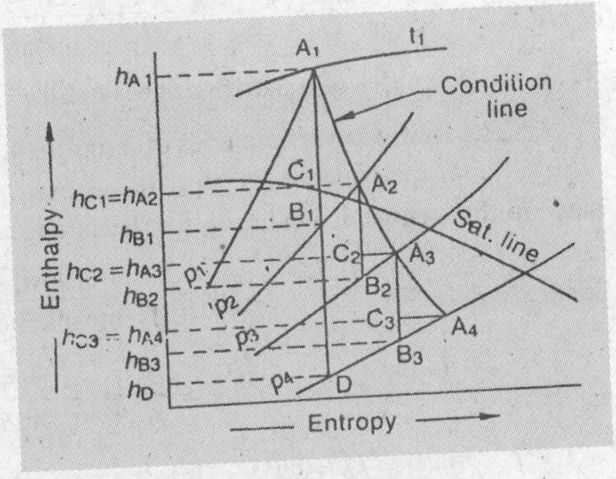

Fig. 25.5. Reheat factor for a three-stage turbine.

Now let us draw the expansion of steam in three stages on a Mollier chart, as shown in Fig. 25.5, as discussed below :

1. First of all, locate the point A_1 with the help of initial pressure (p_1) and temperature (t_1) of the steam. Now find out the enthalpy of steam at A_1 (*i.e.* h_{A1}).

2. From A_1, draw a vertical line A_1B_1 meeting the pressure line (p_2) at B_1 representing isentropic expansion of the steam in the first stage. Now find out the enthalpy of steam at B_1 (*i.e.* h_{B1}).

3. Now cut off B_1C_1 equal to the blade friction in the first stage [or in other words, cut off A_1C_1 equal to $\eta \times (h_{A1} - h_{B1})$]. Through C_1 draw a horizontal line C_1A_2 meeting the pressure line (p_2) at A_2, which gives the final condition of the steam discharged from the first stage. Now find out the enthalpy of steam at A_2 (*i.e.* h_{A2}).

4. Similarly, from A_2 draw a vertical line A_2B_2 to meet the pressure line (p_3) at B_2 representing isentropic expansion of steam in the second stage. Now cut off B_2C_2 equal to the blade friction in the second stage. Through C_2 draw a horizontal line C_2A_3 to meet the pressure line (p_3) at A_3, which gives the final condition of the steam discharged from the second stage. Now find out the enthalpy of steam at A_3 (*i.e.* h_{A3}).

5. Similarly, locate the point A_4 on the pressure line (p_4), which gives the final condition of the steam discharged from the third stage and find out the enthalpy of steam at A_4 (*i.e.* h_{A4}).

Now a draw a smooth curve through the points A_1, A_2, A_3 and A_4. Since these points represent the final condition of steam at the end of each stage, therefore the curve is known as *condition curve*. A little consideration will show, that if the friction is neglected, the isentropic heat drop through all these stages will be represented by the vertical line A_1D.

It will be interesting to know, from the Mollier diagram, that the pressure lines diverge from left to right, which shift the isentropic expansion lines at each stage slightly towards the right side in the diagram. Or in other words, enthalpy or heat drop (as represented by the lines A_1B_1, A_2B_2, A_3B_3) is slightly increased. The sum of increased heat drops is known as *cumulative heat drop*.

Now the ratio of cumulative heat drop to the isentropic heat drop is known as *reheat factor*. Mathematically, reheat factor,

$$\text{R.F.} = \frac{\text{Cumulative heat drop}}{\text{Isentropic heat drop}}$$

$$= \frac{A_1B_1 + A_2B_2 + A_3B_3}{A_1D} = \frac{\Sigma AB}{A_1D}$$

25.7. Efficiencies of a Multi-stage Turbine

The following efficiencies of a multi-stage turbine are important from subject point of view :

1. *Stage efficiency.* It is the ratio of the work done on the rotor (or useful heat drop) in a stage to the isentropic heat drop for the same stage. Mathematically, stage efficiency,

$$\eta_s = \frac{\text{Useful heat drop in one stage}}{\text{Isentropic heat drop for the same stage}}$$

$$= \frac{A_1C_1}{A_1B_1} = \frac{A_2C_2}{A_2B_2} = \frac{A_3C_3}{A_3B_3}$$

2. *Internal efficiency.* It is the ratio of the total work done on the rotor (or total useful heat drop) to the total isentropic neat drop. It accounts for all the losses due to friction etc. Mathematically, internal efficiency,

$$\eta_i = \frac{\text{Total useful heat drop}}{\text{Total isentropic heat drop}} = \frac{A_1C_1 + A_2C_2 + A_3C_3}{A_1D}$$

$$= \frac{(h_{A1} - h_{C1}) + (h_{A2} - h_{C2}) + (h_{A3} - h_{C3})}{h_{A1} - h_D} = \frac{h_{A1} - h_{C3}}{h_{A1} - h_D}$$

Note : The internal efficiency is sometimes known as *isentropic efficiency* or *turbine efficiency*.

3. *Efficiency ratio.* When the external losses due to friction at the bearing etc. are also considered with internal losses, the turbine efficiency is termed as *efficiency ratio.* It is defined as the ratio of the total work produced at the driving end of the shaft per kg of steam to the total isentropic heat drop across the turbine. Mathematically, efficiency ratio,

$$\text{E.R.} = \frac{\text{Total work produced at the driving end of shaft}}{\text{Total isentropic heat drop}}$$

$$= \frac{\text{Total useful heat drop}}{\text{Total isentropic heat drop}}$$

4. *Overall thermal efficiency.* It is the ratio of the work delivered at the turbine shaft to the heat supplied. This efficiency covers both the internal and external losses. Mathematically, overall thermal efficiency,

$$\eta_O = \frac{\text{Work delivered at the turbine shaft}}{\text{Heat supplied}} = \frac{\text{Total useful heat drop}}{\text{Heat supplied}}$$

$$= \frac{A_1C_1 + A_2C_2 + A_3C_3}{h_{A1} - h_{fD}}$$

$$= \frac{(h_{A1} - h_{C1}) + (h_{A2} - h_{C2}) + (h_{A3} - h_{C3})}{h_{A1} - h_{fD}} = \frac{h_{A1} - h_{C3}}{h_{A1} - h_{fD}}$$

5. *Rankine efficiency.* It is the ratio of the isentropic heat drop to the heat supplied. Mathematically, Rankine efficiency,

$$\eta_R = \frac{\text{Isentropic heat drop}}{\text{Heat supplied}} = \frac{h_{A1} - h_D}{h_{A1} - h_{fD}}$$

Example 25.3. *The steam is supplied to a three stage turbine at 30 bar and 350° C. The steam leaves the first stage at 7 bar ; second stage at 1 bar ; and finally at 0.1 bar. If each stage has an efficiency of 0.7, determine : 1. Rankine efficiency, 2. The final condition of steam, 3. Reheat factor, and 4. Overall thermal efficiency.*

Solution. Given : $p_1 = 30$ bar ; $T_1 = 350°$ C ; $p_2 = 7$ bar ; $p_3 = 1$ bar ; $p_4 = 0.1$ bar ; $\eta = 0.7$

Now let us draw expansion of the steam in three stages on the Mollier chart, as shown in Fig. 25.6, as discussed below :

1. First of all, locate a point A_1 at the intersection of initial pressure (30 bar) and initial temperature (350° C). We find that enthalpy of steam at A_1, *i.e.* $h_{A1} = 3120$ kJ/kg.

2. From A_1, draw a vertical line $A_1 B_1$ meeting the pressure line of 7 bar at B_1 representing isentropic expansion of steam in the first stage. We find that enthalpy of steam at B_1, *i.e.* $h_{B1} =$ 2790 kJ/kg.

3. Now cut off, A_1C_1 equal to $\eta\,(h_{A1} - h_{B1}) = 0.7\,(3120 - 2790) = 231$ kJ/kg. Therefore, enthalpy of steam at C_1 *i.e.* $h_{C1} = h_{A1} - 231 = 3120 - 231 = 2889$ kJ/kg. Through C_1, draw a horizontal line $C_1 A_2$ meeting the pressure line of 7 bar at A_2. Now the enthalpy of steam at A_2, *i.e.* $h_{A2} = h_{C1} = 2889$ kJ/kg.

4. Similarly from A_2, draw a vertical line A_2B_2 meeting the pressure line of 1 bar at B_2. We find that enthalpy of steam at B_2, *i.e.* $h_{B2} = 2535$ kJ/kg.

5. Now cut off A_2C_2 equal to $\eta\,(h_{A2} - h_{B2}) = 0.7\,(2889 - 2535) = 248$ kJ/kg. Therefore, enthalpy of steam at C_2, *i.e.* $h_{C2} = h_{A2} - 248 = 2889 - 248 = 2641$ kJ/kg. Through C_2, draw a

horizontal line C_2A_3 meeting the pressure line of 1 bar at A_3. Now the enthalpy of steam at A_3, *i.e.* $h_{A3} = h_{C2} = 2641$ kJ/kg.

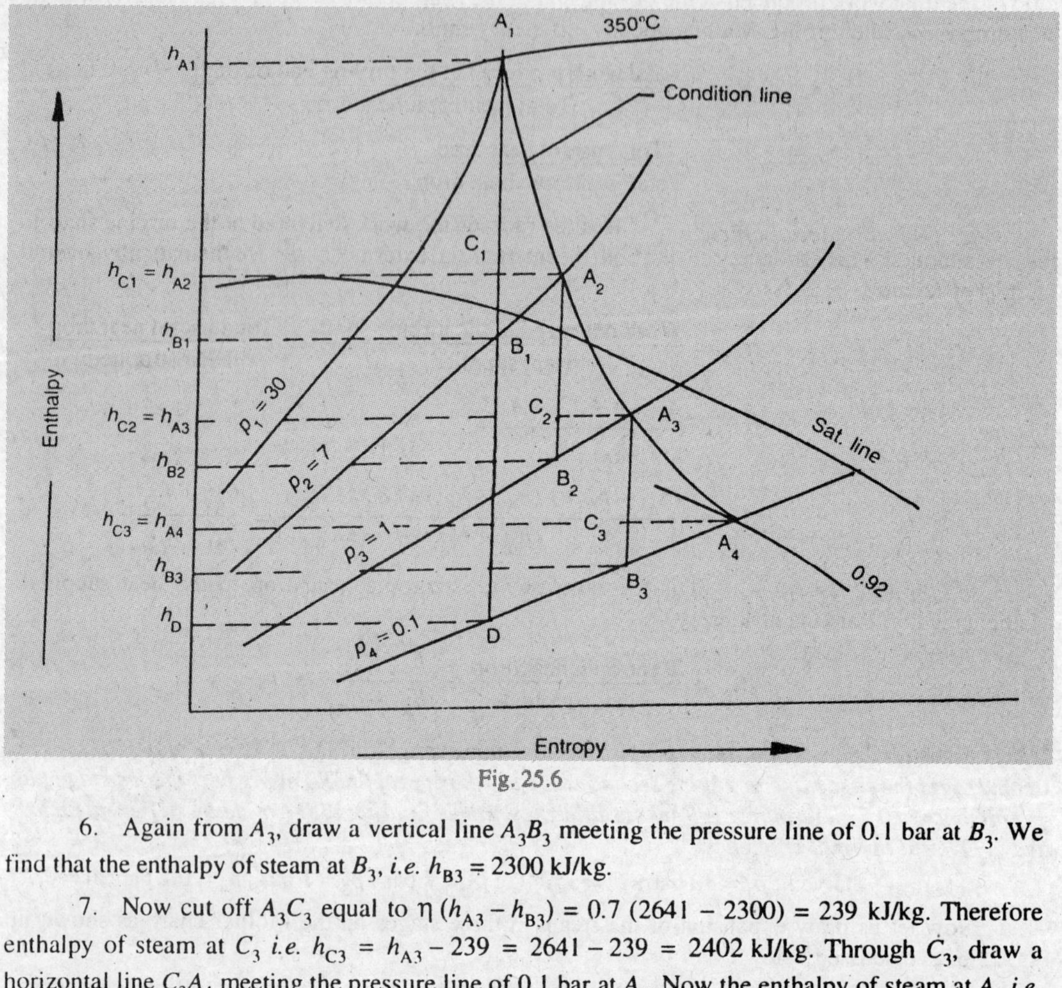

Fig. 25.6

6. Again from A_3, draw a vertical line A_3B_3 meeting the pressure line of 0.1 bar at B_3. We find that the enthalpy of steam at B_3, *i.e.* $h_{B3} = 2300$ kJ/kg.

7. Now cut off A_3C_3 equal to $\eta\,(h_{A3} - h_{B3}) = 0.7\,(2641 - 2300) = 239$ kJ/kg. Therefore enthalpy of steam at C_3 *i.e.* $h_{C3} = h_{A3} - 239 = 2641 - 239 = 2402$ kJ/kg. Through C_3, draw a horizontal line C_3A_4 meeting the pressure line of 0.1 bar at A_4. Now the enthalpy of steam at A_4 *i.e.* $h_{A4} = h_{C3} = 2402$ kJ/kg.

Now let us tabulate the values of enthalpy of steam as found above from the Mollier diagram for convenience.

$h_{A1} = 3120$ kJ/kg ; $h_{C1} = h_{A2} = 2889$ kJ/kg ; $h_{B1} = 2790$ kJ/kg ; $h_{C2} = h_{A3} = 2641$ kJ/kg ;

$h_{B2} = 2535$ kJ/kg ; $h_{C3} = h_{A4} = 2402$ kJ/kg ; $h_{B3} = 2300$ kJ/kg ; $h_D = 2140$ kJ/kg

From steam tables, corresponding to a pressure of 0.1 bar, we find that the enthalpy of water,

$$h_{fD} = 191.8 \text{ kJ/kg}$$

1. *Rankine efficiency*

We know that Rankine efficiency,

$$\eta_R = \frac{h_{A1} - h_D}{h_{A1} - h_{fD}} = \frac{3120 - 2140}{3120 - 191.8} = 0.335 \text{ or } 33.5\% \textbf{ Ans.}$$

2. *Final condition of steam*

From Mollier diagram, we find that the final condition or dryness fraction of steam at A_4 is 0.92. **Ans.**

3. *Reheat factor*

We know that reheat factor,

$$\text{R.F.} = \frac{A_1B_1 + A_2B_2 + A_3B_3}{A_1D} = \frac{(h_{A1} - h_{B1}) + (h_{A2} - h_{B2}) + (h_{A3} - h_{B3})}{h_{A1} - h_D}$$

$$= \frac{(3120 - 2790) + (2889 - 2535) + (2641 - 2300)}{3120 - 2140} = \frac{1025}{980}$$

$$= 1.046 \textbf{ Ans.}$$

4. *Overall thermal efficiency*

We know that overall thermal efficiency,

$$\eta_O = \frac{h_{A1} - h_{C3}}{h_{A1} - h_{fD}} = \frac{3120 - 2402}{3120 - 191.8} = 0.245 \text{ or } 24.5\ \% \textbf{ Ans.}$$

Example 25.4. *In a four stage impulse turbine, the steam is supplied at 350° C and at a pressure of 20 bar. The exhaust pressure is 0.05 bar and the overall turbine efficiency is 80%. Assuming that the work is shared equally between the stages and the condition line to be straight, determine (a) stage steam pressures ; (b) efficiency of each stage ; and (c) reheat factor.*

Solution. Given : $T_1 = 350°$ C ; $p_1 = 20$ bar ; $p_5 = 0.05$ bar ; $\eta_0 = 80\% = 0.8$

Stage steam pressures

Let p_2, p_3, p_4 = Steam pressures leaving the first, second and third stage respectively.

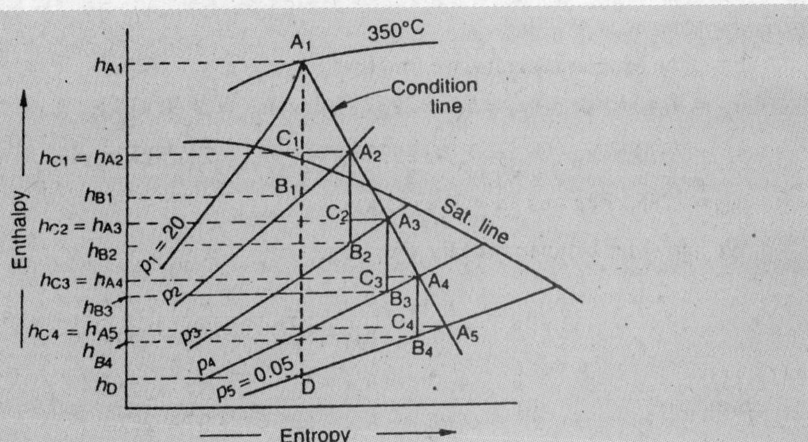

Fig. 25.7

Now let us draw the expansion of steam in four stages on the Mollier chart, as shown in Fig. 25.7, as discussed below :

1. First of all, locate the point A_1 at the intersection of initial pressure (20 bar) and initial temperature (350° C). We find that the enthalpy of steam at A_1, *i.e.* $h_{A1} = 3140$ kJ/kg.

2. From A_1, draw a vertical line A_1D meeting the pressure line of 0.05 bar at D representing the total isentropic expansion of the steam in all the four stages. We find that enthalpy of steam at D, *i.e.* $h_D = 2125$ kJ/kg. Thus we see that total enthalpy or heat drop,

$$h_{A1} - h_D = 3140 - 2125 = 1015 \text{ kJ/kg}$$

We know that overall thermal efficiency,

$$0.8 = \frac{h_{A1} - h_{C4}}{h_{A1} - h_D} = \frac{3140 - h_{C_4}}{1015}$$

$$\therefore \quad h_{C4} = 2328 \text{ kJ/kg}$$

and total work done in all the four stages

$$= h_{A1} - h_{C4} = 3140 - 2328 = 812 \text{ kJ/kg}$$

$\therefore$ Work done in each stage $= 812/4 = 203$ kJ/kg

3. Now locate the point A_5 at the intersection of $h_{C4} = 2328$ kJ/kg and final pressure line of 0.05 bar. Now draw the condition line by joining A_1 and A_5 by a straight line (as given).

4. Now locate the point C_1 by cutting off A_1C_1 equal to 203 kJ/kg. Through C_1, draw a horizontal line meeting the condition line at A_2. From the Mollier diagram, we find that $p_2 = 9$ bar. **Ans.**

5. Similarly, locate the point C_2 by cutting off A_2C_2 equal to 203 kJ/kg. Through C_2, draw a horizontal line meeting the condition line at A_3. From the Mollier diagram, we find that p_3 = 2.4 bar. **Ans.**

6. Again locate the point C_3 by cutting off A_3C_3 equal to 203 kJ/kg. Through C_3, draw a horizontal line meeting the condition line at A_4. From the Mollier diagram, we find that p_4 = 0.5 bar. **Ans.**

Efficiency of each stage

From the Mollier diagram, we find that

$h_{A1} = 3140$ kJ/kg ; $h_{C1} = h_{A2} = 2937$ kJ/kg ; $h_{B1} = 2860$ kJ/kg ; $h_{C2} = h_{A3} = 2734$ kJ/kg ;

$h_{B2} = 2670$ kJ/kg ; $h_{C3} = h_{A4} = 2531$ kJ/kg ; $h_{B3} = 2475$ kJ/kg ; $h_{C4} = h_{A5} = 2328$ kJ/kg ;

$h_{B4} = 2285$ kJ/kg and $h_D = 2125$ kJ/kg

We know that efficiency of the first stage,

$$\eta_1 = \frac{A_1C_1}{A_1B_1} = \frac{h_{A1} - h_{C1}}{h_{A1} - h_{B1}} = \frac{3140 - 2937}{3140 - 2860} = 0.725 = 72.5\% \textbf{ Ans.}$$

Similarly,

$$\eta_2 = \frac{A_2C_2}{A_2B_2} = \frac{h_{A2} - h_{C2}}{h_{A2} - h_{B2}} = \frac{2937 - 2734}{2937 - 2670} = 0.76 = 76\% \textbf{ Ans.}$$

$$\eta_3 = \frac{A_3C_3}{A_3B_3} = \frac{h_{A3} - h_{C3}}{h_{A3} - h_{B3}} = \frac{2734 - 2531}{2734 - 2475} = 0.784 = 78.4\% \textbf{ Ans.}$$

and

$$\eta_4 = \frac{A_4C_4}{A_4B_4} = \frac{h_{A4} - h_{C4}}{h_{A4} - h_{B4}} = \frac{2531 - 2328}{2531 - 2285} = 0.825 = 82.5\% \textbf{ Ans.}$$

Reheat factor

We know that reheat factor,

$$R.F. = \frac{A_1B_1 + A_2B_2 + A_3B_3 + A_4B_4}{A_1D}$$

$$= \frac{(h_{A1} - h_{B1}) + (h_{A2} - h_{B2}) + (h_{A3} - h_{B3}) + (h_{A4} - h_{B5})}{h_{A1} - h_D}$$

$$= \frac{(3140 - 2860) + (2937 - 2670) + (2734 - 2475) + (2531 - 2285)}{3140 - 2125}$$

$$= \frac{1052}{1015} = 1.04 \text{ Ans.}$$

25.8. Regenerative Cycle

We have already discussed in chapter 10, Rankine and Carnot vapour cycles. The efficiency of Rankine cycle is less than that of Carnot cycle, because in the Rankine cycle, all the heat is not added at the higher temperature as is done in the case of Carnot cycle. Moreover, a large amount of additional heat is rejected to the condenser (shown by area *DEFG* in Fig. 25.8). This rejected heat is not compensated by the additional work represented by the area *ADE*. If, however, the working fluid enters the boiler, at some state, between *E* and *A*, the average temperature of heat supplied is increased. Thus the cycle becomes as efficient as Carnot cycle.

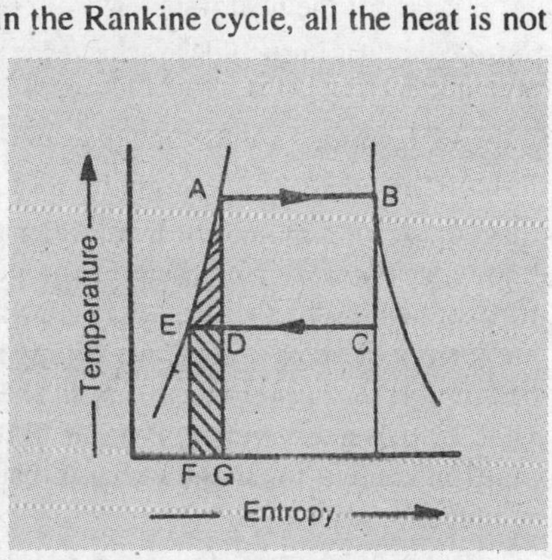

Fig. 25.8. Comparison of Rankine cycle and Carnot cycle.

This is achieved during an ideal regenerative cycle as shown in Fig. 25.9. The dry saturated steam, from the boiler, enters the turbine at a higher temperature, and then expands isentropically to a lower temperature in the same way as that in the Rankine and Carnot cycle. Now the condensate, from the condenser, is pumped back and circulated around the turbine casing, in the direction opposite to the steam flow in the turbine. The steam is thus heated before entering into the boiler. Such a system of heating is known as *regenerative heating*, as the steam is used to heat the steam itself.

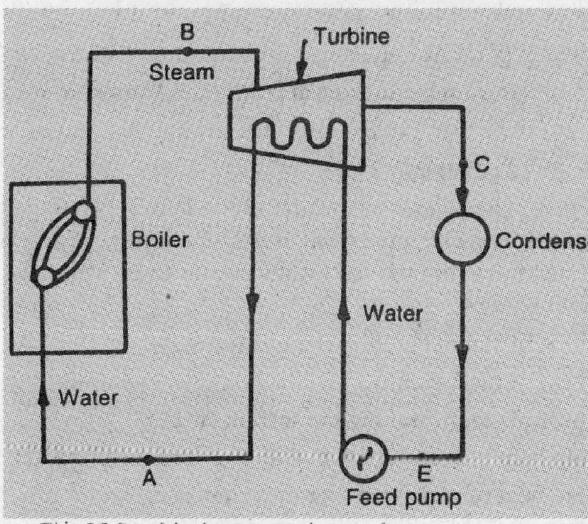

Fig. 25.9. Ideal regenerative cycle.

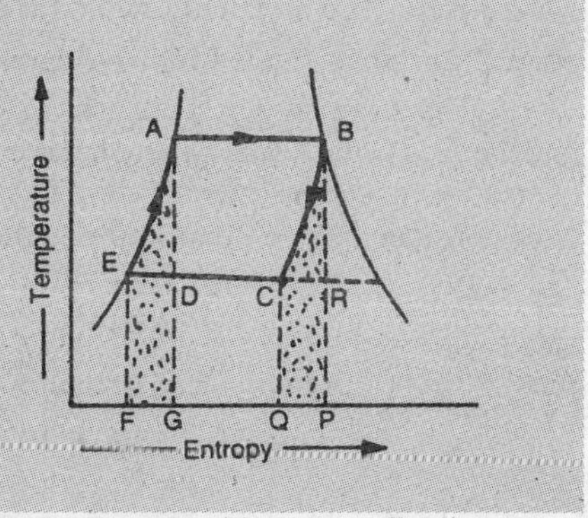

Fig. 25.10. Regenerative cycle on *T-s* diagram.

It may be noted that due to loss of heat, the expansion in the steam turbine is no more isentropic. It follows the path *BC*, which is exactly parallel to *EA*, as shown in Fig. 25.10.

A little consideration will show, that the heat transferred to the liquid (area *EAGF*) is equal to the heat transferred from the steam (area *BPQC*). Moreover, the heat is supplied to the working fluid at constant temperature in the process shown by curve *AB*. This is represented by the area *ABPG*. The heat is rejected from the working fluid at constant temperature shown by curve *CE*. This is represented by the area *CQFE* which is equal to the area *RPGD*. The area *RPGD* represents, to some scale, the heat rejected in the Carnot cycle. Thus, the ideal regenerative cycle has an efficiency equal to the efficiency of Carnot cycle with the same heat supply and heat rejection temperatures.

It will be interesting to know that the ideal regenerative cycle is only of academic interest as it is impossible to achieve the cycle, in actual practice, due to the following two reasons :

1. It is not possible to effect the necessary heat transfer from the steam in the turbine to the liquid feed water.
2. The moisture content of the steam leaving the turbine is considerably increased as a result of the heat transfer.

Note : Since it is impossible to achieve the ideal regenerative cycle, in practice, therefore some advantage of this principle is taken in the form of bleeding a part of the steam, which is discussed below :

25.9. Bleeding

The process of draining steam from the turbine, at certain points during its expansion and using this steam for heating the feed water (in feed water heaters) and then supplying it to the boiler is known as *bleeding*, and the corresponding steam is said to be bled. A feed water heater is a simple form of heat exchanger. It, usually, consists of tubes through which the feed water flows. These tubes are surrounded by a casing, containing the heating steam. The steam condenses and transfers its latent heat to the feed water in the tubes. There is no theoretical limit to the number of heaters, which will yield increased efficiency. But due to the cost of each heater and complicated and costly net work of pipe lines and fittings, its number is restricted to six in the larger modern turbo-generators. The effects of bleeding are :

1. It increases the thermodynamic efficiency of the turbine.
2. The boiler is supplied with a hot water.
3. A small amount of work is lost by the turbine, which decreases the power developed.

25.10. Regenerative Cycle with Single Feed Water Heater

Consider a regenerative cycle with single feed water heater as shown in Fig. 25.11.

The steam (at pressure p_1) enters the turbine at point *A*. Let a small amount of wet steam (say *m* kg) after partial expansion (at pressure p_2) be drained from the turbine at point *B* and enter the feed water heater. The remaining steam (at pressure p_3) is further expanded in the turbine and leaves at point *C* as shown on the Mollier diagram in Fig. 25.12 (Example 25.5).

This steam is then condensed in the condenser. The condensate from the condenser, is pumped into the feed water heater, where it mixes up with the steam extracted from the turbine. The proportion of steam extracted is just sufficient to cause the steam leaving the feed water heater to be saturated. Now consider 1 kg of steam entering the turbine at point *A*.

Let h_1 = Enthalpy or total heat of steam entering the turbine at *A*,

h_2 = Enthalpy or total heat of bled steam,

h_3 = Enthalpy or total heat of steam leaving the turbine at *C*,

h_{f2} = Enthalpy or sensible heat of feed water leaving the feed water heater,

h_{f3} = Enthalpy or sensible heat of steam leaving the condenser, and

m = Amount of bled steam per kg of steam supplied.

We know that heat lost by bled steam

$$= \text{Heat gained by feed water}$$

or
$$m\,(h_2 - h_{f2}) = (1 - m)\,(h_{f2} - h_{f3})$$

$$mh_2 - m\,h_{f2} = h_{f2} - h_{f3} - m\,h_{f2} + m\,h_{f3}$$

$$\therefore \quad m = \frac{h_{f2} - h_{f3}}{h_2 - h_{f3}} \qquad \ldots (i)$$

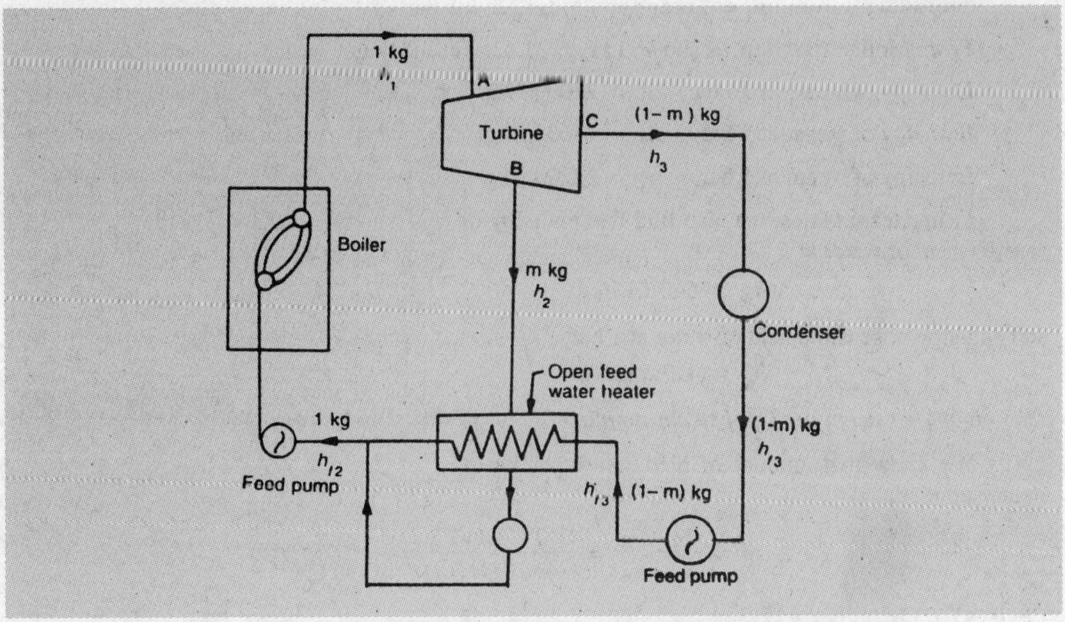

Fig. 25.11. Regenerative cycle with single feed water heater.

We know that mass of steam in the turbine, per kg of feed water, between *A* and *B* is 1 kg. Therefore work done in the turbine per kg of feed water between *A* and *B*

$$= (h_1 - h_2)$$

Mass of steam in the turbine per kg of feed water between *B* and *C*

$$= (1 - m)\text{ kg}$$

$\therefore$ Work done in the turbine between *B* and *C*

$$= (1 - m)\,(h_2 - h_3)$$

Total work done $= (h_1 - h_2) + (1 - m)\,(h_2 - h_3)$

and total heat supplied per kg of feed water

$$= h_1 - h_{f2}$$

$\therefore$ Efficiency of the cycle including the effect of bleeding,

$$\eta = \frac{\text{Total work done}}{\text{Total heat supplied}} = \frac{(h_1 - h_2) + (1 - m)\,(h_2 - h_3)}{h_1 - h_{f2}}$$

Note : If there had been no regenerative feed heating (or in other words, $m = 1$), then the efficiency of the cycle will be the same, as that of Rankine cycle. In this case, Rankine efficiency,

$$\eta = \frac{h_1 - h_3}{h_1 - h_{f3}}$$

Example 25.5. *In a regenerative cycle, having one feed water heater, the dry saturated steam is supplied from the boiler at a pressure of 30 bar and the condenser pressure is 1 bar. The steam is bled at a pressure of 5 bar. Determine the amount of bled steam per kg of steam supplied and the efficiency of the cycle. What would be the efficiency without regenerative feed heating ? Also determine the percentage increase in efficiency due to regeneration.*

Solution. Given : $p_1 = 30$ bar ; $p_3 = 1$ bar ; $p_2 = 5$ bar

From Mollier diagram, as shown in Fig. 25.12, we find that

Enthalpy of steam at 30 bar, $h_1 = 2800$ kJ/kg

Enthalpy of steam at 5 bar, $h_2 = 2460$ kJ/kg

Enthalpy of steam at 1 bar, $h_3 = 2220$ kJ/kg

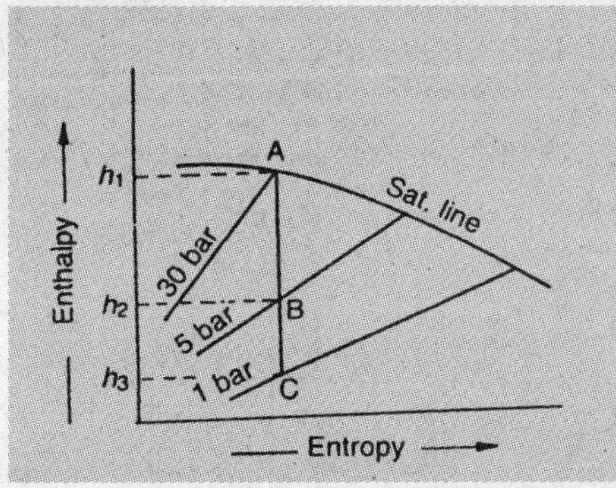

Fig. 25.12

From steam tables, we also find that enthalpy or sensible heat of water at 5 bar,

$$h_{f2} = 640.1 \text{ kJ/kg}$$

and enthalpy or sensible heat of water at 1 bar,

$$h_{f3} = 417.5 \text{ kJ/kg}$$

Amount of bled steam per kg of steam supplied

We know that amount of bled steam per kg of steam supplied,

$$m = \frac{h_{f2} - h_{f3}}{h_2 - h_{f3}} = \frac{640.1 - 417.5}{2460 - 417.5} = 0.109 \text{ kg } \textbf{Ans.}$$

Efficiency of the cycle

We know that efficiency of the cycle,

$$\eta_1 = \frac{(h_1 - h_2) + (1 - m)(h_2 - h_3)}{h_1 - h_{f2}}$$

$$= \frac{(2800 - 2460) + (1 - 0.109)(2460 - 2220)}{2800 - 640.1} = \frac{553.84}{2159.9}$$

$$= 0.256 \text{ or } 25.6\% \textbf{ Ans.}$$

Efficiency of the cycle without regenerative feed heating

We know that efficiency of the cycle,

$$\eta_2 = \frac{h_1 - h_3}{h_1 - h_{f3}} = \frac{2800 - 2220}{2800 - 417.5} = \frac{580}{2382.5}$$

$$= 0.243 \text{ or } 24.3\% \textbf{ Ans.}$$

Percentage increase in efficiency due to regeneration

We know that percentage increase in efficiency due to regeneration

$$= \frac{0.256 - 0.243}{0.243} = 0.0535 \text{ or } 5.35\% \textbf{ Ans.}$$

25.11. Regenerative Cycle with Two Feed Water Heaters

In this case, the steam is removed from the turbine at two points B and B_1. It is then fed into two open feed water heaters 1 and 2 as shown in Fig. 25.13.

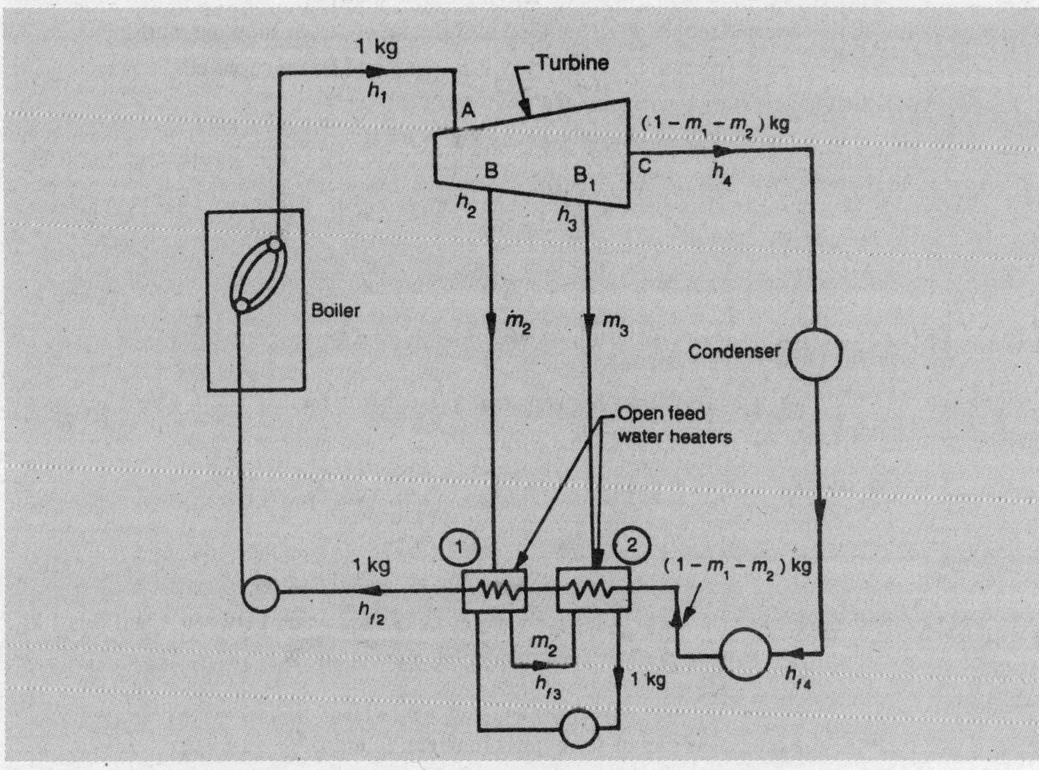

Fig. 25.13. Regenerative cycle with two feed water heaters.

The steam (at pressure p_1) enters the turbine at point A. Let a small amount of steam (say m kg) after partial expansion (at pressure p_2) be drained from the turbine at point B and enter the feed water heater 1. Similarly, let another small amount of steam (say m_2 kg) after further expansion (at pressure p_3) be drained from the turbine at point B_1 and enter the feed water heater 2. The remaining steam equal to $(1 - m_1 - m_2)$ kg (at pressure p_4) is further expanded in turbine, and leaves it at point C as shown on the Mollier diagram in Fig. 25.14.

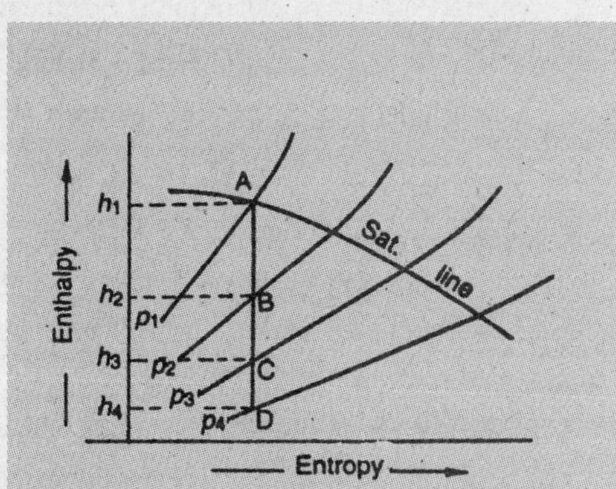

Fig. 25.14

The steam is then condensed in the condenser. The condensate from the condenser is pumped into the feed water heater, where it mixes up with the steam extracted from the turbine. Now consider 1 kg of steam entering into the turbine at A as shown in Fig. 25.13.

Let h_1 = Enthalpy of steam entering the turbine at A,

h_2 = Enthalpy of steam bled at B,

h_3 = Enthalpy of steam bled at B_1,

h_4 = Enthalpy of steam leaving the turbine at C,

h_{f2} = Enthalpy of feed water leaving the feed water heater 1,

h_{f3} = Enthalpy of feed water leaving the feed water heater 2,

h_{f4} = Enthalpy of steam leaving the condenser,

m_1 = Amount of steam bled at B per kg of steam supplied, and

m_2 = Amount of steam bled at B_1 per kg of steam supplied.

We know that heat lost by bled steam at B

= Heat gained by feed water

$$m_1 (h_2 - h_{f2}) = (1 - m_1)(h_{f2} - h_{f3})$$

$$m_1 h_2 - m_1 h_{f2} = h_{f2} - h_{f3} - m_1 h_{f2} + m_1 h_{f3}$$

$$\therefore \quad m_1 = \frac{h_{f2} - h_{f3}}{h_2 - h_{f3}}$$

Similarly, heat lost by bled steam at B_1

= Heat gained by feed water

$$\therefore \quad m_2 (h_3 - h_{f3}) = (1 - m_1 - m_2)(h_{f3} - h_{f4})$$

$$m_2 h_3 - m_2 h_{f3} = h_{f3} - h_{f4} - m_1 h_{f3} + m_1 h_{f4} - m_2 h_{f3} + m_2 h_{f4}$$

or $$m_2 = \frac{(1 - m_1)(h_{f3} - h_{f4})}{h_3 - h_{f4}}$$

We know that the mass of steam in the turbine per kg of feed water between A and B is 1 kg.

$\therefore$ Work done in the turbine per kg of feed water between A and B

$$= h_1 - h_2 \quad \ldots (i)$$

and mass of steam in the turbine per kg of feed water between B and B_1

$$= (1 - m_1) \text{ kg}$$

$\therefore$ Work done in the turbine between B and B_1

$$= (1 - m_1)(h_2 - h_3) \quad \ldots (ii)$$

Similarly, mass of steam in the turbine per kg of feed water between B_1 and C

$$= (1 - m_1 - m_2) \text{ kg}$$

$\therefore$ Work done in the turbine between B_1 and C

$$= (1 - m_1 - m_2)(h_3 - h_4) \quad \ldots (iii)$$

Thus total work done per kg of feed water

$$= (h_1 - h_2) + (1 - m_1)(h_2 - h_3) + (1 - m_1 - m_2)(h_3 - h_4)$$

and total heat supplied per kg of feed water

$$= h_1 - h_{f2}$$

$\therefore$ Efficiency of the plant including the effect of bleeding,

$$\eta = \frac{\text{Total work done}}{\text{Total heat supplied}}$$

Note : When the bleeding takes place at more than two points, the efficiency of the plant may be obtained by proceeding in the same way as explained above.

Example 25.6. *In a steam turbine plant, the steam is generated and supplied to the turbine at 50 bar and 370° C. The condenser pressure is 0.1 bar and the steam enters the condenser with dryness fraction of 0.9. Two feed heaters are used, the steam in the heaters being bled at 5 bar and 0.5 bar. In each heater, the feed water is heated to saturation temperature of the bled steam. The condensate is also pumped at this temperature into the feed line immediately after the heater. Find the masses of the steam bled in the turbine per one kg of steam entering the turbine. Assuming the condition line for the turbine to be straight, calculate the thermal efficiency of the cycle.*

Solution. Given : $p_1 = 50$ bar ; $T_1 = 370°$ C ; $p_4 = 0.1$ bar ; $x_4 = 0.9$; $p_2 = 5$ bar ; $p_3 = 0.5$ bar

First of all, let us draw the Mollier diagram and condition line for the cycle, as shown in Fig. 25.15. From this diagram, we find that

$$h_1 = 3110 \text{ kJ/kg} \; ; \; h_2 = 2780 \text{ kJ/kg} \; ;$$

$$h_3 = 2510 \text{ kJ/kg} \; ; \; h_4 = 2350 \text{ kJ/kg}$$

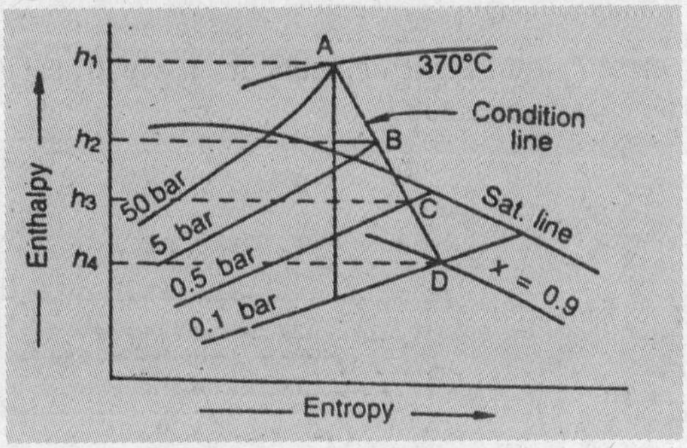

Fig. 25.15

From steam tables, we also find that

$$h_{f2} = 640.1 \text{ kJ/kg (at 5 bar)}$$

$$h_{f3} = 340.6 \text{ kJ/kg (at 0.5 bar)}$$

$$h_{f4} = 191.8 \text{ kJ/kg (at 0.1 bar)}$$

Mass of steam bled in the turbine

We know that mass of steam bled at B,

$$m_1 = \frac{h_{f2} - h_{f3}}{h_2 - h_{f3}} = \frac{640.1 - 340.6}{2780 - 340.6} = 0.123 \text{ kg } \textbf{Ans.}$$

and mass of steam bled at B_1,

$$m_2 = \frac{(1 - m_1)(h_{f3} - h_{f4})}{h_3 - h_{f4}} = \frac{(1 - 0.123)(340.6 - 191.8)}{2510 - 191.8}$$

$$= 0.056 \text{ kg } \textbf{Ans.}$$

Thermal efficiency of the cycle

We know that work done from A to B per kg of feed water

$$= h_1 - h_2 = 3110 - 2780 = 330 \text{ kJ/kg} \quad \ldots (i)$$

Similarly, work done from B to B_1 per kg of feed water

$$= (1 - m_1)(h_2 - h_3) = (1 - 0.123)(2780 - 2510) \text{ kJ/kg}$$

$$= 236.8 \text{ kJ/kg} \quad \ldots (ii)$$

and work done from B_1 to C per kg of feed water

$$= (1 - m_1 - m_2)(h_3 - h_4) = (1 - 0.123 - 0.056)(2510 - 2350) \text{ kJ/kg}$$

$$= 131.4 \text{ kJ/kg} \quad \ldots (iii)$$

$\therefore$ Total work done $= 330 + 236.8 + 131.4 = 698.2$ kJ/kg

Heat supplied $= h_1 - h_{f2} = 3110 - 640.1 = 2469.9$ kJ/kg

∴ Thermal efficiency of the cycle,

$$\eta = \frac{\text{Work done}}{\text{Heat supplied}} = \frac{698.2}{2469.9} = 0.283 \quad \text{or} \quad 28.3\% \ \textbf{Ans.}$$

25.12. Binary Vapour Plants

We have already discussed that the maximum possible efficiency of any steam engine (known as Carnot efficiency) is given by the equation :

$$\eta = \frac{T_1 - T_3}{T_1} = 1 - \frac{T_3}{T_1}$$

where T_1 is the higher temperature at which heat is absorbed and T_3 is the lower temperature at which the heat is rejected.

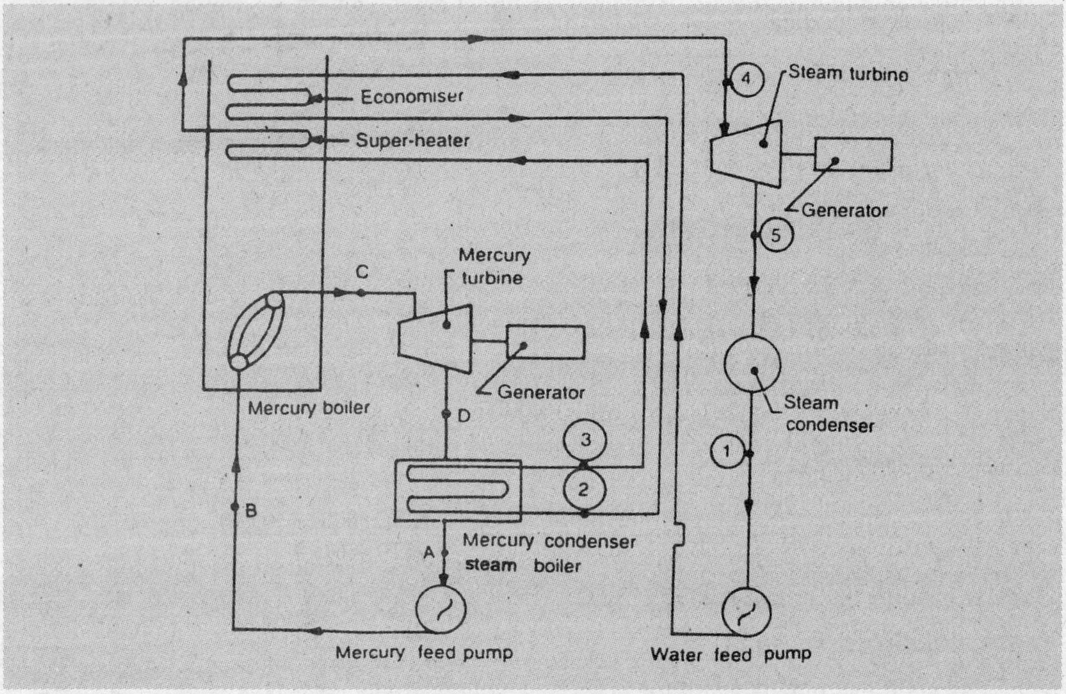

Fig. 25.16. Binary vapour plant.

We know that the value of lower temperature T_3 is fixed by atmospheric conditions. Thus the thermal efficiency of a steam plant can only be improved by increasing the value of T_1. In a steam plant, if T_1 is increased, it will also increase corresponding pressure, which is one of the limiting factors in its design. Thus a substance, other than steam, is used in the high *temperature range. By using mercury vapour, instead of steam, in the high temperature range of the cycle, an increased value of T_1 is obtained without any increase in maximum pressure. The heat rejected by the mercury, in condensing, is utilised in raising superheated steam for the lower temperature range of the cycle. A power plant using binary vapour (mercury and steam) is known as *binary vapour plant*.

* The critical temperature of mercury vapour is 588.4° C at a critical pressure of 21 bar. Moreover, the critical temperature of water vapour is 374.15° C at a critical pressure of 225.65 bar.

A diagrammatic view of a binary vapour plant is shown in Fig. 25.16. The liquid mercury passes from the mercury liquid heater to the mercury vapour boiler, where it is evaporated. It then flows to the mercury turbine through which it is expanded to its low pressure limit. It, now, exhausts to the mercury condenser steam boiler, where its latent heat is given out to the hot feed water. This operation condenses the mercury, while the feed water is evaporated into steam. The mercury is then returned to the mercury liquid heater, thus completing its cycle.

The feed water from the economiser is evaporated into steam in the mercury condenser steam boiler. It then passes to the superheater, where it is superheated by the hot flue gases. Now the superheated steam passes to the steam turbine in which it expands to the condenser pressure. It now passes to the economiser, thus completing the cycle of water and steam.

25.13. Binary Vapour Cycle

The binary vapour cycle on a *T-s* diagram is shown in Fig. 25.17. The line *AB* represents the evaporation of liquid mercury plotted to the same temperature scale as that of steam. But the scale for the corresponding pressures for the mercury is lower. The mercury vapour at *B* has a much higher temperature than the steam at the same pressure. The mercury vapours are now expanded isentropically in a mercury turbine as represented by the line *BC* in Fig. 25.17. The condensation of mercury is shown by the line *CD*. During condensation, the latent heat is utilized for evaporating a corresponding amount of steam. The line *DA* represents the heating of mercury. Thus the mercury has completed a cycle *ABCD*.

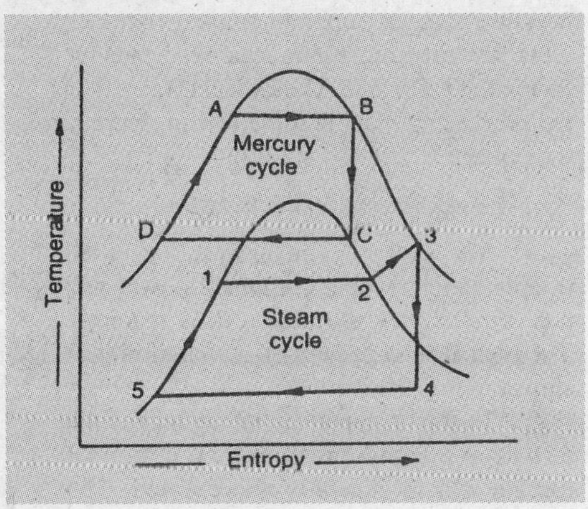

Fig. 25.17. Binary vapour cycle on *T-s* diagram.

The steam cycle is represented by 1-2-3-4-5 as shown in Fig. 25.17. The line 1-2 represents the evaporation by the condensing mercury. The line 2-3 represents the superheating of the steam by the flue gases. The steam is now expanded isentropically through a steam turbine as shown by the line 3-4. The condensation of the exhaust steam is represented by the line 4-5. The heating of feed water is represented by the line 5-1. This completes the steam cycle.

Notes : 1. It is found that 8.196 kg of mercury is required per kg of steam to obtain the correct amount of heat from the condensing mercury for evaporating the steam. It may be seen from the areas of *T-s* diagram that the use of mercury for the high temperature range of the cycle gives a higher efficiency than that obtained from steam with the same addition of heat.

2. The relation between the two cycles is that the heat lost by mercury in *CD* is equal to the heat gained by steam in 1-2. Mathematically, heat lost by mercury at *C* is equal to heat gained by steam at 1.

25.14. Some Special Turbines

The following are some special turbines commonly used in industry :

1. Pass-out or extraction turbine ;
2. Back pressure turbine ; and
3. Exhaust or low pressure turbine.

The function of these turbines, aside from power generation, is to supply steam for manufacturing processes or heating or to utilise steam that would otherwise be wasted.

The above mentioned turbines are discussed, in detail, in the following pages.

25.15. Pass-out or Extraction Turbine

A pass-out or extraction turbine is of the type used in central stations, in which steam is extracted at different stages and used in heating the feed water for the boiler or processing work such as paper making, textile, dying, sugar refining etc. Thus, a pass-out or extraction turbine supplies the required power and also the low pressure steam needed for the purpose.

The typical arrangement of a pass-out or extraction turbine is shown in Fig. 25.18. The high pressure steam from the boiler enters the H.P. stage of the turbine where it expands and the pressure is reduced to such a value as required for processing work. A part of this low pressure steam leaving the high pressure stage is supplied to the processing work while the remaining steam expands further in the L.P. stage. The exhaust steam from the processing plant and the low pressure turbine is condensed in a condenser and pumped back to the boiler.

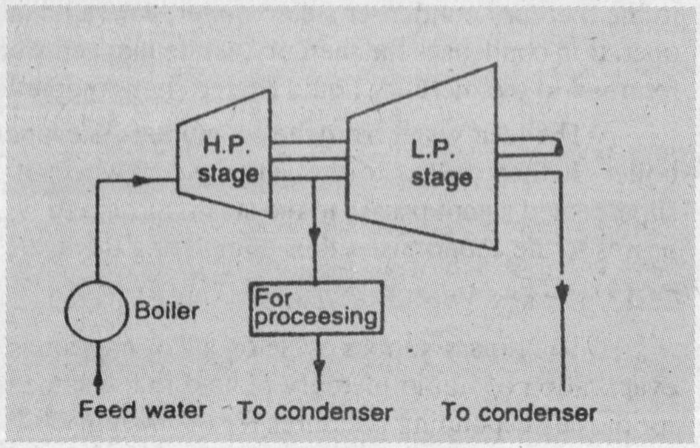

Fig. 25.18. Pass-out or extraction turbine.

25.16. Back Pressure Turbine

The back pressure turbine is also used in applications where combined power and heat in steam for process work is required. The typical arrangement of such a turbine is shown in Fig. 25.19.

In this turbine, the steam leaves the turbine at a higher pressure than in normal turbine and is generally superheated. The superheated steam is not suitable for process work due to the following reasons :

1. The control of its temperature is impossible, and
2. The rate of heat transfer from the superheated steam to the heating surface is lower than that of saturated steam.

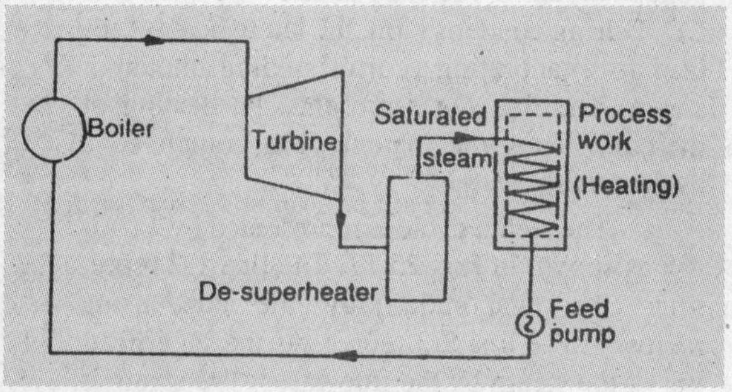

Fig. 25.19. Back pressure turbine.

Thus, the exhaust is passed through a de-superheater to make the steam saturated. The saturated steam is then passed through a heater where it is fully condensed.

25.17. Exhaust or Low Pressure Turbine

The exhaust or low pressure turbine is chiefly used where there are several reciprocating steam engines which work intermittently and are non-condensing such as rolling mill and colliery engines. The arrangement of such a turbine is shown in Fig. 25.20.

In this turbine, the exhaust steam from the engine is expanded in a exhaust or low pressure turbine and then condensed in a condenser. In this turbine, some form of heat accumulator is required to collect the more

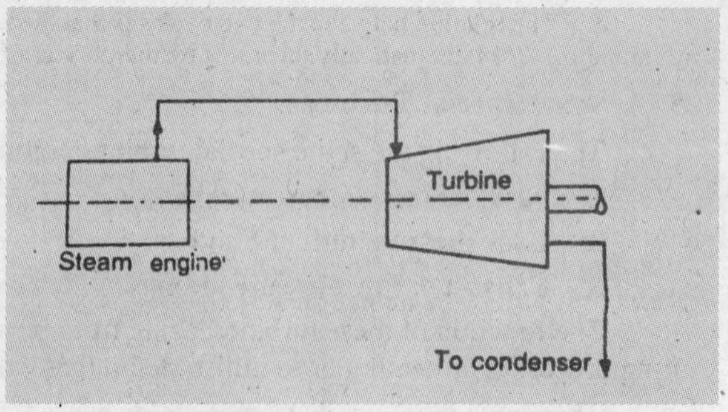

Fig. 25.20. Exhaust or low pressure turbine.

or less irregular supply of low pressure steam for the non-condensing steam engines and deliver it to the turbine at the required rate.

25.18. Future Power Plants

The depleting resources of oil, gas and coal (the conventional fuels used in power plants) along with atmospheric pollution problems have drawn the attention of the scientists and engineers all over the world to find out other sources for the generation of electric power. These sources of energy are going to attain the nerve centre of the future power plants. Though atomic and nuclear power plants have been developed on conventional lines, but lot of work is yet to be done. Efforts are being made to convert atomic and nuclear energy directly into electric power with the help of magneto-hydrodynamic generator and other equipments.

At certain places research is being conducted to use solar energy and gases for the generation of power. It is hoped that our scientists will be able to save the humanity from the power crisis in the near future.

EXERCISES

1. In a power plant, steam at a pressure of 31.5 bar and temperature 370° C is supplied by a boiler. The steam is removed after expansion to 5.6 bar in high pressure turbine, reheated at 370° C and is used in the low pressure turbine. Exhaust is at 0.035 bar. Calculate the gain in efficiency by reheating. [Ans. 1.78%]

2. Steam at a pressure of 150 bar and 550° C is expanded through a turbine to a pressure of 40 bar. It is then reheated to a temperature of 550° C after which it completes its expansion through the turbine to an exhaust pressure of 0.1 bar. Calculate the ideal efficiency of the plant and the work done,

   (*a*) taking the reheating into account, and

   (*b*) if the steam was expanded direct to the exhaust pressure without reheating.

   [Ans. (*a*) 43.75%, 1640 kJ/kg ; (*b*) 41.8%, 1370 kJ/kg]

3. Steam supplied to a three stage turbine is at 20 bar and 350° C. The steam leaves the first stage at 6 bar and the second stage at 1 bar. The steam finally leaves at 0.1 bar. Each stage has an efficiency of 0.8. Find : (*i*) Rankine efficiency, (*ii*) Final condition of steam, (*iii*) Overall thermal efficiency, (*iv*) Efficiency ratio, and (*v*) Reheat factor. [Ans. 35.2% ; 0.89 ; 28.3% ; 0.804 ; 1.008]

4. A three stage pressure compounded impulse turbine operates under initial steam conditions of 28 bar and 350° C and condenser pressure of 50 mm of Hg. Assuming an internal efficiency of 0.78 and equal work is done in all the stages, estimate the probable stage efficiencies and the reheat factor of the turbine.

   [Ans. 71.7%, 74.7%, 78.1% ; 1.043]

5. Steam is supplied at 17.5 bar and 300° C to a 5-stage turbine and exhausts at 0.04 bar. The enthalpies after actual expansion and isentropic expansion in kJ/kg are as follows :

Stage No.	1	2	3	4	5
Actual expansion	2901	2746	2592	2416	2232
Isentropic expansion	2842	2700	2550	2374	2207

   If the reheat factor for the turbine is 1.04, determine the stage efficiencies and the overall efficiency.

   [Ans. 68% ; 77% ; 78.6 %, 80.7%, 88% ; 89.2 %]

6. A steam power plant equipped with regenerative as well as reheat arrangement is supplied with steam to the high pressure turbine at 80 bar and 470° C. A part of the steam is extracted at 7 bar for feed heating and the remaining steam is reheated to 350° C in a reheater and expanded in low pressure turbine to a pressure of 0.035 bar. Determine : 1. The amount of steam bled for feed heating ; 2. The heat supplied in boiler and reheater ; 3. Output of the turbine ; and 4. Overall thermal efficiency.

   [Ans. 0.224 kg/kg of steam ; 2956.6 kJ/kg ; 1303.9 kJ/kg 44.1%]

7. A turbine receives steam at 42 bar and 371°C and exhausts at 0.106 bar. At the actual state of 10.5 bar and 204° C, the steam is withdrawn and part is used for feed water heating, while the remainder passes through a reheater. The reheated steam re-enters the turbine at 9.8 bar and 371° C. A second extraction for feed water

heating occurs at 1.4 bar and 177° C. A third extraction occurs at 0.35 bar and 93° C. The actual exhaust is dry and saturated steam. The actual temperatures of the feed water leaving the heaters are 71° C, 107° C and 167°C respectively. Assuming no losses in and between the heaters, find the actual percentage extracted at each point and the thermal efficiency. Represent the processes on *T-s* and *h-s* plane.

[**Ans.** 0.105 kg, 0.0533 kg, 0.0347 kg ; 30%]

8. A steam power plant using regenerative feed heating generates 27 MW. The steam at 60 bar and 450° C is supplied to the steam turbine. The condenser pressure is 0.07 bar. The steam is bled from the steam turbine at 3 bar. The heating of feed water is done in the direct contact heater and the condensate temperature is raised to 110° C.

The pumps absorb 9 percent of the alternator output, the boiler efficiency is 87 percent, the efficiency ratio of each section of turbine is 85 percent and the alternator efficiency is 97 percent. Ignoring all other losses, calculate : 1. the mass tapped to the feed heater per kg of steam to the condenser ; 2. the steam to be generated per hour, and 3. the overall thermal efficiency of the plant, neglecting the boiler feed pump work in calculating input to the boiler.

[**Ans.** 0.13 kg ; 100 630 kg/h ; 26.94%]

QUESTIONS

1. Explain the process and purpose of reheating steam in steam turbine application.
2. What is reheat factor ? Explain it with the *h-s* diagram.
3. Describe regenerative feed heating as used in thermal power plants. List its advantages.
4. Explain the purpose of bleeding steam turbines, in detail.
5. Describe, with the help of diagram the binary vapour cycle of a thermal power plant ? What are its advantages.
6. Explain with suitable schematic diagram, the following :

   (*a*) Pass out turbine ; (*b*) Back pressure turbine ; and (*c*) Exhaust turbine.

OBJECTIVE TYPE QUESTIONS

1. The efficiency of steam turbines may be improved by

   (*a*) reheating of steam (*b*) regenerative feed heating

   (*c*) binary vapour plant (*d*) any one of these

2. The reheating of steam in a turbine

   (*a*) increases the workdone through the turbine

   (*b*) increases the efficiency of the turbine

   (*c*) reduces wear on the blades

   (*d*) all of the above

3. The ratio of isentropic heat drop to the heat supplied is called

   (*a*) Rankine efficiency (*b*) stage efficiency

   (*c*) reheat factor (*d*) internal efficiency

4. The ratio of total useful heat drop to the total isentropic heat drop is called

   (*a*) internal efficiency (*b*) efficiency ratio

   (*c*) Rankine efficiency (*d*) stage efficiency

   (*e*) both (*a*) and (*b*) (*f*) both (*c*) and (*d*)

5. The reheat factor is the ratio of

   (*a*) cumulative heat drop to the isentropic heat drop

   (*b*) isentropic heat drop to the heat supplied

   (*c*) total useful heat drop to the total isentropic heat drop

   (*d*) none of the above

6. The reheat factor depends upon

(*a*) initial pressure and superheat (*b*) exit pressure
(*c*) turbine stage efficiency (*d*) all of these

7. The value of reheat factor varies from

(*a*) 1.02 to 1.06 (*b*) 1.08 to 1.10 (*c*) 1.2 to 1.6 (*d*) 1.6 to 2.0

8. The process of draining steam from the turbine, at certain points during its expansion and using this steam for heating the feed water in feed water heaters and then supplying it to the boiler, is known as

(*a*) regenerative feed heating (*b*) reheating of steam
(*c*) bleeding (*d*) none of these

9. The effect of bleeding is that

(*a*) it increases the thermodynamic efficiency of the turbine
(*b*) boiler is supplied with hot water
(*c*) it decreases the power developed by the turbine
(*d*) all of the above

10. A binary vapour plant consists of

(*a*) steam turbine (*b*) steam condenser
(*c*) mercury boiler (*d*) economiser
(*e*) superheater (*f*) all of these

ANSWERS

1. (*d*)	2. (*d*)	3. (*a*)	4. (*e*)	5. (*a*)
6. (*d*)	7. (*a*)	8. (*c*)	9. (*d*)	10. (*f*)

26

Internal Combustion Engines

1. Introduction. 2. Comparison of Steam Engines and Internal Combustion Engines. 3. Classification of I.C. Engines. 4. Main Components of I.C. Engines. 5. Sequence of Operations in a Cycle. 6. Two-stroke and Four-stroke Cycle Engines. 7. Advantages and Disadvantages of Two-stroke over Four-stroke Cycle Engines. 8. Valve Timing Diagrams. 9. Four-stroke Cycle Petrol Engine. 10. Actual Indicator Diagram for a Four-stroke Cycle Petrol Engine. 11. Valve Timing Diagram for a Four-stroke Cycle Petrol Engine. 12. Four-stroke Cycle Diesel Engine. 13. Actual Indicator Diagram for a Four-stroke Cycle Diesel Engine. 14. Valve Timing Diagram for a Four-stroke Cycle Diesel Engine. 15. Four-stroke Cycle Gas Engines. 16. Comparison of Petrol and Diesel Engines. 17. Two-stroke Cycle Petrol Engine. 18. Actual Indicator Diagram for a Two-stroke Cycle Petrol Engine. 19. Valve Timing Diagram for a Two-stroke Cycle Petrol Engine. 20. Two-stroke Cycle Diesel Engine. 21. Actual Indicator Diagram for a Two-stroke Cycle Diesel Engine. 22. Valve Timing Diagram for a Two-stroke Cycle Diesel Engine. 23. Scavenging. 24. Types of Scavenging. 25. Detonation in I.C. Engines. 26. Rating of S.I. Engine Fuels-Octane Number. 27. Rating of C.I. Engine Fuels-Cetane Number. 28. Ignition Systems for Petrol Engines. 29. Coil Ignition System. 30. Magneto Ignition System. 31. Fuel Injection System for Diesel Engines. 32. Cooling of I.C. Engines. 33. Cooling Systems for I.C. Engines. 34. Comparison of Air Cooling and Water Cooling Systems. 35. Supercharging of I.C. Engines. 36. Methods of Supercharging. 37. Lubrication of I.C. Engines. 38. Lubrication Systems for I.C. Engines. 39. Governing of I.C. Engines. 40. Methods of Governing I.C. Engines. 41. Carburettor. 42. Spark Plug. 43. Fuel Pump. 44. Injector or Atomiser.

26.1. Introduction

As the name implies or suggests, the internal combustion engines (briefly written as I.C. engines) are those engines in which the combustion of fuel takes place inside the engine cylinder. These are petrol, diesel, and gas engines. We have seen in steam engines or steam turbines that the fuel, fed into the cylinder, is in the form of steam which is already heated (or superheated), and is ready for working in the combustion cycle of the engine. But, in case of internal combustion engines, the combustion of fuel takes place inside the engine cylinder by a spark and produces very high temperature as compared to steam engines.

The high temperature produced may ruin the metal of cylinder, valves, etc. It is, therefore, necessary to abstract some of heat from the engine cylinder. The abstraction of heat or the cooling of cylinder may be effected by the surrounding air as in case of a motor cycle or aeroplane engine; or by circulating water through jackets surrounding the cylinder barrel and cylinder head. The water cooling is mostly adopted for large pistons.

26.2. Comparison of Steam Engines and Internal Combustion Engines

Following points are important for the comparison of steam engines and internal combustion engines.

S. No	Steam Engines	I.C. Engines
1.	The combustion of fuel takes place outside the engine cylinder (*i.e.* in a boiler)	The combustion of fuel takes place inside the engine cylinder.
2.	Since combustion of fuel takes place outside the engine cylinder, therefore these engines are smooth and silent running.	Since combustion of fuel takes place inside the engine cylinder, these engines are very noisy.
3.	The working pressure and temperature inside the engine cylinder is low.	The working pressure and temperature inside the cylinder is very high.
4.	Because of low pressure and temperature, ordinary alloys are used for the manufacture of engine cylinder and its parts.	Because of very high pressure and temperature, special alloys are used for the manufacture of engine cylinder and its parts.
5.	A steam engine requires a boiler and other components to transfer energy. Thus, it is heavy and cumbersome.	An I.C. engine does not require a boiler or other components. Thus, it is light and compact.
6.	The steam engines have efficiency about 15-20 percent.	The I.C. engines have efficiency about 35-40 percent.
7.	It can not be started instantaneously.	It can be started instantaneously.

26.3. Classification of I.C. Engines

The internal combustion engines may be classified in many ways, but the following are important from the subject point of view :

1. *According to the type of fuel used*
   (*a*) Petrol engines, (*b*) Diesel engines or oil engines, and (*c*) Gas engines.
2. *According to the method of igniting the fuel*
   (*a*) Spark ignition engines (briefly written as S.I. engines), (*b*) Compression ignition engines (briefly written as C.I. engines), and (*c*) Hot spot ignition engines.
3. *According to the number of strokes per cycle*
   (*a*) Four stroke cycle engines, and (*b*) Two stroke cycle engines.
4. *According to the cycle of operation*
   (*a*) Otto cycle (also known as constant volume cycle) engines, (*b*) Diesel cycle (also known as constant pressure cycle) engines, and (*c*) Dual combustion cycle (also known as semi-diesel cycle) engines.
5. *According to the speed of the engine*
   (*a*) Slow speed engines, (*b*) Medium speed engines, and (*c*) High speed engines.
6. *According to the cooling system*
   (*a*) Air-cooled engines, (*b*) Water-cooled engines, and (*c*) Evaporative cooling engines.
7. *According to the method of fuel injection*
   (*a*) Carburettor engines, (*b*) Air injection engines, and (*c*) Airless or solid injection engines.
8. *According to the number of cylinders*
   (*a*) Single cylinder engines, and (*b*) Multi-cylinder engines.
9. *According to the arrangement of cylinders*
   (*a*) Vertical engines, (*b*) Horizontal engines, (*c*) Radial engines, (*d*) In-line multi-cylinder engines, (*e*)V-type multi-cylinder engines, (*f*) Opposite-cylinder engines, and (*g*) Opposite-piston engines.

10. *According to the valve mechanism*

    (*a*) Overhead valve engines, and (*b*) Side valve engines.

11. *According to the method of governing*

    (*a*) Hit and miss governed engines, (*b*) Quantitatively governed engines, and (*c*) Qualitatively governed engines.

26.4 Main Components of I.C. Engines

As a matter of fact, an I.C. engine consists of hundreds of different parts, which are important for its proper working. The description of all these parts is beyond the scope of this book. However, the main components, which are important from academic point of view, are shown in Fig. 26.1 and are discussed below :

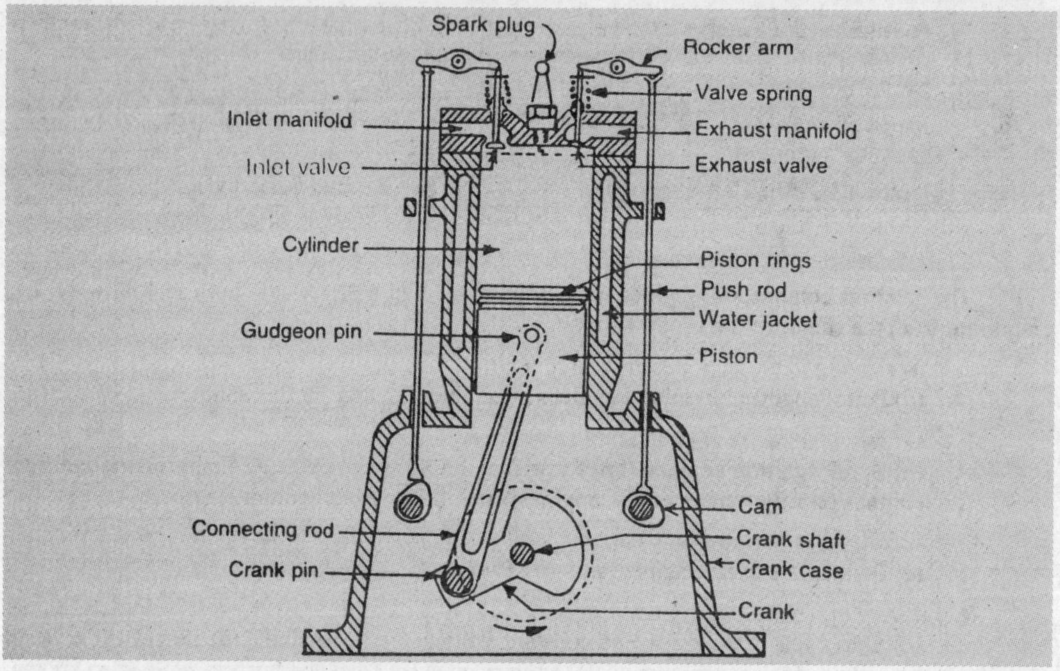

Fig. 26.1. Main components of I.C. engines.

1. *Cylinder.* It is one of the most important part of the engine, in which the piston moves to and fro in order to develop power. Generally, the engine cylinder has to withstand a high pressure (more than 50 bar) and temperature (more than 2000°C). Thus the materials for an engine cylinder should be such that it can retain sufficient strength at such a high pressure and temperature. For ordinary engines, the cylinder is made of ordinary cast iron. But for heavy duty engines, it is made of steel alloys or aluminium alloys. In case of multiple cylinder engines, the cylinders are cast in one block known as cylinder block.

Sometimes, a liner or sleeve is inserted into the cylinder, which can be replaced when worn out. As the material required for liner is comparatively small, it can be made of alloy cast iron having long life and sufficient resistance to rapid wear and tear to the fast moving reciprocating parts.

2. *Cylinder head.* It is fitted on one end of the cylinder, and acts as a cover to close the cylinder bore. Generally, the cylinder head contains inlet and exit valves for admitting fresh charge and exhausting the burnt gases. In petrol engines, the cylinder head also contains a spark plug for igniting the fuel-air mixture, towards the end of compression stroke. But in diesel engines, the cylinder head contains nozzle (*i.e.* fuel valve) for injecting the fuel into the cylinder.

The cylinder head is, usually, cast as one piece and bolted to one end of the cylinder. Generally, the cylinder block and cylinder head are made from the same material. A copper or asbestos gasket is provided between the engine cylinder and cylinder head to make an air-tight joint.

3. *Piston.* It is considered as the heart of an I.C. engine, whose main function is to transmit the force exerted by the burning of charge to the connecting rod. The pistons are generally made of aluminium alloys which are light in weight. They have good heat conducting property and also greater strength at higher temperatures.

4. *Piston rings.* These are circular rings and made of special steel alloys which retain elastic properties even at high temperatures. The piston rings are housed in the circumferential grooves provided on the outer surface of the piston. Generally, there are two sets of rings mounted for the piston. The function of the upper rings is to provide air tight seal to prevent leakage of the burnt gases into the lower portion. Similarly, the function of the lower rings is to provide effective seal to prevent leakage of the oil into the engine cylinder.

5. *Connecting rod.* It is a link between the piston and crankshaft, whose main function is to transmit force from the piston to the crankshaft. Moreover, it converts reciprocating motion of the piston into circular motion of the crankshaft, in the working stroke. The upper (*i.e.* smaller) end of the connecting rod is fitted to the piston and the lower (*i.e.* bigger) end to the crank.

The special steel alloys or aluminium alloys are used for the manufacture of connecting rods. A special care is required for the design and manufacture of connecting rod, as it is subjected to alternatively compressive and tensile stresses as well as bending stresses.

6. *Crankshaft.* It is considered as the backbone of an I.C. engine whose function is to convert the reciprocating motion of the piston into the rotary motion with the help of connecting rod. This shaft contains one or more eccentric portions called cranks. That part of the crank, to which bigger end of the connecting rod is fitted, is called crank pin.

It has been experienced that too many main bearings create difficulty of correct alignment. Special steel alloys are used for the manufacture of crankshaft. A special care is required for the design and manufacture of crankshaft.

7. *Crank case.* It is a cast iron case, which holds the cylinder and crankshaft of an I.C. engine. It also serves as a sump for the lubricating oil. The lower portion of the crank case is known as bed plate, which is fixed with the help of bolts.

8. *Flywheel.* It is a big wheel, mounted on the crankshaft, whose function is to maintain its speed constant. It is done by storing excess energy during the power stroke, which is returned during other strokes.

26.5. Sequence of Operations in a Cycle

Strictly speaking, when an engine is working continuously, we may consider a cycle starting from any stroke. We know that when the engine returns back to the stroke where it started we say that one cycle has been completed.

The readers will find different sequence of operations in different books. But in this chapter, we shall consider the following sequence of operation in a cycle, which is widely used.

1. *Suction stroke.* In this stroke, the fuel vapour in correct proportion, is supplied to the engine cylinder.
2. *Compression stroke.* In this stroke, the fuel vapour is compressed in the engine cylinder.
3. *Expansion or working stroke.* In this stroke, the fuel vapour is fired just before the compression is complete. It results in the sudden rise of pressure, due to expansion of the combustion products in the engine cylinder. This sudden rise of the pressure pushes the piston with a great force, and rotates the crankshaft. The crankshaft, in turn, drives the machine connected to it.
4. *Exhaust stroke.* In this stroke, the burnt gases (or combustion products) are exhausted from the engine cylinder, so as to make space available for the fresh fuel vapour.

Note : The above mentioned strokes are meant for gas and petrol engines. But in case of diesel engines, pure air is sucked in suction stroke which is compressed during the compression stroke. The diesel oil is admitted into the engine cylinder (just before the beginning of the expansion stroke) and it is ignited by the hot air present in the cylinder. The expansion and exhaust strokes are similar to the gas and petrol engines.

26.6. Two-stroke and Four-stroke Cycle Engines

In a two-stroke engine, the working cycle is completed in two strokes of the piston or one revolution of the crankshaft. This is achieved by carrying out the suction and compression processes in one stroke (or more precisely in inward stroke), expansion and exhaust processes in the second stroke (or more precisely in outward stroke). In a four-stroke engine, the working cycle is completed in four-strokes of the piston or two-revolutions of the crankshaft. This is achieved by carrying out suction, compression, expansion and exhaust processes in each stroke.

It will be interesting to know that from the thermodynamic point of view, there is no difference between two-stroke and four-stroke cycle engines. The difference is purely mechanical.

26.7. Advantages and Disadvantage of Two-stroke over Four-stroke Cycle Engines

Following are the advantages and disadvantages of two-stroke cycle engines over four-stroke cycle engines :

Advantages

1. A two stroke cycle engine gives twice the number of power strokes than the four stroke cycle engine at the same engine speed. Theoretically, a two-stroke cycle engine should develop twice the power as that of a four-stroke cycle engine. But in actual practice, a two-stroke cycle engine develops 1.7 to 1.8 times (greater value for slow speed engines) the power developed by four-stroke cycle engine of the same dimensions and speed. This is due to lower compression ratio and effective stroke being less than the theoretical stroke.

2. For the same power developed, a two-stroke cycle engine is lighter, less bulky and occupies less floor area. Thus it makes, a two-stroke cycle engine suitable for marine engines and other light vehicles.

3. As the number of working strokes in a two-stroke cycle engine are twice than the four-stroke cycle engine, so the turning moment of a two-stroke cycle engine is more uniform. Thus it makes a two-stroke cycle engine to have a lighter flywheel and foundations. This also leads to a higher mechanical efficiency of a two-stroke cycle engine.

4. The initial cost of a two-stroke cycle engine is considerably less than a four-stroke cycle engine.

5. The mechanism of a two-stroke cycle engine is much simpler than a four-stroke cycle engine.

6. The two-stroke cycle engines are much easier to start.

Disadvantages

1. Thermal efficiency of a two-stroke cycle engine is less than that a four-stroke cycle engine, because a two-stroke cycle engine has less compression ratio than that of a four-stroke cycle engine.

2. Overall efficiency of a two-stroke cycle engine is also less than that of a four-stroke cycle engine because in a two-stroke cycle, inlet and exhaust ports remain open simultaneously for some time. Inspite of careful design, a small quantity of charge is lost from the engine cylinder.

3. In case of a two-stroke cycle engine, the number of power strokes are twice as those of a four-stroke cycle engine. Thus the capacity of the cooling system must be higher. Beyond a certain limit, the cooling capacity offers a considerable difficulty. Moreover, there is a greater wear and tear in a two-stroke cycle engine.

4. The consumption of lubricating oil is large in a two-stroke cycle engine because of high operating temperature.

5. The exhaust gases in a two-stroke cycle engine creates noise, because of short time available for their exhaust.

26.8. Valve Timing Diagrams

A valve timing diagram is a graphical representation of the exact moments, in the sequence of operations, at which the two valves (*i.e.* inlet and exhaust valves) open and close as well as firing of the fuel. It is, generally, expressed in terms of angular positions of the crankshaft. Here we shall discuss theoretical valve timing diagrams for four stroke and two stroke cycle engines.

1. *Theoretical valve timing diagram for four stroke cycle engine.* The theoretical valve timing diagram for a four-stroke cycle engine is shown in Fig. 26.2. In this diagram, the inlet valve opens at *A* and the suction takes place from *A* to *B*. The crankshaft revolves through 180° and the piston moves from *T.D.C.* to *B.D.C.* At *B*, the inlet valve closes and the compression takes place from *B* to *C*. The crankshaft revolves through 180° and the piston moves from *B.D.C.* to *T.D.C.* At *C*, the fuel is fired and the expansion takes place from *C* to *D*. The crankshaft revolves through 180° and the piston again moves from *T.D.C.* to *B.D.C.* At *D*, the exhaust valve opens and the exhaust takes place from *D* to *E*. The crankshaft again revolves through 180° and the piston moves back to *T.D.C.*

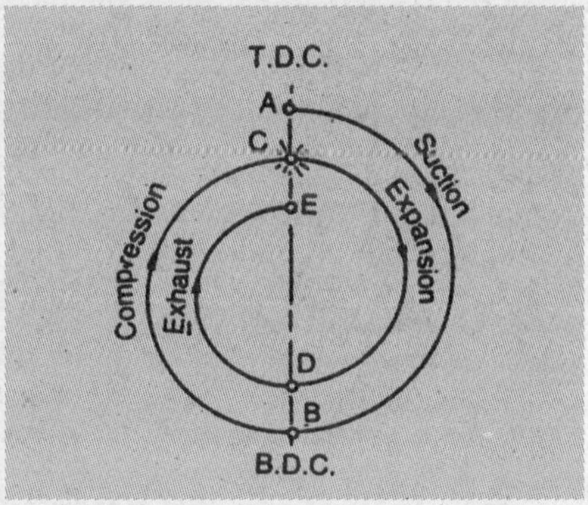

Fig. 26.2. Four-stroke cycle engine.

Note : In four-stroke cycle, the crank revolves through two revolutions.

2. *Theoretical valve timing diagram for two-stroke cycle engine.* The theoretical valve timing diagram for a two-stroke cycle engine is shown in Fig. 26.3. In this diagram, the fuel is fired at *A* and the expansion of gases takes place from *A* to *B*. The crankshaft revolves through approximately 120° and the piston moves from *T.D.C.* towards *B.D.C.* At *B*, both the valves open and suction as well as exhaust take place from *B* to *C*. The crankshaft revolves through approximately 120° and the piston moves first to *BDC* and then little upwards. At *C*, both the valves close and compression takes place from *C* to *A*. The crankshaft revolves through approximately 120° and the piston moves to *T.D.C.*

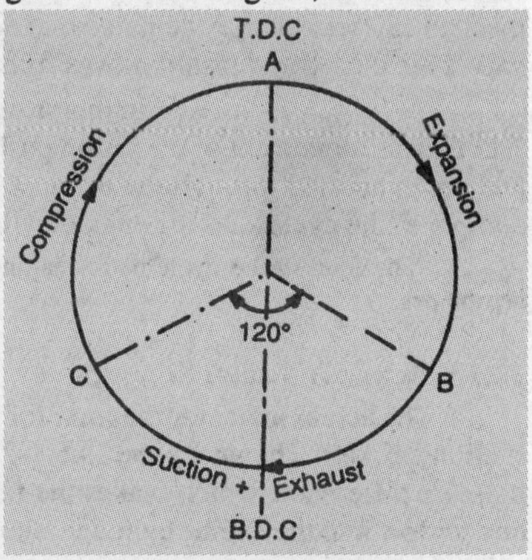

Fig. 26.3. Two-stroke cycle engine.

Notes : 1. In a two stroke cycle, the crank revolves through one revolution.

2. The readers will find valve timing diagram, drawn in other books having slight difference in the angles. This is due to the different engine speed and the manufacturer's design. However, in this book, the authors have given the diagrams which are widely used.

26.9. Four-stroke Cycle Petrol Engine

It is also known as Otto cycle*. It requires four strokes of the piston to complete one cycle of operation in the engine cylinder. The four strokes of a petrol engine sucking fuel-air mixture (petrol mixed with proportionate quantity of air in the carburettor known as charge) are described below :

1. *Suction or charging stroke.* In this stroke, the inlet valve opens and charge is sucked into the cylinder as the piston moves downward from top dead centre (*T.D.C.*). It continues till the piston reaches its bottom dead centre (*B.D.C.*) as shown in Fig. 26.4 (*a*).

---

* In 1978, Doctor Otto, a German engineer devised an engine working on this cycle. The determination of air standard efficiency of Otto cycle has been discussed in Chapter 6.

2. *Compression stroke.* In this stroke, both the inlet and exhaust valves are closed and the charge is compressed as the piston moves upwards from *B.D.C.* to *T.D.C.* As a result of compression, the pressure and temperature of the charge increases considerably (the actual values depend upon the compression ratio). This completes one revolution of the crankshaft. The compression stroke is shown in Fig. 26.4 (*b*).

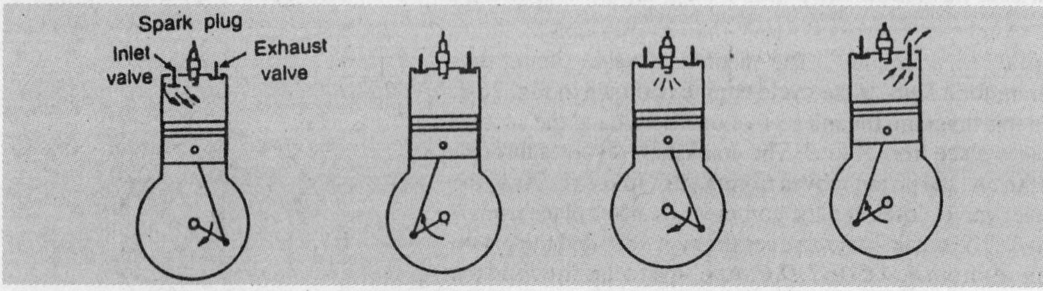

(*a*) Suction or charging stroke. (*b*) Compression stroke. (*c*) Expansion or working stroke. (*d*) Exhaust stroke.

Fig. 26.4. Four-stroke cycle petrol engine.

3. *Expansion or working stroke.* Shortly before the piston reaches *T.D.C.* (during compression stroke), the charge is ignited with the help of a spark plug. It suddenly increases the pressure and temperature of the products of combustion but the volume, practically, remains constant. Due to the rise in pressure, the piston is pushed down with a great force. The hot burnt gases expand due to high speed of the piston. During this expansion, some of the heat energy produced is transformed into mechanical work. It may be noted that during this working stroke, as shown in Fig. 26.4 (*c*), both the valves are closed and piston moves from *T.D.C.* to *B.D.C.*

4. *Exhaust stroke.* In this stroke, the exhaust valve is open as piston moves from *B.D.C.* to *T.D.C.* This movement of the piston pushes out the products of combustion, from the engine cylinder and are exhausted through the exhaust valve into the atmosphere, as shown in Fig. 26.4 (*d*). This completes the cycle, and the engine cylinder is ready to suck the charge again.

Note : The four stroke cycle petrol engine are usually employed in light vehicles such as cars, jeeps and aeroplanes.

26.10. Actual Indicator Diagram for a Four-stroke Cycle Petrol Engine

The actual indicator diagram for a four stroke cycle petrol engine is shown in Fig. 26.5. The suction stroke is shown by the line 1-2, which lies below the atmospheric pressure line. It is this pressure difference, which makes the fuel-air mixture to flow into the engine cylinder. The inlet valve offers some resistance to the incoming charge. That is why, the charge can not enter suddenly into the engine cylinder. As a result of this, pressure inside the cylinder remains somewhat below the atmospheric pressure during the suction stroke. The compression stroke is shown by the line 2-3, which shows that the inlet valve closes (*IVC*) a little beyond 2 (*i.e. BDC*). At the end of this stroke, there is an increase in the pressure inside the engine cylinder. Shortly before the end of compression stroke (*i.e. TDC*), the charge is ignited (*IGN*) with the help of spark plug as shown in the figure. The sparking suddenly increases pressure and temperature of the

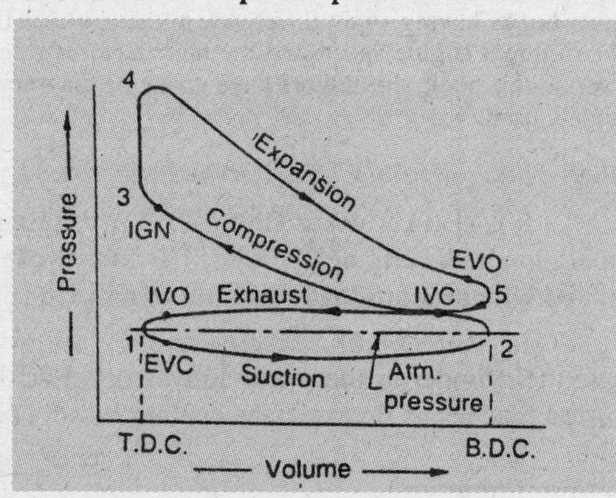

Fig. 26.5. Actual indicator diagram for a four stroke cycle petrol engine.

products of combustion. But the volume, practically, remains constant as shown by the line 3-4. The expansion stroke is shown by the line 4-5, in which the exit valve opens (*EVO*) a little before 5 (*i.e. BDC*). Now the burnt gases are exhausted into the atmosphere through the exit valve. The exhaust stroke is shown by the line 5-1, which lies above the atmospheric pressure line. It is this pressure difference, which makes the burnt gases to flow out of the engine cylinder. The exit valve offers some resistance to the outgoing burnt gases. That is why the burnt gases can not escape suddenly from the engine cylinder. As a result of this, pressure inside the cylinder remains somewhat above the atmospheric pressure line during the exhaust stroke.

26.11. Valve Timing Diagram for a Four-stroke Cycle Petrol Engine

In the valve timing diagram, as shown in Fig. 26.6, we see that the inlet valve opens before the piston reaches *TDC* ; or in other words, while the piston is still moving up before the beginning of the suction stroke. Now the piston reaches the *TDC* and the suction stroke starts. The piston reaches the *BDC* and then starts moving up. The inlet valve closes, when the crank has moved a little beyond the *BDC*. This is done as the incoming charge continues to flow into the cylinder although the piston is moving upwards from *BDC*. Now the charge is compressed (with both valves closed) and then ignited with the help of a spark plug before the end of compression stroke. This is done as the charge

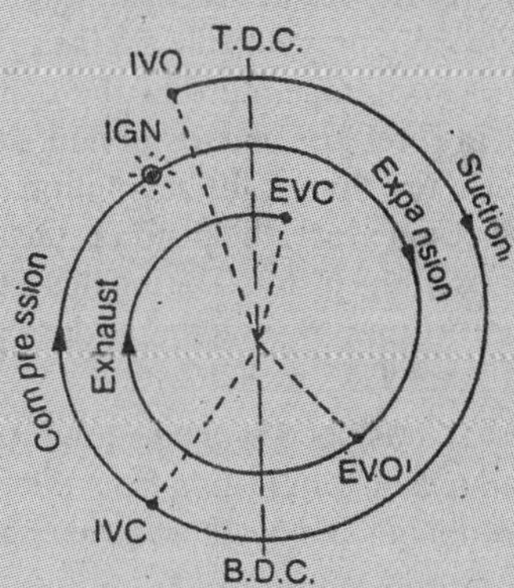

TDC : Top dead centre

BDC : Bottom dead centre

IVO : Inlet valve opens (10°-20° before *TDC*)

IVC : Inlet valve closes (30°- 40° after *BDC*)

IGN : Ignition (20°-30° before *TDC*)

EVO : Exit valve opens (30°-50° before *BDC*)

EVC : Exit valve closes (10°-15° after *TDC*)

Fig. 26.6. Valve timing diagram for a four stroke cycle petrol engine.

requires some time to ignite. By the time, the piston reaches *TDC*, the burnt gases (under high pressure and temperature) push the piston downwards with full force and the expansion or working stroke takes place. Now the exhaust valve opens before the piston again reaches *BDC* and the burnt gases start leaving the engine cylinder. Now the piston reaches *BDC* and then starts moving up, thus performing the exhaust stroke. The inlet valve opens before the piston reaches *TDC* to start suction stroke. This is done as the fresh incoming charge helps in pushing out the burnt gases. Now the piston again reaches *TDC*, and the suction stroke starts. The exit valve closes after the crank has moved a little beyond the *TDC*. This is done as the burnt gases continue to leave the engine cylinder although the piston is moving downwards. It may be noted that for a small fraction of a crank revolution, both the inlet and outlet valves are open. This is known as valve overlap.

26.12. Four-stroke Cycle Diesel Engine

It is also known as *compression ignition engine* because the ignition takes place due to the heat produced in the engine cylinder at the end of compression stroke. The four strokes of a diesel engine sucking pure air are described below :

1. *Suction or charging stroke.* In this stroke, the inlet valve opens and pure air is sucked into the cylinder as the piston moves downwards from the top dead centre (*TDC*). It continues till the piston reaches its bottom dead centre (*BDC*) as shown in Fig. 26.7 (*a*).

2. *Compression stroke.* In this stroke, both the valves are closed and the air is compressed as the piston moves upwards from *BDC* to *TDC*. As a result of compression, pressure and temperature of the air increases considerably (the actual value depends upon the compression ratio). This completes one revolution of the crank shaft. The compression stroke is shown in Fig. 26.7 (*b*).

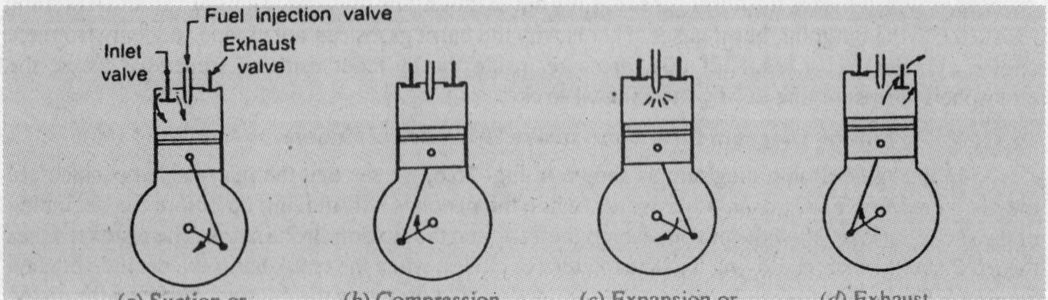

(*a*) Suction or charging stroke. (*b*) Compression stroke. (*c*) Expansion or working stroke. (*d*) Exhaust stroke.

Fig. 26.7. Four stroke cycle diesel engine.

3. *Expansion or working stroke.* Shortly before the piston reaches the *TDC* (during the compression stroke), fuel oil is injected in the form of very fine spray into the engine cylinder, through the nozzle, known as fuel injection valve. At this moment, temperature of the compressed air is sufficiently high to ignite the fuel. It suddenly increases the pressure and temperature of the products of combustion. The fuel oil is continuously injected for a fraction of the revolution. The fuel oil is assumed to be burnt at constant pressure. Due to increased pressure, the piston is pushed down with a great force. The hot burnt gases expand due to high speed of the piston. During this expansion, some of the heat energy is transformed into mechanical work. It may be noted that during this working stroke, both the valves are closed and the piston moves from *TDC* to *BDC*.

4. *Exhaust stroke.* In this stroke, the exhaust valve is open as the piston moves from *BDC* to *TDC*. This movement of the piston pushes out the products of combustion from the engine cylinder through the exhaust valve into the atmosphere. This completes the cycle and the engine cylinder is ready to suck the fresh air again.

Note : The four stroke cycle diesel engines are generally employed in heavy vehicles such as buses, trucks, tractors, pumping sets, diesel locomotives and in earth moving machinery.

26.13. Actual Indicator Diagram for a Four Stroke Cycle Diesel Engine

The actual indicator diagram for a four-stroke cycle diesel engine is shown in Fig. 26.8. The suction stroke is shown by the line 1-2 which lies below the atmospheric pressure line. It is this pressure difference, which makes the fresh air to flow into the engine cylinder. The inlet valve offers some resistance to the incoming air. That is why, the air can not enter suddenly into the engine cylinder. As a result of this, pressure inside the cylinder remains somewhat below the atmospheric pressure during the suction stroke. The compression stroke is shown by the line 2-3, which shows that the inlet valves closes (*IVC*) a little beyond 2 (*i.e. BDC*). At the end of this stroke, there is an increase of pressure inside the engine cylinder. Shortly before the end of compression stroke (*i.e. TDC*), fuel valve opens (*FVO*) and the fuel is injected into the engine cylinder. The fuel is ignited by high temperature of the compressed air. The ignition

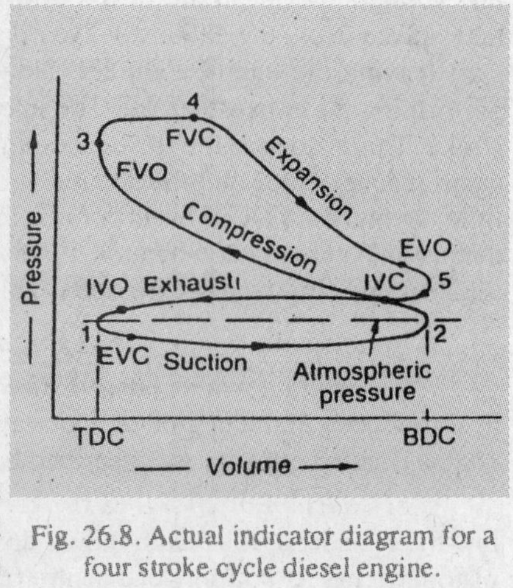

Fig. 26.8. Actual indicator diagram for a four stroke cycle diesel engine.

suddenly increases volume and temperature of the products of combustion. But the pressure, practically, remains constant as shown by the line 3-4. The expansion stroke is shown by the line 4-5, in which the exit valve opens a little before 5 (*i.e. BDC*). Now the burnt gases are exhausted into the atmosphere through the exhaust valve. The exhaust stroke is shown by the line 5-1, which lies above the atmospheric pressure line. It is this pressure difference, which makes the burnt gases to flow out of the engine cylinder. The exhaust valve offers some resistance to the outgoing burnt gases. That is why, the burnt gases can not escape suddenly from the engine cylinder. As a result of this, pressure inside the cylinder remains somewhat above the atmospheric pressure during the exhaust stroke.

26.14. Valve Timing Diagram for a Four-stroke Cycle Diesel Engine

In the valve timing diagram as shown in Fig. 26.9, we see that the inlet valve opens before the piston reaches *TDC* ; or in other words while the piston is still moving up before the beginning of the suction stroke. Now the piston reaches the *TDC* and the suction stroke starts. The piston reaches the *BDC* and then starts moving up. The inlet valve closes, when the crank has moved a little beyond the

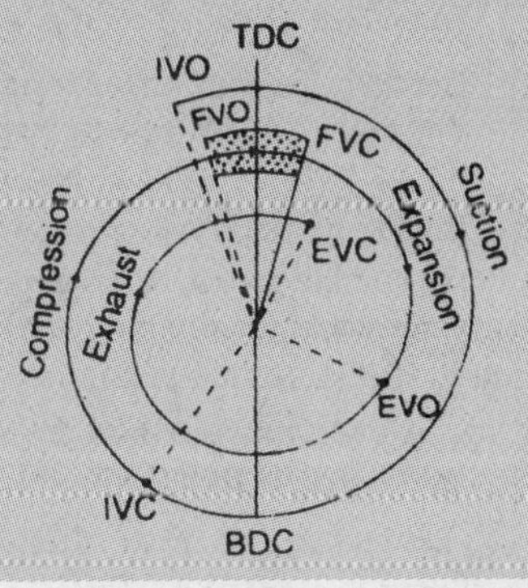

TDC : Top dead centre

BDC : Bottom dead centre

IVO : Inlet valve opens (10° - 20° before *TDC*)

IVC : Inlet valve closes (25° - 40° after *BDC*)

FVO : Fuel valve opens (10° - 15° before *TDC*)

FVC : Fuel valve closes (15° - 20° after *TDC*)

EVO : Exhaust valve opens (39° - 50° before *BDC*)

EVC : Exhaust valve closes (10° - 15° after *TDC*)

Fig. 26.9. Valve timing diagram for a four stroke cycle diesel engine.

BDC. This is done as the incoming air continues to flow into the cylinder although the piston is moving upwards from *BDC*. Now the air is compressed with both valves closed. Fuel valve opens a little before the piston reaches the *TDC*. Now the fuel is injected in the form of very fine spray, into the engine cylinder, which gets ignited due to high temperature of the compressed air. The fuel valve closes after the piston has come down a little from the *TDC*. This is done as the required quantity of fuel is injected into the engine cylinder. The burnt gases (under high pressure and temperature) push the piston downwards, and the expansion or working stroke takes place. Now the exhaust valve opens before the piston again reaches *BDC* and the burnt gases start leaving the engine cylinder. Now the piston reaches *BDC* and then starts moving up thus performing the exhaust stroke. The inlet valve opens before the piston reaches *TDC* to start suction stroke. This is done as the fresh air helps in pushing out the burnt gases. Now the piston again reaches *TDC*, and the suction starts. The exhaust valve closes when the crank has moved a little beyond the *TDC*. This is done as the burnt gases continue to leave the engine cylinder although the piston is moving downwards.

26.15. Four-stroke Cycle Gas Engines

A four stroke cycle gas engine, as the name indicates, uses natural or manufactured gas as the working fuel and works on Otto cycle. All the mechanical features of a gas engine are the same as those of a petrol engine. The ignition system, which usually consists of spark plug, is the same in both the gas and petrol engines. The working of a gas engine on *p-v* diagram is exactly similar to that of the petrol engine. The valve timing diagram for a four stroke cycle engine is also exactly similar to that of a petrol engine.

The only difference between a gas engine and a petrol engine is fuel and the way it is supplied. In a petrol engine, a mixture of petrol and air is sucked inside the engine cylinder during the suction stroke. But in a gas engine, gas and air are supplied to the engine. In most of the cases, the density and calorific value of gas is considerably less than that of the petrol. As a result of this, the cylinder of a gas engine is made larger than that of the petrol engine.

26.16. Comparison of Petrol and Diesel Engines

Following points are important for the comparison of petrol engines and diesel engines :

S. No	Petrol Engines	*Diesel Engines*
1.	A petrol engine draws a mixture of petrol and air during suction stroke.	A diesel engine draws only air during suction stroke.
2.	The carburettor is employed to mix air and petrol in the required proportion and to supply it to the engine during suction stroke.	The injector or atomiser is employed to inject the fuel at the end of compression stroke.
3.	Pressure at the end of compression is about 10 bar.	Pressure at the end of compression is about 35 bar.
4.	The charge (*i.e.* petrol and air mixture) is ignited with the help of spark plug.	The fuel is injected in the form of fine spray. The temperature of the compressed air (about 600°C at a pressure of about 35 bar) is sufficiently high to ignite the fuel.
5.	The combustion of fuel takes place approximately at constant volume. In other words, it works on Otto cycle.	The combustion of fuel takes place approximately at constant pressure. In other words, it works on Diesel cycle.
6.	A petrol engine has compression ratio approximately from 6 to 10.	A diesel engine has compression ratio approximately from 15 to 25.
7.	The starting is easy due to low compression ratio.	The starting is little difficult due to high compression ratio.
8.	As the compression ratio is low, the petrol engines are lighter and cheaper.	As the compression ratio is high, the diesel engines are heavier and costlier.
9.	The running cost of a petrol engine is high because of the higher cost of petrol.	The running cost of diesel engine is low because of the lower cost of diesel.
10.	The maintenance cost is less.	The maintenance cost is more.
11.	The thermal efficiency is upto about 26%.	The thermal efficiency is upto about 40%.
12.	Oveheating trouble is more due to low thermal efficiency.	Oveheating trouble is less due to high thermal efficiency.
13.	These are high speed engines.	These are relatively low speed engines.
14.	The petrol engines are generally employed in light duty vehicles such as scooters, motorcycles, cars. These are also used in aeroplanes.	The diesel engines are generally employed in heavy duty vehicles like buses, trucks, and earth moving machines etc.

26.17. Two-stroke Cycle Petrol Engine

A two-stroke cycle petrol engine was devised by Duglad Clerk in 1880. In this cycle, the suction, compression, expansion and exhaust takes place during two strokes of the piston. It means that there is one working stroke after every revolution of the crank shaft. A two stroke engine has ports instead of valves. All the four stages of a two stroke petrol engine are described below :

1. *Suction stage.* In this stage, the piston, while going down towards *BDC*, uncovers both the transfer port and the exhaust port. The fresh fuel-air mixture flows into the engine cylinder from the crank case, as shown in Fig. 26.10 (*a*).

2. *Compression stage.* In this stage, the piston, while moving up, first covers the transfer port and then exhaust port. After that the fuel is compressed as the piston moves upwards as shown in Fig. 26.10 (*b*). In this stage, the inlet port opens and fresh fuel-air mixture enters into the crank case.

3. *Expansion stage.* Shortly before this piston reaches the *TDC* (during compression stroke), the charge is ignited with the help of a spark plug. It suddenly increases the pressure and temperature of the products of combustion. But the volume, practically, remains constant. Due to rise in the pressure, the piston is pushed downwards with a great force as shown in Fig. 26.10 (*c*). The hot burnt gases expand due to high speed of the piston. During this expansion, some of the heat energy produced is transformed into mechanical work.

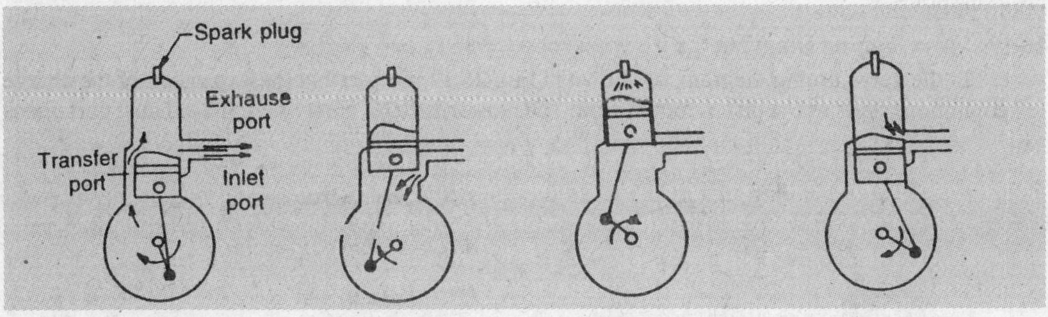

(*a*) Suction. (*b*) Compression. (*c*) Expansion. (*d*) Exhaust.

Fig. 26.10. Two-stroke cycle petrol engine.

4. *Exhaust stage.* In this stage, the exhaust port is opened as the piston moves downwards. The products of combustion, from the engine cylinder are exhausted through the exhaust port into the atmosphere, as shown in Fig. 26.10 (*d*). This completes the cycle and the engine cylinder is ready to suck the charge again.

Note : The two stroke petrol engines are generally employed in very light vehicles such as scooters, motor cycles, three wheelers and sprayers.

26.18. Actual Indicator Diagram for a Two Stroke Cycle Petrol Engine

The actual indicator diagram for a two-stroke cycle petrol engine is shown in Fig 26.11. The suction is shown by the line 1-2-3, *i.e.* from the instant transfer port opens (*TPO*) and transfer port closes (*TPC*). We know that during the suction stage, the exhaust port is also open. In the first half of suction stage, the volume of fuel-air mixture and burnt gases increases. This happens as the piston moves from 1 to 2 (*i.e. BDC*). In the second half of the suction stage, the volume of charge and burnt gases decreases. This happens as the piston moves upwards from 2 to 3. A little beyond 3, the exhaust port closes (*EPC*) at 4. Now the charge inside the engine cylinder is compressed which is shown by the line 4-5. At the end of the compression, there is an increase in the pressure inside the engine cylinder. Shortly before the end of compression (*i.e. TDC*) the charge is ignited (*IGN*) with the help of spark plug as shown in the figure. The sparking suddenly increases pressure and temperature of the products of combustion. But the volume, practically, remains constant as shown by the line 5-6. The expansion is shown by the line 6-7. Now the exhaust port opens (*EPO*) at 7, and the burnt gases are exhausted

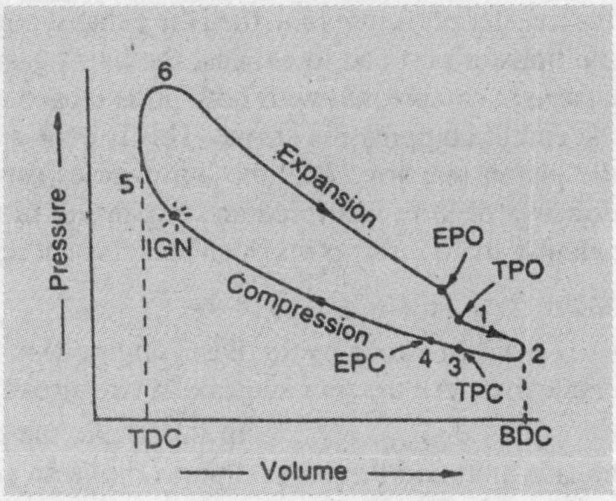

Fig. 26.11. Actual indicator diagram for a two stroke cycle petrol engine.

into the atmosphere through the exhaust port. It reduces the pressure. As the piston is moving towards *BDC*, therefore volume of burnt gases increases from 7 to 1. At 1, the transfer port opens (*TPO*) and the suction starts.

Notes : 1. The piston top is shaped in such a way that the fresh charge will move up towards spark plug, and it will also push out the remaining burnt gases through the exhaust port.

2. As soon as the exhaust port is uncovered by the piston (at this moment the transfer port is still covered), the burnt gases will be exhausted with a great force till the pressure in the cylinder is reduced, approximately to that of atmosphere.

3. Due to downward movement of the piston, the charge below the piston will be compressed. As the transfer port is uncovered, it will flow into cylinder due to pressure difference between the spaces below and above the piston.

26.19. Valve Timing Diagram for a Two-stroke Cycle Petrol Engine

In the valve timing diagram, as shown in Fig. 26.12, we see that the expansion of the charge (after ignition) starts as the piston moves from *TDC* towards *BDC*. First of all, the exhaust port opens

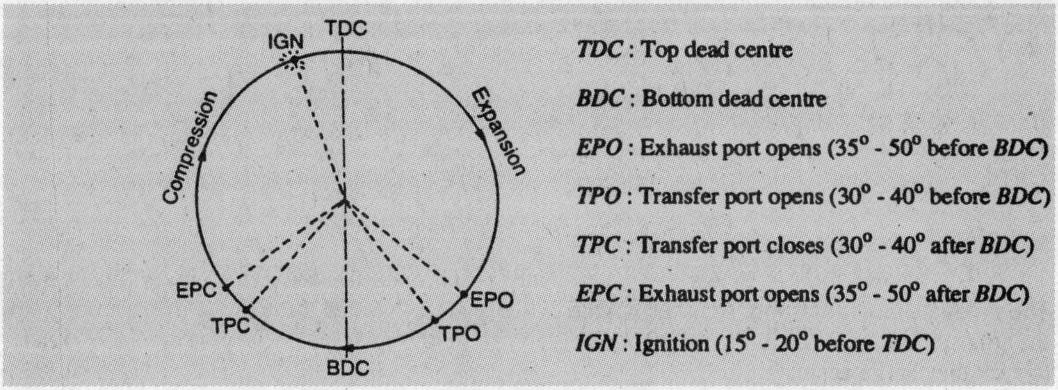

Fig. 26.12. Valve timing diagram for a two-stroke cycle petrol engine.

before the piston reaches *BDC* and the burnt gases start leaving the cylinder. After a small fraction of the crank revolution, the transfer port also opens and the fresh fuel-air mixture enters into the engine cylinder. This is done as the fresh incoming charge helps in pushing out the burnt gases. Now the piston reaches *BDC* and then starts moving upwards. As the crank moves a little beyond *BDC*, first the transfer port closes and then the exhaust port also closes. This is done to suck fresh charge through the transfer port and to exhaust the burnt gases through the exhaust port simultaneously. Now the charge is compressed with both ports closed, and then ignited with the help of a spark plug before the end of compression stroke. This is done as the charge requires some time to ignite. By the time the piston reaches *TDC*, the burnt gases (under high pressure and temperature) push the piston downwards with full force and expansion of the burnt gases takes place. It may be noted that the exhaust and transfer ports open and close at equal angles on either side of the *BDC* position.

26.20. Two-stroke Cycle Diesel Engine

A two-stroke cycle diesel engine also has one working stroke after every revolution of the crank shaft. All the four stages of a two stroke cycle diesel engine are described below :

1. *Suction stage.* In this stage, the piston while going down towards *BDC* uncovers the transfer port and the exhaust port. The fresh air flows into the engine cylinder from the crank case, as shown in Fig. 26.13 (*a*).

2. *Compression stage.* In this stage, the piston while moving up, first covers the transfer port and then exhaust port. After that the air is compressed as the piston moves upwards as shown in Fig. 26.13 (*b*). In this stage, the inlet port opens and the fresh air enters into the crank case.

3. *Expansion stage.* Shortly before the piston reaches the *TDC* (during compression stroke), the fuel oil is injected in the form of very fine spray into the engine cylinder through the nozzle known as fuel injection valve, as shown in fig. 26.13 (*c*). At this moment, temperature of the compressed air is sufficiently high to ignite the fuel. It suddenly increases the pressure and temperature of the products of combustion. The fuel oil is continuously injected for a fraction of the crank revolution. The fuel oil is assumed to be burnt at constant pressure. Due to increased pressure, the piston is pushed with a great force. The hot burnt gases expand due to high speed of the piston. During the expansion, some of the heat energy produced is transformed into mechanical work.

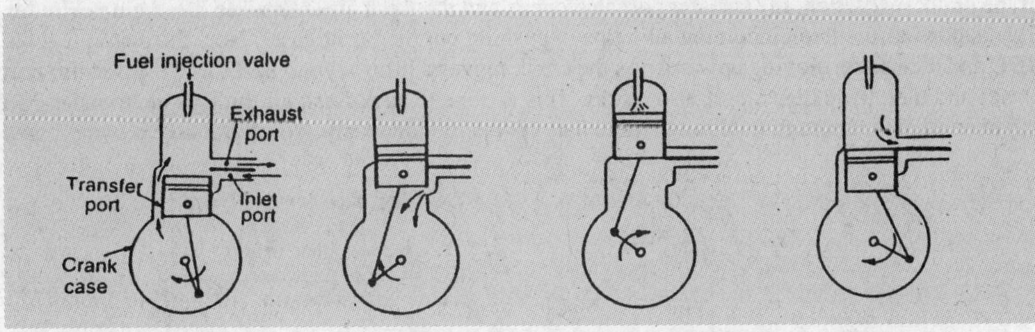

(*a*) Suction. (*b*) Compression. (*c*) Expansion. (*d*) Exhaust.

Fig. 26.13. Two-stroke cycle diesel engine.

4. *Exhaust stage.* In this stage, the exhaust port is opened and the piston moves downwards. The products of combustion from the engine cylinder are exhausted through the exhaust port into the atmosphere as shown in Fig. 26.13 (*d*). This completes the cycle, and the engine cylinder is ready to suck the air again.

Note : The two stroke diesel engines are mainly used in marine propulsion where space and lightness are the main considerations.

26.21. Actual Indicator Diagram for a Two Stroke Cycle Diesel Engine

The actual indicator diagram for a two-stroke cycle diesel engine is shown in Fig. 26.14. The suction is shown by the line 1-2-3 *i.e.* from the instant transfer port opens (*TPO*) and transfer port closes (*TPC*). We know that during the suction stage, the exhaust port is also open. In the first half of suction stage, the volume of air and burnt gases increases. This happens as the piston moves from 1-2 (*i.e. BDC*). In the second half of the suction stage, the volume of air and burnt gases decreases. This happens as the piston moves upwards from 2-3. A little beyond 3, the exhaust port closes (*EPC*) at 4. Now the air inside the engine cylinder is compressed which is shown by the line 4-5. At the end of compression, there is an increase in the pressure inside the engine cylinder. Shortly before the end of compression (*i.e. TDC*), fuel valve opens (*FVO*) and the fuel is injected into the engine cylinder. The fuel is ignited by high temperature of the compressed air. The ignition suddenly increases volume and temperature of the products of combustion. But the pressure, practically, remains constant as shown by the line 5-6. The expansion is shown by the line 6-7, Now the exhaust port opens (*EPO*) at 7 and the

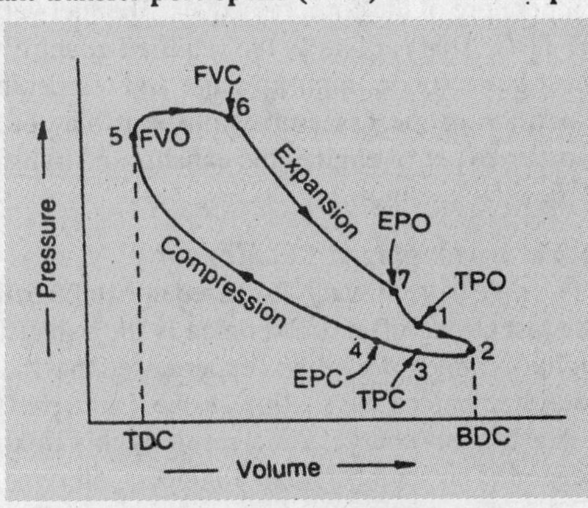

Fig. 26.14. Actual indicator diagram for a two stroke cycle diesel engine.

burnt gases are exhausted into the atmosphere through the exhaust port. It reduces the pressure. As the piston is moving towards *BDC*, therefore volume of burnt gases increases from 7 to 1. At 1, the transfer port opens (*TPO*) and the suction starts.

Note : All the notes given at the end of two-stroke cycle petrol engine are valid for this cycle also.

26.22. Valve Timing Diagram for a Two-stroke Cycle Diesel Engine

In the valve timing diagram, as shown in Fig. 26.15, we see that the expansion of the charge (after ignition) starts as the piston moves from *TDC* towards *BDC*. First of all, the exhaust port opens before the piston reaches *BDC* and the burnt gases start leaving the cylinder. After a small fraction of the crank revolution, the transfer port also opens and the fresh air enters into the engine cylinder. This is done as the fresh incoming air helps in pushing out the burnt gases. Now the piston reaches *BDC* and then starts moving upwards. As the crank moves a little beyond *BDC*, first the transfer port closes and then the exhaust port also closes. This is done to suck fresh air through the transfer port and to exhaust the burnt gases through the exhaust port simultaneously. Now the charge is compressed

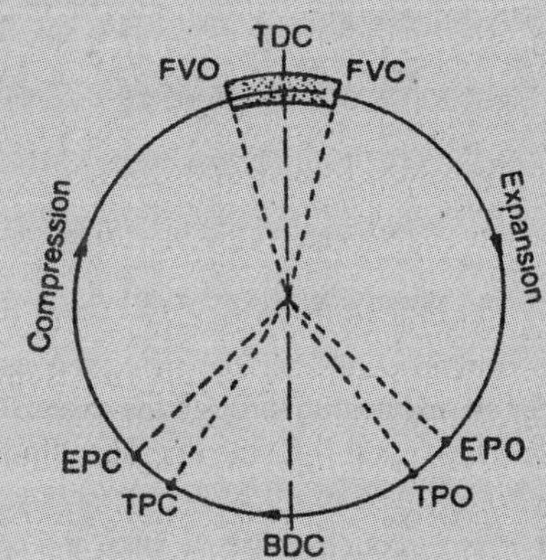

TDC : Top dead centre

BDC : Bottom dead centre

FVO : Fuel valve opens (10° - 15° before *TDC*)

FVC : Fuel valve closes (15° - 20° after *TDC*)

EPO : Exhaust port opens (35° - 50° before *BDC*)

TPO : Transfer port opens (30° - 40° before *BDC*)

TPC : Transfer port closes (30° - 40° after *BDC*)

EPC : Exhaust port closes (35° - 50° after *BDC*)

Fig. 26.15. Valve timing diagram for a two stroke cycle diesel engine.

with both the ports closed. Fuel valve opens a little before the piston reaches the *TDC*. Now the fuel is injected in the form of very fine spray into the engine cylinder, which gets ignited due to high temperature of the compressed air. The fuel valve closes after the piston has come down a little from the *TDC*. This is done as the required quantity of fuel is injected into the engine cylinder. Now the burnt gases (under high pressure and temperature) push the piston downwards with full force and expansion of the gases takes place. It may be noted that in a two-stroke cycle diesel engine, like two-stroke petrol engine, the exhaust and transfer ports open and close at equal angles on either side of the *BDC* position.

26.23. Scavenging of I.C. Engines

We have already discussed in Art. 26.5, the sequence of operations in a cycle of an I.C. engine. The last stroke of an I.C. engine is the exhaust, which means the removal of burnt gases from the engine cylinder. It has been experienced that the burnt gases in the engine cylinder are not completely exhausted before the suction stroke. But a part of the gases still remain inside the cylinder and mix with the fresh charge. As a result of this mixing, the fresh charge gets diluted and its strength is reduced. The scientists and engineers, all over the world, have concentrated on the design of their I.C. engines so that the burnt gases are completely exhausted from the engine cylinder before the suction starts. The process of removing burnt gases, from the combustion chamber of the engine cylinder, is known as scavenging. Now we shall discuss the scavenging in four-stroke and two-stroke cycle engines.

1. *Four-stroke cycle engines.* In a four-stroke cycle engine, the scavenging is very effective, as the piston during the exhaust stroke, pushes out the burnt gases from the engine cylinder. It may be noted that a small quantity of burnt gases remain in the engine cylinder in the clearance space.

2. *Two-stroke cycle engines.* In a two-stroke cycle engine, the scavenging is less effective as the exhaust port is open for a small fraction of the crank revolution. Moreover, as the transfer and exhaust port are open simultaneously during a part of the crank revolution, therefore fresh charge also escapes out along with the burnt gases. This difficulty is overcome by designing the piston crown of a particular shape.

26.24. Types of Scavenging

Though there are many types of scavenging, yet the following are important from the subject point of view :

1. *Crossflow scavenging.* In this method, the transfer port (or inlet port for the engine cylinder) and exhaust port are situated on the opposite sides of the engine cylinder (as is done in case of two-stroke cycle engines). The piston crown is designed into a particular shape, so that the fresh charge moves upwards and pushes out the burnt gases in the form of cross flow as shown in Fig. 26.16 (*a*).

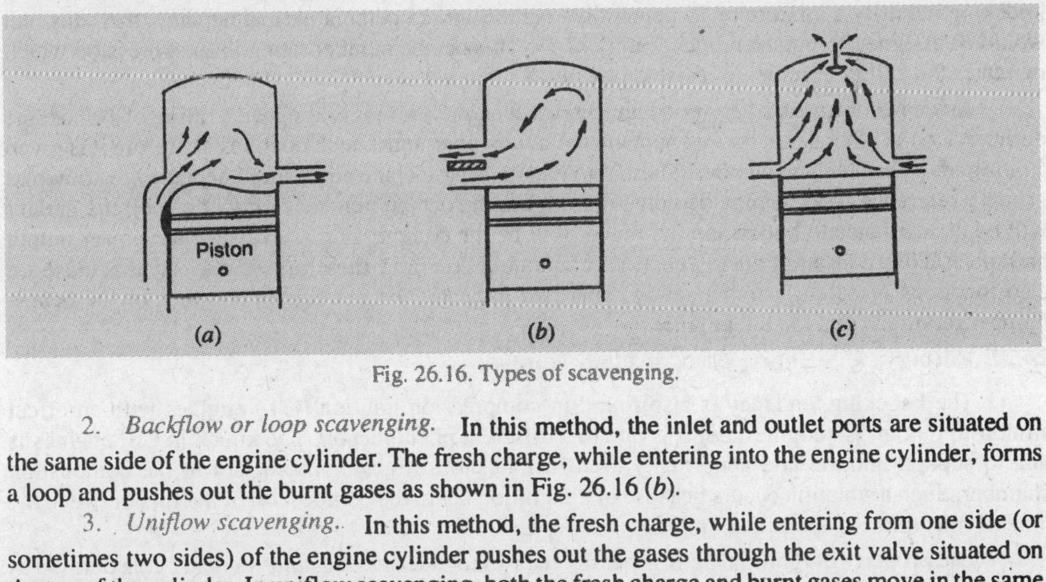

Fig. 26.16. Types of scavenging.

2. *Backflow or loop scavenging.* In this method, the inlet and outlet ports are situated on the same side of the engine cylinder. The fresh charge, while entering into the engine cylinder, forms a loop and pushes out the burnt gases as shown in Fig. 26.16 (*b*).

3. *Uniflow scavenging.* In this method, the fresh charge, while entering from one side (or sometimes two sides) of the engine cylinder pushes out the gases through the exit valve situated on the top of the cylinder. In uniflow scavenging, both the fresh charge and burnt gases move in the same upward direction as shown in Fig. 26.16 (*c*).

26.25. Detonation in I.C. Engines

The loud pulsating noise heard within the engine cylinder is known as *detonation* (also called *knocking* or *pinking*). It is caused due to the propagation of a high speed pressure wave created by the auto-ignition of end portion of unburnt fuel. The blow of this pressure wave may be of sufficient strength to break the piston. Thus, the detonation is harmful to the engine and must be avoided. The following are certain factors which causes detonation :

1. The shape of the combustion chamber,
2. The relative position of the sparking plugs in case of petrol engines,
3. The chemical nature of the fuel,
4. The initial temperature and pressure of the fuel, and

5. The rate of combustion of that portion of the fuel which is the first to ignite. This portion of the fuel in heating up, compresses the remaining unburnt fuel, thus producing the conditions for auto-ignition to occur.

The detonation in petrol engines can be suppressed or reduced by the addition of a small amount of lead ethide or ethyl fluid to the fuel. This is called *doping*.

The following are the chief effects due to detonation :

1. A loud pulsating noise which may be accompanied by a vibration of the engine.
2. An increase in the heat lost to the surface of combustion chamber.
3. An increase in carbon deposits.

26.26. Rating of S.I. Engine Fuels--Octane Number

The hydrocarbon fuels used in spark ignition (S.I.) engine have a tendency to cause engine knock when the engine operating conditions become severe. The knocking tendency of a fuel in S.I. engines is generally expressed by its *octane number*. The percentage, by volume, of iso-octane in a mixture of iso-octane and normal heptane, which exactly matches the knocking intensity of a given fuel, in a standard engine, under given standard operating conditions, is termed as the *octane number rating* of that fuel. Thus, if a mixture of 50 percent iso-octane and 50 percent normal heptane matches the fuel under test, then this fuel is assigned an octane number rating of 50. If a fuel matches in knocking intensity a mixture of 75 percent iso-octane and 25 percent normal heptane, then this fuel would be assigned an octane number rating of 75. This octane number rating is an expression which indicates the ability of a fuel to resist knock in a S.I. engine.

Since iso-octane is a very good anti-knock fuel, therefore it is assigned a rating of 100 octane number. On the other hand, normal heptane has a very poor anti-knock qualities, therefore it is given a rating of 0 (zero) octane number. These two fuels, *i.e.* iso-octane and normal heptane are known as primary reference fuels. It may be noted that higher the octane number rating of a fuel, the greater will be its resistance to knock and the higher will be the compression ratio. Since the power output and specific fuel consumption are functions of compression ratio, therefore we may say that these are also functions of octane number rating. This fact indicates the extreme importance of the octane number rating in fuels for S.I. engines.

26.27. Rating of C.I. Engine Fuels--Cetane Number

The knocking tendency is also found in compression ignition (C.I.) engines with an effect similar to that of S.I. engines, but it is due to a different phenomenon. The knock in C.I. engines is due to sudden ignition and abnormally rapid combustion of accumulated fuel in the combustion chamber. Such a situation occurs because of an ignition lag in the combustion of the fuel between the time of injection and the actual burning.

The property of ignition lag is generally measured in terms of *cetane number* . It is defined as the percentage, by volume, of cetane in a mixture of cetane and alpha-methyl-naphthalene that produces the same ignition lag as the fuel being tested, in the same engine and under the same operating conditions. For example, a fuel of cetane number 50 has the same ignition quality as a mixture of 50 percent cetane and 50 percent alpha-methyl-naphthalene.

The cetane which is a straight chain paraffin with good ignition quality is assigned a cetane number of 100 and alpha-methyl-naphthalene which is a hydrocarbon with poor ignition quality, is assigned a 0 (zero) cetane number.

Note : The knocking in C.I. engines may be controlled by decreasing ignition lag. The shorter the ignition lag, the less is the tendency to knock.

26.28. Ignition Systems of Petrol Engines

We have already discussed that the ignition in a petrol engine, takes place by means of a spark plug at the end of the compression stroke. The voltage required to produce a spark across the gap between the sparking points of a plug, is about 8000 volts. Thus, the ignition system in a petrol engine has to transform the normal battery voltage (6 to 12 volts) to 8000 volts. In addition to this, the ignition system has to provide spark in each cylinder at the appropriate time. Following two ignition systems of petrol engines are important from the subject point of view :

1. Coil ignition system, and 2. Magneto ignition system.

These ignition systems are discussed, in detail, in the following pages :

26.29. Coil Ignition System

It is also known as ***battery ignition system***, and has an induction coil, which consists of two coils known as primary and secondary coils wound on a soft iron core, as shown in Fig. 26.17. The primary coil consists of a few hundred turns (about 300 turns) of wire. Over this coil, but insulated from it, are wound several thousand turns (about 20,000 turns) of secondary coil. The one end of the primary coil is connected to a ignition switch, ammeter and battery generally of 6 volts. The other end of the primary coil is connected to a condenser and a contact breaker.

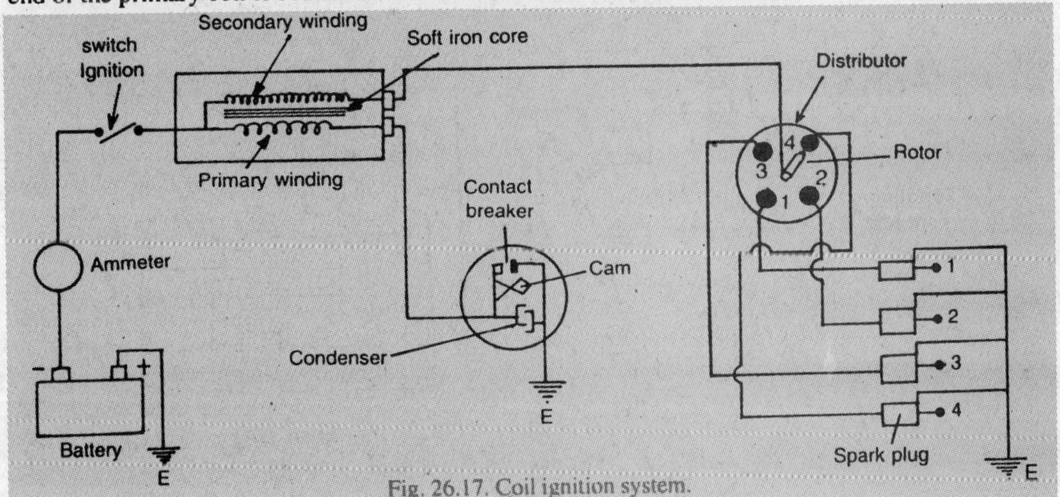

Fig. 26.17. Coil ignition system.

A condenser is connected across the contact-breaker for the following two reasons :

1. It prevents sparking across the gap between the contact breaker points
2. It causes a more rapid break of the primary current, giving a higher voltage in the secondary circuit.

The secondary coil is connected to a distributor (in a multi-cylinder engine) with the central terminal of the sparking plugs. The outer terminals of the sparking plugs are earthed together, and connected to the body of the engine.

When the current flows through the primary coil, it sets up a magnetic field which surrounds both the primary and secondary coils. As the switch is on, the contact-breaker connect the two ends. The magnetic field in coils has tendency to grow from zero to maximum value. Due to this change in the magnetic field, a voltage is generated in both the coils, but opposite to the applied voltage (of battery). Thus the primary coil does not give the final value. The voltage in the secondary coil is, therefore, not sufficient to overcome the resistance of the air gap of the sparking plug, hence no spark occurs.

When the current in the primary coil is switched off by the moving* cam, the magnetic field generated around the coil collapses immediately. The sudden variation of flux, which takes place, gives rise to the voltage generated in each coil. The value of the voltage depends upon the number

* A four lobed cam for four cylinder engine is an essential component of the make and break arrangement. It is rotated at half the engine speed.

of turns in each coil. As a matter of fact, the voltage required to produce a spark across the gap, between the sparking points, is between 10 000 to 20 000 volts. Since the secondary coil has several thousand turns, so it develops a sufficient high voltage to overcome the resistance of the gap of the sparking plug. This high voltage then passes to a distributor. It connects the sparking plugs in rotation depending upon the firing order of the engine. Hence, the ignition of fuel takes place in all the engine cylinders.

The coil ignition system is employed in medium and heavy spark ignition engines such as in cars.

26.30. Magneto Ignition System

The magneto ignition system, as shown in Fig. 26.18, has the same principle of working as that of coil ignition system, except that no battery is required, as the magneto acts as its own generator.

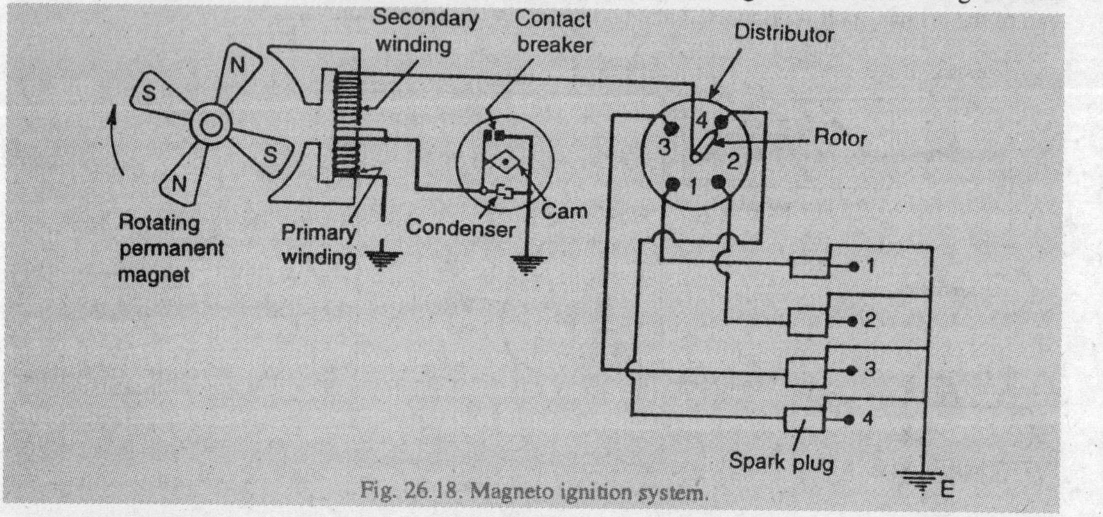

Fig. 26.18. Magneto ignition system.

It consists of either rotating magnets in fixed coils, or rotating coils in fixed magnets. The current produced by the magneto is made to flow to the induction coil which works in the same way as that of coil ignition system. The high voltage current is then made to flow to the distributor, which connects the sparking plugs in rotation depending upon the firing order of the engine.

This type of ignition system is generally employed in small spark ignition engines such as scooters, motor cycles and small motor boat engines.

26.31. Fuel Injection System for Diesel Engines

The following two methods of fuel injection system are generally employed with diesel engines (*i.e.* compression ignition engines) :

1. Air injection method, and 2. Airless or solid injection method.

These methods are discussed, in detail, as follows :

1. *Air injection method.* In this method of fuel injection, a blast of compressed air is used to inject the fuel into the engine cylinder. This method requires the aid of an air compressor which is driven by the engine crankshaft. The air is compressed at a pressure higher than that of engine cylinder at the end of its compression stroke. This method is not used now-a-days because of complicated and expensive system.

2. *Airless or solid injection method.* The most modern compression ignition engines use, now-a-days, the solid injection system. In this method, a separate fuel pump driven by the main crankshaft is used for forcing the fuel. The fuel is compressed in this pump to a pressure higher than that of engine cylinder at the end of compression. This fuel under pressure is directly sprayed into

the combustion chamber of the engine cylinder at the end of compression stroke, with the help of an injector. The solid injection method may be further divided into the following two commonly used systems.

(*a*) Common rail system, and

(*b*) Individual pump system.

These systems are discussed, in detail, as below :

(*a*) *Common rail system.* In the common rail system, as shown in Fig. 26.19, a multi-cylinder high pressure pump is used to supply the fuel at a high pressure to a common rail or header.

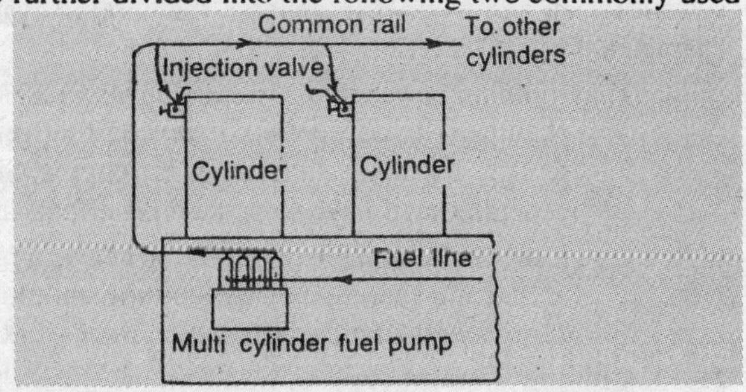

Fig. 26.19. Common rail system.

The high pressure in the common rail forces the fuel to each of the nozzle located in the cylinders. The pressure in this common rail is kept constant with the help of a high pressure relief valve. A metered quantity of fuel is supplied to each cylinder through the nozzle by operating the respective fuel injection valve with the help of cam-mechanism driven by the crankshaft of the engine.

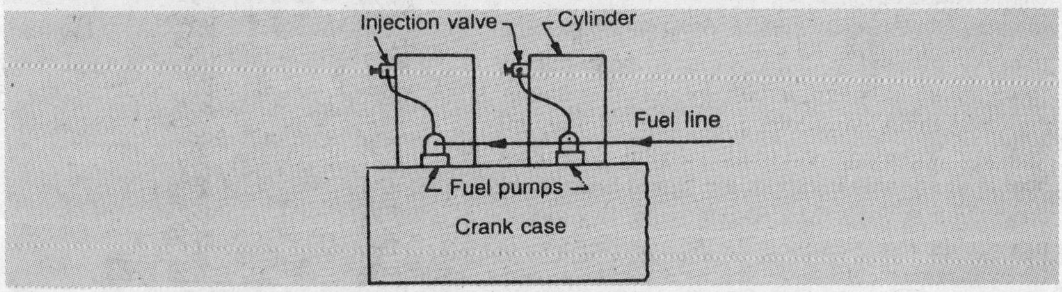

Fig. 26.20. Individual pump system of fuel injection.

(*b*) *Individual pump system.* In the individual pump system, as shown in Fig. 26.20, each cylinder of the engine is provided with an individual injection valve, a high pressure pump and a metering device run by the crankshaft of the engine. The high pressure pump plunger is actuated by a cam and produces the fuel pressure necessary to open the injection valve at the correct time. The amount of fuel injected depends upon the effective stroke of the plunger.

26.32. Cooling of I.C. Engines

We have already discussed that due to combustion of fuel inside the engine cylinder of I.C. engines, intense heat is generated. It has been experimentally found that about 30% of the heat generated is converted into mechanical work. Out of the remaining heat (about 70%) about 40% is carried away by the exhaust gases into the atmosphere. The remaining part of the heat (about 30%), if left un-attended, will be absorbed by engine cylinder, cylinder head piston, and engine valves etc. It has also been found that the overheating of these parts causes the following effects :

1. The overheating causes thermal stresses in the engine parts, which may lead to their distortion.
2. The overheating reduces strength of the piston. The overheating may cause even seizure of the piston.
3. The overheating causes decomposition of the lubricating oil, which may cause carbon deposit on the engine and piston head.
4. The overheating causes burning of valves and valve seats.
5. The overheating reduces volumetric efficiency of the engine.
6. The overheating increases tendency of the detonation.

In order to avoid the adverse effects of overheating, it is very essential to provide some cooling system for an I.C. engine. In general, the cooling system provided should have the following two characteristics for its efficient working :

1. It should be capable of removing about 30% of the total heat generated in the combustion chamber. It has been experienced that removal of more than 30% of heat generated reduces thermal efficiency of the engine. Similarly, removal of less than 30% of the heat generated will have some adverse effects as mentioned above.
2. It should be capable of removing heat at a fast rate, when the engine is hot. But at the time of starting the engine, the cooling should be comparatively slow, so that the various components of the engine attain their working temperature in a short time.

26.33. Cooling Systems for I.C. Engines

We have already discussed, in the last article, the adverse effects of overheating of an I.C. engine and characteristics of the cooling system adopted. The following two systems are used for cooling the I.C. engines these days :

1. *Air cooling system.* The air cooling system, as shown in Fig. 26.21, is used in the engines of motor cycles, scooters, aeroplanes and other stationary installations. In countries with cold climate, this system is also used in car engines. In this system, the heat is dissipated directly to the atmospheric air by conduction through the cylinder walls. In order to increase the rate of cooling, the outer surface area of the cylinder and cylinder head is increased by providing radiating fins and flanges. In bigger units, fans are provided to circulate the air around the cylinder walls and cylinder head.

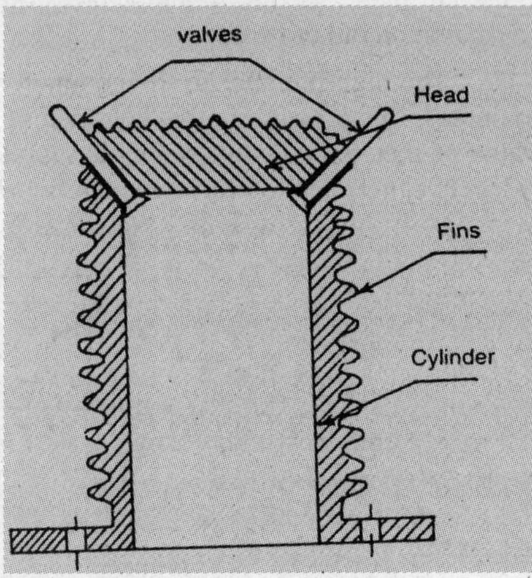

Fig. 26.21. Air cooling system.

2. *Water cooling system (Thermosyphon system of cooling).* The water cooling system as shown in Fig. 26.22, is used in the engines of cars, buses, trucks etc. In this system, the water is circulated through water jackets around each of the combustion chambers, cylinders, valve seats and

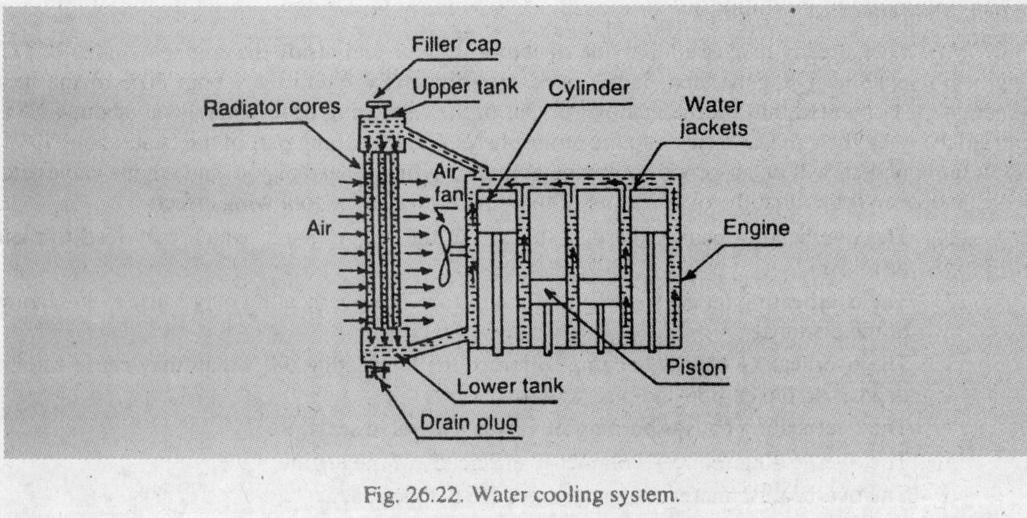

Fig. 26.22. Water cooling system.

valve stems. The water is kept continuously in motion by a centrifugal water pump which is driven by a *V*-belt from the pulley on the engine crank shaft. After passing through the engine jackets in the cylinder block and heads, the water is passed through the radiator. In the radiator, the water is cooled by air drawn through the radiator by a fan. Usually, fan and water pump are mounted and driven on a common shaft. After passing through the radiator, the water is drained and delivered to the water pump through a cylinder inlet passage. The water is again circulated through the engine jackets.

26.34. Comparison of Air Cooling and Water Cooling Systems

The following points are important for the comparison of air cooling and water cooling systems.

S.No.	*Air cooling system*	*Water cooling system*
1.	The design of this system is simple and less costly.	The design of this system is complicated and more costly.
2.	The mass of cooling system (per b.p. of the engine) is very less.	The mass of cooling system (per b.p. of the engine) is much more.
3.	The fuel consumption (per b.p. of the engine) is more.	The fuel consumption (per b.p. of the engine) is less.
4.	Its installation and maintenance is very easy and less costly.	Its installation and maintenance is difficult and more costly.
5.	There is no danger of leakage or freezing of the coolant.	There is a danger of leakage or freezing of the coolant.
6.	It works smoothly and continuously. Moreover it does not depend on any coolant.	If the system fails, it may cause serious damage to the engine within a short time.

26.35. Supercharging of I.C. Engines

It is the process of increasing the mass, or in other words density, of the air-fuel mixture (in spark ignition engine) or air (in compression ignition *i.e.* diesel engines) induced into the engine cylinder. This is, usually, done with the help of compressor or blower known as supercharger.

It has been experimentally found that the supercharging increases the power developed by the engine. It is widely used in aircraft engines, as the mass of air, sucked in the engine cylinder, decreases at very high altitudes. This happens, because atmospheric pressure decreases with the increase in altitude. Now-a-days, supercharging is also used in two-stroke and four-stroke petrol and diesel engines. It will be interesting to know that a supercharged engine is lighter, requires smaller foundations and consumes less lubricating oil as compared to an ordinary engine. Following are the objects of supercharging the engines :

1. To reduce mass of the engine per brake power (as required in aircraft engines).
2. To maintain power of aircraft engines at high altitudes where less oxygen is available for combustion.
3. To reduce space occupied by the engine (as required in marine engines).
4. To reduce the consumption of lubricating oil (as required in all type of engines).
5. To increase the power output of an engine when greater power is required (as required in racing cars and other engines).

26.36. Methods of Supercharging

Strictly speaking, a supercharger is an air pump, which receives air from the atmosphere surrounding the engine, compresses it to a higher pressure and then feeds it into the inlet valve of the engine. Following two methods of supercharging are important from the subject point of view :

1. *Reciprocating type.* It has a piston which moves to and fro inside a cylinder. It is an old method and is not encouraged these days, as it occupies a large space and has lubrication problem.

2. *Rotary type.* It resembles a centrifugal pump in its outward appearance, but differs in action. There are many types of rotary pumps, but gear type, lobe type and vane type are commonly used.

26.37. Lubrication of I.C. Engines

As a matter of fact, the moving parts of an I.C. engine are likely to wear off due to continuous rubbing action of one part with another. In order to avoid an early wearing of the engine parts, a proper lubrication arrangement is provided in I.C. engines. In general, following are the main advantages of lubrication of I.C. engines :

1. It reduces wear and tear of the moving parts.
2. It damps down the vibrations of the engine.
3. It dissipates the heat generated from the moving parts due to friction.
4. It cleans the moving parts.
5. It makes the piston gas-tight.

26.38. Lubrication System for I.C. Engines

The following two lubrication systems of I.C. engines are important from the subject point of view :

1. *Splash lubrication.* This method is generally employed for lubricating small I.C. engines. In this method, an oil sump is fixed to the bottom of the crank case and the pump is immersed, in the lubricating oil, as shown in Fig. 26.23. A small hole is drilled in the crank shaft and the oil is forced through this hole to the bearing. The oil is also forced along the connecting rod either through a hole drilled in the rod or along a small copper pipe to the gudgeon pin and piston.

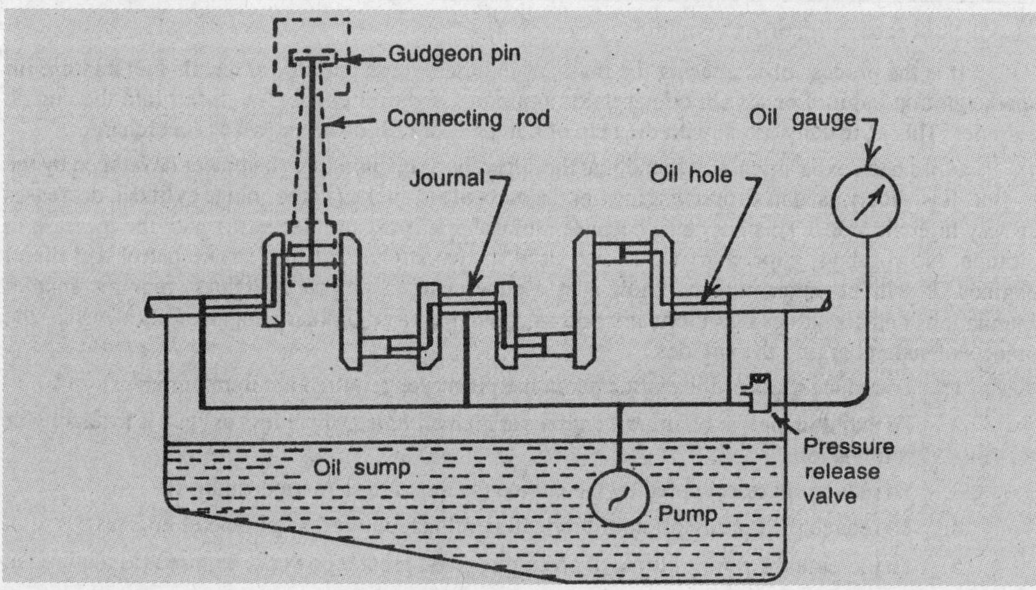

Fig.26.23. Splash lubrication.

The surplus oil is thrown out, in the form of a spray, from the bearings by centrifugal action. The surplus oil lubricates the cams, tappets and valve stems. The whole oil is drained back into the sump.

2. *Forced lubrication.* In this method, the lubricating oil is carried in a separate tank and is pumped at a high pressure to the main bearings. It passes at a lower pressure to the camshaft and

timing gears. As the oil drains with the sump, it is pumped back by a pump known as scavenge pump through an oil cooler to the oil tank, as shown in Fig. 26.24.

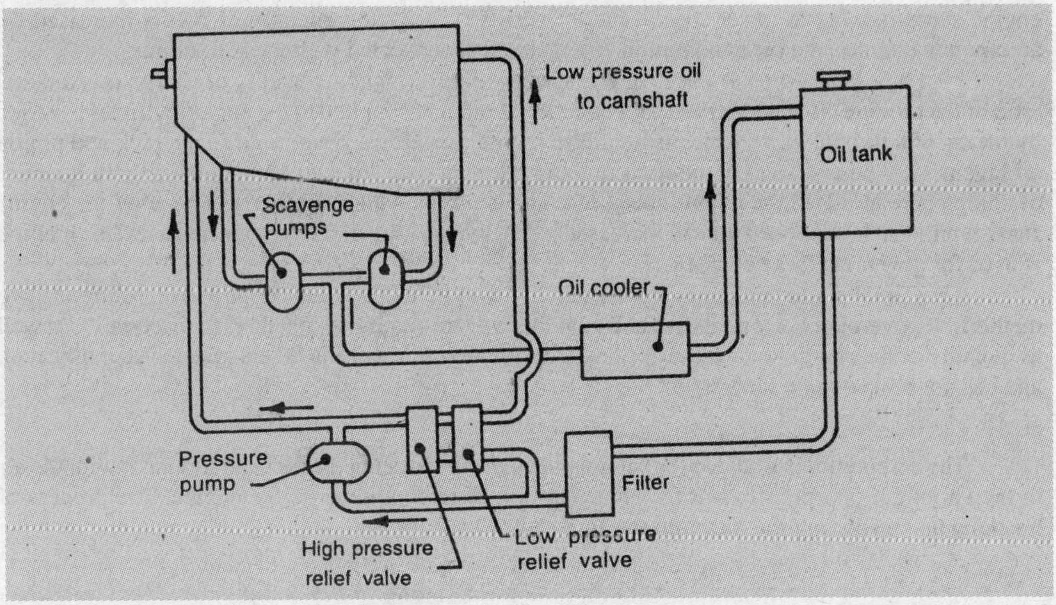

Fig. 26.24. Forced lubrication.

26.39. Governing of I.C. Engines

As a matter of fact, all the I.C. engines like other engines, are always designed to run at a particular speed. But in actual practice, load on the engine keeps on fluctuating from time to time. A little consideration will show, that change of load, on an I.C. engine, is sure to change its speed. It has been observed that if load on an I.C. engine is decreased without changing the quantity of fuel, the engine will run at a higher speed. Similarly, if load on the engine is increased without changing the quantity of fuel, the engine will run at a lower speed.

Now, in order to have a high efficiency of an I.C. engine, at different load conditions, its speed must be kept constant as far as possible. The process of providing any arrangement, which will keep the speed constant (according to the changing load conditions) is known as ***governing of I.C. engines***.

26.40. Methods of Governing I.C. Engines

Through there are many methods for the governing of I.C. engines, yet the following are important from the subject point of view :

1. *Hit and miss governing.* This method of governing is widely used for I.C. engines of smaller capacity or gas engines. This method is most suitable for engines, which are frequently subjected to reduced loads and as a result of this, the engines tend to run at higher speeds. In this system of governing, whenever the engine starts running at higher speed (due to decreased load), some explosions are omitted or missed. This is done with help of centrifugal governor (Art. 24.11) in which the inlet valve of fuel is closed and the explosions are omitted till the engine speed reaches its normal value. The only disadvantage of this method is that there is uneven turning moment due to missing of explosions. As a result of this, it requires a heavy flywheel.

2. *Qualitative governing.* In this system of governing, a control valve is fitted in the fuel delivery pipe, which controls the quantity of fuel to be mixed in the charge. The movement of control valve is regulated by the centrifugal governor through rack and pinion arrangement. It may be noted that in this system, the amount of air used in each cycle remains the same. But with the change in the

quantity of fuel (with quantity of air remaining constant), the quality of charge (*i.e.* air-fuel ratio of mixture) changes. Whenever the engine starts running at higher speed (due to decreased load), the quantity of fuel is reduced till the engine speed reaches its normal value. Similarly, whenever the engine starts running at lower speed (due to increased load), the quantity of fuel is increased. In automobile engines, the rack and pinion arrangement is connected with the accelerator.

3. *Quantitative governing.* In this system of governing, the quality of charge (*i.e.* air-fuel ratio of the mixture) is kept constant. But quantity of mixture supplied to the engine cylinder is varied by means of a throttle valve which is regulated by the centrifugal governor through rack and pinion arrangement. Whenever the engine starts running at higher speed (due to decreased load), the quantity of charge is reduced till the engine speed reaches its normal value. Similarly, whenever the engine starts running at lower speed (due to increased load), the quantity of charge is increased. This method is used for governing large engines.

4. *Combination system of governing.* In this system of governing, the above mentioned two methods of governing (*i.e.* qualitative and quantitative) are combined together, so that quality as well as quantity of the charge is varied according to the changing conditions. This system is complicated, and has not proved to be successful.

26.41. Carburettor

The carburettor is a device for *atomising and **vaporising the fuel and mixing it with the air in the varying proportions to suit the changing operating conditions of the engine. The process of breaking up and mixing the fuel with the air is called *carburation.*

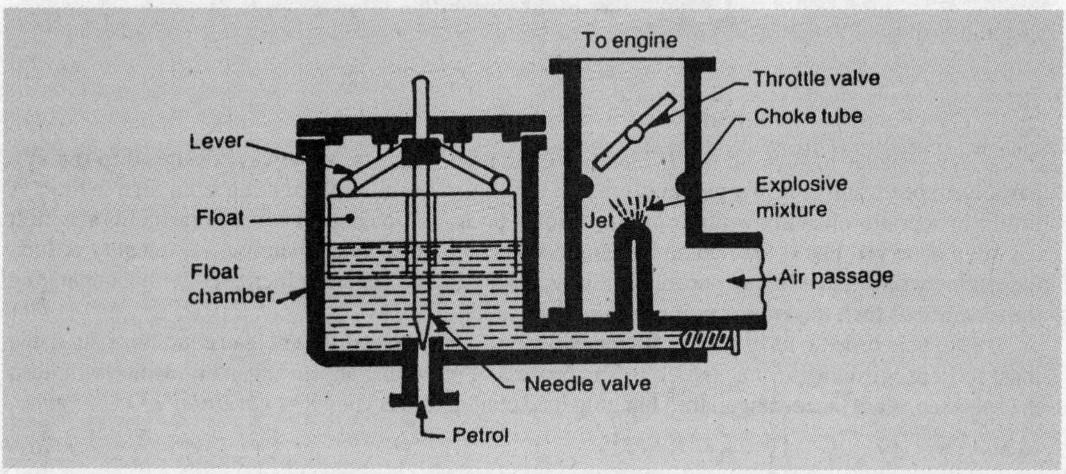

Fig. 26.25. Carburettor.

There are many types of the carburettors in use, but the simplest form of the carburettor is shown in Fig. 26.25. It consists of a fuel jet located in the centre of the choke tube. A float chamber is provided for maintaining the level of the fuel in the jet and is controlled by a float and lever which operates its needle valve. The fuel is pumped into the float chamber and when the correct level of the fuel is reached, the float closes the needle valve, and shuts off the petrol supply.

The suction produced by the engine draws air through the choke tube. The reduced diameter of the choke tube increases the velocity of air and reduces the pressure. The high velocity and low

---

* Atomisation is the mechanical breaking up of the liquid fuel into small particles so that every minute particle of the fuel is surrounded by air.

** Vaporisation is a change of state of fuel from a liquid to vapour.

pressure in the tube facilitates the breaking up of fuel and its admixture with the air. A throttle valve controls the flow of the mixture delivered to the engine cylinder.

26.42. Spark Plug

It is always screwed into the cylinder head for igniting the charge of petrol engines. It is, usually, designed to withstand a pressure up to 35 bar and operate under a current of 10 000 to 30 000 volts.

A spark plug consists of central porcelain insulator, containing an axial electrode length wise and ground elec trode welded to it. The central electrode has an external contact at the top, which is connected to the terminal and communicates with the distributor. A metal tongue is welded to the ground electrode, which bends over to lie across the end of the central electrode. There is a small gap known as spark gap between the end of the central electrode and the metal tongue, as shown in Fig. 26.26. The high tension electric spark jumps over the gap to ignite the charge in the engine cylinder.

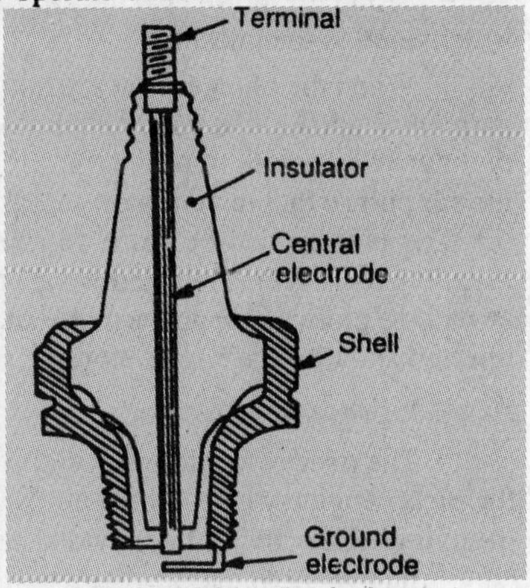

Fig. 26.26. Spark plug.

The electrode material should be such which can withstand corrosiveness, high temperature having good thermal conductivity. The electrodes are generally made from the alloys of platinum, nickle, chromium, barium etc.

Note : The spark plug gap is kept from 0.3 mm to 0.7 mm. The experiments have shown that efficiency of the ignition system is greatly reduced if the gap is too large or too small. Sometimes, foreign materials (such as carbon) gets deposited in the spark gap. It is a source of nuisance, as it permits some of the high voltage current to bypass the gap and reduce the intensity of spark as well as engine efficiency.

26.43. Fuel Pump

The main object of a fuel pump in a diesel engine is to deliver a fuel to the injector which sprays the finely divided particles of the fuel suitable for rapid combustion.

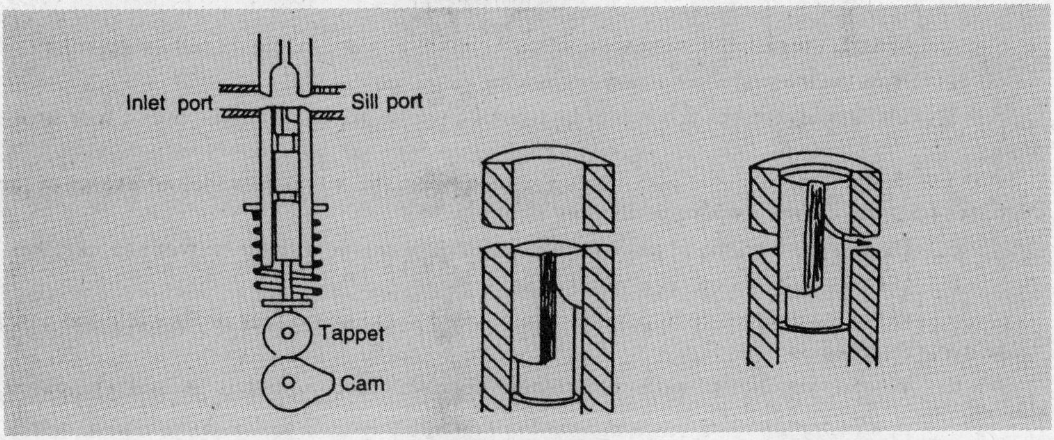

(*a*) Fuel pump. (*b*) Beginning of stroke. (*c*) End of stroke.

Fig. 26.27

The simplified sketch of a fuel pump is shown in Fig. 26.27 (*a*). It consists of a plunger which moves up and down in the barrel by the cam and spring arrangement provided for pushing and lowering the plunger respectively. The fuel oil is highly filtered by means of felt-pack filter before entering the barrel of the pump.

The upper end part of the plunger is cut away in a helix shaped piece forming a groove between the plunger and barrel, which is the most important one. Therefore, the amount of fuel delivered and injected into the engine cylinder depends upon the rotary position of the plunger in the barrel. Fig. 26.27 (*b*) and (*c*) shows how the top part of the plunger is designed so that the correct amount of fuel is delivered to the injector.

When the plunger is at the bottom of its stroke as shown in Fig 26.27 (*b*), the fuel enters the barrel through the inlet port. As the plunger rises, it forces this fuel up into the injector, until the upper part cut away comes opposite the sill port. Then the fuel escapes down the groove and out through the sill port so that injection ceases, as shown in Fig. 26.27 (*c*). The plunger can be made to rotate in the barrel and therefore more fuel is injected. When the plunger is rotated so that the groove is opposite to the sill port, no fuel at all is injected and thus the engine stops.

26.44. Injector or Atomiser

The injector or atomiser is also an important part of the diesel engine which breaks up the fuel and sprays into the cylinder into a very fine divided particles.

Fig. 26.28 shows the type of an injector in which fuel is delivered from the pump along the horizontal pipe connected at *A*. The vertical spindle of the injector is spring loaded at the top which holds the spindle down with a pressure of 140 bar so that the fuel pressure must reach this value before the nozzle will lift to allow fuel to be injected into the engine cylinder. The fuel which leaks past the vertical spindle is taken off by means of an outlet pipe fitted at *B* above the fuel inlet pipe.

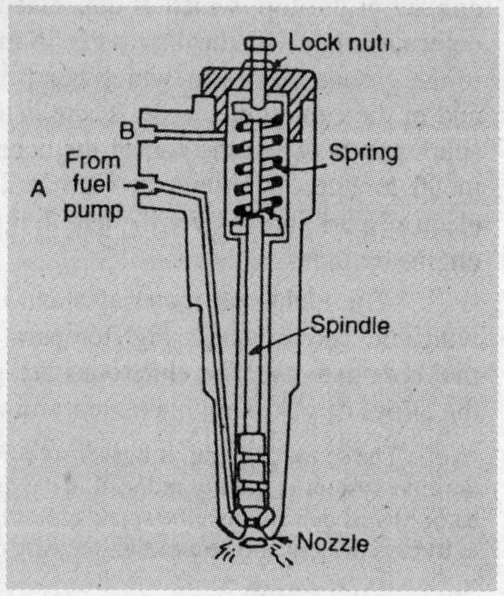

Fig. 26.28. Injector or Atomiser.

QUESTIONS

1. What is the difference between internal combustion and external combustion engines ?
2. How the internal combustion engines are classified ?
3. List the advantages and disadvantages of a two stroke cycle engine over a four stroke one.
4. Describe briefly and with appropriate sketches, the actual sequence of events in the cylinder of a petrol engine working on the four stroke cycle.
5. Discuss the working of a two stroke cycle petrol engine with the help of neat sketches.
6. Differentiate between petrol and diesel engine.
7. Explain with the help of suitable sketches, the working of a four stroke cycle and a two stroke cycle diesel engine.
8. What do you understand by air injection and solid injection system generally employed with the diesel engines ?
9. Describe the phenomenon of detonation in I.C. engines. On what factors does detonation depend ?
10. Explain what do you understand by octane and cetane number rating of a fuel.
11. Draw the electrical circuit used for battery ignition in a four stroke four cylinder engines. Explain the function of each component.

12. Discuss the cooling requirement of an I.C. engine. Describe the different methods of cooling and give specific examples where each method is employed.

13. Explain the need and methods of supercharging in I.C. Engines.

14. Discuss the lubrication system in I.C. engines.

15. What is the function of a carburettor in an S.I. engine ? Briefly explain with a neat sketch the operation of a simple float type carburettor.

16. Write short notes on the following :

(*a*) Scavenging ; (*b*) Fuel pump ; (*c*) Atomiser ; and (*d*) Spark plug.

OBJECTIVE TYPE QUESTIONS

1. In a four stroke engine, the working cycle is completed in
(*a*) one revolution of the crankshaft (*b*) two revolutions of the crankshaft
(*c*) three revolutions of the crankshaft (*d*) four revolutions of the crankshaft

2. A two stroke cycle engine gives...........the number of power strokes as compared to the four stroke cycle engine, at the same engine speed.
(*a*) half (*b*) same (*c*) double (*d*) four times

3. The thermal efficiency of a two stroke cycle engine is..........a four stroke cycle engine.
(*a*) equal to (*b*) less than (*c*) greater than

4. The theoretically correct mixture of air and petrol is
(*a*) 10 : 1 (*b*) 15 : 1 (*c*) 20 : 1 (*d*) 25 : 1

5. The thermodynamic cycle on which the petrol engine works, is
(*a*) Otto cycle (*b*) Joule cycle (*c*) Rankine cycle (*d*) Stirling cycle

6. A diesel engine has
(a) one valve (b) two valves (c) three valves (d) four valves

7. If petrol is used in a diesel engine, then
(*a*) low power will be produced (*b*) efficiency will be low
(*c*) higher knocking will occur (*d*) black smoke will be produced

8. A petrol engine has compression ratio from
(*a*) 6 to 10 (*b*) 10 to 15 (*c*) 15 to 25 (*d*) 25 to 40

9. The function of a distributor in a coil ignition system of I.C. engines is
(*a*) to distribute the spark (*b*) to distribute the power
(*c*) to distribute the current (*d*) to time the spark

10. Supercharging................the power developed by the engine.
(*a*) has no effect on (*b*) increases (*c*) decreases

11. A carburettor is used to supply
(*a*) petrol, air and lubricating oil (*b*) air and diesel
(*c*) petrol and lubricating oil (*d*) petrol and air

12. A spark plug gap is kept from
(*a*) 0.3 to 0.7 mm (*b*) 0.2 to 0.8 mm
(*c*) 0.4 to 0.9 mm (*d*) 0.6 to 1.0 mm

13. The knocking tendency in spark ignition engines may be decreased by
(*a*) controlling the air fuel mixture (*b*) controlling the ignition timing
(*c*) reducing the compression ratio (*d*) all of these

14. The violent sound pulsations within the cylinder of an I.C. engine are due to

(*a*) detonation (*b*) turbulence (*c*) pre-ignition (*d*) none of these

15. Which of the following does not relate to a spark ignition engine ?

(*a*) Ignition coil (*b*) Spark plug (*c*) Distributor (*d*) Fuel injector

ANSWERS

1. (*b*)	2. (*c*)	3. (*b*)	4. (*b*)	5. (*a*)
6. (*c*)	7. (*c*)	8. (*a*)	9. (*d*)	10. (*b*)
11. (*d*)	12. (*a*)	13. (*c*)	14. (*a*)	15. (*d*)

27

Testing of Internal Combustion Engines

1. Introduction. 2. Thermodynamic Tests for I.C. Engines. 3. Indicated Mean Effective Pressure. 4. Indicated Power. 5. Morse Test. 6. Brake Power. 7. Efficiencies of an I.C. Engine. 8. Air Consumption 9. Heat Balance Sheet.

27.1. Introduction

In the last chapter, we have discussed the working *i.e.* operation of the I.C. engines. As a matter of fact, when an I.C. engine is designed and manufactured, then it is tested in a laboratory. The purpose of testing are :

1. To determine the information, which can not be obtained by calculations.
2. To confirm the data used in design, the validity of which may be doubtful.
3. To satisfy the customer regarding the performance of the engine.

Note : By performance, we mean the operation of all variables relating to the working of the engine. These variables are power, fuel consumption etc.

27.2. Thermodynamic Tests for I.C. Engines

An internal combustion engine is put to the thermodynamic tests, so as to determine the following quantities :

1. Indicated mean effective pressure ; 2. Indicated power ; 3.Speed of the engine ; 4.Brake torque ; 5. Brake Power ; 6. Mechanical Losses (Motoring test) ; 7. Mechanical efficiency ; 8. Fuel consumption ; 9. Thermal efficiency ; 10. Air consumption ; 11. Volumetric efficiency ; 12. Various temperatures ; and 13. Heat balance sheet.

It may be noted that these quantities are measured after the engine has reached the steady conditions.

27.3. Indicated Mean Effective Pressure

The indicated mean effective pressure of an engine is obtained from the indicator diagram drawn with the help of an engine indicator, by any one of the following methods :

1. By drawing the diagram on a squared paper and then finding its area by counting the number of squares.
2. By finding the area of the diagram with the help of a planimeter.
3. By mid-ordinates taken from one end to another.

In all these methods, the aim is to determine the height of a rectangle of an area equal to the area of the indicator diagram. The height of this rectangle gives the mean effective pressure.

Fig. 27.1 (*a*) shows the indicator diagram and the equivalent rectangle, *i.e.* a rectangle having the same area as that of the indicator diagram, whose length is equal to the length of the indicator diagram or card.

Let s = Scale of the pressure, *i.e.* scale of indicator spring* in bar per mm,

a = Area of the diagram or rectangle in mm^2, and

l = Length of the diagram in mm.

∴ Mean effective pressure (in bar)

$$= \frac{\text{Area of indicator card} \times \text{Scale of indicator spring}}{\text{Length of indicator card}} = \frac{a \times s}{l}$$

This relation is helpful for finding out the mean effective pressure by using the first two methods.

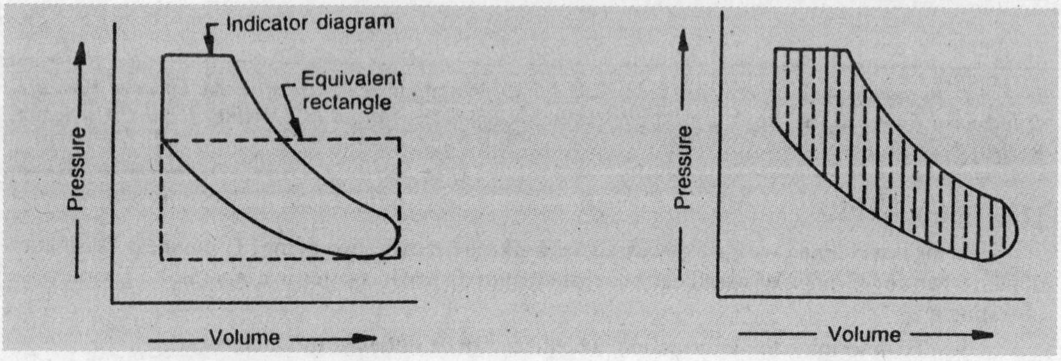

(*a*) Equivalent rectangle method. (*b*) Mid-ordinate method.

Fig. 27.1

In case of mid-ordinate method, the indicator diagram is divided into strips of equal width as shown in Fig. 27.1 (*b*). At the centre of each strip, mid-ordinate (shown dotted) are drawn. All these mid-ordinates are added and the total is divided by number of ordinates to get the mean height of the diagram.

Let m = Mean height of the diagram, and

s = Scale of the indicator spring.

∴ Mean effective pressure

$= m.s$

Note : The mean effective pressure calculated on the basis of the theoretical indicator diagram, is known as ***theoretical mean effective pressure***. If it is based on the actual indicator diagram, then it is called ***actual mean effective pressure***.

27.4. Indicated Power

The indicated power (briefly written as I.P.) is the power actually developed by the engine cylinder. It is based on the information obtained from the indicator diagram of the engine.

Let p_m = Actual mean effective pressure as obtained from the indicator diagram in bar ;

L = Length of stroke in metres ;

A = Area of the piston in m^2 ;

N = Speed of the engine in r.p.m., and

* An engine indicator is, usually, provided with a set of accurately calibrated springs each of which is stamped with a number. This number indicates the pressure (in bar) required to produce a deflection of 1 mm. The number is sometimes known as spring number.

$$n = \text{Number of working strokes per minute}$$

$$= N \quad \text{. . . (For two stroke cycle engine)}$$

$$= N/2 \quad \text{. . . (For four stroke cycle engine)}$$

$$\therefore \text{ Indicated power, I.P.} = \frac{p_m \times 10^5 \times L \times A \times n}{60} \text{ watts}$$

$$= \frac{100\, p_m\, L\, A\, n}{60} \text{ kW} \quad \text{. . . (For single cylinder engine)}$$

$$= \frac{100\, K\, p_m\, L\, A\, n}{60} \text{ kW} \quad \text{. . . (For multi-cylinder engine)}$$

where K = Number of cylinders.

27.5. Morse Test

The Morse test is adopted to find the indicated power of each cylinder of a high speed I.C. engine, without using indicator diagram. The test is carried out as follows :

Consider a four cylinder engine. First of all, the brake power of the engine, when all the cylinders are in operation, is measured accurately (by means of a brake dynamometer) at a constant speed and load. Now, one of the cylinders (say cylinder 1) is cut-off so that it does not develop any power. This is done by short circuiting the spark plug of the cylinder in petrol engines and cutting-off individual fuel supply in diesel engines. The speed of the engine decreases and in order to bring the speed back to the original speed, the load on the engine is reduced. The brake power is now measured in this new condition which gives the brake power of the remaining three cylinders.

In the similar way, each cylinder is cut-off one by one and the brake power of the remaining three cylinders is determined by correcting the engine speed, if necessary.

Let I_1, I_2, I_3 and I_4 = Indicated power of each individual cylinder.

F_1, F_2, F_3 and F_4 = Frictional power of each individual cylinder.

We know that total brake power of the engine when all the cylinders are working is given by

$$B = \text{Total indicated power} - \text{Total frictional power}$$

$$= (I_1 + I_2 + I_3 + I_4) - (F_1 + F_2 + F_3 + F_4) \quad \text{. . . (i)}$$

When the cylinder No. 1 is cut-off, then $I_1 = 0$, but the frictional losses of the cylinder remain the same.

$\therefore$ Brake power of the remaining three cylinders,

$$B_1 = (0 + I_2 + I_3 + I_4) - (F_1 + F_2 + F_3 + F_4) \quad \text{. . . (ii)}$$

Subtracting equation (*ii*) from equation (*i*),

$$B - B_1 = I_1$$

or Indicated power of the first cylinder,

$$I_1 = B - B_1$$

Similarly, indicated power of the second cylinder,

$$I_2 = B - B_2$$

Indicated power of the third cylinder,

$$I_3 = B - B_3$$

Indicated power of the fourth cylinder,

$$I_4 = B - B_4$$

and total indicated power of the engine,

$$I = I_1 + I_2 + I_3 + I_4$$

Example 27.1. *A single cylinder, two stroke petrol engine develops 4 kW indicated power. Find the average speed of the piston, if the mean effective pressure is 6.5 bar and piston diameter is 100 mm.*

Solution. Given : I.P. = 4 kW ; $p_m = 6.5$ bar ; $D_p = 100$ mm = 0.1 m

Let L = Length of stroke in metres, and

N = Speed of the engine in r.p.m.

∴ Number of working strokes per minute,

$$n = N \quad \text{... (∵ Engine works on two stroke cycle)}$$

We know that area of piston,

$$A = \frac{\pi}{4}(0.1)^2 = 7.855 \times 10^{-3} \text{ m}^2$$

∴ Indicated power (I.P.),

$$4 = \frac{100\, p_m L A n}{60} = \frac{100 \times 6.5 \times L \times 7.855 \times 10^{-3} \times N}{60} = 0.0851\, L N$$

or

$$L N = 47$$

∴ Average speed of the piston

$$= 2LN = 2 \times 47 = 94 \text{ m/s} \textbf{ Ans.}$$

Example 27.2. *In a laboratory experiment, the following observations were noted during the test of a four stroke Diesel engine :*

Area of indicator diagram = 420 mm² ; Length of indicator diagram = 62 mm ; Spring number = 1.1 bar/mm ; Diameter of piston = 100 mm ; Length of stroke = 150 mm ; Engine speed = 450 r.p.m. Determine : 1. Indicated mean effective pressure, and 2. Indicated power.

Solution. Given : $a = 420$ mm² ; $l = 62$ mm ; $s = 1.1$ bar/mm ; $D_p = 100$ mm = 0.1 m ; $L = 150$ mm = 0.15 m ; $N = 450$ r.p.m.

We know that area of piston,

$$A = \frac{\pi}{4}(0.1)^2 = 7.855 \times 10^{-3} \text{ m}^2$$

and number of working strokes per minute,

$$n = N/2 = 450/2 = 225 \quad \text{... (∵ Engine works on four stroke cycle)}$$

1. *Indicated mean effective pressure*

We know that indicated mean effective pressure,

$$p_m = \frac{a.s}{l} = \frac{420 \times 1.1}{62} = 7.45 \text{ bar } \textbf{Ans.}$$

2. *Indicated power*

We know that indicated power,

$$\text{I.P.} = \frac{100\, p_m L A n}{60} = \frac{100 \times 7.45 \times 0.15 \times 7.855 \times 10^{-3} \times 225}{60} \text{ kW}$$

$$= 3.29 \text{ kW } \textbf{Ans.}$$

27.6. Brake Power

The brake power (briefly written as B.P.) is the power available at the crank shaft. The brake power of an I.C. engine is, usually, measured by means of brake mechanism (prony brake or rope brake).

In case of prony brake, let

W = Brake load in newtons,

l = Length of arm in metres, and

N = Speed of the engine in r.p.m.

∴ Brake power of the engine,

$$\text{B.P.} = \frac{\text{Torque in N-m} \times \text{Angle turned in radians through 1 revolution}}{60} \times \text{R.P.M. watts}$$

$$= \frac{T \times 2\pi N}{60} = \frac{W\,l \times 2\pi\,N}{60} \text{ watts}$$

In case of rope brake, let

W = Dead load in newtons,

S = Spring balance reading in newtons,

D = Diameter of brake drum in metres,

d = Diameter of the rope in metres, and

N = Speed of the engine in r.p.m.

∴ Brake power of the engine,

$$\text{B.P.} = \frac{(W - S)\,\pi\,DN}{60} \text{ watts} \quad \text{... [without considering diameter } (d) \text{ of the rope]}$$

$$= \frac{(W - S)\,\pi\,(D + d)\,N}{60} \text{ watts} \quad \text{... [Considering diameter (d) of the rope]}$$

Example 27.3. *The following data were recorded during a test on an oil engine :*

Speed of the engine = 1000 r.p.m. ; Load on the brake = 1000 N ; Length of the brake = 750 mm.

Determine : 1. Brake torque ; and 2. Brake power of the engine.

Solution. Given : N = 1000 r.p.m. ; W = 1000 N ; l = 750 mm = 0.75 m

1. *Brake torque*

We know that brake torque,

$$T = W\,l = 1000 \times 0.75 = 750 \text{ N-m } \textbf{Ans.}$$

2. *Brake power of the engine*

We know that brake power of the engine,

$$\text{B.P.} = \frac{T \times 2\,\pi N}{60} = \frac{750 \times 2\,\pi \times 1000}{60} = 78\,550 \text{ W} = 78.55 \text{ kW } \textbf{Ans.}$$

Example 27.4. *A rope brake has brake wheel diameter of 600 mm and the diameter of rope is 5 mm. The dead load on the brake is 210 N and spring balance reads 30 N. If the engine makes 450 r.p.m., find the brake power developed.*

Solution. Given : D = 600 mm = 0.6 m ; d = 5 mm = 0.005 m ; W = 210 N ; S = 30 N ; N = 450 r.p.m.

We know that brake power developed,

$$\text{B.P.} = \frac{(W-S)\,\pi\,(D+d)\,N}{60} = \frac{(210-30)\,\pi\,(0.6+0.005)\,450}{60}\ \text{W}$$

$$= 2570\ \text{W} = 2.57\ \text{kW}\ \textbf{Ans.}$$

27.7. Efficiencies of I.C. Engine

The efficiency of an engine is defined as the ratio of work done to the energy supplied to an engine. The following efficiencies of an I.C. engine are important from the subject point of view :

1. *Mechanical efficiency.* It is the ratio of brake power (B.P.) to the indicated power (I.P.). Mathematically, mechanical efficiency,

$$\eta_m = \frac{\text{B.P.}}{\text{I.P.}}$$

It may be observed that the mechanical efficiency is always less than unity (*i.e.* 100 %) because some power is lost in overcoming the engine friction. In other words, the indicated power is always greater than brake power. The power, which is lost in overcoming the engine friction, is known as *frictional power*. Therefore frictional power,

$$\text{F.P.} = \text{I.P.} - \text{B.P.}$$

2. *Overall efficiency.* It is the ratio of work obtained at the crankshaft in a given time to the energy supplied by the fuel during the same time.

Let m_f = Mass of fuel consumed in kg per hour, and

C = Calorific value of fuel in kJ/kg of fuel.

∴ Energy supplied by the fuel per minute

$$= \frac{m_f \times C}{60}\ \text{kJ}$$

and work obtained at the crankshaft per minute

$$= \text{B.P.} \times 60\ \text{kJ} \qquad \ldots (\because \text{B.P. is in kW and } 1\ \text{kW} = 1\ \text{kJ/s})$$

∴ Overall efficiency, $$\eta_o = \frac{\text{B.P.} \times 60 \times 60}{m_f \times C} = \frac{\text{B.P.} \times 3600}{m_f \times C}$$

3. *Indicated thermal efficiency.* It is the ratio of the heat equivalent to one kW hour to the heat in the fuel per I.P. hour. Mathematically, thermal efficiency,

$$\eta_t = \frac{\text{Heat equivalent to one kW hour}}{\text{Heat in fuel per I.P. hour}}$$

$$= \frac{3600}{\dfrac{m_f \times C}{\text{I.P.}}} = \frac{\text{I.P.} \times 3600}{m_f \times C}$$

Note : The ratio $\dfrac{m_f}{\text{I.P.}}$ is known as specific fuel consumption per I.P. per hour.

4. *Brake thermal efficiency.* It is the ratio of the heat equivalent to one kW hour to the heat in fuel per B. P. hour. It is also known as *overall thermal efficiency* of the engine. Mathematically, brake thermal efficiency,

$$\eta_b = \frac{\text{Heat equivalent to one kW hour}}{\text{Heat in fuel per B.P. hour}}$$

$$= \frac{3600}{\dfrac{m_f \times C}{\text{B.P.}}} = \frac{\text{B.P.} \times 3600}{m_f \times C}$$

Note : The ratio $\dfrac{m_f}{\text{B.P.}}$ is known as specific fuel consumption per B.P. per hour.

5. *Air standard efficiency.* The air standard efficiency of an I.C. engine may also be obtained mathematically from the general expression for the air standard efficiency *i.e.*

$$\eta_{ase} = 1 - \frac{1}{r^{\gamma - 1}} \qquad \text{. . . (For petrol engines)}$$

$$= 1 - \frac{1}{r^{\gamma - 1}} \left[\frac{\rho^{\gamma} - 1}{\gamma(\rho - 1)} \right] \qquad \text{. . . (For Diesel engines)}$$

where r = Compression ratio,

γ = Ratio of specific heats, and

ρ = Cut-off ratio.

6. *Relative efficiency.* It is also known as efficiency ratio. The relative efficiency of an I.C. engine is the ratio of indicated thermal efficiency to the air standard efficiency. Mathematically,

$$\text{Relative efficiency} = \frac{\text{Indicated thermal efficiency}}{\text{Air standard efficiency}}$$

7. *Volumetric efficiency.* It is the ratio of actual volume of charge admitted during the suction stroke at N.T.P. to the swept volume of the piston. Mathematically, volumetric efficiency,

$$\eta_v = \frac{\text{Volume of charge admitted at N.T.P.}}{\text{Swept volume of the piston}^*} = \frac{v_a}{v_s}$$

Note : The volumetric efficiency may also be defined as the ratio of the mass of actual charge admitted to the swept mass of the charge at N.T.P.

Example 27.5. *A gas engine has piston diameter of 150 mm, length of stroke 400 mm and mean effective pressure 5.5 bar. The engine makes 120 explosions per minute. Determine the mechanical efficiency of the engine, if its B.P. is 5 kW.*

Solution. Given : D_p = 150 mm = 0.15 m ; L = 400 mm = 0.4 m ; p_m = 5.5 bar ; n = 120 ; B.P. = 5 kW

We know that area of the piston,

$$A = \frac{\pi}{4}(0.15)^2 = 0.0177 \text{ m}^2$$

and indicated power,

$$\text{I.P.} = \frac{100\, p_m\, L\, A\, n}{60} = \frac{100 \times 5.5 \times 0.4 \times 0.0177 \times 120}{60} = 7.79 \text{ kW}$$

We know that mechanical efficiency of the engine,

$$\eta_m = \frac{\text{B.P.}}{\text{I.P.}} = \frac{5}{7.79} = 0.642 \text{ or } 64.2\%$$ **Ans.**

Example 27.6. *A four cylinder two stroke cycle petrol engine develops 23.5 kW brake power at 2500 r.p.m. The mean effective pressure on each piston is 8.5 bar and the mechanical efficiency is 85%. Calculate the diameter and stroke of each cylinder, assuming the length of stroke equal to 1.5 times the diameter of cylinder.*

Solution. Given : $K = 4$; B.P. = 23.5 kW ; $N = 2500$ r.p.m. ; $p_m = 8.5$ bar ; $\eta_m = 85\% = 0.85$

Let D_c = Diameter of cylinder, and

L = Length of stroke = $1.5\ D_c$ . . . (Given)

We know that area of the cylinder,

$$A = \frac{\pi}{4}(D_c)^2 = 0.7855(D_c)^2$$

and number of working strokes per minute,

$$n = N = 2500 \quad \text{. . . (}\because \text{ Engine works on two stroke cycle)}$$

We know that indicated power,

$$\text{I.P.} = \frac{\text{B.P.}}{\eta_m} = \frac{23.5}{0.85} = 27.65 \text{ kW}$$

We also know that indicated power (I.P.),

$$27.65 = \frac{100\ K\ p_m\ L\ A\ n}{60}$$

$$= \frac{100 \times 4 \times 8.5 \times 1.5\ D_c \times 0.7855\ (D_c)^2\ 2500}{60} = 166\ 920\ (D_c)^3$$

$\therefore$ $(D_c)^3 = 0.000\ 165$ or $D_c = 0.055\text{ m} = 55$ mm **Ans.**

and $L = 1.5 \times 55 = 82.5$ mm **Ans.**

Example 27.7. *During the test on single cylinder oil engine, working on the four stroke cycle and fitted with a rope brake, the following readings are taken :*

Effective diameter of brake wheel = 630 mm ; Dead load on brake = 200 N ; Spring balance reading = 30 N ; Speed = 450 r.p.m. ; Area of indicator diagram = 420 mm² ; Length of indicator diagram = 60 mm ; Spring scale = 1.1 bar per mm ; Diameter of cylinder = 100 mm ; Stroke = 150 mm ; Quantity of oil used = 0.815 kg/h ; Calorific value of oil = 42 000 kJ/kg.

Calculate brake power, indicated power, mechanical efficiency, brake thermal efficiency and brake specific fuel consumption.

Solution. Given : $K = 1$; $D = 630$ mm = 0.63 m ; $W = 200$ N ; $S = 30$ N ; $N = 450$ r.p.m. ; $a = 420$ mm^2 ; $l = 60$ mm ; $s = 1.1$ bar/mm ; $D_c = 100$ mm = 0.1 m ; $L = 150$ mm = 0.15 m ; $m_f = 0.815$ kg/h ; $C = 42\ 000$ kJ/kg

Brake power

We know that brake power,

$$\text{B.P.} = \frac{(W - S)\ \pi DN}{60} = \frac{(200 - 30)\ \pi \times 0.63 \times 450}{60} = 2520 \text{ W}$$

$$= 2.52 \text{ kW} \quad \textbf{Ans.}$$

Indicated power

We know that indicated mean effective pressure,

$$p_m = \frac{a \times s}{l} = \frac{420 \times 1.1}{60} = 7.7 \text{ bar}$$

Area of the cylinder $A = \frac{\pi}{4}(D_c)^2 = \frac{\pi}{4}(0.1)^2 = 7.855 \times 10^{-3}\ \text{m}^2$

and number of working strokes per min,

$$n = N/2 = 450/2 = 225 \quad \ldots (\because \text{Engine works on four stroke cycle})$$

We know that indicated power,

$$\text{I.P.} = \frac{100\, K\, p_m\, L\, A\, n}{60} = \frac{100 \times 1 \times 7.7 \times 0.15 \times 7.855 \times 10^{-3} \times 225}{60} \text{ kW}$$

$$= 3.4 \text{ kW } \textbf{Ans.}$$

Mechanical efficiency

We know that mechanical efficiency,

$$\eta_m = \frac{\text{B.P.}}{\text{I.P.}} = \frac{2.52}{3.4} = 0.7418 \text{ or } 74.18\% \textbf{ Ans.}$$

Brake thermal efficiency

We know that brake thermal efficiency,

$$\eta_b = \frac{\text{B.P.} \times 3600}{m_f \times C} = \frac{2.52 \times 3600}{0.815 \times 42\,000} = 0.265 \text{ or } 26.5\% \textbf{ Ans.}$$

Brake specific fuel consumption

We know that brake specific fuel consumption

$$= \frac{m_f}{\text{B.P.}} = \frac{0.815}{2.52} = 0.323 \text{ kg / B.P. / h } \textbf{Ans.}$$

Example 27.8. *An engine uses 6.5 kg of oil per hour of calorific value 30 000 kJ/kg. If the B.P. of the engine is 22 kW and mechanical efficiency 85%, calculate : 1. Indicated thermal efficiency ; 2. Brake thermal efficiency ; and 3. Specific fuel consumption in kg/B.P./h.*

Solution. Given : m_f = 6.5 kg/h ; C = 30 000 kJ/kg ; B.P. = 22 kW ; η_m = 85% = 0.85

1. *Indicated thermal efficiency*

We know that indicated power,

$$\text{I.P.} = \frac{\text{B.P.}}{\eta_m} = \frac{22}{0.85} = 25.88 \text{ kW}$$

∴ Indicated thermal efficiency,

$$\eta_t = \frac{\text{I.P.} \times 3600}{m_f \times C} = \frac{25.88 \times 3600}{6.5 \times 30\,000} = 0.48 \text{ or } 48\% \textbf{ Ans.}$$

2. *Brake thermal efficiency*

We know that brake thermal efficiency,

$$\eta_b = \frac{\text{B.P.} \times 3600}{m_f \times C} = \frac{22 \times 3600}{6.5 \times 30\,000} = 0.406 \text{ or } 40.6\% \textbf{ Ans.}$$

3. *Specific fuel consumption*

We know that specific fuel consumption

$$= \frac{m_f}{\text{B.P.}} = \frac{6.5}{22} = 0.295 \text{ kg / B.P. / h } \textbf{Ans.}$$

Example 27.9. *A four cylinder engine running at 1200 r.p.m. gave 18.6 kW brake power. The average torque when one cylinder was cut out was 105 N-m. Determine the indicated thermal*

efficiency if the calorific value of the fuel is 42 000 kJ/kg and the engine uses 0.34 kg of petrol per brake power hour.

Solution. Given : $K = 4$; $N = 1200$ r.p.m. ; B.P. = 18.6 kW ; $T = 105$ N-m ; $C = 42\ 000$ kJ/kg ; $m_f = 0.34 \text{ kg / B.P. / h} = 0.34 \times 18.6 = 6.324 \text{ kg / h}$

We know that brake power per cylinder

$$= \frac{18.6}{4} = 4.65 \text{ kW}$$

∴ Brake power for three cylinders (*i.e.* when one cylinder is cut-out)

$$= 4.65 \times 3 = 13.95 \text{ kW}$$

Since the average torque (T) for three cylinders (*i.e.* when one cylinder is cut-out) is given as $T = 105$ N-m, therefore

Brake power for the three cylinders,

$$= \frac{T \times 2\,\pi\,N}{60} = \frac{105 \times 2\,\pi \times 1200}{60} = 13\ 200 \text{ W} = 13.2 \text{ kW}$$

and frictional power per cylinder

$$= 13.95 - 13.2 = 0.75 \text{ kW}$$

∴ Total frictional power for four cylinders,

$$\text{F.P.} = 0.75 \times 4 = 3 \text{ kW}$$

We know that indicated power,

$$\text{I.P.} = \text{B.P.} + \text{F.P.} = 18.6 + 3 = 21.6 \text{ kW}$$

∴ Indicated thermal efficiency,

$$\eta_i = \frac{\text{I.P.} \times 3600}{m_f \times C} = \frac{21.6 \times 3600}{6.324 \times 42\ 000} = 0.293 \quad \text{or} \quad 29.3\% \text{ **Ans.**}$$

Example 27.10. *A four stroke, six cylinder gas engine with a stroke volume of 1.75 litres develops 25 kW at 480 r.p.m.. The mean effective pressure is 6 bar. Find the average number of times each cylinder misfired in one minute.*

Solution. Given $K = 6$; $v_s = LA = 1.75 \text{ litres} = 1.75 \times 10^{-3} \text{ m}^3$; I.P = 25 kW ; $N = 480$ r.p.m ; $p_m = 6$ bar

Let n_1 = Number of working strokes produced per minute.

We know that indicated power (I.P.),

$$25 = \frac{100\,K\,p_m\,L\,A\,n_1}{60} = \frac{100 \times 6 \times 6 \times 1.75 \times 10^{-3} \times n_1}{60} = 0.105\,n_1$$

$$\therefore \quad n_1 = 238$$

But the number of working strokes per minute are given as

$$n = N/2 = 480/2 = 240 \quad \text{...(∵ Engine works on four stroke cycle)}$$

∴ Number of times each cylinder misfired in one minute

$$= n - n_1 = 240 - 238 = 2 \text{ **Ans.**}$$

Example 27.11. *The diameter and stroke length of a single cylinder two stroke gas engine, working on the constant volume cycle, are 200 mm and 300 mm respectively with clearance volume 2.78 litres.*

When the engine is running at 135 r.p.m., the indicated mean effective pressure was 5.2 bar and the gas consumption 8.8 m³/hour. If the calorific value of the gas used is 16 350 kJ/m³, find 1. air standard efficiency ; 2. indicated power developed by the engine ; and 3. indicated thermal efficiency of the engine.

Solution. Given : $D_c = 200$ mm $= 0.2$ m ; $L = 300$ mm $= 0.3$ m ; $v_c = 2.78$ litres $= 0.002\,78$ m^3 ; $N = 135$ r.p.m.; $p_m = 5.2$ bar ; $m_f = 8.8$ m^3/ h ; $C = 16\,350$ kJ/m^3

1. *Air standard efficiency*

We know that area of the cylinder,

$$A = \frac{\pi}{4}(0.2)^2 = 0.031\,42 \text{ m}^2$$

$\therefore$ Stroke volume, $v_s = A\,L = 0.031\,42 \times 0.3 = 0.009\,426 \text{ m}^3$

and compression ratio,
$$r = \frac{\text{Total volume of cylinder}}{\text{Clearance volume}} = \frac{v_c + v_s}{v_c}$$

$$= \frac{0.002\,78 + 0.009\,426}{0.002\,78} = 4.4$$

We know that air standard efficiency,

$$\eta_{ase} = 1 - \frac{1}{r^{\gamma - 1}} = 1 - \frac{1}{(4.4)^{1.4-1}} = 1 - 0.553 = 0.447 \text{ or } 44.7\% \textbf{ Ans.}$$

2. *Indicated power developed by the engine*

We know that indicated power developed by the engine,

$$\text{I.P.} = \frac{100\, p_m\, L\, A\, n}{60} = \frac{100 \times 5.2 \times 0.3 \times 0.031\,42 \times 135}{60} \text{ kW}$$

... ($\because n = N$, for two stroke cycle engine)

$$= 11.03 \text{ kW } \textbf{Ans.}$$

3. *Indicated thermal efficiency*

We know that indicated thermal efficiency,

$$\eta_i = \frac{\text{I.P.} \times 3600}{m_f \times C} = \frac{11.03 \times 3600}{8.8 \times 16\,350} = 0.276 \text{ or } 27.6\% \textbf{ Ans.}$$

Example 27.12. *A four stroke petrol engine 80 mm bore, 100 mm stroke, is tested at full throttle at constant speed. The fuel supply is fixed at 0.068 kg/min and the plugs of the four cylinders are successively short circuited without change of speed, brake torque being correspondingly adjusted. The brake power measurements are the following :*

With all cylinders firing	=	*12.5 kW*
With cylinder No. 1 cut off	=	*9 kW*
With cylinder No. 2 cut off	=	*9.15 kW*
With cylinder No. 3 cut off	=	*9.2 kW*
With cylinder No. 4 cut off	=	*9.1 kW*

Determine I.P. of the engine under these conditions. Also determine the indicated thermal efficiency. Calorific value of the fuel is 44 100 kJ/kg. Compare this efficiency with the air standard value. Clearance volume of one cylinder is 70×10^3 mm^3.

Solution. Given : D_c = 80 mm = 0.08 m ; L = 100 mm = 0.1 m ; m_f = 0.068 kg/min = 4.08 kg/h ; B = 12.5 kW ; B_1 = 9 kW ; B_2 = 9.15 kW ; B_3 = 9.2 kW ; B_4 = 9.1 kW ; C = 44100 kJ/kg ; $v_c = 70 \times 10^3$ mm^3

I.P. of the engine

We know that indicated power produced in cylinder 1,

$$I_1 = B - B_1 = 12.5 - 9 = 3.5 \text{ kW}$$

Indicated power produced in cylinder 2,

$$I_2 = B - B_2 = 12.5 - 9.15 = 3.35 \text{ kW}$$

Indicated power produced in cylinder 3,

$$I_3 = B - B_3 = 12.5 - 9.2 = 3.3 \text{ kW}$$

and indicated power produced in cylinder 4,

$$I_4 = B - B_4 = 12.5 - 9.1 = 3.4 \text{ kW}$$

$\therefore$ Indicated power of the engine,

$$\text{I.P.} = I_1 + I_2 + I_3 + I_4 = 3.5 + 3.35 + 3.3 + 3.4 = 13.55 \text{ kW} \textbf{ Ans.}$$

Indicated thermal efficiency

We know that indicated thermal efficiency,

$$\eta_i = \frac{\text{I.P.} \times 3600}{m_f \times C} = \frac{13.55 \times 3600}{4.08 \times 44\,100} = 0.271 \text{ or } 27.1\% \textbf{ Ans.}$$

Air standard efficiency

We know that swept volume

$$v_s = \frac{\pi}{4}(D_c)^2 L = \frac{\pi}{4}(80)^2\, 100 = 503 \times 10^3 \text{ mm}^3$$

$\therefore$ Compression ratio,

$$r = \frac{\text{Total cylinder volume}}{\text{Clearance volume}} = \frac{v_c + v_s}{v_c}$$

$$= \frac{70 \times 10^3 + 503 \times 10^3}{70 \times 10^3} = 8.18$$

We know that air standard efficiency,

$$\eta_{ase} = 1 - \frac{1}{r^{\gamma - 1}} = 1 - \frac{1}{(8.18)^{1.4-1}} \quad \text{. . .(Taking } \gamma = 1.4 \text{ for air)}$$

$$= 1 - 0.431 = 0.569 \text{ or } 56.9\% \textbf{ Ans.}$$

Ratio of air standard efficiency to indicated thermal efficiency

$$= \frac{0.569}{0.271} = 2.1$$

Thus air standard efficiency is 2.1 times the indicated thermal efficiency. **Ans.**

Example 27.13. *A four stroke diesel engine has a cylinder bore of 150 mm and a stroke of 250 mm. The crankshaft speed is 300 r.p.m. and fuel consumption is 1.2 kg/h, having a calorific value of 39 900 kJ/kg. The indicated mean effective pressure is 5.5 bar. If the compression ratio is 15 and cut-off ratio is 1.8, calculate the relative efficiency, taking $\gamma = 1.4$.*

Solution. Given : $D_c = 150$ mm $= 0.15$ m; $L = 250$ mm $= 0.25$ m ; $N = 300$ r.p.m. ; $m_f = 1.2$ kg/h ; $C = 39\,900$ kJ/kg ; $p_m = 5.5$ bar ; $r = 15$; $\rho = 1.8$; $\gamma = 1.4$

We know that area of the cylinder,

$$A = \frac{\pi}{4}(0.15)^2 = 0.0177 \text{ m}^2$$

and number of working stroke per minute,

$$n = N/2 = 300/2 = 150 \quad \ldots (\because \text{Engine works on four stroke cycle})$$

$$\therefore \text{ Indicated power, } \quad \text{I.P.} = \frac{100\, p_m\, L\, A\, n}{60}$$

$$= \frac{100 \times 5.5 \times 0.25 \times 0.0177 \times 150}{60} = 6.1 \text{ kW}$$

and indicated thermal efficiency,

$$\eta_i = \frac{\text{I.P.} \times 3600}{m_f \times C} = \frac{6.1 \times 3600}{1.2 \times 39\,900} = 0.4586 \text{ or } 45.86\%$$

We know that air standard efficiency for diesel engine,

$$\eta_{ase} = 1 - \frac{1}{r^{\gamma - 1}}\left[\frac{\rho^{\gamma} - 1}{\gamma(\rho - 1)}\right]$$

$$= 1 - \frac{1}{(15)^{1.4-1}}\left[\frac{(1.8)^{1.4} - 1}{1.4\,(1.8 - 1)}\right] = 1 - 0.386 = 0.614 \text{ or } 61.4\%$$

$$\therefore \text{ Relative efficiency, } \eta_r = \frac{\text{Indicated thermal efficiency}}{\text{Air standard efficiency}} = \frac{0.4586}{0.614}$$

$$= 0.747 \text{ or } 74.7\% \textbf{ Ans.}$$

Example 27.14. *A petrol engine has a cylinder diameter of 60 mm and stroke 100 mm. If the mass of the charge admitted per cycle is 0.0002 kg, find the volumetric efficiency of the engine. Assume characteristic constant for the charge as 287 J/kg K.*

Solution. Given : $D_c = 60$ mm $= 0.06$ m ; $L = 100$ mm $= 0.1$ m ; $m = 0.0002$ kg

We know that swept volume of the piston,

$$v_s = \frac{\pi}{4}(0.06)^2\, 0.1 = 0.283 \times 10^{-3} \text{ m}^3$$

Let $\quad v_a$ = Volume of charge admitted at *N.T.P.

We know that according to characteristic gas equations,

$$v_a = \frac{m\,R\,T}{p} = \frac{0.0002 \times 287 \times 273}{1.013 \times 10^5} \text{ m}^3 \quad \ldots (\because pv = mRT)$$

$$= 0.155 \times 10^{-3} \text{ m}^3$$

* N.T.P means normal temperature and pressure, *i.e.* temperature (T) of 0° C or 273 K and pressure 1.013 bar or 1.013×10^5 N/m².

∴ Volumetric efficiency,

$$\eta_v = \frac{v_a}{v_s} = \frac{0.155 \times 10^{-3}}{0.283 \times 10^{-3}} = 0.548 \text{ or } 54.8\% \textbf{ Ans.}$$

Example 27.15. *Find the engine dimensions of a two cylinder, two stroke I. C. engine from the following data :*

Engine speed = 4000 r.p.m. ; Volumetric efficiency = 0.77 ; Mechanical efficiency = 0.75 ; Fuel consumption = 10 litres/h (specific gravity = 0.73) ; Air-fuel ratio = 18:1 ; Piston speed = 600 m/min ; Indicated mean effective pressure = 5 bar.

Find also the brake power. Take R for gas mixture as 281 J/kg K at S.T.P.

Solution. Given : $K = 2$; $N = 4000$ r.p.m. ; $\eta_v = 0.77$; $\eta_m = 0.75$; $m_f = 10 \times 0.73 = 7.3$ kg / h ; $m_a / m_f = 18$; $2LN = 600$ m/min ; $p_m = 5$ bar ; $R = 281$ J/kg K

Engine dimensions

Let D_c = Diameter of the cylinder, and

L = Length of the stroke.

We know that piston speed,

$$2LN = 600$$

or $$L = 600/2N = 600/2 \times 4000 = 0.075 \text{ m or } 75 \text{ mm } \textbf{Ans.}$$

Mass of air required $m_a = m_f \times 18 = 7.3 \times 18 = 131.4$ kg / h ... ($\because m_a/m_f = 18$)

and corresponding volume of air required at *S.T.P.,

$$v_a = \frac{m_a \, R\,T}{p} = \frac{131.4 \times 281 \times 288}{1.013 \times 10^5} = 105 \text{ m}^3\text{/h} = 1.75 \text{ m}^3\text{/min}$$

We know that swept volume of the piston per minute,

$$v_s = \frac{\pi}{4}(D_c)^2 L \times n \times K$$

$$= \frac{\pi}{4}(D_c)^2\, 0.075 \times 4000 \times 2 = 471.3\,(D_c)^2 \text{ m}^3\text{/min}$$

... ($\because n = N$, for two stroke cycle engine)

and volumetric efficiency (η_v),

$$0.77 = \frac{v_a}{v_s} = \frac{1.75}{4.71.3\,(D_c)^2}$$

$$\therefore \quad (D_c)^2 = 4.82 \times 10^{-3} \text{ or } D_c = 0.0694 \text{ m or } 69.4 \text{ mm } \textbf{Ans.}$$

Brake power

We know that area of the cylinder,

$$A = \frac{\pi}{4}(D_c)^2 = \frac{\pi}{4}(0.0694)^2 = 0.003\,78 \text{ m}^2$$

---

* S.T.P. means standard temperature and pressure, *i.e.* temperature (T) of 15° C or 288 K and pressure 1.013 bar or 1.013×10^5 N/m².

and indicated power, $$\text{I.P.} = \frac{100\,K\,p_m\,L\,A\,n}{60} = \frac{100 \times 2 \times 5 \times 0.075 \times 0.003\,78 \times 4000}{60}\ \text{kW}$$

$$= 18.9\ \text{kW}$$

$\therefore$ Brake power, $$\text{B.P.} = \text{I.P.} \times \eta_m = 18.9 \times 0.75 = 14.17\ \text{kW}\ \textbf{Ans.}$$

Example 27.16. *A four stroke petrol engine with a compression ratio of 6.5 to 1 and total piston displacement of 5.2×10^{-3} m³ develops 100 kW brake power and consumes 33 kg of petrol per hour of calorific value 44 300 kJ/kg at 3000 r.p.m.*

Find : 1. Brake mean effective pressure ; 2. Brake thermal efficiency ; 3. Air standard efficiency ($\gamma = 1.4$) ; and 4. Air-fuel ratio by mass.

Assume a volumetric efficiency of 80 %. One kg of petrol vapour occupies 0.26 m³ at 1.013 bar and 15° C. Take R for air 287 J/kg K.

Solution. Given : $r = 6.5$; $v_s = LA = 5.2 \times 10^{-3}$ m³ ; B.P. = 100 kW ; $m_f = 33$ kg/h ; $C = 44\,300$ kJ/kg ; $N = 3000$ r.p.m. ; $\gamma = 1.4$; $\eta_v = 80\% = 0.8$

1. *Brake mean effective pressure*

Let p_{mb} = Brake mean effective pressure in bar.

We know that brake power (B.P.),

$$100 = \frac{100\,p_{mb}\,L\,A\,n}{60} = \frac{100 \times p_{mb} \times 5.2 \times 10^{-3} \times 1500}{60} = 13\,p_{mb}$$

. . . ($\because$ $n = N/2$, for four stroke engine)

$$\therefore \quad p_{mb} = 7.7\ \text{bar}\ \textbf{Ans.}$$

2. *Brake thermal efficiency*

We know that brake thermal efficiency,

$$\eta_b = \frac{\text{B.P.} \times 3600}{m_f \times C} = \frac{100 \times 3600}{33 \times 44\,300} = 0.246 \quad \text{or} \quad 24.6\%\ \textbf{Ans.}$$

3. *Air standard efficiency*

We know that air standard efficiency,

$$\eta_{ase} = 1 - \frac{1}{r^{\gamma - 1}} = 1 - \frac{1}{(6.5)^{1.4-1}} = 1 - 0.473$$

$$= 0.527 \quad \text{or} \quad 52.7\%\ \textbf{Ans.}$$

4. *Air-fuel ratio by mass*

We know that actual volume of charge admitted during the suction stroke at N. T.P.,

$$v_a = v_s \times \eta_v = 5.2 \times 10^{-3} \times 0.8 = 4.16 \times 10^{-3}\ \text{m}^3 \quad \text{. . . (i)}$$

Let v_1 = Specific volume of petrol (*i.e.* volume of 1 kg of petrol) at N.T.P. conditions, *i.e.* at T_1 = 0° C or 273 K and p_1 = 1.013 bar, and

v_2 = Specific volume of petrol (*i.e.* volume of 1 kg of petrol) at T_2 = 15° C or 288 K and p_2 = 1.013 bar

= 0.26 m³ . . . (Given)

We know that $$\frac{p_1\,v_1}{T_1} = \frac{p_2\,v_2}{T_2}$$

$$\therefore \quad v_1 = \frac{p_2 v_2 T_1}{p_1 T_2} = \frac{1.013 \times 0.26 \times 273}{1.013 \times 288} = 0.246 \text{ m}^3/\text{ kg of petrol}$$

Mass of petrol consumed per cycle,

$$m_1 = \frac{\text{Mass of petrol consumed per minute}}{\text{Number of working cycles per minute}}$$

$$= \frac{33/60}{3000/2} = 0.367 \times 10^{-3} \text{kg}$$

$\therefore$ Total volume of petrol at N.T.P.

$$= 0.246 \times 0.367 \times 10^{-3} = 0.0903 \times 10^{-3} \text{ m}^3 \quad \ldots (ii)$$

We know that specific volume of air (*i.e.* volume of 1 kg of air) at N.T.P. (*i.e.* at $T = 0°$ C or 273 K and $p = 1.013$ bar or 1.013×10^5 N/m²),

$$= \frac{m\,R\,T}{p} = \frac{1 \times 287 \times 273}{1.013 \times 10^5} = 0.773 \text{ m}^3/\text{kg of air}$$

Let m kg of air is admitted per cycle, then total volume of air admitted,

$$v = 0.773 \times m \text{ m}^3 \quad \ldots (iii)$$

We know that volume of charge admitted per cycle at N.T.P.,

v_a = Volume of petrol per cycle at N.T.P. + Volume of air per cycle at N.T.P.

$$4.16 \times 10^{-3} = 0.0903 \times 10^{-3} + 0.773 \times m$$

or

$$m = 5.265 \times 10^{-3} \text{ kg}$$

$$\therefore \text{ Air-fuel ratio,} \quad = \frac{m}{m_1} = \frac{5.265 \times 10^{-3}}{0.367 \times 10^{-3}} = 14.35 \text{ Ans.}$$

27.8. Air Consumption

The supply of air to an I.C. engine may be measured experimentally by passing the air through a sharp edged orifice into a large tank (the volume being 500 times the swept volume of the engine). The air is then passed to the engine. It is a cheap and simple method of estimating the air supply to an engine.

The air is drawn into a large tank through an orifice whose diameter and coefficient of discharge are known. The engine now draws air from this tank as shown in Fig. 27.2. The pressure of air in the tank is less than the atmospheric pressure due to the powerful engine suction. Since the tank is relatively large, the air pressure may be assumed to remain constant.

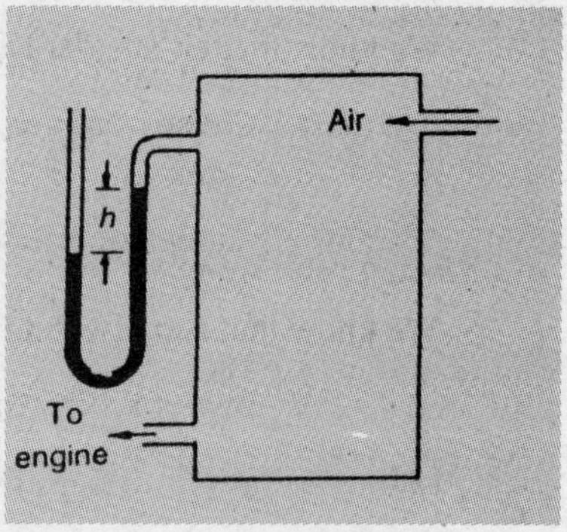

Fig. 27.2. Air consumption.

The outside air is assumed to flow continuously, through the orifice, with a constant velocity. This velocity depends upon the difference of pressure between the air in the tank and the atmospheric air. This pressure difference is measured by a U-tube containing water, whose one limb is connected to the inside of the chamber while the other end is open to the atmosphere. The temperature of atmosphere and barometer reading is also taken.

Let a = Area of orifice in m²,

C_d = Coefficient of discharge of orifice,

H = Head causing flow of air through orifice in metres of air,

ρ_a = Density of air under atmospheric conditions in kg/m^3,

ρ_w = Density of water in kg/m^3 = 1000 kg/m^3,

h = Pressure difference measured in U-tube in metres of water.

Now head causing flow of air through orifice,

$$H = h \times \frac{\rho_w}{\rho_a} = \frac{1000\,h}{\rho_a} \text{ metres of air} \quad \ldots (i)$$

We know that velocity of air,

$$V = \sqrt{2gH} \text{ m/s}$$

∴ Quantity or volume of air passing through the orifice,

$$Q = C_d\, a\, V = C_d\, a\sqrt{2gH} \text{ m}^3\text{/s} \quad \ldots (ii)$$

We know that mass of air supplied,

$$m_a = \text{Volume} \times \text{Density} = v_a\, \rho_a = Q\, \rho_a$$

$$= C_d\, a\sqrt{2gH} \times \rho_a$$

$$= C_d\, a\sqrt{2 \times 9.81 \times \frac{1000\,h}{\rho_a}} \times \rho_a$$

$$= 140\, C_d\, a\sqrt{h\,\rho_a} \text{ kg/s}$$

Note : The density of air (ρ_a) at pressure p bar and absolute temperature T K may be obtained by applying the characteristic equation of a gas, *i.e.*

$$p = \rho_a RT \quad \text{or} \quad \rho_a = p/RT$$

Example 27.17. *Following readings were obtained during the test on a single cylinder, 4-stroke I.C. engine :*

Engine speed = 300 r.p.m. ; Diameter of orifice of the air tank = 20 mm ; Pressure causing air flow through the orifice = 100 mm of water.

Find the quantity of air consumed per second, if its density under atmospheric conditions is 1.15 kg/m^3. Take coefficient of discharge for the orifice as 0.7.

Solution. Given : *N = 300 r.p.m. ; d = 20 mm = 0.02 m ; h = 100 mm of water = 0.1 m of water ; ρ_a = 1.15 kg/m^3 ; C_d = 0.7

We know that area of orifice,

$$a = \frac{\pi}{4}(0.02)^2 = 0.3142 \times 10^{-3} \text{ m}^2$$

and head causing flow of air through orifice,

$$H = h \times \frac{\rho_w}{\rho_a} = 0.1 \times \frac{1000}{1.15} = 86.96 \text{ m of air} \quad \ldots (\because \rho_w = 1000 \text{ kg/m}^3)$$

∴ Quantity of air flow,

$$Q = C_d\, a\sqrt{2gH} = 0.7 \times 0.3142 \times 10^{-3}\sqrt{2 \times 9.81 \times 86.96} \text{ m}^3\text{/s}$$

$$= 0.0091 \text{ m}^3\text{/s} \quad \textbf{Ans.}$$

* *Superfluous data*

27.9. Heat Balance Sheet

The complete record of heat supplied and heat rejected during a certain time (say one minute) by an I.C. engine is entered in a tabulated form known as heat balance sheet. The following values are required to complete the heat balance sheet of an I.C. engine :

1. *Heat supplied by the fuel*

Let m_f = Mass of fuel supplied in kg/min, and

C = Lower calorific value of the fuel in kJ/kg.

We know that the heat in fuel supplied

$$= m_f \times C \text{ kJ/min} \qquad \ldots (i)$$

Note : In case of a gas engine, the volume of the gas supplied is first converted to N.T.P. conditions. It is then multiplied by its lower calorific value to get the heat supplied by the fuel.

Let v_o = Volume of gas supplied in m³/min at N.T.P., and

C = Lower calorific value of the gas in kJ/m³ at N.T.P.

∴ Heat supplied by fuel $= v_o \times C$ kJ/min

2. *Heat absorbed in I.P. produced*

We know that the indicated power produced by I.C. engine,

$$\text{I.P.} = \frac{100\, p_m\, L\, A\, n}{60} \text{ kW} \qquad \ldots \text{(For single cylinder engine)}$$

$$\therefore \text{Heat absorbed in I.P./min} = 100\, p_m\, L\, A\, n \text{ kJ/min} \qquad \ldots (\because 1 \text{ kW} = 1 \text{ kJ/s}) \quad \ldots (ii)$$

3. *Heat rejected to the cooling water*

The mass of cooling water, circulating through the cylinder jackets, as well as its inlet and outlet temperatures are measured in order to determine the heat rejected to the cooling water.

Let m_w = Mass of cooling water supplied in kg/min,

c_w = Specific heat of water which may be taken as 4.2 kJ/kg K,

t_1 = Inlet temperature, and

t_2 = Outlet temperature.

Then heat rejected to cooling water

$$= m_w\, c_w\, (t_1 - t_2) \text{ kJ/min} \qquad \ldots (iii)$$

4. *Heat carried away by exhaust gases*

The mass of exhaust gases may be obtained by adding together the mass of fuel supplied and the mass of air supplied. The mass of air supplied may be measured by an orifice or it may be calculated from the analysis of the exhaust gases. The temperature of the exhaust gases is also measured.

Let m_g = Mass of exhaust gases produced in kg/min,

c_g = Specific heat of exhaust gases, and

t = Rise in temperature.

∴ Heat carried away by exhaust gases

$$= m_g\, c_g\, t \text{ kJ/min} \qquad \ldots (iv)$$

5. *Unaccounted heat*

There is always some loss of heat due to friction, leakage, radiation etc. which cannot be determined experimentally. In order to complete the heat balance sheet, this loss is obtained by the difference of heat supplied by the fuel and heat absorbed in I.P., cooling water and exhaust gases.

Finally, the heat balance sheet is prepared as given below :

S.No.	*Particulars*	*Heat in*	
		kJ	*%*
	Total heat supplied	...	100
1.	Heat absorbed in I.P.	...	...
2.	Heat rejected to the cooling water	...	...
3.	Heat carried away by exhaust gases	...	...
4.	Unaccounted heat	...	...

Example 27.18. *An I.C. engine uses 6 kg of fuel having calorific value 44 000 kJ/kg in one hour. The I.P. developed is 18 kW. The temperature of 11.5 kg of cooling water was found to rise through 25° C per minute. The temperature of 4.2 kg of exhaust gas with specific heat 1 kJ/kg K was found to rise through 220° C. Draw the heat balance sheet for the engine.*

Solution. Given : $m_f = 6$ kg/h = 0.1 kg/min ; $C = 44\,000$ kJ/kg ; I.P. = 18 kW ; $m_w = 1.5$ kg/min ; $t_2 - t_1 = 25°$ C ; $m_g = 4.2$ kg ; $c_g = 1$ kJ/kg K ; $t = 220°$ C

We know that heat supplied by the fuel

$$= m_f \times C = 0.1 \times 44\,000 = 4400 \text{ kJ/min}$$

Heat absorbed in I.P. produced $= 18 \text{ kW} = 18 \text{ kJ/s} = 1080 \text{ kJ/min}$

Heat rejected to cooling water $= m_w\, c_w\, (t_2 - t_1) = 11.5 \times 4.2 \times 25 = 1207.5 \text{ kJ/min}$

Heat lost to exhaust gases $= m_g\, c_g\, t = 4.2 \times 1 \times 220 = 924 \text{ kJ/min}$

and unaccounted heat $= 4400 - (1080 + 1207.5 + 924) = 1188.5 \text{ kJ/min}$

Now prepare the heat balance sheet as given below :

S.No.	*Particulars*	*Heat in*	
		kJ	*%*
	Total heat supplied	4400	100
1.	Heat absorbed in I.P.	1080	24.55
2.	Heat rejected to cooling water	1207.5	27.44
3.	Heat carried away by exhaust gases	924	21.00
4.	Unaccounted heat	1188.5	27.01
	Total	4400	100

Example 27.19. *A gas engine, working on four stroke constant volume cycle, gave the following results when loaded by friction brake during a test of an hour's duration :*

Cylinder diameter 240 mm ; Stroke length 480 mm ; Clearance volume 4450×10^{-6} m³ ; Effective circumference of the brake wheel 3.86 m ; Net load on brake 1260 N at overall speed of 226.7 r.p.m. ; Average explosions/min 77 ; m.e.p. of indicator card 7.5 bar ; Gas used 13 m³/h at 15° C and 771 mm of Hg ; Lower calorific value of gas 49 350 kJ/m³ at N.T.P. ; Cooling jacket water

660 kg raised to 34.2° C ; Heat lost to exhaust gases 8%. Calculate : 1. I.P. ; 2. B.P. ; 3. Indicated thermal efficiency ; and 4. Efficiency ratio. Also draw a heat balance sheet for the engine.

Solution. Given : $D_c = 240$ mm = 0.24 m ; $L = 480$ mm = 0.48 m ; $v_c = 4450 \times 10^{-6} \text{m}^3$; $\pi D = 3.86$ m ; $(W - S) = 1260$ N ; $N = 226.7$ r.p.m. ; $n = 77$; $p_m = 7.5$ bar ; $v_1 = 13 \text{ m}^3\text{/h}$; $T_1 = 15° \text{C} = 288$ K ; $p_1 = 771$ mm of Hg ; $C = 49\ 350 \text{ kJ/m}^3$; $m_w = 660$ kg/h = 11 kg/min ; $t_2 - t_1 = 34.2°$ C ; Heat lost to exhaust gases = 8%

1. *I.P.*

We know that area of the cylinder,

$$A = \frac{\pi}{4}(D_c)^2 = \frac{\pi}{4}(0.24)^2 = 0.045 \text{ m}^2$$

$$\therefore \quad \text{I.P.} = \frac{100\, p_m\, L\, A\, n}{60} = \frac{100 \times 7.5 \times 0.48 \times 0.045 \times 77}{60} = 20.8 \text{ kW} \quad \textbf{Ans.}$$

2. *B.P.*

$$\text{We know that B.P.} = \frac{(W - S)\,\pi DN}{60} = \frac{1260 \times 3.86 \times 226.7}{60} = 18\ 400 \text{ W}$$

$$= 18.4 \text{ kW} \quad \textbf{Ans.}$$

3. *Indicated thermal efficiency*

First of all, let us find the volume of gas at N.T.P. (*i.e.* temperature 0° C and pressure 760 mm of Hg).

Let v_0 = Volume of gas at N.T.P.

T_0 = Absolute temperature of gas = 273 K

p_0 = Pressure at N.T.P. = 760 mm of Hg

According to the gas equation,

$$\frac{p_0 v_0}{T_0} = \frac{p_1 v_1}{T_1}$$

or

$$v_0 = \frac{p_1 v_1}{T_1} \times \frac{T_0}{p_0} = \frac{771 \times 13}{288} \times \frac{273}{760} = 12.5 \text{ m}^3$$

$\therefore$ Indicated thermal efficiency,

$$\eta_i = \frac{\text{I.P.} \times 3600}{v_0 \times C} = \frac{20.8 \times 3600}{12.5 \times 49\ 350} = 0.121 \quad \text{or} \quad 12.1\% \quad \textbf{Ans.}$$

4. *Efficiency ratio**

First of all, let us find out the air standard efficiency (η_{ase}). We know that swept volume,

$$v_s = \frac{\pi}{4}(D_c)^2 L = \frac{\pi}{4}(0.24)^2\, 0.48 = 0.0217 \text{ m}^3$$

and clearance volume, $v_c = 4450 \times 10^{-6} \text{ m}^3$

$\therefore$ Total volume $= v_s + v_c = 0.0217 + 4450 \times 10^{-6} = 0.026\ 15 \text{ m}^3$

---

* It is also known as relative efficiency.

and compression ratio, $r = \dfrac{\text{Total volume}}{\text{Clearance volume}} = \dfrac{0.026\ 15}{4450 \times 10^{-6}} = 5.876$

We know that air standard efficiency,

$$\eta_{ase} = 1 - \frac{1}{r^{\gamma-1}} = 1 - \frac{1}{(5.876)^{1.4-1}} = 1 - 0.492 = 0.518$$

... (For air, $\gamma = 1.4$)

$$\therefore \quad \text{Efficiency ratio} = \frac{\text{Indicated thermal efficiency}}{\text{Air standard efficiency}} = \frac{0.121}{0.518} = 0.234 \quad \text{or} \quad 23.4\% \ \textbf{Ans.}$$

Heat balance sheet

We know that heat supplied by the fuel

$= \dot{v}_0 \times C$ ... (v_0 = Vol. of gas used for one hour at N.T.P.)

$= \dfrac{12.5 \times 49\ 350}{60} = 10\ 280$ kJ/min

Heat absorbed in I.P. produced

$= 20.8$ kW $= 20.8$ kJ/s $= 1248$ kJ/min

Heat rejected to cooling water

$= m_w\, c_w\, (t_2 - t_1) = 11 \times 4.2 \times 34.2 = 1580$ kJ/min

Heat lost to exhaust gases (8% given)

$= 0.08 \times 10\ 280 = 822.4$ kJ/min

and unaccounted heat $= 10\ 280 - (1248 + 1580 + 822.4) = 6629.6$ kJ/min

Now prepare the heat balance sheet as given below :

S.No.	*Particulars*	*Heat in*	
		kJ	*%*
	Total heat supplied	10 280	100
1.	Heat absorbed in I.P. produced	1248	12.14
2.	Heat rejected to cooling water	1580	15.36
3.	Heat lost to exhaust gases	822.4	8.00
4.	Unaccounted heat	6629.6	64.50
	Total	10 280	100

Example 27.20. *A six cylinder, four stroke diesel engine has bore 360 mm and stroke 500 mm. A trial on the engine provided the following data :*

Mean area of indicator diagram = 780 mm² ; Length of the indicator diagram = 75 mm ; Spring number = 0.7 bar per mm of compression ; Brake torque = 14 000 N-m ; Speed = 500 r.p.m. ; Fuel consumption = 240 kg/h ; Calorific value of fuel oil = 44 000 kJ/kg ; Jacket cooling water = 320 kg/min ; Rise in temperature of cooling water = 40° C ; Piston cooling oil = 140 kg/min ; Specific heat = 2.1 kJ/kg K ; Temperature rise of oil = 28° C ; Circulating water in gas calorimeter = 300 kg/min ; Rise in temperature of this water = 42° C.

All heat of the exhaust gases is absorbed in the calorimeter. Estimate the specific fuel consumption and mechanical efficiency of the engine. Draw up a heat balance sheet of the engine on 1 kg of fuel oil basis.

Solution. Given : $K = 6$; $D_c = 360$ mm $= 0.36$ m ; $L = 500$ mm $= 0.5$ m ; $a = 780$ mm^2 ; $l = 75$ mm ; $s = 0.7$ bar/mm ; $T = 14\ 000$ N-m ; $N = 500$ r.p.m. ; $m_f = 240$ kg / h $= 4$ kg / min ; $C = 44\ 000$ kJ/kg ; $m_w = 320$ kg/min ; $t_2 - t_1 = 40°$ C ; $m_o = 140$ kg/min ; $c_o = 2.1$ kJ/kg K ; $t_o = 28°$ C ; $m_c = 300$ kg/min ; $t_c = 42°$ C

Specific fuel consumption

We know that indicated mean effective pressure,

$$p_m = \frac{a \times s}{l} = \frac{780 \times 0.7}{75} = 7.28 \text{ bar}$$

Area of cylinder, $A = \frac{\pi}{4}(D_c)^2 = \frac{\pi}{4}(0.36)^2 = 0.102 \text{ m}^2$

and number of working strokes per minute,

$$n = N/2 = 500/2 = 250 \quad \text{... (}\because \text{ Engine works on four stroke cycle)}$$

We know that indicated power,

$$\text{I.P.} = \frac{100\, K\, p_m\, L\, A\, n}{60} = \frac{100 \times 6 \times 7.28 \times 0.5 \times 0.102 \times 250}{60} \text{ kW}$$

$$= 928 \text{ kW}$$

$\therefore$ Specific fuel consumption

$$= \frac{m_f}{\text{I.P.}} = \frac{240}{928} = 0.258 \text{ kg / kWh} \textbf{ Ans.}$$

Mechanical efficiency

We know that brake power,

$$\text{B.P.} = \frac{T \times 2\pi N}{60} = \frac{14\ 000 \times 2\pi \times 500}{60} = 733\ 000 \text{ W} = 733 \text{ kW}$$

and mechanical efficiency, $\eta_m = \frac{\text{B.P.}}{\text{I.P.}} = \frac{733}{928} = 0.79$ or *79%* **Ans.**

Heat balance sheet for 1 kg of fuel oil

We know that heat supplied by the fuel

$$= 44\ 000 \text{ kJ/kg of fuel}$$

Since the fuel consumption is 240 kg / h, therefore time for 1 kg of fuel consumption

$$= \frac{1}{240} \text{ h} = \frac{3600}{240} = 15 \text{ s}$$

Heat absorbed in I.P. produced

$$= 928 \text{ kW} = 928 \text{ kJ/s}$$

$$= 928 \times 15 = 13\ 920 \text{ kJ/kg of fuel}$$

... ($\because$ 1 kg of fuel consumption takes 15 s)

Heat rejected to cooling water

$$= m_w \times c_w (t_2 - t_1) = 320 \times 4.2 \times 40 = 53\ 760 \text{ kJ/min} = 896 \text{ kJ/s}$$

$$= 896 \times 15 = 13\ 440 \text{ kJ/kg of fuel}$$

Heat lost to piston cooling oil

$$= m_o\, c_o\, t_o = 140 \times 2.1 \times 28 = 8232 \text{ kJ/min} = 137.2 \text{ kJ/s}$$

$$= 137.2 \times 15 = 2058 \text{ kJ/kg of fuel}$$

Heat lost to water in calorimeter

$$= m_c\, c_c\, t_c = 300 \times 4.2 \times 42 = 52\,920 \text{ kJ/min} = 882 \text{ kJ/s}$$

$$= 882 \times 15 = 13\,230 \text{ kJ/kg of fuel}$$

Unaccounted heat $= 44\,000 - (13\,920 + 13\,440 + 2058 + 13\,230) = 1352$ kJ/kg of fuel

Now prepare the heat balance sheel on 1 kg of fuel basis, as given below :

S.No.	*Particulars*	*Heat in*	
		kJ	*%*
	Total heat supplied	44 000	100
1.	Heat absorbed in I.P. produced	13 920	31.64
2.	Heat rejected to cooling water	13 440	30.54
3.	Heat lost to piston cooling oil	2058	4.68
4.	Heat lost to water in calorimeter	13 230	30.07
5.	Unaccounted heat (by difference)	1352	3.07
	Total	44 000	100

EXERCISES

1. The following data were recorded during testing of a four stroke cycle gas engine :

Area of indicator diagram = 900 mm^2 ; Length of indicator diagram = 70 mm ; Spring scale = 0.3 bar/mm ; Diameter of piston = 200 mm ; Length of stroke = 250 mm ; Speed = 300 r.p.m. Determine :

1. Indicated mean effective pressure ; and 2. Indicated power. [**Ans.** 3.86 bar ; 7.58 kW]

2. A two stroke cycle internal combustion engine has a mean effective pressure of 6 bar. The speed of the engine is 1000 r.p.m, If the diameter of piston and stroke are 110 mm and 140 mm respectively, find the indicated power. [**Ans.** 13.3 kW]

3. A gas engine has a piston diameter of 150 mm and stroke 250 mm. The speed of the engine is 250 r.p.m. and the average number of explosions are 90 per minute. The mean effective pressure is 7 bar. If the average torque on the brake is 140 N-m, find indicated power, brake power and mechanical efficiency. [**Ans.** 4.65 kW ; 3.67 kW ; 78.8%]

4. During a trial of a single cylinder four stroke I.C. engine, the following observations were recorded:

Mean effective pressure = 4 bar ; Speed = 200 r.p.m. ; Brake power = 7.5 kW ; Length of stroke = 1.5 times diameter of piston.

If the mechanical efficiency is 70% ; find the dimensions of the engine. [**Ans.** 240 mm ; 360 mm]

5. A constant speed four stroke cycle compression-ignition engine has a bore of 100 mm, stroke 150 mm and runs at 450 r.p.m. The following data refer to a test on this engine:

Brake wheel diameter = 600 mm ; Band thickness = 5 mm ; Load on band = 210 N ; Spring balance reading = 30 N ; Area of indicator diagram = 415 mm^2 ; Length of indicator diagram = 62.5 mm ; Spring scale = 1.1 bar per mm ; Specific fuel consumption. = 0.3 kg /b.p./h ; Calorific value of fuel = 42 000 kJ/kg.

Calculate : 1. the mechanical efficiency, and 2. the indicated thermal efficiency. [**Ans.** 80.3% ; 35.6%]

6. The following data refer to a test on a petrol engine:

Indicated power = 30 kW ; Brake power = 26 kW ; Engine speed = 1800 r.p.m. ; Fuel per brake power hour = 0.35 kg ; Calorific value of the fuel used = 44 100 kJ/kg.

Calculate : 1. the mechanical efficiency ; 2. the indicated thermal efficiency ; and 3. the brake thermal efficiency. **[Ans.** 26.9% ; 23.3%]

7. Following observations were taken during the trial of a single cylinder, four stroke, oil engine, running at full load :

Area of indicator diagram = 300 mm^2 ; Length of the diagram = 40 mm ; Spring stiffness = 1 bar/mm ; Speed of the engine = 400 r.p.m ; Brake load = 400 N ; Spring balance reading = 50 N ; Diameter of the brake drum = 1.2 m ; Fuel consumption per hour = 3 kg ; Calorific value of fuel = 42 000 kJ/kg ; Cylinder diameter = 160 mm ; Stroke = 200 mm.

Find the indicated power, brake power, mechanical efficiency and brake thermal efficiency.

[Ans. 10 kW ; 8.8 kW ; 88% ; 25.14%]

8. The output of an I.C engine is measured by a rope brake dynamometer. The diameter of the brake pulley is 750 mm and rope diameter is 50 mm. The dead load on the tight side of the rope is 410 N and the spring balance reading is 50 N. The engine consumes 4 kg/h of fuel at rated speed of 1000 r.p.m. The calorific value of fuel is 44 100 kJ/kg. Calculate brake specific fuel consumption and the brake thermal efficiency.

[Ans. 0.265 kg/B.P./h ; 30.8%]

9. A compression ignition engine at rated condition develops 7.5 kW brake power. The mechanical losses are 1.5 kW. If the indicated thermal efficiency is 42% ; air fuel ratio 22 and calorific value of fuel 43 260 kJ/kg, determine : 1. fuel consumption is kg/h ; 2 air intake in kg/h ; and 3. brake thermal efficiency.

[Ans. 1.783 kg/h ; 39.23 kg/h ; 35%]

10. A four cylinder, two stroke cycle petrol engine develops 30 kW brake power at 2500 r.p.m. The mean effective pressure on each piston is 8 bar and the mechanical efficiency is 80%. Calculate the diameter and stroke of each cylinder if the stroke to bore ratio is 1.5. Also calculate the brake specific fuel consumption of the engine, if brake thermal efficiency is 28%. The calorific value of the fuel is 44 100 kJ/kg.

[Ans. 62 mm ; 93 mm ; 0.29 kg/B.P./h]

11. An engine is used on a job requiring 110 kW B.P., the mechanical efficiency of the engine is 80 percent and the engine uses 50 kg fuel per hour under the conditions of operation. A design improvement is made which reduces the engine friction by 5 kW. Assuming the indicated thermal efficiency remains the same, how many kg of fuel per hour will be saved ? **[Ans.** 1.8 kg/h]

12. The following data relates to a four cylinder four stroke petrol engine :

Diameter of the piston = 80 mm ; Length of the stroke = 120 mm ; Clearance volume = 100 × 10^3 mm^3 ; Fuel supply = 4.8 kg/h ; Calorific value = 44 100 kJ/kg.

When the Morse test was performed on the engine, the following data were obtained :

B.P. with all the cylinders working = 14.5 kW
B.P. with cylinder 1 cut-off = 9.8 kW
B.P. with cylinder 2 cut-off = 10.3 kW
B.P. with cylinder 3 cut-off = 10.14 kW
B.P. with cylinder 4 cut-off = 10 kW

Find I.P. of the engine and also calculate indicated thermal efficiency, brake thermal efficiency and relative efficiency. **[Ans.** 17.76 kW ; 30.2% ; 24.66% ; 55.7%]

13. A petrol engine uses per brake power hour 0.36 kg of fuel of calorific value 44 100 kJ/kg. The mechanical efficiency is 78 percent and compression ratio is 5.6. Calculate : 1. Brake thermal efficiency ; 2. Indicated thermal efficiency ; and 3. Ideal air standard efficiency. Take $\gamma = 1.4$. **[Ans.** 22.7% ; 29.1% ; 49.8%]

14. A four cylinder four-stroke petrol engine produces 56 kW indicated power when running at 4400 r.p.m. with a volumetric efficiency of 81.5%. The air-fuel ratio is 16 : 1 and the thermal efficiency is 35%. The fuel used has a calorific value of 44 100 kJ/kg. If the bore to stroke ratio is 1 : 1.04, calculate the cylinder dimensions. Assume the charge to have the density of air equal to 1.293 kg/m^3 at N.T.P.

[Ans. 77.2 mm ; 80.3 mm]

15. A six cylinder, four stroke S.I. engine, having a piston displacement of 700 × 10^{-6} m^3 per cylinder developed 78 kW at 3200 r.p.m. and consumed 27 kg of petrol per hour. The calorific value of petrol is 44 MJ/kg.

Estimate 1. the volumetric efficiency of the engine if the air fuel ratio is 12 and the intake air is at 0.9 bar, and 32° C ; 2. the brake thermal efficiency ; and 3. the brake torque. For air, R = 0.287 kJ/kg K.

[Ans. 78.15% ; 23.64% ; 232.7 N-m]

16. Calculate the bore and stroke of a four stroke single cylinder oil engine designed to the following particulars.

Brake power 18 kW at 250 r.p.m. when running on oil having composition by mass C 85%, H 15% and a lower calorific value of 42 000 kJ/kg. The oil is burnt with 25% excess air. The volumetric efficiency reckoned on atmospheric conditions of 1.013 bar and 10° C is 0.8. The mechanical efficiency is 0.9 and indicated thermal efficiency is 0.35. Take R = 0.287 kJ/kg K and bore-stroke ratio as 1 : 1.2. [Ans. 236 mm ; 283.2 mm]

17. In a test of one hour duration on a single cylinder oil engine performing on a four stroke cycle, 8.08 kg of oil of calorific value 42 000 kJ/kg were used. The jacket water was 658 kg and its temperature rise is 22° C. The average speed was 200 r.p.m. and the m.e.p. in the cylinder is 5.95 bar. The cylinder diameter is 300 mm ; stroke 450 mm, and brake friction load 1900 N applied at the periphery of a flywheel of 1.2 m diameter. Show by heat balance chart, how the heat supplied is apportioned between the several items concerned and estimate the brake thermal efficiency and mechanical efficiency of the engine. [Ans. 25.3% ; 75.8%]

18. Calculate the brake specific fuel consumption, indicated thermal efficiency and obtain a heat balance sheet on minute basis from the following test data obtained in a four stroke two cylinder diesel engine :

Duration of test = 1 hour ; Brake power = 15 kW ; Total indicated power = 17.8 kW ; Fuel consumption = 4.24 litres of specific gravity 0.875 ; Lower calorific value of fuel = 43 340 kJ/kg ; Jacket cooling water circulated = 215 kg ; Inlet and outlet cooling water temperature = 30° C and 80° C.

The heat in exhaust gases is measured by an exhaust gas calorimeter as 808 kJ/minute.

[Ans. 0.247 kg/B.P./h ; 39.8%]

19. Draw a heat balance sheet for a two stroke diesel engine run for 20 minutes at full load, from the data given below :

R.P.M. = 350 ; M.E.P. = 3 bar ; Net brake load = 650 N ; Fuel consumption = 1.5 kg ; Cooling water = 160 kg ; Water inlet temperature = 35° C ; Water outlet temperature = 60° C ; Air used per kg of fuel = 30 kg ; Room temperature = 20° C ; Exhaust temperature = 300° C ; Cylinder bore = 200 mm ; Cylinder stroke = 280 mm ; Brake diameter = 1 m ; Calorific value of fuel = 44 100 kJ/kg ; Steam formed per kg of fuel in the exhaust = 1.35 kg ; Specific heat of steam in exhaust = 2.1 kJ/kg K ; Specific heat of dry exhaust gas = 1.008 kJ/kg K.

20. The following particulars refer to the full load test of a single cylinder, petrol engine working on the four stroke cycle :

Speed = 2500 r.p.m. ; Brake power = 118 kW ; Cylinder bore = 110 mm ; Cylinder diameter = 120 mm ; Lower calorific value of fuel = 41150 kJ/kg ; Petrol consumption = 40 kg/h ; Jacket water rate = 2800 kg/h ; Jacket water inlet temperature = 20° C ; Jacket water outlet temperature = 65.5° C ; Fuel-air ratio = 1 : 16 ; Room temperature = 29° C ; Exhaust temperature = 399° C ; Hydrogen in fuel = 15% by mass ; Sp. heat of dry exhaust gases = 0.9945 kJ/kg K ; Sp. heat of water vapour = 1.838 kJ/kg K.

Draw up a heat balance sheet. Calculate the brake thermal efficiency and volumetric efficiency of the engine. [Ans. 25.8% ; 73.7%]

QUESTIONS

1. What is the purpose of engine testing ?
2. Name the various measurements which are to be taken in a test of an I.C. engine ?
3. Describe the method of measuring mean effective pressure of an I.C. engine.
4. Explain the method for determining the indicated power of a multi-cylinder engine without using an indicator.
5. Define volumetric efficiency for an I.C. engine. What is the effect of volumetric efficiency on (*i*) engine power ; and (*ii*) specific fuel consumption.
6. What is the use of heat balance sheet of an engine ? Mention the various items to be determined to complete the heat balance sheet.

OBJECTIVE TYPE QUESTIONS

1. The power actually developed by the engine cylinder of an I.C. engine is known as
(*a*) brake power (*b*) indicated power (*c*) actual power

2. The number of working strokes per minute for a four stroke cycle engine are....the speed of the engine in r.p.m.
(*a*) equal to (*b*) one-half (*c*) twice (*d*) four times

3. If the speed of the engine is increased, the indicated power will
(*a*) increase (*b*) decrease (*c*) remain same

4. The brake power of the engine is the power available
(*a*) at the crank pin (*b*) in the engine cylinder
(*c*) at the crankshaft (*d*) none of these

5. The brake power of an engine is always.....the indicated power.
(*a*) equal to (*b*) less than (*c*) greater than

6. The ratio of the indicated thermal efficiency to the air standard efficiency is called
(*a*) mechanical efficiency (*b*) overall efficiency
(*c*) volumetric efficiency (*d*) relative efficiency

7. The ratio of the volume of charge admitted at N.T.P. to the swept volume of the piston is called
(*a*) mechanical efficiency (*b*) overall efficiency
(*c*) volumetric efficiency (*d*) relative efficiency

8. The thermal efficiency of petrol engines is about
(*a*) 15% (*b*) 30% (*c*) 50% (*d*) 70%

9. The volumetric efficiency of a well designed engine may be
(*a*) 30 to 40% (*b*) 40 to 60% (*c*) 60 to 70% (*d*) 75 to 90%

10. The Morse test is used to find the indicated power of a
(*a*) single cylinder petrol engine (*b*) single cylinder diesel engine
(*c*) multi-cylinder engine (*d*) none of these

ANSWERS

1. (*b*)	2. (*b*)	3. (*a*)	4. (*c*)	5. (*b*)
6. (*d*)	7. (*c*)	8. (*b*)	9. (*d*)	10. (*c*)

28

Reciprocating Air Compressors

1. Introduction. 2. Classification of Air Compressors. 3. Technical Terms. 4. Working of Single Stage Reciprocating Air Compressor. 6. Workdone by a Single Stage Reciprocating Air Compressor without Clearance Volume. 7. Power Required to Drive a Single Stage Reciprocating Air Compressor. 8. Workdone by Reciprocating Air Compressor with Clearance Volume. 9. Multistage Compression. 10. Advantages of Multistage Compression. 11. Two-stage Reciprocating Air Compressor with Intercooler. 12. Assumptions in Two-stage Compressor with Intercooler. 13. Intercooling of Air in a Two-stage Reciprocating Air Compressor. 14. Workdone by a Two-stage Reciprocating Air Compressor with Intercooler. 15. Power Required to Drive a Two-stage Reciprocating Air Compressor. 16. Minimum Work Required for a Two-stage Reciprocating Air Compressor. 17. Heat Rejected in a Reciprocating Air Compressor. 18. Ratio of Cylinder Diameters.

28.1. Introduction

An air compressor, as the name indicates, is a machine to compress the air and to raise its pressure. The air compressor sucks air from the atmosphere, compresses it and then delivers the same under a high pressure to a storage vessel. From the storage vessel, it may be conveyed by the pipeline to a placc where the supply of compressed air is required. Since the compression of air requires some work to be done on it, therefore a compressor must be driven by some prime mover.

The compressed air is used for many purposes such as for operating pneumatic drills, riveters, road drills, paint spraying, in starting and supercharging of internal combustion engines, in gas turbine plants, jet engines and air motors, etc. It is also utilised in the operation of lifts, rams, pumps and a variety of other devices. In industry, compressed air is used for producing blast of air in blast furnaces and bessemer converters.

28.2. Classification of Air Compressors

The air compressors may be classified in many ways, but the following are important from the subject point of view :

1. *According to working*

(*a*) Reciprocating compressors, and (*b*) Rotary compressors.

2. *According to action*

(*a*) Single acting compressors, and (*b*) Double acting compressors.

3. *According to number of stages*

(*a*) Single stage compressors, and (*b*) Multi-stage compressors.

28.3. Technical Terms

The following technical terms, which will be frequently used in this chapter, should be clearly understood at this stage :

1. *Inlet pressure.* **It is the absolute pressure of air at the inlet of a compressor.**

2. *Discharge pressure.* It is the absolute pressure of air at the outlet of a compressor.

3. *Compression ratio (or pressure ratio).* It is the ratio of discharge pressure to the inlet pressure. Since the discharge pressure is always more than the inlet pressure, therefore the value of compression ratio is more than unity.

4. *Compressor capacity.* It is the volume of air delivered by the compressor, and is expressed in m^3/min or m^3/s.

5. *Free air delivery.* It is the actual volume delivered by a compressor when reduced to the normal temperature and pressure condition. The capacity of a compressor is generally given in terms of free air delivery.

6. *Swept volume.* It is the volume of air sucked by the compressor during its suction stroke. Mathematically, the swept volume or displacement of a single acting air compressor is given by

$$v_s = \frac{\pi}{4} \times D^2 \times L$$

where D = Diameter of cylinder bore, and

L = Length of piston stroke.

7. *Mean effective pressure.* As a matter of fact, air pressure on the compressor piston keeps on changing with the movement of the piston in the cylinder. The mean effective pressure of the compressor is found out mathematically by dividing the work done per cycle to the stroke volume.

28.4. Working of Single Stage Reciprocating Air Compressor

A single stage reciprocating air compressor, in its simplest form, consists of a cylinder, piston, inlet and discharge valves, as shown in Fig. 28.1. From the geometry of the compressor, we find that when the piston moves downwards (or in other words, during outward or suction stroke), the pressure

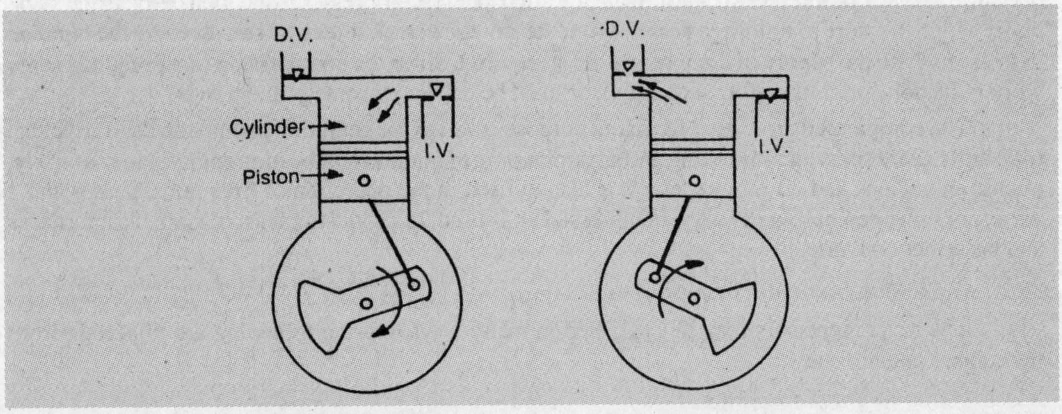

(*a*) Suction stroke. (*b*) Delivery stroke.

Fig. 28.1. Single stage reciprocating air compressor.

inside the cylinder falls below the atmospheric pressure. Due to this pressure difference, the inlet valve (I.V.) gets opened and air is sucked into the cylinder, at inlet pressure until the piston completes the outward stroke. Now when the piston moves upwards (or in other words, during inward or delivery stroke), the pressure inside the cylinder goes on increasing till it reaches the discharge pressure. At this stage, the discharge valve (D.V.) gets opened and air is delivered to the container. At the end of delivery stroke, a small quantity of air, at high pressure, is left in the clearance space. As the piston starts its suction stroke, the air contained in the clearance space expands till its pressure falls below the atmospheric pressure. At this stage, the inlet valve gets opened as a result of which fresh air is sucked into the cylinder, and the cycle is repeated.

It may be noted that in a single acting reciprocating air compressor, the suction, compression and delivery of air takes place in two strokes of the piston or one revolution of the crankshaft.

Note : In a double acting reciprocating compressor, the suction, compression and delivery of air takes place on both sides of the piston. It is thus obvious, that such a compressor will supply double the volume of air than a single acting reciprocating compressor (neglecting volume of piston rod).

28.5. Workdone by a Single Stage Reciprocating Air Compressor

We have already discussed that in a reciprocating air compressor, the air is first sucked, compressed and then delivered. So there are three different operations of the compressor. Thus we see that work is done on the piston during the suction of the air. Similarly, work is done by the piston during compression as well as delivery of the air. A little consideration will show, that the work done by a reciprocating air compressor is mathematically equal to the work done by the compressor during suction. Here we shall discuss the following two important cases of work done :

1. when there is no clearance volume in the cylinder, and
2. when there is some clearance volume.

28.6. Workdone by a Single Stage Reciprocating Air Compressor without Clearance Volume

Consider a single stage reciprocating air compressor without clearance volume delivering air from one side of the piston only.

Let p_1 = Initial pressure of air (before compression),

v_1 = Initial volume of air (before compression),

T_1 = Initial temperature of air (before compression),

p_2, v_2, T_2 = Corresponding values for the final conditions (*i.e.* at the delivery point), and

r = Pressure ratio (*i.e.* p_2 / p_1).

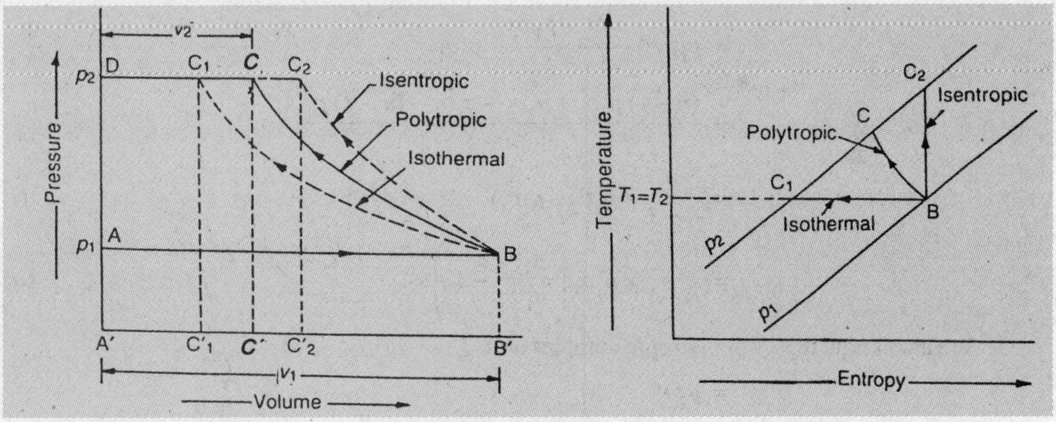

(*a*) *p-v* diagram without clearance volume. (*b*) *T-s* diagram.

Fig. 28.2. *p-v* and *T-s* diagrams for a single stage reciprocating air compressor.

The *p-v* and *T-s* diagrams of a single acting single stage reciprocating air compressor without clearance volume is shown in Fig. 28.2. We know that during return stroke, the air is compressed by its major part (*i.e.* compression stroke *BC*) at constant temperature. The compression continues till the pressure (p_2) in the cylinder is sufficient to force open the delivery valve at *C*. After that no more compression takes place with the inward movement of the piston. Now during the remaining part of compression stroke, the compressed air is delivered till the piston head reaches the cylinder end. After that, the air is sucked from the atmosphere during the suction stroke AB at pressure p_1.

As a matter fact, the compression of air may be isothermal, polytropic or isentropic (reversible adiabatic). Now, we shall find out the amount of work done in compressing the air in all the above mentioned three cases.

1. *Work done during isothermal compression*

The isothermal compression and delivery of air is shown by the graphs BC_1 and C_1D respectively. Now C_1D represents the volume of air delivered. We know that work done by the compressor per cycle,

$$W = \text{Area } ABC_1D$$

$$= \text{Area } A'DC_1C_1' + \text{Area } C_1BB'C_1' - \text{Area } A'ABB'$$

$$= p_2v_2 + 2.3\, p_2v_2 \log\left(\frac{v_1}{v_2}\right) - p_1v_1$$

$$= 2.3\, p_1v_1 \log\left(\frac{v_1}{v_2}\right) = 2.3\, p_1v_1 \log\left(\frac{p_2}{p_1}\right) \qquad \ldots (\because p_1v_1 = p_2v_2)$$

$$= 2.3\, p_1v_1 \log r = 2.3\, m\,R\,T_1 \log r \qquad \ldots (\because p_1v_1 = m\,R\,T_1)$$

2. *Work done during polytropic compression (pv^n = Constant)*

The polytropic compression is shown by the line BC in Fig. 28.2. Now CD represents the volume of air delivered, *i.e.* v_2. We know that work done on the air per cycle,

$$W = \text{Area } ABCD$$

$$= \text{Area } A'DCC' + \text{Area } CBB'C' - \text{Area } A'ABB'$$

$$= p_2v_2 + \frac{p_2v_2 - p_1v_1}{n-1} - p_1v_1$$

$$= \frac{(n-1)\,p_2v_2 + p_2v_2 - p_1v_1 - (n-1)\,p_1v_1}{n-1}$$

$$= \frac{n}{n-1}\,(p_2v_2 - p_1v_1) \qquad \ldots (i)$$

$$= \frac{n}{n-1} \times p_1v_1\left(\frac{p_2v_2}{p_1\,v_1} - 1\right) \qquad \ldots (ii)$$

We also know that for polytropic compression,

$$p_1v_1^n = p_2v_2^n$$

$$\therefore \quad \frac{v_2}{v_1} = \left(\frac{p_1}{p_2}\right)^{\frac{1}{n}} \quad \text{or} \quad \frac{v_1}{v_2} = \left(\frac{p_2}{p_1}\right)^{\frac{1}{n}}$$

Substituting the value of v_2 / v_1 in equation (*ii*),

$$W = \frac{n}{n-1} \times p_1v_1\left[\frac{p_2}{p_1}\left(\frac{p_1}{p_2}\right)^{\frac{1}{n}} - 1\right]$$

$$= \frac{n}{n-1} \times p_1 v_1 \left[\left(\frac{p_2}{p_1} \right)^{\frac{n-1}{n}} - 1 \right]$$

$$= \frac{n}{n-1} \times m\,R\,T_1 \left[\left(\frac{p_2}{p_1} \right)^{\frac{n-1}{n}} - 1 \right] \qquad \ldots (iii)$$

The equation (*i*) may also be written as :

$$W = \frac{n}{n-1} \times p_2 v_2 \left(1 - \frac{p_1 v_1}{p_2 v_2} \right)$$

$$= \frac{n}{n-1} \times p_2 v_2 \left[1 - \frac{p_1}{p_2} \left(\frac{p_2}{p_1} \right)^{\frac{1}{n}} \right]$$

$$= \frac{n}{n-1} \times p_2 v_2 \left[1 - \left(\frac{p_1}{p_2} \right)^{\frac{n-1}{n}} \right]$$

$$= \frac{n}{n-1} \times m\,R\,T_2 \left(1 - \frac{T_1}{T_2} \right) = \frac{n}{n-1} \times m\,R\,(T_2 - T_1) \qquad \ldots (iv)$$

3. *Work done during isentropic compression*

The isentropic compression is shown by the curve BC_2 in Fig. 28.2. In this case, the volume of air delivered v_2 is represented by the line C_2D.

The work done on the air per cycle during isentropic compression may be worked out in the similar way as polytropic compression. The polytropic index n is changed to isentropic index γ in the previous results.

∴ Work done on the air per cycle,

$$W = \frac{\gamma}{\gamma-1} \times p_1 v_1 \left[\left(\frac{p_2}{p_1} \right)^{\frac{\gamma-1}{\gamma}} - 1 \right]$$

$$= \frac{\gamma}{\gamma-1} \times m\,R\,T_1 \left[\left(\frac{p_2}{p_1} \right)^{\frac{\gamma-1}{\gamma}} - 1 \right]$$

$$= \frac{\gamma}{\gamma-1} \times m\,R\,(T_2 - T_1)$$

We know that the ratio of specific heats,

$$\frac{c_p}{c_v} = \gamma; \text{ and } c_p - c_v = R$$

or

$$R = c_p \left(1 - \frac{1}{\gamma} \right) = c_p \left(\frac{\gamma-1}{\gamma} \right)$$

Now work done,

$$W = \frac{\gamma}{\gamma-1} \times m\,R\,(T_2 - T_1)$$

$$= \frac{\gamma}{\gamma-1} \times m\,c_p \left(\frac{\gamma-1}{\gamma} \right) (T_2 - T_1) = m\,c_p\,(T_2 - T_1)$$

We see that the work done on the air during isentropic compression is equal to the heat required to raise the temperature of air from T_1 to T_2 at a constant pressure.

Note : The work done on the air is minimum when the compression is isothermal (*i.e.* when $n = 1$) and it is maximum when the compression is isentropic (*i.e.* when $n = \gamma$) because isothermal line has less slope than isentropic line. It may be noted that in order to perform isothermal process, the compression should be very slow so that the temperature is maintained constant, which is not possible in actual practice. However, the isothermal compression may be approached, if

1. the air or water cooling is done during the compression,
2. the cold water is sprayed (injected) in the cylinder during the compression, and
3. in multi-stage compressors, intercooling is done.

28.7. Power Required to Drive a Single-stage Reciprocating Air Compressor

We have already obtained in the last article the expressions for the work done (W) per cycle during isothermal, polytropic and isentropic compression. The power required to drive the compressor may be obtained from the usual relation,

$$P = \frac{W N_w}{60} \text{ watts}$$

If N is the speed of the compressor in r.p.m., then number of working strokes per minute,

$$N_w = N \qquad \text{. . . (For single acting compressor)}$$

$$= 2N \qquad \text{. . . (For double acting compressor)}$$

Note : Since the compression takes in three different ways, therefore power obtained from different works done will be different. In general, following are the three values of power obtained :

1. Isothermal power $= \dfrac{W \text{ (in isothermal compression) } N_w}{60}$ watts
2. Isentropic power $= \dfrac{W \text{ (in isentropic compression) } N_w}{60}$ watts
3. Indicated power $= \dfrac{W \text{ (in polytropic compression) } N_w}{60}$ watts

The indicated power is also known as *air power* of the compressor.

Example 28.1. *A single stage reciprocating air compressor is required to compress 1 kg of air from 1 bar to 4 bar. The initial temperature is 27° C. Compare the work requirement in the following cases :*

1. Isothermal compression ; 2. Compression with $p\,v^{1.2}$ = constant ; and 3. Isentropic compression.

Solution. Given : $m = 1$ kg ; $p_1 = 1$ bar ; $p_2 = 4$ bar ; $T_1 = 27°\text{ C} = 27 + 273 = 300$ K ; $n = 1.2$

1. *Work required for isothermal compression*

We know that work required by the compressor,

$$W = 2.3\, p_1 v_1 \log\left(\frac{p_2}{p_1}\right) = 2.3\, m R T_1 \log\left(\frac{p_2}{p_1}\right) \qquad \text{. . . } (\because p_1 v_1 = m R T_1)$$

$$= 2.3 \times 1 \times 287 \times 300 \log\left(\frac{4}{1}\right) = 119\,230 \text{ J} = \textbf{119.23 kJ Ans.}$$

$$\text{. . . } (\because R \text{ for air} = 287 \text{ J/kg K})$$

2. *Work required for polytropic compression (i.e. $pv^{1.2}$ = constant)*

We know that work required by the compressor,

$$W = \frac{n}{n-1} \times m\,R\,T_1 \left[\left(\frac{p_2}{p_1} \right)^{\frac{n-1}{n}} - 1 \right]$$

$$= \frac{1.2}{1.2-1} \times 1 \times 287 \times 300 \left[\left(\frac{4}{1} \right)^{\frac{1.2-1}{1.2}} - 1 \right] = 134\,320 \text{ J}$$

$$= 134.32 \text{ kJ} \quad \textbf{Ans.}$$

3. *Work required for isentropic compression*

We know that work required by the compressor,

$$W = \frac{\gamma}{\gamma-1} \times m\,R\,T_1 \left[\left(\frac{p_2}{p_1} \right)^{\frac{\gamma-1}{\gamma}} - 1 \right]$$

$$= \frac{1.4}{1.4-1} \times 1 \times 287 \times 300 \left[\left(\frac{4}{1} \right)^{\frac{1.4-1}{1.4}} - 1 \right] = 146\,630 \text{ J}$$

$$= 146.63 \text{ kJ} \quad \textbf{Ans.}$$

Example 28.2. *Determine the size of the cylinder for a double acting air compressor of 40 kW indicated power, in which air is drawn in at 1 bar and 15° C and compressed according to the law $p\,v^{1.2}$ = constant, to 6 bar. The compressor runs at 100 r.p.m. with average piston speed of 152.5 m/min. Neglect clearance.*

Solution. Given : I.P. = 40 kW = 40×10^3 W ; p_1 = 1 bar = 1×10^5 N/m² ; T_1 = 15° C = 15 + 273 = 288 K ; n = 1.2 ; p_2 = 6 bar ; N = 100 r.p.m. ; Average piston speed = 152.5 m/min

Let D = Diameter of the cylinder in metres, and

L = Length of the stroke in metres.

We know that average piston speed,

$$2\,LN = 152.5$$

$$\therefore \quad L = 152.5 / 2N = 152.5 / 2 \times 100 = 0.7625 \text{ m} \quad \textbf{Ans.}$$

Volume of air before compression,

$$v_1 = \frac{\pi}{4} \times D^2 L = \frac{\pi}{4} \times D^2 \times 0.7625 = 0.6\,D^2 \text{ m}^3$$

and workdone by the compressor,

$$W = \frac{n}{n-1} \times p_1\,v_1 \left[\left(\frac{p_2}{p_1} \right)^{\frac{n-1}{n}} - 1 \right]$$

$$= \frac{1.2}{1.2-1} \times 1 \times 10^5 \times 0.6\,D^2 \left[\left(\frac{6}{1} \right)^{\frac{1.2-1}{1.2}} - 1 \right] \text{N-m}$$

$$= 125\,310\,D^2 \text{ N-m}$$

Since the compressor is double acting, therefore number of working strokes per minute,

$$N_w = 2\,N = 2 \times 100 = 200$$

We know that indicated power (I.P.),

$$40 \times 10^3 = \frac{W \times N_w}{60} = \frac{125\,310\,D^2 \times 200}{60} = 417.7 \times 10^3\,D^2$$

$$\therefore \quad D^2 = 0.096 \text{ or } D = 0.31 \text{ m or } 310 \text{ mm } \textbf{Ans.}$$

Example 28.3. *A single acting reciprocating air compressor has cylinder diameter and stroke of 200 mm and 300 mm respectively. The compressor sucks air at 1 bar and 27° C and delivers at 8 bar while running at 100 r.p.m. Find : 1. Indicated power of the compressor ; 2. Mass of air delivered by the compressor per minute ; and 3. Temperature of the air delivered by the compressor. The compression follows the law* $pv^{1.25} = C$. *Take R as 287 J/kg K.*

Solution. Given : $D = 200$ mm $= 0.2$ m ; $L = 300$ mm $= 0.3$ m ; $p_1 = 1$ bar $= 1 \times 10^5$ N/m² ; $T_1 = 27°$ C $= 27 + 273 = 300$ K ; $p_2 = 8$ bar ; $N = 100$ r.p.m. ; $n = 1.25$; $R = 287$ J/kg K

We know that volume of air before compression,

$$v_1 = \frac{\pi}{4} \times D^2 \times L = \frac{\pi}{4}(0.2)^2\,0.3 = 0.0094 \text{ m}^3$$

1. *Indicated power of the compressor*

We know that workdone by the compressor for polytropic compression of air,

$$W = \frac{n}{n-1} \times p_1 v_1 \left[\left(\frac{p_2}{p_1}\right)^{\frac{n-1}{n}} - 1\right]$$

$$= \frac{1.25}{1.25 - 1} \times 1 \times 10^5 \times 0.0094 \left[\left(\frac{8}{1}\right)^{\frac{1.25-1}{1.25}} - 1\right] \text{ N-m}$$

$$= 4700\,(1.516 - 1) = 2425 \text{ N-m}$$

Since the compressor is single acting, therefore number of working strokes per minute,

$$N_w = N = 100$$

∴ Indicated power of the compressor

$$= \frac{W \times N_w}{60} = \frac{2425 \times 100}{60} = 4042 \text{ W} = 4.042 \text{ kW } \textbf{Ans.}$$

2. *Mass of air delivered by the compressor per minute*

Let m = Mass of air delivered by the compressor per stroke.

We know that $p_1 v_1 = m\,R\,T_1$

$$\therefore \quad m = \frac{p_1 v_1}{R\,T_1} = \frac{1 \times 10^5 \times 0.0094}{287 \times 300} = 0.0109 \text{ kg per stroke}$$

and mass delivered minute $= m \times N_w = 0.0109 \times 100 = 1.09$ kg **Ans.**

3. *Temperature of air delivered by the compressor*

Let T_2 = Temperature of air delivered by the compressor.

We know that

$$\frac{T_2}{T_1} = \left(\frac{p_2}{p_1}\right)^{\frac{n-1}{n}} = \left(\frac{8}{1}\right)^{\frac{1.25-1}{1.25}} = 8^{0.2} = 1.516$$

$$\therefore \quad T_2 = 1.516 \times T_1 = 1.516 \times 300 = 454.8 \text{ K} = 181.8° \text{ C. } \textbf{Ans.}$$

28.8. Workdone by Reciprocating Air Compressor with Clearance Volume

In the previous articles, we have assumed that there is no clearance volume in the compressor cylinder. In other words, the entire volume of air, in the compressor cylinder, is compressed by the inward stroke of the piston. But in actual practice, it is not possible to reduce the clearance volume to zero, for mechanical reasons. Moreover, it is not desirable to allow the piston head to come in contact with the cylinder head. In addition to this, the passage leading to the inlet and outlet valves always contribute to clearance volume. In general, the clearance volume is expressed as some percentage of the piston displacement.

Now consider a reciprocating air compressor with clearance volume, as shown in Fig. 28.3.

Let p_1 = Initial pressure of air (before compression),

v_1 = Initial volume of air (before compression),

T_1 = Initial temperature of air (before compression),

p_2, v_2, T_2 = Corresponding values for the final conditions (*i.e.* at the delivery points),

r = Pressure ratio (*i.e.* p_2/p_1),

v_c = Clearance volume (*i.e.*, volume at point 3),

v_s = Stroke volume = $v_1 - v_c$, and

n = Polytropic index for compression and expansion.

The *p-v* diagram of a single stage single acting reciprocating air compressor with clearance volume (v_c) is shown in Fig. 28.3. We know that during return stroke, the air is compressed by its major part *i.e.* compression stroke 1-2. This compression continues, till the pressure p_2 in the cylinder is sufficient to force open the delivery valve at 2. After that, no more compression takes place with the inward movement of the piston. Now during the remaining part of compression stroke, compressed air is delivered till the piston reaches at 3. At this stage, there will be some air (equal to clearance volume) left in the clearance space of the cylinder at pressure p_2. After that air in the clearance space will expand during some part of outward stroke of the piston *i.e.* expansion stroke 3-4. This expansion continues till the pressure p_1 in the cylinder is sufficient to force open the inlet valve at 4. After that the air is sucked from the atmosphere during the suction stoke 4-1 at pressure p_1.

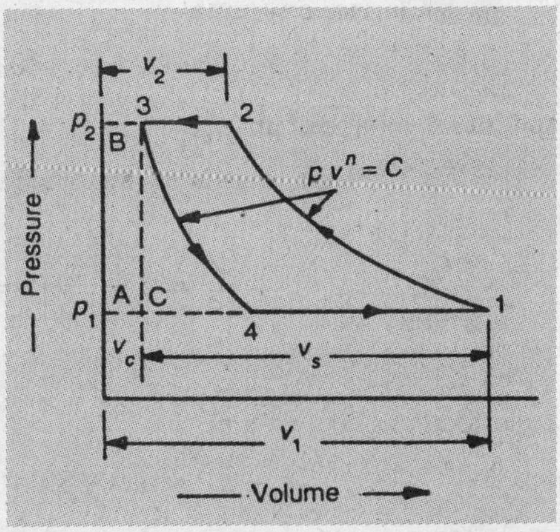

Fig. 28.3. *p-v* diagram with clearance volume.

Though the compression and expansion of air may be isothermal, isentropic or polytropic, yet for all calculation purposes, it is assumed to be polytropic. We know that work done by the compressor per cycle,

$$W = \text{Area 1-2-3-4} = \text{Area A-1-2-B} - \text{Area A-4-3-B}$$

$$= \frac{n}{n-1} \times p_1 v_1 \left[\left(\frac{p_2}{p_1}\right)^{\frac{n-1}{n}} - 1\right] - \frac{n}{n-1} p_1 v_4 \left[\left(\frac{p_2}{p_1}\right)^{\frac{n-1}{n}} - 1\right]$$

$$= \frac{n}{n-1} \times p_1 (v_1 - v_4) \left[\left(\frac{p_2}{p_1}\right)^{\frac{n-1}{n}} - 1\right]$$

$$= \frac{n}{n-1} \times m R T_1 \left[\left(\frac{p_2}{p_1} \right)^{\frac{n-1}{n}} - 1 \right]$$

where $(v_1 - v_4)$ and m is equal to the actual volume and mass of air sucked by the piston per cycle respectively.

We see that the clearance volume does not effect the work done on the air and the power required for compressing the air. This is due to the reason that the work required to compress the clearance volume air is theoretically regained during its expansion from 3 to 4.

Note : The terms v_4 and $(v_1 - v_4)$ are known as expanded clearance volume and effective swept volume respectively.

Example 28.4. *A single stage, single acting reciprocating air compressor has a bore of 200 mm and a stroke of 300 mm. It receives air at 1 bar and 20° C and delivers it at 5.5 bar. If the compression follows the law $pv^{1.3} = C$ and clearance volume is 5 percent of the stroke volume, determine : 1. the mean effective pressure ; and 2. the power required to drive the compressor, if it runs at 500 r.p.m.*

Solution. Given : $D = 200$ mm $= 0.2$ m ; $L = 300$ mm $= 0.3$ m ; $p_1 = 1$ bar $= 1 \times 10^5$ N/m² ; $T_1 = 20°\text{C} = 20 + 273 = 293$ K ; $p_2 = 5.5$ bar ; $n = 1.3$; $v_c = 5\%\ v_s$; $N = 500$ r.p.m

We know that stroke volume,

$$v_s = \frac{\pi}{4} \times D^2 \times L = \frac{\pi}{4} (0.2)^2\, 0.3 = 0.009\,42 \text{ m}^3$$

∴ Clearance volume,

$$v_c = 5\%\ v_s = 0.05 \times 0.009\,42 = 0.000\,47 \text{ m}^3$$

and initial volume of air, $v_1 = v_c + v_s = 0.000\,47 + 0.009\,42 = 0.009\,89 \text{ m}^3$

We know that expanded clearance volume,

$$v_4 = v_c \left(\frac{p_2}{p_1} \right)^{\frac{1}{n}} = 0.000\,47 \left(\frac{5.5}{1} \right)^{\frac{1}{1.3}} = 0.001\,74 \text{ m}^3$$

$$\ldots (\because p_1 v_4^n = p_2 v_c^n)$$

and effective swept volume,

$$v_1 - v_4 = 0.009\,89 - 0.001\,74 = 0.008\,15 \text{ m}^3$$

We know that work done by the compressor per cycle,

$$W = \frac{n}{n-1} \times p_1 (v_1 - v_4) \left[\left(\frac{p_2}{p_1} \right)^{\frac{n-1}{n}} - 1 \right]$$

$$= \frac{1.3}{1.3-1} \times 1 \times 10^5 \times 0.008\,15 \left[\left(\frac{5.5}{1} \right)^{\frac{1.3-1}{1.3}} - 1 \right] = 1702 \text{ N-m}$$

1. *Mean effective pressure*

We know that mean effective pressure,

$$p_m = \frac{\text{Work done}}{\text{Stroke volume}} = \frac{1702}{0.009\,42} = 180\,700 \text{ N/m}^2$$

$$= 1.807 \text{ bar} \quad \textbf{Ans.}$$

2. *Power required to drive the compressor*

Since the compressor is single acting, therefore number of working strokes per minute,

$$N_w = N = 500$$

∴ Power required to drive the compressor,

$$P = \frac{W \times N_w}{60} = \frac{1702 \times 500}{60} = 14\ 183 \text{ W} = 14183 \text{ kW} \quad \textbf{Ans.}$$

28.9. Multistage Compression

In the previous articles, we have been taking into consideration the compression of air in single stage. In other words, air is sucked, compressed in the cylinder and then delivered at a higher pressure. But sometimes, the air is required at a high pressure. In such cases, either we employ a large pressure ratio (in single cylinder) or compress the air in two or more cylinders in series. It has been experienced that if we employ single stage compression for producing high pressure air (say 8 to 10 bar), it suffers the following drawbacks :

1. The size of the cylinder will be too large.
2. Due to compression, there is a rise in temperature of the air. It is difficult to reject heat from the air in the small time available during compression.
3. Sometimes, the temperature of air, at the end of compression, is too high. It may heat up the cylinder head or burn the lubricating oil.

In order to overcome the above mentioned difficulties, two or more cylinders are provided in series with intercooling arrangement between them. Such an arrangement is known as *multistage compression.*

28.10. Advantages of Multistage Compression

Following are the main advantages of multistage compression over single stage compression :

1. The work done per kg of air is reduced in multistage compression with intercooler as compared to single stage compression for the same delivery pressure.
2. It improves the volumetric efficiency for the given pressure ratio.
3. The sizes of the two cylinders (*i.e.* high pressure and low pressure) may be adjusted to suit the volume and pressure of the air.
4. It reduces the leakage loss considerably.
5. It gives more uniform torque, and hence a smaller size flywheel is required.
6. It provides effective lubrication because of lower temperature range.
7. It reduces the cost of compressor.

28.11. Two-stage Reciprocating Air Compressor with Intercooler

A schematic arrangement for a two-stage reciprocating air compressor with water cooled intercooler is shown in Fig. 28.4.

First of all, the fresh air is sucked from the atmosphere in the low pressure (L.P.) cylinder during its suction stroke at intake pressure p_1 and temperature T_1. The air, after compression in the L.P. cylinder (*i.e.* first stage) from 1 to 2, is delivered to the intercooler at pressure p_2 and temperature T_2. Now the air is cooled in the intercooler from 2 to 3 at constant pressure p_2 and from temperature T_2 to T_3. After that, the air is sucked in the high pressure (H.P.) cylinder during its suction stroke.

Finally, the air, after further compression in the H.P. cylinder (*i.e.* second stage) from 3 to 4, is delivered by the compressor at pressure p_3 and temperature T_4.

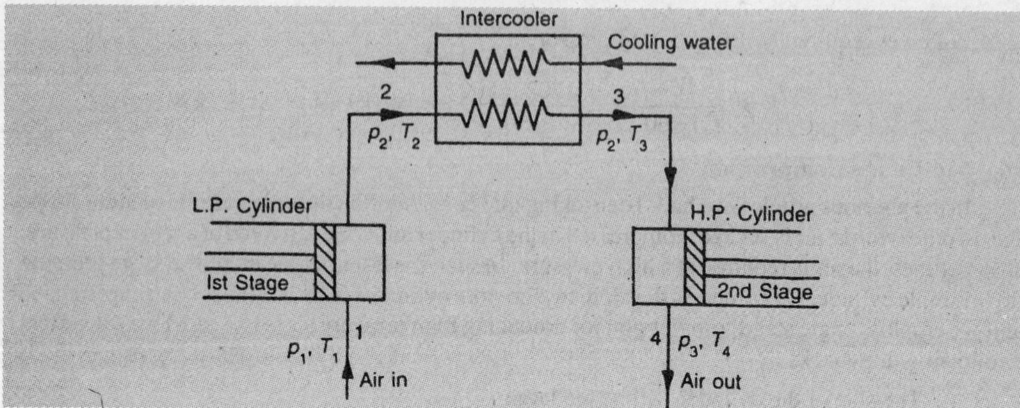

Fig. 28.4. Two-stage reciprocating air compressor with intercooler.

28.12. Assumptions in Two-stage Compression with Intercooler

The following simplifying assumptions are made in case of two stage compression with intercooler :

1. The effect of clearance is neglected.
2. There is no pressure drop in the intercooler.
3. The compression in both the cylinders (*i.e.* L.P. and H.P.) is polytropic (*i.e.* $p\,v^n = C$).
4. The suction and delivery of air takes place at constant pressure.

28.13. Intercooling of Air in a Two-stage Reciprocating Air Compressor

In the previous article, we have discussed the working of a two-stage reciprocating air compressor with an intercooler in between the two stages. As a matter of fact, efficiency of the intercooler plays an important role in the working of a two-stage reciprocating air compressor. Following two types of intercooling are important from the subject point of view :

1. *Complete or perfect intercooling.* When the temperature of the air leaving the intercooler (*i.e.* T_3) is equal to the original atmospheric air temperature (*i.e.* T_1), then the intercooling is known as complete or perfect intercooling. In this case, the point 3 lies on the isothermal curve as shown in Fig. 28.5 (*a*) and (*b*).

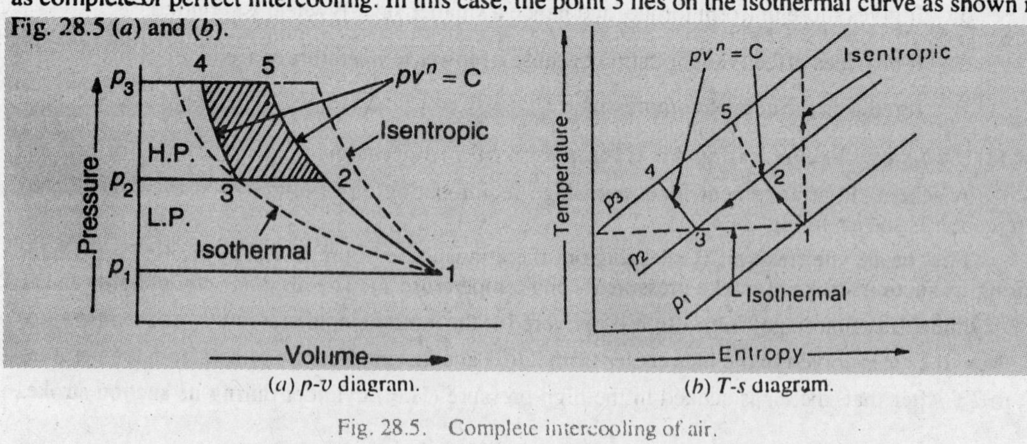

(*a*) *p-v* diagram. (*b*) *T-s* diagram.

Fig. 28.5. Complete intercooling of air.

2. *Incomplete or imperfect intercooling.* When the temperature of the air leaving the intercooler (*i.e.* T_3) is more than the original atmospheric air temperature (*i.e.* T_1), then the intercooling is known as incomplete or imperfect intercooling. In this case, the point 3 lies on the right side of the isothermal curve as shown in Fig. 28.6 (*a*) and (*b*).

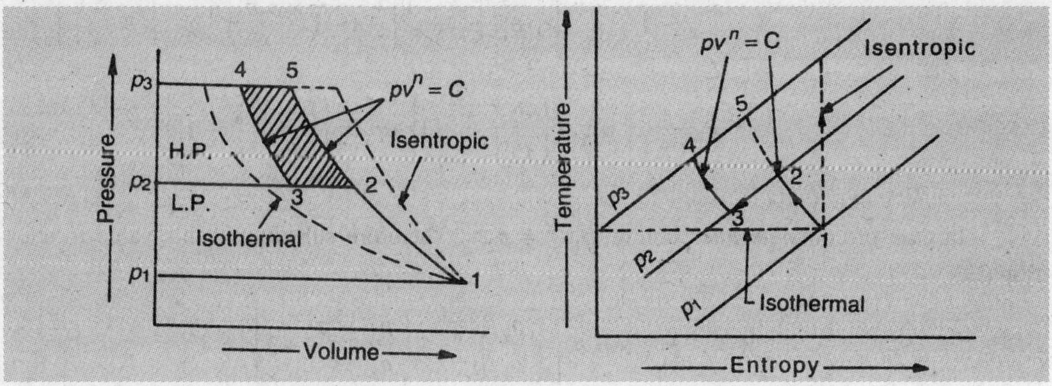

(*a*) *p-v* diagram. (*b*) *T-s* diagram.

Fig. 28.6. Incomplete intercooling of air.

Note : The amount of work saved due to intercooling is shown by the shaded area 2-3-4-5 in both the cases, to some scale. The amount of work saved with incomplete intercooling is less than that in case of complete intercooling.

28.14. Workdone by a Two-stage Reciprocating Air Compressor with Intercooler

Consider a two-stage reciprocating air compressor with intercooler compressing air in its L.P. and H.P. cylinders.

Let p_1 = Pressure of air entering the L.P. cylinder,

v_1 = Volume of the L.P. cylinder,

p_2 = Pressure of air leaving the L.P. cylinder or entering the H.P. cylinder,

v_2 = Volume of H.P. cylinder,

p_3 = Pressure of air leaving the H.P. cylinder, and

n = Polytropic index for both the cylinders.

Now we shall consider both the cases of incomplete intercooling as well as complete intercooling one by one.

1. *When the intercooling is incomplete*

We know that work done per cycle in L.P. cylinder,

$$W_1 = \frac{n}{n-1} \times p_1 v_1 \left[\left(\frac{p_2}{p_1} \right)^{\frac{n-1}{n}} - 1 \right] \qquad \ldots (i)$$

Similarly, work done per cycle in compressing air in H.P. cylinder,

$$W_2 = \frac{n}{n-1} \times p_2 v_2 \left[\left(\frac{p_3}{p_2} \right)^{\frac{n-1}{n}} - 1 \right] \qquad \ldots (ii)$$

$\therefore$ Total work done per cycle,

$$W = W_1 + W_2$$

$$= \frac{n}{n-1} \times p_1 v_1 \left[\left(\frac{p_2}{p_1} \right)^{\frac{n-1}{n}} - 1 \right] + \frac{n}{n-1} \times p_2 v_2 \left[\left(\frac{p_3}{p_2} \right)^{\frac{n-1}{n}} - 1 \right]$$

$$= \frac{n}{n-1} \left[p_1 v_1 \left\{ \left(\frac{p_2}{p_1} \right)^{\frac{n-1}{n}} - 1 \right\} + p_2 v_2 \left\{ \left(\frac{p_3}{p_2} \right)^{\frac{n-1}{n}} - 1 \right\} \right] \quad \ldots (iii)$$

2. *When the intercooling is complete*

In case of complete intercooling, $p_1 v_1 = p_2 v_2$. Therefore substituting this value in equation (*iii*),

$$W = \frac{n}{n-1} \times p_1 v_1 \left[\left(\frac{p_2}{p_1} \right)^{\frac{n-1}{n}} + \left(\frac{p_3}{p_2} \right)^{\frac{n-1}{n}} - 2 \right]$$

$$= \frac{n}{n-1} \times m R T_1 \left[\left(\frac{p_2}{p_1} \right)^{\frac{n-1}{n}} + \left(\frac{p_3}{p_2} \right)^{\frac{n-1}{n}} - 2 \right]$$

Example 28.5. *Estimate the work done by a two stage reciprocating single acting air compressor to compress 2.8 m³ of air per min at 1.05 bar and 10° C to a final pressure of 35 bar. The intermediate receiver cools the air to 30° C and 5.6 bar pressure. For air, take n = 1.4.*

Solution. Given : $v_1 = 2.8$ m³/min ; $p_1 = 1.05$ bar $= 1.05 \times 10^5$ N/m² ; $T_1 = 10°$ C $= 10 + 273 = 283$ K ; $p_3 = 35$ bar ; $T_3 = 30°$ C $= 30 + 273 = 303$ K ; $p_2 = 5.6$ bar $= 5.6 \times 10^5$ N/m² ; $n = 1.4$

Let v_2 = Volume of the high pressure cylinder.

We know that $\dfrac{p_1 v_1}{T_1} = \dfrac{p_2 v_2}{T_3}$

or
$$v_2 = \frac{p_1 v_1 T_3}{p_2 T_1} = \frac{1.05 \times 10^5 \times 2.8 \times 303}{5.6 \times 10^5 \times 283} = 0.562 \text{ m}^3/\text{min}$$

$\therefore$ Work done by the compressor,

$$W = \frac{n}{n-1} \left[p_1 v_1 \left\{ \left(\frac{p_2}{p_1} \right)^{\frac{n-1}{n}} - 1 \right\} + p_2 v_2 \left\{ \left(\frac{p_3}{p_2} \right)^{\frac{n-1}{n}} - 1 \right\} \right]$$

$$= \frac{1.4}{1.4-1} \left[1.05 \times 10^5 \times 2.8 \left\{ \left(\frac{5.6}{1.05} \right)^{\frac{1.4-1}{1.4}} - 1 \right\} \right.$$

$$\left. + 5.6 \times 10^5 \times 0.562 \times \left\{ \left(\frac{35}{5.6} \right)^{\frac{1.4-1}{1.4}} - 1 \right\} \right]$$

$$= 3.5 \, [1.803 \times 10^5 + 2.166 \times 10^5] = 13.9 \times 10^5 \text{ N-m/min} \quad \textbf{Ans.}$$

28.15. Power Required to Drive a Two-stage Reciprocating Air Compressor

We have already obtained in the last article the expressions for the work done (W) per cycle of a two-stage reciprocating air compressor with incomplete and complete intercooling. The power required to drive the compressor may be obtained from the usual relation :

$$P = \frac{W \times N_w}{60} \text{ watts}$$

where N_w = Number of working strokes per minute.

Example 28.6. *A two-stage single acting reciprocating air compressor draws in air at a pressure of 1 bar and 17° C and compresses it to a pressure of 60 bar. After compression in the L.P. cylinder, the air is cooled at constant pressure of 8 bar to a temperature of 37° C. The low pressure cylinder has a diameter of 150 mm and both the cylinders have 200 mm stroke. If the law of compression is $pv^{1.35} = C$, find the power of the compressor, when it runs at 200 r.p.m. Take R = 287 J/kg K.*

Solution. Given : p_1 = 1 bar = 1×10^5 N/m² ; T_1 = 17° C = 17 + 273 = 290 K ; p_3 = 60 bar ; p_2 = 8 bar = 8×10^5 N/m² ; T_3 = 37° C = 37 + 273 = 310 K ; D = 150 mm = 0.15 m ; L = 200 mm = 0.2 m ; n = 1.35 ; N = 200 r.p.m. ; R = 287 J/kg K

We know that volume of L.P. cylinder,

$$v_1 = \frac{\pi}{4} \times D^2 \times L = \frac{\pi}{4} (0.15)^2\, 0.2 = 0.0035 \text{ m}^3$$

Let v_2 = Volume of H.P. cylinder.

We know that $\dfrac{p_1 v_1}{T_1} = \dfrac{p_2 v_2}{T_3}$

or $$v_2 = \frac{p_1 v_1 T_3}{p_2 T_1} = \frac{1 \times 10^5 \times 0.0035 \times 310}{8 \times 10^5 \times 290} = 0.000\,47 \text{ m}^3$$

∴ Workdone by the compressor per stroke,

$$*W = \frac{n}{n-1}\left[p_1 v_1 \left\{\left(\frac{p_2}{p_1}\right)^{\frac{n-1}{n}} - 1\right\} + p_2 v_2 \left\{\left(\frac{p_3}{p_2}\right)^{\frac{n-1}{n}} - 1\right\}\right]$$

$$= \frac{1.35}{1.35-1}\left[1 \times 10^5 \times 0.0035 \left\{\left(\frac{8}{1}\right)^{\frac{1.35-1}{1.35}} - 1\right\}\right.$$

$$\left. + \left[8 \times 10^5 \times 0.000\,47 \left\{\left(\frac{60}{8}\right)^{\frac{1.35-1}{1.35}} - 1\right\}\right]\right] \text{ N-m}$$

$$= 3.86\,(250 + 258) = 1961 \text{ N-m}$$

Since the compressor is single acting, therefore number of working strokes per minute,

$$N_w = N = 200$$

We know that power of the compressor,

$$P = \frac{W \times N_w}{60} = \frac{1961 \times 200}{60} = 6540 \text{ W} = 6.54 \text{ kW} \textbf{ Ans.}$$

* Since T_3 is more than T_1, therefore it a case of incomplete intercooling.

28.16. Minimum Work Required for a Two-stage Reciprocating Air Compressor

We have already discussed in Art. 28.14 that maximum work is saved in a two-stage reciprocating air compressor with complete intercooling. We have also obtained a relation that work required to be done by a two-stage reciprocating air compressor with complete intercooling,

$$W = \frac{n}{n-1} \times p_1 v_1 \left[\left(\frac{p_2}{p_1} \right)^{\frac{n-1}{n}} + \left(\frac{p_3}{p_2} \right)^{\frac{n-1}{n}} - 2 \right] \qquad \ldots (i)$$

If the intake pressure p_1 and the delivery pressure p_3 are fixed, then least value of the intermediate or intercooler pressure p_2 may be obtained by differentiating the above equation with respect to p_2. It may be noted that value of p_2 thus obtained denotes the pressure of the intercooler at which the work required to drive the compressor is minimum. Thus work required is minimum, when

$$\frac{dW}{dp_2} = 0$$

$$\frac{d}{dp_2} \left[\frac{n}{n-1} \times p_1 v_1 \left\{ \left(\frac{p_2}{p_1} \right)^{\frac{n-1}{n}} + \left(\frac{p_3}{p_2} \right)^{\frac{n-1}{n}} - 2 \right\} \right] = 0$$

Now substituting $\left(\frac{n}{n-1} \right)$ as a (a constant) in the above equation,

$$\frac{d}{dp_2} \left[a\, p_1 v_1 \left\{ \left(\frac{p_2}{p_1} \right)^a + \left(\frac{p_3}{p_2} \right)^a - 2 \right\} \right] = 0$$

$$a\, p_1 v_1 \left[\left(\frac{1}{p_1} \right)^a (a) (p_2)^{a-1} + (p_3)^a (-a) (p_2)^{-a-1} \right] = 0$$

$$a (p_1)^{-a} (p_2)^{a-1} - a (p_3)^a (p_2)^{-a-1} = 0$$

$$a (p_1)^{-a} (p_2)^{a-1} = a (p_3)^a (p_2)^{-a-1}$$

$$\frac{(p_2)^{a-1}}{(p_2)^{-a-1}} = \frac{(p_3)^a}{(p_1)^{-a}}$$

$$(p_2)^{a-1} (p_2)^{a+1} = (p_3)^a (p_1)^a$$

$$(p_2)^{2a} = (p_3 p_1)^a \quad \text{or} \quad p_2^2 = p_3 p_1$$

$$\therefore \qquad p_2 = \sqrt{p_3 p_1} \qquad \ldots (ii)$$

or in other words,

$$\frac{p_2}{p_1} = \frac{p_3}{p_2} = \left(\frac{p_3}{p_1} \right)^{1/2} \qquad \ldots (iii)$$

Now substituting the value of $\frac{p_3}{p_2} = \frac{p_2}{p_1}$ in equation (*i*), we have minimum work required for a two stage reciprocating air compressor,

$$W = \frac{n}{n-1} \times p_1 v_1 \left[\left(\frac{p_2}{p_1} \right)^{\frac{n-1}{n}} + \left(\frac{p_2}{p_1} \right)^{\frac{n-1}{n}} - 2 \right]$$

$$= 2 \times \frac{n}{n-1} \times p_1 v_1 \left[\left(\frac{p_2}{p_1} \right)^{\frac{n-1}{n}} - 1 \right] \qquad \ldots (iv)$$

$$= 2 \times \text{Work required for each stage}$$

Now substituting $\dfrac{p_2}{p_1} = \left(\dfrac{p_3}{p_1} \right)^{1/2}$ in equation (*iv*),

$$W = 2 \times \frac{n}{n-1} \times p_1 v_1 \left[\left(\frac{p_3}{p_1} \right)^{\frac{n-1}{2n}} - 1 \right]$$

Similarly, it can be proved that for a three stage compressor

$$\frac{p_2}{p_1} = \frac{p_3}{p_2} = \frac{p_4}{p_3} = \left(\frac{p_4}{p_1} \right)^{1/3}$$

and minimum work required for a three stage compressor,

$$W = 3 \times \frac{n}{n-1} \times p_1 v_1 \left[\left(\frac{p_4}{p_1} \right)^{\frac{n-1}{3n}} - 1 \right]$$

$$= 3 \times \frac{n}{n-1} \times m\, R\, T_1 \left[\left(\frac{p_4}{p_1} \right)^{\frac{n-1}{3n}} - 1 \right]$$

Note: In general, for a compressor having q number of stages,

$$\frac{p_2}{p_1} = \frac{p_3}{p_2} = \frac{p_4}{p_3} = \ldots\ldots = \frac{p_{q+1}}{p_q} = \left(\frac{p_{q+1}}{p_1} \right)^{1/q}$$

and minimum work required for compression,

$$W = q \times \frac{n}{n-1} \times p_1\, v_1 \left[\left(\frac{p_{q+1}}{p_1} \right)^{\frac{n-1}{qn}} - 1 \right]$$

Example 28.7. *Estimate the minimum work required to compress 1 kg of air from 1 bar 27° C to 16 bar in two stages, if the law of compression is $pv^{1.25}$ = constant and the intercooling is perfect. Take R = 287 J/kg K.*

Solution. Given : $m = 1$ kg ; $p_1 = 1$ bar $= 1 \times 10^5$ N/m^2 ; $T_1 = 27°$ C $= 27 + 273 = 300$ K ; $p_3 = 16$ bar ; $n = 1.25$; $R = 287$ J/kg K

We know that for perfect intercooling, the intercooler pressure,

$$p_2 = \sqrt{p_1\, p_3} = \sqrt{1 \times 16} = 4 \text{ bar}$$

∴ Minimum work required to compress 1 kg of air,

$$W = 2 \times \frac{n}{n-1} \times mRT_1 \left[\left(\frac{p_2}{p_1} \right)^{\frac{n-1}{n}} - 1 \right]$$

$$= 2 \times \frac{1.25}{1.25-1} \times 1 \times 287 \times 300 \left[\left(\frac{4}{1} \right)^{\frac{1.25-1}{1.25}} - 1 \right] \text{ N-m}$$

$$= 861\,000 \times 0.3195 = 275\,090 \text{ N-m } \textbf{Ans.}$$

Example 28.8. *A two stage compressor takes in 2.82 m^3 of air per minute at a pressure of 1.05 bar and temperature of 22°C. It delivers the air at 8.44 bar. The compression is carried out in each cylinder according to the law $p\,v^{1.2}$ = Constant. The air is cooled to its initial temperature in intercooler. Neglecting clearance, find the minimum power required to drive the compressor.*

Solution. $v_1 = 2.82\ \text{m}^3/\text{min}$; $p_1 = 1.05\ \text{bar} = 1.05 \times 10^5\ \text{N/m}^2$; $T_1 = 22°\text{C} = 22 + 273 = 295\ \text{K}$; $p_3 = 8.44$ bar ; $n = 1.2$

Since the air is cooled to its initial temperature, therefore the intercooling is perfect. We know that for perfect intercooling, the intercooler pressure,

$$p_2 = \sqrt{p_1\,p_3} = \sqrt{1.05 \times 8.44} = 2.977 \text{ bar}$$

and minimum work required to drive the compressor,

$$W = 2 \times \frac{n}{n-1} \times p_1\,v_1 \left[\left(\frac{p_2}{p_1} \right)^{\frac{n-1}{n}} - 1 \right]$$

$$= 2 \times \frac{1.2}{1.2-1} \times 1.05 \times 10^5 \times 2.82 \left[\left(\frac{2.977}{1.05} \right)^{\frac{1.2-1}{1.2}} - 1 \right]$$

$$= 35.5 \times 10^5 (1.19 - 1) = 674\,500 \text{ N-m/min}$$

∴ Power required to drive the compressor

$$= \frac{674\,500}{60} = 11\,242 \text{ W} = 11.242 \text{ kW } \textbf{Ans.} \qquad \ldots (\because 1 \text{ N-m/s} = 1 \text{ W})$$

Example 28.9. *A two stage air compressor compresses air from 1 bar and 20° C to 42 bar. If the law of compression is $p\,v^{1.35}$ = constant and the intercooling is complete to 20° C, find per kg of air : 1. The work done is compressing ; and 2. The mass of water necessary for abstracting the heat in the intercooler, if the temperature rise of the cooling water is 25° C.*

Take R = 287 J/kg K and c_p = 1 kJ/kg K.

Solution. Given : $p_1 = 1\ \text{bar} = 1 \times 10^5\ \text{N/m}^2$; $T_1 = 20°\text{C} = 20 + 273 = 293\ \text{K}$; $p_3 = 42$ bar $= 42 \times 10^5\ \text{N/m}^2$; $n = 1.35$; $T_3 = 20°\text{C} = 20 + 273 = 293\ \text{K}$; $m = 1$ kg ; Rise in temperature of cooling water = 25° C ; $R = 287$ J/kg K ; $c_p = 1$ kJ/kg K

We know that for complete intercooling, the intercooler pressure,

$$p_2 = \sqrt{p_1\,p_3} = \sqrt{1 \times 42} = 6.48 \text{ bar}$$

and volume of air admitted for compression,

$$v_1 = \frac{mRT_1}{p_1} = \frac{1 \times 287 \times 293}{1 \times 10^5} = 0.84 \text{ m}^3\text{/kg of air} \qquad \ldots (\because p_1 v_1 = mRT_1)$$

1. *Work done in compressing the air*

We know that work done in compresing the air,

$$W = \frac{n}{n-1} \times p_1 v_1 \left[\left(\frac{p_2}{p_1} \right)^{\frac{n-1}{n}} + \left(\frac{p_3}{p_2} \right)^{\frac{n-1}{n}} - 2 \right]$$

$$= \frac{1.35}{1.35-1} \times 1 \times 10^5 \times 0.84 \left[\left(\frac{6.48}{1} \right)^{\frac{1.35-1}{1.35}} + \left(\frac{42}{6.48} \right)^{\frac{1.35-1}{1.35}} - 2 \right] \text{ N-m}$$

$$= 3.24 \times 10^5 \, (1.62 + 1.62 - 2) = 4.017 \times 10^5 \text{ N-m} \quad \textbf{Ans.}$$

2. *Mass of water necessary for abstracting the heat in the intercooler*

Let m_w = Mass of water necessary/kg of air, and

T_2 = Temperature of the air entering the intercooler.

We know that
$$\frac{T_2}{T_1} = \left(\frac{p_2}{p_1} \right)^{\frac{n-1}{n}} = \left(\frac{6.48}{1} \right)^{\frac{1.35-1}{1.35}} = 1.622$$

$$\therefore \quad T_2 = T_1 \times 1.622 = 293 \times 1.622 = 475.6 \text{ K}$$

We also know that heat gained by water

= Heat lost by air

$\therefore \; m_w \times c_w \times$ Rise in temperature

$$= m \, c_p \, (T_2 - T_3)$$

$$m_w \times 4.2 \times 25 = 1 \times 1 \, (475.6 - 293) = 182.6$$

$$\therefore \quad m_w = 1.74 \text{ kg} \quad \textbf{Ans.}$$

Example 28.10. *Find the percentage saving in work by compressing air in two stages from 1 bar to 7 bar instead of in one stage. Assume compression index 1.35 in both the cases and optimum pressure and complete intercooling in two stage compressor.*

Solution. Given : p_1 = 1 bar = 1×10^5 N/m^2 ; p_2 = 7 bar = 7×10^5 N/m^2 ; n = 1.35

We know that workdone in compresing air in one stage,

$$W_1 = \frac{n}{n-1} \times p_1 v_1 \left[\left(\frac{p_2}{p_1} \right)^{\frac{n-1}{n}} - 1 \right]$$

$$= \frac{1.35}{1.35-1} \times 1 \times 10^5 \, v_1 \left[(7)^{\frac{1.35-1}{1.35}} - 1 \right] = 253\,100 \, v_1 \text{ N-m} \quad \ldots (i)$$

and workdone in compressing air in two stages

$$W_2 = 2 \times \frac{n}{n-1} \times p_1 v_1 \left[\left(\frac{p_2}{p_1} \right)^{\frac{n-1}{2n}} - 1 \right]$$

$$= 2 \times \frac{1.35}{1.35-1} \times 1 \times 10^5 \, v_1 \left[(7)^{\frac{1.35-1}{2 \times 1.35}} - 1 \right] \text{N-m}$$

$$= 221\,300 \, v_1 \text{ N-m} \quad \ldots (ii)$$

$$\therefore \text{ Saving in work done } = \frac{253\,100\,v_1 - 221\,300\,v_1}{253\,100\,v_1} = 0.126 \text{ or } 12.6\% \textbf{ Ans.}$$

28.17. Heat Rejected in a Reciprocating Air Compressor

The total heat rejected in a reciprocating air compressor is the sum of the heat rejected during polytropic compression per kg of air and heat rejected in the intercooler per kg of air.

We know that heat rejected during polytropic compression per kg of air,

$$q_1 = \frac{\gamma - n}{\gamma - 1} \times \text{Work done} = \frac{\gamma - n}{\gamma - 1} \times \frac{R\,(T_2 - T_1)}{(n-1)}$$

$$= \frac{c_v\,(\gamma - n)\,(T_2 - T_1)}{n-1} \qquad \ldots [\because R = c_v\,(\gamma - 1)]$$

and heat rejected in the intercooler per kg of air,

$$q_2 = c_p\,(T_2 - T_3)$$

∴ Total heat rejected per kg of air,

$$q = q_1 + q_2 = \frac{c_v\,(\gamma - n)\,(T_2 - T_1)}{(n-1)} + c_p\,(T_2 - T_3)$$

Note : For complete intercooling $T_1 = T_3$. Therefore substituting this value in case of intercooling,

$$q = \frac{c_v\,(\gamma - n)\,(T_2 - T_1)}{n-1} + c_p\,(T_2 - T_1)$$

$$= (T_2 - T_1)\left[\frac{c_v\,(\gamma - n)}{n-1} + c_p\right]$$

Example 28.11. *A two-stage single acting reciprocating compressor takes in air at the rate of 0.2 m³/s. The intake pressure and temperature of air are 0.1 MPa and 16° C. The air is compressed to a final pressure of 0.7 MPa. The intermediate pressure is ideal and intercooling is perfect. The compression index in both the stages is 1.25 and the compressor runs at 600 r.p.m. Neglecting clearance, determine 1. the intermediate pressure, 2. the total volume of each cylinder, 3. the power required to drive the compressor, and 4. the rate of heat rejection in the intercooler. Take C_p = 1.005 kJ/kg K and R = 287 J/kg K.*

Solution. Given : $v_1 = 0.2$ m³/s ; $p_1 = 0.1$ MPa $= 0.1 \times 10^6$ N/m² ; $T_1 = 16°$ C $= 16 + 273 = 289$ K ; $p_3 = 0.7$ MPa $= 0.7 \times 10^6$ N/m² ; $n = 1.25$; $N = 600$ r.p.m. ; $c_p = 1.005$ kJ/kg K ; $R = 287$ J/kg K

1. *Intermediate pressure*

We know that for perfect intercooling, the intermediate pressure (*i.e.* intercooler pressure),

$$p_2 = \sqrt{p_1\,p_3} = \sqrt{0.1 \times 0.7} = 0.2646 \text{ MPa } \textbf{Ans.}$$

2. *Total volume of each cylinder*

Let v_L = Volume of L.P. cylinder, and

v_H = Volume of H.P. cylinder.

Since the compressor is a single acting, therefore number of working strokes/min,

$$N_w = N = 600$$

$$\therefore \quad v_L = \frac{v_1 \times 60}{600} = \frac{0.2 \times 60}{600} = 0.02 \text{ m}^3 \textbf{ Ans.}$$

We know that $p_1 v_L = p_2 v_H$

$$\therefore \quad v_H = v_L \times \frac{p_1}{p_2} = 0.02 \times \frac{0.1}{0.2646} = 0.007\,56 \text{ m}^3 \textbf{ Ans.}$$

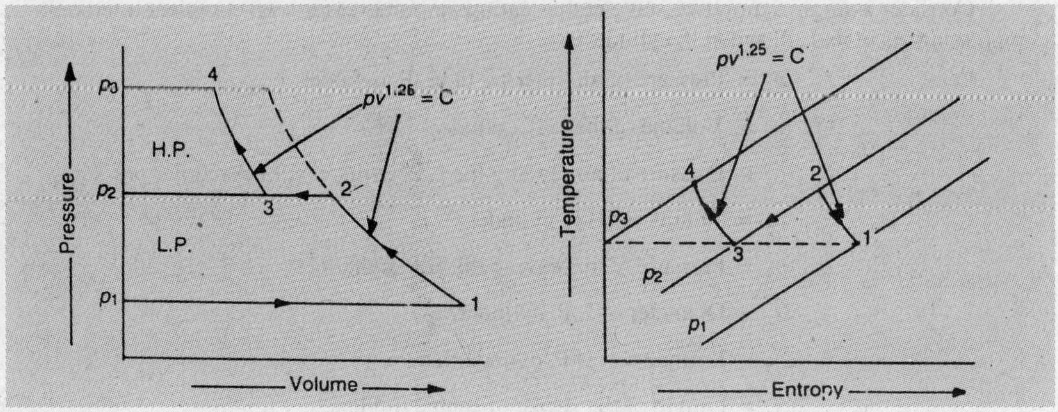

Fig. 28.7

3. *Power required to drive the compressor*

We know that work required to compress the air,

$$W = 2 \times \frac{n}{n-1} \times p_1 v_1 \left[\left(\frac{p_2}{p_1} \right)^{\frac{n-1}{n}} - 1 \right]$$

$$= 2 \times \frac{1.25}{1.25 - 1} \times 0.1 \times 10^6 \times 0.2 \left[\left(\frac{0.2646}{0.1} \right)^{\frac{1.25-1}{1.25}} - 1 \right] \text{ J/s}$$

$$= 42\,970 \text{ J/s} = 42.97 \text{ kJ/s}$$

∴ Power required to drive the compressor,

$$P = 42.97 \text{ kW } \textbf{Ans.} \qquad \ldots (\because 1 \text{ kW} = 1 \text{ kJ/s})$$

4. *Rate of heat rejection in the intercooler*

Let m = Mass of air admitted for compression, and

T_2 = Temperature of the air entering the intercooler.

We know that $p_1 v_1 = mRT_1$

$$\therefore \quad m = \frac{p_1 v_1}{R\,T_1} = \frac{0.1 \times 10^6 \times 0.2}{287 \times 289} = 0.241 \text{ kg/s}$$

We also know that

$$\frac{T_2}{T_1} = \left(\frac{p_2}{p_1} \right)^{\frac{n-1}{n}} = \left(\frac{0.2646}{0.1} \right)^{\frac{1.25-1}{1.25}} = 1.215$$

$$\therefore \quad T_2 = T_1 \times 1.215 = 289 \times 1.215 = 351.1 \text{ K}$$

We know that heat rejected in the intercooler

$$= m\, c_p\, (T_2 - T_3)$$

$$= 0.241 \times 1.005\,(351.1 - 289) = 15 \text{ kJ/s } \textbf{Ans.} \qquad \ldots (\because T_3 = T_1)$$

28.18. Ratio of Cylinder Diameters

Consider a single acting two stage reciprocating air compressor with complete intercooling, compressing air in its L.P. and H.P. cylinders.

Let p_1 = Pressure of air entering the L.P. cylinder,

v_1 = Volume of the L.P. cylinder,

p_2 = Pressure of air leaving the L.P. cylinder (or intercooler pressure),

v_2 = Volume of H.P. cylinder,

p_3 = Pressure of air leaving the H.P. cylinder,

D_1 = Diameter of L.P. cylinder,

D_2 = Diameter of H.P. cylinder, and

L = Length of the stroke of each cylinder.

We know that for complete intercooling,

$$p_1 v_1 = p_2 v_2$$

$$\therefore \quad \frac{p_2}{p_1} = \frac{v_1}{v_2} = \frac{\frac{\pi}{4}(D_1)^2 L}{\frac{\pi}{4}(D_2)^2 L} = \frac{(D_1)^2}{(D_2)^2}$$

or

$$\frac{D_1}{D_2} = \left(\frac{p_2}{p_1}\right)^{1/2} \qquad \ldots (i)$$

We also know that for two stage compression with complete intercooling,

$$\frac{p_2}{p_1} = \left(\frac{p_3}{p_1}\right)^{1/2}$$

Substituting the value of p_2/p_1 in equation (i),

$$\frac{D_1}{D_2} = \left(\frac{p_3}{p_1}\right)^{1/4}$$

Example 28.12. *A two stage air compressor, with complete intercooling, delivers air to the mains at a pressure of 30 bar, the suction conditions being 1 bar and 27° C. If both cylinders have the same stroke, find the ratio of the cylinder diameters, for the efficiency of compression to be maximum. Assume the index of compression to be 1.3.*

Solution. Given : $p_3 = 30$ bar ; $p_1 = 1$ bar, $^*T_1 = 27°$ C ; $n = 1.3$

We know that for complete intercooling, the intercooler pressure,

$$p_2 = \sqrt{p_1\, p_3} = \sqrt{1 \times 30} = 5.48 \text{ bar}$$

---

* Superfluous data

∴ Ratio of cylinder diameters,

$$\frac{D_1}{D_2} = \left(\frac{p_2}{p_1}\right)^{1/2} = \left(\frac{5.48}{1}\right)^{1/2} = 2.34 \textbf{ Ans.}$$

Note : The ratio of cylinder diameters is also given by

$$\frac{D_1}{D_2} = \left(\frac{p_3}{p_1}\right)^{1/4} = \left(\frac{30}{1}\right)^{1/4} = 2.34 \textbf{ Ans.}$$

Example 28.13. *A multi-stage air compressor is to be designed to elevate the pressure from 1 bar to 100 bar such that the stage pressure ratio will not exceed 4. Determine : 1. Number of stages, 2. Exact stage pressure ratio, and 3. Intermediate pressure.*

Solution. Given : $p_1 = 1$ bar ; $p_{q+1} = 100$ bar ; $\dfrac{p_{q+1}}{p_q} = 4$

1. *Number of stages*

Let q = Number of stages.

We know that $$\frac{p_{q+1}}{p_q} = \left(\frac{p_{q+1}}{p_1}\right)^{1/q} \quad \text{or} \quad 4 = \left(\frac{100}{1}\right)^{1/q}$$

Taking log on both sides,

$$\log 4 = \frac{1}{q}\log 100 \quad \text{or} \quad 0.6021 = \frac{1}{q}\times 2$$

$$\therefore \quad q = 2/0.6021 = 3.32 \text{ say } 4 \textbf{ Ans.}$$

2. *Exact stage pressure ratio*

We know that $$\frac{p_{q+1}}{p_q} = \left(\frac{p_{q+1}}{p_1}\right)^{1/q} = \left(\frac{100}{1}\right)^{1/4} = 3.162 \textbf{ Ans.}$$

3. *Intermediate pressure*

Let $^*p_3$ = Intermediate pressure.

We know that $$\frac{p_5}{p_4} = 3.162$$

$$\therefore \quad p_4 = \frac{p_5}{3.162} = \frac{100}{3.162} = 31.62 \text{ bar}$$

Similarly $$\frac{p_4}{p_3} = \frac{p_5}{p_4} = 3.162 \quad \text{or} \quad p_3 = \frac{p_4}{3.162} = \frac{31.62}{3.162} = 10 \text{ bar } \textbf{Ans.}$$

Example 28.14. *A three stage compressor compresses air from 1 bar to 35 bar and delivers it at the higher pressure to a receiver. The initial temperature is 17° C. The law of compression is $pv^{1.25}$ = constant, and is the same for each stage. Assuming conditions of minimum work, perfect intercooling and that the effect of cylinder clearance and valve resistance etc., may be neglected, find the power required to deliver 14 m³/ min air, measured at the suction conditions. Find also the intermediate pressures.*

* The intermediate pressure of an air compressor having four stages is p_3.

Solution. Given : $p_1 = 1$ bar $= 1 \times 10^5$ N/m^2 ; $p_4 = 35$ bar ; $T_1 = 17°$ C $= 17 + 273 = 290$ K ; $v_1 = 14$ m^3/min

Power required

We know that minimum work required per min.,

$$W = \frac{3n}{n-1} \times p_1 v_1 \left[\left(\frac{p_4}{p_1}\right)^{\frac{n-1}{3n}} - 1\right]$$

$$= \frac{3 \times 1.25}{1.25 - 1} \times 1 \times 10^5 \times 14 \left[\left(\frac{35}{1}\right)^{\frac{1.25-1}{3 \times 1.25}} - 1\right] \text{ N-m/min}$$

$$= 5.65 \times 10^6 \text{ N-m/min} = 5650 \text{ kJ/min}$$

$$\therefore \text{ Power required} = \frac{5650}{60} = 94.2 \text{ kW } \textbf{Ans.}$$

Intermediate pressures

Let p_2 and p_3 = Intermediate pressures.

We know that for minimum work,

$$\frac{p_2}{p_1} = \frac{p_3}{p_2} = \frac{p_4}{p_3} = \left(\frac{p_4}{p_1}\right)^{1/3}$$

$$\therefore \quad \frac{p_2}{p_1} = \left(\frac{p_4}{p_1}\right)^{1/3} = (35)^{1/3} = 3.27$$

or $$p_2 = p_1 \times 3.27 = 1 \times 3.27 = 3.27 \text{ bar } \textbf{Ans.}$$

Similarly $$\frac{p_3}{p_2} = \left(\frac{p_4}{p_1}\right)^{1/3} = 3.27$$

$$\therefore \quad p_3 = p_2 \times 3.27 = 3.27 \times 3.27 = 10.7 \text{ bar } \textbf{Ans.}$$

Example 28.15. *A three stage compressor delivers air at 70 bar from an atmospheric pressure of 1 bar and 30° C. Assuming the intercooling complete, estimate the amount of minimum work required to deal with 1 kg of air. Also find the amount of heat rejected in each intercooler. The index of compression is 1.2 throughout. Take c_p for air = 1.005 kJ/kg K.*

Solution. Given : $p_4 = 70$ bar ; $p_1 = 1$ bar ; $T_1 = 30°$ C $= 30 + 273 = 303$ K ; $m = 1$ kg ; $n = 1.2$; $c_p = 1.005$ kJ/kg K

Amount of minimum work required

We know that amount of minimum work required,

$$W = \frac{3n}{n-1} \times mRT_1 \left[\left(\frac{p_4}{p_1}\right)^{\frac{n-1}{3n}} - 1\right]$$

$$= \frac{3 \times 1.2}{1.2 - 1} \times 1 \times 287 \times 303 \left[\left(\frac{70}{1}\right)^{\frac{1.2-1}{3 \times 1.2}} - 1\right] = 417\,070 \text{ J}$$

$$= 417.07 \text{ kJ } \textbf{Ans.}$$

Heat rejected in each intercooler

Let T_2 = Temperature of air at the end of first stage.

Since the air compressor is three stage, therefore

$$\frac{p_2}{p_1} = \left(\frac{p_4}{p_1}\right)^{1/3} = \left(\frac{70}{1}\right)^{1/3} = 4.12$$

We know that $$\frac{T_2}{T_1} = \left(\frac{p_2}{p_1}\right)^{\frac{n-1}{n}} = (4.12)^{\frac{1.2-1}{1.2}} = 1.266$$

$\therefore$ $$T_2 = T_1 \times 1.266 = 303 \times 1.266 = 383.6 \text{ K}$$

Since the pressure ratio and index of compression is same for each stage, and intercooling is complete, therefore temperature of air will be same on entering each intercooler.

$\therefore$ Heat rejected in each intercooler

$$= c_p (T_2 - T_3) = 1.005 (383.6 - 303) = 81 \text{ kJ } \textbf{Ans.}$$

Example 28.16. *A three stage single acting reciprocating compressor has perfect intercooling. The pressure and temperature at the end of suction stroke in L.P. cylinder is 1.013 bar and 15° C respectively. If 8.4 m³ of free air is delivered by the compressor at 70 bar per minute and the work done is minimum, calculate :*

1. L.P. and I.P. delivery pressures ; 2. Ratio of cylinder volumes ; and 3. Total indicated power. Neglect clearance and assume n = 1.2.

Solution. Given : p_1 = 1.013 bar = 1.013×10^5 N/m² ; T_1 = 15° C = 15 + 273 = 288 K ; v_1 = 8.4 m³/min ; p_4 = 70 bar ; n = 1.2

1. *L.P. and I.P. delivery pressures*

Let p_2 = L.P. delivery pressure, and

p_3 = I.P. delivery pressure.

We know that for minimum work,

$$\frac{p_2}{p_1} = \frac{p_3}{p_2} = \frac{p_4}{p_3} = \left(\frac{p_4}{p_1}\right)^{1/3}$$

$\therefore$ $$\frac{p_2}{p_1} = \left(\frac{p_4}{p_1}\right)^{1/3} = \left(\frac{70}{1.013}\right)^{1/3} = 4.1$$

or $$p_2 = p_1 \times 4.1 = 1.013 \times 4.1 = 4.153 \text{ bar } \textbf{Ans.}$$

Similarly, $$p_3 = p_2 \times 4.1 = 4.153 \times 4.1 = 17.03 \text{ bar } \textbf{Ans.}$$

2. *Ratio of cylinder volumes*

Let $v_1 : v_2 : v_3$ = Ratio of cylinder volumes.

We know that for minimum work,

$$p_1 v_1 = p_2 v_2 = p_3 v_3$$

or $$\frac{v_1}{v_2} = \frac{p_2}{p_1} = \frac{4.153}{1.013} = 4.1 \quad \ldots (i)$$

$$\therefore \qquad v_1 = 4.1\, v_2$$

Also
$$\frac{v_2}{v_3} = \frac{p_3}{p_2} = \frac{17.03}{4.157} = 4.1$$

$$\therefore \qquad v_2 = 4.1\, v_3 \qquad \ldots (ii)$$

Now taking v_3 as 1 m³, we find that ratio of cylinder volumes,

$$v_1 : v_2 : v_3 = 16.81 : 4.1 : 1 \text{ Ans.}$$

3. *Total indicated power*

We know that minimum work required,

$$W = \frac{3\,n}{n-1} \times p_1\, v_1 \left[\left(\frac{p_4}{p_1}\right)^{\frac{n-1}{3n}} - 1\right]$$

$$= \frac{3 \times 1.2}{1.2 - 1} \times 1.013 \times 10^5 \times 8.4 \left[\left(\frac{70}{1.013}\right)^{\frac{1.2-1}{3\times 1.2}} - 1\right] \text{J/min}$$

$$= 4067 \times 10^3 \text{ J/min} = 4067 \text{ kJ/min}$$

∴ Total indicated power,

$$P = \frac{4067}{60} = 67.8 \text{ kW Ans.}$$

EXERCISES

1. A single stage reciprocating air compressor is required to compress 60 cubic metres of air from 1 bar to 8 bar at 22° C. Find work done by the compressor, if the compression of air is 1. isothermal, 2. isentropic with isentropic index as 1.4 ; and 3. polytropic with polytropic index as 1.25. [**Ans.** 12.5 MJ ; 17 MJ ; 15.5 MJ]

2. Find the power required to compress and deliver 2 kg of air per minute from 1 bar and 20° C to a delivery pressure 7 bar, when the compression is carried out in a single stage compressor. The compression of air follows the law $p\,v^{1.4}$ = constant. Neglect clearance. Take R = 287 J/kg K. [**Ans.** 7.3 kW)

3. Determine the cylinder dimensions of a double acting, 11 kW indicated power, air compressor which compresses air from 1 bar to 7 bar according to the law $pv^{1.2}$ = constant. The average piston speed is 150 m/s. Assume stroke to diameter ratio of 1.5 and neglect clearance. [**Ans.** 156 mm ; 234 mm]

4. A small single acting compressor has a bore and stroke both of 100 mm and is driven at 400 r.p.m. The clearance volume is 80 × 10³ mm³ and the index of compression and expansion is 1.2. The suction pressure is 0.95 bar and the delivery pressure is 8 bar. Calculate

1. the volume of free air at 1.013 bar and 20° C dealt with per minute if the temperature at the start of compression is 30° C ; 2. the mean effective pressure of the indicator diagram assuming constant suction and delivery pressure ; and 3. the power required to drive the compressor.

[**Ans.** 0.1426 m³/min ; 1.22 bar ; 0.637 kW]

5. An air compressor takes in air at 0.98 bar and 20° C and compresses it according to the law $pv^{1.2}$ = C. It is then delivered to a receiver at constant pressure of 9.8 bar. Determine : 1. the temperature at the end of compression ; 2. the workdone per kg of air ; 3. the heat transferred during the compression ; and 4. the workdone during delivery. Take R = 287 J/kg K and γ = 1.4. [**Ans.** 150° C ; 236 kJ ; 118 kJ ; 123.7 kJ]

6. A two stage air compressor with perfect intercooling takes in air at 1 bar pressure and 27° C. The law of compression in both the stages is $pv^{1.3}$ = constant. The compressed air is delivered at 9 bar from the H.P. cylinder to an air receiver. Calculate per kg of air 1. the minimum workdone, and 2. the heat rejected to intercooler.

[**Ans.** 215 kJ ; 86.6 kJ]

7. A two stage air compressor takes in 22.5 kg of air per minute at 15°C and 1 bar and delivers it at 16.5 bar. At the intermediate pressure, it is cooled to initial temperature. Assuming an ideal diagram with no clearance and compression according to $pv^{1.2}$ = constant, determine the intermediate pressure that gives least work. Also find the heat rejected in the intercooler per minute and minimum power required to run the compressor. Take $c_p = 1$ kJ/kg K; and $R = 287$ J/kg K **[Ans.** 4.06 bar ; 1701 kJ/min ; 97.8 kW**]**

8. A single acting compressor is required to deliver air at 70 bar from a suction pressure of 1 bar at the rate of 2.3 m³/min. measured at free air conditions of 1.013 bar and 15° C. The temperature at the end of the suction stroke is 32° C. Calculate the indicated power required if the compression is carried out in two stages with an ideal intermediate pressure and complete intercooling. The index of compression and expansion for both stages is 1.25. Also calculate the heat rejected per minute to the intercooler and the saving in power over single stage compression. For air, $c_p = 1$ kJ/kg K and $c_v = 0.718$ kJ/kg K. Neglect clearance volume.

[Ans. 22 kW ; 461.2 kJ/min ; 5.84 kW**]**

9. A two stage reciprocating air compressor delivers 40.5 kg/min at 9.8 bar. The intake pressure is 1 bar and the intake temperature is 15.5° C. The compression follows $p\,v^{1.31}$ = constant and the intercooler cools the air back to the intake temperature. Neglecting clearance, calculate 1. the optimum intermediate pressure ; 2. the power to be delivered to each cylinder ; and 3. the rate of heat transfer from the cylinders and intercooler.

[Ans. 3.13 bar ; 73.2 kW ; 3625 kJ/min**]**

10. A three stage reciprocating compressor compresses air from 1 bar and 26° C to 36 bar. The law of compression is $pv^{1.3}$ = constant and is same for all the three stages of compression. Assuming perfect intercooling and neglecting clearance, find the minimum power required to compress 0.25 m³/s of free air. Also find the intermediate pressures. **[Ans.** 103.12 kW ; 3.302 bar ; 10.903 bar**]**

QUESTIONS

1. Classify air compressors. Describe the working of a single stage reciprocating air compressor.

2. Draw *p-v* and *T-s* diagram for a single stage reciprocating air compressor, without clearance. Derive the expression for the workdone when compression is (*a*) isothermal, and (*b*) isentropic.

3. Sketch the theoretical indicator diagram for a single stage, single cylinder reciprocating compressor with clearance volume showing the various processes. For such a compressor, derive the expression for workdone in terms of mass rate of flow of air, initial temperature, pressure ratio and index of compression.

4. When is multi-stage compression used for air ? What are its advantages ?

5. Explain the effect of intercooling in a multistage reciprocating compressor.

6. Discuss briefly a two stage air compressor with intercooler. Draw the ideal *p-v* diagram. Derive the expression for work done per unit mass of air. Establish that the workdone is minimum when the pressure ratio for each stage is the same and there is complete intercooling.

7. In a two stage air compressor, in which intercooling is perfect, prove that the work done in compression is a minimum when the pressure in the intercooler is the geometric mean between the initial and final pressures. Draw the indicator diagram for two stage compression.

8. Discuss the procedure for obtaining the ratio of cylinder dimensions in an air compressor.

OBJECTIVE TYPE QUESTIONS

1. The volume of air delivered by the compressor is called

(*a*) free air delivery (*b*) compressor capacity
(*c*) swept volume (*d*) none of these

2. The volume of air sucked by the compressor during its suction stroke is called

(*a*) free air delivery (*b*) compressor capacity
(*c*) swept volume (*d*) none of these

3. The ratio of workdone per cycle to the stroke volume of the compressor is known as

(*a*) compressor capacity (*b*) compression ratio

(*c*) compressor efficiency (*d*) mean effective pressure

4. In a single stage, single acting reciprocating air compressor without clearance volume, the workdone is minimum during

(*a*) isothermal compression (*b*) isentropic compression

(*c*) polytropic compression (*d*) none of these

5. The pressure of air at the beginning of the compression stroke is ....atmospheric pressure.

(*a*) equal to (*b*) less than (*c*) more than

6. A 3 m^3/min compressor means that it

(*a*) compresses 3 m^3/min of standard air (*b*) compresses 3 m^3/min of free air

(*c*) delivers 3 m^3/min of compressed air

(*d*) delivers 3 m^3/min of compressed air at delivery pressure.

7. The multi-stage compression of air as compared to single stage compression

(*a*) improves volumetric efficiency for the given pressure ratio

(*b*) reduces workdone per kg of air (*c*) gives more uniform torque

(*d*) reduces cost of compressor (*e*) all of the above

8. The intercooling in multi-stage compressors is done

(*a*) to cool the air during compression (*b*) to cool the air at delivery

(*c*) to enable compression in two stages (*d*) to minimise the work of compression

9. The intercooler pressure (p_2) for minimum work required, for a two stage reciprocating air compressor, is given by

(*a*) $p_2 = p_1\, p_3$ (*b*) $p_2 = p_1/p_3$ (*c*) $p_2 = \sqrt{p_1\, p_3}$ (*d*) $p_2 = p_3 / p_1$

where p_1 = Intake pressure of air, and

p_3 = Delivery pressure of air.

10. The ratio of cylinder diameters for a single acting, two stage reciprocating air compressor with complete intercooling, is given by

(*a*) $D_1/D_2 = \sqrt{p_1\, p_2}$ (*b*) $D_1/D_2 = \sqrt{p_1/p_2}$

(*c*) $D_1/D_2 = \sqrt{p_2/p_1}$ (*d*) none of these

where D_1 = Diameter of L.P. cylinder, and

D_2 = Diameter of H.P. cylinder.

ANSWERS

1. (*b*)	2. (*c*)	3. (*d*)	4. (*a*)	5. (*b*)
6. (*b*)	7. (*e*)	8. (*d*)	9. (*c*)	10. (*c*)

29

Rotary Air Compressors

1. Introduction. 2. Comparison of Reciprocating and Rotary Air Compressors. 3. Types of Rotary Air Compressors. 4. Roots Blower Compressor. 5. Vane Blower Compressor. 6. Backflow in Positive Displacement Air Compressors. 7. Centrifugal Compressor. 8. Workdone by a Centrifugal Compressor. 9. Velocity Triangle for Moving Blades of a Centrifugal Compressor. 10. Width of Impeller Blades. 11. Prewhirl. 12. Axial Flow Compressors. 13. Comparison of Centrifugal and Axial Flow Air Compressors. 14. Velocity Diagrams for Axial Flow Air Compressors. 15. Degree of Reaction.

29.1. Introduction

In the previous chapter, we have discussed the reciprocating compressors, in which the pressure of the air is increased in its cylinder with the help of a moving piston. But in a rotary air compressor, the air is entrapped between two sets of engaging surface and the pressure of air is increased by squeezing action or back flow of the air.

29.2. Comparison of Reciprocating and Rotary Air Compressors

Following are the main points of comparison of reciprocating and rotary air compressors :

S.No.	*Reciprocating air compressors*	*Rotary air compressors*
1.	The maximum delivery pressure may be as high as 1000 bar.	The maximum delivery pressure is 10 bar only.
2.	The maximum free air discharge is about 300 m^3/min.	The maximum free air discharge is as high as 3000 m^3/min.
3.	They are suitable for low discharge of air at very high pressure.	They are suitable for large discharge of air at low pressure.
4.	The speed of air compressor is low.	The speed of air compressor is high.
5.	The air supply is intermittent.	The air supply is continuous.
6.	The size of air compressor is large for the given discharge.	The size of air compressor is small for the same discharge.
7.	The balancing is a major problem.	There is no balancing problem.
8.	The lubricating system is complicated.	The lubricating system is simple.
9.	The air delivered is less clean, as it comes in contact with the lubricating oil.	The air delivered is more clean, as it does not come in contact with the lubricating oil.
10.	Isothermal efficiency is used for all sorts of calculations.	Isentropic efficiency is used for all sorts of calculations.

29.3. Types of Rotary Air Compressors

Though there are many types of rotary air compressors, yet the following are important from the subject point of view :

1. Roots blower compressor ; 2. Vane blower compressor ; 3. Centrifugal blower compressor and 4. Axial flow compressor.

The first two compressors are popularly known as positive displacement compressors, whereas the last two as non-positive displacement. We shall discuss all the above mentioned rotary compressors one by one.

Note : The positive displacement compressors (*i.e.* roots blower and vane blower) are not very popular from the practical point of view. However, they have some academic importance. The only important rotary compressor is the centrifugal blower compressor.

29.4. Roots Blower Compressor

A roots blower compressor, in its simplest form, consists of two rotors with lobes rotating in an air tight casing which has inlet and outlet ports. Its action resembles with that of a gear pump. There are many designs of wheels, but they generally have two or three lobes (and sometimes even more). In all cases, their action remains the same as shown in Fig. 29.1 (*a*) and (*b*). The lobes are so designed that they provide an air tight joint at the point of their contact.

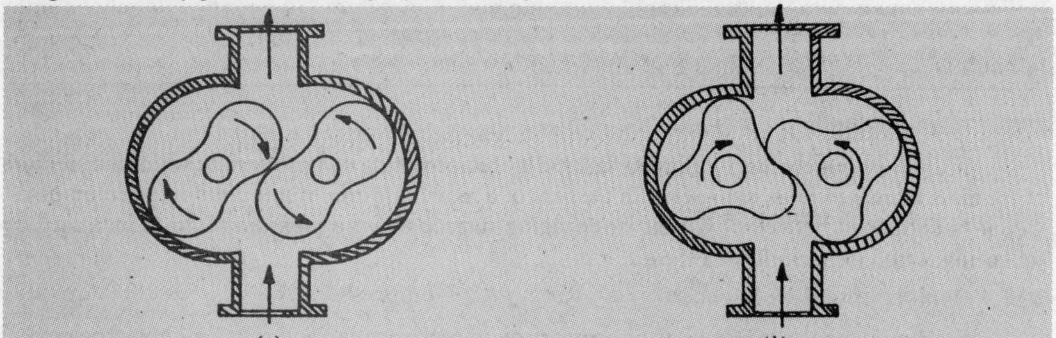

(*a*) (*b*)

Fig. 29.1. Roots blower compressor.

The mechanical energy is provided to one of the rotors from some external source, while the other is gear driven from the first. As the rotors rotate, the air, at atmospheric pressure, is trapped in the pockets formed between the lobes and casing. The rotary motion of the lobes delivers the entrapped air into the receiver. Thus more and more flow of air into the receiver increases its pressure. Finally, the air at a higher pressure is delivered from the receiver.

It will be interesting to know that when the rotating lobe uncovers the exit port, some air (under high pressure) flows back into the pocket from the receiver. It is known as backflow process. The air, which flows from the receiver to the pocket, gets mixed up with the entrapped air. The backflow of air continues, till the pressure in the pocket and receiver is equalised. Thus the pressure of air entrapped in the pocket is increased at constant volume entirely by the backflow of air. The backflow process is shown in Fig. 29.3(*a*). Now the air is delivered to the receiver by the rotation of the lobes. Finally, the air at a higher pressure is delivered from the receiver.

Let p_1 = Intake pressure of air,

p_2 = Discharge pressure of air,

γ = Isentropic index for air, and

v_1 = Volume of air compressed.

We know that theoretical work done in compressing the air,

$$W = \frac{\gamma}{\gamma - 1} \times p_1 v_1 \left[\left(\frac{p_2}{p_1} \right)^{\frac{\gamma - 1}{\gamma}} - 1 \right] \qquad \ldots (i)$$

and actual work $= v_1 (p_2 - p_1)$ .... (*ii*)

∴ Efficiency of roots blower (also known as roots efficiency),

$$\eta = \frac{\frac{\gamma}{\gamma - 1} \times p_1 v_1 \left[\left(\frac{p_2}{p_1} \right)^{\frac{\gamma - 1}{\gamma}} - 1 \right]}{v_1 (p_2 - p_1)}$$

$$= \frac{\gamma}{\gamma - 1} \times \frac{\left[r^{\frac{\gamma - 1}{\gamma}} - 1 \right]}{(r - 1)}$$

where r is the pressure ratio (*i.e.* p_2 / p_1). Now the power required to drive the compressor may be found out from the work done as usual. Thus we see that the efficiency of roots blower decreases with the increase in pressure ratio.

Notes : 1. Sometimes, air at high pressure it obtained by placing two or more roots blower in series, and having intercoolers between each stage.

2. The air is delivered four times in one revolution in case of two-lobbed rotor. Similarly, the air is delivered six times in one revolution in case of three-lobbed rotor.

Example 29.1. *A Roots blower compressor compresses 0.05 m³ of air from 1 bar to 1.5 bar per revolution. Find the compressor efficiency.*

Solution. Given : $v_1 = 0.05\ m^3$; $p_1 = 1\ bar = 1 \times 10^5\ N/m^2$; $p_2 = 1.5\ bar = 1.5 \times 10^5\ N/m^2$

We know that actual work done per revolution,

$$W_1 = v_1 (p_2 - p_1) = 0.05 (1.5 \times 10^5 - 1 \times 10^5) = 2500\ N\text{-}m$$

and ideal work done per revolution.

$$W_2 = \frac{\gamma}{\gamma - 1} \times p_1 v_1 \left[\left(\frac{p_2}{p_1} \right)^{\frac{\gamma - 1}{\gamma}} - 1 \right]$$

$$= \frac{1.4}{1.4 - 1} \times 1 \times 10^5 \times 0.05 \left[\left(\frac{1.5}{1} \right)^{\frac{1.4 - 1}{1.4}} - 1 \right] = 2150\ N\text{-}m$$

∴ Compressor efficiency,

$$\eta = \frac{W_2}{W_1} = \frac{2150}{2500} = 0.86 \quad \text{or} \quad 86\ \% \quad \textbf{Ans.}$$

29.5. Vane Blower Compressor

A vane blower, in its simplest form, as shown in Fig. 29.2, consists of a disc rotating eccentrically in an air tight casing with inlet and outlet ports. The disc has a number of slots (generally 4 to 8) containing vanes. When the rotor rotates the disc, the vanes are pressed against the casing, due to centrifugal force, and form air tight pockets.

Fig. 29.2. Vane blower compressor.

The mechanical energy is provided to the disc from some external source. As the disc rotates, the air is trapped in the pockets formed between the vanes and casing. First of all, the rotary motion of the vanes compresses the air. When the rotating vane uncovers the exit port, some air (under high pressure) flows back into the pocket in the same way as discussed in the case of roots blower compressor. Thus the pressure of air, entrapped in the pocket, is increased first by decreasing the volume and then by the backflow of air as shown in Fig. 29.3(*b*). Now the air is delivered to the receiver by the rotation of the vanes. Finally, the air at a high pressure is delivered from the receiver.

29.6. Backflow in Positive Displacement Air Compressors

We have discussed two important types of positive displacement air compressors *viz*., Roots blower compressor and vane blower compressor in the last articles. In both the cases, the air is delivered to the receiver by the rotating lobes or vanes. It will be interesting to know, that when the rotating lobe (in case of Roots blower) or vane (in case of vane blower) uncovers the exit port, some air (under high pressure) from the receiver flows back into the pockets formed between lobes and casing or vanes and casing. This backflow of air mixes up with the entrapped air, and continues until the pressure in the pockets and receiver are equalised. Thus the pressure of air delivered from the pocket to the receiver is taken to be equal to the receiver pressure. The process of backflow of air is an irreversible process, and called irreversible compression.

It may be noted that the increase of pressure in a Roots blower is entirely due to backflow, and this process is explained on *p-v* diagram as shown in Fig. 29.3 (*a*).

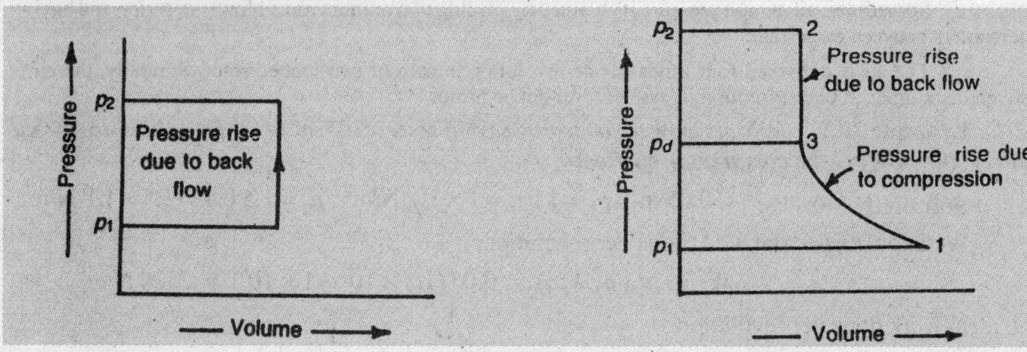

(*a*) Roots blower compressor. (*b*) Vane blower compressor.

Fig. 29.3. *p-v* diagram of air compressor.

The increase of pressure in a vane blower takes place first due to compression and then due to backflow as shown in Fig. 29.3 (*b*). Strictly speaking, the Roots blower compressor is of academic interest only. but vane blower compressor has been used, but with little success. Now consider a vane blower compressor compressing air as shown in Fig. 29.3 (*b*).

Let p_1 = Intake pressure of air,

p_2 = Discharge pressure of air,

p_d = Pressure at point 3.

γ = Isentropic index for air, and

v_1 = Volume of air compressed.

We know that work done due to compression (1-3),

$$W_1 = \frac{\gamma}{\gamma-1} \times p_1 v_1 \left[\left(\frac{p_d}{p_1}\right)^{\frac{\gamma-1}{\gamma}} - 1\right] \quad \ldots (i)$$

and work done due to backflow (3-2),

$$W_2 = v_2 (p_2 - p_d) \quad \ldots (ii)$$

$\therefore$ Total work done, $W = W_1 + W_2$

∴ Efficiency of the vane blower (also known as vane blower efficiency)

$$\eta = \frac{W_2}{W_1 + W_2} \quad \ldots (iii)$$

Now the power required to drive to compressor may be found out from the work done as usual.

Note : The value of v_2 or p_d in equation (*ii*) may be found out from the relation,

$$v_2 = v_1 \left(\frac{p_1}{p_d}\right)^{1/\gamma}$$

Example 29.2. *A rotary vane compressor compresses 4.5 m³ of air per minute from 1 bar to 2 bar when running at 450 r.p.m. Find the power required to drive the compressor when 1. the ports are so placed that there is no internal compression ; and 2. the ports are so placed that there is 50% increase in pressure due to compression before the backflow occurs.*

Solution. Given : $v_1 = 4.5$ m³/min ; $p_1 = 1$ bar $= 1 \times 10^5$ N/m² ; $p_2 = 2$ bar $= 2 \times 10^5$ N/m² ; $N = 450$ r.p.m.

1. *Power required to drive the compressor when there is no internal compression*

We know that work done without internal compression,

$$W = v_1 (p_2 - p_1) = 4.5 (2 \times 10^5 - 1 \times 10^5) = 450\,000 \text{ N-m/min}$$

$$= 450 \text{ kN-m/min}$$

∴ Power required to drive the compressor,

$$P = 450/60 = 7.5 \text{ kW} \textbf{ Ans.}$$

2. *Power required to drive the compressor when there is 50% increase in pressure due to compression*

Since there is 50% increase in the pressure due to compression, therefore delivery pressure before backflow,

$$p_d = 1 + 0.5 (2 - 1) = 1.5 \text{ bar} = 1.5 \times 10^5 \text{ N/m}^2$$

$$\therefore \quad v_2 = v_1 \times \left(\frac{p_1}{p_d}\right)^{1/\gamma} = 4.5 \left(\frac{1}{1.5}\right)^{1/1.4} = 3.37 \text{ m}^3\text{/min}$$

We know that theoretical work done in compressing the air from 1 bar to 1.5 bar,

$$W_1 = \frac{\gamma}{\gamma - 1} \times p_1 v_1 \left[\left(\frac{p_d}{p_1}\right)^{\frac{\gamma - 1}{\gamma}} - 1\right]$$

$$= \frac{1.4}{1.4 - 1} \times 1 \times 10^5 \times 4.5 \left[\left(\frac{1.5}{1}\right)^{\frac{1.4 - 1}{1.4}} - 1\right]$$

$$= 193\,500 \text{ N-m/min} = 193.5 \text{ kN-m/min}$$

and work done in backflow

$$W_2 = v_2 (p_2 - p_d) = 3.37 (2 \times 10^5 - 1.5 \times 10^5)$$

$$= 168\,500 \text{ N-m/min} = 168.5 \text{ kN-m/min}$$

∴ Total work done,

$$W = W_1 + W_2 = 193.5 + 168.5 = 362 \text{ kN-m/min}$$

and power required to drive the compressor,

$$P = 362/60 = 6.03 \text{ kW} \textbf{ Ans.}$$

29.7. Centrifugal Compressor

A centrifugal blower compressor, in its simplest form, consists of a rotor (or impeller) to which a number of curved vanes are fitted symmetrically. The rotor rotates in an air tight volute casing with inlet and outlet points. The casing for the compressor is so designed that the kinetic energy of the air is converted into pressure energy before it leaves the casing as shown in Fig. 29.4.

The mechanical energy is provided to the rotor from some external source. As the rotor rotates, it sucks air through its eye, increases its pressure due to centrifugal force and forces the air to flow over the diffuser. The pressure of air is further increased during its flow over the diffuser.

Finally, the air at a high pressure is delivered to the receiver. It will be interesting to know that the air enters the impeller radially and leaves the vanes axially.

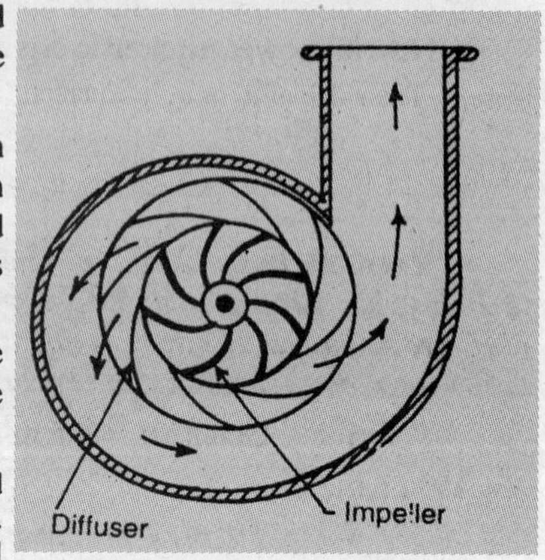

Fig. 29.4. Centrifugal compressor.

Notes : 1. The curved vanes as well as the diffuser are so designed that the air enters and leaves their tips tangentially *i.e.* without shock. Their surface is made very smooth in order to minimise the frictional losses.

2. The workdone by a centrifugal compressor (or power required to drive it) may be found out either by the velocity triangles or otherwise.

29.8. Workdone by a Centrifugal Air Compressor

We have already discussed in Art. 28.6 the work done by a single acting reciprocating air compressor. The equations for work done or power required to drive the reciprocating compressor are applicable for the work done or power required by a rotary compressor also. Thus workdone by a rotary compressor,

$$W = 2.3\, p_1 v_1 \log\left(\frac{v_1}{v_2}\right) \quad \text{... (For isothermal compression)}$$

$$= 2.3\, m\, R\, T_1 \log r \quad \text{... (where } r = v_1 / v_2 \text{ or } p_2 / p_1)$$

$$= \frac{n}{n-1} \times p_1 v_1 \left[\left(\frac{p_2}{p_1}\right)^{\frac{n-1}{n}} - 1\right] \quad \text{... (For polytropic compression)}$$

$$= \frac{n}{n-1} \times m\, R\, T_1 \left[\left(\frac{p_2}{p_1}\right)^{\frac{n-1}{n}} - 1\right] \quad \text{... } (\because pv = mRT)$$

$$= \frac{\gamma}{\gamma - 1} \times p_1 v_1 \left[\left(\frac{p_2}{p_1}\right)^{\frac{\gamma-1}{\gamma}} - 1\right] \quad \text{... (For isentropic compression)}$$

$$= m\, c_p\, (T_2 - T_1)$$

where p_1 = Initial pressure of air,

v_1 = Initial volume of air,

T_1 = Initial temperature of air,

p_2, v_2, T_2 = Corresponding values for the final condition,

m = Mass of air compressed per minute,

n = Polytropic index,

γ = Isentropic index, and

c_p = Specific heat at constant pressure.

Example 29.3. *A centrifugal compressor delivers 50 kg of air per minute at a pressure of 2 bar and 97° C. The intake pressure and temperature of the air is 1 bar and 15° C. If no heat is lost to the surrounding, find : 1. index of compression ; and 2. power required, if the compression is isothermal. Take R = 287 J/ kg K.*

Solution. Given : $m = 50$ kg/min ; $p_2 = 2$ bar ; $T_2 = 97°\text{ C} = 97 + 273 = 370$ K ; $p_1 = 1$ bar ; $T_1 = 15°\text{ C} = 15 + 273 = 288$ K ; $R = 287$ J/kg K

1. *Index of compression*

Let n = Index of compression.

We know that
$$\frac{T_2}{T_1} = \left(\frac{p_2}{p_1}\right)^{\frac{n-1}{n}}$$

$$\frac{370}{288} = \left(\frac{2}{1}\right)^{\frac{n-1}{n}} = (2)^{\frac{n-1}{n}}$$

$$1.285 = (2)^{\frac{n-1}{n}}$$

$$\log 1.285 = \frac{n-1}{n} \times \log 2 \qquad \text{. . . (Taking log of both sides)}$$

$$0.1088 = \frac{n-1}{n} \times 0.301$$

$$0.1088\,n = 0.301\,n - 0.301$$

$$0.1922\,n = 0.301 \quad \text{or} \quad n = 1.57 \textbf{ Ans.}$$

2. *Power required if the compression is isothermal*

We know that work done by the compressor if the compression is isothermal,

$$W = 2.3\,mRT_1 \log r$$

$$= 2.3 \times 50 \times 287 \times 288 \log 2 \text{ J/min} \qquad \text{. . . } (\because r = p_2/p_1)$$

$$= 9\,505\,440 \times 0.301 = 2\,861\,140 \text{ J/min} = 2861.14 \text{ kJ/min}$$

$\therefore$ Power required, $P = 2861.14 / 60 = 47.7$ kW **Ans.**

Example 29.4. *A centrifugal air compressor having a pressure compression ratio of 5 compresses air at the rate of 10 kg/s. If the initial pressure and temperature of the air is 1 bar and 20° C, find : 1. the final temperature of the gas, and 2. power required to drive the compressor. Take γ = 1.4 and c_p = 1 kJ/kg K.*

Solution. Given : $p_2 / p_1 = 5$; $m = 10$ kg/s ; $p_1 = 1$ bar ; $T_1 = 20°\text{ C} = 20 + 273 = 293$ K ; $\gamma = 1.4$; $c_p = 1$ kJ/kg K

1. *Final temperature of the gas*

Let T_2 = Final temperature of the gas.

We know that $\frac{T_2}{T_1} = \left(\frac{p_2}{p_1}\right)^{\frac{\gamma-1}{\gamma}} = (5)^{\frac{1.4-1}{1.4}} = 1.584$

$\therefore \quad T_2 = T_1 \times 1.584 = 293 \times 1.584 = 464 \text{ K} = 191°\text{C}$ **Ans.**

2. *Power required by the compressor*

We know that the workdone by the compressor,

$$W = m\,c_p\,(T_2 - T_1) = 10 \times 1\,(464 - 293) = 1710 \text{ kJ/s}$$

∴ Power required to drive the compressor,

$$P = 1710 \text{ kW} \quad \textbf{Ans.} \qquad \ldots (\because 1 \text{ kJ/s} = 1 \text{ kW})$$

Example 29.5. *A rotary air compressor receives air at a pressure of 1 bar and 17° C, and delivers it at a pressure of 6 bar. Determine, per kg of air delivered, work done by the compressor and heat exchanged with the jacket water when the compression is isothermal, isentropic and by the relation* $pv^{1.6}$ *= Constant.*

Solution. Given : $p_1 = 1$ bar ; $T_1 = 17°\text{C} = 17 + 273 = 290$ K ; $p_2 = 6$ bar ; $m = 1$ kg

Isothermal compression

We know that work done by the compressor,

$$W = 2.3\,mRT_1 \log\left(\frac{p_2}{p_1}\right) = 2.3 \times 1 \times 287 \times 290 \log\left(\frac{6}{1}\right) \text{J}$$

$$= 148\,970 \text{ J} = 148.97 \text{ kJ} \quad \textbf{Ans.}$$

We also know that in isothermal compression, the temperature of air during the process remains constant. Thus the entire work done is carried away by the jacket water in the form of heat. Therefore, heat exchanged with the jacket water

$$= 148.97 \text{ kJ} \quad \textbf{Ans.}$$

Isentropic compression

Let T_2 = Final temperature of the air.

We know that $\frac{T_2}{T_1} = \left(\frac{p_2}{p_1}\right)^{\frac{\gamma-1}{\gamma}} = \left(\frac{6}{1}\right)^{\frac{1.4-1}{1.4}} = 1.668$

$\therefore \quad T_2 = T_1 \times 1.668 = 290 \times 1.668 = 484 \text{ K}$

We know that work done by the compressor

$$= m\,c_p\,(T_2 - T_1) = 1 \times 1\,(484 - 290) = 194 \text{ kJ} \quad \textbf{Ans.}$$

We also know that in isentropic compression, heat exchanged with the jacket water is zero. **Ans.**

Compression by the relation $pv^{1.6}$ *= Constant*

We know that $\frac{T_2}{T_1} = \left(\frac{p_2}{p_1}\right)^{\frac{n-1}{n}} = \left(\frac{6}{1}\right)^{\frac{1.6-1}{1.6}} = 1.958$

$\therefore \quad T_2 = T_1 \times 1.958 = 290 \times 1.958 = 568 \text{ K}$

We know that work done by the compressor

$$= m\,c_p\,(T_2 - T_1) = 1 \times 1\,(568 - 290) = 278 \text{ kJ } \textbf{Ans.}$$

We also know that in polytropic process, heat exchanged

$$= \frac{\gamma - n}{\gamma - 1} \times \text{Work done} = \frac{1.4 - 1.6}{1.4 - 1} \times 278 = -139 \text{ kJ } \textbf{Ans.}$$

The minus sign means that heat is taken by the air from the jacket water

29.9. Velocity Triangles for Moving Blades of a Centrifugal Compressor

We have already discussed that in a centrifugal compressor, the air enters radially and leaves axially. Moreover, the blades and diffuser are so designed that the air enters and leaves them tangentially for the shockless entry and exit.

Consider a stream of air, entering the curved blade at *C*, and leaving it at *D* as shown in Fig. 29.5 (*a*) and (*b*).

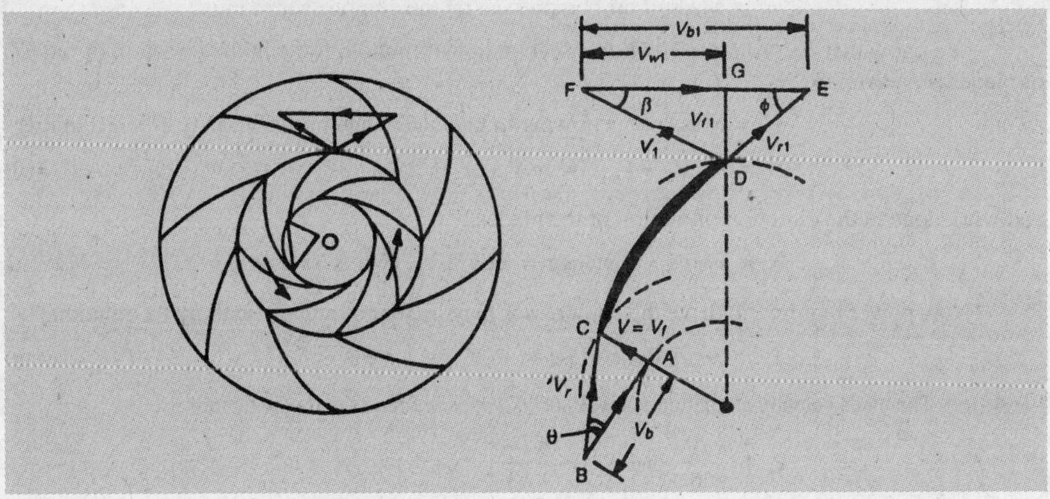

(*a*) (*b*)

Fig. 29.5. Velocity triangles for a centrifugal compressor.

Now let us draw the velocity triangles at the inlet and outlet tips of the blades as shown in Fig. 29.5 (*a*) and (*b*).

Let V_b = Linear velocity of the moving blade at inlet (*BA*),

V = Absolute velocity of the air entering the blade (*AC*),

V_r = Relative velocity of air to the moving blade at inlet (*BC*). It is vectorial difference between V_b and V,

V_f = Velocity of flow at inlet,

θ = Angle which the relative velocity (V_r) makes with the direction of motion of the blade, and

$V_{b1}, V_1, V_{r1}, V_{f1}, \phi$ = Corresponding values at outlet.

It may be seen from the above, that the original notations (*i.e.* V_b, V, V_r, V_f) stand for the inlet triangle. The notations with suffix 1 (*i.e.* V_{b1}, V_1, V_{r1}, V_{f1}) stand for the outlet triangle. A little consideration will show, that as the air enters and leaves the blades without any shock (or in other words tangentially), therefore shape of the blades will be such that V_r and V_{r1} will be along the tangents to the blades at inlet and outlet respectively.

The air enters the blades along AC with a velocity (V). Since the air enters the blades at right angle (*i.e.* radially) to the direction of motion of the blade, therefore velocity of flow (V_f) will be equal to the air velocity (V). Moreover, velocity of whirl at inlet (V_w) will be zero. The linear velocity or mean velocity of blades (V_b) is represented by BA in magnitude and direction. The length BC represents the relative velocity (V_r) of the air with respect to the blade. The air now glides over and leaves the blade with a relative velocity (V_{r1}) which is represented by DE.

The absolute velocity of air (V_1) as it leaves the blade is represented by DF inclined at an angle β with the direction of the blade motion. The tangential component of V_1 (represented by FG) is known as velocity of whirl at exit (V_{w1}). The axial component of V_1 (represented by DG) is known as velocity of flow at exit (V_{f1}).

Let m = Mass of air compressed by the compressor in kg/s.

We know that according to Newton's second law of motion, force in the direction of motion of blades (in newtons),

$$F = \text{Mass of air flowing in kg/s} \times \text{Change in the velocity of whirl in m/s}$$

$$= m\,(V_w + V_{w1}) = m\,V_{w1} \qquad \ldots (\because V_w = 0)$$

and work done in the direction of motion of the blades,

$$W = \text{Force} \times \text{Distance} = m\,V_{w1}\,V_{b1} \text{ N-m/s or J/s}$$

Now power required to drive the compressor may be found out, as usual, by the relation,

$$P = \text{Work done in J/s} = m\,V_{w1}\,V_{b1} \text{ watts} \qquad \ldots (1 \text{ J/s} = 1 \text{ watt})$$

Notes : 1. The blade velocity at inlet or outlet (V_b or V_{b1}) may be found out by the relation,

$$V_b = \frac{\pi DN}{60} \quad \text{and} \quad V_{b1} = \frac{\pi D_1 N}{60}$$

where D and D_1 are the internal and external diameters of the impeller.

2. Under ideal conditions (or in other words for maximum work) $V_{w1} = V_{b1}$.

Therefore ideal work done

$$= m\,(V_{w1})^2 = m\,(V_{b1})^2 \text{ J/s}$$

Example 29.6. *A centrifugal compressor running at 2000 r.p.m. receives air at 17° C. If the outer diameter of the blade tip is 750 mm ; find the temperature of the air leaving the compressor. Take c_p = 1 kJ/kg K.*

Solution. Given : N = 2000 r.p.m. ; T_1 = 17° C = 17 + 273 = 290 K ; D_1 = 750 mm = 0.75 m ; c_p = 1 kJ/kg K

Temperature of the air leaving the compressor

Let T_2 = Temperature of the air leaving the compressor

We know that tangential velocity of the outer blade tip,

$$V_{b1} = \frac{\pi D_1 N}{60} = \frac{\pi \times 0.75 \times 2000}{60} = 78.6 \text{ m/s}$$

∴ Work done per kg of air,

$$W = m\,(V_{b1})^2 = 1\,(78.6)^2 = 6178 \text{ J/s} = 6.178 \text{ kJ/s}$$

We also know that workdone (W),

$$6.178 = m\,c_p\,(T_2 - T_1) = 1 \times 1\,(T_2 - 290)$$

$$\therefore \quad T_2 = 6.178 + 290 = 296.178 \text{ K} = 23.178^\circ \text{C} \textbf{ Ans.}$$

Example 29.7. *A rotary air compressor working between 1 bar and 2.5 bar has internal and external diameters of impeller as 300 mm and 600 mm respectively. The vane angle at inlet and outlet are 30° and 45° respectively. If the air enters the impeller at 15 m/s, find : 1. speed of the impeller in r.p.m. ; and 2. workdone by the compressor per kg of air.*

Solution. Given : p_1 = 1 bar ; *p_2 = 2.5 bar ; D = 300 mm = 0.3 m ; D_1 = 600 mm = 0.6 m ; θ = 30° ; ϕ = 45° ; V = 15 m/s

1. *Speed of the impeller*

Let N = Speed of the impeller in r.p.m.

From the inlet velocity triangle, as shown in Fig. 29.6, we find that the blade velocity,

$$V_b = \frac{V}{\tan 30^\circ} = \frac{15}{0.5774} = 25.98 \text{ m/s}$$

We know that speed of the impeller (V_b),

$$25.98 = \frac{\pi DN}{60} = \frac{\pi \times 0.3\,N}{60} = 0.0157\,N$$

$$\therefore \quad N = 1655 \text{ r.p.m.} \textbf{ Ans.}$$

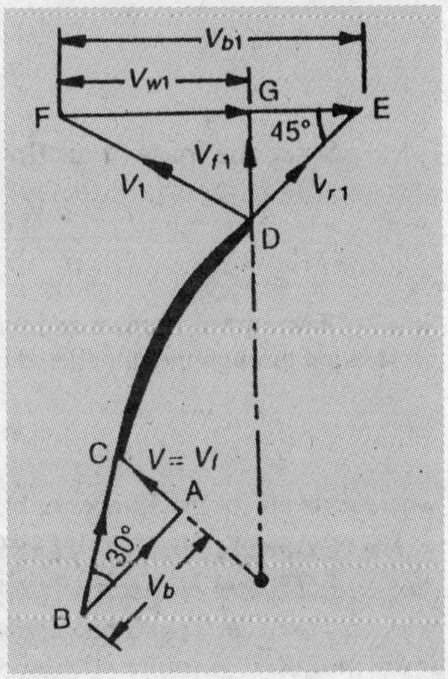

Fig. 29.6

2. *Work done by the compressor per kg of air*

From the outlet triangle, as shown in Fig. 29.6, we find that the blade velocity at outlet,

$$V_{b1} = V_b \times \frac{D_1}{D} = V_b \times \frac{0.6}{0.3} = 25.98 \times \frac{0.6}{0.3} = 51.96 \text{ m/s}$$

and velocity of whirl at outlet,

$$V_{w1} = V_{b1} - \frac{V_{f1}}{\tan 45^\circ} = 51.96 - \frac{15}{1} = 36.96 \text{ m/s} \quad \ldots (\because V_{f1} = V_f = V)$$

Since the velocity of blade at outlet (51.96 m/s) is more than velocity of whirl at outlet (36.96 m/s), therefore shape of the outlet triangle will be as shown in Fig. 29.6.

We know that workdone by the compressor per kg of air,

$$W = m\,V_{w1}\,V_{b1} = 1 \times 36.96 \times 51.96 = 1920.44 \text{ J/s} \textbf{ Ans.}$$

29.10. Width of Impeller Blades

The width of impeller blades at inlet or outlet of a rotary air compressor is found out from the fact that mass of air flowing through the blades at inlet and outlet is constant. Now consider a rotary air compressor compressing the air, whose blade widths at inlet or outlet is required to be found out

---

* Superfluous data

Let b = Width of the impeller blades at inlet.

D = Diameter of impeller at inlet,

V_f = Velocity of flow at inlet,

v_s = Specific volume of air at inlet,

b_1, D_1, V_{f1}, v_{s1} = Corresponding values at outlet, and

m = Mass of the air flowing through the impeller.

We know that the mass of air flowing through the impeller at inlet,

$$m = \frac{\pi D b V_f}{v_s} \quad \ldots (i)$$

Similarly, mass of air flowing through the impeller at outlet,

$$m = \frac{\pi D_1 b_1 V_{f1}}{v_{s1}} \quad \ldots (ii)$$

Since the mass of air flowing through the impeller is constant, therefore,

$$\frac{\pi D b V_{f1}}{v_s} = \frac{\pi D_1 b_1 V_{f1}}{v_{s1}} \quad \ldots (iii)$$

Note : Sometimes, number and thickness of the blades is also taken into consideration. In such a case, mass of air flowing through the impeller at inlet,

$$m = \frac{(\pi D - nb)\, V_f}{v_s}$$

where n stands for the number of blades.

Example 29.8. *A centrifugal air compressor having internal and external diameters of 250 mm and 500 mm respectively compresses 30 kg of air per minute while running at 4000 r.p.m. The vane angles at inlet and outlet are 30° and 40° respectively. Find the necessary thickness of the blade, if the impeller contains 40 blades. Take specific volume of air as 0.8 m³/kg.*

Solution. Given : $D = 250$ mm $= 0.25$ m ; $D_1 = 500$ mm $= 0.5$ m ; $m = 30$ kg/min $= 0.5$ kg/s ; $N = 4000$ r.p.m. ; $\theta = 30°$; $\phi = 40°$; $n = 40$; $v_s = 0.8\ \text{m}^3/\text{kg}$

We know that impeller velocity at inlet,

$$V_b = \frac{\pi D N}{60} = \frac{\pi \times 0.25 \times 4000}{60} = 52.4 \text{ m/s}$$

and velocity of flow at inlet, $V_f = V_b \tan\theta = 52.4 \tan 30° = 30.2$ m/s

Let b = Thickness of the blades.

We know that mass of air flowing through the impeller (m),

$$0.5 = \frac{(\pi D - nb)\, V_f}{v_s} = \frac{(\pi \times 0.25 - 40\, b)\, 30.2}{0.8}$$

or $0.0132 = 0.7855 - 40b$

$\therefore$ $b = 0.0193 \text{ m} = 19.3$ mm **Ans.**

29.11. Prewhirl

It has been observed that tangential velocity of the inlet impeller end is very high due to its exceedingly high revolutions per minute (sometimes, as high as 20 000 r.p m.). At this point, there is always a tendency for the air stream to break away from the trailing face of the curved part of the impeller vane. This phenomenon, under certain set of conditions*causes the shock waves to form. The shock waves increase the loss of energy.

In order to eliminate (or reduce) the shock waves, the air is made to rotate before it enters the impeller blades. This process, which causes the air to enter the impeller blades at a reduced velocity (without effecting the mass of air to flow and velocity of flow), is known as pre-rotation or *prewhirl*.

29.12. Axial Flow Compressors

An axial flow compressor, in its simplest form, consists of a number of rotating blade rows fixed to a rotating drum. The drum rotates inside an air tight casing to which are fixed stator blade rows, as shown in Fig. 29.7. The blades are made of aerofoil section to reduce the loss caused by turbulence and boundary separation.

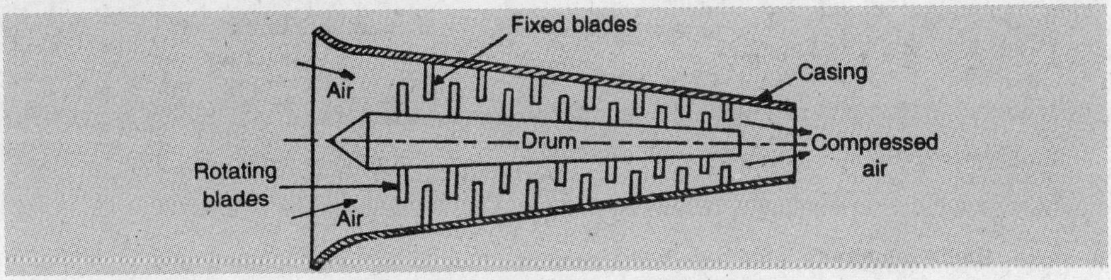

Fig. 29.7. Axial flow compressor.

The mechanical energy is provided to the rotating shaft, which rotates the drum. The air enters from the left side of the compressor. As the drum rotates, the air flows through the alternately arranged stator and rotor. As the air flows from one set of stator and rotor to another, it gets compressed. Thus successive compression of the air, in all the sets of stator and rotor, the air is delivered at a high pressure at the outlet point.

29.13. Comparison of Centrifugal and Axial Flow Air Compressors

Following are the main points of comparison of the centrifugal and axial flow air compressors :

S.No.	*Centrifugal compressor*	*Axial flow compressor*
1.	The flow of air is perpendicular to the axis of compressor.	The flow of air is parallel to the axis of compressor.
2.	It has low manufacturing and running cost.	It has high manufacturing and running cost.
3.	It requires low starting torque.	It requires high starting torque.
4.	It is not suitable for multi-staging.	It is suitable for multi-staging.
5.	It requires large frontal area for a given rate of flow.	It requires less frontal area for a given rate of flow. It makes the compressor suitable for air crafts.

* This condition is popularly defined in terms of Mach number. If the Mach number is less than unity, the flow is known as subsonic. But if it is equal to unity, the flow is known as sonic. Similarly, if the Mach number is more than unity, the flow is known as supersonic. The shock waves are formed when the Mach number exceeds 0.90. The value of Mach number is mathematically given by the relation.

$$M_N = \frac{V_b}{\sqrt{VgRT}}$$

29.14. Velocity Diagrams for Axial Flow Air Compressors

We have already discussed in Art. 29.12 that in an axial flow compressor, the drum with rotor blades, rotates inside a casing with a fixed or stator blades. The inlet and outlet velocity triangles for the rotor blades are shown in Fig. 29.8 (*a*) and (*b*). The general relations between the inlet and outlet velocity triangles are as below :

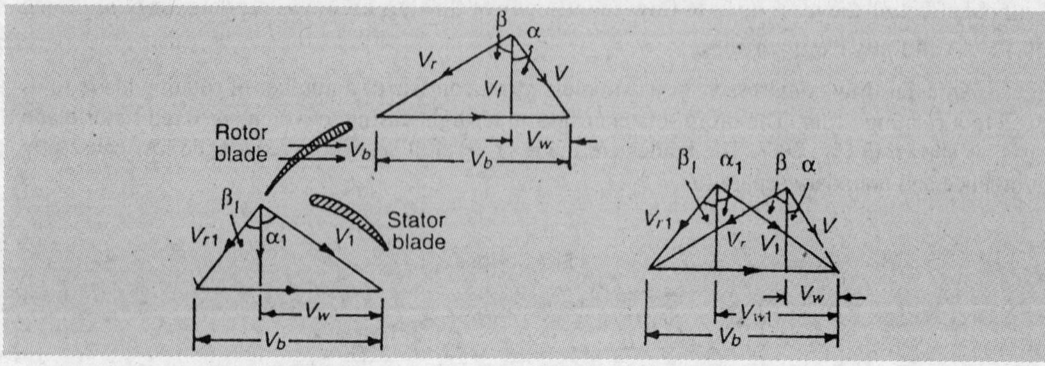

(*a*) Separate velocity diagrams. (*b*) Combined velocity diagram.

Fig. 29.8. Velocity diagrams for axial flow compressor.

1. Blade velocity (V_b) for both the triangles is equal.
2. Velocity of flow (V_f) for both the triangles is also equal.
3. Relative velocity in outlet triangle (V_{r1}) is less than that in inlet triangle (V_r) due to friction.

Notes : 1. Work done by the compressor per kg of air,

$$w = V_b(V_{w1} - V_w) \text{ in N-m or J}$$

2. Sometimes, work factor or work input factor is also given. In such a case, work done by the compressor per kg of air,

$$w = V_b(V_{w1} - V_w) \times \text{Work factor}$$

29.15. Degree of Reaction

It is an important term in the field of axial flow compressor which may be defined as the ratio of pressure rise in the rotor blades to the pressure rise in the compressor in one stage.

As a matter of fact, the degree of reaction is usually kept as 50% or 0.5 for all types of axial flow compressors.

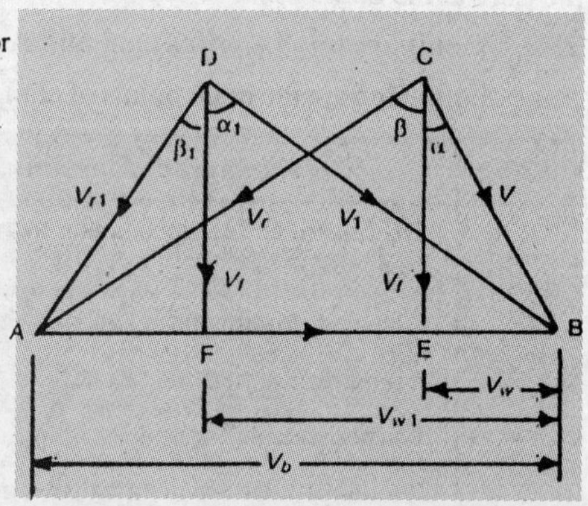

Fig. 29.9. Combined velocity diagram with 50% degree of reaction

Mathematically, degree of reaction,

$$= \frac{\text{Pressure rise in rotor blades}}{\text{Pressure rise in compressor}}$$

$$= \frac{\dfrac{(V_r)^2 - (V_{r1})^2}{2}}{V_b(V_{w1} - V_w)} = \frac{(V_r)^2 - (V_{r1})^2}{2V_b(V_{w1} - V_w)} \quad \ldots (i)$$

First of all, let us draw a combined velocity diagram for an axial flow compressor (with degree of reaction as 0.5), as shown in Fig. 29.9.

From the geometry of the figure, we find that

$$V_w = AB - AE = V_b - V_f \tan\beta \; ; \text{ and } V_{w1} = AB - AF = V_b - V_f \tan\beta_1$$

$$\therefore \quad (V_{w1} - V_w) = (V_b - V_f \tan\beta_1) - (V_b - V_f \tan\beta)$$

$$= V_f(\tan\beta - \tan\beta_1) \qquad \ldots (ii)$$

Moreover, from the geometry of the figure, we also find that

$$(V_r)^2 = (V_f)^2 + (V_f \tan\beta)^2$$

and

$$(V_{r1})^2 = (V_f)^2 + (V_f \tan\beta_1)^2$$

$$\therefore \quad (V_r)^2 - (V_{r1})^2 = [(V_f)^2 + (V_f \tan\beta)^2] - [(V_f)^2 + (V_f \tan\beta_1)^2]$$

$$= (V_f \tan\beta)^2 - (V_f \tan\beta_1)^2 = (V_f)^2 (\tan^2\beta - \tan^2\beta_1) \qquad \ldots (iii)$$

Now substituting the values of $(V_{w1} - V_w)$ and $[(V_r)^2 - (V_{r1})^2]$ from equations (*ii*) and (*iii*) in equation (*i*), we have degree of reaction,

$$R = \frac{(V_f)^2(\tan^2\beta - \tan^2\beta_1)}{2\,V_b\,V_f(\tan\beta - \tan\beta_1)} = \frac{V_f(\tan\beta + \tan\beta_1)}{2\,V_b}$$

Now substituting the value of degree of reaction as 0.5, we have

$$0.5 = \frac{V_f(\tan\beta + \tan\beta_1)}{2\,V_b}$$

$$\therefore \quad \frac{V_b}{V_f} = \tan\beta + \tan\beta_1$$

From the geometry of the figure, we find that

$$\frac{V_b}{V_f} = \tan\alpha + \tan\alpha_1 = \tan\alpha + \tan\beta = \tan\alpha_1 + \tan\beta_1$$

$$\therefore \quad \angle\beta = \angle\alpha_1 \text{ and } \angle\beta_1 = \angle\alpha$$

It is thus obvious, that for 50% reaction, the compressor will have symmetrical blades.

Example 29.9. *An axial flow compressor, with compression ratio as 5, draws air at 20° C delivers it at 50° C. Assuming 50% degree of reaction, find the velocity of flow if the blade velocity is 100 m/s. Also find the number of stages. Take work factor = 0.85 ; α = 10° ; β = 40° and c_p = 1 kJ/kg K.*

Solution. Given : *$p_2/p_1 = 5$; $T_1 = 20°\text{ C} = 20 + 273 = 293\text{ K}$; $T_2 = 50°\text{ C} = 323\text{ K}$; $R = 50\% = 0.5$; $V_b = 100$ m/s ; Work factor = 0.85 ; $\alpha = 10°$; $\beta = 40°$; $c_p = 1$ kJ/kg K

Velocity of flow

Let $\quad V_f$ = Velocity of flow.

From the geometry of the velocity triangle (Fig. 29.9), we know that

$$\frac{V_b}{V_f} = \tan\alpha + \tan\beta = \tan 10° + \tan 40°$$

* Superfluous data

$$\frac{100}{V_f} = 0.1763 + 0.8391 = 1.0154$$

$$\therefore \quad V_f = 98.5 \text{ m/s} \quad \textbf{Ans.}$$

Number of stages

We know that total work required per kg of air

$$= c_p (T_2 - T_1) = 1\,(323 - 293) = 30 \text{ kJ/kg}$$

From the geometry of the velocity triangles, we also know that

$$V_w = V_f \tan \alpha = 98.5 \times \tan 10^\circ = 98.5 \times 0.1763 = 17.4 \text{ m/s}$$

and $$V_{w1} = V_f \tan \alpha_1 = 98.5 \tan 40^\circ = 98.5 \times 0.8391 = 82.7 \text{ m/s}$$

. : . (with 50% reaction, $\angle \alpha_1 = \angle \beta$)

We know that workdone per kg of air per stage,

$$= V_b (V_{w1} - V_w) \times \text{Work factor}$$

$$= 100\,(82.7 - 17.4) \times 0.85 = 5550 \text{ J} = 5.55 \text{ kJ/kg}$$

$$\therefore \text{Number of stages} = \frac{\text{Total work required}}{\text{Work done per stage}} = \frac{30}{5.55} = 5.4 \text{ say } 6 \quad \textbf{Ans.}$$

EXERCISES

1. Air at 1 bar and 30° C is to be compressed to 1.2 bar at the rate of 50 m³/min. Find the power required by a Root's blower. **[Ans. 15.45 kW]**

2. Compare the work inputs required for a Roots blower and a vane type compressor having the same induced volume of 0.03 m^3 per revolution, the inlet pressure being 1.013 bar and the pressure ratio 1.5 to 1. For the vane type, assume that internal compression takes place through half the pressure range. **[Ans. 1.52 kJ, 1.352 kJ]**

3. A centrifugal compressor having compression ratio of 2 delivers air at the rate of 1.5 kg/s. Find the power required to drive the compressor with isothermal compression, if the intake temperature is 300 K. **[Ans. 89.5 kW]**

4. A centrifugal air compressor receives air at 1 bar and deliver it at 3.5 bar. Find the final temperature of air, if the initial temperature of air is 310 K. The compressor compresses 2 kg of air per second. Take γ as 1.4. **[Ans. 443 K]**

5. A rotary air compressor compresses 100 kg of air per minute from 1.2 bar and 20° C to 4.8 bar. Find the power required by the compressor, if the compression is isentropic and by the relation $pv^{1.5} = C$. Take $c_p = 1.008$ kJ/kg K. **[Ans. 247 kW ; 300 kW]**

6. An axial flow compressor, with compression ratio as 4, draws air at 20° C and delivers it at 197° C. The mean blade speed and flow velocity are constant throughout the compressor. Assuming 50 percent reaction blading and taking blade velocity as 180 m/s; find the flow velocity and the number of stages. Take work factor = 0.82 ; $\alpha = 12^\circ$; $\beta = 42^\circ$ and $c_p = 1.005$ kJ/kg K. **[Ans. 162 m/s ; 10]**

QUESTIONS

1. What is the difference between rotary and reciprocating compressor ?

2. Derive an expression for efficiency of a Roots blower in terms of pressure ratio and ratio of specific heats.

3. Describe with a neat sketch, the working of a vane blower compressor and show its *p-v* diagram. For what applications, it is used.

4. Explain, with a neat sketch, the working of a centrifugal compressor and obtain an expression for the workdone.
5. Discuss the method of finding the width of impeller blades in a rotary air compressor.
6. Define 'prewhirl'. Explain its effect on the impeller of a centrifugal pump.
7. Explain, with a neat sketch, the working of an axial flow compressor.
8. Differentiate between centrifugal compressor and axial flow compressor.

OBJECTIVE TYPE QUESTIONS

1. The positive displacement compressor is
   (*a*) roots blower compressor (*b*) vane blower compressor
   (*c*) centrifugal compressor (*d*) axial flow compressor
   (*e*) both (*a*) and (*b*) (*f*) both (*c*) and (*d*)
2. The rotary compressors are used for delivering
   (*a*) small quantities of air at high pressures (*b*) large quantities of air at high pressures
   (*c*) small quantities of air at low pressures (*d*) large quantities of air at low pressures
3. The maximum delivery pressure in a rotary air compressor is
   (*a*) 10 bar (*b*) 20 bar (*c*) 30 bar (*d*) 40 bar
4. The speed of a rotary compressor is ..... as compared to reciprocating air compressor.
   (*a*) high (*b*) low
5. The type of rotary compressor used in gas turbines is of
   (*a*) centrifugal type (*b*) axial flow type (*c*) radial flow type (*d*) none of these
6. If the flow of air through the compressor is perpendicular to its axis, then it is a
   (*a*) reciprocating compressor (*b*) centrifugal compressor
   (*c*) axial flow compressor (*d*) turbo compressor
7. In a centrifugal compressor, an increase in speed at a given pressure ratio causes
   (*a*) increase in flow (*b*) decrease in flow
   (*c*) increase in efficiency (*d*) decrease in efficiency
   (*e*) increase in flow and decrease in efficiency
8. In an axial flow compressor, the ratio of pressure in the rotor blades to the pressure rise in the compressor in one stage is known as
   (*a*) work factor (*b*) slip factor (*c*) degree of reaction (*d*) pressure coefficient
9. A compressor mostly used for supercharging of I.C. engines is
   (*a*) radial flow compressor (*b*) axial flow compressor
   (*c*) roots blower (*d*) reciprocating compressor
10. Which of the following statement is correct as regard to centrifugal compressors ?
    (*a*) The flow of air is parallel to the axis of the compressor.
    (*b*) The static pressure of air in the impeller increases in order to provide centripetal force on the air.
    (*c*) The impeller rotates at high speeds.
    (*d*) The maximum efficiency is higher than multi-stage axial flow compressors.

ANSWERS

1. (*e*)	2. (*d*)	3. (*a*)	4. (*a*)	5. (*b*)
6. (*b*)	7. (*e*)	8. (*c*)	9. (*a*)	10. (*b*)

30

Performance of Air Compressors

1. Introduction. 2. Efficiencies of Reciprocating and Centrifugal Air Compressors. 3. Efficiencies of Reciprocating Air Compressor. 4. Volumetric Efficiency of a Reciprocating Air Compressor with Clearance Volume. 5. Thermodynamic Cycle for a Rotary Air Compressor. 6. Efficiencies of a Centrifugal Air Compressor. 7. Static and Total Head Quantities. 8. Slip Factor. 9. Comparison of Turbine and Centrifugal Compressor Blades.

30.1. Introduction

In the last two chapters, we have discussed reciprocating and rotary air compressors. Now in this chapter, we shall discuss their efficiencies and other important performances.

30.2. Efficiencies of Reciprocating and Centrifugal Air Compressors

The efficiency of any machine is the general term used for the ratio of work done to the energy supplied. The criterion for the thermodynamic efficiency of the reciprocating air compressor is isothermal; whereas that for the centrifugal air compressor is isentropic. The reason for the same is that in case of reciprocating air compressors, due to slow speed of the piston and cooling of the cylinder, the compression of air is approximately isothermal. But in case of centrifugal air compressor, due to high speed of the rotor and without any cooling arrangement, the compression of air is approximately isentropic. Now we shall discuss the efficiencies of both the compressors, in the following pages.

30.3. Efficiencies of Reciprocating Air Compressor

We have already discussed in the last article that the criterion for thermodynamic efficiency of a reciprocating air compressor is isothermal. But in general, the following efficiencies of reciprocating air compressor are important from the subject point of view :

1. *Isothermal efficiency (or compressor efficiency).* It is the ratio of work (or power) required to compress the air isothermally to the actual work required to compress the air for the same pressure ratio. Mathematically, isothermal efficiency or compressor efficiency,

$$\eta_c = \frac{\text{Isothermal power}}{\text{Indicated power}} = \frac{\text{Isothermal work done}}{\text{Indicated work done}}$$

We have already discussed for a single stage reciprocating compressor that the isothermal work done

$$= 2.3\, p_1\, v_1 \log\left(\frac{p_2}{p_1}\right)$$

and indicated work done by the compressor

$$= \text{Work done during polytropic compression}$$

$$= \frac{n}{n-1} \times p_1 v_1 \left[\left(\frac{p_2}{p_1} \right)^{\frac{n-1}{n}} - 1 \right]$$

∴ Isothermal efficiency for a single stage reciprocating compressor

$$= \frac{2.3 \log \left(\frac{p_2}{p_1} \right)}{\frac{n}{n-1} \left[\left(\frac{p_2}{p_1} \right)^{\frac{n-1}{n}} - 1 \right]}$$

2. *Overall isothermal efficiency.* It is the ratio of the isothermal power to the shaft power or brake power of the motor or engine required to drive the compressor. Mathematically, overall isothermal efficiency,

$$\eta_0 = \frac{\text{Isothermal power}}{\text{Shaft power or B.P. of motor}}$$

3. *Mechanical efficiency.* It is the ratio of the indicated power to the shaft power or brake power of the motor or engine required to drive the compressor. Mathematically, mechanical efficiency,

$$\eta_m = \frac{\text{Indicated power}}{\text{Shaft power or B.P. of motor}}$$

Note : The shaft power or brake power of the motor or engine

$$= \frac{\text{Indicated power}}{\text{Mechanical efficiency}}$$

4. *Isentropic efficiency.* It is the ratio of the isentropic power to the brake power required to drive the compressor. Mathematically, isentropic efficiency,

$$\eta_i = \frac{\text{Isentropic power}}{\text{B.P. required to drive the compressor}}$$

5. *Volumetric efficiency.* It is the ratio of volume of free air delivery per stroke to the swept volume of the piston. The volumetric efficiency of a reciprocating air compressor is different when it is with or without clearance volume.

Note : Since it is difficult to visualise the N.T.P. conditions of the swept air, therefore the general trend is to define the volumetric efficiency as the ratio of actual volume of air sucked by the compressor to the swept volume of the piston.

Example 30.1. *A reciprocating air compressor draws in 6 kg of air per minute at 25° C. It compresses the air polytropically and delivers it at 105° C. Find the air power. If the shaft power is 14 kW ; find the mechanical efficiency. Assume R = 0.287 kJ/kg K and n = 1.3.*

Given : $m = 6$ kg/min ; $T_1 = 25° \text{C} = 25 + 273 = 298$ K ; $T_2 = 105° \text{C} = 105 + 273 = 378$ K ; Shaft power = 14 kW ; $R = 0.287$ kJ/kg K ; $n = 1.3$

Air Power

We know that workdone by the compressor,

$$W = \frac{n}{n-1} \times mR\,(T_2 - T_1) = \frac{1.3}{1.3-1} \times 6 \times 0.287\,(378 - 298) \text{ kJ/min}$$

$$= 597 \text{ kJ/min}$$

$$\therefore \text{Air power} = 597/60 = 9.95 \text{ kW} \quad \textbf{Ans.}$$

Mechanical efficiency

We know that mechanical efficiency,

$$\eta_m = \frac{\text{Air power}}{\text{Shaft power}} = \frac{9.95}{14} = 0.7107 \text{ or } 71.07\% \textbf{ Ans.}$$

Example 30.2. *A compressor draws 42.5 m³ of air per minute into the cylinder at a pressure of 1.05 bar. It is compressed polytropically ($pv^{1.3} = C$) to a pressure of 4.2 bar before being delivered to a receiver. Assuming a mechanical efficiency of 80%, find ;*

1. Indicated power ; 2. Shaft power ; and 3. Overall isothermal efficiency.

Solution. Given : $v_1 = 42.5$ m³/min ; $p_1 = 1.05$ bar $= 1.05 \times 10^5$ N/m² ; $n = 1.3$; $p_2 = 4.2$ bar $= 4.2 \times 10^5$ N/m² ; $\eta_m = 80\% = 0.8$

1. *Indicated power*

We know that indicated workdone by the compressor,

$$W = \frac{n}{n-1} \times p_1 v_1 \left[\left(\frac{p_2}{p_1} \right)^{\frac{n-1}{n}} - 1 \right]$$

$$= \frac{1.3}{1.3-1} \times 1.05 \times 10^5 \times 42.5 \left[\left(\frac{4.2}{1.05} \right)^{\frac{1.3-1}{1.3}} - 1 \right] \text{ J/min}$$

$$= 7287 \times 10^3 \text{ J/min} = 7287 \text{ kJ/min}$$

$$\therefore \text{ Indicated power} = 7287/60 = 121.5 \text{ kW } \textbf{Ans.}$$

2. *Shaft power*

We know that shaft power,

$$= \frac{\text{Indicated power}}{\text{Mechanical efficiency}} = \frac{121.5}{0.8} = 151.87 \text{ kW } \textbf{Ans.}$$

3. *Overall isothermal efficiency*

We know that isothermal work done/min,

$$W = 2.3\, p_1 v_1 \log \left(\frac{p_2}{p_1} \right) = 2.3 \times 1.05 \times 10^5 \times 42.5 \log \left(\frac{4.2}{1.05} \right) \text{ J/min}$$

$$= 6180 \times 10^3 \text{ J/min} = 6180 \text{ kJ/min}$$

$$\therefore \text{ Isothermal power} = 6180/60 = 103 \text{ kW}$$

We also know that overall isothermal efficiency,

$$\eta_0 = \frac{\text{Isothermal power}}{\text{Shaft power}} = \frac{103}{151.87} = 0.678 \text{ or } 67.8\% \textbf{ Ans.}$$

30.4. Volumetric Efficiency of a Reciprocating Air Compressor with Clearance Volume

We have already discussed in Art. 30.3, that the volumetric efficiency of a reciprocating air compressor is given by

$$\eta_v = \frac{\text{Volume of free air delivery per stroke}}{\text{Swept volume of the piston}}$$

$$= \frac{\text{Actual volume of air sucked referred to ambient conditions}}{\text{Swept volume of the piston}}$$

Now let us derive an expression for it when the air compressor has clearance volume. Consider a *p-v* diagram of a single acting reciprocating air compressor with clearance volume as shown in Fig. 30.1.

Let p_1 = Initial pressure of air (before compression),

v_1 = Initial volume of air (before compression),

T_1 = Initial temperature of air (before compression),

p_2, v_2, T_2 = Corresponding values for the final conditions (*i.e.* at the delivery point),

p_a, v_a, T_a = Corresponding values for the ambient (*i.e.* N.T.P.) conditions

v_c = Clearance volume,

v_s = Swept volume of the piston, and

n = Polytropic index.

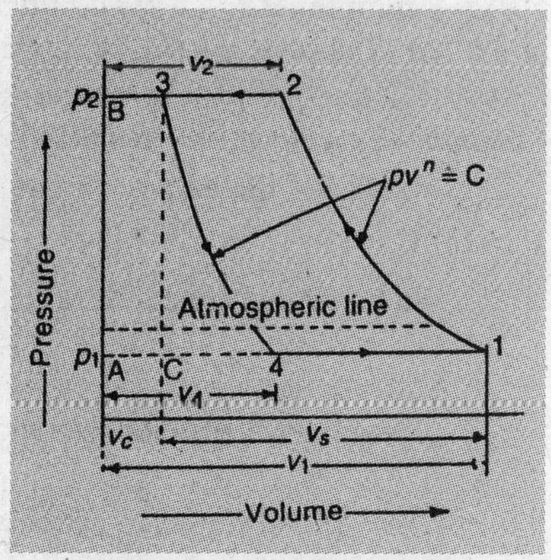

Fig. 30.1. *p-v* diagram with clearance volume.

In actual practice, the temperature at the end of suction *i.e.* at point 1 is not atmospheric because the fresh air passes over hot valves and mixes with the residual air. Also, the pressure at point 1 is not atmospheric as there are obstructions in suction of fresh air. Applying general gas equation to the atmospheric condition of air and the condition of air before compression, we have

$$\frac{p_a v_a}{T_a} = \frac{p_1 (v_1 - v_4)}{T_1}$$

$\therefore$ Volume of air sucked referred to ambient conditions,

$$v_a = \frac{p_1 T_a}{p_a T_1} (v_1 - v_4)$$

We know that volumetric efficiency,

$$\eta_v = \frac{v_a}{v_s} = \frac{p_1 T_a}{p_a T_1}\left(\frac{v_1 - v_4}{v_s}\right)$$

$$= \frac{p_1 T_a}{p_a T_1}\left(\frac{v_s + v_c - v_4}{v_s}\right) \qquad \ldots (\because v_1 = v_s + v_c)$$

$$= \frac{p_1 T_a}{p_a T_1}\left(1 + \frac{v_c}{v_s} - \frac{v_4}{v_c} \times \frac{v_c}{v_s}\right)$$

$$= \frac{p_1 T_a}{p_a T_1}\left(1 + K - K \times \frac{v_4}{v_c}\right) \qquad \ldots (i)$$

where K = Clearance ratio = v_c / v_s

Now for the polytropic expansion process 3-4,

$$p_2 v_c^n = p_1 v_4^n$$

$$\therefore \quad \frac{v_4}{v_c} = \left(\frac{p_2}{p_1}\right)^{1/n}$$

Substituting the value of v_4 / v_c in equation (*i*), we have volumetric efficiency referred to ambient conditions,

$$\eta_v = \frac{p_1 T_a}{p_a T_1}\left[1 + K - K\left(\frac{p_2}{p_1}\right)^{1/n}\right] \qquad \ldots (ii)$$

When the ambient and suction conditions are same, then $p_a = p_1$ and $T_a = T_1$. In such a case,

$$\eta_v = 1 + K - K\left(\frac{p_2}{p_1}\right)^{1/n} \qquad \ldots (iii)$$

Example 30.3. *A single stage air compressor receives air at 1 bar and 27° C and delivers at 6.5 bar. The atmospheric pressure and temperature are 1.013 bar and 15° C. The compression follows the law $p\,v^{1.25}$ = constant and the clearance volume is 5 percent of the stroke volume. Calculate the volumetric efficiency referred to the atmospheric condition.*

Solution. Given : p_1 = 1 bar ; T_1 = 27° C = 27 + 273 = 300 K ; p_2 = 6.5 bar ; p_a = 1.013 bar ; T_a = 15° C = 15 + 273 = 288 K ; n = 1.25 ; $v_c = 5\%\ v_s = 0.05\ v_s$

We know that clearance ratio,

$$K = v_c / v_s = 0.05\ v_s / v_s = 0.05$$

∴ Volumetric efficiency referred to the atmospheric condition,

$$\eta_v = \frac{p_1 T_a}{p_a T_1}\left[1 + K - K\left(\frac{p_2}{p_1}\right)^{1/n}\right]$$

$$= \frac{1 \times 288}{1.013 \times 300}\left[1 + 0.05 - 0.05\left(\frac{6.5}{1}\right)^{1/1.25}\right]$$

$$= 0.783 \text{ or } 78.3\% \text{ Ans.}$$

Example 30.4. *A single stage reciprocating air compressor takes in 7.5 m³/min of air at 1 bar and 30° C and delivers it at 5 bar. The clearance is 5 percent of the stroke. The expansion and compression are polytropic, n = 1.3. Calculate : 1. the temperature of delivered air ; 2. volumetric efficiency, and 3. power of the compressor.*

Solution. Given : $v_1 - v_4 = 7.5$ m³/min ; p_1 = 1 bar = 1×10^5 N/m² ; T_1 = 30° C = 30 + 273 = 303 K ; p_2 = 5 bar = 5×10^5 N/m² ; $v_c = 5\%\ v_s = 0.05\ v_s$; n = 1.3

1. *Temperature of delivered air*

Let T_2 = Temperature of the delivered air.

We know that
$$\frac{T_2}{T_1} = \left(\frac{p_2}{p_1}\right)^{\frac{n-1}{n}} = \left(\frac{5}{1}\right)^{\frac{1.3-1}{1.3}} = 1.45$$

$$\therefore \quad T_2 = T_1 \times 1.45 = 303 \times 1.45 = 439.3 \text{ K} = 166.3^\circ \text{C Ans.}$$

2. *Volumetric efficiency*

We know that clearance ratio,

$$K = \frac{v_c}{v_s} = \frac{0.05\ v_s}{v_s} = 0.05$$

∴ Volumetric efficiency,

$$\eta_v = 1 + K - K\left(\frac{p_2}{p_1}\right)^{1/n} = 1 + 0.05 - 0.05\left(\frac{5}{1}\right)^{1/1.3}$$

$$= 1.05 - 0.172 = 0.878 \text{ or } 87.8\% \textbf{ Ans.}$$

3. *Power of the compressor*

We know that workdone by the compressor,

$$W = \frac{n}{n-1} \times p_1 (v_1 - v_4)\left[\left(\frac{p_2}{p_1}\right)^{\frac{n-1}{n}} - 1\right]$$

$$= \frac{1.3}{1.3-1} \times 1 \times 10^5 \times 7.5\left[\left(\frac{5}{1}\right)^{\frac{1.3-1}{1.3}} - 1\right] \text{ J/min}$$

$$= 1462.5 \times 10^3 \text{ J/min} = 1462.5 \text{ kJ/min}$$

∴ Power of the compressor

$$= 1462.5/60 = 24.4 \text{ kW } \textbf{Ans.}$$

Example 30.5. *A single stage single acting reciprocating air compressor is required to handle 30 m³ of free air per hour measured at 1 bar. The delivery pressure is 6.5 bar and the speed is 450 r.p.m.*

Allowing a volumetric efficiency of 75% ; an isothermal efficiency of 76% and a mechanical efficiency of 80% ; Calculate the indicated mean effective pressure and the power required to drive the compressor.

Solution. Given : $v_1 = 30 \text{ m}^3/\text{h}$; $p_1 = 1 \text{ bar} = 1 \times 10^5 \text{ N/m}^2$; $p_2 = 6.5 \text{ bar} = 6.5 \times 10^5 \text{ N/m}^2$; $N = 450$ r.p.m. ; $\eta_v = 75\% = 0.75$; $\eta_i = 76\% = 0.76$; $\eta_m = 80\% = 0.8$

Indicated mean effective pressure

We know that isothermal work done

$$= 2.3\, p_1 v_1 \log\left(\frac{p_2}{p_1}\right) = 2.3 \times 1 \times 10^5 \times 30 \log\left(\frac{6.5}{1}\right) \text{ J/h}$$

$$= 5609 \times 10^3 \text{ J/h} = 5609 \text{ kJ/h}$$

and indicated workdone

$$= \frac{\text{Isothermal work done}}{\text{Isothermal efficiency}} = \frac{5609}{0.76} = 7380 \text{ kJ/h}$$

We know that swept volume of the piston,

$$v_s = \frac{\text{Volume of free air}}{\text{Volumetric efficiency}} = \frac{30}{0.75} = 40 \text{ m}^3/\text{h}$$

∴ Indicated mean effective pressure,

$$p_m = \frac{\text{Indicated work done}}{\text{Swept volume}} = \frac{7380}{40} = 184.5 \text{ kJ/m}^3$$

$$= 184.5 \text{ kN/m}^2 \qquad \ldots \left(\because \frac{1 \text{ kJ}}{\text{m}^3} = \frac{1 \text{ kN-m}}{\text{m}^3} = 1 \text{ kN/m}^2\right)$$

$$= 1.845 \text{ bar } \textbf{Ans.}$$

Power required to drive the compressor

We know that work done by the compressor

$$= \frac{\text{Indicated work done}}{\text{Mechanical efficiency}} = \frac{7380}{0.8} = 9225 \text{ kJ/h}$$

∴ Power required to drive the compressor

$$= \frac{9225}{3600} = 2.56 \text{ kW} \quad \textbf{Ans.}$$

Example 30.6. *A single stage double acting air compressor delivers 3 m³ of free air per minute at 1.013 bar and 20° C to 8 bar with the following data ;*

R.P.M. = 300 ; Mechanical efficiency = 0.9 ; Pressure loss in passing through intake valves = 0.04 bar ; Temperature rise of air during suction stroke = 12° C ; Clearance volume = 5% of stroke volume ; Index of compression and expansion, n = 1.35 ; Length of the stroke = 1.2 times the cylinder diameter.

Calculate : 1. power input to the shaft ; 2. the volumetric efficiency ; and 3. the cylinder diameter.

Solution. Given : $v_a = 3$ m³/min ; $p_a = 1.013$ bar ; $T_a = 20°$ C = 20 + 273 = 293 K ; $p_2 =$ 8 bar ; $N = 300$ r.p.m. ; $\eta_m = 0.9$; Pressure loss = 0.04 bar ; Temperature rise = 12° C ; $v_c = 5\%\ v_s = 0.05\ v_s$; $n = 1.35$; $L = 1.2\ D$

Let v_1 = Volume of free air at the suction conditions.

Since there is a pressure loss of 0.04 bar in passing through intake valves, therefore suction pressure,

$$p_1 = p_a - 0.04 = 1.013 - 0.04 = 0.973 \text{ bar} = 0.973 \times 10^5 \text{ N/m}^2$$

Also there is a temperature rise of air of 12° C during suction stroke, therefore temperature of air at the beginning of compression,

$$T_1 = T_a + 12 = 20 + 12 = 32° \text{C} = 32 + 273 = 305 \text{ K}$$

We know that $$\frac{p_a v_a}{T_a} = \frac{p_1 v_1}{T_1}$$

$$\therefore \quad v_1 = \frac{p_a v_a T_1}{p_1 T_a} = \frac{1.013 \times 3 \times 305}{0.973 \times 293} = 3.25 \text{ m}^3/\text{min}$$

1. *Power input to the shaft*

We know that indicated workdone

$$= \frac{n}{n-1} \times p_1 v_1 \left[\left(\frac{p_2}{p_1}\right)^{\frac{n-1}{n}} - 1\right]$$

$$= \frac{1.35}{1.35 - 1} \times 0.973 \times 10^5 \times 3.25 \left[\left(\frac{8}{0.973}\right)^{\frac{1.35-1}{1.35}} - 1\right] \text{ J/min}$$

$$= 885 \times 10^3 \text{ J/min} = 885 \text{ kJ/min}$$

and indicated power $= 885/60 = 14.75$ kW

We know that power input to the shaft

$$= \frac{\text{Indicated power}}{\eta_m} = \frac{14.75}{0.9} = 16.4 \text{ kW } \textbf{Ans.}$$

2. *Volumetric efficiency*

We know that clearance ratio,

$$K = v_c / v_s = 0.05\, v_s / v_s = 0.05$$

∴ Volumetric efficiency,

$$\eta_v = 1 + K - K\left(\frac{p_2}{p_1}\right)^{1/n} = 1 + 0.05 - 0.05\left(\frac{8}{0.973}\right)^{1/1.35}$$

$$= 0.812 \text{ or } 81.2\% \textbf{ Ans.}$$

3. *Cylinder diameter*

Let D = Cylinder diameter, and

L = Stroke length = 1.2 D ... (Given)

We know that swept volume per stroke

$$= \frac{\pi}{4} \times D^2 \times L = \frac{\pi}{4} \times D^2 \times 1.2\, D = 0.9426\, D^3$$

Since the compressor is double acting, therefore number of working strokes per minute

$$= 2N = 2 \times 300 = 600$$

and swept volume per minute

$$v_s = 0.9426\, D^3 \times 600 = 565.56\, D^3$$

We know that volumetric efficiency (η_v),

$$0.812 = \frac{v_1}{v_s} = \frac{3.25}{565.56\, D^3}$$

∴ $D^3 = 0.00707$ or $D = 0.192$ m = 192 mm **Ans.**

and $L = 1.2\, D = 1.2 \times 192 = 230.4$ mm **Ans.**

Example 30.7. *A single acting two-stage air compressor deals with air measured at atmospheric conditions of 1.013 bar and 15° C. At suction, the pressure is 1 bar and the temperature is 30° C. The final delivery pressure is 17 bar, the interstage pressure is 4 bar and perfect intercooling is to be assumed. If the L.P. cylinder bore is 230 mm, the common stroke is 150 mm and the speed of the compressor is 350 r.p.m. ; calculate 1. the volumetric efficiency of the compressor ; 2. the volume of atmospheric air dealt with per minute ; and 3. the power of the driving motor required. Assume the clearance volume of L.P. cylinder to be 5% and the indices of compression and expansion in the L.P. and H.P. cylinder to be 1.25 ; the mechanical efficiency being 85%.*

Solution. Given : p_a = 1.013 bar ; T_a = 15° C = 15 + 273 = 288 K ; p_1 = 1 bar = 1 × 10^5 N/m^2 ; T_1 = 30° C = 30 + 273 = 303 K ; p_3 = 17 bar ; p_2 = 4 bar ; D_1 = 230 mm = 0.23 m ; L = 150 mm = 0.15 m ; N = 350 r.p.m. ; $K = v_{c1} / v_{s1}$ = 5% = 0.05 ; n = 1.25 ; η_m = 85% = 0.85

1. *Volumetric efficiency of the compressor*

We know that volumetric efficiency of the compressor,

$$\eta_v = \frac{p_1 T_a}{p_a T_1}\left[1 + K - K\left(\frac{p_2}{p_1}\right)^{1/n}\right]$$

$$= \frac{1 \times 288}{1.013 \times 303}\left[1 + 0.05 - 0.05\left(\frac{4}{1}\right)^{1/1.25}\right]$$

$$= 0.843 \text{ or } 84.3\ \% \textbf{ Ans.}$$

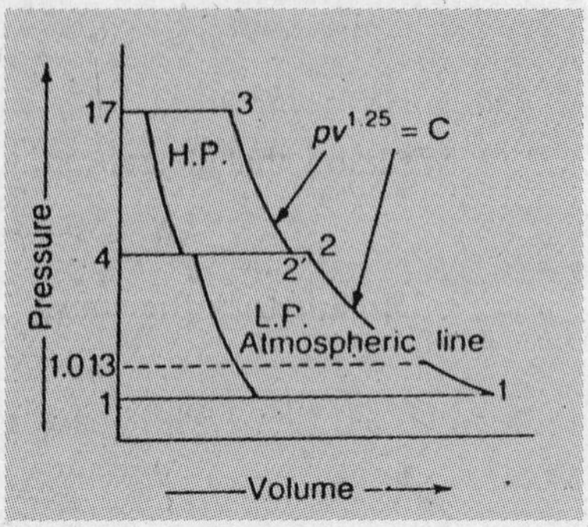

Fig. 30.2

2. *Volume of atmospheric air dealt with per minute*

We know that swept volume of the L.P. cylinder per minute,

$$v_s = \text{Swept volume per stroke} \times \text{No. of working strokes/min}$$

$$= \frac{\pi}{4}(D_1)^2 L \times N_w$$

$$= \frac{\pi}{4}(0.23)^2\, 0.15 \times 350 = 2.18 \text{ m}^3\text{/min} \qquad \ldots [\because \text{For single acting, } N_w = N]$$

$\therefore$ Volume of atmospheric air dealt with per minute,

$$v_a = v_s \times \eta_v = 2.18 \times 0.843 = 1.838 \text{ m}^3\text{/min } \textbf{Ans.} \qquad \ldots (\because \eta_v = v_a / v_s)$$

3. *Power of driving motor required*

Let T_2 = Temperature of air leaving the L.P. cylinder or entering the intercooler,

T_2' = Temperature of air leaving the intercooler or entering the H.P. cylinder, and

T_3 = Temperature of air delivered by the H.P. cylinder

= T_1, for perfect intercooling

We know that for polytropic compression 1-2 in the L.P. cylinder,

$$\frac{T_2}{T_1} = \left(\frac{p_2}{p_1}\right)^{\frac{n-1}{n}} = \left(\frac{4}{1}\right)^{\frac{1.25-1}{1.25}} = 1.3195$$

$$\therefore \quad T_2 = T_1 \times 1.3195 = 303 \times 1.3195 = 400 \text{ K}$$

Similarly, for polytropic compression 2′-3 in the H.P. cylinder,

$$\frac{T_3}{T_2'} = \left(\frac{p_3}{p_2'}\right)^{\frac{n-1}{n}} = \left(\frac{17}{4}\right)^{\frac{1.25-1}{1.25}} = 1.3356 \qquad \ldots (\because p_2' = p_2)$$

$$\therefore \quad T_3 = T_2' \times 1.3356 = 303 \times 1.3356 = 405 \text{ K} \qquad \ldots (\because T_2' = T_1)$$

We know that mass of air dealt with per minute,

$$m = \frac{p_a v_a}{R\, T_a} = \frac{1.013 \times 10^5 \times 1.838}{287 \times 288} = 2.25 \text{ kg/min}$$

∴ Indicated workdone by L.P. compressor,

$$W_L = \frac{n}{n-1} \times mR\,(T_2 - T_1) = \frac{1.25}{1.25-1} \times 2.25 \times 287\,(400 - 303) \text{ J/min}$$

$$= 313\,190 \text{ J/min} = 313.19 \text{ kJ/min}$$

and indicated workdone by H.P. compressor,

$$W_H = \frac{n}{n-1} \times mR\,(T_3 - T_2') = \frac{1.25}{1.25-1} \times 2.25 \times 287\,(405 - 303) \text{ J/min}$$

$$= 329\,330 \text{ J/min} = 329.33 \text{ kJ/min} \qquad \ldots (T_2' = T_1)$$

We know that total indicated work done,

$$W = W_L + W_H = 313.19 + 329.33 = 642.52 \text{ kJ/min}$$

∴ Power of the driving motor required,

$$P = 642.52 / 60 = 10.7 \text{ kW} \quad \textbf{Ans.}$$

Note : The total workdone (W) by a two-stage air compressor with perfect intercooling may also be determined by using the relation,

$$W = \frac{n}{n-1} \times mRT_1 \left[\left(\frac{p_2}{p_1}\right)^{\frac{n-1}{n}} + \left(\frac{p_3}{p_2}\right)^{\frac{n-1}{n}} - 2 \right]$$

Example 30.8. *A two stage, single acting air compressor compresses air to 20 bar. The air enters the L.P. cylinder at 1 bar and 27° C and leaves it at 4.7 bar. The air enters the H.P. cylinder at 4.5 bar and 27° C. The size of L.P. cylinder is 400 mm diameter and 500 mm stroke. The clearance volume in both cylinders is 4% of the respective stroke volume. The compressor runs at 200 r.p.m. Taking index of compression and expansion in the two cylinders as 1.3, estimate 1. the indicated power required to run the compressor ; and 2. the heat rejected in the intercooler per minute.*

Solution. Given : $p_4 = 20$ bar ; $p_1 = 1$ bar $= 1 \times 10^5$ N/m² ; $T_1 = 27°\text{ C} = 27 + 273 = 300$ K ; $p_2 = 4.7$ bar ; $p_3 = 4.5$ bar $= 4.5 \times 10^5$ N/m² ; $T_3 = 27°\text{ C} = 27 + 273 = 300$ K ; $D_1 = 400$ mm $= 0.4$ m ; $L_1 = 500$ mm $= 0.5$ m ; $K = v_{c1} / v_{s1} = v_{c3} / v_{s3} = 4\% = 0.04$; $N = 200$ r.p.m. ; $n = 1.3$

1. *Indicated power required to run the compressor*

We know that swept volume of L.P. cylinder,

$$v_{s1} = \frac{\pi}{4}(D_1)^2 L_1 = \frac{\pi}{4}(0.4)^2\, 0.5 \text{ m}^3$$

$$= 0.062\,84 \text{ m}^3$$

and volumetric efficiency,

$$\eta_v = 1 + K - K\left(\frac{p_2}{p_1}\right)^{1/n}$$

$$= 1 + 0.04 - 0.04\left(\frac{4.7}{1}\right)^{1/1.3}$$

$$= 0.9085 \text{ or } 90.85\ \%$$

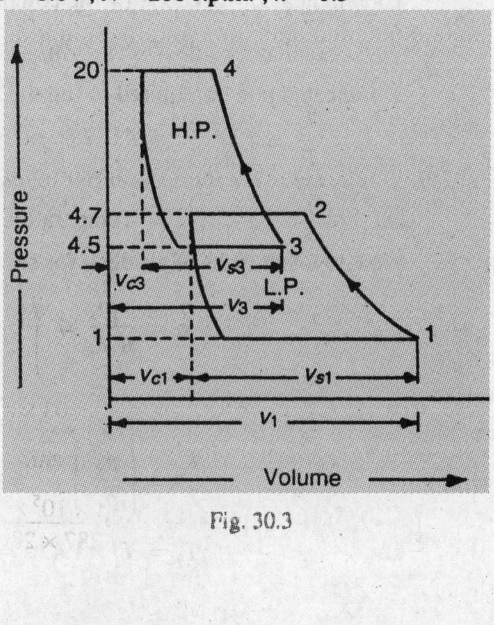

Fig. 30.3

∴ Volume of air sucked by L.P. compressor,

$$v_1 = v_{s1} \times \eta_v = 0.062\,84 \times 0.9085 = 0.0571 \text{ m}^3\text{/stroke}$$

$$= 0.0571 \times N_w = 0.0571 \times 200 = 11.42 \text{ m}^3\text{/min}$$

. . . (∵ For single acting, no. of working strokes per min, $N_w = N = 200$)

and volume of air sucked by H.P compressor,

$$v_3 = \frac{p_1 v_1}{p_3} = \frac{1 \times 11.42}{4.5} = 2.54 \text{ m}^3\text{/min} \qquad \ldots (\because p_1 v_1 = p_3 v_3)$$

We know that indicated workdone by L.P. compressor,

$$W_L = \frac{n}{n-1} \times p_1 v_1 \left[\left(\frac{p_2}{p_1} \right)^{\frac{n-1}{n}} - 1 \right]$$

$$= \frac{1.3}{1.3-1} \times 1 \times 10^5 \times 11.42 \left[\left(\frac{4.7}{1} \right)^{\frac{1.3-1}{1.3}} - 1 \right] \text{ J/min}$$

$$= 2123.3 \times 10^3 \text{ J/min} = 2123.3 \text{ kJ/min}$$

and indicated workdone by H.P. compressor,

$$W_H = \frac{n}{n-1} \times p_3 v_3 \left[\left(\frac{p_4}{p_3} \right)^{\frac{n-1}{n}} - 1 \right]$$

$$= \frac{1.3}{1.3-1} \times 4.5 \times 10^5 \times 2.54 \left[\left(\frac{20}{4.5} \right)^{\frac{1.3-1}{1.3}} - 1 \right] \text{ J/min}$$

$$= 2034.5 \times 10^3 \text{ J/min} = 2034.5 \text{ kJ/min}$$

Total indicated workdone by the compressor,

$$W = W_L + W_H = 2123.3 + 2034.5 = 4157.8 \text{ kJ/min}$$

∴ Indicated power required to run the compressor

$$= 4157.8 / 60 = 69.3 \text{ kW} \textbf{ Ans.}$$

2. *Heat rejected in the intercooler per minute*

Let T_2 = Temperature of air after compression in the L.P. cylinder.

We know that mass of air dealt for compression in the L.P. cylinder,

$$m = \frac{p_1 v_1}{R T_1} = \frac{1 \times 10^5 \times 11.42}{287 \times 300} = 13.26 \text{ kg/min}$$

. . . (∵ R for air = 287 J/kg K)

and
$$\frac{T_2}{T_1} = \left(\frac{p_2}{p_1} \right)^{\frac{n-1}{n}} = \left(\frac{4.7}{1} \right)^{\frac{1.3-1}{1.3}} = 1.429$$

∴
$$T_2 = T_1 \times 1.429 = 300 \times 1.429 = 428.7 \text{ K}$$

We know that heat rejected in the intercooler

$$= m\,c_p\,(T_2 - T_1)$$

$$= 13.26 \times 1\,(428.7 - 300) = 1706.6 \text{ kJ/min } \textbf{Ans.}$$

$$\ldots (\because c_p \text{ for air} = 1 \text{ kJ/kg K})$$

30.5. Thermodynamic Cycle for a Rotary Air Compressor

As a matter of fact, the ideal compression in a rotary compressor is isentropic (*i.e.* frictionless adiabatic) which is shown by the graph 1-2′ in Fig. 30.4 (*a*) and (*b*).

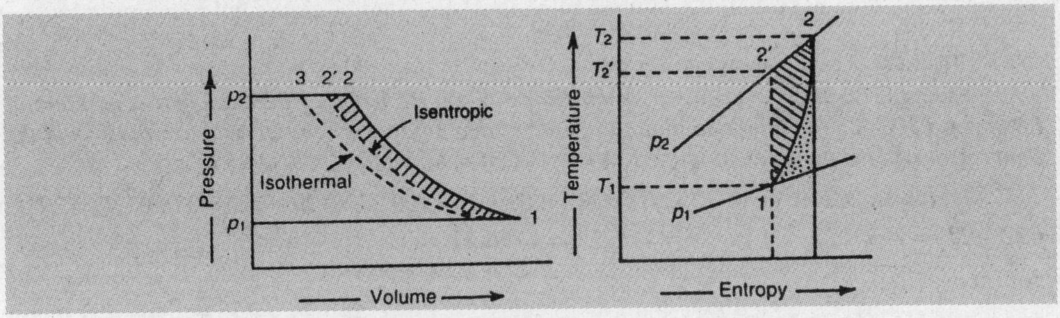

(*a*) *p-v* diagram. (*b*) *T-s* diagram.

Fig. 30.4. Thermodynamic cycle for a rotary air compressor.

But in actual practice, there is always some friction among the air molecules as well as between the air and the compressor casing. Moreover, there is always some shock at the entry and exit of air. It results in the formation of eddies at the entry and exit of the air. The above factors cause an increase in the temperature of the air at the exit without increasing its pressure. As a result of this, the temperature of air coming out of the compressor is more than that if it would have been compressed isentropically. A little consideration will show, that increase in the air temperature causes increase in its volume. Thus the amount of work done is also increased.

In Fig. 30.4 (*a*) and (*b*), the graph 1-2′ shows the ideal isentropic compression from pressure p_1 to p_2 (with an increase in temperature from T_1 to T_2'). The graph 1-2 shows the actual polytropic process (*i.e.* pv^n = constant).

Note : In actual polytropic process, the value of index n is greater than γ (about 1.7).

30.6. Efficiencies of a Centrifugal Air Compressor

We have already discussed in Art. 30.2 that the criterion for thermodynamic efficiency of a centrifugal air compressor is isentropic. But in general, the following efficiencies of a centrifugal air compressor are important from the subject point of view :

1. *Isentropic efficiency (or compressor efficiency).* It is the ratio of work (or power) required to compress the air isentropically to the actual work required to compress the air for the same pressure ratio. Mathematically, isentropic efficiency,

$$\eta_i = \frac{h_2' - h_1}{h_2 - h_1} = \frac{T_2' - T_1}{T_2 - T_1}$$

where h_2' = Enthalpy of air at exit for isentropic compression,

h_2 = Enthalpy of air at exit for actual compression,

h_1 = Enthalpy of air at inlet, and

T_2', T_2, T_1 = Temperatures at corresponding points.

2. *Polytropic efficiency.* It is the ratio of work (or power) required to compress the air polytropically to the actual work required to compress the air for the same pressure ratio. Mathematically, polytropic efficiency,

$$\eta_p = \left(\frac{\gamma - 1}{\gamma}\right)\left(\frac{n}{n-1}\right)$$

where γ = Ratio of specific heats, and

n = Polytropic index.

Note : The value of an n is always greater than γ.

Example 30.9. *A centrifugal compressor delivers 0.5 kg of air per second at a pressure of 1.8 bar and 100° C. The intake conditions are 20° C and 1 bar. Find the isentropic efficiency of the compressor and the power required to drive it. Take n = 1.65 and c_p = 1 kJ/kg K.*

Solution. Given : $m = 0.5$ kg/s ; $p_2 = 1.8$ bar ; $T_2 = 100°$ C = 100 + 273 = 373 K ; $T_1 = 20°$ C = 20 + 273 = 293 K ; $p_1 = 1$ bar ; $n = 1.65$; $c_p = 1$ kJ/kg K

Isentropic efficiency

Let T_2' = Temperature of air at exit for isentropic compression.

We know that
$$\frac{T_2'}{T_1} = \left(\frac{p_2}{p_1}\right)^{\frac{n-1}{n}} = \left(\frac{1.8}{1}\right)^{\frac{1.65-1}{1.65}} = (1.8)^{0.391} = 1.261$$

$$\therefore \quad T_2' = T_1 \times 1.261 = 293 \times 1.261 = 369 \text{ K}$$

We know that isentropic efficiency,

$$\eta_i = \frac{T_2' - T_1}{T_2 - T_1} = \frac{369 - 293}{373 - 293} = 0.95 \quad \text{or} \quad 95\% \textbf{ Ans.}$$

Power required to drive the compressor

We know that work done in compressing the air isentropically,

$$W = m\, c_p (T_2 - T_1) = 0.5 \times 1\,(373 - 293) = 40 \text{ kJ/s}$$

$\therefore$ Power required to drive the compressor,

= 40 kW **Ans.** . . . ($\because$ 1 kJ/s = 1 kW)

Example 30.10. *A centrifugal compressor with 70% isentropic efficiency delivers 20 kg of air per minute at a pressure of 3 bar. If the compressor receives air at 20° C and at a pressure of 1 bar, find the actual temperature of the air at exit. Also find the power required to run the compressor, if its mechanical efficiency is 95%. Take γ and c_p for air as 1.4 and 1 kJ/kg K respectively.*

Solution. Given : $\eta_i = 70\% = 0.7$; $m = 20$ kg/min ; $p_2 = 3$ bar ; $T_1 = 20°$ C = 20 + 273 = 293 K ; $p_1 = 1$ bar ; $\eta_m = 95\% = 0.95$; $\gamma = 1.4$; $c_p = 1$ kJ/kg K

Actual temperature of the air at exit

Let T_2 = Actual temperature of the air at exit, and

T_2' = Temperature of the air at exit for isentropic compression.

We know that $\dfrac{T_2'}{T_1} = \left(\dfrac{p_2}{p_1}\right)^{\frac{n-1}{n}} = \left(\dfrac{3}{1}\right)^{\frac{1.4-1}{1.4}} = (3)^{0.286} = 1.369$

$\therefore \quad T_2' = T_1 \times 1.369 = 293 \times 1.369 = 401.1 \text{ K}$

We also know that isentropic efficiency (η_i),

$$0.7 = \frac{T_2' - T_1}{T_2 - T_1} = \frac{401.1 - 293}{T_2 - 293}$$

$$0.7\,T_2 - 205.1 = 401.1 - 293 = 108.1$$

$$\therefore \quad T_2 = 447.4 \text{ K} = 174.4^\circ \text{C} \quad \textbf{Ans.}$$

Power required to run the compressor

We know that work done in compressing the air isentropically,

$$W = m\,c_p\,(T_2 - T_1) = 20 \times 1\,(447.4 - 293) = 3088 \text{ kJ/min}$$
$$= 51.47 \text{ kJ/s}$$

$\therefore$ Power required to run the compressor

$$= \frac{51.47}{\eta_m} = \frac{51.47}{0.95} = 54.25 \text{ kW} \quad \textbf{Ans.}$$

Example 30.11. *A centrifugal compressor having compression ratio of 2.4 compresses the air polytropically according to law $pv^{1.6}$ = constant. Find the polytropic efficiency of the compressor, if c_p = 0.995 kJ/kg K and c_v = 0.71 kJ/kg K.*

Solution. Given : $^*p_2 / p_1 = 2.4$; $n = 1.6$; $c_p = 0.995$ kJ/kg K ; $c_v = 0.71$ kJ/kg K

We know that ratio of specific heats,

$$\gamma = c_p / c_v = 0.995 / 0.71 = 1.4$$

$\therefore$ Polytropic efficiency of the compressor,

$$\eta_p = \left(\frac{\gamma - 1}{\gamma}\right)\left(\frac{n}{n-1}\right) = \frac{1.4 - 1}{1.4} \times \frac{1.6}{1.6 - 1} = 0.762 \text{ or } 76.2\% \quad \textbf{Ans.}$$

30.7. Static and Total Head Quantities

As a matter of fact, the velocities encountered in a centrifugal compressor are very large as compared with those of a reciprocating air compressor. It is, therefore, very essential to take into consideration the velocities for the analysis purpose.

Now consider a horizontal passage of varying area through which 1 kg of air is flowing as shown in Fig. 30.5.

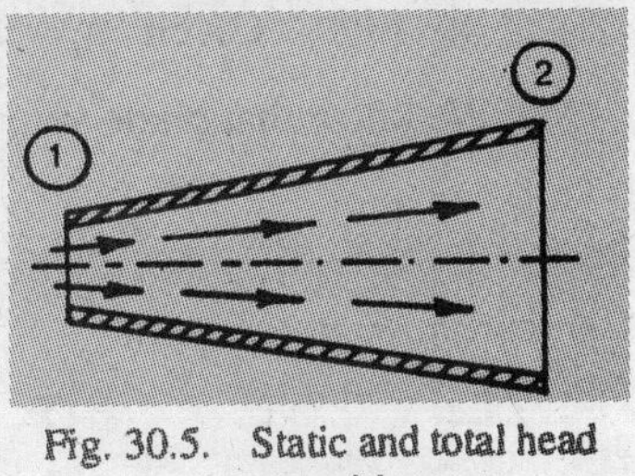

Fig. 30.5. Static and total head quantities.

Let T_1 = Temperature at section 1 in K,

h_1 = Enthalpy at section 1 in kJ/kg,

p_1 = Pressure of air at section 1 in bar,

V_1 = Velocity of air at section 1 in m/s, and

T_2, h_2, p_2, V_2 = Corresponding values at section 2.

---

* Superfluous data

Now let us assume a steady flow of air from section 1 to 2, so that no heat is transferred as well as no work is done, as the air flows through the passage. Applying the steady flow energy equation to the system,

$$h_1 + \frac{10^5 p_1 v_1}{1000} + \frac{V_1^2}{2000} = h_2 + \frac{10^5 p_2 v_2}{1000} + \frac{V_2^2}{2000}$$

$$c_p T_1 + \frac{V_1^2}{2000} = c_p T_2 + \frac{V_2^2}{2000} \qquad \ldots \text{(Assuming } p_1 v_1 = p_2 v_2\text{)}$$

or in other words,

$$c_p T + \frac{V^2}{2000} = \text{Constant}$$

In the above general expression, the temperature (T) stands for the actual temperature of the air recorded by a thermometer which is also moving in the air with the same velocity as that of air. A little consideration will show, that if the moving air is brought to rest under reversible adiabatic conditions, the total kinetic energy will be converted into heat energy, which will increase its temperature and pressure. The new temperature and pressure of the air are called total heat or stagnation temperature (T_0) and pressure (p_0) respectively.

$$\therefore \quad c_p T + \frac{V^2}{2000} = c_p T_0 \quad \text{or} \quad T_0 - T = \frac{V^2}{2000\, c_p}$$

and

$$h_0 - h = \frac{V^2}{2000} \qquad \ldots (\because h = c_p T)$$

where h_0 is the stagnation enthalpy.

The total head pressure may be obtained from the equation,

$$\frac{p_0}{p} = \left(\frac{T_0}{T}\right)^{\frac{\gamma}{\gamma-1}} \quad \text{or} \quad \frac{T_0}{T} = \left(\frac{p_0}{p}\right)^{\frac{\gamma-1}{\gamma}}$$

where γ is the usual ratio of specific heats.

Notes : 1. The term p_0 / p is called static pressure ratio.

2. This relation may be used for any two sections in a centrifugal compressor also.

Example 30.12. *A centrifugal air compressor having isentropic efficiency of 70% receives air at 17° C. If the outer diameter of the blade tip is 1 m and the compressor runs at 5000 r.p.m., find : 1. the temperature rise of the air ; and 2. the static pressure ratio.*

Solution. Given : $\eta_i = 70\% = 0.7$; $T_1 = 17°\text{C} = 17 + 273 = 290$ K ; $D = 1$ m ; $N = 5000$ r.p.m.

We know that blade velocity,

$$V_b = \frac{\pi D N}{60} = \frac{\pi \times 1 \times 5000}{60} = 261.8 \text{ m/s}$$

Temperature rise of the air

Let $(T_2 - T_1)$ = Temperature rise of the air.

We know that work done by the compressor per kg of air,

$$w = \frac{V_b^2}{1000} = \frac{(261.8)^2}{1000} = 68.5 \text{ kJ}$$

We also know that work done by the compressor per kg of air (w),

$$68.5 = c_p(T_2 - T_1) = 1(T_2 - T_1) \qquad \text{... (For air , } c_p = 1 \text{ kJ/kg K)}$$

$$\therefore \qquad (T_2 - T_1) = 68.5^\circ \text{ C or K} \quad \textbf{Ans.}$$

Static pressure ratio

Let $\frac{p_2}{p_1}$ = Static pressure ratio, and

T_2' = Temperature of air at exit for isentropic compression.

We know that isentropic efficiency (η_i),

$$0.7 = \frac{T_2' - T_1}{T_2 - T_1} = \frac{T_2' - 290}{68.5}$$

$$\therefore \qquad T_2' = (0.7 \times 68.5) + 290 = 337.95 \text{ K}$$

and

$$\frac{p_2}{p_1} = \left(\frac{T_2'}{T_1}\right)^{\frac{\gamma}{\gamma-1}} = \left(\frac{337.95}{290}\right)^{\frac{1.4}{1.4-1}} = (1.165)^{3.5} = 1.71 \quad \textbf{Ans.}$$

30.8. Slip Factor

We have already discussed in Art. 29.9 that the ideal or maximum work done by a centrifugal rotary compressor

$$= m(V_{w1})^2 = m(V_b)^2$$

The above relation has been derived under the assumption that $V_{w1} = V_b$. But in actual practice, V_{w1} is a always less than V_b. The difference between V_h and V_{w1} (*i.e.* $V_b - V_{w1}$) is known as slip and the ratio of V_{w1} to V_b (*i.e.* V_{w1}/V_b) is known as *slip factor.*

30.9. Comparison of Turbine and Centrifugal Compressor Blades

Following are the main points of comparison of the turbine and centrifugal compressor blades.

S. No.	Turbine blades	Centrifugal compressor blades
1.	Passage between the blades is converging.	Passage between the blades is diverging.
2.	Due to converging passage, the flow gets accelerated. But the pressure decreases.	Due to diverging passage, the flow gets diffused or decelerated. But the pressure increases.
3.	The flow is more stable.	The flow is less stable.
4.	The flow always takes place in one direction only.	Sometimes, the flow breaks away and reverses its direction.
5.	The blades are simple in design and construction, as their profile consists of circular arc and straight line.	The blades are complicated in design and construction, as their profile consists of aerofile section based on aerodynamic theory.

EXERCISES

1. A single stage reciprocating air compressor takes in air at 1 bar and 15° C. The conditions at the end of suction are 0.97 bar and 30° C. The discharge pressure is 6 bar. The clearance is 5% of the stroke. The compression and expansion follows $pv^{1.3}$ = Constant. Find the volumetric efficiency of the compressor.

[Ans. 78.1%]

2. A single acting single stage reciprocating air compressor with 5% clearance volume compresses air from 1 bar to 5 bar. Find the change in volumetric efficiency of the compressor, if the exponents of expansion process change from 1.25 to 1.4. [**Ans.** 2.5%]

3. Air is compressed by a reciprocating compressor from 1.05 bar and 27° C to 7.9 bar. During the suction and discharge, there are inlet and outlet pressure losses at the valves of 0.05 bar and 0.1 bar respectively and the atmospheric air is heated up after induction to 37° C. Determine the volumetric efficiency of the compressor. Assume law of compression and expansion to be the same, $pv^{1.3}$ = Constant and percentage of clearance volume 4%. [**Ans.** 84.2%]

4. A compressor has 150 mm bore and 200 mm stroke and the linear clearance is 10 mm. Calculate the theoretical volume of air taken in m³ per stroke when working between 1 bar and 7 bar. Take n = 1.25. [**Ans.** 2.87×10^{-3} m³]

5. A compressor is used to compress air from a pressure of 1.013 bar to 7.21 bar. The polytropic exponent for both compression and expansion is n = 1.35. The clearance volume of the compressor is 200×10^{-6} m³. If the volumetric efficiency of the compressor is 80 percent and the stroke is 250 mm, determine the cylinder diameter of the compressor. [**Ans.** 129.2 mm]

6. A single stage reciprocating air compressor takes in air at 1 bar and 15° C. The conditions at the end of suction are 0.97 bar and 30° C. The discharge pressure is 6 bar. The clearance is 5 percent of stroke. The compression and expansion follow $p\,v^{1.3}$ = Constant and the mass of air handled is 1 kg/min. Estimate the stroke volume and power needed in kW. Assume compressor speed as 1000 r.p.m. [**Ans.** 1.147×10^{-3} m³ ; 3.28 kW]

7. A single stage, double acting air compressor delivers 15 m³ of air per minute measured at 1.013 bar and temperature 300 K and delivers at 7 bar. The conditions at the end of suction stroke are pressure 0.98 bar and temperature 313 K. The clearance volume is 4 percent of the swept volume and the stroke/bore ratio is 1.3/1. The compressor runs at 300 r.p.m. Calculate 1. volumetric efficiency ; 2. cylinder dimensions ; 3. indicated power ; and 4. isothermal efficiency of this compressor. Take the index of compression and expansion as 1.3.

[**Ans.** 79.6% ; 313 mm, 407 mm ; 65.8 kW ; 78.8%]

8. A single acting two-stage air compressor deals with 4 m³/min of air under atmospheric conditions of 1.013 bar and 15° C with a speed of 250 r.p.m. The delivery pressure is 80 bar. Assuming complete intercooling, find the minimum power required by the compressor and the bore and stroke of the compressor. Assume a piston speed of 3 m/s, mechanical efficiency of 75 percent, and volumetric efficiency of 80 percent per stage. Assume the polytropic index of compression in both the stages to be n = 1.25 and neglect clearance. [**Ans.** 50 kW]

9. A single stage double acting air compressor delivers air at 7 bar. The pressure and temperature at the end of suction stroke are 1 bar and 27° C. It delivers 2 m³ of free air per minute when the compressor is running at 300 r.p.m. The clearance volume is 5 percent of the stroke volume. The pressure and temperature of ambient air are 1.013 bar and 20° C. The index of compression is 1.3 and index of expansion is 1.35. Find : 1. Volumetric efficiency of the compressor ; and 2. Diameter and stroke of the cylinder if both are equal.

[**Ans.** 83,8% ; 175 mm]

QUESTIONS

1. Explain the following terms :

   (*a*) Isothermal efficiency ; (*b*) Isentropic efficiency ; and (*c*) Volumetric efficiency.

2. Discuss the effect of clearance volume on the volumetric efficiency of a reciprocating air compressor.

3. Prove that the volumetric efficiency of a compressor is given by

$$\frac{p_1 T_a}{p_a T_1}\left[1 + K - K\left(\frac{p_2}{p_1}\right)^{1/n}\right]$$

where suffix a and 1 represent ambient and before compression conditions respectively and K is the ratio of clearance volume to the swept volume.

4. Describe 'thermodynamic cycle for a rotary air compressor'.

5. Define static and total head quantities.

6. What do you understand by the term 'slip factor'?

OBJECTIVE TYPE QUESTIONS

1. The ratio of the indicated power to the shaft power or brake power of the motor or engine required to drive the compressor, is called

   (*a*) compressor efficiency (*b*) volumetric efficiency
   (*c*) isentropic efficiency (*d*) mechanical efficiency

2. The ratio of the volume of free air delivery per stroke to the swept volume of the piston, is known as

   (*a*) compressor efficiency (*b*) volumetric efficiency
   (*c*) isentropic efficiency (*d*) mechanical efficiency

3. If the clearance ratio for a reciprocating air compressor is *K*, then its volumetric efficiency is given by

   (*a*) $1 - K + K\left(\frac{p_1}{p_2}\right)^{1/n}$ (*b*) $1 + K - K\left(\frac{p_1}{p_2}\right)^{1/n}$
   (*c*) $1 + K - K\left(\frac{p_2}{p_1}\right)^{1/n}$ (*d*) $1 - K + K\left(\frac{p_2}{p_1}\right)^{1/n}$

4. The volumetric efficiency of a compressor

   (*a*) increases with decrease in compression ratio
   (*b*) decreases with decrease in compression ratio
   (*c*) increases with increase in compression ratio
   (*d*) decreases with increase in compression ratio

5. The volumetric efficiency for reciprocating air compressors is about

   (*a*) 10 to 40% (*b*) 40 to 60% (*c*) 60 to 70% (*d*) 70 to 90%

ANSWERS

1. (*d*)	2. (*b*)	3. (*c*)	4. (*d*)	5. (*d*)

31

Air Motors

1. Introduction. 2. Workdone by Air in an Air Motor. 3. Combined Air Motor and Air Compressor (Compressed Air System). 4. Efficiency of Compressed Air System. 5. Preheating of Compressed Air.

31.1. Introduction

In the last three chapters, we have discussed air compressors and their performance. Though the air compressors are used for innumerable purposes these days in various fields, yet one of their uses is in air motors attached to portable tools. An air motor is used as an alternative to an electric motor, especially when sparks from the electric motor (or cables) might prove dangerous *e.g.* in explosive factories and mines.

The operation of a reciprocating air motor is similar to that of a reciprocating steam engine ; but reverse to that of reciprocating air compressor.

31.2. Workdone by Air in an Air Motor

As a matter of fact, the compressed air (from an air compressor) is made to enter the cylinder of an air motor which pushes its piston forward in the same way as of a reciprocating steam engine. Now the actual work is done by the movement of the piston. Now consider an air motor working with the help of compressed air.

Let

p_1 = Pressure of the compressed air, and

v_1 = Volume of the compressed air.

The theoretical indicator diagram of a reciprocating air motor without clearance, compression and pressure drop at release is shown in Fig. 31.1.

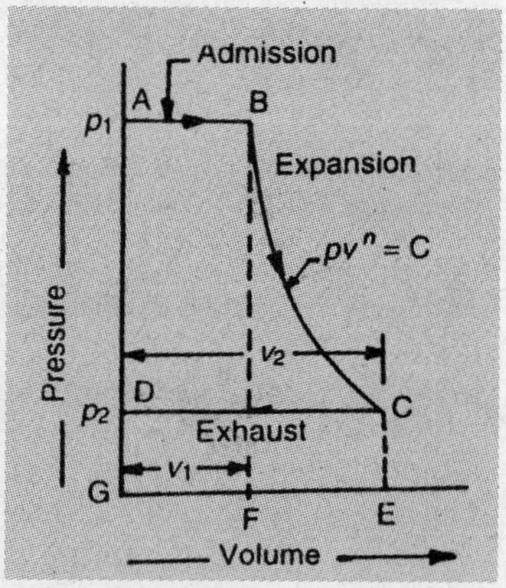

Fig. 31.1. Workdone by air motor.

The compressed air from the compressor is admitted into an air motor at A with pressure p_1. It drives the piston forward. But after a part stroke is performed, the air supply is cut-off at B and the expansion occurs from B to C. After the stroke is completed, the air which has done some work is exhausted into the atmosphere at a constant pressure p_2.

We know that work done by the air per cycle,

$$W = \text{Area } ABCD$$

$$= \text{Area } ABFG + \text{Area } BCEF - \text{Area } CEGD$$

$$= p_1 v_1 + \frac{p_1 v_1 - p_2 v_2}{n-1} - p_2 v_2$$

$$= \frac{n}{n-1}(p_1 v_1 - p_2 v_2)$$

$$= \frac{n}{n-1} \times p_1 v_1 \left[1 - \left(\frac{p_2}{p_1}\right)^{\frac{n-1}{n}}\right]$$

$$= \frac{n}{n-1} \times mRT_1 \left[1 - \left(\frac{p_2}{p_1}\right)^{\frac{n-1}{n}}\right] \quad \ldots (\because pv = mRT)$$

Example 31.1. *An air motor receives air at 3.5 bar and 425 K, and exhaust it at 1 bar. Find the amount of work done per kg of air if the air expands according to the law $pv^{1.35}$ = constant.*

Solution. Given : $p_1 = 3.5$ bar ; $T_1 = 425$ K ; $p_2 = 1$ bar ; $m = 1$ kg ; $n = 1.35$

We know that work done per kg of air,

$$W = \frac{n}{n-1} \times mRT_1 \left[1 - \left(\frac{p_2}{p_1}\right)^{\frac{n-1}{n}}\right]$$

$$= \frac{1.35}{1.35-1} \times 1 \times 287 \times 425 \left[1 - \left(\frac{1}{3.5}\right)^{\frac{1.35-1}{1.35}}\right] \text{J}$$

$$= 130\,320 \text{ J} = 130.32 \text{ kJ} \quad \textbf{Ans.}$$

Example 31.2. *An air motor is supplied with compressed air at 6.5 bar and 157° C. It is expanded to 1.04 bar and then exhausted at constant pressure. Determine the amount of work done by 1 kg of air and the temperature of air at the end of expansion. Assume the expansion according to $pv^{1.3}$ = constant and neglect clearance.*

Solution. Given : $p_1 = 6.5$ bar ; $T_1 = 157°\text{ C} = 157 + 273 = 430$ K ; $p_2 = 1.04$ bar ; $m = 1$ kg ; $n = 1.3$

Work done by 1 kg of air

We know that work done by 1 kg of air,

$$W = \frac{n}{n-1} \times mRT_1 \left[1 - \left(\frac{p_2}{p_1}\right)^{\frac{n-1}{n}}\right]$$

$$= \frac{1.3}{1.3-1} \times 1 \times 287 \times 430 \left[1 - \left(\frac{1.04}{6.5}\right)^{\frac{1.3-1}{1.3}}\right] \text{J}$$

$$= 184\,500 \text{ J} = 184.5 \text{ kJ} \quad \textbf{Ans.}$$

Temperature at the end of the expansion

Let T_1 = Temperature at the end of expansion.

We know that $$\frac{T_2}{T_1} = \left(\frac{p_1}{p_2}\right)^{\frac{n-1}{n}} = \left(\frac{1.04}{6.5}\right)^{\frac{1.3-1}{1.3}} = (0.16)^{\frac{0.3}{1.3}} = 0.655$$

$$T_2 = 430 \times 0.655 = 281.6 \text{ K} = 8.6° \text{ C} \quad \textbf{Ans.}$$

31.3. Combined Air Compressor and Air Motor (Compressed Air System)

In the last article, we have discussed that the compressed air is carried, from the air compressor, to the air motor. Sometimes, air compressor and air motor are installed as two separate units. But sometimes they are installed as one unit. A little consideration will show, that if the air compressor and air motor are installed at some distance apart, the hot* compressed air, flowing through the duct, will get cooled to some extent. But if they are installed as one unit, there is no time for the air to get cooled. In such a case, some cooling arrangement is provided between the two units. *i.e.* after the air compressor or in other words before the motor. Such a system is known as compressed air system.

In a compressed air system, the air is first compressed in an air compressor from pressure p_1 to p_2 with a corresponding rise in its temperature. The hot air, leaving the compressor, is now cooled to the initial compressor temperature. The air is then made to expand in the air motor cylinder from pressure p_2 to p_1 with a corresponding fall in its temperature. Thus the temperature of air discharged from the air motor is less than the initial compressor intake temperature.

31.4. Efficiency of Compressed Air System

The theoretical indicator diagram of a compressed air system is shown in Fig. 31.2. The compression of air, in a compressor cylinder from pressure p_1 to p_2 is represented by the curve 1-2. The hot air leaving the compressor is cooled down in an air cooler to original compressor intake temperature.

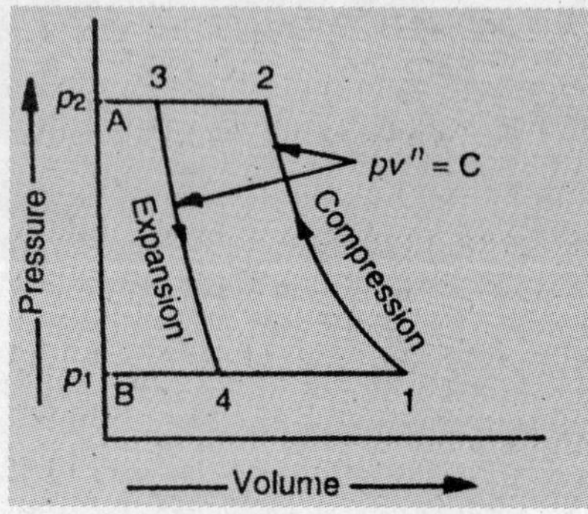

Fig. 31.2. Compressed air system.

The air now enters the air motor cylinder, and expands from pressure p_2 to p_1 as shown by the curve 3-4 in Fig. 31.2. Now let us assume the compression and expansion according to pv^n = constant and neglect clearance.

∴ Work done on the air compressor

$$= \text{Area } B\,12\,A = \frac{n}{n-1}(p_2v_2 - p_1v_1)$$

$$= \frac{n}{n-1} \times mR\,(T_2 - T_1)$$

and work done by the air in the motor

$$= \text{Area } A\,34\,B = \frac{n}{n-1}(p_3v_3 - p_4v_4)$$

$$= \frac{n}{n-1} \times mR\,(T_3 - T_4)$$

Now, let η_m = Efficiency of the air motor, and

η_c = Efficiency of the compressor.

∴ Shaft output of the air motor

$$= \text{Work done by the air} \times \eta_m$$

$$= \frac{n}{n-1} \times mR\,(T_3 - T_4)\,\eta_m \qquad \ldots (i)$$

and shaft input to the compressor

$$= \frac{n}{n-1} \times mR\,(T_2 - T_1)/\eta_c \qquad \ldots (ii)$$

* When the air is compressed, its temperature is increased.

The overall efficiency of the compressed air system is the ratio of the shaft output of the air motor to the shaft input to the compressor. Mathematically, overall efficiency of the compressed air system,

$$= \frac{\text{Shaft output of the air motor}}{\text{Shaft input to the compressor}}$$

$$= \frac{\frac{n}{n-1} \times mR\,(T_3 - T_4)\,\eta_m}{\frac{n}{n-1} \times mR\,(T_2 - T_1)/\eta_c} = \frac{(T_3 - T_4)\,\eta_m}{(T_2 - T_1)/\eta_c}$$

$$= \frac{T_3\left[1 - \frac{T_4}{T_3}\right]\eta_m}{T_1\left[\frac{T_2}{T_1} - 1\right]\eta_c} = \frac{1 - \frac{T_4}{T_3}}{\frac{T_2}{T_1} - 1} \times \eta_m \times \eta_c \qquad \ldots (\because T_3 = T_1) \ldots (iii)$$

We know that $\frac{T_4}{T_3} = \left(\frac{p_1}{p_2}\right)^{\frac{n-1}{n}}$ and $\frac{T_2}{T_1} = \left(\frac{p_2}{p_1}\right)^{\frac{n-1}{n}}$

∴ Equation (*iii*) may also be written as,

$$\eta_0 = \frac{1 - \left(\frac{p_1}{p_2}\right)^{\frac{n-1}{n}}}{\left(\frac{p_2}{p_1}\right)^{\frac{n-1}{n}} - 1} \times \eta_m \times \eta_c$$

Example 31.3. *A compressed air system consists of a single stage compressor of efficiency 75 % and air motor of efficiency 65%. If the compression and expansion follows the law $pv^{1.25}$ = constant, find the overall efficiency of the system. Take pressure ratio for both the machines as 3.5.*

Solution. Given : $\eta_c = 75\,\% = 0.75$; $\eta_m = 65\% = 0.65$; $n = 1.25$; $p_2/p_1 = 3.5$ or $p_1/p_2 = 1/3.5$

We know that overall efficiency of the system,

$$\eta_0 = \frac{1 - \left(\frac{p_1}{p_2}\right)^{\frac{n-1}{n}}}{\left(\frac{p_2}{p_1}\right)^{\frac{n-1}{n}} - 1} \times \eta_m \times \eta_c = \frac{1 - \left(\frac{1}{3.5}\right)^{\frac{1.25-1}{1.25}}}{(3.5)^{\frac{1.25-1}{1.25}} - 1} \times 0.65 \times 0.75$$

$$= \frac{1 - 0.778}{1.285 - 1} \times 0.65 \times 0.75 = 0.38 \quad \text{or} \quad 38\%\ \textbf{Ans.}$$

Example 31.4. *A system using compressed air for power transmission consists of a single stage compressor and air motor both having mechanical efficiency of 80%. The compression and expansion takes place according to $pv^{1.2}$ = constant. The higher and lower pressure for both compressor and air motor are 5 bar and 1 bar respectively. The air is cooled during its passage from the compressor to the motor to the initial temperature of 15° C. Calculate :*

1. Work done in compressor and motor cylinders ; and 2. Overall efficiency of the system.

Solution. Given : $\eta_m = \eta_c = 80\% = 0.8$; $n = 1.2$; $p_2 = p_3 = 5$ bar $= 5 \times 10^5$ N/m^2 ; $p_1 = p_4 = 1$ bar $= 1 \times 10^5$ N/m^2 ; $T_3 = T_1 = 15°$C $= 15 + 273 = 288$ K

Let T_2 = Temperature of air at the end of compression ; and

T_4 = Temperature of air at the end of expansion.

We know that
$$\frac{T_2}{T_1} = \left(\frac{p_2}{p_1}\right)^{\frac{n-1}{n}} = \left(\frac{5}{1}\right)^{\frac{1.2-1}{1.2}} = 1.308$$

$\therefore$
$$T_2 = T_1 \times 1.308 = 288 \times 1.308 = 376.7 \text{ K}$$

Similarly
$$\frac{T_4}{T_3} = \left(\frac{p_1}{p_2}\right)^{\frac{n-1}{n}} = \left(\frac{1}{5}\right)^{\frac{1.2-1}{1.2}} = 0.765$$

$\therefore$
$$T_4 = T_3 \times 0.765 = 288 \times 0.765 = 220.3 \text{ K}$$

1. *Work done in compressor and motor cylinder*

We know that mass of air,

$$m = \frac{p_1 v_1}{R T_1} = \frac{1 \times 10^5 \times 1}{287 \times 288} = 1.21 \text{ kg} \qquad \ldots (\text{Taking } v_1 = 1 \text{ m}^3)$$

$\therefore$ Work done in compressor cylinder,

$$W_{1-2} = \frac{n}{n-1} \times m\,R\,(T_2 - T_1)$$

$$= \frac{1.2}{1.2-1} \times 1.21 \times 287\,(376.7 - 288) = 184\,820 \text{ J}$$

$= 184.82$ kJ **Ans.**

and work done in motor cylinder,

$$W_{3-4} = \frac{n}{n-1} \times m\,R\,(T_3 - T_4)$$

$$= \frac{1.2}{1.2-1} \times 1.21 \times 287\,(288 - 220.3) = 141\,060 \text{ J}$$

$= 141.06$ kJ **Ans.**

2. *Overall efficiency of the system*

We know that overall efficiency of the system,

$$\eta_0 = \frac{(T_3 - T_4)\eta_m}{(T_2 - T_1)/\eta_c} = \frac{(288 - 220.3)\,0.8}{(376.7 - 288)/0.8} = 0.488 \text{ or } 48.8\% \textbf{ Ans.}$$

Note : The overall efficiency may also be found out from the relation,

$$\eta_0 = \frac{\text{Work done in air motor cylinder} \times \eta_m}{\text{Work done in air compressor cylinder}/\eta_c}$$

$$= \frac{141.06 \times 0.8}{184.82/0.8} = 0.488 \text{ or } 48.8\%$$

31.5. Preheating of Compressed Air

We have already discussed in the previous articles that compressed air (from the air compressor) is supplied to an air motor. A little consideration will show, that when the air expands in the motor cylinder, its temperature decreases. It will be interesting to know that if proper care in the design of compressed air system is not taken, then the temperature of exhaust air (from the air motor) may be below the freezing point. As a result of this, if there is any moisture present in the air, the same will be deposited in the form of ice, which will block exhaust valves of the motor. In order to prevent ice formation in the motor cylinder, the air is warmed by the steam in a preheater, at a constant pressure, before admission to the motor. Since the volume of air is proportional to the absolute temperature, a part of this heat energy is converted into additional work in the motor cylinder.

Let W = Work done in the motor cylinder for the given mass of air without preheating, and

W' = Work done for the same mass of air after preheating.

Then assuming that the work done is proportional to the volume of air used, therefore

$$\frac{W'}{W} = \frac{T_1'}{T_1}$$

Now the temperature, at which the air must be heated (T_1') in order to avoid freezing may be found out from the relation :

$$\frac{T_2}{T_1'} = \left(\frac{p_2}{p_1}\right)^{\frac{n-1}{n}}$$

where T_2 = Temperature of the exhaust air from the motor.

Example 31.5. *The initial pressure of the air in air motor is 5 bar and final pressure is 1 bar. Find the temperature at which the air must be preheated in order that the temperature after expansion may be 2° C. Assume the expansion to be according to $pv^{1.3}$ = Constant.*

Solution. Given : $p_1 = 5$ bar ; $p_2 = 1$ bar ; $T_2 = 2°\text{ C} = 2 + 273 = 275$ K

Let T_1' = Temperature at which the air must be heated.

We know that
$$\frac{T_2}{T_1'} = \left(\frac{p_2}{p_1}\right)^{\frac{n-1}{n}} = \left(\frac{1}{5}\right)^{\frac{1.3-1}{1.3}} = (0.2)^{\frac{0.3}{1.3}} = 0.69$$

$$\therefore \quad T_1' = T_2 / 0.69 = 275 / 0.69 = 398.6 \text{ K} = 125.6° \text{ C} \quad \textbf{Ans.}$$

Example 31.6. *An air motor receives air at a pressure of 4 bar and 50° C and exhausts at 1 bar. Determine the temperature to which the air should be preheated, so that exhaust temperature may be 5° C. Also find the ratio of work done due to preheating and otherwise. Take polytropic index as 1.35.*

Solution. Given : $p_1 = 4$ bar ; $T_1 = 50°\text{ C} = 50 + 273 = 323$ K ; $p_2 = 1$ bar ; $T_2 = 5°$ C $= 5 + 273 = 278$ K ; $n = 1.35$

Temperature to which the air should be preheated

Let T_1' = Temperature to which the air should be preheated.

We know that
$$\frac{T_2}{T_1'} = \left(\frac{p_2}{p_1}\right)^{\frac{n-1}{n}} = \left(\frac{1}{4}\right)^{\frac{1.35-1}{1.35}} = (0.25)^{\frac{0.35}{1.35}} = 0.698$$

$$\therefore \quad T_1' = T_2 / 0.698 = 278 / 0.698 = 398.3 \text{ K} = 125.3° \text{ C} \quad \textbf{Ans.}$$

Ratio of work done due to preheating and otherwise

Let W = Work done in the motor cylinder without preheating, and

W' = Work done in the motor cylinder after preheating.

We know that the ratio of work done due to preheating and otherwise,

$$\frac{W'}{W} = \frac{T_1'}{T_1} = \frac{398.3}{323} = 1.233 \text{ Ans.}$$

EXERCISES

1. An air motor receives air at a pressure of 5.5 bar and delivers it at 1 bar. Find the amount of work done per kg of air, if the expansion follows the law $pv^{1.3}$ = Constant. Take temperature of inlet air as 100° C. **[Ans. 155 kJ]**

2. In a compressed air system, the compression and expansion of air takes place according to the law $pv^{1.5}$ = Constant. Find its overall efficiency if both the machines work within pressure limits of 4.5 bar and 1 bar. Take efficiencies of compressor and motor alike. **[Ans. 74%]**

3. The initial pressure of the air in an air motor is 5.25 bar and final pressure is 1.05 bar. Find the temperature at which the air must be preheated in order that the temperature after expansion may be 3° C. Assuming the expansion to be according to $pv^{1.3}$ = Constant. **[Ans. 127.2° C]**

4. A single acting motor works on compressed air at 10.5 bar and 37° C, supplied at the rate of 1 kg/min. The cut-off takes place at 20% of the stroke and the expansion follows adiabatic and frictionless down to 1.03 bar. Determine the cylinder volume, mean effective pressure and indicated power, if the machine runs at 300 r.p.m. Neglect clearance.

QUESTIONS

1. What is air motor ? On what principle does it work ?
2. Obtain expression for the workdone by air in an air motor.
3. Explain the working of compressed air system.
4. Derive an expression for the overall efficiency of a compressed air system.
5. What is preheating of air ? Explain its uses.

OBJECTIVE TYPE QUESTIONS

1. The operation of a reciprocating air motor is similar to that of

(*a*) reciprocating steam engine (*b*) reciprocating air compressor

(*c*) both (*a*) and (*b*) (*d*) none of these

2. Air motors work on the cycle which is the ..... of the reciprocating air compressor cycle.

(*a*) same as that (*b*) reverse

3. In a compressed air system, the temperature of air discharged from the air motor is ..... than the initial compressor intake temperature.

(*a*) more (*b*) less

4. The overall efficiency of the compressed air system is

(*a*) the ratio of shaft output of the air motor to the shaft input to the compressor

(*b*) the ratio of shaft input to the compressor to the shaft output of the air motor

(*c*) the product of the shaft output of the air motor and the shaft input to the compressor

(*d*) none of the above

ANSWERS

1. (*a*) 2. (*b*) 3. (*b*) 4. (*a*)

32

Gas Turbines

1. Introduction. 2. Comparison of Gas Turbines and Steam Turbines. 3. Comparison of Gas Turbines and I.C. Engines. 4. Classification of Gas Turbines. 5. Closed Cycle Gas Turbines. 6. Gas Turbines with Intercooling. 7. Gas Turbines with Reheating. 8. Open Cycle Gas Turbines. 9. Comparison of Closed Cycle and Open Cycle Gas Turbines. 10. Semi-closed Cycle Gas Turbines. 11. Constant Pressure Gas Turbines. 12. Constant Volume Gas Turbines.

32.1. Introduction

The idea of gas turbine is the oldest one, and its working principle is an improved version of the wind mill, which was used several centuries back. In order to achieve an efficient working of the turbine, the movement of gas (or air) is properly controlled and then directed on the blades fixed to the turbine runner. The air, under pressure, is supplied to the turbine by an air compressor, which is run by the turbine itself.

In a gas turbine, first of all, the air is obtained from the atmosphere and compressed in an air compressor. The compressed air is then passed into the combustion chamber, where it is heated considerably. The hot air is then made to flow over the moving blades of the gas turbine, which imparts rotational motion to the runner. During this process, the air gets expanded and finally it is exhausted into the atmosphere. A major part of the power developed by the turbine is consumed for driving the compressor (which supplies compressed air to the combustion chamber). The remaining power is utilised for doing some external work.

32.2. Comparison of Gas Turbines and Steam Turbines

Following are the points of comparison between gas turbines and steam turbines :

S.No.	Gas turbines	Steam turbines
1.	The important components are compressor and combustion chamber.	The important components are steam boiler and accessories.
2.	The mass of gas turbine per kW developed is less.	The mass of steam turbine per kW developed is more.
3.	It requires less space for installation.	It requires more space for installation.
4.	The installation and running cost is less.	The installation and running cost is more.
5.	The starting of gas turbine is very easy and quick.	The starting of steam turbine is difficult and takes long time.
6.	Its control, with the changing load conditions, is easy.	Its control, with the changing load conditions, is difficult.
7.	A gas turbine does not depend on water supply.	A steam turbine depends on water supply.
8.	Its efficiency is less.	Its efficiency is higher.

32.3. Comparison of Gas Turbines and I.C. Engines

Following are the points of comparison between gas turbines and I.C. engines :

S.No.	Gas turbines	I.C. engines
1.	The mass of gas turbine per kW developed is less.	The mass of an I.C. engine per kW deveoped is more.
2.	The installation and running cost is less.	The installation and running cost is more.
3.	Its efficiency is higher.	Its efficiency is less.
4.	The balancing of a gas turbine is perfect.	The balancing of an I.C. engine is not perfect.
5.	The torque produced is uniform. Thus no flywheel is required.	The torque produced is not uniform. Thus flywheel is necessary.
6.	The lubrication and ignition systems are simple.	The lubrication and ignition systems are difficult.
7.	It can be driven at a very high speed.	It can not be driven at a very high speed.
8.	The pressures used are very low (about 5 bar).	The pressures used are high (above 60 bar).
9.	The exhaust of a gas turbine is free from smoke and less polluting.	The exhaust of an I.C. engine is more polluting.
10.	They are very suitable for air crafts.	They are less suitable for air crafts.
11.	The starting of a gas turbine is not simple.	The starting of an I.C. engine is simple.

32.4. Classification of Gas Turbines

Though the gas turbines may be classified in many ways,yet the following are important from the subject point of view :

1. *According to path of the working substance*

   (*a*) Closed cycle gas turbines, (*b*) Open cycle gas turbines, and

   (c) Semi-closed gas turbines.

2. *According to process of heat absorption*

   (*a*) Constant pressure gas turbines, and (*b*) Constant volume gas turbines.

In the following pages, we shall discuss all the above mentioned gas turbines one by one.

32.5. Closed Cycle Gas Turbines

A closed cycle gas turbine, in its simplest form, consists of a compressor, heating chamber, gas turbine which drives the generator and compressor, and a cooling chamber.

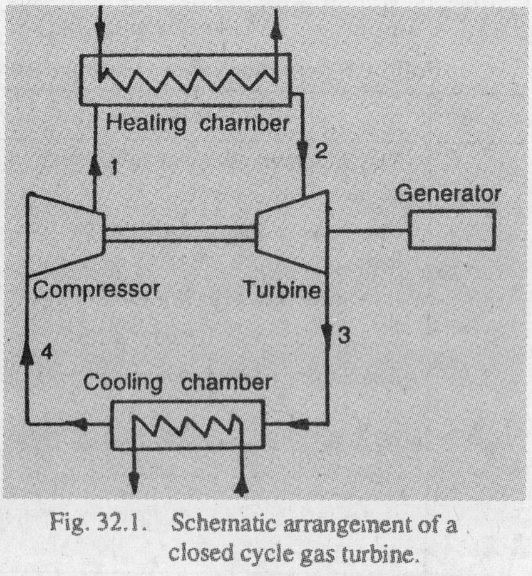

Fig. 32.1. Schematic arrangement of a closed cycle gas turbine.

The schematic arrangement of a closed cycle gas turbine is shown in Fig. 32.1. In this turbine, the air is compressed isentropically (generally in rotary compressor) and then passed into the heating chamber. The compressed air is heated with the help of some external source, and made to flow over the turbine blades (generally reaction type). The gas, while flowing over the blades, gets expanded. From the turbine, the gas is passed to the cooling chamber where it is cooled at constant pressure with the help of circulating water to its original temperature. Now the air is made to flow into the compressor again. It is thus obvious, that in a closed cycle gas turbine, the air is continuously circulated within the turbine.

A closed cycle gas turbine works on Joule's or Brayton's cycle as shown in Fig. 32.2.

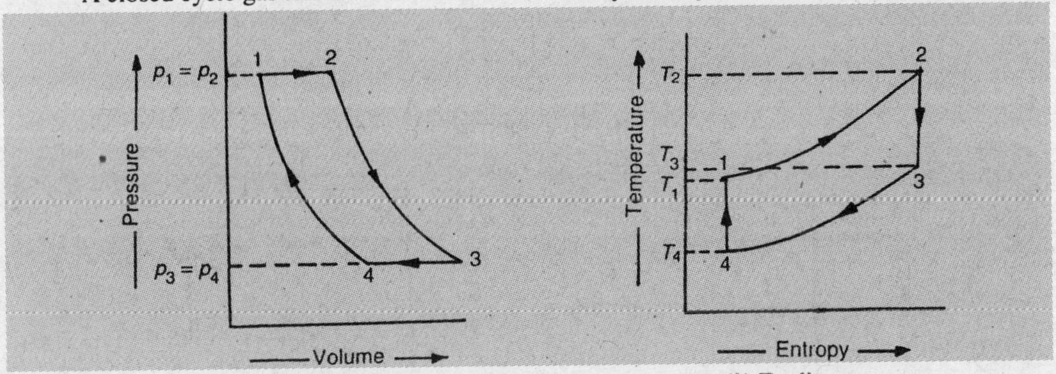

(*a*) *p-v* diagram. (*b*) *T-s* diagram.

Fig. 32.2 Constant pressure closed cycle gas turbine.

The process 1-2 shows heating of the air in heating chamber at constant pressure. The process 2-3 shows isentropic expansion of air in the turbine. Similarly, the process 3-4 shows cooling of the air at constant pressure in cooling chamber. Finally, the process 4-1 shows isentropic compression of the air in the compressor.

∴ Work done by the turbine per kg of air,

$$W_T = c_p\,(T_2 - T_3) \quad \ldots (i)$$

and work required by the compressor per kg of air,

$$W_C = c_p\,(T_1 - T_4) \quad \ldots (ii)$$

Now the net work available,

$$W = W_T - W_C$$

Notes : 1. In the above expressions, c_p is taken in kJ/kg K.

2. The power available (or net power of the installation) may be found out from the work available as usual.

Example 32.1. *A simple closed cycle gas turbine plant receives air at 1 bar and 15° C, and compresses it to 5 bar and then heats it to 800° C in the heating chamber. The hot air expands in a turbine back to 1 bar. Calculate the power developed per kg of air supplied per second. Take c_p for the air as 1 kJ/kg K.*

Solution. Given : $p_3 = p_4 = 1$ bar ; $T_4 = 15°\,C = 15 + 273 = 288$ K ; $p_1 = p_2 = 5$ bar ; $T_2 = 800°\,C = 800 + 273 = 1073$ K ; $c_p = 1$ kJ/kg K

The *p-v* and *T-s* diagram for the closed cycle gas turbine is shown in Fig. 32.3.

Let T_1 and T_3 = Temperature of air after isentropic compression and expansion (*i.e.* at points 1 and 3 respectively).

We know that for isentropic expansion 2-3,

$$\frac{T_3}{T_2} = \left(\frac{p_3}{p_2}\right)^{\frac{\gamma-1}{\gamma}} = \left(\frac{1}{5}\right)^{\frac{1.4-1}{1.4}} = (0.2)^{0.286} = 0.631$$

$$\therefore \quad T_3 = T_2 \times 0.631 = 1073 \times 0.631 = 677 \text{ K}$$

Similarly, for isentropic compression 4-1,

$$\frac{T_4}{T_1} = \left(\frac{p_4}{p_1}\right)^{\frac{\gamma-1}{\gamma}} = \left(\frac{1}{5}\right)^{\frac{1.4-1}{1.4}} = 0.631$$

$$\therefore \qquad T_1 = T_4 / 0.631 = 288 / 0.631 = 456 \text{ K}$$

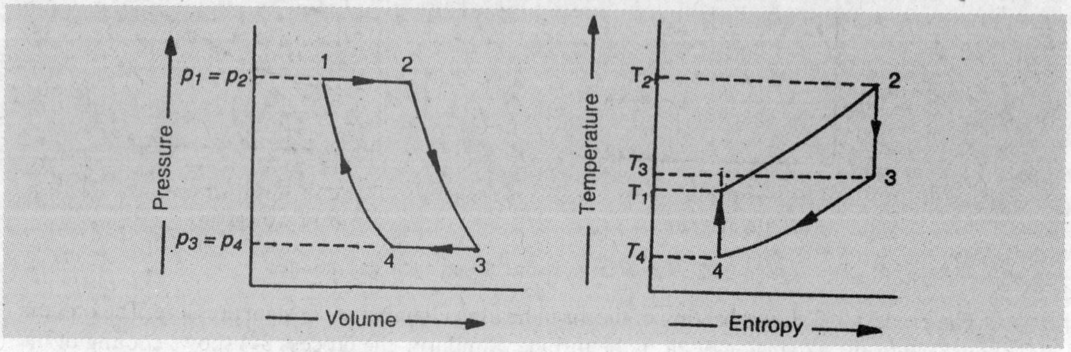

(*a*) *p-v* diagram. (*b*) *T-s* diagram.

Fig. 32.3

We know that work developed by the turbine,

$$W_T = c_p (T_2 - T_3) = 1 (1073 - 677) = 396 \text{ kJ/s}$$

and work required by the compressor,

$$W_C = c_p (T_1 - T_4) = 1 (456 - 288) = 168 \text{ kJ/s}$$

$\therefore$ Net work done by the turbine,

$$W = W_T + W_C = 396 - 168 = 228 \text{ kJ/s}$$

and power developed, $P = 228 \text{ kW}$ **Ans.** ... ($\because$ 1 kJ/s = 1 kW)

Example 32.2. *In an oil-gas turbine installation, it is taken at pressure of 1 bar and 27° C and compressed to a pressure of 4 bar. The oil with a calorific value of 42 000 kJ/kg is burnt in the combustion chamber to raise the temperature of air to 550° C. If the air flows at the rate of 1.2 kg/s ; find the net power of the installation. Also find air fuel ratio. Take c_p = 1.05 kJ/kg K.*

Solution. Given : $p_3 = p_4 = 1$ bar ; $T_4 = 27°$ C = 27 + 273 = 300 K ; $p_1 = p_4 = 4$ bar ; $C = 42\,000$ kJ/kg ; $T_2 = 550°$ C = 550 + 273 = 823 K ; $m = 1.2$ kg/s ; $c_p = 1.05$ kJ/kg K

Net power of the Installation

Let T_1 and T_3 = Temperature of air after isentropic compression and expansion respectively.

We know that for isentropic expansion 2-3 (Refer Fig. 32.3),

$$\frac{T_3}{T_2} = \left(\frac{p_3}{p_2}\right)^{\frac{\gamma-1}{\gamma}} = \left(\frac{1}{4}\right)^{\frac{1.4-1}{1.4}} = (0.25)^{0.286} = 0.673$$

$$\therefore \qquad T_3 = T_2 \times 0.673 = 823 \times 0.673 = 553.9 \text{ K}$$

Similarly, for isentropic compression 4-1,

$$\frac{T_4}{T_1} = \left(\frac{p_4}{p_1}\right)^{\frac{\gamma-1}{\gamma}} = \left(\frac{1}{4}\right)^{\frac{1.4-1}{1.4}} = 0.673$$

$$\therefore \quad T_1 = T_4 / 0.673 = 300 / 0.673 = 445.8 \text{ K}$$

We know that work done by the turbine,

$$W_T = m\, c_p\, (T_2 - T_3) = 1.2 \times 1.05\, (823 - 553.9) = 339.1 \text{ kJ/s}$$

and work done by the compressor,

$$W_C = m\, c_p\, (T_1 - T_4)$$

$$= 1.2 \times 1.05\, (445.8 - 300) = 183.7 \text{ kJ/s}$$

∴ Net power of the installation,

$$= 339.1 - 183.7 = 154.4 \text{ kJ/s} = 154.4 \text{ kW} \quad \textbf{Ans.}$$

Air-fuel ratio

We know that heat supplied by the oil

$$= m\, c_p\, (T_2 - T_1)$$

$$= 1.2 \times 1.05 (823 - 445.8) = 475.3 \text{ kJ/s}$$

∴ Mass of fuel burnt per second

$$= \frac{\text{Heat supplied}}{\text{Calorific value}} = \frac{475.3}{42\,000} = 0.011 \text{ kg}$$

and air-fuel ratio

$$= \frac{\text{Mass of air}}{\text{Mass of fuel}} = \frac{1.2}{0.011} = 109.1 \quad \textbf{Ans.}$$

32.6. Gas Turbine with Intercooling

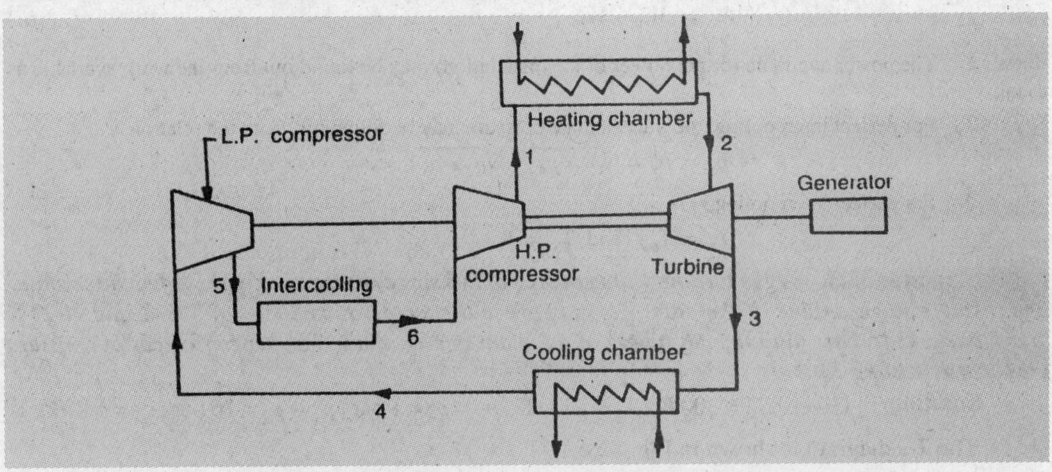

Fig. 32.4. **Schematic arrangement of a closed cycle gas turbine with intercooler.**

We have already discussed that a major portion of the power developed by the gas turbine is utilised by the compressor. It can be reduced by compressing the air in two stages with an intercooler between the two. This improves the efficiency of the gas turbine. The schematic arrangement of a closed cycle gas turbine with an intercooler is shown in Fig. 32.4.

In this arrangement*, first of all, the air is compressed in the first compressor, known as low pressure (L.P.) compressor. We know that as a result of this compression, the pressure and temperature of the air is increased. Now the air is passed to an intercooler which reduces the temperature of the compressed air to its original temperature, but keeping the pressure constant. After that, the compressed air is once again compressed in the second compressor known as high pressure (H.P.) compressor. Now the compressed air is passed through the heating chamber and then through the turbine. Finally, the air is cooled in the cooling chamber and again passed into the low pressure compressor as shown in Fig. 32.4.

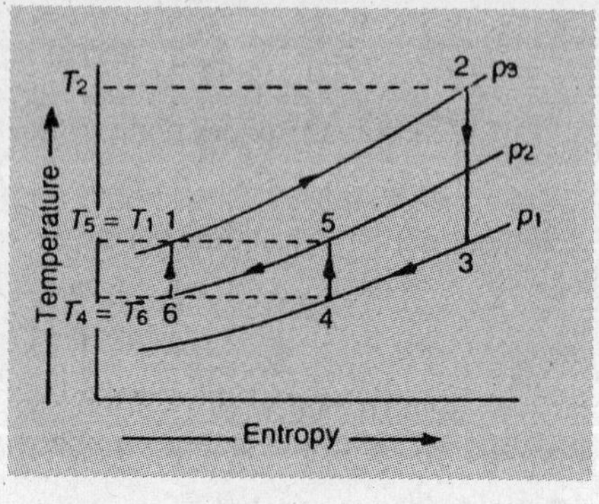

Fig. 32.5. *T-s* diagram for intercooling.

The process of intercooling the air in two stages of compression is shown on *T-s* diagram in Fig. 32.5. The process 1-2 shows heating of the air in heating chamber at constant pressure. The process 2-3 shows isentropic expansion of air in the turbine. The process 3-4 shows cooling of the air in the cooling chamber at constant pressure. The process 4-5 shows compression of air in the L.P. compressor. The process 5-6 shows cooling of the air in the intercooler at constant pressure. Finally, the process 6-1 shows compression of air in the H.P. compressor.

∴ Work done by the compressor per kg of air,

$$W_T = c_p (T_2 - T_3) \quad \ldots (i)$$

and work done by the compressor per kg of air,

$$W_C = c_p [(T_1 - T_6) + (T_5 - T_4)] \quad \ldots (ii)$$

Now the net work available,

$$W = W_T - W_C \quad \ldots (iii)$$

Notes: 1. The power available (or net power of the installation) may be found out from the work available as usual.

2. For perfect intercooling, the intermediate pressure may be found out from the relation,

$$p_6 = p_5 = \sqrt{p_1 \times p_4} = \sqrt{p_2 \times p_3}$$

3. For perfect intercooling,

$$T_4 = T_6 \,; \text{ and } T_5 = T_1$$

Example 32.3. *A gas turbine plant consists of two stage compressor with perfect intercooler and a single stage turbine. If the plant works between the temperature limits of 300 K and 1000 K and 1 bar and 16 bar ; find the net power of the plant per kg of air. Take specific heat at constant pressure as 1 kJ/kg K.*

Solution. Given : $T_4 = 300$ K ; $T_2 = 1000$ K ; $p_3 = p_4 = 1$ bar ; $p_1 = p_2 = 16$ bar ; $c_p = 1$ kJ/kg K

The *T-s* diagram is shown in Fig. 32.6.

Let T_1, T_3, T_5 and T_6 = Temperature of air at corresponding points.

We know that for perfect intercooling, the intermediate pressure,

$$p_5 = p_6 = \sqrt{p_1 \times p_4} = \sqrt{16 \times 1} = 4 \text{ bar}$$

* Please refer Chapter 28 (Art. 28.14) also.

Now for the isentropic process 2-3,

$$\frac{T_3}{T_2} = \left(\frac{p_3}{p_2}\right)^{\frac{\gamma-1}{\gamma}}$$

$$= \left(\frac{1}{16}\right)^{\frac{1.4-1}{1.4}} = 0.453$$

$$\therefore \quad T_3 = T_2 \times 0.453$$

$$= 1000 \times 0.453 = 453 \text{ K}$$

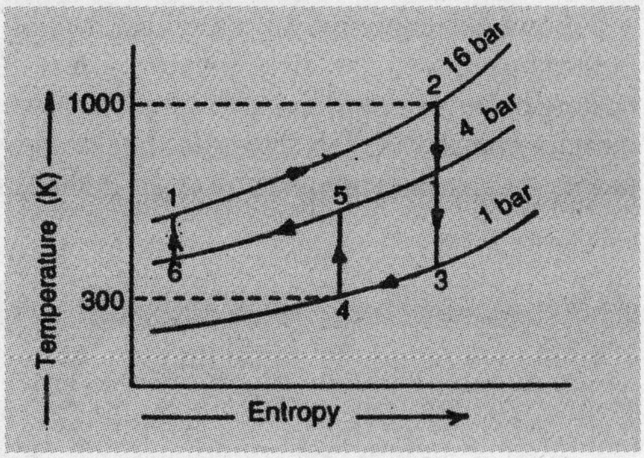

Fig. 32.6

Similarly for the isentropic process 4-5,

$$\frac{T_4}{T_5} = \left(\frac{p_4}{p_5}\right)^{\frac{\gamma-1}{\gamma}}$$

$$= \left(\frac{1}{4}\right)^{\frac{1.4-1}{1.4}} = 0.673$$

$$\therefore \quad T_5 = T_4 / 0.673 = 300 / 0.673 = 446 \text{ K}$$

We know that tor perfect inter cooling,

$$T_1 = T_5 = 446 \text{ K}$$

and for isentropic process 6-1, $\frac{T_6}{T_1} = \left(\frac{p_6}{p_1}\right)^{\frac{\gamma-1}{\gamma}} = \left(\frac{4}{16}\right)^{\frac{1.4-1}{1.4}} = 0.673$

$$\therefore \quad T_6 = T_1 \times 0.673 = 446 \times 0.673 = 300 \text{ K}$$

Now work done by the turbine per kg of air,

$$W_T = c_p (T_2 - T_3) = 1\,(1000 - 453) = 547 \text{ kJ/s}$$

and work absorbed by the compressor per kg of air,

$$W_C = c_p [(T_1 - T_6) + (T_5 - T_4)]$$

$$= 1\,[(446 - 300) + (446 - 300)] = 292 \text{ kJ/s}$$

We know that work done by the plant per kg of air,

$$W = W_T - W_C = 547 - 292 = 255 \text{ kJ/s}$$

$\therefore$ Net power of the plant, $P = 255$ kW **Ans.**

32.7. Gas Turbine with Reheating

The output of a gas turbine can be considerably improved by expanding the hot air in two stages with a reheater between the two. The schematic arrangement of a closed cycle gas turbine with reheating is shown in Fig. 32.7.

In this arrangement, the air is first compressed in the compressor, passed into the heating chamber, and then to the first turbine. The air is once again passed on to another heating chamber and then to the second turbine. Finally, the air is cooled in the cooling chamber and again passed into the compressor as shown in Fig. 32.7.

The process of reheating in the two turbines is shown on *T-s* diagram in Fig. 32.8. The process 1-2 shows heating of the air in the first heating chamber at constant pressure. The process2-3 shows isentropic expansion of air in the first turbine. The process 3-4 shows heating of the air in the second heating chamber at constant pressure. The process 4-5 shows isentropic expansion of air in the second turbine. The process 5-6 shows cooling of the air in the intercooler at constant pressure. Finally, the process 6-1 shows compression of air in the compressor.

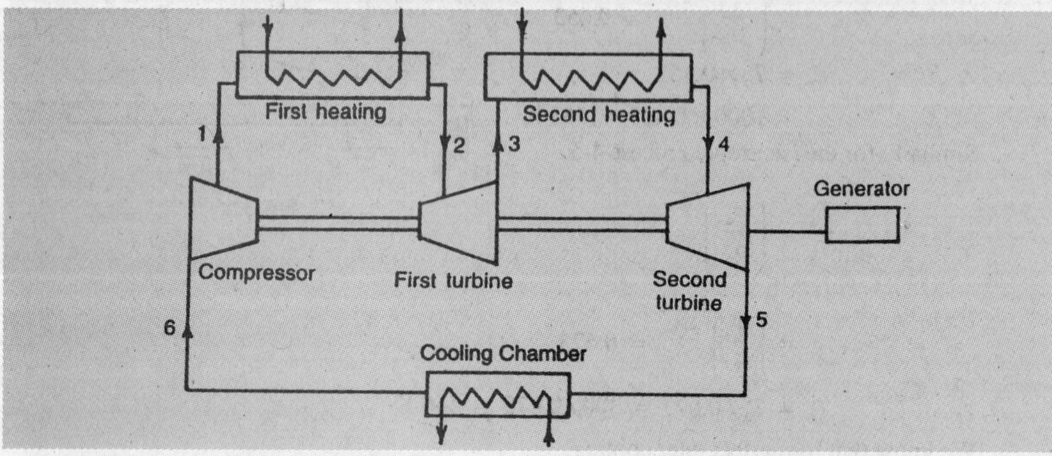

Fig. 32.7. Schematic arrangement of a closed cycle gas turbine with reheating.

$\therefore$ Work done by the turbine per kg of air,

$$W_T = c_p\,[(T_2 - T_3) + (T_4 - T_5)] \quad \ldots (i)$$

and work done by the compressor per kg of air,

$$W_C = c_p\,[(T_1 - T_6)] \quad \ldots (ii)$$

Now net work available,

$$W = W_T - W_C \quad \ldots (iii)$$

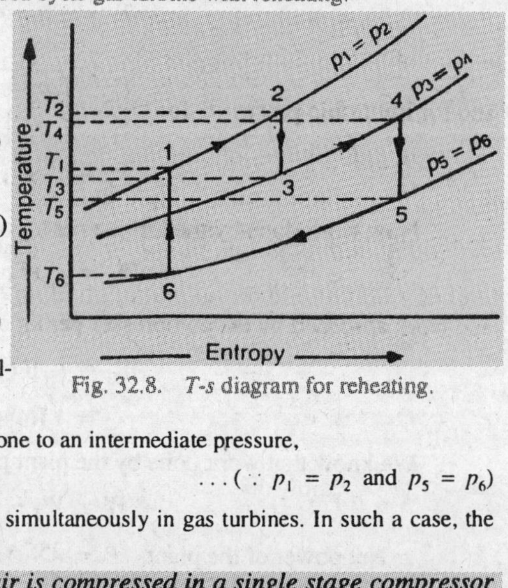

Fig. 32.8. *T-s* diagram for reheating.

Notes : 1. The power available (or net power of the installation may be found out from the work available as usual.

2. . For maximum work, the reheating should be done to an intermediate pressure,

$$p_3 = p_4 = \sqrt{p_2 \times p_5} = \sqrt{p_1 \times p_6} \quad \ldots (\because p_1 = p_2 \text{ and } p_5 = p_6)$$

3. Sometimes, cooling and reheating is provided simultaneously in gas turbines. In such a case, the corresponding values should be used.

Example 32.4. *In a gas turbine plant, the air is compressed in a single stage compressor from 1 bar to 9 bar and from an initial temperature of 300 K. The same air is then heated to a temperature of 800 K and then expanded in the turbine. The air is then reheated to a temperature of 800 K and then expanded in the second turbine. Find the maximum power that can be obtained from the installation, if the mass of air circulated per second is 2 kg. Take* $c_p = 1$ *kJ/kg K.*

Solution. Given : $p_6 = p_5 = 1$ bar ; $p_1 = p_2 = 9$ bar ; $T_6 = 300$ K ; $T_2 = T_4 = 800$ K ; $m = 2$ kg/s ; $c_p = 1$ kJ/kg K

The *T-s* diagram of the reheat cycle is shown in Fig. 32.9.

Let T_1, T_3, T_5 = Temperature of air at the corresponding points.

We know that for maximum power (or work), the intermediate pressure,

$$p_3 = p_4 = \sqrt{p_1 \times p_6}$$
$$= \sqrt{9 \times 1} = 3 \text{ bar}$$

We also know that for isentropic compression of air in the compressor (process 6-1),

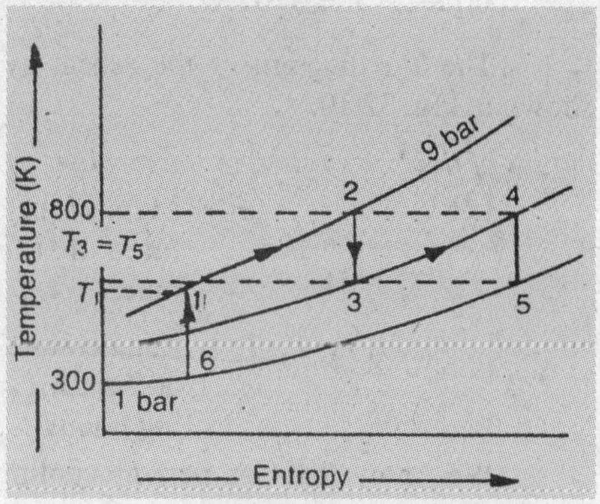

Fig. 32.9

$$\frac{T_1}{T_6} = \left(\frac{p_1}{p_6}\right)^{\frac{\gamma-1}{\gamma}} = \left(\frac{9}{1}\right)^{\frac{1.4-1}{1.4}}$$
$$= (9)^{0.286} = 1.873$$
$$\therefore \quad T_1 = T_6 \times 1.873$$
$$= 300 \times 1.873 = 562 \text{ K}$$

For isentropic expansion of air in the first turbine (process 2-3),

$$\frac{T_2}{T_3} = \left(\frac{p_2}{p_3}\right)^{\frac{\gamma-1}{\gamma}} = \left(\frac{9}{3}\right)^{\frac{1.4-1}{1.4}} = (3)^{0.286} = 1.369$$
$$\therefore \quad T_3 = T_2 / 1.369 = 800 / 1.369 = 584 \text{ K}$$

Similarly, for isentropic expansion of air in the second turbine (process 4-5),

$$\frac{T_4}{T_5} = \left(\frac{p_4}{p_5}\right)^{\frac{\gamma-1}{\gamma}} = \left(\frac{3}{1}\right)^{\frac{1.4-1}{1.4}} = (3)^{0.286} = 1.369$$
$$\therefore \quad T_5 = T_4 / 1.369 = 800 / 1.369 = 584 \text{ K}$$

We know that work done by the turbine,

$$W_T = m\, c_p\, [(T_2 - T_3) + (T_4 - T_5)]$$
$$= 2 \times 1\, [(800 - 584) + (800 - 584)] = 864 \text{ kJ/s}$$

and work absorbed by the compressor,

$$W_C = m\, c_p\, (T_1 - T_6) = 2 \times 1\, (562 - 300) = 524 \text{ kJ/s}$$

We also know that net work available,

$$W = W_T - W_C = 864 - 524 = 340 \text{ kJ/s}$$

$\therefore$ Power that can be obtained from the installation,

$$P = 340 \text{ kJ/s} = 340 \text{ kW} \quad \textbf{Ans.}$$

Example 32.5 *A closed cycle gas turbine consists of a two stage compressor with perfect intercooler and a two stage turbine, with a reheater. All the components are mounted on the same shaft. The pressure and temperature at the inlet of the low pressure compressor are 2 bar and 300 K. The maximum pressure and temperature are limited to 8 bar and 1000 K. The gases are heated in the reheater to 1000 K. Calculate mass of fluid circulated in the turbine, if the net power developed by the turbine is 370 kW. Also find the amount of heat supplied per second from the external source.*

Solution. Given : $p_6 = p_5 = 2$ bar ; $T_6 = 300$ K ; $p_1 = p_2 = 8$ bar ; $T_2 = 1000$ K ; $T_4 = 1000$ K ; $P = 370$ kW

The *T-s* diagram of the reheat cycle is shown in Fig. 32.10.

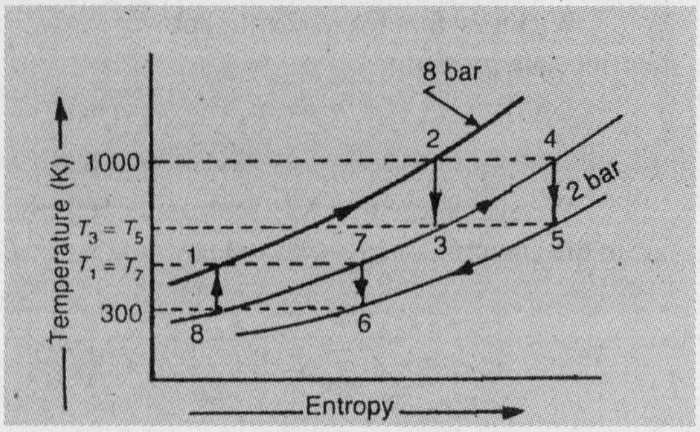

Fig. 32.10

Mass of fluid circulated in the turbine

Let m = Mass of air circulated in the turbine,

T_1, T_3, T_5, T_7, T_8 = Temperature of air at the corresponding points.

We know that for perfect cooling, the intermediate pressure,

$$p_8 = p_7 = p_3 = p_4 = \sqrt{p_1 \times p_6} = \sqrt{8 \times 2} = 4 \text{ bar}$$

Now for the isentropic process 6-7,

$$\frac{T_6}{T_7} = \left(\frac{p_6}{p_7}\right)^{\frac{\gamma-1}{\gamma}} = \left(\frac{2}{4}\right)^{\frac{1.4-1}{1.4}} = (0.5)^{0.286} = 0.82$$

$$\therefore \quad T_7 = T_6 / 0.82 = 300 / 0.82 = 366 \text{ K}$$

We know that for perfect cooling,

$$T_1 = T_7 = 366 \text{ K}$$

Again, for the isentropic process 8-1,

$$\frac{T_8}{T_1} = \left(\frac{p_8}{p_1}\right)^{\frac{\gamma-1}{\gamma}} = \left(\frac{4}{8}\right)^{\frac{1.4-1}{1.4}} = (0.5)^{0.286} = 0.82$$

$$\therefore \quad T_8 = T_1 \times 0.82 = 366 \times 0.82 = 300 \text{ K}$$

and for the isentropic process 2-3,

$$\frac{T_3}{T_2} = \left(\frac{p_3}{p_2}\right)^{\frac{\gamma-1}{\gamma}} = \left(\frac{4}{8}\right)^{\frac{1.4-1}{1.4}} = 0.82$$

$$\therefore \quad T_3 = T_2 \times 0.82 = 1000 \times 0.82 = 820 \text{ K}$$

Similarly, for the isentropic process 4-5,

$$\frac{T_5}{T_4} = \left(\frac{p_5}{p_4}\right)^{\frac{\gamma-1}{\gamma}} = \left(\frac{2}{4}\right)^{\frac{1.4-1}{1.4}} = 0.82$$

$$\therefore \quad T_5 = T_4 \times 0.82 = 1000 \times 0.82 = 820 \text{ K}$$

We know that work done by the turbine,

$$W_T = m\, c_p\, [(T_2 - T_3) + (T_4 - T_5)]$$

$$= m \times 1\, [(1000 - 820) + (1000 - 820)] = 360\, m \text{ kJ/s}$$

and work absorbed by the compressor,

$$W_C = m\,c_p\,[(T_1 - T_8) + (T_7 - T_6)]$$

$$= m \times 1\,[(366 - 300) + (366 - 300)] = 132\,m \text{ kJ/s}$$

$\therefore$ Net work done by the turbine,

$$W = W_T - W_C = 360\,m - 132\,m = 228\,m \text{ kJ/s}$$

We also know that power developed by the turbine (P),

$$370 = 228\,m \quad \text{or} \quad m = 1.62 \text{ kg/s} \quad \textbf{Ans.}$$

Heat supplied from the external source

We know that heat supplied from the external source

$$= m\,c_p\,[(T_2 - T_1) + (T_4 - T_3)]$$

$$= 1.62 \times 1\,[(1000 - 366) + (1000 - 820)] \text{ kJ/s}$$

$$= 1318.7 \text{ kJ/s} \quad \textbf{Ans.}$$

32.8. Open Cycle Gas Turbines

An open cycle gas turbine, in its simplest form, consists of a compressor, combustion chamber and a gas turbine which drives the generator and compressor.

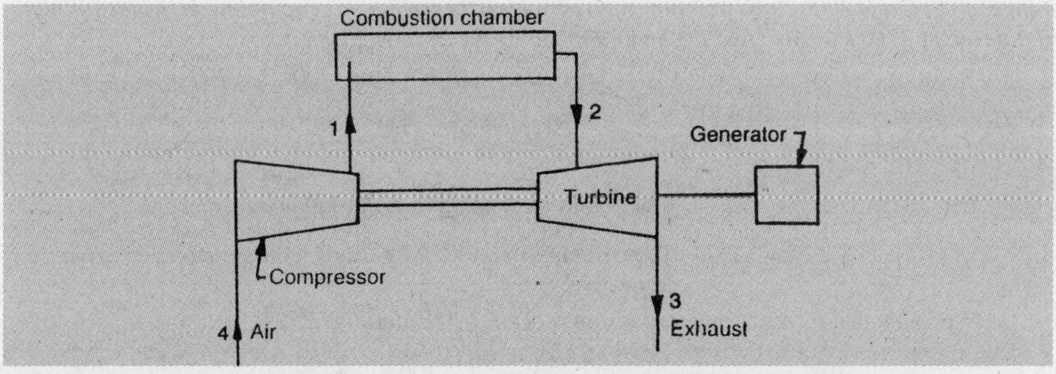

Fig. 32.11. Schematic arrangement of an open cycle gas turbine.

The schematic arrangement of an open cycle gas turbine is shown in Fig. 32.11. In this turbine, the air is first sucked from the atmosphere and then compressed isentropically (generally in a rotary compressor) and then passed into the combustion chamber. The compressed air is heated by the combustion of fuel and the products of combustion (*i.e.* hot gases formed by the combustion of fuel) also get mixed up with the compressed air, thus increasing the mass of compressed air. The hot gas is then made to flow over the turbine blades (generally of reaction type). The gas, while flowing over the blades, gets expanded and finally exhausted into the atmosphere.

An open cycle gas turbine is also called a continuous combustion gas turbine as the combustion of fuel takes place continuously. This turbine also works on Joule's cycle. The relations for work done by the compressor and turbine are same as those of closed cycle gas turbine.

Note In an open cycle gas turbine, the process 3-4 has no practical importance, as the air is exhausted into the atmosphere at point 3 and fresh air is sucked in the compressor at point 4.

32.9. Comparison of Closed Cycle and Open Cycle Gas Turbines

Following are the points of comparison between closed and open cycle gas turbines.

S.No.	Closed cycle gas turbine	Open cycle gas turbine
1.	The compressed air is heated in a heating chamber. Since the gas is heated by an external source, so the amount of gas remains the same.	The compressed air is heated in a combustion chamber. The products of combustion get mixed up in the heated air.
2.	The gas from the turbine is passed into the cooling chamber.	The gas from the turbine is exhausted into the atmosphere.
3.	The working fluid is circulated continuously.	The working fluid is replaced continuously.
4.	Any fluid with better thermodynamic properties can be used.	Only air can be used as the working fluid.
5.	The turbine blades do not wear away earlier, as the enclosed gas does not get contaminated while flowing through the heating chamber.	The turbine blades wear away earlier, as the air from the atmosphere gets contaminated while flowing through the combustion chamber.
6.	Since the air, from the turbine, is cooled by circulating water, it is best suited for stationary installation or marine uses.	Since the air, from the turbine, is discharged into the atmosphere, it is best suited for moving vehicle.
7.	Its maintenance cost is high.	Its maintenance cost is low.
8.	The mass of installation per kW is more.	The mass of installation per kW is less.

Example 32.6. *A constant pressure open cycle gas turbine plant works between temperature range of 15° C and 700° C and pressure ratio of 6. Find the mass of air circulating in the installation, if it develops 1100 kW. Also find the heat supplied by the heating chamber.*

Solution. Given : T_4 = 15° C = 15 + 273 = 288 K ; T_2 = 700° C = 700 + 273 = 973 K ; $p_2 / p_3 = p_1 / p_4 = 6$; P = 1100 kW

Mass of air circulating in the installation

Let m = Mass of air circulating in the installation.

T_1 and T_3 = Temperatures of air after isentropic compression and expansion respectively.

We know that for isentropic expansion process 2-3 (Refer Fig. 32.2),

$$\frac{T_3}{T_2} = \left(\frac{p_3}{p_2}\right)^{\frac{\gamma-1}{\gamma}} = \left(\frac{1}{6}\right)^{\frac{1.4-1}{1.4}} = \left(\frac{1}{6}\right)^{0.286} = 0.599$$

$$\therefore \quad T_3 = T_2 \times 0.599 = 973 \times 0.599 = 583 \text{ K}$$

Similarly, for isentropic compression process 4-1,

$$\frac{T_4}{T_1} = \left(\frac{p_4}{p_1}\right)^{\frac{\gamma-1}{\gamma}} = \left(\frac{1}{6}\right)^{\frac{1.4-1}{1.4}} = \left(\frac{1}{6}\right)^{0.286} = 0.599$$

$$\therefore \quad T_1 = T_4 / 0.599 = 288 / 0.599 = 481 \text{ K}$$

Now work done by the turbine per kg of air,

$$W_T = m\,c_p\,(T_2 - T_3) = m \times 1\,(973 - 583) = 390\,m \text{ kJ/s}$$

and work absorbed by the compressor per kg of air,

$$W_C = m\,c_p\,(T_1 - T_4) = m \times 1\,(481 - 288) = 193\,m \text{ kJ/s}$$

∴ Net work done by the turbine per kg of air,

$$W = W_T - W_C = 390\,m - 193\,m = 197\,m \text{ kJ/s}$$

and power developed by the installation (P),

$$1100 = 197\,m \quad \text{or} \quad m = 1100/197 = \mathbf{5.58 \text{ kg/s}} \text{ Ans.}$$

Heat supplied by the heating chamber

We know that heat supplied by the heating chamber

$$= m\,c_p\,(T_2 - T_1)$$

$$= 5.58 \times 1\,(973 - 481) = \mathbf{2745.4 \text{ kJ/s}} \text{ Ans.}$$

32.10. Semi-closed Cycle Gas Turbines

A semi-closed cycle gas turbine, as the name indicates, is a turbine which is a combination of two turbines, one working on open cycle and the other on closed cycle. The open cycle turbine is used to drive the main generator and works within the pressure limits of atmospheric and about 16 bar. The closed cycle turbine is used to drive the air compressor and works within the pressure limits of about 2 bar and 16 bar.

Strictly speaking, the semi-closed cycle gas turbines are not used on commercial basis, though they are important from academic point of view only.

32.11. Constant Pressure Gas Turbines

A turbine in which the air is heated in the combustion (or heating) chamber at constant pressure, is known as *constant pressure gas turbine*. Almost all the turbines, manufactured today, are constant pressure gas turbines.

32.12. Constant Volume Gas Turbines

A turbine in which the air is heated in combustion (or heating) chamber at constant volume is known as *constant volume gas turbine*. These turbines are not used on commercial basis, though they have academic importance only.

EXERCISES

1. A simple closed cycle gas turbine installation works between the temperature limits of 300 K and 1000 K and pressure limits of 1 bar and 5 bar. If 1.25 kg of air is circulated per second, determine the power developed by the turbine. For air, take c_p = 1.008 kJ/kg K and γ = 1.4. **[Ans. 680.5 kW]**

2. In a gas turbine plant, operating on Brayton cycle, air enters the compressor at 1 bar and 27° C. The pressure ratio in the cycle is 6. Calculate the maximum temperature in the cycle and the power developed by the turbine. Assume the turbine work as 2.5 times the compressor work. Take γ = 1.4 **[Ans. 703.5 kW]**

3. A gas turbine plant consists of two stage compressor (with perfect intercooler) and a single stage turbine. The plant receives air at 1 bar and 290 K. If the maximum pressure and temperature of air in the plant is 12.25 bar and 950 K, find the power developed by the plant per kg of air. Take specific heat at constant pressure as 1 kJ/kg K. **[Ans. 235 kW]**

4. In a gas turbine installation, the air is compressed in a single stage compressor from 1 bar to 6.25 bar and from an initial temperature of 20° C. The air after compression is heated in a chamber to a temperature of 750° C. The hot air is expanded in the turbine and then reheated to a temperature of 750° C. The hot air is once again expanded in the second turbine. Find the power that can be developed per kg of air.

[Ans. 269 kW]

QUESTIONS

1. What is a gas turbine ? How does it differ from a steam turbine ?
2. How does a gas turbine compare with the internal combustion engine power plant ?
3. List the methods of improving the efficiency and specific output of a simple gas turbine.

4. Draw the layout of a gas turbine plant which has two stage compression with complete intercooling. The high pressure turbine develops power enough only to drive the high pressure compressor. The L.P. turbine drives both the L.P. compressor and the load. Indicate the ideal process of this plant on a *T-s* diagram.

5. What are the essential components of a simple open cycle gas turbine plant ?

6. Differentiate clearly between a closed cycle gas turbine and an open cycle gas turbine.

7. Write a short note on semi-closed cycle gas turbine.

OBJECTIVE TYPE QUESTIONS

1. A closed cycle gas turbine works on
   (*a*) Carnot cycle (*b*) Rankine cycle
   (*c*) Ericsson cycle (*d*) Joule cycle
2. In a closed cycle gas turbine, the air is compressed
   (*a*) isothermally (*b*) isentropically
   (*c*) polytropically (*d*) none of these
3. The gas in cooling chamber of a closed cycle gas turbine is cooled at
   (*a*) constant volume (*b*) constant temperature
   (*c*) constant pressure (*d*) none of these
4. A closed cycle gas turbine gives....efficiency as compared to an open cycle gas turbine.
   (*a*) same (*b*) lower (*c*) higher
5. Reheating in a gas turbine
   (*a*) increases the thermal efficiency (*b*) increases the compressor work
   (*c*) increases the turbine work (*d*) decreases the thermal efficiency

ANSWERS

1. (*d*)	2. (*b*)	3. (*c*)	4. (*c*)	5. (*a*)

33

Performance of Gas Turbines

1. Introduction. 2. Assumptions for Overall (Thermal) Efficiency of an Ideal Gas Turbine Plant. 3. Overall (Thermal) Efficiency of an Ideal Gas Turbine Plant. 4. Actual Gas Turbine. 5. Heat Exchanger. 6. Efficiency of Heat Exchanger. 7. Uses of Gas Turbines. 8. Recent Trends in Gas Turbines.

33.1. Introduction

In the last chapter, we have discussed the working of gas turbines under ideal conditions *i.e.* with isentropic expansion and compression. But in actual practice, the ideal conditions do not prevail. Now, in this chapter, we shall discuss the performance and uses of gas turbines under various sets of conditions.

33.2. Assumptions for Overall (Thermal) Efficiency of an Ideal Gas Turbine Plant

Following simplifying assumptions are made to obtain an expression for the overall (thermal) efficiency of an ideal gas turbine installation plant :

1. The compression and expansion processes are considered as isentropic.
2. The specific heat of the working fluid remains constant throughout the cycle.
3. The pressure losses in the cycle are neglected.
4. The heat losses due to various factors are neglected.
5. The kinetic energy of the working fluid, while entering the compressor and leaving the turbine, is equal.

33.3. Overall (Thermal) Efficiency of an Ideal Gas Turbine Plant

It is the ratio of net work done by a turbine (*i.e.* work developed by turbine – work required by compressor) to the heat supplied. Mathematically, thermal efficiency,

$$\eta_{th} = \frac{\text{Net work done}}{\text{Heat supplied}} \quad \ldots (i)$$

Now consider a gas turbine comprising of air compressor and turbine working on Joule's or Brayton's cycle as shown in Fig. 33.1 (*a*) and (*b*).

We know that net work done by the turbine

$$= c_p\,[(T_2 - T_3) - (T_1 - T_4)] \quad \ldots (ii)$$

and heat supplied

$$= c_p\,(T_2 - T_1) \quad \ldots (ii)$$

$$\therefore \quad \eta_{th} = \frac{c_p\,[(T_2 - T_3) - (T_1 - T_4)]}{c_p\,(T_2 - T_1)} \quad \ldots (iv)$$

$$= \frac{(T_2 - T_3) - (T_1 - T_4)}{(T_2 - T_1)} \quad \ldots (v)$$

We know that pressure ratio,

$$r = \frac{p_1}{p_4} = \frac{p_2}{p_3}$$

and for isentropic compression (4-1),

$$\frac{T_1}{T_4} = \left(\frac{p_1}{p_4}\right)^{\frac{\gamma-1}{\gamma}} = (r)^{\frac{\gamma-1}{\gamma}}$$

$$\therefore \quad T_1 = T_4\,(r)^{\frac{\gamma-1}{\gamma}}$$

Similarly, for isentropic expansion (2-3),

$$\frac{T_2}{T_3} = \left(\frac{p_2}{p_3}\right)^{\frac{\gamma-1}{\gamma}} = (r)^{\frac{\gamma-1}{\gamma}}$$

$$\therefore \quad T_2 = T_3\,(r)^{\frac{\gamma-1}{\gamma}}$$

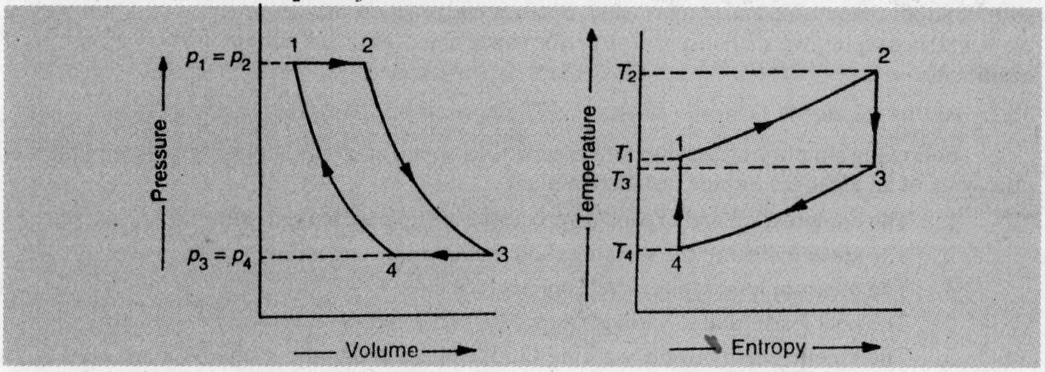

(a) p-v diagram. (b) T-s diagram.

Fig. 33.1. Ideal gas turbine.

Now substituting the values of T_1 and T_2 in equation (v),

$$\eta_{th} = \frac{\left[T_3\,(r)^{\frac{\gamma-1}{\gamma}} - T_3\right] - \left[T_4\,(r)^{\frac{\gamma-1}{\gamma}} - T_4\right]}{\left[T_3\,(r)^{\frac{\gamma-1}{\gamma}} - T_4\,(r)^{\frac{\gamma-1}{\gamma}}\right]}$$

$$= \frac{T_3\left[(r)^{\frac{\gamma-1}{\gamma}} - 1\right] - T_4\left[(r)^{\frac{\gamma-1}{\gamma}} - 1\right]}{(T_3 - T_4)\,(r)^{\frac{\gamma-1}{\gamma}}}$$

$$= \frac{[T_3 - T_4]\left[(r)^{\frac{\gamma-1}{\gamma}} - 1\right]}{(T_3 - T_4)\,(r)^{\frac{\gamma-1}{\gamma}}} = \frac{\left[r^{\frac{\gamma-1}{\gamma}} - 1\right]}{(r)^{\frac{\gamma-1}{\gamma}}} = 1 - \left(\frac{1}{r}\right)^{\frac{\gamma-1}{\gamma}}$$

Notes : 1. The thermal efficiency may also be obtained by substituting the values in equations (*i*) or (*iv*).

2. If the heat supplied is obtained from the mass of fuel × calorific value, then the corresponding efficiency will be overall efficiency.

Example 33.1. *A gas turbine plant receives air at 1 bar and 290 K and compresses it to 5 bar. If the temperature of air after compression is 1000 K ; find the thermal efficiency of the turbine. Take γ for the air as 1.4.*

Solution. Given : $p_3 = p_4 = 1$ bar ; $^*T_4 = 290$ K ; $p_1 = p_2 = 5$ bar ; $^*T_2 = 1000$ K ; $\gamma = 1.4$

We know that pressure ratio,

$$r = \frac{p_1}{p_4} = \frac{p_2}{p_3} = \frac{5}{1} = 5$$

∴ Thermal efficiency of the turbine,

$$\eta_{th} = 1 - \left(\frac{1}{r}\right)^{\frac{\gamma-1}{\gamma}} = 1 - \left(\frac{1}{5}\right)^{\frac{1.4-1}{1.4}} = 1 - (0.2)^{0.286} = 1 - 0.631$$

$$= 0.369 \quad \text{or} \quad 36.9\ \% \quad \textbf{Ans.}$$

Example 33.2. *Air enters the compressor of a gas turbine plant operating on Brayton cycle at 1 bar and 27° C. The pressure ratio in the cycle is 6. Calculate the maximum temperature in the cycle and the cycle efficiency. Assume the turbine work as 2.5 times the compressor work. Take γ = 1.4.*

Solution. Given : $p_3 = p_4 = 1$ bar ; $T_4 = 27°\text{ C} = 27 + 273 = 300$ K ; $r = p_1 / p_4 = p_2 / p_3 = 6$; $W_T = 2.5\ W_C$; $\gamma = 1.4$

The *T-s* diagram of the Brayton cycle is shown in Fig. 33.2.

1. *Maximum temperature in the cycle*

Let T_2 = Maximum temperature in the cycle.

We know that for isentropic compression 4-1,

$$\frac{T_1}{T_4} = \left(\frac{p_1}{p_4}\right)^{\frac{\gamma-1}{\gamma}} = (6)^{\frac{1.4-1}{1.4}} = 1.67$$

$$\therefore \quad T_1 = T_4 \times 1.67 = 300 \times 1.67 = 501 \text{ K}$$

and for isentropic expansion 2-3,

$$\frac{T_2}{T_3} = \left(\frac{p_2}{p_3}\right)^{\frac{\gamma-1}{\gamma}} = (6)^{\frac{1.4-1}{1.4}} = 1.67$$

$$\therefore \quad T_3 = T_2 / 1.67 \qquad \dots (i)$$

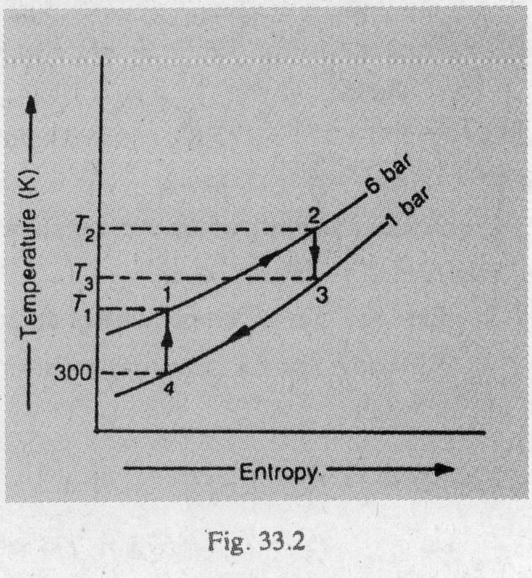

Fig. 33.2

We also know that turbine work per kg of air,

$$W_T = c_p (T_2 - T_3)$$

and compressor work per kg of air, $W_C = c_p (T_1 - T_4)$

Since $W_T = 2.5\ W_C$, therefore

$$c_p (T_2 - T_3) = 2.5\ c_p (T_1 - T_4)$$

* Superfluous data

or $$T_2 - T_3 = 2.5\,(T_1 - T_4)$$

$$= 2.5\,(501 - 300) = 502.5 \text{ K} \qquad \ldots (ii)$$

From equations (*i*) and (*ii*),

$$T_2 = 1252.5 \text{ K} = 979.5°\text{C} \textbf{ Ans.}$$

and $$T_3 = 750 \text{ K}$$

Cycle efficiency

We know that cycle efficiency,

$$\eta = \frac{(T_2 - T_3) - (T_1 - T_4)}{T_2 - T_1}$$

$$= \frac{(1252.5 - 750) - (501 - 300)}{1252.5 - 501} = \frac{301.5}{751.5}$$

$$= 0.40 \text{ or } 40\% \textbf{ Ans.}$$

Note : The cycle efficiency may also be obtained by using the relation,

$$\eta = 1 - \left(\frac{1}{r}\right)^{\frac{\gamma - 1}{\gamma}} = 1 - \left(\frac{1}{6}\right)^{\frac{1.4 - 1}{1.4}} = 1 - \frac{1}{1.67} = 1 - 0.6$$

$$= 0.40 \text{ or } 40\% \textbf{ Ans.}$$

Example 33.3. *In an oil gas turbine installation, air is taken at 1 bar and 30° C. The air is compressed to 4 bar and then heated by burning the oil to a temperature of 500° C. If the air flows at the rate of 90 kg/minute, find the power developed by the plant. Take γ for air as 1.4 and c_p as 1 kJ/kg K.*

If 2.4 kg of oil having calorific value of 40 000 kJ/kg is burnt in the combustion chamber per minute, find the overall efficiency of the plant.

Solution. Given : $p_4 = p_3 = 1$ bar ; $T_4 = 30°\text{C} = 30 + 273 = 303$ K ; $p_1 = p_2 = 4$ bar ; $T_2 = 500°\text{C} = 500 + 273 = 773$ K ; $m_a = 90$ kg/min $= 1.5$ kg/s ; $\gamma = 1.4$; $c_p = 1$ kJ/kg K ; $m_f = 2.4$ kg/min $= 0.04$ kg/s ; $C = 40\,000$ kJ/kg

The *T-s* diagram of the cycle is shown in Fig. 33.3.

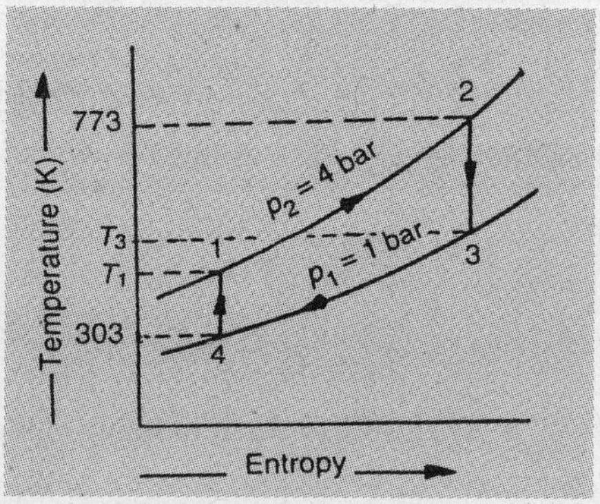

Fig. 33.3

Power developed by the plant

Let T_1, T_3 = Temperature of air at points 1 and 3.

We know that for isentropic expansion 2-3,

$$\frac{T_3}{T_2} = \left(\frac{p_3}{p_2}\right)^{\frac{\gamma - 1}{\gamma}} = \left(\frac{1}{4}\right)^{\frac{1.4 - 1}{1.4}} = 0.673$$

$$\therefore \quad T_3 = T_2 \times 0.673 = 773 \times 0.673 = 520 \text{ K}$$

Similarly, for isentropic compression 4-1,

$$\frac{T_4}{T_1} = \left(\frac{p_4}{p_1}\right)^{\frac{\gamma - 1}{\gamma}} = \left(\frac{1}{4}\right)^{\frac{1.4 - 1}{1.4}} = 0.673$$

$$\therefore \quad T_1 = T_4 / 0.673 = 303 / 0.673 = 450 \text{ K}$$

We know that work developed by the turbine,

$$W_T = m\,c_p\,(T_2 - T_3) = 1.5 \times 1\,(773 - 520) = 379.5 \text{ kJ/s}$$

and work required by compressor,

$$W_C = m\,c_p\,(T_1 - T_4) = 1.5 \times 1\,(450 - 303) = 220.5 \text{ kJ/s}$$

∴ Net work or power of the turbine,

$$P = W_T - W_C = 379.5 - 220.5 = 159 \text{ kJ/s} = 159 \text{ kW} \quad \textbf{Ans.}$$

Overall efficiency of the plant

We know that heat supplied per second

$$= m_f \times C = 0.04 \times 40\,000 = 1600 \text{ kJ/s}$$

∴ Overall efficiency of the plant,

$$\eta_0 = \frac{159}{1600} = 0.099 \text{ or } 9.9\% \quad \textbf{Ans.}$$

33.4. Actual Gas Turbine

We have already discussed in Art. 32.5, an ideal closed cycle gas turbine. In this article, we have shown isentropic expansion of air in the turbine by the curve 2-3′ and isentropic compression of air in the compressor by the curve 4-1′ as shown in Fig. 33.4 (*a*) and (*b*).

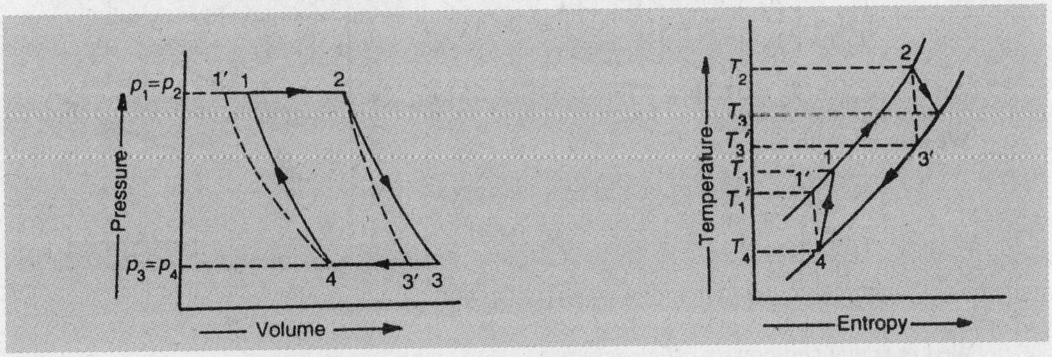

(*a*) *p-v* diagram. (*b*) *T-s* diagram.

Fig. 33.4. Actual and ideal gas turbine.

But in actual practice, the air is expanded adiabatically in the turbine (shown by the curve 2-3) and compressed adiabatically in the compressor (shown by the curve 4-1) in Fig. 33.4 (*a*) and (*b*).

∴ Work required by compressor per kg of air,

$$W_C = c_p\,(T_1 - T_4) \quad \ldots (i)$$

and work done by turbine per kg of air,

$$W_T = c_p\,(T_2 - T_3) \quad \ldots (ii)$$

Now net work available,

$$W = W_T - W_C \quad \ldots (iii)$$

Notes : 1. The power developed by the turbine may be found out from the work done as usual.

2. In this case, isentropic efficiency of the compressor,

$$\eta_C = \frac{T_1' - T_4}{T_1 - T_4}$$

and isentropic efficiency of the turbine (alone),

$$\eta_T = \frac{T_2 - T_3}{T_2 - T_3'}$$

Example 33.4. *A gas turbine plant with a pressure ratio of 1 : 5 takes in air at 15° C. The maximum temperature is 600° C and develops 2200 kW. The turbine and compressor efficiencies are equal to 0.85. Taking $c_p = 1$ kJ/kg K and $c_v = 0.714$ kJ/kg K; determine 1. Actual overall efficiency of the turbine ; and 2. Mass of air circulated by the turbine.*

Solution. Given : $r = p_1 / p_4 = p_2 / p_3 = 5$; $T_4 = 15° \text{ C} = 288 \text{ K}$; $T_2 = 600° \text{ C} = 873 \text{ K}$; $P = 2200 \text{ kW}$; $\eta_T = \eta_C = 0.85$; $c_p = 1$ kJ/kg K ; $c_v = 0.714$ kJ/kg K

We know that ratio of specific heats,

$$\gamma = c_p / c_v = 1 / 0.714 = 1.4$$

Overall efficiency of the turbine

Let T_1', T_1, T_3, T_3' = Temperature of air at corresponding points.

We know that

$$\frac{T_3'}{T_2} = \left(\frac{p_3}{p_2}\right)^{\frac{\gamma - 1}{\gamma}} = \left(\frac{1}{5}\right)^{\frac{1.4-1}{1.4}} = 0.631$$

$$\therefore \quad T_3' = T_2 \times 0.631 = 873 \times 0.631 = 551 \text{ K}$$

We also know that

$$\frac{T_4}{T_1'} = \left(\frac{p_4}{p_1}\right)^{\frac{\gamma - 1}{\gamma}} = \left(\frac{1}{5}\right)^{\frac{1.4-1}{1.4}} = 0.631$$

$$\therefore \quad T_1' = T_4 / 0.631 = 288 / 0.631 = 456 \text{ K}$$

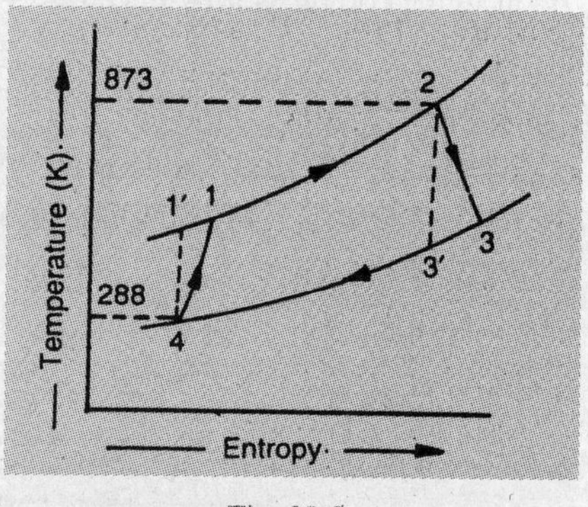

Fig. 33.5

We know that compressor efficiency (η_C),

$$0.85 = \frac{T_1' - T_4}{T_1 - T_4} = \frac{456 - 288}{T_1 - 288} = \frac{168}{T_1 - 288}$$

$$\therefore \quad T_1 = 486 \text{ K}$$

and turbine efficiency (η_T),

$$0.85 = \frac{T_2 - T_3}{T_2 - T_3'} = \frac{873 - T_3}{873 - 551} = \frac{873 - T_3}{322}$$

$$\therefore \quad T_3 = 599 \text{ K}$$

We know that work done by the turbine per kg of air,

$$W_T = c_p (T_2 - T_3) = 1\,(873 - 599) = 274 \text{ kJ}$$

and work required by the compressor per kg of air,

$$W_C = c_p (T_1 - T_4) = 1\,(486 - 288) = 198 \text{ kJ}$$

$\therefore$ Net work done by the turbine per kg of air,

$$W = W_T - W_C = 274 - 198 = 76 \text{ kJ}$$

and heat supplied per kg of air $= c_p (T_2 - T_1) = 1\,(873 - 486) = 387$ kJ

$\therefore$ Overall efficiency of the turbine,

$$\eta_0 = \frac{\text{Work done}}{\text{Heat supplied}} = \frac{76}{387} = 0.1964 \text{ or } 19.64\ \% \textbf{ Ans.}$$

Mass of air circulated by the turbine

Let m_a = Mass of air circulated by the turbine in kg/s.

We know that net workdone by the turbine,

$$W = 76 \text{ kJ/kg of air} = 76 \times m_a \text{ kJ/s} = 76\, m_a \text{ kW}$$

We also know that power developed (P),

$$2200 = 76\, m_a \text{ or } m_a = 2200/76 = 28.95 \text{ kg/s } \textbf{Ans.}$$

Example 33.5. *A gas turbine unit receives air at 100 kPa and 300 K and compresses it adiabatically to 620 kPa with efficiency of the compressor 88%. The fuel has a heating value of 44 180 kJ/kg and the fuel/air ratio is 0.017 kg fuel/kg air. The turbine internal efficiency is 90%. Calculate the compressor work ; turbine work and thermal efficiency.*

Solution. Given : $p_3 = p_4 = 100$ kPa = 1 bar ; $T_4 = 300$ K ; $p_1 = p_2 = 620$ kPa = 6.2 bar ; $\eta_C = 88\% = 0.88$; $C = 44\ 180$ kJ/kg ; $\eta_T = 90\ \% = 0.9$

The T-s diagram is shown in Fig. 33.6.

Let T_1, T_1', T_2, T_3 and T_3' = Temperature at corresponding points.

We know that

$$\frac{T_4}{T_1'} = \left(\frac{p_4}{p_1}\right)^{\frac{\gamma-1}{\gamma}} = \left(\frac{1}{6.2}\right)^{\frac{1.4-1}{1.4}} = 0.593$$

$$\therefore \quad T_1' = T_4 / 0.593 = 300 / 0.593 = 506 \text{ K}$$

and efficiency of the compressor (η_C),

$$0.88 = \frac{T_1' - T_4}{T_1 - T_4} = \frac{506 - 300}{T_1 - 300} = \frac{206}{T_1 - 300}$$

$$\therefore \quad T_1 = 534 \text{ K}$$

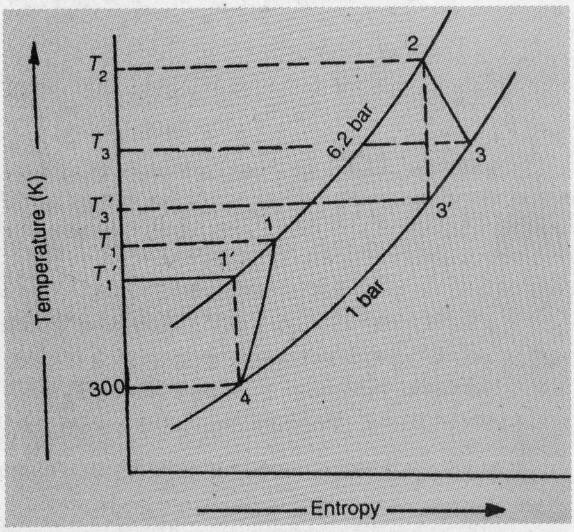

Fig. 33.6

We know that heat supplied by the fuel

$$= m_f \times C = 0.017 \times 44\ 180 = 751 \text{ kJ}$$

We also know that heat supplied per kg of air,

$$751 = m\,c_p\,(T_2 - T_1) = (1 + 0.017)\,1.005\,(T_2 - 534)$$

... (Taking c_p = 1.005 kJ/kg K)

$$\therefore \quad T_2 = 735 + 534 = 1269 \text{ K}$$

Now for isentropic process 2-3',

$$\frac{T_3'}{T_2} = \left(\frac{p_3}{p_2}\right)^{\frac{\gamma-1}{\gamma}} = \left(\frac{1}{6.2}\right)^{\frac{1.4-1}{1.4}} = 0.593$$

$$\therefore \quad T_3' = 0.593\,T_2 = 0.593 \times 1269 = 752 \text{ K}$$

and turbine efficiency (η_T),

$$0.9 = \frac{T_2 - T_3}{T_2 - T_3'} = \frac{1269 - T_3}{1269 - 752}$$

$$\therefore \quad T_3 = 804 \text{ K}$$

Compressor work

We know that compressor work per kg of air

$$W_C = m\,c_p\,(T_1 - T_4) = (1 + 0.017)\,1.005\,(534 - 300) = \textbf{239 kJ Ans.}$$

Turbine work

We know that turbine work per kg of air,

$$W_T = m\,c_p\,(T_2 - T_3) = (1 + 0.017)\,1.005\,(1269 - 804) = \textbf{475 kJ Ans.}$$

Thermal efficiency

We know that net workdone by the turbine per kg of air,

$$W = W_T - W_C = 475 - 239 = 236 \text{ kJ}$$

∴ Thermal efficiency,

$$\eta_{th} = \frac{\text{Net work done}}{\text{Heat supplied}} = \frac{236}{751} = 0.314 \text{ or } \textbf{31.4\% Ans.}$$

Example 33.6. *In a gas turbine plant, the intake temperature and pressure are 1 bar and 18° C respectively. The air is then compressed to a pressure of 4.1 bar by a compressor, whose isentropic efficiency is 80%. The temperature of the gas, whose properties may be assumed to resemble with those of air, is raised to 645° C in the combustion chamber where there is a pressure drop of 0.1 bar. Expansion to atmospheric pressure then occurs.*

If the thermal efficiency of the plant is to be 19%, what must be the isentropic efficiency of the turbine ? The mass of fuel may be neglected. Take γ = 1.4.

Solution. Given : $p_4 = p_3 = 1$ bar ; $T_4 = 18°$ C = 291 K ; $p_1 = 4.1$ bar ; $\eta_C = 80\% = 0.8$; $T_2 = 645°$ C = 918 K ; Pressure drop = 0.1 bar ; $\eta_{th} = 19\% = 0.19$; $\gamma = 1.4$

Since the pressure drop in the combustion chamber is 0.1 bar, therefore

$$p_2 = 4.1 - 0.1 = 4 \text{ bar}$$

The *T-s* diagram is shown in Fig. 33.7.

Let T_1, T_1', T_3, T_3' = Temperature at corresponding points.

We know that for isentropic process 2-3′,

$$\frac{T_3'}{T_2} = \left(\frac{p_3}{p_2}\right)^{\frac{\gamma-1}{\gamma}} = \left(\frac{1}{4}\right)^{\frac{1.4-1}{1.4}} = 0.673$$

$$\therefore \quad T_3' = T_2 \times 0.673 = 918 \times 0.673 = 618 \text{ K}$$

and for isentropic process 4-1′,

$$\frac{T_4}{T_1'} = \left(\frac{p_4}{p_1}\right)^{\frac{\gamma-1}{\gamma}} = \left(\frac{1}{4.1}\right)^{\frac{1.4-1}{1.4}} = 0.668$$

$$\therefore \quad T_1' = T_4 / 0.668 = 291 / 0.668 = 436 \text{ K}$$

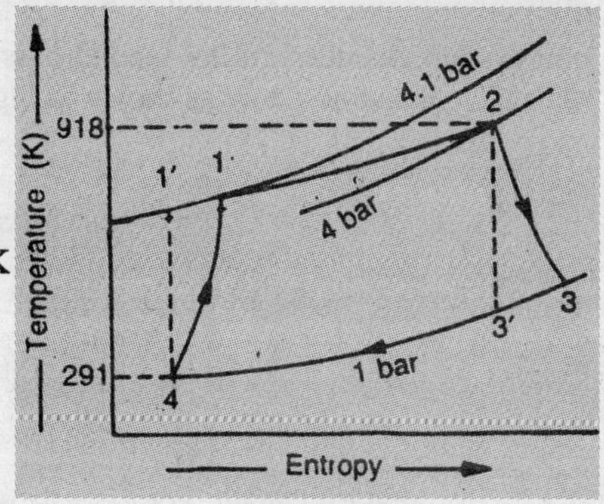

Fig. 33.7

We know that isentropic efficiency of compressor (η_C),

$$0.8 = \frac{T_1' - T_4}{T_1 - T_4} = \frac{436 - 291}{T_1 - 291} = \frac{145}{T_1 - 291}$$

$$\therefore \quad T_1 = 472 \text{ K}$$

We also know that work done by turbine per kg of air,

$$W_T = c_p (T_2 - T_3) = c_p (918 - T_3)$$

and work required by compressor per kg of air,

$$W_C = c_p (T_1 - T_4) = c_p (472 - 291) = 181 \, c_p$$

∴ Net work done by turbine per kg of air

$$= c_p (918 - T_3) - 181 \, c_p = c_p (737 - T_3)$$

and heat supplied per kg of air $= c_p (T_2 - T_1) = c_p (918 - 472) = 446 \, c_p$

We know that thermal efficiency of the turbine (η_{th}),

$$0.19 = \frac{c_p (737 - T_3)}{446 \, c_p} = \frac{737 - T_3}{446}$$

$$\therefore \quad T_3 = 652 \text{ K}$$

and isentropic efficiency of the turbine,

$$\eta_T = \frac{T_2 - T_3}{T_2 - T_3'} = \frac{918 - 652}{918 - 618} = 0.887 \text{ or } \mathbf{88.7\%} \textbf{ Ans.}$$

33.5. Heat Exchanger

As a matter of fact, the exhaust gases, leaving the turbine at the end of expansion, are still at high temperature. If these gases are allowed to pass into the atmosphere, then it amounts to the loss of available heat energy.

In order to achieve fuel economy, some of the available heat energy is recovered by passing the gases from the turbine through a heat exchanger. A heat exchanger, in its simplest form, consists of a chamber having two passages. In one of the passages, air from the compressor flows to the combustion chamber. In the second passage, hot gases from the turbine are made to flow before exhaust to the atmosphere as shown in Fig. 33.8.

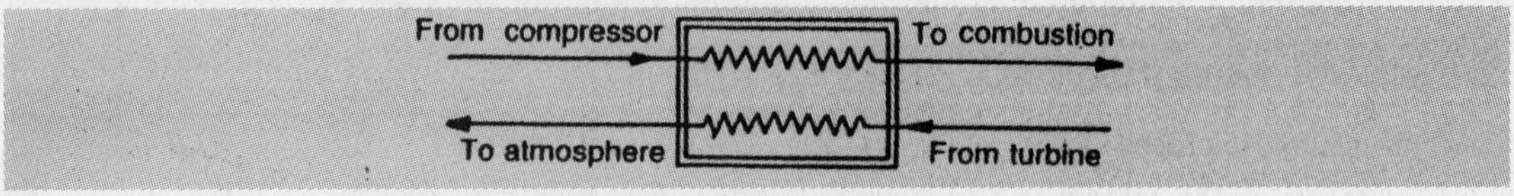

Fig. 33.8. Heat exchanger.

In a heat exchanger, hot gases from the turbine loose some energy and gets cooler, whereas air from compressor gets heated up before entering the combustion chamber.

33.6. Efficiency of Heat Exchanger

We have already discussed in the last article that hot gases from the turbine loose some heat energy in the heat exchanger. This heat energy is taken up by the air flowing from the compressor to the combustion chamber. Both these processes are shown on *T-s* diagram in Fig. 33.9.

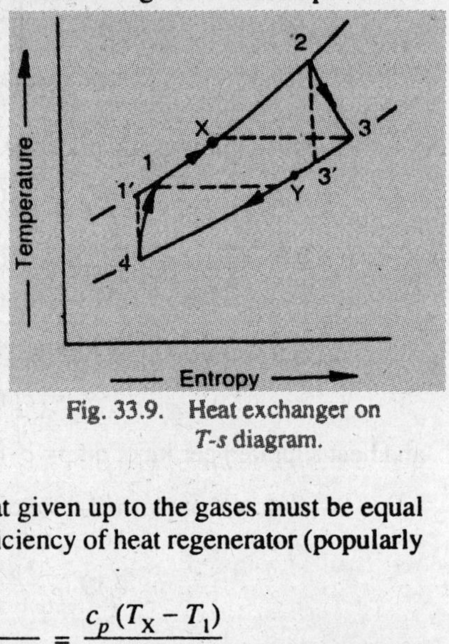

Fig. 33.9. Heat exchanger on *T-s* diagram.

Let T_X be the temperature of the compressed air and T_Y be the temperature of hot gases after passing through the heat exchanger.

In an ideal heat exchanger, the compressed air would be heated from T_1 to T_X (equal to T_3). Similarly, the hot gases will be cooled from T_3 to T_Y (equal to T_1). But in actual practice, it is not possible, as a finite temperature difference is required at all points in the heat exchanger for the heat to flow. In this case, this finite temperature difference is equal to $(T_3 - T_X)$ or $(T_Y - T_1)$.

If there is no loss of heat in the exchanger, then the heat given up to the gases must be equal to the heat taken up by the air. But it is rarely possible. The efficiency of heat regenerator (popularly known as effectiveness) is given by the relation,

$$\eta_e = \frac{\text{Actual heat transfer}}{\text{Maximum possible heat transfer}} = \frac{c_p (T_X - T_1)}{c_p (T_3 - T_1)}$$

Example 33.7. *A gas turbine plant receives air at a pressure of 1 bar and 290 K. The air is then compressed in a rotary compressor to a pressure of 4 bar and then heated to a temperature of 840 K. The efficiencies of compressor and turbine are 82% and 86% respectively. Neglecting the pressure drop, find overall efficiency of the plant (i) without heat exchanger ; and (ii) with heat exchanger of 70% effectiveness. Take* γ *and* c_p *for air and hot gases as 1.4 and 1 kJ/kg K respectively.*

Solution. Given : $p_4 = p_3 = 1$ bar ; $T_4 = 290$ K ; $p_1 = p_2 = 4$ bar ; $T_2 = 840$ K ; $\eta_C = 82\% = 0.82$; $\eta_T = 86\% = 0.86$; $\eta_e = 70\% = 0.7$; $\gamma = 1.4$; $c_p = 1$ kJ/kg K

The *T-s* diagram without heat exchanger and with heat exchanger is shown in Fig. 33.10 (*a*) and (*b*) respectively.

Efficiency of the plant without heat exchanger

Let T_1', T_1, T_3, T_3' = Temperature of air at correspondingpoints.

We know that $\frac{T_3'}{T_2} = \left(\frac{p_3}{p_2}\right)^{\frac{\gamma-1}{\gamma}} = \left(\frac{1}{4}\right)^{\frac{1.4-1}{1.4}} = 0.673$

$\therefore \quad T_3' = T_2 \times 0.673 = 840 \times 0.673 = 565.3 \text{ K}$

and $\frac{T_4}{T_1'} = \left(\frac{p_4}{p_1}\right)^{\frac{\gamma-1}{\gamma}} = \left(\frac{1}{4}\right)^{\frac{1.4-1}{1.4}} = 0.673$

$\therefore \quad T_1' = T_4 / 0.673 = 290 / 0.673 = 430.9 \text{ K}$

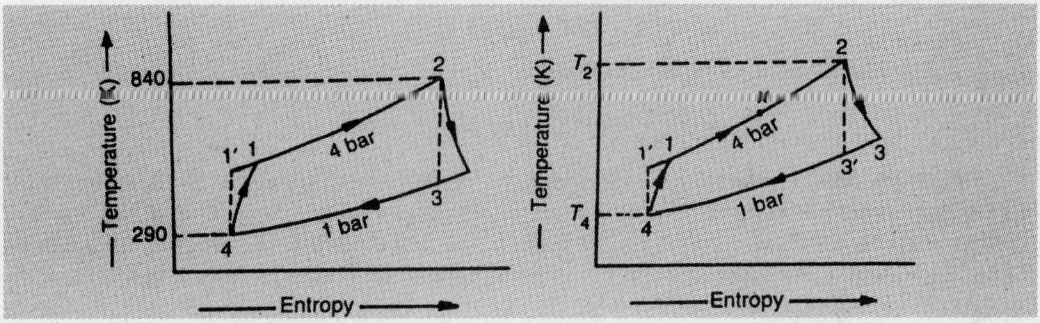

(a) Without heat exchanger. (b) With heat exchanger.

Fig. 33.10

We know that compressor efficiency (η_C),

$$0.82 = \frac{T_1' - T_4}{T_1 - T_4} = \frac{430.9 - 290}{T_1 - 290} = \frac{140.9}{T_1 - 290}$$

$\therefore \quad T_1 = 461.8 \text{ K}$

and turbine efficiency (η_T),

$$0.86 = \frac{T_2 - T_3}{T_2 - T_3'} = \frac{840 - T_3}{840 - 565.3} = \frac{840 - T_3}{274.7}$$

$\therefore \quad T_3 = 603.8 \text{ K}$

We also know that work done by the turbine per kg of air,

$$W_T = c_p (T_2 - T_3) = 1\,(840 - 603.8) = 236.2 \text{ kJ/s}$$

and work required by the compressor,

$$W_C = c_p (T_1 - T_4) = 1\,(461.8 - 290) = 171.8 \text{ kJ/s}$$

$\therefore$ Net work done by the turbine,

$$W = W_T - W_C = 236.2 - 171.8 = 64.4 \text{ kJ/s}$$

and heat supplied per kg of air

$$= c_p (T_2 - T_1) = 1\,(840 - 461.8) = 378.2 \text{ kJ/s}$$

$\therefore$ Thermal efficiency,

$$\eta_{th} = \frac{\text{Net work done}}{\text{Heat supplied}} = \frac{64.4}{378.2} = 0.17 \text{ or } 17\% \text{ **Ans.**}$$

Thermal efficiency of the plant with heat exchanger

Given. Effectiveness of heat exchanger,

$$\eta_e = 70\% = 0.7$$

We know that heat available in exhaust gases

$$= c_p (T_3 - T_1) = 1\,(603.8 - 461.8) = 142 \text{ kJ/s}$$

∴ Actual heat recovered from exhaust gases

$$= 0.7 \times 142 = 99.4 \text{ kJ/s}$$

and heat supplied by the combustion chamber per kg of air

$$= 378.2 - 99.4 = 278.8 \text{ kJ/s}$$

∴ Thermal efficiency with heat exchanger,

$$\eta_{th} = \frac{64.4}{278.8} = 0.231 \text{ or } 23.1\ \% \textbf{ Ans.}$$

Example 33.8. *Determine the efficiency of a gas turbine plant fitted with a heat exchanger of 75% effectiveness. The pressure ratio is 4 : 1 and the compression is carried out in two stages of equal pressure ratio with intercooling back to initial temperature of 290 K. The maximum temperature is 925 K. The turbine isentropic efficiency is 88% and each compressor isentropic efficiency is 85%. For air, $\gamma = 1.4$ and $c_p = 1.005$ kJ/kg K.*

Solution. Given : $\eta_e = 75\% = 0.75$; $r = p_1 / p_4 = p_2 / p_3 = 4$; $T_4 = 290$ K ; $T_2 = 925$ K ; $\eta_T = 88\% = 0.88$; $\eta_C = 85\% = 0.85$; $\gamma = 1.4$; $c_p = 1.005$ kJ/kg K

The *T-s* diagram is shown in Fig. 33.11.

Let $T_1, T_3, T_3', T_5, T_5', T_6$ = Temperature of air at corresponding points.

We know that for equal pressure ratio, the intermediate pressure,

$$p_6 = p_5 = p_5' = \sqrt{p_1 \times p_4} = \sqrt{4 \times 1} = 2 \text{ bar}$$

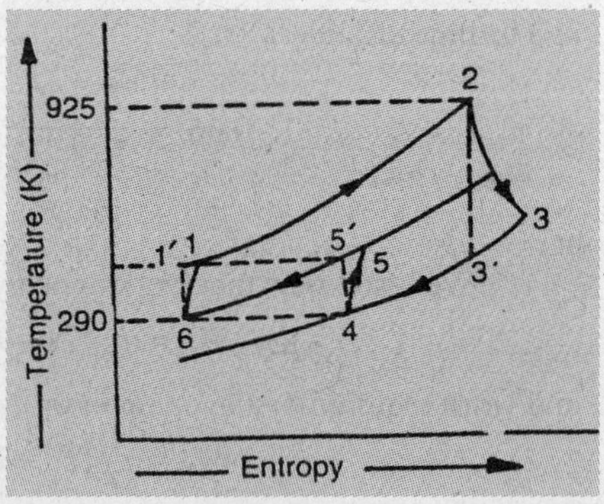

Fig. 33.11

We also know that

$$\frac{T_3'}{T_2} = \left(\frac{p_3}{p_2}\right)^{\frac{\gamma-1}{\gamma}} = \left(\frac{1}{4}\right)^{\frac{1.4-1}{1.4}} = 0.673$$

$$\therefore \quad T_3' = T_2 \times 0.673 = 925 \times 0.673 = 622.6 \text{ K}$$

and

$$\frac{T_4}{T_5'} = \left(\frac{p_4}{p_5}\right)^{\frac{\gamma-1}{\gamma}} = \left(\frac{1}{2}\right)^{\frac{1.4-1}{1.4}} = 0.82$$

$$\therefore \quad T_5' = T_4 / 0.82 = 290 / 0.82 = 353.7 \text{ K}$$

We know that for equal pressure ratio,

$$T_5' = T_1' = 353.7 \text{ K}$$

and

$$T_6 = T_4 = 290 \text{ K}$$

We also know that compressor efficiency (η_C),

$$0.85 = \frac{T_1' - T_6}{T_1 - T_6} = \frac{353.7 - 290}{T_1 - 290} = \frac{63.7}{T_1 - 290}$$

$$\therefore \quad T_1 = 364.9 \text{ K}$$

and turbine efficiency (η_T), $$0.88 = \frac{T_2 - T_3}{T_2 - T_3'} = \frac{925 - T_3}{925 - 622.6} = \frac{925 - T_3}{302.4}$$

$$\therefore \quad T_3 = 658.9 \text{ K}$$

We know that for equal pressure ratio,

$$T_5 = T_1 = 364.9 \text{ K}$$

$\therefore$ Work done by the turbine per kg of air,

$$W_T = c_p (T_2 - T_3) = 1.005 (925 - 658.9) = 267.4 \text{ kJ}$$

and work required by the compressor,

$$W_C = c_p [(T_1 - T_6) + (T_5 - T_4)]$$

$$= 1.005 [(364.9 - 290) + (364.9 - 290)] = 150.5 \text{ kJ}$$

$\therefore$ Net work done by the turbine,

$$W = W_T - W_C = 267.4 - 150.5 = 116.9 \text{ kJ}$$

and heat supplied per kg of air $$= c_p (T_2 - T_1) = 1.005 (925 - 364.9) = 562.9 \text{ kJ}$$

We also know that heat available in exhaust gases

$$= c_p (T_3 - T_1) = 1.005 (658.9 - 364.6) = 295.8 \text{ kJ}$$

$\therefore$ Actual heat recovered from exhaust gases

$$= 0.75 \times 295.8 = 221.8 \text{ kJ}$$

and heat supplied by the combustion chamber pĕr kg of air

$$= 562.9 - 221.8 = 341.1 \text{ kJ}$$

$\therefore$ Efficiency of the turbine plant,

$$\eta_{th} = \frac{116.9}{341.1} = 0.343 \text{ or } 34.3\% \quad \textbf{Ans.}$$

33.7. Uses of Gas Turbines

Though there are innumerable uses of gas turbines these days, yet the following are important from the subject point of view :

1. *Generation of electric power.* The gas turbines are extensively used in the generation of electric power The largest gas turbine power plant is installed in Switzerland. It consists of two turbines of total capacity 40 MW.

2. *Turbojet and turbo propeller engines.* The gas turbines are used to drive air compressors in turbojets. They are also used to drive air compressor and turbines in turbo propeller engines. The gas turbines in turbojet and turbo propeller engines are operated in the temperature range of 800° C to 1000° C.

3. *Supercharger.* The gas turbines are also used to drive superchargers fitted in the aviation gasoline engines as well as for heavy duty diesel engines.

4. *Marine engines.* The gas turbines are used in marine engines. Unlike steam turbine or steam engines, the gas turbines do not require water storage tanks or distillation plants.

5. *Railway engines.* The gas turbines have also entered the field of railway engines. The first gas turbine locomotive was put into service in 1941 in Switzerland. The turbine developed 5800 kW out of which 4200 kW is consumed to drive the air compressor.

33.8. Recent Trends in Gas Turbines

The earlier gas turbines were designed to work on closed cycle with air as the working fluid. In recent years, lot of research is being done all over the world to improve the working of gas turbines.

The main focus of the research is aimed to generate more power from the plant as well as to effect economy in the installation and maintenance expenses. It has lead to the usage of various working fluids such as carbon dioxide, nitrogen, argon etc. Helium has proved to be the most suitable working fluid because of its lower density and higher (about five times) specific heat at constant pressure. Moreover, it requires about 1/3 surface area for heating in combustion chamber and heat exchanger than the turbine using air as the working fluid for the same temperature range and pressure ratio. The second important aspect of the research is aimed to heat the air in the combustion chambers. The scientists have tried many fuels under variable sets of conditions, but the recent trend is to use the heat generated by nuclear fusion in a reactor.

EXERCISES

1. A gas turbine plant receives air at 1 bar and 20° C. Find the thermal efficiency of the plant, if the compression ratio is 4 and specific heat at constant pressure for the working fluid is 1 kJ/kg K. **[Ans.** 32.7%]

2. Find the actual overall efficiency of a simple combustion turbine, the following data of which are available :

Compression ratio = 5 ; Initial temperature of air = 15° C ; γ for air = 1.4 ; Compressor and turbine efficiency = 85%. Maximum temperature of combustion = 600° C and c_p = 1.005 kJ/kg K. **[Ans.** 20.5%]

3. A gas turbine plant works in temperature limits of 300 K and 900 K and the pressure limits are 1 bar and 4 bar. The internal efficiency of the compressor is 0.8 and that of the turbine is 0.85. Estimate the thermal efficiency of the plant and the power available in kilowatts, if the air consumption is 1 kg/s. The heating value of the fuel is 42 000 kJ/kg. **[Ans.** 16.2% ; 71 kW]

4. In a gas turbine plant comprising a single stage compressor, combustion chamber and turbine, the compressor takes in air at 15°C and compresses it to 4 times the initial pressure with an isentropic efficiency of 85 percent. The fuel-air ratio is 0.0125 and the calorific value of the fuel is 42 000 kJ/kg. If the isentropic efficiency of the turbine is 82 percent, find the overall thermal efficiency and the air intake for a power output of 260 kW.

Take the mass of the fuel into account. The turbine inlet temperature is 1000 K.

[Ans. 21.5% ; 2.28 kg/s]

5. In a gas turbine plant, air at 10° C and atmospheric pressure is compressed through a pressure ratio of 4 : 1. In a heat exchanger and combustion chamber, the air is heated to 700° C. After expansion through the turbine, the air passes through a heat exchanger which cools the air through 75% of the maximum range possible, while the pressure drops 0.14 bar and the air is finally exhausted to atmosphere. The isentropic efficiency of the compressor is 0.80 and that of turbine 0.85. Find the efficiency of the plant. **[Ans.** 19%]

6. A gas turbine takes in air at 27° C and 1 bar. The pressure ratio is 4 and the maximum temperature in the cycle is 560° C. The compressor and turbine efficiencies are 0.83 and 0.85 respectively. Determine the overall efficiency if the regenerator effectiveness is 0.75. **[Ans.** 21.2%]

7. In a gas turbine plant, air is compressed from 1 bar and 15° C through a pressure ratio of 4 : 1. It is then heated to 650° C in a combustion chamber and expanded back to a pressure of 1 bar in a turbine. Calculate the cycle efficiency and work ratio, if a perfect heat exchanger is used. Assume isentropic efficiency of the turbine and compressor as 85% and 80% respectively. **[Ans.** 31.9% ; 31.9%]

8. A gas turbine plant consists of two stage compressor with perfect intercooler and a set of H.P. and L.P. turbines. The exhaust from L.P. turbine passes through a heat exchanger which heats up the air leaving the H.P. compressor. The overall pressure ratio is 10 and temperature range is 20° C to 600° C. Assuming isentropic efficiencies for compressors, turbines and heat exchanger as 80%, 85% and 70% respectively, find the power developed and thermal efficiency of the plant. Take mass of air flow as 1.15 kg/s and 2% work of each turbine is lost in overcoming friction. **[Ans.** 144.6 kW ; 25.7%]

QUESTIONS

1. State the assumptions made for thermal efficiency of a gas turbine plant.
2. Derive an expression for the thermal efficiency of a gas turbine plant, and show that it is independent of the mass of air circulated in it.
3. Describe the difference between an ideal gas turbine plant and an actual gas turbine plant. Give relations for the isentropic efficiencies of compressor and turbine.
4. What is heat exchanger ? Describe its utility.
5. Obtain an expression for the effectiveness of a heat exchanger.
6. Write briefly the uses of gas turbines.

OBJECTIVE TYPE QUESTIONS

1. In an ideal gas turbine plant, it is assumed that the compression and expansion processes are

   (*a*) isothermal (*b*) isentropic (*c*) polytropic

2. The thermal efficiency of an ideal gas turbine plant is given by

   (*a*) $r^{\gamma-1}$ (*b*) $1 - r^{\gamma-1}$ (*c*) $1 - \left(\frac{1}{r}\right)^{\frac{\gamma}{\gamma-1}}$ (*d*) $1 - \left(\frac{1}{r}\right)^{\frac{\gamma-1}{\gamma}}$

3. The gas turbine cycle with regenerator improves

   (*a*) thermal efficiency (*b*) work ratio

   (*c*) avoids pollution (*d*) none of these

4. High air-fuel ratio is gas turbines

   (*a*) increases power output (*b*) improves thermal efficiency

   (*c*) reduces exhaust temperature (*d*) do not damage turbine blades

ANSWERS

1. (*b*)	2. (*d*)	3. (*a*)	4. (*c*)

34

Introduction to Heat Transfer

1. Introduction. 2. Methods of Heat Transfer. 3. Newton's Law of Cooling. 4. Fourier's Law of Heat Conduction. 5. Heat Transfer by Conduction through a Slab. 6. Thermal Conductivity. 7. Temperature Gradient. 8. Heat Transfer by Conduction through a Composite Wall. 9. Radial Heat Transfer by Conduction through a Thick Cylinder. 10. Heat Transfer by Conduction through a Thick Sphere. 11. Overall Coefficient of Heat Transfer.

34.1. Introduction

As a matter of fact, the subject of heat transfer is getting more and more importance in the sphere of science and technology these days. The mechanical engineers deal with the problems of heat transfer in the design of their I.C. engines, refrigeration and air-conditioning system as well as various types of steam generation plants. The electrical engineers deal with the problems of heat transfer in the cooling systems of power generation machines, transformers and various types of huge electrical installations. Similarly, the civil engineers require the knowledge of heat transfer in the design of dams, tunnels, cold storages, cinema halls and other important structures.

The subject of heat transfer is very vast, and its detailed study is beyond the scope of this book. However, its basic principles will be discussed in this book.

34.2. Methods of Heat Transfer

The heat transfer may be broadly defined as the transmission of heat energy from one region to another due to the temperature difference between these two regions.

The following methods, of heat transfer, from one body to another, are important from the subject point of view :

1. *Conduction.* It is a process of heat transfer from one particle of the body to another in the direction of fall of temperature. The particles themselves remain in fixed position relative to each other.

2. *Convection.* It is a process of heat transfer from one particle of the body to another by convection current. In this case, the particles of the body move relative to each other.

3. *Radiation.* It is a process of heat transfer from a hot body to a cold body, in a straight line, without affecting the intervening medium.

In this chapter, we shall discuss the heat transfer by conduction only.

34.3. Newton's Law of Cooling

It is an important law in the field of heat transfer which states, *"Heat transfer from a hot body to a cold body is directly proportional to the surface area and difference of temperatures between the two bodies".*

It is a general law, for the heat transfer which can not be applied to all sets of conditions. But it paved the way for other laws dealing in the heat loss.

34.4. Fourier's* Law of Heat Conduction

It is also an important law in heat conduction, which is represented by the equation,

$$Q \propto A \times \frac{dT}{dx} = k\,A \times \frac{dT}{dx}$$

where Q = Amount of heat flow through the body in a unit time.

A = Surface area of heat flow. It is taken at right angles to the direction of flow.

dT = Temperature difference on the two faces of the body.

dx = Thickness of the body through which the heat flows. It is taken along the direction of heat flow.

k = Constant of proportionality known as *thermal conductivity* of the body.

34.5. Heat Transfer by Conduction through a Slab

Consider a solid slab having one of its face (say left) at a higher temperature and the other (say right) at a lower temperature as shown in Fig. 34.1.

Let T_1 = Temperature of the left face (*i.e.* higher temperature) in K,

T_2 = Temperature of the right face (*i.e.* lower temperature) in K,

x = Thickness of the slab,

A = Area of the slab.

k = Thermal conductivity of the body,

t = Time through which the heat flow has taken place.

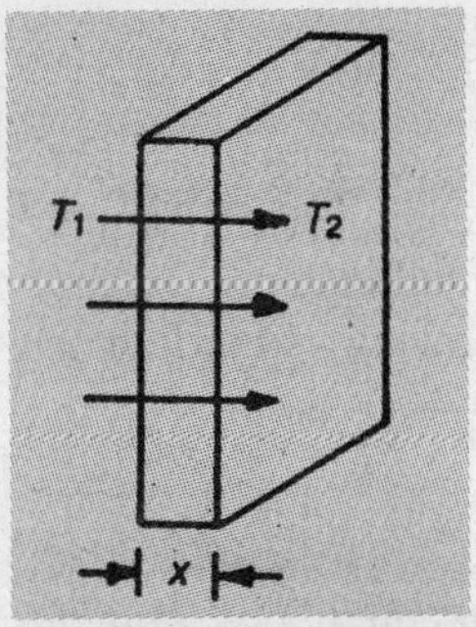

Fig. 34.1. Heat transfer through a slab.

As per the Fourier's law of heat conduction, the heat flow (assuming no loss of heat from the sides) through the slab,

$$Q = k\,A \times \frac{dT}{dx} = \frac{k\,A\,(T_1 - T_2)}{dx}$$

Now the total amount of heat flow in time t may be found out by the equation,

$$Q = \frac{k\,A\,(T_1 - T_2)\,t}{x}$$

Notes : 1. Since the temperature of the slab decreases as x increases, therefore sometimes negative sign is put on the right hand side of the above equation.

2. The rate of heat flow per second is given by the relation,

$$Q = \frac{k\,A\,(T_1 - T_2)}{x}$$

34.6. Thermal Conductivity

We have discussed in the previous article that the amount of heat flow through a body.

$$Q = \frac{k\,A\,(T_1 - T_2)\,t}{x}$$

---

* The fundamental law of heat conduction was proposed by Biot in 1804. But is could not get proper recognition until 1822, when Fourier confirmed it by his outstanding contribution in the field. After confirmation, the law is popularly known as Fourier's Law (or Fourier's equation) of heat conduction.

In the above equation, if we substitute $A = 1\ m^2$; $(T_1 - T_2) = 1$ K ; $t = 1$ s and $x = 1$ m, then $Q = k$.

It is thus obvious, that the thermal conductivity of a material is numerically equal to the quantity of heat (in joules) which flows in one second through a slab of the material of area 1 m^2 and thickness 1 m when its faces differ in temperature by 1 K. It may also be defined as the quantity of heat in joules that flows in one second through one metre cube of a material when opposite faces are maintained at a temperature difference of 1 K.

Notes : 1. The unit of thermal conductivity depends upon the units of the quantities on the right side of the above equation. In S.I. units, the unit of thermal conductivity is J/m/K/s = W/m K ($\because$ 1 J/s = 1W).

2. The above expression for rate of heat flow may also be written as :

$$Q = \frac{T_1 - T_2}{x/kA}$$

The term x/kA is known as *thermal resistance* and corresponds to resistance '*R*' in Ohm's law.

34.7. Temperature Gradient

We know that the rate of heat flow through a body,

$$Q = \frac{kA(T_1 - T_2)}{x}$$

The term $(T_1 - T_2)/x$ is known as *temperature gradient*. If the temperature falls by dT over a small distance dx, then temperature gradient may be written as dT/dx. Hence the quantity of heat flowing in a differential form may be written as :

$$Q = kA \times \frac{dT}{dx}$$

Example 34.1. *The glass windows of a room have a total area of 10 m^2 and the glass is 4 mm thick. Calculate the quantity of heat that escapes from the room by conduction per second when the inside surfaces of windows are at 25° C and the outside surfaces at 10° C. The value of k is 0.84 W/m K.*

Solution. Given : $A = 10\ m^2$; $x = 4$ mm = 0.004 m ; $T_1 = 25°$ C = 298 K ; $T_2 = 10°$ C = 283 K ; $k = 0.84$ W/m K

We know that the quantity of heat that escapes from the room per second,

$$Q = \frac{kA(T_1 - T_2)}{x} = \frac{0.84 \times 10\,(298 - 283)}{0.004} = 31\,500\ \text{J}$$

$= 31.5$ kJ **Ans.**

Example 34.2. *A boiler is made of iron plates 12 mm thick. If the temperature of the outside surface be 120° C and that of the inner 100° C, calculate the mass of water evaporated per hour. Assume that the area of heating surface is 5 m^2, and k for iron as 84 W/m K.*

Solution. Given : $x = 12$ mm = 0.012 m ; $T_1 = 120°$ C = 393 K ; $T_2 = 100°$ C = 373 K ; $t = 1$h = 3600 s ; $A = 5\ m^2$; $k = 84$ W/m K

We know that amount of heat transferred,

$$Q = \frac{kA(T_1 - T_2)\,t}{x} = \frac{84 \times 5\,(393 - 373)\,3600}{0.012} = 2520 \times 10^6\ \text{J/h}$$

$= 2520 \times 10^3$ kJ/h

We know that the heat required to evaporate 1 kg of water at 100° C is equal to its latent heat, *i.e.* 2260 kJ.

∴ Mass of water evaporated per hour

$$= \frac{\text{Total amount of heat transferred}}{\text{Heat reqd. to evaporate 1 kg of water}}$$

$$= \frac{2520 \times 10^3}{2260} = 1115 \text{ kg } \textbf{Ans.}$$

34.8. Heat Transfer by Conduction through a Composite Wall

Consider a composite wall consisting of two different materials through which the heat is being transferred by conduction, as shown in Fig. 34.2.

Let x_1 = Thickness of first material,

k_1 = Thermal conductivity of first material,

x_2, k_2 = Corresponding values for the second material,

T_1, T_3 = Temperatures of the two outer surfaces,

T_2 = Temperature at junction point, and

A = Surface area of the wall.

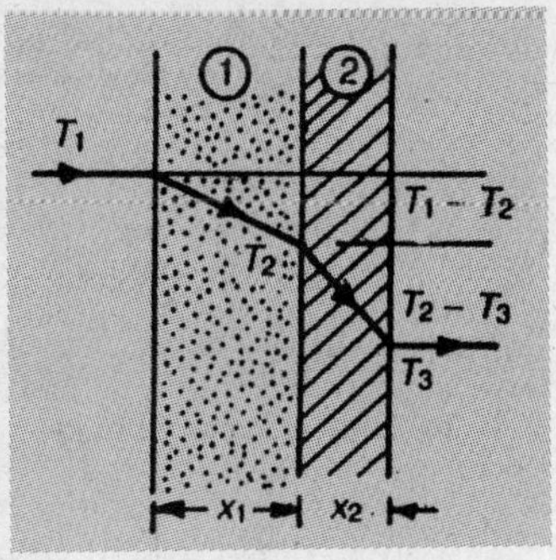

Fig. 34.2. Heat transfer through a composite wall.

Now assuming T_1 to be higher than T_2, the heat will flow from left to right as shown in the figure. Under steady conditions, the rate of heat flow through section 1 is equal to that through section 2. We know that heat flowing through section 1,

$$Q = \frac{k_1 A (T_1 - T_2)}{x_1}$$

or $$(T_1 - T_2) = \frac{Q}{A} \times \frac{x_1}{k_1} \qquad \ldots (i)$$

Similarly for section 2,

$$(T_2 - T_3) = \frac{Q}{A} \times \frac{x_2}{k_2} \qquad \ldots (ii)$$

Adding equations (*i*) and (*ii*),

$$(T_1 - T_3) = \frac{Q}{A}\left(\frac{x_1}{k_1} + \frac{x_2}{k_2}\right)$$

or $$Q = \frac{A(T_1 - T_3)}{\dfrac{x_1}{k_1} + \dfrac{x_2}{k_2}} = \frac{(T_1 - T_3)}{\dfrac{x_1}{k_1 A} + \dfrac{x_2}{k_2 A}} = \frac{(T_1 - T_3)}{\sum \dfrac{x}{kA}}$$

Notes : 1. We have taken the composite wall consisting of two different materials for simplicity. But this relation may be extended for any number of materials.

2. Now the total heat flow in any time (t) may be found out by the equation :

$$Q = \frac{(T_1 - T_3)\, t}{\sum \dfrac{x}{kA}}$$

3. The term $\Sigma \dfrac{x}{kA}$ is known as thermal resistance of the wall.

Example 34.3. *A furnace wall is made up of refractory bricks of 300 mm thick. The inner and outer surfaces of the wall have temperature of 1000° C and 150° C. Find the heat loss per square metre per hour.*

If the outside temperature becomes 50° C, the furnace wall is covered with insulating bricks of 200 mm thickness. Find the reduction in heat loss. Take thermal conductivities of refractory and insulating bricks as 4.5 and 0.5 W/m K.

Solution. Given : x_1 = 300 mm = 0.3 m ; T_1 = 1000° C = 1273 K ; T_2 = 150° C = 423 K ; t = 1 h = 3600 s ; T_3 = 50° C = 323 K ; x_2 = 200 mm = 0.2 m ; k_1 = 4.5 W/m K ; k_2 = 0.5 W/mK

(a) Only refractory bricks. (b) Lined with insulating bricks.

Fig. 34.3

Heat loss when the furnace wall is made of refractory bricks

We know that heat loss per square metre per hour,

$$Q_1 = \frac{k_1 A\,(T_1 - T_2)\,t}{x_1} = \frac{4.5 \times 1\,(1273 - 423)\,3600}{0.3}\ \text{J/m}^2\ \text{h}$$

$$= 45.9 \times 10^6\ \text{J/m}^2\ \text{h} = 45.9\ \text{MJ/m}^2\ \text{h}\ \textbf{Ans.}$$

Reduction in heat loss when the furnace wall is covered with insulating bricks

We know that heat loss per square metre per hour,

$$Q_2 = \frac{A\,(T_1 - T_3)\,t}{\dfrac{x_1}{k_1} + \dfrac{x_2}{k_2}} = \frac{1\,(1273 - 323)\,3600}{\dfrac{0.3}{4.5} + \dfrac{0.2}{0.5}}\ \text{J/m}^2\ \text{h}$$

$$= 7.3 \times 10^6\ \text{J/m}^2\ \text{h} = 7.3\ \text{MJ/m}^2\text{h}$$

$\therefore$ Reduction in heat loss $= Q_1 - Q_2 = 45.9 - 7.3 = 38.6$ MJ/m²h **Ans.**

Example 34.4. *Heat is conducted through a compound plate composed of two parallel plates of different materials A and B of conductivities 134 W/mK and 60 W/mK and each of thickness 36 and 42 mm respectively. If the temperature of the outer face of the slab A and that of B are found to be steady at 96° C and 8° C respectively, find the temperature of the interface A/B.*

Solution. Given : k_1 = 134 W/mK ; k_2 = 60 W/mK ; x_1 = 36 mm = 0.036 m ; x_2 = 42 mm = 0.042 m ; T_1 = 96° C = 369 K ; T_3 = 8° C = 281 K

Let T_2 = Temperature of the interface *A*/*B*, and

A = Area of plates *A* and *B*.

We know that the rate of heat flow through the compound plate,

$$Q = \frac{k\,A\,(T_1 - T_2)}{x}$$

$$\therefore \qquad Q_A = \frac{134\,A\,(369 - T_2)}{0.036} \qquad \ldots (i)$$

Similarly,
$$Q_B = \frac{60\,A\,(T_2 - 281)}{0.042} \qquad \ldots (ii)$$

We know that under steady conditions, the rate of heat flow through *A* and *B* is same, therefore equating (*i*) and (*ii*),

$$\frac{134\,A\,(369 - T_2)}{0.036} = \frac{60\,A\,(T_2 - 281)}{0.042}$$

$$3722\,(369 - T_2) = 1429\,(T_2 - 281)$$

$$2.6\,(369 - T_2) = T_2 - 281$$

$$\therefore \qquad T_2 = 344.5 \text{ K} = 71.5^\circ \text{ C} \quad \textbf{Ans.}$$

Example 34.5. *The walls of a room consist of parallel layers in contact of cement, brick and wood of thickness 20 mm, 300 mm and 10 mm respectively. Find the quantity of heat that passes through each m² of wall per minute, if the temperature of air in contact with the wall is 5° C and 30° C inside. The values of k for cement, brick and wood are 0.294, 0.252 and 0.168 W/mK respectively.*

Solution. Given : $x_1 = 20$ mm = 0.02 m ; $x_2 = 300$ mm = 0.3 m ; $x_3 = 10$ mm = 0.01 m ; $T_1 = 30^\circ$ C = 303 K ; $T_2 = 5^\circ$ C = 278 K ; $k_1 = 0.294$ W/mK ; $k_2 = 0.252$ W/m K ; $k_3 = 0.168$ W/m K ; $A = 1\text{ m}^2$; $t = 1$ min = 60 s

We know that heat passing through the wall,

$$Q = \frac{A\,(T_1 - T_2)\,t}{\dfrac{x_1}{k_1} + \dfrac{x_2}{k_2} + \dfrac{x_3}{k_3}} = \frac{1\,(303 - 278)\,60}{\dfrac{0.02}{0.294} + \dfrac{0.3}{0.252} + \dfrac{0.01}{0.168}}$$

$$= \frac{1500}{0.068 + 1.19 + 0.059} = 1139 \text{ J} \quad \textbf{Ans.}$$

34.9. Radial Heat Transfer by Conduction through a Thick Cylinder

The heat transfer through boiler tubes or refrigerator pipings are the examples of conduction of heat transferred radially through the walls of hollow thick cylindrical pipes.

Consider a thick pipe of length *l* carrying steam or a hot liquid at a higher temperature as shown in Fig 34.4.

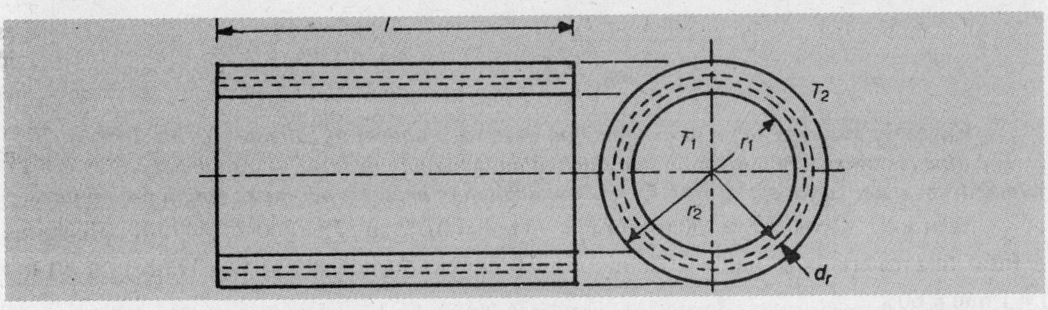

Fig. 34.4. Heat transfer through a thick cylinder.

Let T_1 = Inside (higher) temperaturcof liquid,

T_2 = Outside (lower) temperatureof the surroundings,

$$r_1 = \text{Inside radius of the pipe, and}$$

$$r_2 = \text{Outside radius of the pipe.}$$

$$\therefore \quad (r_2 - r_1) = \text{Thickness of the pipe.}$$

This thick pipe may be imagined to consist of a large number of thin concentric cylinders of increasing radii. Now consider any thin imaginary cylinder of infinitesimal thickness (dr) at a distance (r) from the axis of the pipe as shown in Fig. 34.4. Let the temperature drop across the thickness be dT.

We know that the surface area of this imaginary cylinder,

$$A = 2\pi rl$$

and heat conduction through this elementary cylinder,

$$Q = kA\left(\frac{-dT^*}{dr}\right) = -k \times 2\pi rl\left(\frac{dT}{dr}\right)$$

or

$$\frac{dr}{r} = \left(\frac{-2\pi lk}{Q}\right)dT$$

Integrating the above expression,

$$\int_{r_1}^{r_2} \frac{dr}{r} = \frac{-2\pi lk}{Q}\int_{T_1}^{T_2} dT$$

$$\left[\log_e r\right]_{r_1}^{r_2} = \frac{-2\pi lk}{Q}\left[T\right]_{T_1}^{T_2}$$

$$\therefore \quad \log_e\left(\frac{r_2}{r_1}\right) = \frac{-2\pi lk}{Q}(T_2 - T_1) = \frac{2\pi lk}{Q}(T_1 - T_2)$$

or

$$Q = \frac{2\pi lk\,(T_1 - T_2)}{\log_e\left(\frac{r_2}{r_1}\right)} = \frac{2\pi lk\,(T_1 - T_2)}{2.3\log\left(\frac{r_2}{r_1}\right)}$$

Notes : 1. The above relation also holds good when the outside temperature is higher than the inside temperature and the heat flows from outside to inside of the pipe.

2. In case of a composite cylinder, the heat transfer,

$$Q = \frac{2\pi l\,(T_1 - T_2)}{\sum \frac{2.3}{k}\log\left(\frac{r_2}{r_1}\right)}$$

Example 34.6. *A metal pipe having an external diameter of 150 mm carries steam at 200° C. The pipe is covered by a layer 25 mm thick of an insulating material whose conductivity is 0.21 W/mK. If the outer surface is at 100° C, find the amount of heat lost per metre length per minute.*

Solution. Given : $d_1 = 150$ mm or $r_1 = 75$ mm $= 0.075$ m ; $T_1 = 200°$ C $= 473$ K ; Thickness of insulating material $= 25$ mm $= 0.025$ m ; $k = 0.21$ W/mK ; $T_2 = 100°$ C $= 373$ K ; $l = 1$ m ; $t = 1$ min $= 60$ s

* The negative sign indicates that as the distance (r) increases from the centre outwards, the temperature decreases.

We know that amount of heat lost per minute,

$$Q = \frac{2\pi lk\,(T_1 - T_2)\,t}{2.3 \log\left(\dfrac{r_2}{r_1}\right)} = \frac{2\pi \times 1 \times 0.21\,(473 - 373)\,60}{2.3 \log\left(\dfrac{0.075 + 0.025}{0.075}\right)} \text{ J}$$

$$= 27\,600 \text{ J} = 27.6 \text{ kJ } \textbf{Ans.} \qquad \ldots (\because r_2 = r_1 + 0.025)$$

Example 34.7. *Water is pumped through an iron pipe (k = 67.2 W/mK), 2 metres long at the rate of 1000 kg/min. The inner and outer diameters of the tube are 50 mm and 60 mm respectively. Calculate the rise in temperature of water when the outside of the tube is heated to a temperature of 600° C. The initial temperature of the water is 30° C.*

Solution. Given : $k = 67.2$ W/mK ; $l = 2$ m ; $m = 1000$ kg/min ; $d_1 = 50$ mm or $r_1 = 25$ mm $= 0.025$ m ; $d_2 = 60$ mm or $r_2 = 30$ mm $= 0.03$ m ; $T_1 = 600°$ C $= 873$ K ; $T_{w1} = 30°$ C $= 303$ K

Let T_{w2} = Final temperature of water in K.

We know that heat transferred through the tube per second,

$$Q = \text{Mass} \times \text{Sp. heat} \times \text{Rise in temp.}$$

$$= \frac{1000 \times 4.2\,(T_{w2} - 303)}{60} = 70\,(T_{w2} - 303) \text{ kJ} \qquad \ldots (i)$$

We also know that

$$Q = \frac{2\pi lk\,(T_1 - T_2)\,t}{2.3 \log\left(\dfrac{r_2}{r_1}\right)} = \frac{2\pi \times 2 \times 67.2\left[873 - \dfrac{(303 + T_{w2})}{2}\right]}{2.3 \log\left(\dfrac{0.03}{0.025}\right)} \text{ J}$$

$$\ldots (\because t = 1 \text{ s})$$

$$= 4640\left[873 - \frac{(303 + T_{w2})}{2}\right] = 2320\,(1443 - T_{w2}) \text{ J}$$

$$= 2.32\,(1443 - T_{w2}) \text{ kJ} \qquad \ldots (ii)$$

Equating equations (*i*) and (*ii*),

$$70\,(T_{w2} - 303) = 2.32\,(1443 - T_{w2})$$

or

$$T_{w2} = 339.6 \text{ K} = 66.6° \text{C}$$

$\therefore$ Rise in temperature $= T_{w2} - T_{w1} = 66.6 - 30 = 36.6°$ C **Ans.**

Example 34.8. *A steam pipe 20 m long, 100 mm internal diameter and 40 mm thick is covered by a layer of lagging of 25 mm thick. The coefficient of thermal conductivities for the pipe material and lagging are 0.07 W/m K and 0.1 W/mK respectively. If the steam is conveyed at a pressure of 17 bar with 30° C superheat and the outside temperature of the lagging is 24° C, determine : 1. the heat lost per hour ; and 2. the interface temperature. Neglect the pressure drop across the steam pipe.*

Solution. Given : $l = 20$ m ; $d_1 = 100$ mm or $r_1 = 50$ mm $= 0.05$ m ; $r_2 = r_1$ + Pipe thickness $= 50 + 40 = 90$ mm $= 0.09$ m ; $r_3 = r_2$ + Lagging thickness $= 90 + 25 = 115$ mm $= 0.115$ m ; $k_1 = 0.07$ W/mK ; $k_2 = 0.1$ W/mK ; $p = 17$ bar ; Degree of superheat $= 30°$ C ; $T_3 = 24°$ C $= 297$ K

1. *Total heat lost per hour*

From steam tables, we find that the temperature of steam corresponding to pressure of 17 bar is 204.3° C. Therefore, temperature of steam,

$$T_1 = 204.3° + 30° = 234.3° \text{C} = 507.3 \text{ K}$$

We know that total heat lost per second,

$$Q = \frac{2\pi l\,(T_1 - T_3)}{\frac{1}{k_1} \times 2.3 \log\left(\frac{r_2}{r_1}\right) + \frac{1}{k_2} \times 2.3 \log\left(\frac{r_3}{r_2}\right)}$$

$$= \frac{2\pi \times 20\,(507.3 - 297)}{\left[\frac{1}{0.07} \times 2.3 \log\left(\frac{0.09}{0.05}\right)\right] + \left[\frac{1}{0.1} \times 2.3 \log\left(\frac{0.115}{0.09}\right)\right]}$$

$$= \frac{2\pi \times 20 \times 210.3}{8.39 + 2.45} = 2438 \text{ J/s}$$

$\therefore$ Heat lost per hour $= 2438 \times 3600 = 8\,776\,800$ J/h $= 8776.8$ kJ/h **Ans.**

2. *Interface temperature*

Let T_2 = Interface temperature in K.

Now we shall consider, the flow of heat (*i.e.* 2438 J/s) through the steam pipe*. We know that

$$Q = \frac{2\pi\, lk\,(T_1 - T_2)}{2.3 \log\left(\frac{r_2}{r_1}\right)}$$

$$2438 = \frac{2\pi \times 20 \times 0.07\,(T_1 - T_2)}{2.3 \log\left(\frac{0.09}{0.05}\right)} = \frac{8.8\,(507.3 - T_2)}{0.587}$$

$$1431 = 4464 - 8.8\,T_2 \quad \text{or} \quad T_2 = 344.6 \text{ K} = 71.6^\circ \text{ C} \;\textbf{Ans.}$$

34.10. Heat Transfer by Conduction through a Thick Sphere

Consider a hollow thick spherical shell containing liquid at a higher temperature as shown in Fig. 34.5.

Let T_1 = Inside (higher) temperature of the liquid,

T_2 = Outside (lower) temperature of the surroundings,

r_1 = Inside radius of the sphere, and

r_2 = Outside radius of the sphere.

Fig. 34.5. Heat transfer through a thick sphere.

This thick sphere may be imagined to consist of a large number of thin concentric spheres of increasing radii. Now consider any thin imaginary sphere of infinitesimal thickness (dr) at a distance (r) from the centre of the sphere. Let the temperature across the thickness be dT.

---

* It may also be found by considering the flow of heat through the pipe lagging as discussed below :

$$Q = \frac{2\pi lk\,(T_2 - T_3)}{2.3 \log\left(\frac{r_3}{r_2}\right)}$$

$$2438 = \frac{2\pi \times 20 \times 0.1\,(T_2 - 297)}{2.3 \log\left(\frac{0.115}{0.09}\right)} = \frac{12.57\,(T_2 - 297)}{0.245}$$

$$597.3 = 12.57\,T_2 - 3733 \quad \text{or} \quad T_2 = 344.6 \text{ K} = 71.6^\circ \text{C} \;\textbf{Ans.}$$

We know that surface area of the imaginary sphere,

$$A = 4\pi r^2$$

and heat conduction through the imaginary sphere,

$$* \qquad Q = kA\left(\frac{-dT}{dr}\right)^* = -k \times 4\pi^2\left(\frac{dT}{dr}\right)$$

$$\frac{dr}{r^2} = \left(\frac{-4\pi k}{Q}\right)dT$$

Integrating the above expression,

$$\int_{r_1}^{r_2} \frac{dr}{r^2} = \frac{-4\pi k}{Q}\int_{T_1}^{T_2} dT$$

$$\left[\frac{r^{-1}}{-1}\right]_{r_1}^{r_2} = \frac{-4\pi k}{Q}[T]_{T_1}^{T_2}$$

$$\frac{1}{r_1} - \frac{1}{r_2} = \frac{-4\pi k}{Q}(T_2 - T_1)$$

$$\frac{r_2 - r_1}{r_1 r_2} = \frac{4\pi k}{Q}(T_1 - T_2)$$

$$\therefore \qquad Q = \frac{4\pi k r_1 r_2 (T_1 - T_2)}{(r_2 - r_1)}$$

Notes : 1. The above relation also holds good when the outside temperature is higher than the inside temperature, and the heat flows from outside to inside of the sphere.

2. In case of a composite sphere, the heat transfer,

$$Q = \frac{4\pi (T_1 - T_2)}{\sum \frac{r_2 - r_1}{k_1 r_1 r_2}}$$

Example. 34.9. *A spherical shaped vessel of 1 m outside diameter is 100 mm thick. Find the rate of heat leakage, if the temperature difference between the inner and outer surface is 150 K. Take k for the vessel material as 0.2 kJ/mh K.*

Solution. Given : $d_2 = 1$ m or $r_2 = 0.5$ m ; $r_1 = r_2 -$ Thickness $= 0.5 - 0.1 = 0.4$ m ; $T_1 - T_2 = 150$ K ; $k = 0.2$ kJ/mh K

We know that rate of heat leakage,

$$Q = \frac{4\pi k r_1 r_2 (T_1 - T_2)}{(r_2 - r_1)} = \frac{4\pi \times 0.2 \times 0.4 \times 0.5 \times 150}{(0.5 - 0.4)} = 754 \text{ kJ/h } \textbf{Ans.}$$

Example 34.10. *An air-conditioned hall has one of its walls 0.7 m thick. The inside space is required to be maintained at 20° C. The thermal conductivity of the wall is variable, and is given by the relation :*

$$k = 0.93 + 1.163 \times 10^{-4} T^2$$

* The negative sign indicates that as the distance (r) increases from the centre outwards, the temperature decreases.

where T is in degrees centigrade and k in W/m °C. Find the heat lost per hour through one of its walls 10 m × 5 m when the outside temperature is 40° C.

Solution. Given : $x = 0.7$ m ; $T_2 = 20°$ C ; $k = 0.93 + 1.163 \times 10^{-4} T^2$; $A = 10 \times 5 = 50$ m^2 ; $T_1 = 40°$ C

We know that heat lost,

$$Q = -kA \times \frac{dT}{dx}$$

$$\therefore \quad Q\,dx = -kA\,dT = -(0.93 + 1.163 \times 10^{-4} T^2)\,50 \times dT$$

Integrating this expression for the wall thickness,

$$Q \int_0^{0.7} dx = -50 \int_{T_1}^{T_2} (0.93 + 1.163 \times 10^{-4} T^2)\,dT$$

$$Q \times 0.7 = -50 \left[0.93\,T + \frac{1.163 \times 10^{-4} \times T^3}{3} \right]_{T_1}^{T_2}$$

$$= -50 \left[0.93\,(T_2 - T_1) + \frac{1.163 \times 10^{-4}}{3} (T_2^3 - T_1^3) \right]$$

$$= -50 \left[0.93\,(20 - 40) + \frac{1.163 \times 10^{-4}}{3} (20^3 - 40^3) \right]$$

$$= -50\,[-18.6 - 2.17] = 1038.5$$

$$\therefore \quad Q = 1038.5 / 0.7 = 1483.6 \text{ J/s } \textbf{Ans.}$$

34.11. Overall Coefficient of Heat Transfer

In the previous articles, we have discussed only the theoretical cases of heat transfer through conduction. But in actual practice, the heat from a hot body is transferred to the cold body by the combined effect of conduction and convection.

Consider a wall through which heat is transferred from a hot surface to a cold surface as shown in Fig. 34.6.

Let $(T_1 - T_2)$ = Difference of temperatures,

A = Surface area of the wall,

x = Thickness of the wall, and

k = Thermal conductivity of the wall material.

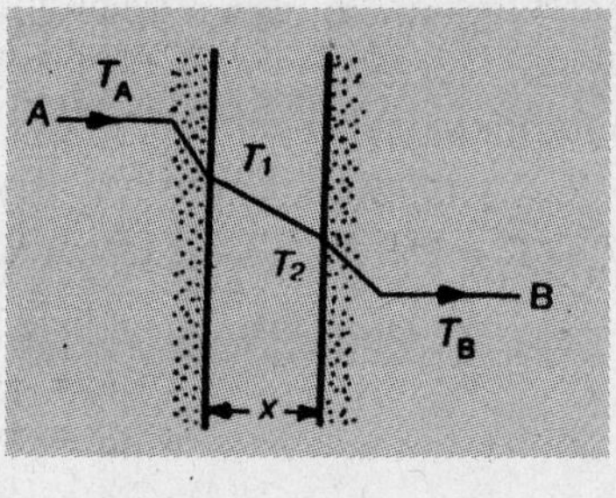

Fig. 34.6. Overall coefficient of heat transfer.

As a matter of fact, there will be a thin film of air on both the hot as well as cold faces of the wall, which will act as transition layers adjacent to the wall surface, and through which the heat also has to flow in addition to the wall as shown in Fig. 34.6. Let A and B be the effective film of air for the heat flow.

Let T_A and T_B = Temperatures at ends of two thin films of air A and B respectively.

h_A and h_B = Coefficients of heat transfer for A and B respectively.

U = Overall coefficient of heat transfer.

We know that the rate of heat flow through air film A,

$$Q = h_A A (T_A - T_1) \text{ or } (T_A - T_1) = \frac{Q}{h_A A} \quad \ldots (i)$$

Similarly, rate of heat flow through the wall,

$$Q = \frac{kA (T_1 - T_2)}{x} \text{ or } (T_1 - T_2) = \frac{Q\,x}{k\,A} \quad \ldots (ii)$$

and rate of heat flow through the film B,

$$Q = h_B A (T_2 - T_B) \text{ or } (T_2 - T_B) = \frac{Q}{h_B A} \quad \ldots (iii)$$

Adding equations (*i*), (*ii*) and (*iii*), we get

$$(T_A - T_B) = \frac{Q}{A}\left[\frac{1}{h_A} + \frac{x}{k} + \frac{1}{h_B}\right]$$

$$\therefore \quad Q = \frac{A (T_A - T_B)}{\left(\frac{1}{h_A} + \frac{x}{k} + \frac{1}{h_B}\right)} \quad \ldots (iv)$$

We know that the rate of heat flow,

$$Q = UA (T_A - T_B) \quad \ldots (v)$$

Now equating the equations (*iv*) and (*v*),

$$U A (T_A - T_B) = \frac{A (T_A - T_B)}{\left(\frac{1}{h_A} + \frac{x}{k} + \frac{1}{h_B}\right)}$$

$$\therefore \quad U = \frac{1}{\left(\frac{1}{h_A} + \frac{x}{k} + \frac{1}{h_B}\right)}$$

Notes : 1. For the sake of simplicity, we have considered a simple wall of one material only. But this relation may be extended to all the cases discussed earlier, *e.g.* for a composite wall,

$$U = \frac{1}{\left(\frac{1}{h_A} + \frac{x_1}{k_1} + \frac{x_2}{k_2} + \ldots. + \frac{1}{h_B}\right)}$$

Similarly, for a simple cylinder,

$$U = \frac{1}{\left\{\frac{1}{r_1 h_A} + 2.3 \log\left(\frac{r_2}{r_1}\right) + \frac{1}{r_2 h_B}\right\}}$$

2. If ambient temperature (*i.e.* atmospheric temperature) is existing on one side of the wall or cylinder, then thickness of air film is zero on that side.

Example 34.11. *A furnace wall 200 mm thick is made of a material having thermal conductivity of 1.45 W/m² K. The inner and outer surfaces are exposed to average temperatures of 350° C and 40° C respectively. If the gas and air film coefficients are 58 and 11.63 W/m² K respectively, find the rate of heat transfer through a wall of 2.5 square metre. Also, find the temperatures on the two sides of the wall.*

Solution. Given : x = 200 mm = 0.2 m ; k = 1.45 W/m² K ; T_A = 350° C = 623 K ; T_B = 4° C = 313 K ; h_A = 58 W/m² K ; h_B = 11.63 W/m² K ; A = 2.5 m²

Rate of heat transfer

We know that rate of heat transfer,

$$Q = \frac{A(T_A - T_B)}{\frac{1}{h_A}+\frac{x}{k}+\frac{1}{h_B}} = \frac{2.5(623-313)}{\frac{1}{58}+\frac{0.2}{1.45}+\frac{1}{11.63}} = \frac{775}{0.017+0.138+0.086} \text{ J/s}$$

$$= 3216 \text{ J/s } \textbf{Ans.}$$

Temperature on two sides of the wall

Let T_1 = Temperature on the inner side of the wall, and

T_2 = Temperature on the outside of the wall.

We know that the rate of heat transfer through the gas film, wall and air film is the same. So we shall first consider the flow of heat through the gas film and then through the air film.

We know that $Q = h_A A (T_A - T_1)$

$$\therefore \quad 3216 = 58 \times 2.5 (623 - T_1) = 145 (623 - T_1)$$

$$623 - T_1 = 3216/145 = 22.2 \quad \text{or} \quad T_1 = 600.8 \text{ K} = 327.8° \text{ C } \textbf{Ans.}$$

Similarly $Q = h_B A (T_2 - T_B)$

$$3216 = 11.63 \times 2.5 (T_2 - 313) = 29 (T_2 - 313)$$

$$\therefore \quad T_2 = \frac{3216}{29} + 313 = 423.9 \text{ K} = 150.9° \text{ C } \textbf{Ans.}$$

Example 34.12. *A composite wall is made up of brickwork, fiberglass and insulating board with thickness and thermal conductivities as given below :*

Brick work	*110 mm thick*	*1.15 W/mK*
Fiberglass	*75 mm thick*	*0.04 W/mK*
Board	*25 mm thick*	*0.06 W/mK*

Find the overall coefficient of heat transfer, if the coefficient of heat transfer for the outside and inside walls are 3.1 W/m² K and 2.5 W/m² K respectively. Also find the heat lost per hour through such a wall of 20 square metre, when the temperature difference is 27 K.

Solution. Given : For brick work : x_1 = 110 mm = 0.11 m ; k_1 = 1.15 W/mK ; For fibre glass : x_2 = 75 mm = 0.075 m ; k_2 = 0.04 W/m K ; For board : x_3 = 25 mm = 0.025 m ; k_3 = 0.06 W/mK ; h_A = 3.1 W/m² K ; h_B = 2.5 W/m² K ; A = 20 m² ; $T_A - T_B$ = 27 K

Overall coefficient of heat transfer

We know that overall coefficient of heat transfer,

$$U = \frac{1}{\left(\frac{1}{h_A}+\frac{x_1}{k_1}+\frac{x_2}{k_2}+\frac{x_3}{k_3}+\frac{1}{h_B}\right)}$$

$$= \frac{1}{\left(\frac{1}{3.1}+\frac{0.11}{1.15}+\frac{0.075}{0.04}+\frac{0.025}{0.06}+\frac{1}{2.5}\right)} = 0.322 \text{ W/m}^2 \text{ K } \textbf{Ans.}$$

Heat loss per hour through the wall

We know that heat loss per hour through the wall,

$$Q = UA\,(T_A - T_B)$$

$$= 0.322 \times 20 \times 27 = 173.9 \text{ J/s} \textbf{ Ans.}$$

Example 34.13. *A 200 mm diameter steel pipe, conveying saturated steam at a pressure of 9.8 bar, is covered by a layer of lagging material of thickness 60 mm and thermal conductivity 0.116 W/mK. If the heat transfer coefficient of steam is 9.3 W/m² K and ambient temperature is 25° C, find the loss of heat per metre length of the pipe.*

Solution. Given : d_1 = 200 mm or r_1 = 100 mm = 0.1 m ; p = 9.8 bar ; $r_2 = r_1$ + Lagging thickness = 100 + 60 = 160 mm = 0.16 m ; k = 0.116 W/m K ; h = 9.3 W/m² K ; T_B = 25° C = 298 K

From steam tables, we find that the temperature of steam at 9.8 bar,

$$t_A = 179° \text{ C} = 452 \text{ K}$$

We know that loss of heat per metre length of the pipe,

$$Q = \frac{2\pi l\,(T_A - T_B)}{\left[\frac{2.3}{k}\log\left(\frac{r_2}{r_1}\right) + \frac{1}{r_2 h}\right]}$$

$$= \frac{2\pi \times 1\,(452 - 298)}{\left[\frac{2.3}{0.116}\log\left(\frac{0.16}{0.10}\right) + \frac{1}{0.16 \times 9.3}\right]} \text{ J/s}$$

$$= \frac{967.74}{4.05 + 0.672} = 205 \text{ J/s} \textbf{ Ans.}$$

Note : In this example, thickness and thermal conductivity of pipe as well as heat transfer coefficient of steam is not given. So the example has to be solved ignoring these values.

EXERCISES

1. The temperatures inside and outside of a room are 5° C and 18° C respectively. Calculate the rate at which heat is being transferred through a glass window of area 1 m² and thickness 5 mm. Take k for glass as 0.84 W/mK. **[Ans. 2.18 kJ/s]**

2. The glass windows of a room have total area of 10 square metres, while thickness is 4 mm. Find the rate at which the heat escapes from the room per second by conduction when the temperature difference is 20° C. Take k for glass as 3 W/mK. **[Ans. 41.67 kJ/s]**

3. An iron boiler 12.5 mm thick contains water at atmospheric pressure. The heated surface is 2 m² in area and the temperature of the inner side is 120° C. If the thermal conductivity of iron is 84 W/mK and the latent heat of evaporation of water is 22.5 J, find the mass of water evaporated. **[Ans. 370 kg]**

4. An aluminium plate (k = 193.2) and a gold plate (k = 294) are joined with faces in contact. The outer face of the aluminium plate is kept at a temperature of 100° C and the gold face is at 60° C. Compare the thickness required, if the temperature drop through aluminium plate is three times that of gold plate. The surface areas of the two plates is same. **[Ans. 197 : 1]**

5. A tube has internal radius 20 mm and external radius 25 mm. The inside of the tube is maintained at 100° C and the outside at 20° C. Calculate the quantity of heat conducted through unit length of the tube per second. Take the value of k as 380 W/mK **[Ans. 8.6 kJ]**

6. A spherical storage vessel of 3 m internal diameter and 150 mm thick has a liquid at a temperature of 120° C. If the thermal conductivity of the vessel material is 10.5 W/m² K, find the heat transfer rate. **[Ans. 2395 J/h]**

7. A furnace wall is built up with 200 mm thick refractory bricks and 150 mm insulating bricks. The temperature of the surrounding is 40° C, whereas that inside the furnace is 1000° C. The thermal conductivities of the refractory bricks and insulating bricks are 5 W/m K and 0.5 W/m K respectively. If the coefficients of heat transfer for the furnace gas and air is 80 and 40 W/m^2 K, determine the rate of heat flow per square metre.

[Ans. 2543 kJ/h]

QUESTIONS

1. What are the three methods of heat transfer ?
2. How do you define the thermal conductivity of a material ?
3. Deduce an expression for the quantity of heat flow through solid bodies when joined together.
4. Derive an expression for the quantity of heat flow through a thick cylindrical pipe.
5. Define the overall coefficient of heat transfer.

OBJECTIVE TYPE QUESTIONS

1. The process of heat transfer from one particle of the body to another is called conduction, when the particles of the body

   (*a*) move actually (*b*) do not move actually
   (*c*) affect the intervening medium (*d*) does not affect the intervening medium

2. The heat transfer takes place according to

   (*a*) zeroth law of thermodynamics (*b*) first law of thermodynamics
   (*c*) second law of thermodynamics (*d*) Stefan's law

3. The rate of heat flow through a body is $Q = \dfrac{k A (T_1 - T_2)}{x}$. The term *x*/*k A* is known as

   (*a*) thermal coefficient (*b*) thermal resistance
   (*c*) thermal conductivity (*d*) none of these

4. The thermal conductivity of sold metals ..... with rise in temperature.

   (*a*) remains same (*b*) decreases (*c*) increases

5. The overall coefficient of heat transfer is used in problems of

   (*a*) conduction (*b*) convection
   (*c*) radiation (*d*) conduction and convection
   (*e*) conduction and radiation.

ANSWERS

1. (*b*)	2. (*c*)	3. (*b*)	4. (*b*)	5. (*d*)

35

Air Refrigeration Cycles

1. Introduction. 2. Air Refrigeration Cycle. 3. Units of Refrigeration. 4. Coefficient of Performance of a Refrigerator. 5. Difference between a Heat Engine, Refrigerator and Heat Pump. 6. Open Air Refrigeration Cycle. 7. Closed or Dense Air Refrigeration Cycle. 8. Air Refrigerator Working on Reversed Carnot Cycle. 9. Temperature Limitations for Reversed Carnot Cycle. 10. Air Refrigerator Working on a Bell-Coleman Cycle (or Reversed Joule or Brayton Cycle).

35.1. Introduction

The term 'refrigeration' in a broad sense is used for the process of removing heat (*i.e.* cooling) from a substance. It also includes the process of reducing and maintaining the temperature of a body below the general temperature of its surroundings. In other words, the refrigeration means a continued extraction of heat from a body, whose temperature is already below the temperature of its surroundings.

For example, if some space (say in cold storage) is to be kept at $-2°$ C, we must continuously extract heat which flows into it due to leakage through the walls and also the heat, which is brought into it with the articles stored after the temperature is once reduced to $-2°$ C. Thus in a refrigerator, heat is virtually being pumped from a lower temperature to a higher temperature. According to second* law of thermodynamics, this process can only be performed with the aid of some external work. It is thus obvious, that supply of power (say electric motor) is regularly required to drive a refrigerator. Theoretically, the refrigerator is a reversed heat engine, or a heat pump which pumps heat from a cold body and delivers it to a hot body. The substance which works in a heat pump to extract heat from a cold body and to deliver it to a hot body is called *refrigerant.*

The refrigeration system is known to the man, since the middle of nineteenth century. The scientists, of the time, developed a few stray machines to achieve some pleasure. But it paved the way by inviting the attention of scientists for proper studies and research. They were able to build a reasonably reliable machine by the end of nineteenth century for the refrigeration jobs. But with the advent of efficient rotary compressors and gas turbines, the science of refrigeration reached its present height. Today, it is used for the manufacture of ice and other similar products. It is also widely used for the cooling of storage chambers in which perishable food, drinks and medicines are stored. The refrigeration also has wide applications in sub-marine ships, rockets and aircrafts.

35.2. Air Refrigeration Cycle

In an air refrigeration cycle, the air is used as a refrigerant. In olden days, air was widely used in commercial applications because of its availability at free of cost. Since air does not change its phase *i.e.* remains gaseous throughout the cycle, therefore its heat carrying capacity per kg of air is

* Second law of thermodynamics states, "*It is impossible for a self-acting machine, working in a cyclic process, to transfer heat from a body at a lower temperature to a body at a higher temperature, without the aid of an external agency.*" For more details, please refer Chapter 1.

very small as compared to vapour absorbing systems. The air-cycle refrigeration systems, as originally designed and installed, are now practically obsolete because of their low coefficient of performance and high power requirements. However, this system continues to be favoured for aircraft refrigeration because of the low weight and volume of the equipment. The basic elements of an air cycle refrigeration system are the compressor, the cooler or heat exchanger, the expander and the refrigerator.

Before discussing the air refrigeration cycles, we should first know about the unit of refrigeration, coefficient of performance of a refrigerator and the difference between the heat engine, a refrigerator and a heat pump.

35.3. Units of Refrigeration

The practical unit of refrigeration is expressed in terms of 'tonne of refrigeration' (briefly written as TR). A tonne of refrigeration is defined as the amount of refrigeration effect produced by the uniform melting of one tonne (1000 kg) of ice from and at 0° C in 24 hours. Since the latent heat of ice is 335 kJ/kg, therefore one tonne of refrigeration,

$$1 \text{ TR} = 1000 \times 335 \text{ kJ in 24 hours}$$

$$= \frac{1000 \times 335}{24 \times 60} = 232.6 \text{ kJ/min}$$

In actual practice, one tonne of refrigeration is taken as equivalent to 210 kJ/min or 3.5 kW (*i.e.* 3.5 kJ/s).

35.4. Coefficient of Performance of a Refrigerator

The coefficient of performance (briefly written as C.O.P.) is the ratio of heat extracted in the refrigerator to the work done on the refrigerant. It is also known as theoretical coefficient of performance. Mathematically,

$$\text{Theoretical C.O.P.} = \frac{Q}{W}$$

where Q = Amount of heat extracted in the refrigerator (or the amount of refrigeration effect produced, or the capacity of a refrigerator), and

W = Amount of work done.

Notes : 1. The coefficient of performance is the reciprocal of the efficiency (*i.e.* $1/\eta$) of a heat engine. It is thus obvious, that the value of C.O.P. is always greater than unity.

2. The ratio of the actual C.O.P. to the theoretical C.O.P. is known as relative coefficient of performance. Mathematically,

$$\text{Relative C.O.P.} = \frac{\text{Actual C.O.P.}}{\text{Theoretical C.O.P.}}$$

Example 35.1. *An ice plant produces 10 tonnes of ice per day at 0° C using water at room temperature of 20° C. Estimate the power rating of the compressor-motor, if the C.O.P. of the plant is 2.5 and overall electro-mechanical efficiency is 90%.*

Solution. Given : m = 10 t/day = 10 × 1000 / 24 × 60 = 6.94 kg/min ; T_1 = 0° C = 273 K ; T_2 = 20° C = 293 K ; C.O.P. = 2.5 ; η_0 = 90 % = 0.9

Let W = Work required to drive the compressor / min.

We know that heat extracted from 1 kg of water at 20° C to produce 1 kg of ice at 0° C

$$= 1 \times 4.187\,(20 - 0) + 335 = 418.74 \text{ kJ/kg}$$

... ($\because$ Latent heat of ice = 335 kJ/kg)

$\therefore$ Total heat extracted,

$$Q = 418.74 \times 6.94 = 2906 \text{ kJ/min}$$

We know that C.O.P. of the plant,

$$2.5 = \frac{Q}{W} = \frac{2906}{W}$$

$$\therefore \quad W = 2906 / 2.5 = 1162.4 \text{ kJ/min}$$

and power, $$P = \frac{1162.4}{60 \times \eta} = \frac{1162.4}{60 \times 0.9} = 21.5 \text{ kW} \quad \textbf{Ans.}$$

Example 35.2. *Five hundred kg of fruits are supplied to a cold storage at 20° C. The cold storage is maintained at –5° C and the fruits get cooled to the storage temperature in 10 hours. The latent heat of freezing is 105 kJ/kg and specific heat of fruit is 1.26. Find the refrigeration capacity of the plant.*

Solution. Given : $m = 500$ kg ; $T_2 = 20°$ C = 293 K ; $T_1 = -5°$ C = 268 K ; $h_{fg} = 105$ kJ/kg ; $c_F = 1.26$

We know that heat removed from the fruit in 10 hrs,

$$Q_1 = m\, c_F (T_2 - T_1) = 500 \times 1.26\,(293 - 268) = 15\,750 \text{ kJ}$$

and total latent heat of freezing,

$$Q_2 = m\, h_{fg} = 500 \times 105 = 52\,500 \text{ kJ}$$

∴ Total heat removed in 10 hrs,

$$Q = Q_1 + Q_2 = 15\,750 + 52\,500 = 68\,250 \text{ kJ}$$

and total heat removed in one minute,

$$= \frac{68\,250}{10 \times 60} = 113.75 \text{ kJ/min}$$

∴ Refrigeration capacity of the plant

$$= \frac{113.75}{210} = 0.542 \text{ TR} \quad \textbf{Ans.} \qquad \ldots(\because 1 \text{ TR} = 210 \text{ kJ/min})$$

35.5. Difference between a Heat Engine, Refrigerator and Heat Pump

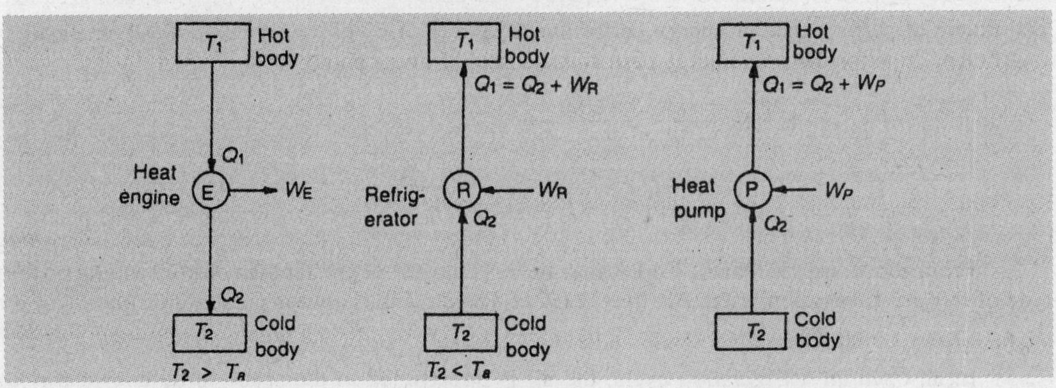

(*a*) Heat-engine. (*b*) Refrigerator. (*c*) Heat pump.

Fig. 35.1. Difference between a heat engine, refrigerator and heat pump.

In a heat engine, as shown in Fig. 35.1 (*a*), the heat supplied to the engine is converted into useful work. If Q_1 is the heat supplied to the engine and Q_2 is the heat rejected from the engine, then the net work done by the engine is given by

$$W_E = Q_1 - Q_2$$

The performance of a heat engine is expressed by its efficiency. We know that the efficiency or coefficient of performance of a heat engine,

$$\eta_E \text{ or } (\text{C.O.P.})_E = \frac{\text{Work done}}{\text{Heat supplied}} = \frac{W_E}{Q_1} = \frac{Q_1 - Q_2}{Q_1}$$

A refrigerator, as shown in Fig. 35.1 (*b*), is a reversed heat engine which either cool or maintain the temperature of a body (T_2) lower than the atmospheric temperature (T_a). This is done by extracting the heat (Q_2) from a cold body and delivering it to a hot body (Q_1). In doing so, work W_R is required to be done on the system. According to first law of thermodynamics,

$$W_R = Q_1 - Q_2$$

The performance of a refrigerator is expressed by the ratio of amount of heat taken from the cold body (Q_2) to the amount of work required to be done on the system (W_R). This ratio is called coefficient of performance. Mathematically, coefficient of performance of a refrigerator,

$$(\text{C.O.P.})_R = \frac{Q_2}{W_R} = \frac{Q_2}{Q_1 - Q_2}$$

Any refrigerating system is a heat pump, as shown in Fig. 35.1 (*c*), which extracts heat (Q_2) from a cold body and delivers it to a hot body. Thus there is no difference between the cycle of operations of a heat pump and a refrigerator. The main difference between the two is in their operating temperatures. A refrigerator works between the cold body temperature (T_2) and the atmospheric temperature (T_a) whereas the heat pump operates between the hot body temperature (T_1) and the atmospheric temperature (T_a). A refrigerator used for cooling in summer can be used as a heat pump for heating in winter.

In the similar way, as discussed for refrigerator, we have

$$W_P = Q_1 - Q_2$$

The performance of a heat pump is expressed by the ratio of the amount of heat delivered to the hot body (Q_1) to the amount of work required to be done on the system (W_P). This ratio is called coefficient of performance or energy performance ratio (E.P.R.) of a heat pump. Mathematically, coefficient of performance or energy performance ratio of a heat pump,

$$(\text{C.O.P.})_P \text{ or E.P.R.} = \frac{Q_1}{W_P} = \frac{Q_1}{Q_1 - Q_2}$$

$$= \frac{Q_2}{Q_1 - Q_2} + 1 = (\text{C.O.P.})_R + 1$$

From above, we see that the C.O.P. may be less than one or greater than one depending on the type of refrigeration system used. But the C.O.P. of a heat pump is always greater than one.

35.6. Open Air Refrigeration Cycle

In an open air refrigeration cycle, the air is directly led to the space to be cooled (*i.e.* a refrigerator), allowed to circulate through the cooler and then returned to the compressor to start another cycle. Since the air is supplied to the refrigerator at atmospheric pressure, therefore, volume of air handled by the compressor and expander is large. Thus the size of compressor and expander should be large. Another disadvantage of the open cycle system is that the moisture is regularly carried away by the air circulated through the cooled space. This leads to the formation of frost at the end of expansion process and clog the line. Thus in an open cycle system, a drier should be used.

35.7. Closed or Dense Air Refrigeration Cycle

In a closed or dense air refrigeration cycle, the air is passed through the pipes and component parts of the system at all times. The air, in this system, is used for absorbing heat from the other fluid (say brine) and this cooled brine is circulated into the space to be cooled. The air in the closed system does not come in contact directly with the space to be cooled.

The closed air refrigeration cycle has the following thermodynamic advantages :

1. Since it can work at a suction pressure higher than that of atmospheric pressure, therefore the volume of air handled by the compressor and expander are smaller as compared to an open air refrigeration cycle system.
2. The operating pressure ratio can be reduced, which results in higher coefficient of performance.

35.8. Air Refrigerator Working on Reversed Carnot Cycle

In refrigerating systems, the Carnot cycle considered is the reversed Carnot cycle. We know that a heat engine working on Carnot cycle has the highest possible efficiency. Similarly, a refrigerating system working on the reversed Carnot cycle, will have the maximum possible coefficient of performance. We also know that it is not possible to make an engine working on the Carnot cycle. Similarly, it is also not possible to make a refrigerating machine working on the reversed Carnot cycle. However, it is used as the ultimate standard of comparison.

A reversed Carnot cycle, using air as working medium (or refrigerant) is shown on *p-v* and *T-s* diagram in Fig. 35.2 (*a*) and (*b*) respectively. At point 1, let p_1, v_1 and T_1 be the pressure, volume and temperature of air respectively.

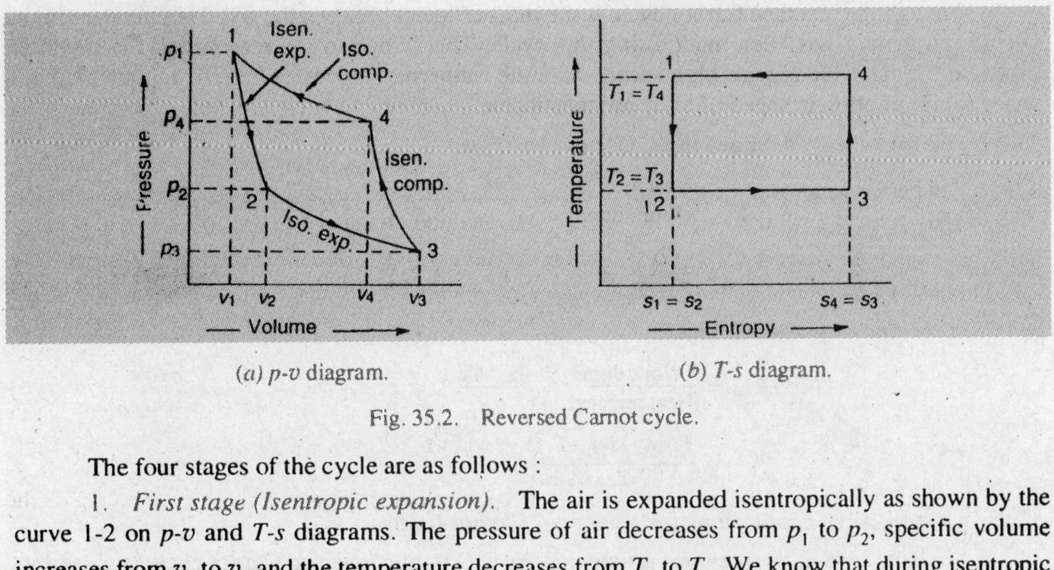

(*a*) *p-v* diagram. (*b*) *T-s* diagram.

Fig. 35.2. Reversed Carnot cycle.

The four stages of the cycle are as follows :

1. *First stage (Isentropic expansion).* The air is expanded isentropically as shown by the curve 1-2 on *p-v* and *T-s* diagrams. The pressure of air decreases from p_1 to p_2, specific volume increases from v_1 to v_2 and the temperature decreases from T_1 to T_2. We know that during isentropic expansion, no heat is absorbed or rejected by the air.

2. *Second stage (Isothermal expansion).* The air is now expanded isothermally (*i.e.* at constant temperature, $T_2 = T_3$) as shown by the curve 2-3 on *p-v* and *T-s* diagrams. The pressure of air decreases from p_2 to p_3 and the specific volume increases from v_2 to v_3. We know that the heat absorbed by the air (or heat extracted from the cold body) during isothermal expansion per kg of air,

$$q_A = q_{2-3} = T_2 (s_3 - s_2) = T_3 (s_3 - s_2)$$

3. *Third stage (Isentropic compression).* The air is compressed isentropically as shown by the curve 3-4 on p-v and T-s diagrams. During this process, the pressure of air increases from p_3 to p_4, specific volume decreases from v_3 to v_4 and temperature increases from T_3 to T_4. We know that during isentropic compression, no heat is absorbed or rejected by the air.

4. *Fourth stage (Isothermal compression).* The air is now compressed isothermally (*i.e.* at constant temperature, $T_4 = T_1$) as shown by the curve 4-1 on p-v and T-s diagrams. During this process, the pressure of air increases from p_4 to p_1 and specific volume decreases from v_4 to v_1. We know that the heat rejected by the air during isothermal compression per kg of air,

$$q_R = q_{4-1} = T_4 (s_4 - s_1) = T_1 (s_3 - s_2)$$

We know that work done during the cycle per kg of air,

$$w_R = {}^*\text{Heat rejected} - \text{Heat absorbed} = q_R - q_A$$

$$= T_1 (s_3 - s_2) - T_3 (s_3 - s_2) = (T_1 - T_3)(s_3 - s_2)$$

∴ Coefficient of performance of the refrigeration system working on reversed Carnot cycle,

$$(\text{C.O.P.})_R = \frac{\text{Heat absorbed}}{\text{Work done}} = \frac{q_A}{q_R - q_A}$$

$$= \frac{T_3 (s_3 - s_2)}{(T_1 - T_3)(s_3 - s_2)} = \frac{T_3}{T_1 - T_3} = \frac{T_2}{T_1 - T_2} \qquad \ldots (\because T_3 = T_2)$$

Though the reversed Carnot cycle is the most efficient between the fixed temperature limits, yet no refrigerator has been made using this cycle. This is due to the reason that the isentropic processes of the cycle require high speed while the isothermal processes require an extremely low speed. This variation in speed of air is not practicable.

Note : We have already discussed that C.O.P. of a heat pump,

$$(\text{C.O.P.})_P = \frac{\text{Heat rejected}}{\text{Work done}} = \frac{q_R}{q_R - q_A} = \frac{T_1 (s_3 - s_2)}{T_1 (s_3 - s_2) - T_2 (s_3 - s_2)}$$

$$= \frac{T_1}{T_1 - T_2} = \frac{T_2}{T_1 - T_2} + 1 = (\text{C.O.P.})_R + 1$$

and C.O.P. or efficiency of a heat engine,

$$(\text{C.O.P.})_E = \frac{\text{Work done}}{\text{Heat rejected}} = \frac{q_R - q_A}{q_R}$$

$$= \frac{T_1 (s_3 - s_2) - T_2 (s_3 - s_2)}{T_1 (s_3 - s_2)} = \frac{T_1 - T_2}{T_1}$$

35.9. Temperature Limitations for Reversed Carnot Cycle

We have seen in the previous article that the C.O.P. of a reversed Carnot cycle is given by

$$\text{C.O.P} = \frac{T_2}{T_1 - T_2}$$

where T_1 = Higher temperature, and

T_2 = Lower temperature.

---

* In a refrigerating machine, heat rejected is more than heat absorbed.

The C.O.P. of the reversed Carnot cycle may be improved by

1. decreasing the higher temperature (*i.e.* temperature of hot body, T_1), or
2. increasing the lower temperature (*i.e.* temperature of cold body, T_2).

This applies to all refrigerating machines, both theoretical and practical. It may be noted that temperatures T_1 and T_2 cannot be varied at will, due to certain functional limitations. It should be kept in mind that the higher temperature T_1 is the temperature of cooling water or air available for rejection of heat and the lower temperature T_2 is the temperature to be maintained in the refrigerator. The heat transfer will take place in the right direction only when the higher temperature is more than the temperature of cooling water or air to which heat is to be rejected, while the lower temperature must be less than the temperature of substance to be cooled.

Thus if the temperature of cooling water or air (*i.e.* T_1) available for heat rejection is low, the C.O.P. of the Carnot refrigerator will be high. Since T_1 in winter is less than T_1 in summer, therefore, C.O.P. in winter will be higher than C.O.P. in summer. In other words, the Carnot refrigerators work more efficiently in winter than in summer. Similarly, if the lower temperature fixed by the refrigeration application is high, the C.O.P. of the Carnot refrigerator will be high. Thus a Carnot refrigerator used for making ice at 0° C (273 K) will have less C.O.P. than a Carnot refrigerator used for air-conditioned plant in summer at 20° C when the atmospheric temperature is 40° C. In other words, we can say that the Carnot C.O.P. of a domestic refrigerator is less than the Carnot C.O.P. of a domestic air-conditioner.

Example 35.3. *A machine working on a Carnot cycle operates between 305 K and 260 K. Determine the C.O.P. when it is operated as 1. a refrigerating machine, 2. a heat pump, and 3. a heat engine.*

Solution. Given : $T_1 = 305$ K ; $T_2 = 260$ K

1. *C.O.P. of a refrigerating machine*

We know that C.O.P. of a refrigerating machine,

$$(\text{C.O.P.})_R = \frac{T_2}{T_1 - T_2} = \frac{260}{305 - 260} = 5.78 \textbf{ Ans.}$$

2. *C.O.P. of a heat pump*

We know that C.O.P. of a *heat pump

$$(\text{C.O.P.})_P = \frac{T_1}{T_1 - T_2} = \frac{305}{305 - 260} = 6.78 \textbf{ Ans.}$$

3. *C.O.P. of a heat engine*

We know that C.O.P. of a heat engine,

$$(\text{C.O.P.})_E = \frac{T_1 - T_2}{T_1} = \frac{305 - 260}{305} = 0.147 \textbf{ Ans.}$$

Example 35.4. *A Carnot refrigeration cycle absorbs heat at – 3° C and rejects it at 27° C.*

1. Calculate the coefficient of performance of this refrigeration cycle.

2. If the cycle is absorbing 1130 kJ/min at – 3° C, how many kJ of work is required per second ?

3. If the Carnot heat pump operates between the same temperatures as the above refrigeration cycle, what is the coefficient of performance ?

4. How many kJ/min will the heat pump deliver at 27° C if it absorbs 1130 kJ/min at – 3° C ?

---

* We know that C.O.P. of a heat pump,

$$(\text{C.O.P.})_P = (\text{C.O.P})_R + 1 = 5.78 + 1 = 6.78 \textbf{ Ans.}$$

Solution. Given : $T_2 = -3° C = 270 K$; $T_1 = 27° C = 300 K$

1. *Coefficient of performance of Carnot refrigeration cycle*

We know that coefficient of performance of Carnot refrigeration cycle,

$$(C.O.P.)_R = \frac{T_2}{T_1 - T_2} = \frac{270}{300 - 270} = 9 \textbf{ Ans.}$$

2. *Work required per second*

Let W_R = Work required per second.

Heat absorbed at $-3°$ C (*i.e.* T_2),

$$Q_2 = 1130 \text{ kJ/min} = 18.83 \text{ kJ/s} \qquad \text{... (Given)}$$

We know that $(C.O.P.)_R = \dfrac{Q_2}{W_R}$

$$\therefore \quad 9 = \frac{18.83}{W_R} \quad \text{or} \quad W_R = 2.1 \text{ kJ/s } \textbf{Ans.}$$

3. *Coefficient of performance of Carnot heat pump*

We know that coefficient of performance of a Carnot heat pump,

$$(C.O.P.)_P = \frac{T_1}{T_1 - T_2} = \frac{300}{300 - 270} = 10 \textbf{ Ans.}$$

4. *Heat delivered by heat pump at 27° C*

Let Q_1 = Heat delivered by heat pump at 27° C.

Heat absorbed at $-3°$ C (*i.e.* T_2),

$$Q_2 = 1130 \text{ kJ/min} \qquad \text{... (Given)}$$

We know that $(C.O.P.)_P = \dfrac{Q_1}{Q_1 - Q_2}$

$$\therefore \quad 10 = \frac{Q_1}{Q_1 - 1130}$$

$$10\,Q_1 - 11\,300 = Q_1$$

$$\therefore \quad Q_1 = 1256 \text{ kJ/min } \textbf{Ans.}$$

Example 35.5. *1.5 kW per tonne of refrigeration is required to maintain the temperature of − 40° C in the refrigerator. If the refrigeration cycle works on Carnot cycle, determine the following :*

1. C.O.P. of the cycle, 2. Temperature of the sink, 3. Heat rejected to the sink per tonne of refrigeration, and 4. Heat supplied and E.P.R., if the cycle is used as a heat pump.

Solution. Given : $T_2 = -40° C = 233 K$

1. *C.O.P. of the cycle*

Since 1.5 kW per tonne of refrigeration is required to maintain the temperature in the refrigerator, therefore amount of work required to be done,

$$W_R = 1.5 \text{ kW} = 1.5 \text{ kJ/s} = 1.5 \times 60 = 90 \text{ kJ/min}$$

and heat extracted from the cold body,

$$Q_2 = 1 \text{ TR} = 210 \text{ kJ/min}$$

We know that $(\text{C.O.P.})_R = \dfrac{Q_2}{W_R} = \dfrac{210}{90} = 2.33$ **Ans.**

2. *Temperature of the sink*

Let T_1 = Temperature of the sink.

We know that $(\text{C.O.P.})_R = \dfrac{T_2}{T_1 - T_2}$ or $2.33 = \dfrac{233}{T_1 - 233}$

$\therefore \quad T_1 = \dfrac{233}{2.33} + 233 = 333 \text{ K} = 60° \text{ C}$ **Ans.**

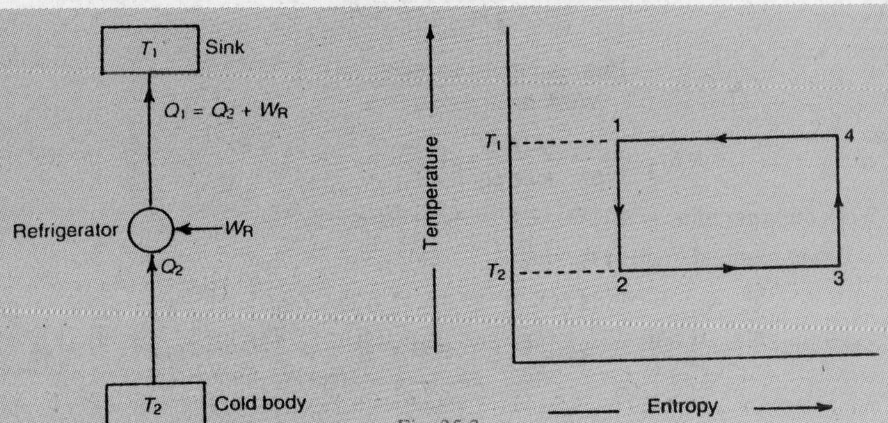

Fig. 35.3

3. *Heat rejected to the sink per tonne of refrigeration*

We know that heat rejected to the sink,

$$Q_1 = Q_2 + W_R = 210 + 90 = 300 \text{ kJ/min} \textbf{ Ans.}$$

4. *Heat supplied and E.P.R., if the cycle is used as a heat pump*

We know that heat supplied when the cycle is used as a heat pump is Q_1 = 300 kJ/min **Ans.**

and $\text{E.P.R.} = (\text{C.O.P.})_R + 1 = 2.33 + 1 = 3.33$ **Ans.**

Example 35.6. *The capacity of a refrigerator is 200 TR when working between – 6° C and 25° C. Determine the mass of ice produced per day from water at 25° C. Also find the power required to drive the unit. Assume that the cycle operates on reversed Carnot cycle and latent heat of ice is 335 kJ/kg.*

Solution. Given : Q = 200 TR ; T_2 = – 6° C = 267 K ; T_1 = 25° C = 298 K, T_w = 25° C ; $h_{fg\,(ice)}$ = 335 kJ/kg

Mass of ice produced per day

We know that heat extraction capacity of the refrigerator

$$= 200 \times 210 = 42\,000 \text{ kJ/min} \quad \ldots (\because 1 \text{ TR} = 210 \text{ kJ/min})$$

and heat removed from 1 kg of water at 25° C (298 K) to form ice at 0° C (273 K)

$$= \text{Mass} \times \text{Sp. heat} \times \text{Rise in temperature} + h_{fg\,(ice)}$$

$$= 1 \times 4.187\,(298 - 273) + 335 = 439.7 \text{ kJ/kg}$$

∴ Mass of ice produced per min

$$= \frac{42\,000}{439.7} = 95.52 \text{ kg/min}$$

and mass of ice produced per day

$$= 95.52 \times 60 \times 24 = 137\,550 \text{ kg} = 137.55 \text{ tonnes } \textbf{Ans.}$$

Power required to drive the unit

We know that C.O.P. of the reversed Carnot cycle,

$$= \frac{T_2}{T_1 - T_2} = \frac{267}{298 - 267} = 8.6$$

Also $\quad$ $\text{C.O.P.} = \frac{\text{Heat extraction capacity}}{\text{Work done per min}}$

$$\therefore \quad 8.6 = \frac{42\,000}{\text{Work done per min}}$$

or $\quad$ Work done per min = 42 000 / 8.6 = 4884 kJ/min

∴ Power required to drive the unit

$$= 4884 / 60 = 81.4 \text{ kW } \textbf{Ans.}$$

Example 35.7. *A cold storage plant is required to store 20 tonnes of fish. The fish is supplied at a temperature of 30° C. The specific heat of fish above freezing point is 2.93 kJ/kg K. The specific heat of fish below freezing point is 1.26 kJ/kg K. The fish is stored in cold storage which is maintained at – 8° C. The freezing point of fish is – 4° C. The latent heat of fish is 235 kJ/kg. If the plant requires 75 kW to drive it, find*

1. The capacity of the plant, and 2. Time taken to achieve cooling.

Assume actual C.O.P. of the plant as 0.3 of the Carnot C.O.P.

Solution. Given : $m = 20$ t = 20 000 kg ; $T_1 = 30°$ C = 303 K ; $c_{AF} = 2.93$ kJ/kg K ; $c_{BF} =$ 1.26 kJ/kg K ; $T_2 = -8°$ C = 265 K ; $T_3 = -4°$ C = 269 K ; $h_{fg\,(fish)} = 235$ kJ/kg ; $P = 75$ kW = 75 kJ/s

1. *Capacity of the plant*

We know that Carnot C.O.P.

$$= \frac{T_2}{T_1 - T_2} = \frac{265}{303 - 265} = 6.97$$

$$\therefore \quad \text{Actual C.O.P.} = 0.3 \times 6.97 = 2.091$$

and heat removed by the plant

$$= \text{Actual C.O.P.} \times \text{Work required}$$

$$= 2.091 \times 75 = 156.8 \text{ kJ/s} = 9408 \text{ kJ/min}$$

∴ Capacity of the plant

$$= 9408 / 210 = 44.8 \text{ TR } \textbf{Ans.} \qquad \ldots (\because 1 \text{ TR} = 210 \text{ kJ/min})$$

2. *Time taken to achieve cooling*

We know that heat removed from the fish above freezing point,

$$Q_1 = m \times c_{AF} (T_1 - T_3)$$

$$= 20\,000 \times 2.93\,(303 - 269) = 1.992 \times 10^6 \text{ kJ}$$

Similarly, heat removed from the fish below freezing point,

$$Q_2 = m \times c_{BF}(T_3 - T_2)$$

$$= 20\,000 \times 1.26\,(269 - 265) = 0.101 \times 10^6 \text{ kJ}$$

and total latent heat of fish,

$$Q_3 = m \times h_{fg\,(fish)} = 20\,000 \times 235 = 4.7 \times 10^6 \text{ kJ}$$

∴ Total heat removed by the plant

$$= Q_1 + Q_2 + Q_3$$

$$= 1.992 \times 10^6 + 0.101 \times 10^6 + 4.7 \times 10^6 = 6.793 \times 10^6 \text{ kJ}$$

and time taken to achieve cooling

$$= \frac{\text{Total heat removed by the plant}}{\text{Heat removed by the plant per min}}$$

$$= \frac{6.793 \times 10^6}{9408} = 722 \text{ min} = 12.03 \text{ h} \quad \textbf{Ans.}$$

35.10. Air Refrigerator Working on a Bell-Coleman Cycle (or Reversed Joule or Brayton Cycle)

A Bell-Coleman air refrigerating machine was developed by Bell-Coleman and Light Foot by reversing the Joule's air cycle. It was one of the earliest types of refrigerators used in ships carrying frozen meat. Fig. 35.4 shows a schematic diagram of such a machine which consists of a compressor, a cooler, an expander and a refrigerator.

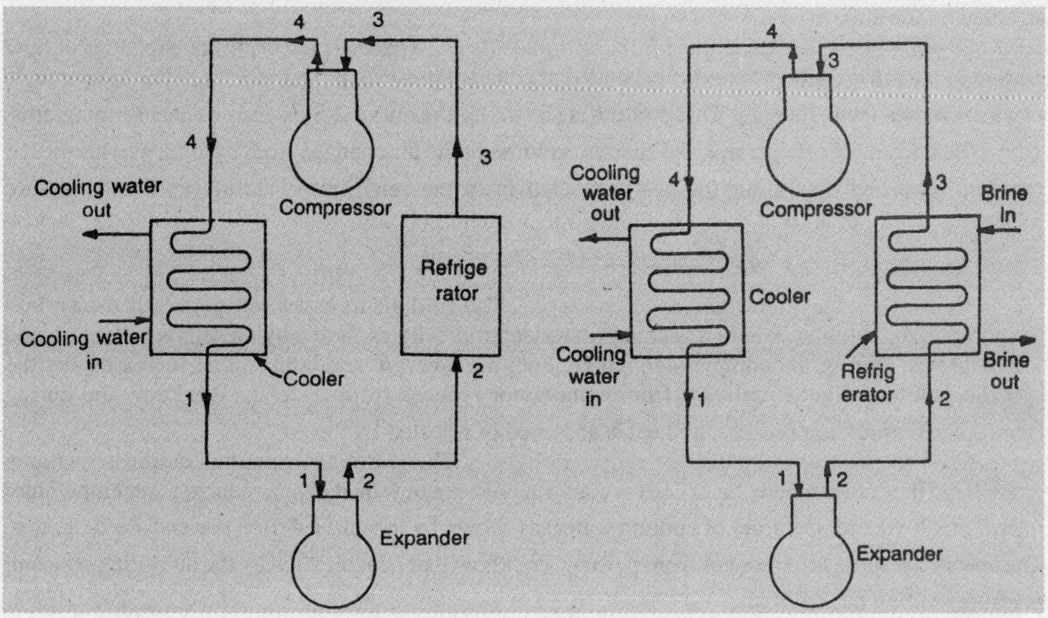

Fig. 35.4. Open cycle air Bell-Coleman refrigerator.

Fig. 35.5. Closed cycle or dense air Bell-Coleman refrigerator.

The Bell-Coleman cycle (also known as *reversed Joule* or *Brayton cycle*) is a modification of reversed Carnot cycle. The cycle is shown on *p-v* and *T-s* diagrams in Fig.35.6 (*a*) and (*b*). At point

1, let p_1, v_1, and T_1 be the pressure, volume and temperature of air respectively. The four stages of the cycle are as follows :

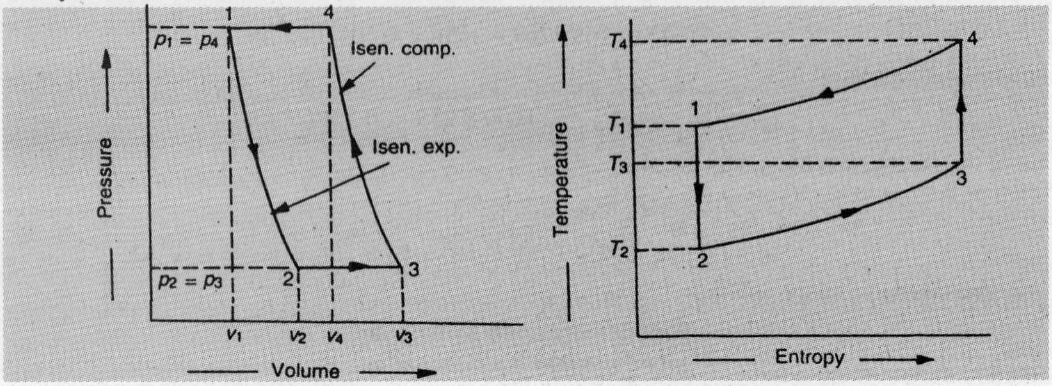

(*a*) *p-v* diagram. (*b*) *T-s* diagram..

Fig. 35.6. Bell-Coleman cycle.

1. *First stage (Isentropic expansion).* The air from the cooler is drawn into the expander cylinder where it is expanded isentropically from pressure p_1 to the refrigerator pressure p_2 which is equal to the atmospheric pressure.The temperature of air during expansion falls from T_1 to T_2 (*i.e.* the temperature much below the temperature of cooling water, T_1). The expansion process is shown by the curve 1-2 on the *p-v* and *T-s* diagrams. The specific volume of air at entry to the refrigerator increases from v_1 to v_2. We know that during isentropic expansion of air, no heat is absorbed or rejected by the air.

2. *Second stage (Constant pressure expansion).* The cold air from the expander is now passed to the refrigerator where it is expanded at constant pressure p_3 (equal to p_2). The temperature of air increases from T_2 to T_3. This process is shown by the curve 2-3 on the *p-v* and *T-s* diagrams. Due to heat from the refrigerator, the specific volume of the air changes from v_2 to v_3. We know that the heat absorbed by the air (or heat extracted from the refrigerator) during constant pressure expansion per kg of air is

$$q_A = q_{2-3} = c_p (T_3 - T_2)$$

3. *Third stage (Isentropic compression).* The cold air from the refrigerator is drawn into the compressor cylinder, where it is compressed isentropically as shown by the curve 3-4 on *p-v* and *T-s* diagrams. During the compression stroke, both the pressure and temperature increases and the specific volume of air at delivery from compressor reduces from v_3 to v_4. We know the during isentropic compression process, no heat is absorbed or rejected by the air.

4. *Fourth stage (Constant pressure cooling).* The warm air from the compressor is now passed into the cooler where it is cooled at constant pressure p_4 (equal to p_1), reducing the temperature from T_4 to T_1 (the temperature of cooling water) as shown by the curve 4-1 on *p-v* and *T-s* diagrams. The specific volume also reduces from v_4 to v_1. We know that heat rejected by the air during constant pressure cooling per kg of air,

$$q_R = q_{4-1} = c_p (T_4 - T_1)$$

We know that work done during the cycle per kg of air

$$w = \text{Heat rejected} - \text{Heat absorbed} = q_R - q_A$$

$$= c_p (T_4 - T_1) - c_p (T_3 - T_2)$$

$\therefore$ Coefficient of performance,

$$\text{C.O.P.} = \frac{\text{Heat absorbed}}{\text{Work done}} = \frac{q_A}{q_R - q_A}$$

$$= \frac{c_p (T_3 - T_2)}{c_p (T_4 - T_1) - c_p (T_3 - T_2)} = \frac{T_3 - T_2}{(T_4 - T_1) - (T_3 - T_2)}$$

$$= \frac{T_2\left(\dfrac{T_3}{T_2} - 1\right)}{T_1\left(\dfrac{T_4}{T_1} - 1\right) - T_2\left(\dfrac{T_3}{T_2} - 1\right)} \qquad \ldots (i)$$

We know that for isentropic expansion process 1-2,

$$\frac{T_1}{T_2} = \left(\frac{p_1}{p_2}\right)^{\frac{\gamma - 1}{\gamma}} \qquad \ldots (ii)$$

Similarly, for isentropic compression process 3-4,

$$\frac{T_4}{T_3} = \left(\frac{p_4}{p_3}\right)^{\frac{\gamma - 1}{\gamma}} \qquad \ldots (iii)$$

Since $p_1 = p_4$ and $p_2 = p_3$, therefore, from equations (*ii*) and (*iii*),

$$\frac{T_1}{T_2} = \frac{T_4}{T_3} \quad \text{or} \quad \frac{T_3}{T_2} = \frac{T_4}{T_1} \qquad \ldots (iv)$$

Now substituting these values in equation (*i*), we get

$$\text{C.O.P.} = \frac{T_2}{T_1 - T_2} = \frac{1}{\dfrac{T_1}{T_2} - 1}$$

$$= \frac{1}{\left(\dfrac{p_1}{p_2}\right)^{\frac{\gamma - 1}{\gamma}} - 1} = \frac{1}{\left(\dfrac{p_4}{p_3}\right)^{\frac{\gamma - 1}{\gamma}} - 1}$$

$$= \frac{1}{(r_p)^{\frac{\gamma - 1}{\gamma}} - 1} \qquad \ldots (v)$$

where $\quad r_p$ = Expansion or compression ratio $= \dfrac{p_1}{p_2} = \dfrac{p_4}{p_3}$

Sometimes, the compression and expansion processes take place according to the law pv^n = constant. In such a case, the C.O.P. is obtained from the fundamentals as discussed below :

We know that work done by the expander during the process 1-2 per kg of air,

$$w_E = w_{1\text{-}2} = \frac{n}{n - 1}(p_1 v_1 - p_2 v_2) = \frac{n}{n - 1}(RT_1 - RT_2) \qquad \ldots (\because pv = RT)$$

and work done by the compressor during the process 3-4 per kg of air,

$$w_C = w_{3-4} = \frac{n}{n-1}(p_4 v_4 - p_3 v_3) = \frac{n}{n-1}(RT_4 - RT_3)$$

∴ Net work done during the cycle per kg of air,

$$w = w_C - w_E = \frac{n}{n-1} \times R[(T_4 - T_3) - (T_1 - T_2)]$$

We also know that heat absorbed during constant pressure process 2-3

$$= c_p (T_3 - T_2)$$

$$\therefore \quad \text{C.O.P.} = \frac{\text{Heat absorbed}}{\text{Work done}} = \frac{q_A}{w}$$

$$= \frac{c_p (T_3 - T_2)}{\frac{n}{n-1} \times R[(T_4 - T_3) - (T_1 - T_2)]} \quad \ldots (vi)$$

We know that $R = c_p - c_v = c_v (\gamma - 1)$

Substituting the value of R in equation (*vi*),

$$\text{C.O.P.} = \frac{c_p (T_3 - T_2)}{\frac{n}{n-1} \times c_v (\gamma - 1)[(T_4 - T_3) - (T_1 - T_2)]}$$

$$= \frac{T_3 - T_2}{\frac{n}{n-1} \times \frac{\gamma - 1}{\gamma}[(T_4 - T_3) - (T_1 - T_2)]}$$

$$= \frac{T_3 - T_2}{\frac{n}{n-1} \times \frac{\gamma - 1}{\gamma}[(T_4 - T_1) - (T_3 - T_2)]} \quad \ldots (vii)$$

Notes : 1. In this case, the values of T_2 and T_4 are to be obtained from the relations

$$\frac{T_1}{T_2} = \left(\frac{p_1}{p_2}\right)^{\frac{n-1}{n}}, \quad \text{and} \quad \frac{T_4}{T_3} = \left(\frac{p_4}{p_3}\right)^{\frac{n-1}{n}}$$

2. For isentropic expansion or compression $n = \gamma$. Therefore, the equation (*vii*) may be written as

$$\text{C.O.P.} = \frac{T_3 - T_2}{(T_4 - T_1) - (T_3 - T_2)} \quad \ldots \text{(same as before)}$$

3. We have already discussed that the main drawback of the open cycle air refrigerator is freezing of the moisture in the air during expansion stroke which is liable to choke up the valves. Due to this reason, a closed cycle or dense air Bell-Coleman refrigerator, as shown in Fig. 35.5, is preferred. In this case, the cold air does not come in direct contact of the refrigerator. The cold air is passed through the pipes and it is used for absorbing heat from the brine and this cooled brine is circulated into the refrigerated space. The term dense air system, is derived from the fact that the suction to the compressor is at higher pressure than the open cycle system (which is atmospheric).

Example 35.8. *A refrigerator working on Bell-Coleman cycle operates between pressure limits of 1.05 bar and 8.5 bar. Air is drawn from the cold chamber at 10° C, compressed and then it is cooled to 30° C before entering the expansion cylinder. The expansion and compression follows the law $pv^{1.3}$ = constant. Determine the theoretical C.O.P. of the system.*

Solution. Given : $p_3 = p_2 = 1.05$ bar ; $p_4 = p_1 = 8.5$ bar ; $T_3 = 10° \text{ C} = 283 \text{ K}$; $T_1 = 30° \text{ C} = 303 \text{ K}$; $n = 1.3$

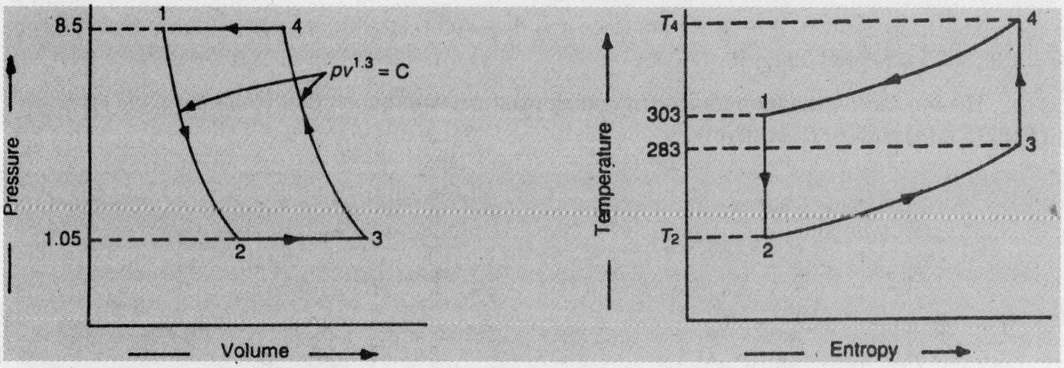

(*a*) *p-v* diagram. (*b*) *T-s* diagram.

Fig. 35.7

The *p-v* and *T-s* diagram for a refrigerator working on the Bell-Coleman cycle is shown in Fig. 35.7 (*a*) and (*b*) respectively.

Let T_2 and T_4 = Temperature of air at the end of expansion and compression respectively.

Since the expansion and compression follows the law $pv^{1.3} = C$, therefore

$$\frac{T_1}{T_2} = \left(\frac{p_1}{p_2}\right)^{\frac{n-1}{n}} = \left(\frac{8.5}{1.05}\right)^{\frac{1.3-1}{1.3}} = (8.1)^{0.231} = 1.62$$

$$\therefore \qquad T_2 = T_1 / 1.62 = 303 / 1.62 = 187 \text{ K}$$

Similarly,
$$\frac{T_4}{T_3} = \left(\frac{p_4}{p_3}\right)^{\frac{n-1}{n}} = \left(\frac{8.5}{1.05}\right)^{\frac{1.3-1}{1.3}} = 1.62$$

$$\therefore \qquad T_4 = T_3 \times 1.62 = 283 \times 1.62 = 458.5 \text{ K}$$

We know that theoretical coefficient of performance,

$$\text{C.O.P.} = \frac{T_3 - T_2}{\frac{n}{n-1} \times \frac{\gamma - 1}{\gamma}\left[(T_4 - T_1) - (T_3 - T_2)\right]}$$

$$= \frac{283 - 187}{\frac{1.3}{1.3-1} \times \frac{1.4-1}{1.4}\left[(458.5 - 303) - (283 - 187)\right]} \qquad \ldots (\text{Taking } \gamma = 1.4)$$

$$= \frac{96}{1.24 \times 59.5} = 1.3 \textbf{ Ans.}$$

Example 35.9. *The atmospheric air at pressure 1 bar and temperature – 5°C is drawn in the cylinder of the compressor of a Bell-Coleman refrigeration machine. It is compressed isentropically to a pressure of 5 bar. In the cooler, the compressed air is cooled to 15° C, pressure remaining the same. It is then expanded to a pressure of 1 bar in an expansion cylinder, from where it is passed to the cold chamber. Find 1. the work done per kg of air ; and 2. the C.O.P. of the plant.*

For air, assume law for expansion $pv^{1.2}$ = constant, law for compression $pv^{1.4}$ = constant and specific heat of air at constant pressure = 1 kJ/kg K.

Solution. Given : $p_3 = p_2 = 1$ bar ; $T_3 = -5°$ C = 268 K ; $p_4 = p_1 = 5$ bar ; $T_1 = 15°$ C = 288 K ; $n = 1.2$; $\gamma = 1.4$; $c_p = 1$ kJ/kg K

The *p-v* and *T-s* diagram for a refrigerating machine working on Bell-Coleman cycle is shown in Fig. 35.8 (*a*) and (*b*) respectively.

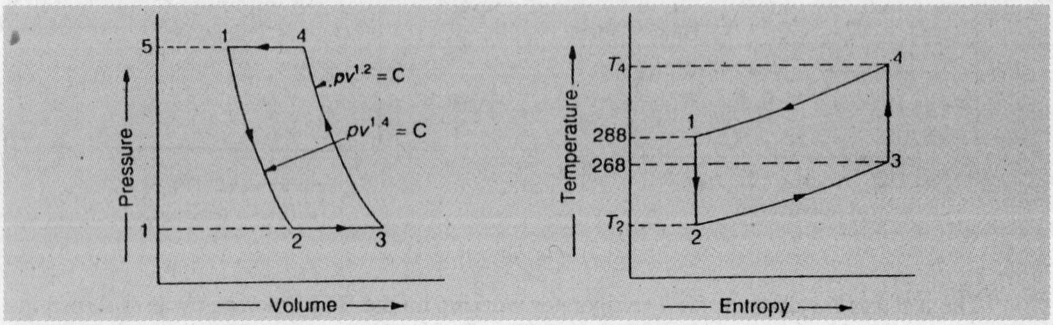

(*a*) *p-v* diagram. (*b*) *T-s* diagram.

Fig. 35.8

1. *Work done per kg of air*

Let T_2 and T_4 = Temperatures at the end of expansion and compression respectively.

The expansion process 1-2 follows the law $pv^{1.2}$ = constant.

$$\therefore \quad \frac{T_1}{T_2} = \left(\frac{p_1}{p_2}\right)^{\frac{n-1}{n}} = \left(\frac{5}{1}\right)^{\frac{1.2-1}{1.2}} = (5)^{0.167} = 1.31$$

or $$T_2 = T_1 / 1.31 = 288 / 1.31 = 220 \text{ K}$$

The compression process 3-4 is isentropic and follows the law $pv^{1.4}$ = constant.

$$\therefore \quad \frac{T_4}{T_3} = \left(\frac{p_4}{p_3}\right)^{\frac{\gamma-1}{\gamma}} = \left(\frac{5}{1}\right)^{\frac{1.4-1}{1.4}} = (5)^{0.286} = 1.585$$

or $$T_4 = T_3 \times 1.585 = 268 \times 1.585 = 424.8 \text{ K}$$

We know that work done by the expander during the process 1-2 per kg of air,

$$w_E = w_{1-2} = \frac{n}{n-1} \times R\,(T_1 - T_2)$$

$$= \frac{1.2}{1.2-1} \times 0.29\,(288 - 220) = 118.3 \text{ kJ/kg}$$

... (Taking R = 0.29 kJ/kg K)

and work done by the compressor during the isentropic process 3-4 per kg of air,

$$w_C = w_{3-4} = \frac{\gamma}{\gamma-1} \times R\,(T_4 - T_3)$$

$$= \frac{1.4}{1.4-1} \times 0.29\,(424.8 - 268) = 159 \text{ kJ/kg}$$

∴ Net work done per kg of air,

$$w = w_C - w_E = 159 - 118.3 = 40.7 \text{ kJ/kg } \textbf{Ans.}$$

2. C.O.P. of the plant

We know that heat absorbed during constant pressure process 2-3 per kg of air,

$$q_A = c_p (T_3 - T_2) = 1 (268 - 220) = 48 \text{ kJ/kg}$$

$$\therefore \text{ C.O.P. of the plant } = \frac{\text{Heat absorbed}}{\text{Work done}} = \frac{q_A}{w} = \frac{48}{40.7} = 1.18 \textbf{ Ans.}$$

Example 35.10. *In an open cycle air refrigeration machine, air is drawn from a cold chamber at – 2° C and 1 bar and compressed to 11 bar. It is then cooled, at this pressure, to the cooler temperature of 20° C and then expanded in expansion cylinder and returned to the cold room. The compression and expansion are isentropic, and follows the law $pv^{1.4}$ = constant. Sketch the p-v and T-s diagrams of the cycle and for a refrigeration of 15 tonnes, find : 1. theoretical C.O.P. ; 2. rate of circulation of the air in kg/min ; 3. piston displacement per minute in the compressor and expander; and 4. theoretical power per tonne of refrigeration.*

Solution. Given : $T_3 = -2°\text{ C} = 271 \text{ K}$; $p_3 = p_2 = 1 \text{ bar} = 1 \times 10^5 \text{ N/m}^2$; $p_4 = p_1 = 11$ bar ; $T_1 = 20°\text{ C} = 293 \text{ K}$; $\gamma = 1.4$; $Q = 15$ TR

The *p-v* and *T-s* diagrams of the cycle are shown in Fig. 35.9 (*a*) and (*b*) respectively.

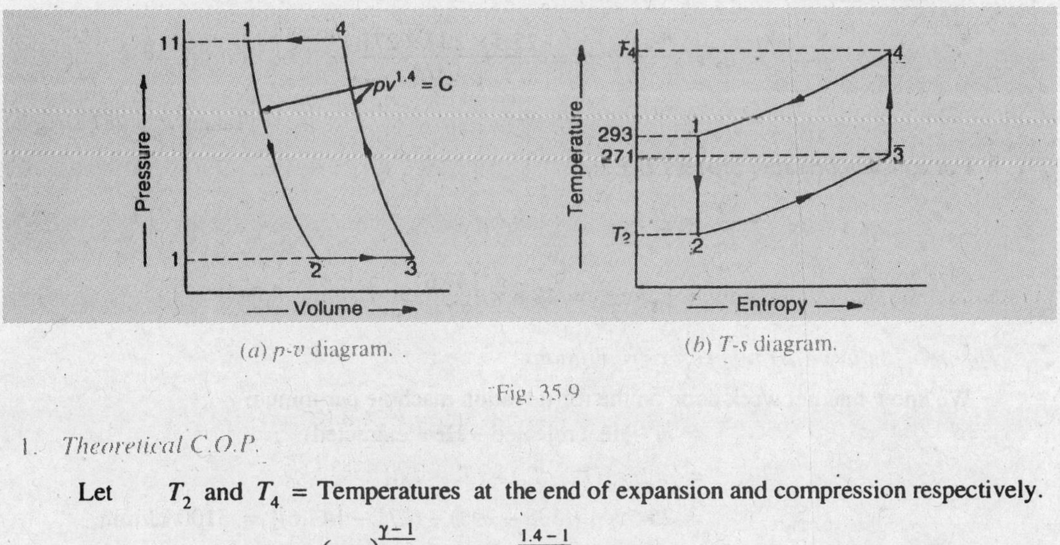

(*a*) *p-v* diagram. (*b*) *T-s* diagram.

Fig. 35.9

1. Theoretical C.O.P.

Let T_2 and T_4 = Temperatures at the end of expansion and compression respectively.

We know that
$$\frac{T_1}{T_2} = \left(\frac{p_1}{p_2}\right)^{\frac{\gamma-1}{\gamma}} = \left(\frac{11}{1}\right)^{\frac{1.4-1}{1.4}} = 1.985$$

$$\therefore \quad T_2 = T_1 / 1.985 = 293 / 1.985 = 147.6 \text{ K}$$

Similarly,
$$\frac{T_4}{T_3} = \left(\frac{p_4}{p_3}\right)^{\frac{\gamma-1}{\gamma}} = \left(\frac{11}{1}\right)^{\frac{1.4-1}{1.4}} = (11)^{0.286} = 1.985$$

$$\therefore \quad T_4 = T_3 \times 1.985 = 271 \times 1.985 = 538 \text{ K}$$

We also know that theoretical C.O.P.

$$= \frac{T_3 - T_2}{(T_4 - T_1) - (T_3 - T_2)}$$

$$= \frac{271 - 147.6}{(538 - 293) - (271 - 147.6)} = \frac{123.4}{121.6} = 1.015 \text{ Ans.}$$

2. *Rate of circulation of the air in kg/min*

Refrigeration capacity $= 15$ TR ... (Given)

$\therefore$ Heat extracted/min $= 15 \times 210 = 3150$ kJ/min ... ($\because$ 1 TR = 210 kJ/min)

We know that heat extracted from cold chamber per kg

$$= c_p (T_3 - T_2) = 1\,(271 - 147.6) = 123.4 \text{ kJ/kg}$$

... ($\because$ c_p for air = 1 kJ/kg K)

$\therefore$ Rate of circulation of air,

$$m_a = \frac{\text{Heat extracted /min}}{\text{Heat extracted /kg}} = \frac{3150}{123.4} = 25.5 \text{ kg/min Ans.}$$

3. *Piston displacement per minute in the compressor and expander*

Let v_3 and v_2 = Piston displacement per minute in the compressor and expander respectively.

We know that $p_3 v_3 = m_a R_a T_3$

$$\therefore \quad v_3 = \frac{m_a R_a T_3}{p_3} = \frac{25.5 \times 287 \times 271}{1 \times 10^5} = 19.8 \text{ m}^3\text{/min Ans.}$$

... (Taking R_a = 287 kJ/kg K)

For constant pressure process 2-3,

$$\frac{v_2}{T_2} = \frac{v_3}{T_3}$$

$$\therefore \quad v_2 = v_3 \times \frac{T_2}{T_3} = 19.8 \times \frac{147.6}{271} = 10.8 \text{ m}^3 \text{ Ans.}$$

4. *Theoretical power per tonne of refrigeration*

We know that net work done on the refrigeration machine per minute

$$= m_a \text{ (Heat rejected – Heat extracted)}$$

$$= m_a c_p [(T_4 - T_1) - (T_3 - T_2)]$$

$$= 25.5 \times 1 [(538 - 293) - (271 - 147.6)] = 3100 \text{ kJ/min}$$

$\therefore$ Theoretical power of the refrigerating machine

$$= 3100/60 = 51.57 \text{ kW}$$

and theoretical power per tonne of refrigeration

$$\doteq 51.67/15 = 3.44 \text{ kW / TR Ans.}$$

EXERCISES

1. A refrigerating plant is required to produce 2.5 tonnes of ice per day at – 4° C from water at 20° C. If the temperature range in the compressor is between 25° C and – 6° C, calculate power required to drive the compressor. Latent heat of ice = 335 kJ/kg and specific heat of ice = 2.1 kJ/kg K. [Ans. 1.437 kW]

2. A refrigeration system has working temperature of – 30° C and 40° C. What is the maximum C.O.P. possible? If the actual C.O.P. is 75% of the maximum, calculate the actual refrigerating effect produced per kW hour. [Ans. 0.743 TR]

3. A Carnot cycle machine operates between the temperature limits of 47° C and – 30°C. Determine the C.O.P. when it operates as 1. a refrigerating machine ; 2. a heat pump ; and 3. a heat engine.

[Ans. 3.16 : 4.16 : 0.24]

4. A refrigerator using Carnot cycle requires 1.25 kW per tonne of refrigeration to maintain a temperature of – 30° C. Find : 1. C.O.P. of Carnot refrigerator ; 2. Temperature at which heat is rejected ; and 3. heat rejected per tonne of refrigeration. [Ans. 2.8 ; 55.4°C ; 284 kJ/min]

5. A refrigerator storage is supplied with 30 tonnes of fish at a temperature of 27° C. The fish has to be cooled to – 9° C for preserving it for long period without deterioration. The cooling takes place in 10 hours. The specific heat of fish is 2.93 kJ/kg K above freezing point of fish and 1.26 kJ/kg K below freezing point of fish which is – 3° C. The latent heat of freezing is 232 kJ/kg. What is the capacity of the plant in tonne of refrigeration for cooling the fish ? What would be the ideal C.O.P. between this temperature range ? If the actual C.O.P. is 40% of the ideal, find the power required to run the cooling plant. [Ans. 78 TR ; 7.33 ; 93.3 kW]

6. A refrigerating system working on Bell-Coleman cycle receives air from cold chamber at – 5° C and compresses it from 1 bar to 4.5 bar. The compressed air is then cooled to a temperature of 37° C before it is expanded in the expander. Calculate the C.O.P. of the system when compression and expansion are (*i*) isentropic, and (*ii*) follow the law $pv^{1.25}$ = constant. [Ans. 1.86 : 1.98]

7. A Bell-Coleman refrigerator works between 4 bar and 1 bar pressure limits. After compression, the cooling water reduces the air temperature to 17° C. What is the lowest temperature produced by the ideal machine ?

Compare the coefficient of performance of this machine with that of the ideal Carnot cycle machine working between the same pressure limits, the temperature at the beginning of compression being – 13° C.

[Ans. – 78°C ; 2.07, 1.02]

8. A dense air refrigerating system operating between pressures of 17.5 bar and 3.5 bar is to produce 10 tonne of refrigeration. Air leaves the refrigerating coils at – 7° C and it leaves the air cooler at 15.5° C. Neglecting losses and clearance, calculate the net work done per minute and the coefficient of performance. For air, c_p = 1.005 kJ/kg K and γ = 1.4. [Ans. 1237 kJ/min ; 1.7]

9. In a refrigeration plant using Bell-Coleman cycle, air at 0.8 bar, 6° C enters the compressor. The conditions at entry to the air turbine are 3.2 bar and 30° C. Assuming the isentropic efficiencies of compressor and turbine to be 83% and 85%, estimate the C.O.P. and air flow rate per tonne of refrigeration.

[Ans. 0.76 ; 3.46 kg/min]

10. An air refrigeration used for food storage provides 25 TR. The temperature of air entering the compressor is 7° C and the temperature at exit of cooler is 27° C. Find (*a*) C.O.P. of the cycle, and (*b*) power per tonne of refrigeration required by the compressor. The quantity of air circulated in the system is 3000 kg/h. The compression and expansion both follows the law $pv^{1.3}$ = constant and take γ = 1.4 and c_p = 1 for air.

[Ans. 1.13 ; 3.1 kW/TR]

11. A dense air machine operates on reversed Brayton cycle and is required for a capacity of 10 TR. The cooler pressure is 4.2 bar and the refrigerator pressure is 1.4 bar. The air is cooled in the cooler to a temperature of 50° C and the temperature of air at inlet to compressor is – 20° C. Determine for the ideal cycle (*a*) C O.P. (*b*) mass of air circulated per minute, (*c*) theoretical piston displacement of compressor, (*d*) theoretical piston displacement of expander, and (*e*) net power per tonne of refrigeration.Show the cycle on *p-v* and *T-s* planes.

[Ans. 2.83 ; 123.5 kg/min ; 64 m³ ; 60 m³ ; 1.235 kW/TR]

QUESTIONS

1. Define the following terms :

(*a*) Coefficient of performance ; (*b*) tonne of refrigeration.

2. Discuss the advantages of the dense air refrigerating system over an open air refrigeration system.

3. What is the difference between a refrigerator and a heat pump ? Derive an expression for the performance factor for both if they are running on reversed Carnot cycle.

4. Describe the Bell-Coleman cycle and obtain an expression for the C.O.P. of the cycle.

5. Show that C.O.P. of a Bell-Coleman cycle is given by the expression :

$$\text{C.O.P.} = \frac{1}{\left(\frac{1}{r_p}\right)^{\frac{\gamma-1}{\gamma}} - 1}$$

where r_p is the compression ratio and γ is the usual ratio of specific heats.

OBJECTIVE TYPE QUESTIONS

1. One tonne of refrigeration is equal to
   (*a*) 21 kJ/min (*b*) 210 kJ/min (*c*) 420 kJ/min (*d*) 620 kJ/min
2. One tonne refrigerating machine means that
   (*a*) one tonne is the total mass of the machine
   (*b*) one tonne of refrigerant is used
   (*c*) one tonne of water can be converted into ice
   (*d*) one tonne of ice when melts from and at 0° C in 24 hours, the refrigeration effect produced is equivalent to 210 kJ/min
3. The coefficient of performance is always .... one.
   (*a*) equal to (*b*) less than (*c*) greater than
4. The relative coefficient of performance is equal to
   (*a*) $\frac{\text{Theoretical C.O.P.}}{\text{Actual C.O.P.}}$ (*b*) $\frac{\text{Actual C.O.P.}}{\text{Theoretical C.O.P.}}$
   (*c*) Theoretical C.O.P. × Actual C.O.P.
5. In a closed or dense air refrigeration cycle, the operating pressure ratio can be reduced, which results in .... coefficient of performance.
   (*a*) lower (*b*) higher
6. Air refrigeration cycle is used in
   (*a*) commercial refrigerators (*b*) domestic refrigerators
   (*c*) air-conditioning (*d*) gas liquefaction
7. In a refrigerating machine, heat rejected is .... heat absorbed.
   (*a*) equal to (*b*) less than (*c*) greater than
8. Air refrigerator works on
   (*a*) Carnot cycle (*b*) Rankine cycle
   (*c*) reversed Carnot cycle (*d*) Bell-Coleman cycle
   (*e*) both (*a*) and (*b*) (*f*) both (*c*) and (*d*)
9. In air-conditioning of aeroplanes, using air as a refrigerant, the cycle used is
   (*a*) reversed Carnot cycle (*b*) reversed Joule cycle
   (*c*) reversed Brayton cycle (*d*) reversed Otto cycle

ANSWERS

1. (*b*)	2. (*d*)	3. (*c*)	4. (*b*)	5. (*b*)
6. (*d*)	7. (*c*)	8. (*f*)	9. (*c*)	

36

Vapour Compression Refrigeration Systems

1. Introduction. 2. Advantages and Disadvantages of Vapour Compression Refrigeration System over Air Refrigeration System. 3. Mechanism of a Simple Vapour Compression Refrigeration System. 4. Pressure-Enthalpy (p-h) Chart. 5. Types of Vapour Compression Cycles. 6. Theoretical Vapour Compression Cycle with Dry Saturated Vapour after Compression. 7. Theoretical Vapour Compression Cycle with Wet Vapour after Compression. 8. Theoretical Vapour Compression Cycle with Superheated Vapour after Compression. 9. Theoretical Vapour Compression Cycle with Superheated Vapour before Compression. 10. Theoretical Vapour Compression Cycle with Under-cooling or Sub-cooling of Refrigerant. 11. Actual Vapour Compression Cycle. 12. Vapour Absorption Refrigeration System. 13. Advantages of Vapour Absorption Refrigeration System over Vapour Compression Refrigeration System. 14. Ammonia-Hydrogen (Electrolux) Refrigerator. 15. Properties of a Refrigerant. 16. Refrigerants Commonly Used in Practice.

36.1. Introduction

A vapour compression refrigeration system* is an improved type of air refrigeration system in which a suitable working substance, termed as refrigerant, is used. It condenses and evaporates at temperature and pressures close to the atmospheric conditions. The refrigerants, usually, used for this purpose are ammonia, carbon dioxide and sulphur dioxide. The refrigerant used, does not leave the system, but is circulated throughout the system alternately condensing and evaporating. In evaporating, the refrigerant absorbs its latent heat from the brine** (salt water) which is used for circulating it around the cold chamber. While condensing, it gives out its latent heat to the circulating water of the cooler. The vapour compression refrigeration system is, therefore, a latent heat pump, as it pumps its latent heat from the brine and delivers it to the cooler.

The vapour compression refrigeration system is now-a-days used for all purpose refrigeration. It is generally used for all industrial purposes from a small domestic refrigerator to a big air conditioning plant.

36.2. Advantages and Disadvantages of Vapour Compression Refrigeration System over Air Refrigeration System

Following are the advantages and disadvantages of the vapour compression refrigeration system over air refrigeration system :

Advantages

1. It has smaller size for the given capacity of refrigeration.
2. It has less running cost.

* Since low pressure vapour refrigerant from the evaporator is changed into high pressure vapour refrigerant in the compressor, therefore it is named as vapour compression refrigeration system.

** Brine is used as it has a very low freezing temperature.

3. It can be employed over a large range of temperatures.
4. The coefficient of performance is quite high.

Disadvantages

1. The initial cost is high.
2. The prevention of leakage of the refrigerant is the major problem in vapour compression system.

36.3. Mechanism of a Simple Vapour Compression Refrigeration System

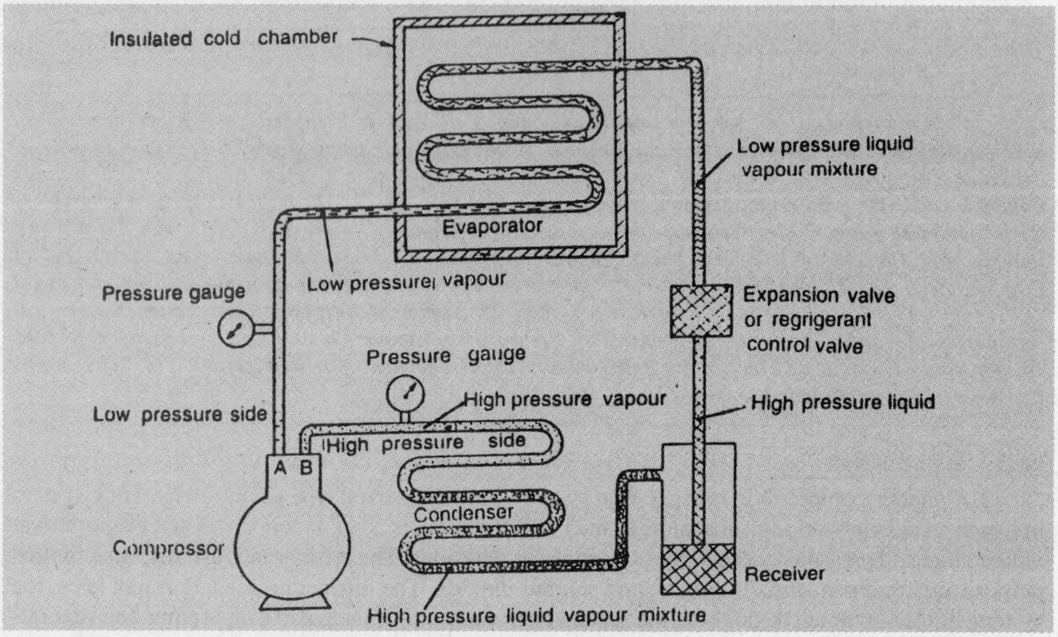

Fig. 36.1. Simple vapour compression refrigeration system.

Fig. 36.1 shows the schematic diagram of a simple vapour compression refrigeration system. It consists of the following five essential parts :

1. *Compressor.* The low pressure and temperature vapour refrigerant from evaporator is drawn into the compressor through the inlet or suction valve *A*, where it is compressed to a high pressure and temperature. This high pressure and temperature vapour refrigerant is discharged into the condenser through the delivery or discharge valve *B*.

2. *Condenser.* The condenser or cooler consists of coils of pipe in which the high pressure and temperature vapour refrigerant is cooled and condensed. The refrigerant, while passing through the condenser, gives up its latent heat to the surrounding condensing medium which is normally air or water.

3. *Receiver.* The condensed liquid refrigerant from the condenser is stored in a vessel known as receiver from where it is supplied to the evaporator through the expansion valve or refrigerant control valve.

4. *Expansion valve.* It is also called throttle valve or refrigerant control valve. The function of the expansion valve is to allow the liquid refrigerant under high pressure and temperature to pass at a controlled rate after reducing its pressure and temperature. Some of the liquid refrigerant evaporates as it passes through the expansion valve, but the greater portion is vaporised in the evaporator at the low pressure and temperature.

5. *Evaporator.* An evaporator consists of coils of pipe in which the liquid-vapour refrigerant at low pressure and temperature is evaporated and changed into vapour refrigerant at low pressure and temperature. In evaporating, the liquid vapour refrigerant absorbs its latent heat of vaporisation from the medium (air, water or brine) which is to be cooled.

Note : In any compression refrigeration system, there are two different pressure conditions. One is called the ***high pressure side*** and the other is known as ***low pressure side***. The high pressure side includes the discharge line (*i.e.* piping from delivery valve *B* to the condenser), receiver and expansion valve. The low pressure side includes the evaporatior, piping from the expansion valve to the evaporator and the suction line (*i.e.* piping from the evaporator to the suction valve *A*).

36.4. Pressure - Enthalpy (*p-h*) Chart

The most convenient chart for studying the behavior of a refrigerant is the *p-h* chart in which the vertical ordinates represent pressure and horizontal ordinates represent enthalpy (*i.e.* total heat). A typical chart is shown in Fig. 36.2, in which a few important lines of the complete chart are drawn. The saturated liquid line and the saturated vapour line merge into one another at the critical point. A

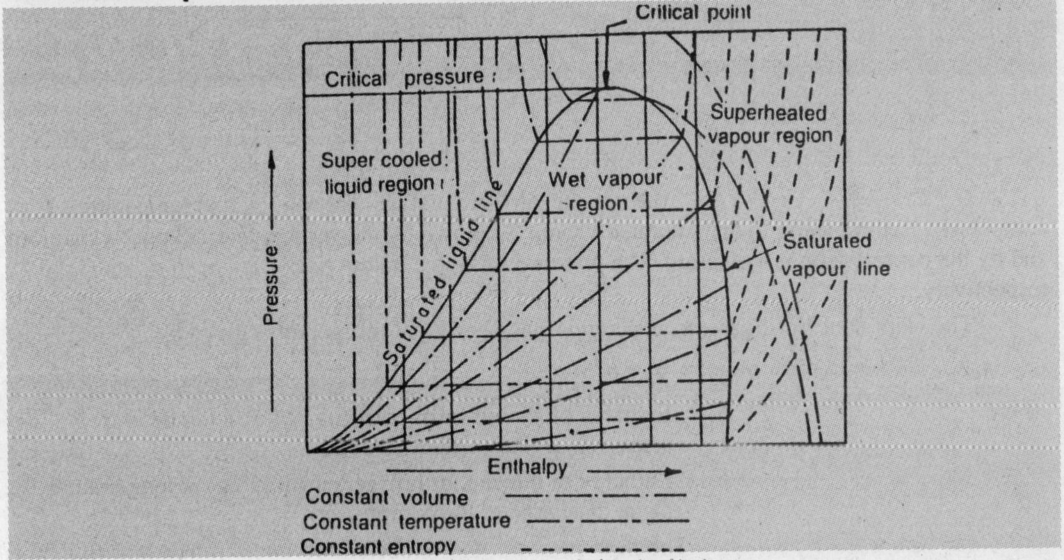

Fig.36.2. Pressure-enthalpy (*p-h*) chart.

saturated liquid is one which has a temperature equal to the saturation temperature corresponding to its pressure. The space to the left of the saturated liquid line will, therefore, be sub-cooled liquid region. The space between the liquid and the vapour lines is called wet vapour region and to the right of the saturated vapour line is a superheated vapour region.

In the following pages, we shall draw the *p-h* chart along with the *T-s* diagram of the cycles.

36.5. Types of Vapour Compression Cycles

We have already discussed that a vapour compression cycle essentially consists of compression, condensation, throttling and evaporation. Many scientists have focussed their attention to increase the coefficient of performance of the cycle. Though there are many cycles, yet the following are important from the subject point of view :

1. Cycle with dry saturated vapour after compression,
2. Cycle with wet vapour after compression,
3. Cycle with superheated vapour after compression,
4. Cycle with superheated vapour before compression, and
5. Cycle with undercooling or subcooling of refrigerant.

Now we shall discuss all the above mentioned cycles, one by one, in the following pages.

36.6. Theoretical Vapour Compression Cycle with Dry Saturated Vapour after Compression

A vapour compression cycle with dry saturated vapour after compression is shown on *T-s* and *p-h* diagrams in Fig. 36.3 (*a*) and (*b*) respectively. At point 1, let T_1, p_1 and s_1 be the temperature, pressure and entropy of the vapour refrigerant respectively. The four processes of the cycle are as follows :

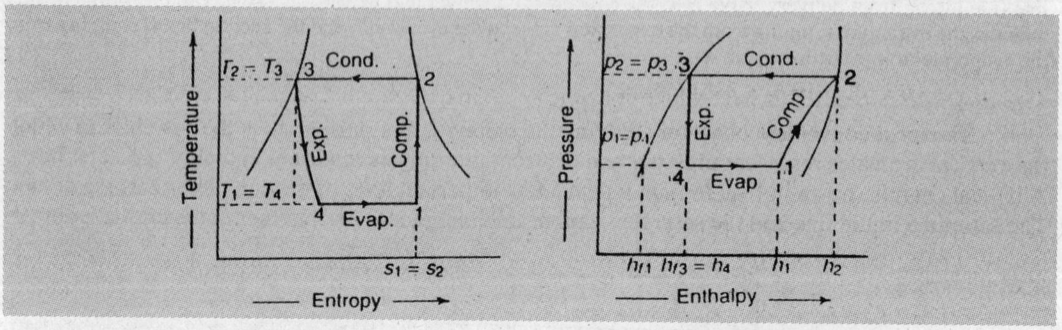

(*a*) *T-s* diagram. (*b*) *p-h* diagram.

Fig. 36.3. Theoretical vapour compression cycle with dry saturated vapour after compression.

1. *Compression process.* The vapour refrigerant at low pressure p_1 and temperature T_1 is compressed isentropically to dry saturated vapour as shown by the vertical line 1-2 on *T-s* diagram and by the curve 1-2 on *p-h* diagram. The pressure and temperature rises from p_1 to p_2 and T_1 to T_2 respectively.

The work done during isentropic compression per kg of refrigerant is given by

$$w_{1-2} = h_2 - h_1$$

where h_1 = Enthalpy of vapour refrigerant (in kJ/kg) at temperature T_1, *i.e.* at suction of the compressor, and

h_2 = Enthalpy of the vapour refrigerant (in kJ/kg) at temperature T_2, *i.e.* at discharge of the compressor.

2. *Condensing process.* The high pressure and temperature vapour refrigerant from the compressor is passed through the condenser where it is completely condensed at constant pressure p_2 and temperature T_2, as shown by the horizontal line 2-3 on *T-s* and *p-h* diagrams. The vapour refrigerant is changed into liquid refrigerant. The refrigerant, while passing through the condenser, gives its latent heat to the surrounding condensing medium.

3. *Expansion process.* The liquid refrigerant at pressure $p_3 = p_2$ and temperature $T_3 = T_2$ is expanded by throttling process* through the expansion valve to a low pressure $p_4 = p_1$ and temperature $T_4 = T_1$, as shown by the curve 3-4 on *T-s* diagram and by the vertical line 3-4 on *p-h* diagram. We have already discussed that some of the liquid refrigerant evaporates as it passes through the expansion valve, but the greater portion is vaporised in the evaporator. We know that during the throttling process, no heat is absorbed or rejected by the liquid refrigerant.

Notes : (*a*) In case an expansion cylinder is used in place of throttle or expansion valve to expand the liquid refrigerant, then the refrigerant will expand isentropically as shown by dotted vertical line on *T-s* diagram in Fig. 36.3 (*a*). The isentropic expansion reduces the external work being expanded in running the compressor and increases the refrigerating effect. Thus, the net result of using the expansion cylinder is to increase the coefficient of performance.

* The throttling process is an irreversible process.

Since the expansion cylinder system of expanding the liquid refrigerant is quite complicated and involves greater initial cost, therefore its use is not justified for small gain in cooling capacity. Moreover, the flow rate of the refrigerant can be controlled with throttle valve which is not possible in case of expansion cylinder which has a fixed cylinder volume.

(*b*) In modern domestic refrigerators, a capillary (small bore tube) is used in place of an expansion valve.

4. *Vaporising process.* The liquid-vapour mixture of the refrigerant at pressure $p_4 = p_1$ and temperature $T_4 = T_1$ is evaporated and changed into vapour refrigerant at constant pressure and temperature, as shown by the horizontal line 4-1 on *T-s* and *p-h* diagrams. During evaporation, the liquid-vapour refrigerant absorbs its latent heat of vaporisation from the medium (air, water or brine) which is to be cooled. This heat which is absorbed by the refrigerant is called *refrigerating effect.* The process of vaporisation continues upto point 1 which is the starting point and thus the cycle is completed.

We know that the refrigerating effect or the heat absorbed or extracted by the liquid-vapour refrigerant during evaporation per kg of refrigerant is given by

$$R_E = h_1 - h_4 = h_1 - h_{f3} \qquad \ldots (\because h_{f3} = h_4)$$

where h_{f3} = Sensible heat at temperature T_3.

It may be noticed from the cycle that the liquid-vapour refrigerant has extracted heat during evaporation and the work will be done by the compressor for isentropic compression of the high pressure and temperature vapour refrigerant.

∴ Coefficient of performance,

$$\text{C.O.P.} = \frac{\text{Refrigerating effect}}{\text{Work done}} = \frac{h_1 - h_{f3}}{h_2 - h_1}$$

Example 36.1. *The temperature limits of an ammonia refrigerating system are 25° C and – 10° C. If the gas is dry at the end of compression, calculate the coefficient of performance of the cycle assuming no undercooling of the liquid ammonia. Use the following table for properties of ammonia :*

Temperature (°C)	*Liquid heat (kJ/kg)*	*Latent heat (kJ/kg)*	*Liquid entropy (kJ/kg K)*
25	298.9	1166.94	1.1242
– 10	135.37	1297.68	0.5443

Solution. Given : $T_2 = T_3 = 25°\text{ C} = 298\text{ K}$; $T_1 = T_4 = -10°\text{ C} = 263\text{ K}$; $h_{f3} = h_{f2} = 298.9$ kJ/kg ; $h_{fg2} = 1166.94$ kJ/kg ; $s_{f2} = 1.1242$ kJ/kg K ; $h_{f1} = 135.37$ kJ/kg ; $h_{fg1} = 1297.68$ kJ/kg ; $s_{f1} = 0.5443$ kJ/kg K

The *T-s* and *p-h* diagrams are shown in Fig. 36.4 (*a*) and (*b*) respectively.

Let x_1 = Dryness fraction at point 1.

We know that entropy at point 1,

$$s_1 = s_{f1} + \frac{x_1 h_{fg1}}{T_1} = 0.5443 + \frac{x_1 \times 1297.68}{263}$$

$$= 0.5443 + 4.934\, x_1 \qquad \ldots (i)$$

Similarly, entropy at point 2,

$$s_2 = s_{f2} + \frac{h_{fg2}}{T_2} = 1.1242 + \frac{1166.94}{298} = 5.04 \qquad \ldots (ii)$$

Since the entropy at point 1 is equal to entropy at point 2, therefore equating equations (*i*) and (*ii*),

$$0.5443 + 4.934\, x_1 = 5.04 \quad \text{or} \quad x_1 = 0.91$$

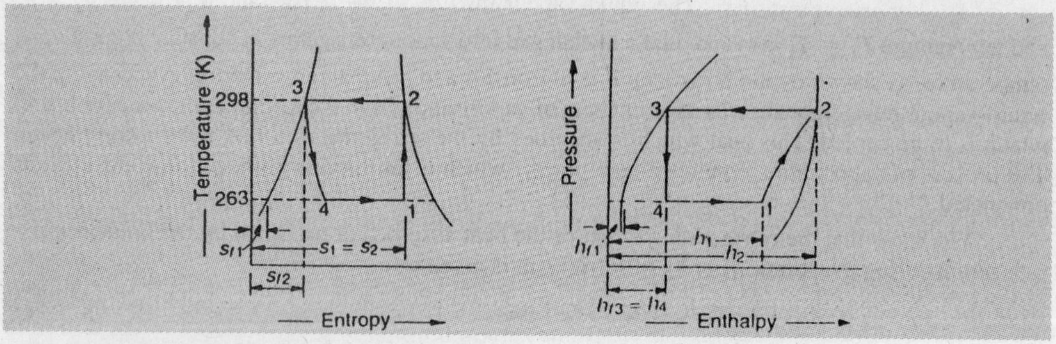

(*a*) *T-s* diagram. (*b*) *p-h* diagram.

Fig. 36.4

We know that enthalpy at point 1,

$$h_1 = h_{f1} + x_1\, h_{fg1} = 135.37 + 0.91 \times 1297.68 = 1316.26 \text{ kJ/kg}$$

and enthalpy at point 2,

$$h_2 = h_{f2} + h_{fg2} = 298.9 + 1166.94 = 1465.84 \text{ kJ/kg}$$

∴ Coefficient of performance of the cycle

$$= \frac{h_1 - h_{f3}}{h_2 - h_1} = \frac{1316.26 - 298.9}{1465.84 - 1316.26} = 6.8 \textbf{ Ans.}$$

Example 36.2. *A vapour compression refrigerator works between the pressure limits of 60 bar and 25 bar. The working fluid is just dry at the end of compression and there is no under cooling of the liquid before the expansion valve. Determine : 1. C.O.P. of the cycle; and 2. Capacity of the refrigerator if the fluid flow is at the rate of 5 kg/min. Data :*

Pressure, bar	*Saturation temperature, K*	*Enthalpy, kJ/kg*		*Entropy, kJ/kg K*	
		Liquid	*Vapour*	*Liquid*	*Vapour*
60	295	61.9	208.1	0.197	0.703
25	261	– 18.4	234.5	– 0.075	0.896

Solution. Given : $p_2 = p_3 = 60$ bar ; $p_1 = p_4 = 25$ bar ; $T_2 = T_3 = 295$ K ; $T_1 = T_4 = 261$ K ; $h_{f3} = h_4 = 61.9$ kJ/kg ; $h_{f1} = -18.4$ kJ/kg ; $h_{g2} = h_2 = 208.1$ kJ/kg ; $h_{g1} = 234.5$ kJ/kg ; *$s_{f2} = 0.197$ kJ/kg K ; $s_{f1} = -0.075$ kJ/kg K ; $s_{g1} = 0.896$ kJ/kg K ; $s_{g2} = s_2 = 0.703$ kJ/kg K

1. *C.O.P. of the cycle*

The *T-s* and *p-h* diagrams are shown in Fig. 36.5 (*a*) and (*b*) respectively.

Let x_1 = Dryness fraction of the vapour refrigerant entering the compressor at point 1.

* Superfluous data

We know that entropy at point 1 (s_1)

$$= \text{Entropy at point 2 } (s_2)$$

$$s_{f1} + x_1 s_{fg1} = s_2$$

$$s_{f1} + x_1 (s_{g1} - s_{f1}) = s_{g2} \qquad \ldots (\because s_{g1} = s_{f1} + s_{fg1} \text{ and } s_{g2} = s_2)$$

$$-0.075 + x_1 [0.896 - (-0.075)] = 0.703$$

$$\therefore \qquad 0.971 x_1 = 0.778 \text{ or } x_1 = 0.8$$

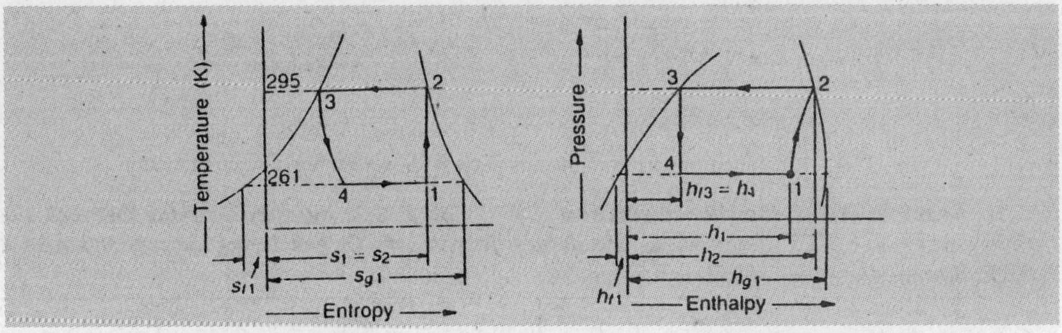

(a) T-s diagram. (b) p-h diagram.

Fig. 36.5

We know that enthalpy at point 1,

$$h_1 = h_{f1} + x_1 h_{fg1} = h_{f1} + x_1 (h_{g1} - h_{f1}) \qquad \ldots (h_{g1} = h_{f1} + h_{fg1})$$

$$= -18.4 + 0.8 [234.5 - (-18.4)] = 183.9 \text{ kJ/kg}$$

$$\therefore \text{ C.O.P. of the cycle} = \frac{h_1 - h_{f3}}{h_2 - h_1} = \frac{183.9 - 61.9}{208.1 - 183.9} = 5.04 \textbf{ Ans.}$$

2. *Capacity of the refrigerator*

We know that the heat extracted or refrigerating effect produced per kg of refrigerant

$$= h_1 - h_{f3} = 183.9 - 61.9 = 122 \text{ kJ/kg}$$

Since the fluid flow is at the rate of 5 kg/min, therefore total heat extracted

$$= 5 \times 122 = 610 \text{ kJ/min}$$

$$\therefore \text{ Capacity of the refrigerator} = \frac{610}{210} = 2.9 \text{ TR } \textbf{Ans.} \qquad \ldots (\because 1 \text{ TR} = 210 \text{ kJ/min})$$

36.7. Theoretical Vapour Compression Cycle with Wet Vapour after Compression

A vapour compression cycle with wet vapour after compression is shown on *T-s* and *p-h* diagrams in Fig 36.6 (*a*) and (*b*) respectively. In this cycle, the enthalpy or total heat at point 2 is found out with the help of dryness fraction at this point. The dryness fraction at point 1 and 2 may be obtained by equating entropies at points 1 and 2.

Now the coefficient of performance may be found out as usual from the relation,

$$\text{C.O.P.} = \frac{\text{Refrigerating effect}}{\text{Work done}} = \frac{h_1 - h_{f3}}{h_2 - h_1}$$

The remaining cycle is the same as discussed in the last article.

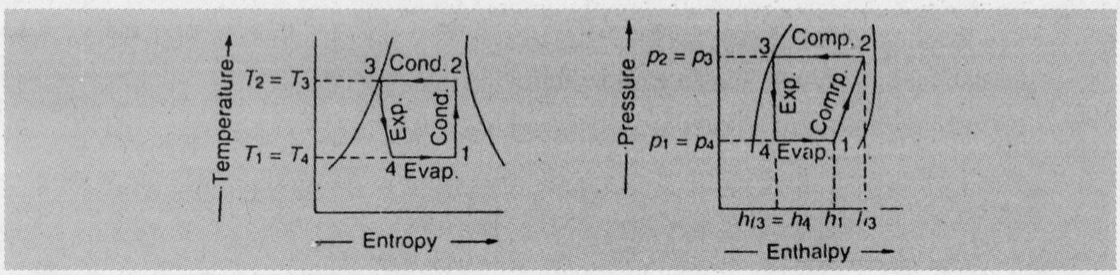

(a) T-s diagram. (b) p-h diagram.

Fig. 36.6. Theoretical vapour compression cycle with wet vapour after compression.

Example 36.3. *Find the theoretical C.O.P. for a CO_2 machine working between the temperature range of 25° C and −5° C. The dryness fraction of CO_2 gas during the suction stroke is 0.6. Following properties of CO_2 are given :*

Temperature °C	*Liquid*		*Vapour*		*Latent heat kJ/kg*
	Enthalpy kJkg	*Entropy kJ/kg K*	*Enthalpy kJ/kg*	*Entropy kJ/kg K*	
25	81.3	0.251	202.6	0.63	121.4
−5	−7.54	−0.042	237	0.84	245.3

Solution. Given : $T_2 = T_3 = 25°$ C = 298 K ; $T_1 = T_4 = -5°$C = 268 K ; x_1 = 0.6 ; $h_{f3} = h_{f2} = 81.3$ kJ/kg ; $h_{f1} = h_{f4} = -7.54$ kJ/kg ; $s_{f2} = 0.251$ kJ/kg K ; $s_{f1} = -0.042$ kJ/kg K ; $h_2' = 202.6$ kJ/kg ; $*h_1' = 237$ kJ/kg ; $*s_2' = 0.63$ kJ/kg K ; $*s_1' = 0.84$ kJ/kg K ; $h_{fg2} = 121.4$ kJ/kg ; $h_{fg1} = 245.3$ kJ/kg

The *T-s* and *p-h* diagrams are shown in Fig. 36.7 (*a*) and (*b*) respectively.

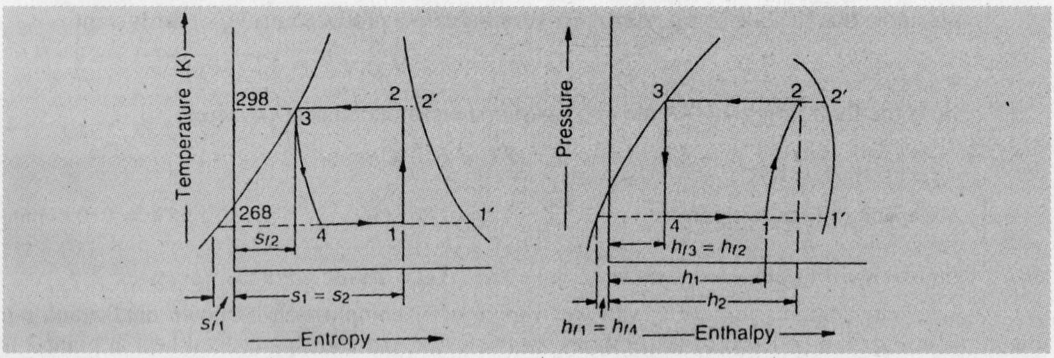

(*a*) *T-s* diagram. (*b*) *p-h* diagram.

Fig.36.7

First of all, let us find the dryness fraction at point 2, *i.e.* x_2. We know that the entropy at point 1,

* Superfluous data

$$s_1 = s_{f1} + \frac{x_1\, h_{fg1}}{T_1} = -0.042 + \frac{0.6 \times 245.3}{268} = 0.507 \quad \ldots (i)$$

Similarly, entropy at point 2,

$$s_2 = s_{f2} + \frac{x_2\, h_{fg2}}{T_2} = 0.251 + \frac{x_2 \times 121.4}{298} = 0.251 + 0.407\, x_2 \quad \ldots (ii)$$

Since the entropy at point 1 (s_1) is equal to entropy at point 2 (s_2), therefore equating equations (*i*) and (*ii*),

$$0.507 = 0.251 + 0.407\, x_2 \quad \text{or} \quad x_2 = 0.629$$

$\therefore$ Enthalpy at point 1, $h_1 = h_{f1} + x_1\, h_{fg1} = -7.54 + 0.6 \times 245.3 = 139.64$ kJ/kg

and enthalpy at point 2, $h_2 = h_{f2} + x_2\, h_{fg2} = 81.3 + 0.629 \times 121.4 = 157.66$ kJ/kg

$\therefore$ Theoretical C.O.P. $= \dfrac{h_1 - h_{f3}}{h_2 - h_1} = \dfrac{139.64 - 81.3}{157.66 - 139.64} = 3.24$ **Ans.**

Example 36.4. *An ammonia refrigerating machine fitted with an expansion valve works between the temperature limits of $-10°$ C and $30°$ C. The vapour is 95% dry at the end of isentropic compression and the fluid leaving the condenser is at $30°$ C. Assuming actual C.O.P. as 60% of the theoretical, calculate the kilograms of ice produced per kW hour at $0°$ C from water at $10°$ C. Latent heat of ice is 335 kJ/kg. Ammonia has the following properties :*

Temperature °C	*Liquid heat kJ/kg*	*Latent heat kJ/kg*	*Liquid entropy*	*Total entropy of dry saturated vapour*
30	*323.08*	*1145.80*	*1.2037*	*4.9842*
–10	*135.37*	*1297.68*	*0.5443*	*5.4770*

Solution. Given : $T_1 = T_4 = -10°$ C = 263 K ; $T_2 = T_3 = 30°$ C = 303° K ; $x_2 = 0.95$; $h_{f3} = h_{f2} = 323.08$ kJ/kg ; $h_{f1} = h_{f4} = 135.37$ kJ/kg ; $h_{fg2} = 1145.8$ kJ/kg ; $h_{fg1} = 1297.68$ kJ/kg ; $s_{f2} = 1.2037$; $s_{f1} = 0.5443$; $*s_2' = 4.9842$; $*s_1' = 5.4770$

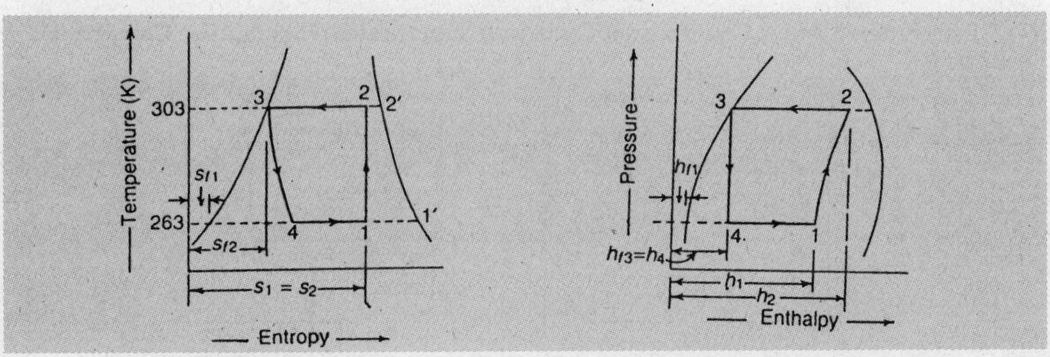

(*a*) *T-s* diagram. (*b*) *p-h* diagram.

Fig.36.8

The *T-s* and *p-h* diagrams are shown in Fig. 36.8 (*a*) and (*b*) respectively.

* Superfluous data

Let x_1 = Dryness fraction at point 1.

We know that entropy at point 1,

$$s_1 = s_{f1} + \frac{x_1 h_{fg1}}{T_1} = 0.5443 + \frac{x_1 \times 1297.68}{263} = 0.5443 + 4.934\, x_1 \quad \ldots (i)$$

Similarly, entropy at point 2,

$$s_2 = s_{f2} + \frac{x_2 h_{fg2}}{T_2} = 1.2037 + \frac{0.95 \times 1145.8}{303} = 4.796 \quad \ldots (ii)$$

Since the entropy at point 1 (s_1) is equal to entropy at point 2 (s_2), therefore equating equations (*i*) and (*ii*),

$$0.5443 + 4.934\, x_1 = 4.796 \quad \text{or} \quad x_1 = 0.86$$

∴ Enthalpy at point 1, $h_1 = h_{f1} + x_1 h_{fg1} = 135.37 + 0.86 \times 1297.68 = 1251.4$ kJ/kg

and enthalpy at point 2, $h_2 = h_{f2} + x_2 h_{fg2} = 323.08 + 0.95 \times 1145.8 = 1411.6$ kJ/kg

We know that theoretical C.O.P.

$$= \frac{h_1 - h_{f3}}{h_2 - h_1} = \frac{1251.4 - 323.08}{1411.6 - 1251.4} = 5.8$$

∴ Actual C.O.P. $= 0.6 \times 5.8 = 3.48$

Work to be spent corresponding to 1 kW hour,

$$W = 3600 \text{ kJ}$$

∴ Actual heat extracted or refrigeration effect produced per kW hour

$$= W \times \text{Actual C.O.P.} = 3600 \times 3.48 = 12\,528 \text{ kJ}$$

We know that heat extracted from 1 kg of water at 10°C for the formation of 1 kg of ice at 0°C

$$= 1 \times 4.187 \times 10 + 335 = 376.87 \text{ kJ}$$

∴ Amount of ice produced $= \dfrac{12\,528}{376.87} = 33.2$ kg/kW hour **Ans.**

36.8. Theoretical Vapour Compression Cycle with Superheated Vapour after Compression

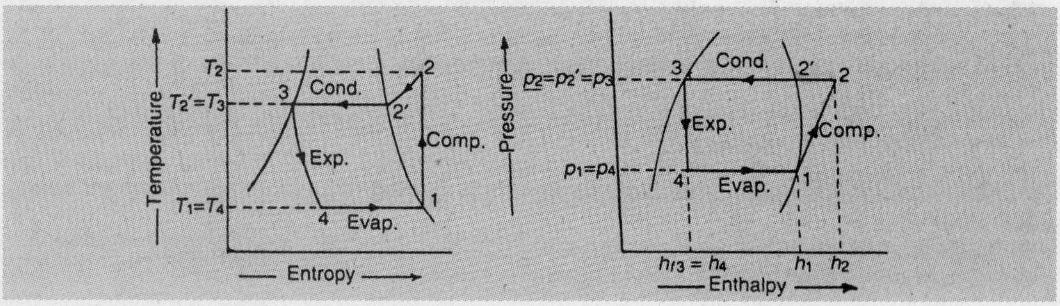

(*a*) *T-s* diagram. (*b*) *p-h* diagram.

Fig. 36.9. Theoretical vapour compression cycle with superheated vapour after compression.

A vapour compression cycle with superheated vapour after compression is shown on *T-s* and *p-h* diagrams in Fig. 36.9 (*a*) and (*b*) respectively. In this cycle, the enthalpy or total heat at point 2 is found out with the help of degree of superheat. The degree of superheat may be found out by equating the entropies at points 1 and 2.

Now the coefficient of performance may be found out as usual from the relation,

$$\text{C.O.P.} = \frac{\text{Refrigerating effect}}{\text{Work done}} = \frac{h_1 - h_{f3}}{h_2 - h_1}$$

A little consideration will show that the superheating increases the refrigerating effect and the amount of work done in the compressor. Since the increase in refrigerating effect is less as compared to the increase in work done, therefore, the net effect of superheating is to have low coefficient of performance.

Note : In this cycle, the cooling of superheated vapour will take place in two stages. Firstly, it will be condensed to dry saturated state at constant pressure (shown by graph 2-2′) and secondly it will be condensed at constant temperature (shown by graph 2′-3). The remaining cycle is same as discussed in the last article.

Example 36.5. *A vapour compression refrigerator uses methyl chloride (R-40) and operates between temperature limits of – 10° C and 45° C. At entry to the compressor, the refrigerant is dry saturated and after compression it acquires a temperature of 60° C. Find the C.O.P. of the refrigerator. The relevant properties of methyl chloride are as follows :*

Saturation Temperature in °C	*Enthalpy in kJ/kg*		*Entropy in kJ/kg K*	
	Liquid	*Vapour*	*Liquid*	*Vapour*
–10	*45.4*	*460.7*	*0.183*	*1.637*
45	*133.0*	*483.6*	*0.485*	*1.587*

Solution. Given : $T_1 = T_4 = -10°\text{ C} = 263\text{ K}$; $T_2' = T_3 = 45°\text{ C} = 318\text{ K}$; $T_2 = 60°\text{ C} = 333\text{ K}$; $^*h_{f1} = 45.4$ kJ/kg ; $h_{f3} = 133$ kJ/kg ; $h_1 = 460.7$ kJ/kg ; $h_2' = 483.6$ kJ/kg ; $^*s_{f1} = 0.183$ kJ/kg K ; $^*s_{f3} = 0.485$ kJ/kg K ; $s_2 = 1.637$ kJ/kg K ; $s_2' = 1.587$ kJ/kg

The *T-s* and *p-h* diagrams are shown in Fig. 36.10 (*a*) and (*b*) respectively.

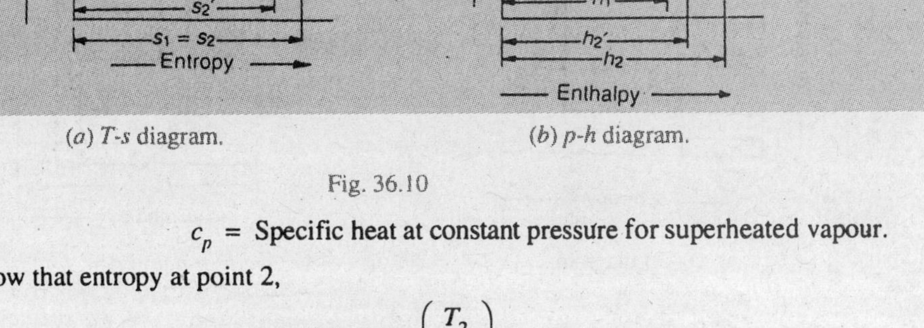

(*a*) *T-s* diagram. (*b*) *p-h* diagram.

Fig. 36.10

Let c_p = Specific heat at constant pressure for superheated vapour.

We know that entropy at point 2,

$$s_2 = s_2' + 2.3\, c_p \log\left(\frac{T_2}{T_2'}\right)$$

* Superfluous data

$$1.637 = 1.587 + 2.3\, c_p \log\left(\frac{333}{318}\right)$$
$$= 1.587 + 2.3\, c_p \times 0.02 = 1.587 + 0.046\, c_p$$

$\therefore \quad c_p = 1.09$

and enthalpy at point 2, $h_2 = h_2' + c_p \times$ Degree of superheat $= h_2' + c_p\,(T_2 - T_2')$

$$= 483.6 + 1.09\,(333 - 318) = 500 \text{ kJ/kg}$$

$\therefore$ C.O.P. of the refrigera tor $= \dfrac{h_1 - h_{f3}}{h_2 - h_1} = \dfrac{460.7 - 133}{500 - 460.7} = 8.34$ **Ans.**

Example 36.6. *A refrigeration machine using R-12 as refrigerant operates between the pressures 2.5 bar and 9 bar. The compression is isentropic and there is no undercooling in the condenser.*

The vapour is in dry saturated condition at the beginning of the compression. Estimate the theoretical coefficient of performance. If the actual coefficient of performance is 0.65 of theoretical value, calculate the net cooling produced per hour. The refrigerant flow is 5 kg per minute. Properties of refrigerant are :

Pressure, bar	*Saturation temperature, °C*	*Enthalpy, kJ/kg*		*Entropy of saturated vapour, kJ/kg K*
		Liquid	*Vapour*	
9.0	*36*	*456.4*	*585.3*	*4.74*
2.5	*–7*	*412.4*	*570.3*	*4.76*

Take c_p for superheated vapour at 9 bar as 0.67 kJ/kg K.

Solution. Given : $T_2' = T_3 = 36°\text{ C} = 309\text{ K}$; $T_1 = T_4 = -7°\text{ C} = 266\text{ K}$; $(\text{C.O.P.})_{actual} = 0.65\,(\text{C.O.P.})_{th}$; $m = 5$ kg/min $= 300$ kg/h ; $h_{f3} = h_4 = 456.4$ kJ/kg ; $*h_{f1} = h_{f4} = 412.4$ kJ/kg ; $h_2' = 585.3$ kJ/kg ; $h_1 = 570.3$ kJ/kg ; $s_2' = 4.74$ kJ/kg K ; $s_1 = s_2 = 4.76$ kJ/kg K ; $c_p = 0.67$ kJ/kg K

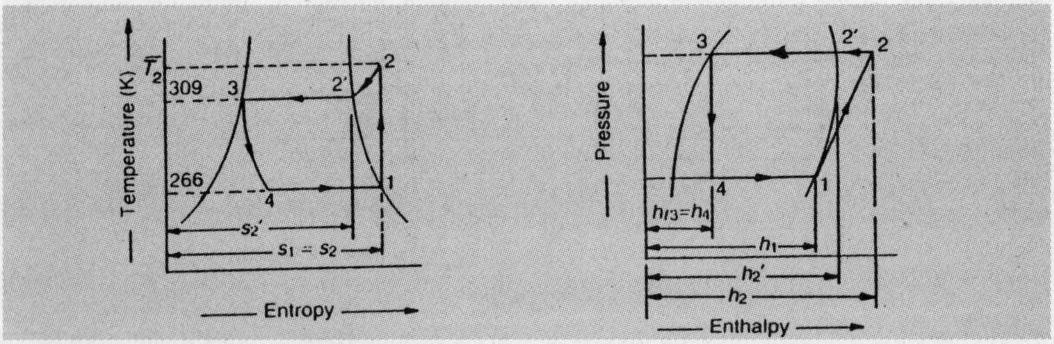

(*a*) *T-s* diagram. (*b*) *p-h* diagram.

Fig. 36.11

The *T-s* and *p-h* diagrams are shown in Fig. 36.11 (*a*) and (*b*) respectively.

* Superfluous data

Theoretical coefficient of performance

First of all, let us find the temperature at point 2 (T_2).

We know that entropy at point 2,

$$s_2 = s_2' + 2.3\, c_p \log\left(\frac{T_2}{T_2'}\right)$$

$$4.76 = 4.74 + 2.3 \times 0.67 \log\left(\frac{T_2}{309}\right)$$

$$\therefore \quad \log\left(\frac{T_2}{309}\right) = \frac{4.76 - 4.74}{2.3 \times 0.67} = 0.013$$

$$\frac{T_2}{309} = 1.03 \qquad \text{...(Taking antilog of 0.013)}$$

$$\therefore \quad T_2 = 309 \times 1.03 = 318.3 \text{ K}$$

We know that enthalpy of superheated vapour at point 2,

$$h_2 = h_2' + c_p(T_2 - T_2')$$

$$= 585.3 + 0.67\,(318.3 - 309) = 591.5 \text{ kJ/kg}$$

$\therefore$ Theoretical coefficient of performance,

$$(\text{C.O.P.})_{th} = \frac{h_1 - h_{f3}}{h_2 - h_1} = \frac{570.3 - 456.4}{591.5 - 570.3} = 5.37 \textbf{ Ans.}$$

Net cooling produced per hour

We also know that actual C.O.P. of the machine,

$$(\text{C.O.P.})_{actual} = 0.65 \times (\text{C.O.P.})_{th} = 0.65 \times 5.37 = 3.49$$

and actual work done,

$$W_{actual} = h_2 - h_1 = 591.5 - 570.3 = 21.2 \text{ kJ/kg}$$

We know that net cooling (or refrigerating effect) produced per kg of refrigerant

$$= W_{actual} \times (\text{C.O.P.})_{actual} = 21.2 \times 3.49 = 74 \text{ kJ/kg}$$

$\therefore$ Net cooling produced per hour

$$= m \times 74 = 300 \times 74 = 22\,200 \text{ kJ/h}$$

$$= \frac{22\,200}{210 \times 60} = 1.76 \text{ TR } \textbf{Ans.} \qquad \text{...}(\because 1 \text{ TR} = 210 \text{ kJ/min})$$

36.9. Theoretical Vapour Compression Cycle with Superheated Vapour before Compression

A vapour compression cycle with superheated vapour before compression is shown on *T-s* and *p-h* diagrams in Fig. 36.12 (*a*) and (*b*) respectively. In this cycle, the evaporation starts at point 4 and continues upto point 1′, when it is dry saturated. The vapour is now superheated before entering the compressor upto the point 1.

The coefficient of performance may be found out as usual from the relation,

$$\text{C.O.P.} = \frac{\text{Refrigerating effect}}{\text{Work done}} = \frac{h_1 - h_{f3}}{h_2 - h_1}$$

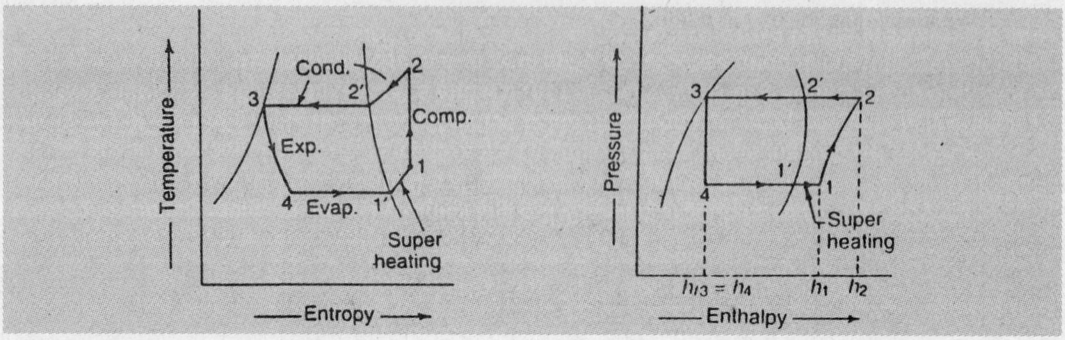

(a) T-s diagram. (b) p-h diagram.

Fig. 36.12. Theoretical vapour compression cycle with superheated vapour before compression.

Note : In this cycle, the heat is absorbed (or extracted) in two stages. Firstly from point 4 to point 1′ and from point 1′ to point 1. The remaining cycle is same as discussed in the previous article.

Example 36.7. *A vapour compression refrigeration plant works between pressure limits of 5.3 bar and 2.1 bar. The vapour is superheated at the end of compression, its temperature being 37° C. The vapour is superheated by 5° C before entering the compressor.*

If the specific heat of superheated vapour is 0.63 kJ/kg K, find the coefficient of performance of the plant. Use the data given below :

Pressure, bar	*Saturation temperature, °C*	*Liquid heat, kJ/kg*	*Latent heat, kJ/kg*
5.3	15.5	56.15	144.9
2.1	– 14.0	25.12	158.7

Solution. Given : $p_2 = 5.3$ bar ; $p_1 = 2.1$ bar ; $T_2 = 37° \text{ C} = 310 \text{ K}$; $T_1 - T_1' = 5° \text{ C}$; $c_{pv} = 0.63$ kJ/kg K ; $T_2' = 15.5° \text{ C} = 288.5 \text{ K}$; $T_1' = -14° \text{ C} = 259 \text{ K}$; $h_{f3} = h_{f2}' = 56.15$ kJ/kg ; $h_{f1}' = 25.12$ kJ/kg ; $h_{fg2}' = 144.9$ kJ/kg ; $h_{fg1}' = 158.7$ kJ/kg

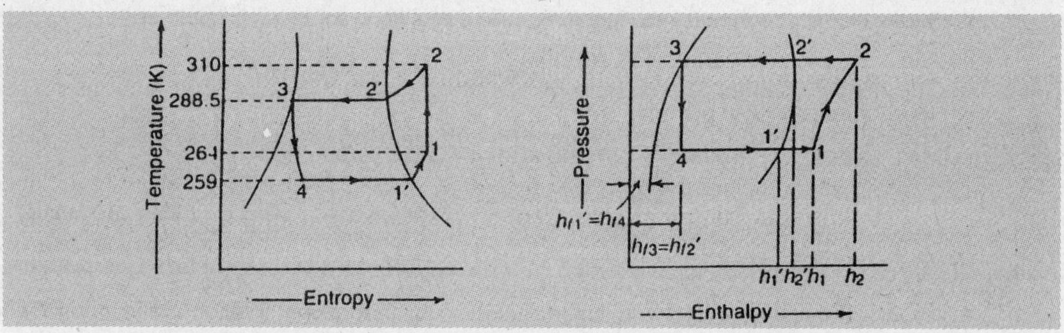

(a) T-s diagram. (b) p-h diagram.

Fig. 36.13

The *T-s* and *p-h* diagrams are shown in Fig. 36.13 (*a*) and (*b*) respectively.

We know that enthalpy of vapour at point 1,

$$h_1 = h_1' + c_{pv}(T_1 - T_1') = (h_{f1}' + h_{fg1}') + c_{pv}(T_1 - T_1')$$

$$= (25.12 + 158.7) + 0.63 \times 5 = 186.97 \text{ kJ/kg}$$

Similarly, enthalpy of vapour at point 2,

$$h_2 = h_2' + c_{pv}(T_2 - T_2') = (h_{f2}' + h_{fg2}') + c_{pv}(T_2 - T_2')$$

$$= (56.15 + 144.9) + 0.63\,(310 - 288.5) = 214.6 \text{ kJ/kg}$$

∴ Coefficient of performance of the plant,

$$\text{C.O.P.} = \frac{h_1 - h_{f3}}{h_2 - h_1} = \frac{186.97 - 56.15}{214.6 - 186.97} = \frac{130.82}{27.63} = 4.735 \textbf{ Ans.}$$

36.10. Theoretical Vapour Compression Cycle with Undercooling or Sub-cooling of Refrigerant

Sometimes, the refrigerant, after condensation process 2′-3′, is cooled below the saturation temperature (T_3') before expansion by throttling. Such a process is called *undercooling* or *subcooling* of the refrigerant and is generally done along the liquid line as shown in Fig. 36.14 (*a*) and (*b*). The ultimate effect of the undercooling is to increase the value of coefficient of performance under the same set of conditions.

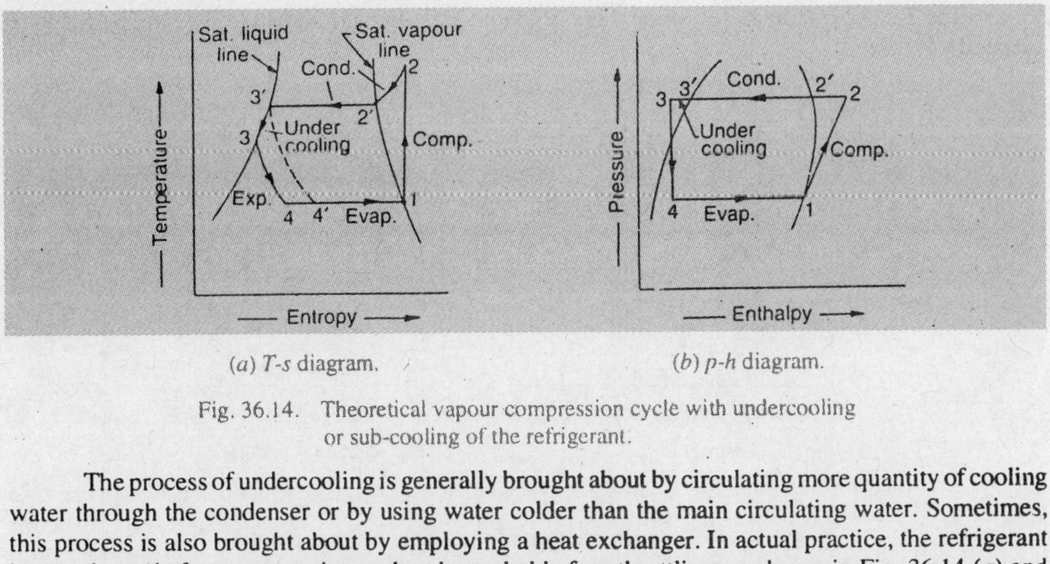

Fig. 36.14. Theoretical vapour compression cycle with undercooling or sub-cooling of the refrigerant.

The process of undercooling is generally brought about by circulating more quantity of cooling water through the condenser or by using water colder than the main circulating water. Sometimes, this process is also brought about by employing a heat exchanger. In actual practice, the refrigerant is superheated after compression and undercooled before throttling, as shown in Fig. 36.14 (*a*) and (*b*). A little consideration will show, that the refrigerating effect is increased by adopting both the superheating and undercooling process as compared to a cycle without them, which is shown by dotted lines in Fig. 36.14 (*a*).

In this case, the refrigerating effect or heat absorbed or extracted,

$$R_E = h_1 - h_4 = h_1 - h_{f3} \quad \ldots (\because h_4 = h_{f3})$$

and work done, $W = h_2 - h_1$

$$\therefore \quad \text{C.O.P.} = \frac{\text{Refrigerating effect}}{\text{Work done}} = \frac{h_1 - h_{f3}}{h_2 - h_1}$$

Note : The value of h_{f3} may be found out from the relation,

$$h_{f3} = h_{f3}' - c_p \times \text{Degree of undercooling}$$

Example 36.8. *A vapour compression refrigerator uses R-12 as refrigerant and the liquid evaporates in the evaporator at – 15°C. The temperature of this refrigerant at the delivery from the compressor is 15° C when the vapour is condensed at 10° C. Find the coefficient of performance if 1. there is no undercooling ; and 2. the liquid is cooled by 5° C before expansion by throttling.*

Take specific heat at constant pressure for the superheated vapour as 0.64 kJ/kg K and that for liquid as 0.94 kJ/kg K. The other properties of refrigerant are as follows :

Temperature in °C	*Enthalpy in kJ/kg*		*Entropy in kJ/kg K*	
	Liquid	*Vapour*	*Liquid*	*Vapour*
–15	*22.3*	*180.88*	*0.0904*	*0.7051*
+10	*45.4*	*191.76*	*0.1750*	*0.6921*

Solution. Given : $T_1 = T_4 = -15°\text{C} = 258\text{ K}$; $T_2 = 15°\text{C} = 288\text{ K}$; $T_2' = 10°\text{C} = 283\text{ K}$; $c_{pv} = 0.64$ kJ/kg K ; $c_{pl} = 0.94$ kJ/kg K ; $h_{f1} = 22.3$ kJ/kg ; $h_{f3}' = 45.4$ kJ/kg ; $h_1' = 180.88$ kJ/kg ; $h_2' = 191.76$ kJ/kg ; $s_{f1} = 0.0904$ kJ/kg K ; $*s_{f2} = 0.1750$ kJ/kg K ; $s_{g1} = 0.7051$ kJ/kg K ; $s_2' = 0.6921$ kJ/kg K

1. *Coefficient of performance if there is no undercooling*

The *T-s* and *p-h* diagrams, when there is no undercooling, are shown in Fig. 36.15 (*a*) and (*b*) respectively.

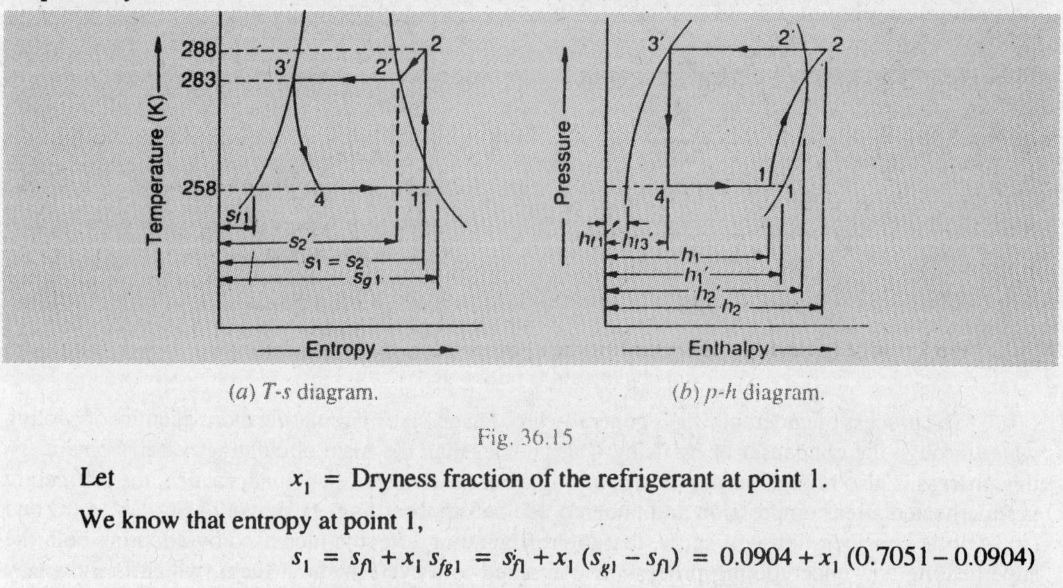

(*a*) *T-s* diagram. (*b*) *p-h* diagram.

Fig. 36.15

Let x_1 = Dryness fraction of the refrigerant at point 1.

We know that entropy at point 1,

$$s_1 = s_{f1} + x_1 s_{fg1} = s_{f1} + x_1 (s_{g1} - s_{f1}) = 0.0904 + x_1 (0.7051 - 0.0904)$$

$$= 0.0904 + 0.6147 x_1 \quad \ldots (i)$$

and entropy at point 2,

$$s_2 = s_2' + 2.3 c_{pv} \log\left(\frac{T_2}{T_2'}\right) = 0.6921 + 2.3 \times 0.64 \log\left(\frac{288}{283}\right)$$

$$= 0.6921 + 2.3 \times 0.64 \times 0.0077 = 0.7034 \quad \ldots (ii)$$

* Superfluous data

Since the entropy at point 1 is equal to entropy at point 2, therefore equating equations (*i*) and (*ii*),

$$0.0904 + 0.6147\, x_1 = 0.7034 \quad \text{or} \quad x_1 = 0.997$$

We know that the enthalpy at point 1,

$$h_1 = h_{f1} + x_1\, h_{fg1} = h_{f1} + x_1\,(h_1' - h_{f1}) \qquad \ldots (\because h_{fg1} = h_1' - h_{f1})$$

$$= 22.3 + 0.997\,(180.88 - 22.3) = 180.4 \text{ kJ/kg}$$

and enthalpy at point 2,

$$h_2 = h_2' + c_{pv}\,(T_2 - T_2')$$

$$= 191.76 + 0.64\,(288 - 283) = 194.96 \text{ kJ/kg}$$

$$\therefore \quad \text{C.O.P.} = \frac{h_1 - h_{f3}'}{h_2 - h_1} = \frac{180.4 - 45.4}{194.96 - 180.4} = 9.27 \ \textbf{Ans.}$$

2. *Coefficient of performance when there is an undercooling of 5° C*

The *T-s* and *p-h* diagrams, when there is an undercooling of 5° C, are shown in Fig. 36.16 (*a*) and (*b*) respectively.

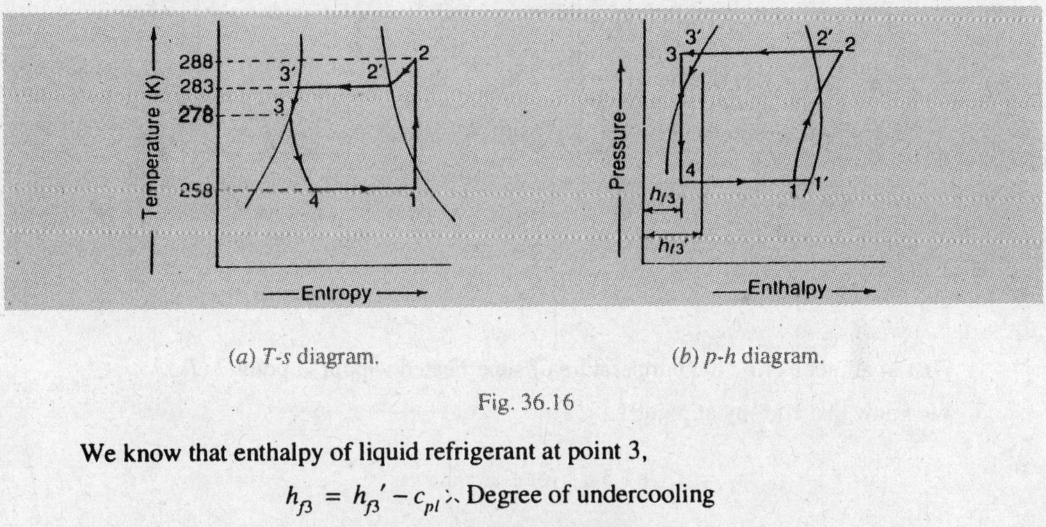

(*a*) *T-s* diagram. (*b*) *p-h* diagram.

Fig. 36.16

We know that enthalpy of liquid refrigerant at point 3,

$$h_{f3} = h_{f3}' - c_{pl} \times \text{Degree of undercooling}$$

$$= 45.4 - 0.94 \times 5 = 40.7 \text{ kJ/kg}$$

$$\therefore \quad \text{C.O.P.} = \frac{h_1 - h_{f3}}{h_2 - h_1} = \frac{180.4 - 40.7}{194.96 - 180.4} = 9.59 \ \textbf{Ans.}$$

Example 36.9. *A food storage locker requires a refrigeration capacity of 12 TR and works between the evaporating temperature of – 8° C and condensing temperature of 30° C. The refrigerant R-12 is subcooled by 5° C before entry to expansion valve and the vapour is superheated to – 2° C before leaving the evaporator coils. Determine : 1. coefficient of performance ; and 2. theoretical power per tonne of refrigeration.*

Use the following data for R-12

Saturation temperature, °C	Pressure, bar	Enthalpy, kJ/kg		Entropy, kJ/kg K	
		Liquid	Vapour	Liquid	Vapour
–8	2.354	28.72	184.07	0.1149	0.7007
30	7.451	64.59	199.62	0.2400	0.6853

The specific heat of liquid R-12 is 1.235 kJ/kg K, and of vapour R-12 is 0.733 kJ/kg K.

Solution. Given : Q = 12 TR ; $T_1' = -8°$C = 265 K; $T_2' = 30°$C = 303 K; $T_3' - T_3 = 5°$C ; $T_1 = -2°$C = 271 K ; h_{f1} = 28.72 kJ/kg ; h_{f3}' = 64.59 kJ/kg ; h_1' = 184.07 kJ/kg ; h_2' = 199.62 kJ/kg ; *s_{f1} = 0.1149 kJ/kg K ; *s_{f3} = 0.2400 kJ/kg K ; s_1' = 0.7007 kJ/kg K ; s_2' = 0.6853 kJ/kg K ; c_{pl} = 1.235 kJ/kg K ; c_{pv} = 0.733 kJ/kg K

The *T-s* and *p-h* diagrams are shown in Fig. 36.17 (*a*) and (*b*) respectively,

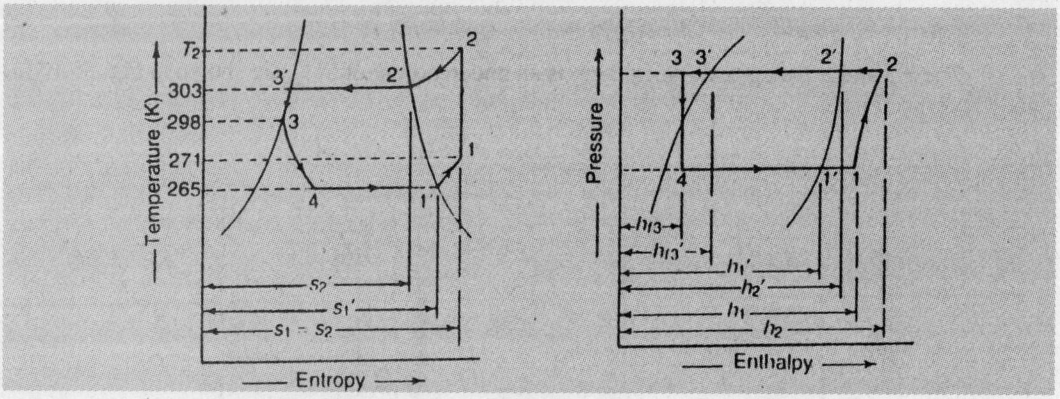

(*a*) *T-s* diagram. (*b*) *p-h* diagram.

Fig. 36.17

1. *Coefficient of performance*

First of all, let us find the temperature of superheated vapour at point 2 (T_2).

We know that entropy at point 1,

$$s_1 = s_1' + 2.3\, c_{pv} \log\left(\frac{T_1}{T_1'}\right)$$

$$= 0.7007 + 2.3 \times 0.733 \log\left(\frac{271}{265}\right) = 0.7171 \quad \ldots (i)$$

and entropy at point 2,

$$s_2 = s_2' + 2.3\, c_{pv} \log\left(\frac{T_2}{T_2'}\right)$$

$$= 0.6853 + 2.3 \times 0.733 \log\left(\frac{T_2}{303}\right)$$

$$= 0.6853 + 1.686 \log\left(\frac{T_2}{303}\right) \quad \ldots (ii)$$

* Superfluous data

Since the entropy at point 1 is equal to entropy at point 2, therefore equating equations (*i*) and (*ii*),

$$0.7171 = 0.6853 + 1.686 \log \left(\frac{T_2}{303} \right)$$

or $$\log \left(\frac{T_2}{303} \right) = \frac{0.7171 - 0.6853}{1.686} = 0.0188$$

$$\frac{T_2}{303} = 1.0444 \qquad \text{...(Taking antilog of 0.0188)}$$

$$\therefore \quad T_2 = 316.4 \text{ K or } 43.4\,^\circ\text{C}$$

We know that enthalpy at point 1,

$$h_1 = h_1' + c_{pv}(T_1 - T_1')$$

$$= 184.07 + 0.733\,(271 - 265) = 188.47 \text{ kJ/kg}$$

Enthalpy at point 2, $$h_2 = h_2' + c_{pv}(T_2 - T_2')$$

$$= 199.62 + 0.733\,(316.4 - 303) = 209.44 \text{ kJ/kg}$$

and enthalpy of liquid refrigerant at point 3,

$$h_{f3} = h_{f3}' - c_{pl}(T_3' - T_3)$$

$$= 64.59 - 1.235 \times 5 = 58.42 \text{ kJ/kg}$$

$$\therefore \quad \text{C.O.P.} = \frac{h_1 - h_{f3}}{h_2 - h_1} = \frac{188.47 - 58.42}{209.44 - 188.47} = \frac{130.05}{20.97} = 6.2 \text{ Ans.}$$

2. *Theoretical power per tonne of refrigeration*

We know that the heat extracted or refrigerating effect per kg of the refrigerant,

$$R_E = h_1 - h_{f3} = 188.47 - 58.42 = 130.05 \text{ kJ/kg}$$

and the refrigerating capacity of the system,

$$Q = 12 \text{ TR} = 12 \times 210 = 2520 \text{ kJ/min} \qquad \text{...(Given)}$$

∴ Mass flow of the refrigerant,

$$m_R = \frac{Q}{R_E} = \frac{2520}{130.05} = 19.4 \text{ kg/min}$$

Work done during compression of the refrigerant

$$= m_R(h_2 - h_1) = 19.4\,(209.44 - 188.47) = 406.82 \text{ kJ/min}$$

∴ Theoretical power per tonne of refrigeration

$$= \frac{406.82}{60 \times 12} = 0.565 \text{ kW/TR Ans.}$$

36.11. Actual Vapour Compression Cycle

The actual vapour compression cycle differs from the theoretical vapour compression cycle in many ways, some of which are unavoidable and cause losses. The main deviations between the theoretical cycle and actual cycle are as follows :

1. The vapour refrigerant leaving the evaporator is in superheated state.

2. The compression of refrigerant is neither isentropic nor polytropic.
3. The liquid refrigerant before entering the expansion valve is sub-cooled in the condenser.
4. The pressure drops in the evaporator and condenser.

The actual vapour compression cycle on *T-s* diagram is shown in Fig. 36.18. The various processes are discussed below :

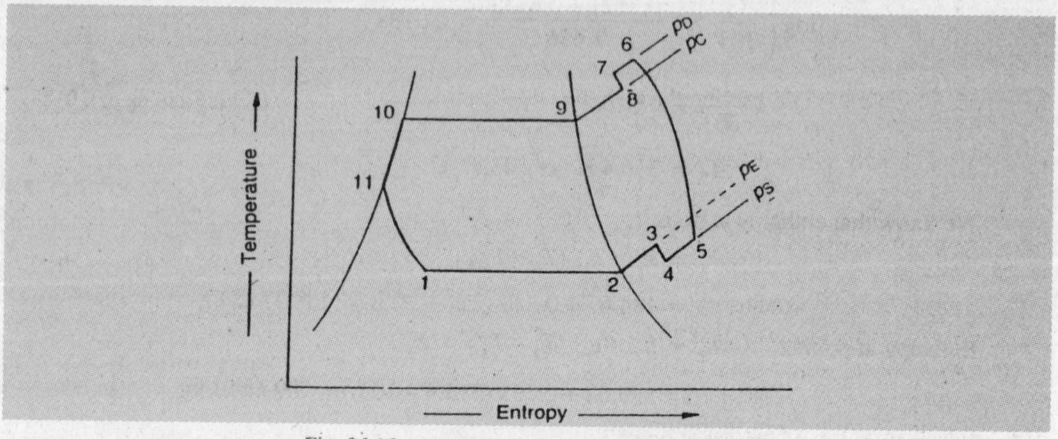

Fig. 36.18. Actual vapour compression cycle.

(*a*) *Process 1-2-3.* This process shows the flow of refrigerant in the evaporator. The point 1 represents the entry of refrigerant into the evaporator and the point 3 represents the exit of refrigerant from the evaporator in a superheated state. The point 3 also represents the entry of refrigerant into the compressor in a superheated condition. The superheating of vapour refrigerant from point 2 to point 3 may be due to :

(*i*) automatic control of expansion valve so that the refrigerant leaves the evaporator as the superheated vapour.

(*ii*) picking up of larger amount of heat from the evaporator through pipes located within the cooled space.

(*iii*) picking up of heat from the suction pipe, *i.e.* the pipe connecting the evaporator delivery and the compressor suction valve.

In the first and second case of superheating the vapour refrigerant, the refrigerating effect as well as the compressor work is increased. The coefficient of performance, as compared to saturation cycle at the same suction pressure may be greater, less or unchanged.

The superheating also causes increase in the required displacement of compressor and load on the compressor and condenser. This is indicated by 2-3 on *T-s* diagram as shown in Fig. 36.18.

(*b*) *Process 3-4-5-6-7-8.* This process represents the flow of refrigerant through the compressor. When the refrigerant enters the compressor through the suction valve at point 3, the pressure falls to point 4 due to frictional resistance to flow. Thus the actual suction pressure (p_S) is lower than the evaporator pressure (p_E). During suction and prior to compression, the temperature of the cold refrigerant vapour rises to point 5 when it comes in contact with the compressor cylinder walls. The actual compression of the refrigerant is shown by 5-6 in Fig. 36.18, which is neither isentropic nor polytropic. This is due to the heat transfer between the cylinder walls and the vapour refrigerant. The temperature of the cylinder walls is some-what in between the temperature of cold suction vapour refrigerant and hot discharge vapour refrigerant. It may be assumed that the heat absorbed by the vapour refrigerant from the cylinder walls during the first part of the compression stroke is equal to heat rejected by the vapour refrigerant to the cylinder walls. Like the heating effect at suction given by 4-5 in Fig. 36.18, there is a cooling effect at discharge as given by 6-7. These heating and cooling effects take place at constant pressure. Due to the frictional resistance of flow, there is a pressure drop *i.e.* the actual discharge pressure (p_D) is more than the condenser pressure (p_C).

(c) Process 8-9-10-11. This process represents the flow of refrigerant through the condenser. The process 8-9 represents the cooling of superheated vapour refrigerant to the dry saturated state. The process 9-10 shows the removal of latent heat which changes the dry saturated refrigerant into liquid refrigerant. The process 10-11 represents the sub-cooling of liquid refrigerant in the condenser before passing through the expansion valve. This is desirable as it increases the refrigerating effect per kg of the refrigerant flow. It also reduces the volume of the refrigerant partially evaporated from the liquid refrigerant while passing through the expansion valve. The increase in refrigerating effect can be obtained by large quantities of circulating cooling water which should be at a temperature much lower than the condensing temperatures.

(d) Process 11-1 This process represents the expansion of sub-cooled liquid refrigerant by throttling from the condenser pressure to the evaporator pressure.

36.12. Vapour Absorption Refrigeration System

The idea of a vapour absorption refrigeration system is to avoid compression of the refrigerant. In this type of refrigeration system, the vapour produced by the evaporation of the refrigerant, in the cold chamber, passes into a vessel containing a homogeneous mixture of ammonia and water (known as aqua-ammonia). In this chamber, the vapour is absorbed, which maintains constant low pressure, thus facilitating its further evaporation.

The refrigerant is liberated in the vapour state subsequently by the direct application of heat, and at such a pressure that condensation can be effected at the temperature of the air or by cold water. Fig. 36.19 shows a schematic arrangement of the essential elements of such a system.

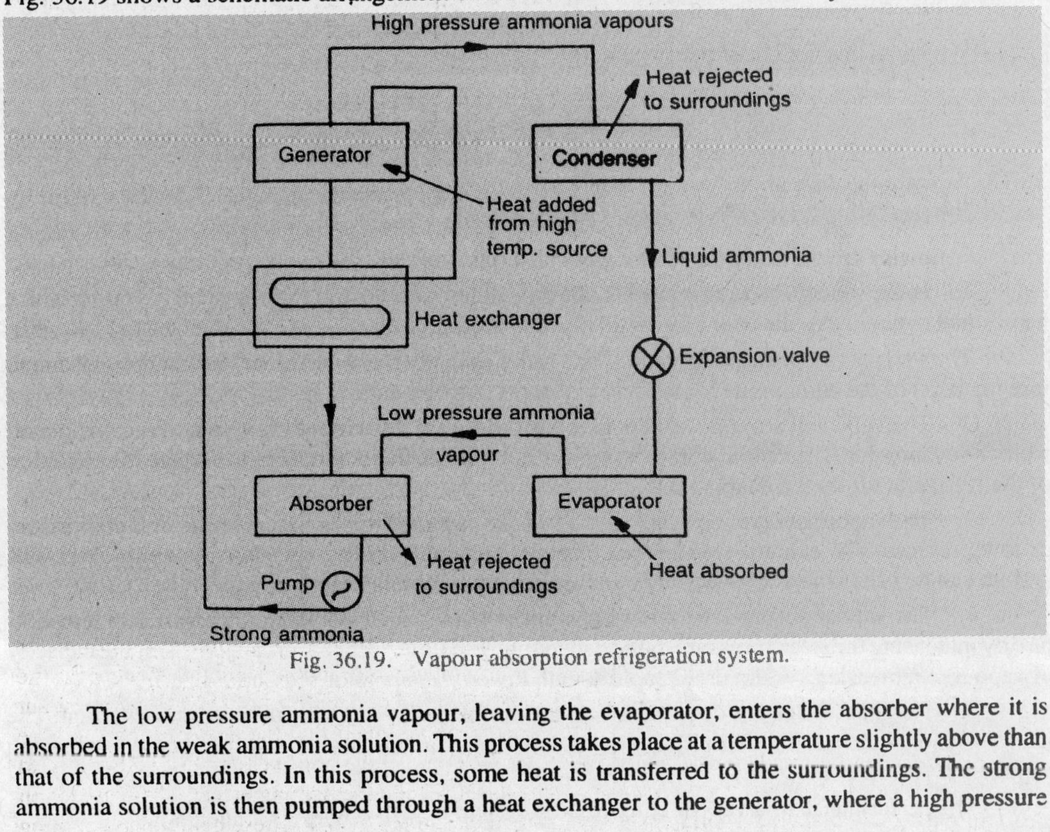

Fig. 36.19. Vapour absorption refrigeration system.

The low pressure ammonia vapour, leaving the evaporator, enters the absorber where it is absorbed in the weak ammonia solution. This process takes place at a temperature slightly above than that of the surroundings. In this process, some heat is transferred to the surroundings. The strong ammonia solution is then pumped through a heat exchanger to the generator, where a high pressure and temperature is maintained. Under these conditions, the ammonia vapour is driven from the solution. This happens because of the heat transfer from a high temperature source. The ammonia vapour enters into the condenser, where it gets condensed, in the same way as in the vapour compression system. The weak ammonia solution returns back to the absorber through a heat exchanger.

The equipment used in a vapour absorption system is somewhat complicated than in a vapour compression system. It can be economically justified only in those cases where a suitable source of heat is available which would otherwise be wasted.

The coefficient of performance of this refrigerator is given by :

$$\text{C.O.P.} = \frac{\text{Heat absorbed during evaporation}}{\text{Work done by pump + Heat supplied in heat exchanger}}$$

Mathematically,
$$\text{C.O.P.} = \frac{T_3\,(T_1 - T_2)}{T_1\,(T_2 - T_3)}$$

where T_1 = Temperature at which the working substance receives heat,

T_2 = Temperature of the cooling water, and

T_3 = Evaporator temperature.

Example 36.10. *In an absorption type refrigerator, heating, cooling and refrigeration takes place at the temperature of 100° C ; 20° C and – 5° C respectively. Find the theoretical C.O.P. of the system.*

Solution. Given : T_1 = 100° C = 373 K ; T_2 = 20° C = 293 K ; T_3 = – 5° C = 268 K

We know that C.O.P. of the system

$$= \frac{T_3\,(T_1 - T_2)}{T_1\,(T_2 - T_3)} = \frac{268\,(373 - 293)}{373\,(293 - 268)} = 2.3 \text{ Ans.}$$

36.13. Advantages of Vapour Absorption Refrigeration System over Vapour Compression Refrigeration System

Following are the advantages of vapour absorption system over vapour compression system :

1. In the vapour absorption system, the only moving part of the entire system is a pump which has a small motor. Thus, the operation of this system is essentially quiet and is subjected to little wear.

The vapour compression system of the same capacity has more wear, tear and noise due to moving parts of the compressor.

2. The vapour absorption system uses heat energy to change the condition of the refrigerant from the evaporator. The vapour compression system uses mechanical energy to change the condition of the refrigerant from the evaporator.

3. The vapour absorption systems are usually designed to use steam, either at high pressure or low pressure. The exhaust steam from furnaces and solar energy may also be used. Thus, this system can be used where the electric power is difficult to obtain or is very expensive.

4. The vapour absorption systems can operate at reduced evaporator pressure and temperature by increasing the steam pressure to the generator, with little decrease in capacity. But the capacity of vapour compression system drops rapidly with lowered evaporator pressure.

5. The load variations does not effect the performance of a vapour absorption system. The load variations are met by controlling the quantity of aqua circulated and the quantity of steam supplied to the generator.

The performance of a vapour compression system at partial loads is poor.

6. In the vapour absorption system, the liquid refrigerant leaving the evaporator has no bad effect on the system except that of reducing the refrigerating effect. In the vapour compression system, it is essential to superheat the vapour refrigerant leaving the evaporator so that no liquid may enter the compressor.

7. The vapour absorption systems can be built in capacities well above 1000 tonne of refrigeration each which is the largest size for single compressor units.

8. The space requirements and automatic control requirements favour the absorption system more and more as the desired evaporator temperature drops.

36.14. Ammonia-Hydrogen (Electrolux) Refrigerator

In small domestic installations, working on the ammonia absorption process, the pump may be omitted by the introduction of hydrogen into the low pressure side. The ammonia acts, normally, under its partial pressure. The total pressure is arranged to be practically uniform throughout the system. Thus the weak solution, passing from the boiler to the absorber, moves under gravity. The flow of strong solution, in the opposite direction, is assisted by a vertical pipe between the boiler and absorber, which is heated at its lower end by a small heating coil or gas jet. The electrolux refrigerator, as shown in Fig. 36.20, makes use of this principle.

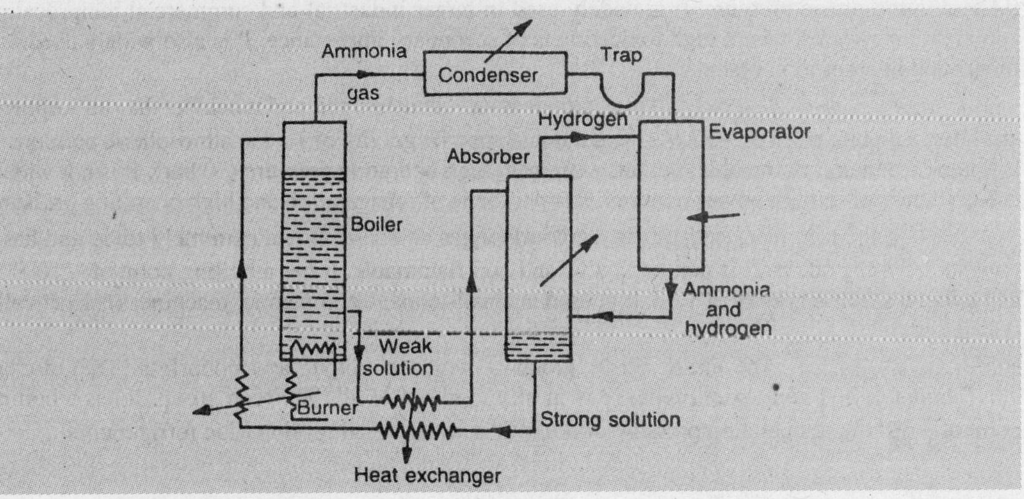

Fig. 36.20. Ammonia-Hydrogen refrigerator.

The chief advantage of this type of refrigerator is that no compressor, pump or fan is required in it. Therefore, there is no noise due to moving parts. Moreover, there is no machinery to give mechanical trouble. The coefficient of performance of this refrigerator is given by :

$$\text{C.O.P.} = \frac{\text{Heat absorbed by evaporator}}{\text{Heat supplied by burner}}$$

36.15. Properties of a Refrigerant

A substance which absorbs heat through expansion or vaporisation is termed as a refrigerant. An ideal refrigerant should possess chemical, physical and thermodynamic properties which permit its efficient application in the refrigerating system. An ideal refrigerant should have the following properties :

1. Low boiling point.
2. High critical temperature.

3. High latent heat of vaporisation.
4. Low specific heat of liquid.
5. Low specific volume of vapour.
6. Non-corrosive to metal.
7. Non-flammable and non-explosive.
8. Non-toxic.
9. Easy to liquify at moderate pressure and temperature.
10. Easy of locating leaks by odour or suitable indicator.
11. Low cost.
12. Mixes well with oil.

36.16. Refrigerants Commonly Used in Practice

Through there are many refrigerants which are commonly used, yet the following are important from the subject point of view :

1. *Ammonia (NH_3).* It is one of the oldest and the most commonly used of all the refrigerants. It is highly toxic and flammable. It has a boiling point of – 333° C and a liquid specific gravity of 0.684 at atmospheric pressure. It is widely used in larger industrial and commercial reciprocating compression systems, where high toxicating is of secondary importance. It is also widely used as a refrigerant in absorption systems.

2. *Carbon dioxide (CO_2).* It is a colourless and odourless gas, and is heavier than atmospheric air. It has a boiling point of – 77.6° C and a liquid specific gravity of 1.56 at atmospheric pressure. It is nontoxic and non-flammable, but has extremely high operating pressure (70 bar). It is not widely used, because of its high power requirements per tonne of refrigeration and high operating pressure.

3. *Sulphur dioxide (SO_2).* It is a colourless gas or liquid. It is a extremely toxic and has a pungent irritating odour. It is non-explosive and non-flammable. It has a boiling point of – 10.5° C and a liquid specific gravity of 1.36. It is used in small-tonnage commercial machines (hermetically sealed units).

4. *Freon-12*.* The entire Freon group is white in colour and odourless. They are all non-flammable and non-toxic. Freon-12 is mostly used out of all the Freon group. It has a boiling point of – 30° C and operating pressure of 8 bar. It is widely used for domestic refrigerators.

EXERCISES

1. A refrigerator works between – 7° C and – 27° C. The vapour is dry at the end of isentropic compression. There is no undercooling, and the evaporation is by throttle valve. Find : 1. C.O.P. ; and 2. power of the compressor to remove 175 kJ/min.

The properties of the refrigerant are as under ;

Temperature °C	*Sensible heat kJ/kg*	*Latent heat kJ/kg*	*Entropy Liquid kJ/kg K*	*Entropy of dry saturated vapour, kJ/kg K*
– 7	– 29.4	1298	– 0.1088	4.748
27	124.8	1172.4	0.427	4.334

[**Ans.** 7.51 ; 0.55]

2. An ammonia vapour compression refrigerator works between the temperature limits of – 6.7° C and 26.7° C. The vapour is dry at the end of compression, and there is no undercooling of the liquid, which is

* Other types of refrigerants are Freon-11, Freon-13, Freon-21, and Freon-22, etc. (abbreviated as F-11, F-13 and so on). These are also abbreviated as Refrigerant-11, Refrigerant-13, etc. or simply R-11, R-13, etc.

throttled to the lower temperature. Estimate the C.O.P. of the machine. Properties of ammonia given below should be used :

Temperature, °C	*Enthalpy, kJ/kg*			*Entropy, kJ/kg K*	
	Sensible (h_f)	*Latent* (h_{fg})	*Vapour* (h)	*Liquid* (s_f)	*Vapour* (s_g)
– 6.7	–29.3	1293.8	1264.5	– 0.113	4.752
26.7	125.6	1172.4	1297.9	0.427	4.334

[**Ans.** 7.1]

3. Determine the theoretical coefficient of performance for CO_2 refrigerating machine working between the limits of pressures of 65.1 bar and 30.8 bar. The CO_2 during the suction stroke has a dryness fraction of 0.6. Properties of refrigerant are :

Pressure, bar	*Temperature, °C*	*Liquid heat, kJ/kg*	*Latent heat, kJ/kg*	*Entropy of liquid, kJ/kg K*
65.1	25	81.23	121.42	0.2512
30.8	– 5	– 7.54	245.36	– 0.042

How many tonnes of ice would a machine, working between the same limits and having a relative coefficient of performance of 40%, make in 24 hours ? The water for the ice is supplied at 10° C and the compressor takes 6.8 kg of CO_2 per minute. Latent heat of ice is 335 kJ/kg. [**Ans.** 3.3 ; 0.606 tonnes]

4. An ammonia refrigerator produces 30 tonnes of ice from and at 0° C in 24 hours. The temperature range of the compressor is from 25° C to – 15° C. The vapour is dry saturated at the end of compression and an expansion valve is used. Assume a coefficient of performance to be 60% of the theoretical value. Calculate the power required to drive the compressor. Latent heat of ice = 335 kJ/kg.

Properties of ammonia are :

Temperature, °C	*Enthalpy, kJ/kg,*		*Entropy, kJ/kg K*	
	Liquid	*Vapour*	*Liquid*	*Vapour*
25	298.9	1465.84	1.1242	5.0391
– 15	112.34	1426.54	0.4572	5.5490

[**Ans.** 33.24 kW]

5. An ammonia refrigerating machine fitted with an expansion valve works between the temperature limits of – 10° C and 30° C. The vapour is 95% dry at the end of isentropic compression and the fluid leaving the condenser is at 30° C. If the actual coefficient of performance is 60% of the theoretical, find the ice produced per kW hour at 0° C from water at 10° C. The latent heat of ice is 335 kJ/kg. The ammonia has the following properties :

Temperature, °C	*Liquid heat kJ/kg*	*Latent heat kJ/kg*	*Entropy, kJ/kg K*	
			Liquid	*Vapour*
30	323.08	1145.79	1.2037	4.9842
– 10	135.37	1297.68	0.5443	5.4770

[**Ans.** 33.24 kg/kWh]

6. A R-12 refrigerating machine works on vapour compression cycle. The temperature of refrigerant in the evaporator is – 20° C. The vapour is dry saturated when it enters the compressor and leaves it in a superheated condition. The condenser temperature is 30° C. Assuming specific heat at constant pressure for R-12 in the superheated condition as 1.884 kJ/kg K, determine :

1. condition of vapour at the entrance to condenser,

2. condition of vapour at the entrance to the evaporator, and
3. theoretical C.O.P. of the machine.

The properties of R-12 are :

Temperature, °C	*Enthalpy, kJ/kg*		*Entropy, kJ/kg K*	
	Liquid	*Vapour*	*Liquid*	*Vapour*
– 20	17.82	178.73	0.0731	0.7087
30	64.59	199.62	0.2400	0.6853

[**Ans.** 33.8° C, 29% dry ; 4.07]

7. A CO_2 refrigerating plant fitted with an expansion valve, works between the pressure limits of 56.25 bar and 21.2 bar. The vapour is compressed isentropically and leaves the compressor cylinder at 32° C and with a total heat of 246.2 kJ/kg. The condensation takes place at 18.5° C in the condenser and there is no undercooling of the liquid. Determine the theoretical coefficient of performance of the plant. The properties of CO_2 are :-

Pressure bar	*Saturation temperature °C*	*Enthalpy, kJ/kg*		*Entropy, kJ/kg K*	
		Liquid	*Vapour*	*Liquid*	*Vapour*
56.25	18.5	52.75	214.37	0.1675	0.7244
21.2	−18.0	−37.68	234.90	−0.1507	0.9170

[**Ans.** 4.84]

8. A vapour compression refrigerator works between the temperature limits of – 20° C and 25° C. The refrigerant leaves the compressor in dry saturated condition. If the liquid refrigerant is undercooled to 20° C before entering the throttle valve, determine :

1. work required to drive the compressor,
2. refrigerating effect produced per kg of the refrigerant, and
3. theoretical C.O.P.

Assume specific heat of the refrigerant as 1.15. The properties of the refrigerant are :

Temperature, °C	*Enthalpy, kJ/kg*		*Entropy, kJ/kg K*	
	Liquid	*Vapour*	*Liquid*	*Vapour*
−20	327.4	1655.9	3.8416	9.09
25	536.3	1703.3	4.5956	8.50

[**Ans.** 189.7 kJ/kg ; 990.2 kJ/kg ; 5.01]

9. A food storage chamber requires a refrigeration system of 12 TR capacity with an evaporator temperature of – 8° C and condenser temperature of 30° C. The refrigerant R-12 is subcooled by 5° C before entering the throttle valve, and the vapour is superheated by 6° C before entering the compressor. If the liquid and vapour specific heats are 1.235 and 0.733 kJ/kg K respectively, find : 1. refrigerating effect per kg ; 2. mass of refrigerant circulated per minute ; and 3. coefficient of performance.

The relevant properties of the refrigerant R-12 are given below :

Saturation temperature, °C	*Enthalpy, kJ/kg*		*Entropy, kJ/ K*	
	Liquid	*Vapour*	*Liquid*	*Vapour*
−8	28.72	184.07	0.1149	0.7007
30	64.59	199.62	0.2400	0.6853

[**Ans.** 130.05 kJ/kg ; 19.4 kg ; 6.2]

QUESTIONS

1. Mention the advantages of vapour compression refrigeration system over air refrigeration system.

2. Draw the layouts of a vapour compression refrigerating system. State the function of each of the component and show the thermodynamic processes on a pressure-enthalpy diagram.

3. Sketch the *T-s* and *p-h* diagrams for the vapour compression cycles when the vapour after compression is (*i*) dry saturated, and (*ii*) wet.

4. Why in practice a throttle valve is used in vapour compression refrigerator rather than an expansion cylinder to reduce pressure between the condenser and the evaporator ?

5. What is sub-cooling and superheating ? Explain with the help of diagram. Why is superheating considered to be good in certain cases ?

6. Establish how an actual cycle differs from a theoretical vapour compression cycle.

7. Describe briefly with the help of a diagram, the vapour absorption system of refrigeration. In what way this system is advantageous over the vapour compression system ?

8. State the properties of a good refrigerant. What are the normal refrigerants used.

OBJECTIVE TYPE QUESTIONS

1. During a refrigeration cycle, heat is rejected by the refrigerant in a
   (*a*) compressor (*b*) condenser (*c*) evaporator (*d*) expansion valve
2. In a vapour compression system, the condition of refrigerant before entering the compressor is
   (*a*) saturated liquid (*b*) wet vapour
   (*c*) dry saturated liquid (*d*) superheated vapour
3. The highest temperature during the cycle, in a vapour compression system, occurs after
   (*a*) compression (*b*) condensation (*c*) expansion (*d*) evaporation
4. In a vapour compression system, the lowest temperature during the cycle occurs after
   (*a*) compression (*b*) condensation (*c*) expansion (*d*) evaporation
5. The sub-cooling in a refrigeration cycle
   (*a*) does not alter C.O.P. (*b*) increases C.O.P. (*c*) decreases C.O.P.
6. The refrigerant, commonly used in vapour absorption refrigeration systems, is
   (*a*) sulphur dioxide (*b*) ammonia (*c*) freon (*d*) aqua-ammonia
7. In ammonia-hydrogen refrigerator,
   (*a*) ammonia is absorbed in hydrogen (*b*) ammonia is absorbed in water
   (*c*) ammonia evaporates in hydrogen (*d*) hydrogen evaporates in ammonia
8. The boiling point of ammonia is
   (*a*) – 10.5° C (*b*) – 30° C (*c*) – 33.3° C (*d*) – 77.6° C
9. Which of the following refrigerant has the lowest boiling point ?
   (*a*) Ammonia (*b*) Carbon dioxide (*c*) Sulphur dioxide (d) Freon-12
10. Which of the following refrigerant is highly toxic and flammable ?
    (*a*) Ammonia (*b*) Carbon dioxide (*c*) Sulphur dioxide (*d*) Freon-12

ANSWERS

1. (*b*)	2. (*d*)	3. (*a*)	4. (*d*)	5. (*b*)
6. (*d*)	7. (*c*)	8. (*c*)	9. (*b*)	10. (*a*)

37

Psychrometry

1. Introduction. 2. Psychrometric Terms. 3. Dalton's Law of Partial Pressures. 4. Psychrometric Relations. 5. Enthalpy (Total Heat) of Moist Air. 6. Thermodynamic Wet Bulb Temperature or Adiabatic Saturation Temperature. 7. Psychrometric Chart. 8. Psychrometric Processes. 9. Sensible Cooling. 10. Sensible Heating. 11. Humidification and Dehumidification. 12. Sensible Heat Factor. 13. Cooling and Dehumidification. 12. Heating and Humidification. 15. Adiabatic Mixing of Two Air Streams.

37.1. Introduction

The psychrometry is that branch of engineering science, which deals with the study of moist air *i.e.* dry air mixed with water vapour or humidity. It also includes the study of behaviour of dry air and water vapour mixture under various sets of conditions. Though the earth's atmosphere is a mixture of gases including nitrogen (N_2), oxygen (O_2), argon (Ar) and carbon dioxide (CO_2), yet for the purpose of psychrometry, it is considered to be a mixture of dry air and water vapour only.

37.2. Psychrometric Terms

Though there are many psychrometric terms, yet the following are important from the subject point of view :

1. *Dry air.* The pure dry air is a mixture of number of gases such as nitrogen, oxygen, carbon-dioxide, hydrogen, argon, neon, helium etc. But the nitrogen and oxygen have the major portion of the combination.

The dry air is considered to have the composition as given in the following table.

Table 37.1. Composition of dry air.

S.No.	*Constituent*	*By volume*	*By mass*
1.	Nitrogen (N_2)	78.03%	75.47%
2.	Oxygen (O_2)	20.99%	23.19%
3.	Argon (Ar)	0.94%	1.29%
4.	Carbon-dioxide (CO_2)	0.03%	0.05%
5.	Hydrogen (H_2)	0.01%	--

The molecular mass of dry air is taken as 28.966 and the gas constant for air (R_a) is equal to 0.287 kJ/kg K.

Notes : (*a*) The pure dry air does not, ordinarily, exist in nature because it always contain some water vapour.

(*b*) The term air, wherever used in this text, means dry air containing moisture in the vapour form.

(*c*) Both dry air and water vapour can be considered as perfect gases because both exist in the atmosphere at low pressure. Thus all the perfect gas terms can be applied to them individually.

(*d*) The density of dry air is taken as 1.293 kg/m^3 at pressure 1.0135 bar or 101.35 kN/m^2 and at temperature 0° C (273 K).

2. *Moist air.* It is a mixture of dry air and water vapour. The amount of water vapour, present in the air, depends upon the absolute pressure and temperature of the mixture.

3. *Saturated air.* It is a mixture of dry air and water vapour, when the air has diffused the maximum amount of water vapour into it. The water vapours, usually, occur in the form of superheated steam as an invisible gas. However, when the saturated air is cooled, the water vapour in the air starts condensing, and the same may be visible in the form of moist, fog or condensation on cold surfaces.

4. *Degree of saturation.* It is the ratio of actual mass of water vapour in a unit mass of dry air to the mass of water vapour in the same mass and pressure of dry air when it is saturated at the same temperature.

5. *Humidity.* It is the mass of water vapour present in 1 kg of dry air, and is generally expressed in terms of gm per kg of dry air. It is also called ***specific humidity*** or ***humidity ratio***.

6. *Absolute humidity.* It is the mass of water vapour present in 1 m^3 of dry air, and is generally expressed in terms of gm per cubic metre of dry air. It is also expressed in terms of grains per cubic metre of dry air. Mathematically, one kg of water vapour is equal to 15 430 grains.

7. *Relative humidity.* It is the ratio of actual mass of water vapour in a given volume of moist air to the mass of water vapour in the same volume of saturated air at the same temperature and pressure.

8. *Dry bulb temperature.* It is the temperature of air recorded by a thermometer, when it is not affected by the moisture present in the air. The dry bulb temperature is generally denoted by t_d or t_{db}.

9. *Wet bulb temperature.* It is the temperature of air recorded by a thermometer, when its bulb is surrounded by a wet cloth exposed to the air. Such a thermometer is called *Wet bulb thermometer. The wet bulb temperature is generally denoted by t_w or t_{wb}.

10. *Wet bulb depression.* It is the difference between dry bulb temperature and wet bulb temperature at any point. The wet bulb depression indicates relative humidity of the air.

11. *Dew point temperature.* It is the temperature of air recorded by a thermometer, when the moisture (water vapour) present in it begins to condense. In other words, the dew point temperature is the saturation temperature (t_{sat}) corresponding to the partial pressure of water vapour (p_v). It is, usually, denoted by t_{dp}.

Note : For saturated air, the dry bulb temperature, wet bulb temperature and dew point temperature is same.

12. *Dew point depression.* It is the difference between the dry bulb temperature and dew point temperature of air..

13. *Psychrometer.* There are many types of psychrometers, but the sling psychrometer, as shown in Fig. 37.1, is widely used. It consists of a dry bulb thermometer and a wet

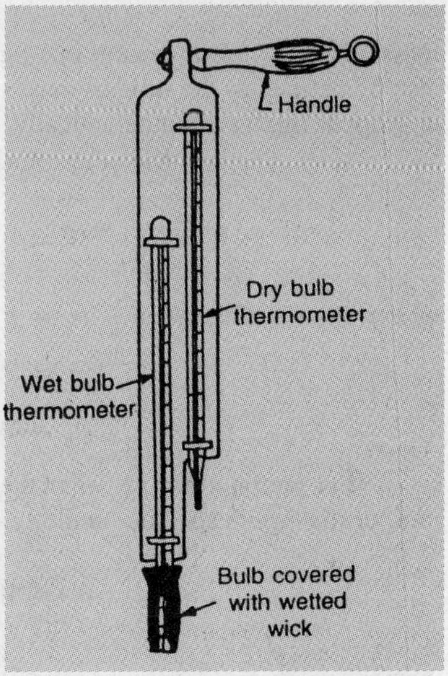

Fig. 37.1. Sling psychrometer.

* A wet bulb thermometer has its bulb covered with a piece of soft cloth (or silk wick) which is exposed to the air. The lower part of this cloth is dipped in a small basin of water. The water from the basin rises up in the cloth by the capillary action, and then gets evaporated. It may be noted that if relative humidity of air is high (*i.e.* the air contains more water vapour), there will be little evaporation and thus there will be a small cooling effect. On the other hand, if relative humidity of air is low (*i.e.* the air contains less water vapour), there will be more evaporation and thus there will be more cooling effect.

bulb thermometer mounted side by side in a protective case that is attached to a handle by a swivel connection so that the case can be easily rotated. The dry bulb thermometer is directly exposed to air and measures the actual temperature of the air. The bulb of the wet bulb thermometer is covered by a wick thoroughly wetted by distilled water. The temperature measured by this wick covered bulb of a thermometer is the temperature of liquid water in the wick and is called wet bulb temperature.

The sling psychrometer is rotated in the air for approximately one minute after which the readings from both the thermometers are taken. This process is repeated several times to assure that the lowest possible wet bulb temperature is recorded.

37.3. Dalton's Law of Partial Pressures

It states, "*The total pressure exerted by the mixture of air and water vapour is equal to the sum of the pressures, which each constituent would exert, if it occupied the same space by itself*". In other words, the total pressure exerted by air and water vapour mixture is equal to the barometric pressure. Mathematically, barometric pressure of the mixture,

$$p_b = p_a + p_v$$

where p_a = Partial pressure of dry air, and

p_v = Partial pressure of water vapour.

37.4. Psychrometric Relations

We have already discussed some psychrometric terms in Art. 37.2. These terms have some relations between one another. The following psychrometric relations are important from the subject point of view :

1. *Humidity, specific or absolute humidity, humidity ratio or moisture content.* It is the mass of water vapour present in 1 kg of dry air and is generally expresed in kg/kg of dry air. It may also be defined as the ratio of mass of water vapour to the mass of dry air in a given volume of the air-vapour mixture. Mathematically,

Humidity, specific or absolute humidity, humidity ratio or moisture content,

$$W = 0.622 \times \frac{p_v}{p_a} = \frac{0.622\, p_v}{p_b - p_v}$$

where p_v = Partial pressure of water vapour,

p_a = Partial pressure of dry air, and

p_b = Barometric pressure.

For saturated air (*i.e.* when the air is holding maximum amount of water vapour), the humidity ratio or maximum specific humidity,

$$W_s = W_{max} = 0.622 \times \frac{p_s}{p_b - p_s}$$

where p_s = Partial pressure of air corresponding to saturation temperature (*i.e.* dry bulb temperature, t_d).

2. *Degree of saturation or percentage humidity.* We have already discussed that the degree of saturation is the ratio of actual mass of water vapour in a unit mass of dry air to the mass of water vapour in the same mass of dry air when it is saturated at the same temperature (dry bulb temperature). In other words, it may be defined as the ratio of actual specific humidity to the specific humidity of saturated air at the same dry bulb temperature. It is, usually, denoted by μ.

Mathematically, degree of saturation,

$$\mu = \frac{W}{W_s} = \frac{\dfrac{0.622\,p_v}{p_b - p_v}}{\dfrac{0.622\,p_s}{p_b - p_s}}$$

$$= \frac{p_v}{p_s}\left(\frac{p_b - p_s}{p_b - p_v}\right) = \frac{p_v}{p_s}\left[\frac{1 - \dfrac{p_s}{p_b}}{1 - \dfrac{p_v}{p_b}}\right]$$

Notes : (*a*) The partial pressure of saturated air (p_s) is obtained from the steam tables corresponding to dry bulb temperature t_d.

(*b*) If the relative humidity, $\phi = p_v / p_s$ is equal to zero, then the humidity ratio, $W = 0$, *i.e.*, for dry air, degree of saturation, $\mu = 0$.

(*c*) If the relative humidity, $\phi = p_v / p_s$ is equal to 1, then $W = W_s$, and $\mu = 1$. Thus, degree of saturation, μ varies between 0 and 1.

3. *Relative humidity.* We have already discussed that the relative humidity is the ratio of actual mass of water vapour (m_v) in a given volume of moist air to the mass of water vapour (m_s) in the same volume of saturated air at the same temperature and pressure.

The relative humidity may also be defined as the ratio of actual partial pressure of water vapour in moist air (p_v) at a given temperature (dry bulb temperature) to the saturation pressure (p_s) of water vapour (or partial pressure of water vapour in saturated air) at the same temperature. It is usually denoted by ϕ. Mathematically, relative humidity,

$$\phi = \frac{m_v}{m_s} = \frac{p_v}{p_s}$$

The relative humidity may also be obtained as discussed below :

We know that degree of saturation,

$$\mu = \frac{p_v}{p_s}\left[\frac{1 - \dfrac{p_s}{p_b}}{1 - \dfrac{p_v}{p_b}}\right] = \phi\left[\frac{1 - \dfrac{p_s}{p_b}}{1 - \phi \times \dfrac{p_s}{p_b}}\right] \qquad \ldots\left(\because \phi = \frac{p_v}{p_s}\right)$$

$$\therefore \qquad \phi = \frac{\mu}{1 - (1 - \mu)\dfrac{p_s}{p_b}}$$

Note : For saturated air, the relative humidity is 100%.

4. *Pressure of water vapour.* According to Carrier's equation, the partial pressure of water vapour,

$$p_v = p_w - \frac{(p_b - p_w)(t_d - t_w)}{1544 - 1.44\,t_w}$$

where p_w = Saturation pressure corresponding to wet bulb temperature (from steam tables),

p_b = Barometric pressure,

t_d = Dry bulb temperature, and

t_w = Wet bulb temperature.

5. *Vapour density or absolute humidity.* We have already discussed that the vapour density or absolute humidity is the mass of water vapour present in 1 m³ of dry air. Mathematically, vapour density or absolute humidity,

$$\rho_v = \frac{W\,p_a}{R_a\,T_d} = \frac{W\,(p_b - p_v)}{R_a\,T_d} \qquad \ldots (\because p_b = p_a + p_v)$$

where p_a = Pressure of air in kN/m²,

W = Humidity ratio,

R_a = Gas constant for air = 0.287 kJ/kg K, and

T_d = Dry bulb temperature in K.

Example 37.1. *The atmospheric air has a dry bulb temperature of 21° C and wet bulb temperature 18° C. If the barometer reads 750 mm of Hg, determine : 1. partial pressure of water vapour ; 2. Relative humidity ; and 3. Dew point temperature.*

Solution. Given : t_d = 21° C ; t_w = 18° C ; p_b = 750 mm of Hg

1. *Partial pressure of water vapour*

From steam tables, we find that saturation pressure of vapour corresponding to wet bulb temperature of 18° C, is

$$p_w = 0.020\,62 \text{ bar} = \frac{0.020\,62}{0.001\,33} = 15.5 \text{ mm of Hg}$$

$$\ldots (\because 1 \text{ mm of Hg} = 0.001\,33 \text{ bar})$$

We know that partial pressure of water vapour,

$$p_v = p_w - \frac{(p_b - p_w)(t_d - t_w)}{1544 - 1.44\,t_w}$$

$$= 15.5 - \frac{(750 - 15.5)(21 - 18)}{1544 - 1.44 \times 18} = 14 \text{ mm of Hg } \textbf{Ans.}$$

2. *Relative humidity*

From steam tables, we find that saturation pressure of vapour, corresponding to dry bulb temperature of 21° C, is

$$p_s = 0.024\,85 \text{ bar} = \frac{0.024\,85}{0.001\,33} = 18.7 \text{ mm of Hg}$$

We know that relative humidity,

$$\phi = \frac{p_v}{p_s} = \frac{14}{18.7} = 0.748 \text{ or } 74.8\ \% \textbf{ Ans.}$$

3. *Dew point temperature*

We know that dew point temperature has a saturation pressure equal to the pressure of water vapour in the air. From steam tables, we find that the temperature corresponding to 14 mm of Hg is 16.4° C. Therefore dew point temperature,

$$t_{dp} = 16.4° \text{ C } \textbf{Ans.}$$

Example 37.2. *A sling psychrometer gives reading of 25° C dry bulb temperature and 15° C wet bulb temperature. The barometer indicates 760 mm of Hg. Assuming partial pressure of the vapour as 10 mm of Hg, determine : 1. Specific humidity, and 2. Saturation ratio.*

Solution. Given : $t_d = 25°$ C ; $t_w = 15°$ C ; $p_b = 760$ mm of Hg ; $p_v = 10$ mm of Hg

1. *Specific humidity*

We know that specific humidity,

$$W = \frac{0.622\, p_v}{p_b - p_v} = \frac{0.622 \times 10}{760 - 10} = 0.0083 \text{ kg/kg of dry air} \quad \textbf{Ans.}$$

2. *Saturation ratio*

From steam tables, corresponding to a dry bulb temperature of 25° C, we find that partial pressure,

$$p_s = 0.031\,66 \text{ bar} = \frac{0.031\,66}{0.001\,33} = 23.8 \text{ mm of Hg} \quad (\because 1 \text{ mm of Hg} = 0.001\,33 \text{ bar})$$

We know that saturation ratio,

$$\mu = \frac{p_v\,(p_b - p_s)}{p_s\,(p_b - p_v)} = \frac{10\,(760 - 23.8)}{23.8\,(760 - 10)} = 0.41 \quad \textbf{Ans.}$$

37.5. Enthalpy (Total heat) of Moist Air

The enthalpy of moist air is numerically equal to the enthalpy of dry air *plus* the enthalpy of water vapour associated with dry air. Let us consider one kg of dry air. We know that enthalpy of 1 kg of dry air,

$$h_a = c_{pa} \times t_d \qquad \ldots (i)$$

where c_{pa} = Specific heat of dry air which is normally taken as 1.005 kJ/kg K, and

t_d = Dry bulb temperature.

Enthalpy of water vapour associated with 1 kg of dry air,

$$h_v = W\,h_s \qquad \ldots (ii)$$

where W = Specific humidity (*i.e.* mass of water vapour) in kg per kg of dry air, and

h_s = Enthalpy of water vapour in kJ per kg of dry air at dew point temperature (t_{dp}).

If the moist air is superheated, then the enthalpy of water vapour

$$= W\,c_{ps}\,(t_d - t_{dp}) \qquad \ldots (iii)$$

where c_{ps} = Specific heat of superheated water vapour which is normally taken as 1.9 kJ/kg K, and

$t_d - t_{dp}$ = Degree of superheat of the water vapour.

∴ Total enthalpy of superheated water vapour,

$$h = c_{pa}\,t_d + W\,h_s + W\,c_{ps}\,(t_d - t_{dp})$$

$$= c_{pa}\,t_d + W\,[h_{fdp} + h_{fgdp} + c_{ps}\,(t_d - t_{dp})] \qquad \ldots (\because h_s = h_{fdp} + h_{fgdp})$$

$$= c_{pa}\,t_d + W\,[4.2 \times t_{dp} + h_{fgdp} + c_{ps}\,(t_d - t_{dp})] \qquad \ldots (\because h_{fdp} = 4.2\,t_{dp})$$

$$= (c_{pa} + W\,c_{ps})\,t_d + W\,[h_{fgdp} + (4.2 - c_{ps})\,t_{dp}]$$

The term $(c_{pa} + W c_{ps})$ is called *humid specific heat* (c_{vm}). It is the specific heat or heat capacity of moist air, *i.e.* $(1 + W)$ kg/kg of dry air. At low temperature of air conditioning range, the value of W is very small. The general value of humid specific heat in air conditioning range is taken as 1.022 kJ/kg K.

$$\therefore \quad h = 1.022\, t_d + W\,[h_{fgdp} + (4.2 - 1.9)\, t_{dp}] \qquad \ldots (\because c_{ps} = 1.9 \text{ kJ/kg K})$$

$$= 1.022\, t_d + W\,[h_{fgdp} + 2.3\, t_{dp}] \text{ kJ}$$

where h_{fgdp} = Latent heat of vaporisation of water corresponding to dew point temperature (from steam tables).

Example 37.3. *A sleeve psychrometer reads 40° C dry bulb temperature and 28° C wet bulb temperature. Assuming the barometric pressure as 1.013 bar, determine : 1. Humidity ratio ; 2. Relative humidity ; 3. Dew point temperature ; and 4. Enthalpy of the mixture per kg of dry air.*

Solution. Given : $t_d = 40°$ C ; $t_w = 28°$ C ; $p_b = 1.013$ bar

1. *Humidity ratio*

First of all, let us find the partial pressure of vapour (p_v). From steam tables, we find that the saturation pressure corresponding to wet bulb temperature of 28° C is

$$p_w = 0.037\,78 \text{ bar}$$

$$\therefore \quad p_v = p_w - \frac{(p_b - p_w)(t_d - t_w)}{1544 - 1.44\, t_w}$$

$$= 0.037\,78 - \frac{(1.013 - 0.037\,78)(40 - 28)}{1544 - 1.44 \times 28}$$

$$= 0.037\,78 - 0.007\,78 = 0.03 \text{ bar } \textbf{Ans.}$$

We know that humidity ratio,

$$W = \frac{0.622\, p_v}{p_b - p_v} = \frac{0.622 \times 0.03}{1.013 - 0.03} = 0.019 \text{ kg / kg of dry air } \textbf{Ans.}$$

2. *Relative humidity*

From steam tables, we find that the saturation pressure of vapour, corresponding to dry bulb temperature of 40° C is

$$p_s = 0.073\,75 \text{ bar}$$

∴ Relative humidity,

$$\phi = \frac{p_v}{p_s} = \frac{0.03}{0.073\,75} = 0.407 \text{ or } 40.7\,\% \textbf{ Ans.}$$

3. *Dew point temperature*

Since the dew point temperature is the saturation temperature corresponding to the partial pressure of water vapour (p_v), therefore from steam tables, corresponding to 0.03 bar, dew point temperature is

$$t_{dp} = 24° \text{ C } \textbf{Ans.}$$

4. *Enthalpy of mixture per kg of dry air*

From steam tables, we find that the latent heat of vaporisation of water at dew point temperature of 24° C is

$$h_{fgdp} = 2445 \text{ kJ/kg}$$

$\therefore$ Enthalpy of mixture per kg of dry air,

$$= 1.022\, t_d + W\,[h_{fgdp} + 2.3\, t_{dp}]$$

$$= 1.022 \times 40 + 0.019\,[2445 + 2.3 \times 24] \text{ kJ/kg of dry air}$$

$$= 88.4 \text{ kJ/kg of dry air} \quad \textbf{Ans.}$$

Example 37.4. *In a laboratory test, a sling psychrometer recorded dry bulb and wet bulb temperatures as 30° C and 25° C respectively. Calculate : 1. vapour pressure, 2. relative humidity, 3. specific humidity, 4. degree of saturation, 5. dew point temperature, and 6. enthalpy of the mixture.*

Solution. Given : $t_d = 30°$ C ; $t_w = 25°$ C

1. *Vapour pressure*

From steam tables, we find that saturation pressure of vapour corresponding to wet bulb temperature of 25° C is

$$p_w = 0.031\,66 \text{ bar}$$

We know that vapour pressure,

$$p_v = p_w - \frac{(p_b - p_w)(t_d - t_w)}{1544 - 1.44\, t_w}$$

$$= 0.031\,66 - \frac{(1.0133 - 0.031\,66)(30 - 25)}{1544 - 1.44 \times 25}$$

$$= 0.031\,66 - 0.003\,26 = 0.0284 \text{ bar} \quad \textbf{Ans.}$$

2. *Relative humidity*

From steam tables, we also find that the saturation pressure of vapour at 30° C,

$$p_s = 0.042\,42 \text{ bar}$$

We know that relative humidity,

$$\phi = \frac{p_v}{p_s} = \frac{0.0284}{0.042\,42} = 0.66 \text{ or } 66\% \quad \textbf{Ans.}$$

3. *Specific humidity*

We know that specific humidity,

$$W = \frac{0.622\, p_v}{(p_b - p_v)} = \frac{0.622 \times 0.0284}{1.0133 - 0.0284} \text{ kg / kg of dry air}$$

$$= 0.018 \text{ kg / kg of dry air} \quad \textbf{Ans.}$$

4. *Degree of saturation*

We know that degree of saturation,

$$\mu = \frac{p_v\,(p_b - p_s)}{p_s\,(p_b - p_v)}$$

$$= \frac{0.0284\,(1.0133 - 0.042\,42)}{0.042\,42\,(1.0133 - 0.0284))} = 0.651 \quad \textbf{Ans.}$$

5. *Dew point temperature*

We know that dew point temperature has a saturation pressure equal to the pressure of water vapour in the air. From steam tables, we find that the temperature corresponding to 0.0284 bar is 23° C. Therefore dew point temperature,

$$t_{dp} = 23° \text{ C} \quad \textbf{Ans.}$$

6. *Enthalpy of the mixture*

From steam tables, we find that the latent heat of vaporisation of water at dew point temperature of 23° C, is

$$h_{fgdp} = 2447.2 \text{ kJ/kg}$$

∴ Enthalpy of the mixture

$$= 1.022\, t_d + W\,(h_{fgdp} + 2.3\, t_{dp})$$

$$= 1.022 \times 30 + 0.018\,(2447.2 + 2.3 \times 23) = 75.66 \text{ kJ} \quad \textbf{Ans.}$$

37.6. Thermodynamic Wet Bulb Temperature or Adiabatic Saturation Temperature

The thermodynamic wet bulb temperature or adiabatic saturation temperature is the temperature at which the air can be brought to saturation state, adiabatically, by the evaporation of water into the flowing air.

The equipment used for the adiabatic saturation of air, in its simplest form, consists of an insulated chamber containing adequate quantity of water. There is also an arrangement for extra water (known as make-up water) to flow into the chamber from its top, as shown in Fig. 37.2 .

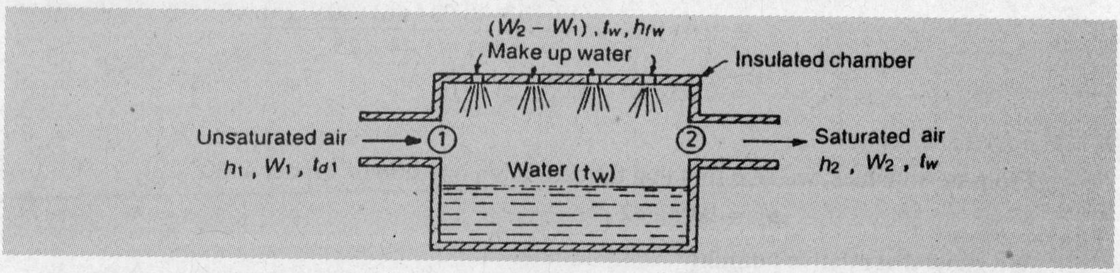

Fig. 37.2. Adiabatic saturation of air.

Let the unsaturated air enters the chamber at section 1. As the air passes through the chamber over a long sheet of water, the water evaporates which is carried with the flowing stream of air, and the specific humidity of the air increases. The make up water is added to the chamber at this temperature to make the water level constant. Both the air and water are cooled as the evaporation takes place. This process continues until the energy transferred from the air to the water is equal to the energy required to vaporise the water. When steady conditions are reached, the air flowing at section 2 is saturated with water vapour. The temperature of the saturated air at section 2 is known as *thermodynamic wet bulb temperature* or *adiabatic saturation temperature.*

The adiabatic saturation process can be represented on *T-s* diagram as shown by the curve 1-2 in Fig. 37.3. During the adiabatic saturation process, the partial pressure of vapour increases, although the total pressure of the air-vapour mixture remains constant. The unsaturated air initially at dry bulb temperature t_{d1} is cooled adiabatically to dry bulb temperature t_{d2} which is equal to the adiabatic saturation temperature t_w. It may be noted that the adiabatic saturation temperature is taken equal to the wet bulb temperature for all practical purposes.

Let h_1 = Enthalpy of unsaturated air at section 1,

W_1 = Specific humidity of air at section 1,

h_2, W_2 = Corresponding values of saturated air at section 2, and

h_{fw} = Sensible heat of water at adiabatic saturation temperature.

Balancing the enthalpies of air at inlet and outlet (*i.e.* at sections 1 and 2),

$$h_1 + (W_2 - W_1)\, h_{fw} = h_2 \quad \ldots (i)$$

or

$$h_1 - W_1\, h_{fw} = h_2 - W_2\, h_{fw} \quad \ldots (ii)$$

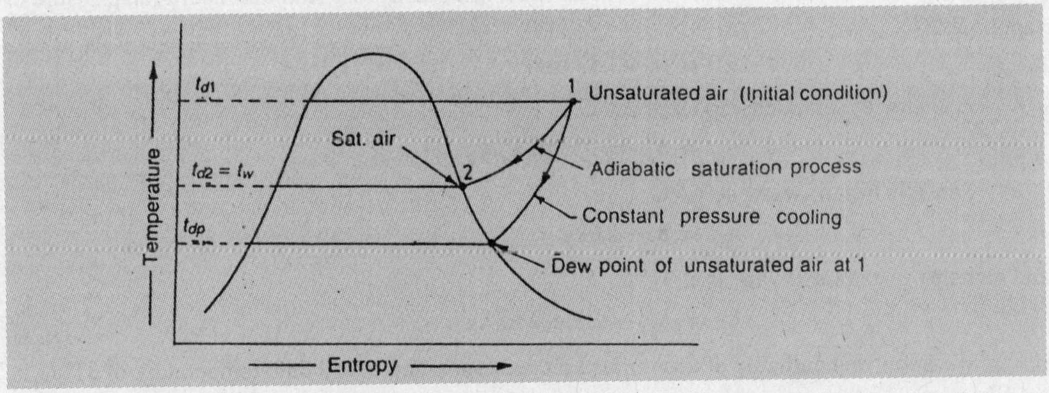

Fig. 37.3. *T-s* diagram for adiabatic saturation process.

The term $(h_2 - W_2\, h_{fw})$ is known as *sigma heat* and remains constant during the adiabatic process.

We know that $h_1 = h_{a1} + W_1\, h_{s1}$

and $h_2 = h_{a2} + W_2\, h_{s2}$

where h_{a1} = Enthalpy of 1 kg of dry air at dry bulb temperature t_{d1},

*h_{s1} = Enthalpy of superheated vapour at t_{d1} per kg of vapour,

h_{a2} = Enthalpy of 1 kg of air at wet bulb temperature t_w, and

h_{s2} = Enthalpy of saturated vapour at wet bulb temperature t_w per kg of vapour.

Now the equation (*ii*) may be written as :

$$(h_{a1} + W_1\, h_{s1}) - W_1\, h_{fw} = (h_{a2} + W_2\, h_{s2}) - W_2\, h_{fw}$$

$$(h_{a1} + W_1\, h_{s1}) + W_2\, h_{fw} - W_1\, h_{fw} = h_{a2} + W_2\, h_{s2}$$

$$W_1\,(h_{s1} - h_{fw}) = W_2\,(h_{s2} - h_{fw}) + h_{a2} - h_{a1}$$

$$\therefore \quad W_1 = \frac{W_2\,(h_{s2} - h_{fw}) + h_{a2} - h_{a1}}{h_{s1} - h_{fw}}$$

Example 37.5. *Atmospheric air at 0.965 bar enters the adiabatic saturator. The wet bulb temperature is 20° C and dry bulb temperature is 31° C during adiabatic saturation process. Determine 1. humidity ratio of the entering air, 2. vapour pressure and relative humidity at 31° C, and 3. dew point temperature.*

Solution. Given : $p_b = 0.965$ bar ; $t_w = 20°C$; $t_d = 31°C$

---

* In psychrometry, the enthalpy of superheated vapour at dry bulb temperature t_{d1} is taken equal to the enthalpy of saturated vapour corresponding to dry bulb temperature t_{d1}.

1. *Humidity ratio of the entering air*

Let W_1 = Humidity ratio of the entering air, and

W_2 = Humidity ratio of the saturated air.

First of all, let us find the value of W_2. From steam tables, we find that saturation pressure of vapour at 20° C,

$$p_{v2} = 0.023\ 37 \text{ bar}$$

Enthalpy of saturated vapour at 20° C,

$$h_{s2} = h_{g2} = 2538.2 \text{ kJ/kg}$$

Sensible heat of water at 20° C,

$$h_{fw} = 83.9 \text{ kJ/kg}$$

and enthalpy of saturated vapour at 31° C,

$$h_{s1} = h_{g1} = 2558.2 \text{ kJ/kg}$$

We know that enthalpy of unsaturated air corresponding to dry bulb temperature of 31° C,

$$h_{a1} = m\,c_p\,t_d = 1 \times 1.005 \times 31 = 31.155 \text{ kJ/kg}$$

... (Taking c_p for air = 1.005 kJ/kg K)

Enthalpy of 1 kg of saturated air corresponding to wet bulb temperature of 20° C,

$$h_{a2} = m\,c_p\,t_w = 1 \times 1.005 \times 20 = 20.1 \text{ kJ/kg}$$

We know that

$$W_2 = \frac{0.622\,p_{v2}}{p_b - p_{v2}} = \frac{0.622 \times 0.023\ 37}{0.965 - 0.023\ 37} = 0.0154 \text{ kg / kg of dry air}$$

$$\therefore \quad W_1 = \frac{W_2\,(h_{s2} - h_{fw}) + h_{a2} - h_{a1}}{h_{s1} - h_{fw}}$$

$$= \frac{0.0154\,(2538.2 - 83.9) + 20.1 - 31.155}{2558.2 - 83.9}$$

$$= 0.0108 \text{ kg / kg of dry air} \quad \textbf{Ans.}$$

2. *Vapour pressure and relative humidity at 31° C*

Let p_{v1} = Vapour pressure at 31° C.

We know that

$$W_1 = \frac{0.622\,p_{v1}}{p_b - p_{v1}}$$

$$0.0108 = \frac{0.622\,p_{v1}}{0.965 - p_{v1}}$$

$$0.0104 - 0.0108\,p_{v1} = 0.622\,p_{v1}$$

$$\therefore \quad p_{v1} = 0.0164 \text{ bar} \quad \textbf{Ans.}$$

From steam tables, we find that the saturation pressure corresponding to dry bulb temperature of 31° C is

$$p_s = 0.044\ 91 \text{ bar}$$

$$\therefore \text{ Relative humidity, } \quad \phi = \frac{p_{v1}}{p_s} = \frac{0.0164}{0.044\ 91} = 0.365 \text{ or } 36.5\% \quad \textbf{Ans.}$$

3. *Dew point temperature*

Since the dew point temperature (t_{dp}) is the saturation temperature corresponding to the partial pressure of water vapour (p_{v1}), therefore from steam tables, corresponding to 0.0164 bar, we find that

$$t_{dp} = 14.5°\text{ C} \quad \textbf{Ans.}$$

37.7. Psychrometric Chart

It is a graphical representation of the various thermodynamic properties of moist air. The psychrometric chart is very useful for finding out the properties of air (which are required in the field of air conditioning) and eliminate lot of calculations. There is a slight variation in the charts prepared by different air-conditioning manufactures but basically they are all alike. The psychrometric chart is normally drawn for standard atmospheric pressure of 760 mm of Hg (or 1.01325 bar).

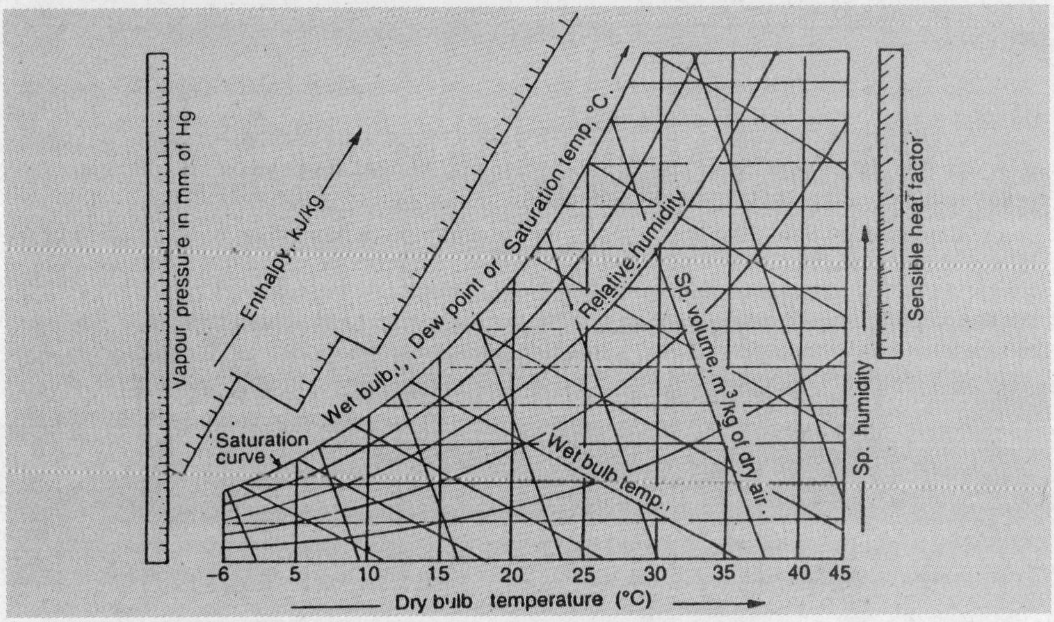

Fig. 37.4. Psychrometric chart.

In a psychrometric chart, dry bulb temperature is taken as abscissa and specific humidity *i.e.* moisture contents as ordinate as shown in Fig. 37.4. Now the saturation curve is drawn by plotting the various saturation points at corresponding dry bulb temperatures. The saturation curve represents 100% relative humidity at various dry bulb temperatures. It also represents the wet bulb and dew point temperatures.

Though the psychrometric chart has a number of details, yet the following lines are important from the subject point of view :

1. *Dry bulb temperature lines.* The dry bulb temperature lines are vertical *i.e.* parallel to the ordinate and uniformly spaced as shown in Fig. 37.5. Generally, the temperature range of these lines on psychrometric chart is from – 6° C to 45° C. The dry bulb temperature lines are drawn with difference of every 5° C and up to the saturation curve as shown in the figure. The values of dry bulb temperature are also shown on the saturation curve.

2. *Specific humidity or moisture content lines.* The specific humidity (moisture content) lines are horizontal *i.e.* parallel to the abscissa and are also uniformly spaced as shown in Fig. 37.6. Generally, moisture content range of these lines on psychrometric chart is from 0 to 30 g / kg of dry

air (or from 0 to 0.030 kg / kg of dry air). The moisture content lines are drawn with a difference of every 1 g (or 0.001 kg) and up to the saturation curve as shown in the figure.

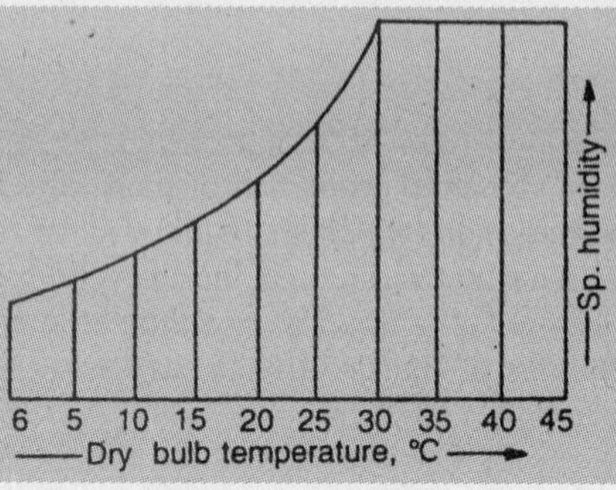

Fig. 37.5. Dry bulb temperature lines.

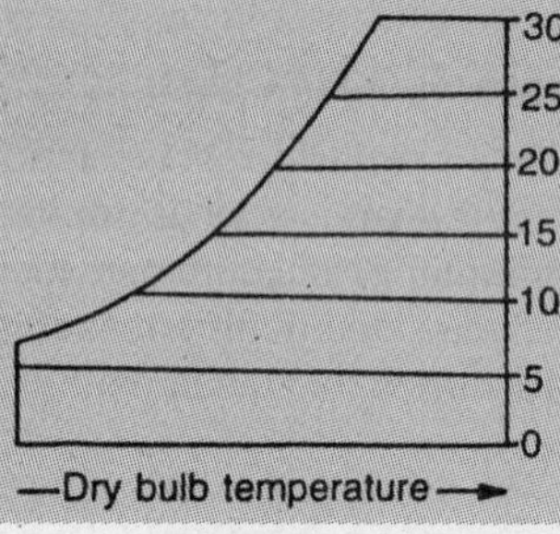

Fig.37.6. Specific humidity lines.

3. *Dew point temperature lines.* The dew point temperature lines are horizontal *i.e.* parallel to the abscissa and non-uniformly spaced as shown in Fig. 37.7. At any point on the saturation curve, the dry bulb and dew point temperatures are equal.

The values of dew point temperatures are generally given along the saturation curve of the chart as shown in the figure.

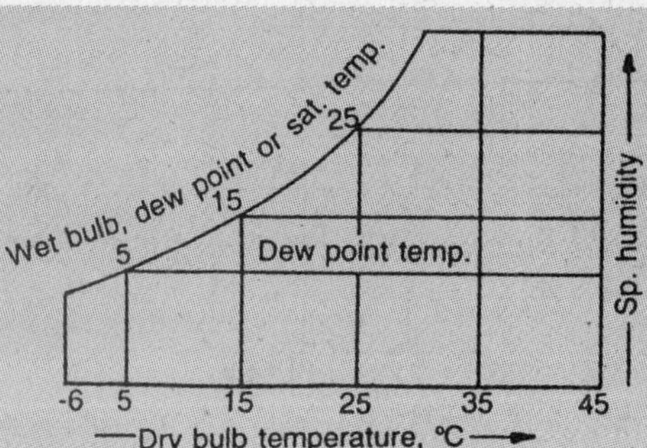

Fig. 37.7. Dew point temperature lines.

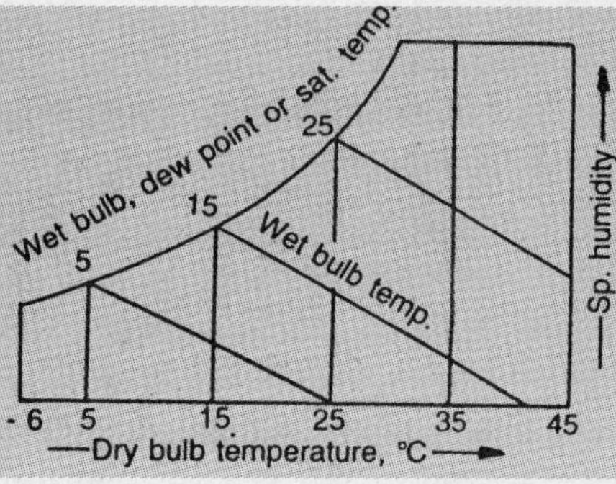

Fig. 37.8. Wet bulb temperature lines.

4. *Wet bulb temperatures lines.* The wet bulb temperature lines are inclined straight lines and non-uniformly spaced as shown in Fig. 37.8. At any point on the saturation curve, the dry bulb and wet bulb temperatures are equal.

The values of wet bulb temperatures are generally given along the saturation curve of the chart as shown in the figure.

5. *Enthalpy (total heat) lines.* The enthalpy (or total heat) lines are inclined straight lines and uniformly spaced as shown in Fig. 37.9. These lines are parallel to the wet bulb temperature lines, and are drawn up to the saturation curve. Some of these lines coincide with the wet bulb temperature lines also.

The values of total enthalpy are given on a scale above the saturation curve as shown in the figure.

6. *Specific volume lines.* The specific volume lines are obliquely inclined straight lines and uniformly spaced as shown in Fig. 37.10. These lines are drawn up to the saturation curve.

The values of volume lines are generally given at the base of the chart.

7. *Vapour pressure lines.* The vapour pressure lines are horizontal and uniformly spaced. Generally, the vapour pressure lines are not drawn in the main chart. But a scale showing vapour pressure in mm of Hg is given on the extreme left side of the chart as shown in Fig. 37.11.

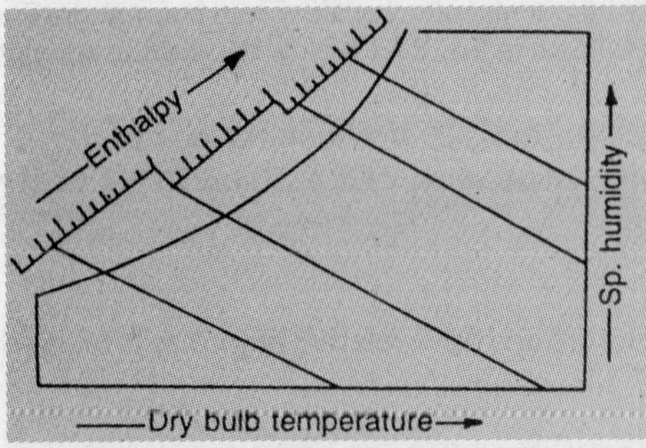

Fig. 37.9. Enthalpy lines.

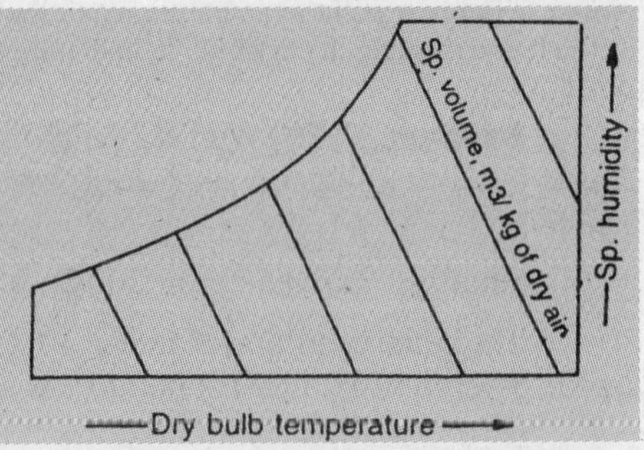

Fig. 37.10. Specific volume lines.

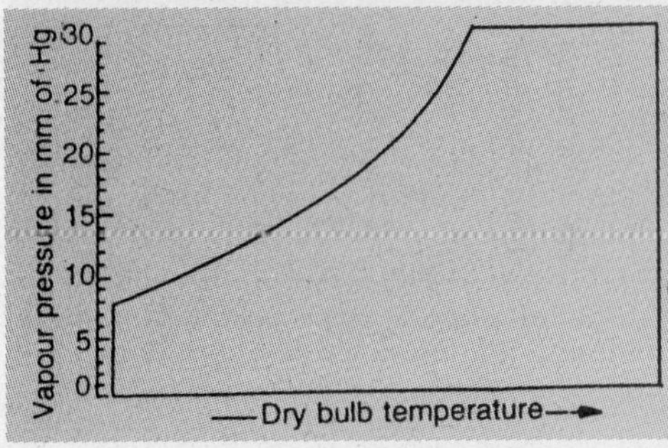

Fig. 37.11. Vapour pressure lines.

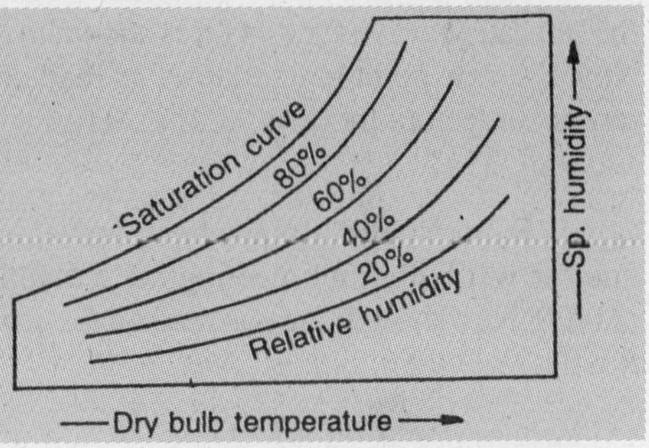

Fig. 37.12. Relative humidity lines.

8. *Relative humidity lines.* The relative humidity lines are curved lines and follows the saturation curve. Generally, these lines are drawn with values 10%, 20%, 30% etc., and upto 100%. The saturation curve represents 100% relative humidity. The values of relative humidity lines are generally given along the lines themselves as shown in Fig. 37.12.

Example 37.6. *Atmospheric air at 760 mm of Hg barometric pressure has 25° C dry bulb temperature and 15° C wet bulb temperature. By using psychrometric chart, determine : 1. Relative humidity, 2. Humidity ratio, and 3. Dew point temperature.*

Solution. $^*p_b = 760$ mm of Hg ; $t_d = 25°$ C ; $t_w = 15°$ C

The initial condition of the air, *i.e.* 25° C dry bulb temperature and 15° C wet bulb temperature is marked on the psychrometric chart at point P as shown in Fig. 37.13.

1. *Relative humidity*

The relative humidity, as read along the relative humidity curve, is 35%. **Ans.**

2. *Humidity ratio*

From point P, draw a horizontal line meeting the humidity ratio line at point Q, as shown in Fig. 37.13. Now, the humidity ratio, as read on the scale at point Q, is 6.7 g / kg of dry air. **Ans.**

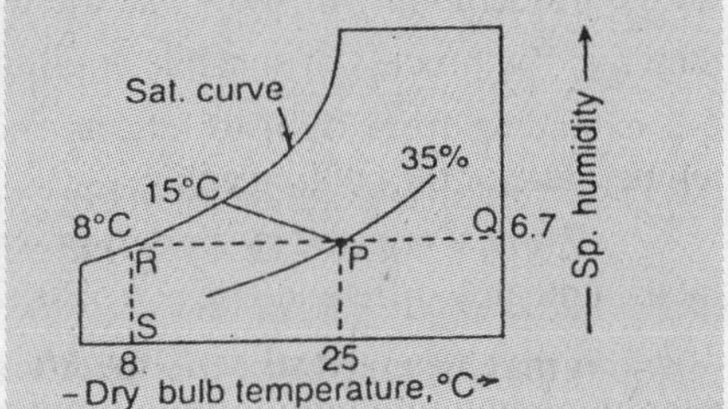

Fig. 37.13

* Superfluous data.

3. *Dew point temperature*

From point *P*, draw a horizontal line meeting the saturation curve at point *R*, as shown in Fig. 37.13. Now, the dew point temperature, as read on the saturation curve at point *R* is 8° C.

Note : The dew point temperature may also be read on the dry bulb temperature scale. From point *R*, draw a vertical line meeting the dry bulb temperature scale at point *S*, as shown in Fig. 37.13. Now the reading at point *S* is 8° C.

Example 37.7. *With the help of psychrometric chart, find : 1. dew point temperature ; 2. Enthalpy ; and 3. Vapour pressure of air having a dry bulb temperature of 30° C and 50% relative humidity.*

Solution. Given : t_d = 30° C ; ϕ = 50%

The initial condition of air, *i.e.* 30° C dry bulb temperature and 50% relative humidity is marked on the psychrometric chart at point *P*, as shown in Fig. 37.14.

1. *Dew point temperature*

From point *P*, draw a horizontal line meeting the saturation curve at point *Q*, as shown in Fig. 37.14. Now the dew point temperature, as read on the saturation curve at point *Q*, is 18.5° C. **Ans.**

2. *Enthalpy*

From point *P*, draw a line parallel to enthalpy line (or wet bulb line) meeting the enthalpy scale at point *R*. Now the enthalpy as read on the scale at point *R* is 64.7 kJ / kg of dry air. **Ans.**

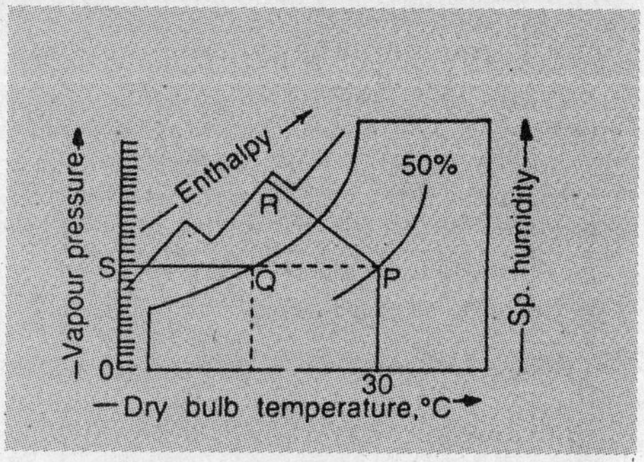

Fig. 37.14

3. *Vapour pressure*

Extend the horizontal line *PQ* to meet the vapour pressure scale at point *S*. Now the vapour pressure, as read on the scale at point *S*, is 16 mm of Hg. **Ans.**

37.8. Psychrometric Processes

Though there are many psychrometric processes, yet the following are important from the subject point of view :

1. Sensible cooling. 2. Sensible heating. 3. Humidification and dehumidification.

We shall now discuss these processes, in detail, in the following pages.

37.9. Sensible Cooling

The cooling of air, without any change in its specific humidity, is known as ***sensible cooling***. Let air at temperature t_{d1} passes over a cooling coil of temperature t_{d3}, as shown in Fig. 37.15 (*a*). A little consideration will show that the temperature of air leaving the cooling coil (t_{d2}) will be more than t_{d3}. The process of sensible cooling, on the psychrometric chart, is shown by a horizontal line 1-2 extending from right to left, as shown in Fig. 37.15 (*b*). The point 3 represents the surface temperature of the cooling coil.

The heat rejected by air during sensible cooling may be obtained from the psychrometric chart by the enthalpy difference ($h_1 - h_2$), as shown in Fig. 37.15 (*b*).

It may be noted that the specific humidity during the sensible cooling remains constant (*i.e.* $W_1 = W_2$). The dry bulb temperature reduces from t_{d1} to t_{d2} and relative humidity increases from ϕ_1 to ϕ_2, as shown in Fig. 37.15 (*b*). The amount of heat rejected during sensible cooling may also be obtained from the relation :

Heat rejected, $h = h_1 - h_2 = c_{pa}(t_{d1} - t_{d2}) + W c_{ps}(t_{d1} - t_{d2})$

$$= (c_{pa} + W c_{ps})(t_{d1} - t_{d2}) = c_{pm}(t_{d1} - t_{d2})$$

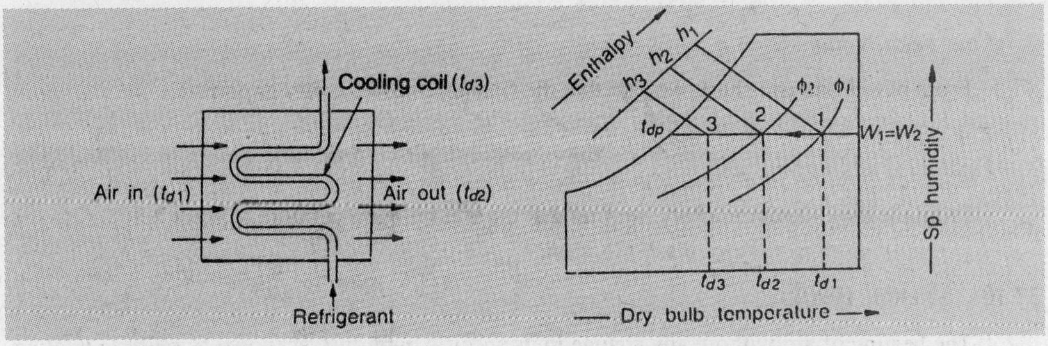

(*a*) Psychrometric process. (*b*) Psychrometric chart.

Fig. 37.15. Sensible cooling.

The term $(c_{pa} + W c_{ps})$ is called ***humid specific heat*** (c_{pm}) and its value is taken as 1.022 kJ/kg K.

$\therefore$ Heat rejected, $h = 1.022\,(t_{d1} - t_{d2})$ kJ/kg

We know that *by pass factor in sensible cooling,

$$\text{B.P.F.} = \frac{t_{d2} - t_{d3}}{t_{d1} - t_{d3}}$$

and coil efficiency, $\eta_c = 1 - \text{B.P.F.} = 1 - \dfrac{t_{d2} - t_{d3}}{t_{d1} - t_{d3}}$

Notes : 1. For sensible cooling, the cooling coil may have refrigerant, cooling water or cool gas flowing through it.

2. The sensible cooling can be done only upto the dew point temperature (t_{dp}), as shown in Fig. 37.15 (*b*). The cooling below this temperature will result in the condensation of moisture.

Example 37.8. ***Atmospheric air with dry bulb temperature of 28° C and a wet bulb temperature of 17° C is cooled to 15° C without changing its moisture content. Find : 1. Original relative humidity ; 2. Final relative humidity ; and 3. Final wet bulb temperature.***

Solution. Given : $t_{d1} = 28°$ C ; $t_{w1} = 17°$ C ; $t_{d2} = 15°$ C

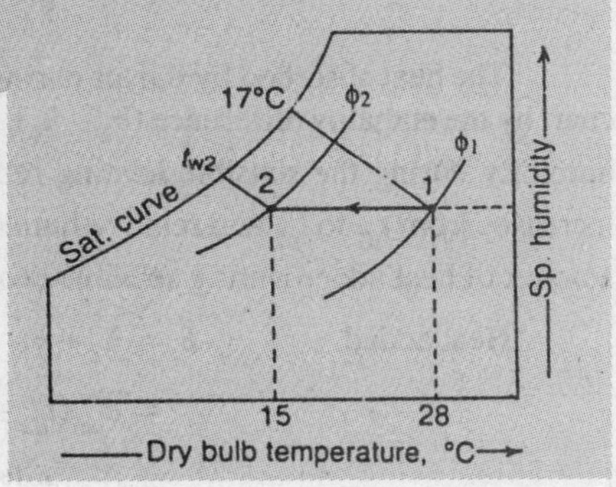

Fig. 37.16

The initial condition of air, *i.e.* 28° C dry bulb temperature and 17° C wet bulb temperature is marked on the psychrometric chart at point 1, as shown in Fig. 37.16. Now mark the final condition of air by drawing a horizontal line through point 1 (because there is no change in moisture content of the air) to meet the 15° C dry bulb temperature line at point 2, as shown in Fig. 37.16.

* Under ideal conditions, the dry bulb temperature of the air leaving the apparatus (t_{d2}) should be equal to that of the cooling coil (t_{d3}). But it is not so, because of the inefficiency of the cooling coil. This phenomenon is known as by pass factor.

1. *Original relative humidity*

From the psychrometric chart, we find that the original relative humidity at point 1,

$$\phi_1 = 34\% \quad \textbf{Ans.}$$

2. *Final relative humidity*

From psychrometric chart, we find that the final relative humidity at point 2,

$$\phi_2 = 73\% \quad \textbf{Ans.}$$

3. *Final wet bulb temperature*

From the psychrometric chart, we find that the wet bulb temperature at point 2,

$$t_{w2} = 12.2^\circ\text{C} \quad \textbf{Ans.}$$

37.10. Sensible Heating

The heating of air, without any change in its specific humidity, is known as *sensible heating*. Let air at temperature t_{d1} passes over a heating coil of temperature t_{d3}, as shown in Fig. 37.17 (*a*). A little consideration will show that the temperature of air leaving the heating coil (*i.e.* t_{d2}) will be less than t_{d3}. The process of sensible heating, on the psychrometric chart, is shown by a horizontal line 1-2 extending from left to right, as shown in Fig. 37.17 (*b*). The point 3 represents the surface temperature of the heating coil.

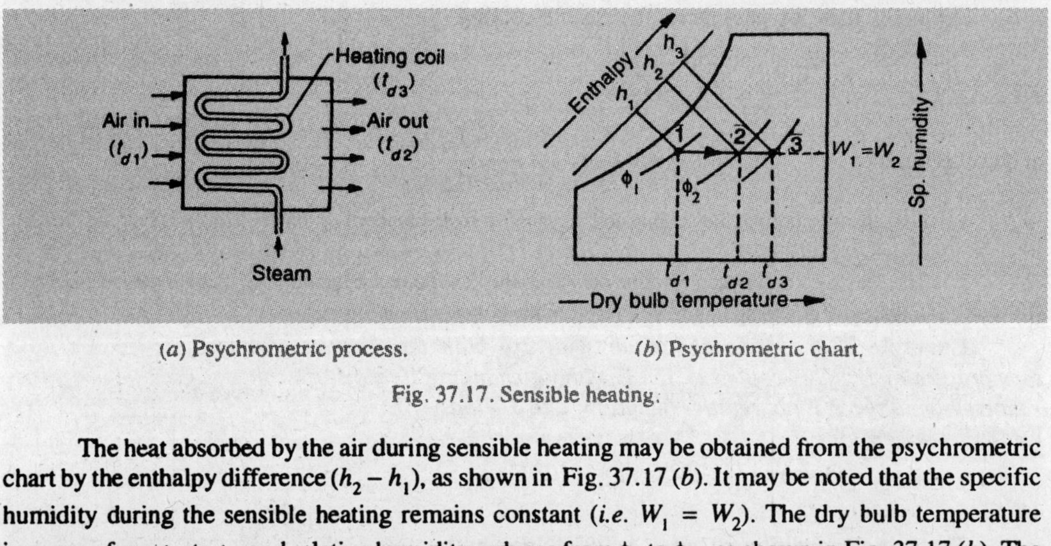

(*a*) Psychrometric process. (*b*) Psychrometric chart.

Fig. 37.17. Sensible heating.

The heat absorbed by the air during sensible heating may be obtained from the psychrometric chart by the enthalpy difference $(h_2 - h_1)$, as shown in Fig. 37.17 (*b*). It may be noted that the specific humidity during the sensible heating remains constant (*i.e.* $W_1 = W_2$). The dry bulb temperature increases from t_{d1} to t_{d2} and relative humidity reduces from ϕ_1 to ϕ_2, as shown in Fig. 37.17 (*b*). The amount of heat added during sensible heating may also be obtained from the relation :

$$\text{Heat added,} \quad h = h_2 - h_1$$

$$= c_{pa}(t_{d2} - t_{d1}) + W\,c_{ps}(t_{d2} - t_{d1})$$

$$= (c_{pa} + W\,c_{ps})(t_{d2} - t_{d1}) = c_{pm}(t_{d2} - t_{d1})$$

The term $(c_{pa} + W\,c_{ps})$ is called *humid specific heat* (c_{pm}) and its value is taken as 1.022 kJ/kg K.

$$\therefore \text{Heat added,} \quad h = 1.022\,(t_{d2} - t_{d1}) \text{ kJ/kg}$$

We know that by pass factor in sensible heating,

$$\text{B.P.F.} = \frac{t_{d3} - t_{d2}}{t_{d3} - t_{d1}}$$

and coil efficiency, $$\eta_c = 1 - \text{B.P.F.} = 1 - \frac{t_{d3} - t_{d2}}{t_{d3} - t_{d1}}$$

Notes : 1 For sensible heating, steam or hot water is passed through the heating coil. The heating coil may be electric resistance coil.

2 The sensible heating of moist air can be done to any desired temperature.

Example 37.9. *The atmospheric air at 760 mm of Hg, dry bulb temperature 15° C and wet bulb temperature 11° C enters a heating coil whose temperature is 41° C. Assuming by-pass factor of heating coil as 0.5, determine dry bulb temperature, wet bulb temperature and relative humidity of the air leaving the coil. Also determine the sensible heat added to the air per kg of dry air.*

Solution. Given : p_b = 760 mm of Hg ; t_{d1} = 15° C ; t_{w1} = 11° C ; t_{d3} = 41° C ; BPF = 0.5

The initial condition of air entering the coil at dry bulb temperature of 15° C, and wet bulb temperature of 11° C, is marked by point 1 on the psychrometric chart as shown in Fig. 37.18. Now draw a constant specific humidity line from point 1 to interest the vertical line drawn through 41° C at point 3. The point 2 lies on the line 1-3.

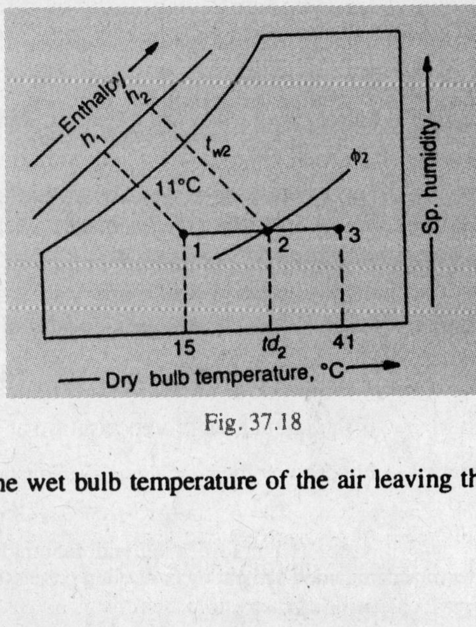

Fig. 37.18

Dry bulb temperature of the air leaving the coil

t_{d2} = Dry bulb temperature of the air leaving the coil.

We know that by pass factor (BPF),

$$0.5 = \frac{t_{d3} - t_{d2}}{t_{d3} - t_{d1}} = \frac{41 - t_{d2}}{41 - 15} = \frac{41 - t_{d2}}{26}$$

$$\therefore \quad t_{d2} = 28° \text{ C} \quad \textbf{Ans.}$$

Wet bulb temperature of the air leaving the coil

From the psychrometric chart, we find that the wet bulb temperature of the air leaving the coil at point 2 is

$$t_{w2} = 16.1° \text{ C} \quad \textbf{Ans.}$$

Relative humidity of the air leaving the coil

From the psychrometric chart, we find that the relative humidity of the air leaving the coil at point 2 is,

$$\phi_2 = 29\% \quad \textbf{Ans.}$$

Sensible heat added to the air per kg of dry air

From the psychrometric chart, we find that enthalpy of air at point 2,

$$h_2 = 46 \text{ kJ/kg of dry air}$$

and enthalpy of air at point 1, h_1 = 31.8 kJ/kg of dry air

∴ Sensible heat added to the air per kg of dry air,

$$h = h_2 - h_1 = 46 - 31.8 = 14.2 \text{ kJ/kg of dry air} \quad \textbf{Ans.}$$

37.11. Humidification and Dehumidification

The addition of moisture to the air, without change in its dry bulb temperature, is known as *humidification*. Similarly, removal of moisture from the air, without change in its dry bulb temperature is known as *dehumidification*. The heat added during humidification process and heat removed during dehumidification process is shown on the psychrometric chart in Fig. 37.19 and 37.20 respectively.

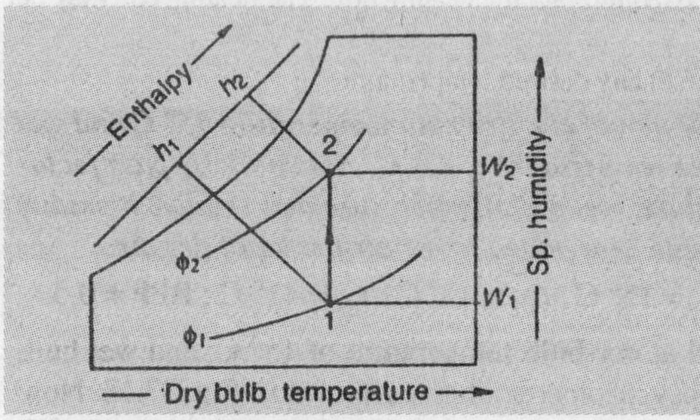

Fig. 37.19. Humidification.

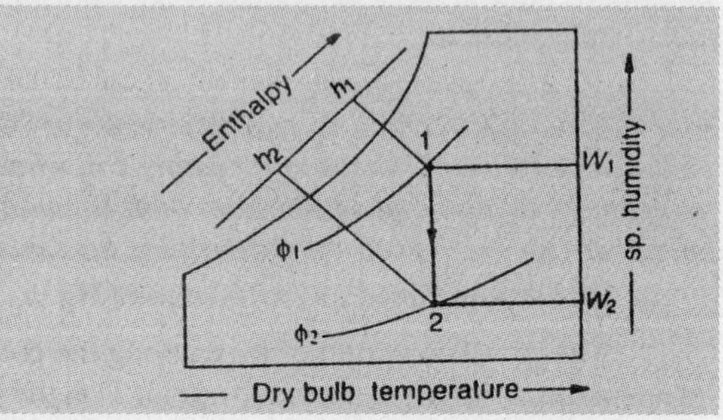

Fig. 37.20. Dehumidification.

It may be noted that in humidification, the relative humidity increases from ϕ_1 to ϕ_2 and specific humidity also increases from W_1 to W_2, as shown in Fig. 37.19. Similarly, in dehumidification, the relative humidity decreases from ϕ_1 to ϕ_2 and specific humidity also decreases from W_1 to W_2, as shown in Fig. 37.20.

A little consideration will show, that in humidification, change in enthalpy is shown by the intercept $(h_2 - h_1)$ on the psychrometric chart. Since the dry bulb temperature of air during the humidification remains constant, therefore its sensible heat also remains constant. It is thus obvious that the change in enthalpy per kg of dry air is the latent heat of vaporisation of the increased moisture content equal to $(W_2 - W_1)$ kg per kg of dry air. Mathematically,

$$(h_2 - h_1) = h_{fgd}(W_2 - W_1)$$

where h_{fgd} is the latent heat of vaporisation at dry bulb temperature (t_d).

Notes : 1. For dehumidification, the above equation may be written as :

$$(h_1 - h_2) = h_{fgd}(W_1 - W_2)$$

2. Absolute humidification and dehumidification processes are rarely found in practice. These are always accompanied by heating or cooling processes.

37.12. Sensible Heat Factor

As a matter of fact, the heat added during a psychrometric process may be split up into sensible heat and latent heat. The ratio of the sensible heat to the total heat is known as *sensible heat factor* (briefly written as SHF) or *sensible heat ratio* (briefly written as SHR). Mathematically,

$$\text{SHF} = \frac{\text{Sensible heat}}{\text{Total heat}} = \frac{SH}{SH + LH}$$

where SH = Sensible heat, and

LH = Latent heat.

The sensible heat factor scale is shown on the right hand side of the psychrometric chart.

37.13. Cooling and Dehumidification

This process is generally used in summer air conditioning to cool and dehumidify the air. The process of cooling and dehumidification is shown by line 1-2 on the psychrometric chart, as shown in Fig. 37.21. The air enters at condition 1 and leaves at condition 2. In this process, the dry bulb temperature of air decreases from t_{d1} to t_{d2} (*i.e.* cooling) and specific humidity of air decreases from W_1 to W_2 (*i.e.* dehumidification). The final relative humidity of the air is generally higher than that of the entering air.

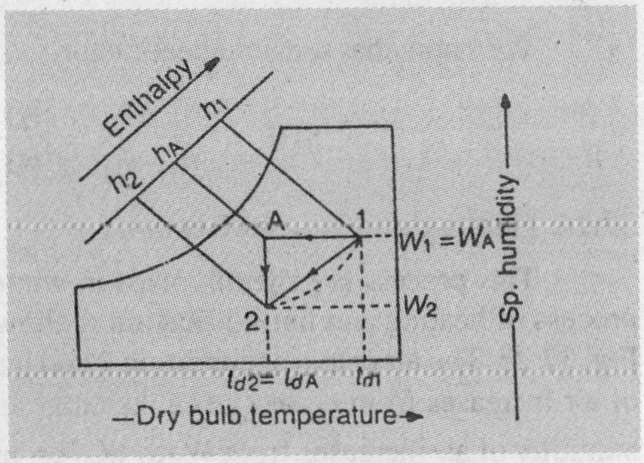

Fig. 37.21. Cooling and dehumidification.

Actually, the cooling and dehumidification process follows the path as shown by a dotted curve in Fig. 37.21, but for the calculation of psychrometric properties, only end points are important. Thus the cooling and dehumidification process shown by a line 1-2 may be assumed to have followed a path 1-*A* (*i.e.* sensible cooling) and *A*-2 (*i.e.* dehumidification) as shown in Fig. 37.21. We see that the total heat removed from the air during the cooling and dehumidification process is

$$h = h_1 - h_2$$

$$= (h_1 - h_A) + (h_A - h_2) = SH + LH$$

where $SH = h_1 - h_A$ = Sensible heat removed, and

$LH = h_A - h_2$ = Latent heat removed due to condensation of vapour of the reduced moisture content $(W_1 - W_2)$.

We know that sensible heat factor,

$$\text{SHF} = \frac{\text{Sensible heat}}{\text{Total heat}} = \frac{SH}{SH + LH} = \frac{h_1 - h_A}{h_1 - h_2}$$

Example 37.10. *Atmospheric air at 30° C dry bulb temperature and 45% R.H. is to be conditioned to 17° C dry bulb temperature and 15° C wet bulb temperature. Find the amount of heat rejected by the air. Also find the sensible heat factor of the process.*

Solution. Given : $t_{d1} = 30°\text{ C}$; $\phi_1 = 45\%$; $t_{d2} = 17°\text{ C}$; $t_{w2} = 15°\text{ C}$

Amount of heat rejected by the air

First of all, mark the initial condition of air, *i.e.* at 30° C dry bulb temperature and 45% relative humidity on the psychrometric chart as point 1. Then mark the final condition of air at 17° C dry bulb temperature and 15° C wet bulb temperature on the psychrometric chart as point 2 shown in Fig 37.22. Now locate the point *A* by drawing horizontal line through point 1 and vertical line through point 2.

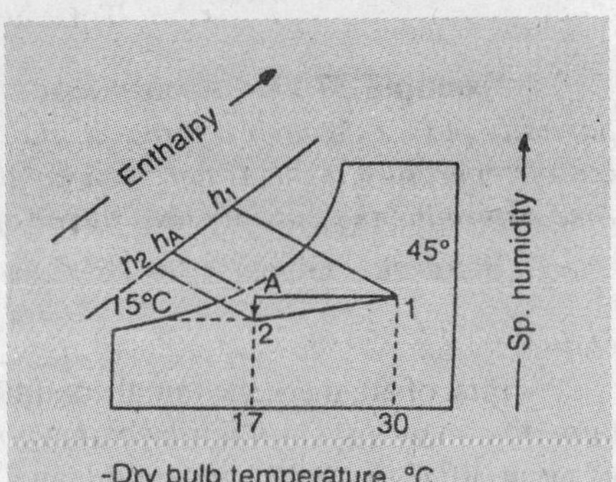

Fig. 37.22

From the psychrometric chart, we find that enthalpy of air at point 1,

$$h_1 = 60 \text{ kJ/kg of dry air}$$

Similarly $h_A = 49$ kJ/kg of dry air

and $h_2 = 41$ kJ/kg of dry air

We know that amount of heat rejected by the air

$$= h_1 - h_2 = 60 - 41 = 19 \text{ kJ/kg of dry air } \textbf{Ans.}$$

Sensible heat factor

We know that sensible heat factor,

$$SHF = \frac{h_1 - h_A}{h_1 - h_2} = \frac{60 - 49}{60 - 41} = 0.58 \textbf{ Ans.}$$

37.14. Heating and Humidification

This process is generally used in winter air conditioning to warm and humidify the air The process of heating and humidification is shown by line 1-2 on the psychrometric chart, as shown in Fig. 37.23. The air enters at condition 1 and leaves at condition 2. In this process, the dry temperature of air increases from t_{d1} to t_{d2} (*i.e.* heating) and specific humidity of air increases from W_1 to W_2 (*i.e.* humidification). The final relative humidity of the air can be lower or higher than that of the entering air.

Actually, the heating and humidification process follows the path as shown by dotted curve in Fig. 37.23, but for the calculation of psychrometric properties, only the end points are important. Thus, the heating and humidification process shown by a line 1-2 on the psychrometric chart may be assumed to have followed the path 1-*A* (*i.e.* sensible heating) and *A*-2 (*i.e.* humidification), as shown in Fig. 37.23. We see that the total heat added to the air during heating and humidification is

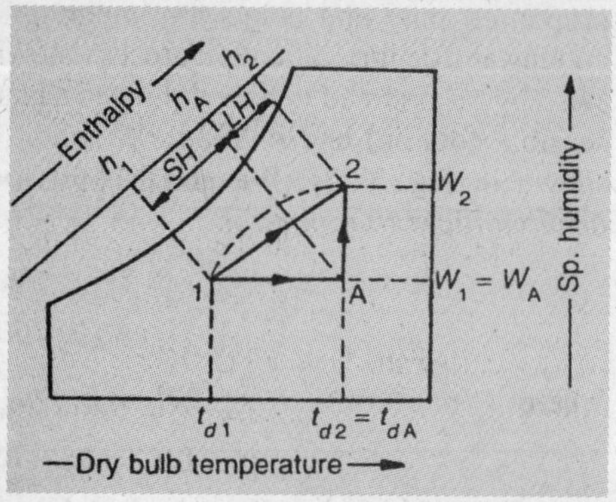

Fig. 37.23. Heating and humidification.

$$h = h_2 - h_1$$

$$= (h_2 - h_A) + (h_A - h_1) = LH + SH$$

where $LH = h_2 - h_A$ = Latent of vaporisation of the increased moisture content $(W_2 - W_1)$, and

$SH = h_A - h_1$ = Sensible heat added.

We know that sensible heat factor,

$$SHF = \frac{\text{Sensible heat}}{\text{Total heat}} = \frac{SH}{SH + LH} = \frac{h_A - h_1}{h_2 - h_1}$$

Example 37.11. *Atmospheric air at a dry bulb temperature of 16° C and 25% relative humidity passes through a furnace and then through a humidifier, in such a way that the final dry bulb temperature is 30° C and 50% relative humidity. Find the heat and moisture added to the air. Also determine the sensible heat factor of the process.*

Solution. Given : $t_{d1} = 16°$ C ; $\phi_1 = 25\%$; $t_{d2} = 30°$ C ; $\phi_2 = 50\%$

Heat added to the air

First of all, mark the initial condition of air *i.e.* at 16° C dry bulb temperature and 25% relative humidity on the psychrometric chart at point 1, as shown in Fig. 37.24. Then mark the final condition of air at 30° C dry bulb temperature and 50% relative humidity on the psychrometric chart as point

2. Now locate the point *A* by drawing horizontal line through point 1 and vertical line through point 2. From the psychrometric chart, we find that enthalpy of air at point 1,

$$h_1 = 23 \text{ kJ/kg of dry air}$$

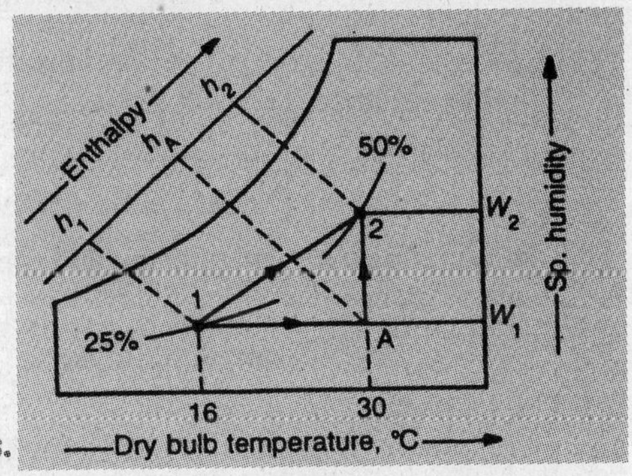

Fig. 37.24

Enthalpy of air at point *A*,

$$h_A = 37.5 \text{ kJ/kg of dry air}$$

and enthalpy of air at point 2,

$$h_2 = 64.5 \text{ kJ/kg of dry air}$$

∴ Heat added to the air

$$= h_2 - h_1 = 64.5 - 23$$

$$= 41.5 \text{ kJ/kg of dry air} \quad \textbf{Ans.}$$

Moisture added to the air

From the psychrometric chart, we find that the specific humidity in the air at point 1,

$$W_1 = 0.0028 \text{ kg / kg of dry air}$$

and specific humidity in the air at point 2,

$$W_2 - 0.0134 \text{ kg / kg of dry air}$$

∴ Moisture added to the air

$$= W_2 - W_1 = (0.0134 - 0.0028) = 0.0106 \text{ kg / kg of dry air} \quad \textbf{Ans.}$$

Sensible heat factor of the process

We know that sensible heat factor of the process,

$$\text{SHF} = \frac{h_A - h_1}{h_2 - h_1} = \frac{37.5 - 23}{64.5 - 23} = 0.35 \quad \textbf{Ans.}$$

37.15. Adiabatic Mixing of Two Air Streams

When two quantities of air having different enthalpies and different specific humidities are mixed, the final condition of the air mixture depends upon the masses involved, and on the enthalpy and specific humidity of each of the constituent masses which enter the mixture.

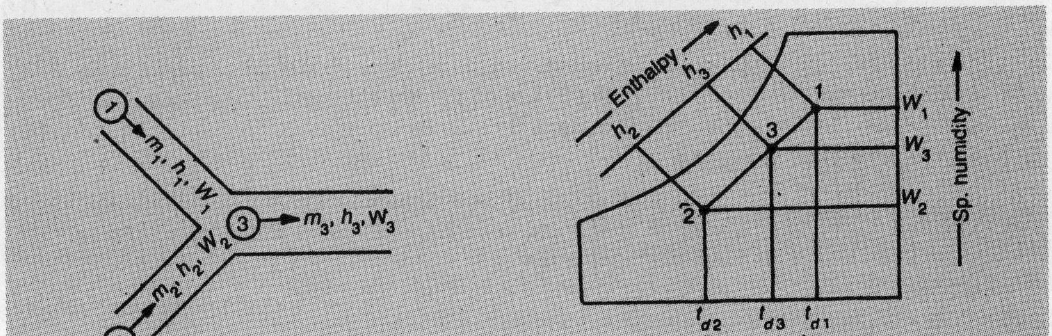

Fig. 37.25. Adiabatic mixing of two air streams

Now consider two air streams 1 and 2 mixing adiabatically, as shown in Fig. 37.25 (*a*).

Let m_1 = Mass of air entering at 1,

h_1 = Enthalpy of air entering at 1,

W_1 = Specific humidity of air entering at 1,

m_2, h_2, W_2 = Corresponding values of air entering at 2, and

m_3, h_3, W_3 = Corresponding values of the mixture leaving at 3.

Assuming no loss of enthalpy and specific humidity during the air mixing process, we have for the mass balance,

$$m_1 + m_2 = m_3 \quad \ldots (i)$$

For the energy balance,

$$m_1 h_1 + m_2 h_2 = m_3 h_3 \quad \ldots (ii)$$

and for the mass balance of water vapour,

$$m_1 W_1 + m_2 W_2 = m_3 W_3 \quad \ldots (iii)$$

From equations (*i*) and (*ii*),

$$m_1 h_1 + m_2 h_2 = (m_1 + m_2) h_3 = m_1 h_3 + m_2 h_3$$

or

$$m_1 h_1 - m_1 h_3 = m_2 h_3 - m_2 h_2$$

$$m_1 (h_1 - h_3) = m_2 (h_3 - h_2)$$

$$\therefore \quad \frac{m_1}{m_2} = \frac{h_3 - h_2}{h_1 - h_3} \quad \ldots (iv)$$

Similarly, from equations (*i*) and (*iii*),

$$\frac{m_1}{m_2} = \frac{W_3 - W_2}{W_1 - W_3} \quad \ldots (v)$$

Now from equations (*iv*) and (*v*),

$$\frac{m_1}{m_2} = \frac{h_3 - h_2}{h_1 - h_3} = \frac{W_3 - W_2}{W_1 - W_3} \quad \ldots (vi)$$

The adiabatic mixing process is represented on the psychrometric chart, as shown in Fig. 37.25 (*b*). The final condition of the mixture (point 3) lies on the straight line 1-2. The point 3 divides the line 1-2 in the inverse ratio of the mixing masses. By calculating the value of W_3 from equation (*vi*), the point 3 is plotted on the line 1-2.

Example 37.12. ***One kg of air at 40° C dry bulb temperature and 50% relative humidity is mixed with 2 kg of air at 20° C dry bulb temperature and 20° C dew point temperature. Calculate the temperature and specific humidity of the mixture.***

Solution. Given : $m_1 = 1$ kg ; $t_{d1} = 40°$ C ; $\phi_1 = 50\%$; $m_2 = 2$ kg ; $t_{d2} = 20°$ C ; $t_{dp} = 20°$ C

Specific humidity of the mixture

Let W_3 = Specific humidity of the mixture.

The condition of first mass of air at 40° C dry bulb temperature and 50% relative humidity is marked on the psychrometric chart at point 1, as shown in Fig. 37.26. Now mark the condition of second mass of air at 20° C dry bulb temperature and 20° C dew point temperature at point 2, as shown in the figure. This point lies on the saturation curve. Join the points 1 and 2. From the psychrometric chart, we find that specific humidity of the first mass of air,

$$W_1 = 0.0238 \text{ kg / kg of dry air}$$

and specific humidity of the second mass of air,

$$W_2 = 0.0148 \text{ kg / kg of dry air}$$

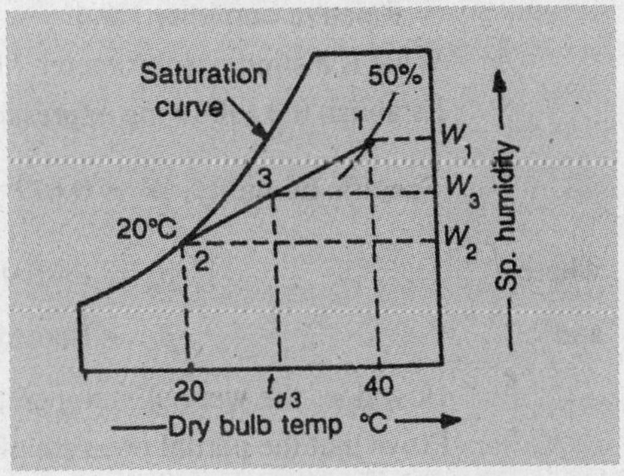

Fig. 37.26

We know that

$$\frac{m_1}{m_2} = \frac{W_3 - W_2}{W_1 - W_3}$$

$$\frac{1}{2} = \frac{W_3 - 0.0148}{0.0238 - W_3}$$

or $$0.0238 - W_3 = 2W_3 - 0.0296$$

$$\therefore \quad W_3 = 0.0178 \text{ kg / kg of dry air } \textbf{Ans.}$$

Temperature of the mixture

Now plot point 3 on the line joining the points 1 and 2 corresponding to specific humidity $W_3 = 0.0178$ kg / kg of dry air as shown in Fig. 37.26. We find that at point 3, the dry bulb temperature of the mixture is

$$t_{d3} = 26.8° \text{ C } \textbf{Ans.}$$

EXERCISES

1. A sling psychrometer reads 44° C dry bulb temperature and 30° C wet bulb temperature. Calculate 1. vapour pressure ; 2. specific humidity ; 3. relative humidity ; 4. dew point temperature ; and 5. enthalpy of the mixture per kg of dry air. [**Ans.** 0.033 bar ; 0.021 kg/kg of dry air ; 36.7% ; 25.9° C ; 99.3 kJ]

2. The humidity ratio of atmospheric air at 28° C dry bulb temperature and 760 mm of mercury is 0.016 kg/kg of dry air. Determine : 1. partial pressure of water vapour, 2. relative humidity, 3. dew point temperature, and 4. specific enthalpy. [**Ans.** 0.0253 bar ; 67% ; 21.1° C ; 68.6 kJ/kg of dry air]

3. The atmospheric air enters the adiabatic saturator at 33° C dry bulb temperature and 23° C wet bulb temperature. The barometric pressure is 740 mm of Hg. Determine the specific humidity and vapour pressure at 33° C. [**Ans.** 0.012 kg/kg of dry air ; 13 mm of Hg]

4. A sample of air has 22° C DBT, relative humidity 30 percent at barometric pressure of 760 mm of Hg. Calculate : 1. Vapour pressure, 2. Humidity ratio, 3. Vapour density, and 4. Enthalpy.

Verify your results by psychrometric chart.

[**Ans.** 0.008 bar ; 0.005 kg/kg of dry air ; 0.005 82 kg/m³ of dry air ; 34.8 kJ/kg of dry air]

5. A quantity of air having a volume of 300 m³ at 30° C dry bulb temperature and 25° C wet bulb temperature is heated to 40° C dry bulb temperature. Estimate the amount of heat added, final relative humidity and wet bulb temperature. The air pressure is 1.013 25 bar.

[**Ans.** 3533 .4 kJ/kg ; 39% ; 27.5° C]

6. 2 kg of air at 40° C dry bulb temperature and 50% relative humidity is mixed with 3 kg of air at 20° C dry bulb temperature and 12° C dew point temperature. Calculate specific humidity and the dry bulb temperature of the mixture. [**Ans.** 14.5 g/kg of dry air ; 28° C]

7. 800 m³/min of recirlculated air at 22° C DBT and 10° C dew point temperature is to be mixed with 300 m³/min of fresh ait at 30° C DBT and 50% RH. Determine the enthalpy, specific volume, humidity ratio and dew point temperature of the mixture. [**Ans.** 47.6 kJ/kg ; 0.856 m³/kg ; 9.2 g/kg of dry air ; 13° C]

QUESTIONS

1. What do you understand by the term 'psychrometry'.
2. Define the following :

   1. Specific humidity ; 2. Absolute humidity ;
   3. Relative humidity ; and 4. Dew point temperature.
3. What is a sling psychrometer ? Make a neat sketch and explain its use.
4. Establish the following expression for air-vapour mixture :

$$\text{Specific humidity, } W = 0.622 \times \frac{p_v}{p_b - p_v}$$

where p_v = Partial pressure of water vapour,

and p_b = Barometric pressure.

5. How does the wet bulb temperature differ from thermodynamic wet bulb temperature ?
6. Prove that the partial pressure of water vapour in the atmospheric air remains constant as long as the specific humidity remains constant.
7. Prove that the enthalpy of the humid air remains constant along a wet bulb temperature line on the psychrometric chart.
8. When is dehumidification of air necessary and how it is achieved.
9. Define sensible heat factor.
10. Show the following processes on the skeleton psychrometric chart :

    (*a*) Dehumidification of moist air by cooling ; and

    (*b*) Adiabatic mixing of two streams.

OBJECTIVE TYPE QUESTIONS

1. A mixture of dry air and water vapour, when the air has diffused the maximum amount of water vapour into it, is called

   (*a*) dry air (*b*) moist air (*c*) saturated air (*d*) specific humidity
2. The temperature of air recorded by a thermometer, when it is not effected by the moisture present in it, is called

   (*a*) wet bulb temperature (*b*) dry bulb temperature
   (*c*) dew point temperature (*d*) none of these
3. For unsaturated air, the dew point temperature is ..... wet bulb temperature.

   (*a*) equal to (*b*) less than (*c*) more than
4. The difference between dry bulb temperature and wet bulb temperature, is called

   (*a*) dry bulb depression (*b*) wet bulb depression
   (*c*) dew point depression (*d*) degree of saturation
5. The vertical and uniformly spaced linės on a psychrometric chart indicates

   (*a*) dry bulb temperature (*b*) wet bulb temperature
   (*c*) dew point temperature (*d*) specific humidity
6. The curved lines on a psychrometric chart indicates

   (*a*) dry bulb temperature (*b*) wet bulb temperature
   (*c*) specific humidity (*d*) relative humidity
7. During sensible cooling of air, the specific humidity

   (*a*) increases (*b*) decreases (*c*) remains constant

8. During sensible cooling of air, the dry bulb temperature
   (*a*) increases (*b*) decreases (*c*) remains constant
9. During sensible cooling of air, the wet bulb temperature
   (*a*) increases (*b*) decreases (*c*) remains constant
10. The process generally used in winter air conditioning to warm and humidify the air, is called
    (*a*) humidification (*b*) dehumidification
    (*c*) heating and humidification (*d*) cooling and dehumidification

ANSWERS

1. (*c*)	2. (*b*)	3. (*b*)	4. (*b*)	5. (*a*)
6. (*d*)	7. (*c*)	8. (*b*)	9. (*b*)	10. (*c*)

38

Air Conditioning Systems

1. Introduction. 2. Factors Affecting Comfort Air Conditioning. 3. Air Conditioning System. 4. Equipments used in an Air Conditioning System. 5. Classifications of Air Conditioning Systems. 6. Comfort Air Conditioning System. 7. Industrial Air Conditioning System. 8. Winter Air Conditioning System. 9. Summer Air Conditioning System. 10. Year-round Air Conditioning System. 11. Unitary Air Conditioning System. 12. Central Air Conditioning System.

38.1. Introduction

The air conditioning is that branch of engineering science which deals with the study of conditioning of air for *human comfort. This subject, in its broad sense, also deals with the conditioning of air for industrial purposes, food processing, storage of food and other materials.

38.2. Factors Affecting Comfort Air Conditioning

The four important factors for comfort air conditioning are discussed as below :

1. *Temperature of air.* In air conditioning, the control of temperature means the maintenance of any desired temperature within an enclosed space even though the temperature of the outside air is above or below the desired room temperature. This is accomplished either by the addition or removal of heat from the enclosed space as and when demanded. It may be noted that a human being feels comfortable when the air is at 21° C with 56% relative humidity.

2. *Humidity of air.* The control of humidity of air means the increasing or decreasing of moisture contents of air during summer or winter respectively in order to produce comfortable and healthy conditions. The control of humidity is not only necessary for human comfort but it also increases the efficiency of the workers. In general, for summer air conditioning, the relative humidity should not be less than 60% whereas for winter air conditioning it should not be more than 40%.

3. *Purity of air.* It is an important factor for the comfort of a human body. It has been noticed that people do not feel comfortable when breathing contaminated air, even if it is within acceptable temperature and humidity ranges. It is thus obvious that proper filtration, cleaning and purification of air is essential to keep it free from dust and other impurities.

4. *Motion of air.* The motion or circulation of air is another important factor which should be controlled, in order to keep constant temperature throughout the conditioned space. It is, therefore, necessary that there should be equi-distribution of air throughout the space to be air conditioned.

* Strictly speaking, the human comfort depends upon physiological and psychological conditions. Thus it is difficult to define the term 'human comfort'. There are many definitions given for this term by different bodies. But the most accepted definition, from the subject point of view, is given by the American Society of Heating, Refrigeration and Air conditioning Engineers (ASHRAE) which states "*human comfort is that condition of mind, which expresses satisfaction with the thermal environment.*"

38.3. Air Conditioning System

We have already discussed in Art. 38.2, the four important factors which effect the human comfort. The system which effectively controls these conditions to produce the desired effects upon the occupants of the space, is known as an *air conditioning system*.

38.4. Equipments Used in an Air Conditioning System

Following are the main equipments or parts used in an air conditioning system :

1. *Circulation fan.* The main function of this fan is to move air to and from the room.

2. *Air conditioning units.* It is a unit, which consists of cooling and dehumidifying processes for summer air conditioning or heating and humidification processes for winter air conditioning.

3. *Supply duct.* It directs the conditioned air from the circulating fan to the space to be air conditioned at proper point.

4. *Supply outlets.* These are grills, which distribute the conditioned air evenly in the room.

5. *Return outlets.* These are the openings in a room surface which allow the room air to enter the return duct.

6. *Filters.* The main function of the filters is to remove dust, dirt and other harmful bacteria from the air.

38.5. Classifications of Air Conditioning Systems

The air conditioning systems may be broadly classified as follows :

1. *According to the purpose*

   (*a*) Comfort air conditioning system, and

   (*b*) Industrial air conditioning system.

2. *According to season of the year*

   (*a*) Winter air conditioning system,

   (*b*) Summer air conditioning system, and

   (*c*) Year-round air conditioning system.

3. *According to the arrangement of equipment*

   (*a*) Unitary air conditioning system, and

   (*b*) Central air conditioning system.

In this chapter, we shall discuss all the above mentioned air conditioning systems, one by one.

38.6. Comfort Air Conditioning System

In comfort air conditioning, the air is brought to the required dry bulb temperature and relative humidity. If sufficient data of the required condition is not given, then it is assumed to be 21° C dry bulb temperature and 50% relative humidity. The sensible heat factor is, generally, kept as following :

For residence or private office	=	0.9
For restaurant or busy office	=	0.8
Auditorium or cinema hall	=	0.7
Ball room dance hall etc.	=	0.6

The comfort air conditioning may be adopted for small cabins, office halls or big halls like cinema houses.

Example 38.1. *An air conditioning plant is required to supply 60 m^3 of air per minute at a DBT of 21° C and 55% R.H. The outside air is at DBT of 28° C and 60% RH. Determine the mass of water drained and capacity of the cooling coil. Assume the air conditioning plant first to dehumidify and then to cool the air.*

Solution. Given : $v_2 = 60\ m^3/min$; $t_{d2} = 21°\ C$; $\phi_2 = 55\%$; $t_{d1} = 28°\ C$; $\phi_1 = 60\%$

Mass of water drained

First of all, mark the initial condition of air at 28° C dry bulb temperature and 60% relative humidity on the psychrometric chart as point 1, as shown in Fig. 38.1. Now mark the final condition of air at 21° C dry bulb temperature and 55% relative humidity as point 2. From the psychrometric chart, we find that

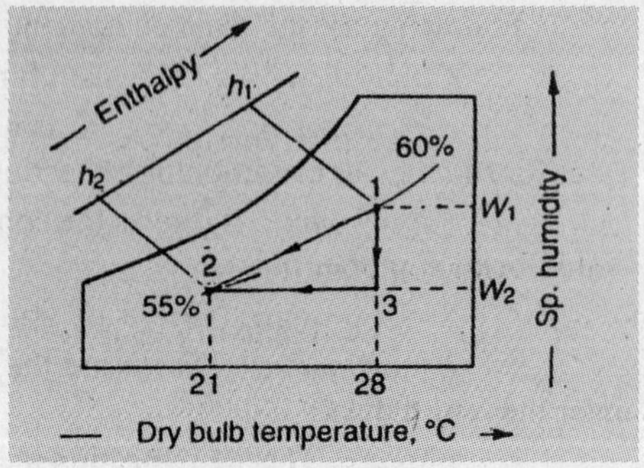

Fig. 38.1

Specific humidity of air at point 1,

$$W_1 = 0.0142\ kg\ /\ kg\ of\ dry\ air$$

Specific humidity of air at point 2,

$$W_2 = 0.0088\ kg\ /\ kg\ of\ dry\ air$$

and specific volume of air at point 2,

$$v_{s2} = 0.845\ m^3/\ kg\ of\ dry\ air$$

We know that mass of air circulated,

$$m_a = \frac{v_2}{v_{s2}} = \frac{60}{0.845} = 71\ kg\ /\ min$$

∴ Mass of water drained

$$= m_a (W_1 - W_2) = 71\ (0.0142 - 0.0088) = 0.3834\ kg\ /\ min$$

$$= 23\ kg\ /\ h \quad \textbf{Ans.}$$

Capacity of the cooling coil

From the psychrometric chart, we find that

Enthalpy of air at point 1,

$$h_1 = 63.8\ kJ\ /\ kg\ of\ dry\ air$$

and enthalpy of air at point 2,

$$h_2 = 44.5\ kJ\ /\ kg\ of\ dry\ air$$

∴ Capacity of the cooling coil

$$= m_a (h_1 - h_2) = 71\ (63.8 - 44.5) = 1370.3\ kJ\ /\ min \quad \textbf{Ans.}$$

38.7. Industrial Air Conditioning System

It is an important system of air conditioning these days in which the inside dry bulb temperature and relative humidity of the air is kept constant for proper working of the machines. Some of the sophisticated electronic and other machines need a particular dry bulb temperature and relative humidity. Sometimes, these machines also require a particular method of psychrometric processes.

Example 38.2. *Following data refers to an air conditioning system to be designed for an industrial process for hot and wet climate :*

Outside conditions	*= 30° C DBT and 75% RH*
Required inside conditions	*= 20° C DBT and 60% RH*

The required condition is to be achieved first by cooling and dehumidifying and then by heating. If 20 m³ of air is absorbed by the plant every minute, find : 1. capacity of the cooling coil in tonne of refrigeration ; 2. capacity of the heating coil in kW ; and 3. amount of water removed per hour.

Solution. Given : $t_{d1} = 30°\ C$; $\phi_1 = 75\%$; $t_{d2} = 20°\ C$; $\phi_2 = 60\%$; $v_1 = 20\ m^3/min$

1. *Capacity of the cooling coil in tonne of refrigeration*

First of all, mark the initial condition of air at 30° C dry bulb temperature and 75% relative humidity, on the psychrometric chart as point 1, as shown in Fig. 38.2. Then mark the final condition of air at 20° C dry bulb temperature and 60% relative humidity on the chart as point 3.

Now locate the points 2′ and 2 on the saturation curve by drawing horizontal lines through points 1 and 3, as shown in Fig. 38.2. On the chart, the process 1-2′-2 represents the cooling and dehumidifying process and 2-3 represents the heating process. From the psychrometric chart, we find that the specific volume of air at point 1,

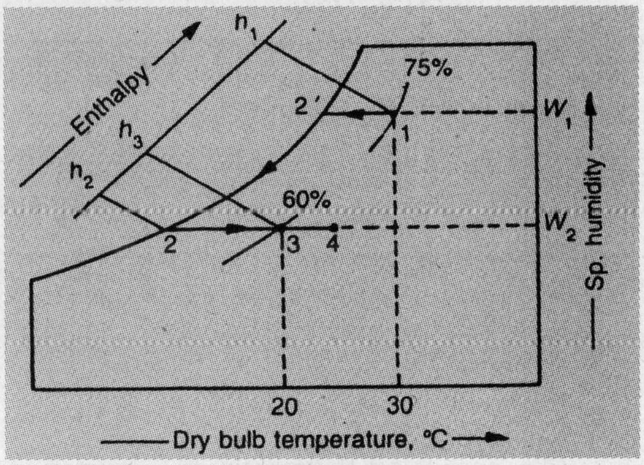

Fig 38.2

$$v_{s1} = 0.886 \text{ m}^3/\text{ kg of dry air}$$

Enthalpy of air at point 1,

$$h_1 = 81.8 \text{ kJ / kg of dry air}$$

and enthalpy of air at point 2,

$$h_2 = 34.2 \text{ kJ / kg of dry air}$$

We know that mass of air absorbed by the plant,

$$m_a = \frac{v_1}{v_{s1}} = \frac{20}{0.886} = 22.6 \text{ kg / min}$$

$\therefore$ Capacity of the cooling coil

$$= m_a(h_1 - h_2) = 22.6\,(81.8 - 34.2) = 1075.76 \text{ kJ / min}$$

$$= \frac{1075.76}{210} = 5.1 \text{ TR } \textbf{Ans.}$$

2. *Capacity of the heating coil in kW*

From the psychrometric chart, we find that enthalpy of air at point 3,

$$h_3 = 42.6 \text{ kJ / kg of dry air}$$

$\therefore$ Capacity of the heating coil

$$= m_a(h_3 - h_2) = 22.6\,(42.6 - 34.2) = 189.84 \text{ kJ / min}$$

$$= \frac{189.84}{60} = 3.16 \text{ kW } \textbf{Ans.}$$

3. *Amount of water removed per hour*

From the psychrometric chart, we find that specific humidity of air at point 1,

$$W_1 = 0.0202 \text{ kg / kg of dry air}$$

and specific humidity of air at point 2,

$$W_2 = 0.0088 \text{ kg / kg of dry air}$$

$\therefore$ Amount of water removed per hour

$$= m_a(W_1 - W_2) = 22.6\,(0.0202 - 0.0088) = 0.258 \text{ kg / min}$$

$$= 15.48 \text{ kg / h } \textbf{Ans.}$$

38.8. Winter Air Conditioning System

In winter air conditioning, the air is heated, which is generally accompanied by humidification. The schematic arrangement of the system is shown in Fig. 38.3.

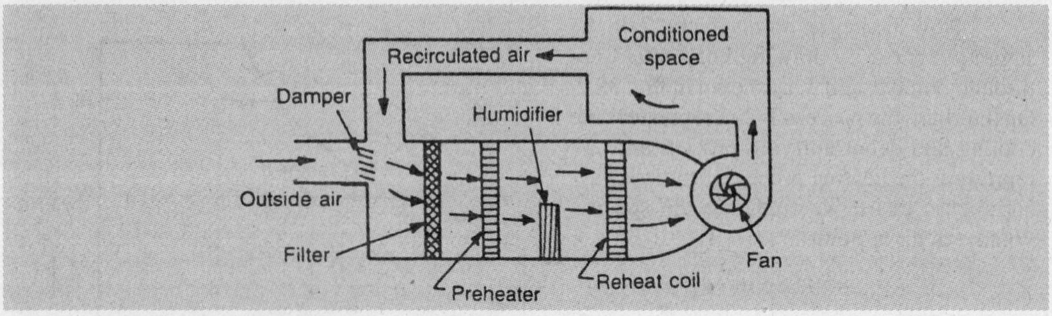

Fig. 38.3. Winter air conditioning system.

The outside air flows through a damper and mixes up with the recirculated air (which is obtained from the conditioned space). The mixed air passes through a filter to remove dirt, dust and other impurities. The air now passes through a preheat coil in order to prevent the possible freezing of water and to control the evaporation of water in the humidifier. After that, the air is made to pass through a reheat coil to bring the air to the designed dry bulb temperature. Now, the conditioned air is supplied to the conditioned space by a fan. From the conditioned space, a part of the used air is exhausted to the atmosphere by the exhaust fans or ventilators. The remaining part of the used air (known as recirculated air) is again conditioned, as shown in Fig. 38.3.

The outside air is sucked and made to mix with the recirculated air, in order to make up for the loss of conditioned (or used) air through exhaust fans or ventilation from the conditioned space.

Example 38.3. *In a winter air conditioning system, 100 m³ of air per minute at 15° C dry bulb temperature and 80% RH is heated until its dry bulb temperature is 22° C. Find heat added to the air per minute.*

Solution. Given : $v_1 = 100$ m³/min ; $t_{d1} = 15°$ C ; $\phi_1 = 80\%$; $t_{d2} = 22°$ C

First of all, mark the initial condition of air at 15° C dry bulb temperature and 80% relative humidity on the psychrometric chart as point 1. Now mark final condition of air by drawing a horizontal line through point 1 up to 22° C dry bulb temperature, as shown in Fig. 38.4. From the psychrometric chart, we find that enthalpy of air at point 1,

$$h_1 = 36.5 \text{ kJ / kg of dry air}$$

Similarly, $h_2 = 43.5$ kJ / kg of dry air

Moreover, specific volume of air at point 1,

$$v_{s1} = 0.827 \text{ m}^3\text{/ kg}$$

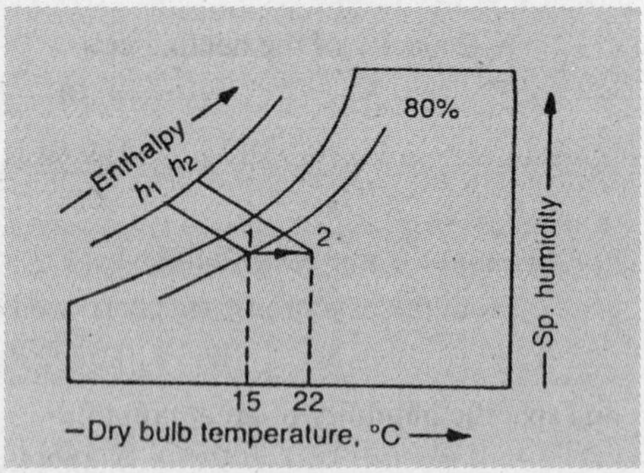

Fig. 38.4

∴ Mass of air heated per minute,

$$m_a = \frac{v_1}{v_{s1}} = \frac{100}{0.827} = 120.9 \text{ kg}$$

and heat added to the air per minute

$$= m_a (h_2 - h_1) = 120.9 (43.5 - 36.5) = 846.3 \text{ kJ} \quad \textbf{Ans.}$$

38.9. Summer Air Conditioning System

It is the most important type of air conditioning, in which the air is cooled and generally dehumidified. The schematic arrangement of a typical summer air conditioning system is shown in Fig. 38.5.

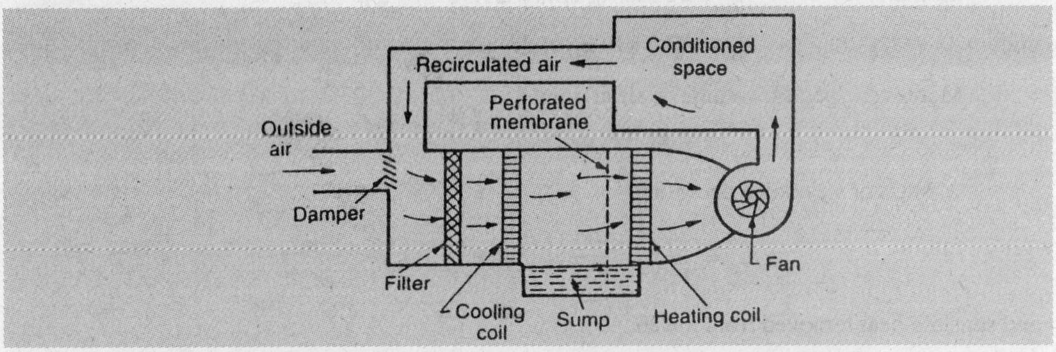

Fig. 38.5. Summer air conditioning system.

The outside air flows through the damper, and mixes up with recirculated air (which is obtained from the conditioned space). The mixed air passes through a filter to remove dirt, dust and other impurities. The air now passes through a cooling coil. The coil has a temperature much below the required dry bulb temperature of the air in the conditioned space. The cooled air passes through a perforated membrane and looses its moisture in the condensed form which is collected in a sump. After that, the air is made to pass through a heating coil which heats up the air slightly. This is done to bring the air to the designed dry bulb temperature and relative humidity.

Now the conditioned air is supplied to the conditioned space by a fan. From the conditioned space a part of the used air is exhausted to the atmosphere by the exhaust fans or ventilators. The remaining part of the used air (known as recirculated air) is again conditioned, as shown in Fig. 38.5. The outside air is sucked and made to mix with the recirculated air in order to make up for the loss of conditioned (or used) air through exhaust fans or ventilation from the conditioned space.

Example 38.4. *A theatre of 1200 seating capacity is to be air conditioned for summer conditions with the following data :*

Outdoor conditions	*30° C DBT and 55% RH*
Required conditions	*20° C DBT and 60% RH*
Amount of air supplied	*0.25 m³/min/person*

Find the sensible heat, latent heat removed from the air per minute and sensible heat factor for the system.

Solution. Given : Seating capacity = 1200 persons ; t_{d1} = 30° C ; ϕ_1 = 55% ; t_{d2} = 22° C ; ϕ_2 = 60% ; v_1 = 0.25 m³/min / person = 0.25 × 1200 = 300 m³/min

Sensible heat removed from the air

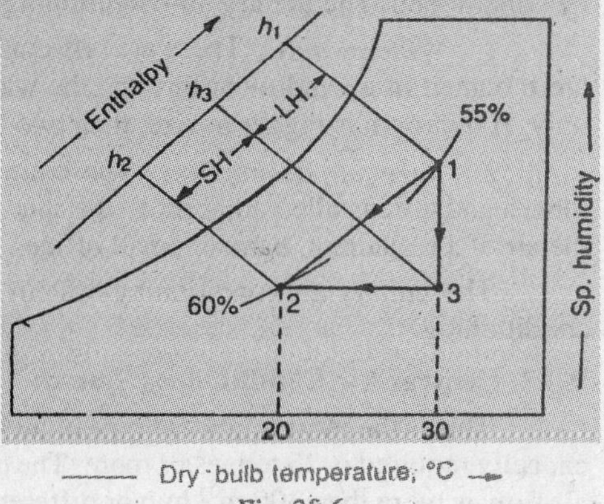

Fig. 38.6

First of all, mark the initial condition of air at 30° C dry bulb temperature and 55% relative humidity on the psychrometric chart as point 1, as shown in Fig. 38.6. Now mark the final condition of air at 20° C dry bulb temperature and 60% relative humidity on the chart as point 2. Now locate point 3 on the chart by drawing horizontal line through point 2 and vertical line through point 1. From the psychrometric chart, we find that enthalpy of air at point 1,

$$h_1 = 68 \text{ kJ / kg of dry air}$$

Similarly $$h_2 = 42 \text{ kJ / kg of dry air}$$

and $$h_3 = 52.5 \text{ kJ / kg of dry air}$$

Moreover, specific volume of air at point 1,

$$v_{s1} = 0.88 \text{ m}^3\text{/ kg}$$

∴ Mass of air cooled per minute

$$m_a = \frac{v_1}{v_{s1}} = \frac{300}{0.88} = 340.9 \text{ kg}$$

and sensible heat removed from the air,

$$\text{SH} = m_a (h_3 - h_2) = 340.9 (52.5 - 42) = 3579.5 \text{ kJ / min} \quad \textbf{Ans.}$$

Latent heat removed from the air

We know that the latent heat removed from the air,

$$\text{LH} = m_a (h_1 - h_3) = 340.9 (68 - 52.5) = 5284 \text{ kJ / min} \quad \textbf{Ans.}$$

Sensible heat factor for the system

We also know that sensible heat factor for the system,

$$= \frac{\text{SH}}{\text{SH} + \text{LH}} = \frac{3579.5}{3579.5 + 5284} = 0.4 \quad \textbf{Ans.}$$

38.10. Year-round Air Conditioning System

An year-round air conditioning system should have equipment for both the winter and summer air conditioning. There can be so many combinations for providing year-round air conditioning. The arrangement of the combination should be such that one part is working in winter and the other in summer. For example, heating and humidification is needed in winter, whereas cooling and dehumidification is needed in summer.

38.11. Unitary Air Conditioning System

In this system, factory assembled air conditioners are installed in or adjacent to the space to be conditioned. The unitary air conditioning systems are of the following two types :

1. *Window units.* These are self-contained units of small capacity of 1 tonne to 3 tonnes, and are mounted in a window or through the wall. They are employed to condition the air of one room only. If the room is bigger in size, then two or more units are installed.

2. *Vertical packed units.* These are also self-contained units of bigger capacity of 5 to 10 tonnes and are installed adjacent to the space to be conditioned. This is very useful for conditioning the air of a restaurant, bank or small office.

The unitary air conditioning system may be adopted for winter, summer or year-round air conditioning.

38.12. Central Air Conditioning System

This is the most important type of air conditioning system, which is adopted when the cooling capacity required is 25 tonnes or more. The central air conditioning system is also adopted when the air flow is more than 300 m³/ min or different zones in a building are to be conditioned.

Example 38.5. *An air conditioning plant is to be designed for a small office for winter conditions :*

Out-door conditions	*= 10° C DBT 8° C WBT*
Required indoor conditions	*= 20° C DBT and 60% RH*
Amount of air circulation	*= 0.3 m³/min/person*
Seating capacity of the office	*= 50 persons*

The required conditions is achieved first by heating and then by adiabatic humidifying. Find : 1. heating capacity of the coil in kW ; and 2. capacity of the humidifier.

Solution. Given : $t_{d1} = 10°$ C ; $t_{w1} = 8°$ C ; $t_{d2} = 20°$ C ; $\phi_2 = 60\%$; Seating capacity = 50 persons ; $v_1 = 0.3$ m³/min / person $= 0.3 \times 50 = 15$ m³/min

First of all, mark the initial condition of air at 10° C dry bulb temperature and 8° C wet bulb temperature on the psychrometric chart as point 1, as shown in Fig. 38.7. Now mark the final condition of air at 20° C dry bulb temperature and 60% relative humidity on the chart as point 2. Now locate point 3 on the chart by drawing horizontal line through point 1 and constant enthalpy line through point 2. From the psychrometric chart, we find that the specific volume at point 1,

$$v_{s1} = 0.81 \text{ m}^3/\text{ kg of dry air}$$

∴ Mass of air supplied per minute,

$$m_a = \frac{v_1}{v_{s1}} = \frac{15}{0.81} = 18.52 \text{ kg / min}$$

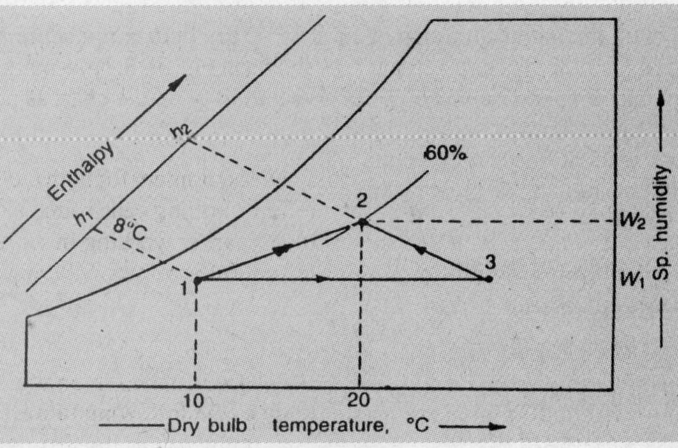

Fig. 38.7

1. *Heating capacity of the coil in kW*

From the psychrometric chart, we find that enthalpy at point 1,

$$h_1 = 24.8 \text{ kJ / kg of dry air}$$

and enthalpy at point 2, $\quad h_2 = 42.6$ kJ / kg of dry air

We know that heating capacity of the coil

$$= m_a(h_2 - h_1) = 18.52(42.6 - 24.8) = 329.6 \text{ kJ / min}$$

$$= \frac{329.6}{60} = 5.49 \text{ kW } \textbf{Ans.}$$

2. *Capacity of the humidifier*

From the psychrometric chart, we find that specific humidity at point 1,

$$W_1 = 0.0058 \text{ kg / kg of dry air}$$

and specific humidity at point 2,

$$W_2 = 0.0088 \text{ kg / kg of dry air}$$

We know that capacity of the humidifier,

$$= m_a (W_2 - W_1) = 18.52 (0.0088 - 0.0058) = 0.055 \text{ kg / min}$$

$$= 3.3 \text{ kg / h} \quad \textbf{Ans.}$$

Example 38.6. *A small office hall of 25 persons capacity is provided with summer air conditioning system with the following data :*

Outside conditions	*= 34° C DBT and 28° C WBT*
Inside conditions	*= 24° C DBT and 50% RH*
Volume of air supplied	*= 0.4 m³/min/person*
Sensible heat load in room	*= 125 600 kJ/h*
Latent heat load in the room	*= 42 000 kJ/h*

Find the sensible heat factor of the plant.

Solution. Given : Seating capacity = 25 persons ; t_{d1} = 34° C ; t_{w1} = 28° C ; t_{d2} = 24° C ; ϕ_2 = 50% ; v_1 = 0.4 m³/min / person = 0.4 × 25 = 10 m³/ h ; SH load = 125 600 kJ/h ; LH load = 42 000 kJ/h

First of all, mark the initial condition of air at 34° C dry bulb temperature and 28° C wet bulb temperature on the psychrometric chart as point 1, as shown in Fig. 38.8. Now mark the final condition of air at 24° C dry bulb temperature and 50% relative humidity on the chart as point 2. Now locate point 3 on the chart by drawing horizontal line through point 2 and vertical line through point 1. From the psychrometric chart, we find that the specific volume at point 1,

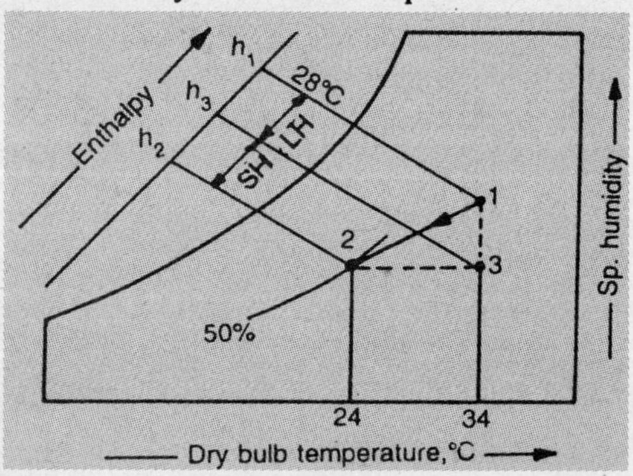

Fig. 38.8

$$v_{s1} = 0.9 \text{ m}^3\text{/ kg of dry air}$$

Enthalpy of air at point 1,

$$h_1 = 90 \text{ kJ / kg of dry air}$$

Enthalpy of air at point 2,

$$h_2 = 48.2 \text{ kJ / kg of dry air}$$

and enthalpy of air at point 3,

$$h_3 = 59 \text{ kJ / kg of dry air}$$

We know that mass of air supplied per min,

$$m_a = \frac{v_1}{v_{s1}} = \frac{10}{0.9} = 11.1 \text{ kg / min}$$

and sensible heat removed from the air

$$= m_a (h_3 - h_2) = 11.1 (59 - 48.2) = 119.88 \text{ kJ / min}$$

$$= 7192.8 \text{ kJ/h}$$

∴ Total sensible heat of the room,

$$SH = 7192.8 + 125\,600 = 132\,792.8 \text{ kJ/h}$$

We know that latent heat removed from the air

$$= m_a (h_1 - h_3) = 11.1\,(90 - 59) = 344.1 \text{ kJ / min} = 20\,646 \text{ kJ/h}$$

∴ Total latent heat of the room

$$LH = 20\,646 + 42\,000 = 62\,646 \text{ kJ/h}$$

We know that sensible heat factor,

$$SHF = \frac{SH}{SH + LH} = \frac{132\,792.8}{132\,792.8 + 62\,646} = 0.679 \textbf{ Ans.}$$

EXERCISES

1. An air conditioning plant is handling 30 m^3 of air per minute at 32° C dry bulb temperature and 22° C wet bulb temperature. If the final conditions of air are 22° C dry bulb temperature and 50% relative humidity, find the heating capacity of the dehumidifier. [**Ans.** 7.3 kg/h]

2. A cinema hall of seating capacity 1500 persons has been provided with an air conditioned plant with the following data :

   Outdoor conditions = 40° C DBT and 20° C WBT

   Required conditions = 20° C DBT and 60% RH

   Amount of air supplied = 0.3 m^3/min/person

   If the required condition is achieved first by adiabatic humidifying and then cooling, find the capacity of cooling coil and capacity of the humidifier. [**Ans.** 37.6 TR ; 54.2 kg/h]

3. An office hall of 70 persons capacity is to be provided with summer air conditioning plant with the following data :

   Outside conditions = 35° C DBT and 26° WBT

   Inside conditions = 20° C DBT and 45% RH

   Volume of air supplied = 0.3 m^3/min/person

   Find the sensible heat factor for the plant. [**Ans.** 0.35]

4. The amount of air supplied to an air conditioned hall is 300 m^3/min. The atmospheric conditions are 35° C DBT and 55% RH. The required conditions are 20° C DBT and 60% RH. Find out the sensible heat and latent heat removed from the air per minute. Also find sensible heat factor for the system. [**Ans.** 5040 kJ/min ; 9517 kJ/min ; 0.346]

5. An auditorium of 300 seating capacity is to be provided with air conditioning with the following data :

   Outdoor conditions = 12° C DBT and 70% RH

   Indoor conditions = 20° C DBT and 50% RH

   Amount of air supplied = 0.25 m^3/min/person

   Find the heating capacity of the coil. [**Ans.** 13.6 kW]

6. A restaurant with a capacity of 100 persons is to be air-conditioned with the following conditions :

   Outside conditions = 30° C DBT, 70% RH

   Desired inside conditions = 23° C DBT, 55% RH

   Quantity of air supplied = 0.5 m^3/min/person

   The desired conditions are achieved by cooling, dehumidifying and then heating. Determine : 1. Capacity of cooling coils in tonnes of refrigeration ; 2. Capacity of heating coil : and 3. Amount of water removed by dehumidifier. [**Ans.** 9.3 TR ; 8.7 kW ; 31.5 kg/h]

QUESTIONS

1. Define comfort. What are the factors which affect comfort air conditioning ?
2. How does industrial air conditioning is different from comfort air conditioning ?
3. Draw a line diagram of air conditioning system required in winter season. Explain the working of different components in the circuit.
4. Draw a block diagram of a simple year round air conditioning system and briefly explain the process on a psychrometric chart.
5. Describe unitary and central air conditioning systems.

OBJECTIVE TYPE QUESTIONS

1. For summer air conditioning, the relative humidity should not be less than
   (*a*) 40% (*b*) 60% (*c*) 75% (*d*) 90%
2. For winter air conditioning, the relative humidity should not be more than
   (*a*) 40% (*b*) 60% (*c*) 75% (*d*) 90%
3. The sensible heat factor for auditorium or cinema hall is generally kept as
   (*a*) 0.6 (*b*) 0.7 (*c*) 0.8 (*d*) 0.9
4. The conditioned air supplied to the room must have the capacity to take up
   (*a*) room sensible heat load only
   (*b*) room latent heat load only
   (*c*) both room sensible heat and latent heat loads

ANSWERS

1. (*b*) 2. (*a*) 3. (*b*) 4. (*c*)

39

General Thermodynamic Relations

1. Introduction. 2. Helmholtz Function. 3. Gibbs Function. 4. Mathematical Fundamentals. 5. Maxwell's Equations. 6. Co-efficient of Volume Expansion. 7. Isothermal Compressibility. 8. Adiabatic Compressibility. 9. Ratio of Co-efficient of Volume Expansion and Isothermal Compressibility. 10. General Equations for Change in Internal Energy. 11. T dS Equations. 12. Specific Heats. 13. Specific Heat Relations. 14. Joule Thomson Co-efficient. 15. Equations of State. 16. Compressibility Factor.

39.1. Introduction

We have already discussed, in the previous chapters, the six thermodynamic properties, *i.e.* pressure (p), volume (v), temperature (T), internal energy (U), enthalpy (H) and entropy (S). There are other two important properties also which are known as Helmholtz function (A) and Gibbs function (G). In this chapter, we shall discuss some important thermodynamic relations between different properties. It may be noted that only the three thermodynamic properties (*i.e.* pressure, volume and temperature) are directly measurable while the other properties are evaluated in terms of these measurable properties. Thus, it is very important to develop the thermodynamic relations for the properties which cannot be measured directly in terms of the combination of properties which can be measured directly or evaluated experimentally. These all combinations are defined as the rate of change of one property with respect to another while the third property is held constant. The thermodynamic relations have been derived with the help of first and second laws of thermodynamics and the mathematical concepts of differentiation, particularly partial differentiation.

39.2. Helmholtz Function

Helmholtz function (A) is a property of a system and is given by substracting the product of absolute temperature (T) and entropy (S) from the internal energy (U). Mathematically,

Helmholtz function , $A = U - TS$ ... (*i*)

Since ($U - TS$) is made up entirely of properties, therefore Helmholtz function (A) is also a property.

For per unit mass, specific Helmholtz function,

$$a = u - Ts$$

where u = Specific internal energy, and

s = Specific entropy.

The general gas energy equation is

$$\delta Q = dU + \delta W \quad \text{or} \quad \delta W = \delta Q - dU \qquad \ldots (ii)$$

The equation (*ii*) implies that the energy converted into work is provided partly by the heat reservoir with which the system is in contact and which gives up a quantity of heat δQ and partly by the system whose internal energy decreases by $-dU$.

Let us now, derive the relationship for the *maximum work* that can be done when a system undergoes a process between two equilibrium states. Assume that the system exchanges heat only with a single heat reservoir at a temperature T_0.

Let dS = Increase in entropy of the system,

dS_0 = Increase in entropy of the surroundings (reservoir),

δQ = Heat absorbed by the system from the reservoir, and

T_0 = Temperature of the reservoir.

From the *principle of increase of entropy, it is known that the sum of the increase in entropy of the system and that of the surroundings is equal to or greater than zero.

$$\therefore \quad dS + dS_0 \geq 0 \qquad \ldots (iii)$$

We know that

$$dS_0 = \frac{-\delta Q}{T_0} \qquad \ldots (\because \text{ Reservoir has given out heat, therefore } \delta Q \text{ is } -\text{ve})$$

Substituting this value of dS_0 in equation (*iii*), we get

$$dS - \frac{\delta Q}{T_0} \geq 0$$

or

$$T_0\, dS \geq \delta Q \qquad \ldots (iv)$$

From equations (*ii*) and (*iv*), we get

$$\delta W \leq T_0\, dS - dU \qquad \ldots (v)$$

On integrating the above equation from state 1 to state 2,

$$\int_1^2 \delta W \leq T_0 \int_1^2 dS - \int_1^2 dU \qquad \ldots (\because T_0 \text{ is a constant})$$

or

$$W_{1-2} \leq T_0\ (S_2 - S_1) - (U_2 - U_1)$$

$$W_{1-2} \leq (U_1 - U_2) - T_0\ (S_1 - S_2) \qquad \ldots (vi)$$

When initial and final temperatures are equal and are the same as that of the heat reservoir, then

$$T = T_0$$

where T = Temperature of the system.

Now, equation (*vi*) can be written as

$$(W_T)_{1-2} \leq (U_1 - U_2)_T - T\,(S_1 - S_2)_T \qquad \ldots (vii)$$

* Refer Chapter 4, Art. 4.7.

We know that, Helmholtz function,

$$A = U - T S$$

For a small change, the Helmholtz function can be written as

$$dA = dU - d(T S) \qquad \ldots (viii)$$

On integrating the above equation from state 1 to state 2, keeping temperature (T) constant. we have

$$\int_1^2 dA = \int_1^2 dU - T\int_1^2 dS$$

$$(A_2 - A_1)_T = (U_2 - U_1)_T - T(S_2 - S_1)_T$$

or
$$(A_1 - A_2)_T = (U_1 - U_2)_T - T(S_1 - S_2)_T \qquad \ldots (ix)$$

From equations (*vii*) and (*ix*), we have

$$(W_T)_{1-2} \leq (A_1 - A_2)_T \qquad \ldots (x)$$

Thus, the work done in any process between two equilibrium states keeping the temperature constant is equal to the decrease in Helmholtz function of the system but the energy converted into work is provided partly by the system and the remaining heat is taken from the heat reservoir.

The work done is maximum when the process is reversible. If the process is irreversible, the work done is less than the maximum. Therefore, for maximum work, equation (x) may be written as :

$$(W_T)_{max} = (W_T)_{rev} = (A_1 - A_2)_T.$$

Note: Helmholtz function (A) is useful when considering chemical reactions and is of fundamental importance in statistical thermodynamics.

39.3. Gibbs Function

The Gibbs function (also known as thermodynamic potential) is also a property of a system and is denoted by G. Mathematically, Gibbs function,

$$G = U - T S + p\,v$$

$$= H - T S \qquad \ldots (\because H = U + p\,v)$$

Since ($H - TS$) is made up entirely of properties, therefore Gibbs function (G) is also a property.

For per unit mass, specific Gibbs function,

$$g = h - T s$$

where h = Specific enthalpy, and

s = Specific entropy.

Consider a system that can do other forms of work, in addition to ($p\,dv$) like electrical work, magnetic work etc.

Let δW_0 = Some other form of work in addition to $p\,dv$.

$\therefore$ $\delta W = p\,dv + \delta W_0$

On integrating the above equation from state 1 to state 2, keeping the pressure p (equal to the external pressure, p_0) constant, we have

$$\int_1^2 \delta W = p \int_1^2 dv + \int_1^2 \delta W_0$$

or $$W_{1-2} = p_0 (v_2 - v_1) + W_0^* \quad \ldots (\because p = p_0) \quad \ldots (i)$$

We have already discussed that when a system exchanges heat with a reservoir whose temperature is T_0, then work done,

$$W_{1-2} \le (U_1 - U_2) - T_0 (S_1 - S_2) \quad \ldots (ii)$$

Substituting the value of W_{1-2} from equation (*i*) in equation (*ii*), we get

$$p_0 (v_2 - v_1) + W_0 \le (U_1 - U_2) - T_0 (S_1 - S_2)$$

or $$W_0 \le (U_1 - U_2) - T_0 (S_1 - S_2) - p_0 (v_2 - v_1) \quad \ldots (iii)$$

When the initial and final equilibrium states are at the same temperature and pressure, then

$$T_0 = T \quad \text{and} \quad p_0 = p$$

Thus, equation (*iii*) can be written as

$$(W_0)_{p,\,\mathrm{T}} \le (U_1 - U_2)_{p,\,\mathrm{T}} - \mathrm{T}\,(S_1 - S_2)_{p,\,\mathrm{T}} + p\,(v_1 - v_2)_{p,\,\mathrm{T}} \quad \ldots (iv)$$

We know that, Gibbs function,

$$G = U - TS + pv$$

For a small change, the Gibbs function can be written as

$$dG = dU - d(TS) + d(pv)$$

On integrating from state 1 to state 2, keeping both pressure (p) and temperature (T) constant, we have

$$\int_1^2 dG = \int_1^2 dU - T\int_1^2 dS + p\int_1^2 dv \quad \ldots (\because p \text{ and } T \text{ are constant})$$

$$G_2 - G_1 = (U_2 - U_1) - T(S_2 - S_1) + p(v_2 - v_1)$$

or $$(G_1 - G_2)_{p,\mathrm{T}} = (U_1 - U_2)_{p,\mathrm{T}} - T(S_1 - S_2)_{p,\mathrm{T}} + p(v_1 - v_2)_{p,\mathrm{T}} \quad \ldots (v)$$

From equations (*iv*) and (*v*), we have

$$(W_0)_{p,\mathrm{T}} \le (G_1 - G_2)_{p,\mathrm{T}} \quad \ldots (vi)$$

Thus, the difference between Gibbs function of a system between two equilibrium states sets the maximum limit to the work that can be performed in addition to *pdv* work, provided that two equilibrium states are at the same pressure and temperature and the system exchange heat with a single heat reservoir.

The work done is maximum when the process is reversible. If the process is irreversible,

* W_0 may be electrical work as in case of voltaic cell which is equal to $-E\,dI$ or for a magnetic material, it may be magnetic work which is equal to $-mdH$.

the work done is less than the maximum. Therefore, for maximum work, equation (*vi*) can be written as :

$$\left((W_0)_{p,T}\right)_{max} = \left((W_0)_{p,T}\right)_{rev} = \left(G_1 - G_2\right)_{p,T}$$

Note : The Gibbs function (G) is also useful when considering chemical reactions. It is also useful when considering processes involving a change of phase.

39.4. Mathematical Fundamentals

The following mathematical fundamentals are important from the subject point of view :

1. *Exact differential.* Suppose z is a function of two independent properties x and y, such that

$$z = f(x, y) \qquad \ldots (i)$$

According to the condition of exact differentials,

$$dz = \left(\frac{\partial z}{\partial x}\right)_y dx + \left(\frac{\partial z}{\partial y}\right)_x dy \qquad \ldots (ii)$$

where $\left(\frac{\partial z}{\partial x}\right)_y$ = Partial derivative of z with respect to x, when y is held constant, and

$\left(\frac{\partial z}{\partial y}\right)_x$ = Partial derivative of z with respect to y, when x is held constant.

Now, the above equation (*ii*) can be written as

$$dz = Mdx + Ndy \qquad \ldots (iii)$$

where $M = \left(\frac{\partial z}{\partial x}\right)_y$ and $N = \left(\frac{\partial z}{\partial y}\right)_x$

We know that,

$$\left(\frac{\partial M}{\partial y}\right)_x = \left[\frac{\partial}{\partial y}\left(\frac{\partial z}{\partial x}\right)_y\right]_x = \left(\frac{\partial^2 z}{\partial x \times \partial y}\right) \qquad \ldots (iv)$$

and

$$\left(\frac{\partial N}{\partial x}\right)_y = \left[\frac{\partial}{\partial x}\left(\frac{\partial z}{\partial y}\right)_x\right]_y = \left(\frac{\partial^2 z}{\partial y \times \partial x}\right) \qquad \ldots (v)$$

According to calculus, the magnitude of second mixed partial derivatives of a function are independent of the order of differentiation and hence are equal.

$$\therefore \quad \left(\frac{\partial M}{\partial y}\right)_x = \left(\frac{\partial N}{\partial x}\right)_y$$

2. *Cyclic relation.* We know that, when z is a function of two independent variables x and y, then we can write

$$z = f(x, y) \qquad \ldots (i)$$

It may be noted that from above equation, x can be similarly expressed explicitly in terms of other two variables *i.e.* y and z, such that

$$x = f(y, z) \qquad \ldots (ii)$$

According to the condition of exact differentials,

$$dx = \left(\frac{\partial x}{\partial y}\right)_z dy + \left(\frac{\partial x}{\partial z}\right)_y dz \qquad \ldots (iii)$$

Similarly y can be expressed explicitly in terms of x and z, so that

$$y = f(x, z) \qquad \ldots (iv)$$

According to the condition of exact differentials,

$$dy = \left(\frac{\partial y}{\partial x}\right)_z dx + \left(\frac{\partial y}{\partial z}\right)_x dz \qquad \ldots (v)$$

Now from equation (*iii*), we have

$$dx = \left(\frac{\partial x}{\partial z}\right)_y dz + \left(\frac{\partial x}{\partial y}\right)_z dy$$

Substituting the value of dy from equation (v) in the above equation, we have

$$dx = \left(\frac{\partial x}{\partial z}\right)_y dz + \left(\frac{\partial x}{\partial y}\right)_z \left[\left(\frac{\partial y}{\partial x}\right)_z dx + \left(\frac{\partial y}{\partial z}\right)_x dz\right]$$

$$= \left(\frac{\partial x}{\partial z}\right)_y dz + \left(\frac{\partial x}{\partial y}\right)_z \left(\frac{\partial y}{\partial x}\right)_z dx + \left(\frac{\partial x}{\partial y}\right)_z \left(\frac{\partial y}{\partial z}\right)_x dz$$

$$= \left[\left(\frac{\partial x}{\partial z}\right)_y + \left(\frac{\partial x}{\partial y}\right)_z \left(\frac{\partial y}{\partial z}\right)_x\right] dz + dx$$

or $$\left(\frac{\partial x}{\partial y}\right)_z \left(\frac{\partial y}{\partial z}\right)_x + \left(\frac{\partial x}{\partial z}\right)_y = 0$$

or $$\left(\frac{\partial x}{\partial y}\right)_z \left(\frac{\partial y}{\partial z}\right)_x = -\left(\frac{\partial x}{\partial z}\right)_y$$

or $$\left(\frac{\partial x}{\partial y}\right)_z \left(\frac{\partial y}{\partial z}\right)_x \left(\frac{\partial z}{\partial x}\right)_y = -1$$

The above relation is known as *cyclic relation.*

Note : For thermodynamic properties p, v and T, the cyclic relation is

$$\left(\frac{\partial p}{\partial v}\right)_T \left(\frac{\partial v}{\partial T}\right)_p \left(\frac{\partial T}{\partial p}\right)_v = -1$$

39.5. Maxwell's Equations

Now let us derive the Maxwell's equation.

We know that for a system undergoing an infinitesimal reversible process from one equilibrium state to another :

1. Internal energy, $dU = \delta Q - \delta W$ . . . (General gas energy equation)

$$= T\,dS - p\,dv \qquad \ldots (\because \delta Q = T\,dS \text{ and } \delta W = p\,dv) \qquad \ldots (i)$$

The above equation (*i*) is of the form

$$dz = M\,dx + N\,dy$$

where $M = T,\ N = -p,\ x = S$ and $y = v$

$$\therefore \quad \left(\frac{\partial T}{\partial v}\right)_S = -\left(\frac{\partial p}{\partial S}\right)_v \quad \ldots (ii)$$

2. Enthalpy, $dH = dU + d(pv)$

$$= dU + p\,dv + v\,dp$$

$$= (T\,dS - p\,dv) + p\,dv + v\,dp \quad \ldots (\because dU = T\,dS - p\,dv)$$

$$= T\,ds + v\,dp \quad \ldots (iii)$$

The above equation (*iii*) is of the form

$$dz = M\,dx + N\,dy$$

where $M = T,\ N = v,\ x = S$ and $y = p$

$$\therefore \quad \left(\frac{\partial T}{\partial p}\right)_S = \left(\frac{\partial v}{\partial S}\right)_p \quad \ldots (iv)$$

3. Helmholtz function (A),

$$dA = dU - d(TS)$$

$$= dU - T\,dS - S\,dT$$

$$= (T\,dS - p\,dv) - T\,dS - S\,dT \quad \ldots (\because dU = T\,dS - p\,dv)$$

$$= -p\,dv - S\,dT \quad \ldots (v)$$

The above equation (v) is of the form

$$dz = M\,dx + N\,dy$$

where $M = -p, N = -S, x = v, y = T$

$$\therefore \quad -\left(\frac{\partial p}{\partial T}\right)_v = -\left(\frac{\partial S}{\partial v}\right)_T$$

or

$$\left(\frac{\partial p}{\partial T}\right)_v = \left(\frac{\partial S}{\partial v}\right)_T \quad \ldots (vi)$$

4. Gibbs function (G), $dG = dH - d(TS)$

$$= dH - T\,dS - S\,dT$$

$$= (T\,dS + v\,dp) - T\,dS - S\,dT \quad \ldots (\because dH = T\,dS + v\,dp)$$

$$= v\,dp - S\,dT \quad \ldots (vii)$$

The above equation (*vii*) is of the form

$$dz = M\,dx + N\,dy$$

where $M = v, N = -S,\ x = p, y = T$

$$\therefore \quad \left(\frac{\partial v}{\partial T}\right)_p = -\left(\frac{\partial S}{\partial p}\right)_T \quad \ldots (viii)$$

The equations (*ii*), (*iv*), (*vi*) and (*viii*) are known as *Maxwell's equations* in thermodynamics.

39.6. Co-efficient of Volume Expansion

The co-efficient of volume expansion is defined as the change in volume with change in temperature per unit volume keeping the pressure constant. It is denoted by β. Mathematically, co-efficient of volume expansion,

$$\beta = \frac{1}{v}\left(\frac{\partial v}{\partial T}\right)_p$$

For specific volume, the coefficient of volume expansion is written as,

$$\beta = \frac{1}{v_s}\left(\frac{\partial v_s}{\partial T}\right)_p$$

where v_s = Specific volume in m^3/kg.

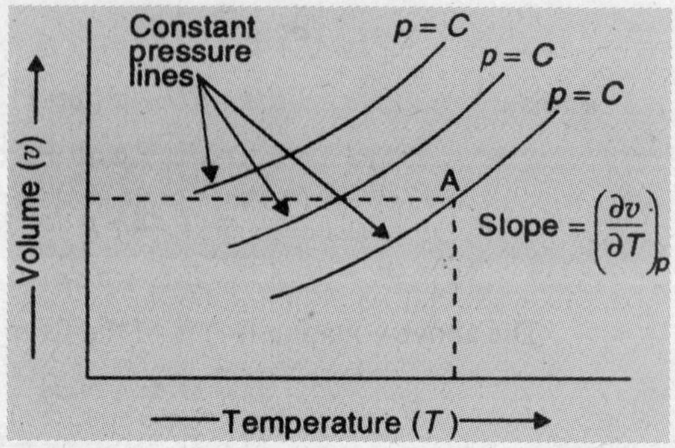

Fig 39.1. Coefficient of volume expansion.

The volume-temperature (v–T) diagram at constant pressure is shown in Fig. 39.1. The slope at any point A is given by $\left(\frac{\partial v}{\partial T}\right)_p$. If this slope is divided by the volume at that point, then we get co-efficient of volume expansion (β).

39.7. Isothermal Compressibility

The isothermal compressibility of a substance is defined as the change in volume with the change in pressure per unit volume, keeping the temperature constant. It is denoted by K. Mathematically, isothermal compressibility,

$$K = -\frac{1}{v}\left(\frac{\partial v}{\partial p}\right)_T^*$$

For specific volume, the isothermal compressibility is written as

$$K = -\frac{1}{v_s}\left(\frac{\partial v_s}{\partial p}\right)_T$$

where v_s = Specific volume in m^3/kg.

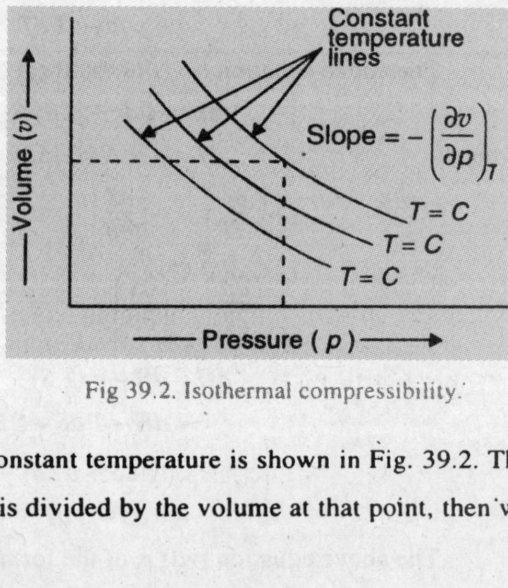

Fig 39.2. Isothermal compressibility.

The pressure-volume ($p - v$) diagram at constant temperature is shown in Fig. 39.2. The slope at a point is given by $-\left(\frac{\partial v}{\partial p}\right)_T$. If this slope is divided by the volume at that point, then we get isothermal compressibility (K).

39.8. Adiabatic Compressibility

The adiabatic compressibility is defined as the change in volume with change in pressure per unit volume when entropy is kept constant. It is denoted by K_S. Mathematically, adiabatic compressibility,

$$K_S = -\frac{1}{v}\left(\frac{\partial v}{\partial p}\right)_S$$

* The minus sign in the equation indicates that the slope is decreasing.

For specific volume, the adiabatic compressibility is written as

$$K_S = -\frac{1}{v_s}\left(\frac{\partial v_s}{\partial p}\right)_S$$

where v_s = Specific volume in m^3/kg.

Note : It may be noted that the ratio of specific heats (*i.e.* c_p/c_v) is equal to the ratio of isothermal compressibility and adiabatic compressibility. Mathematically, adiabatic index,

$$\gamma = \frac{c_p}{c_v} = \frac{K}{K_S} = \frac{-(1/v)(\partial v/\partial p)_T}{-(1/v)(\partial v/\partial p)_S} = \frac{(\partial p/\partial v)_S}{(\partial p/\partial v)_T}$$

For specific volume, the adiabatic index is written as

$$\gamma = \frac{(\partial p/\partial v_s)_S}{(\partial p/\partial v_s)_T}$$

Since γ is always greater than unity, therefore K is always greater than K_S.

39.9. Ratio of Co-efficient of Volume Expansion and Isothermal Compressibility

We know that the co-efficient of volume expansion,

$$\beta = \frac{1}{v}\left(\frac{\partial v}{\partial T}\right)_p \quad \text{or} \quad \left(\frac{\partial v}{\partial T}\right)_p = \beta\, v \qquad \ldots (i)$$

and isothermal compressibility,

$$K = -\frac{1}{v}\left(\frac{\partial v}{\partial p}\right)_T \quad \text{or} \quad \left(\frac{\partial v}{\partial p}\right)_T = -K\, v \qquad \ldots (ii)$$

We also know by cyclic property of differentials,

$$\left(\frac{\partial p}{\partial T}\right)_v \left(\frac{\partial T}{\partial v}\right)_p \left(\frac{\partial v}{\partial p}\right)_T = -1$$

Substituting the value of $\left(\frac{\partial v}{\partial T}\right)_p$ and $\left(\frac{\partial v}{\partial p}\right)_T$ from equations (*i*) and (*ii*), we get

$$\left(\frac{\partial p}{\partial T}\right)_v \left(\frac{1}{\beta v}\right)\left(-Kv\right) = -1$$

or

$$\left(\frac{\partial p}{\partial T}\right)_v = \frac{\beta\, v}{K\, v} = \frac{\beta}{K} \qquad \ldots (iii)$$

The ratio of co-efficient of volume expansion and isothermal compressibility is defined as the change in pressure with temperature when volume is kept constant. Thus

$$\frac{\beta}{K} = \left(\frac{\partial p}{\partial T}\right)_v$$

39.10. General Equations for Change in Internal Energy

We have already discussed that internal energy (U) is a property of a system which cannot be measured directly or experimentally. Hence thermodynamic general equations are derived in terms of pressure (p), volume (v) and temperature (T) and specific heats to measure change in internal energy (dU).

Let us now derive the following equations for the change in internal energy.

1. *When pressure (p) and volume (v) are independent.* **When p and v are taken as independent variables, then change in internal energy,**

$$dU = \left(\frac{\partial U}{\partial p}\right)_v dp + \left(\frac{\partial U}{\partial v}\right)_p dv \quad \ldots (i)$$

It may be noted that

$$\left(\frac{\partial U}{\partial p}\right)_v = \left(\frac{\partial U}{\partial T}\right)_v \left(\frac{\partial T}{\partial p}\right)_v = m\, c_v \times \frac{K}{\beta} \quad \ldots (ii)$$

$$\ldots \left[\because \left(\frac{\partial U}{\partial T}\right)_v = m\, c_v, \text{ and } \left(\frac{\partial T}{\partial p}\right)_v = \frac{K}{\beta}\right]$$

We know that $H = U + p\,v$ ... *(iii)*

Taking the partial derivative of equation (*iii*) with respect to volume (v) keeping the pressure (p) constant, we get

$$\left(\frac{\partial H}{\partial v}\right)_p = \left(\frac{\partial U}{\partial v}\right)_p + \frac{\partial}{\partial v}(pv) = \left(\frac{\partial U}{\partial v}\right)_p + p$$

or

$$\left(\frac{\partial U}{\partial v}\right)_p = \left(\frac{\partial H}{\partial v}\right)_p - p \quad \ldots (iv)$$

It may be noted that

$$\left(\frac{\partial H}{\partial v}\right)_p = \left(\frac{\partial H}{\partial T}\right)_p \left(\frac{\partial T}{\partial v}\right)_p = \frac{m\, c_p}{\beta\, v} \quad \ldots (v)$$

$$\ldots \left[\left(\frac{\partial H}{\partial T}\right)_p = m\, c_p \text{ and } \left(\frac{\partial T}{\partial v}\right)_p = \frac{1}{\beta v}\right]$$

Substituting this value of $\left(\frac{\partial H}{\partial v}\right)_p$ in equation (*iv*), we get

$$\left(\frac{\partial U}{\partial v}\right)_p = \frac{m\, c_p}{\beta\, v} - p \quad \ldots (vi)$$

Now substituting the values of $\left(\frac{\partial U}{\partial p}\right)_v$ from equation (*ii*) and $\left(\frac{\partial U}{\partial v}\right)_p$ from equation (*vi*), in equation (*i*), we get

$$dU = \left(m\, c_v \times \frac{K}{\beta}\right) dp + \left(\frac{m\, c_p}{\beta v} - p\right) dv$$

Note : The change is specific internal energy (du) is given as

$$du = \left(c_v \times \frac{K}{\beta}\right) dp + \left(\frac{c_p}{v_s\, \beta} - p\right) dv_s$$

where v_s = Specific volume of the gas in m^3/kg.

2. *When volume (v) and temperature (T) are independent.* **When v and T are taken as independent variables, then change in internal energy,**

$$dU = \left(\frac{\partial U}{\partial v}\right)_T dv + \left(\frac{\partial U}{\partial T}\right)_v dT \quad \ldots (i)$$

By definition, we know that ,

$$\left(\frac{\partial U}{\partial T}\right)_v = m\, c_v$$

$$\therefore \qquad dU = \left(\frac{\partial U}{\partial v}\right)_T dv + m\, c_v\, dT \qquad \ldots (ii)$$

From the first law of thermodynamics,

$$\delta Q = dU + \delta W = dU + pdv \qquad \ldots (iii)$$

and from the second law of thermodynamics,

$$dS = \frac{\delta Q}{T} \quad \text{or} \quad \delta Q = TdS \qquad \ldots (iv)$$

From equations (*iii*) and (*iv*),

$$TdS = dU + pdv \qquad \ldots (v)$$

We know that enthalpy,

$$H = U + p\, v$$

Differentiating the above equation, we get

$$dH = dU + d\,(p\, v) = dU + p\, dv + v\, dp = TdS + vdp$$

$$\ldots (\because TdS = dU + pdv)$$

$$\text{or} \qquad TdS = dH - vdp \qquad \ldots (vi)$$

When T and p are taken as independent variables, then change in enthalpy,

$$dH = \left(\frac{\partial H}{\partial T}\right)_p dT + \left(\frac{\partial H}{\partial p}\right)_T dp = m\, c_p\, dT + \left(\frac{\partial H}{\partial p}\right)_T dp \qquad \ldots (vii)$$

$$\ldots \left[\because \left(\frac{\partial H}{\partial T}\right)_p = m\, c_p\right]$$

From equations (*v*) and (*vi*),

$$dU + p\, dv = dH - v\, dp$$

Substituting the values of dU and dH in above equation, we get

$$\left(\frac{\partial U}{\partial v}\right)_T dv + m\, c_v\, dT + p\, dv = m\, c_p\, dT + \left(\frac{\partial H}{\partial p}\right)_T dp - v\, dp$$

$$\text{or} \qquad m\, c_v\, dT + \left[p + \left(\frac{\partial U}{\partial v}\right)_T\right] dv = m\, c_p\, dT - \left[v - \left(\frac{\partial H}{\partial p}\right)_T\right] dp$$

When pressure is assumed to be constant, then $dp = 0$.

$$\therefore \qquad m\,(c_p - c_v)\,(dT)_p = \left[p + \left(\frac{\partial U}{\partial v}\right)_T\right](dv)_p$$

$$\text{or} \qquad m\,(c_p - c_v) = \left[p + \left(\frac{\partial U}{\partial v}\right)_T\right]\left(\frac{dv}{dT}\right)_p$$

or $$m\,(c_p - c_v) = \left[p + \left(\frac{\partial U}{\partial v}\right)_T \right] \beta\, v \qquad \ldots \left[\because \left(\frac{\partial v}{\partial T}\right)_p = \beta\, v \right]$$

$$\frac{m\,(c_p - c_v)}{\beta\, v} = p + \left(\frac{\partial U}{\partial v}\right)_T$$

$$\therefore \quad \left(\frac{\partial U}{\partial v}\right)_T = \frac{m\,(c_p - c_v)}{\beta\, v} - p \qquad \ldots (viii)$$

Substituting this value in equation (*ii*), we get

$$dU = \left[\frac{m\,(c_p - c_v)}{\beta\, v} - p \right] dv + m\, c_v\, dT$$

Note : The change in specific internal energy (du) is given as

$$du = \left[\frac{(c_p - c_v)}{\beta\, v_s} - p \right] dv_s + c_v\, dT$$

where v_s = Specific volume of the gas in m³/kg.

3. *When temperature (T) and pressure (p) are independent.* When T and p are taken as independent variables, then change in internal energy,

$$dU = \left(\frac{\partial U}{\partial T}\right)_p dT + \left(\frac{\partial U}{\partial p}\right)_T dp \qquad \ldots (i)$$

By definition, we know that,

$$m\, c_p = \left(\frac{\partial H}{\partial T}\right)_p = \left[\frac{\partial\,(U + pv)}{\partial T} \right]_p = \left(\frac{\partial U}{\partial T}\right)_p + v \left(\frac{\partial p}{\partial T}\right)_p + p \left(\frac{\partial v}{\partial T}\right)_p$$

$$\ldots (\because H = U + pv)$$

$$= \left(\frac{\partial U}{\partial T}\right)_p + p \left(\frac{\partial v}{\partial T}\right)_p \qquad \ldots (\because \text{Pressure is constant})$$

$$\therefore \quad \left(\frac{\partial U}{\partial T}\right)_p = m\, c_p - p \left(\frac{\partial v}{\partial T}\right)_p = m\, c_p - p\, \beta\, v \qquad \ldots \left[\because \left(\frac{\partial v}{\partial T}\right)_p = \beta\, v \right] \qquad \ldots (ii)$$

Also, from the first law of thermodynamics,

$$\delta Q = dU + \delta W = dU + p dv$$

If T and p are taken as independent variables, then we can say

$$\delta Q = \left(\frac{\partial U}{\partial T}\right)_p dT + \left(\frac{\partial U}{\partial p}\right)_T dp + p \left[\left(\frac{\partial v}{\partial T}\right)_p dT + \left(\frac{\partial v}{\partial p}\right)_T dp \right]$$

$$= \left[\left(\frac{\partial U}{\partial T}\right)_p + p \left(\frac{\partial v}{\partial T}\right)_p \right] dT + \left[\left(\frac{\partial U}{\partial p}\right)_T + p \left(\frac{\partial v}{\partial p}\right)_T \right] dp$$

We know that

$$m\, c_p = \left(\frac{\partial U}{\partial T}\right)_p + p \left(\frac{\partial v}{\partial T}\right)_p$$

$$\therefore \qquad \delta Q = c_p\, dT + \left[\left(\frac{\partial U}{\partial p}\right)_T + p\left(\frac{\partial v}{\partial p}\right)_T\right] dp$$

If volume is assumed to be constant, then

$$(\delta Q)_v = m\, c_v\, (dT)_v = m\, c_p\, (dT)_v + \left[\left(\frac{\partial U}{\partial p}\right)_T + p\left(\frac{\partial v}{\partial p}\right)_T\right] (dp)_v$$

$$\therefore \qquad m\, c_v = m\, c_p + \left[\left(\frac{\partial U}{\partial p}\right)_T + p\left(\frac{\partial v}{\partial p}\right)_T\right]\left(\frac{\partial p}{\partial T}\right)_v$$

But $\left(\frac{\partial v}{\partial p}\right)_T = -K\,v$ and $\left(\frac{\partial p}{\partial T}\right)_v = \frac{\beta}{K}$

$$\therefore \qquad m\, c_v = m\, c_p + \left[\left(\frac{\partial U}{\partial p}\right)_T - pK\,v\right]\frac{\beta}{K}$$

Thus

$$\left(\frac{\partial U}{\partial p}\right)_T = p\,K\,v - \frac{m\,(c_p - c_v)\,K}{\beta} \qquad \ldots (iii)$$

From equations (*ii*) and (*iii*),

$$dU = (mc_p - p\,\beta\, v)\, dT + \left[p\,K\,v - \frac{m(c_p - c_v)\,K}{\beta}\right] dp$$

Note : For per unit mass, the change in specific internal energy (*du*) is given as

$$du = (c_p - p\beta v_s)\, dT + \left[pKv_s - \frac{(c_p - c_v)\,K}{\beta}\right] dp$$

where v_s = Specific volume of the gas in m^3/kg.

39.11. *T dS* Equations

The First Law of Thermodynamics as applied to a closed system undergoing a reversible process is given by

$$\delta Q = dU + \delta W = dU + p\, dv \qquad \ldots (\because \delta W = pdv) \qquad \ldots (i)$$

According to Second Law of Thermodynamics, we know that

$$dS = \frac{\delta Q}{T} \quad \text{or} \quad \delta Q = T\, dS \qquad \ldots (ii)$$

From equations (*i*) and (*ii*),

$$TdS = dU + p\, dv \qquad \ldots (iii)$$

The above equation (*iii*) is sometimes called as ***general TdS equation***. Let us derive the following useful relationships for the *TdS* equation.

1. *When temperature (T) and volume (v) are independent.* When *T* and *v* are taken as independent variables, then change in internal energy,

$$dU = \left(\frac{\partial U}{\partial T}\right)_v dT + \left(\frac{\partial U}{\partial v}\right)_T dv \qquad \ldots (i)$$

We know that

$$T\,dS = dU + p\, dv$$

Substituting the value of dU in the above equation, we get

$$TdS = \left[\left(\frac{\partial U}{\partial T}\right)_v dT + \left(\frac{\partial U}{\partial v}\right)_T dv\right] + pdv$$

or
$$dS = \frac{1}{T}\left[\left(\frac{\partial U}{\partial T}\right)_v dT + \left(\frac{\partial U}{\partial v}\right)_T dv\right] + \frac{p}{T}dv$$

$$= \frac{1}{T}\left(\frac{\partial U}{\partial T}\right)_v dT + \frac{1}{T}\left[p + \left(\frac{\partial U}{\partial v}\right)_T\right]dv \qquad \ldots (ii)$$

But, we know that for dS, taking T and v as independent variables, we can write

$$dS = \left(\frac{\partial S}{\partial T}\right)_v dT + \left(\frac{\partial S}{\partial v}\right)_T dv \qquad \ldots (iii)$$

Comparing equations (*ii*) and (*iii*), we get

$$\left(\frac{\partial S}{\partial T}\right)_v = \frac{1}{T}\left(\frac{\partial U}{\partial T}\right)_v \qquad \ldots (iv)$$

and
$$\left(\frac{\partial S}{\partial v}\right)_T = \frac{1}{T}\left[p + \left(\frac{\partial U}{\partial v}\right)_T\right] \qquad \ldots (v)$$

Taking the partial derivative of equation (*iv*) with respect to volume (v), we get

$$\frac{\partial^2 S}{\partial v.\partial T} = \frac{1}{T}\left[\frac{\partial^2 U}{\partial v.\partial T}\right] \qquad \ldots (vi)$$

Taking the partial derivative of equation (v) with respect to temperature (T), we get

$$\frac{\partial^2 S}{\partial T.\partial v} = -\frac{1}{T^2}\left[p + \left(\frac{\partial U}{\partial v}\right)_T\right] + \frac{1}{T}\left[\left(\frac{\partial p}{\partial T}\right)_v + \frac{\partial^2 U}{\partial T.\partial v}\right] \qquad \ldots (vii)$$

We have already derived that the magnitude of second mixed partial derivative of a function are independent of the order of differentiation and hence are equal.

$$\therefore \qquad \frac{\partial^2 S}{\partial v.\partial T} = \frac{\partial^2 S}{\partial T.\partial v}$$

or
$$\frac{1}{T}\left[\frac{\partial^2 U}{\partial v.\partial T}\right] = -\frac{1}{T^2}\left[p + \left(\frac{\partial U}{\partial v}\right)_T\right] + \frac{1}{T}\left[\left(\frac{\partial p}{\partial T}\right)_v + \frac{\partial^2 U}{\partial T.\partial v}\right]$$

or
$$\frac{1}{T^2}\left[p + \left(\frac{\partial U}{\partial v}\right)_T\right] = \frac{1}{T}\left(\frac{\partial p}{\partial T}\right)_v$$

or
$$\frac{1}{T}\left[p + \left(\frac{\partial U}{\partial v}\right)_T\right] = \left(\frac{\partial p}{\partial T}\right)_v = \frac{\beta}{K} \qquad \ldots \left[\because \frac{\beta}{K} = \left(\frac{\partial p}{\partial T}\right)_v\right] \qquad \ldots (viii)$$

Also, we know that

$$m\,c_v = T\left(\frac{\partial S}{\partial T}\right)_v \quad \text{or} \quad \left(\frac{\partial S}{\partial T}\right)_v = \frac{m\,c_v}{T}$$

where $m =$ Mass of the gas in kg.

From equation (*iv*), we know that

$$\left(\frac{\partial S}{\partial T}\right)_v = \frac{1}{T}\left(\frac{\partial U}{\partial T}\right)_v$$

or

$$\frac{1}{T}\left(\frac{\partial U}{\partial T}\right)_v = \frac{m\,c_v}{T} \qquad \ldots\left(\because \left(\frac{\partial S}{\partial T}\right)_v = \frac{m\,c_v}{T}\right) \qquad \ldots(ix)$$

Substituting the values from equation (*viii*) and (*ix*) in equation (*ii*), we get

$$dS = \frac{m\,c_v}{T}\,dT + \frac{\beta}{K}\,dv$$

or

$$TdS = m\,c_v\,dT + \frac{T\beta}{K}\,dv \qquad \ldots(x)$$

The above equation (*x*) is called the ***first TdS equation.***

2. *When temperature (T) and pressure (p) are independent.* When T and p are taken as independent variables, then change in internal energy,

$$dU = \left(\frac{\partial U}{\partial T}\right)_p dT + \left(\frac{\partial U}{\partial p}\right)_T dp \qquad \ldots(i)$$

and change in volume,

$$dv = \left(\frac{\partial v}{\partial T}\right)_p dT + \left(\frac{\partial v}{\partial p}\right)_T dp \qquad \ldots(ii)$$

But, we have discussed when T and v are taken as independent variables, then change in internal energy (dU) can be expressed as

$$dU = (m\,c_p - p\,\beta\,v)\,dT + \left[p\,K\,v - \frac{m\,(c_p - c_v)\,K}{\beta}\right]dp \qquad \ldots(iii)$$

We also know that

$$m\,(c_p - c_v) = \left[p + \left(\frac{\partial U}{\partial v}\right)_T\right]\left(\frac{\partial v}{\partial T}\right)_p = \frac{T\beta}{K}\left(\frac{\partial v}{\partial T}\right)_p = \frac{vT\beta^2}{K} \qquad \ldots(iv)$$

$$\ldots\left[\because p + \left(\frac{\partial U}{\partial v}\right)_T = \frac{T\beta}{K} \text{ and } \left(\frac{\partial v}{\partial T}\right)_p = \beta\,v\right]$$

Substituting this value of $m\,(c_p - c_v)$ in equation (*iii*) we get

$$dU = (m\,c_p - p\,\beta\,v)\,dT + (p\,K\,v - vT\,\beta)\,dp \qquad \ldots(v)$$

We also know that

$$\left(\frac{\partial v}{\partial T}\right)_p = \beta\,v \text{ and } \left(\frac{\partial v}{\partial p}\right)_T = -K\,v$$

Substituting these values in equation (*ii*), we get

$$dv = \beta\,v\,dT - K\,v\,dp \qquad \ldots(vi)$$

We know that

$$T\,dS = dU + p\,dv$$

Substituting the values of dU and dv in the above equation, we get

$$T\,dS = (m\,c_p - p\,\beta\,v)\,dT + (p\,K\,v - v\,T\,\beta)\,dp + p\,(\beta\,v\,dT - K\,v\,dp)$$

$$= m\,c_p\,dT - v\,T\,\beta\,dp \quad \dots (vii)$$

The above equation (*vii*) is called the *second TdS equation.*

3. *When pressure (p) and volume (v) are independent.* When p and v are taken as independent variables, then change in internal energy,

$$dU = \left(\frac{\partial U}{\partial p}\right)_v dp + \left(\frac{\partial U}{\partial v}\right)_p dv \quad \dots (i)$$

But we have already discussed that when p and v are taken as independent variables, then change in internal energy,

$$dU = \frac{K\,m\,c_v}{\beta}\,dp + \left(\frac{m\,c_p}{\beta\,v} - p\right) dv \quad \dots (ii)$$

We know that

$$TdS = dU + pdv$$

$$= \frac{K\,m\,c_v}{\beta}\,dp + \left(\frac{m\,c_p}{\beta\,v} - p\right) dv + pdv \quad \dots \text{[From equation (ii)]}$$

$$= \frac{K\,m\,c_v}{\beta}\,dp + \frac{m\,c_p}{\beta\,v}\,dv \quad \dots (iii)$$

The above equation (*iii*) is called the *third TdS equation.*

39.12. Specific Heat

We know that, there are two specific heats of a gas, *i.e.* specific heat at a constant volume (c_v) and specific heat at a constant pressure (c_p). These two specific heats can be measured experimentally. *Joly's differential steam calorimeter* is used to determine specific heat of a gas at constant volume and *Regnault's method* is used to determine specific heat of a gas at constant pressure experimentally*. Now let us discuss the two specific heats as follows:

1. *Specific heat at constant volume (c_v).* The specific heat at constant volume is given by

$$c_v = \left(\frac{\partial u}{\partial T}\right)_v \quad \dots (i)$$

$$= \left(\frac{\partial u}{\partial s}\right)_v \left(\frac{\partial s}{\partial T}\right)_v = T\left(\frac{\partial s}{\partial T}\right)_v \quad \dots \left[\because T = \left(\frac{\partial u}{\partial s}\right)_v\right] \quad \dots (ii)$$

2. *Specific heat at constant pressure (c_p).* The specific heat at constant pressure is given by

$$c_p = \left(\frac{\partial u}{\partial T}\right)_p \quad \dots (i)$$

$$= \left(\frac{\partial h}{\partial s}\right)_p \left(\frac{\partial s}{\partial T}\right)_p = T\left(\frac{\partial s}{\partial T}\right)_p \quad \dots \left[\because T = \left(\frac{\partial h}{\partial s}\right)_p\right] \quad \dots (ii)$$

---

* The detailed study of these experiments are beyond the scope of this book.

39.13. Specific Heat Relations

We have already derived the results of first and second TdS equations in Art.39.11.

The first TdS equation is given by

$$TdS = m\,c_v\,dT + \frac{T\beta}{K}\,dv$$

$$= m\,c_v\,dT + T\left(\frac{\partial p}{\partial T}\right)_v dv \qquad \ldots\left[\because \frac{\beta}{K} = \left(\frac{\partial p}{\partial T}\right)_v\right] \qquad \ldots (i)$$

and the second TdS equation is given by

$$TdS = m\,c_p\,dT - vT\beta\,dp$$

$$= m\,c_p\,dT - T\left(\frac{\partial v}{\partial T}\right)_p dp \qquad \ldots\left[\because \beta\,v = \left(\frac{\partial v}{\partial T}\right)_p\right] \qquad \ldots (ii)$$

Equating the equations (i) and (ii), we get

$$m\,c_v\,dT + T\left(\frac{\partial p}{\partial T}\right)_v dv = m\,c_p\,dT - T\left(\frac{\partial v}{\partial T}\right)_p dp$$

or
$$m\,(c_p - c_v)\,dT = T\left(\frac{\partial p}{\partial T}\right)_v dv + T\left(\frac{\partial v}{\partial T}\right)_p dp$$

$$\therefore \qquad dT = \frac{T\left(\frac{\partial p}{\partial T}\right)_v}{m\,(c_p - c_v)}\,dv + \frac{T\left(\frac{\partial v}{\partial T}\right)_p}{m\,(c_p - c_v)}\,dp \qquad \ldots (iii)$$

Taking p and v as independent variables, the change in temperature (dT) can be expressed as

$$dT = \left(\frac{\partial T}{\partial v}\right)_p dv + \left(\frac{\partial T}{\partial p}\right)_v dp \qquad \ldots (iv)$$

Comparing equations (iv) and (iii), we get

$$\left(\frac{\partial T}{\partial v}\right)_p = \frac{T\left(\frac{\partial p}{\partial T}\right)_v}{m\,(c_p - c_v)} \quad \text{or} \quad m\,(c_p - c_v) = T\left(\frac{\partial p}{\partial T}\right)_v \left(\frac{\partial v}{\partial T}\right)_p \qquad \ldots (v)$$

and
$$\left(\frac{\partial T}{\partial p}\right)_v = \frac{T\left(\frac{\partial v}{\partial T}\right)_p}{m\,(c_p - c_v)} \quad \text{or} \quad m\,(c_p - c_v) = T\left(\frac{\partial v}{\partial T}\right)_p \left(\frac{\partial p}{\partial T}\right)_v \qquad \ldots (vi)$$

Thus from both equations (v) and (vi), it may be concluded that

$$m\,(c_p - c_v) = T\left(\frac{\partial p}{\partial T}\right)_v \left(\frac{\partial v}{\partial T}\right)_p \qquad \ldots (vii)$$

We know that by cyclic property of differentials,

$$\left(\frac{\partial p}{\partial T}\right)_v \left(\frac{\partial T}{\partial v}\right)_p \left(\frac{\partial v}{\partial p}\right)_T = -1$$

or
$$\left(\frac{\partial p}{\partial T}\right)_v = \frac{-1}{\left(\frac{\partial T}{\partial v}\right)_p \left(\frac{\partial v}{\partial p}\right)_T} = -\left(\frac{\partial v}{\partial T}\right)_p \left(\frac{\partial p}{\partial v}\right)_T$$

Substituting this value of $\left(\frac{\partial p}{\partial T}\right)_v$ in equation (*vii*), we get

$$m\,(c_p - c_v) = T\left[-\left(\frac{\partial v}{\partial T}\right)_p \left(\frac{\partial p}{\partial v}\right)_T\right]\left(\frac{\partial v}{\partial T}\right)_p = -T\left(\frac{\partial v}{\partial T}\right)_p^2 \left(\frac{\partial p}{\partial v}\right)_T$$

or
$$c_p - c_v = -\frac{T}{m}\left(\beta^2 v^2\right)\times\frac{-1}{Kv} = \frac{T\beta^2 v}{mK} = \frac{T\beta^2 v_s}{K}$$
$$\ldots\left[\because \left(\frac{\partial v}{\partial T}\right)_p = \beta\, v \text{ and } \left(\frac{\partial p}{\partial v}\right)_T = K\, v\right]$$

where $v_s = \frac{v}{m}$ = Specific volume of gas in m³/kg.

39.14. Joule Thomson Co-efficient

The change in temperature with drop in pressure at constant enthalpy is termed as *Joule-Thomson coefficient* (μ). Mathematically,

$$^*\mu = \left(\frac{dT}{dp}\right)_h \qquad \ldots(i)$$

It varies with both the temperature and pressure of the gas.

The magnitude of the Joule-Thomson coefficient is a measure of the imperfection of a gas or its deviation from perfect gas behaviour. For real gases, μ may be either positive or negative depending upon the thermodynamic state of the gas. When μ is zero, the temperature of the gas remains constant with throttling. The temperature at which $\mu = 0$, is called the *inversion temperature* for a given pressure. If μ is greater than zero, the temperature of the gas decreases with throttling and when μ is less than zero, the temperature of the gas increases with throttling. Thus, in cooling of a gas by throttling, we require that the gas shows a large positive value of μ.

The Joule-Thomson coefficient is not a constant but is a function of both pressure and temperature. We shall now derive the functional relationship for the coefficient.

We know that enthalpy,

$$H = U + p\,v \qquad \ldots(ii)$$

Differentiating equation (*ii*), we get

$$dH = dU + d\,(p\,v) = dU + p\,dv + v\,dp \qquad \ldots(iii)$$

We also know that

$$T\,dS = dU + p\,dv \quad \text{or} \quad dU = T\,dS - p\,dv \qquad \ldots(iv)$$

Substituting the value of dU in equation (*iii*), we get

$$dH = T\,dS - p\,dv + p\,dv + v\,dp = T\,dS + v\,dp$$

When T and p are taken as independent variables, then we can write

---

* For details, refer Chapter 3, Art. 3.21.

$$dH = T\left[\left(\frac{\partial S}{\partial T}\right)_p dT + \left(\frac{\partial S}{\partial p}\right)_T dp\right] + v\,dp$$

$$= T\left(\frac{\partial S}{\partial T}\right)_p dT + \left[T\left(\frac{\partial S}{\partial p}\right)_T + v\right] dp \quad \ldots (v)$$

We know that

$$T\left(\frac{\partial S}{\partial T}\right)_p = m\,c_p \quad \ldots (vi)$$

where $m =$ Mass of the gas in kg.

Also, from Maxwell relations,

$$\left(\frac{\partial S}{\partial p}\right)_T = -\left(\frac{\partial v}{\partial T}\right)_p \quad \ldots (vii)$$

Substituting the value from equations (*vi*) and (*vii*) in equation (*v*), we get

$$dH = m\,c_p\,dT + \left[v - T\left(\frac{\partial v}{\partial T}\right)_p\right] dp \quad \ldots (viii)$$

Solving the above equation (*viii*) for dT, we have

$$dT = \frac{dH}{m\,c_p} + \frac{T\left(\frac{\partial v}{\partial T}\right)_p - v}{m\,c_p}\,dp$$

$$= \frac{dh}{c_p} + \frac{T\left(\frac{\partial v_s}{\partial T}\right)_p - v_s}{c_p}\,dp \quad \ldots (ix)$$

where $dh = dH/m =$ Change in specific enthalpy in kJ/kg,

$\frac{\partial v_s}{\partial T} = \frac{\partial}{\partial T}\left(\frac{v}{m}\right) =$ Change in specific volume with temperature, and

$v_s = v/m =$ Specific volume of gas in m^3/kg.

When specific enthalpy (h) and pressure (p) are taken as independent variables, then for dT we can write

$$dT = \left(\frac{\partial T}{\partial h}\right)_p dh + \left(\frac{\partial T}{\partial p}\right)_h dp \quad \ldots (x)$$

Comparing equations (*ix*) and (*x*), we get

$$\left(\frac{\partial T}{\partial h}\right)_p = \frac{1}{c_p}$$

and

$$\left(\frac{\partial T}{\partial p}\right)_h = \frac{T\left(\frac{\partial v_s}{\partial T}\right)_p - v_s}{c_p}$$

We know that

$$\left(\frac{\partial T}{\partial p}\right)_h = \text{Joule-Thomson coefficient, } \mu$$

$$\therefore \qquad \mu = \left(\frac{\partial T}{\partial p}\right)_h = \frac{T^2}{c_p}\left[\frac{T\left(\frac{\partial v_s}{\partial T}\right)_p - v_s}{T^2}\right] = \frac{T^2}{c_p}\left[\frac{\partial}{\partial T}\left(\frac{v_s}{T}\right)\right]_p$$

... (Multiplying the numerator and denominator by T^2)

Note : We have derived above that

$$\mu = \frac{T^2}{c_p}\left[\frac{\partial}{\partial T}\left(\frac{v_s}{T}\right)\right]_p$$

For an ideal gas,

$$p\,v_s = R\,T \quad \text{or} \quad \frac{v_s}{T} = \frac{R}{p}$$

Substituting this value of (v_s/T) in the above equation,

$$\mu = \frac{T^2}{c_p}\left[\frac{\partial}{\partial T}\left(\frac{R}{p}\right)\right]_p = 0$$

... ($\because$ R and p are constants)

Thus, the Joule Thomson coefficient for an ideal gas is zero and the ideal gas passing through the porous gas does not show any change in temperature.

39.15. Equations of State

Practically there is no real gas which obeys the gas laws perfectly. The deviation is small at ordinary temperatures and pressures, but it is large at high pressure and low temperatures. In deriving the perfect gas laws on the basis of kinetic theory of gases, the following assumptions were made which do not hold true for real gases :

1. The molecules of gas are mere mass-points occupying no space; and
2. There is no attraction or repulsion between the molecules.

In actual practice, the molecules of all actual or real gases do occupy some space and do attract each other. Hence no real gas equation conforms to the perfect gas equation $p\,v_s = RT$. Some of the equations used to correct the perfect gas equation are defined below.

(*a*) *Van der Waal's equation**. The Dutch physicist J.D. Van der Waal was the first scientist to correct the perfect gas equation $pv_s = R\,T$. Van der Waal proposed an equation of state, to account for non ideal gas behaviour. Thus the equation given by Van der Waal has the form

$$\left(p + \frac{a}{v_s^2}\right)(v_s - b) \doteq R\,T$$

where a is a constant of proportionality which takes into account the molecular force-field interactions and b is a constant of proportionality which takes account of the volume occupied by all molecules per kg of gas.

(*b*) *Clausius equation.* The Clausius equation is a modification of Van der Waal's equation. It neglects molecular force-field interactions, a. Thus, the equation has the form

$$p\,(v_s - b) = R\,T$$

* For details, refer Chapter 5, Art. 5.11.

where b = Volume occupied by all molecules.

(c) Beattie Bridgeman equation. Beattie Bridgeman in 1928 proposed an equation of state which gives accurate results. Thus, the equation has the form

$$p = \frac{RT(1-\varepsilon)}{v_s}(v_s + B) - \frac{A}{v_s}$$

where

$$A = A_0\left(1 - \frac{a}{v_s}\right),$$

$$B = B_0\left(1 - \frac{b}{v_s}\right),$$

$$\varepsilon = \left(\frac{C}{v_s T^3}\right),$$

and A_0, a, B_0, b and c are constants whose values are different for different gases.

Beattie Bridgeman equation of state gives properties accurately (within 2 percent) in regions where the density is less than 0.8 times the critical density.

(d) Bertholet equation. Bertholet equation is similar to the Van der Waals equation except for the inclusion of the temperature in the denominator of second term. Thus, the equation has the form

$$\left(p - \frac{a}{T v_s^2}\right)(v_s - b) = RT$$

where a and b are constants, as in Van-der Waals equation.

Bertholet equation gives accurate properties within 1 percent.

(e) Dieterici equation. Dieterici equation of state gives properties very near to exact values of the properties in the neighbourhood of the critical point and on the critical isotherm (constant temperature). This equation does not give satisfactory results in other region and may produce large errors if applied away from critical region. Thus, the equation has the form

$$p = \frac{RT}{v_s - b} e^{-a/RTv_s}$$

where a and b are constants.

(f) Redlich-Kwong equation. The equation given by Redlich-Kwong has the form

$$\left(p - \frac{a}{\sqrt{T}\, v_s (v_s + b)}\right)(v_s - b) = RT$$

where a and b are constants.

(g) Benedict-Webb-Rubin equation. Benedict-Webb-Rubin equation of state is used for high density regions upto approximately two times the critical density. Thus, the equation involves eight experimentally determined constants and has the form

$$p = \frac{RT}{v_s} + \frac{B_0 RT - A_0 - C_0/T^2}{v_s^2} + \frac{bRT - a}{v_s^3} + \frac{a\alpha}{v_s^6} + \frac{C}{v_s^3 T^2}\left(1 + \frac{\gamma}{v_s^2}\right) e^{-\gamma/v_s^2}$$

where A_0, B_0, C_0, a, b, c, α and γ are constants.

(h) Viral equation. The viral equation of state was first introduced by the Dutch physicist Kammerlingh Onnes, and is only applicable to gases of low and medium densities. Basically, the viral equation of state represents an expansion of the $p\,v_s$ product in infinite series form. It is given as

$$p\,v_s = R\,T\left(1 + \frac{B}{v_s} + \frac{C}{v_s^2} + \frac{D}{v_s^3} + \ldots\right) \qquad \ldots (i)$$

It is sometimes convenient to express the viral equation of state with pressure as the independent variable, *i.e.*

$$p\,v_s = R\,T\left(1 + B'p + C'p^2 + D'p^3 + \ldots\right) \qquad \ldots (ii)$$

In the above equations, B, B', C, C', D, D' etc. are called viral coefficients. B and B' are called second viral coefficients, C and C' are called third viral coefficients, D and D' are called fourth viral coefficients and so on. Also, there exists a relation between the viral coefficients in equation (*i*) to the corresponding viral coefficients in equation (*ii*), such that

$$B' = \frac{B}{RT},\quad C' = \frac{C - B^2}{(RT)^2},\quad D' = \frac{D - 3BC + 2B^3}{(RT)^3}, \text{ and so on.}$$

39.16. Compressibility Factor

We have discussed that the viral equation of state is given by

$$p\,v_s = R\,T\left(1 + \frac{B}{v_s} + \frac{C}{v_s^2} + \frac{D}{v_s^3} + \ldots\right)$$

or

$$\frac{p\,v_s}{R\,T} = \left(1 + \frac{B}{v_s} + \frac{C}{v_s^2} + \frac{D}{v_s^3} + \ldots\right)$$

The ratio of $p\,v_s / R\,T$ is called the *compressibility factor*. It is denoted by Z. For the ideal gas, $Z = 1$ and for real gases, Z is determined from experimental data. The magnitude of Z for a certain gas at particular pressure and temperature is a measure of deviation of the real gas from the ideal gas behaviour.

QUESTIONS

1. Define the following :

   (*a*) Co-efficient of volume expansion ; (*b*) Isothermal compressibility; and

   (*c*) Adiabatic compressibility.

2. Derive the Maxwell's thermodynamic relations and explain their importance in thermodynamics.

3. Show from the consideration of Maxwell's thermodynamic relations that

   (*a*) For a perfect gas, $c_p - c_v = R$

   (*b*) For a gas obeying Van der Waal's equation,

$$c_p - c_v = \frac{R}{1 - 2a\,(v_s - b)^2 / RT\,v_s^3}$$

4. Show that for specific helmholtz function, $a = u - T_s$

(a) $\left(\frac{\partial a}{\partial v_s}\right)_T = -p$ (b) $\left(\frac{\partial a}{\partial T}\right)_{v_s} = -s$

(c) $u = a - T\left(\frac{\partial a}{\partial T}\right)_{v_s}$ (d) $c_v = -T\left(\frac{\partial^2 a}{\partial T^2}\right)_{v_s}$

5. Show that for specific Gibbs function, $g = h - Ts$

(a) $\left(\frac{\partial g}{\partial T}\right)_p = -s$ (b) $\left(\frac{\partial g}{\partial p}\right)_T = -v_s$

(c) $u = g - T\left(\frac{\partial g}{\partial T}\right)_p - p\left(\frac{\partial g}{\partial p}\right)_T$ (d) $c_v = -T\left(\frac{\partial^2 g}{\partial T^2}\right)_{v_s}$

6. Show that the equation of state of a substance may be written in the form

$$\frac{dv}{v} = -K\,dp + \beta\,dT$$

7. A substance has the volume expansivity (β) and isothermal compressibility (*K*) as

$$\beta = \frac{1}{T} \quad \text{and} \quad K = \frac{1}{p}$$

Show that the equation of state is $p\,v_s/T$ = constant

8. Using the cyclic relation, prove that

$$\left(\frac{\partial p}{\partial T}\right)_v = \frac{\beta}{KT}$$

9. Derive the following thermodynamic relations

(a) $\left(\frac{\partial h}{\partial p}\right)_T = -T\left(\frac{\partial v_s}{\partial T}\right)_{v_s} + v_s = -c_p\left(\frac{\partial T}{\partial p}\right)_h$ (b) $\left(\frac{\partial h}{\partial v_s}\right)_T = T\left(\frac{\partial p}{\partial T}\right)_{v_s} + v_s\left(\frac{\partial p}{\partial v_s}\right)_T$

(c) $\left(\frac{\partial h}{\partial s}\right)_{v_s} = T - v_s\left(\frac{\partial T}{\partial v_s}\right)_s$ (d) $\left(\frac{\partial u}{\partial v_s}\right) = T\left(\frac{\partial p}{\partial T}\right)_{v_s} - p$

10. Show that for a Van der Waal's gas

$$\left(\frac{\partial c_v}{\partial v}\right)_T = 0$$

OBJECTIVE TYPE QUESTIONS

1. Helmholtz function (*A*) is given as

(a) $A = U - TS$ (b) $A = U + TS$

(c) $A = TS - U$ (d) $A = U - T/S$

2. Gibbs function (G) is given as

(a) $G = H + TS$ (b) $G = H - TS$

(c) $G = TS - H$ (d) $G = H + T/S$

3. Difference in specific heat at constant pressure (c_p) and specific heat at constant volume (c_v) *i.e.* ($c_p - c_v$) is

(a) $\frac{T\beta^2}{K v_s}$ (b) $\frac{T\beta^2 K}{v_s}$

(c) $\frac{T v_s}{\beta^2 K}$ (d) $\frac{T\beta^2 v_s}{k}$

4. TdS equation taking temperature (T) and volume (v) as independent variables is

(a) $m\, c_v\, dT + T\frac{K}{\beta}\, dv$ (b) $m\, c_v\, dT + T\beta K\, dv$

(c) $m\, c_v\, dT + T\frac{\beta}{K}\, dv$ (d) $m\, c_v\, dT + T\beta\, dv$

5. Van der Waal's equation of state is

(a) $p\, v_s = R\, T$ (b) $p\, v_s = 1 + \frac{B}{v_s} + \frac{C}{v_s^2}$

(c) $p = \frac{R\,T}{v_s - b} - \frac{a}{v_s^2}$ (d) $v_s = \frac{R\,T}{p} - \frac{K}{T^3}$

ANSWERS

1. (a)	2. (b)	3. (d)	4. (c)	5. (c)

40

Variable Specific Heat

1. Introduction. 2. Molar Specific Heats of a Gas. 3. Variation of Specific Heat with Temperature. 4. Atmospheric Air. 5. Heat Transfer during a Process with Variable Specific Heat. 6. Heat Transfer during Polytropic Expansion or Compression Process. 7. Change in Thermodynamic Properties with Variable Specific Heat. 8. Isentropic Expansion with Variable Specific Heats. 9. Effect of Variable Specific Heat in I.C. Engines. 10. Effect of Variable Specific Heat on Otto Cycle. 11. Effect of Variable Specific Heat on Diesel Cycle.

40.1. Introduction

We have already discussed in Chapter 1 (Art. 1.40), that specific heat of a substance may be broadly defined as the amount of heat required to raise the temperature of a unit mass of any substance through one degree. It is generally denoted by c. In S.I. system of units, the unit of specific heat (c) is taken as kJ/kg K.

Since the solids and liquids does not have appreciable change in their volume on heating therefore they have only one specific heat. But the gases have the following two *specific heats depending upon the process adopted for heating the gas.

1. Specific heat at constant pressure (c_p), and
2. Specific heat at constant volume (c_v).

In general, specific heat is not constant. The specific heat of any gas increases significantly with rise in temperature, but it does not increase significantly with pressure, except at high pressure. Thus, for simple calculations, the change in specific heat with pressure is neglected.

40.2. Molar Specific Heats of a Gas

The molar or volumetric specific heat of a gas may be defined as the amount of heat required to raise the temperature of unit mole of gas through one degree. It is generally denoted by c_m. In S.I. system of units, the unit of molar specific heat (c_m) is kJ/kg-mol K. Mathematically, molar specific heat (c_m) is given as

$$c_m = M \times c$$

where M = Molar mass of the gas, and

c = Specific heat of the gas.

---

* For further details, refer Chapter 2 (Art. 2.11).

In the similar way, the molar specific heat at constant volume

$$c_{mv} = M\,c_v$$

and molar specific heat at constant pressure,

$$c_{mp} = M\,c_p$$

40.3. Variation of Specific Heat with Temperature

We have already discussed that the specific heat of any gas increases with increase in temperature. The specific heat increases because of increase in energy of vibration of molecules at high temperature. The vibrations are caused because of collisions among molecules which are significant at higher temperature. Therefore, the energy of vibration of a polyatomic gas will undergo a considerable change at ordinary temperature, thus increasing the value of specific heat.

It is assumed for gases that in a temperature range of 300 K to 1500 K, the specific heat is a linear function of temperature and may be expressed as

$$c_p = a + K\,T \text{ , and}$$

$$c_v = b + K\,T$$

where *a*, *b* and *K* are constants.

We know that characteristic gas constant,

$$^{*}R = c_p - c_v$$

$$= (a + K\,T) - (b + K\,T) = a - b$$

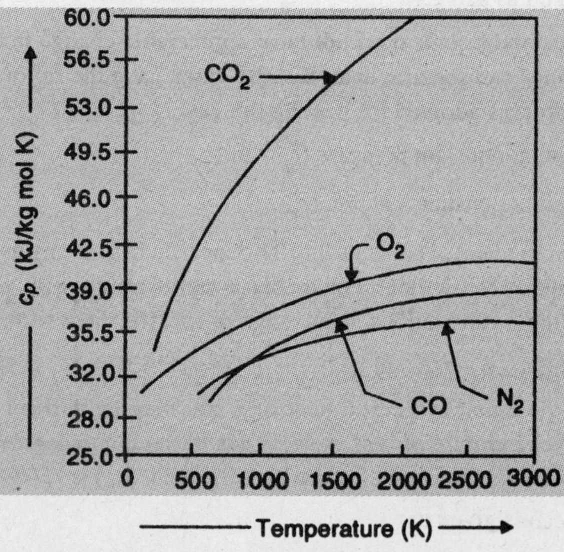

Fig. 40.1. Variation of molar specific heat at constant pressure with temperature.

Above 1500 K, the specific heats increases very rapidly and is given by the following expressions :

$$c_p = a + K_1 T + K_2 T^2 \text{, and}$$

$$c_v = b + K_1 T + K_2 T^2$$

---

* For details, refer Chapter 2 (Art. 2.17).

In the above equations, if the term involving T^2 is neglected, then the expression for c_p and c_v becomes same as for the temperature range of 300 K to 1500 K. Fig. 40.1 shows the variation of molar specific heat (c_{mp}) for CO_2, O_2, CO and N_2 with temperature.

The following table shows the expressions for variable molar specific heat at low pressure.

Table 40.1. Variable molar specific heat at low pressure.

S.No.	*Name of gas*	*Molecular mass (M)*	*Temp. range (K)*	c_{mp} *(kJ/kg-mol K)*
1.	Air	29	280-1500	$26.63 + 7473.44 \times 10^{-4} T - 115.97 \times 10^{-8} T^2$
2.	Carbon dioxide (CO_2)	44	280-3500	$67.83 - 15156.22 T^{-1} + 181.71 \times 10^4 T^{-2}$
3.	Oxygen (O_2)	32	280-2800	$48.21 - 535.917 T^{-0.5} + 3516.91 T^{-1}$
4.	Nitrogen (N_2)	28	280-5000	$39.65 - 8080.52 T^{-1} + 1494.69 T^{-2}$
5.	Ammonia (NH_3)	17	280-1000	$25.92 + 330.76 \times 10^{-4} T - 305.64 \times 10^{-8} T^2$
6.	Carbon monoxide (CO)	28	280-2800	$39.61 - 7640.91 T^{-1} + 138.16 \times 10^4 T^{-2}$
7.	Hydrogen (H_2)	2	300-2200	$24.12 + 43.54 \times 10^{-4} T + 62.38 T^{-0.5}$
8.	Sulphur dioxide (SO_2)	64	300-1900	$49.78 + 46.05 \times 10^{-4} T + 110.11 \times 10^{-4} T^{-2}$
9.	Methane (CH_4)	16	280-1500	$14.15 + 753.62 \times 10^{-4} T - 1808.70 \times 10^{-8} T^2$
10.	Water vapour	18	280-3000	$83.15 - 1863.13 T^{-0.5} - 173.75 T^{-1}$

40.4. Atmospheric Air

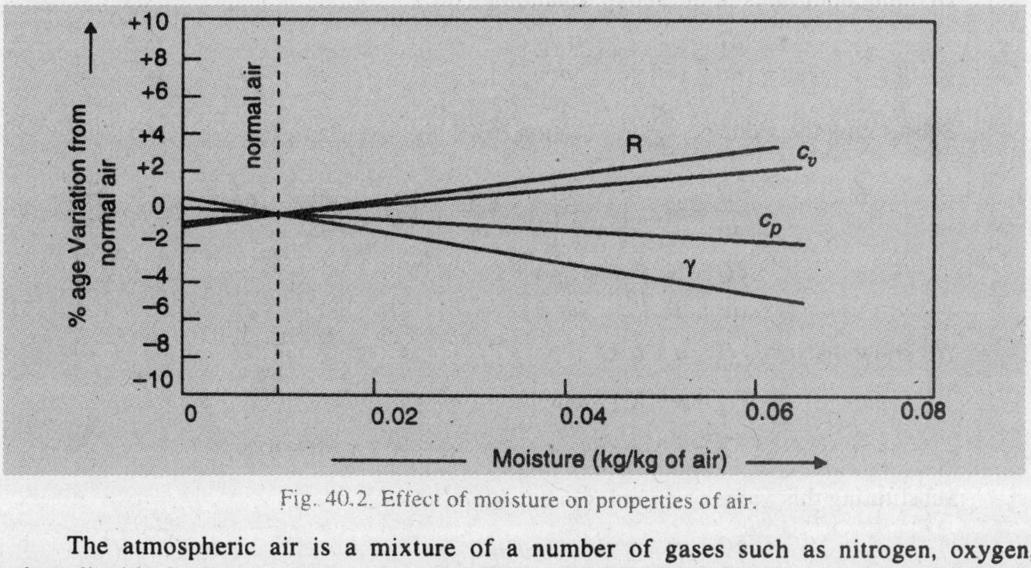

Fig. 40.2. Effect of moisture on properties of air.

The atmospheric air is a mixture of a number of gases such as nitrogen, oxygen, carbon-dioxide, hydrogen, argon, neon, helium etc. But the nitrogen and oxygen have the major portion of the combination. The atmospheric air consists of 23.19% oxygen and 75.47% nitrogen by mass.

However, the atmospheric air also contains water vapours known as *moisture*. The amount of water vapour, present in the air, depends upon the absolute pressure and temperature of the mixture. Fig. 40.2 shows the effect of moisture on the properties of air, like characteristic gas constant (R), specific heat at constant volume (c_v), specific heat at constant pressure (c_p), and ratio of specific heats (γ).

40.5. Heat Transfer during a Process with Variable Specific Heat

We know that for a small change in the state of a working substance, the general gas energy equation is

$$\delta Q = dU + \delta W$$

or $$\delta Q = m\,c_v\,dT + p\,dv \qquad \ldots (i)$$

where dT = Change in temperature, and

dv = Change in volume.

Dividing both sides of equation (*i*) by dv,

$$\frac{\delta Q}{dv} = m\,c_v \times \frac{dT}{dv} + p \qquad \ldots (ii)$$

We know that according to general gas equation,

$$p\,v = m\,R\,T$$

or $$T = \frac{p\,v}{m\,R} \qquad \ldots (iii)$$

Differentiating equation (*iii*),

$$dT = \frac{1}{mR}\,[p\,dv + v\,dp]$$

Dividing both sides of the above equation by dv,

$$\frac{dT}{dv} = \frac{1}{mR}\left(p + \frac{v\,dp}{dv}\right) \qquad \ldots (iv)$$

Substituting the value of $\frac{dT}{dv}$ in equation (*ii*),

$$\frac{\delta Q}{dv} = m\,c_v\left[\frac{1}{mR}\left(p + \frac{v\,dp}{dv}\right)\right] + p$$

or $$\frac{\delta Q}{dv} = \frac{c_v}{R}\left(p + \frac{v\,dp}{dv}\right) + p \qquad \ldots (v)$$

We know that $c_p = a + K\,T$,

$c_v = b + K\,T$, and

$$R = c_p - c_v = (a + K\,T) - (b + K\,T) = a - b$$

Substituting this value of c_v and R in equation (*v*),

$$\frac{\delta Q}{dv} = \frac{b + K\,T}{a - b}\left(p + \frac{v\,dp}{dv}\right) + p$$

$$= \frac{1}{a-b}\left(b\,p + \frac{b\,v\,dp}{dv} + K\,T\,p + \frac{K\,T\,v\,dp}{dv} + a\,p - b\,p\right)$$

$$= \frac{b}{a-b}\left(\frac{v\,dp}{dv} + \frac{K\,T\,p}{b} + \frac{K\,T\,v}{b}\frac{dp}{dv} + \frac{a\,p}{b}\right)$$

$$= \frac{1}{\frac{a}{b}-1}\left(\frac{v\,dp}{dv} + \frac{K\,T\,p}{b} + \frac{K\,T\,v}{b}\frac{dp}{dv} + \frac{a\,p}{b}\right)$$

$$= \frac{1}{\frac{a}{b}-1}\left(\frac{a\,p}{b} + \frac{v\,dp}{dv}\right) + \frac{KT}{a-b}\left(p + \frac{v\,dp}{dv}\right)$$

Let $\frac{a^*}{b} = \frac{c_p - K\,T}{c_v - K\,T} = \gamma'$

$$\therefore \quad \frac{\delta Q}{dv} = \frac{1}{\gamma'-1}\left(\gamma' p + \frac{v\,dp}{dv}\right) + \frac{K\,T}{a-b}\left(p + \frac{v\,dp}{dv}\right) \qquad \ldots (vi)$$

Note: For constant specific heats, $K = 0$. Thus the above equation (*vi*) becomes

$$\frac{\delta Q}{dv} = \frac{1}{\gamma'-1}\left(\gamma' p + \frac{v\,dp}{dv}\right)$$

40.6. Heat Transfer during Polytropic Expansion or Compression Process

We know that for a polytropic process,

$$p\,v^n = \text{Constant}$$

Differentiating the above equation with respect to v,

$$n\,p\,v^{n-1} + v^n \times \frac{dp}{dv} = 0 \qquad \ldots \left[\because \frac{d}{dx}(u \times v) = u \times \frac{dv}{dx} + v \times \frac{du}{dx}\right]$$

$$v^n \times \frac{dp}{dv} = -\,n\,p\,v^{n-1}$$

$$v \times \frac{dp}{dv} = -\,n\,p$$

We have already derived that

$$\frac{\delta Q}{dv} = \frac{1}{\gamma'-1}\left(\gamma' p + \frac{v\,dp}{dv}\right) + \frac{KT}{a-b}\left(p + \frac{v\,dp}{dv}\right)$$

Substituting the value of $v \times \frac{dp}{dv}$ in the above equation, we get

$$\frac{\delta Q}{dv} = \frac{1}{\gamma'-1}\left(\gamma' p - n\,p\right) + \frac{K\,T}{a-b}\left(p - n\,p\right)$$

$$= \frac{\gamma'-n}{\gamma'-1} \times p + \frac{K\,T}{a-b}\left(1-n\right)p$$

---

* We know that

$$c_p = a + K\,T \quad \text{or} \quad a = c_p - K\,T$$

$$c_v = b + K\,T \quad \text{or} \quad b = c_v - K\,T$$

$$\therefore \quad \frac{a}{b} = \frac{c_p - K\,T}{c_v - K\,T}$$

$$= \left(\frac{\gamma' - n}{\gamma' - 1} - K\,T\,\frac{n-1}{a-b} \right) p \qquad \ldots (vii)$$

Note : Since for constant specific heats, $K = 0$, therefore the above equation (*vii*) becomes

$$\frac{\delta Q}{dv} = \frac{\gamma' - n}{\gamma' - 1} \times p$$

40.7. Change in Thermodynamic Properties with Variable Specific Heat

We have already discussed that when a gas is heated, its specific heat increases. Since the thermodynamic properties like internal energy, enthalpy and entropy depends upon specific heat also, thus there is a considerable change in these properties with change in temperature. Consider m kg of gas being heated from the initial state 1 to the final state 2.

Let m = Mass of the gas,

p_1, v_1 and T_1 = Pressure, volume and temperature respectively at the initial state 1,

p_2, v_2 and T_2 = Pressure, volume and temperature respectively at the final state 2,

c_p = Specific heat at constant pressure = $a + K\,T$, and

c_v = Specific heat at constant volume = $b + K\,T$

Now, let us derive the following relations for the change in thermodynamic properties with variable specifc heat.

1. *Change in internal energy*

We know that change in internal energy of gas,

$$dU = mc_v\, dT \qquad \ldots \text{(Joule's law)}$$

$$= m\,(b + K\,T)\, dT \qquad \ldots (\because c_v = b + K\,T)$$

$$= m\,b\, dT + m\,K\,T\, dT$$

On integrating from state 1 to state 2,

$$\int_1^2 dU = m\,b \int_1^2 dT + m\,K \int_1^2 T\, dT$$

or
$$U_2 - U_1 = m\,b\,(T_2 - T_1) + mK \times \frac{(T_2^2 - T_1^2)}{2}$$

$$= m\,(T_2 - T_1) \left[b + K \times \frac{(T_2 + T_1)}{2} \right] = m\,(T_2 - T_1)\,(b + K\,T_m)$$

$$= m\,c_{vm}\,(T_2 - T_1)$$

where T_m = Mean temperature = $\dfrac{T_2 + T_1}{2}$, and

c_{vm} = Mean specific heat at constant volume = $b + K\,T_m$

Note : For per unit mass of gas, the expression for change in internal energy may be written as

$$u_2 - u_1 = c_{vm}\,(T_2 - T_1)$$

where $u_2 - u_1$ = Change in specific internal energy in kJ/kg.

2. *Change in enthalpy*

We know that change in enthalpy of gas,

$$dH = mc_p\,dT = m\,(a + K\,T)\,dT \qquad \ldots (\because c_p = a + K\,T)$$

$$= m\,a\,dT + m\,K\,T\,dT$$

On integrating from state 1 to state 2,

$$\int_1^2 dH = m\,a\int_1^2 dT + m\,K\int_1^2 T\,dT$$

or
$$H_2 - H_1 = m\,a(T_2 - T_1) + mK \times \frac{(T_2^2 - T_1^2)}{2}$$

$$= m\,(T_2 - T_1)\left[a + K \times \frac{(T_2 + T_1)}{2}\right]$$

$$= m\,(T_2 - T_1)\,(a + KT_m)$$

$$= m\,c_{pm}\,(T_2 - T_1)$$

where T_m = Mean temperature $= \dfrac{T_2 + T_1}{2}$, and

c_{pm} = Mean specific heat at constant pressure $= a + K\,T_m$.

Note For per unit mass of gas, the expression for change in enthalpy can be written as

$$h_2 - h_1 = c_{pm}\,(T_2 - T_1)$$

where $h_2 - h_1$ = Change in specific enthalpy in kJ/kg.

3. *Change in entropy*

We know that change in entropy of the gas,

$$dS = \frac{\delta Q}{T} \qquad \ldots (i)$$

where δQ = Heat transferred in kJ, and

T = Absolute temperature in K.

Now the relation for the change of entropy for variable specific heat can be expressed in the following three ways :

(*a*) *In terms of volume and absoute temperature*

We know that for a small change in the state of a working substance, the general gas energy equation is

$$\delta Q = dU + \delta W \qquad \ldots \text{(First Law of Thermodynamics)}$$

$$= m\,c_v\,dT + p\,dv \qquad \ldots (ii)$$

where dT = Change in temperature, and

dv = Change in volume.

Dividing throughout equation (*ii*) by T,

$$\frac{\delta Q}{T} = \frac{m\, c_v\, dT}{T} + \frac{p\, dv}{T}$$

Since $p\, v = m\, R\, T$ or $\frac{p}{T} = \frac{mR}{v}$ and $\frac{\delta Q}{T} = dS$, therefore

$$dS = \frac{m\, c_v\, dT}{T} + \frac{m\, R\, dv}{v} \qquad \ldots (iii)$$

$$= m\,(b + KT)\,\frac{dT}{T} + m\,(a - b)\,\frac{dv}{v}$$

$$\ldots (\because c_v = b + KT \text{ and } R^* = a - b)$$

$$= m\, b \times \frac{dT}{T} + m\, K\, T \times \frac{dT}{T} + m\,(a - b)\,\frac{dv}{v}$$

$$= m\, b \times \frac{dT}{T} + m\, K \times dT + m(a - b)\,\frac{dv}{v} \qquad \ldots (iv)$$

On integrating equation (*iv*) from state 1 to state 2,

$$\int_1^2 dS = m\, b \int_1^2 \frac{dT}{T} + m\, K \int_1^2 dT + m\,(a - b) \int_1^2 \frac{dv}{v}$$

or
$$S_2 - S_1 = m\, b \log_e \left(\frac{T_2}{T_1}\right) + m\, K\,(T_2 - T_1) + m\,(a - b) \log_e \left(\frac{v_2}{v_1}\right) \qquad \ldots (v)$$

$$= 2.3\, m\, b \log \left(\frac{T_2}{T_1}\right) + m\, K\,(T_2 - T_1) + 2.3\, m\,(a - b) \log \left(\frac{v_2}{v_1}\right)$$

Note : For per unit mass, the above expression may be written as

$$s_2 - s_1 = 2.3\, b \log \left(\frac{T_2}{T_1}\right) + K\,(T_2 - T_1) + 2.3\,(a - b) \log \left(\frac{v_{s2}}{v_{s1}}\right)$$

where $s_2 - s_1$ = Change in specific entropy in kJ/kg K, and

v_{s1} and v_{s2} = Specific volume at states 1 and 2 in m^3/kg of gas.

(*b*) *In terms of pressure and absolute temperature*

We know from the general gas equation,

$$\frac{p_1\, v_1}{T_1} = \frac{p_2\, v_2}{T_2} \quad \text{or} \quad \frac{v_2}{v_1} = \frac{p_1}{p_2} \times \frac{T_2}{T_1}$$

or
$$\log_e \left(\frac{v_2}{v_1}\right) = \log_e \left(\frac{p_1}{p_2}\right) + \log_e \left(\frac{T_2}{T_1}\right)$$

Substituting the value of $\log_e \left(\frac{v_2}{v_1}\right)$ in equation (*v*),

$$S_2 - S_1 = m\, b \log_e \left(\frac{T_2}{T_1}\right) + m\, K\,(T_2 - T_1) + m\,(a - b) \left[\log_e \left(\frac{p_1}{p_2}\right) + \log_e \left(\frac{T_2}{T_1}\right)\right]$$

* For details, refer Art. 40.5.

$$= m \log_e \left(\frac{T_2}{T_1}\right)\left(b + a - b\right) + m\,K\,(T_2 - T_1) + m\,(a-b) \log_e \left(\frac{p_1}{p_2}\right)$$

$$= m\,a \log_e \left(\frac{T_2}{T_1}\right) + m\,K\,(T_2 - T_1) + m\,(a-b) \log_e \left(\frac{p_1}{p_2}\right)$$

$$= 2.3\,m\,a \log \left(\frac{T_2}{T_1}\right) + m\,K\,(T_2 - T_1) + 2.3\,m\,(a-b) \log \left(\frac{p_1}{p_2}\right)$$

Note For per unit mass, the above expression may be written as :

$$s_2 - s_1 = 2.3\,a \log \left(\frac{T_2}{T_1}\right) + K\,(T_2 - T_1) + 2.3\,(a-b) \log \left(\frac{p_1}{p_2}\right)$$

where $s_2 - s_1 =$ Change in specific entropy in kJ/kg K.

(*c*) *In terms of pressure and volume*

We know from the general gas equation,

$$\frac{p_1 v_1}{T_1} = \frac{p_2 v_2}{T_2} \quad \text{or} \quad \frac{T_2}{T_1} = \frac{p_2}{p_1} \times \frac{v_2}{v_1}$$

or

$$\log_e \left(\frac{T_2}{T_1}\right) = \log_e \left(\frac{p_2}{p_1}\right) + \log_e \left(\frac{v_2}{v_1}\right)$$

Substituting the value of $\log_e \left(\frac{T_2}{T_1}\right)$ in equation (*v*),

$$S_2 - S_1 = m\,b \left[\log_e \left(\frac{p_2}{p_1}\right) + \log_e \left(\frac{v_2}{v_1}\right)\right] + m\,K\,(T_2 - T_1) + m\,(a-b) \log_e \left(\frac{v_2}{v_1}\right)$$

$$= m\,b \log_e \left(\frac{p_2}{p_1}\right) + m \log_e \left(\frac{v_2}{v_1}\right)\left(b + a - b\right) + m\,K\,(T_2 - T_1)$$

$$= m\,b \log_e \left(\frac{p_2}{p_1}\right) + m\,a \log_e \left(\frac{v_2}{v_1}\right) + m\,K\,(T_2 - T_1)$$

$$= 2.3\,m\,b \log \left(\frac{p_2}{p_1}\right) + 2.3\,m\,a \log \left(\frac{v_2}{v_1}\right) + m\,K\,(T_2 - T_1)$$

Note For per unit mass, the above expression may be written as

$$s_2 - s_1 = 2.3\,b \log \left(\frac{p_2}{p_1}\right) + 2.3\,a \log \left(\frac{v_{s2}}{v_{s1}}\right) + K\,(T_2 - T_1)$$

where $s_2 - s_1 =$ Change in specific entropy in kJ/kg K, and

v_{s1} and $v_{s2} =$ Change in specific volume at states 1 and 2 in m^3/kg of gas.

Example 40.1. *2 kg of a certain gas is heated from 400 K to 1000 K. Calculate the change in internal energy and enthalpy.*

Take $c_p = 0.946 + 0.184 \times 10^{-3}\,T$ kJ/kg K, and $c_v = 0.653 + 0.184 \times 10^{-3}\,T$ kJ/kg K

Solution. Given : $m = 2$ kg ; $T_1 = 400$ K ; $T_2 = 1000$ K ; $a = 0.946$ kJ/kg K ; $b = 0.653$ kJ/kg K ; $K = 0.184 \times 10^{-3}$ kJ/kg K^2

We know that mean temperature,

$$T_m = \frac{T_1 + T_2}{2} = \frac{400 + 1000}{2} = 700 \text{ K}$$

∴ Mean specific heat at constant volume,

$$c_{vm} = b + K\,T_m = 0.653 + 0.184 \times 10^{-3} \times 700 = 0.782 \text{ kJ/kg K}$$

and mean specific heat at constant pressure,

$$c_{pm} = a + K\,T_m = 0.946 + 0.184 \times 10^{-3} \times 700 = 1.075 \text{ kJ/kg K}$$

Change in internal energy

We know that change in internal energy,

$$dU = U_2 - U_1 = m\,c_{vm}\,(T_2 - T_1)$$

$$= 2 \times 0.782\,(1000 - 400) = 938.4 \text{ kJ} \quad \textbf{Ans.}$$

Change in enthalpy

We know that change in enthalpy,

$$dH = H_2 - H_1 = m\,c_{pm}\,(T_2 - T_1)$$

$$= 2 \times 1.075\,(1000 - 400) = 1290 \text{ kJ} \quad \textbf{Ans.}$$

Example 40.2. *A certain quantity of gas having a volume of 5 m^3 at 1 bar and 27°C is heated at constant pressure to 55°C. Determine : 1. mass of the gas, 2. heat transfer during the process, and 3. change of entropy.*

Take $c_p = 0.946 + 0.184 \times 10^{-3}\,T$ kJ/kg K, and $c_v = 0.653 + 0.184 \times 10^{-3}\,T$ kJ/kg K

Solution. Given : $v_1 = 5 \text{ m}^3$; $p_1 = 1 \text{ bar} = 100 \text{ kN/m}^2$; $T_1 = 27°\text{C} = 27 + 273 = 300$ K ; $T_2 = 55°\text{C} = 55 + 273 = 328$ K ; $a = 0.946$ kJ/kg K ; $b = 0.653$ kJ/kg K ; $K = 0.184 \times 10^{-3}$ kJ/kg K^2

In Fig. 40.3, 1-2 represents the constant pressure process.

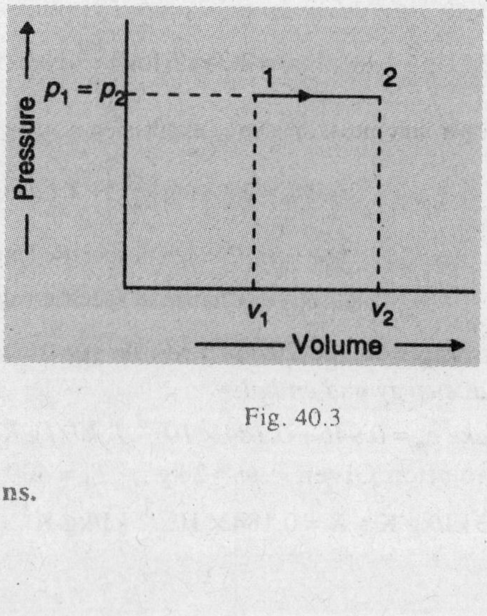

Fig. 40.3

1. *Mass of the gas*

Let m = Mass of the gas.

First of all, let us find the characteristic gas constant (R). We know that

$$R = c_p - c_v = a - b = 0.946 - 0.653$$

$$= 0.293 \text{ kJ/kg K}$$

We know that

$$p_1\,v_1 = m\,R\,T_1$$

$$\therefore \quad m = \frac{p_1\,v_1}{R\,T_1} = \frac{100 \times 5}{0.293 \times 300} = 5.69 \text{ kg} \quad \textbf{Ans.}$$

2. *Heat transfer during the process*

First of all, let us find the value of mean temperature (T_m). We know that

$$T_m = \frac{T_1 + T_2}{2} = \frac{300 + 328}{2} = 314 \text{ K}$$

∴ Mean specific heat at constant pressure,

$$c_{pm} = a + K\,T_m = 0.946 + 0.184 \times 10^{-3} \times 314 = 1 \text{ kJ/kg K}$$

We know that heat transfer during the process,

$$Q_{1-2} = m\,c_{pm}\,(T_2 - T_1)$$

$$= 5.69 \times 1\,(328 - 300) = 159.32 \text{ kJ } \textbf{Ans.}$$

3. *Change in entropy*

We know that change in entropy,

$$S_2 - S_1 = 2.3\,m\,a\log\left(\frac{T_2}{T_1}\right) + m\,K\,(T_2 - T_1) + 2.3\,m\,(a - b)\log\left(\frac{p_1}{p_2}\right)^{*}$$

$$= 2.3 \times 5.69 \times 0.946 \log\left(\frac{328}{300}\right) + 5.69 \times 0.184 \times 10^{-3}\,(328 - 300)$$

$$= 0.48 + 0.029 = 0.509 \text{ kJ/K } \textbf{Ans.}$$

40.8. Isentropic Expansion with Variable Specific Heat

Isentropic expansion is a reversible adiabatic process, in which no heat enters or leaves the gas. Thus

$$\delta Q = 0$$

We know that change in entropy,

$$dS = \frac{\delta Q}{T}$$

Since $\delta Q = 0$, therefore $dS = 0$.

The expression for isentropic expansion with variable specific heat may be expressed in the following ways :

(*a*) *In terms of volume and absolute temperature*

We know that for a small change in the state of a working substance, the general gas energy equation is,

$$\delta Q = dU + \delta W \qquad \text{. . . (First Law of Thermodynamics)}$$

$$= m\,c_v\,dT + p\,dv \qquad \text{. . . }(i)$$

where dT = Small change in temperature, and

dv = Small change in volume.

---

* Since $p_1 = p_2$, therefore $\log\left(\frac{p_1}{p_2}\right) = 0$

Dividing throughout equation (*i*) by *T*,

$$\frac{\delta Q}{T} = m\, c_v \times \frac{dT}{T} + \frac{p\, dv}{T}$$

Since $p\,v = m\,R\,T$ or $\frac{p}{T} = \frac{mR}{v}$ and $\frac{\delta Q}{T} = dS = 0$, therefore

$$m\, c_v \times \frac{dT}{T} + mR \times \frac{dv}{v} = 0$$

or $$m\,(b + K\,T)\,\frac{dT}{T} + m\,(a - b)\,\frac{dv}{v} = 0 \quad \ldots (\because\ c_v = b + KT \text{ and } R = a - b)$$

$$m\,b \times \frac{dT}{T} + m\,K\,dT + m\,(a - b)\,\frac{dv}{v} = 0 \quad \ldots (ii)$$

On integrating equation (*ii*),

$$m\,b \int \frac{dT}{T} + mK \int dT + m\,(a - b) \int \frac{dv}{v} = \int 0$$

$$mb \log_e T + mKT + m\,(a - b) \log_e v = \text{Constant}$$

Dividing throughout by *mb*, we get

$$\log_e T + \frac{K\,T}{b} + \left(\frac{a}{b} - 1\right) \log_e v = \text{Constant}$$

or $$\log_e T + \log_e e^{\frac{K\,T}{b}} + \log_e v^{\left(\frac{a}{b} - 1\right)} = \text{Constant}$$

or $$T\, e^{\frac{K\,T}{b}}\, v^{\left(\frac{a}{b} - 1\right)} = \text{Constant} \quad \ldots (iii)$$

$$\therefore \quad \frac{T}{v}\, e^{\frac{K\,T}{b}}\, v^{\frac{a}{b}} = \text{Constant} \quad \ldots (iv)$$

The more useful form of above equation is

$$\frac{T_1}{v_1}\, e^{\frac{K\,T_1}{b}}\, v_1^{\frac{a}{b}} = \frac{T_2}{v_2}\, e^{\frac{K\,T_2}{b}}\, v_2^{\frac{a}{b}} = \frac{T_3}{v_3}\, e^{\frac{K\,T_3}{b}}\, v_3^{\frac{a}{b}} = \ldots = \text{Constant}$$

where suffixes $_{1,\,2}$ and $_3$ . . . refer to different sets of condition.

Note : For per unit mass, the above expression may be written as

$$\frac{T_1}{v_{s1}}\, e^{\frac{K\,T_1}{b}}\, v_{s1}^{\frac{a}{b}} = \frac{T_2}{v_{s2}}\, e^{\frac{K\,T_2}{b}}\, v_{s2}^{\frac{a}{b}} = \frac{T_3}{v_{s3}}\, e^{\frac{K\,T_3}{b}}\, v_{s3}^{\frac{a}{b}} = \ldots = \text{Constant}$$

where v_{s1}, v_{s2} and v_{s3} = Specific volumes at states 1, 2 and 3 respectively in m^3/kg of gas.

(*b*) *In terms of pressure and volume*

We know that the general gas equation is

$$p\,v = m\,R\,T \quad \text{or} \quad \frac{T}{v} = \frac{p}{mR} = \frac{p}{m\,(a - b)}$$

Substituting this value of T/v in equation (*iv*), we get

$$\frac{p}{m(a-b)} e^{\frac{KT}{b}} v^{\frac{a}{b}} = \text{Constant}$$

or $$p\, e^{\frac{KT}{b}} v^{\frac{a}{b}} = \text{Constant} \quad \ldots [\because m(a-b) \text{ is constant for a particular gas}]$$

The more useful form of above equation is

$$p_1 e^{\frac{KT_1}{b}} v_1^{\frac{a}{b}} = p_2 e^{\frac{KT_2}{b}} v_2^{\frac{a}{b}} = p_3 e^{\frac{KT_3}{b}} v_3^{\frac{a}{b}} = \ldots = \text{Constant}$$

where suffixes $_1$, $_2$ and $_3$ refer to different sets of condition.

Note : For per unit mass, the above expression may be written as

$$p_1 e^{\frac{KT_1}{b}} v_{s1}^{\frac{a}{b}} = p_2 e^{\frac{KT_2}{b}} v_{s2}^{\frac{a}{b}} = p_3 e^{\frac{KT_3}{b}} v_{s3}^{\frac{a}{b}} = \ldots = \text{Constant}$$

where v_{s1}, v_{s2} and v_{s3} = Specific volumes at states 1, 2 and 3 respcetively in m^3/kg of gas.

Example 40.3. *A mass of 1.5 kg and volume of 0.14 m^3 of certain gas at 40 bar is expanded isentropically such that temperature falls to 500 K. Determine : 1. Initial temperature of the gas, 2. Work done during the process, and 3. Pressure at the end of expansion.*

Take $c_p = 0.946 + 0.000\,184\,T$ kJ/kg K, and $c_v = 0.653 + 0.000\,184\,T$ kJ/kg K.

Solution. Given : $m = 1.5$ kg ; $v_1 = 0.14\ m^3$; $p_1 = 40$ bar $= 4 \times 10^6\ N/m^2 = 4 \times 10^3\ kN/m^2$, $T_2 = 500$ K ; $a = 0.946$ kJ/kg K ; $b = 0.653$ kJ/kg K ; $K = 0.000\,184$ kJ/kg K^2

In Fig. 40.4, the process 1-2 represents the isentropic expansion of the gas.

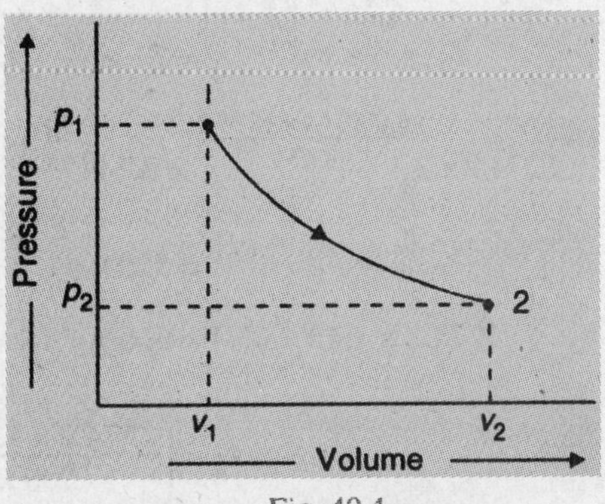

Fig. 40.4

1. *Initial temperature of the gas*

Let T_1 = Initial temperature of the gas.

First of all, let us find the characteristic gas constant (R). We know that

$$R = c_p - c_v$$
$$= a - b = 0.946 - 0.653$$
$$= 0.293 \text{ kJ/kg K}$$

We know that

$$p_1 v_1 = m R T_1$$

$$\therefore \quad T_1 = \frac{p_1 v_1}{m R} = \frac{4 \times 10^3 \times 0.14}{1.5 \times 0.293} = 1274.2 \text{ K} \textbf{ Ans.}$$

2. *Work done by the gas*

Let W_{1-2} = Work done by the gas.

We know that

$$\delta Q = dU + \delta W \quad \ldots \text{(First Law of Thermodynamics)}$$

In case of isentropic expansion, no heat enters or leaves the gas.

$$\therefore \qquad \delta Q = 0 \quad \text{or} \quad \delta W = -dU$$

On integrating from state 1 to state 2,

$$W_{1-2} = -(U_2 - U_1)$$

Now, let us find the change in internal energy $(U_2 - U_1)$.

We know that mean temperature

$$T_m = \frac{T_1 + T_2}{2} = \frac{1274.2 + 500}{2} = 887.1 \text{ K}$$

and mean specific heat at constant volume,

$$c_{vm} = b + K\,T_m = 0.653 + 0.000\,184 \times 887.1 = 0.816 \text{ kJ kg K}$$

We know that change in internal energy,

$$dU = U_2 - U_1 = m\,c_{vm}\,(T_2 - T_1)$$

$$= 1.5 \times 0.816 \times (500 - 1274.2) = -947.6 \text{ kJ}$$

$\therefore$ Work done by the gas,

$$W_{1-2} = -(U_2 - U_1) = 947.6 \text{ kJ} \quad \textbf{Ans.}$$

3. *Pressure at the end of expansion*

Let $\qquad p_2 =$ Pressure at the end of expansion.

First of all, let us find out the volume at the end of expansion.

We know that for isentropic process,

$$T_1\, e^{\frac{K T_1}{b}}\, v_1^{\left(\frac{a}{b} - 1\right)} = T_2\, e^{\frac{K T_2}{b}}\, v_2^{\left(\frac{a}{b} - 1\right)}$$

$$1274.2 \times e^{\frac{0.000184 \times 1274.2}{0.653}} \times 0.14^{\frac{0.946}{0.653} - 1}$$

$$= 500 \times e^{\frac{0.000184 \times 500}{0.653}} \times v_2^{\frac{0.946}{0.653} - 1}$$

$$1274.2 \times 1.43 \times 0.41 = 500 \times 1.15 \times v_2^{0.45}$$

or

$$v_2 = \left(\frac{1274.2 \times 1.43 \times 0.41}{500 \times 1.5}\right)^{\frac{1}{0.45}} = 0.99 \text{ m}^3$$

We know that $\qquad p_2\, v_2 = m\,R\,T_2$

$$\therefore \qquad p_2 = \frac{m\,R\,T_2}{v_2} = \frac{1.5 \times 0.293 \times 500}{0.99} = 221.97 \text{ kN/m}^2 = 2.2197 \text{ bar} \quad \textbf{Ans.}$$

40.9. Effect of Variable Specific Heat in I.C. Engines

We have already discussed that with the increase in temperature, the specific heat of any gas also increases. In an I.C. engine, the pressure and temperature developed during combustion are much less than that calculated by assuming specific heat to be constant. This results in lower efficiency of the cycle. This effect of variable specific heat on air standard efficiencies of Otto and Diesel cycles is discussed in the following pages.

40.10. Effect of Variable Specific Heat on Otto Cycle

The effect of variable specific heat on Otto cycle is shown on p-v and T-s diagram in Fig. 40.5. The ideal Otto cycle is represented by 1-2-3-4. When the variation in specific heat is taken into account, the cycle is represented by 1′-2′- 3-4′ (*i.e.* by dotted lines).

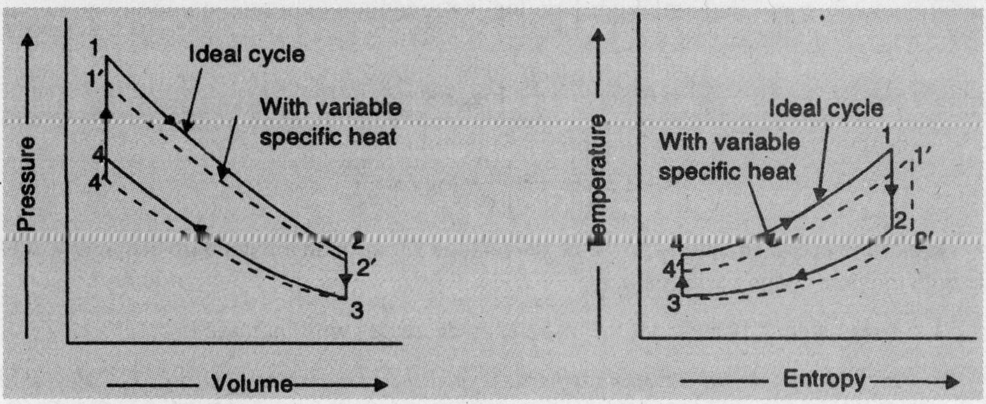

(*a*) p-v diagram. (*b*) T-s diagram.

Fig. 40.5. Effect of variable specific heat on Otto cycle.

Let us derive the relation for the variation in air standard efficiency of Otto cycle with percentage variation in c_v. We have already discussed in Chapter 6 (Art. 6.16), that air standard efficiency of Otto cycle is

$$\eta = 1 - \frac{1}{(r)^{\gamma - 1}} = 1 - (r)^{-(\gamma - 1)} = 1 - (r)^{-R^{*}/c_v}$$

where $r =$ Compression ratio, and

$\gamma =$ Ratio of specific heats $= c_p/c_v$

$$\therefore \quad 1 - \eta = (r)^{\frac{-R}{C_v}}$$

Taking $\log_e$ on both sides, we get

$$\log_e \left(1 - \eta\right) = \frac{-R}{c_v} \log_e r$$

Differentiating the above equation, we get

$$-\frac{1}{1-\eta}\, d\eta = -R \log_e r \left(\frac{-1}{c_v^2}\, dc_v\right) = R \log_e r \times \frac{dc_v}{c_v^2}$$

---

* We know that

$$c_p - c_v = R$$

Dividing throughout by c_v,

$$\frac{c_p}{c_v} - 1 = \frac{R}{c_v} \quad \text{or} \quad \gamma - 1 = \frac{R}{c_v} \qquad \ldots \left(\because \gamma = c_p/c_v\right)$$

or $$-d\eta = (1-\eta)\,\frac{R}{c_v}\,\log_e r \times \frac{dc_v}{c_v}$$

Dividing both sides by η, we get

$$-\frac{d\eta}{\eta} = \left(\frac{1-\eta}{\eta}\right)\frac{R}{c_v}\,\log_e r \times \frac{dc_v}{c_v}$$

or $$\frac{d\eta}{\eta} = -\left(\frac{1-\eta}{\eta}\right)\frac{R}{c_v}\,\log_e r \times \frac{dc_v}{c_v}$$

$$= -2.3\left(\frac{1-\eta}{\eta}\right)\frac{R}{c_v}\,\log r \times \frac{dc_v}{c_v}$$

Hence the above equation shows the percentage variation in air standard efficiency of Otto cycle with the percentage variation in c_v.

The *minus* sign indicates that the efficiency decreases with increase in c_v.

Example 40.4. *In an air standard Otto cycle, the compression ratio is 7. Calculate the change in efficiency if the specific heat at constant volume increases by 2%. Take $\gamma = 1.4$.*

Solution. Given : $r = 7$; $dc_v/c_v = 2\% = 0.02$; $\gamma = 1.4$

We know that air standard efficiency for Otto cycle,

$$\eta = 1 - \frac{1}{(r)^{\gamma-1}} = 1 - \frac{1}{(7)^{1.4-1}} = 1 - \frac{1}{2.18}$$

$$= 1 - 0.46 = 0.54 \quad \text{or} \quad 54\%$$

$\therefore$ Change in efficiency,

$$\frac{d\eta}{\eta} = -2.3\left(\frac{1-\eta}{\eta}\right)\frac{R}{c_v}\,\log r \times \frac{dc_v}{c_v}$$

$$= -2.3\left(\frac{1-0.54}{0.54}\right)(1.4-1)\,\log 7 \times 0.02 \qquad \ldots\left(\because \frac{R}{c_v} = \frac{c_p - c_v}{c_v} = \gamma - 1\right)$$

$$= -0.0133 = -1.33\% \quad \textbf{Ans.}$$

The *minus* sign indicates that there is a decrease in the efficiency of Otto cycle.

Example 40.5. *In an Otto cycle, the compression ratio is 7 and the compression begins at 1 bar and 325 K. The air fuel ratio is 15 : 1 and the calorific value of fuel is 46 000 kJ/kg. Determine the maximum pressure in the cycle, if the mean index of compression is 1.37 and specific heat at constant volume is given by*

$$c_v = 0.7116 + 0.21 \times 10^{-3} \text{ kJ/kg K}$$

If value of c_v remains constant as 0.7116 kJ/kg, and the value of compression index is unaltered, what will be the maximum pressure ?

Solution. Given : $r = v_3/v_4 = v_2/v_1 = 7$; $p_3 = 1\text{bar} = 100 \text{ kN/m}^2$; $T_3 = 325$ K ; Air-fuel ratio = 15 : 1 ; $C.V = 46\,000$ kJ/ kg ; $n = 1.37$; $b = 0.7116$ kJ/kg K ; $K = 0.21 \times 10^{-3}$ kJ/kg K^2.

The *p-v* diagram for the Otto cycle with variable specific heat (c_v) is shown in Fig. 40.6.

First of all, let us find the pressure and temperature at the end of compression.

Let p_4 = Pressure at the end of compression, and

T_4 = Temperature at the end of compression.

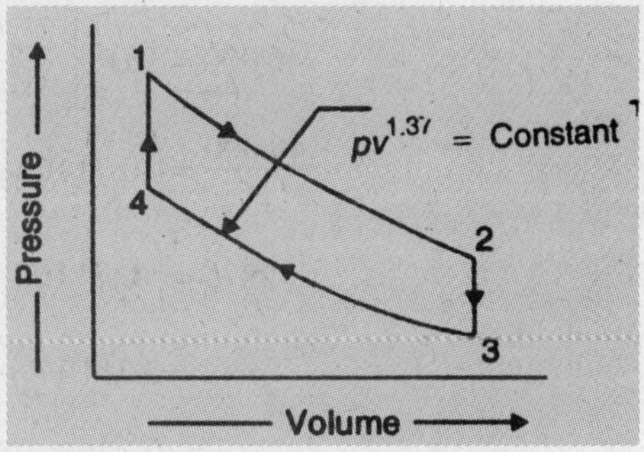

Fig. 40.6

We know that, during compression process 3-4,

$$p_3 (v_3)^n = p_4 (v_4)^n$$

or $$p_4 = p_3 \left(\frac{v_3}{v_4}\right)^n = p_3 \times r^n = 100\,(7)^{1.37} = 1438 \text{ kN/m}^2$$

$\ldots (\because v_3/v_4 = r)$

Also, $$\frac{T_4}{T_3} = \left(\frac{p_4}{p_3}\right)^{\frac{n-1}{n}}$$

$\therefore$ $$T_4 = T_3 \left(\frac{p_4}{p_3}\right)^{\frac{n-1}{n}} = 325 \left(\frac{1438}{100}\right)^{\frac{1.37-1}{1.37}} = 667.54 \text{ K}$$

Now, let us find the temperature at the end of constant volume heat addition process 4-1.

Let T_1 = Temperature at the end of constant volume heat addition process.

We know that mean temperature,

$$T_m = \frac{T_4 + T_1}{2} = \frac{667.54 + T_1}{2}$$

$\therefore$ Mean specific heat at constant volume,

$$c_{vm} = b + K\,T_m = 0.7716 + 0.21 \times 10^{-3} \left(\frac{667.54 + T_1}{2}\right) \text{ kJ/kg K}$$

Now, let us find out the amount of heat added during the process 4-1,

Let Q_{4-1} = Amount of heat added in kJ.

We know that, Air fuel ratio = 15 : 1 ... (Given)

i.e. One kg of fuel mixes with 15 kg of air and thus total mass of mixture (m) = 15 + 1 = 16 kg.

Since the calorific value of fuel is 46 000 kJ/kg of fuel therefore,

$$Q_{4-1} = \text{Mass of fuel} \times C.V. = 1 \times 46\,000 = 46\,000 \text{ kJ}$$

We know that in constant volume process, the amount of heat added is equal to change in internal energy, *i.e.*

$$Q_{4-1} = U_1 - U_4$$

$$\therefore \quad Q_{4-1} = m\, c_{vm}\,(T_1 - T_4) \qquad \ldots [\because U_1 - U_4 = m\, c_{vm}\,(T_1 - T_4)]$$

$$46\,000 = 16 \times \left[0.7116 + 0.21 \times 10^{-3}\left(\frac{667.54 + T_1}{2}\right)\right]\left(T_1 - 667.54\right)$$

$$\frac{46\,000}{16} = \left(0.7116 + 0.105 \times 10^{-3} \times 667.54 + 0.105 \times 10^{-3} \times T_1\right)\left(T_1 - 667.54\right)$$

$$2875 = \left(0.7116 + 0.0701 + 0.105 \times 10^{-3} \times T_1\right)\left(T_1 - 667.54\right)$$

$$= 0.7116\,T_1 + 0.0701\,T_1 + 0.105 \times 10^{-3}\,(T_1)^2 - 475.02 - 46.79 - 0.0701\,T_1$$

$$= 0.105 \times 10^{-3}\,(T_1)^2 + 0.7116\,T_1 - 521.81$$

or $\quad 0.105 \times 10^{-3}\,(T_1)^2 + 0.7116\,T_1 - 3396.81 = 0$

$$(T_1)^2 + 6.78 \times 10^3\,T_1 - 32.34 \times 10^6 = 0$$

$$\therefore \quad T_1 = -\frac{6.78 \times 10^3 \pm \sqrt{(6.78 \times 10^3)^2 - 4 \times (-32.34 \times 10^6)}}{2}$$

$$= \frac{-6.78 \times 10^3 \pm \sqrt{(45.97 \times 10^6) + (129.36 \times 10^6)}}{2}$$

$$= \frac{-6.78 \times 10^3 \pm 13.24 \times 10^3}{2}$$

$$\therefore \quad T_1 = 3230\text{ K} \quad \text{or} \quad -10\,010\text{ K}$$

Taking the positive value of T_1, we get

$$T_1 = 3230\text{ K}$$

Maximum pressure in the cycle

Let $\quad p_3 =$ Maximum pressure in the cycle.

We know that for the process 4-1,

$$\frac{p_1}{T_1} = \frac{p_4}{T_4} \qquad \ldots (\because \text{Volume is constant})$$

$$\therefore \quad p_1 = p_4\left(\frac{T_1}{T_4}\right) = 1438\left(\frac{3230}{667.54}\right) = 6958\text{ kN/m}^2 = 69.58\text{ bar} \quad \textbf{Ans.}$$

Maximum pressure in the cycle if c_v remains constant

Let $\quad p_1' =$ Maximum pressure in the cycle if c_v remains constant.

First of all, let us find temperature at the end of compression again with c_v as constant.

Let $\quad T_1' =$ Temperature at the end of compression.

We know that

$$Q_{4-1} = m\, c_v\,(T_1' - T_4)$$

$$46\,000 = 16 \times 0.7116 \left(T_1' - 667.54 \right)$$

$$T_1' - 667.54 = \frac{46\,000}{16 \times 0.7116} = 4040.2$$

or $$T_1' = 4040.2 + 667.54 = 4707.74 \text{ K}$$

$\therefore$ **Maximum pressure air the cycle,**

$$p_1' = p_4 \left(\frac{T_1'}{T_4} \right) = 1438 \left(\frac{4707.74}{667.54} \right) = 10.141 \times 10^3 \text{ kN/m}^2$$

$$= 101.41 \text{ bar} \textbf{ Ans.}$$

Example 40.6. *An engine working on Otto cycle (constant volume cycle) has a compression ratio of 10. The temperature rise during compression is from 320 K to 820 K and at the beginning and end of expansion, the temperatures are 1720 K and 688 K, both the compression and expansion being adiabatic and occupying the whole stroke.*

Assuming that the specific heat at constant volume has the form b + KT where b and K are constants and T is the absolute temperature, and the value of R = 287 J/kg K, find : 1. The values of b and K, and 2. the theoretical efficiency.

Solution. Given : $r = v_3/v_4 = v_2/v_1 = 10$; $T_3 = 320$ K ; $T_4 = 820$ K ; $T_1 = 1720$ K ; $T_2 = 688$ K ; $R = 287$ J/kg K

1. The values of b and K

The *p-v* diagram for Otto cycle with variable specific heat (c_v) is shown in Fig. 40.7.

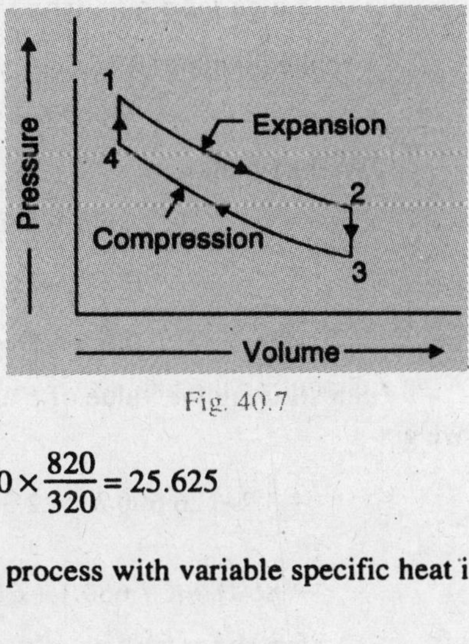

Fig. 40.7

We know that during the expansion process 1-2,

$$\frac{p_1 v_1}{T_1} = \frac{p_2 v_2}{T_2}$$

or $$\frac{p_1}{p_2} = \left(\frac{v_2}{v_1} \right) \times \left(\frac{T_1}{T_2} \right) = 10 \times \frac{1720}{688} = 25$$

We also know that for compression process 3-4,

$$\frac{p_3 v_3}{T_3} = \frac{p_4 v_4}{T_4} \quad \text{or} \quad \frac{p_4}{p_3} = \left(\frac{v_3}{v_4} \right) \times \left(\frac{T_4}{T_3} \right) = 10 \times \frac{820}{320} = 25.625$$

We know that the equation for reversible adiabatic process with variable specific heat is

$$p \; e^{\frac{KT}{b}} \; v^{\frac{a}{b}} = \text{Constant}$$

Hence for the expansion process 1-2,

$$p_1 \; e^{\frac{KT_1}{b}} \; v_1^{\frac{a}{b}} = p_2 \; e^{\frac{KT_2}{b}} \; v_2^{\frac{a}{b}}$$

or $$\log_e p_1 + \frac{KT_1}{b} + \frac{a}{b} \log_e v_1 = \log_e p_2 + \frac{KT_2}{b} + \frac{a}{b} \log_e v_2$$

$$-2.3\log\left(\frac{p_1}{p_2}\right)+2.3\times\frac{a}{b}\log\left(\frac{v_2}{v_1}\right)=\frac{K}{b}\left(T_1-T_2\right)$$

$$-2.3\log\left(25\right)+2.3\times\frac{a}{b}\log\left(10\right)=\frac{K}{b}\left(1720-688\right)$$

$$-3.22\,b+2.3\times a=1032\,K \qquad \ldots(i)$$

Similarly for the compression process 3-4,

$$p_3\,e^{\frac{K\,T_3}{b}}\,v_3^{\frac{a}{b}}=p_4\,e^{\frac{K\,T_4}{b}}\,v_4^{\frac{a}{b}}$$

or $$\log_e p_3+\frac{K\,T_3}{b}+\frac{a}{b}\times\log_e v_3=\log_e p_4+\frac{K\,T_4}{b}+\frac{a}{b}\log_e v_4$$

or $$-2.3\log\left(\frac{p_4}{p_3}\right)+2.3\times\frac{a}{b}\log\left(\frac{v_3}{v_4}\right)=\frac{K}{b}\left(T_4-T_1\right)$$

$$-2.3\log\left(25.625\right)+2.3\times\frac{a}{b}\log\left(10\right)=\frac{K}{b}\left(820-320\right)$$

$$-3.24+2.3\times\frac{a}{b}=500\times\frac{K}{b}$$

or $$-3.24\,b+2.3\,a=500\,K \qquad \ldots(ii)$$

From equations (*i*) and (*ii*), we get

$$0.02\,b=532\,K \quad \text{or} \quad b=26\,600\ K \qquad \ldots(iii)$$

We know that

$$c_v=b+K\,T;\ \text{and}\ c_p=a+K\,T$$

$$\therefore \quad R=c_p-c_v=a-b$$

or $$a=R+b=287+26\,600\ K \qquad \ldots(iv)$$

Substituting the value of b and a from equations (*iii*) and (*iv*) respectively in equation (*ii*), we get

$$-3.24\left(26\,600\,K\right)+2.3\left(287+26\,600\,K\right)=500\,K$$

$$-86\,184\,K+660.1+61\,180\,K=500\,K$$

or $$25\,504\,K=660.1 \quad \text{or} \quad K=0.0259\ \text{J/kg K}^2$$ **Ans.**

$$\therefore \quad b=26\,600\,K=26\,600\times0.0259=688.9\ \text{J/kg}$$ **Ans.** ... [From equation (*iii*)]

2. *Theoretical efficiency*

We know that theoretical efficiency,

$$\eta=1-\frac{1}{(r)^{\gamma-1}}=1-\frac{1}{(10)^{1.4-1}}=1-0.40=0.60 \text{ or } 60\%$$ **Ans.**

40.11. Effect of Variable Specific Heat on Diesel Cycle

The effect of variable specific heat on Diesel cycle is shown on p-v and T-s diagram in Fig.40.8. The ideal Diesel cycle is represented by 1-2-3-4. When the variation in specific heat is taken into account, the cycle is represented by 1′-2′-3′-4.

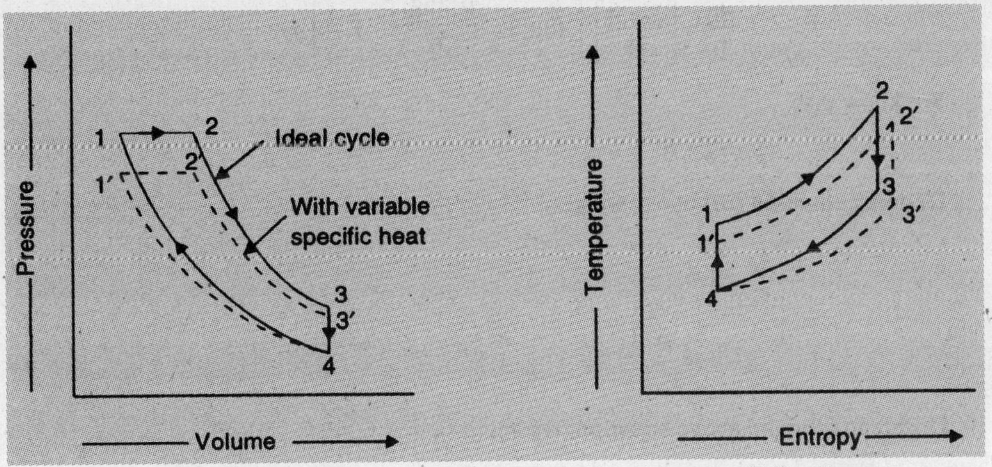

(a) $p-v$ diagram. (b) $T-s$ diagram.

Fig. 40.8. Effect of variable specific heat on Diesel cycle.

Let us derive the relation for the percentage variation in air standard efficiency of Diesel cycle with the percentage variation of c_v.

We have already discussed in Chapter 6 (Art. 6.17) that the efficiency of Diesel cycle is

$$\eta = 1 - \frac{1}{r^{\gamma-1}}\left[\frac{\rho^{\gamma}-1}{\gamma(\rho-1)}\right] \quad \ldots (i)$$

where r = Compression ratio,

ρ = Cut-off ratio, and

γ = Ratio of specific heats $= c_p/c_v$.

$$\therefore \quad 1-\eta = \frac{1}{r^{\gamma-1}}\left[\frac{\rho^{\gamma}-1}{\gamma(\rho-1)}\right]$$

or

$$1-\eta = r^{-(\gamma-1)}\left[\frac{\rho^{\gamma}-1}{\gamma(\rho-1)}\right]$$

Taking $\log_e$ on both sides, we get

$$\log_e\left(1-\eta\right) = -\left(\gamma-1\right)\log_e r + \log_e\left(\rho^{\gamma}-1\right) - \log_e \gamma - \log_e\left(\rho-1\right)$$

Now, differentiate the above equation with respect to γ,

$$-\frac{1}{1-\eta}\times\frac{d\eta}{d\gamma} = -\log_e r + \frac{\rho^{\gamma}\log_e \rho}{\rho^{\gamma}-1} - \frac{1}{\gamma}$$

or $$d\eta = \left(1-\eta\right)\left[\log_e r - \frac{\rho^{\gamma}\log_e \rho}{\rho^{\gamma}-1} + \frac{1}{\gamma}\right] d\gamma$$

Dividing both sides by η , we get

$$\frac{d\eta}{\eta} = \left[\frac{1-\eta}{\eta}\right]\left[\log_e r - \frac{\rho^{\gamma}\log_e \rho}{\rho^{\gamma}-1} + \frac{1}{\gamma}\right] d\gamma \quad \ldots (ii)$$

We know that

$$c_p - c_v = R \quad \ldots (iii)$$

Dividing equation (*iii*) by c_v, we get

$$\frac{c_p}{c_v} - 1 = \frac{R}{c_v}$$

or $$\gamma - 1 = \frac{R}{c_v} \quad \ldots \left(\because \gamma = \frac{c_p}{c_v}\right)$$

Differentiating the above equation, we get

$$d\gamma = -\frac{R}{c_v^2}\, dc_v = \frac{-R}{c_v} \times \frac{dc_v}{c_v}$$

or $$d\gamma = -\left(\gamma - 1\right)\frac{dc_v}{c_v} \quad \ldots\left(\because \gamma - 1 = \frac{R}{c_v}\right)$$

Substituting this value of $d\gamma$ in equation (*ii*), we get

$$\frac{d\eta}{\eta} = -\left[\frac{1-\eta}{\eta}\right]\left[\log_e r - \frac{\rho^{\gamma}\log_e \rho}{\rho^{\gamma}-1} + \frac{1}{\gamma}\right]\left(\gamma - 1\right)\frac{dc_v}{c_v}$$

or $$\frac{d\eta}{\eta} = -\left[\frac{1-\eta}{\eta}\right]\left(\gamma - 1\right)\left[\log_e r - \frac{\rho^{\gamma}\log_e \rho}{\rho^{\gamma}-1} + \frac{1}{\gamma}\right]\frac{dc_v}{c_v}$$

$$= -\left[\frac{1-\eta}{\eta}\right]\left(\gamma - 1\right)\left[2.3\log_e r - \frac{2.3\,\rho^{\gamma}\log \rho}{\rho^{\gamma}-1} + \frac{1}{\gamma}\right]\frac{dc_v}{c_v}$$

Hence the above equation shows the percentage variation in air standard efficiency of Diesel cycle with the percentage variation in c_v.

The *minus* sign indicates that there is a decrease in efficiency with increase in c_v.

Example 40.7. *In an air standard Diesel cycle, the compression ratio is 13 and the fuel is cut-off at 8% of the stroke. Calculate the change in efficiency if the specific heat at constant volume increases by 2%. Take $\gamma = 1.4$.*

Solution. Given : $r = 13$; Cut- off = 8% of stroke volume ; $dc_v/c_v = 2\% = 0.02$; $\gamma = 1.4$

First of all, let us find out the cut-off ratio. We know that cut-off ratio,

$$^*\rho = 1 + \frac{\text{Cut–off \%}}{100}\left(r - 1\right)$$

$$= 1 + \frac{8}{100}\left(13 - 1\right) = 1 + 0.08 \times 12 = 1.96$$

∴ Air standard efficiency for Diesel cycle,

$$\eta = 1 - \frac{1}{(r)^{\gamma - 1}}\left[\frac{\rho^{\gamma} - 1}{\gamma(\rho - 1)}\right] = 1 - \frac{1}{(13)^{1.4-1}}\left[\frac{(1.96)^{1.4} - 1}{1.4\,(1.96 - 1)}\right]$$

$$= 1 - 0.417 = 0.583 \quad \text{or} \quad 58.3\%$$

We know that change in efficiency for Diesel cycle,

$$\frac{d\eta}{\eta} = -\left[\frac{1-\eta}{\eta}\right]\left(\gamma - 1\right)\left[2.3 \log r - \frac{2.3\,\rho^{\gamma} \log \rho}{\rho^{\gamma} - 1} + \frac{1}{\gamma}\right]\frac{dc_v}{c_v}$$

$$= -\left[\frac{1 - 0.583}{0.583}\right]\left(1.4 - 1\right)\left[2.3 \log 13 - \frac{2.3 \times (1.96)^{1.4} \log (1.96)}{(1.96)^{1.4} - 1} + \frac{1}{1.4}\right] 0.02$$

$$= -0.0124 = -1.24\%$$ **Ans.**

The *minus* sign indicates that there is a decrease in the efficiency of Diesel cycle.

Example 40.8. *In an Diesel cycle, the compression ratio is 15 and the compression begins at 1 bar and 420 K. The air-fuel raio is 25 : 1 and the calorific value of fuel is 42 000 kJ/kg. The law of compression follows the relation $p\,v^{1.35}$ = constant. Determine the percentage at which the constant pressure combustion stops. Neglect molecular changes during combustion.*

Take $c_v = 0.7116 + 0.21 \times 10^{-3}\,T$ kJ/kg K, and $R = 0.287$ kJ/kg K.

Solution. Given : $r = v_4/v_1 = 15$; $p_4 = 1$ bar $= 100$ kN/m^2 ; $T_4 = 420$ K ; Air-fuel ratio = 25 : 1 ; $C.V. = 42\,000$ kJ/kg ; $n = 1.35$; $b = 0.7116$ kJ/kg K ; $K = 0.21 \times 10^{-3}$ kJ/kg K^2 ; $R = 0.287$ kJ/kg K

The *p-v* diagram for Diesel cycle with variable specific heat is shown in Fig. 40.9.

First of all, let us find out the temperature at the end of compression process 4-1.

Let T_1 = Temperature at the end of compression process 4-1.

---

* We know that volume at cut-off,

$$v_2 = v_1 + \frac{\text{Cut–off \%}}{100} \times \text{Stroke volume}$$

Dividing throughout by v_1,

$$\frac{v_2}{v_1} = 1 + \frac{\text{Cut–off \%}}{100}\left(\frac{v_4}{v_1} - 1\right) \qquad \ldots (\because \text{Stroke volume} = v_4 - v_1)$$

or

$$\rho = 1 + \frac{\text{Cut–off \%}}{100}\left(r - 1\right) \qquad \ldots \left(\because \frac{v_2}{v_1} = \rho \text{ and } \frac{v_4}{v_1} = r\right)$$

We know that during compression process 4-1 ,

$$\frac{T_1}{T_4} = \left(\frac{v_4}{v_1}\right)^{n-1}$$

$$\therefore \quad T_1 = T_4\left(\frac{v_4}{v_1}\right)^{n-1}$$

$$= 420\left(15\right)^{1.35-1} = 1083.63 \text{ K}$$

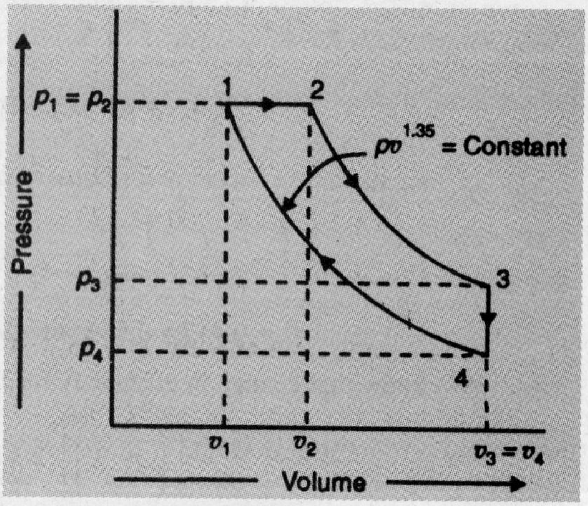

Fig. 40.9

Now let us find out the amount of heat added during the constant pressure heat addition process 1-2.

Let Q_{1-2} = Amount of heat added in kJ.

We know that air fuel ratio = 25 : 1 , *i.e.* one kg of fuel mixes with 25 kg of air and thus the total mass of the mixture $(m) = 25 + 1 = 26$ kg.

Since the calorific value of fuel is 42 000 kJ/kg of fuel, therefore

$$Q_{1-2} = \text{Mass of fuel} \times C.V. = 1 \times 42\,000 = 42\,000 \text{ kJ}$$

Now, let us find out the change in internal energy during the constant pressure heat addition process.

Let $U_2 - U_1$ = Change in internal energy.

We know that mean temperature (T_m) during process 1-2,

$$T_m = \frac{T_1 + T_2}{2} = \frac{1083.63 + T_2}{2}$$

$\therefore$ Mean specific heat at constant volume,

$$c_{vm} = b + K\,T_m = 0.7116 + 0.21 \times 10^{-3}\left(\frac{1083.63 + T_2}{2}\right) \text{ kJ/kg K}$$

$$= 0.7116 + 0.105 \times 10^{-3}\,(1083.63 + T_2)$$

We also know that change in internal energy,

$$U_2 - U_1 = m\,c_{vm}\,(T_2 - T_1)$$

$$= 26\left[0.7116 + 0.105 \times 10^{-3}\left(1083.63 + T_2\right)\right]\left(T_2 - 1083.63\right)$$

$$= \left(21.46 + 2.73 \times 10^{-3}\,T_2\right)\left(T_2 - 1083.63\right)$$

$$= 2.73 \times 10^{-3}\,(T_2)^2 + 18.5\,T_2 - 23254.7$$

Now let us find out the work done during the process 1-2. We know that for constant pressure process 1-2, work done

$$W_{1-2} = p\,(v_2 - v_1) = m\,R\,(T_2 - T_1) = 16 \times 0.287\;(T_2 - 1083.63)$$

$$= 4.6\,T_2 - 4984.7 \text{ kJ}$$

Now, let us find out the temperature at the end of constant pressure process 1-2.

Let T_2 = Temperature at the end of process 1-2.

We know that for a process, the general gas energy equation is

$$Q_{1-2} = U_2 - U_1 + W_{1-2}$$

Substituting the values, in the above equation, we get

$$42\,000 = (2.73 \times 10^{-3}\,(T_2)^2 + 18.5\,T_2 - 23\,254.7) + (4.6\,T_2 - 4984.7)$$

$$2.73 \times 10^{-3}\,(T_2)^2 + 23.1\,T_2 - 70239.4 = 0$$

$$(T_2)^2 + 8.46 \times 10^3\,T_2 - 25.73 \times 10^6 = 0$$

$$\therefore \quad T_2 = \frac{-8.46 \times 10^3 \pm \sqrt{(8.46 \times 10^3)^2 - 4 \times 1 \times (-25.73 \times 10^6)}}{2}$$

$$= \frac{-8.46 \times 10^3 \pm \sqrt{71.57 \times 10^6 + 102.92 \times 10^6}}{2}$$

$$= \frac{-8.46\ 10^3 \pm 13.21 \times 10^3}{2}$$

$$\therefore \quad T_2 = 2375 \text{ K} \quad \text{or} \quad T_2 = -10\,830 \text{ K}$$

Taking positive value of T_2, we get

$$T_2 = 2375 \text{ K}$$

We know that for constant pressure process 1-2,

$$\frac{v_2}{v_1} = \frac{T_2}{T_1} = \frac{2375}{1083.63} = 2.19$$

Percentage at which the constant pressure combustion stops

We know that the percentage at which the constant pressure combustion stops

$$= \frac{v_2 - v_1}{v_4 - v_1} \times 100 = \frac{\dfrac{v_2}{v_1} - 1}{\dfrac{v_4}{v_1} - 1} \times 100$$

$$= \frac{2.19 - 1}{15 - 1} \times 100 = 8.5\% \quad \textbf{Ans.}$$

EXERCISES

1. One kg of a certain gas is heated from 500 K to 1300 K. Calculate the change in internal energy and enthalpy.

Take $c_p = 0.946 + 0.000\,184\,T$ kJ/kg K and $c_v = 0.653 + 0.000\,184\,T$ kJ/kg K.

[**Ans.** 655.2 kJ ; 889.6 kJ]

2. A certain quantity of gas having a volume of 4 m^3 at 1.2 bar and 30°C is heated at constant pressure to 50°C. Determine : 1. Mass of the gas ; 2. Heat transfer during the process and 3. Change of entropy.

Take $c_p = 0.946 + 0.184 \times 10^{-3} T$ kJ/kg K and $c_v = 0.946 + 0.184 \times 10^{-3} T$ kJ/kg K

[Ans. 5.41 kg ; 108.2 kJ]

3. A mass of 1.8 kg and volume of 0.15 m^3 of certain gas at 45 bar is expanded isentropically such that the temperature falls to 600 K. Determine : 1. Initial temperature of the gas, 2. Work done during the process, and 3. Pressure at the end of expansion.

Take $c_p = 0.946 + 0.000\,184\,T$ kJ/kg K and $c_v = 0.653 + 0.000\,184\,T$ kJ/kg K

[Ans. 1279.86 K ; 1010.82 kJ ; 2.532 bar]

4. In an air standard Otto cycle, the compression ratio is 7.5. Calculate the change in efficiency if the specific heat at constant volume increases by 3%. Take $\gamma = 1.4$ [Ans. – 1.95%]

5. In an air standard Diesel cycle, the compression ratio is 18 and the fuel is cut-off at 5% of the stroke. Calculate the change in efficiency if the specific heat at constant volume increases by 2.5%. Take $\gamma = 1.4$. [Ans. – 1.44%]

QUESTIONS

1. What do you understand by molar specific heat?
2. Explain the effect of variation of specific heat with temperature.
3. Describe how moisture content of air affects the specific heats of air.
4. Derive the relations for the change in internal energy and enthalpy during a process with variable specific heats.
5. The specific heats of a gas are of the form $c_p = a + KT$ and $c_v = b + KT$, where a, b and K are constants and T is in K. Derive the formula $T^b\, v_s^{a-b}\, e^{KT} = \text{constant}$, for the adiabatic expansion of the gas.
6. Explain the effect of variable specific heat on air standard efficiencies of Otto and Diesel cycle.
7. Derive the following relation for Otto cycle, by taking variation of specific heat in account :

$$\frac{d\eta}{\eta} = -\left[\frac{1-\eta}{\eta}\right]\frac{R}{c_v}\log_e r \times \frac{dc_v}{c_v}$$

OBJECTIVE TYPE QUESTIONS

1. Gases have

(a) one specific heat (b) two specific heats

(c) three specific heats (d) four specific heats

2. The value of specific heat at constant pressure (c_p) .... with increase in temperature.

(a) increases (b) decreases (c) remains same

3. Molar specific heat at constant pressure (c_{mp}) is

   (*a*) $M + c_p$ (*b*) $M - c_p$ (*c*) $M\,c_p$ (*d*) M/c_p

where M = Molecular mass of gas, and

c_p = Specific heat at constant pressure.

4. Molar specific heat at constant volume (c_{mv}) is

   (*a*) $M + c_v$ (*b*) $M - c_v$ (*c*) $M\,c_v$ (*d*) M/c_v

where M = Molecular mass of gas, and

c_v = Specific heat at constant volume.

5. The ratio of specific heats $\left(\gamma = \frac{c_p}{c_v}\right)$ for air

   (*a*) increases with increase in moisture content in air

   (*b*) decreases with increase in moisture content in air

   (*c*) remains constant irrespective of the increase in moisture content of air

   (*d*) increases first and then decreases later with moisture content in air

6. The ratio of specific heats $\left(\gamma = \frac{c_p}{c_v}\right)$ for real gas

   (*a*) increases with increase in temperature

   (*b*) decreases with increase in temperature

   (*c*) remains same irrespective of increase in temperature

   (*d*) increases first with increase in temperature and then decreases with further increase in temperature

7. The gas constant (R) is equal to

   (*a*) ratio of two specific heats

   (*b*) sum of two specific heats

   (*c*) difference of two specific heats

   (*d*) product of two specific heats

8. The change in internal energy during a process with variable specific heats is equal to

   (*a*) $m\,c_v\,(T_2 - T_1)$ (*b*) $m\,c_{vm}/(T_2 - T_1)$

   (*c*) $m\,c_v/(T_2 - T_1)$ (*d*) $m\,c_{vm}\,(T_2 - T_1)$

9. The efficiency of Otto cycle

   (*a*) increases with increase in specific heat

   (*b*) decreases with increase in specific heat

   (*c*) remains same with increase in specific heat

   (*d*) first increases then decreases with increase in specific heat

10. The efficiency of Diesel cycle

(*a*) increases with increase in specific heat

(*b*) decreases with increase in specific heat

(*c*) remains same with increase in specific heat

(*d*) first increases and then decreases with increase in specific heat

ANSWERS

1. (*b*)	2. (*a*)	3. (*c*)	4. (*c*)	5. (*b*)
6. (*b*)	7. (*c*)	8. (*d*)	9. (*b*)	10. (*b*)

Index

A

Absolute humidity, 799, 800, 801
— pressure, 12
— temperature, 11
— units of force, 7
Actual gas turbine, 725
— indicator diagram, 381, 588, 590, 593, 595
— vapour compression cycle, 789
Adiabatic compressibility, 842
— process, 63
— mixing of two air streams, 819
— saturation temperature, 806
Advantages of compounding of steam engines, 395
— condensers in a steam power plant, 446
— mechanical draught, 361
— multistage compression, 647
— reheating of steam, 561
— steam turbines over reciprocating steam engines, 501
— superheating of steam, 206
— two stroke cycle over four stroke cycle engines, 586
— vapour absorption refrigeration system over vapour compression refrigeration system, 792
— vapour compression refrigeration sys tem over air refrigeration system, 771
Air conditioning system, 825
— consumption, 626
— cooling system, 602
— injection method, 600
— preheater, 340
— pump, 464
— refrigeration cycle, 751
— refrigerator working on Bell Coleman cycle, 761
— on reversed Carnot cycle, 755
— standard efficiency, 617
Ammonia hydrogen refrigerator, 793
Anthracite coal, 290
Application of first law of thermodynamics to a nonflow process, 51
— to a steady flow process, 86
— of kinetic energy to laws of perfect gases, 146
— of steady flow energy equation to engineering system, 92
Approximate method for heat absorbed, 135
Artificial draught, 360
Arrangement of cylinders in compound steam engines, 394
Assumptions for overall (thermal) efficiency of an ideal gas turbine cycle, 721
— in thermodynamic cycle, 153
— in kinetic theory of gases, 140
— in two stage compressor with inter cooler, 648
Atomic mass, 301
Atomiser, 608
Atmospheric air, 861
Atoms and molecules, 301
Axial flow compressors, 677
Available heat energy, 105
Average piston speed, 377
Avogadro's law, 35

B

Babcock and Wilcox boiler, 329
Back flow in positive displacement air compressors, 668
Back pressure turbine, 577
Balanced draught, 362
Barometric jet condensers, 449
Barrel calorimeter, 211
Benson boiler, 331
Binary vapour cycle, 577
— plant, 567
Bituminous coal, 290
Blading efficiency, 535
Blast furnace gas, 292
Bleeding, 570
Blow off cock, 338
Boiler accessories, 323, 340
— efficiency, 346
— mountings, 323, 334
— shell, 323
— trial, 351
Bomb calorimeter, 295
Boyle's law, 30
Boy's gas calorimeter, 296
Brake power, 389, 615
— thermal efficiency, 423, 616
Brayton cycle, 761
Briquetted coal, 290
Brown coal, 290
By-pass factor, 813, 815

C

Caking bituminous coal, 290
Calorific value of fuels, 293
Carburettor, 606
Carnot cycle, 157, 264
Celsius or centigrade scale, 10
— flow surface condenser, 451
Centrifugal compressor, 670
Cetane number, 598
C.G.S. units, 2
Change of entropy of a perfect gas, 110
— during reversible adiabatic process, 124
— constant pressure process, 115
— constant temperature process, 120
— constant volume process, 110
— in thermodynamic properties with variable specific heat, 864
— isentropic process, 124
— polytropic process, 126
Characteristic equation of a gas, 33
Charles' law, 31
Chimney draught, 360
Classification of air compressors, 637
— air conditioning systems, 825
— compound steam engines, 395
— of draughts, 360
— fuels, 289
— gas turbines, 708
— I.C. engines, 583
— properties of a system, 8
— steam boilers, 324
— steam condensers, 447
— steam engines, 374
— steam turbines, 502
— thermodynamic cycles, 158
— thermodynamic process, 51
— thermodynamic systems, 7
Clausius statement, 22
— inequality, 106
Clearance volume, 156, 377
Closed air refrigeration cycle, 755
— cycle gas turbines, 708
— system of thermodynamics, 7
Coal gas, 291
Cochran boiler, 326
Coefficient of performance, 752
— volume expansion, 841
Coil ignition system, 599
Coke, 290
— oven gas, 292
Combined air compressor and air motor, 701
— indicator diagram for a compound steam engine, 397
— separating and throttling calorimeter, 214
— velocity triangle for moving blades of an impulse turbine, 505
— of a reaction turbine, 524
Combustion chamber, 323
— equations of gaseous fuels, 303
— of solid fuels, 302
Comfort air conditioning system, 825
Comparison between air cooling and water cooling-system, 603
— centrifugal and axial flow air compres sors, 677
— closed and open cycle gas turbines, 718
— forced draught and induced draught, 361
— gas turbines and I.C. engines, 708
— impulse turbines and reaction turbines, 522
— jet and surface condensers, 452
— petrol and diesel engines, 592
— reciprocating and centrifugal air com pressor, 665
— steam engines and I.C. engines, 582
— turbine and centrifugal compressor blades, 697
— water tube and fire tube boilers, 332
Components of I.C. engine, 584
Compounds, 301
Compound steam engine, 394
Compounding of impulse steam turbines, 552
Compressed air system, 701
Compressibility factor, 856
Compression ratio, 156, 538
Compressor capacity, 638
— efficiency, 682
Condenser efficiency, 458
Condition for maximum discharge through the nozzle, 476
— efficiency of an impulse turbine, 545
— efficiency of a reaction turbine, 548
— maximum discharge through the chim ney, 366
connecting rod, 376, 585
Constant enthalpy process, 84, 92
— entropy process, 63, 247
— internal energy process, 84
— pressure gas turbines, 719
— pressure lines, 225
— pressure process, 57, 238
— temperature process, 61, 242
— total heat process, 91, 257
— volume gas turbines, 719
— volume lines, 225
— volume process, 54, 231

Convergent nozzle, 469
— divergent nozzle, 469
Conversion of volumetric analysis into mass analysis, 307
— mass analysis into volumetric analysis, 307
Cooling and duhumidification, 817
— of I.C. engines, 601
— systems of I.C. engines, 602
— towers, 465
Cornish boiler, 328
Counterflow jet condensers, 448
Crank case, 585
— shaft, 376, 585
Critical pressure ratio, 476
— Physical significance of, 479
— Values of, 479
Cross head, 376
Cyclic process, 9
Cylinder bore, 156, 377
— dimensions for compound steam engine, 398
— head of I.C. engines, 584
— of steam engines, 375
Cushion steam, 429
Cut off governing of compound steam engines, 440
— of simple steam engines, 439
— volume, 377

D

Dalton's law of partial pressure, 452, 800
Dead centres, 377
— weight safety valve, 336
Degree of reaction, 525, 678
— saturation, 799, 800
Degrees of freedom, 148
Dehumidification, 816, 817
De-Laval impulse turbine, 502
Demerits of gaseous fuels, 292
— liquid fuels, 291
Dense air regrigeration cycle, 755
Derived units, 2
Detonation in I.C. engines, 597
Dew point temperature, 799
— depression, 799
— lines, 810
Deagram efficiency, 535
— factor, 382
Diameters of throat and exit of a nozzle for maximum discharge, 481
Diesel cycle, 178
Difference between a heat engine, refrigerator and heat pump, 753
Disadvantages of mechanical draught, 361
— two stroke over four stroke cycle en gine, 586
—vapour compression refrigeration system over air refrigeration system, 771
Discharge pressure, 638
Divergent nozzle, 469
Down flow surface condenser, 450
D'slide valve, 375
Dry air, 798
— bulb temperature, 799
— bulb temperature lines, 809
— saturated steam, 200
Dryness fraction lines, 224
— of steam, 201
Dual combustion cycle, 188

E

Eccentric, 376
— rod, 376
Economiser, 340
Edward's air pump, 464
Effect of variable specific heat in I.C. engines, 872
— on Otto cycle, 872
— on Diesel cycle, 879
Effects of air leakage, 464
— friction on the velocity diagram of an impulse turbine, 507
— piston rod in a double acting steam en gine, 387
— supersaturation, 491
Efficiencies of a compressor, 682, 693
— multistage turbine, 564
— steam engine, 422
— steam turbine, 564
Efficiency of compressed air system, 702
— chimney, 370
— cycle, 157
— heat exchanger, 730
— modified Rankine cycle, 277
— ratio, 267, 565
Ejector condensers, 449
Electrolux refrigerator, 793
Elements, 301
— Symbols for, 302
Equations of state, 854
Energy, 13
— Law of conservation of, 15
Enthalpy of gas, 39
— entropy (h-s) diagram for water and steam, 223
— lines, 810

— moist air, 803
— steam, 202
Entropy, 103
— importance of, 104
— increase during evaporation, 220
— of superheated steam, 220
— of water, 219
— of wet and dry steam, 220
— Principle of increase of, 107
— Units of, 106
Equipment used in air conditioning system, 825
Equaility of temperature, 11
Equivalence of Kelvin-Plank and Clausius statements, 23
Equivalent evaporation, 345
Ericsson cycle, 166
Essentials of a good steam boiler, 324
Evaporative condenser, 451
Excess air, 310
— Mass of, 310
Exhaust ports, 375
— turbine, 578
Experimental determination of higher calorific value, 294
Extensive properties of thermodynamics, 9
External work done during evaporation, 207
Extraction turbine, 578

F

Factors affecting comfort air conditioning, 824
Fahrenheit scale, 11
Fan draught, 360
Feed check valve, 339
— pump, 340
First law of thermodynamics, 19
— limitations of, 20
Flow of steam through nozzles, 470
— process, 86
Flue gas analysis by Orsat apparatus, 319
Flywheel, 376, 585
Force, 6
Forced draught, 361
— lubrication, 604
Formation of steam, 199
Fourier's law of heat conduction, 737
Four stroke cycle diesel engine, 589
— gas engine, 591
— petrol engine, 587
F.P.S. units, 2
Free air delivery, 638
— expansion process, 83
Frame of steam engine, 375
Friction in a nozzle, 470
Fuel pump, 607
Fundamental units, 1
Furance, 323
Fusible plug, 339
Future power plants, 579

G

Gas turbine with intercooling, 711
— with reheating, 713
— uses of, 733
Gauge pressure, 12
Gaseous fuels, 291
Gasoline, 291
Gay-Lussac law, 31
General equations of change in internal energy, 843
General expression for the change of entropy of a perfect gas, 108
— gas equation, 31
— law for expansion and compression of perfect gas, 85
Gibbs function, 837
Governor, 376
Governing of I.C. engines, 605
— steam engines, 437
— steam turbines, 55
— simple steam engines, 438
Grate, 323
Gravimetric analysis, 307
Gravitational units of force, 7
Gross calorific value, 293
— efficiency, 536
Guide mechanism, 521

H

Heat, 15, 17
— absorbed or rejected during polytropic process, 61
— balance sheet, 352, 433, 628
— capacity, 16
— exchanger, 729
— losses in a boiler, 351
— rejected in a reciprocating air compres sor, 656
— transfer by conduction through a slab, 737
— transfer during a process with variable specific heat, 862
— during polytropic expansion or com pression process, 863
— through a composite wall, 739
— through thick cylinders, 741
— through thick sphere, 744
Heating and humidification, 818
Heavy fuel oils, 291

Height of blades for reaction turbine, 527
— chimney, 362
Helmholtz function, 835
Higher calorific value, 293
— Experimental determination of 294
High level jet condensers, 449
— steam law water safety valve, 336
Humidification, 816, 818
Humidity, 799, 800
— lines, 883
— ratio, 800
Hyperbolic process, 60, 244
Hypothetical indicator diagram, 378
— mean effective pressure, 379

I

Ignition system of petrol engines, 598
Indicated mean effective pressure, 611
— power, 383, 612
— thermal efficiency, 423, 616
Indicator diagram of a simple steam engine, 378
Induced draught, 361
Industrial air conditioning system, 826
Injector, 608
Inlet ports, 375
— pressure, 637
Inner dead centre, 377
Intensive properties of thermodynamics, 9
Intercooling, 760
— of air in a two-stage reciprocating air compressor, 648
Internal energy, 14
— of steam, 208
— efficiency, 564
— losses in turbine, 555
International system of units, 2
Importance of entropy, 104
Important terms used in steam, 200
— thermodynamic cycles, 156
— for steam boilers, 323
Impulse turbine, 502
Irreversible cycle, 154
— process, 50
Isentropic process, 63, 247
— efficiency, 683
— expansion with variable specific heat, 869
— lines, 223, 226
Isobaric process, 57, 238
Isochoric process, 54, 231
Isothermal compressibility, 842
— lines 223, 226
— efficiency, 682
— process, 61, 242

J

Jet condensers, 447
Joule's cycle, 168
— law, 32
Joule Thomson coefficient, 92, 852

K

Kelvin, 3
Kerosene, 291
Kilogram, 3
Kinetic energy, 14
— per kg molecule of a gas, 143

L

La-Mont boiler, 330
Lancashire boiler, 327
Latent heat of vaporisation, 202
Law of conservation of energy, 14
— equipartition of energy, 14
— perfect gases, 30
— thermodyamics, 19
Lever safety valve, 335
Lignite, 290
Limitations of first law of thermodynamics, 20
Liquid fuels, 290
Locomotive boiler, 328
Loeffler boiler, 331
Lower calorific value, 293
Low level jet condenser, 448
—pressure turbine, 578
Lubrication of I.C. engines, 604

M

Magneto ignition system, 600
Main components of I.C. engines, 584
Mass, 6
— Atomic, 301
— of carbon in flue gases, 308
— of cooling water required for condensation of steam, 460
— of excess air supplied, 310
— fuel gases per kg of fuel burnt, 308
— of steam discharged through a nozzle, 474
— of steam in the engine cylinder, 420
— Molecular, 302
Mathematical Fundamentals, 839
Maximum discharge through nozzle, 477
— through chimney, 366
Maxwell's equation, 840
Mean effective pressure, 156, 378, 638
Measurement of brake power, 389
— dryness fraction of steam, 211
— vacuum in a condenser, 453

Mechanical draught, 360
— efficiency, 422, 616
— equivalent of heat, 16
Mechanism of simple vapour compression refrigeration system, 772
Merits of gaseous fuels, 292
— liquid fuels, 291
Metastable flow through nozzles, 490
Methods of governing I.C. engines, 605
— heat transfer, 736
— reducing missing quantity or cylinder condensation, 431
— reducing rotor speeds, 552
Metre, 3
Minimum mass of air required for complete combustion, 304, 306
— work required for a two stage compressor, 652
Missing quantity, 430
Mixture of air and steam in condensers, 452
M.K.S. units, 2
Modified Rankine cycle, 277
— Efficiency of, 278
Moist air, 799
Moisture content, 800
— lines, 809
Molar constant, 36
— specific heat of a gas, 39, 859
Molecular mass, 302
Molecules and atoms, 301
Mollier chart, 223
Mand gas, 292
Morse text, 613
Mountings, 323, 334
Multistage compression, 647
— turbine, 563

N

Natural draught, 360
Net calorific value, 293
Newton's law of cooling, 736
— motion, 5
Non-caking bituminous coal, 290
Non-flow process, 51
— Application of first law of thermody namics to, 51
Normal temperature and pressure (N.T.P.), 13
Nozzle efficiency, 470, 536

O

Octane number, 598
Open air refrigeration cycle, 754
— cycle gas turbines, 717
— system of thermodynamics, 8
Orsat apparatus, 319
Otto cycle, 171
Outer dead centre, 377
Oven gas, 292
Overall coefficient of heat transfer, 746
— efficiency, 422, 565, 616, 721
— isothermal efficiency, 683

P

Paraffin oil, 291
Parallel flow jet condensers, 448
Parson's reaction turbine, 521
Pass-out turbine, 578
Path of change of state, 9
Peat, 289
Percentage humidity, 800
Performance criteria for thermodynamic vapour cycles, 267
Perpetual motion machine of the first kind (PMM-I), 20
— second kind PMM-II), 21
Physical significance of critical pressure ratio, 479
Piston, 375, 585
— displacement volume, 377
— rings, 585
— rod, 375
Polytropic index, determination of, 82
— process, 72, 253
Potential energy 14
Power, 18
— produced by compound steam engine 398
— impulse turbine, 505
— reaction turbine, 524
— required to drive a compressor, 642, 651
— required to drive a fan, 369
— simple steam engine, 383
Preheating of compressed air, 704
Presentation of units and their values, 3
Pressure, 12
— and velocity of steam in turbine, 503, 522
— compounding of an impulse turbine, 553
— enthalpy chart, 773
— exerted by an ideal gas, 142
— gauge, 335
— of water vapour, 801
— velocity compounding of an impulse turbine, 554
Prewhirl, 676
Principle of increase of entropy, 107
Producer gas, 292

Prony brake dynamometer, 389
Properties of a refrigerant, 793
— system, 8
Psychrometer, 799
Psychrometric chart, 809
— processes, 812
— relations, 800
— terms, 798
Pulverised coal, 290

Q

Quality of wet steam, 201
Quasi-equilibrium or Quasi-static process, 9

R

Rankine cycle, 268
— with incomplete evaporation, 272
— with superheated steam, 274
— efficiency, 565
Rate of heat absorpotion or rejection per unit volume during a polytropic process, 81
Rating of C.I. engine fuels, 598
— S.I. engine fuels, 598
Ratio of coefficient of volume expansion and isothermal compressibility, 843
— cylinder diameters, 658
— specific heats, 43, 149
Receiver type compound steam engine , 397
Recent trends in gas turbines, 734
Refrigerants commonly used in practice, 794
Regenerative cycle, 569
— with single feed water heater, 570
— with two feed water heaters, 573
— surface condenser, 451
Regnault's law, 42
Reheat cycle, 561
— factor, 563
Reheating of steam, 560
— advantages of, 561
Relation between cycle and engine, 155
— heat and entropy, 103
— specific heats, 42
Relative efficiency, 423, 617
— humidity, 799, 800
— humidity lines, 811
Requirements of a steam condensing plant, 446
— good fuel, 292
Reversed Carnot cycle, 755
— Joule cycle, 761
Reversible adiabatic process, 247
— cycle, 154
— process, 50
Reversibility and irreversibility of thermodynamic processes, 155
Roots blower compressor, 666
Rope brake dynamometer, 390
Rules for S.I. units, 4

S

Safety valves, 335
Saturated air, 799
Scavenging of I.C. engines, 596
Scotch marine boiler, 326
Second, 3
— law of thermodynamics, 20
Selection of a steam boiler, 324
Semiclosed cycle gas turbine, 719
Sensible cooling, 812
— heat factor, 816
— heating, 814
— heat of water, 201
Separating calorimeter, 213
Sequence of operations in a cycle, 585
Simple vertical boiler, 325
S.I. Units, 2
Single stage reciprocating air compressor, 638
Slip factor, 697
Solid fueld, 289
— injection method, 600
Spark plug, 607
splash lubrication, 604
Sources of air in the condenser, 464
Specific heat, 15, 37, 850
— at constant pressure, 37
— at constant volume, 38
— humidity, 800
— humidity lines, 809
— ratio of, 43
— relations, 851
— relation between, 42
— steam consumption, 267
— volume lines, 810
— volume of steam, 202
Spring loaded safety valve, 337
State of a system, 9
Standard temperature and pressure (S.T.P.), 13
Static and total head quantities, 695
Steady flow process, 86
— Workdone in a, 89
Steam consumption, 440
— chest, 375
— injector, 493, 494
— jet draught, 361
— stop valve, 338
— tables, 203
Stirling cycle, 164
Stored energy, 13
— Types of, 14

Stroke length, 156
— volume, 377
Summary of formulae of heating and expansion of perfect gases, 85
Summer air conditioning system, 829
Supercharging of I.C. engines, 603
Superheated steam, 201, 206
Supersaturated flow through nozzles, 490
Surface condenser, 449
Swept volume, 156, 377, 638
Symbols of elements and compounds, 302
System of units, 2

T

Tandem type compound steam engine, 395
T ds equations, 847
Temperature, 10
— entropy diagram for water and steam, 222
— gradient, 738
— limitations for reversed Carnot cycle, 756
— vs total heat graph during steam formation, 200
Theroretical indicator diagram of simple steam engine, 378
— mass of air required for complete com bustion, 304, 306
— mean effective pressure, 379
— vapour compression cycle with dry saturated vapour after compression, 774
— with superheated vapour after compres sion, 783
— with undercooling or subcooling of re frigerant, 785
— with wet vapour after compression, 778
Thermal capacity, 16
— conductivity, 737
— equilibrium, 11
Thermodynamic cycle,9
— assumptions in, 153
— equilibrium, 11
— for rotary air compressor, 693
— Important terms used in, 156
— systems, 7
— text of I.C. engines, 611
— wet bulb temperature, 806
Thermosyphon system of cooling, 602
Three cylinder compound steam engine, 414
Throttle governing of compound steam engine, 414
Throttle governing of compound steam engines, 439
— simple steam engines, 438
— steam turbines, 555
Throttling calorimeter, 213
— lines, 226
— process, 91, 257
Total heat lines, 810
— of moist air, 803
— of steam, 202
Transit energy, 13
Two cylinder compound steam engine, 399, 401, 407
— stage reciprocating air compressor, 647
— stroke cycle diesel engine, 594
— stroke cycle petrol engine, 592
Types of cooling towers, 465
— draughts, 360
— jet condensers, 448
— rotary air compressors, 665
— scavenging, 597
— steam nozzles, 469
— stored energy, 14
— surface condensers, 450
— thermodynamic cycles, 157
— vapour compression cycles, 773

U

Unavailable heat energy, 105
Unitary air conditioning system, 830
Units of entropy, 106
— refrigeration, 752
Universal gas constant, 36
Unresisted expansion process, 83
Uses of gas turbines, 733
— steam tables, 203

V

Vacuum efficiency of steam condenser, 453
Values for maximum discharge through a nozzle, 477
— of critical pressure ratio, 479
Valve timing diagrams, 587, 589, 591, 594, 596
Vane blower compressor, 667
Vander Waals' equation of a real gas, 150
Vapour absorption refrigeration system, 791
— compression cycles, 773
— pressure lines, 811
Variation of specific heat with temperature, 860
Velocity compounding of an impulse turbine, 552
— diagram for axial flow air compressor, 678
— diagram for moving blade of an im pulse turbine, 503

— diagram for two stage impulse turbine, 514
— of a molecule, 141
— of steam flowing through a nozzle, 471
— triangle for centrifugal compressor, 673
— triangle for moving blades of a reaction turbine, 522, 524
Vertical multi-tubular boiler, 326
Volumetric efficiency, 617, 683, 684

W

Water cooling system of I.C. engines, 602
— equivalent, 16
— gas, 292
— level indicator, 334
Weight, 6
Wet bulb depression, 799
— temperature, 799
— temperature lines, 810
Wet steam, 200
Width of impeller blades, 675
Willian's law, 440
Winter air conditioning system, 828
Wood, 289
— charcoal, 290
Woolf type compound steam engine, 396
Work, 16, 17
— ratio, 267
Workdone by air in air motor, 700
— compound steam engine, 398
— centrifugal air compressor, 670
— during a non-flow precess, 51
— for various steady flow processes, 90
— single stage reciprocating air compres sor, 639, 645
— steady flow process, 89
— two stage reciprocating air compressor with intercooler, 649
Working of an ideal engine, 155
— single cylinder double acting horizontal reciprocating steam engine, 376
— single stage reciprocating air compres sor, 638

Y

Year round air conditioning, 830

Z

Zeroth law of thermodynamics, 19